Standard Normal Distribution *(continued)*

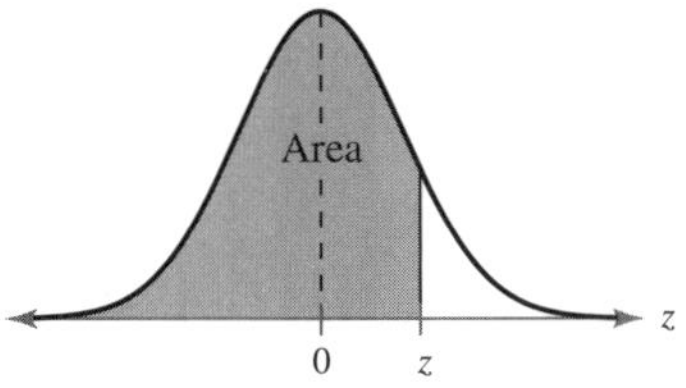

z	.00	.01	.02	.03	.04	.05	.06	.07	.08	.09
0.0	.5000	.5040	.5080	.5120	.5160	.5199	.5239	.5279	.5319	.5359
0.1	.5398	.5438	.5478	.5517	.5557	.5596	.5636	.5675	.5714	.5753
0.2	.5793	.5832	.5871	.5910	.5948	.5987	.6026	.6064	.6103	.6141
0.3	.6179	.6217	.6255	.6293	.6331	.6368	.6406	.6443	.6480	.6517
0.4	.6554	.6591	.6628	.6664	.6700	.6736	.6772	.6808	.6844	.6879
0.5	.6915	.6950	.6985	.7019	.7054	.7088	.7123	.7157	.7190	.7224
0.6	.7257	.7291	.7324	.7357	.7389	.7422	.7454	.7486	.7517	.7549
0.7	.7580	.7611	.7642	.7673	.7704	.7734	.7764	.7794	.7823	.7852
0.8	.7881	.7910	.7939	.7967	.7995	.8023	.8051	.8078	.8106	.8133
0.9	.8159	.8186	.8212	.8238	.8264	.8289	.8315	.8340	.8365	.8389
1.0	.8413	.8438	.8461	.8485	.8508	.8531	.8554	.8577	.8599	.8621
1.1	.8643	.8665	.8686	.8708	.8729	.8749	.8770	.8790	.8810	.8830
1.2	.8849	.8869	.8888	.8907	.8925	.8944	.8962	.8980	.8997	.9015
1.3	.9032	.9049	.9066	.9082	.9099	.9115	.9131	.9147	.9162	.9177
1.4	.9192	.9207	.9222	.9236	.9251	.9265	.9279	.9292	.9306	.9319
1.5	.9332	.9345	.9357	.9370	.9382	.9394	.9406	.9418	.9429	.9441
1.6	.9452	.9463	.9474	.9484	.9495	.9505	.9515	.9525	.9535	.9545
1.7	.9554	.9564	.9573	.9582	.9591	.9599	.9608	.9616	.9625	.9633
1.8	.9641	.9649	.9656	.9664	.9671	.9678	.9686	.9693	.9699	.9706
1.9	.9713	.9719	.9726	.9732	.9738	.9744	.9750	.9756	.9761	.9767
2.0	.9772	.9778	.9783	.9788	.9793	.9798	.9803	.9808	.9812	.9817
2.1	.9821	.9826	.9830	.9834	.9838	.9842	.9846	.9850	.9854	.9857
2.2	.9861	.9864	.9868	.9871	.9875	.9878	.9881	.9884	.9887	.9890
2.3	.9893	.9896	.9898	.9901	.9904	.9906	.9909	.9911	.9913	.9916
2.4	.9918	.9920	.9922	.9925	.9927	.9929	.9931	.9932	.9934	.9936
2.5	.9938	.9940	.9941	.9943	.9945	.9946	.9948	.9949	.9951	.9952
2.6	.9953	.9955	.9956	.9957	.9959	.9960	.9961	.9962	.9963	.9964
2.7	.9965	.9966	.9967	.9968	.9969	.9970	.9971	.9972	.9973	.9974
2.8	.9974	.9975	.9976	.9977	.9977	.9978	.9979	.9979	.9980	.9981
2.9	.9981	.9982	.9982	.9983	.9984	.9984	.9985	.9985	.9986	.9986
3.0	.9987	.9987	.9987	.9988	.9988	.9989	.9989	.9989	.9990	.9990
3.1	.9990	.9991	.9991	.9991	.9992	.9992	.9992	.9992	.9993	.9993
3.2	.9993	.9993	.9994	.9994	.9994	.9994	.9994	.9995	.9995	.9995
3.3	.9995	.9995	.9995	.9996	.9996	.9996	.9996	.9996	.9996	.9997
3.4	.9997	.9997	.9997	.9997	.9997	.9997	.9997	.9997	.9997	.9998

t-Distribution

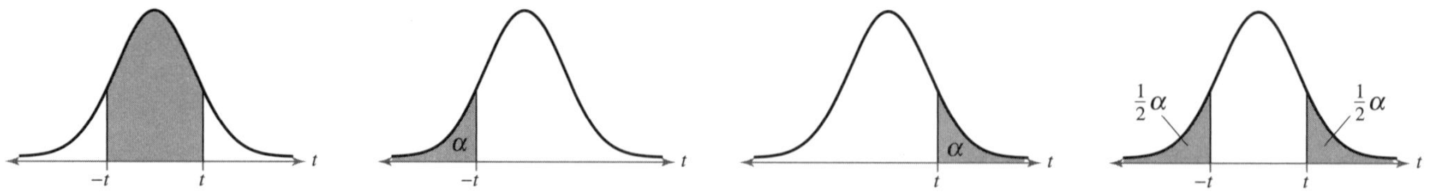

c-confidence interval | Left-tailed test | Right-tailed test | Two-tailed test

d.f.	Level of confidence, *c*	0.50	0.80	0.90	0.95	0.98	0.99
	One tail, α	**0.25**	**0.10**	**0.05**	**0.025**	**0.01**	**0.005**
	Two tails, α	**0.50**	**0.20**	**0.10**	**0.05**	**0.02**	**0.01**
1		1.000	3.078	6.314	12.706	31.821	63.657
2		.816	1.886	2.920	4.303	6.965	9.925
3		.765	1.638	2.353	3.182	4.541	5.841
4		.741	1.533	2.132	2.776	3.747	4.604
5		.727	1.476	2.015	2.571	3.365	4.032
6		.718	1.440	1.943	2.447	3.143	3.707
7		.711	1.415	1.895	2.365	2.998	3.499
8		.706	1.397	1.860	2.306	2.896	3.355
9		.703	1.383	1.833	2.262	2.821	3.250
10		.700	1.372	1.812	2.228	2.764	3.169
11		.697	1.363	1.796	2.201	2.718	3.106
12		.695	1.356	1.782	2.179	2.681	3.055
13		.694	1.350	1.771	2.160	2.650	3.012
14		.692	1.345	1.761	2.145	2.624	2.977
15		.691	1.341	1.753	2.131	2.602	2.947
16		.690	1.337	1.746	2.120	2.583	2.921
17		.689	1.333	1.740	2.110	2.567	2.898
18		.688	1.330	1.734	2.101	2.552	2.878
19		.688	1.328	1.729	2.093	2.539	2.861
20		.687	1.325	1.725	2.086	2.528	2.845
21		.686	1.323	1.721	2.080	2.518	2.831
22		.686	1.321	1.717	2.074	2.508	2.819
23		.685	1.319	1.714	2.069	2.500	2.807
24		.685	1.318	1.711	2.064	2.492	2.797
25		.684	1.316	1.708	2.060	2.485	2.787
26		.684	1.315	1.706	2.056	2.479	2.779
27		.684	1.314	1.703	2.052	2.473	2.771
28		.683	1.313	1.701	2.048	2.467	2.763
29		.683	1.311	1.699	2.045	2.462	2.756
∞		.674	1.282	1.645	1.960	2.326	2.576

Annotated Instructor's Edition

Elementary Statistics

Picturing the World

Third Edition

Ron Larson

The Pennsylvania State University
The Behrend College

Betsy Farber

Bucks County
Community College

PEARSON
Prentice Hall

Upper Saddle River, NJ 07458

Editor-in-Chief: ***Sally Yagan***
Executive Acquisition Editor: ***Petra Recter***
Project Manager: ***Jacquelyn Riotto Zupic***
Production Editor/Assistant Managing Editor: ***Bayani Mendoza de Leon***
Vice President/Director of Production and Manufacturing: ***David W. Riccardi***
Senior Managing Editor: ***Linda Mihatov Behrens***
Executive Managing Editor: ***Kathleen Schiaparelli***
Assistant Manufacturing Manager/Buyer: ***Michael Bell***
Managing Editor, Digital Supplements: ***Nicole M. Jackson***
Marketing Assistant: ***Rebecca Alimena***
Director of Creative Services: ***Paul Belfanti***
Art Director/Cover Designer: ***John Christiana***
Art Editor: ***Thomas Benfatti***
Manager, Cover Visual Research & Permissions: ***Karen Sanatar***
Director, Image Resource Center: ***Melinda Reo***
Manager, Rights and Permissions: ***Zina Arabia***
Manager, Visual Research: ***Beth Brenzel***
Image Permission Coordinator: ***Debbie Latronica***
Photo Researcher: ***Melinda Alexander***
Composition/Art/Interior Design: ***Larson Texts, Inc.***
Editorial Assistant/Supplement Editor: ***Joanne Wendelken***
Cover photos: *Commuter reading newspaper—©Taxi/Getty Images, Inc.; crowd applauding—©Lisette Le Bon/Superstock; research chemist—©Kim Steele/Photodisc/Getty Images, Inc.; soccer players—©Royalty-Free/CORBIS*

Pearson Prentice Hall
Pearson Education, Inc.
Upper Saddle River, New Jersey 07458

Printed in the United States of America

10 9 8 7 6 5 4 3

ISBN 0-13-148317-X

(Student Edition ISBN 0-13-148316-1)

Pearson Education LTD., *London*
Pearson Education Australia PTY, Limited, *Sydney*
Pearson Education Singapore, Pte. Ltd
Pearson Education North Asia Ltd, *Hong Kong*
Pearson Education Canada, Ltd., *Toronto*
Pearson Educación de Mexico, S.A. de C.V.
Pearson Education—Japan, *Tokyo*
Pearson Education Malaysia, Pte. Ltd

About the Authors

Ron Larson
The Pennsylvania State University
The Behrend College

Ron Larson received his Ph.D. in mathematics from the University of Colorado in 1970. At that time he accepted a position with Penn State University, and he currently holds the rank of professor of mathematics at the university. Larson is the lead author of more than two dozen mathematics textbooks that range from sixth grade through calculus levels. Many of his texts, such as the seventh edition of his calculus text, are leaders in their markets. Larson is also one of the pioneers in the use of multimedia and the Internet to enhance the learning of mathematics. He has authored multimedia programs, extending from the elementary school through calculus levels. Larson is a member of several professional groups and is a frequent speaker at national and regional mathematics meetings.

Betsy Farber
Bucks County
Community College

Betsy Farber received her Bachelor's degree in mathematics from Penn State University and Master's degree in mathematics from the College of New Jersey. Since 1976, she has been teaching all levels of mathematics at Bucks County Community College in Newtown, Pennsylvania, where she currently holds the rank of professor. She is particularly interested in developing new ways to make statistics relevant and interesting to her students and has been teaching statistics in many different modes—with TI-83, with MINITAB, and by distance learning as well as in the traditional classroom. A member of the American Mathematical Association of Two-Year Colleges (AMATYC), she is an author of *The Student Edition to MINITAB* and *A Guide to MINITAB.* She served as consulting editor for *Statistics, A First Course* and has written computer tutorials for the CD-ROM correlating to the texts in the Streeter Series in mathematics.

Contents

Preface x
Supplements xiii
Acknowledgments xiv
Features xv
How to Study Statistics xxi
Index of Applications xxiii

Chapter 1 Introduction to Statistics 0

Where You've Been 1
Where You're Going 1
1.1 An Overview of Statistics 2
1.2 Data Classification 9
Case Study: *Rating Television Shows in the United States* 15
1.3 Experimental Design 16
Uses and Abuses 23
Chapter Summary 24
Review Exercises 25
Chapter Quiz 27
Real Statistics–Real Decisions—Putting It All Together 28
History of Statistics—Timeline 29
Technology: *Using Technology in Statistics* 30

Chapter 2 Descriptive Statistics 32

Where You've Been 33
Where You're Going 33
2.1 Frequency Distributions and Their Graphs 34
2.2 More Graphs and Displays 49
2.3 Measures of Central Tendency 60
2.4 Measures of Variation 74
Case Study: *Sunglass Sales in the United States* 92
2.5 Measures of Position 93
Uses and Abuses 105
Chapter Summary 106
Review Exercises 107
Chapter Quiz 111
Real Statistics–Real Decisions—Putting It All Together 112
Technology: *Monthly Milk Production* 113
Using Technology to Determine Descriptive Statistics 114

Chapter 3 Probability 116

Where You've Been 117
Where You're Going 117
3.1 Basic Concepts of Probability 118
3.2 Conditional Probability and the Multiplication Rule 130
3.3 The Addition Rule 140
Case Study: *Probability and Parking Lot Strategies* 149
3.4 Counting Principles 150
Uses and Abuses 161
Chapter Summary 162
Review Exercises 163
Chapter Quiz 167
Real Statistics–Real Decisions—Putting It All Together 168
Technology: *Simulation: Composing Mozart Variations with Dice* 169

Chapter 4 Discrete Probability Distributions 170

Where You've Been 171
Where You're Going 171
4.1 Probability Distributions 172
4.2 Binomial Distributions 184
Case Study: *Binomial Distribution of Airplane Accidents* 197
4.3 More Discrete Probability Distributions 198
Uses and Abuses 205
Chapter Summary 206
Review Exercises 207
Chapter Quiz 211
Real Statistics–Real Decisions—Putting It All Together 212
Technology: *Using Poisson Distributions as Queuing Models* 213

Chapter 5 Normal Probability Distributions 214

Where You've Been 215
Where You're Going 215
5.1 Introduction to Normal Distributions and the Standard Normal Distribution 216
5.2 Normal Distributions: Finding Probabilities 229
5.3 Normal Distributions: Finding Values 237
Case Study: *Birth Weights in America* 245
5.4 Sampling Distributions and the Central Limit Theorem 246
5.5 Normal Approximations to Binomial Distributions 259
Uses and Abuses 269
Chapter Summary 270
Review Exercises 271
Chapter Quiz 275
Real Statistics–Real Decisions—Putting It All Together 276
Technology: *Age Distribution in the United States* 277

Chapter 6 Confidence Intervals 278

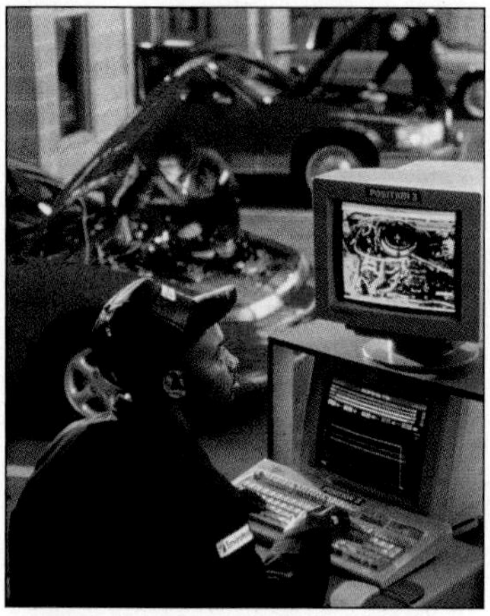

Where You've Been 279
Where You're Going 279
6.1 Confidence Intervals for the Mean (Large Samples) 280
Case Study: *Shoulder Heights of Appalachian Black Bears* 294
6.2 Confidence Intervals for the Mean (Small Samples) 295
6.3 Confidence Intervals for Population Proportions 303
6.4 Confidence Intervals for Variance and Standard Deviation 312
Uses and Abuses 319
Chapter Summary 320
Review Exercises 321
Chapter Quiz 325
Real Statistics–Real Decisions—Putting It All Together 326
Technology: *Most Admired Polls* 327
Using Technology to Construct Confidence Intervals 328

Chapter 7 Hypothesis Testing with One Sample 330

Where You've Been 331
Where You're Going 331
7.1 Introduction to Hypothesis Testing 332
7.2 Hypothesis Testing for the Mean (Large Samples) 347
Case Study: *Human Body Temperature: What's Normal?* 363
7.3 Hypothesis Testing for the Mean (Small Samples) 364
7.4 Hypothesis Testing for Proportions 374
7.5 Hypothesis Testing for Variance and Standard Deviation 380
Uses and Abuses 389
A Summary of Hypothesis Testing 390
Chapter Summary 392
Review Exercises 393
Chapter Quiz 397
Real Statistics–Real Decisions—Putting It All Together 398
Technology: *The Case of the Vanishing Women* 399
Using Technology to Perform Hypothesis Tests 400

Chapter 8 Hypothesis Testing with Two Samples 402

Where You've Been 403
Where You're Going 403
8.1 Testing the Difference Between Means (Large Independent Samples) 404
Case Study: *DASH Diet and Blood Pressure* 415
8.2 Testing the Difference Between Means (Small Independent Samples) 416
8.3 Testing the Difference Between Means (Dependent Samples) 425
8.4 Testing the Difference Between Proportions 437
Uses and Abuses 445
Chapter Summary 446
Review Exercises 447
Chapter Quiz 451
Real Statistics–Real Decisions—Putting It All Together 452
Technology: *Tails Over Heads* 453
Using Technology to Perform Two-Sample Hypothesis Tests 454

Chapter 9 Correlation and Regression 456

Where You've Been		457
Where You're Going		457
9.1	Correlation	458
9.2	Linear Regression	474
	Case Study: *Correlation of Body Measurements*	483
9.3	Measures of Regression and Prediction Intervals	484
9.4	Multiple Regression	494
	Uses and Abuses	499
	Chapter Summary	500
	Review Exercises	501
	Chapter Quiz	505
	Real Statistics–Real Decisions—Putting It All Together	506
	Technology: *Nutrients in Breakfast Cereals*	507

Chapter 10 Chi-Square Tests and the *F*-Distribution 508

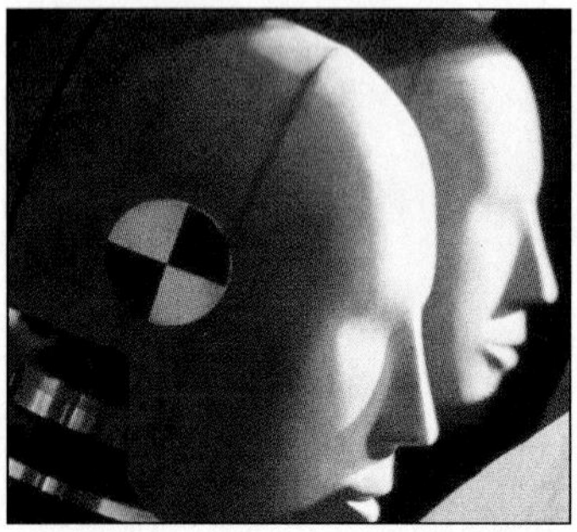

Where You've Been		509
Where You're Going		509
10.1	Goodness of Fit	510
10.2	Independence	521
	Case Study: *Traffic Safety Facts*	532
10.3	Comparing Two Variances	533
10.4	Analysis of Variance	542
	Uses and Abuses	555
	Chapter Summary	556
	Review Exercises	557
	Chapter Quiz	561
	Real Statistics–Real Decisions—Putting It All Together	562
	Technology: *Teacher Salaries*	563

Chapter 11 Nonparametric Tests 564

Where You've Been 565
Where You're Going 565
11.1 The Sign Test 566
11.2 The Wilcoxon Tests 577
Case Study: *Earnings by College Degree* 586
11.3 The Kruskal-Wallis Test 587
11.4 Rank Correlation 593
11.5 The Runs Test 599
Uses and Abuses 607
Chapter Summary 608
Review Exercises 609
Chapter Quiz 613
Real Statistics–Real Decisions—Putting It All Together 614
Technology: *U.S. Income and Economic Research* 615

Appendices

Appendix A Alternative Presentation of the Standard Normal Distribution A1
Standard Normal Distribution Table (0-to-z) A1
Alternative Presentation of the Standard Normal Distribution A2

Appendix B Tables A7
Table 1 *Random Numbers* A7
Table 2 *Binomial Distribution* A8
Table 3 *Poisson Distribution* A11
Table 4 *Standard Normal Distribution* A16
Table 5 *t-Distribution* A18
Table 6 *Chi-Square Distribution* A19
Table 7 *F-Distribution* A20
Table 8 *Critical Values for the Sign Test* A25
Table 9 *Critical Values for the Wilcoxon Signed-Rank Test* A26
Table 10 *Critical Values for the Spearman Rank Correlation* A27
Table 11 *Critical Values for the Pearson Correlation Coefficient* A28
Table 12 *Critical Values for the Number of Runs* A29

Answers to the Try It Yourself Exercises A30
Answers to the Odd-Numbered Exercises A47
Instructor's Answers to Selected Even-Numbered Exercises A89
Index I1

Preface

Welcome to *Elementary Statistics: Picturing the World*, Third Edition. We are grateful for the overwhelming acceptance and support of the First and Second Editions. It is gratifying to know that our vision of combining theory, pedagogy, and design to exemplify how statistics is used to picture and describe the world has helped students learn about statistics and make informed decisions.

New to the Third Edition

Chapter Openers Chapter openers have been updated. Each chapter opener provides a road map for the chapter and begins with a two-page visual description of a real-life problem that is often revisited in the chapter. The chapter opener shows students how the chapter fits into the bigger picture of statistics by connecting it to topics learned in earlier chapters and then gives an overview of the chapter in the context of the real-world data presented.

Interpretation Examples are solved in a step-by-step manner, and students are asked to reflect upon conclusion of the example in an Interpretation step. This added step promotes critical thinking and writing skills.

Exercises The exercise sets in the Third Edition include about 1900 exercises, giving students practice in performing calculations, making decisions, providing explanations, and applying results to a real-life setting. Approximately 35% of these exercises are new or revised. The exercises are divided into three sections:

Building Basic Skills and Vocabulary Each exercise set begins with a group of exercises called *Building Basic Skills and Vocabulary*. These exercises are short answer, true and false, or vocabulary carefully written to nurture student understanding.

NEW! *Using and Interpreting Concepts* Each exercise set contains a group of exercises called *Using and Interpreting Concepts*. These exercises are skill or word problems carefully written to nurture student understanding and proficiency. They move from basic skill development to more challenging and interpretive problems.

Extending Concepts Each exercise set ends with a group of exercises called *Extending Concepts*. These exercises go beyond the material presented in the section—they tend to be more challenging and are not required as prerequisites of subsequent sections.

Round off Rules Round off rules have been included in Study Tips at point of use. These rules guide the student during calculations.

Course Coverage In response to suggestions from statistics instructors, the coverage of topics in Chapters 4, 5, 7, 9, and 11 is revised in the Third Edition. Many of these changes emphasize interpretation of results rather than calculations.

- **In Chapter 4,** additional exercises place more emphasis on mean and standard deviation of binomial distributions. Also, more emphasis is placed on identification of unusual values within this distribution.

- **In Chapter 5,** we combined two sections to form Section 5.1 **Introduction to Normal Distributions and the Standard Normal Distribution.** This section replaces Section 5.1 **Introduction to Normal Distributions** and Section 5.2 **The Standard Normal Distribution** in the Second Edition. The Empirical Rule has been covered in Chapter 2, so there is no need for it in Chapter 5. In Section 5.4, **Sampling Distributions and the Central Limit Theorem,** more emphasis is placed on identifying unusual values within sampling distributions.
- **In Chapter 7,** we have placed more emphasis on the logic behind the hypotheses testing process and interpreting the decision.
- **In Chapter 9,** for instructors who prefer to cover Section 9.1, **Correlation,** immediately after covering graphing paired data in Chapter 2, we present a method for testing a population correlation coefficient that does not involve hypothesis testing. The method is simple and can easily be covered after Chapters 1 and 2.
- **In Chapter 11,** we added a new section—Section 11.5 **Runs Test.** This section provides students with a method—a nonparametric test called the Runs Test—for testing randomness in a set of data.

Continuing Strong Pedagogy from the Second Edition

Versatile Course Coverage The table of contents of the text was developed to give instructors many options. For instance, assigning the *Extending Concepts* exercises and spending time on the chapter projects provide sufficient content to use the text in a two-semester course. More commonly, we expect the text to be used in a three-credit semester course or a four-credit semester course that includes a lab component. In such cases, instructors will have to pare down the text's 46 sections. If you wish to view sample syllabi, you can access these within **MyMathLab** under the Instructor Resource Area.

Graphical Approach As with most introductory statistics texts, we begin the descriptive statistics chapter with a survey of different ways to display data graphically. A difference between this text and many others is that **we continue to incorporate the graphical display of data throughout the text.** For example, see the use of stem-and-leaf plots to display data on pages 362 and 363. This emphasis on graphical displays is beneficial to all students, especially those utilizing visual learning strategies.

Balanced Approach The text strikes a **balance between computation, decision making, and conceptual understanding.** We have provided many Examples, Exercises, and Try It Yourself exercises that go beyond mere computation. For instance, look at Exercises 31–34 on page 257. Students are not just asked to calculate a probability; they also are asked to use the probability to make a decision about a claim.

Variety of Real-Life Applications We have chosen real-life applications that are representative of the majors of students taking introductory statistics courses. These include business, psychology, health sciences, sports, computer science, political science, and many others. Choosing meaningful applications for such a diverse audience is difficult. We want statistics to come alive and appear relevant to students so they understand the importance and rationale for studying statistics. We wanted the applications to be **authentic**—but they also need to be **accessible.** See the Index of Applications on page xxiii.

Data and Source Lines The data sets in the book were chosen for interest, variety, and their ability to illustrate concepts. Most of the **over 200 data sets** contain actual data with source lines. The remaining data sets contain simulated data that are representative of real-life situations. All data sets containing 20 or more entries are available in a variety of electronic forms, including disk and Internet. In the exercise sets, the data sets that are available electronically are indicated by the icon DATA.

Flexible Technology Although most formulas in the book are illustrated with tabular "hand" calculations, we assume that most students who take this course have access to some form of technology tool, such as MINITAB, Excel, the TI-83, or SPSS. Because the use of technology varies widely, we have made the text flexible. **It can be used in courses with no more technology than a scientific calculator—or it can be used in courses that require frequent use of sophisticated technology tools.** For those who want specific instructions on particular technology tools, separate technology manuals are available to augment the text. Whatever your use of technology, we are sure that you agree with us that the goal of the course is not computation. Rather, it is to gain an understanding of the basic concepts and uses of statistics.

Prerequisites Statistics contains many formulas and variables, including radicals, summation notation, Greek letters, and subscripts. So, **some familiarity with algebra and evaluation of algebraic expressions** is a prerequisite. Nevertheless, we have made every effort to keep algebraic manipulations to a minimum—often we display informal versions of formulas using words in place of or in addition to variables. For instance, see the definitions of midpoint and relative frequency on page 36.

Choice of Tables Our experience has shown that students find a **cumulative density function** (CDF) table easier to use than a "0-to-z" table. Using the CDF table to find the area under a normal curve is a topic of Section 5.1 on pages 219–223. Because we realize that many teachers prefer to use the "0-to-z" table, we have provided an alternative presentation of this topic using the "0-to-z" table in Appendix A of the book.

MAA, AMATYC, NCTM Standards This text answers the call for a **student-friendly text that emphasizes the uses of statistics** and not just the computation of its myriad formulas. Our experience indicates that our job as instructors of an introductory course in statistics is not to produce statisticians but to produce informed consumers of statistical reports. For this reason, we have included many exercises that require students to provide written explanations, find patterns, and make decisions.

Page Layout We believe that statistics is more accessible to students when it is carefully formatted on each page with a consistent open layout. This text is the first college-level statistics book to be written to design, which means that its features (Examples, Try It problems, Definitions, or Guidelines) are not split from one page to the next. Although this process requires extra planning and work in the development stage, the result is a presentation that is clean and clear.

Supplements

Student Resources

Student Study Pack

(Stand-alone: 0-13-134367-X; Value Pack: 0-13-136366-1)

Everything a student needs to succeed in one place. Free packaged with the book, or available for purchase stand-alone. The **Student Study Pack** contains:

- **Chapter Quiz Prep Video (CD-ROM)** Provides step-by-step video solutions to every problem in the textbook *Chapter Quizzes*.
- **Student Solutions Manual** Includes complete worked-out solutions to all of the *Try It Yourself* exercises, the odd-numbered exercises, and all of the *Chapter Quiz* exercises.
- **CD Lecture Series** A comprehensive set of CD-ROMs, tied to the textbook, containing short video clips of an instructor working every *Try It Yourself* exercise.
- **Technology Manuals** Tutorial instruction and worked-out examples for the TI-83 Calculator, Excel, and MINITAB.
- **Pearson Tutor Center** Tutors provide one-on-one tutoring for any problem with an answer at the back of the book. Students access the Tutor Center via toll-free phone, fax, or email. (not available in high school channel)

Instructor Resources

Content Distribution Center

All instructor resources can be downloaded from Prentice Hall website *www.prenhall.com* (then search for Larson/Farber, *Elementary Statistics 3e*). This is a password-protected website that requires instructors to set up an account or, alternatively, instructor resources can be ordered individually from your Prentice Hall sales representative:

- **Instructor-to-Instructor CD Lecture Video** (0-13-186162-X) Offers classroom suggestions, time-saving tips, and strategies for engaging students and presenting key topics throughout the course.
- **Instructor Solutions Manual** (0-13-148318-8) Includes complete solutions to all of the exercises, *Try It Yourself* exercises, Case Studies, Technology pages, Uses and Abuses exercises, and Real Statistics-Real Decisions exercises.
- **TestGen** (0-13-148320-X) Test-generating software—create tests from textbook section objectives. Includes more than 1050 additional questions with an answer key. Questions are algorithmically generated, allowing for unlimited test versions. Instructors may also edit problems or create their own.
- **Test Item File** (0-13-148319-6) A printed test bank derived from TestGen.
- **PowerPoint Lecture Slides** Fully editable and printable slides that follow the textbook. Use during lecture or post to a website in an online course. Most slides include notes offering suggestions for how the material may effectively be presented in class.

Annotated Instructor's Edition

Includes:

- *Notes to Instructors* appear in the margin of the text to suggest activities that correspond to the example or concept, additional ways to present the material, common pitfalls students encounter, alternative formulas or approaches that may be used, and other helpful teaching tips for instructors.
- All answers to the section and review exercises are provided with short answers (numerical, tabular, and/or graphical) appearing in the margin next to the exercise.

Internet Resources

MyMathLab

(Student: 0-13-147894-X; Instructor: 0-13-147898-2)

MyMathLab is a text-specific, online course. MyMathLab contains MathXL (Internet) at its core, providing all of the tutorial, homework, testing, and diagnostic power of MathXL. Beyond the power of MathXL, you also get access to

- **Course Management Tools** MyMathLab is a fully functioning course management system. Upload documents (e.g., syllabi, lecture notes, etc.) utilize communication tools (e.g., email, chat rooms, virtual classroom), create and post assignments, and more.
- **Additional Resources for Instructors and Students** Access a multimedia textbook, online lecture videos, solutions manuals, online graphing calculator manual, sample syllabi, a video on overcoming math anxiety, PowerPoint lecture slides, and more.

MathXL® (Internet)

A powerful online homework, tutorial, and assessment system. MathXL (Internet) provides powerful tutorial capabilities and allows instructors to create, edit, and assign online homework and tests (built from provided text-specific question banks) and to track student work in a gradebook. MathXL creates study plans personalized for students on the basis of test results.

Acknowledgments

We owe a debt of gratitude to the many reviewers who helped us shape and refine *Elementary Statistics: Picturing the World*, Third Edition.

Previous Edition Reviewers and/or Contributors

John J. Avioli, Christopher Newport University
John Bernard, University of Texas–Pan American
G. Andy Chang, Youngstown State University
Keith J. Craswell, Western Washington University
Gary Egan, Monroe Community College
Charles Ehler, Anne Arundel Community College
Jill Fanter, Walters State Community College
Douglas Frank, Indiana University of Pennsylvania
Frieda Ganter, California State University–Fresno
Sonja Hensler, St. Petersbury Jr. College
David Kay, Moorpark College
Hyune-Ju Kim, Syracuse University
Rita Kolb, Catonsville Community College
Rowan Lindley, Westchester Community College
Jeffrey Linek, St. Petersburg Junior College
Diane Long, College of DuPage
Rhonda Magel, North Dakota State University
Mike McGann, Ventura Community College
Vicki McMillian, Ocean County College
Julie Norton, California State University–Hayward
Lynn Onken, San Juan College
Lindsay Packer, College of Charleston
Neal Rogness, Grand Valley State University
Jean Sells, Sacred Heart University
Carol Shapero, Oakton Community College
Aileen Solomon, Trident Technical College
Sandra L. Spain, Thomas Nelson Community College
Michelle Strager-McCarney, Penn State Erie, The Behrend College
Agnes Tuska, California State University–Fresno
Ting-Xiu Wang, Oakton Community College
Dex Whittinghill, Rowan University

Third Edition Reviewers

Olcay Akman, College of Charleston
Cara Delong, Fayetteville Technical Community College
Ginger Dewey, York Technical College
Nancy Johnson, Manatee Community College
Martin Jones, College of Charleston
Mohammad Kazemi, University of North Carolina–Charlotte
Jane Keller, Metropolitan Community College
Susan Kellicut, Seminole Community College
Benny Lo, DeVry University, Freemont
Lynn Meslinsky, Erie Community College
Eric Preibisius, Cuyamaca Community College
Elisabeth Schuster, Benedictine University
Jean Sells, Sacred Heart University
Aileen Solomon, Trident Technical College
Kathleen Suchora, University of Central Florida

Accuracy Checker

Sarah Streett

We also give special thanks to the people at Prentice Hall who worked with us in the development of *Elementary Statistics: Picturing the World*, Third Edition: Sally Yagan, Petra Recter, Joanne Wendelken, Bayani Mendoza de Leon, Donna Crilly, Linda Behrens, John Christiana, Michael Bell, and Jacquelyn Riotto Zupic.

We also thank the staff of Larson Texts, Inc., who assisted with the development and production of the book.

On a personal level, we are grateful to our spouses, Deanna Gilbert Larson and Richard Farber, for their love, patience, and support. Also, a special thanks goes to R. Scott O'Neil.

We have worked hard to make *Elementary Statistics: Picturing the World*, Third Edition, a clean, clear, and enjoyable text from which to teach and learn statistics. Despite our best efforts to ensure accuracy and ease of use, many users will undoubtedly have suggestions for improvement. We welcome your suggestions.

Ron Larson, odx@psu.edu

Betsy Farber, farberb@bucks.edu

Features

Chapter Openers

Where You've Been

Each chapter begins with a two-page photographic description of a real-life problem. The feature called *Where You've Been* shows students how the chapter fits into the bigger picture of statistics, by connecting it to topics learned in earlier chapters.

Where You're Going

The feature called *Where You're Going* gives students an overview of the chapter, exploring concepts in the context of real-world settings.

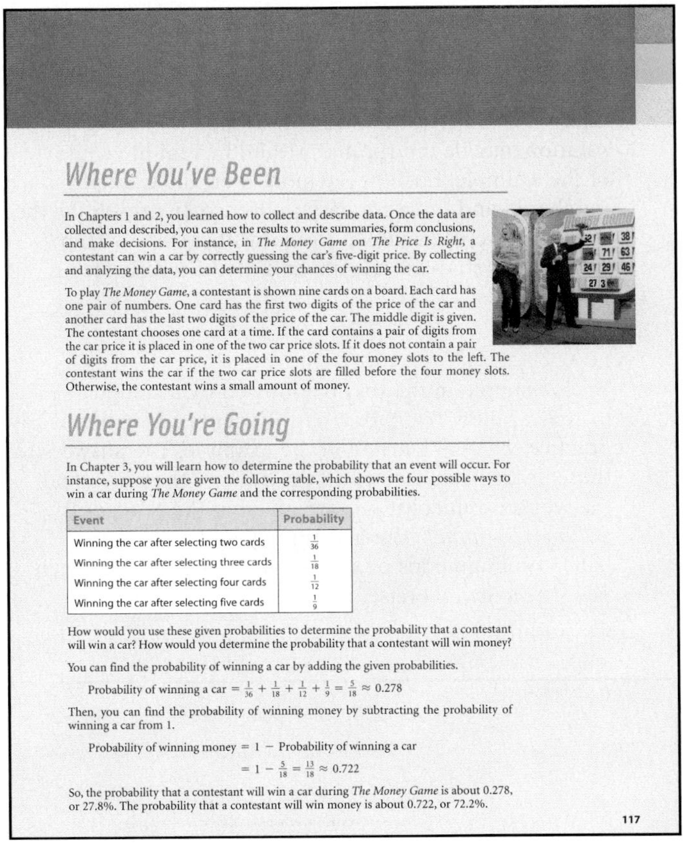

Where You've Been

In Chapters 1 and 2, you learned how to collect and describe data. Once the data are collected and described, you can use the results to write summaries, form conclusions, and make decisions. For instance, in *The Money Game* on *The Price Is Right*, a contestant can win a car by correctly guessing the car's five-digit price. By collecting and analyzing the data, you can determine your chances of winning the car.

To play *The Money Game*, a contestant is shown nine cards on a board. Each card has one pair of numbers. One card has the first two digits of the price of the car and another card has the last two digits of the price of the car. The middle digit is given. The contestant chooses one card at a time. If the card contains a pair of digits from the car price it is placed in one of the two car price slots. If it does not contain a pair of digits from the car price, it is placed in one of the four money slots to the left. The contestant wins the car if the two car price slots are filled before the four money slots. Otherwise, the contestant wins a small amount of money.

Where You're Going

In Chapter 3, you will learn how to determine the probability that an event will occur. For instance, suppose you are given the following table, which shows the four possible ways to win a car during *The Money Game* and the corresponding probabilities.

Event	Probability
Winning the car after selecting two cards	$\frac{1}{36}$
Winning the car after selecting three cards	$\frac{1}{18}$
Winning the car after selecting four cards	$\frac{1}{12}$
Winning the car after selecting five cards	$\frac{1}{9}$

How would you use these given probabilities to determine the probability that a contestant will win a car? How would you determine the probability that a contestant will win money?

You can find the probability of winning a car by adding the given probabilities.

$$\text{Probability of winning a car} = \frac{1}{36} + \frac{1}{18} + \frac{1}{12} + \frac{1}{9} = \frac{5}{18} \approx 0.278$$

Then, you can find the probability of winning money by subtracting the probability of winning a car from 1.

$$\text{Probability of winning money} = 1 - \text{Probability of winning a car}$$
$$= 1 - \frac{5}{18} = \frac{13}{18} \approx 0.722$$

So, the probability that a contestant will win a car during *The Money Game* is about 0.278, or 27.8%. The probability that a contestant will win money is about 0.722, or 72.2%.

117

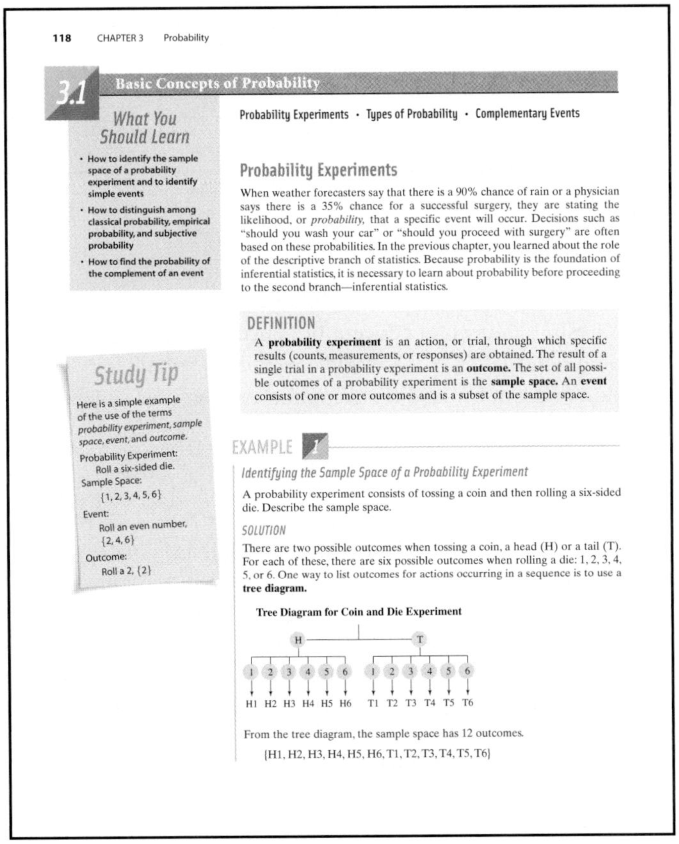

118 CHAPTER 3 Probability

3.1 Basic Concepts of Probability

What You Should Learn

- How to identify the sample space of a probability experiment and to identify simple events
- How to distinguish among classical probability, empirical probability, and subjective probability
- How to find the probability of the complement of an event

Probability Experiments • Types of Probability • Complementary Events

Probability Experiments

When weather forecasters say that there is a 90% chance of rain or a physician says there is a 35% chance for a successful surgery, they are stating the likelihood, or *probability*, that a specific event will occur. Decisions such as "should you wash your car" or "should you proceed with surgery" are often based on these probabilities. In the previous chapter, you learned about the role of the descriptive branch of statistics. Because probability is the foundation of inferential statistics, it is necessary to learn about probability before proceeding to the second branch—inferential statistics.

DEFINITION

A **probability experiment** is an action, or trial, through which specific results (counts, measurements, or responses) are obtained. The result of a single trial in a probability experiment is an **outcome.** The set of all possible outcomes of a probability experiment is the **sample space.** An **event** consists of one or more outcomes and is a subset of the sample space.

Study Tip

Here is a simple example of the use of the terms *probability experiment, sample space, event,* and *outcome.*

Probability Experiment: Roll a six-sided die.
Sample Space: {1, 2, 3, 4, 5, 6}
Event: Roll an even number, {2, 4, 6}
Outcome: Roll a 2, {2}

EXAMPLE 1

Identifying the Sample Space of a Probability Experiment

A probability experiment consists of tossing a coin and then rolling a six-sided die. Describe the sample space.

SOLUTION

There are two possible outcomes when tossing a coin, a head (H) or a tail (T). For each of these, there are six possible outcomes when rolling a die: 1, 2, 3, 4, 5, or 6. One way to list outcomes for actions occurring in a sequence is to use a **tree diagram.**

Tree Diagram for Coin and Die Experiment

H T
1 2 3 4 5 6 1 2 3 4 5 6
H1 H2 H3 H4 H5 H6 T1 T2 T3 T4 T5 T6

From the tree diagram, the sample space has 12 outcomes.

{H1, H2, H3, H4, H5, H6, T1, T2, T3, T4, T5, T6}

Section Organization

Each section is organized by learning objectives. These objectives are presented in everyday language in a margin feature called *What You Should Learn*. The same objectives are then used as subsection titles throughout the section.

Study Tips

Most sections contain one or more study tips placed on "sticky notes" in the margin. These tend to be informal learning aids, which show how to read a table, use technology, or interpret a result or a graph.

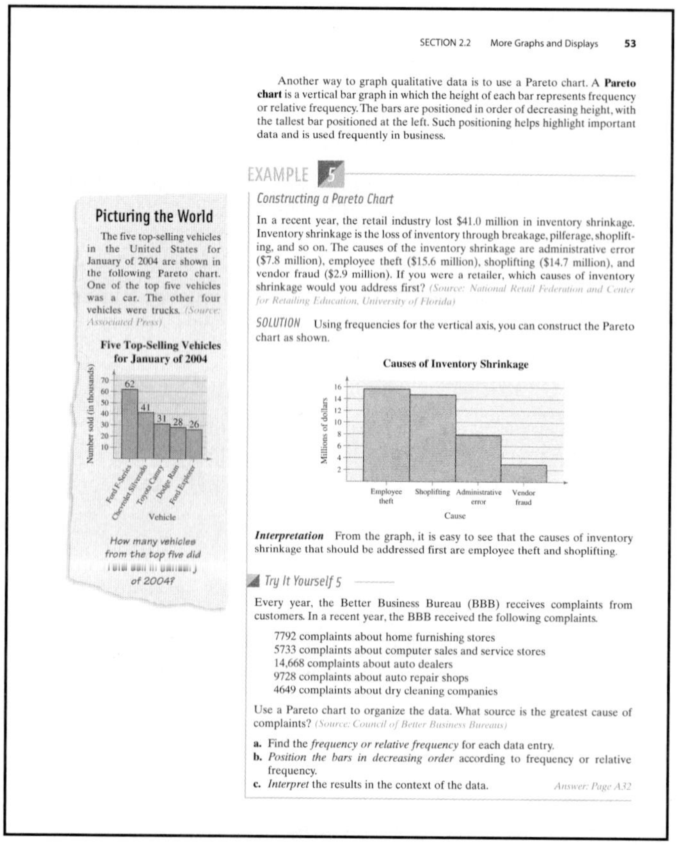

SECTION 2.2 More Graphs and Displays 53

Another way to graph qualitative data is to use a Pareto chart. A **Pareto chart** is a vertical bar graph in which the height of each bar represents frequency or relative frequency. The bars are positioned in order of decreasing height, with the tallest bar positioned at the left. Such positioning helps highlight important data and is used frequently in business.

EXAMPLE 5

Constructing a Pareto Chart

In a recent year, the retail industry lost $41.0 million in inventory shrinkage. Inventory shrinkage is the loss of inventory through breakage, pilferage, shoplifting, and so on. The causes of the inventory shrinkage are administrative error ($7.8 million), employee theft ($15.6 million), shoplifting ($14.7 million), and vendor fraud ($2.9 million). If you were a retailer, which causes of inventory shrinkage would you address first? *(Source: National Retail Federation and Center for Retailing Education, University of Florida)*

SOLUTION Using frequencies for the vertical axis, you can construct the Pareto chart as shown.

Interpretation From the graph, it is easy to see that the causes of inventory shrinkage that should be addressed first are employee theft and shoplifting.

Try It Yourself 5

Every year, the Better Business Bureau (BBB) receives complaints from customers. In a recent year, the BBB received the following complaints.

7792 complaints about home furnishing stores
5733 complaints about computer sales and service stores
14,668 complaints about auto dealers
9728 complaints about auto repair shops
4649 complaints about dry cleaning companies

Use a Pareto chart to organize the data. What source is the greatest cause of complaints? *(Source: Council of Better Business Bureaus)*

a. Find the *frequency or relative frequency* for each data entry.
b. *Position the bars in decreasing order* according to frequency or relative frequency.
c. *Interpret* the results in the context of the data. *Answer: Page A32*

Picturing the World

The five top-selling vehicles in the United States for January of 2004 are shown in the following Pareto chart. One of the top five vehicles was a car. The other four vehicles were trucks. *(Source: Associated Press)*

How many vehicles from the top five did [illegible] of 2004?

Titled Examples

Every concept in the text is clearly illustrated with one or more step-by-step examples. Many examples have an interpretation step that shows the student how the solution may be interpreted within the real-life context of the example. Each of the more than 200 examples is numbered and titled for easy reference. In presenting the examples, we use an open format with a step-by-step display that students can use as a model when solving the exercises.

Try It Yourself

Each example in the text is followed by a similar exercise called *Try It Yourself.* Students can immediately practice the skill learned in the example. The answers to these exercises are given in the back of the book, and the worked-out solutions are given in the *Student's Solutions Manual.* The accompanying CD Lecture Videos contain clips of an instructor working out each *Try It Yourself* exercise.

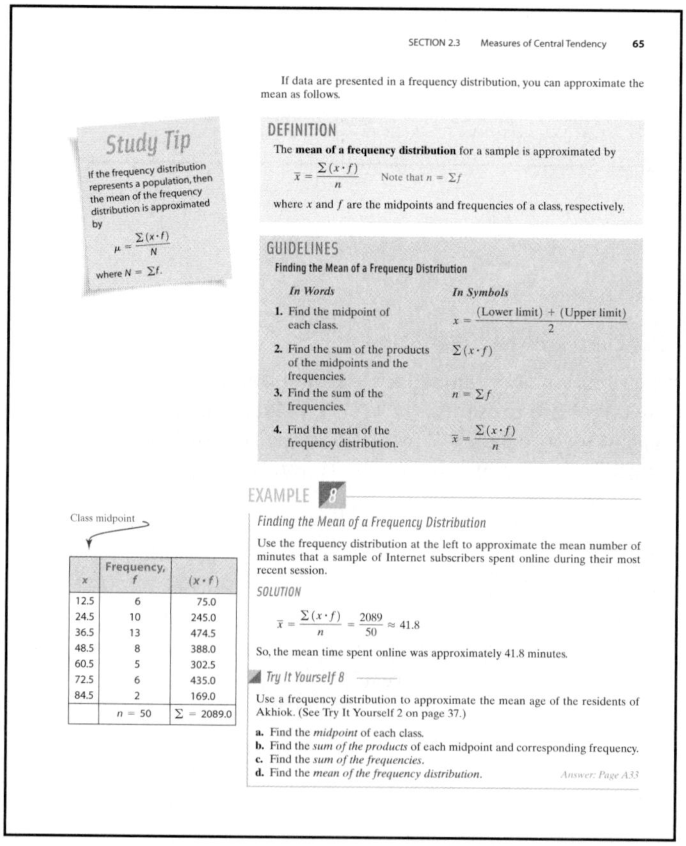

SECTION 2.3 Measures of Central Tendency 65

If data are presented in a frequency distribution, you can approximate the mean as follows.

DEFINITION

The **mean of a frequency distribution** for a sample is approximated by

$$\bar{x} = \frac{\Sigma(x \cdot f)}{n} \qquad \text{Note that } n = \Sigma f$$

where x and f are the midpoints and frequencies of a class, respectively.

GUIDELINES

Finding the Mean of a Frequency Distribution

In Words	*In Symbols*
1. Find the midpoint of each class.	$x = \frac{(\text{Lower limit}) + (\text{Upper limit})}{2}$
2. Find the sum of the products of the midpoints and the frequencies.	$\Sigma(x \cdot f)$
3. Find the sum of the frequencies.	$n = \Sigma f$
4. Find the mean of the frequency distribution.	$\bar{x} = \frac{\Sigma(x \cdot f)}{n}$

Study Tip

If the frequency distribution represents a population, then the mean of the frequency distribution is approximated by

$$\mu = \frac{\Sigma(x \cdot f)}{N}$$

where $N = \Sigma f$.

EXAMPLE 8

Finding the Mean of a Frequency Distribution

Use the frequency distribution at the left to approximate the mean number of minutes that a sample of Internet subscribers spent online during their most recent session.

Class midpoint

x	Frequency, f	$(x \cdot f)$
12.5	6	75.0
24.5	10	245.0
36.5	13	474.5
48.5	8	388.0
60.5	5	302.5
72.5	6	435.0
84.5	2	169.0
	$n = 50$	$\Sigma = 2089.0$

SOLUTION

$$\bar{x} = \frac{\Sigma(x \cdot f)}{n} = \frac{2089}{50} \approx 41.8$$

So, the mean time spent online was approximately 41.8 minutes.

Try It Yourself 8

Use a frequency distribution to approximate the mean age of the residents of Akhiok. (See Try It Yourself 2 on page 37.)

a. Find the *midpoint* of each class.
b. Find the *sum of the products* of each midpoint and corresponding frequency.
c. Find the *sum of the frequencies.*
d. Find the *mean of the frequency distribution.* *Answer: Page A33*

Picturing the World

Each section contains a real-life "mini case study" that illustrates the important concept or concepts of the section. Each *Picturing the World* concludes with a question and can be used for general class discussion or group work.

Definitions

The critical statistics definitions are set off with screens. Formal definitions are often followed by guidelines that explain, in everyday English, how to apply the definition.

Guidelines

Throughout the book, the presentation of a statistical formula is followed by a set of step-by-step guidelines for applying the formula. The guidelines are divided into two columns titled *In Words* and *In Symbols.*

Technology Examples

Many sections contain a worked example that shows how technology can be used to calculate formulas, perform tests, or display data. Screen displays from MINITAB, Excel, and TI-83 are given. Additional screen displays are given at the ends of selected chapters, and detailed instructions are given in separate technology manuals available with the book.

Insights

Most sections also contain one or more insights placed on blue "sticky notes" in the margin. The purpose of each insight is to help drive home an important interpretation or help connect different concepts.

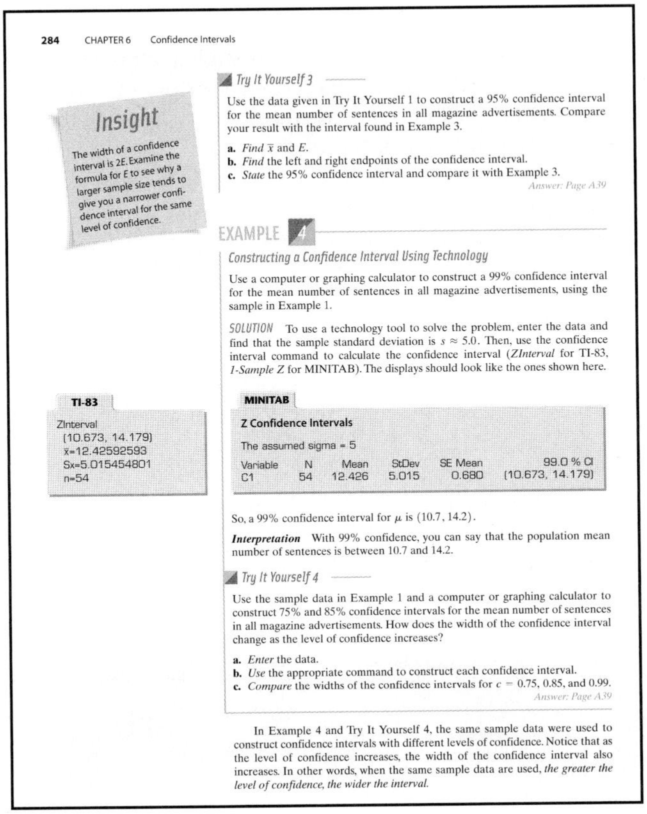

284 CHAPTER 6 Confidence Intervals

Insight

The width of a confidence interval is 2*E*. Examine the formula for *E* to see why a larger sample size tends to give you a narrower confidence interval for the same level of confidence.

Try It Yourself 3

Use the data given in Try It Yourself 1 to construct a 95% confidence interval for the mean number of sentences in all magazine advertisements. Compare your result with the interval found in Example 3.

a. *Find* $\bar{x}$ and E.
b. *Find* the left and right endpoints of the confidence interval.
c. *State* the 95% confidence interval and compare it with Example 3.

Answer: Page A39

EXAMPLE 4

Constructing a Confidence Interval Using Technology

Use a computer or graphing calculator to construct a 99% confidence interval for the mean number of sentences in all magazine advertisements, using the sample in Example 1.

SOLUTION To use a technology tool to solve the problem, enter the data and find that the sample standard deviation is $s \approx 5.0$. Then, use the confidence interval command to calculate the confidence interval (*ZInterval* for TI-83, *1-Sample Z* for MINITAB). The displays should look like the ones shown here.

TI-83

```
ZInterval
 (10.673, 14.179)
 x̄=12.42592593
 Sx=5.015454801
 n=54
```

MINITAB

Z Confidence Intervals

The assumed sigma = 5

Variable	N	Mean	StDev	SE Mean	99.0 % CI
C1	54	12.426	5.015	0.680	(10.673, 14.179)

So, a 99% confidence interval for μ is $(10.7, 14.2)$.

Interpretation With 99% confidence, you can say that the population mean number of sentences is between 10.7 and 14.2.

Try It Yourself 4

Use the sample data in Example 1 and a computer or graphing calculator to construct 75% and 85% confidence intervals for the mean number of sentences in all magazine advertisements. How does the width of the confidence interval change as the level of confidence increases?

a. *Enter* the data.
b. *Use* the appropriate command to construct each confidence interval.
c. *Compare* the widths of the confidence intervals for $c = 0.75, 0.85$, and 0.99.

Answer: Page A39

In Example 4 and Try It Yourself 4, the same sample data were used to construct confidence intervals with different levels of confidence. Notice that as the level of confidence increases, the width of the confidence interval also increases. In other words, when the same sample data are used, *the greater the level of confidence, the wider the interval.*

Section Exercise Sets

The section exercises are split into three categories, listed below. The exercises build from basic skill development to more challenging and interpretive problems. Most exercises are labeled for easy reference. Exercises labeled *Graphical Analysis* ask students to use the graphs provided to answer the questions. Almost all exercises are given in "paired format" so that the odd-numbered exercise, whose answer is given in the back of the book, is paired with an even-numbered exercise, whose answer is not given.

Building Basic Skills and Vocabulary These exercises are short answer, true and false, or vocabulary for students to test their knowledge of vocabulary and concepts from the section.

Using and Interpreting Concepts These exercises are skill or word problems carefully written to nurture student understanding and proficiency.

Extending Concepts These exercises go beyond the material presented in the section—they tend to be more challenging and are not required as prerequisites of subsequent sections.

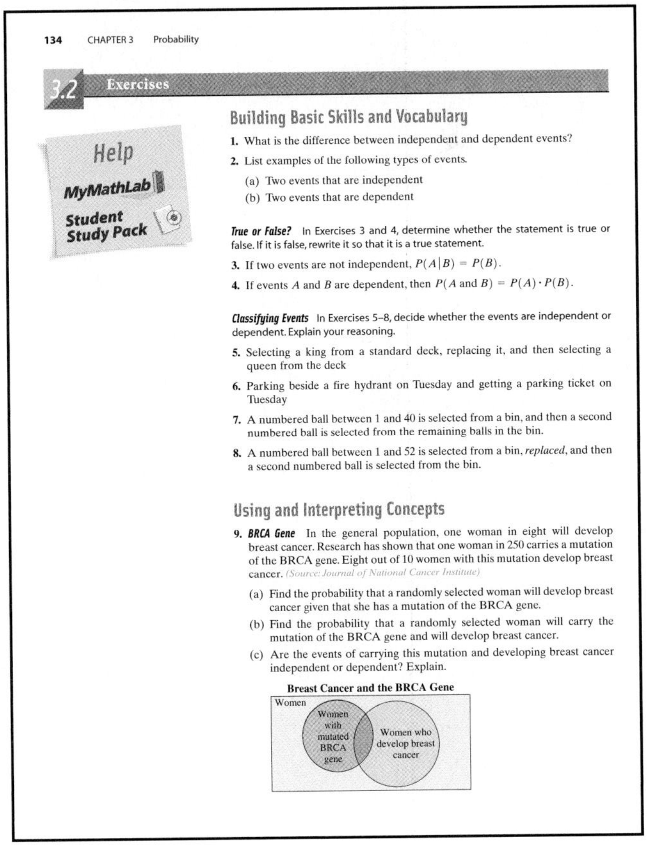

134 CHAPTER 3 Probability

3.2 Exercises

Help
MyMathLab
Student Study Pack

Building Basic Skills and Vocabulary

1. What is the difference between independent and dependent events?
2. List examples of the following types of events.
   (a) Two events that are independent
   (b) Two events that are dependent

True or False? In Exercises 3 and 4, determine whether the statement is true or false. If it is false, rewrite it so that it is a true statement.

3. If two events are not independent, $P(A|B) = P(B)$.
4. If events A and B are dependent, then $P(A \text{ and } B) = P(A) \cdot P(B)$.

Classifying Events In Exercises 5–8, decide whether the events are independent or dependent. Explain your reasoning.

5. Selecting a king from a standard deck, replacing it, and then selecting a queen from the deck
6. Parking beside a fire hydrant on Tuesday and getting a parking ticket on Tuesday
7. A numbered ball between 1 and 40 is selected from a bin, and then a second numbered ball is selected from the remaining balls in the bin.
8. A numbered ball between 1 and 52 is selected from a bin, *replaced*, and then a second numbered ball is selected from the bin.

Using and Interpreting Concepts

9. ***BRCA Gene*** In the general population, one woman in eight will develop breast cancer. Research has shown that one woman in 250 carries a mutation of the BRCA gene. Eight out of 10 women with this mutation develop breast cancer. *(Source: Journal of National Cancer Institute)*
   (a) Find the probability that a randomly selected woman will develop breast cancer given that she has a mutation of the BRCA gene.
   (b) Find the probability that a randomly selected woman will carry the mutation of the BRCA gene and will develop breast cancer.
   (c) Are the events of carrying this mutation and developing breast cancer independent or dependent? Explain.

Breast Cancer and the BRCA Gene

Chapter Case Study

Each chapter has a full-page Case Study featuring actual data from a real-world context and a series of thought-provoking questions that are designed to illustrate the important concepts of the chapter.

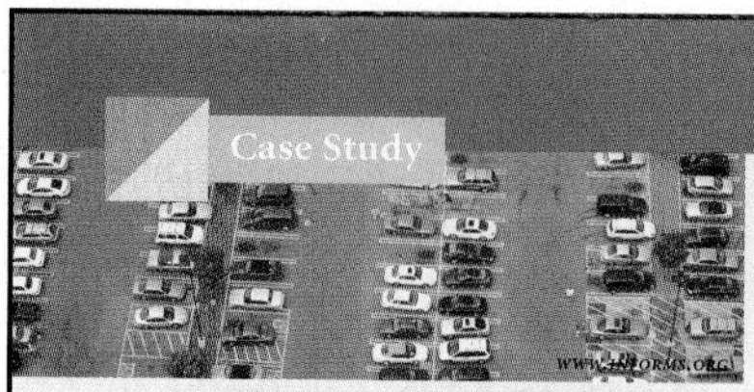

Probability and Parking Lot Strategies

The Institute for Operations Research and the Management Sciences (INFORMS) is an international scientific society with over 12,000 members. It is dedicated to the application of scientific methods to improve decision making, management, and operations. Members of the institute work primarily in business, government, and education. They represent fields as diverse as airlines, health care, law enforcement, the military, the stock market, and telecommunications.

One study published by INFORMS was the result of research conducted by Dr. C. Richard Cassady of Mississippi State University and Dr. John Kobza of Virginia Polytechnic Institute. The parking space study was conducted at a mall that has four entrances, seven rows with 72 spaces each, and directional restrictions. The researchers compared several parking lot strategies to see which strategy saves the most time. The two best strategies are called *Pick a Row* and *Cycling*. The results are shown in the table.

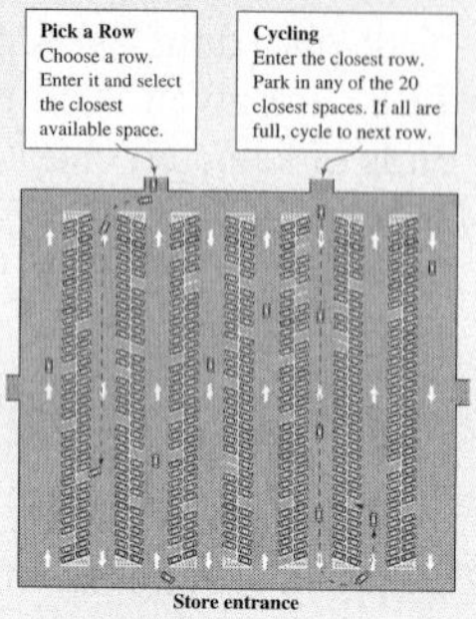

Time or Distance	Pick a Row	Cycling
Time from lot entrance to parking space	37.7 seconds	52.5 seconds
Time from lot entrance to store's door	61.3 seconds	70.7 seconds
Average walking distance to store	257 feet	208 feet

Exercises

1. In a parking lot study, is each parking space equally likely to be empty? Explain your reasoning.
2. According to the results of the study, are you more likely to spend less time using the Pick-a-Row strategy or the Cycling strategy? Explain.
3. According to the results of the study, are you more likely to walk less using the Pick-a-Row strategy or the Cycling strategy? Explain.
4. A key assumption in the study was that the drivers can see which spaces are available as soon as they enter a lane. Why is that important?
5. The parking lot is completely full, and one car leaves. What is the probability that the car was in the first row? Explain your reasoning.
6. A person is leaving from a row that is full. What is the probability that the person was parked in one of the 20 spaces that are closest to the store?
7. Draw a diagram of the parking lot. Color code the parking spaces into three categories of 168 spaces each: most desirable, moderately desirable, and least desirable. Assume that the parking lot is half full. Estimate the probability that you can find a parking space in the most desirable category. Explain your reasoning.

149

Uses and Abuses: Statistics in the Real World

Each chapter features an expanded discussion on how statistical techniques should be used, while cautioning students about common abuses. Exercises help students to apply their knowledge.

Uses and Abuses

Statistics in the Real World

Uses

Probability affects decisions when the weather is forecast, when marketing strategies are determined, when medications are selected, and even when players are selected for professional sports teams. Although intuition is often used for determining probabilities, you will be better able to assess the likelihood that an event will occur by applying the rules of classical probability and empirical probability.

For instance, suppose you work for a real estate company and are asked to estimate the likelihood that a particular house will sell for a particular price within the next 90 days. You could use your intuition, but you could better assess the probability by looking at sales records for similar houses.

Abuses

Assuming Probability Has a "Memory" One common abuse of probability is thinking that probabilities have "memories." For example, if a coin is tossed eight times, the probability that it will land heads up all eight times is only about 0.004. However, if the coin has already been tossed seven times and has landed heads up each time, the probability that it will land heads up on the eighth time is 0.5. Each toss is independent of all other tosses. The coin does not "remember" that it has already landed heads up seven times.

Adding Probabilities Incorrectly Another common abuse is adding probabilities incorrectly. For instance, suppose a company has 100 employees. Of these, 40 are women and 20 are minorities. If one employee is selected at random, the probability that the employee is a woman is 0.4 and the probability that the employee is a minority is 0.2. This does not mean, however, that the probability that a randomly selected employee is a woman or a minority is $0.4 + 0.2$ or 0.6. To determine this probability, you need to know how many employees belong to both groups, women and minorities.

Exercises

1. ***Assuming Probability Has a "Memory"*** A "Daily Number" lottery has a three-digit number from 000 to 999. You buy one ticket each day. Your number is 389.
   a. What is the probability of winning next Tuesday and Wednesday?
   b. You won on Tuesday. What's the probability of winning on Wednesday?
   c. You didn't win on Tuesday. What's the probability of winning on Wednesday?
2. ***Adding Probabilities Incorrectly*** A town has a population of 500 people. Suppose that the probability that a randomly chosen person owns a pickup is 0.25 and the probability that a randomly chosen person owns an SUV is 0.30. What can you say about the probability that a randomly chosen person owns a pickup or an SUV? Could this probability be 0.55? Could it be 0.60? Explain your reasoning.

161

Chapter Summary

Each chapter concludes with a Chapter Summary that answers the question *What did you learn?* This can be used as a study aid in conjunction with the chapter Review Exercises.

Chapter Review Exercises

A set of Review Exercises follows each Chapter Summary. The order of the exercises follows the chapter organization. Answers to all odd-numbered exercises are given in the back of the book. Worked-out solutions are available in the *Student's Solutions Manual.*

Chapter Quizzes

Each chapter ends with a Chapter Quiz. The answers to all quiz questions are provided in the back of the book. For additional help, see the step-by-step video solutions on the Chapter Quiz Prep Video.

---

162 CHAPTER 3 Probability

3 Chapter Summary

What did you learn?	*Review Exercises*
Section 3.1	
◆ How to identify the sample space of a probability experiment and to identify simple events	1–4
◆ How to distinguish among classical probability, empirical probability, and subjective probability	5–10
◆ How to identify and use several properties of probability	11, 12
Section 3.2	
◆ How to find conditional probabilities	13, 14
◆ How to distinguish between independent and dependent events	15, 16
◆ How to use the Multiplication Rule to find the probability of two events occurring in sequence $P(A \text{ and } B) = P(A)\cdot P(B\mid A)$ if events are dependent $P(A \text{ and } B) = P(A)\cdot P(B)$ if events are independent	17, 18
Section 3.3	
◆ How to determine if two events are mutually exclusive	19, 20
◆ How to use the Addition Rule to find the probability of two events $P(A \text{ or } B) = P(A) + P(B) - P(A \text{ and } B)$ $P(A \text{ or } B) = P(A) + P(B)$ if events are mutually exclusive	21–30
Section 3.4	
◆ How to use the Fundamental Counting Principle to find the number of ways two or more events can occur	31, 32
◆ How to find the number of ways a group of objects can be arranged in order and the number of ways to choose several objects from a group without regard to order $_nP_r = \frac{n!}{(n-r)!}, \frac{n!}{n_1!\cdot n_2!\cdot n_3!\cdots n_k!}, {_nC_r} = \frac{n!}{(n-r)!r!}$	33–36
◆ How to use counting principles to find probabilities	37–40

Review Exercises 163

3 Review Exercises

Section 3.1

In Exercises 1–4, identify the sample space of the probability experiment and list the outcomes of the event.

1. *Experiment:* Tossing four coins
   Event: Getting three heads
2. *Experiment:* Rolling 2 six-sided dice
   Event: Getting a sum of 4 or 5
3. *Experiment:* Choosing a one-digit number
   Event: Choosing a number less than 2
4. *Experiment:* Guessing the gender of the three children in a family
   Event: The family has two boys

In Exercises 5–10, classify the statement as an example of classical probability, empirical probability, or subjective probability.

5. On the basis of prior counts, a quality control officer says there is a 0.05 probability that a randomly chosen part is defective.
6. The probability of randomly selecting five cards of the same suit (a flush) from a standard deck is about 0.0005.
7. The chance that Corporation A's stock price will fall today is 75%.
8. The probability of a person from the United States being left handed is 11%.
9. The probability of rolling 2 six-sided dice and getting a sum greater than nine is $\frac{1}{6}$.
10. The chance that a randomly selected person in the United States is between 15 and 24 years old is about 14%. *(Source: U.S. Census Bureau)*

In Exercises 11 and 12, the table shows the approximate U.S. age distribution for the year 2005. Use the table to determine the probability of the event. *(Source: U.S. Census Bureau)*

Age	19 and under	20–34	35–59	60–84	85 and over
Population	28%	20%	35%	15%	2%

11. What is the probability that a randomly selected person in the United States will be at least 20 years old?
12. What is the probability that a randomly selected person in the United States will be less than 60 years old?

Chapter Quiz 167

3 Chapter Quiz

Take this quiz as you would take a quiz in class. After you are done, check your work against the answers given in the back of the book.

1. The table shows the estimated number (in thousands) of earned degrees conferred in the United States in the year 2004 by level and gender. *(Source: National Center for Education Statistics)*

		Gender		
		Male	Female	Total
Level of Degree	Associate	231	401	632
	Bachelor's	553	769	1322
	Master's	197	270	467
	Doctorate	25	20	45
	Total	1006	1460	2466

A person who earned a degree in the year 2004 is randomly selected. Find the probability of selecting someone who

(a) earned a bachelor's degree.
(b) earned a bachelor's degree given that the person is a female.
(c) earned a bachelor's degree given that the person is not a female.
(d) earned an associate degree or a bachelor's degree.
(e) earned a doctorate given that the person is a male.
(f) earned a master's degree or is a female.
(g) earned an associate degree and is a male.
(h) is a female given that the person earned a bachelor's degree.

2. Decide if the events are mutually exclusive. Then decide if the events are independent or dependent. Explain your reasoning.
   Event *A*: A golfer scoring the best round in a four-round tournament
   Event *B*: Losing the golf tournament
3. A shipment of 150 television sets contains three defective units. Determine how many ways a vending company can buy three of these units and receive (a) no defective units, (b) all defective units, and (c) at least one good unit.
4. In Exercise 3, find the probability of the vending company receiving (a) no defective units, (b) all defective units, and (c) at least one good unit.
5. The access code for a car's security system consists of four digits. The first digit cannot be 0 and the last digit must be even. How many different codes are available?
6. From a pool of 25 candidates, the offices of president, vice-president, secretary, and treasurer will be filled. In how many different ways can the offices be filled?

Real Statistics–Real Decisions

This feature encourages students to think critically and make informed decisions about real-world data. Exercises guide students from interpretation to drawing conclusions.

PUTTING IT ALL TOGETHER

Real Statistics ■ Real Decisions

You work for the company that runs the Powerball lottery. Powerball is a lottery game in which five white balls are chosen from a drum containing 53 balls and one red ball is chosen from a drum containing 42 balls. To win the jackpot, a player must match all five white balls and the red ball. Other winners and their prizes are also shown in the table.

Working in the public relations department, you handle many inquiries from the media and from lottery players. You receive the following e-mail.

> *You list the probability of matching only the red ball as* 1/70. *I know from my statistics class that the probability of winning is the ratio of the number of successful outcomes to the total number of outcomes. Could you please explain why the probability of matching only the red ball is* 1/70?

Your job is to answer this question, using the probability techniques you have learned in this chapter to justify your answer. In answering the question, assume only one ticket is purchased.

www.musl.com

Powerball Winners and Prizes

Match	Prize	Approximate Probability
5 white, 1 red	Jackpot	1/120,526,770
5 white	$100,000	1/2,939,677
4 white, 1 red	$5,000	1/502,195
4 white	$100	1/12,249
3 white, 1 red	$100	1/10,685
3 white	$7	1/261
2 white, 1 red	$7	1/697
1 white, 1 red	$4	1/124
1 red	$3	1/70

(Source: Multi-State Lottery Association)

Exercises

1. ***How Would You Do It?***
   (a) How would you investigate the question about the probability of matching only the red ball?
   (b) What statistical methods taught in this chapter would you use?
2. [illegible]
   Write an explanation that answers the question about the probability of matching only the red ball. Include in your explanation any probability formulas that justify your explanation.
3. ***Another Question***
   You receive another question asking how the overall probability of winning a prize in the Powerball lottery is determined. The overall probability of winning a prize in the Powerball lottery is 1/36. Write an explanation that answers the question and include any probability formulas that justify your explanation.

Where Is Powerball Played?
Powerball is played in 26 states, Washington, D.C., and the U.S. Virgin Islands

(Source: Multi-State Lottery Association)

Chapter Technology Project

Each chapter has a full-page Technology project using tools from MINITAB, Excel, and TI-83 that gives students additional insight into the way technology is used to handle large data sets or complex, real-life questions.

TECHNOLOGY MINITAB EXCEL TI-83

Simulation: Composing Mozart Variations with Dice

Wolfgang Mozart (1756–1791) composed a wide variety of musical pieces. In his Musical Dice Game, he wrote a Wiener minuet with an almost endless number of variations. Each minuet has 16 bars. In the eighth and sixteenth bars, the player has a choice of two musical phrases. In each of the other 14 bars, the player has a choice of 11 phrases.

To create a minuet, Mozart suggested that the player toss 2 six-sided dice 16 times. For the eighth and sixteenth bars, choose Option 1 if the dice total is odd and Option 2 if it is even. For each of the other 14 bars, subtract 1 from the dice total. The following minuet is the result of the following sequence of numbers.

5	7	1	6	4	10	5	1
6	6	2	4	6	8	8	2

Exercises

1. How many phrases did Mozart write to create the Musical Dice Game minuet? Explain.
2. How many possible variations are there in Mozart's Musical Dice Game minuet? Explain.
3. Use technology to select randomly a number from 1 to 11.
   (a) What is the theoretical probability of each number from 1 to 11 occurring?
   (b) Use this procedure to select 100 integers between 1 and 11. Tally your results and compare them with the probabilities in part (a).
4. What is the probability of randomly selecting options 6, 7, or 8 for the first bar? For all 14 bars? Find each probability using (a) theoretical probability and (b) the results of Exercise 3(b).
5. Use technology to select randomly two numbers from 1, 2, 3, 4, 5, and 6. Find the sum and subtract 1 to obtain a total.
   (a) What is the theoretical probability of each total from 1 to 11?
   (b) Use this procedure to select 100 totals between 1 and 11. Tally your results and compare them with the probabilities in part (a).
6. What is the probability of randomly selecting options 6, 7, or 8 for the first bar? For all 14 bars? Find each probability using (a) theoretical probability and (b) the results of Exercise 5(b).

Extended solutions are given in the *Technology Supplement*.
Technical instruction is provided for MINITAB, Excel, and the TI-83.

How to Study Statistics

Study Strategies

Congratulations! You are about to begin your study of statistics. As you progress through the course, you should discover how to use statistics in your everyday life and in your career. The prerequisites for this course are two years of algebra, an open mind, and a willingness to study. When you are studying statistics, the material you learn each day builds on material you learned previously. There are no shortcuts—you must keep up with your studies every day. Before you begin, read through the following hints that will help you succeed.

Making a Plan Make your own course plan right now! A good rule of thumb is to study at least two hours for every hour in class. After your first major exam, you will know if your efforts were sufficient. If you did not get the grade you wanted, then you should increase your study time, improve your study efficiency, or both.

Preparing for Class Before every class, review your notes from the previous class and read the portion of the text that is to be covered. Pay special attention to the definitions and rules that are highlighted. Read the examples and work through the Try It Yourself exercises that accompany each example. These steps take self-discipline, but they pay off because you will benefit much more from your instructor's presentation.

Attending Class Attend every class. Arrive on time with your text, materials for taking notes, and your calculator. If you must miss a class, get the notes from another student, go to a tutor or your instructor for help, or view the appropriate CD Lecture Video. Try to learn the material that was covered in the missed class before attending the next class.

Participating in Class When reading the text before class, reviewing your notes from a previous class, or working on your homework, write down any questions you have about the material. Ask your instructor these questions during class. Doing so will help you (and others in your class) understand the material better.

Taking Notes During class, be sure to take notes on definitions, examples, concepts, and rules. Focus on the instructor's cues to identify important material. Then, as soon after class as possible, review your notes and add any explanations that will help to make your notes more understandable to you.

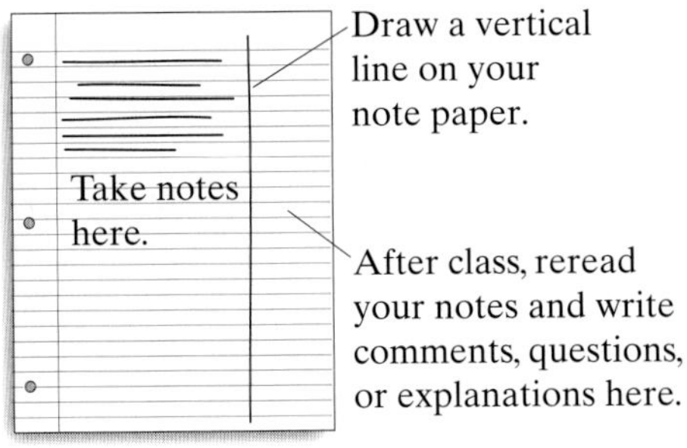

(continued on next page)

Doing the Homework Learning statistics is like learning to play the piano or basketball. You cannot develop skills just by watching someone do it; you must do it yourself. The best time to do your homework is right after class, when the concepts are still fresh in your mind. Doing homework at this time increases your chances of retaining the information in long-term memory.

Finding a Study Partner When you get stuck on a problem, you may find that it helps to work with a partner. Even if you feel you are giving more help than you are getting, you will find that teaching others is an excellent way to learn.

Keeping Up with the Work Don't let yourself fall behind in this course. If you are having trouble, seek help immediately—from your instructor, a statistics tutor, your study partner, or additional study aids such as videotapes and software tutorials. Remember: If you have trouble with one section of your statistics text, there's a good chance that you will have trouble with later sections unless you take steps to improve your understanding.

Getting Stuck Every statistics student has had this experience: You work a problem and cannot solve it, or the answer you get does not agree with the one given in the text. When this happens, consider asking for help or taking a break to clear your thoughts. You might even want to sleep on it, or rework the problem, or reread the section in the text. Avoid getting frustrated or spending too much time on a single problem.

Preparing for Tests Cramming for a statistics test seldom works. If you keep up with the work and follow the suggestions given here, you should be almost ready for the test. To prepare for the chapter test, review the Chapter Summary and work the Review Exercises. Then set aside some time to take the sample Chapter Quiz. Analyze the results of your Chapter Quiz to locate and correct test-taking errors.

Taking a Test Most instructors do not recommend studying right up to the minute the test begins. Doing so tends to make people anxious. The best cure for test-taking anxiety is to prepare well in advance. Once the test begins, read the directions carefully and work at a reasonable pace. (You might want to read the entire test first, then work the problems in the order in which you feel most comfortable.) Don't rush! People who hurry tend to make careless errors. If you finish early, take a few moments to clear your thoughts and then go over your work.

Learning from Mistakes After your test is returned to you, go over any errors you might have made. Doing so will help you avoid repeating some systematic or conceptual errors. Don't dismiss any error as just a "dumb mistake." Take advantage of any mistakes by hunting for ways to improve your test-taking skills.

Index of Applications

Biology and Life Sciences

Air quality, 68
Air pollution, 326
Atlantic croaker fish, 45, 233
Birth weights in America, 245
Black bears, 294
BRCA gene, 134
Breastfeeding, 584
Brown trout, 200
Butterfly wingspan, 101
Cats, 89, 164, 181
Child delivery, 143
Chrysanthemums, plant hormone, 26
Citrus growers, 397
Conscious consumers, 378
Cow milk, 501
Crawling, 485, 506
Cuckoos, 546
DNA testing, 275
Dogs, 69, 108, 127, 137, 164, 181, 183
Elephants, 57
Farm acreage, 47
Farm values, 87, 88, 273
Fish, 13, 121, 124, 235, 496
Fisher's iris data set, 54
Fruit flies, 103
Gray whale, 314
Green turtle migration, 272, 273
House flies, 58, 255
Kangaroos, 26
Kitti's hog-nosed bat, 272, 273
Lamb, 594
Mice, 448, 449
Newborn babies, 606
Nitrogen dioxide, 361
Oat, 595
Pets, 6, 88, 304
Plants, 196
Pregnancy study, Cebu Philippines, 26
Rabbits, 200
Rhesus monkey, 44, 233, 235
Salmon, 123, 131–133
Seawater, 318
Seedling, 288
Snapdragon flowers, 127
Soil, 538
Soybean, harvest, 21
Sunflower yield, 505
Storing fish, 4
Trees, 44, 47, 154, 158, 218, 250, 255, 289, 480, 497
Waste, 301, 366, 371
Water, 228, 318
  conductivity, 368
  consumption, 89
  pH level, 368
  pollution, 27
Weight of newborns, 223, A6
Wheat, 288, 559, 595

Business

Advertisements, 57, 487, 553
Advertising expense, 459, 462, 467, 489
Advertising and sales, 475, 477, 485
Beverage company, 129
Board of directors, 153
Bookbinding defects, 138
Book store sales, 179
Business failures, 203
Clothing store purchases, 194
Consumer ratings, 435
Defects parts, 137, 146, 159, 163, 167, 606
Free samples, 379
Internet advertisers, 557
Inventory shrinkage, 53
Magazine advertisement, 280, 282–284, 286
Manufacturer
  claims, 226
  earnings, 101, 102
  performance ratings, 435
Online shopping, 439
Product assembly, 158
Quality control, 21, 27, 30, 31, 119, 181
Sales, 27, 46, 59, 110, 111, 142, 172, 174, 176, 177, 198, 199, 203, 351, 491, 492, 498, 546
Salesperson, 46, 100
Shipping errors, 344
Small business
  managers, 25
  websites, 190
Software piracy, 330, 331
Wal-Mart shareholder's equity, 498
Warehouses, 138
Website costs, 318

Combinatorics

Answers guessing, 194
Area code, 159, 166
Letters, 156, 158
License plate, 151, 158, 166
Number generator, 606
Password, 158
Security code, 152, 157, 166, 167

Computers

Calculators, 166
Computer, 7, 8, 181, 183, 191, 196, 223, 229, 233–235, 289, 300, 309, 344, 371,
Digital cameras, 255
Disk drive, 554
Email addresses, 207
Internet, 3, 35, 37–42, 65, 136, 164, 165, 209, 210, 309, 323, 345, 404, 526
Microchips, 204
Microsoft Internet Browser, 264
Monitor, 419

Demographics

Age, 5, 25, 27, 56, 71, 141, 145, 469, 514, 613
Bride's age, 84, 576
Children per household, 82, 180
Children's heights, 100
City rent, 274
Drive to work, 69, 179, 190
Home, 267
Education, 558, 561
Employee, 44, 60, 122, 124, 127, 158, 159, 161, 166, 180, 255
Eye color, 141
Groom's age, 85, 576
Gun control, 20, 23
Height
  of family, 502
  of men, 44, 71, 72, 80, 103, 226, 233, 243, 256, 425, 425, 486
  of women, 71, 80, 243, 256
Height and weight, 215
Household, 183, 448, 560
Juror's age, 557
Left handed, 147–148, 163
Marriage, 1, 5, 123, 519, 520
Most admired polls, 327
Moving out, 444
New car, 150
New home prices, 111
Population
  Akhiok, Alaska, 33, 36, 37, 39–41, 50, 51, 61, 62, 65, 93, 95–97, 128
  Alaska, 81
  Florida, 81
  Japan, 90
  U.S., 9, 90
West Ridge County, 18–20
Race relations, 136
Retirement age, 47
Secondary school students, 165
Senior citizens, 17
Shoe size, 45, 461, 480, 486
Sleeping, 502
Student ages, 290
U.S. age distribution, 147, 163, 277
U.S. unemployment rate, 108
Weight
  of adults, 432
  of family, 502
Women's attitudes, 515
Zip codes, 25

Earth Science

Clear days, May, San Diego, CA, 192
Climate conditions, 171
Cloudy days, June, Pittsburgh, PA, 192
Global warming, El Ninos, 14
Hurricane, 182, 203
  relief efforts, 21
NASA
  budget, 58
  expenditures, 58
Old Faithful, Yellowstone National Park, 40, 88, 211, 258, 460, 463, 465, 476, 477, 485
Precipitation, 202, 203,
  Anchorage, AL, 292

Nome, AL, 293
Sacramento, CA, 12
Rain, 613
Rainy days in July, FL, 179
Saffir-Simpson Hurricane Scale, 211
Sea water, 288
Shark attacks, 210
Snowfall, 604
Alaska, 179
Evansville, IN, 204
January average, 14
New York county, 255
Space shuttle
exploration, 25
flights, 110
menu, 157
Sunny and rainy days, 171, 175
Seattle, WA, 129
Temperature, 13, 606
Chicago, IL, 107
Mohave, AZ, 25
Pittsburgh, PA, 572
Sacramento, CA, 12
San Diego, CA, 572
UV Index, 58
Memphis, TN, 59
Water temperatures, 590
Wet or dry, Seattle, WA, 129

Economics and Finance

Amount spent, 553, 560
ATM machine, 25, 46, 48, 182, 537
Audit, 31, 146
Bond, 345, 492, 493
Book spending, 45
Charitable donations, 459, 462, 465, 475
Children's savings accounts, 274
Credit card, 253, 265, 322, 323, 373, 407, 575
Credit card debt, 573, 609
Debt and income, 596
Dividends and earnings, 471, 473
Dow Jones Industrial Average, 173
Financial debt, 573
Forecasting earnings, 5
Economic growth, 322
Emergency savings, 137
Ethical business leaders, 323
Expenditures, 550
Financial shape, 159
Franchise, 351
Home owner income, 7
Honeymoon financing, 195
Income taxes, 396
Investment professionals, 310, 311
Lodging cost, 317, 369, 373, 388, 411, 540
Missing tax deductions, 323
Money managing, 7
Monthly income, 318s
Mortgage, 298
Mutual funds, 282
Paycheck errors, 204
Personal income, 505, 551
Prices of jeans, 56
Primary investor in household, 7
Raising a child, cost, 356
Restaurant spending, 373, 411
Retirement income, 187
Salaries, 4, 6, 7, 25, 59, 67, 70, 74–77, 85, 109, 111, 256, 273, 301, 355, 360, 372, 388, 397, 413, 494, 496, 540, 563, 583, 584, 590–592, 611, 613
Ann Arbor, MI, 561
Chicago, IL, 77
Dallas, TX, 86, 561
Houston, TX, 86
Long Beach, CA, 86
Los Angeles, CA, 86
San Jose, CA, 561
Spending before traveling, 83
Stock, 57, 68, 105, 128, 163, 208, 209, 288, 490, 492, 538, 609
McDonald's, 497
U.S. exports, 72
U.S. income and economic research, 615
Utility bills, 98, 234, 235, 239
Vacation cost, 7, 369, 394, 397, 408

Education

Accounting exam scores, 13
ACT, 8, 100, 227, 233, 302, 411
Ages of students, 285
Astronomy majors, 25
Average starting salary for an MBA, 4
Biology, 102, 301, 325
Books, 288
Business majors, 146
Canadians with college degrees, 450
Class size, 372
Classes, 63
College courses, 145
College freshman grades, 528
College graduates, 267, 334, 339, 340
College major, 7
College students, 344
Completing a test, 173
Continuing, 529
Day care, 413
Degrees, 586
Degrees and gender, 167, 595
Doctorate degree, 573, 610, 611
Dormitory room prices, 109
Economics degrees, 301
Educational attainment, 519
Engineering, 444
Enrollment, 207
Extracurricular activities, 182
Faculty hours, 372, 373, 395
Final exam, 425
Final grade, 70, 495, 496
Genders of students, 602
GPA, 69, 301, 468, 547, 554, 606
Grading public schools, 274
Heights of students, 110
Internship by graduation, 2
Life science assessment tests, 387
Literary skills, 344
Majors, 450
Mathematics assessment test, 275, 359
Midterm scores, 425
Musical training, 417
Nobel Prize, 58
Nursing major, 135, 140
Ph.D., 121
Physical science assessment, 387, 540
Plus/minus grading, 164
Preschool, 413
Public schools, 529
Quiz, 157, 182, 185
Reliability of testing, 139
Safety, 528
SAT scores, 45, 48, 86, 97, 227, 232, 234, 258, 301, 397, 431, 432, 451, 499, 559, 574, 575
School board, 23
Science assessment, 451
Social science, 444, 573
Student ID numbers, 27
Student loans, 469
Student sleep habits, 395
Study habits, 26, 412, 470, 472, 479
Supplies, 372
Survey of students, 6
Tardiness, 606
Teaching methods, 22, 423, 448
Test grades/scores, 27, 47, 64, 67, 70, 102, 108, 110, 244, 425, 461, 471, 472, 520, 540
Test scores and GNP, 597
Textbooks, 254
Tuition, 68, 94, 95
Undergraduate program, 146
Vocabulary, 470, 479

Engineering

Ball bearings, 226
Beverage cans liquid content, 324
Bolts, 302, 317, 396
Brick mortar, 46
Building heights, Miami, FL, 107
Building space, 551
Cooling capacity, 478
Gears, 236
Liquid dispenser, 396, 599
Liquid volume of cans, 107
Machine
calibrations, 258
cuts, 276

part supplier, 129
Mold, 560
Nails, 236
Nut, 412
O-ring damage, 565
Plastic sheet cutting, 291
Repairs, 159
Resistors, 271
Steel pipe fitting, 559
Tensile strength, 422
Volume of bottled water, 173
Washers, 412

Entertainment

Best-selling novel, 121
Box office rank, musical artists, 14
Cable, 7, 109, 191
news, 377
Cards, 120, 125, 129, 130–132, 134, 141, 146, 156, 163, 165, 166, 184, 186
CDs, 154
Chess, 344
Coin toss, 31, 118, 122, 131–133, 161, 163, 164, 209, 335, 453, 605
Concert tickets, 157, 14
Daily lottery, 204
Die roll, 31, 67, 72, 118–120, 124, 125, 129, 131–133, 140–142, 146, 163, 165, 186
Games of chance, 183
Guessing, 138
Horse race, 153, 157, 158
Lottery, 125, 156, 158, 159, 161, 194
Magazine, 8, 108, 166, 460
Meal costs, 411
Monopoly game, 131
Motion Picture Association, ratings, 10
Movie ticket prices, 109
Movies, 10, 147, 210, 254
Music preference, 510, 511, 513
Network news broadcast, 377
Networks, Portland, OR, 10
Newspaper, 45, 289
Nielsen Media ratings, 15, 22
Portable CD player, 596
Powerball lottery, 168
Puzzle, 17
Radio stations, 12, 110, 334, 339, 341
Raffle ticket, 123, 166, 178
Reading, 189
Summer vacation, 135
Television, 10, 101, 102, 110, 167, 208, 216, 334, 411, 413, 443, 471, 472, 479, 612
The Price Is Right, 117
The Tonight Show, 121
VCR, 25, 289
Videotape or DVD rentals, 209

Food and Nutrition

Apple, 57, 243
Beef, 594
Breakfast, 378
Caffeine, 89, 360, 468
Calories, 345, 424, 447, 501, 502
Candy, 569
Carbohydrates, 424, 541
Cereal, 244, 318, 324, 334, 339, 361, 393, 507
Cheese, 290
Chicken wings, 209
Coffee, 72, 89, 297, 298, 425, 471, 516
Cookies, 195, 267, 244
Egg prices, 59
Family dining, 2
Fat substitute, 21
Fruit consumption, 273, 442
Genetically modified foods, 378
Grocery store, 110, 302
Hamburger, 395
Hot dogs, 188, 480
Ice cream, 244, 257
M&M's, 209
Meat, 202, 255, 310, 514, 515
Menu, 157
Milk, 383, 461
containers, 256
production, 113
Oranges, 56, 243
Peanuts, 235, 497
Pepper pungencies, 46
Pizza, 158, 350
Protein, 424, 478
Restaurant, 527
Arby's, 424
Long John Silver's, 447
McDonald's, 424, 447
meals purchased, 107
serving, 384
Rice, 497
Safety, 26, 308
Saturated fat intake, 47
Soda, 372
Soft drinks, 210, 255
Soup, 393
Sport drink, 383
Sugar, 211, 501, 502
Supermarket, 89, 230, 568
T-Bone steak, price, 59
Tea drinker, 128, 360
Tuna, 360
Vegetable consumption, 273, 442
Vending machine, 167, 207, 244,
Water, 291, 459, 470

Government

Armed forces, 308
Better Business Bureau, 53
Civil service exams, 240
Congress
ages, 21
gender profile, 14
override veto, 121
Courtroom, 342
Department of Energy, gas prices, 3
Entitlements, 517
Federal employees' salaries, 406
Governor ages, 6
Legal system in U.S., 336
Registered voters, 6, 31, 145
Social Security Administration, 25, 398
State tax collection, 14
Tax cuts, 517
U.S Department of Labor, 345
U.S Post Office, 345
U.S. Census, 4
undercount, 17

Health and Medicine

Allergies, 375, 468
Allergy medicines, 315
Alternative medicines, 438, 441
Antidepressants, 442
Anxiety, 443
Appetite suppressant, 426
Arthritis medication, 445
Assisted reproductive technology, 136, 212
Backaches, 443
Blood, 138, 179
donations, 140, 143, 180, 181
drives, 195
pressure, 216, 307, 435, 470, 479
type, 125, 138, 195, 266
BMI, 70
Body measurements, 483
Body temperature, 11, 342, 363, 430, 469
Calcium supplements, 449, 583, 609
Cholesterol, 6, 68, 231, 232, 234, 235, 241, 290, 424, 440, 443, 541
Chronic fatigue syndrome, 529
Colds, 333, 571
Cough syrup, 316, 318
CPR training, 93, 95, 96
Cyanosis, 194
DASH, 415
Dentist, 225
Diabetic, 16, 443
Diet, 26, 209
Doctor visits, 511, 557
Doctors aid dying patients, 262
Drinking habits, 21
Drug testing, 125, 185, 268, 436
ER visits, 344
Exercise, 21, 111, 202, 273, 431, 526
Flu, 325
Fluorouracil, 414
Generic medicine, 527
Growth of a virus, 21
Headaches, 434, 574, 584, 611

Health club, 234, 395
Healthy foods, 22, 165, 166
Heart, 17, 345
Heart rate, 11, 70, 131, 299, 429, 431
Heart rhythm abnormality, 7
Heart transplant, 244
Herbal medicine, 414, 436, 449, 610
HIV, 396
Hospital beds, 72
Hospital costs, 388
Hospital length of stay, 387, 550, 592
Hospital waiting times, 302, 387, 540
Hospitalization insurance, 71
Intravenous, 537
Irinotecan, 414
Kidney cancer survival rate, 274
Lead levels, 27
Managed health care, 25
Medicine cabinets, 263
Nonprescription pain relievers, 393
No trouble sleeping, 195
Organ donors, 265
Outpatient care, 21
Over-the-counter drugs, 376
Overweight, 308
Pain relievers, 545, 548
Patient care, 240
Physical examination, 5
Placebo, 527
Pneumonia, 452
  and chicken pox, 579
Prescription drugs, 21
Prostate cancer, 2, 265
Pulse rate, 317
Recovery time, 342
Registered nurse salaries, 111, 256
Resting heart rates, 425
Salmonella, 336
Sleep, 8, 27, 70, 471, 480
Smoking, 25, 27, 128, 308, 311, 344, 360, 376, 378, 402, 403, 439, 440, 442, 443, 491, 492, 504, 579
Stroke, increase, 8
Sudden infant death syndrome, 7
Sunscreen, children, 27
Surgery, 194
  procedure, 185
  survival, 136
  technique, 211
Vitamins, 311, 316, 318
Weight, 56, 67, 225, 431, 459
Weight loss, 225, 361, 414, 426, 433, 434, 470

Housing and Construction

City house value, 274
Construction, 299
Home insurance, 591, 613
Home repairs, 308
Housing costs, 216
Prices of homes, 63, 252, 551, 552, 569, 572
Real estate, 161, 379, 537
Residence, rent or own, 22
Room and board, 252
Security system, 151, 157, 344
Square footage, 573
Subdivision, 154
Tacoma Narrows Bridge, 147, 200
Unit size, 573

Law

Case of the vanishing women, 399
Child support, 249
Crime, 68, 471, 473
Fraudulent insurance claims, 121
Jury selection, 156, 158
Radar gun, 26
Repeat offenders, 571
Speeding, 131, 612
Telemarketing fraud, 562
Victim, serious crime, 121

Miscellaneous

Air conditioners, 60–62, 503, 597
Appliances, 387, 539
Ball, numbered, 134
Bank customers, 568
Battery, 31, 235, 344
Birthday, 119, 138, 139
Blackouts, California, 179
Camcorder, 309
CD player, 322
Cell phone, 8, 25, 55, 57, 332, 482
Cheaters, 202
Chemical compound, weight, 179
Chlorine levels in a pool, 383
Clocks, 344
Coauthored articles, 530
Consumption
  energy, 552, 592
  fuel, 539
Cordless phone, 612
Deck treatment, 550
Direct-mail advertising, 308
Dishwashers, 325
Dreams, 345
DVD player, 317
Electricity usage, 362
Extraterrestrial being sightings, 376
Fluorescent lamps, 361
Gas grill, 188, 503, 288
Greeting cards, 207
Lawn mower, 317
Length of phone calls, 325
Light bulbs, 137, 302, 344, 361
Liquid dispenser, 236
Mail-order company, 609
Marbles, 185
Microwave, 300, 362, 371, 397, 410
Mozart, 169
New Year's resolution, 260, 262, 263, 308
Pages, section, 208
Paint cans, 256, 290
Paint sprayer, 291
Parachute assembly, 336
Perfume, 274
Phone numbers, 10, 13
Picture development, 345
Power failure, 68
Product warning label, 21
Queuing models, 213
Random number selection, 125
Refrigerator, 345
Socks, 164
Sound-system receivers, 253
Spray on water repellent, 579
Suitcases, 597
Sunglasses, 92
Telephone 47, 71, 125, 209, 249, 419
Toaster, 302
Toothpaste, 164, 205, 549
Travel concerns, 522, 525
Typographical error, 203, 211
Vacuum cleaner, 410
Waking times, 321
Washing cars, 273
Washing machine, 125, 288
Water faucet, 334, 339
Winning a prize, 203
Writing, 148, 196

Mortality

Airplane accidents, 197
Accidents, 181, 199, 200
Alcohol-related accidents, 308, 451, 530, 558
Blood alcohol level, 530
Cancer survival times, 344
Car accident, 131, 175, 470
Emergency response time, 45, 384
Fatal work injuries, 519
Lightning deaths, 210
Memorial Day fatalities, 2
Motor vehicle casualties, 52, 179, 477, 508, 531, 558
Pedestrian motor vehicle collisions, 558

Motor Vehicles and Transportation

Age of vehicles, 493
Air travel, 147
Airfare, 292
Airplanes, 17, 69, 110
Annual vehicle miles, 179
Auto parts, 203
Automobile insurance, 112, 367, 388, 582
Automobiles, 26, 342, 350
  battery, 316, 334, 339, 359, 549

four-door, 389
Bicycle, 517, 518
Brakes, 257
Braking distance, 240, 272, 340, 410, 418, 431, 454, 472, 481
Bumper, 421
Bus, 612
Car occupancy, 147, 182, 183, 200
Chevrolet Blazer, 232
Compact cars, consumer testing, 73
Convertible, 501
Crash test, 17, 421, 508, 509
Drivers, 57, 306
Driver's license exam, 164
Driving time, 251
Driving to work, 268
Engine, 239, 502, 503, 600
Flights, 139, 547
Footwell intrusion, 421
Fuel additive, 585
Fuel economy ratings, 278, 279, 504
Fuel-efficient cars, 557
Gas prices, 47, 256, 258
Luxury car, 8
Mileage, 229, 345, 355, 362, 373, 387, 397, 431, 433
Motor company departments, 26
New highway, 155
Oil tankers, 202
Parking, 149
Parking ticket, 134
Pickup trucks, 135
Pilot's test, 202
Power boats, 431
Price of a car, 9, 344, 345, 367
Public transportation, 266
Rental car company mileage, 109
Repeat car buyers, 344
Road rage, 195
Seat belts, 344
Speed of vehicles, 6, 26, 68, 69, 98, 350, 612
Sports cars, 68, 302
Stressful travel, 308
SUV, 318
seating, 68
Taxi cab, 345
Tires, 103, 227, 244, 257, 344, 385, 393
Top-selling vehicles, 53
Trade-in value, 569
Traffic safety, 532
Traffic signal, 254
Travel concerns, 522, 525
Travel plans, 309
Vehicle
crashes, 530
manufacturers, 141
owned, 181, 527, 558
purchased, 68
sales, 553

Political Science

108th Congress, 13, 14
American voters, age, 126
Campaign spending, 492, 493
Candidate's approval ratings, 308
Council members, 605
Grading our leaders, 528
Gubernatorial Election, Michigan, 126
Judge donations, Sioux Falls, SD, 26
Officers, 158, 167
Political parties, 62
Political polls, 194
President's approval ratings, 17
Supreme Court justice, ages, 109
U.S Presidents
ages at inauguration, 96
children, 48
greatest, 303, 305
U.S. Senator term, 119
Voters, 307, 429, 492, 493
Voting democrat, 126
Voting republican, 123

Psychology

Depression, 73
Eating disorders, 69
Experimental group, 158
IQ, 130, 461
Mouthing behavior, 16
Obsessive-compulsive disorder, 529, 531
Passive-aggressive traits, 174, 176, 177
Psychological tests, 26, 180
Reaction times, 46

Sports

400-meter dash, women's, 88
Advertisers, 57
Baseball, 152, 158
American League, 49–51
Barry Bonds, 62, 210
home run totals, 11, 173
Major League, 4, 25, 111, 231
National League, 152
New York Yankees World Series victories, 11
retired uniform numbers, 27
Basketball, 100, 166, 194
Chicago Bulls, 26
free throws, 203
heights, weights, 90
NBA draft, 160
NBA, average points, 68
women's basketball, 195
Bass catching, 207
Bicycle race, 166
Boston marathon, 27
Bowling, 46
Daytona 500, 153
Fishing, 189, 385
Football, 299, 305
Arena Football League, 99
College football, 13, 179
defensive player weights, 110
fans, 323
National football league, 89, 99, 144
Super Bowl, 604
Golf, 85, 167, 380, 428, 454
Tiger Woods, 204
Injury, 527
Hockey, 10, 70
Olympics, 456, 457, 460, 463, 467, 476
Soccer, 13, 208, 291, 292
Softball, 157
Sports, 268, 552
Team photo, 158
Tennis, 302
Workout, 578
Wrestling, 324

Work

Annual wage, 422
Arrangements, 523
Career placement, 610
CEO ages, 521, 522, 524
Committee
composition, 205
members, 157
Commute, 321
CPA, 198
Department manager, 155
Earnings, 58, 491, 493, 576, 584
Employee income, 107
Employee tenure, 614
Employee's well-being, 516
Employment, 54, 241
Executives, ages, 104
Genders of recent hires, 603
Happy at work, 308
Hourly earnings, 101, 573, 574, 589
Injuries, 598, 602
Industry workers, 128
Managers, use of digital media, 22
Leaving job, 517
Lumber cutter, 257
Office parties, 575
Office rentals,
Miami, FL, 78
Seattle, WA, 78
Overtime hours, 182
Sick days, 199
Time, 491, 492
Union, 27, 308, 323
Vacation days, 101, 195
Work days, 355
Work time and leisure time, 478
Work weeks, 265, 267
Volunteer, 379

CHAPTER 1

Introduction to Statistics

1.1 An Overview of Statistics
1.2 Data Classification
Case Study
1.3 Experimental Design
Uses and Abuses
Real Statistics–Real Decisions
Technology

New York City, with a population of over 8 million, has more than twice the population of any other American city. The next four most populated cities in the United States are Los Angeles, Chicago, Houston, and Philadelphia.

Where You've Been

You are already familiar with many of the practices of statistics, such as taking surveys, collecting data, and describing populations. What you may not know is that collecting accurate statistical data is often difficult and costly. Consider, for instance, the monumental task of counting and describing the entire population of the United States. If you were in charge of such a census, how would you do it? How would you ensure that your results are accurate? These and many more concerns are the responsibility of the United States Census Bureau.

Where You're Going

In Chapter 1, you will be introduced to the basic concepts and goals of statistics. For instance, statistics was used to construct the following graphs, which show the age and marital status of the United States population, as determined by the 2000 census. When conducting the census, the Bureau sends short forms to the entire population that ask about characteristics such as gender, age, race, and home ownership. A long form, which covers many additional topics, is sent to 17% of the population. This 17% forms a sample. In this course, you will learn how the data collected from a sample are used to infer characteristics about the entire population.

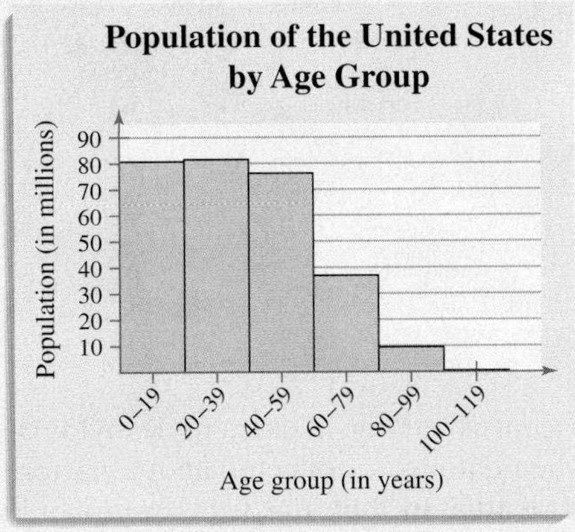

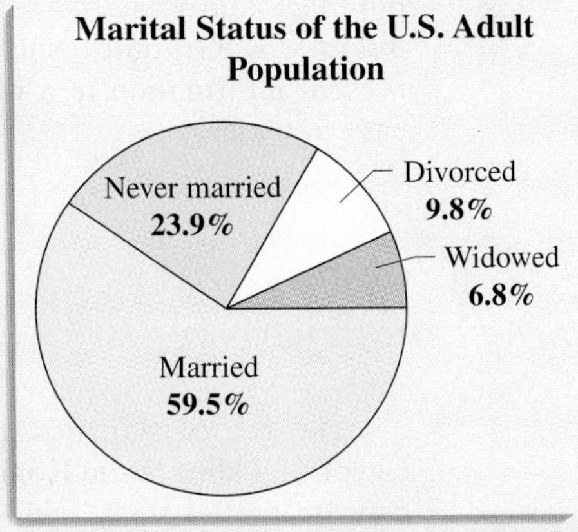

1.1 An Overview of Statistics

What You Should Learn

- The definition of statistics
- How to distinguish between a population and a sample and between a parameter and a statistic
- How to distinguish between descriptive statistics and inferential statistics

A Definition of Statistics • Data Sets • Branches of Statistics

A Definition of Statistics

As you begin this course, you may wonder: *What is statistics? Why should I study statistics? How can studying statistics help me in my profession?* Almost every day you are exposed to statistics. For example, consider the following excerpts from recent newspapers and journals.

- "A . . . survey of traffic deaths during this past Memorial Day weekend shows a 36% decrease in fatalities compared with last year." *(Source: National Safety Council)*
- "Men who eat just two servings of tomatoes a week in raw, sauce, or pizza form have a 34% less risk of developing prostate cancer." *(Source: Journal of the National Cancer Institute)*
- "More than three fourths of all college seniors in the United States complete at least one internship by graduation and 55% participate in two or more." *(Source: UPI)*

The three statements you just read are based on the collection of **data.**

DEFINITION

Data consist of information coming from observations, counts, measurements, or responses. The singular for data is datum.

Note to Instructor*

To initiate a class discussion, bring a recent newspaper article that describes the results of a statistical study.

*Annotations such as this appear in this Annotated Instructor's Edition, but not in the student text.

Sometimes data are presented graphically. If you have ever read *USA TODAY*, you have certainly seen one of that newspaper's most popular features, *USA TODAY Snapshots*. Graphics such as this present information in a way that is easy to understand.

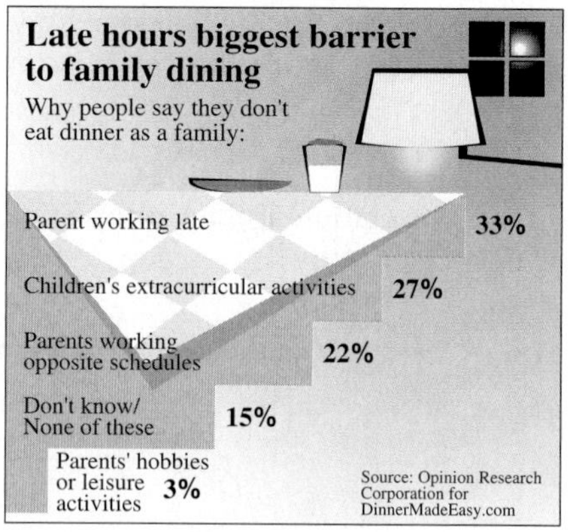

The use of statistics dates back to census taking in ancient Babylonia, Egypt, and later in the Roman Empire, when data were collected about matters concerning the state, such as births and deaths. In fact, the word *statistics* is derived from the Latin word *status*, meaning "state."

But statistics today involves more than collecting data, presenting facts, calculating averages, and drawing graphs. So, what is statistics?

DEFINITION

Statistics is the science of collecting, organizing, analyzing, and interpreting data in order to make decisions.

Data Sets

There are two types of data sets you will use when studying statistics. These data sets are called *populations* and *samples*.

DEFINITION

A **population** is the collection of *all* outcomes, responses, measurements, or counts that are of interest.

A **sample** is a subset of a population.

Insight

A census consists of data from an entire population. But, unless a population is small, it is usually impractical to obtain all the population data. In most studies, information must be obtained from a sample.

EXAMPLE 1

Identifying Data Sets

In a recent survey, 3002 adults in the United States were asked if they read news on the Internet at least once a week. Six hundred of the adults said yes. Identify the population and the sample. Describe the data set. *(Source: Pew Research Center)*

SOLUTION The population consists of the responses of all adults in the United States, and the sample consists of the responses of the 3002 adults in the United States in the survey. The sample is a subset of the responses of all adults in the United States. The data set consists of 600 yes's and 2402 no's.

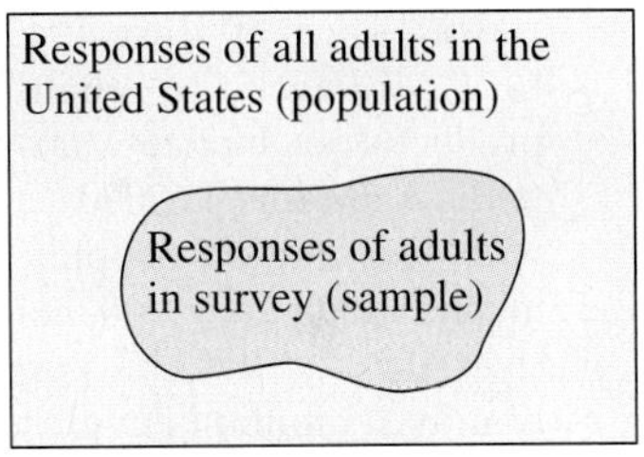

Note to Instructor

Point out to students that Venn diagrams show only which sets are subsets of other sets. They do not show relative sizes of the sets. For instance, the Venn diagram at the right should not be interpreted to mean that the sample size is roughly one-fourth of the population size.

Try It Yourself 1

The U.S. Department of Energy conducts weekly surveys of approximately 900 gasoline stations to determine the average price per gallon of regular gasoline. On December 29, 2003, the average price was $1.478 per gallon. Identify the population and the sample. *(Source: U.S. Department of Energy)*

a. Identify the *population*.
b. Identify the *sample*.
c. What does the data set consist of?

Answer: Page A30

Whether a data set is a population or a sample usually depends on the context of the real-life situation. For instance, in Example 1, the population was the set of responses of all adults in the United States. Depending on the purpose of the survey, the population could have been the set of responses of all adults who live in California or who have telephones or who read a particular newspaper.

Two important terms that are used throughout this course are *parameter* and *statistic.*

Study Tip

The terms *parameter* and *statistic* are easy to remember if you use the mnemonic device of matching the first letters in ***p**opulation **p**arameter* and the first letters in ***s**ample **s**tatistic.*

DEFINITION

A **parameter** is a numerical description of a *population* characteristic.

A **statistic** is a numerical description of a *sample* characteristic.

EXAMPLE 2

Distinguishing between a Parameter and a Statistic

Decide whether the numerical value describes a population parameter or a sample statistic. Explain your reasoning.

1. A recent survey of a sample of MBAs reported that the average starting salary for an MBA is less than $65,000. *(Source: The Washington Post Company)*
2. Starting salaries for the 667 MBA graduates from the University of Chicago Graduate School of Business increased 8.5% from the previous year.
3. In a random check of a sample of retail stores, the Food and Drug Administration found that 34% of the stores were not storing fish at the proper temperature.

SOLUTION

1. Because the average of $65,000 is based on a subset of the population, it is a sample statistic.
2. Because the percent increase of 8.5% is based on all 667 graduates' starting salaries, it is a population parameter.
3. Because the percent of 34% is based on a subset of the population, it is a sample statistic.

Try It Yourself 2

In 2003, major league baseball teams spent a total of $2,130,863,461 on players' salaries. Does this numerical value describe a population parameter or a sample statistic? *(Source: USA TODAY)*

a. Decide whether the numerical value is from a *population* or a *sample.*
b. Specify whether the numerical value is a *parameter* or a *statistic.*

Answer: Page A30

Picturing the World

How accurate is the U.S. census? According to a post-census evaluation conducted by the Census Bureau, the 1990 census undercounted the U.S. population by an estimated 4.0 million people. The 1990 census was the first census since at least 1940 to be less accurate than its predecessor. Notice that the undercount for the 2000 census was −1.3 million people. This means that the 2000 census *overcounted* the U.S. population by 1.3 million people.

U.S. Census Undercount

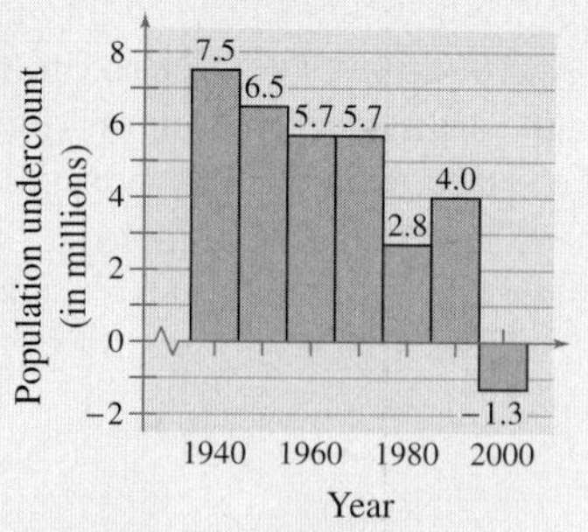

What are some difficulties in collecting population data?

In this course, you will see how the use of statistics can help you make informed decisions that affect your life. Consider the census that the U.S. government takes every decade. When taking the census, the Census Bureau attempts to contact everyone living in the United States. This is an impossible task. It is important that the census be accurate, because public officials make many decisions based on the census information. Data collected in the 2010 census will determine how to assign congressional seats and how to distribute public funds.

Branches of Statistics

The study of statistics has two major branches: **descriptive statistics** and **inferential statistics.**

DEFINITION

Descriptive statistics is the branch of statistics that involves the organization, summarization, and display of data.

Inferential statistics is the branch of statistics that involves using a sample to draw conclusions about a population. A basic tool in the study of inferential statistics is probability.

EXAMPLE 3

Descriptive and Inferential Statistics

Decide which part of the study represents the descriptive branch of statistics. What conclusions might be drawn from the study using inferential statistics?

1. A large sample of men, aged 48, was studied for 18 years. For unmarried men, approximately 70% were alive at age 65. For married men, 90% were alive at age 65. *(Source: The Journal of Family Issues)*
2. In a sample of Wall Street analysts, the percentage who incorrectly forecasted high-tech earnings in a recent year was 44%. *(Source: Bloomberg News)*

Note to Instructor

Ask students to identify the sample and the population in Example 3. Also ask students to suggest other inferences that might be drawn. This would be a good time to tell students that one can never be 100% sure about inferences.

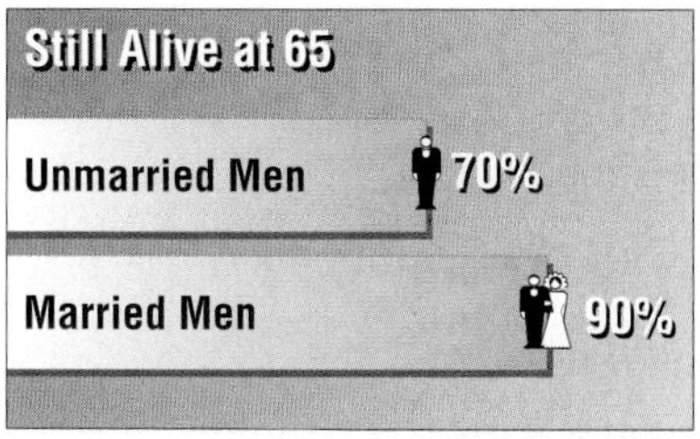

SOLUTION

1. Descriptive statistics involves statements such as "For unmarried men, approximately 70% were alive at age 65" and "For married men, 90% were alive at 65." A possible inference drawn from the study is that being married is associated with a longer life for men.
2. The part of this study that represents the descriptive branch of statistics involves the statement "the percentage of Wall Street analysts who incorrectly forecasted high-tech earnings in a recent year was 44%." A possible inference drawn from the study is that the stock market is difficult to forecast, even for professionals.

Try It Yourself 3

A survey conducted among 1017 men and women by Opinion Research Corporation International found that 76% of women and 60% of men had had a physical examination within the previous year. *(Source: Men's Health)*

a. Identify the descriptive aspect of the survey.
b. What inferences could be drawn from this survey? *Answer: Page A30*

Throughout this course you will see applications of both branches. A major theme in this course will be how to use sample statistics to make inferences about unknown population parameters.

1.1 Exercises

Help

MyMathLab

Student Study Pack

Building Basic Skills and Vocabulary

1. How is a sample related to a population?

2. Why is a sample used more often than a population?

3. What is the difference between a parameter and a statistic?

4. What are the two main branches of statistics?

True or False? In Exercises 5–10, determine whether the statement is true or false. If it is false, rewrite it as a true statement.

5. A statistic is a measure that describes a population characteristic.

6. A sample is a subset of a population.

7. It is impossible for the Census Bureau to obtain all the census data about the population of the United States.

8. Inferential statistics involves using a population to draw a conclusion about a corresponding sample.

9. A population is the collection of some outcomes, responses, measurements, or counts that are of interest.

10. The word *statistics* is derived from the Latin word *status*, meaning "state."

Classifying a Data Set In Exercises 11–16, determine whether the data set is a population or a sample. Explain your reasoning.

11. The age of each state governor

12. The speed of every fifth car passing a police speed trap

13. A survey of 500 students from a university with 10,000 students

14. The annual salary for each employee at a company

15. The cholesterol levels of 20 patients in a hospital with 100 patients

16. The number of pets in each U.S. household

Graphical Analysis In Exercises 17–20, use the Venn diagram to identify the population and the sample.

17.

1. A sample is a subset of a population.
2. It is usually impractical (too expensive and time consuming) to obtain all the population data.
3. A parameter is a numerical description of a population characteristic. A statistic is a numerical description of a sample characteristic.
4. The two main branches of statistics are descriptive statistics and inferential statistics.
5. False. A statistic is a measure that describes a sample characteristic.
6. True
7. True
8. False. Inferential statistics involves using a sample to draw conclusions about a population.
9. False. A population is the collection of *all* outcomes, responses, measurements, or counts that are of interest.
10. True
11. Population, because it is a collection of the ages of all the state governors.
12. Sample, because only the speed of every fifth car is recorded.
13. Sample, because the collection of the 500 students is a subset within the population.
14. Population, because it is a collection of the annual salaries of all employees at a company.

15. Sample, because the collection of the 20 patients is a subset within the population.

16. Population, because it is a collection of the number of pets in all U.S. households.

17. Population: Party of registered voters in Bucks County

Sample: Party of Bucks County voters responding to phone survey

18. Population: Major of college students at Central College

Sample: Major of college students at Central College who take statistics

19. Population: Ages of adults in the U.S. who own computers

Sample: Ages of adults in the U.S. who own Dell computers

20. Population: Income of all home owners in Ohio

Sample: Income of home owners in Ohio with mortgages

21. Population: Collection of all infants in Italy

Sample: Collection of the 33,043 infants in the study

22. Population: Collection of all households from the U.S.

Sample: Collection of the 2104 U.S. households surveyed

23. Population: Collection of all women in the U.S.

Sample: Collection of the 546 U.S. women surveyed

24. Population: Collection of all vacationers from the U.S.

Sample: Collection of the 791 U.S. vacationers surveyed

25. Statistic. The value $68,000 is a numerical description of a sample of annual salaries.

26. Statistic. 43% is a numerical description of a sample of high school students.

18.

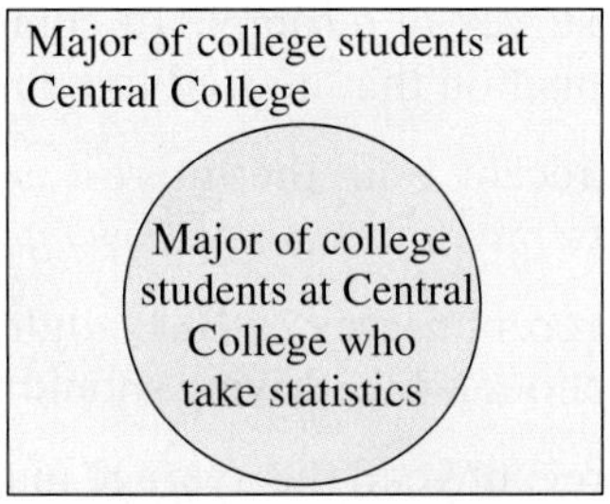

19.

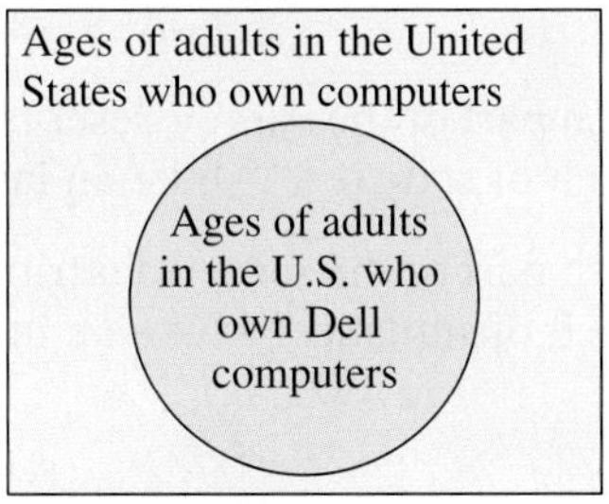

20.

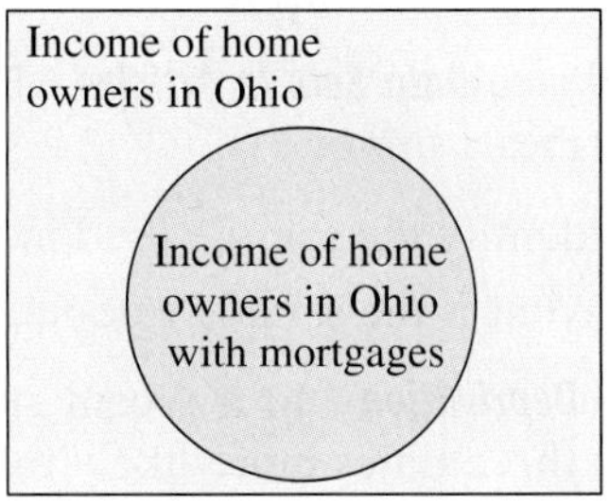

Using and Interpreting Concepts

Identifying Populations and Samples In Exercises 21–24, identify the population and the sample.

21. A study of 33,043 infants in Italy was conducted to find a link between a heart rhythm abnormality and sudden infant death syndrome. *(Source: New England Journal of Medicine)*

22. A survey of 2104 households in the United States found that 65% subscribe to cable television.

23. A survey of 546 women found that more than 56% are the primary investor in their household. *(Adapted from: Roper Starch Worldwide for Intuit)*

24. A survey of 791 vacationers from the United States found that they planned on spending at least $2000 for their next vacation.

Distinguishing between a Parameter and a Statistic In Exercises 25–30, determine whether the numerical value is a parameter or a statistic. Explain your reasoning.

25. The average annual salary for 35 of a company's 1200 accountants is $68,000.

26. In a survey of a sample of high school students, 43% said that their mother has taught them the most about managing money. *(Source: Harris Poll for Girls Incorporated)*

27. Statistic. 8% is a numerical description of a sample of computer users.

28. See Selected Answers, page A89.

29. Statistic. 47% is a numerical description of a sample of people in the United States.

30. See Selected Answers, page A89.

31. The statement "56% are the primary investor in their household" is an application of descriptive statistics.

An inference drawn from the sample is that an association exists between U.S. women and being the primary investor in their household.

32. The statement "spending at least $2000 for their next vacation" is an application of descriptive statistics.

An inference drawn from the sample is that vacationers in the U. S. are associated with spending more than $2000 for their next vacation.

33. Answers will vary.

34. (a) The sample is the volunteers in the study.

(b) The population is the collection of all individuals who completed the math test.

(c) The statement "three times more likely to answer correctly" is an application of descriptive statistics.

(d) An inference drawn from the sample is that individuals who are not sleep deprived will be three times more likely to answer math questions correctly than individuals who are sleep deprived.

35. (a) An inference drawn from the sample is that senior citizens who own luxury cars have more of their natural teeth than senior citizens who do not own luxury cars.

(b) This inference may incorrectly imply that if you purchase a luxury car, you will keep more of your natural teeth.

36. See Selected Answers, page A89.

37. Answers will vary.

27. In a survey of a sample of computer users, 8% said their computer had a malfunction that needed to be repaired by a service technician.

28. In a recent year, the interest category for 12% of all new magazines was sports. *(Source: Oxbridge Communications)*

29. In a recent survey of 1000 adults from the United States, 47% said using a cell phone while driving should be illegal. *(Source: Rasmussen Research)*

30. In a recent year, the average math scores for all graduates on the ACT was 20.7. *(Source: ACT, Inc.)*

31. Which part of the survey described in Exercise 23 represents the descriptive branch of statistics? Make an inference based on the results of the survey.

32. Which part of the survey described in Exercise 24 represents the descriptive branch of statistics? Make an inference based on the results of the survey.

Extending Concepts

33. ***Identifying Data Sets in Articles*** Find a newspaper or magazine article that describes a survey.

(a) Identify the sample used in the survey.

(b) What is the sample's population?

34. ***Sleep Deprivation*** In a recent study, volunteers who had 8 hours of sleep were three times more likely to answer correctly on a math test than were sleep-deprived participants. *(Source: CBS News)*

(a) Identify the sample used in the study.

(b) What is the sample's population?

(c) Which part of the study represents the descriptive branch of statistics?

(d) Make an inference based on the results of the study.

35. ***Owning a Luxury Car*** A study shows that senior citizens who own a luxury car have more of their natural teeth than senior citizens who do not own a luxury car.

(a) Make an inference based on the results of this study.

(b) What is wrong with this type of reasoning?

36. ***Increase in Strokes*** A study shows that the number of people who have a stroke has increased every year for the past 15 years.

(a) Make an inference based on the results of this study.

(b) What is wrong with this type of reasoning?

37. ***Writing*** Write an essay about the importance of statistics for one of the following.

(a) A study on the effectiveness of a new drug

(b) An analysis of a manufacturing process

(c) Making conclusions about voter opinions using surveys

1.2 Data Classification

What You Should Learn

- How to distinguish between qualitative data and quantitative data
- How to classify data with respect to the four levels of measurement: nominal, ordinal, interval, and ratio

Types of Data • Levels of Measurement

Types of Data

When doing a study, it is important to know the kind of data involved. The nature of the data you are working with will determine which statistical procedures can be used. In this section, you will learn how to classify data by type and by level of measurement. Data sets can consist of two types of data: *qualitative data* and *quantitative data.*

DEFINITION

Qualitative data consist of attributes, labels, or nonnumerical entries.

Quantitative data consist of numerical measurements or counts.

EXAMPLE 1

Classifying Data by Type

The base prices of several vehicles are shown in the table. Which data are qualitative data and which are quantitative data? Explain your reasoning. *(Source: Ford Motor Company)*

Model	Base price
ZX2	$13,750
Focus LX	$13,800
Ranger XL	$14,720
Taurus LX	$20,490
Explorer Sport-Trac	$23,840
Crown Victoria	$24,515
Windstar LX	$27,000
Expedition XLT	$34,710

SOLUTION The information shown in the table can be separated into two data sets. One data set contains the names of vehicle models, and the other contains the base prices of vehicle models. The names are nonnumerical entries, so these are qualitative data. The base prices are numerical entries, so these are quantitative data.

Try It Yourself 1

The populations of several U.S. cities are shown in the table. Which data are qualitative data and which are quantitative data? *(Source: U.S. Census Bureau)*

a. Identify the contents of each data set.
b. Decide whether each data set consists of numerical or nonnumerical entries.
c. Specify the qualitative data and the quantitative data.

City	Population
Baltimore, MD	638,614
Boston, MA	589,281
Dallas, TX	1,211,467
Las Vegas, NV	508,604
Lincoln, NE	232,362
Seattle, WA	570,426

Answer: Page A30

Levels of Measurement

Another characteristic of data is its level of measurement. The level of measurement determines which statistical calculations are meaningful. The four levels of measurement, in order from lowest to highest, are *nominal, ordinal, interval,* and *ratio.*

DEFINITION

Data at the **nominal level of measurement** are qualitative only. Data at this level are categorized using names, labels, or qualities. No mathematical computations can be made at this level.

Data at the **ordinal level of measurement** are qualitative or quantitative. Data at this level can be arranged in order, but differences between data entries are not meaningful.

Insight

When numbers are at the nominal level of measurement, they simply represent a label. Examples of numbers used as labels include Social Security numbers and numbers on sports jerseys. For instance, it would not make sense to add the numbers on the players' jerseys for the Chicago Bears.

EXAMPLE 2

Classifying Data by Level

Two data sets are shown. Which data set consists of data at the nominal level? Which data set consists of data at the ordinal level? Explain your reasoning. *(Source: Nielsen Media Research)*

Top 5 TV Programs (from 3/08/04 to 3/14/04)
1. *CSI*
2. *American Idol–Tuesday*
3. *American Idol–Wednesday*
4. *Without a Trace*
5. *Survivor*

Network Affiliates in Portland, Oregon	
KATU	(ABC)
KGW	(NBC)
KOIN	(CBS)
KPTV	(FOX)

SOLUTION The first data set lists the rank of five TV programs. The data consist of the ranks 1, 2, 3, 4, and 5. Because the rankings can be listed in order, these data are at the ordinal level. Note that the difference between a rank of 1 and 5 has no mathematical meaning. The second data set consists of the call letters of each network affiliate in Portland. The call letters are simply the names of network affiliates, so these data are at the nominal level.

Try It Yourself 2

Consider the following data sets. For each data set, decide whether the data are at the nominal level or at the ordinal level.

1. The final standings for the Northeast Division of the National Hockey League
2. A collection of phone numbers

a. *Identify* what each data set represents.
b. Specify the *level of measurement* and justify your answer.

Answer: Page A30

Picturing the World

At the end of the twentieth century, the American Film Institute chose the 100 greatest American movies. The institute started with a population of more than 40,000 movies and chose a sample of 400 to put on a ballot. The ballot was mailed to over 1500 film artists and executives. The films were judged on popularity over time, historical significance, cultural impact, and other factors.

The American Film Institute's Top Five American Movies
1. *Citizen Kane*
2. *Casablanca*
3. *The Godfather*
4. *Gone with the Wind*
5. *Lawrence of Arabia*

In this list, what is the level of measurement?

DEFINITION

Data at the **interval level of measurement** are quantitative. The data can be ordered, and you can calculate meaningful differences between data entries. At the interval level, a zero entry simply represents a position on a scale; the entry is not an inherent zero.

Data at the **ratio level of measurement** are similar to data at the interval level, with the added property that a zero entry is an inherent zero. A ratio of two data values can be formed so one data value can be expressed as a multiple of another.

An *inherent zero* is a zero that implies "none." For instance, the amount of money you have in a savings account could be zero dollars. In this case, the zero represents no money; it is an inherent zero. On the other hand, a temperature of 0°C does not represent a condition in which no heat is present. The 0°C temperature is simply a position on the Celsius scale; it is not an inherent zero.

To distinguish between data at the interval level and at the ratio level, determine whether the expression "twice as much" has any meaning in the context of the data. For instance, \$2 is twice as much as \$1, so these data are at the ratio level. On the other hand, 2°C is not twice as warm as 1°C, so these data are at the interval level.

New York Yankees' World Series Victories (Years)
1923, 1927, 1928, 1932, 1936, 1937, 1938, 1939, 1941, 1943, 1947, 1949, 1950, 1951, 1952, 1953, 1956, 1958, 1961, 1962, 1977, 1978, 1996, 1998, 1999, 2000

2003 National League Home Run Totals (by Team)	
Atlanta	235
Arizona	152
Chicago	172
Cincinnati	182
Colorado	198
Florida	157
Houston	191
Los Angeles	124
Milwaukee	196
Montreal	144
New York	124
Philadelphia	166
Pittsburgh	163
San Diego	128
San Francisco	180
St. Louis	196

EXAMPLE 3

Classifying Data by Level

Two data sets are shown at the left. Which data set consists of data at the interval level? Which data set consists of data at the ratio level? Explain your reasoning. *(Source: Major League Baseball)*

SOLUTION Both of these data sets contain quantitative data. Consider the dates of the Yankees' World Series victories. It makes sense to find differences between specific dates. For instance, the time between the Yankees' first and last World Series victories is

$$2000 - 1923 = 77 \text{ years.}$$

But it does not make sense to write a ratio using these dates. So, these data are at the interval level. Using the home run totals, you can find differences *and* write ratios. From the data, you can see that Milwaukee hit 52 more home runs than Montreal hit and that Atlanta hit about twice as many home runs as Los Angeles hit. So, these data are at the ratio level.

Try It Yourself 3

Decide whether the data are at the interval level or at the ratio level.

1. The body temperatures (in degrees Fahrenheit) of an athlete during an exercise session
2. The heart rates (in beats per minute) of an athlete during an exercise session

a. *Identify* what each data set represents.
b. Specify the *level of measurement* and justify your answer.

Answer: Page A30

The following tables summarize which operations are meaningful at each of the four levels of measurement. When identifying a data set's level of measurement, use the highest level that applies.

Level of measurement	Put data in categories	Arrange data in order	Subtract data values	Determine if one data value is a multiple of another
Nominal	Yes	No	No	No
Ordinal	Yes	Yes	No	No
Interval	Yes	Yes	Yes	No
Ratio	Yes	Yes	Yes	Yes

Summary of Four Levels of Measurement

	Example of a Data Set	Meaningful Calculations
Nominal Level (Qualitative data)	*Types of Music Played by a Radio Station* Pop Modern rock Contemporary jazz Hip hop	*Put in a category.* For instance, a song played by the radio station could be put into one of the four categories shown.
Ordinal Level (Qualitative or quantitative data)	*Motion Picture Association of America Ratings Description* G General Audiences PG Parental Guidance Suggested PG-13 Parents Strongly Cautioned R Restricted NC-17 No One Under 17 Admitted	Put in a category and *put in order.* For instance, a PG rating has a stronger restriction than a G rating.
Interval Level (Quantitative data)	*Average Monthly Temperature (in degrees Fahrenheit) for Sacramento, CA* Jan 46.3 Jul 75.4 Feb 51.2 Aug 74.8 Mar 54.5 Sep 71.7 Apr 58.9 Oct 64.4 May 65.5 Nov 53.3 Jun 71.3 Dec 45.8 *(Source: National Climatic Data Center)*	Put in a category, put in order, and *find differences between values.* For instance, $71.5 - 65.5 = 6°F$. So, June is 6° warmer than May.
Ratio Level (Quantitative data)	*Average Monthly Precipitation (in inches) for Sacramento, CA* Jan 3.8 Jul 0.1 Feb 3.5 Aug 0.1 Mar 2.8 Sep 0.4 Apr 1.0 Oct 0.9 May 0.5 Nov 2.2 Jun 0.2 Dec 2.5 *(Source: National Climatic Data Center)*	Put in a category, put in order, find differences between values, and *find ratios of values.* For instance, $\frac{1.0}{0.5} = 2$. So, there is twice as much rain in April as in May.

1.2 Exercises

Building Basic Skills and Vocabulary

1. Name each level of measurement for which data can be qualitative.
2. Name each level of measurement for which data can be quantitative.

True or False? In Exercises 3–6, determine whether the statement is true or false. If it is false, rewrite it as a true statement.

3. Data at the ordinal level are quantitative only.
4. For data at the interval level, you cannot calculate meaningful differences between data entries.
5. More types of calculations can be performed with data at the nominal level than with data at the interval level.
6. Data at the ratio level cannot be put in order.

Classifying Data by Type In Exercises 7–10, determine whether the data are qualitative or quantitative.

7. The telephone numbers in a telephone directory
8. The daily low temperatures for the month of January
9. The scores of a class on an accounting exam
10. The player numbers for a soccer team

Using and Interpreting Concepts

Classifying Data by Level In Exercises 11–16, identify the data set's level of measurement. Explain your reasoning.

11. ***Football*** The top five teams in the final college football poll released on January 4, 2004 are listed. *(Source: Associated Press)*

    1. USC 2. LSU 3. Oklahoma 4. Ohio 5. Miami (Florida)

12. ***Politics*** The three political parties in the 108th Congress are listed below.

    Republican
    Democrat
    Independent

13. ***Top Salesperson*** The region representing the top salesperson in a corporation for the past six years is given.

Southeast	Northwest	Northeast
Southeast	Southwest	Southwest

14. ***Fish Lengths*** The lengths (in inches) of a sample of striped bass caught in Maryland waters are listed. *(Adapted from National Marine Fisheries Service, Fisheries Statistics and Economics Division)*

    16 17.25 19 18.75 21 20.3 19.8 24 21.82

1. Nominal and ordinal
2. Ordinal, interval, and ratio
3. False. Data at the ordinal level can be qualitative or quantitative.
4. False. For data at the interval level, you can calculate meaningful differences between data entries. You cannot calculate meaningful differences at the nominal or ordinal level.
5. False. More types of calculations can be performed with data at the interval level than with data at the nominal level.
6. False. Data at the ratio level can be placed in a meaningful order.
7. Qualitative
8. Quantitative
9. Quantitative
10. Qualitative
11. Ordinal. Data can be arranged in order, but the differences between data entries make no sense.
12. Nominal. No mathematical computations can be made and data are categorized using names.
13. Nominal. No mathematical computations can be made and data are categorized by region.
14. Ratio. A ratio of two data values can be formed so one data value can be expressed as a multiple of another.

15. Ordinal. Data can be arranged in order, but the differences between data entries are not meaningful.
16. Ratio. A ratio of two data values can be formed so one data value can be expressed as a multiple of another.
17. Ordinal
18. Ratio
19. Nominal
20. Interval
21. An inherent zero is a zero that implies "none." Answers will vary.
22. Answers will vary.

15. *Box Office Rank* The top five musical artists on July 23, 2003 by box office rank are listed. *(Source: New York Times)*

1. The Eagles
2. Dave Matthews Band
3. The Dixie Chicks
4. Fleetwood Mac
5. Cher

16. *Ticket Prices* The average ticket prices for 10 rock concerts in 2003 are listed. *(Source: New York Times)*

$104 $42 $60 $85 $68 $58 $53 $37 $37 $38

Graphical Analysis In Exercises 17–20, identify the level of measurement of the data listed on the horizontal axis in the graph.

17. **Does Global Warming Contribute to More Severe El Niños?**

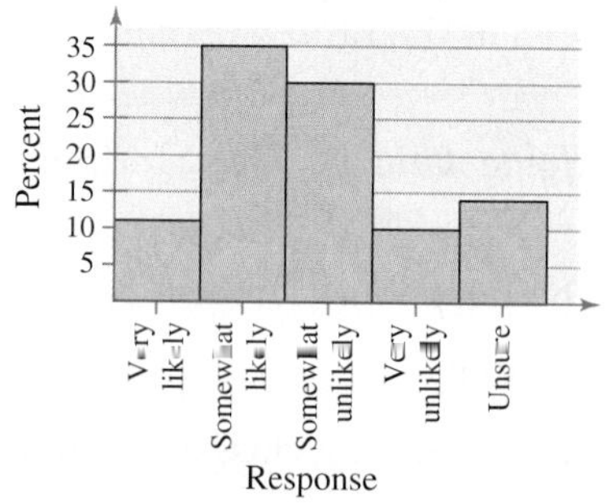

(Source: Yankelovich for the National Representatives Science Foundation, American Meteorological Society)

18. **Average January Snowfall for 15 Cities**

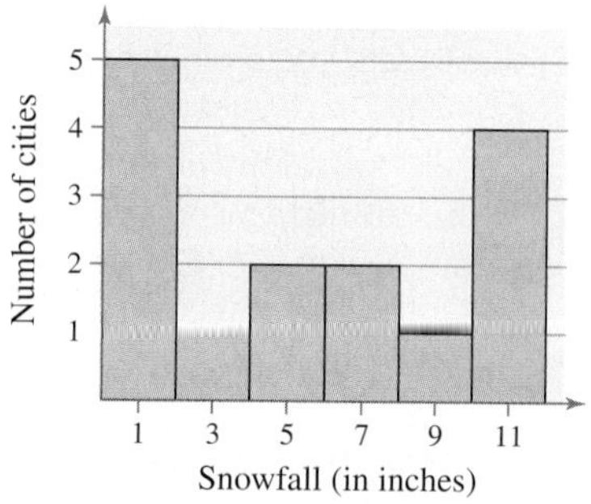

(Source: National Climatic Data Center)

19. **Gender Profile of the 108th Congress**

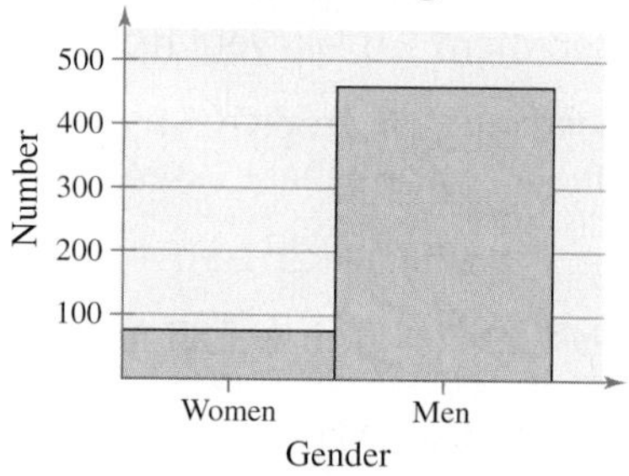

(Source: U.S. House of Representatives, Office of the Clerk)

20. **State Government Tax Collections by Year**

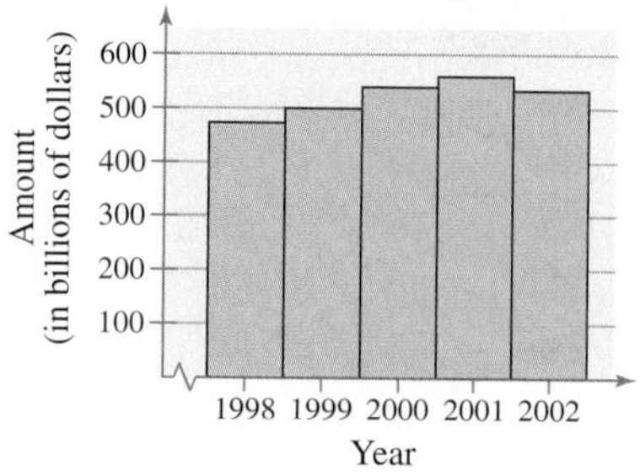

(Source: U.S. Census Bureau)

Extending Concepts

21. *Writing* What is an inherent zero? Describe three examples of data sets that have an inherent zero and three that do not.

22. *Writing* Describe two examples of data sets for each of the four levels of measurement. Justify your answer.

Case Study

Rating Television Shows in the United States

Nielsen Media Research has been rating television programs for nearly 50 years. Nielsen uses several sampling procedures, but its main one is to track the viewing patterns of 5000 households. These contain over 13,000 people and are chosen to form a cross section of the overall population. The households represent various locations, ethnic groups, and income brackets. The data gathered from the Nielsen sample of 5000 households are used to draw inferences about the population of all households in the United States.

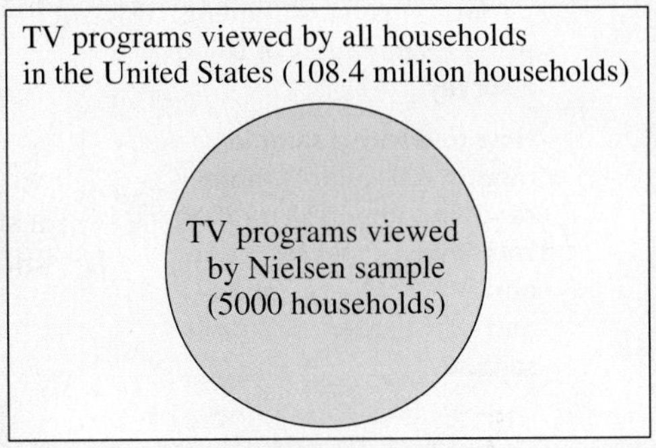

Top-Ranked Programs in Prime Time for the Week of 3/08/04 – 3/14/04

Rank	Rank last week	Program name	Network	Day, Time	Rating	Share	Audience
1	3	*CSI*	CBS	Thu., 9:00 P.M.	16.8	25	27,402,000
2	1	*American Idol–Tuesday*	FOX	Tue., 8:00 P.M.	14.5	22	24,569,000
3	7	*American Idol–Wednesday*	FOX	Wed., 8:30 P.M.	13.0	20	22,023,000
4	9	*Without a Trace*	CBS	Thu., 10:00 P.M.	13.0	22	19,683,000
5	2	*Survivor*	CBS	Thu., 8:00 P.M.	12.6	20	22,630,000
6	4	*Apprentice*	NBC	Thu., 9:00 P.M.	12.3	18	19,207,000
7	8	*Everybody Loves Raymond*	CBS	Mon., 9:00 P.M.	10.8	17	16,692,000
8	11	*Friends*	NBC	Thu., 8:00 P.M.	10.8	17	16,572,000
9	14	*Will & Grace*	NBC	Thu., 8:30 P.M.	10.4	16	15,611,000
10	5	*CSI: Miami*	CBS	Mon., 10:00 P.M.	10.3	17	15,382,000

Exercises

1. ***Rating Points*** Each rating point represents 1,084,000 households, or 1% of the households in the United States. Does a program with a rating of 8.4 have twice the number of households as a program with a rating of 4.2? Explain your reasoning.

2. ***Sampling Percent*** What percentage of the total number of U.S. households is used in the Nielsen sample?

3. ***Nominal Level of Measurement*** Which columns in the table contain data at the nominal level?

4. ***Ordinal Level of Measurement*** Which columns in the table contain data at the ordinal level? Describe two ways that the data can be ordered.

5. ***Interval Level of Measurement*** Which column in the table contains data at the interval level? How can these data be ordered? What is the unit of measure for the difference of two entries in the data set?

6. ***Ratio Level of Measurement*** Which three columns contain data at the ratio level?

7. ***Share*** The column listed as "Share" gives the percentage of televisions in use at a given time. Does the Nielsen rating rank shows by rating or by share? Explain your reasoning.

8. ***Inferences*** What decisions (inferences) can be made on the basis of the Nielsen ratings?

1.3 Experimental Design

What You Should Learn

- How to design a statistical study
- How to collect data by doing an observational study, performing an experiment, using a simulation, or using a survey
- How to create a sample using random sampling, simple random sampling, stratified sampling, cluster sampling, and systematic sampling and how to identify a biased sample

Experimental Design • Data Collection • Sampling Techniques

Experimental Design

The goal of every statistical study is to collect data and then use the data to make a decision. Any decision you make using the results of a statistical study is only as good as the process used to obtain the data. If the process is flawed, then the resulting decision is questionable.

You may never have to develop a statistical study, but it is likely that you will have to interpret the results of one. And before you interpret the results of a study, you should determine whether the results are valid. In other words, you should be familiar with how to design a statistical study.

GUIDELINES

Designing a Statistical Study

1. Identify the variable(s) of interest (the focus) and the population of the study.
2. Develop a detailed plan for collecting data. If you use a sample, make sure the sample is representative of the population.
3. Collect the data.
4. Describe the data, using descriptive statistics techniques.
5. Interpret the data and make decisions about the population using inferential statistics.
6. Identify any possible errors.

Insight

The difference between an observational study and an experiment is that, in an observational study, a researcher does not influence the responses, whereas in an experiment, a researcher deliberately applies a treatment before observing the responses.

Note to Instructor

Discuss the use of a placebo and a control group in an experiment. The placebo effect can be eliminated by making experiments "double blind." That is, neither the researcher nor the subject knows which subjects are receiving the placebo.

Data Collection

There are several ways you can collect data. Often, the focus of the study dictates the best way to collect data. The following is a brief summary of four methods of data collection.

- ***Do an observational study*** In an **observational study,** a researcher observes and measures characteristics of interest of part of a population. For instance, an observational study was performed in which researchers observed and recorded the mouthing behavior on nonfood objects of children up to three years old. *(Source: Pediatrics Magazine)*
- ***Perform an experiment*** In performing an **experiment,** a treatment is applied to part of a population and responses are observed. For instance, an experiment was performed in which diabetics took cinnamon extract daily while a control group took none. After 40 days, the diabetics who had the cinnamon reduced their risk of heart disease while the control group experienced no change. Each experimental unit is called a *block.* Care must be taken to ensure that the blocks are similar. *(Source: Diabetes Care)*

Note to Instructor

Students will be given the opportunity to use simulations when they study probability. Refer to page 169 to see an interesting example of a simulation.

- ***Use a simulation*** A **simulation** is the use of a mathematical or physical model to reproduce the conditions of a situation or process. Collecting data often involves the use of computers. Simulations allow you to study situations that are impractical or even dangerous to create in real life, and often they save time and money. For instance, automobile manufacturers use simulations with dummies to study the effects of crashes on humans.
- ***Use a survey*** A **survey** is an investigation of one or more characteristics of a population. Most often, surveys are carried out on *people* by asking them questions. The most common types of surveys are done by interview, mail, or telephone. In designing a survey, it is important to word the questions so that they do not lead to biased results. Surveys can be used to take a census or a sampling. A **census** is a count or measure of an *entire* population. Taking a census provides complete information, but it is often costly and difficult to perform. A **sampling** is a count or measure of *part* of a population. For instance, every year the U.S. Census Bureau samples the U.S. population to update the most recent census data. Using samples is often more practical than taking a census.

Picturing the World

The Gallup Organization conducts many polls (or surveys) regarding the president, Congress, and political and nonpolitical issues. A commonly cited Gallup poll is the public approval rating of the president. For example, the approval ratings for President George W. Bush from 2002 to 2004 are shown in the following graph. (The rating is from the first poll conducted in January of each year.)

President's Approval Ratings, 2002–2004

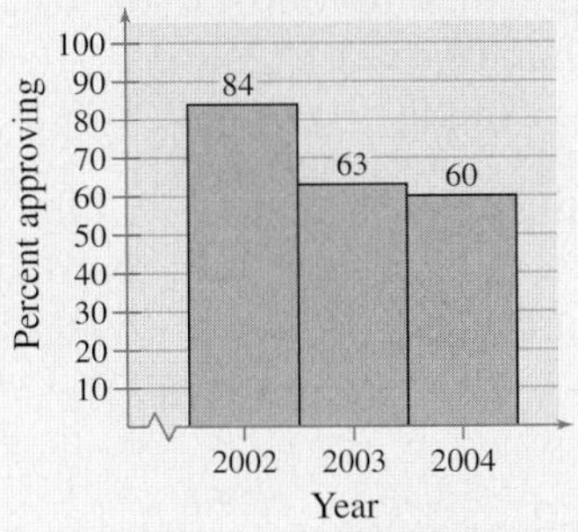

Discuss some ways that Gallup could select a biased sample to conduct a poll. How could Gallup select a sample that is unbiased?

EXAMPLE 1

Deciding on Methods of Data Collection

Consider the following statistical studies. Which method of data collection would you use to collect data for each study? Explain your reasoning.

1. A study of the effect of changing flight patterns on the number of airplane accidents
2. A study of the effect of aspirin on preventing heart attacks
3. A study of how fourth grade students solve a puzzle
4. A study of U.S. residents' approval rating of the U.S. president

SOLUTION

1. Because it is impractical to create this situation, use a simulation.
2. In this study, you want to measure the effect a treatment (taking aspirin) has on patients. So you would want to perform an experiment.
3. Because you want to observe and measure certain characteristics of part of a population, you could do an observational study.
4. You could use a survey that asks "Do you approve of the way the president is handling his job?"

Try It Yourself 1

Consider the following statistical studies. Which method of data collection would you use to collect data for each study?

1. A study of the effect of exercise on senior citizens
2. A study of the effect of radiation fallout on senior citizens

a. Identify the *focus* of the study.
b. Identify the *population* of the study.
c. Choose an appropriate *method of data collection*.

Answer: Page A30

Insight

A **biased sample** is one that is not representative of the population from which it is drawn. For instance, a sample consisting of only 18- to 22-year-old college students would not be representative of the entire 18- to 22-year-old population in the country.

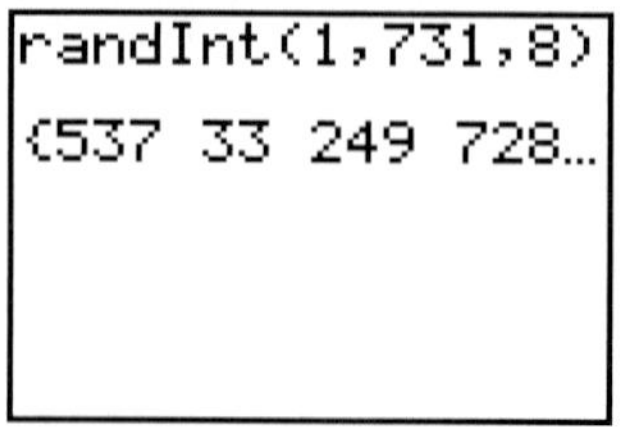

Random numbers generated using TI-83

Sampling Techniques

To collect unbiased data, a researcher must ensure that the sample is representative of the population. Appropriate sampling techniques must be used to ensure that inferences about the population are valid. Remember that when a study is done with faulty data, the results are questionable.

A **random sample** is one in which every member of the population has an equal chance of being selected. A **simple random sample** is a sample in which every possible sample of the same size has the same chance of being selected. One way to collect a simple random sample is to assign a different number to each member of the population and then use a random number table like the one in Appendix B. Responses, counts, or measures from members of the population whose numbers correspond to those generated using the table would be in the sample. Calculators and computer software programs are also used to generate random numbers (see page 30).

Table 1—Random Numbers

92630	78240	19267	95457	53497	23894	37708	79862
79445	78735	71549	44843	26104	67318	00701	34986
59654	71966	27386	50004	05358	94031	29281	18544
31524	49587	76612	39789	13537	48086	59483	60680
06348	76938	90379	51392	55887	71015	09209	79157

Portion of Table 1 found in Appendix B

For instance, to use a simple random sample to count the number of people who live in West Ridge County households, you could assign a different number to each household, use a computer or table of random numbers to generate a sample of numbers, and then count the number of people living in each selected household.

EXAMPLE 2

Using a Simple Random Sample

There are 731 students currently enrolled in statistics at your school. You wish to form a sample of eight students to answer some survey questions. Select the students who will belong to the simple random sample.

SOLUTION Assign numbers 1 to 731 to each student in the course. On the table of random numbers, choose a starting place at random and read the digits in groups of three (because 731 is a three-digit number). For example, if you started in the third row of the table at the beginning of the second column, you would group the numbers as follows:

719|66 2|738|6 50|004| 053|58 9|403|1 29|281| 185|44

Ignoring numbers greater than 731, the first eight numbers are 719, 662, 650, 4, 53, 589, 403, and 129. The students assigned these numbers will make up the sample.

Try It Yourself 2

A company employs 79 people. Choose a simple random sample of five to survey.

a. On the table, randomly choose a *starting place*.
b. *Read the digits* in groups of two.
c. Write the five random numbers.

Answer: Page A30

When you choose members of a sample, you should decide whether it is acceptable to have the same population member selected more than once. If it is acceptable, then the sampling process is said to be *with replacement*. If it is not acceptable, then the sampling process is said to be *without replacement*.

There are several other commonly used sampling techniques. Each has advantages and disadvantages.

- ***Stratified Sample*** When it is important for the sample to have members from each segment of the population, you should use a stratified sample. Depending on the focus of the study, members of the population are divided into two or more subsets, called *strata*, that share a similar characteristic such as age, gender, ethnicity, or even political preference. A sample is then randomly selected from each of the strata. Using a stratified sample ensures that each segment of the population is represented. For example, to collect a stratified sample of the number of people who live in West Ridge County households, you could divide the households into socioeconomic levels, and then randomly select households from each level.

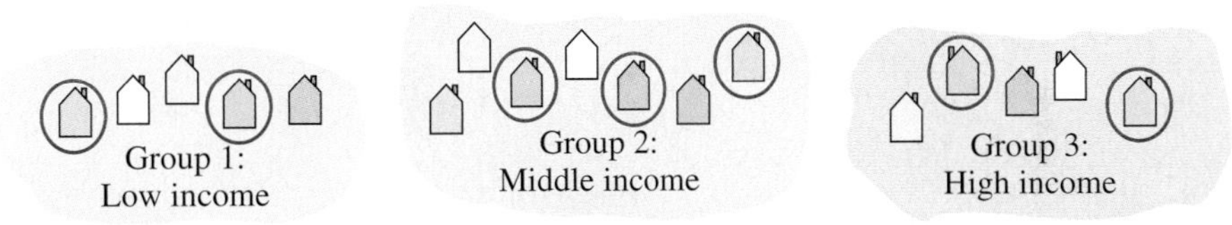

Stratified Sampling

- ***Cluster Sample*** When the population falls into naturally occurring subgroups, each having similar characteristics, a cluster sample may be the most appropriate. To select a cluster sample, divide the population into groups, called *clusters*, and select all of the members in one or more (but not all) of the clusters. Examples of clusters could be different sections of the same course or different branches of a bank. For instance, to collect a cluster sample of the number of people who live in West Ridge County households, divide the households into groups according to zip codes, then select all the households in one or more, but not all, zip codes and count the number of people living in each household. In using a cluster sample, care must be taken to ensure that all clusters have similar characteristics. For example, if one of the zip code clusters has a greater proportion of high-income people, the data might not be representative of the population.

Insight

For stratified sampling, each of the strata contains members with a certain characteristic (for example, a particular age group). In contrast, clusters consist of geographic groupings, and each cluster should consist of members with all of the characteristics (for example, all age groups). With stratified samples, some of the members of each group are used. In a cluster sampling, all of the members of one or more groups are used.

Zip Code Zones in West Ridge County

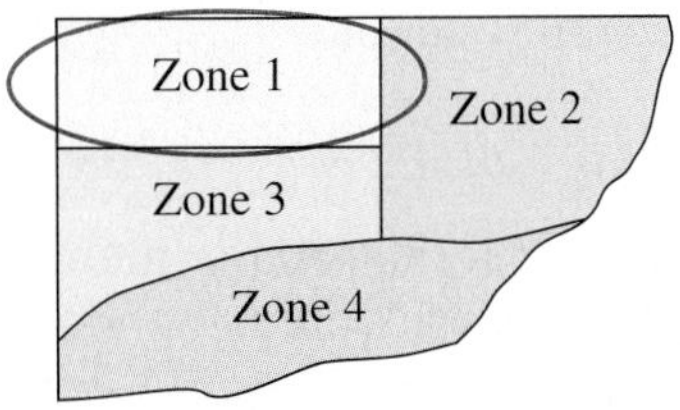

Cluster Sampling

- ***Systematic Sample*** A systematic sample is a sample in which each member of the population is assigned a number. The members of the population are ordered in some way, a starting number is randomly selected, and then sample members are selected at regular intervals from the starting number. (For instance, every 3rd, 5th, or 100th member is selected.) For example, to collect a systematic sample of the number of people who live in West Ridge County households, you could assign a different number to each household, randomly choose a starting number, select every 100th household, and count the number of people living in each.

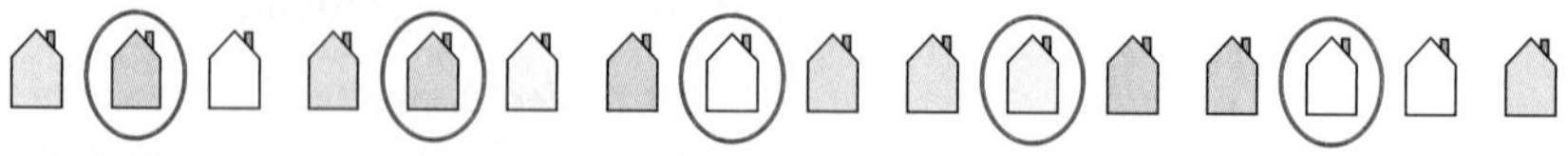

Systematic Sampling

An advantage of systematic sampling is that it is easy to use. In the case of any regularly occurring pattern in the data, however, this type of sampling should be avoided.

A type of sample that often leads to biased studies (so it is not recommended) is a **convenience sample.** A convenience sample consists only of available members of the population.

Note to Instructor

Ask students whether they have ever been part of a sampling process. Discuss the types of samples they have been part of. For students who don't think they have ever participated in a sampling, ask them if they have ever had a blood sample drawn or sampled a food.

EXAMPLE 3

Identifying Sampling Techniques

You are doing a study to determine the opinion of students at your school regarding gun control. Identify the sampling technique you are using if you select the samples listed.

1. You select a class at random and question each student in the class.
2. You divide the student population with respect to majors and randomly select and question some students in each major.
3. You assign each student a number and generate random numbers. You then question each student whose number is randomly selected.

SOLUTION

1. Because each class is a naturally occurring subgroup (a cluster) and you question each student in the class, this is a cluster sample.
2. Because students are divided into strata (majors) and a sample is selected from each major, this is a stratified sample.
3. Each sample of the same size has an equal chance of being selected and each student has an equal chance of being selected, so this is a simple random sample.

Try It Yourself 3

You want to determine the opinion of students at your school regarding gun control. Identify the sampling technique you are using if you select the samples listed.

1. You select students who are in your statistics class.
2. You assign each student a number and, after choosing a starting number, question every 25th student.

a. Determine *how* the sample is *selected.*
b. Identify the corresponding *sampling technique.*

Answer: Page A30

1.3 Exercises

Building Basic Skills and Vocabulary

True or False? In Exercises 1–4, determine whether the statement is true or false. If it is false, rewrite it as a true statement.

1. Using a systematic sample guarantees that members of each group within a population will be sampled.

2. A census is a count of part of a population.

3. The method for selecting a stratified sample is to order a population in some way and then select members of the population at regular intervals.

4. To select a cluster sample, divide a population into groups and then select all of the members in at least one (but not all) of the groups.

Deciding on the Method of Data Collection In Exercises 5–8, decide which method of data collection you would use to collect data for the study. Explain.

5. A study of the effect on the human digestive system of potato chips made with a fat substitute

6. A study of the effect of a product's warning label to determine whether consumers still buy the product

7. A study of how fast a virus would spread in a metropolitan area

8. A study of the ages of the 535 members of the U.S. Congress

Using and Interpreting Concepts

Identifying Sampling Techniques In Exercises 9–18, identify the sampling technique used and discuss potential sources of bias (if any). Explain.

9. Using random digit dialing, researchers called 1599 people and asked what obstacles (such as childcare) kept them from exercising. *(Source: Yankelovich Partners, Inc. for Shape Up America!)*

10. Chosen at random, 500 rural and 500 urban persons age 65 or older were asked about their health and experience with prescription drugs.

11. Questioning students as they left a university library, a researcher asked 358 students about their drinking habits.

12. After a hurricane, a disaster area is divided into 200 equal grids. Thirty of the grids are selected, and every occupied household in the grid is interviewed to help focus relief efforts on what residents require the most.

13. Chosen at random, 1210 hospital outpatients were contacted and asked their opinion of the care they received.

14. For quality assurance, every twentieth engine part is selected from an assembly line and tested for durability.

15. Soybeans are planted on a 48-acre field. The field is divided into one-acre subplots. A sample of plants is taken from each subplot to estimate the harvest.

1. False. Using stratified sampling guarantees that members of each group within a population will be sampled.
2. See Selected Answers, page A89.
3. See Odd Answers, page A47.
4. True
5. Perform an experiment because you want to measure the effect of a treatment on the human digestive system.
6. See Selected Answers, page A89.
7. Use a simulation because the situation is impractical.
8. See Selected Answers, page A89.
9. Simple random sampling is used because each telephone number has an equal chance of being dialed, and all samples of 1599 phone numbers have an equal chance of being selected. The sample may be biased because only homes with telephones will be sampled.
10. See Selected Answers, page A89.
11. See Odd Answers, page A47.
12. See Selected Answers, page A89.
13. Simple random sampling is used because each outpatient had an equal chance of being contacted, and all samples of 1210 outpatients had an equal chance of being selected.
14. See Selected Answers, page A89.
15. Stratified sampling is used because a sample is taken from each one-acre subplot.

16. Questioning teachers as they left a faculty lounge, a researcher asked 32 teachers about their teaching styles and grading methods.

17. A list of managers is compiled and ordered. After a starting number is randomly chosen, every ninth name is selected until 1000 managers are selected. The managers are questioned about the use of digital media.

18. From calls made with randomly generated telephone numbers, 1012 respondents were asked if they rented or owned their residence.

Choosing between a Census and a Sampling In Exercises 19 and 20, determine whether you would take a census or use a sampling. If you would use a sampling, decide what sampling technique you would use. Explain your reasoning.

19. The average salary of the 50 employees of a company

20. The most popular car color among car owners in the United States.

Recognizing a Biased Question In Exercises 21–24, determine whether the survey question is biased. If the question is biased, suggest a better wording.

21. Why is drinking fruit juice good for you?

22. Why are drivers who change lanes several times dangerous?

23. How many hours of sleep do you get on an average night?

24. Do you think the media have a negative effect on teen girls' dieting habits?

25. ***Writing*** Television program ratings by Nielsen Media Research are described on page 15. Discuss the strata used in the sample.

26. ***Writing*** Television program ratings by Nielsen Media Research are described on page 15. Why is it important to have a stratified sample for these ratings?

Extending Concepts

27. ***Open and Closed Questions*** Two types of survey questions are open questions and closed questions. An open question allows for any kind of response; a closed question allows only for a fixed response. For example, an open question and a closed question are given below. List an advantage and a disadvantage of an open question. Then list an advantage and a disadvantage of a closed question.

Open Question What can be done to get students to eat healthier foods?
Closed Question How would you get students to eat healthier foods?

1. Mandatory nutrition course
2. Offer only healthy foods in the cafeteria and remove unhealthy foods
3. Offer more healthy foods in the cafeteria and raise the prices on unhealthy foods

28. ***Who Picked These People?*** Sometimes, instead of selecting a sample to survey, some polling agencies ask people to call a telephone number and give their response to a question. (a) List an advantage and a disadvantage of a survey conducted in this manner. (b) What sampling technique is used in such a survey?

16. Convenience sampling is used because a sample is taken from members of a population that are readily available. The sample may be biased if the teachers sampled are not representative of the population of teachers.

17. Systematic sampling is used because every ninth name on a list is being selected.

18. See Selected Answers, page A89.

19. Census, because it is relatively easy to obtain the salaries of the 50 employees.

20. See Selected Answers, page A89.

21. The question is biased because it already suggests that drinking fruit juice is good for you. The question might be rewritten as "How does drinking fruit juice affect your health?"

22. The question is biased because it already suggests that drivers who change lanes several times are dangerous. The question might be rewritten as "Are drivers who change lanes several times dangerous?"

23. The survey question is unbiased.

24. The question is biased because it already suggests that the media have a negative effect on teen girls' dieting habits. The question might be rewritten as "Do you think the media have an effect on teen girls' dieting habits?"

25. The households sampled represent various locations, ethnic groups, and income brackets. Each of these variables is considered a stratum.

26. See Selected Answers, page A89.

27. Open Question

Advantage: Allows respondent to express some depth and shades of meaning in the answer.

Disadvantage: Not easily quantified and difficult to compare surveys.

Closed Question

Advantage: Easy to analyze results.

Disadvantage: May not provide appropriate alternatives and may influence the opinion of the respondent.

28. See Selected Answers, page A89.

Uses and Abuses

Statistics in the Real World

Uses

Surveys Surveys can be valuable in determining the attitude of a population about a candidate, product, or issue.

If you are working for a political candidate, it is important that you know how the voting population views your candidate. With this information, you might be able to address voter concerns and increase your candidate's chance of winning the election.

If you are working in the market research department of a manufacturing company, it is important that you know how the public will react to a proposed new product *before* the product is produced. Perhaps you will be able to alter the product's design to make it more appealing and capture a greater market share.

If you are working for an activist organization, it is important that you know how the population feels about your organization's issues. If you discover that the population does not support an issue, perhaps you will be able to change the public's attitude through an advertising campaign.

Abuses

Biased Samples The most common abuse (or misuse) of statistics is using a sample that does not represent the entire population of the study. Consider a phone survey of opinions about a candidate for a local school board. The survey consisted of phone calls to numbers listed in the local phone directory. Of the 930 numbers dialed, 543 were answered. Of the calls that were answered, 162 people agreed to take the survey. Of those surveyed, 62% plan on voting for the candidate. It should be clear that you cannot conclude from this survey that 62% of the voters in the locality plan on voting for the candidate. The survey did not include voters with unlisted phone numbers, or who did not happen to be home, or who did not agree to participate in the phone survey. In such cases, we say that the survey is biased toward people who have listed phone numbers, tend to be at home in the evening, and are willing to participate in surveys.

Biased Survey Questions Another common abuse is using survey questions that encourage respondents, either intentionally or unintentionally, to answer in a certain way. For instance, consider a survey about gun control. It seems clear that the following questions would produce very different results. (a) "Do people have the right to own a gun to protect themselves and their families?" (b) "Do people have the right to possess loaded guns in their homes?"

Exercises

1. ***Biased Sample*** Find an example of a survey that used a biased sample. Try hunting in magazines, newspapers, or on the Internet. Explain why the sample is biased. What could have been done to make the sample unbiased?

2. ***Biased Survey Question*** Find an example of a survey that used a biased survey question. Try hunting in magazines, newspapers, or on the Internet. Explain why the question is biased. Rewrite the survey question so that it is unbiased, or at least less biased.

1 Chapter Summary

What did you learn?

What did you learn?	Review Exercises
Section 1.1	
◆ How to distinguish between a population and a sample	*1–4*
◆ How to distinguish between a parameter and a statistic	*5–8*
◆ How to distinguish between descriptive statistics and inferential statistics	*9–10*
Section 1.2	
◆ How to distinguish between qualitative data and quantitative data	*11–14*
◆ How to classify data with respect to the four levels of measurement: nominal, ordinal, interval, and ratio	*15–18*
Section 1.3	
◆ How data are collected: by doing an observational study, performing an experiment, using a simulation, or using a survey	*19–22*
◆ How to create a sample using random sampling, simple random sampling, stratified sampling, cluster sampling, and systematic sampling	*23–28*
◆ How to identify a biased sample	*29–32*

1 Review Exercises

Section 1.1

In Exercises 1–4, identify the population and the sample.

1. A survey of 898 U.S. adult VCR owners found that 16% had VCR clocks that were currently blinking "12:00." *(Source: Wirthlin Worldwide)*

2. Thirty-eight nurses working in the San Francisco area were surveyed concerning their opinions of managed health care.

3. A study of 82,188 U.S. ATMs (automated teller machines) found that the average surcharge for withdrawals from a competing bank was $1.48. *(Source: Dove Consulting Group, Inc.)*

4. A survey of 787 small-business managers found that 72% are optimistic about the future of their company. *(Source: OPEN Small Business Network)*

In Exercises 5–8, determine whether the numerical value describes a parameter or a statistic.

5. The 2003 team payroll of the Baltimore Orioles was $69,452,275. *(Source: Associated Press)*

6. In a survey of a sample of U.S. adults, 62% owned a portable cellular phone. *(Source: Scarborough Research)*

7. In a recent survey at the University of Arizona, 89 students were majoring in astronomy. *(Source: University of Arizona Student Research Office)*

8. Nineteen percent of a sample of Indiana ninth graders surveyed smoked cigarettes daily. *(Source: Indiana University)*

9. Which part of the study described in Exercise 3 represents the descriptive branch of statistics? Make an inference based on the results of the study.

10. Which part of the survey described in Exercise 4 represents the descriptive branch of statistics? Make an inference based on the results of the survey.

Section 1.2

In Exercises 11–14, determine which data are qualitative data and which are quantitative data. Explain your reasoning.

11. The monthly salaries of the employees at an accounting firm

12. The Social Security numbers of the employees at an accounting firm

13. The ages of a sample of 350 residents of nursing homes

14. The zip codes of a sample of 350 residents of nursing homes

In Exercises 15–18, identify the data set's level of measurement. Explain your reasoning.

15. The daily high temperatures (in degrees Fahrenheit) for Mohave, Arizona, for a week in June are listed. *(Source: Arizona Meteorological Network)*

93 91 86 94 103 104 103

---

1. Population: Collection of all U.S. adult VCR owners
   Sample: Collection of the 898 U.S. adult VCR owners that were sampled
2. Population: Collection of all nurses in San Francisco area
   Sample: Collection of 38 nurses in San Francisco area that were sampled
3. Population: Collection of all U.S. ATMs
   Sample: Collection of 82,188 ATMs that were sampled
4. Population: Collection of all small-business managers
   Sample: Collection of 787 small-business managers that were sampled
5. Parameter
6. Statistic
7. Parameter
8. Statistic
9. The statement "the average surcharge for withdrawals from a competing bank was $1.48" is an application of descriptive statistics.
   An inference drawn from the sample is that all ATMs charge an average of $1.48 for withdrawals from a competing bank.
10. The statement "72% are optimistic about the future of their company" is an application of descriptive statistics.
    An inference drawn from the sample is that 72% of all small-business managers are optimistic about the future of their company.
11. Quantitative, because monthly salaries are numerical measurements.
12. Qualitative because Social Security numbers are merely labels for employees.
13. Quantitative, because ages are numerical measurements.
14. Qualitative because zip codes are merely labels for nursing homes.
15. Interval. It makes no sense saying that 100 degrees is twice as hot as 50 degrees.

16. The EPA size classes for a sample of automobiles are listed.

subcompact compact midsize large compact large

17. The four departments of a motor company are listed.

Sales Service Parts Body Shop

18. The heights (in inches) of the 2003–2004 Chicago Bulls are listed. *(Source: National Basketball Association)*

82 76 85 77 83 81 79 80 77 75 80 80 81 74 81

Section 1.3

In Exercises 19–22, decide which method of data collection you would use to collect data for the study. Explain your reasoning.

19. A study of charitable donations of the judges in Sioux Falls, South Dakota

20. A study of the effect of kangaroos on the Florida Everglades ecosystem

21. A study of the effects of a plant hormone on chrysanthemums

22. A study of college students' opinions on space exploration

In Exercises 23–28, identify which sampling technique was used in the study. Explain your reasoning.

23. Calling randomly generated telephone numbers, a study asked 1001 U.S. adults which medical conditions could be prevented by their diet. *(Adapted from Wirthlin Worldwide)*

24. A student asks 18 friends to participate in a psychology experiment.

25. A pregnancy study in Cebu, Philippines, randomly selected 33 communities from the Cebu metropolitan area, then interviewed all available pregnant women in these communities. *(Adapted from Cebu Longitudinal Health and Nutrition Survey)*

26. Law enforcement officials use a radar gun to measure the speed of every tenth vehicle on an interstate.

27. Twenty-five students are randomly selected from each grade level at a high school and surveyed about their study habits.

28. A journalist interviews 123 people after they leave a restaurant and asks them how confident they are that the food is safe.

In Exercises 29–32, identify a bias or error that might occur in the indicated survey or study.

29. The phone survey in Exercise 23

30. The psychology experiment in Exercise 24

31. The pregnancy study in Exercise 25

32. The vehicle speed sampling in Exercise 26

16. Ordinal. The data are categorical but could be arranged in order of car size.

17. Nominal. The data are categorical and cannot be arranged in a meaningful order.

18. Ratio. The data are numerical, and it makes sense to say one player is twice as tall as another player.

19. Take a census because judges keep accurate records of charitable donations.

20. Perform a simulation because it is impractical to create this situation.

21. Perform an experiment because you want to measure the effect of a treatment on chrysanthemums.

22. Take a sample because asking every college student about space exploration would be nearly impossible.

23. Simple random sampling is used because random telephone numbers were generated and called.

24. Convenience sampling is used because the student sampled a convenient group of friends.

25. Cluster sampling is used because each community is considered a cluster and every pregnant woman in a selected community is surveyed.

26. Systematic sampling is used because every tenth car is measured.

27. Stratified sampling is used because 25 students are randomly selected from each grade level.

28. See Selected Answers, page A89.

29. Telephone sampling samples only individuals who have telephones, are available, and are willing to respond.

30. See Selected Answers, page A89.

31. The selected communities may not be representative of the entire area.

32. See Selected Answers, page A89.

1 Chapter Quiz

1. Population: Collection of all individuals with sleep disorders

   Sample: Collection of 254 patients in study
2. (a) Statistic (b) Parameter
3. (a) Qualitative (b) Quantitative
4. (a) Nominal, because no mathematical computations can be made.

   (b) Ratio, because one data value can be expressed as a multiple of another.

   (c) Interval, because meaningful differences between entries can be calculated but a zero entry is not an inherent zero.
5. (a) Perform an experiment because you want to measure the effect of a treatment on lead levels in adults.

   (b) Use sampling because it would be impossible to question everyone in the population.
6. (a) Convenience sampling, because all of the people sampled are in one convenient location.

   (b) Systematic sampling, because every fifth part is sampled.

   (c) Stratified sampling, because the population is first stratified and then a sample is collected from each stratum.
7. Convenience
8. (a) False. A statistic is a numerical measure that describes a sample characteristic.

   (b) False. Ratio data represent the highest level of measurement.

Take this quiz as you would take a quiz in class. After you are done, check your work against the answers given in the back of the book.

1. Identify the population and the sample in the following study.

A study of 254 patients with sleep disorders was conducted to find a link between obesity and sleep disorders.

2. Determine whether the numerical value is a parameter or a statistic.

(a) In a survey of a sample of parents, 53% said they protect their children from sun exposure using sunscreen. *(Source: Morbidity and Mortality Weekly Report)*

(b) In a union's vote, 67% of all union members voted to ratify a contract proposal.

3. Determine whether the data are qualitative or quantitative.

(a) A database of student identification numbers

(b) The test scores in a statistics class

4. Identify each data set's level of measurement. Explain your reasoning.

(a) A list of the uniform numbers retired by each major league baseball team

(b) The number of products sold by a toy manufacturer each quarter for the current fiscal year

(c) The finishing times of the runners in the Boston marathon

5. Decide which method of data collection you would use to gather data for each study. Explain your reasoning.

(a) A study on the effect of low dietary intake of vitamin C and iron on lead levels in adults

(b) The ages of people living within 500 miles of your home

6. Identify which sampling technique was used in each study. Explain your reasoning.

(a) A journalist goes to a beach to ask people how they feel about water pollution.

(b) For quality assurance, every fifth engine part is selected from an assembly line and tested for durability.

(c) A study on attitudes about smoking is conducted at a college. The students are divided by class (freshman, sophomore, junior, and senior). Then a random sample is selected from each class and interviewed.

7. Which sampling technique used in Exercise 6 could lead to a biased study?

8. Determine whether each statement is true or false. If it is false, rewrite it as a true statement.

(a) A parameter is a numerical measure that describes a sample characteristic.

(b) Ordinal data represent the highest level of measurement.

PUTTING IT ALL TOGETHER

Real Statistics ■ Real Decisions

You are a researcher for a professional research firm. Your firm has won a contract on doing a study for a transportation industry publication. The editors of the publication would like to know their readers' thoughts on personal and public transportation in fields such as safety, mobility, economic growth, human and natural environment, and security. They would also like to know the thoughts of adults who use transportation for business as well as for recreation.

The editors have given you their readership database and 20 questions they would like to ask (two sample questions from a previous study are given at the right). You know that it is too expensive to contact all of the readers, so you need to determine a way to contact a representative sample of the entire readership population.

Exercises

1. *How Would You Do It?*

(a) What sampling technique would you use to select the sample for the study? Why?

(b) Will the technique you chose in part (a) give you a sample that is representative of the population?

(c) Describe the method for collecting data.

(d) Identify possible flaws or biases in your study.

2. *Data Classification*

(a) What type of data do you expect to collect: qualitative, quantitative, or both? Why?

(b) What levels of measurement do you think the data in the study will be? Why?

(c) Will the data collected for the study represent a population or a sample?

(d) Will the numerical descriptions of the data be parameters or statistics?

3. *How They Did It*

When the *Bureau of Transportation Statistics*, part of the *Department of Transportation*, did a similar study, they used a telephone survey. They called 3626 numbers and received 1121 complete surveys.

(a) Describe some possible errors in collecting data by telephone surveys.

(b) Compare your method for collecting data in Exercise 1 to this method.

When you travel by personal vehicle, what is most important to you?

Response	Percent
Convenience	44.5%
Security/safety	32.2%
Time	10.7%
Cost	6.7%
Other	1.1%
No response	4.8%

(Source: Bureau of Transportation Statistics)

On a typical day in the past month, how did you get to work?

Method	Percent
Personal vehicle	47.1%
Company car	2.5%
Public transit	1.3%
Walk	1.2%
Carpool	0.5%
Bicycle	0.4%
Other	3.2%
No response	43.8%

(Source: Bureau of Transportation Statistics)

www.bts.gov

HISTORY OF STATISTICS - TIMELINE

CONTRIBUTOR	TIME	CONTRIBUTION
John Graunt (1620–1674)	17th century	Studied records of deaths in London in the early 1600s. The first to make extensive statistical observations from massive amounts of data (Chapter 2), his work laid the foundation for modern statistics.
Blaise Pascal (1623–1662) *Pierre deFermat* (1601–1665)	17th century	Pascal and Fermat corresponded about basic probability problems (Chapter 3)—especially those dealing with gaming and gambling.
Pierre Laplace (1749–1827)	18th century	Studied probability (Chapter 3) and is credited with putting probability on a sure mathematical footing.
Carl Friedrich Gauss (1777–1855)	18th century	Studied regression and the method of least squares (Chapter 9) through astronomy. In his honor, the normal distribution is sometimes called the Gaussian distribution.
Lambert Quetelet (1796–1874)	19th century	Used descriptive statistics (Chapter 2) to analyze crime and mortality data and studied census techniques. Described normal distributions (Chapter 5) in connection with human traits such as height.
Francis Galton (1822–1911)	19th century	Used regression and correlation (Chapter 9) to study genetic variation in humans. He is credited with the discovery of the Central Limit Theorem (Chapter 5).
Karl Pearson (1857–1936) 	20th century	Studied natural selection using correlation (Chapter 9). Formed first academic department of statistics and helped develop chi-square analysis (Chapter 6).
William Gosset (1876–1937)	20th century	Studied process of brewing and developed *t*-test to correct problems connected with small sample sizes (Chapter 6).
Charles Spearman (1863–1945)	20th century	British psychologist who was one of the first to develop intelligence testing using factor analysis (Chapter 10).
Ronald Fisher (1890–1962) 	20th century	Studied biology and natural selection and developed ANOVA (Chapter 10), stressed the importance of experimental design (Chapter 1), and was the first to identify the null and alternative hypotheses (Chapter 7).
Frank Wilcoxon (1892–1965)	20th century (later)	Biochemist who used statistics to study plant pathology. He introduced two-sample tests (Chapter 8), which led the way to the development of nonparametric statistics.
John Tukey (1915–2000) 	20th century (later)	Worked at Princeton during World War II. Introduced exploratory data analysis techniques such as stem-and-leaf plots (Chapter 2). Also, worked at Bell Laboratories and is best known for his work in inferential statistics (Chapters 6–11).
David Kendall (1918–)	20th century (later)	Worked at Princeton and Cambridge. Is a leading authority on applied probability and data analysis (Chapters 2 and 3).

TECHNOLOGY

Note to Instructor

As an instructor, you can decide the extent to which you want to use technology with this text. Supplements with detailed instructions are available for each of the technology tools shown.

Using Technology in Statistics

With large data sets, you will find that calculators or computer software programs can help perform calculations and create graphics. Of the many calculators and statistical software programs that are available, we have chosen to incorporate the TI-83 graphing calculator and MINITAB and Excel software into this text.

The following example shows how to use these three technologies to generate a list of random numbers. This list of random numbers can be used to select sample members or perform simulations.

EXAMPLE

Generating a List of Random Numbers

A quality control department inspects a random sample of 15 of the 167 cars that are assembled at an auto plant. How should the cars be chosen?

SOLUTION

One way to choose the sample is to number the cars from 1 to 167. To form a list of random numbers from 1 to 167, you can use technology. Each of the technology tools shown requires different steps to generate the list. Each, however, does require that you identify the minimum value as 1 and the maximum value as 167. Check your user's manual for specific instructions.

MINITAB

↓	C1
1	167
2	11
3	74
4	160
5	18
6	70
7	80
8	56
9	37
10	6
11	82
12	126
13	98
14	104
15	137

EXCEL

	A
1	41
2	16
3	91
4	58
5	151
6	36
7	96
8	154
9	2
10	113
11	157
12	103
13	64
14	135
15	90

TI-83

```
randInt(1, 167, 15)
{17 42 152 59 5 116
125 64 122 55 58 60
82 152 105}
```

MINITAB EXCEL TI-83

Note to Instructor

In Exercises 6 and 8, students may argue that these results are reasonable because there is a relatively small number of trials in each case. This is a reasonable answer because of the law of large numbers, which will be discussed in Chapter 3.

Recall that when you generate a list of random numbers, you should decide whether it is acceptable to have numbers that repeat. If it is acceptable, then the sampling process is said to be *with replacement.* If it is not acceptable, then the sampling process is said to be *without replacement.*

With each of the three technology tools shown on page 30, you have the capability of sorting the list so that the numbers appear in order. Sorting helps you see whether any of the numbers in the list repeat. If it is not acceptable to have repeats, you should specify that the tool generate more random numbers than you need.

Exercises

1. A CPA is examining the records of a business. The business has 74 major accounts, and the CPA decides to audit a random sample of 8 of the accounts. Describe how this audit could be done. Then use technology to generate a list of 8 random numbers from 1 to 74 and order the list.

2. A quality control department is testing 20 batteries from a shipment of 200 batteries. Describe how this test could be done. Then use technology to generate a list of 20 random numbers from 1 to 200 and order the list.

3. Consider the population of ten digits: 0, 1, 2, 3, 4, 5, 6, 7, 8, and 9. Select three random samples of five digits from this list. Find the average of each sample. Compare your results with the average of the entire population. Comment on your results. (*Hint:* To find the average, sum the data entries and divide the sum by the number of entries.)

4. Consider the population of 41 whole numbers from 0 to 40. What is the average of these numbers? Select three random samples of seven numbers from this list. Find the average of each sample. Compare your results with the average of the entire population. Comment on your results. (*Hint:* To find the average, sum the data entries and divide the sum by the number of entries.)

5. Use random numbers to simulate rolling a six-sided die 60 times. How many times did you obtain each number from 1 to 6? Are the results what you expected?

6. You rolled a six-sided die 60 times and got the following tally.

20 ones	20 twos	15 threes
3 fours	2 fives	0 sixes

Does this seem like a reasonable result? What inference might you draw from the result?

7. Use random numbers to simulate tossing a coin 100 times. Let 0 represent heads, and let 1 represent tails. How many times did you obtain each number? Are the results what you expected?

8. You tossed a coin 100 times and got 77 heads and 23 tails. Does this seem like a reasonable result? What inference might you draw from the result?

9. A political analyst would like to survey a sample of the registered voters in a county. The county has 47 election districts. How could the analyst use random numbers to obtain a cluster sample?

Extended solutions are given in the *Technology Supplement.*
Technical instruction is provided for MINITAB, Excel, and the TI-83.

CHAPTER 2

Descriptive Statistics

2.1 Frequency Distributions and Their Graphs

2.2 More Graphs and Displays

2.3 Measures of Central Tendency

2.4 Measures of Variation

Case Study

2.5 Measures of Position

Uses and Abuses

Real Statistics–Real Decisions

Technology

Akhiok is a small fishing village on Kodiak Island. Akhiok has a population of 80 residents.

Photographs © Roy Corral

Where You've Been

In Chapter 1, you learned that there are many ways to collect data. Usually, researchers must work with sample data in order to analyze populations, but occasionally it is possible to collect all the data for a given population. For instance, the following represents the ages of the entire population of the 80 residents of Akhiok, Alaska, from the 2000 census.

25, 5, 18, 12, 60, 44, 24, 22, 2, 7, 15, 39, 58, 53, 36, 42, 16, 20, 1, 5, 39, 51, 44, 23, 3, 13, 37, 56, 58, 13, 47, 23, 1, 17, 39, 13, 24, 0, 39, 10, 41, 1, 48, 17, 18, 3, 72, 20, 3, 9, 0, 12, 33, 21, 40, 68, 25, 40, 59, 4, 67, 29, 13, 18, 19, 13, 16, 41, 19, 26, 68, 49, 5, 26, 49, 26, 45, 41, 19, 49

Where You're Going

In Chapter 2, you will learn ways to organize and describe data sets. The goal is to make the data easier to understand by describing trends, averages, and variations. For instance, in the raw data showing the ages of the residents of Akhiok, it is not easy to see any patterns or special characteristics. Here are some ways you can organize and describe the data.

Make a frequency distribution table.

Class	Frequency, f
0–9	15
10–19	19
20–29	14
30–39	7
40–49	14
50–59	6
60–69	4
70–79	1

Draw a histogram.

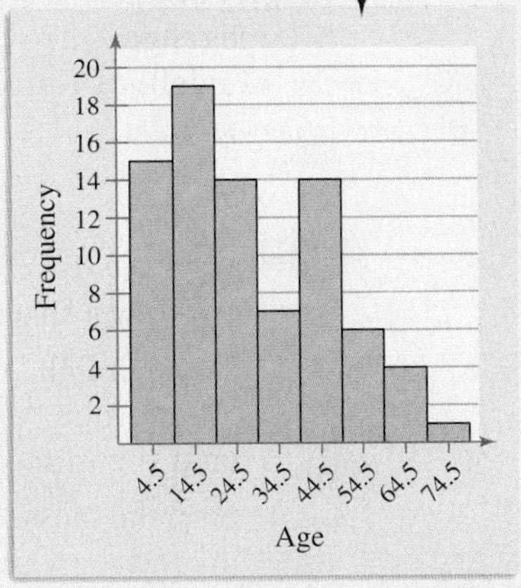

$$\text{Mean} = \frac{0 + 0 + 1 + 1 + 1 + \cdots + 67 + 68 + 68 + 72}{80}$$

$$= \frac{2226}{80}$$

$$\approx 27.8 \text{ years}$$

Find an average.

$$\text{Range} = 72 - 0$$

$$= 72 \text{ years}$$

Find how the data vary.

2.1 Frequency Distributions and Their Graphs

What You Should Learn

- How to construct a frequency distribution including limits, boundaries, midpoints, relative frequencies, and cumulative frequencies
- How to construct frequency histograms, frequency polygons, relative frequency histograms, and ogives

Frequency Distributions • Graphs of Frequency Distributions

Frequency Distributions

When a data set has many entries, it can be difficult to see patterns. In this section, you will learn how to organize data sets by grouping the data into intervals called classes and forming a frequency distribution. You will also learn how to use frequency distributions to construct graphs.

DEFINITION

A **frequency distribution** is a table that shows **classes** or **intervals** of data entries with a count of the number of entries in each class. The **frequency** f of a class is the number of data entries in the class.

Example of a Frequency Distribution

Class	Frequency, f
1–5	5
6–10	8
11–15	6
16–20	8
21–25	5
26–30	4

In the frequency distribution shown there are six classes. The frequencies for each of the six classes are 5, 8, 6, 8, 5, and 4. Each class has a **lower class limit,** which is the least number that can belong to the class, and an **upper class limit,** which is the greatest number that can belong to the class. In the frequency distribution shown, the lower class limits are 1, 6, 11, 16, 21, and 26, and the upper class limits are 5, 10, 15, 20, 25, and 30. The **class width** is the distance between lower (or upper) limits of consecutive classes. For instance, the class width in the frequency distribution shown is $6 - 1 = 5$.

The difference between the maximum and minimum data entries is called the **range.** For instance, if the maximum data entry is 29, and the minimum data entry is 1, the range is $29 - 1 = 28$. You will learn more about the range in Section 2.4.

Guidelines for constructing a frequency distribution from a data set are as follows.

Study Tip

In a frequency distribution, it is best if each class has the same width. Answers shown will use the minimum data value for the lower limit of the first class. Sometimes it may be more convenient to choose a value that is slightly lower than the minimum value. The frequency distribution produced will vary slightly.

GUIDELINES

Constructing a Frequency Distribution from a Data Set

1. Decide on the number of classes to include in the frequency distribution. The number of classes should be between 5 and 20; otherwise, it may be difficult to detect any patterns.
2. Find the class width as follows. Determine the range of the data, divide the range by the number of classes, and *round up to the next convenient number.*
3. Find the class limits. You can use the minimum data entry as the lower limit of the first class. To find the remaining lower limits, add the class width to the lower limit of the preceding class. Then find the upper limit of the first class. Remember that classes cannot overlap. Find the remaining upper class limits.
4. Make a tally mark for each data entry in the row of the appropriate class.
5. Count the tally marks to find the total frequency f for each class.

Note to Instructor

Let students know that there are many correct versions for a frequency distribution. To make it easy to check answers, however, they should follow the conventions shown in the text.

EXAMPLE 1

Constructing a Frequency Distribution from a Data Set

The following sample data set lists the number of minutes 50 Internet subscribers spent on the Internet during their most recent session. Construct a frequency distribution that has seven classes.

50	40	41	17	11	7	22	44	28	21	19	23	37	51	54	42	88
41	78	56	72	56	17	7	69	30	80	56	29	33	46	31	39	20
18	29	34	59	73	77	36	39	30	62	54	67	39	31	53	44	

Insight

If you obtain a whole number when calculating the class width of a frequency distribution, use the next whole number as the class width. Doing this ensures you have enough space in your frequency distribution for all the data values.

SOLUTION

1. The number of classes (7) is stated in the problem.
2. The minimum data entry is 7 and the maximum data entry is 88, so the range is 81. Divide the range by the number of classes and round up to find that the class width is 12.

$$\text{Class width} = \frac{88 - 7}{7} \qquad \frac{\text{Maximum entry} - \text{Minimum entry}}{\text{Number of classes}}$$

$$= \frac{81}{7} \qquad \frac{\text{Range}}{\text{Number of classes}}$$

$$\approx 11.57 \qquad \text{Round up to 12.}$$

3. The minimum data entry is a convenient lower limit for the first class. To find the lower limits of the remaining six classes, add the class width of 12 to the lower limit of each previous class. The upper limit of the first class is 18, which is one less than the lower limit of the second class. The upper limits of the other classes are $18 + 12 = 30$, $30 + 12 = 42$, and so on. The lower and upper limits for all seven classes are shown.
4. Make a tally mark for each data entry in the appropriate class.
5. The number of tally marks for a class is the frequency for that class.

Lower limit	Upper limit
7	18
19	30
31	42
43	54
55	66
67	78
79	90

The frequency distribution is shown in the following table. The first class, 7–18, has six tally marks. So, the frequency for this class is 6. Notice that the sum of the frequencies is 50, which is the number of entries in the sample data set. The sum is denoted by Σf, where Σ is the uppercase Greek letter **sigma.**

Study Tip

The uppercase Greek letter sigma (Σ) is used throughout statistics to indicate a summation of values.

Frequency Distribution for Internet Usage (in minutes)

Minutes online → Class; Number of subscribers → Frequency, f

Class	Tally	Frequency, f
7–18	卌 I	6
19–30	卌 卌	10
31–42	卌 卌 III	13
43–54	卌 III	8
55–66	卌	5
67–78	卌 I	6
79–90	II	2
		$\Sigma f = 50$

Check that the sum of the frequencies equals the number in the sample.

Note to Instructor

Be sure that students interpret the class width correctly as the distance between lower (or upper) limits of consecutive classes. A common error is to use a class width of 11 for the class 7–18. Students should be shown that this class actually has a width of 12.

Try It Yourself 1

Construct a frequency distribution using the Akhiok population data set listed in the Chapter Opener on page 33. Use eight classes.

a. State the *number of classes*.
b. Find the minimum and maximum values and the *class width*.
c. Find the *class limits*.
d. *Tally* the data entries.
e. Write the *frequency f* for each class.

Answer: Page A30

After constructing a standard frequency distribution such as the one in Example 1, you can include several additional features that will help provide a better understanding of the data. These features, the midpoint, relative frequency, and cumulative frequency of each class, can be included as additional columns in your table.

DEFINITION

The **midpoint** of a class is the sum of the lower and upper limits of the class divided by two. The midpoint is sometimes called the *class mark*.

$$\text{Midpoint} = \frac{(\text{Lower class limit}) + (\text{Upper class limit})}{2}$$

The **relative frequency** of a class is the portion or percentage of the data that falls in that class. To find the relative frequency of a class, divide the frequency f by the sample size n.

$$\text{Relative frequency} = \frac{\text{Class frequency}}{\text{Sample size}} = \frac{f}{n}$$

The **cumulative frequency** of a class is the sum of the frequency for that class and all previous classes. The cumulative frequency of the last class is equal to the sample size n.

After finding the first midpoint, you can find the remaining midpoints by adding the class width to the previous midpoint. For instance, if the first midpoint is 12.5 and the class width is 12, then the remaining midpoints are

$$12.5 + 12 = 24.5$$
$$24.5 + 12 = 36.5$$
$$36.5 + 12 = 48.5$$
$$48.5 + 12 = 60.5$$

and so on.

You can write the relative frequency as a fraction, decimal, or percent. The sum of the relative frequencies of all the classes must equal 1 or 100%.

EXAMPLE 2

Midpoints, Relative and Cumulative Frequencies

Using the frequency distribution constructed in Example 1, find the midpoint, relative frequency, and cumulative frequency for each class. Identify any patterns.

SOLUTION The midpoint, relative frequency, and cumulative frequency for the first three classes are calculated as follows.

Class	f	Midpoint	Relative frequency	Cumulative frequency
7–18	6	$\frac{7+18}{2} = 12.5$	$\frac{6}{50} = 0.12$	6
19–30	10	$\frac{19+30}{2} = 24.5$	$\frac{10}{50} = 0.2$	$6 + 10 = 16$
31–42	13	$\frac{31+42}{2} = 36.5$	$\frac{13}{50} = 0.26$	$16 + 13 = 29$

The remaining midpoints, relative frequencies, and cumulative frequencies are shown in the following expanded frequency distribution.

Frequency Distribution for Internet Usage (in minutes)

Minutes online → Class; Number of subscribers → Frequency, f; Portion of subscribers → Relative frequency

Class	Frequency, f	Midpoint	Relative frequency	Cumulative frequency
7–18	6	12.5	0.12	6
19–30	10	24.5	0.2	16
31–42	13	36.5	0.26	29
43–54	8	48.5	0.16	37
55–66	5	60.5	0.1	42
67–78	6	72.5	0.12	48
79–90	2	84.5	0.04	50
	$\Sigma f = 50$		$\Sigma \frac{f}{n} = 1$	

Interpretation There are several patterns in the data set. For instance, the most common time span that users spent online was 31 to 42 minutes.

Try It Yourself 2

Using the frequency distribution constructed in Try It Yourself 1, find the midpoint, relative frequency, and cumulative frequency for each class. Identify any patterns.

a. Use the formulas to *find each midpoint, relative frequency*, and *cumulative frequency*.
b. *Organize your results* in a frequency distribution.
c. *Identify* patterns that emerge from the data.

Answer: Page A30

Graphs of Frequency Distributions

Sometimes it is easier to identify patterns of a data set by looking at a graph of the frequency distribution. One such graph is a frequency histogram.

DEFINITION

A **frequency histogram** is a bar graph that represents the frequency distribution of a data set. A histogram has the following properties.

1. The horizontal scale is quantitative and measures the data values.
2. The vertical scale measures the frequencies of the classes.
3. Consecutive bars must touch.

Study Tip

If data entries are integers, subtract 0.5 from each lower limit to find the lower class boundaries. To find the upper class boundaries, add 0.5 to each upper limit. The upper boundary of a class will equal the lower boundary of the next higher class.

Because consecutive bars of a histogram must touch, bars must begin and end at class boundaries instead of class limits. **Class boundaries** are the numbers that separate classes *without* forming gaps between them. You can mark the horizontal scale either at the midpoints or at the class boundaries, as shown in Example 3.

EXAMPLE 3

Constructing a Frequency Histogram

Draw a frequency histogram for the frequency distribution in Example 2. Describe any patterns.

SOLUTION First, find the class boundaries. The distance from the upper limit of the first class to the lower limit of the second class is $19 - 18 = 1$. Half this distance is 0.5. So, the lower and upper boundaries of the first class are as follows:

First class lower boundary $= 7 - 0.5 = 6.5$

First class upper boundary $= 18 + 0.5 = 18.5$

The boundaries of the remaining classes are shown in the table. Using the class midpoints or class boundaries for the horizontal scale and choosing possible frequency values for the vertical scale, you can construct the histogram.

Class	Class boundaries	Frequency, f
7–18	6.5–18.5	6
19–30	18.5–30.5	10
31–42	30.5–42.5	13
43–54	42.5–54.5	8
55–66	54.5–66.5	5
67–78	66.5–78.5	6
79–90	78.5–90.5	2

Internet Usage (labeled with class midpoints)

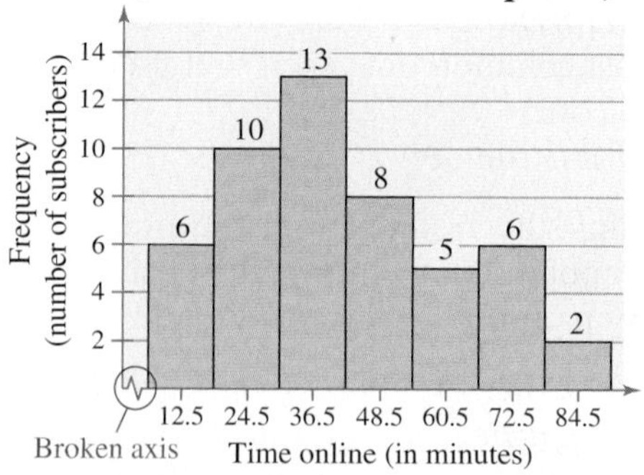

Internet Usage (labeled with class boundaries)

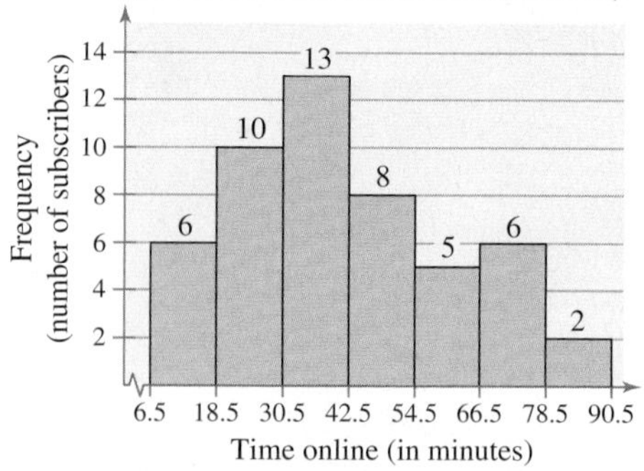

Interpretation From either histogram, you can see that more than half of the subscribers spent between 19 and 54 minutes on the Internet during their most recent session.

Try It Yourself 3

Use the frequency distribution from Try It Yourself 1 to construct a frequency histogram that represents the ages of the residents of Akhiok. Describe any patterns.

a. Find the *class boundaries*.
b. Choose appropriate *horizontal and vertical scales*.
c. Use the frequency distribution to *find the height of each bar*.
d. *Describe* any patterns for the data.

Answer: Page A31

Another way to graph a frequency distribution is to use a frequency polygon. A **frequency polygon** is a line graph that emphasizes the continuous change in frequencies.

EXAMPLE 4

Constructing a Frequency Polygon

Draw a frequency polygon for the frequency distribution in Example 2.

SOLUTION To construct the frequency polygon, use the same horizontal and vertical scales that were used in the histogram labeled with class midpoints in Example 3. Then plot points that represent the midpoint and frequency of each class and connect the points in order from left to right. Because the graph should begin and end on the horizontal axis, extend the left side to one class width before the first class midpoint and extend the right side to one class width after the last class midpoint.

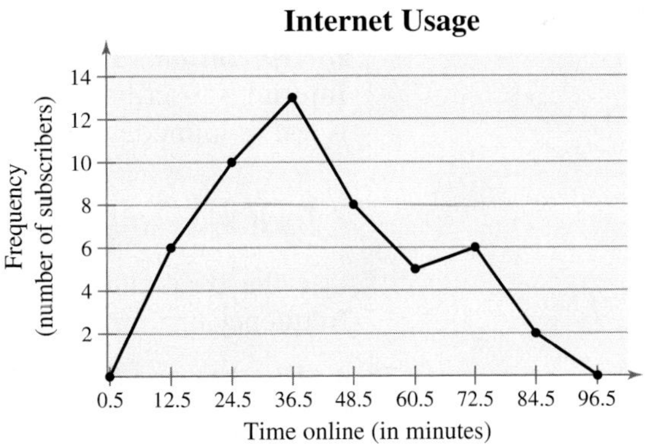

Interpretation You can see that the frequency of subscribers increases up to 36.5 minutes and then decreases.

> **Study Tip**
>
> A histogram and its corresponding frequency polygon are often drawn together. If you have not already constructed the histogram, begin constructing the frequency polygon by choosing appropriate horizontal and vertical scales. The horizontal scale should consist of the class midpoints, and the vertical scale should consist of appropriate frequency values.

Try It Yourself 4

Use the frequency distribution from Try It Yourself 1 to construct a frequency polygon that represents the ages of the residents of Akhiok. Describe any patterns.

a. Choose appropriate *horizontal* and *vertical scales.*
b. *Plot points* that represent the midpoint and frequency for each class.
c. *Connect the points* and extend the sides as necessary.
d. *Describe* any patterns for the data.

Answer: Page A31

A **relative frequency histogram** has the same shape and the same horizontal scale as the corresponding frequency histogram. The difference is that the vertical scale measures the *relative* frequencies, not frequencies.

Picturing the World

Old Faithful, a geyser at Yellowstone National Park, erupts on a regular basis. The time spans of a sample of eruptions are given in the relative frequency histogram. *(Source: Yellowstone National Park)*

Old Faithful Eruptions

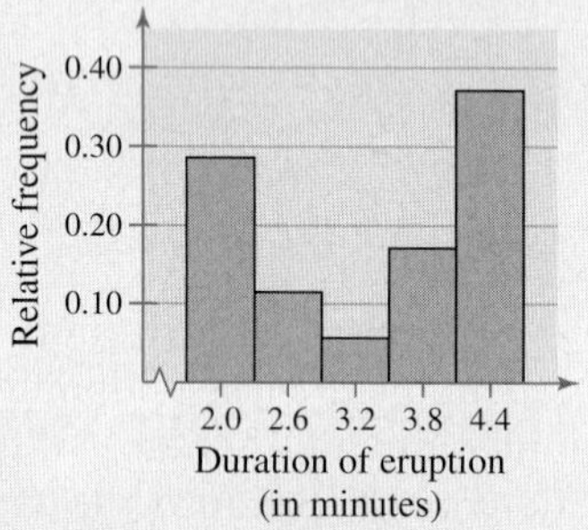

Fifty percent of the eruptions last less than how many minutes?

EXAMPLE 5

Constructing a Relative Frequency Histogram

Draw a relative frequency histogram for the frequency distribution in Example 2.

SOLUTION The relative frequency histogram is shown. Notice that the shape of the histogram is the same as the shape of the frequency histogram constructed in Example 3. The only difference is that the vertical scale measures the relative frequencies.

Internet Usage

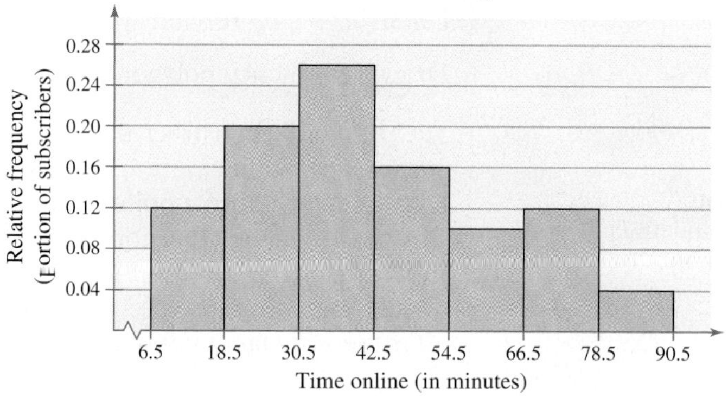

Interpretation From this graph, you can quickly see that 0.20 or 20% of the Internet subscribers spent between 18.5 minutes and 30.5 minutes online, which is not as immediately obvious from the frequency histogram.

Try It Yourself 5

Use the frequency distribution from Try It Yourself 1 to construct a relative frequency histogram that represents the ages of the residents of Akhiok.

a. *Use the same horizontal scale* as used in the frequency histogram.
b. *Revise the vertical scale* to reflect relative frequencies.
c. Use the relative frequencies to *find the height of each bar.* *Answer: Page A31*

If you want to describe the number of data entries that are equal to or below a certain value, you can easily do so by constructing a cumulative frequency graph.

DEFINITION

A **cumulative frequency graph,** or **ogive** (pronounced ō′jīve), is a line graph that displays the cumulative frequency of each class at its upper class boundary. The upper boundaries are marked on the horizontal axis, and the cumulative frequencies are marked on the vertical axis.

GUIDELINES

Constructing an Ogive (Cumulative Frequency Graph)

1. Construct a frequency distribution that includes cumulative frequencies as one of the columns.
2. Specify the horizontal and vertical scales. The horizontal scale consists of upper class boundaries, and the vertical scale measures cumulative frequencies.
3. Plot points that represent the upper class boundaries and their corresponding cumulative frequencies.
4. Connect the points in order from left to right.
5. The graph should start at the lower boundary of the first class (cumulative frequency is zero) and should end at the upper boundary of the last class (cumulative frequency is equal to the sample size).

EXAMPLE 6

Constructing an Ogive

Draw an ogive for the frequency distribution in Example 2. Estimate how many subscribers spent 60 minutes or less online during their last session. Also, use the graph to estimate when the greatest increase in usage occurs.

SOLUTION Using the frequency distribution, you can construct the ogive shown. The upper class boundaries, frequencies, and cumulative frequencies are shown in the table. Notice that the graph starts at 6.5, where the cumulative frequency is 0, and the graph ends at 90.5, where the cumulative frequency is 50.

Upper class boundary	f	Cumulative frequency
18.5	6	6
30.5	10	16
42.5	13	29
54.5	8	37
66.5	5	42
78.5	6	48
90.5	2	50

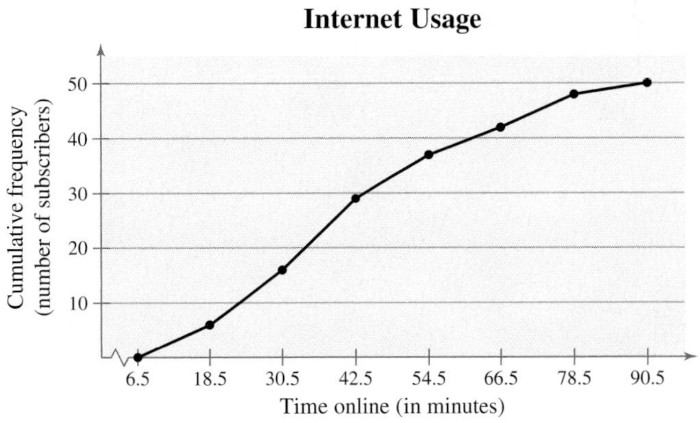

Interpretation From the ogive, you can see that about 40 subscribers spent 60 minutes or less online during their last session. The greatest increase in usage occurs between 30.5 minutes and 42.5 minutes because the line segment is steepest between these two class boundaries.

Another type of ogive uses percent as the vertical axis instead of frequency (see Example 5 in Section 2.5).

Try It Yourself 6

Use the frequency distribution from Try It Yourself 1 to construct an ogive that represents the ages of the residents of Akhiok. Estimate the number of residents who are 49 years old or younger.

a. Specify the *horizontal* and *vertical scales.*
b. *Plot* the points given by the upper class boundaries and the cumulative frequencies.
c. *Construct* the graph.
d. *Estimate* the number of residents who are 49 years old or younger.

Answer: Page A31

EXAMPLE 7

Using Technology to Construct Histograms

Use a calculator or a computer to construct a histogram for the frequency distribution in Example 2.

SOLUTION MINITAB, Excel, and the TI-83 each have features for graphing histograms. Try using this technology to draw the histograms as shown.

MINITAB

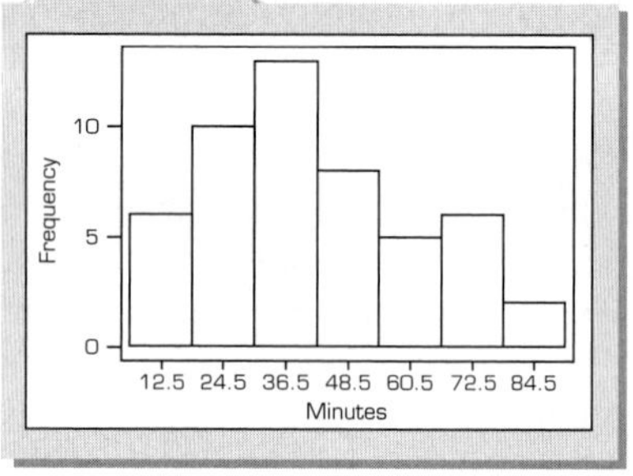

EXCEL

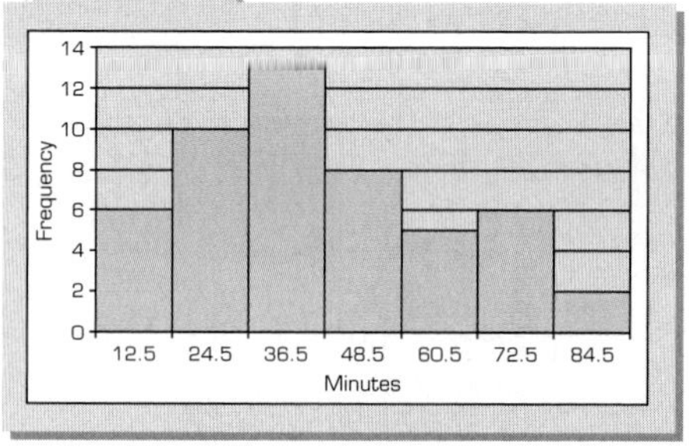

TI-83

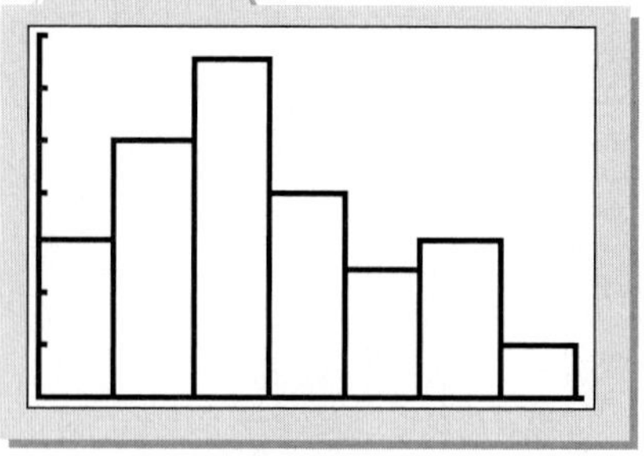

> **Study Tip**
>
> Detailed instructions for using MINITAB, Excel, and the TI-83 are shown in the Technology Guide that accompanies this text. For instance, here are instructions for creating a histogram on a TI-83.
>
> STAT ENTER
>
> Enter midpoints in L1.
> Enter frequencies in L2.
>
> 2nd STATPLOT
>
> Turn on Plot 1.
> Highlight Histogram.
>
> Xlist: L1
> Freq: L2
>
> ZOOM 9
> WINDOW
>
> Xscl=12
>
> GRAPH

Try It Yourself 7

Use a calculator or a computer to construct a frequency histogram that represents the ages of the residents of Akhiok listed in the Chapter Opener on page 33. Use eight classes.

a. *Enter* the data.
b. *Construct* the histogram.

Answer: Page A31

2.1 Exercises

Help

MyMathLab

Student Study Pack

Building Basic Skills and Vocabulary

1. What are some benefits of representing data sets using frequency distributions?

2. What are some benefits of representing data sets using graphs of frequency distributions?

3. What is the difference between class limits and class boundaries?

4. What is the difference between frequency and relative frequency?

True or False? In Exercises 5–8, determine whether the statement is true or false. If it is false, rewrite it as a true statement.

5. The midpoint of a class is the sum of its lower and upper limits.

6. The relative frequency of a class is the sample size divided by the frequency of the class.

7. An ogive is a graph that displays cumulative frequency.

8. Class limits are used to ensure that consecutive bars of a histogram do not touch.

Reading a Frequency Distribution In Exercises 9 and 10, use the given frequency distribution to find the

(a) class width.

(b) class midpoints.

(c) class boundaries.

9. Employee Age

Class	Frequency, f
20–29	10
30–39	132
40–49	284
50–59	300
60–69	175
70–79	65
80–89	25

10. Tree Height

Class	Frequency, f
16–20	100
21–25	122
26–30	900
31–35	207
36–40	795
41–45	568
46–50	322

11. Use the frequency distribution in Exercise 9 to construct an expanded frequency distribution, as shown in Example 2.

12. Use the frequency distribution in Exercise 10 to construct an expanded frequency distribution, as shown in Example 2.

1. Organizing the data into a frequency distribution may make patterns within the data more evident.
2. Sometimes it is easier to identify patterns of a data set by looking at a graph of the frequency distribution.
3. Class limits determine which numbers can belong to that class.

   Class boundaries are the numbers that separate classes without forming gaps between them.
4. Frequency for a class is the number of data entries in each class.

   Relative frequency of a class is the percent of the data that fall in each class.
5. False. The midpoint of a class is the sum of the lower and upper limits of the class divided by two.
6. False. The relative frequency of a class is the frequency of the class divided by the sample size.
7. True
8. False. Class boundaries are used to ensure that consecutive bars of a histogram do not touch.
9. See Odd Answers, page A49.
10. See Selected Answers, page A89.
11. See Odd Answers, page A49.
12. See Selected Answers, page A89.

13. (a) Number of classes = 7
(b) Least frequency ≈ 10
(c) Greatest frequency ≈ 300
(d) Class width = 10

14. (a) Number of classes = 7
(b) Least frequency ≈ 100
(c) Greatest frequency ≈ 900
(d) Class width = 5

15. (a) 50
(b) 12.5–13.5 pounds

16. (a) 50
(b) 68–70 inches

17. (a) 24
(b) 19.5 pounds

18. (a) 44
(b) 70 inches

Graphical Analysis In Exercises 13 and 14, use the frequency histogram to

(a) determine the number of classes.
(b) estimate the frequency of the class with the least frequency.
(c) estimate the frequency of the class with the greatest frequency.
(d) determine the class width.

13.

14.

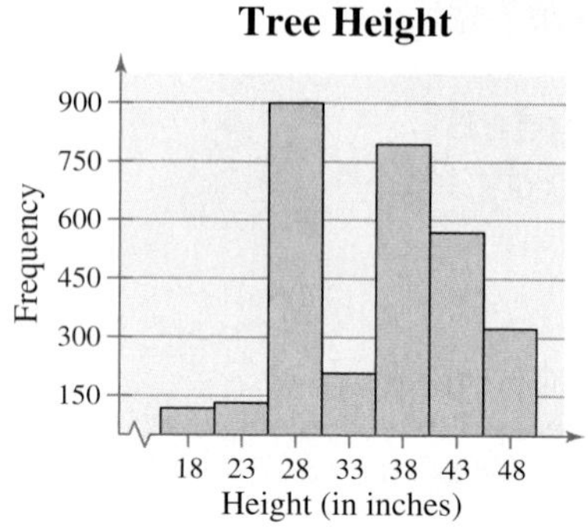

Graphical Analysis In Exercises 15 and 16, use the ogive to approximate

(a) the number in the sample.
(b) the location of the greatest increase in frequency.

15.

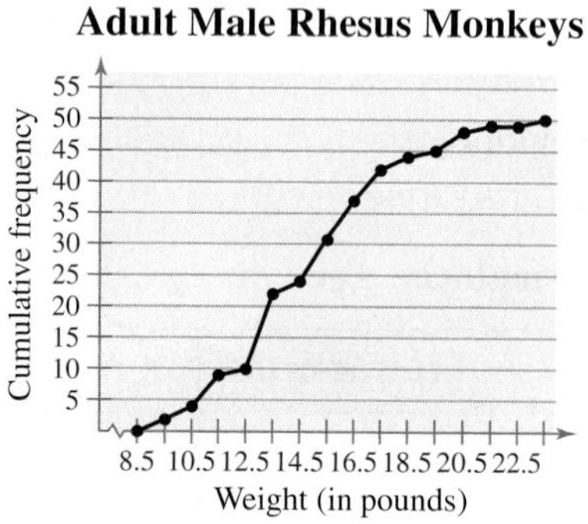

16.

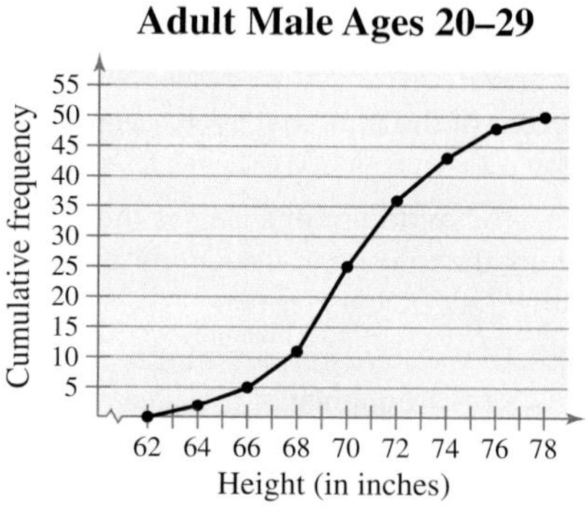

17. Use the ogive in Exercise 15 to approximate

(a) the cumulative frequency for a weight of 14.5 pounds.
(b) the weight for which the cumulative frequency is 45.

18. Use the ogive in Exercise 16 to approximate

(a) the cumulative frequency for a height of 74 inches.
(b) the height for which the cumulative frequency is 25.

Graphical Analysis In Exercises 19 and 20, use the relative frequency histogram to

(a) identify the class with the greatest and the least relative frequency.

(b) approximate the greatest and least relative frequency.

(c) approximate the relative frequency of the second class.

19. **Atlantic Croaker Fish**

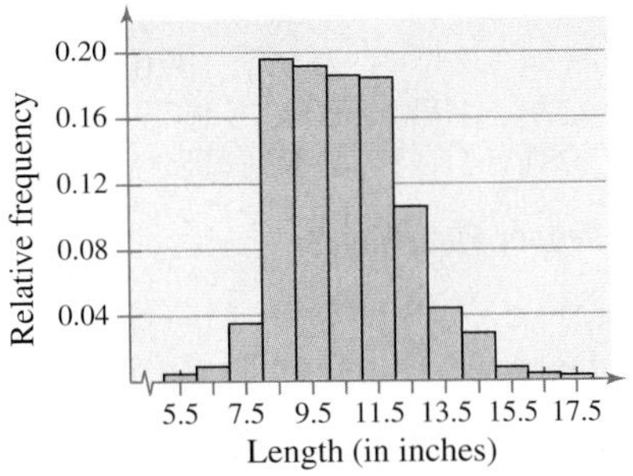

20. **Emergency Response Time**

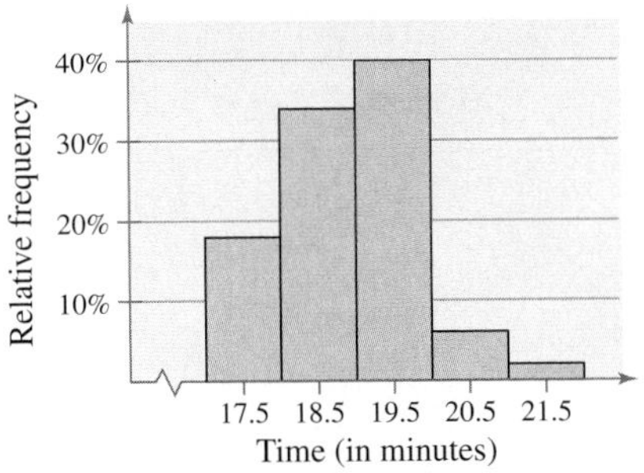

Graphical Analysis In Exercises 21 and 22, use the frequency polygon to identify the class with the greatest and the least frequency.

21. **SAT Scores for 50 Students**

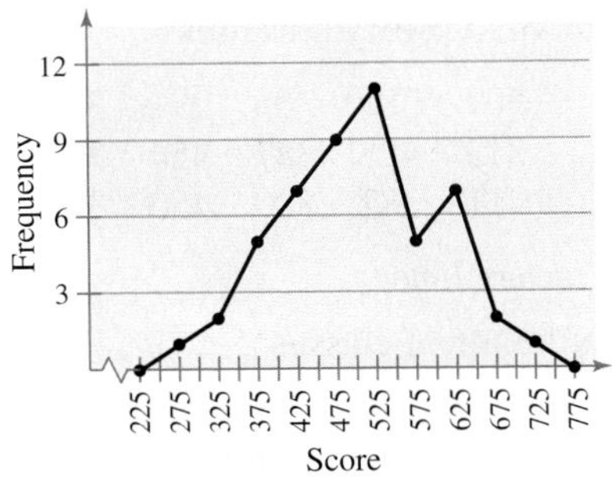

22. **Shoe Sizes for 50 Females**

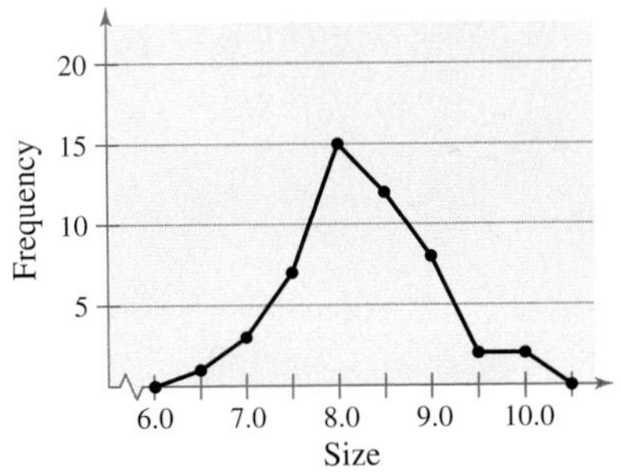

Using and Interpreting Concepts

Constructing a Frequency Distribution In Exercises 23 and 24, construct a frequency distribution for the data set using the indicated number of classes. In the table, include the midpoints, relative frequencies, and cumulative frequencies. Which class has the greatest frequency and which has the least frequency?

DATA **23.** ***Newspaper Reading Times***

Number of classes: 5

Data set: Time (in minutes) spent reading the newspaper in a day

7	39	13	9	25	8	22	0	2	18	2	30	7
35	12	15	8	6	5	29	0	11	39	16	15	

DATA **24.** ***Book Spending***

Number of classes: 6

Data set: Amount (in dollars) spent on books for a semester

91	472	279	249	530	376	188	341	266	199
142	273	189	130	489	266	248	101	375	486
190	398	188	269	43	30	127	354	84	

DATA indicates that the data set for this exercise is available electronically.

19. (a) Class with greatest relative frequency: 8–9 inches

Class with least relative frequency: 17–18 inches

(b) Greatest relative frequency ≈ 0.195

Least relative frequency ≈ 0.005

(c) Approximately 0.015

20. (a) Class with greatest relative frequency: 19–20 minutes

Class with least relative frequency: 21–22 minutes

(b) Greatest relative frequency ≈ 40%

Least relative frequency ≈ 2%

(c) Approximately 33%

21. Class with greatest frequency: 500–550

Classes with least frequency: 250–300 and 700–750

22. Class with greatest frequency: 7.75–8.25

Class with least frequency: 6.25–6.75

23. See Odd Answers, page A49.

24. See Selected Answers, page A90.

25. See Odd Answers, page A49.
26. See Selected Answers, page A90.
27. See Odd Answers, page A49.
28. See Selected Answers, page A90.
29. See Odd Answers, page A50.
30. See Selected Answers, page A90.

Constructing a Frequency Distribution and a Frequency Histogram In Exercises 25–28, construct a frequency distribution and a frequency histogram for the data set using the indicated number of classes. Describe any patterns.

DATA **25.** ***Sales***

Number of classes: 6

Data set: July sales (in dollars) for all sales representatives at a company

2114	2468	7119	1876	4105	3183	1932	1355
4278	1030	2000	1077	5835	1512	1697	2478
3981	1643	1858	1500	4608	1000		

DATA **26.** ***Pepper Pungencies***

Number of classes: 5

Data set: Pungencies (in 1000s of Scoville units) of 24 tabasco peppers

35	51	44	42	37	38	36	39	44	43	40	40
32	39	41	38	42	39	40	46	37	35	41	39

DATA **27.** ***Reaction Times***

Number of classes: 8

Data set: Reaction times (in milliseconds) of a sample of 30 adult females to an auditory stimulus

507	389	305	291	336	310	514	442	307	337
373	428	387	454	323	441	388	426	469	351
411	382	320	450	309	416	359	388	422	413

DATA **28.** ***Fracture Times***

Number of classes: 5

Data set: Amount of pressure (in pounds per square inch) at fracture time for 25 samples of brick mortar

2750	2862	2885	2490	2512	2456	2554	2532	2885
2872	2601	2877	2721	2692	2888	2755	2853	2517
2867	2718	2641	2834	2466	2596	2519		

Constructing a Frequency Distribution and a Relative Frequency Histogram In Exercises 29–32, construct a frequency distribution and a relative frequency histogram for the data set using five classes. Which class has the greatest relative frequency and which has the least relative frequency?

DATA **29.** ***Bowling Scores***

Data set: Bowling scores of a sample of league members

154	257	195	220	182	240	177	228	235
146	174	192	165	207	185	180	264	169
225	239	148	190	182	205	148	188	

DATA **30.** ***ATM Withdrawals***

Data set: A sample of ATM withdrawals (in dollars)

35	10	30	25	75	10	30	20	20	10	40
50	40	30	60	70	25	40	10	60	20	80
40	25	20	10	20	25	30	50	80	20	

31. See Odd Answers, page A50.
32. See Selected Answers, page A90.
33. See Odd Answers, page A50.
34. See Selected Answers, page A90.
35. See Odd Answers, page A50.
36. See Selected Answers, page A91.
37. See Odd Answers, page A51.

DATA **31.** ***Tree Heights***

Data set: Heights (in feet) of a sample of Douglas-fir trees

40	44	35	49	35	43	35	36	39
37	41	41	48	52	37	45	40	36
35	50	42	51	33	34	51	39	

DATA **32.** ***Farm Acreage***

Data set: Number of acres on a sample of small farms

12	7	9	8	9	8	12	10	9
10	6	8	13	12	10	11	7	14
12	9	8	10	9	11	13	8	

Constructing a Cumulative Frequency Distribution and an Ogive In Exercises 33–36, construct a cumulative frequency distribution and an ogive for the data set using six classes. Then describe the location of the greatest increase in frequency.

DATA **33.** ***Retirement Ages***

Data set: Retirement ages for a sample of engineers

60	65	68	63	66	67	69	67
58	65	67	61	63	65	62	64
73	50	61	71	62	69	72	63

DATA **34.** ***Saturated Fat Intakes***

Data set: Daily saturated fat intakes (in grams) of a sample of people

38	32	34	39	40	54	32	17	29	33
57	40	25	36	33	24	42	16	31	33

DATA **35.** ***Gasoline Purchases***

Data set: Gasoline (in gallons) purchased by a sample of drivers during one fill-up

7	4	18	4	9	8	8	7	6	2
9	5	9	12	4	14	15	7	10	2
3	11	4	4	9	12	5	3		

DATA **36.** ***Long-Distance Phone Calls***

Data set: Lengths (in minutes) of a sample of long-distance phone calls

1	20	10	20	13	23	3	7
18	7	4	5	15	7	29	10
18	10	10	23	4	12	8	6

Constructing a Frequency Distribution and a Frequency Polygon In Exercises 37 and 38, construct a frequency distribution and a frequency polygon for the data set. Describe any patterns.

DATA **37.** ***Exam Scores***

Number of classes: 5

Data set: Exam scores for all students in a statistics class

83	92	94	82	73	98	78	85	72	90
89	92	96	89	75	85	63	47	75	82

38. See Selected Answers, page A91.

39. See Odd Answers, page A51.

40. See Selected Answers, page A91.

41. Histogram (5 Classes)

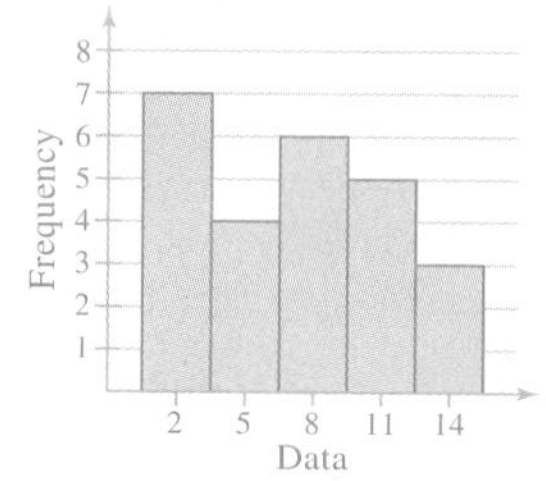

Histogram (10 Classes)

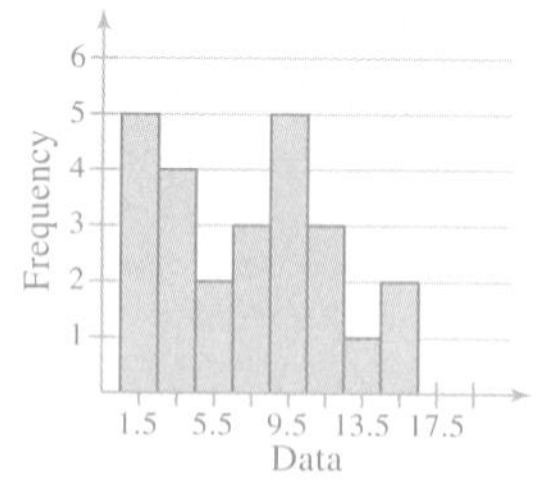

Histogram (20 Classes)

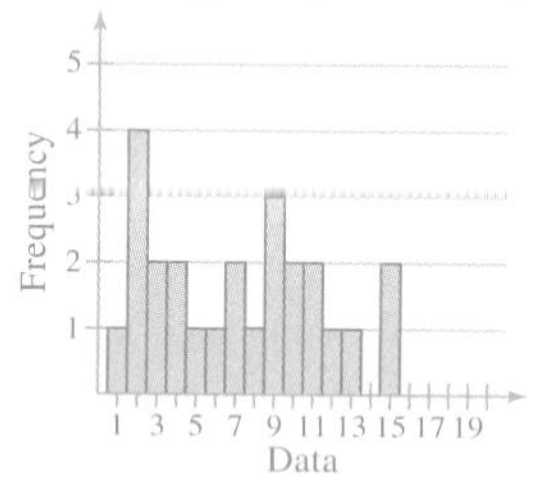

In general, a greater number of classes better preserves the actual values of the data set but is not as helpful for observing general trends and making conclusions. In choosing the number of classes, an important consideration is the size of the data set. For instance, you would not want to use 20 classes if your data set contained 20 entries. In this particular example, as the number of classes increases, the histogram shows more fluctuation. The histograms with 10 and 20 classes have classes with zero frequencies. Not much is gained by using more than five classes. Therefore, it appears that five classes would be best.

DATA **38. *Children of the President***

Number of classes: 6

Data set: Number of children of the U.S. presidents *(Source: infoplease.com)*

0	5	6	0	2	4	0	4	10	15	0	6	2	3
0	4	5	4	8	7	3	5	3	2	6	3	3	0
2	2	6	1	2	3	2	2	4	4	4	6	1	2

Extending Concepts

DATA **39. *What Would You Do?*** You work at a bank and are asked to recommend the amount of cash to put in an ATM each day. You don't want to put in too much (security) or too little (customer irritation). Here are the daily withdrawals (in 100s of dollars) for a period of 30 days.

72	84	61	76	104	76	86	92	80	88
98	76	97	82	84	67	70	81	82	89
74	73	86	81	85	78	82	80	91	83

(a) Construct a relative frequency histogram for the data, using eight classes.

(b) If you put $9000 in the ATM each day, what percent of the days in a month should you expect to run out of cash? Explain your reasoning.

(c) If you are willing to run out of cash for 10% of the days, how much cash, in hundreds of dollars, should you put in the ATM each day? Explain your reasoning.

DATA **40. *What Would You Do?*** You work in the admissions department for a college and are asked to recommend the minimum SAT scores that the college will accept for a position as a full-time student. Here are the SAT scores for a sample of 50 applicants.

1325	1072	982	996	872	849	785	706	669	1049
885	1367	935	980	1188	869	1006	1127	979	1034
1052	1165	1359	667	1264	727	808	955	544	1202
1051	1173	410	1148	1195	1141	1193	768	812	887
1211	1266	830	672	917	988	791	1035	688	700

(a) Construct a relative frequency histogram for the data using 10 classes.

(b) If you set the minimum score at 986, what percent of the applicants will you be accepting? Explain your reasoning.

(c) If you want to accept the top 88% of the applicants, what should the minimum score be? Explain your reasoning.

DATA **41. *Writing*** What happens when the number of classes is increased for a frequency histogram? Use the data set listed and a technology tool to create frequency histograms with 5, 10, and 20 classes. Which graph displays the data best?

2	7	3	2	11	3	15	8	4	9	10	13	9
7	11	10	1	2	12	5	6	4	2	9	15	

2.2 More Graphs and Displays

What You Should Learn

- How to graph and interpret quantitative data sets using stem-and-leaf plots and dot plots
- How to graph and interpret qualitative data sets using pie charts and Pareto charts
- How to graph and interpret paired data sets using scatter plots and time series charts

Graphing Quantitative Data Sets • Graphing Qualitative Data Sets • Graphing Paired Data Sets

Graphing Quantitative Data Sets

In Section 2.1, you learned several traditional ways to display quantitative data graphically. In this section, you will learn a newer way to display quantitative data, called a **stem-and-leaf plot.** Stem-and-leaf plots are examples of **exploratory data analysis (EDA),** which was developed by John Tukey in 1977.

In a stem-and-leaf plot, each number is separated into a **stem** (for instance, the entry's leftmost digits) and a **leaf** (for instance, the rightmost digit). A stem-and-leaf plot is similar to a histogram but has the advantage that the graph still contains the original data values. Another advantage of a stem-and-leaf plot is that it provides an easy way to sort data.

EXAMPLE 1

Constructing a Stem-and-Leaf Plot

The following are the numbers of league-leading runs batted in (RBIs) for baseball's American League during a recent 50-year period. Display the data in a stem-and-leaf plot. What can you conclude? *(Source: Major League Baseball)*

155	159	144	129	105	145	126	116	130	114	122	112	112	142	126
118	118	108	122	121	109	140	126	119	113	117	118	109	109	119
139	139	122	78	133	126	123	145	121	134	124	119	132	133	124
129	112	126	148	147										

SOLUTION Because the data entries go from a low of 78 to a high of 159, you should use stem values from 7 to 15. To construct the plot, list these stems to the left of a vertical line. For each data entry, list a leaf to the right of its stem. For instance, the entry 155 has a stem of 15 and a leaf of 5. The resulting stem-and-leaf plot will be unordered. To obtain an ordered stem-and-leaf plot, rewrite the plot with the leaves in increasing order from left to right. It is important to include a key for the display to identify the values of the data.

RBIs for American League Leaders

Key: 15|5 = 155

Stem	Leaves
7	8
8	
9	
10	5 8 9 9 9
11	6 4 2 2 8 8 9 3 7 8 9 9 2
12	9 6 2 6 2 1 6 2 6 3 1 4 4 9 6
13	0 9 9 3 4 2 3
14	4 5 2 0 5 8 7
15	5 9

Unordered Stem-and-Leaf Plot

RBIs for American League Leaders

Key: 15|5 = 155

Stem	Leaves
7	8
8	
9	
10	5 8 9 9 9
11	2 2 2 3 4 6 7 8 8 8 9 9 9
12	1 1 2 2 2 3 4 4 6 6 6 6 6 9 9
13	0 2 3 3 4 9 9
14	0 2 4 5 5 7 8
15	5 9

Ordered Stem-and-Leaf Plot

Interpretation From the ordered stem-and-leaf plot, you can conclude that more than 50% of the RBI leaders had between 110 and 130 RBIs.

Study Tip

In a stem-and-leaf plot, you should have as many leaves as there are entries in the original data set.

Insight

You can use stem-and-leaf plots to identify unusual data values called **outliers.** In Example 1, the data value 78 is an outlier. You will learn more about outliers in Section 2.3.

Try It Yourself 1

Use a stem-and-leaf plot to organize the Akhiok population data set listed in the Chapter Opener on page 33. What can you conclude?

a. List all possible *stems.*
b. List the leaf of each data entry to the right of its stem and include a key.
c. Rewrite the stem-and-leaf plot so that the leaves are ordered.
d. Use the plot to make a conclusion.

Answer: Page A31

EXAMPLE 2

Constructing Variations of Stem-and-Leaf Plots

Organize the data given in Example 1 using a stem-and-leaf plot that has two lines for each stem. What can you conclude?

SOLUTION Construct the stem-and-leaf plot as described in Example 1, except now list each stem twice. Use the leaves 0, 1, 2, 3, and 4 in the first stem row and the leaves 5, 6, 7, 8, and 9 in the second stem row. The revised stem-and-leaf plot is shown.

RBIs for American League Leaders

Key: 15|5 = 155

7	
7	8
8	
8	
9	
9	
10	
10	5 8 9 9 9
11	4 2 2 3 2
11	6 8 8 9 7 8 9 9
12	2 2 1 2 3 1 4 4
12	9 6 6 6 6 9 6
13	0 3 4 2 3
13	9 9
14	4 2 0
14	5 5 8 7
15	
15	5 9

Unordered Stem-and-Leaf Plot

RBIs for American League Leaders

Key: 15|5 = 155

7	
7	8
8	
8	
9	
9	
10	
10	5 8 9 9 9
11	2 2 2 3 4
11	6 7 8 8 8 9 9 9
12	1 1 2 2 2 3 4 4
12	6 6 6 6 6 9 9
13	0 2 3 3 4
13	9 9
14	0 2 4
14	5 5 7 8
15	
15	5 9

Ordered Stem-and-Leaf Plot

Interpretation From the display, you can conclude that most of the RBI leaders had between 105 and 135 RBIs.

Try It Yourself 2

Using two rows for each stem, revise the stem-and-leaf plot you constructed in Try It Yourself 1.

a. List each stem *twice.*
b. List all leaves *using the appropriate stem row.*

Answer: Page A31

Note to Instructor

If you are using MINITAB or Excel, ask students to use this technology to construct a stem-and-leaf plot.

Insight

Compare Examples 1 and 2. Notice that by using two lines per stem, you obtain a more detailed picture of the data.

You can also use a dot plot to graph quantitative data. In a **dot plot,** each data entry is plotted, using a point, above a horizontal axis. Like a stem-and-leaf plot, a dot plot allows you to see how data are distributed, determine specific data entries, and identify unusual data values.

EXAMPLE 3

Constructing a Dot Plot

Use a dot plot to organize the RBI data given in Example 1.

155	159	144	129	105	145	126	116	130
114	122	112	112	142	126	118	118	108
122	121	109	140	126	119	113	117	118
109	109	119	139	139	122	78	133	126
123	145	121	134	124	119	132	133	124
129	112	126	148	147				

SOLUTION So that each data entry is included in the dot plot, the horizontal axis should include numbers between 70 and 160. To represent a data entry, plot a point above the entry's position on the axis. If an entry is repeated, plot another point above the previous point.

RBIs for American League Leaders

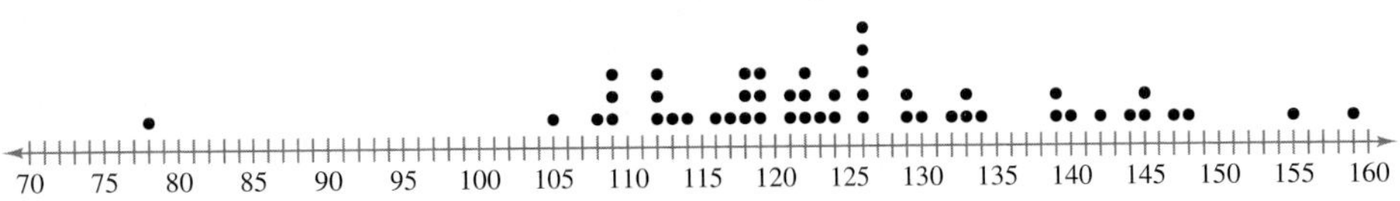

Interpretation From the dot plot, you can see that most values cluster between 105 and 148 and the value that occurs the most is 126. You can also see that 78 is an unusual data value.

Try It Yourself 3

Use a dot plot to organize the Akhiok population data set listed in the Chapter Opener on page 33. What can you conclude from the graph?

a. Choose an appropriate scale for the *horizontal axis.*
b. Represent each data entry by *plotting a point.*
c. *Describe* any patterns for the data.

Answer: Page A32

Technology can be used to construct stem-and-leaf plots and dot plots. For instance, a MINITAB dot plot for the RBI data is shown.

MINITAB

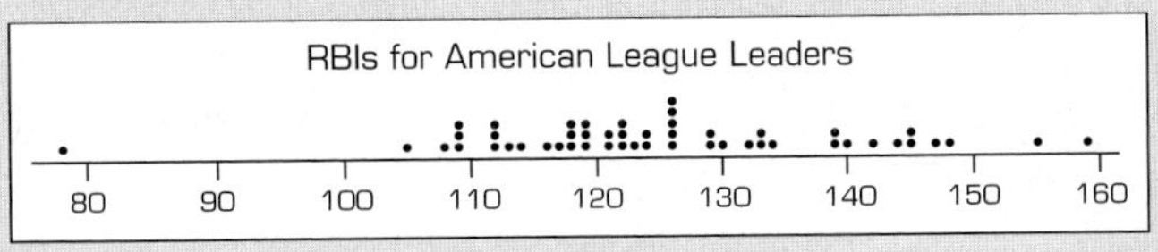

Graphing Qualitative Data Sets

Pie charts provide a convenient way to present qualitative data graphically. A **pie chart** is a circle that is divided into sectors that represent categories. The area of each sector is proportional to the frequency of each category.

EXAMPLE 4

Constructing a Pie Chart

Motor Vehicle Occupants Killed in 2001

Vehicle type	Killed
Cars	20,269
Trucks	12,260
Motorcycles	3,067
Other	612

The numbers of motor vehicle occupants killed in crashes in 2001 are shown in the table. Use a pie chart to organize the data. What can you conclude? *(Source: U.S. Department of Transportation, National Highway Traffic Safety Administration)*

SOLUTION Begin by finding the relative frequency, or percent, of each category. Then construct the pie chart using the central angle that corresponds to each category. To find the central angle, multiply 360° by the category's relative frequency. For example, the central angle for cars is $360°(0.56) \approx 202°$. From the pie chart, you can see that most fatalities in motor vehicle crashes were those involving the occupants of cars.

	f	Relative frequency	Angle
Cars	20,269	0.56	202°
Trucks	12,260	0.34	122°
Motorcycles	3,067	0.08	29°
Other	610	0.02	7°

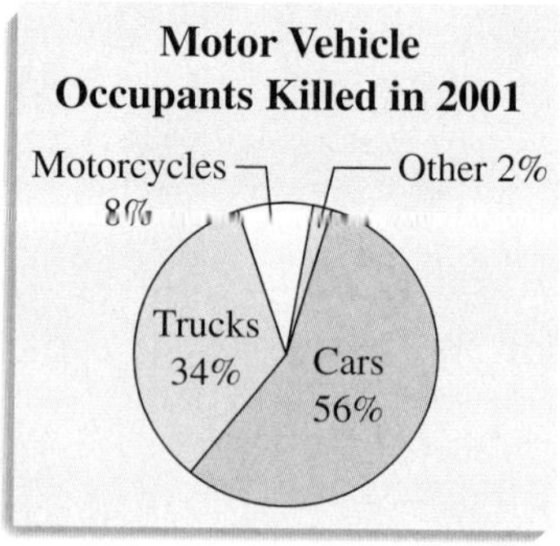

Try It Yourself 4

The numbers of motor vehicle occupants killed in crashes in 1991 are shown in the table. Use a pie chart to organize the data. Compare the 1991 data with the 2001 data. *(Source: U.S. Department of Transportation, National Highway Safety Administration)*

Motor Vehicle Occupants Killed in 1991

Vehicle type	Killed
Cars	22,385
Trucks	8,457
Motorcycles	2,806
Other	497

a. Find the *relative frequency* of each category.
b. Use the *central angle* to find the portion that corresponds to each category.
c. Compare the 1991 data with the 2001 data. *Answer: Page A32*

Technology can be used to construct pie charts. For instance, an Excel pie chart for the data in Example 4 is shown.

EXCEL

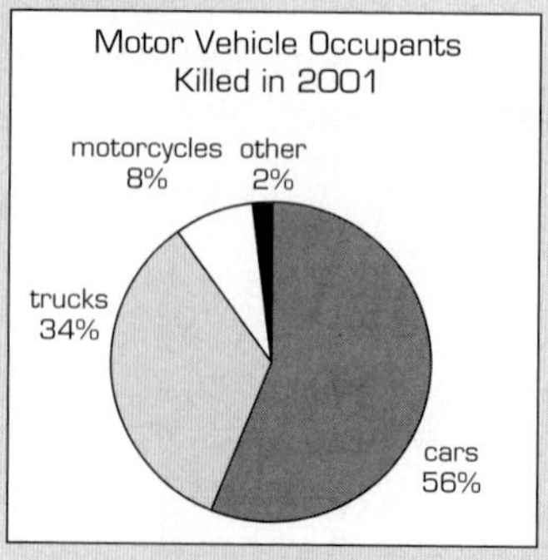

Another way to graph qualitative data is to use a Pareto chart. A **Pareto chart** is a vertical bar graph in which the height of each bar represents frequency or relative frequency. The bars are positioned in order of decreasing height, with the tallest bar positioned at the left. Such positioning helps highlight important data and is used frequently in business.

EXAMPLE 5

Constructing a Pareto Chart

In a recent year, the retail industry lost $41.0 million in inventory shrinkage. Inventory shrinkage is the loss of inventory through breakage, pilferage, shoplifting, and so on. The causes of the inventory shrinkage are administrative error ($7.8 million), employee theft ($15.6 million), shoplifting ($14.7 million), and vendor fraud ($2.9 million). If you were a retailer, which causes of inventory shrinkage would you address first? *(Source: National Retail Federation and Center for Retailing Education, University of Florida)*

SOLUTION Using frequencies for the vertical axis, you can construct the Pareto chart as shown.

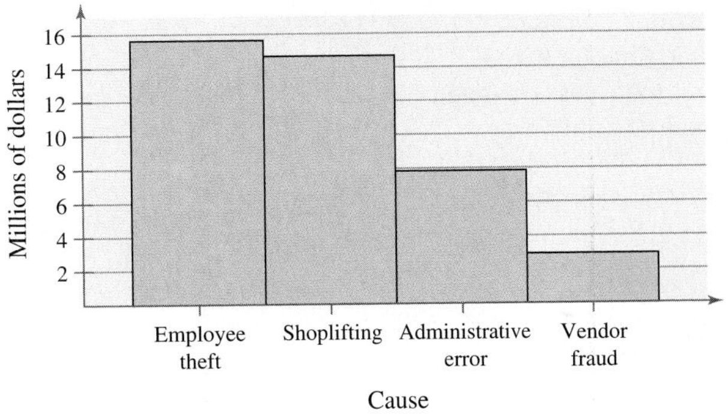

Interpretation From the graph, it is easy to see that the causes of inventory shrinkage that should be addressed first are employee theft and shoplifting.

Try It Yourself 5

Every year, the Better Business Bureau (BBB) receives complaints from customers. In a recent year, the BBB received the following complaints.

7792 complaints about home furnishing stores
5733 complaints about computer sales and service stores
14,668 complaints about auto dealers
9728 complaints about auto repair shops
4649 complaints about dry cleaning companies

Use a Pareto chart to organize the data. What source is the greatest cause of complaints? *(Source: Council of Better Business Bureaus)*

a. Find the *frequency or relative frequency* for each data entry.
b. *Position the bars in decreasing order* according to frequency or relative frequency.
c. *Interpret* the results in the context of the data. *Answer: Page A32*

Picturing the World

The five top-selling vehicles in the United States for January of 2004 are shown in the following Pareto chart. One of the top five vehicles was a car. The other four vehicles were trucks. *(Source: Associated Press)*

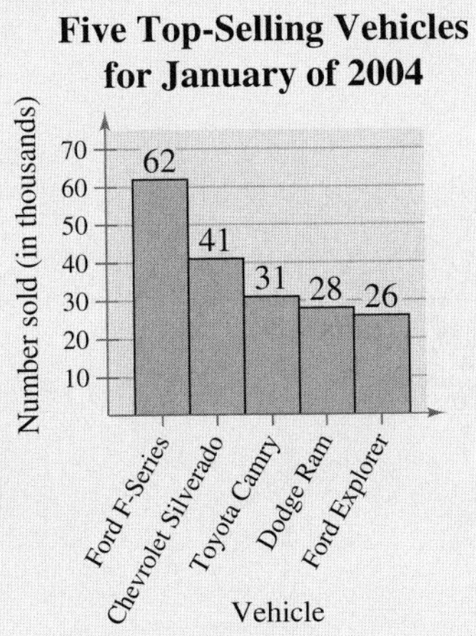

How many vehicles from the top five did Ford sell in January of 2004?

Graphing Paired Data Sets

When each entry in one data set corresponds to one entry in a second data set, the sets are called **paired data sets.** For instance, suppose a data set contains the costs of an item and a second data set contains sales amounts for the item at each cost. Because each cost corresponds to a sales amount, the data sets are paired. One way to graph paired data sets is to use a **scatter plot,** where the ordered pairs are graphed as points in a coordinate plane. A scatter plot is used to show the relationship between two quantitative variables.

EXAMPLE 6

Interpreting a Scatter Plot

The British statistician Ronald Fisher (see page 29) introduced a famous data set called Fisher's Iris data set. This data set describes various physical characteristics, such as petal length and petal width (in millimeters), for three species of iris. In the scatter plot shown, the petal lengths form the first data set and the petal widths form the second data set. As the petal length increases, what tends to happen to the petal width? *(Source: Fisher, R. A., 1936)*

Note to Instructor

A complete discussion of types of correlation occurs in Chapter 9. You may want, however, to discuss positive correlation, negative correlation, and no correlation at this point. Be sure that students do not confuse correlation with causation.

Fisher's Iris Data Set

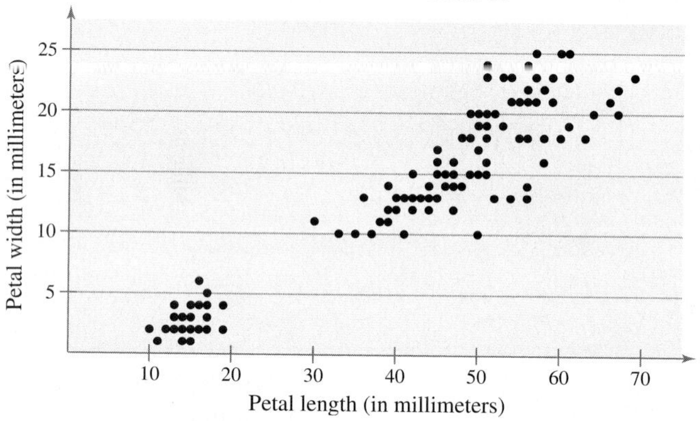

SOLUTION The horizontal axis represents the petal length, and the vertical axis represents the petal width. Each point in the scatter plot represents the petal length and petal width of one flower.

Interpretation From the scatter plot, you can see that as the petal length increases, the petal width also tends to increase.

Try It Yourself 6

The lengths of employment and the salaries of 10 employees are listed in the table at the left. Graph the data using a scatter plot. What can you conclude?

Length of employment (in years)	Salary (in dollars)
5	32,000
4	32,500
8	40,000
4	27,350
2	25,000
10	43,000
7	41,650
6	39,225
9	45,100
3	28,000

a. Label the *horizontal and vertical axes.*
b. *Plot* the paired data.
c. *Describe* any trends.

Answer: Page A32

You will learn more about scatter plots and how to analyze them in Chapter 9.

A data set that is composed of quantitative entries taken at regular intervals over a period of time is a **time series.** For instance, the amount of precipitation measured each day for one month is an example of a time series. You can use a **time series chart** to graph a time series.

See MINITAB and TI-83 steps on pages 114 and 115.

EXAMPLE 7

Constructing a Time Series Chart

The table lists the number of cellular telephone subscribers (in millions) and a subscriber's average local monthly bill for service (in dollars) for the years 1991 through 2001. Construct a time series chart for the number of cellular subscribers. What can you conclude? *(Source: Cellular Telecommunications & Internet Association)*

Year	Subscribers (in millions)	Average bill (in dollars)
1991	7.6	72.74
1992	11.0	68.68
1993	16.0	61.48
1994	24.1	56.21
1995	33.8	51.00
1996	44.0	47.70
1997	55.3	42.78
1998	69.2	39.43
1999	86.0	41.24
2000	109.5	45.27
2001	128.4	47.37

Note to Instructor

Consider asking students to find a time series plot in a magazine or newspaper and bring it to class for discussion.

SOLUTION Let the horizontal axis represent the years and the vertical axis represent the number of subscribers (in millions). Then plot the paired data and connect them with line segments.

Cellular Telephone Subscribers

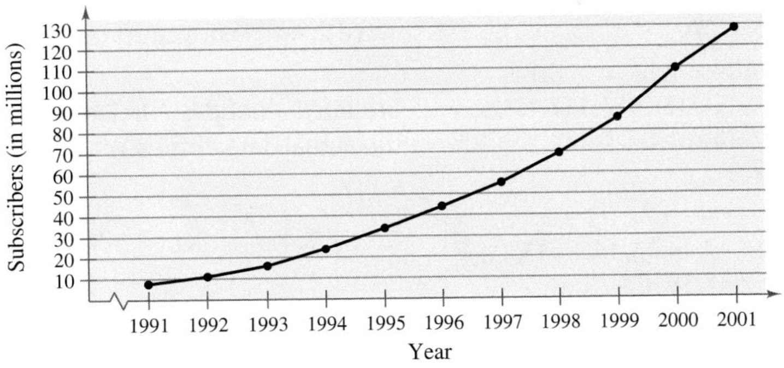

Interpretation The graph shows that the number of subscribers has been increasing since 1991, with greater increases recently.

Try It Yourself 7

Use the table in Example 7 to construct a time series chart for a subscriber's average local monthly cellular telephone bill for the years 1991 through 2001. What can you conclude?

a. Label the *horizontal and vertical axes.*
b. *Plot* the paired data and *connect* them with line segments.
c. *Describe* any patterns you see.

Answer: Page A32

2.2 Exercises

Help
MyMathLab
Student Study Pack

Building Basic Skills and Vocabulary

1. Name some ways to display quantitative data graphically. Name some ways to display qualitative data graphically.

2. What is an advantage of using a stem-and-leaf plot instead of a histogram? What is a disadvantage?

Putting Graphs in Context In Exercises 3–6, match the plot with the description of the sample.

3.

2	8 9	Key: 2|8 = 28
3	2 2 2 3 4 5 7 7 8 9	
4	0 2 4 5	
5	1	
6	5 6	
7	2	

4.

6	7 8	Key: 6|7 = 67
7	4 5 5 8 8 8	
8	1 3 5 5 8 8 9	
9	0 0 0 2 4	

5.

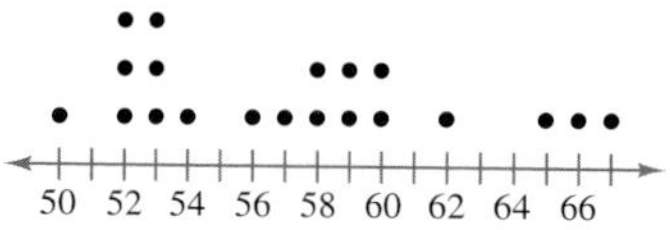

6.

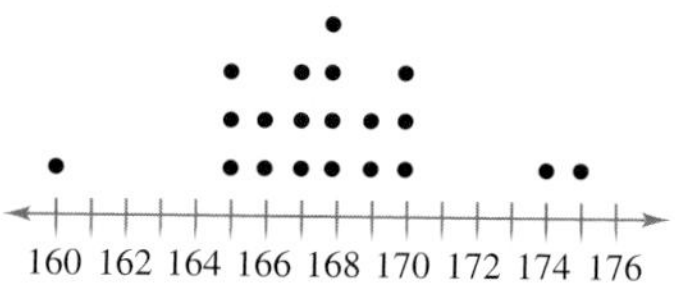

(a) Prices (in dollars) of a sample of 20 brands of jeans
(b) Weights (in pounds) of a sample of 20 first grade students
(c) Volumes (in cubic centimeters) of a sample of 20 oranges
(d) Ages (in years) of a sample of 20 residents of a retirement home

Graphical Analysis In Exercises 7–10, use the stem-and-leaf plot or dot plot to list the actual data entries. What is the maximum data entry? What is the minimum data entry?

7.

2	7	Key: 2|7 = 27
3	2	
4	1 3 3 4 7 7 8	
5	0 1 1 2 3 3 3 4 4 4 4 5 6 6 8 9	
6	8 8 8	
7	3 8 8	
8	5	

8.

12		Key: 12|9 = 12.9
12	9	
13	3	
13	6 7 7	
14	1 1 1 1 3 4 4	
14	6 9 9	
15	0 0 0 1 2 4	
15	6 7 8 8 8 9	
16	1	
1 6	6 7	

9.

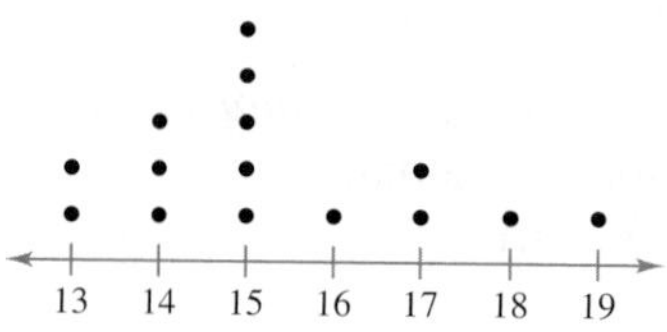

10.

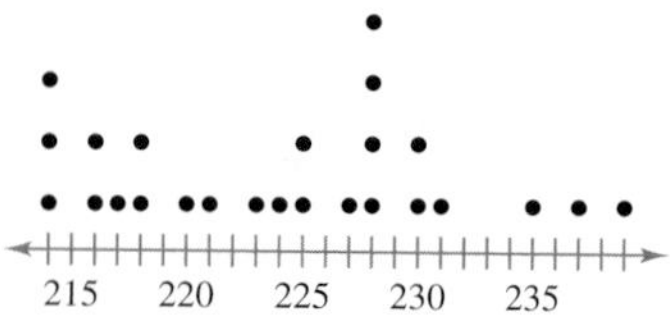

1. Quantitative: stem-and-leaf plot, dot plot, histogram, scatter plot, time series chart
Qualitative: pie chart, Pareto chart
2. Unlike the histogram, the stem-and-leaf plot still contains the original data values. However, some data are difficult to organize in a stem-and-leaf plot.
3. a 4. d 5. b 6. c
7. 27, 32, 41, 43, 43, 44, 47, 47, 48, 50, 51, 51, 52, 53, 53, 53, 54, 54, 54, 54, 55, 56, 56, 58, 59, 68, 68, 68, 73, 78, 78, 85
Max: 85; Min: 27
8. 129, 133, 136, 137, 137, 141, 141, 141, 141, 143, 144, 144, 146, 149, 149, 150, 150, 150, 151, 152, 154, 156, 157, 158, 158, 158, 159, 161, 166, 167
Max: 167; Min: 129
9. 13, 13, 14, 14, 14, 15, 15, 15, 15, 15, 16, 17, 17, 18, 19
Max: 19; Min: 13
10. 214, 214, 214, 216, 216, 217, 218, 218, 220, 221, 223, 224, 225, 225, 227, 228, 228, 228, 228, 230, 230, 231, 235, 237, 239
Max: 239; Min: 214
11. Anheuser-Busch spends the most on advertising and Honda spends the least. (Answers will vary.)
12. Value increased the most between 2000 and 2003. (Answers will vary.)
13. Tailgaters irk drivers the most, and too-cautious drivers irk drivers the least. (Answers will vary.)

14. Twice as many people "sped up" than "cut off a car." (Answers will vary.)

15. Key: 3|3 = 33

3	2 3 3 4 5 9
4	0 1 1 3 4 5 5 6 6 7 8
5	1 3 3
6	0 0 6 9

It appears that most elephants tend to drink less than 55 gallons of water per day. (Answers will vary.)

16. Key: 31|9 = 319

29	8
30	5
31	9
32	7
33	
34	5
35	1
36	
37	
38	
39	0 3
40	3 9
41	0 5 9
42	
43	
44	6 8 9
45	0 5
46	0 5
47	9 9
48	
49	1
50	3

It appears that the majority of the elephants eat between 390 and 480 pounds of hay each day. (Answers will vary.)

17. Key: 17|5 = 17.5

16	4 8
17	1 1 3 4 5 5 6 7 9
18	1 3 4 4 6 6 6 9
19	0 0 2 3 3 5 6
20	1 8

It appears that most farmers charge 17 to 19 cents per pound of apples. (Answers will vary.)

18. See Selected Answers, page A91.

Using and Interpreting Concepts

Graphical Analysis In Exercises 11–14, what can you conclude from the graph?

11. **Top Five Sports Advertisers**

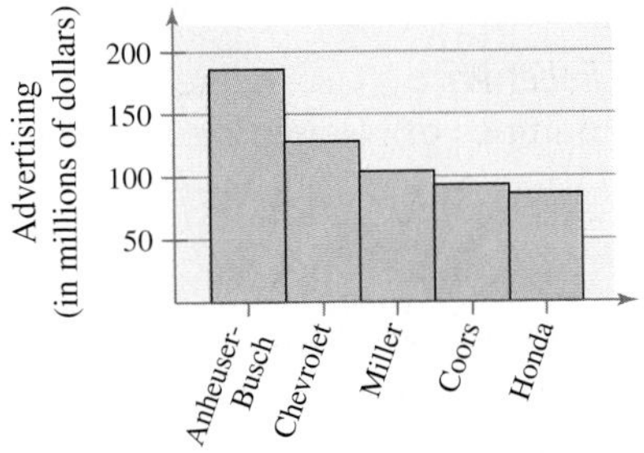

(Source: Nielsen Media Research)

12. **Stock Portfolio**

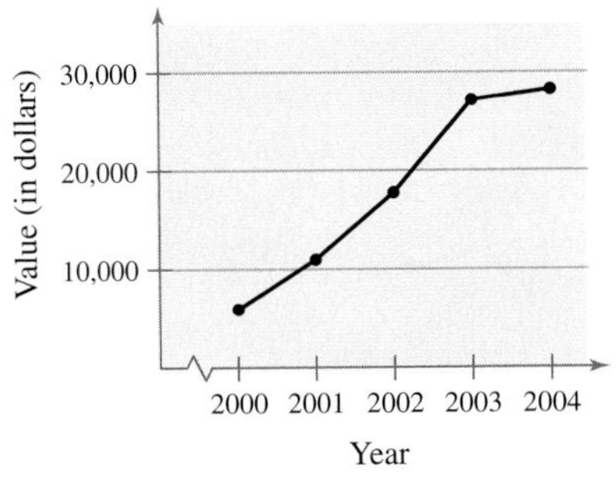

13. **How Other Drivers Irk Us**

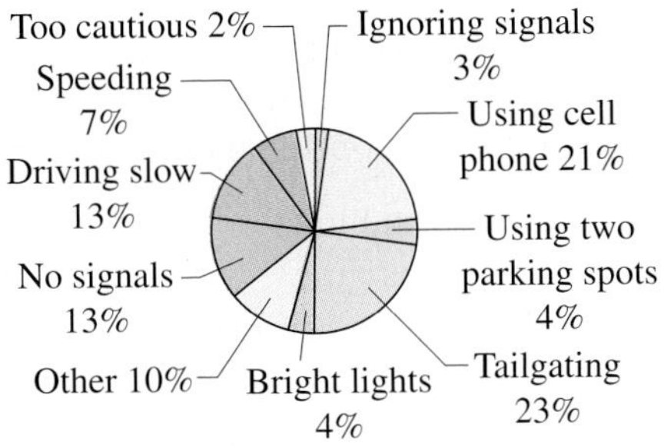

(Adapted from Reuters/Zogby)

14. **Driving and Cell Phone Use**

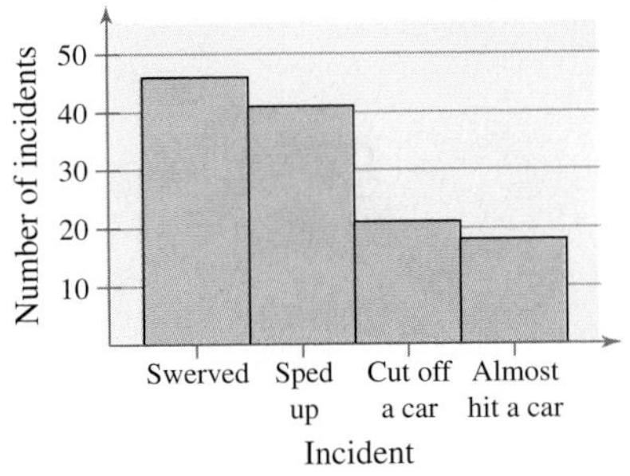

(Adapted from USA TODAY)

Graphing Data Sets In Exercises 15–28, organize the data using the indicated type of graph. What can you conclude about the data?

DATA **15.** ***Elephants: Water Consumed*** Use a stem-and-leaf plot to display the data. The data represent the amount of water (in gallons) consumed by 24 elephants in one day.

33	45	34	47	43	48	35	69	45	60	46	51
41	60	66	41	32	40	44	39	46	33	53	53

DATA **16.** ***Elephants: Hay Eaten*** Use a stem-and-leaf plot to display the data. The data represent the amount of hay (in pounds) eaten daily by 24 elephants.

449	450	419	448	479	410	446	465	415	455	345	305
491	479	390	393	403	298	503	327	460	351	409	319

DATA **17.** ***Apple Prices*** Use a stem-and-leaf plot to display the data. The data represent the price (in cents per pound) paid to 28 farmers for apples.

19.2	19.6	16.4	17.1	19.0	17.4	17.3	20.1	19.0	17.5
17.6	18.6	18.4	17.7	19.5	18.4	18.9	17.5	19.3	20.8
19.3	18.6	18.6	18.3	17.1	18.1	16.8	17.9		

DATA **18.** ***Advertisements*** Use a dot plot to display the data. The data represent the number of advertisements seen or heard in one week by a sample of 30 people from the United States.

598	494	441	595	728	690	684	486	735	808	734	590	673	545	702
481	298	135	846	764	317	649	732	582	637	588	540	727	486	703

19.

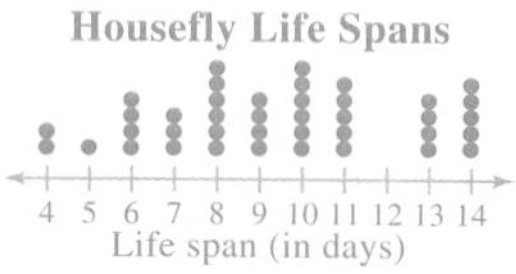

It appears that the life span of a housefly tends to be between 4 and 14 days. (Answers will vary.)

20.

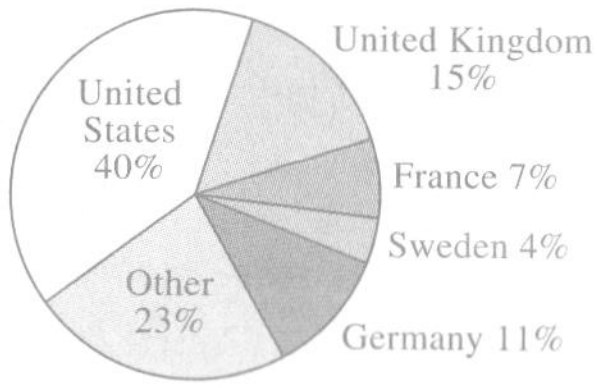

The United States had the greatest number of Nobel Prize laureates during the years 1901–2002.

21.

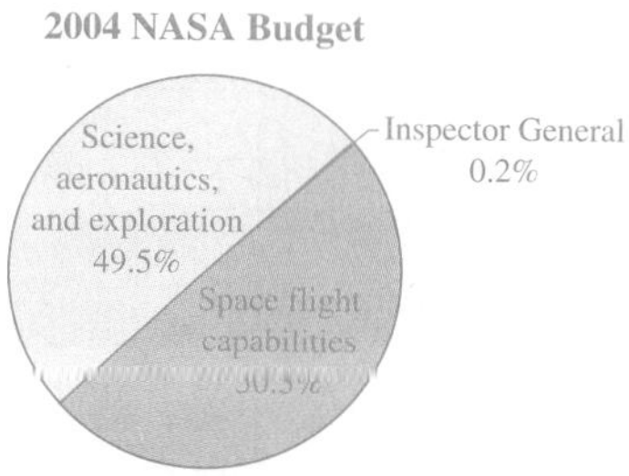

It appears that 50.3% of NASA's budget went to space flight capabilities. (Answers will vary).

22. See Selected Answers, page A91.

23.

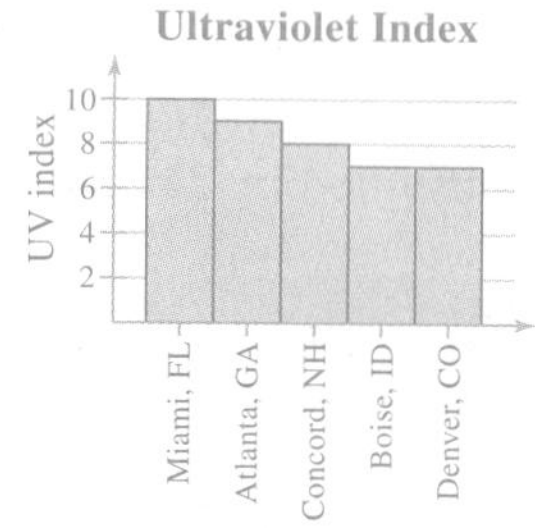

It appears that Boise, ID, and Denver, CO, have the same UV index. (Answers will vary.)

24.

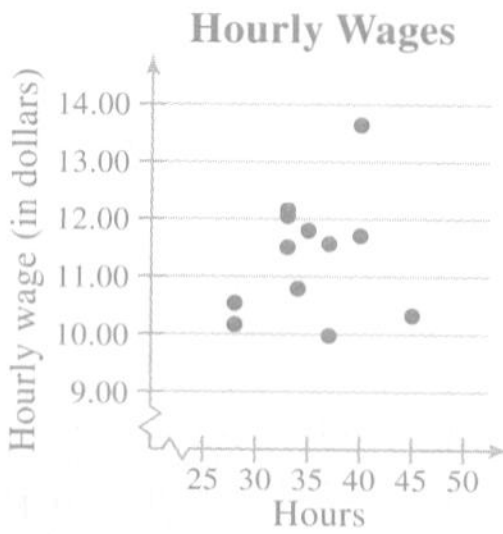

It appears that hourly wage increases as the number of hours worked increases. (Answers will vary.)

DATA **19. *Life Spans of House Flies*** Use a dot plot to display the data. The data represent the life span (in days) of 40 house flies.

9	9	4	4	8	11	10	5	8	13	9	6	7	11
13	11	6	9	8	14	10	6	10	10	8	7	14	11
7	8	6	11	13	10	14	14	8	13	14	10		

20. *Nobel Prize* Use a pie chart to display the data. The data represent the number of Nobel Prize laureates by country during the years 1901–2002.

United States	270	France	49	Germany	77
United Kingdom	100	Sweden	30	Other	157

21. *NASA Budget* Use a pie chart to display the data. The data represent the 2004 NASA budget (in millions of dollars) divided among three categories. *(Source: NASA)*

Science, aeronautics, and exploration	7661
Space flight capabilities	7782
Inspector General	26

22. *NASA Expenditures* Use a Pareto chart to display the data. The data represent the estimated 2003 NASA space shuttle operations expenditures (in millions of dollars). *(Source: NASA)*

External tank	265.4
Main engine	249.0
Reusable solid rocket motor	374.9
Solid rocket booster	156.3
Vehicle and extravehicular activity	636.1
Flight hardware upgrades	162.6

23. *UV Index* Use a Pareto chart to display the data. The data represent the ultraviolet index for five cities at noon on a recent date. *(Source: National Oceanic and Atmospheric Administration)*

Atlanta, GA	Boise, ID	Concord, NH	Denver, CO	Miami, FL
9	7	8	7	10

24. *Hourly Wages* Use a scatter plot to display the data in the table. The data represent the number of hours worked and the hourly wage (in dollars) for a sample of 12 production workers. Describe any trends shown.

Hours	Hourly wage
33	12.16
37	9.98
34	10.79
40	11.71
35	11.80
33	11.51
40	13.65
33	12.05
28	10.54
45	10.33
37	11.57
28	10.17

Table for Exercise 25

Number of students per teacher	Average teacher's salary
17.1	28.7
17.5	47.5
18.9	31.8
17.1	28.1
20.0	40.3
18.6	33.8
14.4	49.8
16.5	37.5
13.3	42.5
18.4	31.9

25. Teachers' Salaries

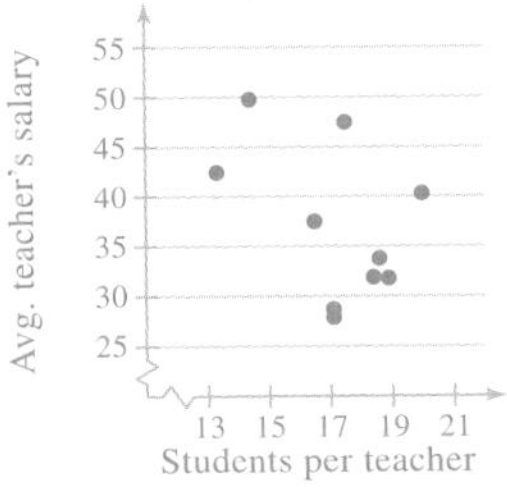

It appears that a teacher's average salary decreases as the number of students per teacher increases. (Answers will vary.)

26. See Selected Answers, page A91.

27. Price of Grade A Eggs

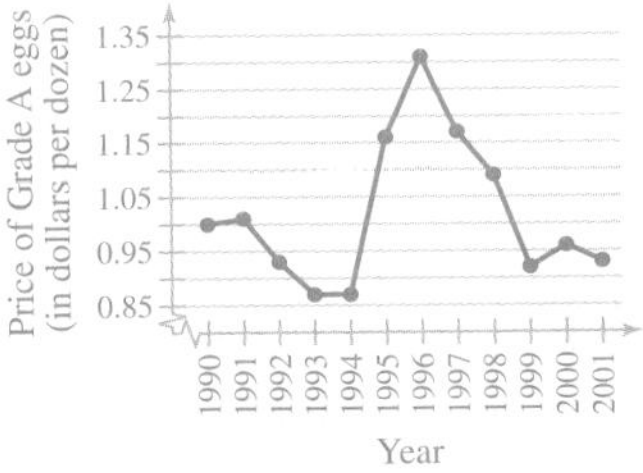

It appears the price of eggs peaked in 1996. (Answers will vary.)

28. See Selected Answers, page A92.

29. See Odd Answers, page A52.

30. See Selected Answers, page A92.

25. *Salaries* Use a scatter plot to display the data shown in the table. The data represent the number of students per teacher and the average teacher salary (in thousands of dollars) for a sample of 10 school districts. Describe any trends shown.

26. *UV Index* Use a time series chart to display the data. The data represent the ultraviolet index for Memphis, TN, on June 14–23 during a recent year. *(Source: Weather Services International)*

June 14	June 15	June 16	June 17	June 18
9	4	10	10	10
June 19	**June 20**	**June 21**	**June 22**	**June 23**
10	10	10	9	9

27. *Egg Prices* Use a time series chart to display the data. The data represent the prices of Grade A eggs (in dollars per dozen) for the indicated years. *(Source: U.S. Bureau of Labor Statistics)*

1990	1991	1992	1993	1994	1995
1.00	1.01	0.93	0.87	0.87	1.16
1996	**1997**	**1998**	**1999**	**2000**	**2001**
1.31	1.17	1.09	0.92	0.96	0.93

28. *T-Bone Steak Prices* Use a time series chart to display the data. The data represent the prices of T-bone steak (in dollars per pound) for the indicated years. *(Source: U.S. Bureau of Labor Statistics)*

1990	1991	1992	1993	1994	1995
5.45	5.21	5.39	5.77	5.86	5.92
1996	**1997**	**1998**	**1999**	**2000**	**2001**
5.87	6.07	6.40	6.71	6.82	7.31

Extending Concepts

A Misleading Graph? In Exercises 29 and 30,

(a) explain why the graph is misleading.

(b) redraw the graph so that it is not misleading.

29.

Sales for Company A

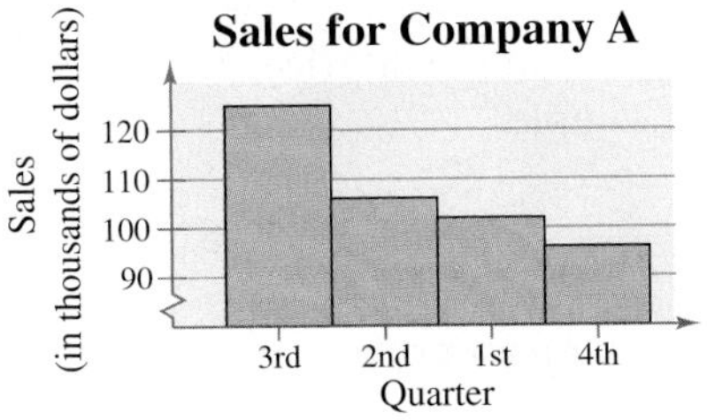

30.

Sales for Company B

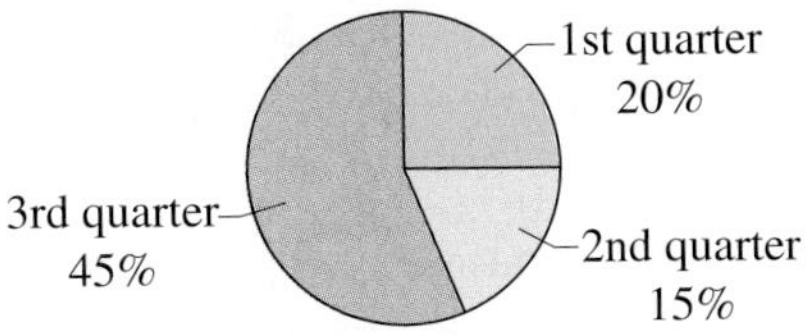

1st quarter	2nd quarter	3rd quarter	4th quarter
20%	15%	45%	20%

2.3 Measures of Central Tendency

What You Should Learn

- How to find the mean, median, and mode of a population and a sample
- How to find a weighted mean of a data set and the mean of a frequency distribution
- How to describe the shape of a distribution as symmetric, uniform, or skewed and how to compare the mean and median for each

Mean, Median, and Mode • Weighted Mean and Mean of Grouped Data • The Shape of Distributions

Mean, Median, and Mode

A **measure of central tendency** is a value that represents a typical, or central, entry of a data set. The three most commonly used measures of central tendency are the mean, the median, and the mode.

DEFINITION

The **mean** of a data set is the sum of the data entries divided by the number of entries. To find the mean of a data set, use one of the following formulas.

Population Mean: $\mu = \dfrac{\Sigma x}{N}$ Sample Mean: $\bar{x} = \dfrac{\Sigma x}{n}$

The lowercase Greek letter μ (pronounced mu) represents the population mean and $\bar{x}$ (read as "x bar") represents the sample mean. Note that N represents the number of entries in a *population* and n represents the number of entries in a *sample*.

EXAMPLE 1

Finding a Sample Mean

The prices (in dollars) for a sample of room air conditioners (10,000 Btus per hour) are listed. What is the mean price of the air conditioners?

500 840 470 480 420 440 440

SOLUTION The sum of the air conditioner prices is

$$\Sigma x = 500 + 840 + 470 + 480 + 420 + 440 + 440 = 3590.$$

To find the mean price, divide the sum of the prices by the number of prices in the sample.

$$\bar{x} = \frac{\Sigma x}{n} = \frac{3590}{7} \approx 512.9$$

So, the mean price of the air conditioners is about $512.90.

Study Tip

Notice that the mean in Example 1 has one more decimal place than the original set of data values. This *round-off rule* will be used throughout the text. Another important *round-off rule* is that rounding should not be done until the final answer of a calculation.

Try It Yourself 1

The ages of employees in a department are listed. What is the mean age?

34 27 50 45 41 37 24
57 40 38 62 44 39 40

a. *Find the sum* of the data entries.
b. *Divide the sum* by the number of data entries.
c. *Interpret* the results in the context of the data.

Answer: Page A32

DEFINITION

The **median** of a data set is the value that lies in the middle of the data when the data set is ordered. If the data set has an odd number of entries, the median is the middle data entry. If the data set has an even number of entries, the median is the mean of the two middle data entries.

Study Tip

In a data set, there are the same number of data values above the median as there are below the median. For instance, in Example 2, three of the prices are below \$470 and three are above \$470.

EXAMPLE 2

Finding the Median

Find the median of the air conditioner prices given in Example 1.

SOLUTION To find the median price, first order the data.

420 440 440 470 480 500 840

Because there are seven entries (an odd number), the median is the middle, or fourth, data entry. So, the median air conditioner price is \$470.

Try It Yourself 2

One of the families of Akhiok is planning to relocate to another city. The ages of the family members are 33, 37, 3, 7, and 59. What will be the median age of the remaining residents of Akhiok after this family relocates?

a. *Order* the data entries.
b. *Find the middle* data entry. *Answer: Page A32*

Akhiok, Alaska is a fishing village on Kodiak Island.
(Photograph © Roy Corral.)

EXAMPLE 3

Finding the Median

The air conditioner priced at \$480 is discontinued. What is the median price of the remaining air conditioners?

SOLUTION The remaining prices, in order, are

420, 440, 440, 470, 500, and 840.

Because there are six entries (an even number), the median is the mean of the two middle entries.

$$\text{Median} = \frac{440 + 470}{2}$$
$$= 455$$

So, the median price of the remaining air conditioners is \$455.

Try It Yourself 3

Find the median age of the residents of Akhiok using the population data set listed in the Chapter Opener on page 33.

a. *Order* the data entries.
b. *Find the mean* of the two middle data entries.
c. *Interpret* the results in the context of the data. *Answer: Page A32*

DEFINITION

The **mode** of a data set is the data entry that occurs with the greatest frequency. If no entry is repeated, the data set has no mode. If two entries occur with the same greatest frequency, each entry is a mode and the data set is called **bimodal.**

Insight

The mode is the only measure of central tendency that can be used to describe data at the nominal level of measurement.

EXAMPLE 4

Finding the Mode

Find the mode of the air conditioner prices given in Example 1.

SOLUTION Ordering the data helps to find the mode.

420 440 440 470 480 500 840

From the ordered data, you can see that the entry of 440 occurs twice, whereas the other data entries occur only once. So, the mode of the air conditioner prices is $440.

Try It Yourself 4

Find the mode of the ages of the Akhiok residents. The data are given below.

25, 5, 18, 12, 60, 44, 24, 22, 2, 7, 15, 39, 58, 53, 36, 42, 16, 20, 1, 5, 39,
51, 44, 23, 3, 13, 37, 56, 58, 13, 47, 23, 1, 17, 39, 13, 24, 0, 39, 10, 41,
1, 48, 17, 18, 3, 72, 20, 3, 9, 0, 12, 33, 21, 40, 68, 25, 40, 59, 4, 67, 29,
13, 18, 19, 13, 16, 41, 19, 26, 68, 49, 5, 26, 49, 26, 45, 41, 19, 49

a. Write the data in *order*.
b. Identify the entry, or entries, that occur with the *greatest frequency*.
c. *Interpret* the results in the context of the data.

Answer: Page A32

EXAMPLE 5

Finding the Mode

At a political debate a sample of audience members was asked to name the political party to which they belong. Their responses are shown in the table. What is the mode of the responses?

Political party	Frequency, *f*
Democrat	34
Republican	56
Other	21
Did not respond	9

SOLUTION The response occurring with the greatest frequency is Republican. So, the mode is Republican.

Interpretation In this sample, there were more Republicans than people of any other single affiliation.

Try It Yourself 5

In a survey, 250 baseball fans were asked if Barry Bonds's home run record would ever be broken. One hundred sixty-nine of the fans responded "yes," 54 responded "no," and 27 "didn't know." What is the mode of the responses?

a. Identify the entry that occurs with the *greatest frequency*.
b. *Interpret* the results in the context of the data.

Answer: Page A32

Although the mean, the median, and the mode each describe a typical entry of a data set, there are advantages and disadvantages of using each, especially when the data set contains outliers.

DEFINITION

An **outlier** is a data entry that is far removed from the other entries in the data set.

Ages in a class						
20	20	20	20	20	20	21
21	21	21	22	22	22	23
23	23	23	24	24	65	

Outlier (65)

EXAMPLE 6

Comparing the Mean, the Median, and the Mode

Find the mean, the median, and the mode of the sample ages of a class shown at the left. Which measure of central tendency best describes a typical entry of this data set? Are there any outliers?

SOLUTION

Mean: $\bar{x} = \frac{\Sigma x}{n} = \frac{475}{20} \approx 23.8$ years

Median: $\text{Median} = \frac{21 + 22}{2} = 21.5$ years

Mode: The entry occurring with the greatest frequency is 20 years.

Interpretation The mean takes every entry into account but is influenced by the outlier of 65. The median also takes every entry into account, and it is not affected by the outlier. In this case the mode exists, but it doesn't appear to represent a typical entry. Sometimes a graphical comparison can help you decide which measure of central tendency best represents a data set. The histogram shows the distribution of the data and the location of the mean, the median, and the mode. In this case, it appears that the median best describes the data set.

Ages of Students in a Class

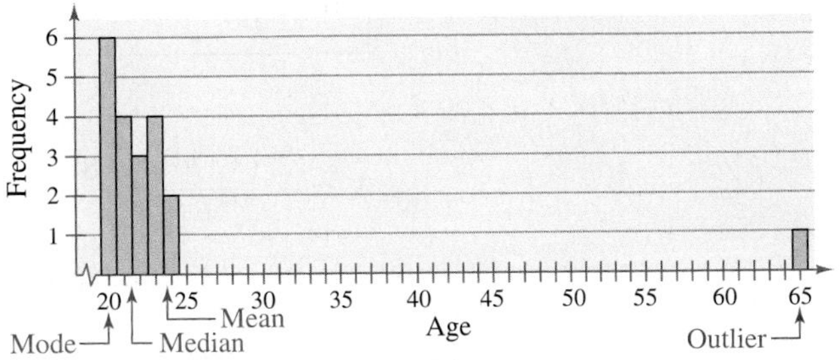

Try It Yourself 6

Remove the data entry of 65 from the preceding data set. Then rework the example. How does the absence of this outlier change each of the measures?

a. Find the *mean*, the *median*, and the *mode*.

b. Compare these measures of central tendency with those found in Example 6.

Answer: Page A32

Picturing the World

The National Association of Realtors keeps a databank of existing-home sales. One list uses the *median* price of existing homes sold and another uses the *mean* price of existing homes sold. The sales for the first quarter of 2003 are shown in the graph. *(Source: National Association of Realtors)*

2003 U.S. Existing-Home Sales

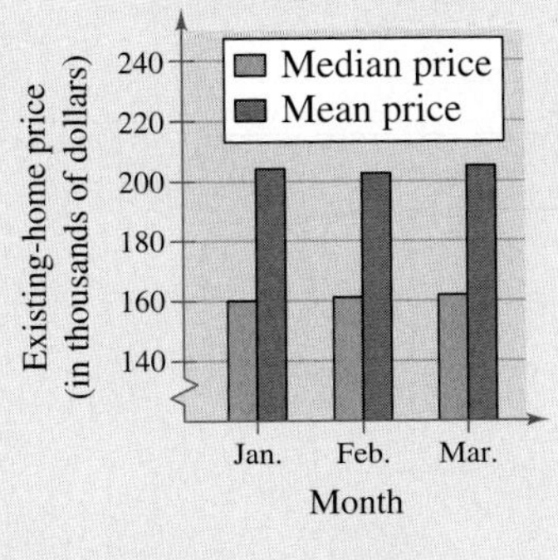

Notice in the graph that each month the mean price is about $40,000 more than the median price. What factors would cause the mean price to be greater than the median price?

Weighted Mean and Mean of Grouped Data

Sometimes data sets contain entries that have a greater effect on the mean than do other entries. To find the mean of such data sets, you must find the weighted mean.

DEFINITION

A **weighted mean** is the mean of a data set whose entries have varying weights. A weighted mean is given by

$$\bar{x} = \frac{\Sigma(x \cdot w)}{\Sigma w}$$

where w is the weight of each entry x.

EXAMPLE 7

Finding a Weighted Mean

You are taking a class in which your grade is determined from five sources: 50% from your test mean, 15% from your midterm, 20% from your final exam, 10% from your computer lab work, and 5% from your homework. Your scores are 86 (test mean), 96 (midterm), 82 (final exam), 98 (computer lab), and 100 (homework). What is the weighted mean of your scores?

SOLUTION Begin by organizing the scores and the weights in a table.

Source	Score, x	Weight, w	xw
Test Mean	86	0.50	43.0
Midterm	96	0.15	14.4
Final Exam	82	0.20	16.4
Computer Lab	98	0.10	9.8
Homework	100	0.05	5.0
		$\Sigma w = 1$	$\Sigma(x \cdot w) = 88.6$

$$\bar{x} = \frac{\Sigma(x \cdot w)}{\Sigma w} = \frac{88.6}{1} = 88.6$$

So, your weighted mean for the course is 88.6.

Try It Yourself 7

An error was made in grading your final exam. Instead of getting 82, you scored 98. What is your new weighted mean?

a. Multiply each score by its weight and *find the sum of these products.*
b. Find the *sum of the weights.*
c. Find the *weighted mean.*
d. *Interpret* the results in the context of the data.

Answer: Page A33

If data are presented in a frequency distribution, you can approximate the mean as follows.

DEFINITION

The **mean of a frequency distribution** for a sample is approximated by

$$\bar{x} = \frac{\Sigma(x \cdot f)}{n} \qquad \text{Note that } n = \Sigma f$$

where x and f are the midpoints and frequencies of a class, respectively.

Study Tip

If the frequency distribution represents a population, then the mean of the frequency distribution is approximated by

$$\mu = \frac{\Sigma(x \cdot f)}{N}$$

where $N = \Sigma f$.

GUIDELINES

Finding the Mean of a Frequency Distribution

In Words	*In Symbols*
1. Find the midpoint of each class.	$x = \frac{\text{(Lower limit)} + \text{(Upper limit)}}{2}$
2. Find the sum of the products of the midpoints and the frequencies.	$\Sigma(x \cdot f)$
3. Find the sum of the frequencies.	$n = \Sigma f$
4. Find the mean of the frequency distribution.	$\bar{x} = \frac{\Sigma(x \cdot f)}{n}$

EXAMPLE 8

Finding the Mean of a Frequency Distribution

Use the frequency distribution at the left to approximate the mean number of minutes that a sample of Internet subscribers spent online during their most recent session.

Class midpoint

x	Frequency, f	$(x \cdot f)$
12.5	6	75.0
24.5	10	245.0
36.5	13	474.5
48.5	8	388.0
60.5	5	302.5
72.5	6	435.0
84.5	2	169.0
	$n = 50$	$\Sigma = 2089.0$

SOLUTION

$$\bar{x} = \frac{\Sigma(x \cdot f)}{n} = \frac{2089}{50} \approx 41.8$$

So, the mean time spent online was approximately 41.8 minutes.

Try It Yourself 8

Use a frequency distribution to approximate the mean age of the residents of Akhiok. (See Try It Yourself 2 on page 37.)

a. Find the *midpoint* of each class.
b. Find the *sum of the products* of each midpoint and corresponding frequency.
c. Find the *sum of the frequencies*.
d. Find the *mean of the frequency distribution*.

Answer: Page A33

The Shape of Distributions

A graph reveals several characteristics of a frequency distribution. One such characteristic is the shape of the distribution.

DEFINITION

A frequency distribution is **symmetric** when a vertical line can be drawn through the middle of a graph of the distribution and the resulting halves are approximately mirror images.

A frequency distribution is **uniform** (or **rectangular**) when all entries, or classes, in the distribution have equal frequencies. A uniform distribution is also symmetric.

A frequency distribution is skewed if the "tail" of the graph elongates more to one side than to the other. A distribution is **skewed left (negatively skewed)** if its tail extends to the left. A distribution is **skewed right (positively skewed)** if its tail extends to the right.

Insight

The mean will always fall in the direction the distribution is skewed. For instance, when a distribution is skewed left, the mean is to the left of the median.

When a distribution is symmetric and unimodal, the mean, median, and mode are equal. If a distribution is skewed left, the mean is less than the median and the median is usually less than the mode. If a distribution is skewed right, the mean is greater than the median and the median is usually greater than the mode. Examples of these commonly occurring distributions are shown.

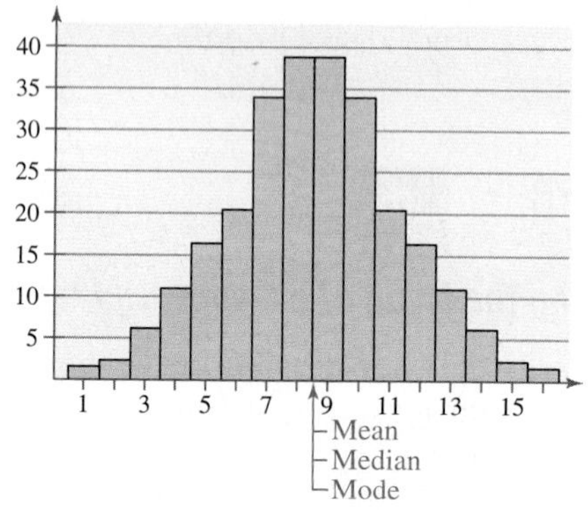

Symmetric Distribution

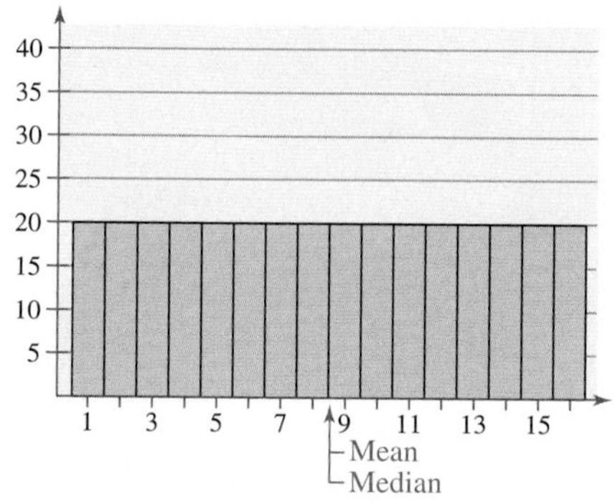

Uniform Distribution

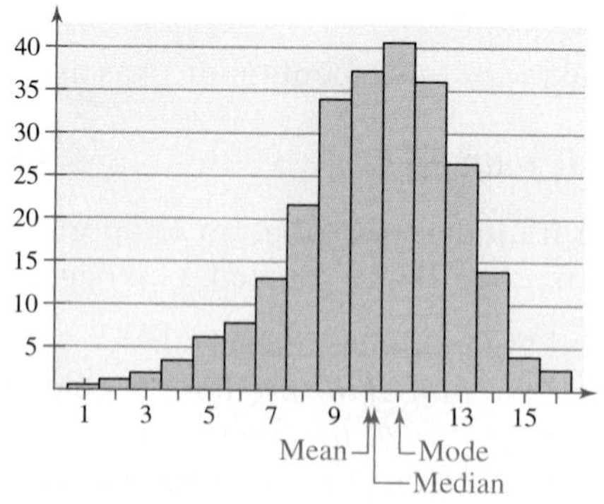

Skewed-Left Distribution

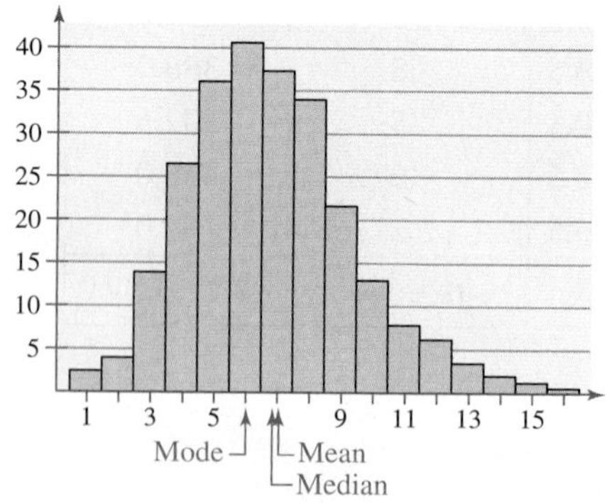

Skewed-Right Distribution

2.3 Exercises

Building Basic Skills and Vocabulary

True or False? In Exercises 1–4, determine whether the statement is true or false. If it is false, rewrite it so it is a true statement.

1. The median is the measure of central tendency most likely to be affected by an extreme value (an outlier).
2. Every data set must have a mode.
3. Some quantitative data sets do not have a median.
4. The mean is the only measure of central tendency that can be used for data at the nominal level of measurement.
5. Give an example in which the mean of a data set is *not* representative of a typical number in the data set.
6. Give an example in which the median and the mode of a data set are the same.

Graphical Analysis In Exercises 7–10, determine whether the approximate shape of the distribution in the histogram is symmetric, uniform, skewed left, skewed right, or none of these. Justify your answer.

7.

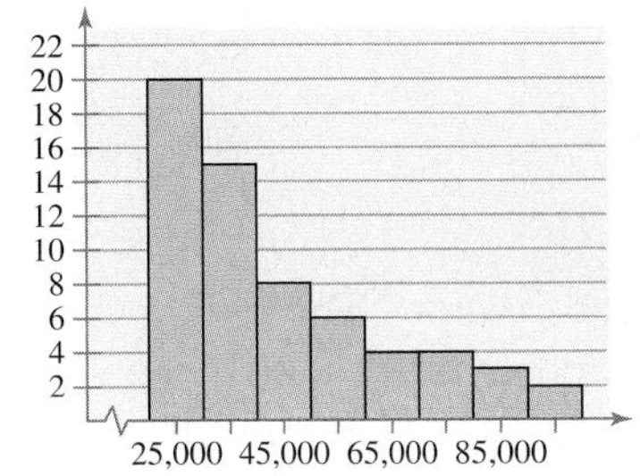

8.

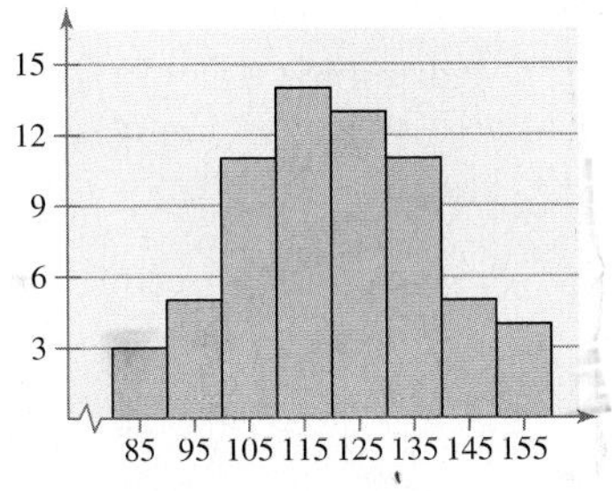

9.

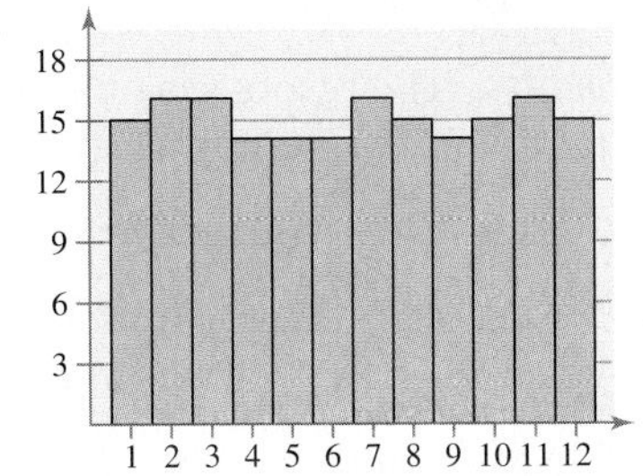

10.

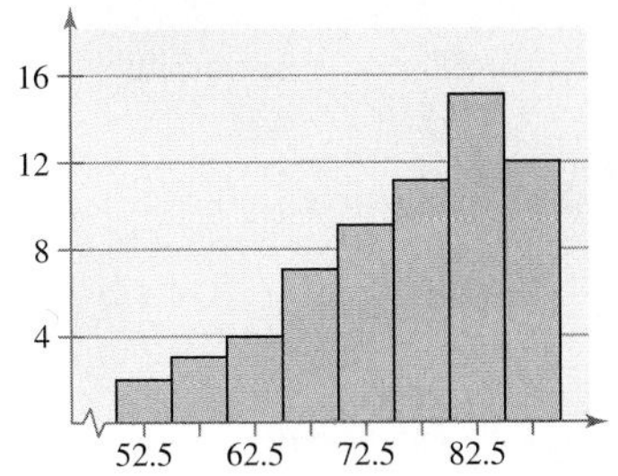

Matching In Exercises 11–14, match the distribution with one of the graphs in Exercises 7–10. Justify your decision.

11. The frequency distribution of 180 rolls of a dodecagon (a 12-sided die)
12. The frequency distribution of salaries at a company where a few executives make much higher salaries than the majority of employees
13. The frequency distribution of scores on a 90-point test where a few students scored much lower than the majority of students
14. The frequency distribution of weights for a sample of seventh grade boys

1. False. The mean is the measure of central tendency most likely to be affected by an extreme value (or outlier).
2. False. Not all data sets must have a mode.
3. False. All quantitative data sets have a median.
4. False. The mode is the only measure of central tendency that can be used for data at the nominal level of measurement.
5. A data set with an outlier within it would be an example. (Answers will vary.)
6. Any data set that is symmetric has the same median and mode.
7. The shape of the distribution is skewed right because the bars have a "tail" to the right.
8. Symmetric. If a vertical line is drawn down the middle, the two halves look approximately the same.
9. The shape of the distribution is uniform because the bars are approximately the same height.
10. See Selected Answers, page A92.
11. (9), because the distribution of values ranges from 1 to 12 and has (approximately) equal frequencies.
12. See Selected Answers, page A92.
13. (10), because the distribution has a maximum value of 90 and is skewed left owing to a few students' scoring much lower than the majority of the students.
14. See Selected Answers, page A92.

15. (a) $\bar{x} \approx 6.2$
median = 6
mode = 5
(b) Median, because the distribution is skewed.

16. (a) $\bar{x} = 19.6$
median = 19.5
mode = 19, 20
(b) Mean, because there are no outliers.

17. (a) $\bar{x} \approx 4.57$
median = 4.8
mode = 4.8
(b) Median, because there are no outliers.

18. (a) $\bar{x} = 184.6$
median = 182.5
mode = none
(b) Mean, because there are no outliers.

19. (a) $\bar{x} \approx 93.81$
median = 92.9
mode = 90.3, 91.8
(b) Median, because the distribution is skewed.

20. (a) $\bar{x} = 61.2$
median = 55
mode = 80, 125
(b) Median, because the distribution is skewed.

21. (a) $\bar{x}$ = not possible
median = not possible
mode = "Worse"
(b) Mode, because the data are at the nominal level of measurement.

22. (a) $\bar{x}$ = not possible
median = not possible
mode = "Watchful"
(b) Mode, because the data are at the nominal level of measurement.

23. (a) $\bar{x} \approx 170.63$
median = 169.3
mode = none
(b) Mean, because there are no outliers.

Using and Interpreting Concepts

Finding and Discussing the Mean, Median, and Mode In Exercises 15–32,

(a) find the mean, median, and mode of the data, if possible. If it is not possible, explain why the measure of central tendency cannot be found.

(b) determine which measure of central tendency best represents the data. Explain your reasoning.

15. ***SUVs*** The maximum number of seats in a sample of 13 sport utility vehicles

6 6 9 9 6 5 5 5 7 5 5 5 8

16. ***Education*** The education cost per student (in thousands of dollars) from a sample of 10 liberal arts colleges

22 26 19 20 20 18 21 17 19 14

17. ***Sports Cars*** The time (in seconds) for a sample of seven sports cars to go from 0 to 60 miles per hour

3.7 4.0 4.8 4.8 4.8 4.8 5.1

18. ***Cholesterol*** The cholesterol level of a sample of 10 female employees

154 216 171 188 229 203 184 173 181 147

DATA **19.** ***NBA*** The average points per game scored by each NBA team during the 2003–2004 regular season *(Source: NBA)*

89.8	88.0	95.3	90.3	92.0	94.0
90.3	91.8	92.8	89.7	103.5	98.0
92.9	85.4	105.2	97.2	94.5	91.5
90.1	96.7	88.7	93.3	98.2	94.2
91.8	94.8	90.7	102.8	97.1	

DATA **20.** ***Power Failures*** The duration (in minutes) of every power failure at a residence in the last 10 years

18	26	45	75	125	80	33	40	44	49
89	80	96	125	12	61	31	63	103	28

21. ***Air Quality*** The responses of a sample of 1040 people who were asked if the air quality in their community is better or worse than it was 10 years ago

Better: 346 Worse: 450 Same: 244

22. ***Crime*** The responses of a sample of 1019 people who were asked how they felt when they thought about crime

Unconcerned: 34 Watchful: 672 Nervous: 125 Afraid: 188

23. ***Top Speeds*** The top speed (in miles per hour) for a sample of seven sports cars

187.3 181.8 180.0 169.3 162.2 158.1 155.7

24. ***Purchase Preference*** The responses of a sample of 1001 people who were asked if their next vehicle purchase will be foreign or domestic

Domestic: 704 Foreign: 253 Don't know: 44

25. ***Stocks*** The recommended prices (in dollars) for several stocks that analysts predict should produce at least 10% annual returns *(Source: Money)*

41 20 22 14 15 25 18 40 17 14

26. *Eating Disorders* The number of weeks it took to reach a target weight for a sample of five patients with eating disorders treated by psychodynamic psychotherapy *(Source: The Journal of Consulting and Clinical Psychology)*

15.0 31.5 10.0 25.5 1.0

27. *Eating Disorders* The number of weeks it took to reach a target weight for a sample of 14 patients with eating disorders treated by psychodynamic psychotherapy and cognitive behavior techniques *(Source: The Journal of Consulting and Clinical Psychology)*

2.5	20.0	11.0	10.5	17.5	16.5	13.0
15.5	26.5	2.5	27.0	28.5	1.5	5.0

28. *Aircraft* The number of aircraft 11 airlines have in their fleets *(Source: Airline Transport Association)*

819	366	573	280	375	567
444	145	102	26	37	

29. Weights (in pounds) of Dogs at a Kennel

Key: 1|0 = 10

1	0 2
2	1 4 7
3	7 8
4	1 5 5
5	0 7
6	5
7	
8	
9	
10	6

30. Grade Point Averages of Students in a Class

Key: 0|8 = 0.8

0	8
1	5 6 8
2	1 3 4 5
3	0 9
4	0 0

31. Time (in minutes) it Takes Employees to Drive to Work

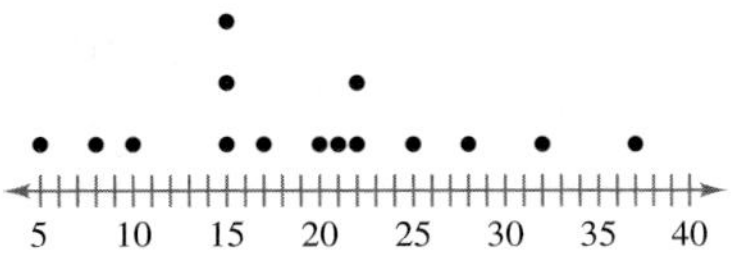

32. Top Speeds (in miles per hour) of High-Performance Sports Cars

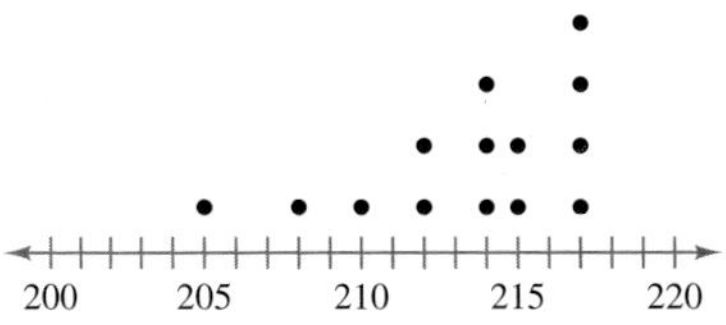

Graphical Analysis In Exercises 33 and 34, the letters A, B, and C are marked on the horizontal axis. Determine which is the mean, which is the median, and which is the mode. Justify your answers.

33. Sick Days Used by Employees

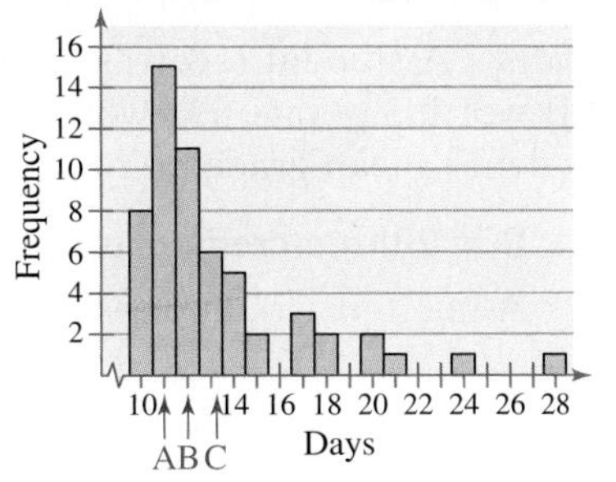

34. Hourly Wages of Employees

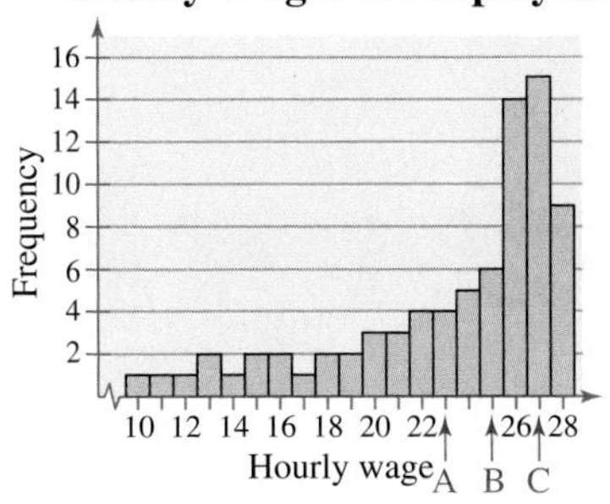

24. (a) $\bar{x}$ = not possible
median = not possible
mode = "Domestic"
(b) Mode, because the data are at the nominal level of measurement.

25. (a) $\bar{x} = 22.6$
median = 19
mode = 14
(b) Median, because the distribution is skewed.

26. (a) $\bar{x} = 16.6$
median = 15
mode = none
(b) Mean, because there are no outliers.

27. (a) $\bar{x} \approx 14.11$
median = 14.25
mode = 2.5
(b) Mean, because there are no outliers.

28. (a) $\bar{x} \approx 339.5$
median = 366
mode = none
(b) Median, because the distribution is skewed.

29. (a) $\bar{x} = 41.3$
median = 39.5
mode = 45
(b) Median, because the distribution is skewed.

30. (a) $\bar{x} \approx 2.5$
median = 2.35
mode = 4.0
(b) Mean, because there are no outliers.

31. (a) $\bar{x} \approx 19.5$
median = 20
mode = 15
(b) Median, because the distribution is skewed.

32. See Selected Answers, page A92.

33. A = mode, because it's the data entry that occurred most often.
B = median, because the distribution is skewed right.
C = mean, because the distribution is skewed right.

34. See Selected Answers, page A92.

35. Mode, because the data are at the nominal level of measurement.

36. Median, because the distribution is skewed.

37. Mean, because there are no outliers.

38. Median, because the distribution is skewed.

39. 89.3

40. \$32,640

41. 2.8

In Exercises 35–38, determine which measure of central tendency best represents the graphed data without performing any calculations. Explain your reasoning.

35. **Are You Getting Enough Sleep?**

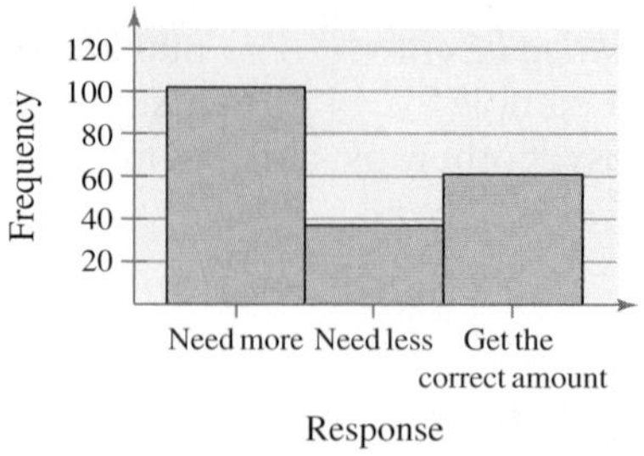

36. **Heights of Players on a Hockey Team**

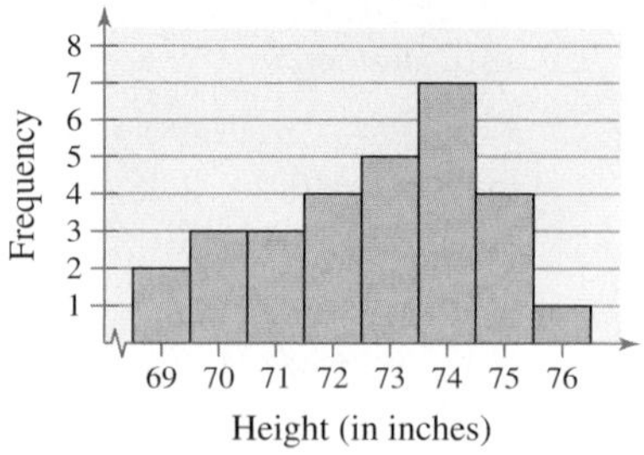

37. **Heart Rate of a Sample of Adults**

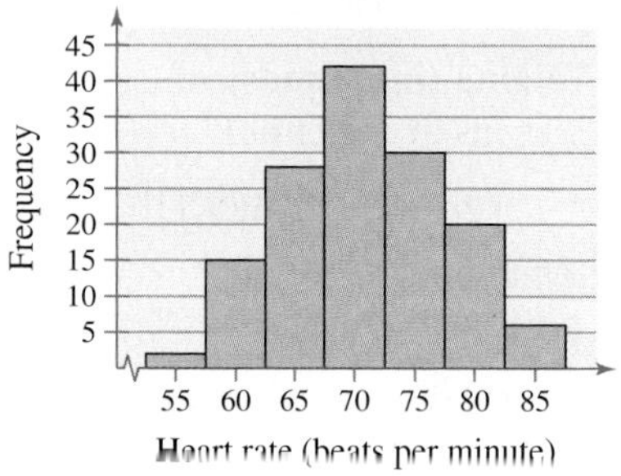

38. **Body Mass Index (BMI) of People in a Gym**

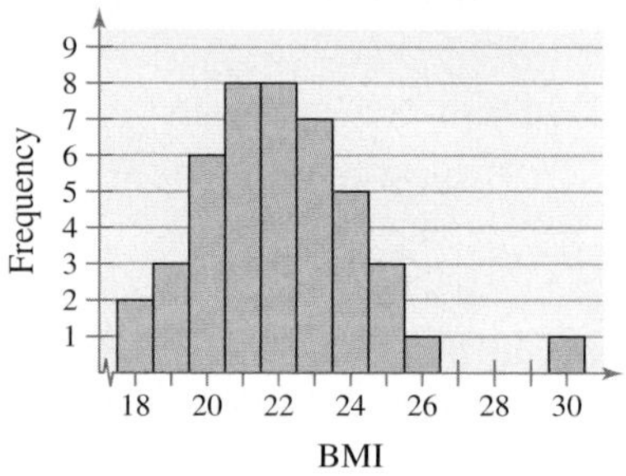

Finding the Weighted Mean In Exercises 39–42, find the weighted mean of the data.

39. ***Final Grade*** The scores and their percent of the final grade for a statistics student are given. What is the student's mean score?

	Score	Percent of final grade
Homework	85	15%
Quiz	80	10%
Quiz	92	10%
Quiz	76	10%
Project	100	15%
Speech	90	15%
Final Exam	93	25%

40. ***Salaries*** The average starting salaries (by degree attained) for 25 employees at a company are given. What is the mean starting salary for these employees?

8 with MBAs: \$42,500

17 with BAs in business: \$28,000

41. ***Grades*** A student receives the following grades, with an A worth 4 points, a B worth 3 points, a C worth 2 points, and a D worth 1 point. What is the student's mean grade point score?

B in 2 three-credit classes

A in 1 four-credit class

D in 1 two-credit class

C in 1 three-credit class

42. 82

43. 65.5

44. 70.1

45. 35.0

46. 15.3

47.

Class	Frequency, f	Midpoint
3–4	3	3.5
5–6	8	5.5
7–8	4	7.5
9–10	2	9.5
11–12	2	11.5
13–14	1	13.5
	$\sum f = 20$	

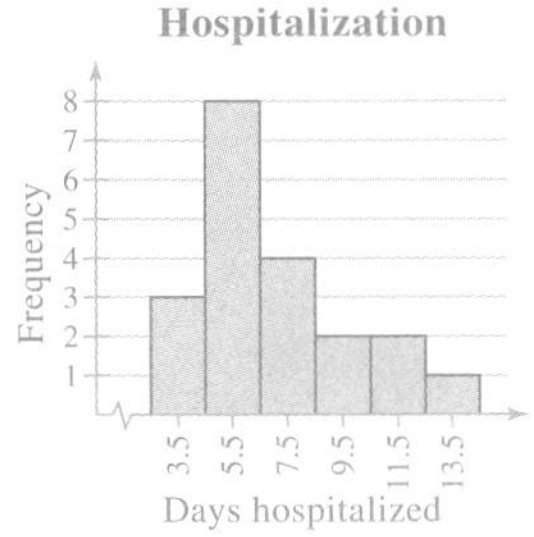

Positively skewed

42. ***Scores*** The mean scores for a statistics course (by major) are given. What is the mean score for the class?

8 engineering majors: 83

5 math majors: 87

11 business majors: 79

Finding the Mean of Grouped Data In Exercises 43–46, approximate the mean of the grouped data.

43. ***Heights of Females*** The heights (in inches) of 16 female students in a physical education class

Height (in inches)	Frequency
60–62	3
63–65	4
66–68	7
69–71	2

44. ***Heights of Males*** The heights (in inches) of 21 male students in a physical education class

Height (in inches)	Frequency
63–65	2
66–68	4
69–71	8
72–74	5
75–77	2

45. ***Ages*** The ages of residents of a town

Age	Frequency
0–9	57
10–19	68
20–29	36
30–39	55
40–49	71
50–59	44
60–69	36
70–79	14
80–89	8

46. ***Phone Calls*** The lengths of long-distance calls (in minutes) made by one person in one year

Length of call	Number of calls
1–5	12
6–10	26
11–15	20
16–20	7
21–25	11
26–30	7
31–35	4
36–40	4
41–45	1

Identifying the Shape of a Distribution In Exercises 47–50, construct a frequency distribution and a frequency histogram of the data using the indicated number of classes. Describe the shape of the histogram as symmetric, uniform, negatively skewed, positively skewed, or none of these.

DATA **47.** ***Hospitalization***

Number of classes: 6

Data set: The number of days 20 patients remained hospitalized

6	9	7	14	4	5	6	8	4	11
10	6	8	6	5	7	6	6	3	11

48.

Class	Frequency, f	Midpoint
127–161	9	144
162–196	8	179
197–231	3	214
232–266	3	249
267–301	1	284
	$\sum f = 24$	

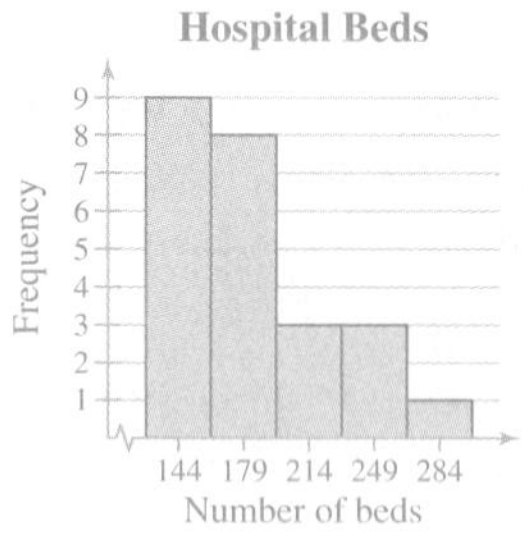

Positively skewed

49.

Class	Frequency, f	Midpoint
62–64	3	63
65–67	7	66
68–70	9	69
71–73	8	72
74–76	3	75
	$\sum f = 30$	

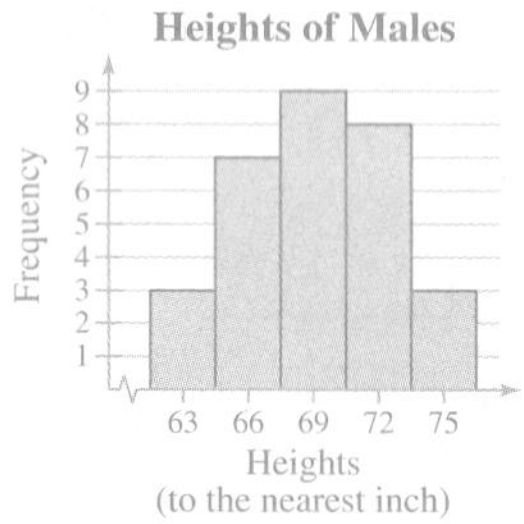

Symmetric

50. See Selected Answers, page A92.

51. (a) $\bar{x} = 6.005$

median $= 6.01$

(b) $\bar{x} = 5.945$

median $= 6.01$

(c) Mean

52. (a) $\bar{x} \approx 29.63$

median $= 18.3$

(b) $\bar{x} \approx 22.34$

median $= 17.25$

(c) Mean

48. *Hospital Beds*

DATA

Number of classes: 5

Data set: The number of beds in a sample of 24 hospitals

149	167	162	127	130	180	160	167
221	145	137	194	207	150	254	262
244	297	137	204	166	174	180	151

49. *Height of Males*

DATA

Number of classes: 5

Data set: The heights (to the nearest inch) of 30 males

67	76	69	68	72	68	65	63	75	69
66	72	67	66	69	73	64	62	71	73
68	72	71	65	69	66	74	72	68	69

50. *Six-Sided Die*

DATA

Number of classes: 6

Data set: The results of rolling a six-sided die 30 times

1	4	6	1	5	3	2	5	4	6
1	2	4	3	5	6	3	2	1	1
5	6	2	4	4	3	1	6	2	4

51. *Coffee Content* During a quality assurance check, the actual coffee content (in ounces) of six jars of instant coffee was recorded as 6.03, 5.59, 6.40, 6.00, 5.99, and 6.02.

(a) Find the mean and the median of the coffee content.

(b) The third value was incorrectly measured and is actually 6.04. Find the mean and median of the coffee content again.

(c) Which measure of central tendency, the mean or the median, was affected more by the data entry error?

52. *U.S. Exports* The following data are the U.S. exports (in billions of dollars) to 19 countries for a recent year. *(Source: U.S. Department of Commerce)*

Canada	160.8	Japan	51.4
Mexico	97.5	United Kingdom	33.3
Germany	26.6	South Korea	22.6
Taiwan	18.4	Singapore	16.2
Netherlands	18.3	France	19.0
China	22.1	Brazil	12.4
Australia	13.1	Belgium	13.3
Malaysia	10.3	Italy	10.1
Switzerland	7.8	Thailand	4.9
Saudi Arabia	4.8		

(a) Find the mean and median.

(b) Find the mean and median without the U.S. exports to Canada.

(c) Which measure of central tendency, the mean or the median, was affected more by the elimination of the Canadian export data?

Extending Concepts

53. *Data Analysis* A consumer testing service obtained the following miles per gallon in five test runs performed with three types of compact cars.

	Run 1	Run 2	Run 3	Run 4	Run 5
Car A:	28	32	28	30	34
Car B:	31	29	31	29	31
Car C:	29	32	28	32	30

(a) The manufacturer of Car A wants to advertise that their car performed best in this test. Which measure of central tendency—mean, median, or mode—should be used for their claim? Explain your reasoning.

(b) The manufacturer of Car B wants to advertise that their car performed best in this test. Which measure of central tendency—mean, median, or mode—should be used for their claim? Explain your reasoning.

(c) The manufacturer of Car C wants to advertise that their car performed best in this test. Which measure of central tendency—mean, median, or mode—should be used for their claim? Explain your reasoning.

54. *Midrange* The **midrange** is

$$\frac{(\text{Maximum data entry}) + (\text{Minimum data entry})}{2}.$$

Which of the manufacturers in Exercise 53 would prefer to use the midrange statistic in their ads? Explain your reasoning.

DATA

55. *Data Analysis* Students in an experimental psychology class did research on depression as a sign of stress. A test was administered to a sample of 30 students. The scores are given.

44	51	11	90	76	36	64	37	43	72	53	62	36	74	51
72	37	28	38	61	47	63	36	41	22	37	51	46	85	13

(a) Find the mean of the data.

(b) Find the median of the data.

(c) Draw a stem-and-leaf plot for the data using one line per stem. Locate the mean and median on the display.

(d) Describe the shape of the distribution.

56. *Trimmed Mean* To find the 10% **trimmed mean** of a data set, order the data, delete the lowest 10% of the entries and the highest 10% of the entries, and find the mean of the remaining entries.

(a) Find the 10% trimmed mean for the data in Exercise 55.

(b) Compare the four measures of central tendency.

(c) What is the benefit of using a trimmed mean versus using a mean found using all data entries? Explain your reasoning.

57. *Writing* The population mean μ and the sample mean $\bar{x}$ have essentially the same formulas. Explain why it is necessary to have two different symbols.

58. *Writing* Describe in words the shape of a distribution that is symmetric but whose mean, median, and mode are not all equal. Then sketch this distribution.

---

53. (a) Mean, because Car A has the highest mean of the three.

(b) Median, because Car B has the highest median of the three.

(c) Mode, because Car C has the highest mode of the three.

54. Car A, because its midrange is the largest.

55. (a) $\bar{x} \approx 49.2$ (b) median $= 46.5$

(c) Key: 3|6 = 36

1	1 3	
2	2 8	
3	6 6 6 7 7 7 8	
4	1 3 4 6 7	mean
5	1 1 1 3	
6	1 2 3 4	median
7	2 2 4 6	
8	5	
9	0	

(d) Positively skewed

56. (a) 49.2

(b) $\bar{x} = 49.2$; median $= 46.5$; mode $= 36, 37, 51$

(c) Using a trimmed mean eliminates potential outliers that may affect the mean of all the entries.

57. Two different symbols are needed because they describe a measure of central tendency for two different sets of data (sample is a subset of the population).

58. A distribution with one data entry in each class would be an example of a rectangular (uniform) distribution whose mean and median are equal and whose mode does not exist.

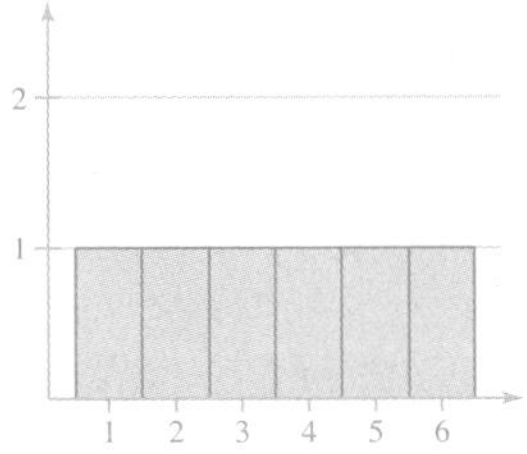

2.4 Measures of Variation

What You Should Learn

- How to find the range of a data set
- How to find the variance and standard deviation of a population and of a sample
- How to use the Empirical Rule and Chebychev's Theorem to interpret standard deviation
- How to approximate the sample standard deviation for grouped data

Range • Deviation, Variance, and Standard Deviation • Interpreting Standard Deviation • Standard Deviation for Grouped Data

Range

In this section, you will learn different ways to measure the variation of a data set. The simplest measure is the range of the set.

DEFINITION

The **range** of a data set is the difference between the maximum and minimum data entries in the set.

Range = (Maximum data entry) − (Minimum data entry)

EXAMPLE 1

Finding the Range of a Data Set

Two corporations each hired 10 graduates. The starting salaries for each are shown. Find the range of the starting salaries for Corporation A.

Starting Salaries for Corporation A (1000s of dollars)

Salary	41	38	39	45	47	41	44	41	37	42

Starting Salaries for Corporation B (1000s of dollars)

Salary	40	23	41	50	49	32	41	29	52	58

SOLUTION Ordering the data helps to find the least and greatest salaries.

37 38 39 41 41 41 42 44 45 47

Minimum (37) — Maximum (47)

Range = (Maximum salary) − (Minimum salary)

= 47 − 37

= 10

So, the range of the starting salaries for Corporation A is 10, or $10,000.

Insight

Both data sets in Example 1 have a mean of 41.5, a median of 41, and a mode of 41. And yet the two sets differ significantly.

The difference is that the entries in the second set have greater variation. Your goal in this section is to learn how to measure the variation of a data set.

Try It Yourself 1

Find the range of the starting salaries for Corporation B.

a. Identify the *minimum* and *maximum* salaries.
b. Find the *range*.
c. Compare your answer with that for Example 1.

Answer: Page A33

Deviation, Variance, and Standard Deviation

As a measure of variation, the range has the advantage of being easy to compute. Its disadvantage, however, is that it uses only two entries from the data set. Two measures of variation that use all the entries in a data set are the variance and the standard deviation. However, before you learn about these measures of variation, you need to know what is meant by the deviation of an entry in a data set.

Note to Instructor

Remind students of the reason for the difference between the symbols μ and $\bar{x}$.

DEFINITION

The **deviation** of an entry x in a population data set is the difference between the entry and the mean μ of the data set.

Deviation of $x = x - \mu$

Deviations of Starting Salaries for Corporation A

Salary (1000s of dollars) x	Deviation (1000s of dollars) $x - \mu$
41	−0.5
38	−3.5
39	−2.5
45	3.5
47	5.5
41	−0.5
44	2.5
41	−0.5
37	−4.5
42	0.5
$\Sigma x = 415$	$\Sigma(x - \mu) = 0$

EXAMPLE 2

Finding the Deviations of a Data Set

Find the deviation of each starting salary for Corporation A given in Example 1.

SOLUTION The mean starting salary is $\mu = 415/10 = 41.5$. To find out how much each salary deviates from the mean, subtract 41.5 from the salary. For instance, the deviation of 41 (or \$41,000) is

$$\underset{x}{41} - \underset{\mu}{41.5} = -0.5 \text{ (or } -\$500\text{)}. \qquad \text{Deviation of } x = x - \mu$$

The table at the left lists the deviations of each of the 10 starting salaries.

Try It Yourself 2

Find the deviation of each starting salary for Corporation B given in Example 1.

a. Find the *mean* of the data set.
b. *Subtract* the mean from each salary.

Answer: Page A33

In Example 2, notice that the sum of the deviations is zero. Because this is true for any data set, it doesn't make sense to find the average of the deviations. To overcome this problem, you can square each deviation. In a population data set, the mean of the squares of the deviations is called the **population variance.**

Study Tip

When you add the squares of the deviations, you compute a quantity called the sum of squares, denoted SS_x.

DEFINITION

The **population variance** of a population data set of N entries is

$$\text{Population variance} = \sigma^2 = \frac{\Sigma(x - \mu)^2}{N}$$

The symbol σ is the lowercase Greek letter sigma.

DEFINITION

The **population standard deviation** of a population data set of N entries is the square root of the population variance.

$$\text{Population standard deviation} = \sigma = \sqrt{\sigma^2} = \sqrt{\frac{\Sigma(x-\mu)^2}{N}}$$

Note to Instructor

We have used the formulas here that are derived from the definition of the population variance and standard deviation because we feel they are easier to remember than the shortcut formula. If you prefer to use the shortcut formula, we have included it on page 91.

GUIDELINES

Finding the Population Variance and Standard Deviation

In Words	*In Symbols*
1. Find the mean of the population data set.	$\mu = \frac{\Sigma x}{N}$
2. Find the deviation of each entry.	$x - \mu$
3. Square each deviation.	$(x-\mu)^2$
4. Add to get the **sum of squares.**	$SS_x = \Sigma(x-\mu)^2$
5. Divide by N to get the **population variance.**	$\sigma^2 = \frac{\Sigma(x-\mu)^2}{N}$
6. Find the square root of the variance to get the **population standard deviation.**	$\sigma = \sqrt{\frac{\Sigma(x-\mu)^2}{N}}$

Sum of Squares of Starting Salaries for Corporation A

Salary x	Deviation $x - \mu$	Squares $(x-\mu)^2$
41	−0.5	0.25
38	−3.5	12.25
39	−2.5	6.25
45	3.5	12.25
47	5.5	30.25
41	−0.5	0.25
44	2.5	6.25
41	−0.5	0.25
37	−4.5	20.25
42	0.5	0.25
	$\Sigma = 0$	$SS_x = 88.5$

EXAMPLE 3

Finding the Population Standard Deviation

Find the population variance and standard deviation of the starting salaries for Corporation A given in Example 1.

SOLUTION The table at the left summarizes the steps used to find SS_x.

$$SS_x = 88.5, \quad N = 10, \quad \sigma^2 = \frac{88.5}{10} \approx 8.9, \quad \sigma = \sqrt{8.85} \approx 3.0$$

So, the population variance is about 8.9, and the population standard deviation is about 3.0, or \$3000.

Study Tip

Notice that the variance and standard deviation in Example 3 have one more decimal place than the original set of data values. This is the same *round-off rule* that was used to calculate the mean.

Try It Yourself 3

Find the population standard deviation of the starting salaries for Corporation B given in Example 1.

a. Find the *mean* and each *deviation,* as you did in Try It Yourself 2.
b. *Square* each deviation and *add* to get the sum of squares.
c. *Divide* by N to get the population variance.
d. Find the *square root* of the population variance.
e. *Interpret* the results by giving the population standard deviation in dollars.

Answer: Page A33

Study Tip

Note that when you find the *population* variance, you divide by N, the number of entries, but when you find the *sample* variance, you divide by $n - 1$, one less than the number of entries.

DEFINITION

The **sample variance** and **sample standard deviation** of a sample data set of n entries are listed below.

$$\text{Sample variance} = s^2 = \frac{\Sigma(x - \bar{x})^2}{n - 1}$$

$$\text{Sample standard deviation} = s = \sqrt{s^2} = \sqrt{\frac{\Sigma(x - \bar{x})^2}{n - 1}}$$

Symbols in Variance and Standard Deviation Formulas

	Population	Sample
Variance	σ^2	s^2
Standard deviation	σ	s
Mean	μ	$\bar{x}$
Number of entries	N	n
Deviation	$x - \mu$	$x - \bar{x}$
Sum of squares	$\Sigma(x - \mu)^2$	$\Sigma(x - \bar{x})^2$

GUIDELINES

Finding the Sample Variance and Standard Deviation

In Words	*In Symbols*
1. Find the mean of the sample data set.	$\bar{x} = \frac{\Sigma x}{n}$
2. Find the deviation of each entry.	$x - \bar{x}$
3. Square each deviation.	$(x - \bar{x})^2$
4. Add to get the **sum of squares.**	$SS_x = \Sigma(x - \bar{x})^2$
5. Divide by $n - 1$ to get the **sample variance.**	$s^2 = \frac{\Sigma(x - \bar{x})^2}{n - 1}$
6. Find the square root of the variance to get the **sample standard deviation.**	$s = \sqrt{\frac{\Sigma(x - \bar{x})^2}{n - 1}}$

EXAMPLE 4

Finding the Sample Standard Deviation

See MINITAB and TI-83 steps on pages 114 and 115.

The starting salaries given in Example 1 are for the Chicago branches of Corporations A and B. Each corporation has several other branches, and you plan to use the starting salaries of the Chicago branches to estimate the starting salaries for the larger populations. Find the *sample* standard deviation of the starting salaries for the Chicago branch of Corporation A.

SOLUTION

$$SS_x = 88.5, \quad n = 10, \quad s^2 = \frac{88.5}{9} \approx 9.8, \quad s = \sqrt{\frac{88.5}{9}} \approx 3.1$$

So, the sample variance is about 9.8, and the sample standard deviation is about 3.1, or \$3100.

Try It Yourself 4

Find the sample standard deviation of the starting salaries for the Chicago branch of Corporation B.

a. Find the *sum of squares*, as you did in Try It Yourself 3.
b. *Divide* by $n - 1$ to get the sample variance.
c. Find the *square root* of the sample variance.

Answer: Page A33

EXAMPLE 5

Office Rental Rates		
35.00	33.50	37.00
23.75	26.50	31.25
36.50	40.00	32.00
39.25	37.50	34.75
37.75	37.25	36.75
27.00	35.75	26.00
37.00	29.00	40.50
24.50	33.00	38.00

Using Technology to Find the Standard Deviation

Sample office rental rates (in dollars per square foot per year) for Miami's central business district are shown in the table. Use a calculator or a computer to find the mean rental rate and the sample standard deviation. *(Adapted from Cushman & Wakefield Inc.)*

SOLUTION MINITAB, Excel, and the TI-83 each have features that automatically calculate the mean and the standard deviation of data sets. Try using this technology to find the mean and the standard deviation of the office rental rates. From the displays, you can see that $\bar{x} \approx 33.73$ and $s \approx 5.09$.

MINITAB

Descriptive Statistics

Variable	N	Mean	Median	TrMean	StDev
Rental Rates	24	33.73	35.38	33.88	5.09

Variable	SE Mean	Minimum	Maximum	Q1	Q3
Rental Rates	1.04	23.75	40.50	29.56	37.44

EXCEL

	A	B
1	Mean	33.72917
2	Standard Error	1.038864
3	Median	35.375
4	Mode	37
5	Standard Deviation	5.089373
6	Sample Variance	25.90172
7	Kurtosis	-0.74282
8	Skewness	-0.70345
9	Range	16.75
10	Minimum	23.75
11	Maximum	40.5
12	Sum	809.5
13	Count	24

TI-83

1-Var Stats
$\bar{x}$=33.72916667
Σx=809.5
Σx^2=27899.5
Sx=5.089373342
σx=4.982216639
n=24

Sample Mean

Sample Standard Deviation

Note to Instructor

The standard deviations reported by MINITAB and Excel represent sample standard deviations. The TI-83 also reports σ, the population standard deviation. Ask students to compare the values of s and σ shown from the same data.

Try It Yourself 5

Sample office rental rates (in dollars per square foot per year) for Seattle's central business district are listed. Use a calculator or a computer to find the mean rental rate and the sample standard deviation. *(Adapted from Cushman & Wakefield Inc.)*

40.00	43.00	46.00	40.50	35.75	39.75	32.75
36.75	35.75	38.75	38.75	36.75	38.75	39.00
29.00	35.00	42.75	32.75	40.75	35.25	

a. *Enter* the data.
b. *Calculate* the sample mean and the sample standard deviation.

Answer: Page A33

Insight

When all data values are equal, the standard deviation is 0. Otherwise, the standard deviation must be positive.

Interpreting Standard Deviation

When interpreting the standard deviation, remember that it is a measure of the typical amount an entry deviates from the mean. The more the entries are spread out, the greater the standard deviation.

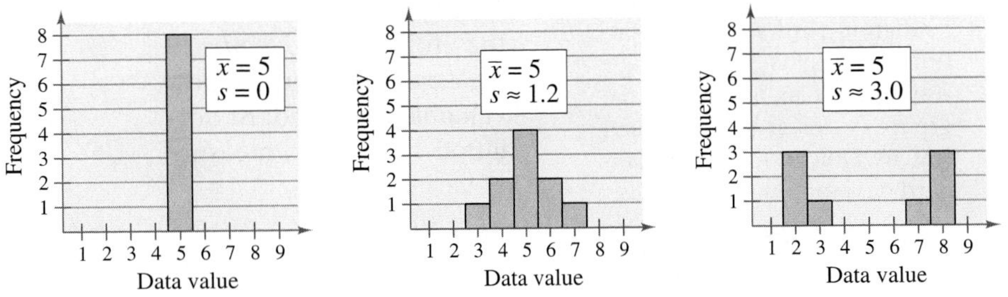

EXAMPLE 6

Estimating Standard Deviation

Without calculating, estimate the population standard deviation of each data set.

1.
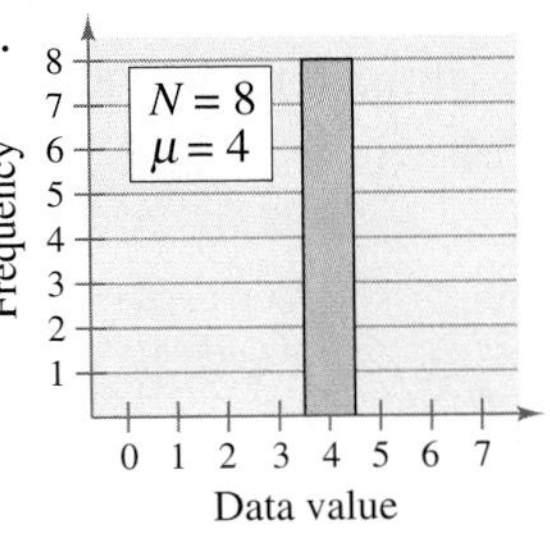

2.
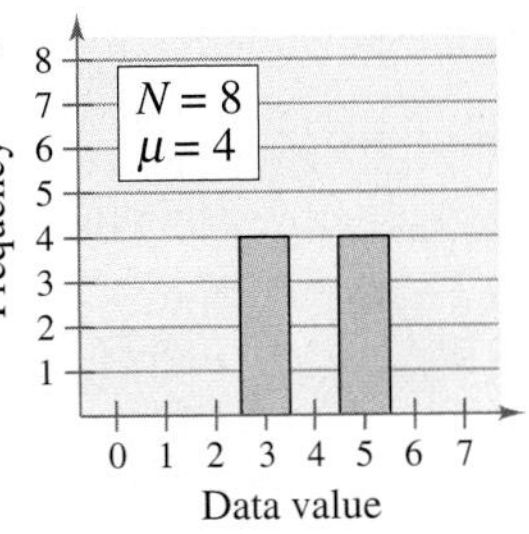

3.
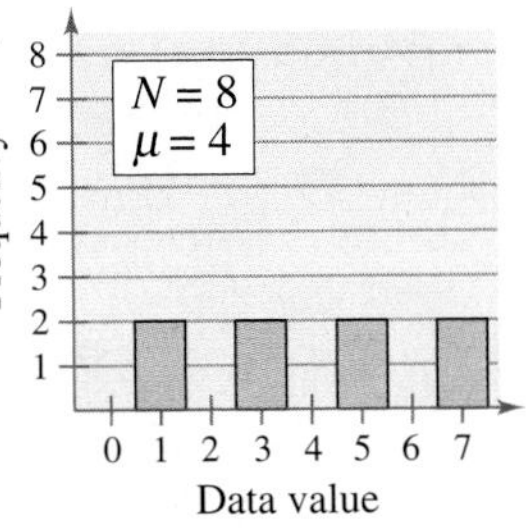

SOLUTION

1. Each of the eight entries is 4. So, each deviation is 0, which implies that

$$\sigma = 0.$$

2. Each of the eight entries has a deviation of ± 1. So, the population standard deviation should be 1. By calculating, you can see that

$$\sigma = 1.$$

3. Each of the eight entries has a deviation of ± 1 or ± 3. So, the population standard deviation should be about 2. By calculating, you can see that

$$\sigma \approx 2.24.$$

Try It Yourself 6

Write a data set that has 10 entries, a mean of 10, and a population standard deviation that is approximately 3. (There are many correct answers.)

a. *Write* a data set that has five entries that are three units less than 10 and five entries that are three units more than 10.

b. *Calculate* the population standard deviation to check that σ is approximately 3.

Answer: Page A33

Picturing the World

A survey was conducted by the National Center for Health Statistics to find the mean height of males in the U.S. The histogram shows the distribution of heights for the 2485 respondents in the 20–29 age group. In this group, the mean was 69.2 inches and the standard deviation was 2.9 inches.

Heights of Men in the U.S. Ages 20–29

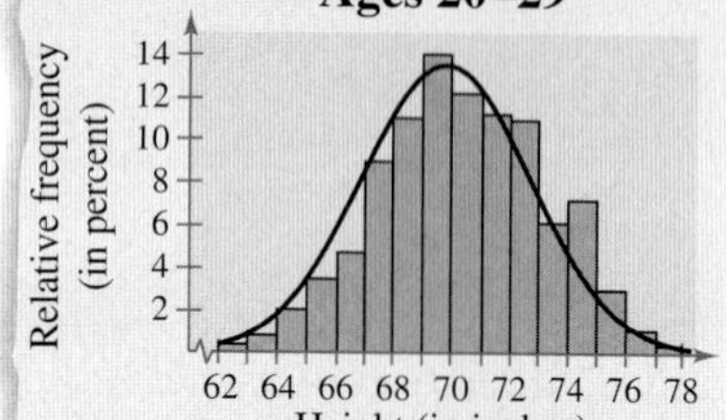

About what percent of the heights lie within two standard deviations of the mean?

Many real-life data sets have distributions that are approximately symmetric and bell shaped. Later in the text, you will study this type of distribution in detail. For now, however, the following *Empirical Rule* can help you see how valuable the standard deviation can be as a measure of variation.

Bell-Shaped Distribution

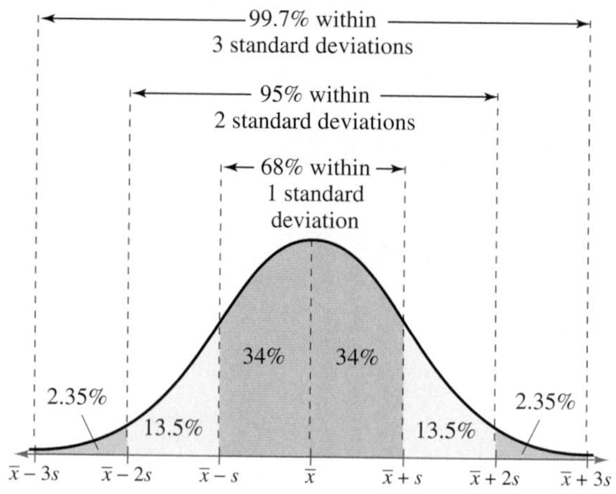

Empirical Rule (or 68-95-99.7 Rule)

For data with a (symmetric) bell-shaped distribution, the standard deviation has the following characteristics.

1. About 68% of the data lie within one standard deviation of the mean.
2. About 95% of the data lie within two standard deviations of the mean.
3. About 99.7% of the data lie within three standard deviations of the mean.

Insight

Data values that lie more than two standard deviations from the mean are considered unusual. Data values that lie more than three standard deviations from the mean are very unusual.

EXAMPLE 7

Using the Empirical Rule

In a survey conducted by the National Center for Health Statistics, the sample mean height of women in the United States (ages 20–29) was 64 inches, with a sample standard deviation of 2.75 inches. Estimate the percent of the women whose heights are between 64 inches and 69.5 inches.

SOLUTION The distribution of the women's heights is shown. Because the distribution is bell shaped, you can use the Empirical Rule. The mean height is 64, so when you add two standard deviations to the mean height, you get

$$\bar{x} + 2s = 64 + 2(2.75) = 69.5.$$

Because 69.5 is two standard deviations above the mean height, the percent of the heights between 64 inches and 69.5 inches is 34% + 13.5% = 47.5%.

Interpretation So, 47.5% of women are between 64 and 69.5 inches tall.

Heights of Women in the U.S. Ages 20–29

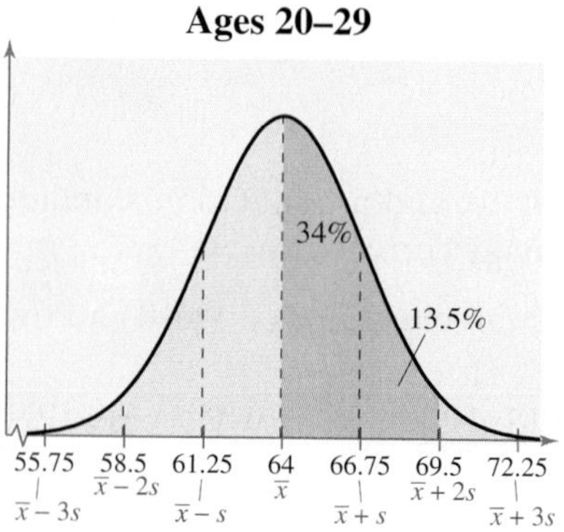

Try It Yourself 7

Estimate the percent of the heights that are between 61.25 and 64 inches.

a. How many standard deviations is 61.25 to the left of 64?

b. Use the Empirical Rule to estimate the percent of the data between $\bar{x} - s$ and $\bar{x}$.

c. *Interpret* the result in the context of the data.

Answer: Page A33

The Empirical Rule applies only to (symmetric) bell-shaped distributions. What if the distribution is not bell-shaped, or what if the shape of the distribution is not known? The following theorem applies to *all* distributions. It is named after the Russian statistician Pafnuti Chebychev (1821–1894).

Note to Instructor

Explain that k represents the number of standard deviations from the mean. Ask students to calculate the percents for $k = 4$ and $k = 5$. Then ask them what happens as k increases. Point out that it is helpful to draw a number line and mark it in units of standard deviations.

Chebychev's Theorem

The portion of any data set lying within k standard deviations $(k > 1)$ of the mean is at least

$$1 - \frac{1}{k^2}.$$

- $k = 2$: In any data set, at least $1 - \frac{1}{2^2} = \frac{3}{4}$, or 75%, of the data lie within 2 standard deviations of the mean.
- $k = 3$: In any data set, at least $1 - \frac{1}{3^2} = \frac{8}{9}$, or 88.9%, of the data lie within 3 standard deviations of the mean.

Insight

In Example 8, Chebychev's Theorem tells you that at least 75% of the population of Florida is under the age of 88.8. This is a true statement, but it is not nearly as strong a statement as could be made from reading the histogram.

In general, Chebychev's Theorem gives cautious estimates of the percent lying within k standard deviations of the mean. Remember, the theorem applies to all distributions.

EXAMPLE 8

Using Chebychev's Theorem

The age distributions for Alaska and Florida are shown in the histograms. Decide which is which. Apply Chebychev's Theorem to the data for Florida using $k = 2$. What can you conclude?

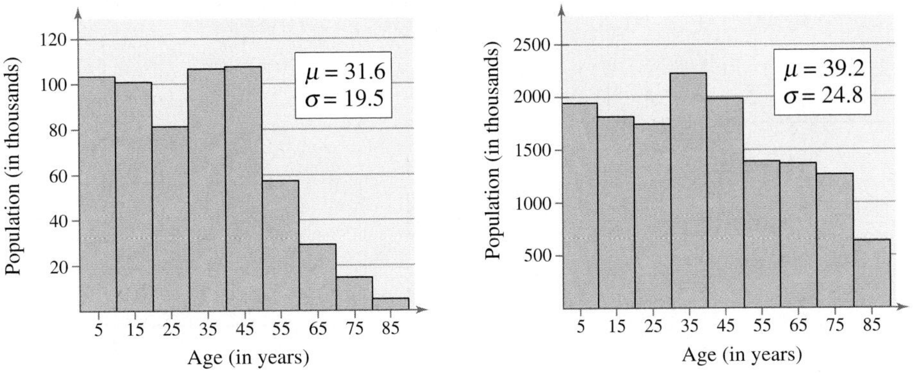

SOLUTION The histogram on the right shows Florida's age distribution. You can tell because the population is greater and older. Moving two standard deviations to the left of the mean puts you below 0, because $\mu - 2\sigma = 39.2 - 2(24.8) = -10.4$. Moving two standard deviations to the right of the mean puts you at $\mu + 2\sigma = 39.2 + 2(24.8) = 88.8$. By Chebychev's Theorem, you can say that at least 75% of the population of Florida is between 0 and 88.8 years old.

Try It Yourself 8

Apply Chebychev's Theorem to the data for Alaska using $k = 2$.

a. *Subtract* two standard deviations from the mean.
b. *Add* two standard deviations to the mean.
c. *Apply* Chebychev's Theorem for $k = 2$ and *interpret* the results.

Answer: Page A33

Standard Deviation for Grouped Data

In Section 2.1, you learned that large data sets are usually best represented by a frequency distribution. The formula for the sample standard deviation for a frequency distribution is

$$\text{Sample standard deviation} = s = \sqrt{\frac{\Sigma(x - \bar{x})^2 f}{n - 1}}$$

where $n = \Sigma f$ is the number of entries in the data set.

Number of Children in 50 Households

1	3	1	1	1
1	2	2	1	0
1	1	0	0	0
1	5	0	3	6
3	0	3	1	1
1	1	6	0	1
3	6	6	1	2
2	3	0	1	1
4	1	1	2	2
0	3	0	2	4

Study Tip

Remember that formulas for grouped data require you to multiply by the frequencies.

EXAMPLE 9

Finding the Standard Deviation for Grouped Data

You collect a random sample of the number of children per household in a region. The results are shown at the left. Find the sample mean and the sample standard deviation of the data set.

SOLUTION These data could be treated as 50 individual entries, and you could use the formulas for mean and standard deviation. Because there are so many repeated numbers, however, it is easier to use a frequency distribution.

x	f	xf
0	10	0
1	19	19
2	7	14
3	7	21
4	2	8
5	1	5
6	4	24
	$\Sigma = 50$	$\Sigma = 91$

$x - \bar{x}$	$(x - \bar{x})^2$	$(x - \bar{x})^2 f$
−1.8	3.24	32.40
−0.8	0.64	12.16
0.2	0.04	0.28
1.2	1.44	10.08
2.2	4.84	9.68
3.2	10.24	10.24
4.2	17.64	70.56
		$\Sigma = 145.40$

$$\bar{x} = \frac{\Sigma xf}{n} = \frac{91}{50} \approx 1.8$$ Sample mean

Use the sum of squares to find the sample standard deviation.

$$s = \sqrt{\frac{\Sigma(x - \bar{x})^2 f}{n - 1}} = \sqrt{\frac{145.4}{49}} \approx 1.7$$ Sample standard deviation

So, the sample mean is 1.8 children, and the standard deviation is 1.7 children.

Try It Yourself 9

Change three of the 6s in the data set to 4s. How does this change affect the sample mean and sample standard deviation?

a. Write the first three columns of a *frequency distribution.*
b. Find the *sample mean.*
c. Complete the *last three columns* of the frequency distribution.
d. Find the *sample standard deviation.*

Answer: Page A33

When a frequency distribution has classes, you can estimate the sample mean and standard deviation by using the midpoint of each class.

EXAMPLE 10

Using Midpoints of Classes

The circle graph at the right shows the results of a survey in which 1000 adults were asked how much they spend in preparation for personal travel each year. Make a frequency distribution for the data. Then use the table to estimate the sample mean and the sample standard deviation of the data set. *(Adapted from Travel Industry Association of America)*

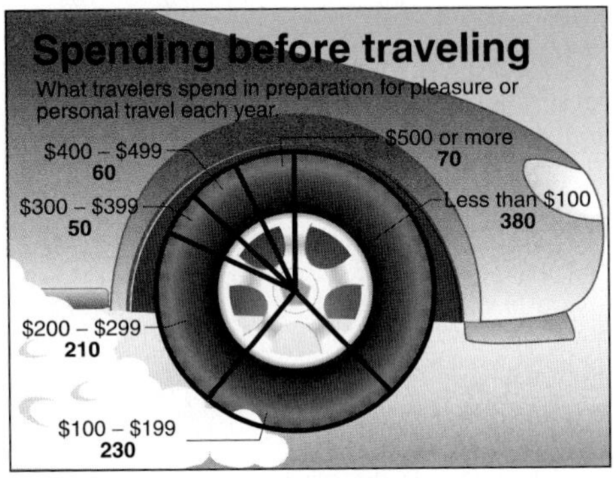

SOLUTION Begin by using a frequency distribution to organize the data.

Class	x	f	xf	$x - \bar{x}$	$(x - \bar{x})^2$	$(x - \bar{x})^2 f$
0–99	49.5	380	18,810	−142.5	20,306.25	7,716,375.0
100–199	149.5	230	34,385	−42.5	1,806.25	415,437.5
200–299	249.5	210	52,395	57.5	3,306.25	694,312.5
300–399	349.5	50	17,475	157.5	24,806.25	1,240,312.5
400–499	449.5	60	26,970	257.5	66,306.25	3,978,375.0
500+	599.5	70	41,965	407.5	166,056.25	11,623,937.5
		$\Sigma = 1{,}000$	$\Sigma = 192{,}000$			$\Sigma = 25{,}668{,}750.0$

Study Tip

When a class is open, as in the last class, you must assign a single value to represent the midpoint. For this example, we selected 599.5.

$$\bar{x} = \frac{\Sigma xf}{n} = \frac{192{,}000}{1{,}000} = 192 \qquad \text{Sample mean}$$

Use the sum of squares to find the sample standard deviation.

$$s = \sqrt{\frac{\Sigma(x - \bar{x})^2 f}{n - 1}} = \sqrt{\frac{25{,}668{,}750}{999}} \approx 160.3 \qquad \text{Sample standard deviation}$$

So, the sample mean is \$192 per year, and the sample standard deviation is about \$160.3 per year.

Try It Yourself 10

In the frequency distribution, 599.5 was chosen to represent the class of \$500 or more. How would the sample mean and standard deviation change if you used 650 to represent this class?

a. Write the first four columns of a *frequency distribution.*
b. Find the *sample mean.*
c. Complete the *last three columns* of the frequency distribution.
d. Find the *sample standard deviation.*

Answer: Page A34

2.4 Exercises

Building Basic Skills and Vocabulary

In Exercises 1 and 2, find the range, mean, variance, and standard deviation of the population data set.

1. 11 10 8 4 6 7 11 6 11 7

2. 13 23 15 13 18 13 15
14 20 20 18 17 20 13

In Exercises 3 and 4, find the range, mean, variance, and standard deviation of the sample data set.

3. 15 8 12 5 19 14 8 6 13

4. 24 26 27 23 9 14 8
8 26 15 15 27 11

Graphical Reasoning In Exercises 5 and 6, find the range of the data set represented by the display or graph.

5. Key: $2|3 = 23$

2	3 9
3	0 0 2 3 6 7
4	0 1 2 3 3 8
5	0 1 1 9
6	1 2 9 9
7	5 9
8	4 8
9	0 2 5 6

6. **Bride's Age at First Marriage**

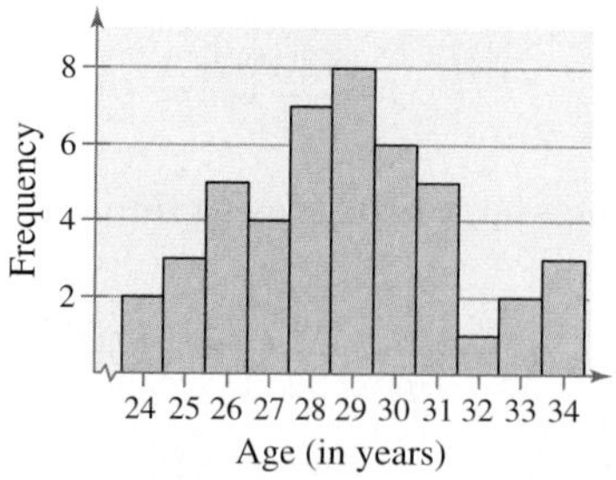

7. Explain how to find the range of a data set. What is an advantage of using the range as a measure of variation? What is a disadvantage?

8. Explain how to find the deviation of an entry in a data set. What is the sum of all the deviations in any data set?

9. Why is the standard deviation used more frequently than the variance? (*Hint:* Consider the units of the variance.)

10. Explain the relationship between variance and standard deviation. Can either of these measures be negative? Explain. Find a data set for which $n = 5$, $\bar{x} = 7$, and $s = 0$.

1. Range = 7, mean = 8.1, variance ≈ 5.7, standard deviation ≈ 2.4

2. Range = 10
Mean ≈ 16.6
Variance ≈ 10.2
Standard deviation ≈ 3.2

3. Range = 14, mean ≈ 11.1, variance ≈ 21.6, standard deviation ≈ 4.6

4. Range = 19
Mean ≈ 17.9
Variance ≈ 59.6
Standard deviation ≈ 7.7

5. 73 **6.** 10

7. The range is the difference between the maximum and minimum values of a data set. The advantage of the range is that it is easy to calculate. The disadvantage is that it uses only two entries from the data set.

8. A deviation $(x - \mu)$ is the difference between an observation x and the mean of the data μ. The sum of the deviations is always zero.

9. The units of variance are squared. Its units are meaningless. (Example: dollars^2)

10. The standard deviation is the positive square root of the variance. The standard deviation and variance can never be negative. Squared deviations can never be negative.
$\{7, 7, 7, 7, 7\}$

11. *Marriage Ages* The ages of 10 grooms at their first marriage are given below.

24.3 46.6 41.6 32.9 26.8 39.8 21.5 45.7 33.9 35.1

(a) Find the range of the data set.

(b) Change 46.6 to 66.6 and find the range of the new data set.

(c) Compare your answer to part (a) with your answer to part (b).

12. Find a population data set that contains six entries, has a mean of 5, and has a standard deviation of 2.

Using and Interpreting Concepts

13. *Graphical Reasoning* Both data sets have a mean of 165. One has a standard deviation of 16, and the other has a standard deviation of 24. Which is which? Explain your reasoning.

(a) Key: 12|8 = 128

12	8 9
13	5 5 8
14	1 2
15	0 0 6 7
16	4 5 9
17	1 3 6 8
18	0 8 9
19	6
20	3 5 7

(b)

12	
13	1
14	2 3 5
15	0 4 5 6 8
16	1 1 2 3 3 3
17	1 5 8 8
18	2 3 4 5
19	0 2
20	

14. *Graphical Reasoning* Both data sets represented below have a mean of 50. One has a standard deviation of 2.4, and the other has a standard deviation of 5. Which is which? Explain your reasoning.

(a)

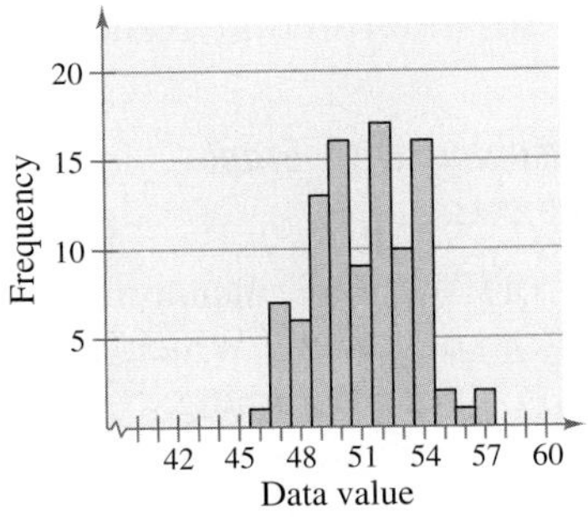

(b)

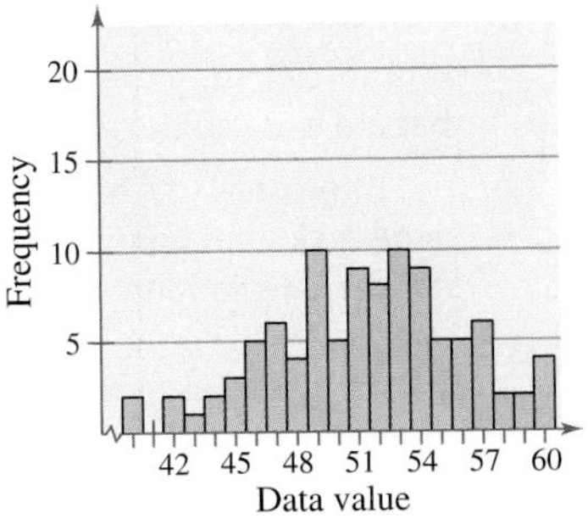

15. *Writing* Describe the difference between the calculation of population standard deviation and sample standard deviation.

16. *Writing* Given a data set, how do you know whether to calculate σ or s?

17. *Salary Offers* You are applying for a job at two companies. Company A offers starting salaries with $\mu = \$31,000$ and $\sigma = \$1000$. Company B offers starting salaries with $\mu = \$31,000$ and $\sigma = \$5000$. From which company are you more likely to get an offer of $33,000 or more?

18. *Golf Strokes* An Internet site compares the strokes per round of two professional golfers. Which golfer is more consistent: Player A with $\mu = 71.5$ strokes and $\sigma = 2.3$ strokes, or Player B with $\mu = 70.1$ strokes and $\sigma = 1.2$ strokes?

11. (a) Range = 25.1
(b) Range = 45.1
(c) Changing the maximum value of the data set greatly affects the range.

12. $\{3, 3, 3, 7, 7, 7\}$

13. (a) has a standard deviation of 24 and (b) has a standard deviation of 16, because the data in (a) have more variability.

14. (a) has a standard deviation of 2.4 and (b) has a standard deviation of 5 because the data in (b) have more variability.

15. When calculating the population standard deviation, you divide the sum of the squared deviations by n, then take the square root of that value. When calculating the sample standard deviation, you divide the sum of the squared deviations by $n - 1$, then take the square root of that value.

16. When given a data set, one would have to determine if it represented the population or was a sample taken from the population. If the data are a population, then σ is calculated. If the data are a sample, then s is calculated.

17. Company B

18. Player B

19. (a) Los Angeles: 17.6, 37.35, 6.11

Long Beach: 8.7, 8.71, 2.95

(b) It appears from the data that the annual salaries in Los Angeles are more variable than the salaries in Long Beach.

20. (a) Dallas: 18.1, 37.33, 6.11

Houston: 13, 12.26, 3.50

(b) It appears from the data that the annual salaries in Dallas are more variable than the salaries in Houston.

21. (a) Males: 405; 16,225.3; 127.4

Females: 552; 34,575.1; 185.9

(b) It appears from the data that the SAT scores for females are more variable than the SAT scores for males.

22. (a) Public teachers: 5.1, 2.95, 1.72

Private teachers: 4.2, 1.99, 1.41

(b) It appears from the data that the annual salaries for public teachers are more variable than the salaries for private teachers.

23. (a) Greatest sample standard deviation: (ii)

Data set (ii) has more entries that are farther away from the mean.

Least sample standard deviation: (iii)

Data set (iii) has more entries that are close to the mean.

(b) The three data sets have the same mean but have different standard deviations.

Comparing Two Data Sets In Exercises 19–22, you are asked to compare two data sets and interpret the results.

19. *Annual Salaries* Sample annual salaries (in thousands of dollars) for municipal employees in Los Angeles and Long Beach are listed.

Los Angeles:	20.2	26.1	20.9	32.1	35.9	23.0	28.2	31.6	18.3
Long Beach:	20.9	18.2	20.8	21.1	26.5	26.9	24.2	25.1	22.2

(a) Find the range, variance, and standard deviation of each data set.

(b) Interpret the results in the context of the real-life setting.

20. *Annual Salaries* Sample annual salaries (in thousands of dollars) for municipal employees in Dallas and Houston are listed.

Dallas:	34.9	25.7	17.3	16.8	26.8	24.7	29.4	32.7	25.5
Houston:	25.6	23.2	26.7	27.7	25.4	26.4	18.3	26.1	31.3

(a) Find the range, variance, and standard deviation of each data set.

(b) Interpret the results in the context of the real-life setting.

21. *SAT Scores* Sample SAT scores for eight males and eight females are listed.

Male SAT scores:	1059	1328	1175	1123	923	1017	1214	1042
Female SAT scores:	1226	965	841	1053	1056	1393	1312	1222

(a) Find the range, variance, and standard deviation of each data set.

(b) Interpret the results in the context of the real-life setting.

22. *Annual Salaries* Sample annual salaries (in thousands of dollars) for public and private elementary school teachers are listed.

Public teachers:	38.6	38.1	38.7	36.8	34.8	35.9	39.9	36.2
Private teachers:	21.8	18.4	20.3	17.6	19.7	18.3	19.4	20.8

(a) Find the range, variance, and standard deviation of each data set.

(b) Interpret the results in the context of the real-life setting.

Reasoning with Graphs In Exercises 23–26, you are asked to compare three data sets.

23. (a) Without calculating, which data set has the greatest sample standard deviation? Which has the least sample standard deviation? Explain your reasoning.

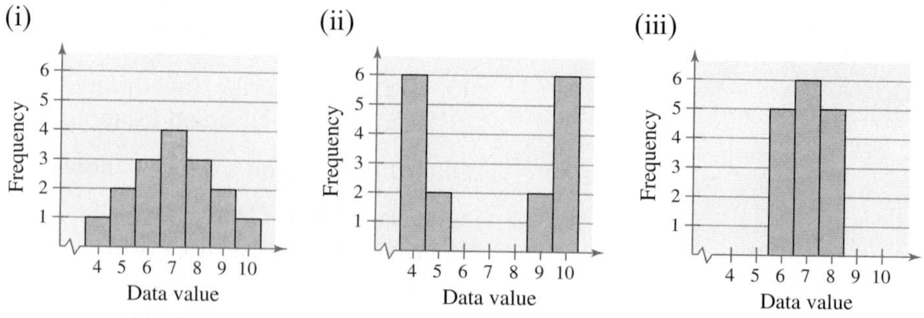

(b) How are the data sets the same? How do they differ?

24. (a) Greatest sample standard deviation: (i)

Data set (i) has more entries that are farther away from the mean.

Least sample standard deviation: (iii)

Data set (iii) has more entries that are close to the mean.

(b) The three data sets have the same mean, median, and mode but have different standard deviations.

25. (a) Greatest sample standard deviation: (ii)

Data set (ii) has more entries that are farther away from the mean.

Least sample standard deviation: (iii)

Data set (iii) has more entries that are close to the mean.

(b) The three data sets have the same mean, median, and mode but have different standard deviations.

26. (a) Greatest sample standard deviation: (iii)

Data set (iii) has more entries that are farther away from the mean.

Least sample standard deviation: (i)

Data set (i) has more entries that are close to the mean.

(b) The three data sets have the same mean and median but have different modes and standard deviations.

27. Similarity: Both estimate proportions of the data contained within k standard deviations of the mean.

Difference: The Empirical Rule assumes the distribution is bell shaped; Chebychev's Theorem makes no such assumption.

28. You must know that the distribution is bell shaped.

29. 68%

24. (a) Without calculating, which data set has the greatest sample standard deviation? Which has the least sample standard deviation? Explain your reasoning.

(i)

0	9
1	5 8
2	3 3 7 7
3	2 5
4	1

Key: 4|1 = 41

(ii)

0	9
1	5
2	3 3 3 7 7 7
3	5
4	1

Key: 4|1 = 41

(iii)

0	
1	5
2	3 3 3 3 7 7 7 7
3	5
4	

Key: 4|1 = 41

(b) How are the data sets the same? How do they differ?

25. (a) Without calculating, which data set has the greatest sample standard deviation? Which has the least sample standard deviation? Explain your reasoning.

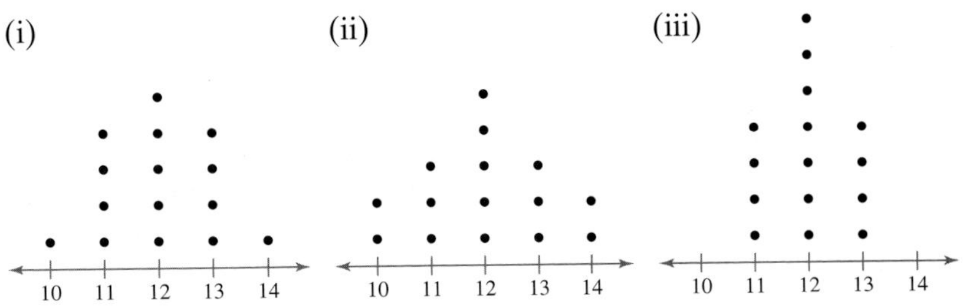

(b) How are the data sets the same? How do they differ?

26. (a) Without calculating, which data set has the greatest sample standard deviation? Which has the least sample standard deviation? Explain your reasoning.

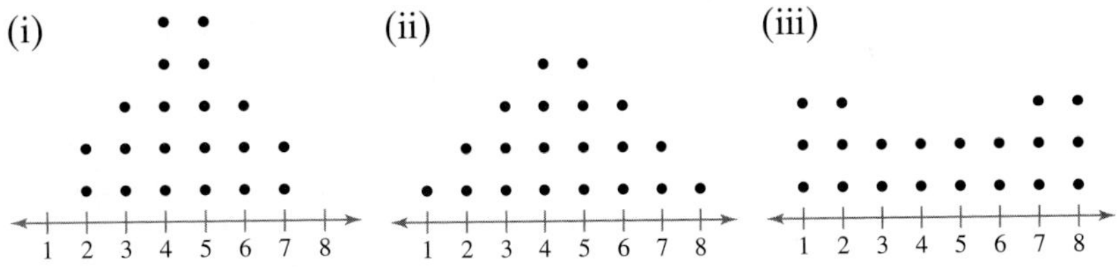

(b) How are the data sets the same? How do they differ?

27. ***Writing*** Discuss the similarities and the differences between the Empirical Rule and Chebychev's Theorem.

28. ***Writing*** What must you know about a data set before you can use the Empirical Rule?

Using the Empirical Rule In Exercises 29–34, you are asked to use the Empirical Rule.

29. The mean value of land and buildings per acre from a sample of farms is \$1000, with a standard deviation of \$200. The data set has a bell-shaped distribution. Estimate the percent of farms whose land and building values per acre are between \$800 and \$1200.

30. Between \$500 and \$1900
31. (a) 51 (b) 17
32. (a) 38 (b) 19
33. \$1250, \$1375, \$1450, \$550
34. \$1950, \$475, \$2050
35. 24
36. (48.07, 56.67); so, at least 75% of the 400-meter dash times lie between 48.07 and 56.67 seconds.
37. Sample mean $\approx$ 2.1

Sample standard deviation $\approx$ 1.3

30. The mean value of land and buildings per acre from a sample of farms is \$1200, with a standard deviation of \$350. Between what two values do about 95% of the data lie? (Assume the data set has a bell-shaped distribution.)

31. Using the sample statistics from Exercise 29, do the following. (Assume the number of farms in the sample is 75.)

(a) Use the Empirical Rule to estimate the number of farms whose land and building values per acre are between \$800 and \$1200.

(b) If 25 additional farms were sampled, about how many of these farms would you expect to have land and building values between \$800 per acre and \$1200 per acre?

32. Using the sample statistics from Exercise 30, do the following. (Assume the number of farms in the sample is 40.)

(a) Use the Empirical Rule to estimate the number of farms whose land and building values per acre are between \$500 and \$1900.

(b) If 20 additional farms were sampled, about how many of these farms would you expect to have land and building values between \$500 per acre and \$1900 per acre?

33. Using the sample statistics from Exercise 29 and the Empirical Rule, determine which of the following farms, whose land and building values per acre are given, are outliers (more than two standard deviations from the mean).

\$1250, \$1375, \$1125, \$1450, \$550, \$800

34. Using the sample statistics from Exercise 30 and the Empirical Rule, determine which of the following farms, whose land and building values per acre are given, are outliers (more than two standard deviations from the mean).

\$1875, \$1950, \$475, \$600, \$2050, \$1600

35. ***Chebychev's Theorem*** Old Faithful is a famous geyser at Yellowstone National Park. From a sample with $n = 32$, the mean duration of Old Faithful's eruptions is 3.32 minutes and the standard deviation is 1.09 minutes. Using Chebychev's Theorem, determine at least how many of the eruptions lasted between 1.14 minutes and 5.5 minutes. *(Source: Yellowstone National Park)*

36. ***Chebychev's Theorem*** The mean time in a women's 400-meter dash is 52.37 seconds, with a standard deviation of 2.15. Apply Chebychev's Theorem to the data using $k = 2$. Interpret the results.

Calculating Using Grouped Data In Exercises 37–44, use the grouped data formulas to find the indicated mean and standard deviation.

37. ***Pets per Household*** The results of a random sample of the number of pets per household in a region are shown in the histogram. Estimate the sample mean and the sample standard deviation of the data set.

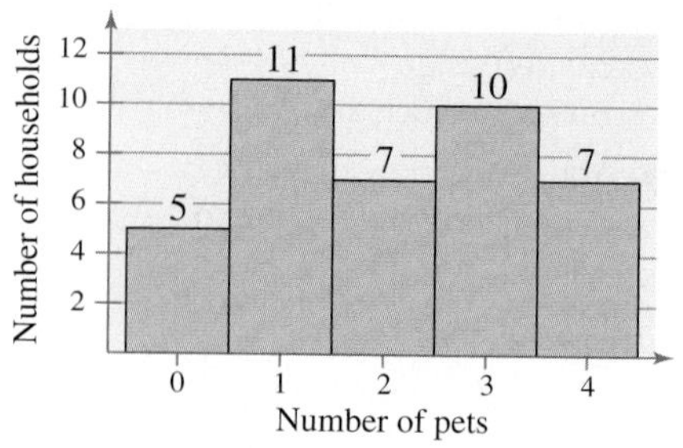

38. Sample mean ≈ 1.7
Sample deviation ≈ 0.8
39. See Odd Answers, page A54.
40. See Selected Answers, page A92.
41. See Odd Answers, page A54.
42. See Selected Answers, page A92.

38. ***Cars per Household*** A random sample of households in a region and the number of cars per household are shown in the histogram. Estimate the sample mean and the sample deviation of the data set.

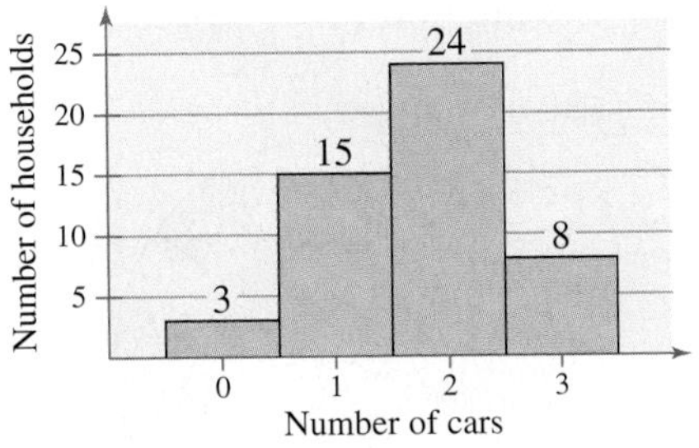

39. ***Football Wins*** The number of wins for each National Football League team in 2003 are listed. Make a frequency distribution (using five classes) for the data set. Then approximate the population mean and the population standard deviation of the data set. *(Source: National Football League)*

DATA

14	10	6	6	10	8	6	5	12	12	5
5	13	10	4	4	12	10	5	4	10	9
7	5	11	8	7	5	12	10	7	4	

40. ***Water Consumption*** The number of gallons of water consumed per day by a small village are listed. Make a frequency distribution (using five classes) for the data set. Then approximate the population mean and the population standard deviation of the data set.

DATA

167	180	192	173	145	151	174
175	178	160	195	224	244	146
162	146	177	163	149	188	

41. ***Amount of Caffeine*** The amount of caffeine in a sample of five-ounce servings of brewed coffee is shown in the histogram. Make a frequency distribution for the data. Then use the table to estimate the sample mean and the sample standard deviation of the data set.

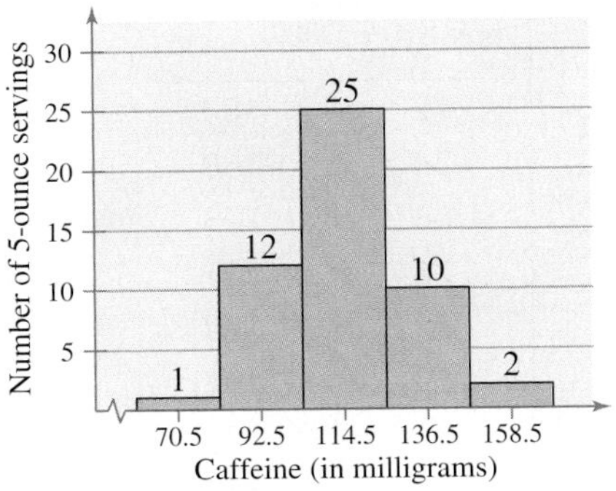

Figure for Exercise 41

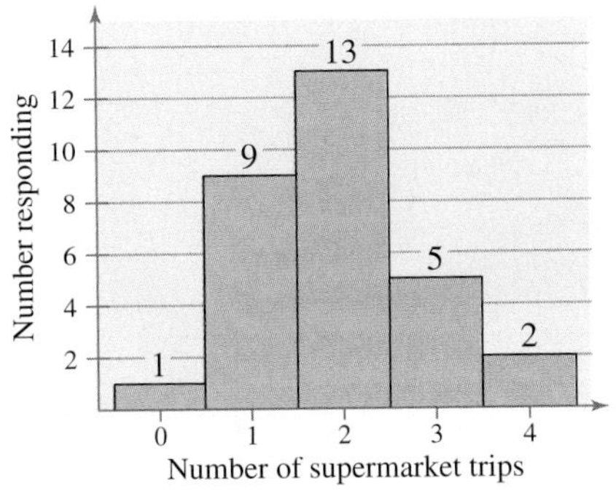

Figure for Exercise 42

42. ***Supermarket Trips*** Thirty people were randomly selected and asked how many trips to the supermarket they made in the past week. The responses are shown in the histogram. Make a frequency distribution for the data. Then use the table to estimate the sample mean and the sample standard deviation of the data set.

43. See Odd Answers, page A54.

44. See Selected Answers, page A93.

45. $CV_{heights} = \frac{3.44}{72.75} \cdot 100 \approx 4.73$

$CV_{weights} = \frac{18.47}{187.83} \cdot 100 \approx 9.83$

It appears that weight is more variable than height.

43. *U.S. Population* The estimated distribution (in millions) of the U.S. population by age for the year 2009 is shown in the circle graph. Make a frequency distribution for the data. Then use the table to estimate the sample mean and the sample standard deviation of the data set. Use 70 as the midpoint for "65 years and over." *(Source: U.S. Census Bureau)*

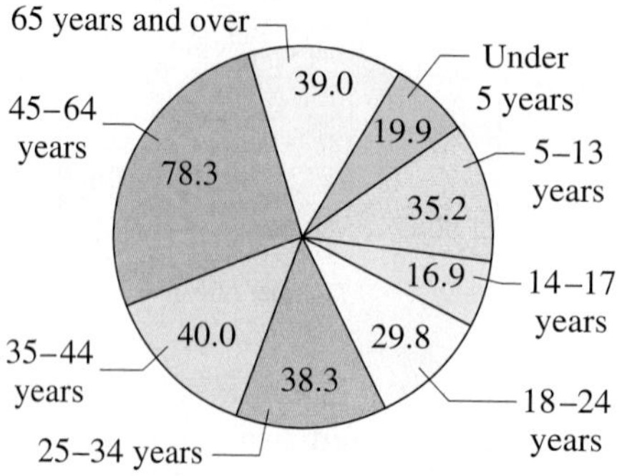

Figure for Exercise 43

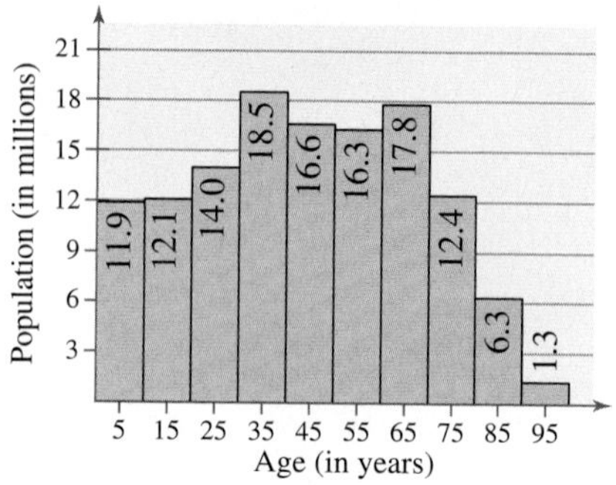

Figure for Exercise 44

44. *Japan's Population* Japan's estimated population for the year 2010 is shown in the bar graph. Make a frequency distribution for the data. Then use the table to estimate the sample mean and the sample standard deviation of the data set. *(Source: U.S. Census Bureau, International Data Base)*

Extending Concepts

DATA **45. *Coefficient of Variation*** The **coefficient of variation** CV describes the standard deviation as a percent of the mean. Because it has no units, you can use the coefficient of variation to compare data with different units.

$$CV = \frac{\text{Standard deviation}}{\text{Mean}} \times 100\%$$

The following table shows the heights (in inches) and weights (in pounds) of the members of a basketball team. Find the coefficient of variation for each data set. What can you conclude?

Heights	Weights
72	180
74	168
68	225
76	201
74	189
69	192
72	197
79	162
70	174
69	171
77	185
73	210

46. (a) Male: 127.4

Female: 185.9

47. (a) $\bar{x} = 550$, $s \approx 302.8$

(b) $\bar{x} = 5500$, $s \approx 3028$

(c) $\bar{x} = 55$, $s \approx 30.28$

(d) When each entry is multiplied by a constant k, the new sample mean is $k \cdot \bar{x}$, and the new sample standard deviation is $k \cdot s$.

48. (a) $\bar{x} = 550$, $s \approx 302.8$

(b) $\bar{x} = 560$, $s \approx 302.8$

(c) $\bar{x} = 540$, $s \approx 302.8$

(d) Adding or subtracting a constant k to each entry makes the new sample mean $\bar{x} + k$ with the sample standard deviation being unaffected.

49. 10

Set $1 - \frac{1}{k^2} = 0.99$ and solve for k.

50. (a) $P \approx -2.61$

The data are skewed left.

(b) $P \approx 4.12$

The data are skewed right.

46. ***Shortcut Formula*** You used $SS_x = \Sigma(x - \bar{x})^2$ when calculating variance and standard deviation. An alternative formula that is sometimes more convenient for hand calculations is

$$SS_x = \Sigma x^2 - \frac{(\Sigma x)^2}{n}.$$

You can find the sample variance by dividing the sum of squares by $n - 1$ and the sample standard deviation by finding the square root of the sample variance.

(a) Use the shortcut formula to calculate the sample standard deviation for the data set given in Exercise 21.

(b) Compare your results with those obtained in Exercise 21.

47. ***Team Project: Scaling Data*** Consider the following sample data set.

100	200	300	400	500
600	700	800	900	1000

(a) Find $\bar{x}$ and s.

(b) Multiply each entry by 10. Find $\bar{x}$ and s for the revised data.

(c) Divide the original data by 10. Find $\bar{x}$ and s for the revised data.

(d) What can you conclude from the results of (a), (b), and (c)?

48. ***Team Project: Shifting Data*** Consider the following sample data set.

100	200	300	400	500
600	700	800	900	1000

(a) Find $\bar{x}$ and s.

(b) Add 10 to each entry. Find $\bar{x}$ and s for the revised data.

(c) Subtract 10 from the original data. Find $\bar{x}$ and s for the revised data.

(d) What can you conclude from the results of (a), (b), and (c)?

49. ***Chebychev's Theorem*** At least 99% of the data in any data set lie within how many standard deviations of the mean? Explain how you obtained your answer.

50. ***Pearson's Index of Skewness*** The English statistician Karl Pearson (1857–1936) introduced a formula for the skewness of a distribution.

$$P = \frac{3(\bar{x} - \text{median})}{s} \qquad \text{Pearson's index of skewness}$$

Most distributions have an index of skewness between -3 and 3. When $P > 0$, the data are skewed right. When $P < 0$, the data are skewed left. When $P = 0$, the data are symmetric. Calculate the coefficient of skewness for each distribution. Describe the shape of each.

(a) $\bar{x} = 17$, $s = 2.3$, median $= 19$

(b) $\bar{x} = 32$, $s = 5.1$, median $= 25$

Case Study

Outlet type	Number of locations
Optical Store	34,043
Sunglass Specialty	2,060
Dept. Store	6,866
Discount Dept. Store	10,376
Catalog Showroom	887
General Merchandise	11,868
Supermarket	21,613
Convenience Store	83,613
Chain Drug Store	31,127
Indep. Drug Store	7,034
Chain Apparel Store	26,831
Chain Sports Store	5,760
Indep. Sports Store	14,683

Sunglass Sales in the United States

The Sunglass Association of America is a not-for-profit association of manufacturers and distributors of sunglasses. Part of the association's mission is to gather and distribute marketing information about the sale of sunglasses. The data presented here are based on surveys administered by Jobson Optical Research International.

Number (in 1000s) of Pairs of Sunglasses Sold

Price	$0–$10	$11–$30	$31–$50	$51–$75	$76–$100	$101–$150	$151+
Optical Store	0	290	3,164	1,240	3,654	842	478
Sunglass Specialty	192	708	2,515	1,697	1,145	805	378
Dept. Store	1,224	1,464	1,527	488	38	16	5
Discount Dept. Store	8,793	5,284	147	67	16	8	0
Catalog Showroom	153	100	65	35	29	9	0
General Merchandise	6,147	495	0	0	0	0	0
Supermarket	14,108	316	0	0	0	0	0
Convenience Store	19,726	2,985	0	0	0	0	0
Chain Drug Store	17,883	3,432	50	0	0	0	0
Indep. Drug Store	1,352	1,110	12	0	0	0	0
Chain Apparel Store	3,464	1,804	186	112	40	17	7
Chain Sports Store	672	526	430	72	45	18	4
Indep. Sports Store	875	1,997	1,320	528	206	85	11

Exercises

1. ***Mean Price*** Estimate the mean price of a pair of sunglasses sold at (a) an optical store, (b) a sunglass specialty store, and (c) a department store. Use $200 as the midpoint for $151+.

2. ***Revenue*** Which type of outlet had the greatest total revenue? Explain your reasoning.

3. ***Revenue*** Which type of outlet had the greatest revenue per location? Explain your reasoning.

4. ***Standard Deviation*** Estimate the standard deviation for the price of a pair of sunglasses sold at (a) optical stores, (b) sunglass specialty stores, and (c) department stores. Use $200 as a midpoint for $151+.

5. ***Standard Deviation*** Of the 13 distributions, which has the greatest standard deviation? Explain your reasoning.

6. ***Bell-Shaped Distribution*** Of the 13 distributions, which is more bell shaped? Explain.

2.5 Measures of Position

What You Should Learn

- How to find the first, second, and third quartiles of a data set
- How to find the interquartile range of a data set
- How to represent a data set graphically using a box-and-whisker plot
- How to interpret other fractiles such as percentiles
- How to find and interpret the standard score (*z*-score)

Quartiles • Percentiles and Other Fractiles • The Standard Score

Quartiles

In this section, you will learn how to use fractiles to specify the position of a data entry within a data set. **Fractiles** are numbers that partition, or divide, an ordered data set into equal parts. For instance, the median is a fractile because it divides an ordered data set into two equal parts.

DEFINITION

The three **quartiles,** Q_1, Q_2, and Q_3, approximately divide an ordered data set into four equal parts. About one quarter of the data fall on or below the **first quartile** Q_1. About one half the data fall on or below the **second quartile** Q_2 (the second quartile is the same as the median of the data set). About three quarters of the data fall on or below the **third quartile** Q_3.

EXAMPLE 1

Finding the Quartiles of a Data Set

The test scores of 15 employees enrolled in a CPR training course are listed. Find the first, second, and third quartiles of the test scores.

13 9 18 15 14 21 7 10 11 20 5 18 37 16 17

SOLUTION First, order the data set and find the median Q_2. Once you find Q_2, divide the data set into two halves. The first and third quartiles are the medians of the lower and upper halves of the data set.

Lower half: 5 7 9 10 11 13 14

15

Upper half: 16 17 18 18 20 21 37

Q_1 = 10, Q_2 = 15, Q_3 = 18

Interpretation About one fourth of the employees scored 10 or less; about one half scored 15 or less; and about three fourths scored 18 or less.

Try It Yourself 1

Find the first, second, and third quartiles for the ages of the Akhiok residents using the population data set listed in the Chapter Opener on page 33.

a. *Order* the data set.
b. Find the median Q_2.
c. Find the first and third quartiles Q_1 and Q_3.

Answer: Page A34

EXAMPLE 2

Using Technology to Find Quartiles

The tuition costs (in thousands of dollars) for 25 liberal arts colleges are listed. Use a calculator or a computer to find the first, second, and third quartiles.

23 25 30 23 20 22 21 15 25 24 30 25 30
20 23 29 20 19 22 23 29 23 28 22 28

SOLUTION MINITAB, Excel, and the TI-83 each have features that automatically calculate quartiles. Try using this technology to find the first, second, and third quartiles of the tuition data. From the displays, you can see that $Q_1 = 21.5$, $Q_2 = 23$, and $Q_3 = 28$.

MINITAB

Descriptive Statistics

Variable	N	Mean	Median	TrMean	StDev
Tuition	25	23.960	23.000	24.087	3.942

Variable	SE Mean	Minimum	Maximum	Q1	Q3
Tuition	0.788	15.000	30.000	21.500	28.000

EXCEL

	A	B	C	D
1	23			
2	25		Quartile(A1:A25,1)	
3	30		22	
4	23			
5	20		Quartile(A1:A25,2)	
6	22		23	
7	21			
8	15		Quartile(A1:A25,3)	
9	25		28	
10	24			
11	30			
12	25			
13	30			
14	20			
15	23			
16	29			
17	20			
18	19			
19	22			
20	23			
21	29			
22	23			
23	28			
24	22			
25	28			

TI-83

1-Var Stats
↑n=25
minX=15
Q_1=21.5
Med=23
Q_3=28
maxX=30

Interpretation About one quarter of these colleges charge tuition of $21,500 or less; one half charge $23,000 or less; and about three quarters charge $28,000 or less.

Study Tip

There are several ways to find the quartiles of a data set. Regardless of how you find the quartiles, the results are rarely off by more than one data entry. For instance, in Example 2, the first quartile, as determined by Excel, is 22 instead of 21.5.

Note to Instructor

For MINITAB and the TI-83, quartiles are found with the following ranks.

Q_1: $\dfrac{1(n + 1)}{4}$

Q_2: $\dfrac{2(n + 1)}{4}$

Q_3: $\dfrac{3(n + 1)}{4}$

Try It Yourself 2

The tuition costs (in thousands of dollars) for 25 universities are listed. Use a calculator or a computer to find the first, second, and third quartiles.

20 26 28 25 31 14 23 15 12 26 29 24 31
19 31 17 15 17 20 31 32 16 21 22 28

a. *Enter* the data.
b. *Calculate* the first, second, and third quartiles.
c. What can you conclude?

Answer: Page A34

Insight

The IQR is a measure of variation that gives you an idea of how much the middle 50% of the data varies. It can also be used to identify outliers. Any data value that lies more than 1.5 IQRs to the left of Q_1 or to the right of Q_3 is an outlier. For instance, 37 is an outlier of the 15 test scores in Example 1.

After finding the quartiles of a data set, you can find the interquartile range.

DEFINITION

The **interquartile range (IQR)** of a data set is the difference between the third and first quartiles.

$$\text{Interquartile range (IQR)} = Q_3 - Q_1$$

EXAMPLE 3

Finding the Interquartile Range

Find the interquartile range of the 15 test scores given in Example 1. What can you conclude from the result?

SOLUTION From Example 1, you know that $Q_1 = 10$ and $Q_3 = 18$. So, the interquartile range is

$$\begin{aligned} \text{IQR} &= Q_3 - Q_1 \\ &= 18 - 10 \\ &= 8. \end{aligned}$$

Interpretation The test scores in the middle portion of the data set vary by at most 8 points.

Try It Yourself 3

Find the interquartile range for the ages of the Akhiok residents listed in the Chapter Opener on page 33.

a. Find the first and third quartiles, Q_1 and Q_3.
b. *Subtract* Q_1 from Q_3.
c. *Interpret* the result in the context of the data.

Answer: Page A34

Another important application of quartiles is to represent data sets using box-and-whisker plots. A **box-and-whisker plot** is an exploratory data analysis tool that highlights the important features of a data set. To graph a box-and-whisker plot, you must know the following values.

Picturing the World

Of the first 43 U.S. presidents, Theodore Roosevelt was the youngest at the time of inauguration, at the age of 42. Ronald Reagan was the oldest president, inaugurated at the age of 69. The box-and-whisker plot summarizes the ages of the first 43 U.S. presidents at inauguration. *(Source: infoplease.com)*

Ages of U.S. Presidents at Inauguration

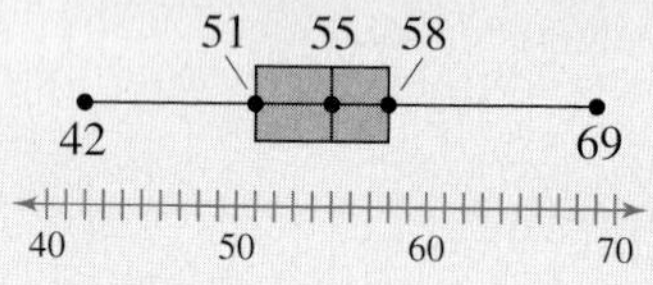

How many U.S. presidents' ages are represented by the box?

1. The minimum entry
2. The first quartile Q_1
3. The median Q_2
4. The third quartile Q_3
5. The maximum entry

These five numbers are called the **five-number summary** of the data set.

GUIDELINES

Drawing a Box-and-Whisker Plot

1. Find the five-number summary of the data set.
2. Construct a horizontal scale that spans the range of the data.
3. Plot the five numbers above the horizontal scale.
4. Draw a box above the horizontal scale from Q_1 to Q_3 and draw a vertical line in the box at Q_2.
5. Draw whiskers from the box to the minimum and maximum entries.

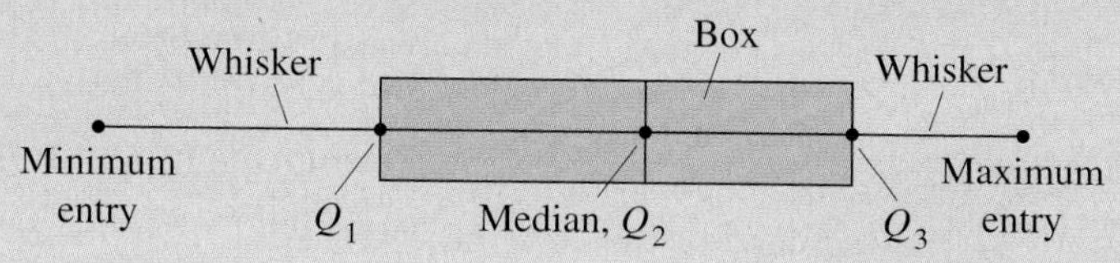

EXAMPLE 4

Drawing a Box-and-Whisker Plot

See MINITAB and TI-83 steps on pages 114 and 115.

Draw a box-and-whisker plot that represents the 15 test scores given in Example 1. What can you conclude from the display?

SOLUTION The five-number summary of the test scores is below. Using these five numbers, you can construct the box-and-whisker plot shown.

Min = 5 $\quad Q_1 = 10 \quad Q_2 = 15 \quad Q_3 = 18 \quad$ Max = 37

Test Scores in CPR Class

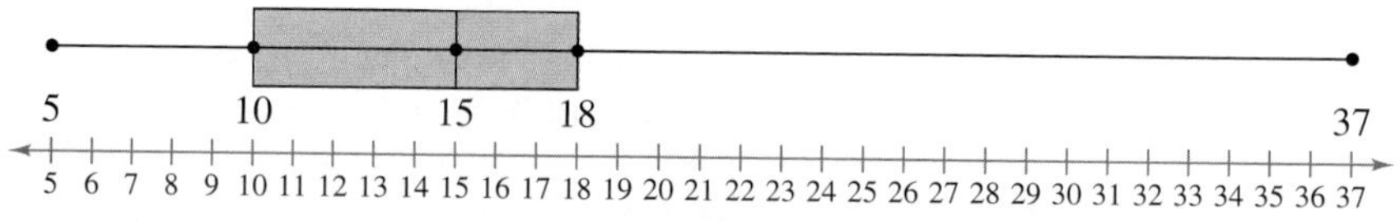

Interpretation You can make several conclusions from the display. One is that about half the scores are between 10 and 18.

Insight

You can use a box-and-whisker plot to determine the shape of a distribution. Notice that the box-and-whisker plot in Example 4 represents a distribution that is skewed right.

Try It Yourself 4

Draw a box-and-whisker plot that represents the ages of the residents of Akhiok listed in the chapter opener on page 33.

a. Find the *five-number summary* of the data set.
b. Construct a *horizontal scale* and *plot* the five numbers above it.
c. Draw the *box*, the *vertical line*, and the *whiskers*.
d. Make some conclusions.

Answer: Page A34

Percentiles and Other Fractiles

In addition to using quartiles to specify a measure of position, you can also use percentiles and deciles. These common fractiles are summarized as follows.

Fractiles	Summary	Symbols
Quartiles	Divide a data set into 4 equal parts.	Q_1, Q_2, Q_3
Deciles	Divide a data set into 10 equal parts.	$D_1, D_2, D_3, \ldots, D_9$
Percentiles	Divide a data set into 100 equal parts.	$P_1, P_2, P_3, \ldots, P_{99}$

Insight

Notice that the 25th percentile is the same as Q_1; the 50th percentile is the same as Q_2, or the median; the 75th percentile is the same as Q_3.

Percentiles are often used in education and health-related fields to indicate how one individual compares with others in a group. They can also be used to identify unusually high or unusually low values. For instance, test scores and children's growth measurements are often expressed in percentiles. Scores or measurements in the 95th percentile and above are unusually high, while those in the 5th percentile and below are unusually low.

Study Tip

It is important that you understand what a percentile means. For instance, if the weight of a six-month-old infant is at the 78th percentile, the infant weighs more than 78% of all six-month-old infants. It does not mean that the infant weighs 78% of some ideal weight.

EXAMPLE 5

Interpreting Percentiles

The ogive represents the cumulative frequency distribution for SAT test scores of college-bound students in a recent year. What test score represents the 64th percentile? How should you interpret this? *(Source: College Board Online)*

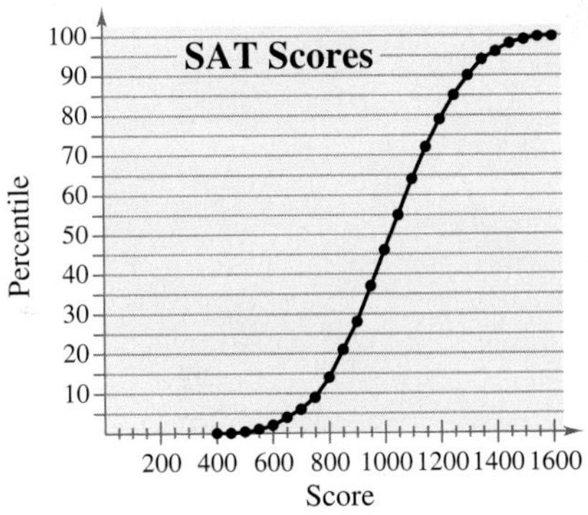

SOLUTION From the ogive, you can see that the 64th percentile corresponds to a test score of 1100.

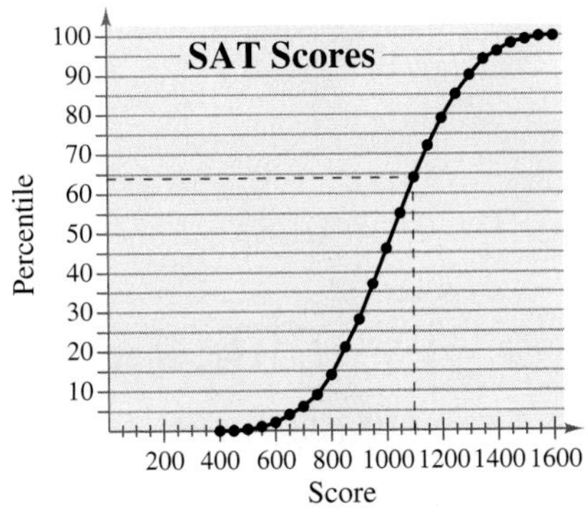

Interpretation This means that 64% of the students had an SAT score of 1100 or less.

Try It Yourself 5

The ages of the residents of Akhiok are represented in the cumulative frequency graph at the left. At what percentile is a resident whose age is 45?

a. *Use the graph* to find the percentile that corresponds to the given age.

b. *Interpret* the results in the context of the data. *Answer: Page A34*

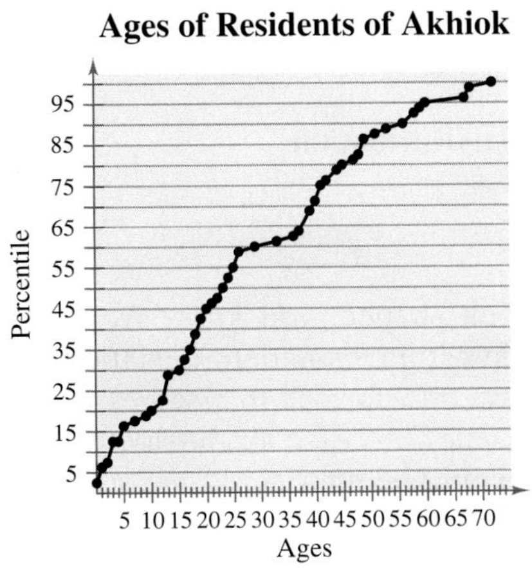

The Standard Score

When you know the mean and standard deviation of a data set, you can measure a data value's position in the data set with a standard score, or z-score.

DEFINITION

The **standard score,** or ***z*-score,** represents the number of standard deviations a given value x falls from the mean μ. To find the z-score for a given value, use the following formula.

$$z = \frac{\text{Value} - \text{Mean}}{\text{Standard deviation}} = \frac{x - \mu}{\sigma}$$

A z-score can be negative, positive, or zero. If z is negative, the corresponding x-value is below the mean. If z is positive, the corresponding x-value is above the mean. And if $z = 0$, the corresponding x-value is equal to the mean.

EXAMPLE 6

Finding z-Scores

The mean speed of vehicles along a stretch of highway is 56 miles per hour with a standard deviation of 4 miles per hour. You measure the speed of three cars traveling along this stretch of highway as 62 miles per hour, 47 miles per hour, and 56 miles per hour. Find the z-score that corresponds to each speed. What can you conclude?

SOLUTION The z-score that corresponds to each speed is calculated below.

$x = 62$ mph

$$z = \frac{62 - 56}{4} = 1.5$$

$x = 47$ mph

$$z = \frac{47 - 56}{4} = -2.25$$

$x = 56$ mph

$$z = \frac{56 - 56}{4} = 0$$

Interpretation From the z-scores, you can conclude that a speed of 62 miles per hour is 1.5 standard deviations above the mean; a speed of 47 miles per hour is 2.25 standard deviations below the mean; and a speed of 56 miles per hour is equal to the mean.

Try It Yourself 6

The monthly utility bills in a city have a mean of \$70 and a standard deviation of \$8. Find the z-scores that correspond to utility bills of \$60, \$71, and \$92. What can you conclude?

a. *Identify* μ and σ of the nonstandard normal distribution.
b. *Transform* each value to a z-score.
c. *Interpret* the results.

Answer: Page A34

Insight

Notice that if the distribution of the speeds in Example 6 is approximately bell shaped, the car going 47 miles per hour is traveling at an unusually slow speed because the speed corresponds to a z-score of -2.25.

When a distribution is approximately bell shaped, you know from the Empirical Rule that about 95% of the data lie within 2 standard deviations of the mean. So, when this distribution's values are transformed to z-scores, about 95% of the z-scores should fall between -2 and 2. A z-score outside of this range will occur about 5% of the time and would be considered unusual. So, according to the Empirical Rule, a z-score less than -3 or greater than 3 would be *very* unusual, with such a score occurring about 0.3% of the time.

In Example 6, you used z-scores to compare data values within the same data set. You can also use z-scores to compare data values from different data sets.

	W	L	T	Pct	PF	PA
Jacksonville	5	11	0	.312	276	331
Houston	5	11	0	.312	255	380
West						
	W	L	T	Pct	PF	PA
yz-Kansas City	13	3	0	.812	484	332
x-Denver	10	6	0	.625	381	301
Oakland	4	12	0	.250	270	379
San Diego	4	12	0	.250	313	441
NATIONAL CONFERENCE						
East						
	W	L	T	Pct	PF	PA
yz-Philadelphia	12	4	0	.750	374	287
x-Dallas	10	6	0	.625	289	260
Washington	5	11	0	.312	287	372
N.Y. Giants	4	12	0	.250	243	387

NATIONAL CONFERENCE

EASTERN DIVISION

Team	Won	Lost	Tie	Pct	PF	PA
x-New York	8	8	0	.500	857	825
y-Detroit	8	8	0	.500	799	819
y-Las Vegas	8	8	0	.500	756	821
Buffalo	5	11	0	.313	554	751

SOUTHERN DIVISION

Team	Won	Lost	Tie	Pct	PF	PA
x-Tampa Bay	12	4	0	.750	849	689
y-Orlando	12	4	0	.750	805	670
y-Georgia	8	8	0	.500	731	701
Carolina	0	16	0	.000	553	886

y--clinched playoff berth, x--clinched division title

EXAMPLE 7

Comparing z-Scores from Different Data Sets

During the 2003 regular season the Kansas City Chiefs, one of 32 teams in the National Football League (NFL), scored 63 touchdowns. During the 2003 regular season the Tampa Bay Storm, one of 16 teams in the Arena Football League (AFL), scored 119 touchdowns. The mean number of touchdowns in the NFL is 37.4, with a standard deviation of 9.3. The mean number of touchdowns in the AFL is 111.7, with a standard deviation of 17.3. Find the z-score that corresponds to the number of touchdowns for each team. Then compare your results. *(Source: The National Football League and the Arena Football League)*

SOLUTION

The z-score that corresponds to the number of touchdowns for each team is calculated below.

Kansas City Chiefs

$$z = \frac{x - \mu}{\sigma} = \frac{63 - 37.4}{9.3} \approx 2.75$$

Tampa Bay Storm

$$z = \frac{x - \mu}{\sigma} = \frac{119 - 111.7}{17.3} \approx 0.42$$

The number of touchdowns scored by the Chiefs is 2.75 standard deviations above the mean, and the number of touchdowns scored by the Storm is 0.42 standard deviations above the mean.

Interpretation The z-score corresponding to the number of touchdowns for the Chiefs is more than two standard deviations from the mean, so it is considered unusual. The Chiefs scored an unusually high number of touchdowns in the NFL, whereas the number of touchdowns scored by the Storm was only slightly higher than the AFL average.

Try It Yourself 7

During the 2003 regular season the Kansas City Chiefs scored 16 field goals. During the 2003 regular season the Tampa Bay Storm scored 12 field goals. The mean number of field goals in the NFL is 23.6, with a standard deviation of 6.0. The mean number of field goals in the AFL is 11.7, with a standard deviation of 4.6. Find the z-score that corresponds to the number of field goals for each team. Then compare your results. *(Source: The National Football League and the Arena Football League)*

a. *Identify* μ and σ of each nonstandard normal distribution.
b. *Transform* each value to a z-score.
c. *Compare* your results.

Answer: Page A34

2.5 Exercises

Building Basic Skills and Vocabulary

In Exercises 1 and 2, (a) find the three quartiles and (b) draw a box-and-whisker plot of the data.

DATA **1.** 4 7 7 5 2 9 7 6 8 5 8 4 1 5 2 8 7 6 6 9

DATA **2.** 2 7 1 3 1 2 8 9 9 2 5 4 7 3 7 5 4 7
2 3 5 9 5 6 3 9 3 4 9 8 8 2 3 9 5

3. The points scored per game by a basketball team represent the third quartile for all teams in a league. What can you conclude about the team's points scored per game?

4. A salesperson at a company sold $6,903,435 of hardware equipment last year, a figure that represented the eighth decile of sales performance at the company. What can you conclude about the salesperson's performance?

5. A student's score on the ACT placement test for college algebra is in the 63rd percentile. What can you conclude about the student's test score?

6. A doctor tells a child's parents that their child's height is in the 87th percentile for the child's age group. What can you conclude about the child's height?

True or False? In Exercises 7–10, determine whether the statement is true or false. If it is false, rewrite it as a true statement.

7. The second quartile is the median of an ordered data set.

8. The five numbers you need to graph a box-and-whisker plot are the minimum, the maximum, Q_1, Q_3, and the mean.

9. The 50th percentile is equivalent to Q_1.

10. It is impossible to have a negative z-score.

Using and Interpreting Concepts

Graphical Analysis In Exercises 11–16, use the box-and-whisker plot to identify

(a) the minimum entry.
(b) the maximum entry.
(c) the first quartile.
(d) the second quartile.
(e) the third quartile.
(f) the interquartile range.

11.

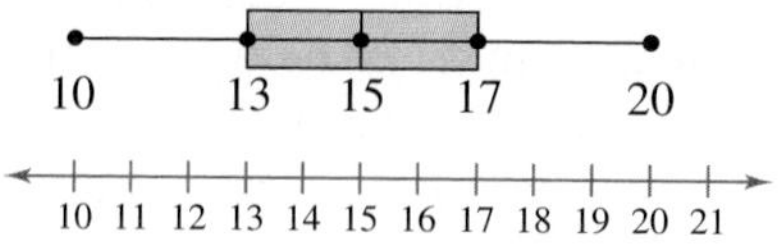

12.

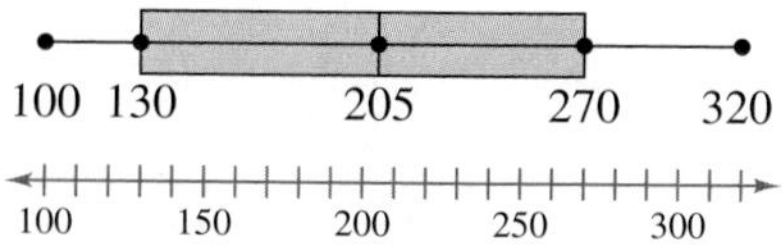

13.

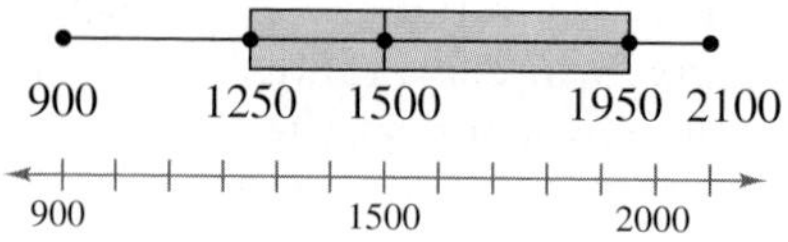

14.

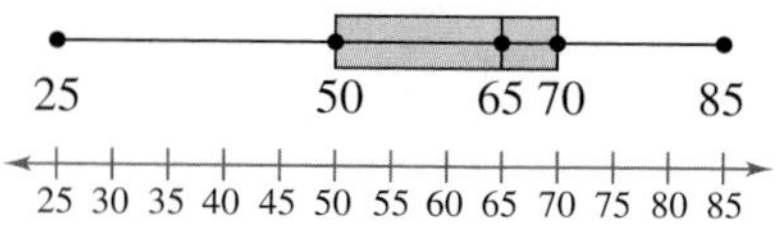

1. (a) $Q_1 = 4.5, Q_2 = 6, Q_3 = 7.5$
(b)

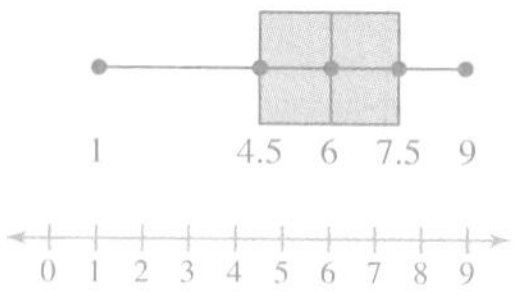

2. (a) $Q_1 = 3, Q_2 = 5, Q_3 = 8$
(b)

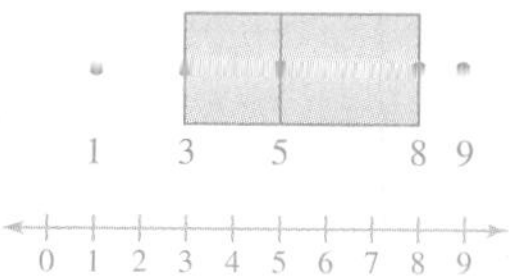

3. The basketball team scored more points per game than 75% of the teams in the league.

4. The salesperson sold more hardware equipment than 80% of the other salespeople.

5. The student scored above 63% of the students who took the ACT placement test.

6. The child is taller than 87% of the other children in the same age group.

7. True

8. False. The five numbers you need to graph a box-and-whisker plot are the minimum, the maximum, Q_1, Q_3, and the median.

9. False. The 50th percentile is equivalent to Q_2.

10. False. The only way to have a negative z-score is if the value is less than the mean.

11. (a) Min = 10 (b) Max = 20
(c) $Q_1 = 13$ (d) $Q_2 = 15$
(e) $Q_3 = 17$ (f) IQR = 4

12. (a) Min = 100 (b) Max = 320
(c) Q_1 = 130 (d) Q_2 = 205
(e) Q_3 = 270 (f) IQR = 140

13. (a) Min = 900 (b) Max = 2100
(c) Q_1 = 1250 (d) Q_2 = 1500
(e) Q_3 = 1950 (f) IQR = 700

14. (a) Min = 25 (b) Max = 85
(c) Q_1 = 50 (d) Q_2 = 65
(e) Q_3 = 70 (f) IQR = 20

15. (a) Min = −1.9 (b) Max = 2.1
(c) Q_1 = −0.5 (d) Q_2 = 0.1
(e) Q_3 = 0.7 (f) IQR = 1.2

16. (a) Min = −1.3 (b) Max = 2.1
(c) Q_1 = −0.3 (d) Q_2 = 0.2
(e) Q_3 = 0.4 (f) IQR = 0.7

17. Q_1 = B, Q_2 = A, Q_3 = C, because about one quarter of the data fall on or below 17, 18.5 is the median of the entire data set, and about three quarters of the data fall on or below 20.

18. P_{10} = T, P_{50} = R, P_{80} = S

Because 10% of the values are below T, 50% of the values are below R, and 80% of the values are below S.

19. (a) Q_1 = 2, Q_2 = 4, Q_3 = 5

(b) Watching Television

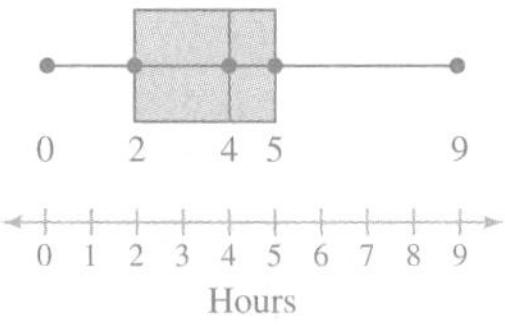

20. (a) Q_1 = 2, Q_2 = 4.5, Q_3 = 6.5

(b) Vacation Days

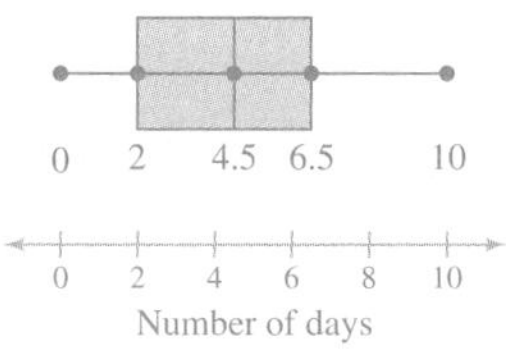

21. (a) Q_1 = 3.2, Q_2 = 3.65, Q_3 = 3.9

(b) Butterfly Wingspans

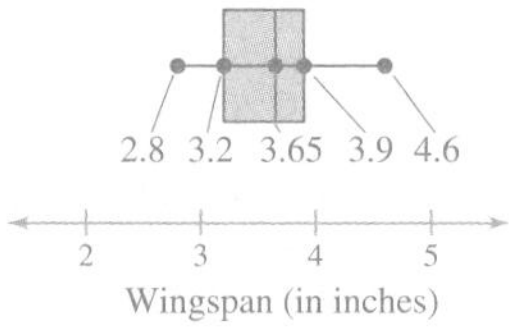

22. See Selected Answers, page A93.

15.

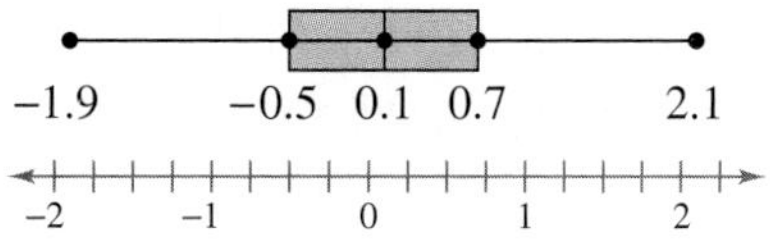

16.

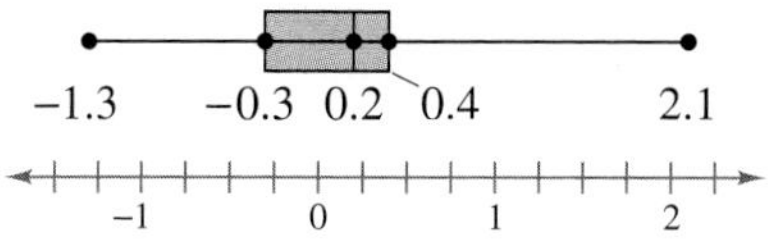

17. *Graphical Analysis* The letters A, B, and C are marked on the histogram. Match them to Q_1, Q_2 (the median), and Q_3. Justify your answer.

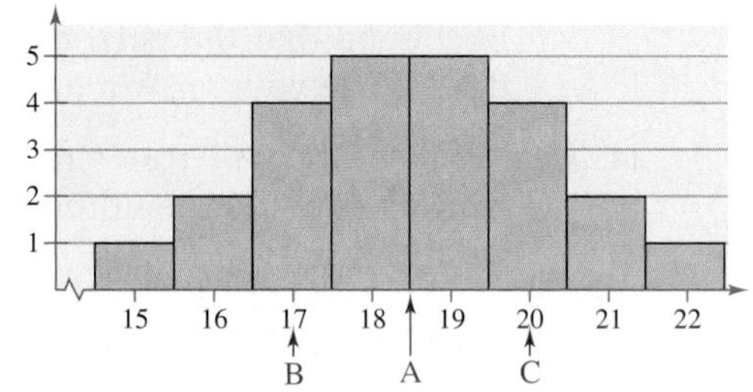

18. *Graphical Analysis* The letters R, S, and T are marked on the histogram. Match them to P_{10}, P_{50}, and P_{80}. Justify your answer.

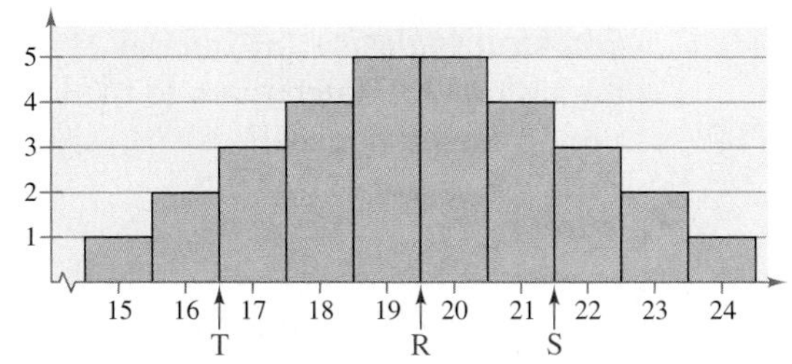

Using Technology to Find Quartiles and Draw Graphs In Exercises 19–22, use a calculator or a computer to (a) find the data set's first, second, and third quartiles, and (b) draw a box-and-whisker plot that represents the data set.

DATA **19. *TV Viewing*** The number of hours of television watched per day by a sample of 28 people

2	4	1	5	7	2	5	4	4	2	3	6	4	3
5	2	0	3	5	9	4	5	2	1	3	6	7	2

DATA **20. *Vacation Days*** The number of vacation days used by a sample of 20 employees in a recent year

3	9	2	1	7	5	3	2	2	6
4	0	10	0	3	5	7	8	6	5

DATA **21. *Butterfly Wingspans*** The lengths (in inches) of a sample of 22 butterfly wingspans

3.2	3.1	2.9	4.6	3.7	3.8	4.0	3.0
2.8	3.3	3.6	3.9	3.7	3.9	4.1	2.9
3.2	3.8	3.9	3.5	3.7	3.3		

DATA **22. *Hourly Earnings*** The hourly earnings (in dollars) of a sample of 25 railroad equipment manufacturers

15.60	18.75	14.60	15.80	14.35	13.90	17.50	17.55	13.80
14.20	19.05	15.35	15.20	19.45	15.95	16.50	16.30	15.25
15.05	19.10	15.20	16.22	17.75	18.40	15.25		

23. (a) 5
(b) 50%
(c) 25%

24. (a) $17.65
(b) 50%
(c) 50%

25. A → $z = -1.43$
B → $z = 0$
C → $z = 2.14$
A z-score of 2.14 would be unusual.

26. B → $z = 0.77$
C → $z = 1.54$
A → $z = -1.54$
None of the z-scores are unusual.

27. (a) Statistics: $z = \frac{73 - 63}{7} \approx 1.43$

Biology: $z = \frac{26 - 23}{3.9} \approx 0.77$

(b) The student did better on the statistics test.

28. (a) Statistics: $z = \frac{60 - 63}{7} \approx -0.43$

Biology: $z = \frac{20 - 23}{3.9} \approx -0.77$

(b) The student did better on the statistics test.

29. (a) Statistics: $z = \frac{78 - 63}{7} \approx 2.14$

Biology: $z = \frac{29 - 23}{3.9} \approx 1.54$

(b) The student did better on the statistics test.

30. (a) Statistics: $z = \frac{63 - 63}{7} = 0$

Biology: $z = \frac{23 - 23}{3.9} = 0$

(b) The student performed equally on both tests.

23. *TV Viewing* Refer to the data set given in Exercise 19 and the box-and-whisker plot you drew that represents the data set.

(a) About 75% of the people watched no more than how many hours of television per day?

(b) What percent of the people watched more than 4 hours of television per day?

(c) If you randomly selected one person from the sample, what is the likelihood that the person watched less than 2 hours of television per day? Write your answer as a percent.

24. *Manufacturer Earnings* Refer to the data set given in Exercise 22 and the box-and-whisker plot you drew that represents the data set.

(a) About 75% of the manufacturers made less than what amount per hour?

(b) What percent of the manufacturers made more than $15.80 per hour?

(c) If you randomly selected one manufacturer from the sample, what is the likelihood that the manufacturer made less than $15.80 per hour? Write your answer as a percent.

Graphical Analysis In Exercises 25 and 26, the midpoints A, B, and C are marked on the histogram. Match them to the indicated z-scores. Which z-scores, if any, would be considered unusual?

25. $z = 0$
$z = 2.14$
$z = -1.43$

Statistics Test Scores

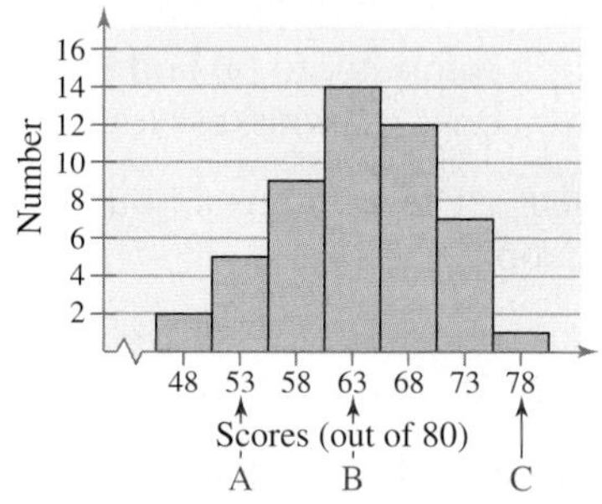

26. $z = 0.77$
$z = 1.54$
$z = -1.54$

Biology Test Scores

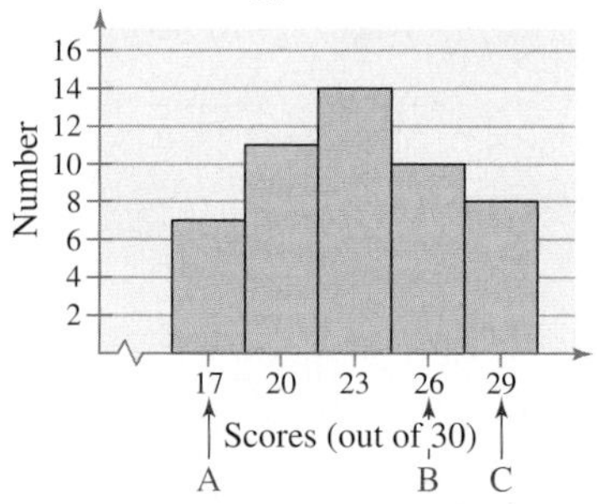

Comparing Test Scores For the statistics test scores in Exercise 25, the mean is 63 and the standard deviation is 7.0, and for the biology test scores in Exercise 26 the mean is 23 and the standard deviation is 3.9. In Exercises 27–30, you are given the test scores of a student who took both tests.

(a) Transform each test score to a z-score.

(b) Determine on which test the student had a better score.

27. A student gets a 73 on the statistics test and a 26 on the biology test.

28. A student gets a 60 on the statistics test and a 20 on the biology test.

29. A student gets a 78 on the statistics test and a 29 on the biology test.

30. A student gets a 63 on the statistics test and a 23 on the biology test.

31. (a) $z_1 = \dfrac{34{,}000 - 35{,}000}{2250}$

≈ -0.44

$z_2 = \dfrac{37{,}000 - 35{,}000}{2250} \approx 0.89$

$z_3 = \dfrac{31{,}000 - 35{,}000}{2250}$

≈ -1.78

None of the selected tires have unusual life spans.

(b) For 30,500, 2.5th percentile

For 37,250, 84th percentile

For 35,000, 50th percentile

32. (a) $z_1 = \dfrac{34 - 33}{4} = 0.25,$

$z_2 = \dfrac{30 - 33}{4} = -0.75,$

$z_3 = \dfrac{42 - 33}{4} = 2.25$

The life span of 42 days is unusual.

(b) For 29, 16th percentile
For 41, 97.5th percentile
For 25, 2.5th percentile

33. About 67 inches; 20% of the heights are below 67 inches.

34. 99th percentile

35. $z_1 = \dfrac{74 - 69.2}{2.9} \approx 1.66$

$z_2 = \dfrac{62 - 69.2}{2.9} \approx -2.48$

$z_3 = \dfrac{80 - 69.2}{2.9} \approx 3.72$

The heights that are 62 and 80 inches are unusual.

36. $z_1 = \dfrac{70 - 69.2}{2.9} \approx 0.28$

$z_2 = \dfrac{66 - 69.2}{2.9} \approx -1.10$

$z_3 = \dfrac{68 - 69.2}{2.9} \approx -0.41$

None of the heights are unusual.

31. ***Life Span of Tires*** A certain brand of automobile tire has a mean life span of 35,000 miles and a standard deviation of 2250 miles. (Assume the life spans of the tires have a bell-shaped distribution.)

(a) The life spans of three randomly selected tires are 34,000 miles, 37,000 miles, and 31,000 miles. Find the z-score that corresponds to each life span. According to the z-scores, would the life spans of any of these tires be considered unusual?

(b) The life spans of three randomly selected tires are 30,500 miles, 37,250 miles, and 35,000 miles. Using the Empirical Rule, find the percentile that corresponds to each life span.

32. ***Life Span of Fruit Flies*** The life spans of a species of fruit fly have a bell-shaped distribution, with a mean of 33 days and a standard deviation of 4 days.

(a) The life spans of three randomly selected fruit flies are 34 days, 30 days, and 42 days. Find the z-score that corresponds to each life span and determine if any of these life spans are unusual.

(b) The life spans of three randomly selected fruit flies are 29 days, 41 days, and 25 days. Using the Empirical Rule, find the percentile that corresponds to each life span.

Interpreting Percentiles In Exercises 33–38, use the cumulative frequency distribution to answer the questions. The cumulative frequency distribution represents the heights of males in the United States in the 20–29 age group. The heights have a bell-shaped distribution (see Picturing the World, page 80) with a mean of 69.2 inches and a standard deviation of 2.9 inches. *(Source: National Center for Health Statistics)*

Adult Males Ages 20–29

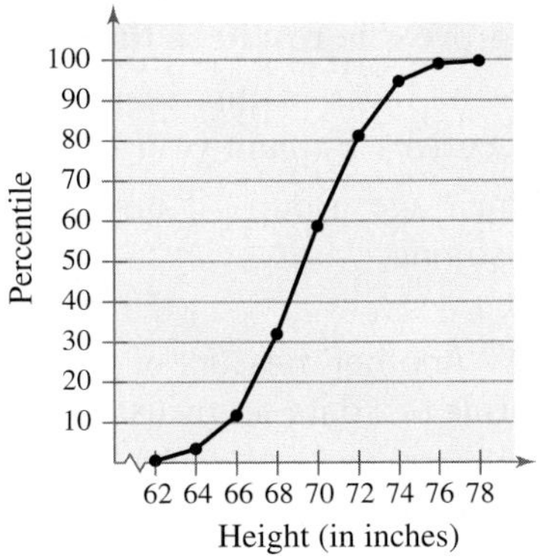

33. What height represents the 20th percentile? How should you interpret this?

34. What percentile is a height of 76 inches? How should you interpret this?

35. Three adult males in the 20–29 age group are randomly selected. Their heights are 74 inches, 62 inches, and 80 inches. Use z-scores to determine which heights, if any, are unusual.

36. Three adult males in the 20–29 age group are randomly selected. Their heights are 70 inches, 66 inches, and 68 inches. Use z-scores to determine which heights, if any, are unusual.

37. $z = \frac{71.1 - 69.2}{2.9} \approx 0.66$

About the 70th percentile

38. $z = \frac{66.3 - 69.2}{2.9} = -1$

About the 11th percentile

39. (a) $Q_1 = 42, Q_2 = 49, Q_3 = 56$

(b)

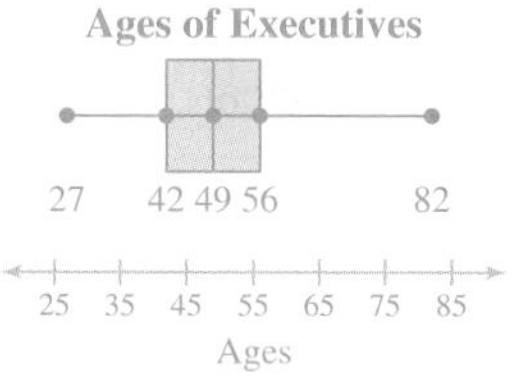

(c) Half of the ages are between 42 and 56 years.

(d) 49, because half of the executives are older and half are younger.

40. 5

41. 33.75

42. 10.975

43. 19.8

37. Find the z-score for a male in the 20–29 age group whose height is 71.1 inches. What percentile is this?

38. Find the z-score for a male in the 20–29 age group whose height is 66.3 inches. What percentile is this?

Extending Concepts

DATA **39.** ***Ages of Executives*** The ages of a sample of 100 executives are listed.

31	62	51	44	61	47	49	45	40	52	60	51	67	47	63	54	59	43	63	52
50	54	61	41	48	49	51	54	39	54	47	52	36	53	74	33	53	68	44	40
60	42	50	48	42	42	36	57	42	48	56	51	54	42	27	43	43	41	54	49
49	47	51	28	54	36	36	41	60	55	42	59	35	65	48	56	82	39	54	49
61	56	57	32	38	48	64	51	45	46	62	63	59	63	32	47	40	37	49	57

(a) Order the data and find the first, second, and third quartiles.

(b) Draw a box-and-whisker plot that represents the data set.

(c) Interpret the results in the context of the data.

(d) On the basis of this sample, at what age would you expect to be an executive? Explain your reasoning.

(e) Which age groups, if any, can be considered unusual? Explain your reasoning.

Midquartile Another measure of position is called the **midquartile.** You can find the midquartile of a data set by using the following formula.

$$\text{Midquartile} = \frac{Q_1 + Q_3}{2}$$

In Exercises 40–43, find the midquartile of the given data set.

40. 5 7 1 2 3 10 8 7 5 3

41. 23 36 47 33 34 40 39 24 32 22 38 41

42. 12.3 9.7 8.0 15.4 16.1 11.8 12.7 13.4
12.2 8.1 7.9 10.3 11.2

43. 21.4 20.8 19.7 15.2 31.9 18.7 15.6 16.7
19.8 13.4 22.9 28.7 19.8 17.2 30.1

Uses and Abuses

Statistics in the Real World

Uses

It can be difficult to see trends or patterns from a set of raw data. Descriptive statistics helps you do so. A good description of a data set consists of three features: (1) the shape of the data, (2) a measure of the center of the data, and (3) a measure of how much variability there is in the data. When you read reports, news items, or advertisements prepared by other people, you are seldom given raw data sets. Instead, you are given graphs, measures of central tendency, and measures of variation. To be a discerning reader, you need to understand the terms and techniques of descriptive statistics.

Abuses

Cropped Vertical Axis Misleading statistical graphs are common in newspapers and magazines. Compare the two time series charts below. The data are the same for each. However, the first graph has a cropped vertical axis, which makes it appear that the stock price has increased greatly over the 10-year period. In the second graph, the scale on the vertical axis begins at zero. This graph correctly shows that stock prices increased only modestly during the 10-year period.

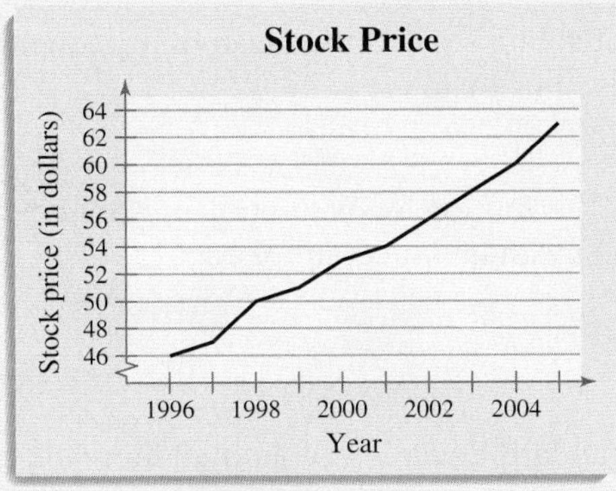

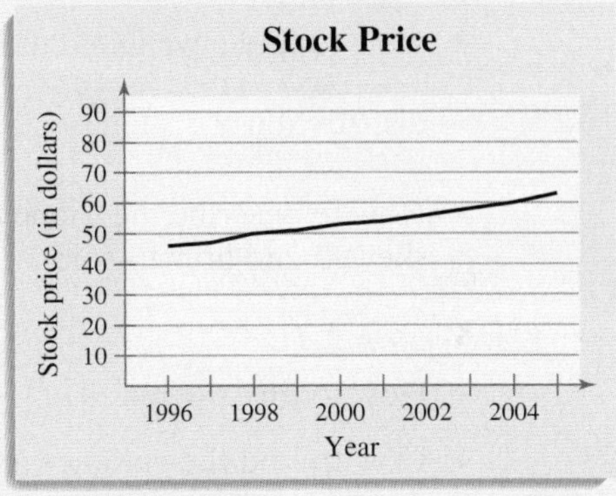

Effect of Outliers on the Mean Outliers, or extreme values, can have significant effects on the mean. Suppose, for example, that in recruiting information, a company stated that the average commission earned by the five people in its salesforce was $60,000 last year. This statement would be misleading if four of the five earned $25,000 and the fifth person earned $200,000.

Exercises

1. ***Cropped Vertical Axis*** In a newspaper or magazine, find an example of a graph that has a cropped vertical axis. Is the graph misleading? Do you think this graph was intended to be misleading? Redraw the graph so that it is not misleading.

2. ***Effect of Outliers on the Mean*** Describe a situation in which an outlier can make the mean misleading. Is the median also affected significantly by outliers? Explain your reasoning.

2 Chapter Summary

What did you learn?

	Review Exercises
Section 2.1	
◆ How to construct a frequency distribution including limits, boundaries, midpoints, relative frequencies, and cumulative frequencies	*1*
◆ How to construct frequency histograms, frequency polygons, relative frequency histograms, and ogives	*2–6*
Section 2.2	
◆ How to graph quantitative data sets using the exploratory data analysis tools of stem-and-leaf plots and dot plots	*7, 8*
◆ How to graph and interpret paired data sets using scatter plots and time series charts	*9, 10*
◆ How to graph qualitative data sets using pie charts and Pareto charts	*11, 12*
Section 2.3	
◆ How to find the mean, median, and mode of a population and a sample $\mu = \frac{\Sigma x}{N}, \bar{x} = \frac{\Sigma x}{n}$	*13, 14*
◆ How to find a weighted mean of a data set and the mean of a frequency distribution $\bar{x} = \frac{\Sigma(x \cdot w)}{\Sigma w}, \bar{x} = \frac{\Sigma(x \cdot f)}{n}$	*15–18*
◆ How to describe the shape of a distribution as symmetric, uniform, or skewed and how to compare the mean and median for each	*19–24*
Section 2.4	
◆ How to find the range of a data set	*25, 26*
◆ How to find the variance and standard deviation of a population and a sample $\sigma = \sqrt{\frac{\Sigma(x-\mu)^2}{N}}, s = \sqrt{\frac{\Sigma(x-\bar{x})^2}{n-1}}$	*27–30*
◆ How to use the Empirical Rule and Chebychev's Theorem to interpret standard deviation	*31–34*
◆ How to approximate the sample standard deviation for grouped data $s = \sqrt{\frac{\Sigma(x-\bar{x})^2 f}{n-1}}$	*35, 36*
Section 2.5	
◆ How to find the quartiles and interquartile range of a data set	*37–39, 41*
◆ How to draw a box-and-whisker plot	*40, 42*
◆ How to interpret other fractiles such as percentiles	*43, 44*
◆ How to find and interpret the standard score (z-score) $z = (x - \mu)/\sigma$	*45–48*

2 Review Exercises

1. See Odd Answers, page A56.
2. See Selected Answers, page A93.
3. Liquid Volume 12-oz Cans

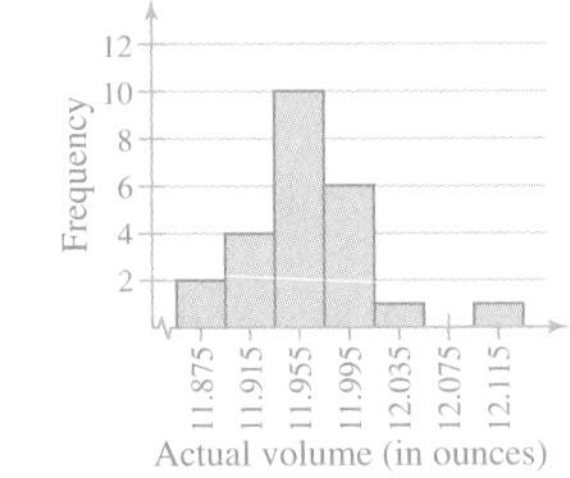

4. See Selected Answers, page A93.
5.

Class	Midpoint	Frequency, f
79–93	86	9
94–108	101	12
109–123	116	5
124–138	131	3
139–153	146	2
154–168	161	1
		$\sum f = 32$

Meals Purchased

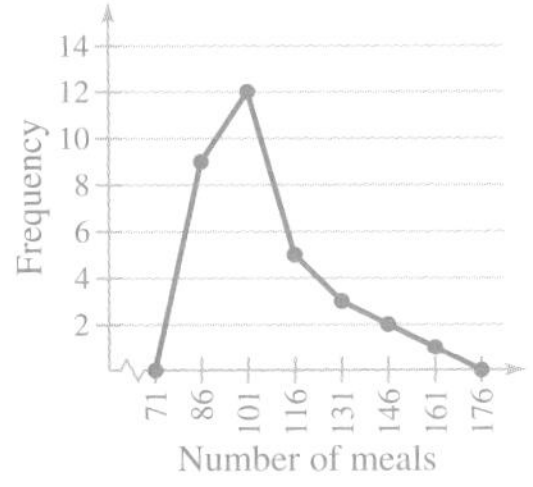

6. See Selected Answers, page A93.
7.

1	3 7 8 9
2	0 1 2 3 3 3 4 4 5 5 5 7 8 8 9
3	1 1 2 3 4 5 7 8
4	3 4 7
5	1

8. See Selected Answers, page A93.
9. Height of Buildings

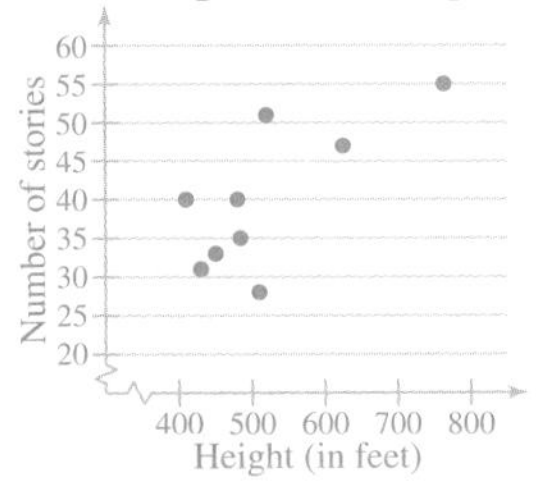

The number of stories appears to increase with height.

Section 2.1

DATA In Exercises 1 and 2, use the following data set. The data set represents the income (in thousands of dollars) of 20 employees at a small business.

30 28 26 39 34 33 20 39 28 33
26 39 32 28 31 39 33 31 33 32

1. Make a frequency distribution of the data set using five classes. Include the class midpoints, limits, boundaries, frequencies, relative frequencies, and cumulative frequencies.

2. Make a relative frequency histogram using the frequency distribution in Exercise 1. Then determine which class has the greatest relative frequency and which has the least relative frequency.

DATA In Exercises 3 and 4, use the following data set. The data represent the actual liquid volume (in ounces) in 24 twelve-ounce cans.

11.95 11.91 11.86 11.94 12.00 11.93 12.00 11.94
12.10 11.95 11.99 11.94 11.89 12.01 11.99 11.94
11.92 11.98 11.88 11.94 11.98 11.92 11.95 11.93

3. Make a frequency histogram using seven classes.

4. Make a relative frequency histogram of the data set using seven classes.

DATA In Exercises 5 and 6, use the following data set. The data represent the number of meals purchased during one night's business at a sample of restaurants.

153 104 118 166 89 104 100 79 93 96 116
94 140 84 81 96 108 111 87 126 101 111
122 108 126 93 108 87 103 95 129 93

5. Make a frequency distribution with six classes and draw a frequency polygon.

6. Make an ogive of the data set using six classes.

Section 2.2

DATA In Exercises 7 and 8, use the following data set. The data represent the average daily high temperature (in degrees Fahrenheit) during the month of January for Chicago, Illinois. *(Source: National Oceanic and Atmospheric Administration)*

33 31 25 22 38 51 32 23 23 34 44 43 47 37 29 25
28 35 21 24 20 19 23 27 24 13 18 28 17 25 31

7. Make a stem-and-leaf plot of the data set. Use one line per stem.

8. Make a dot plot of the data set.

9. The following are the heights (in feet) and the number of stories of nine notable buildings in Miami. Use the data to construct a scatter plot. What type of pattern is shown in the scatter plot? *(Source: Skyscrapers.com)*

Height (in feet)	764	625	520	510	484	480	450	430	410
Number of stories	55	47	51	28	35	40	33	31	40

10. The U.S. unemployment rate over a 12-year period is given. Use the data to construct a time series chart. *(Source: U.S. Bureau of Labor Statistics)*

Year	1992	1993	1994	1995	1996	1997
Unemployment rate	7.5	6.9	6.1	5.6	5.4	4.9
Year	1998	1999	2000	2001	2002	2003
Unemployment rate	4.5	4.2	4.0	4.7	5.8	6.0

In Exercises 11 and 12, use the following data set. The data set represents the top seven American Kennel Club registrations (in thousands) in 2003. *(Source: American Kennel Club)*

Breed	Labrador Retriever	Golden Retriever	Beagle	German Shepherd	Dachshund	Yorkshire Terrier	Boxer
Number registered (in thousands)	145	53	45	44	39	38	34

11. Make a Pareto chart of the data set.

12. Make a pie chart of the data set.

Section 2.3

13. Find the mean, median, and mode of the data set.

9 7 8 6 9 12 11 5 9 10

14. Find the mean, median, and mode of the data set.

28 35 29 29 33 32 29 33 31 29

15. Estimate the mean of the frequency distribution you made in Exercise 1.

16. The following frequency distribution shows the number of magazine subscriptions per household for a sample of 60 households. Find the mean number of subscriptions per household.

Number of magazines	0	1	2	3	4	5	6
Frequency	13	9	19	8	5	2	4

17. Six test scores are given. The first five test scores are 15% of the final grade, and the last test score is 25% of the final grade. Find the weighted mean of the test scores.

65 72 84 89 70 90

18. Four test scores are given. The first three test scores are 20% of the final grade, and the last test score is 40% of the final grade. Find the weighted mean of the test scores.

81 95 89 87

19. Describe the shape of the distribution in the histogram you made in Exercise 3. Is the distribution symmetric, uniform, or skewed?

20. Describe the shape of the distribution in the histogram you made in Exercise 4. Is the distribution symmetric, uniform, or skewed?

10.

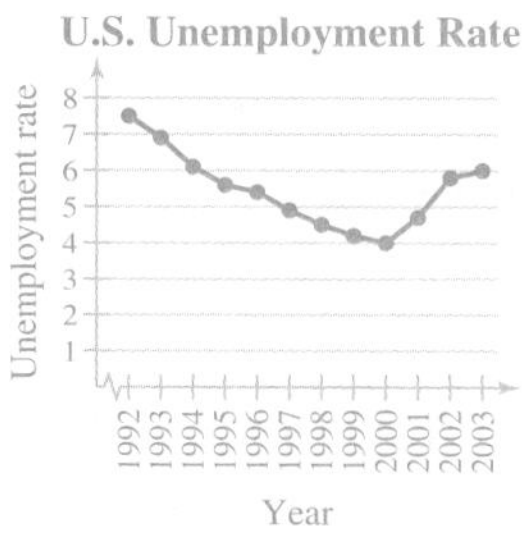

11.

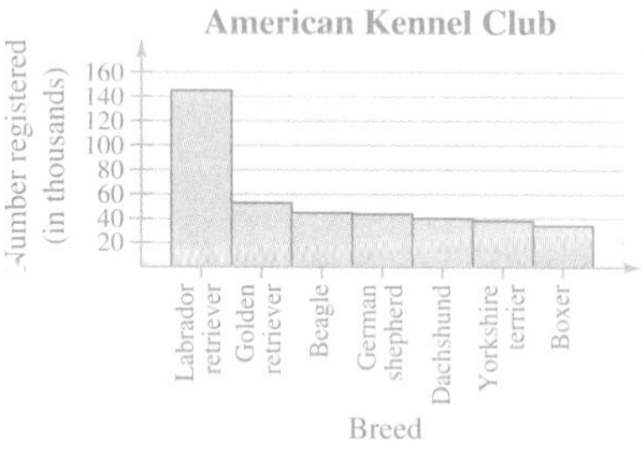

12.

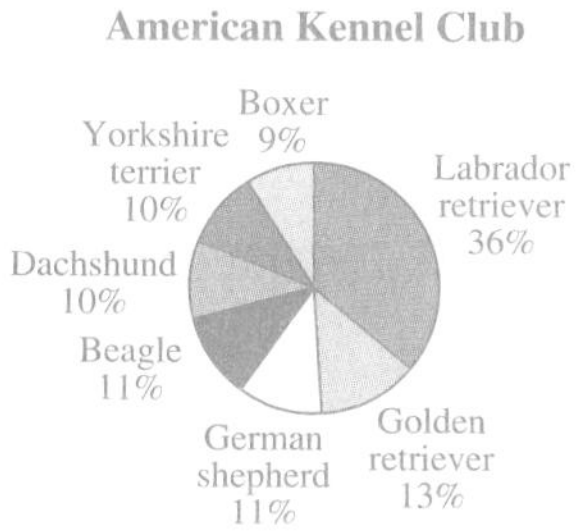

13. Mean = 8.6
Median = 9
Mode = 9

14. Mean = 30.8
Median = 30
Mode = 29

15. 31.7

16. 2.1

17. 79.5

18. 87.8

19. Skewed

20. Skewed

21. Skewed left
22. Skewed right
23. Median
24. Mean
25. 2.8
26. 3.84
27. Population mean = 9
Standard deviation ≈ 3.2
28. Population mean = 69
Standard deviation ≈ 7.8
29. Sample mean = 2453.4
Standard deviation ≈ 306.1
30. Sample mean = 38,653.5
Standard deviation ≈ 6762.6
31. Between \$21.50 and \$36.50
32. 68%

In Exercises 21 and 22, determine whether the approximate shape of the distribution in the histogram is skewed right, skewed left, or symmetric.

21.

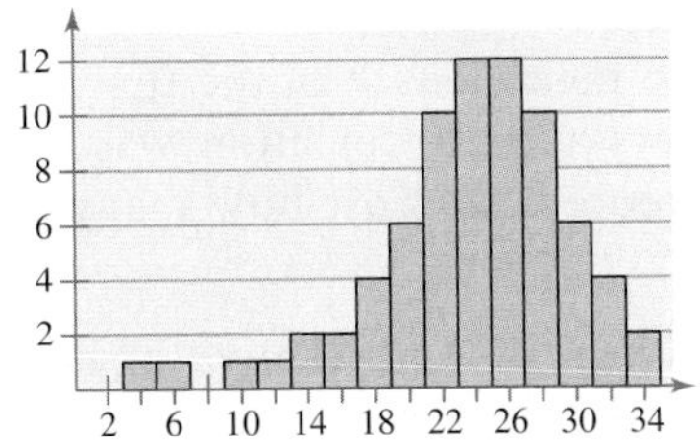

22.

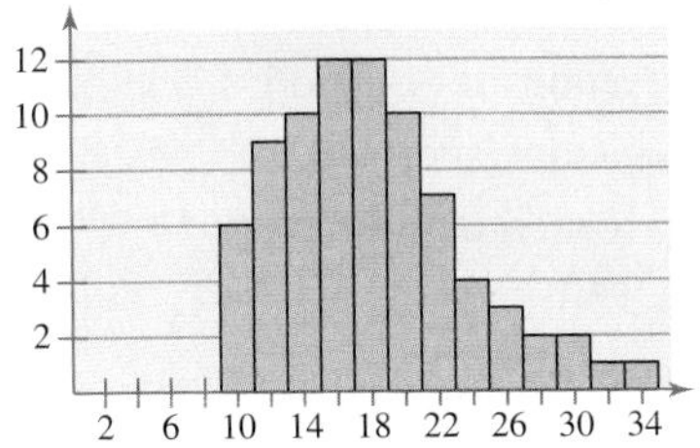

23. For the histogram in Exercise 21, which is greater, the mean or the median?

24. For the histogram in Exercise 22, which is greater, the mean or the median?

Section 2.4

25. The data set represents the mean price of a movie ticket (in U.S. dollars) for a sample of 12 U.S. cities. Find the range of the data set.

7.82 7.38 6.42 6.76 6.34 7.44 6.15 5.46 7.92 6.58 8.26 7.17

26. The data set represents the mean price of a movie ticket (in U.S. dollars) for a sample of 12 Japanese cities. Find the range of the data set.

19.73 16.48 19.10 18.56 17.68 17.19
16.63 15.99 16.66 19.59 15.89 16.49

27. The mileage (in thousands) for a rental car company's fleet is listed. Find the population mean and standard deviation of the data.

6 14 3 7 11 13 8 5 10 9 12 10

28. The age of each Supreme Court justice as of August 20, 2003 is listed. Find the population mean and standard deviation of the data. *(Source: Supreme Court of the United States)*

78 83 73 67 67 63 55 70 65

29. Dormitory room prices (in dollars for one school year) for a sample of four-year universities are listed. Find the sample mean and the sample standard deviation of the data.

2445 2940 2399 1960 2421 2940 2657 2153
2430 2278 1947 2383 2710 2761 2377

30. Sample salaries (in dollars) of public school teachers are listed. Find the sample mean and standard deviation of the data.

46,098 36,259 35,084 38,617 42,690 26,202 47,169 37,109

31. The mean rate for cable television from a sample of households was \$29.00 per month, with a standard deviation of \$2.50 per month. Between what two values do 99.7% of the data lie? (Assume a bell-shaped distribution.)

32. The mean rate for cable television from a sample of households was \$29.50 per month, with a standard deviation of \$2.75 per month. Estimate the percent of cable television rates between \$26.75 and \$32.25. (Assume that the data set has a bell-shaped distribution.)

33. 30

34. 15

35. Sample mean ≈ 2.5
Standard deviation ≈ 1.2

36. Sample mean = 2.4
Standard deviation ≈ 1.7

37. 56

38. 70

39. 14

40.

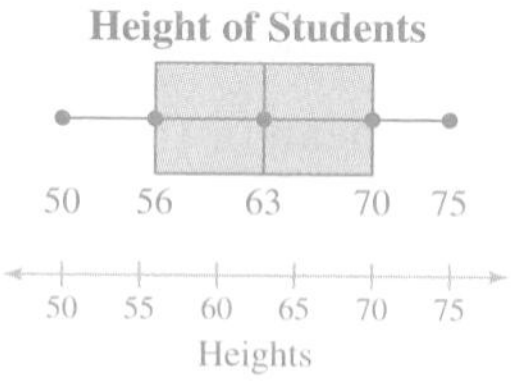

41. 4

42.

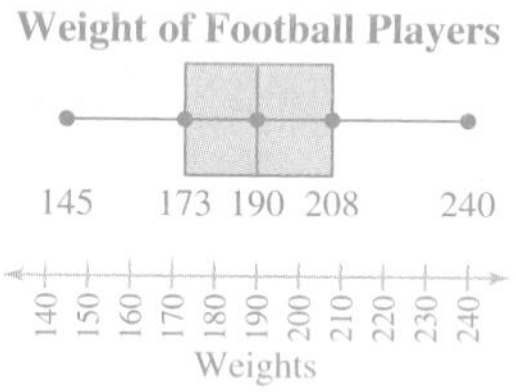

43. 23% scored higher than 68.

44. 88th percentile

45. $z = 2.33$, unusual

46. $z = -1.5$, not unusual

47. $z = 1.25$, not unusual

48. $z = -2.125$, unusual

33. The mean sale per customer for 40 customers at a grocery store is $23.00, with a standard deviation of $6.00. On the basis of Chebychev's Theorem, at least how many of the customers spent between $11.00 and $35.00?

34. The mean length of the first 20 space shuttle flights was about 7 days, and the standard deviation was about 2 days. On the basis of Chebychev's Theorem, at least how many of the flights lasted between 3 days and 11 days? *(Source: NASA)*

35. From a random sample of households, the number of television sets are listed. Find the sample mean and standard deviation of the data.

Number of televisions	0	1	2	3	4	5
Number of households	1	8	13	10	5	3

36. From a random sample of airplanes, the number of defects found in their fuselages are listed. Find the sample mean and standard deviation of the data.

Number of defects	0	1	2	3	4	5	6
Number of airplanes	4	5	2	9	1	3	1

Section 2.5

In Exercises 37–40, use the following data set. The data represent the heights (in inches) of students in a statistics class.

50	51	54	54	56	59	60	61	61	63
64	65	68	69	70	70	71	71	75	

37. Find the height that corresponds to the first quartile.

38. Find the height that corresponds to the third quartile.

39. Find the interquartile range.

40. Make a box-and-whisker plot of the data.

41. Find the interquartile range of the data from Exercise 14.

42. The weights (in pounds) of the defensive players on a high school football team are given. Make a box-and-whisker plot of the data.

173	145	205	192	197	227	156	240	172	185
208	185	190	167	212	228	190	184	195	

43. A student's test grade of 68 represents the 77th percentile of the grades. What percent of students scored higher than 68?

44. In 2004 there were 728 "oldies" radio stations in the United States. If one station finds that 84 stations have a larger daily audience than it does, what percentile does this station come closest to in the daily audience rankings? *(Source: Radioinfo.com)*

In Exercises 45–48, use the following information. The weights of 19 high school football players have a bell-shaped distribution, with a mean of 192 pounds and a standard deviation of 24 pounds. Use *z*-scores to determine if the weights of the following randomly selected football players are unusual.

45. 248 pounds **46.** 156 pounds **47.** 222 pounds **48.** 141 pounds

2 Chapter Quiz

1. See Odd Answers, page A56.
2. 125.2, 13.0
3. (a)

(b)

4. (a) 751.6, 784.5, none
   The mean best describes a typical salary because there are no outliers.
   (b) 575; 48,135.1; 219.4
5. Between \$125,000 and \$185,000
6. (a) $z = 3.0$, unusual
   (b) $z \approx -6.67$, very unusual
   (c) $z \approx 1.33$
   (d) $z = -2.2$, unusual
7. (a) 71, 84.5, 90
   (b) 19
   (c)

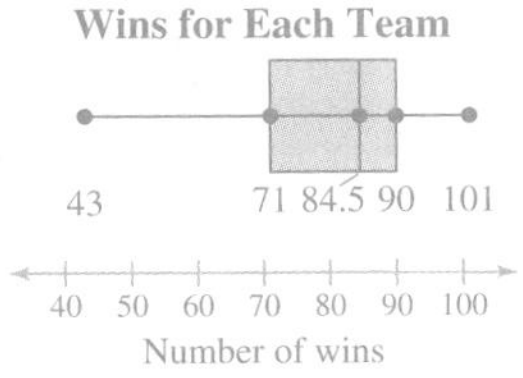

Take this quiz as you would take a quiz in class. After you are done, check your work against the answers given in the back of the book.

DATA **1.** The data set is the number of minutes a sample of 25 people exercise each week.

108	139	120	123	120	132	123	131	131
157	150	124	111	101	135	119	116	117
127	128	139	119	118	114	127		

(a) Make a frequency distribution of the data set using five classes. Include class limits, midpoints, frequencies, boundaries, relative frequencies, and cumulative frequencies.

(b) Display the data using a frequency histogram and a frequency polygon on the same axes.

(c) Display the data using a relative frequency histogram.

(d) Describe the distribution's shape as symmetric, uniform, or skewed.

(e) Display the data using a box-and-whisker plot.

(f) Display the data using an ogive.

2. Use frequency distribution formulas to approximate the sample mean and standard deviation of the data set in Exercise 1.

3. U.S. sporting goods sales (in billions of dollars) can be classified in four areas: clothing (10.0), footwear (14.1), equipment (21.7), and recreational transport (32.1). Display the data using (a) a pie chart and (b) a Pareto chart. *(Source: National Sporting Goods Association)*

4. Weekly salaries (in dollars) for a sample of registered nurses are listed.

774 446 1019 795 908 667 444 960

(a) Find the mean, the median, and the mode of the salaries. Which best describes a typical salary?

(b) Find the range, variance, and standard deviation of the data set. Interpret the results in the context of the real-life setting.

5. The mean price of new homes from a sample of houses is \$155,000 with a standard deviation of \$15,000. The data set has a bell-shaped distribution. Between what two prices do 95% of the houses fall?

6. Refer to the sample statistics from Exercise 5 and use z-scores to determine which, if any, of the following house prices is unusual.

(a) \$200,000 (b) \$55,000 (c) \$175,000 (d) \$122,000

DATA **7.** The number of wins for each Major League Baseball team in 2003 are listed. *(Source: Major League Baseball)*

101	95	86	71	63	90	86	83	68	43
96	93	77	71	101	91	86	83	66	88
87	85	75	69	68	100	85	84	74	64

(a) Find the quartiles of the data set.

(b) Find the interquartile range.

(c) Draw a box-and-whisker plot.

PUTTING IT ALL TOGETHER

Real Statistics ■ Real Decisions

You are a consumer journalist for a newspaper. You have received several letters and emails from readers who are concerned about the cost of their automobile insurance premiums. One of the readers wrote the following:

> *"I think, on the average, a driver in our city pays a higher automobile insurance premium than drivers in other cities like ours in this state."*

Your editor asks you to investigate the costs of insurance premiums and write an article about it. You have gathered the data shown at the right (your city is City A). The data represent the automobile insurance premiums paid annually (in dollars) by a random sample of drivers in your city and three other cities of similar size in your state. (The prices of the premiums from the sample include comprehensive, collision, bodily injury, property damage, and uninsured motorist coverage.)

runzheimer
INTERNATIONAL

The Prices, in Dollars, of Automobile Insurance Premiums Paid by 10 Randomly Selected Drivers in Four Cities

City A	City B	City C	City D
2465	2514	2030	2345
1984	1600	1450	2152
2545	1545	2715	1570
1640	2716	2145	1830
1983	1987	1600	1450
2302	2200	1430	1745
2542	2005	1545	1590
1875	1945	1792	1800
1920	1380	1645	2575
2655	2400	1368	2016

(Adapted from Runzheimer International)

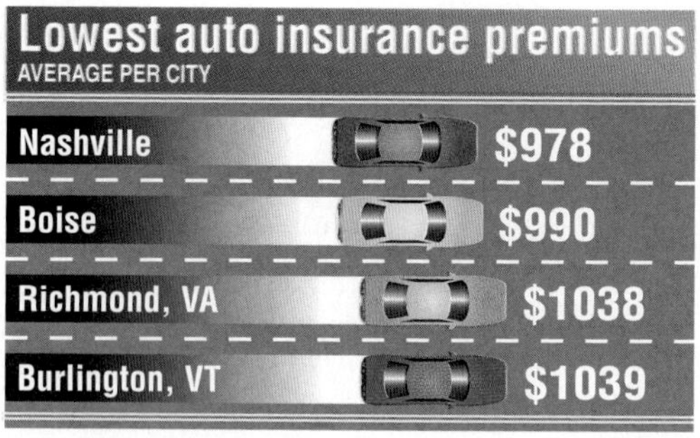

(Source: Runzheimer International)

Exercises

1. ***How Would You Do It?***
   (a) How would you investigate the statement about the price of automobile insurance premiums?
   (b) What statistical measures in this chapter would you use?

2. ***Displaying the Data***
   (a) What type of graph would you choose to display the data? Why?
   (b) Construct the graph from part (a).
   (c) On the basis of what you did in part (b), does it appear that the average automobile insurance premium in your city, City A, is higher than in any of the other cities? Explain.

3. ***Measuring the Data***
   (a) What statistical measures discussed in this chapter would you use to analyze the automobile insurance premium data?
   (b) Calculate the measures from part (a).
   (c) Compare the measures from part (b) with the graphs you made in Exercise 2. Do the measurements support your conclusion in Exercise 2? Explain.

4. ***Discussing the Data***
   (a) What would you tell your readers? Is the average automobile insurance premium in your city more than in the other cities?
   (b) What reasons might you give to your readers as to why the prices of automobile insurance premiums vary from city to city?

TECHNOLOGY

MINITAB | EXCEL | TI-83

DAIRY FARMERS of AMERICA

Dairy Farmers of America is an association that provides help to dairy farmers. Part of this help is gathering and distributing statistics on milk production.

www.dfamilk.com

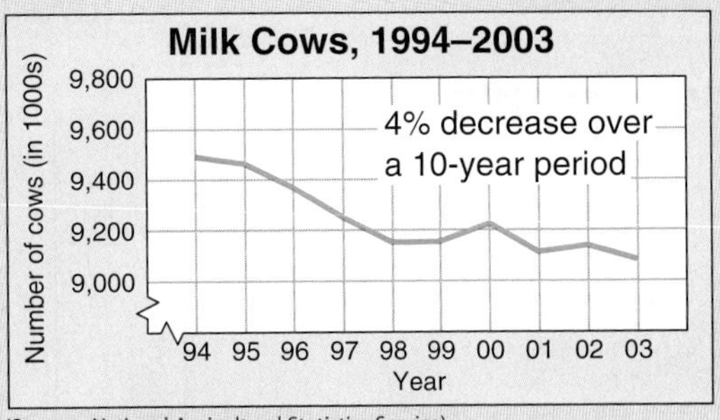

(Source: National Agricultural Statistics Service)

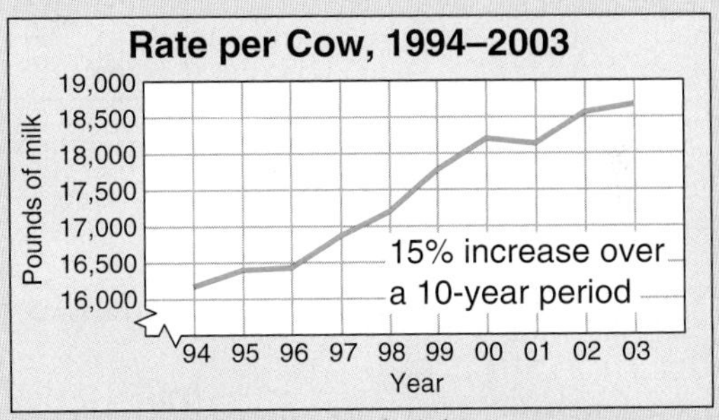

(Source: National Agricultural Statistics Service)

From 1994 to 2003, the number of dairy cows in the United States decreased and the yearly milk production increased.

Monthly Milk Production

The following data set was supplied by a dairy farmer. It lists the monthly milk production (in pounds) for 50 Holstein dairy cows. *(Source: Matlink Dairy, Clymer, NY)*

2825	2072	2733	2069	2484
4285	2862	3353	1449	2029
1258	2982	2045	1677	1619
2597	3512	2444	1773	2284
1884	2359	2046	2364	2669
3109	2804	1658	2207	2159
2207	2882	1647	2051	2202
3223	2383	1732	2230	1147
2711	1874	1979	1319	2923
2281	1230	1665	1294	2936

Exercises

In Exercises 1–4, use a computer or calculator. If possible, print your results.

1. Find the sample mean of the data.

2. Find the sample standard deviation of the data.

3. Make a frequency distribution for the data. Use a class width of 500.

4. Draw a histogram for the data. Does the distribution appear to be bell shaped?

5. What percent of the distribution lies within one standard deviation of the mean? Within two standard deviations of the mean? How do these results agree with the Empirical Rule?

In Exercises 6–8, use the frequency distribution found in Exercise 3.

6. Use the frequency distribution to estimate the sample mean of the data. Compare your results with Exercise 1.

7. Use the frequency distribution to find the sample standard deviation for the data. Compare your results with Exercise 2.

8. ***Writing*** Use the results of Exercises 6 and 7 to write a general statement about the mean and standard deviation for grouped data. Do the formulas for grouped data give results that are as accurate as the individual entry formulas?

Extended solutions are given in the *Technology Supplement.*
Technical instruction is provided for MINITAB, Excel, and the TI-83.

2 Using Technology to Determine Descriptive Statistics

Here are some MINITAB and TI-83 printouts for three examples in this chapter.

(See Example 7, page 55.)

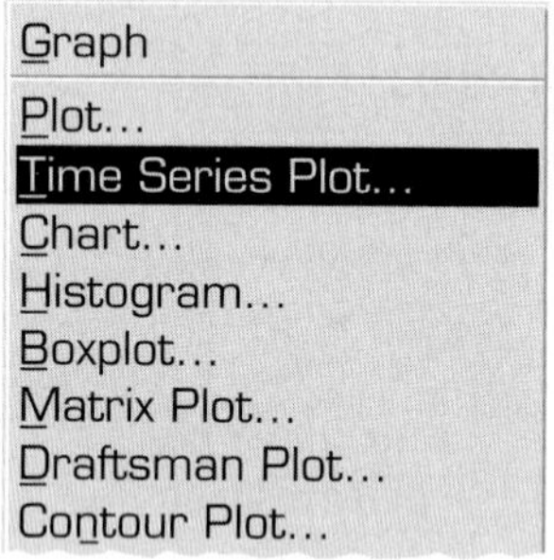

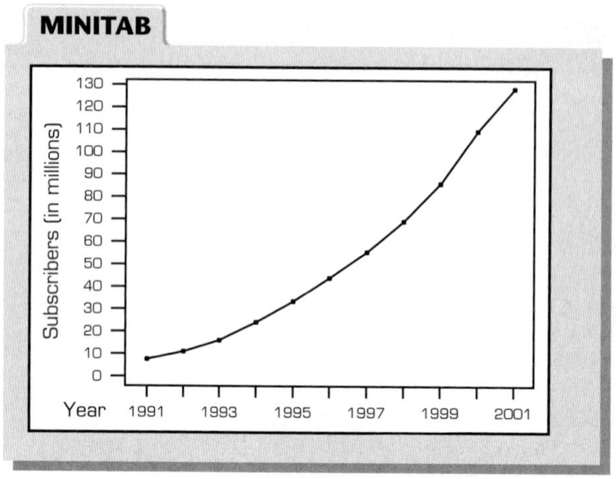

(See Example 4, page 77.)

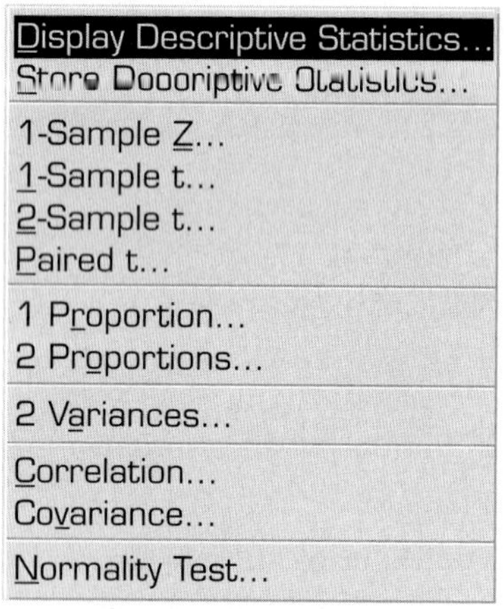

MINITAB

Descriptive Statistics

Variable	N	Mean	Median	TrMean	StDev	SE Mean
Salaries	10	41.500	41.000	41.375	3.136	0.992

Variable	Minimum	Maximum	Q1	Q3
Salaries	37.000	47.000	38.750	44.250

(See Example 4, page 96.)

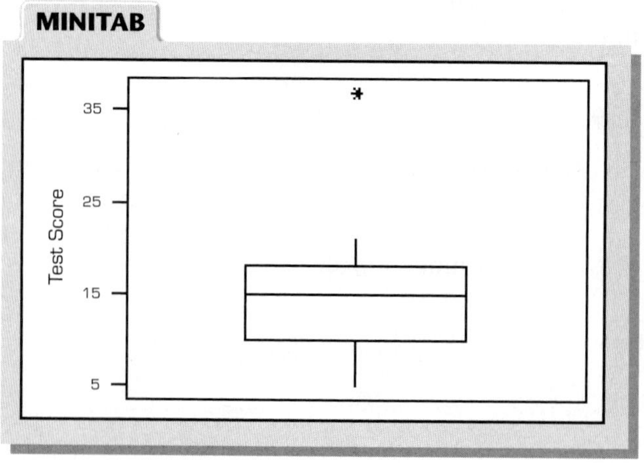

(See Example 7, page 55.)

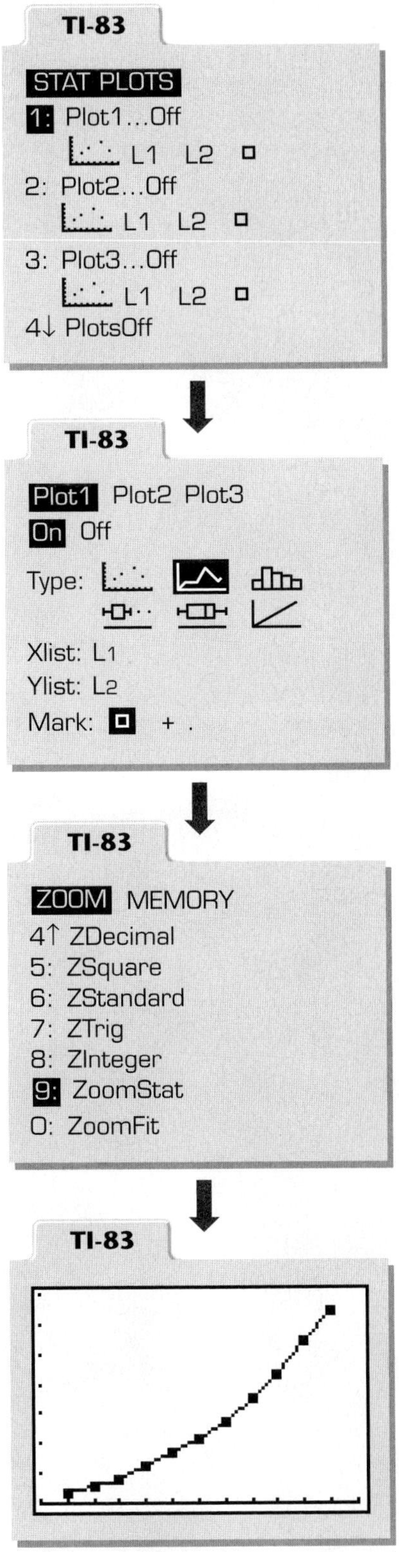

(See Example 4, page 77.)

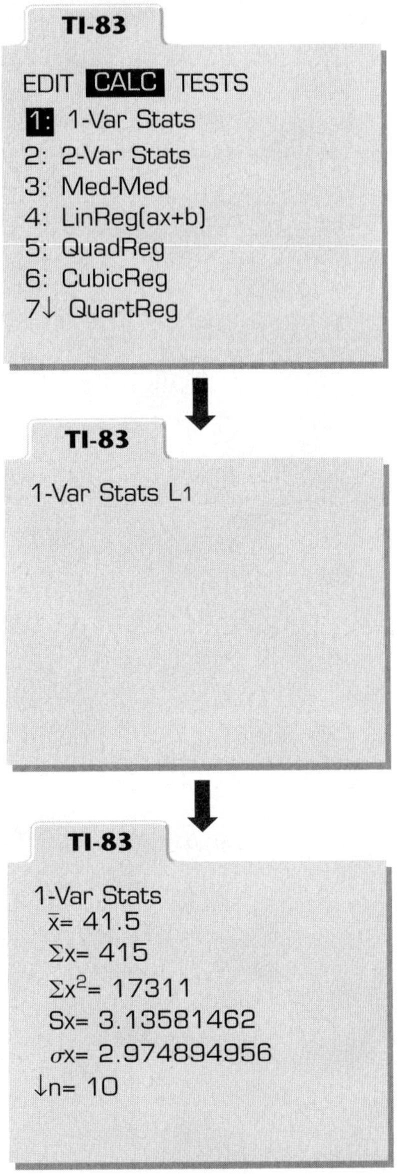

(See Example 4, page 96.)

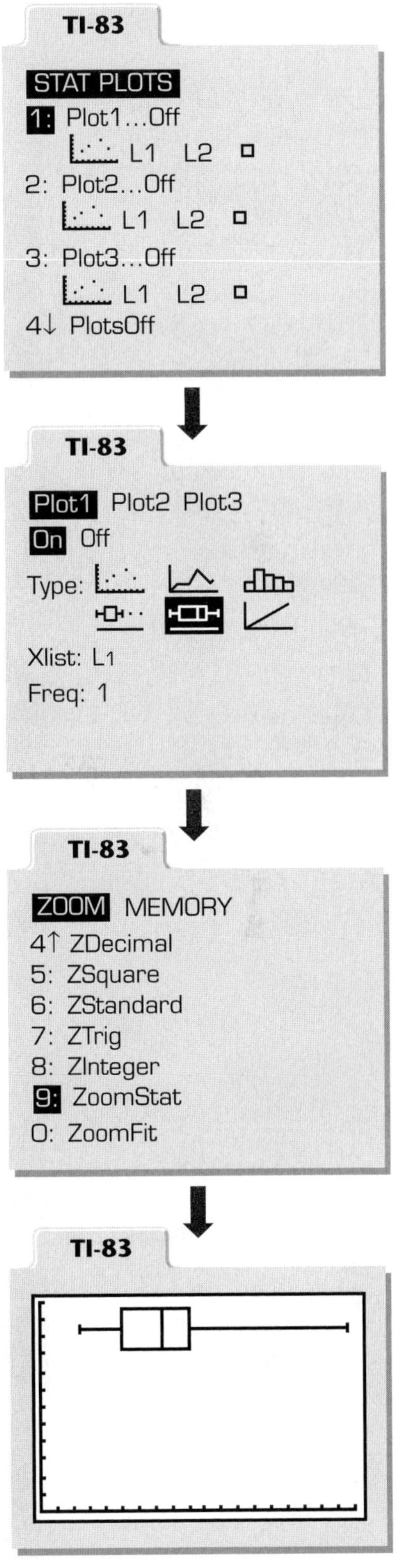

CHAPTER

3 Probability

3.1 Basic Concepts of Probability

3.2 Conditional Probability and the Multiplication Rule

3.3 The Addition Rule

Case Study

3.4 Counting Principles

Uses and Abuses

Real Statistics–Real Decisions

Technology

The television game show *The Price is Right* presents a wide range of pricing games in which contestants compete for prizes using strategy, luck, and their knowledge of prices. In *The Money Game*, a contestant can win a car by correctly guessing the car's five-digit price.

Where You've Been

In Chapters 1 and 2, you learned how to collect and describe data. Once the data are collected and described, you can use the results to write summaries, form conclusions, and make decisions. For instance, in *The Money Game* on *The Price Is Right*, a contestant can win a car by correctly guessing the car's five-digit price. By collecting and analyzing the data, you can determine your chances of winning the car.

To play *The Money Game*, a contestant is shown nine cards on a board. Each card has one pair of numbers. One card has the first two digits of the price of the car and another card has the last two digits of the price of the car. The middle digit is given. The contestant chooses one card at a time. If the card contains a pair of digits from the car price it is placed in one of the two car price slots. If it does not contain a pair of digits from the car price, it is placed in one of the four money slots to the left. The contestant wins the car if the two car price slots are filled before the four money slots. Otherwise, the contestant wins a small amount of money.

Where You're Going

In Chapter 3, you will learn how to determine the probability that an event will occur. For instance, suppose you are given the following table, which shows the four possible ways to win a car during *The Money Game* and the corresponding probabilities.

Event	Probability
Winning the car after selecting two cards	$\frac{1}{36}$
Winning the car after selecting three cards	$\frac{1}{18}$
Winning the car after selecting four cards	$\frac{1}{12}$
Winning the car after selecting five cards	$\frac{1}{9}$

How would you use these given probabilities to determine the probability that a contestant will win a car? How would you determine the probability that a contestant will win money?

You can find the probability of winning a car by adding the given probabilities.

$$\text{Probability of winning a car} = \frac{1}{36} + \frac{1}{18} + \frac{1}{12} + \frac{1}{9} = \frac{5}{18} \approx 0.278$$

Then, you can find the probability of winning money by subtracting the probability of winning a car from 1.

$$\begin{aligned}\text{Probability of winning money} &= 1 - \text{Probability of winning a car}\\ &= 1 - \frac{5}{18} = \frac{13}{18} \approx 0.722\end{aligned}$$

So, the probability that a contestant will win a car during *The Money Game* is about 0.278, or 27.8%. The probability that a contestant will win money is about 0.722, or 72.2%.

Note to Instructor

Asking the students to calculate the probabilities in the table would be a challenge for this course. The calculations are as follows:

Two cards: $\dfrac{({}_2C_2)({}_7C_0)}{{}_9C_2} = \dfrac{1}{36}$

Three cards:

$$\frac{({}_2C_1)({}_7C_1)}{{}_9C_2} \cdot \frac{({}_1C_1)({}_6C_0)}{{}_7C_1} = \frac{1}{18}$$

Four cards:

$$\frac{({}_2C_1)({}_7C_2)}{{}_9C_3} \cdot \frac{({}_1C_1)({}_5C_0)}{{}_6C_1} = \frac{1}{12}$$

Five cards:

$$\frac{({}_2C_1)({}_7C_3)}{{}_9C_4} \cdot \frac{({}_1C_1)({}_4C_0)}{{}_5C_1} = \frac{1}{9}$$

3.1 Basic Concepts of Probability

What You Should Learn

- How to identify the sample space of a probability experiment and to identify simple events
- How to distinguish among classical probability, empirical probability, and subjective probability
- How to find the probability of the complement of an event

Probability Experiments • Types of Probability • Complementary Events

Probability Experiments

When weather forecasters say that there is a 90% chance of rain or a physician says there is a 35% chance for a successful surgery, they are stating the likelihood, or *probability*, that a specific event will occur. Decisions such as "should you wash your car" or "should you proceed with surgery" are often based on these probabilities. In the previous chapter, you learned about the role of the descriptive branch of statistics. Because probability is the foundation of inferential statistics, it is necessary to learn about probability before proceeding to the second branch—inferential statistics.

DEFINITION

A **probability experiment** is an action, or trial, through which specific results (counts, measurements, or responses) are obtained. The result of a single trial in a probability experiment is an **outcome.** The set of all possible outcomes of a probability experiment is the **sample space.** An **event** consists of one or more outcomes and is a subset of the sample space.

Study Tip

Here is a simple example of the use of the terms *probability experiment, sample space, event,* and *outcome.*

Probability Experiment:
Roll a six-sided die.

Sample Space:
{1, 2, 3, 4, 5, 6}

Event:
Roll an even number,
{2, 4, 6}

Outcome:
Roll a 2, {2}

EXAMPLE 1

Identifying the Sample Space of a Probability Experiment

A probability experiment consists of tossing a coin and then rolling a six-sided die. Describe the sample space.

SOLUTION

There are two possible outcomes when tossing a coin, a head (H) or a tail (T). For each of these, there are six possible outcomes when rolling a die: 1, 2, 3, 4, 5, or 6. One way to list outcomes for actions occurring in a sequence is to use a **tree diagram.**

Tree Diagram for Coin and Die Experiment

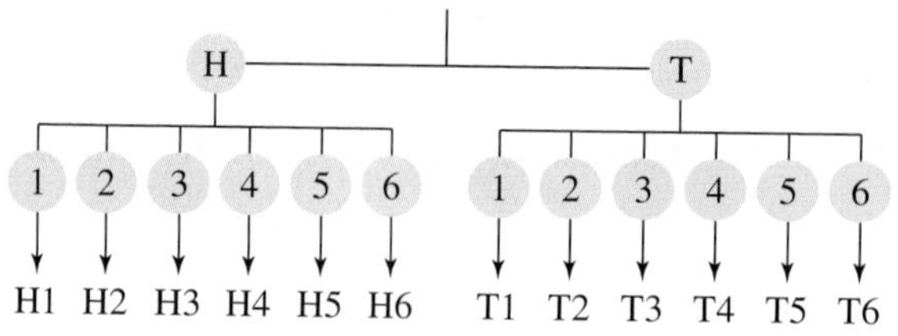

From the tree diagram, the sample space has 12 outcomes.

{H1, H2, H3, H4, H5, H6, T1, T2, T3, T4, T5, T6}

Note to Instructor

Bring dice, a deck of cards, and some coins to class to illustrate examples. Differently colored gaming chips work well also.

Try It Yourself 1

For each probability experiment, identify the sample space.

1. A probability experiment consists of recording a response to the survey statement at the left *and* the gender of the respondent.
2. A probability experiment consists of recording a response to the survey statement at the left *and* the political party (Democrat, Republican, or Other) of the respondent.

a. Start a tree diagram by forming a branch for each possible response to the survey.
b. At the end of each survey response branch, draw a new branch for each possible outcome.
c. Find the *number of outcomes* in the sample space.
d. List the *sample space.*

Answer: Page A34

In the rest of this chapter, you will learn how to calculate the probability or likelihood of an event. Events are often represented by uppercase letters, such as A, B, and C. An event that consists of a single outcome is called a **simple event.** In Example 1, the event "tossing heads and rolling a 3" is a simple event and can be represented as $A = \{H3\}$. In contrast, the event "tossing heads and rolling an even number" is not simple because it consists of three possible outcomes $\{H2, H4, H6\}$.

EXAMPLE 2

Identifying Simple Events

Decide whether each event is simple or not. Explain your reasoning.

1. For quality control, you randomly select a computer chip from a batch that has been manufactured that day. Event A is selecting a specific defective chip.
2. You roll a six-sided die. Event B is rolling at least a 4.

SOLUTION

1. Event A has only one outcome: choosing the specific defective chip. So, the event is a simple event.
2. B has three outcomes: rolling a 4, a 5, or a 6. Because the event has more than one outcome, it is not simple.

Try It Yourself 2

You ask for a student's age at his or her last birthday. Decide whether each event is simple or not.

1. Event C: The student's age is between 18 and 23, inclusive.
2. Event D: The student's age is 20.

a. Decide how many outcomes are in the event.
b. State whether the event is *simple* or not.

Answer: Page A34

Note to Instructor

Ask students to use a tree diagram to develop the sample space that results from rolling two six-sided dice. Emphasize that the outcome $\{3, 2\}$ is different from the outcome $\{2, 3\}$ for a statistician, but not for a player. Point out that the event $\{3, 2\}$ is a simple event. The event "roll a sum of 5" is not simple because it consists of the four outcomes $\{1, 4\}$, $\{2, 3\}$, $\{3, 2\}$, and $\{4, 1\}$.

Types of Probability

The method you will use to calculate a probability depends on the type of probability. There are three types of probability: classical probability, empirical probability, and subjective probability. The probability that event E will occur is written as $P(E)$ and is read "the probability of event E."

Study Tip

Probabilities can be written as fractions, decimals, or percents. In Example 3, the probabilities are written as fractions and decimals, rounded when necessary to three places. This *round-off rule* will be used throughout the text.

DEFINITION

Classical (or **theoretical**) **probability** is used when each outcome in a sample space is equally likely to occur. The classical probability for an event E is given by

$$P(E) = \frac{\text{Number of outcomes in } E}{\text{Total number of outcomes in sample space}}.$$

EXAMPLE 3

Finding Classical Probabilities

You roll a six-sided die. Find the probability of each event.

1. Event A: rolling a 3
2. Event B: rolling a 7
3. Event C: rolling a number less than 5

SOLUTION When a six-sided die is rolled, the sample space consists of six outcomes: $\{1, 2, 3, 4, 5, 6\}$.

1. There is one outcome in event $A = \{3\}$. So,

$$P(\text{rolling a 3}) = \frac{1}{6} \approx 0.167.$$

2. Because 7 is not in the sample space, there are no outcomes in event B. So,

$$P(\text{rolling a 7}) = \frac{0}{6} = 0.$$

3. There are four outcomes in event $C = \{1, 2, 3, 4\}$. So,

$$P(\text{rolling a number less than 5}) = \frac{4}{6} = \frac{2}{3} \approx 0.667.$$

Standard Deck of Playing Cards

Hearts	Diamonds	Spades	Clubs
A ♥	A ♦	A ♠	A ♣
K ♥	K ♦	K ♠	K ♣
Q ♥	Q ♦	Q ♠	Q ♣
J ♥	J ♦	J ♠	J ♣
10 ♥	10 ♦	10 ♠	10 ♣
9 ♥	9 ♦	9 ♠	9 ♣
8 ♥	8 ♦	8 ♠	8 ♣
7 ♥	7 ♦	7 ♠	7 ♣
6 ♥	6 ♦	6 ♠	6 ♣
5 ♥	5 ♦	5 ♠	5 ♣
4 ♥	4 ♦	4 ♠	4 ♣
3 ♥	3 ♦	3 ♠	3 ♣
2 ♥	2 ♦	2 ♠	2 ♣

Try It Yourself 3

You select a card from a standard deck. Find the probability of each event.

1. Event D: Selecting a seven of diamonds
2. Event E: Selecting a diamond
3. Event F: Selecting a diamond, heart, club, or spade

a. Identify the *total number of outcomes* in the sample space.
b. Find the *number of outcomes* in the event.
c. Use the *classical probability formula*.

Answer: Page A34

When an experiment is repeated many times, regular patterns are formed. These patterns make it possible to find empirical probability. Empirical probability can be used even if each outcome of an event is not equally likely to occur.

DEFINITION

Empirical (or **statistical**) **probability** is based on observations obtained from probability experiments. The empirical probability of an event E is the relative frequency of event E.

$$P(E) = \frac{\text{Frequency of event } E}{\text{Total frequency}}$$

$$= \frac{f}{n}.$$

EXAMPLE 4

Finding Empirical Probabilities

A pond contains three types of fish: bluegills, redgills, and crappies. Each fish in the pond is equally likely to be caught. You catch 120 fish and record the type. Each time, you release the fish back into the pond. The frequency distribution shows your results. If you catch another fish, what is the probability that it will be a bluegill?

Fish type	Number of times caught, f
Bluegill	39
Redgill	51
Crappy	30
	$\Sigma f = 120$

SOLUTION The event is "catching a bluegill." In your experiment, the frequency of this event is 39. Because the total of the frequencies is 120, the empirical probability of catching a bluegill is

$$P(\text{bluegill}) = \frac{39}{120}$$

$$= 0.325.$$

Try It Yourself 4

An insurance company determines that in every 100 claims, 4 are fraudulent. What is the probability that the next claim the company processes will be fraudulent?

a. *Identify* the event. Find the *frequency* of the event.
b. *Find the total frequency for the experiment.*
c. Find the *relative frequency* of the event.

Answer: Page A34

Picturing the World

It seems as if no matter how strange an event is, somebody wants to know the probability that it will occur. The following table lists the probability that some intriguing events will happen. *(Source: What Are the Chances?)*

Event	Probability
Appearing on *The Tonight Show*	1 in 490,000
Being a victim of serious crime	5%
Writing a best-selling novel	0.00205
Congress will override a veto	4%
Earning a Ph.D.	0.008

Which of these events is most likely to occur? Least likely?

Note to Instructor

Ask each student in your class to toss a coin 10 times and count the number of heads. Tally the results for the entire class. How close is the total number of heads to 50% of the total number of tosses? Emphasize that variation in experimental results is expected.

As you increase the number of times a probability experiment is repeated, the empirical probability (relative frequency) of an event approaches the theoretical probability of the event. This is known as the **law of large numbers.**

Law of Large Numbers

As an experiment is repeated over and over, the empirical probability of an event approaches the theoretical (actual) probability of the event.

Probability of Tossing a Head

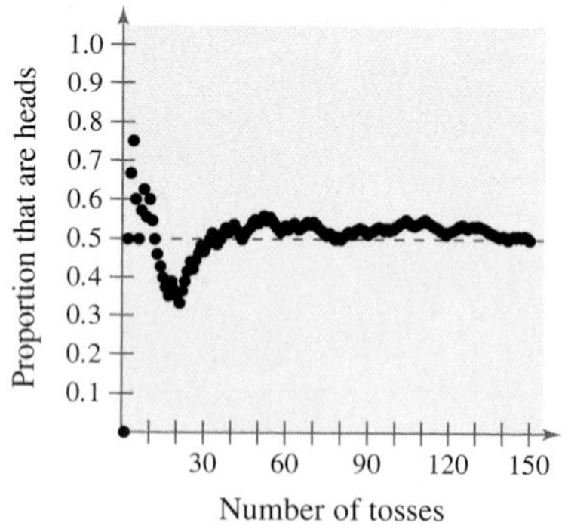

As an example of this law, suppose you want to determine the probability of tossing a head with a fair coin. If you toss the coin 10 times and get only 3 heads, you obtain an empirical probability of $\frac{3}{10}$. Because you tossed the coin only a few times, your empirical probability is not representative of the theoretical probability, which is $\frac{1}{2}$. If, however, you toss the coin several thousand times, then the law of large numbers tells you that the empirical probability will be very close to the theoretical or actual probability.

The scatter plot at the left shows the results of simulating a coin toss 150 times. Notice that, as the number of tosses increases, the probability of tossing a head gets closer and closer to the theoretical probability of 0.5.

EXAMPLE 5

Using Frequency Distributions to Find Probabilities

Employee ages	Frequency, f
15 to 24	54
25 to 34	366
35 to 44	233
45 to 54	180
55 to 64	125
65 and over	42
	$\Sigma f = 1000$

You survey a sample of 1000 employees at a company and record the age of each. The results are shown at the left in the frequency distribution. If you randomly select another employee, what is the probability that the employee will be between 25 and 34 years old?

SOLUTION The event is selecting an employee who is between 25 and 34 years old. In your survey, the frequency of this event is 366. Because the total of the frequencies is 1000, the probability of selecting an employee between the ages of 25 and 34 years old is

$$P(\text{age 25 to 34}) = \frac{366}{1000} = 0.366.$$

Try It Yourself 5

Find the probability that an employee chosen at random will be between 15 and 24 years old.

a. Find the *frequency* of the event.
b. Find the *total of the frequencies.*
c. Find the *relative frequency* of the event.

Answer: Page A34

Note to Instructor

If your students have a technology tool available in class, ask them to perform a simulation of tossing a coin 1000 times. Use 0 for heads and 1 for tails. Discuss the results.

The third type of probability is **subjective probability.** Subjective probabilities result from intuition, educated guesses, and estimates. For instance, given a patient's health and extent of injuries, a doctor may feel that the patient has a 90% chance of a full recovery. Or a business analyst may predict that the chance of the employees of a certain company going on strike is 0.25.

EXAMPLE 6

Classifying Types of Probability

Classify each statement as an example of classical probability, empirical probability, or subjective probability. Explain your reasoning.

1. The probability that you will be married by age 30 is 0.5.
2. The probability that a voter chosen at random will vote Republican is 0.45.
3. The probability of winning a 1000-ticket raffle with one ticket is $\frac{1}{1000}$.

SOLUTION

1. This probability is most likely based on an educated guess. It is an example of subjective probability.
2. This statement is most likely based on a survey of a sample of voters, so it is an example of empirical probability.
3. Because you know the number of outcomes and each is equally likely, this is an example of classical probability.

Try It Yourself 6

Based on previous counts, the probability of a salmon successfully passing through a dam on the Columbia River is 0.85. Is this statement an example of classical probability, empirical probability, or subjective probability? *(Source: Army Corps of Engineers)*

a. Identify the *event.*
b. Decide whether the probability is *determined* by knowing all possible outcomes, whether the probability is *estimated* from the results of an experiment, or whether the probability is an *educated guess.*
c. Make a *conclusion.*

Answer: Page A34

A probability cannot be negative or greater than 1. So, the probability of an event E is between 0 and 1, inclusive, as stated in the following rule.

Range of Probabilities Rule

The probability of an event E is between 0 and 1, inclusive. That is,

$$0 \le P(E) \le 1.$$

If the probability of an event is 1, the event is certain to occur. If the probability of an event is 0, the event is impossible. A probability of 0.5 indicates that an event has an even chance of occurring.

The following graph shows the possible range of probabilities and their meanings.

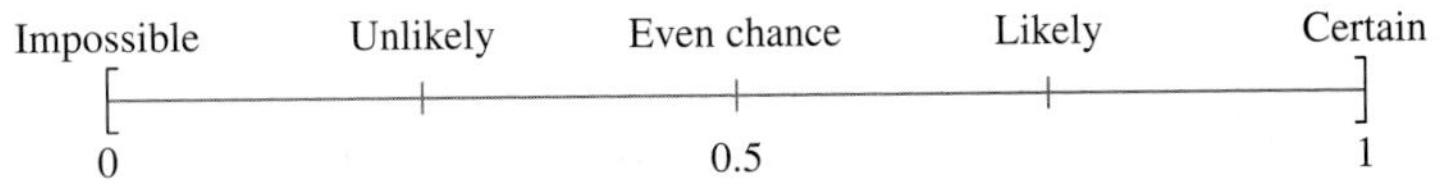

An event that occurs with a probability of 0.05 or less is typically considered unusual. Unusual events are highly unlikely to occur.

Complementary Events

The sum of the probabilities of all outcomes in a sample space is 1 or 100%. An important result of this fact is that if you know the probability of an event E, you can find the probability of the *complement of event* E.

DEFINITION

The **complement of event** $\boldsymbol{E}$ is the set of all outcomes in a sample space that are not included in event E. The complement of event E is denoted by E' and is read as "E prime."

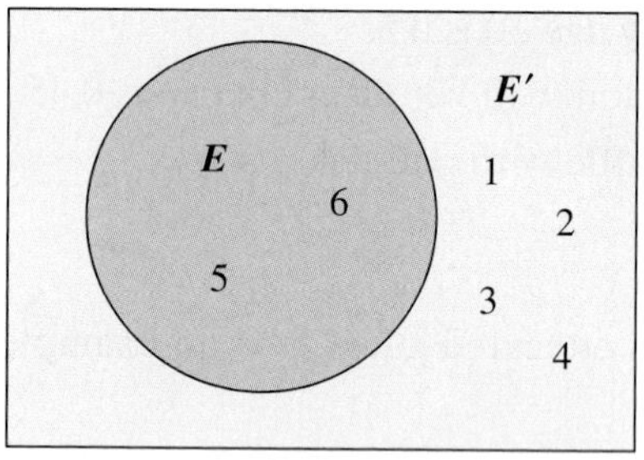

The area of the rectangle represents the total probability of the sample space ($1 = 100\%$). The area of the circle represents the probability of event E, and the area outside the circle represents the probability of the complement of event E.

For instance, if you roll a die and let E be the event "the number is at least 5," then the complement of E is the event "the number is less than 5." In symbols, $E = \{5, 6\}$ and $E' = \{1, 2, 3, 4\}$.

Using the definition of the complement of an event and the fact that the sum of the probabilities of all outcomes is 1, you can determine the following formulas:

$$P(E) + P(E') = 1 \qquad P(E) = 1 - P(E') \qquad P(E') = 1 - P(E)$$

The Venn diagram illustrates the relationship between the sample space, an event E, and its complement E'.

EXAMPLE 7

Finding the Probability of the Complement of an Event

Use the frequency distribution in Example 5 to find the probability of randomly choosing an employee who is not between 25 and 34 years old.

SOLUTION From Example 5, you know that

$$P(\text{age 25 to 34}) = \frac{366}{1000} = 0.366.$$

So, the probability that an employee is not between 25 and 34 years old is

$$P(\text{age is not 25 to 34}) = 1 - \frac{366}{1000} = \frac{634}{1000} = 0.634.$$

Try It Yourself 7

Use the frequency distribution in Example 4 to find the probability that a fish that is caught is *not* a redgill.

a. *Find the probability* that the fish is a redgill.
b. *Subtract* the resulting probability from 1.
c. State the probability as a fraction and as a decimal.

Answer: Page A34

Note to Instructor

Discuss other events (and complements) from the roll of a die. Ask students to calculate probabilities of the event and of the complement to see the relationship.

Event A: Roll a 4.

Event B: Roll an odd number.

Event C: Roll a 7.

Event D: Roll a 2 or greater.

3.1 Exercises

Help

MyMathLab

Student Study Pack

Building Basic Skills and Vocabulary

1. Determine which of the following numbers could represent the probability of an event. Explain your reasoning.

(a) 0 (b) 1.5 (c) -1 (d) 50% (e) $\frac{2}{3}$

2. Explain why the following statement is incorrect:

The probability of rain tomorrow is 120%.

Identifying a Sample Space In Exercises 3–6, identify the sample space of the probability experiment. Draw a tree diagram if it is appropriate.

3. Guessing the last digit in a telephone number

4. Tossing three coins

5. Determining a person's blood type (A, B, AB, O) and Rh-factor (positive, negative)

6. Rolling two dice

Recognizing Simple Events In Exercises 7–10, decide whether the event is a simple event or not. Explain your reasoning.

7. A computer is used to select randomly a number between 1 and 2000. Event A is selecting 359.

8. A computer is used to select randomly a number between 1 and 2000. Event B is selecting a number less than 200.

9. You randomly select one card from a standard deck. Event A is selecting a king.

10. You randomly select one card from a standard deck. Event B is selecting a four of hearts.

Classifying Types of Probability In Exercises 11 and 12, classify the statement as an example of classical probability, empirical probability, or subjective probability. Explain your reasoning.

11. According to company records, the probability that a washing machine will need repairs during a six-year period is 0.10.

12. The probability of choosing six numbers from 1 to 40 that match the six numbers drawn by a state lottery is $1/3{,}838{,}380 \approx 0.00000026$.

Finding Probabilities In Exercises 13–16, consider a company that selects employees for random drug tests. The company uses a computer to select randomly employee numbers that range from 1 to 6296.

13. Find the probability of selecting a number less than 1000.

14. Find the probability of selecting a number greater than 1000.

1. (a) Yes. The probability of an event occurring must be contained in the interval $[0, 1]$ or $[0\%, 100\%]$.
(b) No. The probability of an event occurring cannot be greater than 1.
(c) No. The probability of an event occurring cannot be less than 0.
(d) Yes. The probability of an event occurring must be contained in the interval $[0, 1]$ or $[0\%, 100\%]$.
(e) Yes. The probability of an event occurring must be contained in the interval $[0, 1]$ or $[0\%, 100\%]$.

2. The probability of an event cannot exceed 100%.

3. $\{0, 1, 2, 3, 4, 5, 6, 7, 8, 9\}$

4. See Selected Answers, page A93.

5. See Odd Answers, page A58.

6. See Selected Answers, page A93.

7. Simple event because it is an event that consists of a single outcome.

8. Not a simple event because it is an event that consists of more than a single outcome.

9. Not a simple event because it is an event that consists of more than a single outcome.

10. Simple event because it is an event that consists of a single outcome.

11. See Odd Answers, page A58.

12. See Selected Answers, page A93.

13. 0.159 **14.** 0.841

15. 0.000953
16. 0.999
17. 0.474
18. 0.541
19. 0.056
20. 0.219
21. 0.963
22. 0.842

15. Find the probability of selecting a number divisible by 1000.

16. Find the probability of selecting a number that is not divisible by 1000.

Using and Interpreting Concepts

Graphical Analysis In Exercises 17 and 18, use the diagram to answer the question.

17. What is the probability that a registered voter in Michigan voted in the 2002 gubernatorial election? *(Source: Michigan Department of State)*

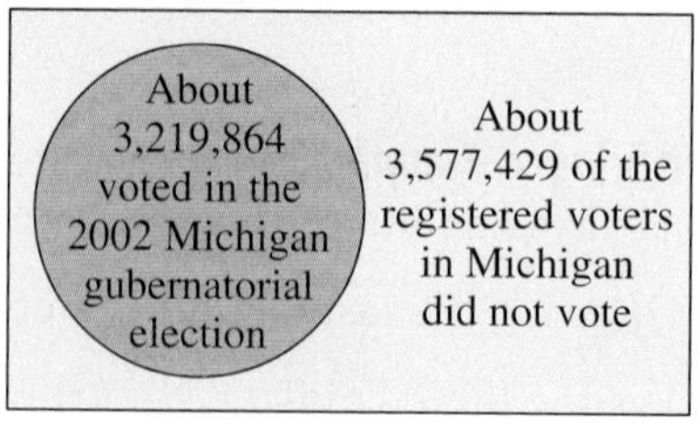

18. What is the probability that a voter chosen at random did not vote for a Democratic representative in the 2002 election? *(Source: Federal Election Commission)*

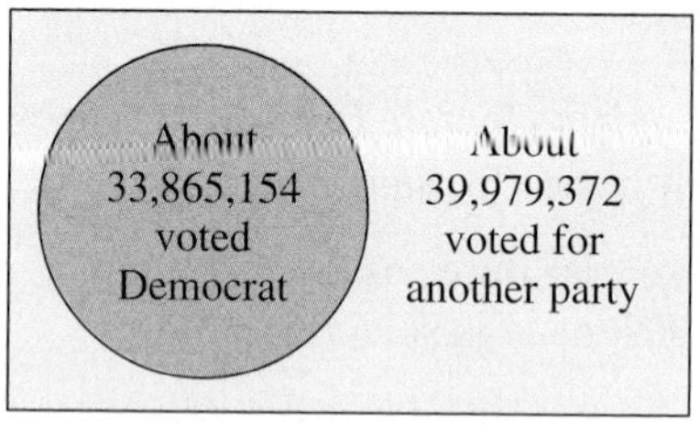

Using a Frequency Distribution to Find Probabilities In Exercises 19–22, use the frequency distribution, which shows the number of American voters (in millions) according to age. *(Source: U.S. Bureau of the Census)*

Ages of voters	Frequency (in millions)
18 to 20 years old	4.8
21 to 24 years old	7.3
25 to 34 years old	20.4
35 to 44 years old	28.4
45 to 64 years old	43.7
65 years old and over	24.9

Find the probability that a voter chosen at random is

19. between 21 and 24 years old.

20. between 35 and 44 years old.

21. not between 18 and 20 years old.

22. not between 25 and 34 years old.

23. 0.090
24. 0.202
25. 0.236
26. 0.371
27. (a) 0.5
 (b) 0.25
 (c) 0.25
28. 0.5

Using a Bar Graph to Find Probabilities In Exercises 23–26, use the following bar graph, which shows the highest level of education received by employees of a company.

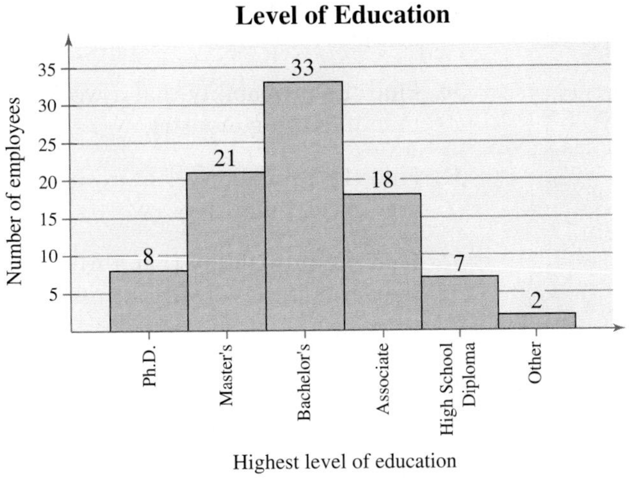

Find the probability that an employee chosen at random has

23. a Ph.D.

24. an Associate degree.

25. a Master's degree.

26. a Bachelor's degree.

27. ***Genetics*** When two pink snapdragon flowers (RW) are crossed, there are four equally likely possible outcomes for the genetic makeup of the offspring: red (RR), pink (RW), pink (WR), and white (WW). If two pink snapdragons are crossed, what is the probability that the offspring will be (a) pink, (b) red, and (c) white?

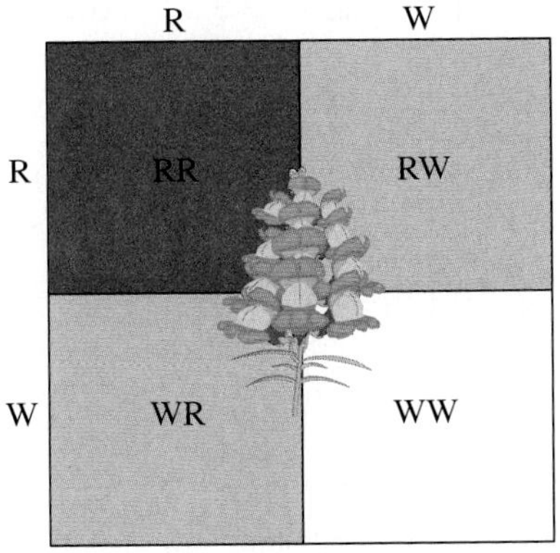

Parents

Ssmm and SsMm

	SM	Sm
Sm	SSMm	SSmm
Sm	SSMm	SSmm
sm	SsMm	Ssmm
sm	SsMm	Ssmm

	sM	sm
Sm	SsMm	Ssmm
Sm	SsMm	Ssmm
sm	ssMm	ssmm
sm	ssMm	ssmm

28. ***Genetics*** There are six basic types of coloring in registered collies: sable (SSmm), tricolor (ssmm), trifactored sable (Ssmm), blue merle (ssMm), sable merle (SSMm), and trifactored sable merle (SsMm). The *Punnett square* at the left shows the possible coloring of the offspring of a trifactored sable merle collie and a trifactored sable collie. What is the probability that the offspring will have the same coloring as one of its parents?

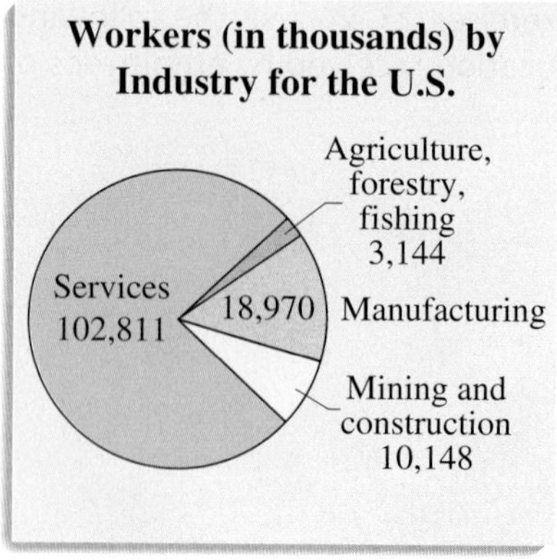

Using a Pie Chart to Find Probabilities In Exercises 29–32, use the pie chart at the left, which shows the number of workers (in thousands) by industry for the United States. *(Source: U.S. Department of Labor)*

29. Find the probability that a worker chosen at random was employed in the services industry.

30. Find the probability that a worker chosen at random was employed in the manufacturing industry.

31. Find the probability that a worker chosen at random was not employed in the services industry.

32. Find the probability that a worker chosen at random was not employed in the agriculture, forestry, or fishing industry.

Writing In Exercises 33 and 34, write a statement that represents the complement of the given probability.

33. The probability of randomly choosing a tea drinker who has a college degree (Assume that you are choosing from the population of all tea drinkers.)

34. The probability of randomly choosing a smoker whose mother also smoked (Assume that you are choosing from the population of all smokers.)

Extending Concepts

DATA **35.** ***Back to Akhiok*** A stem-and-leaf plot for the ages of all 80 residents of Akhiok, Alaska, is shown. If a resident is selected at random, find the probability he or she will be (a) at least 21 years old, (b) between 40 and 50 years old inclusive, and (c) older than 69.

0	0 0 1 1 1 2 3 3 3 4 5 5 5 7 9	Key: 1|0 = 10
1	0 2 2 3 3 3 3 3 5 6 6 7 7 8 8 8 9 9 9	
2	0 0 1 2 3 3 4 4 5 5 6 6 6 9	
3	3 6 7 9 9 9 9	
4	0 0 1 1 1 2 4 4 5 7 8 9 9 9	
5	1 3 6 8 8 9	
6	0 7 8 8	
7	2	

36. ***Individual Stock Price*** An individual stock is selected at random from the portfolio represented by the box-and-whisker plot shown. Find the probability that the stock price is (a) less than $21, (b) between $21 and $50, and (c) $30 or more.

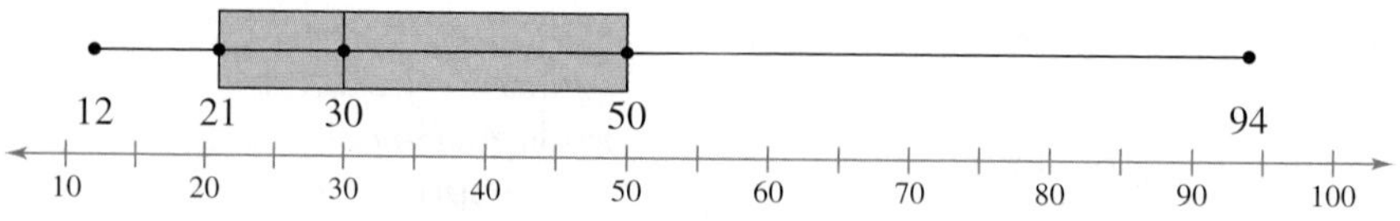

29. 0.761
30. 0.140
31. 0.239
32. 0.977
33. The probability of choosing a tea drinker who does not have a college degree.
34. The probability of choosing a smoker whose mother did not smoke.
35. (a) 0.55
(b) 0.175
(c) 0.013
36. (a) 0.25
(b) 0.50
(c) 0.50

37. *Rolling a Pair of Dice* You roll a pair of dice and record the sum.

(a) List all of the possible sums and determine the probability of rolling each sum.

(b) Use a technology tool to simulate rolling a pair of dice and recording the sum 100 times. Make a tally of the 100 sums and use these results to list the probability of rolling each sum.

(c) Compare the probabilities in part (a) with the probabilities in part (b). Explain any similarities or differences.

38. *Wet or Dry?* You are planning a three-day trip to Seattle, Washington, in October. Use the following tree diagram to answer each question.

(a) List the sample space.

(b) List the outcome(s) of the event "It rains all three days."

(c) List the outcome(s) of the event "It rains on exactly one day."

(d) List the outcome(s) of the event "It rains on at least one day."

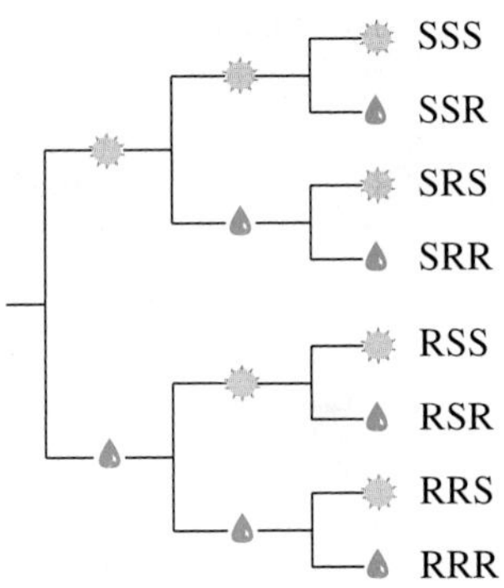

39. *Sunny and Rainy Days* You are planning a four-day trip to Seattle, Washington, in October. Make a sunny day/rainy day tree diagram (see Exercise 38) for your trip.

40. *Machine Part Suppliers* Your company buys machine parts from three different suppliers. Make a tree diagram that shows the three suppliers and whether the parts they supply are defective.

Odds In Exercises 41–44, use the following information. In gambling, the chances of winning are often written in terms of odds rather than probabilities. The **odds of winning** is the ratio of the number of successful outcomes to the number of unsuccessful outcomes. The **odds of losing** is the ratio of the number of unsuccessful outcomes to the number of successful outcomes. For example, if the number of successful outcomes is 2 and the number of unsuccessful outcomes is 3, the odds of winning are 2 : 3 (read "2 to 3") or $\frac{2}{3}$. (Note: The *probability* of success is $\frac{2}{5}$.)

41. A beverage company puts game pieces under the caps of its drinks and claims that one in six game pieces wins a prize. The official rules of the contest state that the odds of winning a prize are 1 : 6. Is the claim "one in six game pieces wins a prize" correct? Why or why not?

42. The odds of an event occurring are 4 : 5. Find (a) the probability that the event will occur and (b) the probability that the event will not occur.

43. A card is picked at random from a standard deck of 52 playing cards. Find the odds that it is a spade.

44. A card is picked at random from a standard deck of 52 playing cards. Find the odds that it is not a spade.

37. (a)

Sum	Probability
2	0.028
3	0.056
4	0.083
5	0.111
6	0.139
7	0.167
8	0.139
9	0.111
10	0.083
11	0.056
12	0.028

(b) Answers will vary.

(c) Answers will vary.

38. (a) {(SSS), (SSR), (SRS), (SRR), (RSS), (RSR), (RRS), (RRR)}

(b) {(RRR)}

(c) {(SSR), (SRS), (RSS)}

(d) {(SSR), (SRS), (SRR), (RSS), (RSR), (RRS), (RRR)}

39.

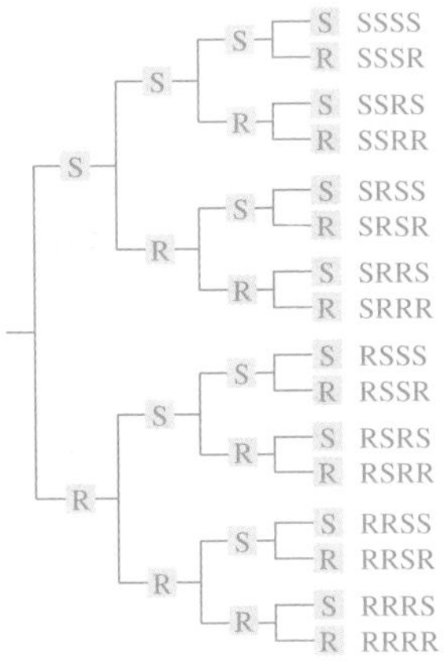

40.

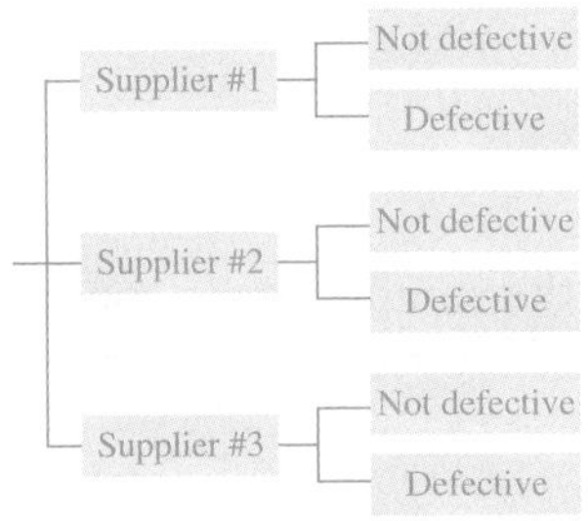

41. No. The odds of winning a prize are 1 : 6 (one winning cap and six losing caps). So, the statement should read, "one in seven game pieces wins a prize."

42. (a) 0.444

(b) 0.556

43. 13 : 39 = 1 : 3

44. 3 : 1

3.2 Conditional Probability and the Multiplication Rule

What You Should Learn

- How to find the probability of an event given that another event has occurred
- How to distinguish between independent and dependent events
- How to use the Multiplication Rule to find the probability of two events occurring in sequence
- How to use the Multiplication Rule to find conditional probabilities

Conditional Probability • Independent and Dependent Events • The Multiplication Rule

Conditional Probability

In this section, you will learn how to find the probability that two events occur in sequence. Before you can find this probability, however, you must know how to find conditional probabilities.

DEFINITION

A **conditional probability** is the probability of an event occurring, given that another event has already occurred. The conditional probability of event B occurring, given that event A has occurred, is denoted by $P(B|A)$ and is read as "probability of B, given A."

EXAMPLE 1

Finding Conditional Probabilities

1. Two cards are selected in sequence from a standard deck. Find the probability that the second card is a queen, given that the first card is a king. (Assume that the king is not replaced.)
2. The table at the left shows the results of a study in which researchers examined a child's IQ and the presence of a specific gene in the child. Find the probability that a child has a high IQ, given that the child has the gene. *(Source: Psychological Science)*

	Gene present	Gene not present	Total
High IQ	33	19	52
Normal IQ	39	11	50
Total	72	30	102

SOLUTION

1. Because the first card is a king and is not replaced, the remaining deck has 51 cards, 4 of which are queens. So,

$$P(B|A) = \frac{4}{51} \approx 0.078.$$

2. There are 72 children who have the gene. So, the sample space consists of these 72 children, as shown at the left. Of these, 33 have a high IQ. So,

$$P(B|A) = \frac{33}{72} \approx 0.458.$$

Sample Space

	Gene present
High IQ	33
Normal IQ	39
Total	72

Try It Yourself 1

1. Find the probability that a child does not have the gene.
2. Find the probability that a child does not have the gene, given that the child has a normal IQ.

a. Find the *number of outcomes* in the event and in the sample space.
b. *Divide* the number of outcomes in the event by the number of outcomes in the sample space.

Answer: Page A35

Picturing the World

Truman Collins, a probability and statistics enthusiast, wrote a program that finds the probability of landing on each square of a Monopoly board during a game. Collins explored various scenarios, including the effects of the Chance and Community Chest cards and the various ways of landing in or getting out of jail. Interestingly, Collins discovered that the length of each jail term affects the probabilities.

Monopoly square	Probability given short jail term	Probability given long jail term
Go	0.0310	0.0291
Chance	0.0087	0.0082
In Jail	0.0395	0.0946
Free Parking	0.0288	0.0283
Park Place	0.0219	0.0206
B&O RR	0.0307	0.0289
Water Works	0.0281	0.0265

Why do the probabilities depend on how long you stay in jail?

Independent and Dependent Events

In some experiments, one event does not affect the probability of another. For instance, if you roll a die and flip a coin, the outcome of the roll of the die does not affect the probability of the coin landing on heads. These two events are independent. The question of the independence of two or more events is important to researchers in fields such as marketing, medicine, and psychology. You can use conditional probabilities to determine whether events are independent.

DEFINITION

Two events are **independent** if the occurrence of one of the events does not affect the probability of the occurrence of the other event. Two events A and B are independent if

$$P(B|A) = P(B) \quad \text{or if} \quad P(A|B) = P(A).$$

Events that are not independent are **dependent.**

To determine if A and B are independent, calculate $P(B)$ and $P(B|A)$. If the values are equal, the events are independent. If $P(B) \neq P(B|A)$, then A and B are dependent events.

EXAMPLE 2

Classifying Events as Independent or Dependent

Decide whether the events are independent or dependent.

1. Selecting a king from a standard deck (A), not replacing it, and then selecting a queen from the deck (B)
2. Tossing a coin and getting a head (A), and then rolling a six-sided die and obtaining a 6 (B)
3. Driving over 85 miles per hour (A), and then getting in a car accident (B)

SOLUTION

1. $P(B|A) = \frac{4}{51}$ and $P(B) = \frac{4}{52}$. The occurrence of A changes the probability of the occurrence of B, so the events are dependent.
2. $P(B|A) = \frac{1}{6}$ and $P(B) = \frac{1}{6}$. The occurrence of A does not change the probability of the occurrence of B, so the events are independent.
3. If you drive over 85 miles per hour, the chances of getting in a car accident are greatly increased, so these events are dependent.

Try It Yourself 2

Decide whether the events are independent or dependent.

1. A salmon swims successfully through a dam (A) and another salmon swims successfully through the same dam (B).
2. Exercising frequently (A) and having a low resting heart rate (B)

a. *Decide* whether the occurrence of the first event affects the probability of the second event.

b. State if the events are *independent* or *dependent*.

Answer: Page A35

Study Tip

In words, to use the Multiplication Rule,

1. find the probability the first event occurs,
2. find the probability the second event occurs given the first event has occurred, and
3. multiply these two probabilities.

The Multiplication Rule

To find the probability of two events occurring in sequence, you can use the Multiplication Rule.

The Multiplication Rule for the Probability of A and B

The probability that two events A and B will occur in sequence is

$$P(A \text{ and } B) = P(A) \cdot P(B|A).$$

If events A and B are independent, then the rule can be simplified to $P(A \text{ and } B) = P(A) \cdot P(B)$. This simplified rule can be extended for any number of independent events.

EXAMPLE 3

Using the Multiplication Rule to Find Probabilities

1. Two cards are selected, without replacing the first card, from a standard deck. Find the probability of selecting a king and then selecting a queen.
2. A coin is tossed and a die is rolled. Find the probability of getting a head and then rolling a 6.

SOLUTION

1. Because the first card is not replaced, the events are dependent.

$$\begin{aligned} P(K \text{ and } Q) &= P(K) \cdot P(Q|K) \\ &= \frac{4}{52} \cdot \frac{4}{51} \\ &= \frac{16}{2652} \approx 0.006 \end{aligned}$$

So, the probability of selecting a king and then a queen is about 0.006.

2. The events are independent.

$$\begin{aligned} P(H \text{ and } 6) &= P(H) \cdot P(6) \\ &= \frac{1}{2} \cdot \frac{1}{6} \\ &= \frac{1}{12} \approx 0.083 \end{aligned}$$

So, the probability of tossing a head and then rolling a 6 is about 0.083.

Try It Yourself 3

1. The probability that a salmon swims successfully through a dam is 0.85. Find the probability that two salmon successfully swim through the dam.
2. Two cards are selected from a standard deck without replacement. Find the probability that they are both hearts.

a. Decide if the events are *independent* or *dependent*.
b. Use the *Multiplication Rule* to find the probability.

Answer: Page A35

EXAMPLE 4

Using the Multiplication Rule to Find Probabilities

1. A coin is tossed and a die is rolled. Find the probability of getting a head and then rolling a 2.
2. The probability that a salmon swims successfully through a dam is 0.85. Find the probability that three salmon swim successfully through the dam.
3. Find the probability that none of the three salmon is successful.
4. Find the probability that at least one of the three salmon is successful in swimming through the dam.

SOLUTION

1. $P(H) = \frac{1}{2}$. Whether or not the coin is a head, $P(2) = \frac{1}{6}$. The events are independent.

$$P(H \text{ and } 2) = P(H) \cdot P(2) = \frac{1}{2} \cdot \frac{1}{6} = \frac{1}{12} \approx 0.083$$

So, the probability of tossing a head and then rolling a 2 is about 0.083.

2. The probability that each salmon is successful is 0.85. One salmon's chance of success is independent of the others' chances.

$$P(3 \text{ salmon are successful}) = (0.85)(0.85)(0.85) \approx 0.614$$

So, the probability that all three are successful is about 0.614.

3. Because the probability of success for one salmon is 0.85, the probability of failure for one salmon is $1 - 0.85 = 0.15$.

$$P(\text{none of the three is successful}) = (0.15)(0.15)(0.15) \approx 0.003$$

So, the probability that none are successful is about 0.003.

4. The phrase "at least one" means one or more. The complement to the event "at least one is successful" is the event "none are successful." Using the rule of complements,

$$\begin{aligned} P(\text{at least 1 is successful}) &= 1 - P(\text{none are successful}) \\ &\approx 1 - 0.003 \\ &= 0.997. \end{aligned}$$

There is about a 0.997 probability that at least one of the three salmon is successful.

Insight

In Example 4, you were asked to find a probability using the phrase "at least one." Notice that it was easier to find the probability of its complement, "none," and then use the Complement Rule.

Try It Yourself 4

The probability of a salmon successfully swimming through a dam has increased to 0.90.

1. Find the probability that three salmon swim successfully through the dam.
2. Find the probability that at least one of three salmon swims successfully through the dam.

a. Determine whether to find the probability of the event or its complement.
b. Use the *Multiplication Rule* to find the probability. If necessary, use the *Complement Rule*.

Answer: Page A35

3.2 Exercises

Help

MyMathLab

Student Study Pack

Building Basic Skills and Vocabulary

1. What is the difference between independent and dependent events?

2. List examples of the following types of events.

(a) Two events that are independent

(b) Two events that are dependent

True or False? In Exercises 3 and 4, determine whether the statement is true or false. If it is false, rewrite it so that it is a true statement.

3. If two events are not independent, $P(A|B) = P(B)$.

4. If events A and B are dependent, then $P(A \text{ and } B) = P(A) \cdot P(B)$.

Classifying Events In Exercises 5–8, decide whether the events are independent or dependent. Explain your reasoning.

5. Selecting a king from a standard deck, replacing it, and then selecting a queen from the deck

6. Parking beside a fire hydrant on Tuesday and getting a parking ticket on Tuesday

7. A numbered ball between 1 and 40 is selected from a bin, and then a second numbered ball is selected from the remaining balls in the bin.

8. A numbered ball between 1 and 52 is selected from a bin, *replaced*, and then a second numbered ball is selected from the bin.

Using and Interpreting Concepts

9. ***BRCA Gene*** In the general population, one woman in eight will develop breast cancer. Research has shown that one woman in 250 carries a mutation of the BRCA gene. Eight out of 10 women with this mutation develop breast cancer. *(Source: Journal of National Cancer Institute)*

(a) Find the probability that a randomly selected woman will develop breast cancer given that she has a mutation of the BRCA gene.

(b) Find the probability that a randomly selected woman will carry the mutation of the BRCA gene and will develop breast cancer.

(c) Are the events of carrying this mutation and developing breast cancer independent or dependent? Explain.

Breast Cancer and the BRCA Gene

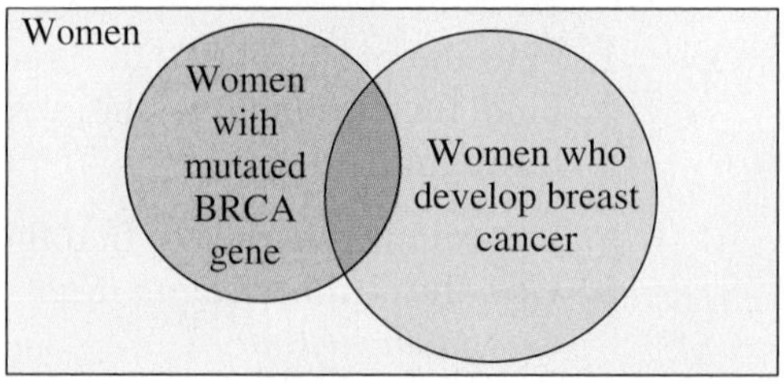

---

1. Two events are independent if the occurrence of one of the events does not affect the probability of the occurrence of the other event. If $P(B|A) = P(B)$ or $P(A|B) = P(A)$, then events A and B are independent.

2. (a) Roll a die twice. The outcome of the 2nd toss is independent of the outcome of the 1st toss.

(b) Draw two cards (without replacement) from a standard 52-card deck. The outcome of the 2nd card is dependent on the outcome of the 1st card.

3. False. If two events are independent, then $P(A|B) = P(A)$.

4. False. If events A and B are independent, then $P(A \text{ and } B) = P(A) \cdot P(B)$.

5. Independent. The outcome of the 1st card drawn does not affect the outcome of the 2nd card drawn.

6. Dependent. Parking beside a fire hydrant affects getting a parking ticket.

7. Dependent. The outcome of the 1st ball drawn affects the outcome of the 2nd ball drawn.

8. Independent. The outcome of the 1st ball drawn does not affect the outcome of the 2nd ball drawn.

9. (a) 0.8

(b) 0.003

(c) Dependent. The probability that a woman develops breast cancer depends on whether she carries the gene mutation.

10. (a) 0.222
(b) 0.067
(c) Dependent. Ford makes pickup trucks.

11. (a) 0.807
(b) 0.315
(c) Dependent. The probability that the family takes a summer vacation depends on whether they own a computer.

12. (a) 0.063
(b) 0.022
(c) Dependent. Some male students are nursing majors.

10. ***Pickup Trucks*** In a survey, 510 adults were asked if they drive a pickup truck and if they drive a Ford. The results showed that one in six adults surveyed drives a pickup truck, and three in ten adults surveyed drives a Ford. Of the adults surveyed that drive Fords, two in nine drive a pickup truck.

(a) Find the probability that a randomly selected adult drives a pickup truck given that he or she drives a Ford.

(b) Find the probability that a randomly selected adult drives a Ford and drives a pickup truck.

What Do You Drive?

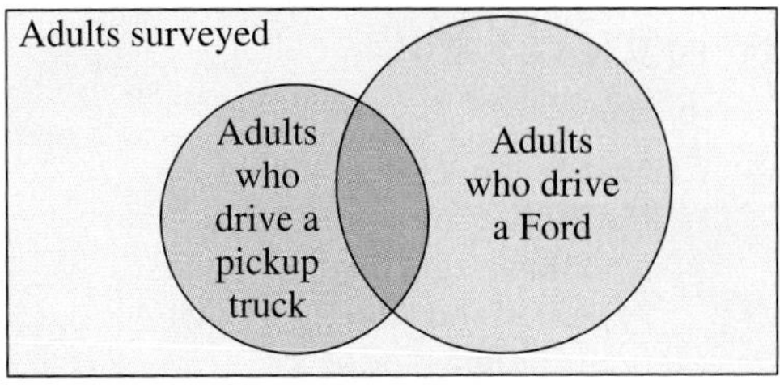

(c) Are the events driving a Ford and driving a pickup truck independent or dependent? Explain.

11. ***Summer Vacation*** The table shows the results of a survey in which 146 families were asked if they own a computer and if they will be taking a summer vacation this year.

		Summer Vacation This Year		
		Yes	No	Total
Own a Computer	Yes	46	11	57
	No	55	34	89
	Total	101	45	146

(a) Find the probability a randomly selected family is taking a summer vacation this year given that they own a computer.

(b) Find the probability a randomly selected family is taking a summer vacation this year and owns a computer.

(c) Are the events of owning a computer and taking a summer vacation this year independent or dependent events? Explain.

12. ***Nursing Majors*** The table shows the number of male and female students enrolled in nursing at the University of Oklahoma Health Sciences Center for a recent semester. *(Source: University of Oklahoma Health Center Office of Admissions and Records)*

	Nursing majors	Non-nursing majors	Total
Males	65	971	1036
Females	571	1393	1964
Total	636	2364	3000

(a) Find the probability that a randomly selected student is a nursing major given that the student is male.

(b) Find the probability that a randomly selected student is a nursing major and male.

(c) Are the events being a male student and being a nursing major independent or dependent events? Explain.

13. (a) 0.017
(b) 0.93
(c) Yes, this would be unusual because the probability of a pregnancy and a multiple birth is only 1.7%.

14. (a) 0.24
(b) 0.6
(c) No, this is not unusual because the probability of an adult who thinks race relations have improved and who thinks civil rights progress is too slow is 24%.

15. 0.291

16. 0.3

13. *Assisted Reproductive Technology* A study found that 24% of the assisted reproductive technology (ART) cycles resulted in a pregnancy. Seven percent of the ART pregnancies resulted in multiple births. *(Source: U.S. Department of Health and Human Services)*

Pregnancies

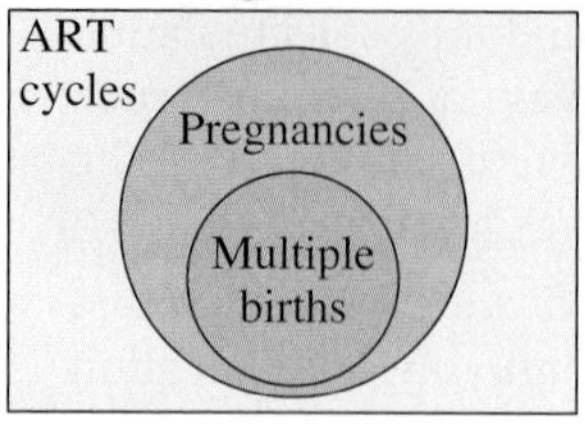

(a) Find the probability that a randomly selected ART cycle resulted in a pregnancy *and* produced a multiple birth.

(b) Find the probability that a randomly selected ART cycle that resulted in a pregnancy did *not* produce a multiple birth.

(c) Would it be unusual for a randomly selected ART cycle to result in a pregnancy and produce a multiple birth? Explain.

14. *Race Relations* In a survey, 60% of adults in the United States think race relations have improved since the death of Martin Luther King Jr. Of these 60%, 4 out of 10 said the rate of civil rights progress is too slow. *(Source: Marist Institute for Public Opinion)*

Race Relations

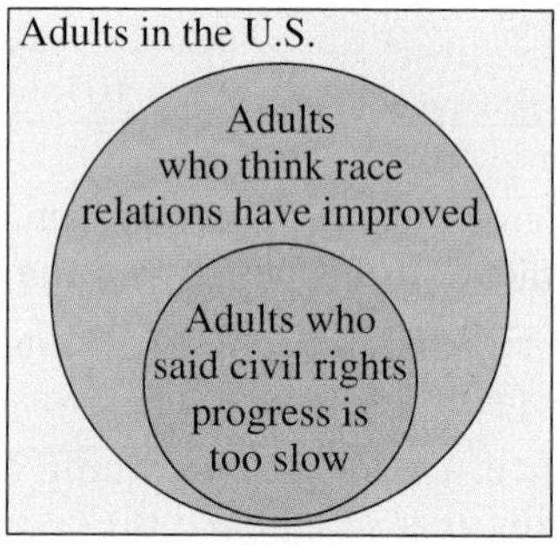

(a) Find the probability that a randomly selected adult thinks race relations have improved since the death of Martin Luther King Jr. *and* thinks the rate of civil rights progress is too slow.

(b) Given that a randomly selected adult thinks race relations have improved since the death of Martin Luther King Jr., find the probability that he or she thinks the rate of civil rights progress is *not* too slow.

(c) Would it be unusual for a randomly selected adult to think race relations have improved since the death of Martin Luther King Jr. and think the rate of civil rights progress is too slow? Explain.

15. *Computers and Internet Access* A study found that 57% of households in the United States have a computer. Of those 57%, 51% have Internet access. Find the probability that a U.S. household selected at random has Internet access. *(U.S. Census Bureau)*

16. *Surviving Surgery* A doctor gives a patient a 60% chance of surviving bypass surgery after a heart attack. If the patient survives the surgery, he has a 50% chance that the heart damage will heal. Find the probability that the patient survives surgery and the heart damage heals.

17. (a) 0.109
(b) 0.382
(c) 0.618

18. (a) 0.045
(b) 0.545
(c) 0.455

19. (a) 0.481
(b) 0.465
(c) 0.449
(d) Dependent. Whether a person has at least one month's income saved depends on whether the person is male.

20. (a) 0.556
(b) 0.525
(c) 0.167
(d) Dependent. The probability that the owner spent $100 or more on health care depends on whether the dog was a mixed breed.

17. *Defective Parts* In a box of 11 parts, four of the parts are defective. Two parts are selected at random without replacement.

(a) Find the probability that both parts are defective.
(b) Find the probability that both parts are *not* defective.
(c) Find the probability that at least one part is defective.

18. *Light Bulbs* Twelve light bulbs are tested to see if they last as long as the manufacturer claims they do. Three light bulbs fail the test. Two light bulbs are selected at random without replacement.

(a) Find the probability that both light bulbs failed the test.
(b) Find the probability that both light bulbs passed the test.
(c) Find the probability that at least one light bulb failed the test.

19. *Emergency Savings* The table shows the results of a survey in which 142 men and 145 women workers ages 25 to 64 were asked if they have at least one month's income set aside for emergencies.

	Men	Women	Total
Less than one month's income	66	83	149
One month's income or more	76	62	138
Total	142	145	287

(a) Find the probability that a randomly selected worker has one month's income or more set aside for emergencies.
(b) Given that a randomly selected worker is a male, find the probability that the worker has less than one month's income.
(c) Given that a randomly selected worker has one month's income or more, find the probability that the worker is a female.
(d) Are the events "having less than one month's income" and "being male" independent or dependent? Explain.

20. *Health Care for Dogs* The table shows the results of a survey in which 90 dog owners were asked how much they have spent in the last year for their dog's health care, and whether their dogs were purebred or mixed breeds.

		Type of Dog		
		Purebred	Mixed breed	Total
Health Care	Less than $100	19	21	40
	$100 or more	35	15	50
	Total	54	36	90

(a) Find the probability that $100 or more was spent on a randomly selected dog's health care in the last year.
(b) Given that a randomly selected dog owner spent less than $100, find the probability that the dog was a mixed breed.
(c) Find the probability that a randomly selected dog owner spent $100 or more on health care and the dog was a mixed breed.
(d) Are the events "spending $100 or more on health care" and "having a mixed breed dog" independent or dependent? Explain.

21. (a) 0.0000000243
(b) 0.859
(c) 0.141

22. (a) 0.055
(b) 0.238
(c) 0.762

23. (a) 0.2
(b) 0.04
(c) 0.008
(d) 0.512
(e) 0.488

24. (a) 0.985
(b) 0.015
(c) 0.000000125

25. 0.011

26. (a) 0.00000751
(b) 0.992

27. 0.444

21. *Blood Types* The probability that a person in the United States has type AB^+ blood is 3%. Five unrelated people in the United States are selected at random. *(Source: American Association of Blood Banks)*

(a) Find the probability that all five have type AB^+ blood.
(b) Find the probability that none of the five has type AB^+ blood.
(c) Find the probability that at least one of the five has type AB^+ blood.

22. *Blood Types* The probability that a person in the United States has type O^+ blood is 38%. Three unrelated people in the United States are selected at random. *(Source: American Association of Blood Banks)*

(a) Find the probability that all three have type O^+ blood.
(b) Find the probability that none of the three has type O^+ blood.
(c) Find the probability that at least one of the three has type O^+ blood.

23. *Guessing* A multiple-choice quiz has three questions, each with five answer choices. Only one of the choices is correct. You have no idea what the answer is to any question and have to guess each answer.

(a) Find the probability of answering the first question correctly.
(b) Find the probability of answering the first two questions correctly.
(c) Find the probability of answering all three questions correctly.
(d) Find the probability of answering none of the questions correctly.
(e) Find the probability of answering at least one of the questions correctly.

24. *Bookbinding Defects* A printing company's bookbinding machine has a probability of 0.005 of producing a defective book. This machine is used to bind three books.

(a) Find the probability that none of the books are defective.
(b) Find the probability that at least one of the books is defective.
(c) Find the probability that all of the books are defective.

25. *Warehouses* A distribution center receives shipments of a product from three different factories in the following quantities: 50, 35, and 25. Three times a product is selected at random, each time without a replacement. Find the probability that all three products came from the third factory.

26. *Birthdays* Three people are selected at random. Find the probability that (a) all three share the same birthday and (b) none of the three share the same birthday. Assume 365 days in a year.

Extending Concepts

According to **Bayes's Theorem,** the probability of event A, given that event B has occurred, is

$$P(A|B) = \frac{P(A) \cdot P(B|A)}{P(A) \cdot P(B|A) + P(A') \cdot P(B|A')}.$$

In Exercises 27–30, use Bayes's Theorem to find $P(A|B)$.

27. $P(A) = \frac{2}{3}$, $P(A') = \frac{1}{3}$, $P(B|A) = \frac{1}{5}$, and $P(B|A') = \frac{1}{2}$

28. 0.4
29. 0.167
30. 0.797
31. (a) 0.074
(b) 0.999
32. (a) 0.462
(b) 0.538
(c) Yes
(d) Answers will vary.
33. 0.954
34. 0.933

28. $P(A) = \frac{3}{8}$, $P(A') = \frac{5}{8}$, $P(B|A) = \frac{2}{3}$, and $P(B|A') = \frac{3}{5}$

29. $P(A) = 0.25$, $P(A') = 0.75$, $P(B|A) = 0.3$, and $P(B|A') = 0.5$

30. $P(A) = 0.62$, $P(A') = 0.38$, $P(B|A) = 0.41$, and $P(B|A') = 0.17$

31. ***Reliability of Testing*** A certain virus infects one in every 200 people. A test used to detect the virus in a person is positive 80% of the time if the person has the virus and 5% of the time if the person does not have the virus. (This 5% result is called a *false positive.*) Let A be the event "the person is infected" and B be the event "the person tests positive."

(a) If a person tests positive, what is the probability that the person is infected?

(b) If a person tests negative, what is the probability that the person is *not* infected?

32. ***Birthday Problem*** You are in a class that has 24 students. You want to find the probability that at least two of the students share the same birthday.

(a) First, find the probability that each student has a different birthday.

$$P(\text{different birthdays}) = \overbrace{\frac{365}{365}\cdot\frac{364}{365}\cdot\frac{363}{365}\cdot\frac{362}{365}\cdots\frac{343}{365}\cdot\frac{342}{365}}^{24 \text{ factors}}$$

(b) The probability that at least two students have the same birthday is the complement of the probability in part (a). What is this probability?

(c) We used a technology tool to generate 24 random numbers between 1 and 365. Each number represents a birthday. Did we get at least two people with the same birthday?

228	348	181	317	81	183
52	346	177	118	315	273
252	168	281	266	285	13
118	360	8	193	57	107

(d) Use a technology tool to simulate the "Birthday Problem." Repeat the simulation 10 times. How many times did you get "at least two people" with the same birthday?

The Multiplication Rule and Conditional Probability By rewriting the formula for the Multiplication Rule, you can write a formula for finding conditional probabilities. The conditional probability of event B occurring, given that event A has occurred, is

$$P(B|A) = \frac{P(A \text{ and } B)}{P(A)}.$$

In Exercises 33 and 34, use the following information.

- The probability that an airplane flight departs on time is 0.89.
- The probability that a flight arrives on time is 0.87.
- The probability that a flight departs and arrives on time is 0.83.

33. Find the probability that a flight departed on time given that it arrives on time.

34. Find the probability that a flight arrives on time given that it departed on time.

3.3 The Addition Rule

What You Should Learn

- How to determine if two events are mutually exclusive
- How to use the Addition Rule to find the probability of two events

Mutually Exclusive Events • The Addition Rule • A Summary of Probability

Mutually Exclusive Events

In Section 3.2, you learned how to find the probability of two events, A and B, occurring in sequence. Such probabilities are denoted by $P(A \text{ and } B)$. In this section, you will learn how to find the probability that at least one of two events will occur. Probabilities such as these are denoted by $P(A \text{ or } B)$ and depend on whether the events are mutually exclusive.

DEFINITION

Two events A and B are **mutually exclusive** if A and B cannot occur at the same time.

The Venn diagrams show the relationship between events that are mutually exclusive and events that are not mutually exclusive.

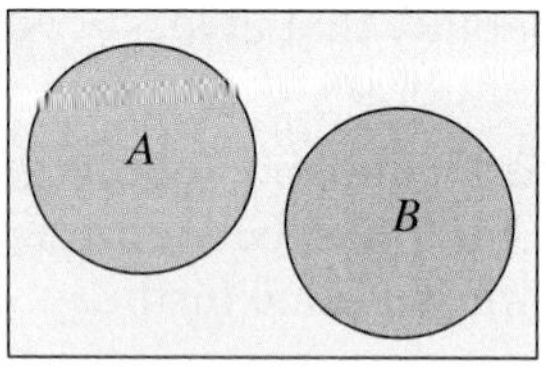

A and *B* are mutually exclusive.

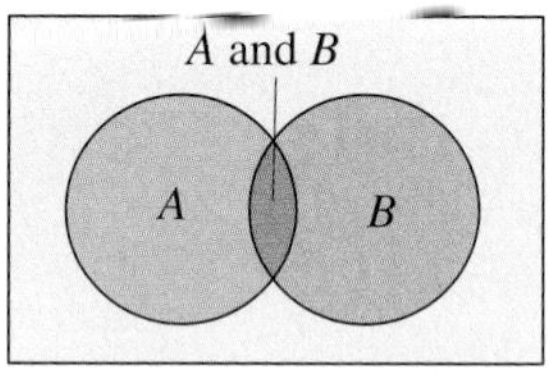

A and *B* are not mutually exclusive.

Study Tip

In probability and statistics, the word *or* is usually used as an "inclusive or" rather than an "exclusive or." For instance, there are three ways for "Event *A* or *B*" to occur.

(1) *A* occurs and *B* does not occur.

(2) *B* occurs and *A* does not occur.

(3) *A* and *B* both occur.

EXAMPLE 1

Mutually Exclusive Events

Decide if the events are mutually exclusive. Explain your reasoning.

1. Event A: Roll a 3 on a die.
   Event B: Roll a 4 on a die.
2. Event A: Randomly select a male student.
   Event B: Randomly select a nursing major.
3. Event A: Randomly select a blood donor with type O blood.
   Event B: Randomly select a female blood donor.

SOLUTION

1. The first event has one outcome, a 3. The second event also has one outcome, a 4. These outcomes cannot occur at the same time, so the events are mutually exclusive.
2. Because the student can be a male nursing major, the events are not mutually exclusive.
3. Because the donor can be a female with type O blood, the events are not mutually exclusive.

Note to Instructor

Students often confuse the concept of independent events with the concept of mutually exclusive events. Ask students whether the pairs of events in Example 1 are independent or dependent.

Try It Yourself 1

Decide if the events are mutually exclusive.

1. Event A: Randomly select a jack from a standard deck of cards.
   Event B: Randomly select a face card from a standard deck of cards.
2. Event A: Randomly select a 20-year-old student.
   Event B: Randomly select a student with blue eyes.
3. Event A: Randomly select a vehicle that is a Ford.
   Event B: Randomly select a vehicle that is a Toyota.

a. Decide if one of the following statements is true.
- Events A and B cannot occur at the same time.
- Events A and B have no outcomes in common.
- $P(A \text{ and } B) = 0$

b. Make a conclusion.

Answer: Page A35

The Addition Rule

Study Tip

By subtracting $P(A \text{ and } B)$ you avoid double counting the probability of outcomes that occur in both A and B.

The Addition Rule for the Probability of *A* or *B*

The probability that events A or B will occur $P(A \text{ or } B)$ is given by

$$P(A \text{ or } B) = P(A) + P(B) - P(A \text{ and } B).$$

If events A and B are mutually exclusive, then the rule can be simplified to $P(A \text{ or } B) = P(A) + P(B)$. This simplified rule can be extended to any number of mutually exclusive events.

EXAMPLE 2

Using the Addition Rule to Find Probabilities

1. You select a card from a standard deck. Find the probability that the card is a 4 or an ace.
2. You roll a die. Find the probability of rolling a number less than three or rolling an odd number.

Deck of 52 Cards

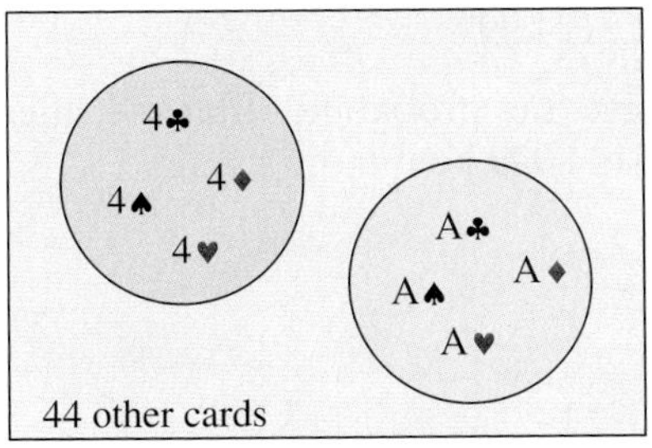

Roll a Die

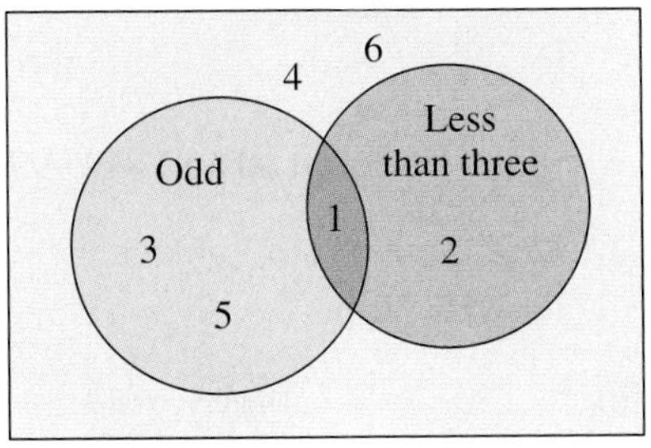

SOLUTION

1. If the card is a 4, it cannot be an ace. So, the events are mutually exclusive. The probability of selecting a 4 or an ace is

$$P(4 \text{ or ace}) = P(4) + P(\text{ace}) = \frac{4}{52} + \frac{4}{52} = \frac{8}{52} = \frac{2}{13} \approx 0.154.$$

2. The events are not mutually exclusive because 1 is an outcome of both events. So, the probability of rolling a number less than 3 or an odd number is

$$P(\text{less than 3 or odd}) = P(\text{less than 3}) + P(\text{odd}) - P(\text{less than 3 and odd})$$

$$= \frac{2}{6} + \frac{3}{6} - \frac{1}{6} = \frac{4}{6} = \frac{2}{3} \approx 0.667.$$

Try It Yourself 2

1. A die is rolled. Find the probability of rolling a 6 or an odd number.
2. A card is selected from a standard deck. Find the probability that the card is a face card or a heart.

a. Decide whether the events are *mutually exclusive.*
b. Find $P(A)$, $P(B)$, and, if necessary, $P(A \text{ and } B)$.
c. Use the *Addition Rule* to find the probability.

Answer: Page A35

EXAMPLE 3

Finding Probabilities of Mutually Exclusive Events

The frequency distribution shows the volume of sales (in dollars) and the number of months a sales representative reached each sales level during the past three years. If this sales pattern continues, what is the probability that the sales representative will sell between \$75,000 and \$124,999 next month?

Sales volume (\$)	Months
0–24,999	3
25,000–49,999	5
50,000–74,999	6
75,000–99,999	7
100,000–124,999	9
125,000–149,999	2
150,000–174,999	3
175,000–199,999	1

SOLUTION To solve this problem, define events A and B as follows.

A = monthly sales between \$75,000 and \$99,999

B = monthly sales between \$100,000 and \$124,999

Because events A and B are mutually exclusive, the probability that the sales representative will sell between \$75,000 and \$124,999 next month is

$$P(A \text{ or } B) = P(A) + P(B)$$
$$= \frac{7}{36} + \frac{9}{36}$$
$$= \frac{16}{36} = \frac{4}{9} \approx 0.444.$$

Try It Yourself 3

Find the probability that the sales representative will sell between \$0 and \$49,999.

a. *Identify* events A and B.
b. Verify that A and B are *mutually exclusive.*
c. Find the *probability* of each event.
d. Use the *Addition Rule* to find the probability.

Answer: Page A35

EXAMPLE 4

Using the Addition Rule to Find Probabilities

A blood bank catalogs the types of blood, including positive or negative Rh-factor, given by donors during the last five days. The number of donors who gave each blood type is shown in the table. A donor is selected at random.

1. Find the probability that the donor has type O or type A blood.
2. Find the probability that the donor has type B blood or is Rh-negative.

		Blood Type				
		O	A	B	AB	Total
Rh-factor	Positive	156	139	37	12	344
	Negative	28	25	8	4	65
	Total	184	164	45	16	409

SOLUTION

1. Because a donor cannot have type O blood and type A blood, these events are mutually exclusive. So, on the basis of the Addition Rule, the probability that a randomly chosen donor has type O or type A blood is

$$\begin{aligned} P(\text{type O or type A}) &= P(\text{type O}) + P(\text{type A}) \\ &= \frac{184}{409} + \frac{164}{409} \\ &= \frac{348}{409} \\ &\approx 0.851. \end{aligned}$$

2. Because a donor can have type B blood and be Rh-negative, these events are not mutually exclusive. So, on the basis of the Addition Rule, the probability that a randomly chosen donor has type B blood or is Rh-negative is

$$\begin{aligned} P(\text{type B or Rh-neg}) &= P(\text{type B}) + P(\text{Rh-neg}) - P(\text{type B and Rh-neg}) \\ &= \frac{45}{409} + \frac{65}{409} - \frac{8}{409} \\ &= \frac{102}{409} \\ &\approx 0.249. \end{aligned}$$

Try It Yourself 4

1. Find the probability that the donor has type B or type AB blood.
2. Find the probability that the donor has type O blood or is Rh-positive.

a. Decide if the events are *mutually exclusive.*
b. Use the *Addition Rule.*

Answer: Page A35

Picturing the World

In a survey conducted by the National Family Organization, new mothers were asked to rate the difficulty of delivering their first child compared with what they expected.

How Difficult Was the Delivery?

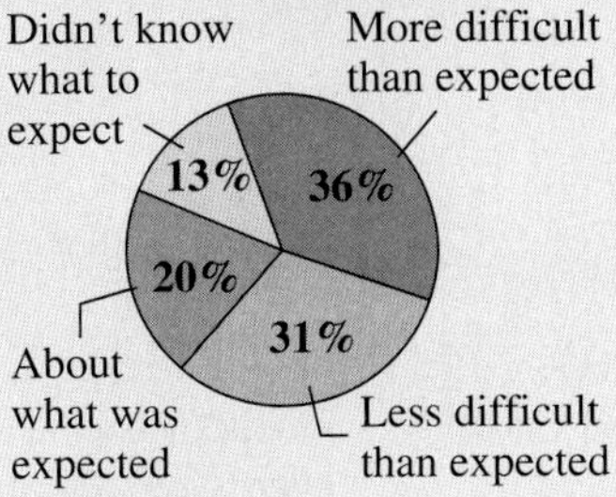

If you selected a new mother at random and asked her to compare the difficulty of her delivery with what she expected, what is the probability that she would say that it was the same or more difficult than what she expected?

A Summary of Probability

Type of Probability and Probability Rules	Summary	Formula
Classical Probability	The number of outcomes in the sample space is known and each outcome is equally likely to occur.	$P(E) = \dfrac{\text{Number of outcomes in event } E}{\text{Number of outcomes in sample space}}$
Empirical Probability	The frequency of outcomes in the sample space is estimated from experimentation.	$P(E) = \dfrac{\text{Frequency of event } E}{\text{Total frequency}} = \dfrac{f}{n}$
Range of Probabilities Rule	The probability of an event is between 0 and 1, inclusive.	$0 \le P(E) \le 1$
Complementary Events	The complement of event E is the set of all outcomes in a sample space that are not included in E, denoted by E'.	$P(E') = 1 - P(E)$
Multiplication Rule	The Multiplication Rule is used to find the probability of two events occurring in a sequence.	$P(A \text{ and } B) = P(A) \cdot P(B \mid A)$ $P(A \text{ and } B) = P(A) \cdot P(B)$ *Independent events*
Addition Rule	The Addition Rule is used to find the probability of at least one of two events occurring.	$P(A \text{ or } B) = P(A) + P(B) - P(A \text{ and } B)$ $P(A \text{ or } B) = P(A) + P(B)$ *Mutually exclusive events*

EXAMPLE 5

Combining Rules to Find Probabilities

Use the graph at the right to find the probability that a randomly selected draft pick is not a running back or a wide receiver.

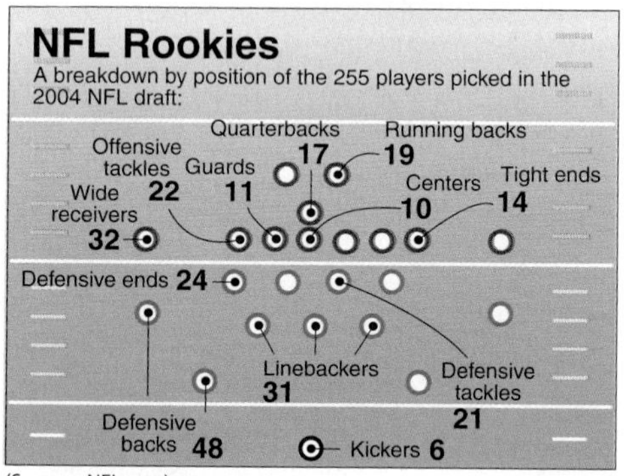

(Source: NFL.com)

SOLUTION Define events A and B.

A: Draft pick is a running back.
B: Draft pick is a wide receiver.

These events are mutually exclusive, so the probability that the draft pick is a running back or wide receiver is

$$P(A \text{ or } B) = P(A) + P(B) = \frac{19}{255} + \frac{32}{255} = \frac{51}{255} = \frac{1}{5}.$$

By taking the complement of $P(A \text{ or } B)$, you can determine that the probability of randomly selecting a draft pick who is not a running back or wide receiver is

$$1 - P(A \text{ or } B) = 1 - \frac{1}{5} = \frac{4}{5} = 0.8.$$

Try It Yourself 5

Find the probability that a randomly selected draft pick is not a linebacker or a quarterback.

a. Find the *probability* that the draft pick is a linebacker or a quarterback.
b. Find the *complement* of the event.

Answer: Page A35

3.3 Exercises

1. $P(A \text{ and } B) = 0$ because A and B cannot occur at the same time.
2. Answers will vary. Sample answers are given.
   (a) Toss coin once: $A = \{\text{head}\}$ and $B = \{\text{tail}\}$
   (b) Draw one card: $A = \{\text{ace}\}$ and $B = \{\text{spade}\}$
3. True
4. False. Two events being independent does not imply they are mutually exclusive.
5. False. The probability that event A or event B will occur is $P(A \text{ or } B) = P(A) + P(B) - P(A \text{ and } B)$.
6. True
7. Not mutually exclusive. The student can be taking both economics and statistics.
8. Mutually exclusive. A person cannot register to vote in more than one state.
9. Not mutually exclusive. The worker can be female and have a college degree.
10. Not mutually exclusive. The worker can be male and be employed part time.
11. Mutually exclusive. The person cannot be in both age classes.
12. Not mutually exclusive. A person can be between 18 and 24 years old and earn between \$20,000 and \$29,999.

Building Basic Skills and Vocabulary

1. If two events are mutually exclusive, why is $P(A \text{ and } B) = 0$?

2. List examples of

(a) two events that are mutually exclusive.

(b) two events that are not mutually exclusive.

True or False? In Exercises 3–6, determine whether the statement is true or false. If it is false, explain why.

3. If two events are mutually exclusive, they have no outcomes in common.

4. If two events are independent, then they are also mutually exclusive.

5. The probability that event A or event B will occur is

$$P(A \text{ or } B) = P(A) + P(B) - P(A \text{ or } B).$$

6. If events A and B are mutually exclusive, then

$$P(A \text{ or } B) = P(A) + P(B).$$

Graphical Analysis In Exercises 7 and 8, decide if the events shown in the Venn diagram are mutually exclusive. Explain your reasoning.

7.

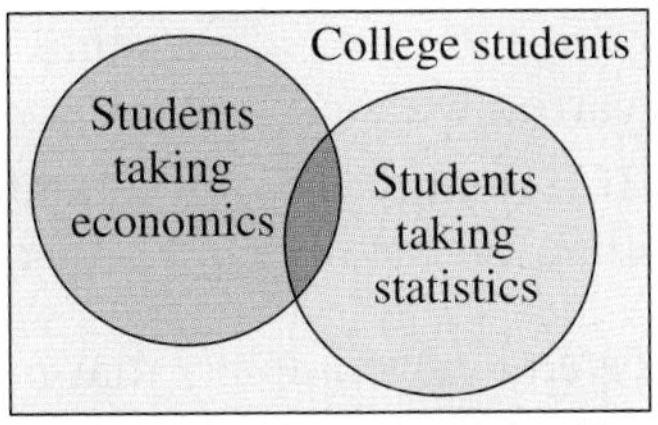

8.

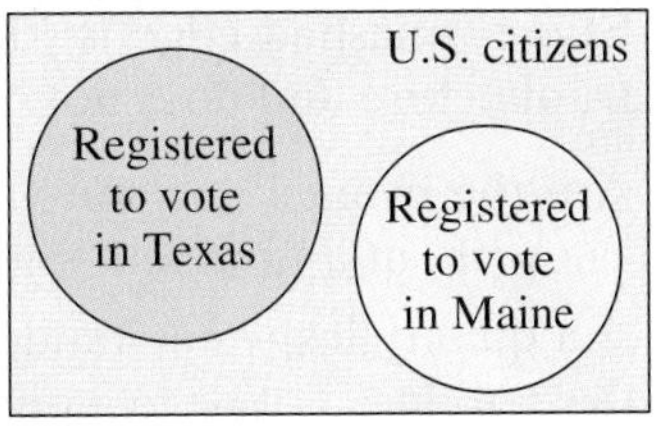

Recognizing Mutually Exclusive Events In Exercises 9–12, decide if the events are mutually exclusive. Explain your reasoning.

9. Event A: Randomly select a female worker.
Event B: Randomly select a worker with a college degree.

10. Event A: Randomly select a male worker.
Event B: Randomly select a worker employed part time.

11. Event A: Randomly select a person between 18 and 24 years old.
Event B: Randomly select a person between 25 and 34 years old.

12. Event A: Randomly select a person between 18 and 24 years old.
Event B: Randomly select a person earning between \$20,000 and \$29,999.

13. (a) Not mutually exclusive. For five weeks the events overlap.
(b) 0.423
14. (a) Not mutually exclusive. A student can be a woman and be a business major.
(b) 0.639
15. (a) Not mutually exclusive. A carton can have a puncture and a smashed corner.
(b) 0.126
16. (a) Not mutually exclusive. A can can have no punctures and no smashed edges.
(b) 0.997
17. (a) 0.308
(b) 0.538
(c) 0.308
18. (a) 0.333
(b) 0.833
(c) 0.667

Using and Interpreting Concepts

13. *Audit* During a 52-week period, a company paid overtime wages for 18 weeks and hired temporary help for 9 weeks. During 5 weeks, the company paid overtime *and* hired temporary help.

(a) Are the events "selecting a week that contained overtime wages" and "selecting a week that contained temporary help wages" mutually exclusive? Explain.

(b) If an auditor randomly examined the payroll records for only one week, what is the probability that the payroll for that week contained overtime wages or temporary help wages?

14. *Newspaper Survey* A college has an undergraduate enrollment of 3500. Of these, 860 are business majors and 1800 are women. Of the business majors, 425 are women.

(a) Are the events "selecting a woman student" and "selecting a business major" mutually exclusive? Explain.

(b) If a college newspaper conducts a poll and selects students at random to answer a survey, find the probability that a selected student is a woman or a business major.

15. *Carton Defects* A company that makes cartons finds the probability of producing a carton with a puncture is 0.05, the probability that a carton has a smashed corner is 0.08, and the probability that a carton has a puncture and has a smashed corner is 0.004.

(a) Are the events "selecting a carton with a puncture" and "selecting a carton with a smashed corner" mutually exclusive? Explain.

(b) If a quality inspector randomly selects a carton, find the probability that the carton has a puncture or has a smashed corner.

16. *Can Defects* A company that makes soda pop cans finds the probability of producing a can without a puncture is 0.96, the probability that a can does not have a smashed edge is 0.93, and the probability that a can does not have a puncture and does not have a smashed edge is 0.893.

(a) Are the events "selecting a can without a puncture" and "selecting a can without a smashed edge" mutually exclusive? Explain.

(b) If a quality inspector randomly selects a can, find the probability that the can does not have a puncture or does not have a smashed edge.

17. *Selecting a Card* A card is selected at random from a standard deck. Find each probability.

(a) Randomly selecting a heart or a 3

(b) Randomly selecting a black suit or a king

(c) Randomly selecting a 5 or a face card

18. *Rolling a Die* You roll a die. Find each probability.

(a) Rolling a 6 or a number greater than 4

(b) Rolling a number less than 5 or an odd number

(c) Rolling a 3 or an even number

19. (a) 0.067
(b) 0.868
(c) 0.231
20. (a) 0.298
(b) 0.445
(c) 0.435
21. (a) 0.110
(b) 0.720
22. (a) 0.170
(b) 0.621
23. (a) 0.014
(b) 0.774
(c) 0.226

19. *U.S. Age Distribution* The estimated percent distribution of the U.S. population for 2010 is shown in the pie chart. Find each probability. *(Source: U.S. Census Bureau)*

(a) Randomly selecting someone under five years old
(b) Randomly selecting someone who is not 65 years or over
(c) Randomly selecting someone who is between 18 and 34 years old

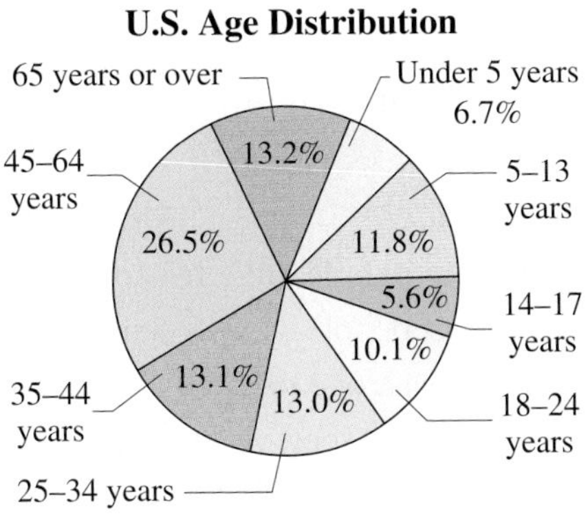

Figure for Exercise 19

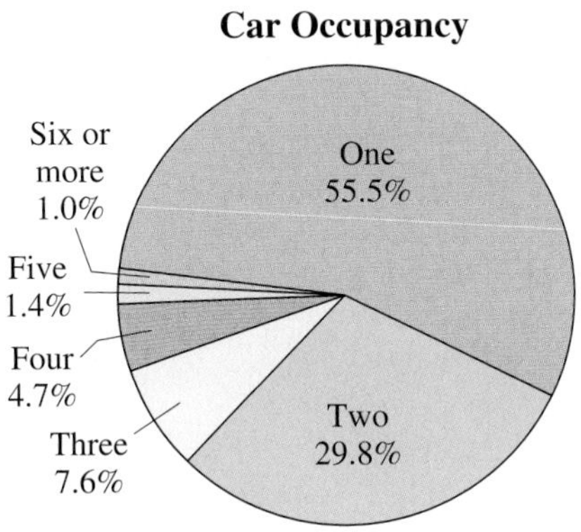

Figure for Exercise 20

20. *Tacoma Narrows Bridge* The percent distribution of the number of occupants in vehicles crossing the Tacoma Narrows Bridge in Washington is shown in the pie chart. Find each probability. *(Source: Washington State Department of Transportation)*

(a) Randomly selecting a car with two occupants
(b) Randomly selecting a car with two or more occupants
(c) Randomly selecting a car with between two and five occupants, inclusive

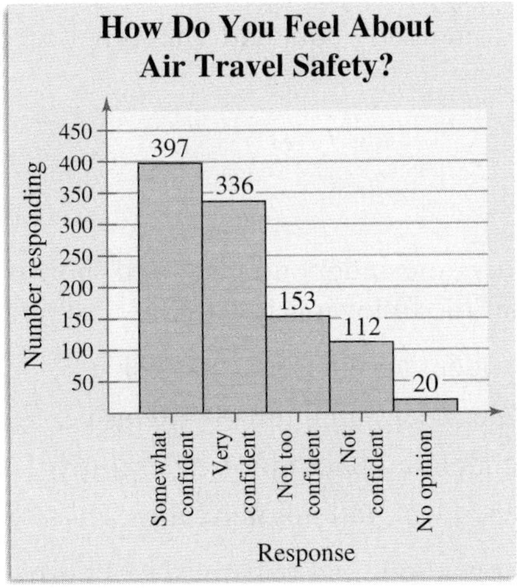

Figure for Exercise 21

21. *Air Travel* The number of responses to a survey are shown in the Pareto chart. The survey asked 1018 U.S. adults how they feel about the safety standards of the major U.S. commercial airlines. Each person gave one response. Find each probability. *(Source: Gallup/CNN/USA TODAY)*

(a) Randomly selecting a person from the sample who is not confident about air travel safety
(b) Randomly selecting a person from the sample who is somewhat confident or very confident about air travel safety

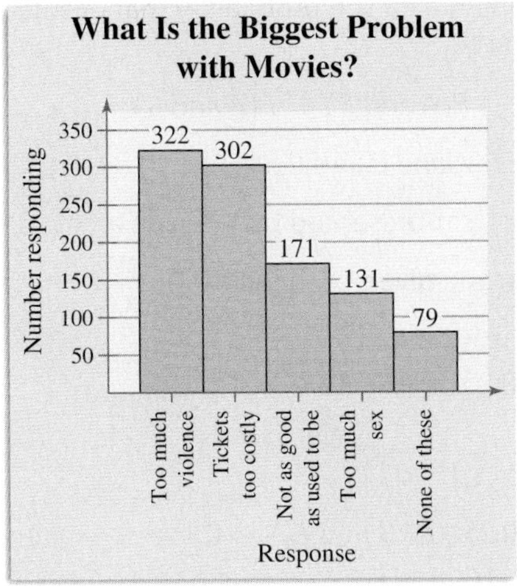

Figure for Exercise 22

22. *Movies* The number of responses to a survey are shown in the Pareto chart. The survey asked 1005 U.S. adults what they feel is the biggest problem with movies. Each person gave one response. Find each probability. *(Source: Associated Press)*

(a) Randomly selecting a person from the sample who feels the biggest problem with movies is that movies are not as good as they used to be
(b) Randomly selecting a person from the sample who feels the biggest problem with movies is that movies have too much violence or that the tickets cost too much

23. *Left-Handed People* In a sample of 1000 people, 120 are left handed. Two unrelated people are selected at random from the sample. Find the probability that (a) both people are left handed, (b) neither person is left handed, and (c) at least one of the two people is left handed.

24. (a) 0.004
(b) 0.003
(c) 0.213
(d) 0.003
(e) 0.006
25. Answers will vary.
26. 0.63
27. 0.54

24. *Left-Handed People* In a sample of 1000 people (525 men and 475 women), 113 are left handed (63 men and 50 women). The results of the sample are shown in the table. Two unrelated people are selected at random from the sample. Find the probability of each event.

		Gender		
		Men	Women	Total
Dominant Hand	Left	63	50	113
	Right	462	425	887
	Total	525	475	1000

(a) Both people are left-handed men.
(b) Both people are left-handed women.
(c) At least one of the two people is left handed.
(d) The first person is a left-handed man and the second person is a left-handed woman.
(e) One of the two people is a left-handed man and the other is a left-handed woman.

Extending Concepts

25. *Writing* Is there a relationship between independence and mutual exclusivity? To decide, find examples of the following, if possible.

(a) Describe two events that are dependent and mutually exclusive.
(b) Describe two events that are independent and mutually exclusive.
(c) Describe two events that are dependent and not mutually exclusive.
(d) Describe two events that are independent and not mutually exclusive.

Use your results to write a conclusion about the relationship between independence and mutual exclusivity.

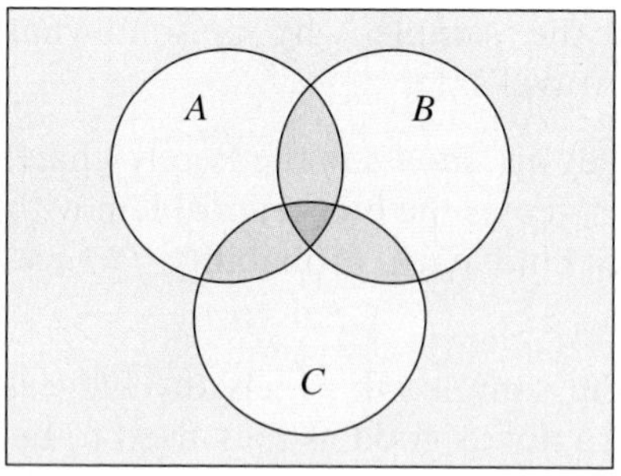

Addition Rule for Three Events The Addition Rule for the probability that events A or B or C will occur $P(A \text{ or } B \text{ or } C)$ is given by

$$P(A \text{ or } B \text{ or } C) = P(A) + P(B) + P(C) - P(A \text{ and } B) - P(A \text{ and } C) - P(B \text{ and } C) + P(A \text{ and } B \text{ and } C).$$

In the Venn diagram at the left, $P(A \text{ or } B \text{ or } C)$ is represented by the blue areas.

In Exercises 26 and 27, find $P(A \text{ or } B \text{ or } C)$ for the given probabilities.

26. $P(A) = 0.40$, $P(B) = 0.10$, $P(C) = 0.50$,
$P(A \text{ and } B) = 0.05$, $P(A \text{ and } C) = 0.25$, $P(B \text{ and } C) = 0.10$,
$P(A \text{ and } B \text{ and } C) = 0.03$

27. $P(A) = 0.35$, $P(B) = 0.35$, $P(C) = 0.30$,
$P(A \text{ and } B) = 0.15$, $P(A \text{ and } C) = 0.25$, $P(B \text{ and } C) = 0.19$,
$P(A \text{ and } B \text{ and } C) = 0.13$

Case Study

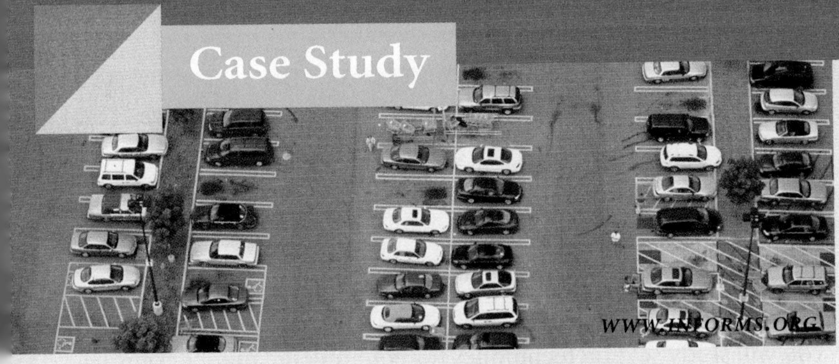

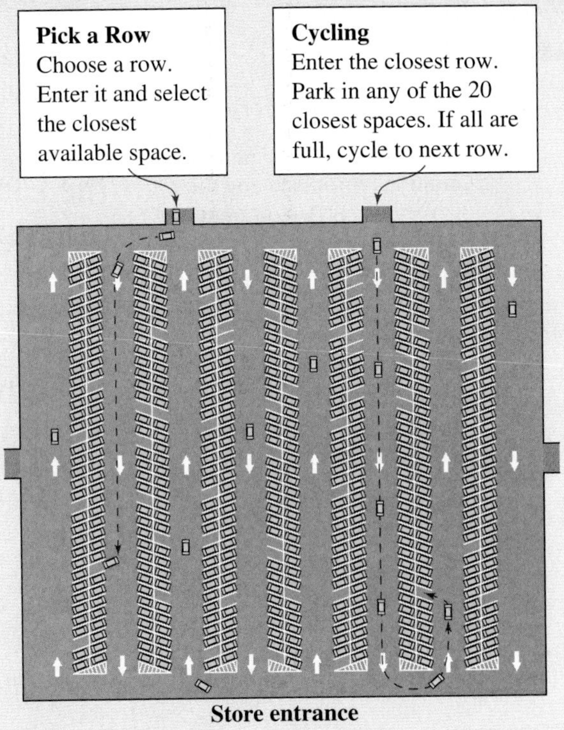

Probability and Parking Lot Strategies

The Institute for Operations Research and the Management Sciences (INFORMS) is an international scientific society with over 12,000 members. It is dedicated to the application of scientific methods to improve decision making, management, and operations. Members of the institute work primarily in business, government, and education. They represent fields as diverse as airlines, health care, law enforcement, the military, the stock market, and telecommunications.

One study published by INFORMS was the result of research conducted by Dr. C. Richard Cassady of Mississippi State University and Dr. John Kobza of Virginia Polytechnic Institute. The parking space study was conducted at a mall that has four entrances, seven rows with 72 spaces each, and directional restrictions. The researchers compared several parking lot strategies to see which strategy saves the most time. The two best strategies are called *Pick a Row* and *Cycling*. The results are shown in the table.

Time or Distance	Pick a Row	Cycling
Time from lot entrance to parking space	37.7 seconds	52.5 seconds
Time from lot entrance to store's door	61.3 seconds	70.7 seconds
Average walking distance to store	257 feet	208 feet

Exercises

1. In a parking lot study, is each parking space equally likely to be empty? Explain your reasoning.
2. According to the results of the study, are you more likely to spend less time using the Pick-a-Row strategy or the Cycling strategy? Explain.
3. According to the results of the study, are you more likely to walk less using the Pick-a-Row strategy or the Cycling strategy? Explain.
4. A key assumption in the study was that the drivers can see which spaces are available as soon as they enter a lane. Why is that important?
5. The parking lot is completely full, and one car leaves. What is the probability that the car was in the first row? Explain your reasoning.
6. A person is leaving from a row that is full. What is the probability that the person was parked in one of the 20 spaces that are closest to the store?
7. Draw a diagram of the parking lot. Color code the parking spaces into three categories of 168 spaces each: most desirable, moderately desirable, and least desirable. Assume that the parking lot is half full. Estimate the probability that you can find a parking space in the most desirable category. Explain your reasoning.

3.4 Counting Principles

What You Should Learn

- How to use the Fundamental Counting Principle to find the number of ways two or more events can occur
- How to find the number of ways a group of objects can be arranged in order
- How to find the number of ways to choose several objects from a group without regard to order
- How to use counting principles to find probabilities

The Fundamental Counting Principle • Permutations • Combinations • Applications of Counting Principles

The Fundamental Counting Principle

In this section, you will study several techniques for counting the number of ways an event can occur. One is the Fundamental Counting Principle. You can use this principle to find the number of ways two or more events can occur in sequence.

The Fundamental Counting Principle

If one event can occur in m ways and a second event can occur in n ways, the number of ways the two events can occur in sequence is $m \cdot n$. This rule can be extended for any number of events occurring in sequence.

Note to Instructor

This section is optional. The material presented here is not a prerequisite to successive sections.

EXAMPLE 1

Using the Fundamental Counting Principle

You are purchasing a new car. Using the following manufacturers, car sizes, and colors, how many different ways can you select one manufacturer, one car size, and one color?

Manufacturer:	Ford, GM, Chrysler
Car size:	small, medium
Color:	white (W), red (R), black (B), green (G)

SOLUTION There are three choices of manufacturers, two car sizes, and four colors. So, the number of ways to select one manufacturer, one car size, and one color is

$$3 \cdot 2 \cdot 4 = 24 \text{ ways}.$$

A tree diagram can help you see why there are 24 options.

Tree Diagram for Car Selections

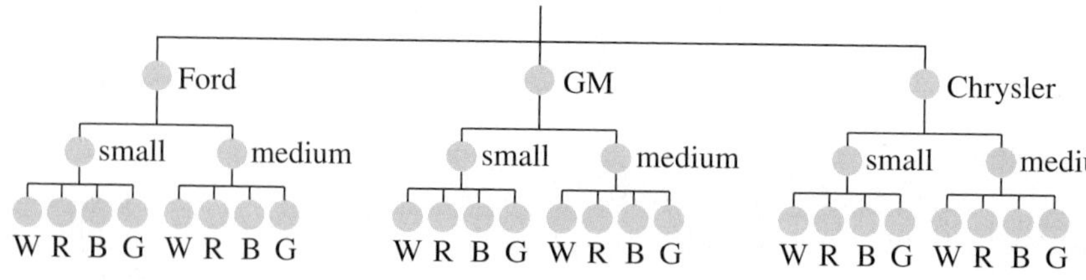

Try It Yourself 1

Your choices now include a Toyota, a large car, or a tan or gray car. How many different ways can you select one manufacturer, one car size, and one color?

a. Find the *number of ways* each event can occur.
b. Use the *Fundamental Counting Principle.*

Answer: Page A35

EXAMPLE 2

Using the Fundamental Counting Principle

The access code for a car's security system consists of four digits. Each digit can be 0 through 9. How many access codes are possible if

1. each digit can be used only once and not repeated?
2. each digit can be repeated?

SOLUTION

1. Because each digit can be used only once, there are 10 choices for the first digit, 9 choices left for the second digit, 8 choices left for the third digit, and 7 choices for the fourth digit. Using the Fundamental Counting Principle, you can conclude that there are

   $$10 \cdot 9 \cdot 8 \cdot 7 = 5040$$

   possible access codes.
2. Because each digit can be repeated, there are 10 choices for each of the four digits. So, there are

   $$\begin{aligned} 10 \cdot 10 \cdot 10 \cdot 10 &= 10^4 \\ &= 10{,}000 \end{aligned}$$

   possible access codes.

Try It Yourself 2

How many license plates can you make if a license plate consists of

1. six (out of 26) alphabetical letters each of which can be repeated?
2. six (out of 26) alphabetical letters each of which cannot be repeated?

a. *Identify* each event and the *number of ways* each event can occur.
b. Use the *Fundamental Counting Principle.* *Answer: Page A35*

Permutations

An important application of the Fundamental Counting Principle is determining the number of ways that n objects can be arranged in order or in a permutation.

Study Tip

The six permutations for the letters A, B, and C are

ABC, ACB, BAC
BCA, CAB, CBA.

DEFINITION

A **permutation** is an ordered arrangement of objects. The number of different permutations of n distinct objects is $n!$.

The expression $n!$ is read as n **factorial** and is defined as follows.

$$n! = n \cdot (n-1) \cdot (n-2) \cdot (n-3) \cdots 3 \cdot 2 \cdot 1$$

As a special case,

$$0! = 1.$$

Study Tip

Notice at the right that as n increases, $n!$ becomes very large. Take some time now to learn how to use the factorial key on your calculator.

Here are several other values of $n!$.

$1! = 1$

$2! = 2 \cdot 1 = 2$

$3! = 3 \cdot 2 \cdot 1 = 6$

$4! = 4 \cdot 3 \cdot 2 \cdot 1 = 24$

$5! = 5 \cdot 4 \cdot 3 \cdot 2 \cdot 1 = 120$

EXAMPLE 3

Finding the Number of Permutations of n Objects

The starting lineup for a baseball team consists of nine players. How many different batting orders are possible using the starting lineup?

SOLUTION

The number of permutations is $9! = 9 \cdot 8 \cdot 7 \cdot 6 \cdot 5 \cdot 4 \cdot 3 \cdot 2 \cdot 1 = 362{,}880$. So, there are 362,880 different batting lineups.

Try It Yourself 3

The teams in the National League Central Division are listed at the right. How many different final standings are possible?

National League Central Division	
Chicago Cubs	Cincinnati Reds
Houston Astros	Milwaukee Brewers
Pittsburgh Pirates	St. Louis Cardinals

a. Determine *how many teams*, n, are in the Central Division.
b. Evaluate $n!$.

Answer: Page A35

Suppose you want to choose some of the objects in a group and put them in order. Such an ordering is called a **permutation of *n* objects taken *r* at a time.**

Permutations of *n* Objects Taken *r* at a Time

The number of permutations of n distinct objects taken r at a time is

$${}_nP_r = \frac{n!}{(n-r)!}, \text{ where } r \le n.$$

EXAMPLE 4

Finding ${}_nP_r$

Find the number of ways of forming three-digit codes in which no digit is repeated.

SOLUTION

To form a three-digit code with no repeating digits, you need to select 3 digits from a group of 10, so $n = 10$ and $r = 3$.

$${}_nP_r = {}_{10}P_3 = \frac{10!}{(10-3)!} = \frac{10!}{7!} = \frac{10 \cdot 9 \cdot 8 \cdot \not{7} \cdot \not{6} \cdot \not{5} \cdot \not{4} \cdot \not{3} \cdot \not{2} \cdot \not{1}}{\not{7} \cdot \not{6} \cdot \not{5} \cdot \not{4} \cdot \not{3} \cdot \not{2} \cdot \not{1}} = 720$$

So, there are 720 possible three-digit codes that do not have repeating digits.

Try It Yourself 4

In a race with eight horses, how many ways can three of the horses finish in first, second, and third place? Assume that there are no ties.

a. Find the *quotient* of $n!$ and $(n - r)!$. (List the factors and divide out.)
b. *Write* the result as a sentence.

Answer: Page A35

EXAMPLE 5

Permutations of n Objects Taken r at a Time

Forty-three race cars started the 2004 Daytona 500. How many ways can the cars finish first, second, and third? *(Source: NASCAR.com)*

SOLUTION You need to select three race cars from a group of 43, so $n = 43$ and $r = 3$. Because the order is important, the number of ways the cars can finish first, second, and third is

$$
\begin{aligned}
{}_nP_r = {}_{43}P_3 &= \frac{43!}{(43-3)!} \\
&= \frac{43!}{40!} \\
&= 43 \cdot 42 \cdot 41 \\
&= 74{,}046.
\end{aligned}
$$

Try It Yourself 5

The board of directors for a company has 12 members. One member is the president, another is the vice-president, another is the secretary, and another is the treasurer. How many ways can can these positions be assigned?

a. *Identify* the total number of objects n and the number of objects r being chosen in order.
b. Evaluate ${}_nP_r$.

Answer: Page A35

> **Study Tip**
>
> The letters *AAAABBC* can be rearranged in 7! orders, but many of these are not distinguishable. The number of distinguishable orders is
>
> $$\frac{7!}{4! \cdot 2! \cdot 1!} = \frac{7 \cdot 6 \cdot 5}{2} = 105.$$

You may want to order a group of n objects in which some of the objects are the same. For instance, consider a group of letters consisting of four As, two Bs, and one C. How many ways can you order such a group? Using the previous formula, you might conclude that there are ${}_7P_7 = 7!$ possible orders. However, because some of the objects are the same, not all of these permutations are *distinguishable.* How many distinguishable permutations are possible? The answer can be found using the following formula.

Distinguishable Permutations

The number of **distinguishable permutations** of n objects, where n_1 are of one type, n_2 are of another type, and so on is

$$\frac{n!}{n_1! \cdot n_2! \cdot n_3! \cdots n_k!}, \text{ where } n_1 + n_2 + n_3 + \cdots + n_k = n.$$

EXAMPLE 6

Distinguishable Permutations

A building contractor is planning to develop a subdivision. The subdivision is to consist of 6 one-story houses, 4 two-story houses, and 2 split-level houses. In how many distinguishable ways can the houses be arranged?

SOLUTION There are to be 12 houses in the subdivision, 6 of which are of one type (one-story), 4 of another type (two-story), and 2 of a third type (split-level). So, there are

$$\frac{12!}{6! \cdot 4! \cdot 2!} = \frac{12 \cdot 11 \cdot 10 \cdot 9 \cdot 8 \cdot 7 \cdot 6!}{6! \cdot 4! \cdot 2!}$$
$$= 13{,}860$$

distinguishable ways to arrange the houses in the subdivision.

Try It Yourself 6

The contractor wants to plant six oak trees, nine maple trees, and five poplar trees along the subdivision street. The trees are to be spaced evenly apart. In how many distinguishable ways can they be planted?

a. *Identify* the total number of objects, n, and the number of each type of object in the group, $n_1, n_2,$ and n_3.

b. *Evaluate* $\dfrac{n!}{n_1! \cdot n_2! \cdots n_k!}$.

Answer: Page A35

Combinations

You want to buy three CDs from a selection of five CDs. There are 10 ways to make your selections.

ABC, ABD, ABE,
ACD, ACE,
ADE,
BCD, BCE,
BDE,
CDE

In each selection, order does not matter (*ABC* is the same set as *BAC*). The number of ways to choose r objects from n objects without regard to order is called the number of **combinations of n objects taken r at a time.**

Insight

You can think of a combination of n objects chosen r at a time as a permutation of n objects in which the r selected objects are alike and the remaining $n - r$ (not selected) objects are alike.

Combination of n Objects Taken r at a Time

A **combination** is a selection of r objects from a group of n objects without regard to order and is denoted by ${}_nC_r$. The number of combinations of r objects selected from a group of n objects is

$${}_nC_r = \frac{n!}{(n-r)!r!}.$$

EXAMPLE 7

Finding the Number of Combinations

A state's department of transportation plans to develop a new section of interstate highway and receives 16 bids for the project. The state plans to hire four of the bidding companies. How many different combinations of four companies can be selected from the 16 bidding companies?

SOLUTION The state is selecting four companies from a group of 16, so $n = 16$ and $r = 4$. Because order is not important, there are

$$
\begin{aligned}
{}_nC_r = {}_{16}C_4 &= \frac{16!}{(16-4)!4!} \\
&= \frac{16!}{12!4!} \\
&= \frac{16 \cdot 15 \cdot 14 \cdot 13 \cdot 12!}{12! \cdot 4!} \\
&= 1820
\end{aligned}
$$

different combinations.

Try It Yourself 7

The manager of an accounting department wants to form a three-person advisory committee from the 16 employees in the department. In how many ways can the manager form this committee?

a. *Identify* the number of objects in the group n and the number of objects to be selected r.
b. *Evaluate* ${}_nC_r$.
c. *Write* the results as a sentence.

Answer: Page A35

The table summarizes the counting principles.

Principle	Description	Formula
Fundamental Counting Principle	If one event can occur in m ways and a second event can occur in n ways, the number of ways the two events can occur in sequence is $m \cdot n$.	$m \cdot n$
Permutation	The number of different ordered arrangements of n distinct objects	$n!$
	The number of permutations of n distinct objects taken r at a time, where $r \leq n$	${}_nP_r = \frac{n!}{(n-r)!}$
	The number of distinguishable permutations of n objects where n_1 are of one type, n_2 are of another type, and so on	$\frac{n!}{n_1! \cdot n_2! \cdots n_k!}$
Combinations	The number of combinations of r objects selected from a group of n objects without regard to order	${}_nC_r = \frac{n!}{(n-r)!r!}$

Picturing the World

The largest lottery jackpot ever, $363 million, was won in the Big Game lottery (now called Mega Millions). When the Big Game jackpot was won, five numbers were chosen from 1 to 50 and one number, the Big Money Ball, was chosen from 1 to 36. The winning numbers are shown below.

Big Money Ball

If you buy one ticket, what is the probability that you will win the Big Game lottery?

Applications of Counting Principles

EXAMPLE 8

Finding Probabilities

A word consists of one M, four Is, four Ss, and two Ps. If the letters are randomly arranged in order, what is the probability that the arrangement spells the word *Mississippi*?

SOLUTION There is one favorable outcome and there are

$$\frac{11!}{1! \cdot 4! \cdot 4! \cdot 2!} = 34{,}650 \qquad \text{11 letters with 1, 4, 4, and 2 like letters}$$

distinguishable permutations of the word *Mississippi*. So, the probability that the arrangement spells the word *Mississippi* is

$$P(\text{Mississippi}) = \frac{1}{34{,}650} \approx 0.000029.$$

Try It Yourself 8

A word consists of one L, two Es, two Ts, and one R. If the letters are randomly arranged in order, what is the probability that the arrangement spells the word *letter*?

a. *Find* the number of favorable outcomes and the number of distinguishable permutations.
b. *Divide* the number of favorable outcomes by the number of distinguishable permutations.

Answer: Page A35

EXAMPLE 9

Finding Probabilities

Find the probability of being dealt five diamonds from a standard deck of playing cards. (In poker, this is a diamond flush.)

SOLUTION The possible number of ways of choosing 5 diamonds out of 13 is ${}_{13}C_5$. The number of possible five-card hands is ${}_{52}C_5$. So, the probability of being dealt 5 diamonds is

$$P(\text{diamond flush}) = \frac{{}_{13}C_5}{{}_{52}C_5} = \frac{1287}{2{,}598{,}960} \approx 0.0005.$$

Try It Yourself 9

A jury consists of five men and seven women. Three are selected at random for an interview. Find the probability that all three are men.

a. *Find* the product of the number of ways to choose three men from five and the number of ways to choose zero women from seven.
b. *Find* the number of ways to choose 3 jury members from 12.
c. *Divide* the result of part a by the result of part b.

Answer: Page A35

3.4 Exercises

Building Basic Skills and Vocabulary

1. When you use the Fundamental Counting Principle, what are you counting?

2. What is the difference between a permutation and a combination?

True or False? In Exercises 3 and 4, determine whether the statement is true or false. If it is false, rewrite it so it is a true statement.

3. A combination is an ordered arrangement of objects.

4. The number of different ordered arrangements of n distinct objects is $n!$.

In Exercises 5–8, perform the indicated calculation.

5. $7!$

6. ${}_{14}P_2$

7. ${}_{7}C_4$

8. $\dfrac{{}_{8}C_3}{{}_{12}C_3}$

In Exercises 9 and 10, decide if the situation involves permutations, combinations, or neither. Explain your reasoning.

9. The number of ways 15 people can line up in a row for concert tickets

10. The number of ways a four-member committee can be chosen from 10 people

Using and Interpreting Concepts

11. ***Space Shuttle Menu*** Space shuttle astronauts each consume an average of 3000 calories per day. One meal normally consists of a main dish, a vegetable dish, and two different desserts. The astronauts can choose from 10 main dishes, 8 vegetable dishes, and 13 desserts. How many different meals are possible? *(Source: NASA)*

12. ***Menu*** A menu has three choices for salad, six main dishes, and four desserts. How many different meals are available if you select a salad, a main dish, and a dessert?

13. ***Security System*** The access code for a car's security system consists of four digits. The first digit cannot be zero and the last digit must be odd. How many different codes are available?

14. ***True or False Quiz*** Assuming that no questions are left unanswered, in how many ways can a six-question true-false quiz be answered?

15. ***Horse Race*** A horse race has eight entries. Assuming that there are no ties, in how many different orders can the horses finish?

16. ***Security Code*** In how many ways can the letters A, B, C, D, E, and F be arranged for a six-letter security code?

17. ***Starting Lineup*** The starting lineup for a softball team consists of 10 players. How many different batting orders are possible using the starting lineup?

1. You are counting the number of ways two or more events can occur in sequence.
2. Permutation: Order matters
   Combination: Order does not matter
3. False. A permutation is an ordered arrangement of objects.
4. True
5. 5040
6. 182
7. 35
8. 0.255
9. Permutation. The order of the 15 people in line matters.
10. Combination. The order of the committee members does not matter.
11. 6240
12. 72
13. 4500
14. 64
15. 40,320
16. 720
17. 3,628,800

18. 24
19. 1320
20. 741,354,768,000
21. 24,024
22. 13! = 6,227,020,800
23. 9,189,180
24. 50,400
25. 5,586,853,480
26. 4845
27. 84
28. 20,358,520
29. (a) 70
(b) 16
(c) 0.086
30. (a) 6,760,000
(b) 5,760,000
(c) 0.5
31. (a) 67,600,000
(b) 19,656,000
(c) 0.000000015

18. ***Assembly Process*** There are four processes involved in assembling a certain product. These processes can be performed in any order. Management wants to find which order is the least time consuming. How many different orders will have to be tested?

19. ***Horse Race*** A horse race has 12 entries. Assuming that there are no ties, in how many different ways can these horses finish first, second, and third?

20. ***Starting Lineup*** The starting lineup for a baseball team consists of nine players. Assuming that each member of a team with 25 players can play each position, in how many different ways can the starting lineup be filled?

21. ***Officers*** From a pool of 14 candidates, the offices of president, vice-president, secretary, and treasurer will be filled. In how many different ways can the offices be filled?

22. ***Team Photo*** In how many ways can a team of 13 players line up in a row to pose for a photo?

23. ***Tree Planting*** A landscaper wants to plant four oak trees, eight maple trees, and six poplar trees along the border of a lawn. The trees are to be spaced evenly apart. In how many distinguishable ways can they be planted?

24. ***Letters*** In how many distinguishable ways can the letters in the word *statistics* be written?

25. ***Jury Selection*** From a group of 40 people, a jury of 12 people is selected. In how many different ways can a jury of 12 people be selected?

26. ***Experimental Group*** In order to conduct an experiment, 4 subjects are randomly selected from a group of 20 subjects. How many different groups of four subjects are possible?

27. ***Pizza Toppings*** A pizza shop offers nine toppings. No topping is used more than once. In how many different ways can a three-topping pizza be formed?

28. ***Lottery Number Selection*** A lottery has 52 numbers. In how many different ways can 6 of the numbers be selected? (Assume that order of selection is not important.)

29. ***Employee Selection*** Four sales representatives for a company are to be chosen to participate in a training program. The company has eight sales representatives, two in each of four regions. In how many ways can the four sales representatives be chosen if (a) there are no restrictions and (b) the selection must include a sales representative from each region? (c) What is the probability that the four sales representatives chosen to participate in the training program will be from only two of the four regions if they are chosen at random?

30. ***License Plates*** In a certain state, each automobile license plate number consists of two letters followed by a four-digit number. How many distinct license plate numbers can be formed if (a) there are no restrictions and (b) the letters O and I are not used? (c) What is the probability of selecting at random a license plate that ends in an even number?

31. ***Password*** A password consists of two letters followed by a five-digit number. How many passwords are possible if (a) there are no restrictions and (b) none of the letters or digits can be repeated? (c) What is the probability of guessing the password in one trial if there are no restrictions?

32. (a) 1000
(b) 800
(c) 0.5
33. (a) 120
(b) 12
(c) 12
(d) 0.4
34. (a) 56
(b) 56
(c) 112
(d) 0.067
35. 0.000012
36. 8.99×10^{-9}
37. 6.00×10^{-20}
38. 5.09×10^{-13}
39. (a) 658,008
(b) 0.00000152
40. (a) 1.4629×10^{22}
(b) 8.5323×10^{19}
(c) 0.00583
(d) Yes, because the probability that no minorities are selected is very small.

32. *Area Code* An area code consists of three digits. How many area codes are possible if (a) there are no restrictions and (b) the first digit cannot be a 1 or a 0? (c) What is the probability of selecting an area code at random that ends in an odd number if the first digit cannot be a 1 or a 0?

33. *Repairs* In how many orders can three broken computers and two broken printers be repaired if (a) there are no restrictions, (b) the printers must be repaired first, and (c) the computers must be repaired first? (d) If the order of repairs has no restrictions and the order of repairs is done at random, what is the probability that a printer will be repaired first?

34. *Defective Units* A shipment of 10 microwave ovens contains two defective units. In how many ways can a restaurant buy three of these units and receive (a) no defective units, (b) one defective unit, and (c) at least two good units? (d) What is the probability of the restaurant buying at least two defective units?

Financial Shape In Exercises 35–38, use the pie chart, which shows how U.S. adults rate their financial shape. *(Source: Pew Research Center)*

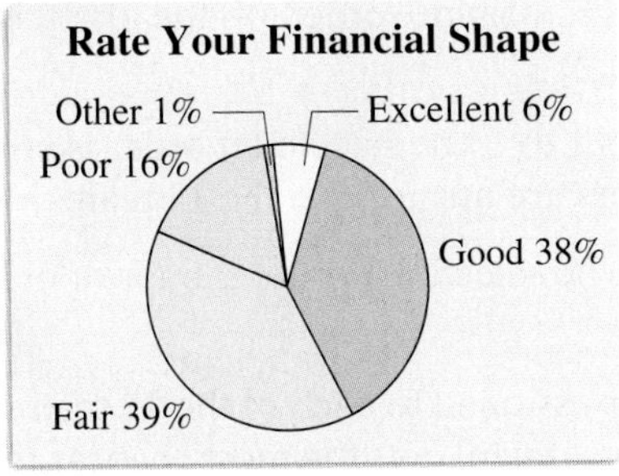

35. Suppose 4 people are chosen at random from a group of 1200. What is the probability that all 4 would rate their financial shape as excellent? (Make the assumption that the 1200 people are represented by the pie chart.)

36. Suppose 10 people are chosen at random from a group of 1200. What is the probability that all 10 would rate their financial shape as poor? (Make the assumption that the 1200 people are represented by the pie chart.)

37. Suppose 80 people are chosen at random from a group of 500. What is the probability that none of the 80 people would rate their financial shape as fair? (Make the assumption that the 500 people are represented by the pie chart.)

38. Suppose 55 people are chosen at random from a group of 500. What is the probability that none of the 55 people would rate their financial shape as good? (Make the assumption that the 500 people are represented by the pie chart.)

39. *Probability* In a state lottery, you must select 5 numbers (in any order) out of 40 correctly to win the top prize.

(a) How many ways can 5 numbers be chosen from 40 numbers?

(b) You purchase one lottery ticket. What is the probability of winning the top prize?

40. *Probability* A company that has 200 employees chooses a committee of 15 to represent employee retirement issues. When the committee was formed, none of the 56 minority employees were selected.

(a) Use a technology tool to find the number of ways 15 employees can be chosen from 200.

(b) Use a technology tool to find the number of ways 15 employees can be chosen from 144 nonminorities.

(c) If the committee was chosen randomly (without bias), what is the probability that it contained no minorities?

(d) Does your answer to part (c) indicate that the committee selection was biased? Explain your reasoning.

Extending Concepts

NBA Draft Lottery In Exercises 41–46, use the following information. The National Basketball Association (NBA) uses a lottery to determine which team gets the first pick in its annual draft. The teams eligible for the lottery are the 13 non-playoff teams. Fourteen Ping-Pong balls numbered 1 through 14 are placed in a drum. Each of the 13 teams is assigned several of the possible four-number combinations that correspond to the numbers on the Ping-Pong balls, such as 3, 8, 10, and 12. Four balls are then drawn out to determine the first pick in the draft. The order in which the balls are drawn is not important. All of the four-number combinations are assigned to the 13 teams by computer except for one four-number combination. When this four-number combination is drawn, the balls are put back in the drum and another drawing takes place. For instance, if Team A has been assigned the four-number combination 3, 8, 10, 12 and the balls shown at the left are drawn, then Team A wins the first pick.

After the first pick of the draft is determined, the process continues to choose the teams that will select second and third. The remaining order of the draft is determined by the number of losses of each team.

41. In how many ways can four of the numbers 1 to 14 be selected if order is *not* important? How many sets of four numbers are assigned to the 13 teams?

42. In how many ways can four of the numbers be selected if order is important?

In the Pareto chart, the number of combinations assigned to each of the 13 teams is shown. The team with the most losses (the worst team) gets the most chances to win the lottery. So, the worst team receives the greatest frequency of four-number combinations, 250. The team with the best record of the 13 non-playoff teams has the fewest chances, with 5 four-number combinations.

Frequency of Four-Number Combinations Assigned in the NBA Draft Lottery

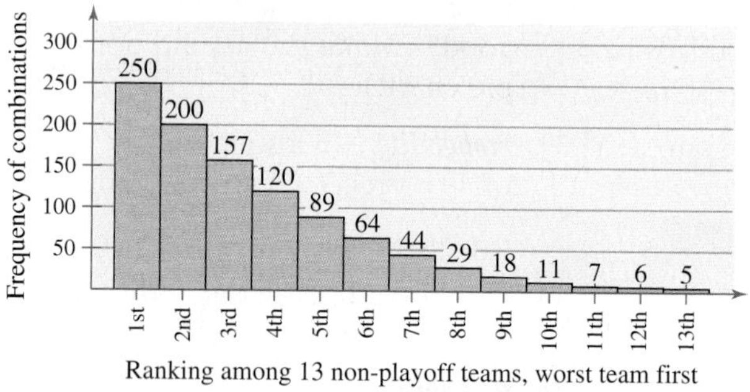

43. For each team, find the probability that the team will win the first pick.

44. What is the probability that the team with the worst record will win the second pick, given that the team with the best record, ranked 13th, wins the first pick?

45. What is the probability that the team with the worst record will win the third pick, given that the team with the best record, ranked 13th, wins the first pick and the team ranked 2nd wins the second pick?

46. What is the probability that neither the first- nor the second-worst team will get the first pick?

41. 1001; 1000

42. 24,024

43.

Team (worst team first)	1	2	3	4
Probability	0.25	0.20	0.157	0.12

Team (worst team first)	5	6	7
Probability	0.089	0.064	0.044

Team (worst team first)	8	9	10
Probability	0.029	0.018	0.011

Team (worst team first)	11	12	13
Probability	0.007	0.006	0.005

44. 0.251

45. 0.566

46. 0.55

Uses and Abuses

Statistics in the Real World

Uses

Probability affects decisions when the weather is forecast, when marketing strategies are determined, when medications are selected, and even when players are selected for professional sports teams. Although intuition is often used for determining probabilities, you will be better able to assess the likelihood that an event will occur by applying the rules of classical probability and empirical probability.

For instance, suppose you work for a real estate company and are asked to estimate the likelihood that a particular house will sell for a particular price within the next 90 days. You could use your intuition, but you could better assess the probability by looking at sales records for similar houses.

Abuses

Assuming Probability Has a "Memory" One common abuse of probability is thinking that probabilities have "memories." For example, if a coin is tossed eight times, the probability that it will land heads up all eight times is only about 0.004. However, if the coin has already been tossed seven times and has landed heads up each time, the probability that it will land heads up on the eighth time is 0.5. Each toss is independent of all other tosses. The coin does not "remember" that it has already landed heads up seven times.

Adding Probabilities Incorrectly Another common abuse is adding probabilities incorrectly. For instance, suppose a company has 100 employees. Of these, 40 are women and 20 are minorities. If one employee is selected at random, the probability that the employee is a woman is 0.4 and the probability that the employee is a minority is 0.2. This does not mean, however, that the probability that a randomly selected employee is a woman or a minority is $0.4 + 0.2$ or 0.6. To determine this probability, you need to know how many employees belong to both groups, women and minorities.

Exercises

1. ***Assuming Probability Has a "Memory"*** A "Daily Number" lottery has a three-digit number from 000 to 999. You buy one ticket each day. Your number is 389.
   a. What is the probability of winning next Tuesday and Wednesday?
   b. You won on Tuesday. What's the probability of winning on Wednesday?
   c. You didn't win on Tuesday. What's the probability of winning on Wednesday?

2. ***Adding Probabilities Incorrectly*** A town has a population of 500 people. Suppose that the probability that a randomly chosen person owns a pickup is 0.25 and the probability that a randomly chosen person owns an SUV is 0.30. What can you say about the probability that a randomly chosen person owns a pickup or an SUV? Could this probability be 0.55? Could it be 0.60? Explain your reasoning.

3 Chapter Summary

What did you learn?

	Review Exercises
Section 3.1	
◆ How to identify the sample space of a probability experiment and to identify simple events	1–4
◆ How to distinguish among classical probability, empirical probability, and subjective probability	5–10
◆ How to identify and use several properties of probability	11, 12
Section 3.2	
◆ How to find conditional probabilities	13, 14
◆ How to distinguish between independent and dependent events	15, 16
◆ How to use the Multiplication Rule to find the probability of two events occurring in sequence	17, 18

$$P(A \text{ and } B) = P(A) \cdot P(B|A) \quad \text{if events are dependent}$$
$$P(A \text{ and } B) = P(A) \cdot P(B) \quad \text{if events are independent}$$

Section 3.3	
◆ How to determine if two events are mutually exclusive	19, 20
◆ How to use the Addition Rule to find the probability of two events	21–30

$$P(A \text{ or } B) = P(A) + P(B) - P(A \text{ and } B)$$
$$P(A \text{ or } B) = P(A) + P(B) \quad \text{if events are mutually exclusive}$$

Section 3.4	
◆ How to use the Fundamental Counting Principle to find the number of ways two or more events can occur	31, 32
◆ How to find the number of ways a group of objects can be arranged in order and the number of ways to choose several objects from a group without regard to order	33–36

$$_nP_r = \frac{n!}{(n-r)!}, \quad \frac{n!}{n_1! \cdot n_2! \cdot n_3! \cdots n_k!}, \quad _nC_r = \frac{n!}{(n-r)!r!}$$

◆ How to use counting principles to find probabilities	37–40

3 Review Exercises

1. Sample space:
{HHHH, HHHT, HHTH, HHTT, HTHH, HTHT, HTTH, HTTT, THHH, THHT, THTH, THTT, TTHH, TTHT, TTTH, TTTT}
Event: Getting three heads
{HHHT, HHTH, HTHH, THHH}
2. Sample space:
{(1, 1), (1, 2), (1, 3), (1, 4), (1, 5), (1, 6), (2, 1), (2, 2), (2, 3), (2, 4), (2, 5), (2, 6), (3, 1), (3, 2), (3, 3), (3, 4), (3, 5), (3, 6), (4, 1), (4, 2), (4, 3), (4, 4), (4, 5), (4, 6), (5, 1), (5, 2), (5, 3), (5, 4), (5, 5), (5, 6), (6, 1), (6, 2), (6, 3), (6, 4), (6, 5), (6, 6)}
Event: Sum of 4 or 5
{(1, 3), (1, 4), (2, 2), (2, 3), (3, 1), (3, 2), (4, 1)}
3. Sample space:
{0, 1, 2, 3, 4, 5, 6, 7, 8, 9}
Event: Choosing a number less than 2
{0, 1}
4. Sample Space:
{GGG, GGB, GBG, GBB, BGG, BGB, BBG, BBB}
Event: Family has two boys
{GBB, BGB, BBG}
5. Empirical probability
6. Classical probability
7. Subjective probability
8. Empirical probability
9. Classical probability
10. Empirical probability
11. 0.72
12. 0.83

Section 3.1

In Exercises 1–4, identify the sample space of the probability experiment and list the outcomes of the event.

1. *Experiment:* Tossing four coins
Event: Getting three heads
2. *Experiment:* Rolling 2 six-sided dice
Event: Getting a sum of 4 or 5
3. *Experiment:* Choosing a one-digit number
Event: Choosing a number less than 2
4. *Experiment:* Guessing the gender of the three children in a family
Event: The family has two boys

In Exercises 5–10, classify the statement as an example of classical probability, empirical probability, or subjective probability.

5. On the basis of prior counts, a quality control officer says there is a 0.05 probability that a randomly chosen part is defective.
6. The probability of randomly selecting five cards of the same suit (a flush) from a standard deck is about 0.0005.
7. The chance that Corporation A's stock price will fall today is 75%.
8. The probability of a person from the United States being left handed is 11%.
9. The probability of rolling 2 six-sided dice and getting a sum greater than nine is $\frac{1}{6}$.
10. The chance that a randomly selected person in the United States is between 15 and 24 years old is about 14%. *(Source: U.S. Census Bureau)*

In Exercises 11 and 12, the table shows the approximate U.S. age distribution for the year 2005. Use the table to determine the probability of the event. *(Source: U.S. Census Bureau)*

Age	19 and under	20–34	35–59	60–84	85 and over
Population	28%	20%	35%	15%	2%

11. What is the probability that a randomly selected person in the United States will be at least 20 years old?
12. What is the probability that a randomly selected person in the United States will be less than 60 years old?

13. 0.92
14. 0.07
15. Independent
16. Dependent
17. 0.025
18. 0.242
19. Mutually exclusive
20. Not mutually exclusive
21. 0.60

Section 3.2

In Exercises 13 and 14, the list shows the results of a study on the use of plus/minus grading at North Carolina State University. It shows the percents of graduate and undergraduate students who received grades with pluses and minuses (for example, C+, A–, etc.). *(Source: North Carolina State University)*

- Of all students who received one or more plus grades, 92% were undergraduates and 8% were graduates.
- Of all students who received one or more minus grades, 93% were undergraduates and 7% were graduates.

13. Find the probability that a student is an undergraduate student, given that the student received a plus grade.

14. Find the probability that a student is a graduate student, given that the student received a minus grade.

In Exercises 15 and 16, decide whether the events are independent or dependent.

15. Tossing a coin four times, getting four heads, and tossing it a fifth time and getting a head

16. Taking a driver's education course and passing the driver's license exam

In Exercises 17 and 18, find the probability of the sequence of events.

17. You are shopping, and your roommate has asked you to pick up toothpaste and dental rinse. However, your roommate did not tell you which brands to get. The store has eight brands of toothpaste and five brands of dental rinse. What is the probability that you will purchase the correct brands of both products?

18. Your sock drawer has 12 folded pairs of socks, with 4 pairs each of white, black, and blue. What is the probability, without looking in the drawer, that you will first select and remove a black pair, then select either a blue or a white pair?

Section 3.3

In Exercises 19 and 20, decide if the events are mutually exclusive.

19. Event A: Randomly select a person who uses the Internet at least twice a week.
Event B: Randomly select a person who has not used the Internet in seven days.

20. Event A: Randomly select a person who loves cats.
Event B: Randomly select a person who owns a dog.

21. A random sample of 250 working adults found that 37% access the Internet at work, 44% access the Internet at home, and 21% access the Internet at both work and home. What is the probability that a person in this sample selected at random accesses the Internet at home or at work?

22. 0.57
23. 0.538
24. 0.538
25. 0.583
26. 0.625
27. 0.463
28. 0.418

22. Another random sample of 225 working adults found that 21% access the Internet at work, 56% access the Internet at home, and 20% access the Internet both at work and home. What is the probability that a randomly selected person from this sample accesses the Internet at home or at work?

In Exercises 23–26, determine the probability.

23. A card is randomly selected from a standard deck. Find the probability that the card is between 4 and 8 (inclusive) or is a club.

24. A card is randomly selected from a standard deck. Find the probability that the card is red or a queen.

25. A 12-sided die, numbered 1–12, is rolled. Find the probability that the roll results in an odd number or a number less than 4.

26. An eight-sided die, numbered 1–8, is rolled. Find the probability that the roll results in an even number or a number greater than six.

In Exercises 27 and 28, use the pie chart, which shows the percent distribution of the number of students in traditional U.S. secondary schools. *(U.S. National Center for Education Statistics)*

Students in Secondary Schools

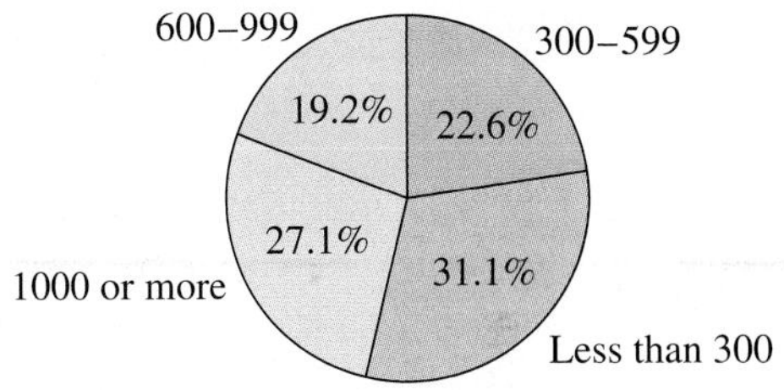

27. Find the probability of randomly selecting a school with 600 or more students.

28. Find the probability of randomly selecting a school with between 300 and 999 students, inclusive.

In Exercises 29 and 30, use the Pareto chart, which shows the results of a survey in which 500 adults were asked why they don't always eat healthy foods.

Why Don't You Eat Healthy Foods?

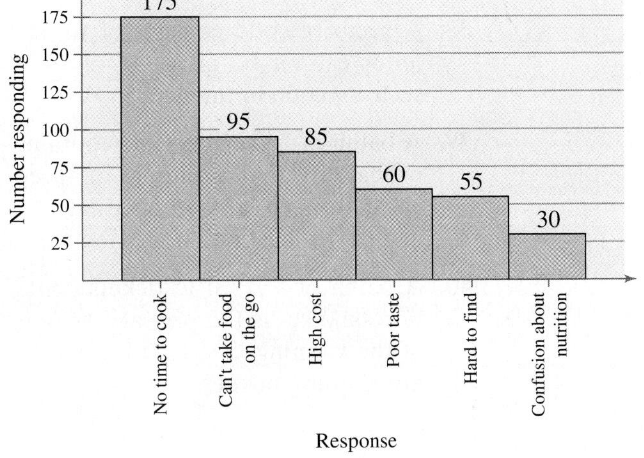

29. 0.23
30. 0.41
31. 144
32. 1,679,616
33. 2730
34. 120
35. 2380
36. 78
37. 0.00000923
38. 0.0000064
39. (a) 0.955
(b) 0.000000761
(c) 0.045
(d) 0.999999239
40. (a) 0.958
(b) 0.00000014
(c) 0.042
(d) 0.9999998589

29. Find the probability of randomly selecting an adult from the sample who feels that healthy foods have poor taste or are hard to find.

30. Find the probability of randomly selecting an adult from the sample who doesn't always eat healthy foods because he or she has no time to cook or is confused about nutrition.

Section 3.4

In Exercises 31 and 32, use the Fundamental Counting Principle.

31. Until recently, with the advent of cellular phones, modems, and pagers, the area codes in the United States and Canada followed a certain system. The first number could not be 0 or 1, the second could only be 0 or 1, and the third could not be 0. Under this system, how many area codes are possible?

32. Assuming that each character on a state's license plate can be either a letter or a digit, how many four-character license plates are possible?

In Exercises 33–36, use combinations and permutations.

33. Fifteen cyclists enter a race. In how many ways can they finish first, second, and third?

34. Five players on a basketball team must choose a player on the opposing team to defend. In how many ways can they choose their defensive assignments?

35. A literary magazine editor must choose 4 short stories for this month's issue from 17 submissions. In how many ways can the editor choose this month's stories?

36. An employer must hire 2 people from a list of 13 applicants. In how many ways can the employer choose to hire 2 people?

In Exercises 37–40, use counting principles to find the probability.

37. In poker, a full house consists of a three-of-a-kind and a two-of-a-kind. Find the probability of a full house consisting of three kings and two queens.

38. A security code consists of three letters followed by one digit. The first letter cannot be an A, B, or C. What is the probability of guessing the security code in one trial?

39. A batch of 200 calculators contains three defective calculators. What is the probability that a sample of three calculators will have (a) no defective calculators, (b) all defective calculators, (c) at least one defective calculator, and (d) at least one nondefective calculator?

40. A batch of 350 raffle tickets contains three winning tickets. You buy five tickets. What is the probability that you have (a) no winning tickets, (b) all of the winning tickets, (c) at least one winning ticket, and (d) at least one nonwinning ticket?

3 Chapter Quiz

Take this quiz as you would take a quiz in class. After you are done, check your work against the answers given in the back of the book.

1. The table shows the estimated number (in thousands) of earned degrees conferred in the United States in the year 2004 by level and gender. *(Source: National Center for Education Statistics)*

		Gender		
		Male	**Female**	**Total**
Level of Degree	**Associate**	231	401	632
	Bachelor's	553	769	1322
	Master's	197	270	467
	Doctorate	25	20	45
	Total	1006	1460	2466

A person who earned a degree in the year 2004 is randomly selected. Find the probability of selecting someone who

(a) earned a bachelor's degree.

(b) earned a bachelor's degree given that the person is a female.

(c) earned a bachelor's degree given that the person is not a female.

(d) earned an associate degree or a bachelor's degree.

(e) earned a doctorate given that the person is a male.

(f) earned a master's degree or is a female.

(g) earned an associate degree and is a male.

(h) is a female given that the person earned a bachelor's degree.

2. Decide if the events are mutually exclusive. Then decide if the events are independent or dependent. Explain your reasoning.

Event A: A golfer scoring the best round in a four-round tournament
Event B: Losing the golf tournament

3. A shipment of 150 television sets contains three defective units. Determine how many ways a vending company can buy three of these units and receive (a) no defective units, (b) all defective units, and (c) at least one good unit.

4. In Exercise 3, find the probability of the vending company receiving (a) no defective units, (b) all defective units, and (c) at least one good unit.

5. The access code for a car's security system consists of four digits. The first digit cannot be 0 and the last digit must be even. How many different codes are available?

6. From a pool of 25 candidates, the offices of president, vice-president, secretary, and treasurer will be filled. In how many different ways can the offices be filled?

1. (a) 0.536
(b) 0.527
(c) 0.550
(d) 0.792
(e) 0.025
(f) 0.672
(g) 0.094
(h) 0.582

2. Not mutually exclusive. A golfer can score the best round in a four-round tournament and still lose the tournament.

Dependent. One event can affect the occurrence of the second event.

3. (a) 518,665
(b) 1
(c) 551,299

4. (a) 0.941
(b) 0.00000181
(c) 0.999998

5. 4500

6. 303,600

PUTTING IT ALL TOGETHER

Real Statistics ■ Real Decisions

You work for the company that runs the Powerball lottery. Powerball is a lottery game in which five white balls are chosen from a drum containing 53 balls and one red ball is chosen from a drum containing 42 balls. To win the jackpot, a player must match all five white balls and the red ball. Other winners and their prizes are also shown in the table.

Working in the public relations department, you handle many inquiries from the media and from lottery players. You receive the following e-mail.

> *You list the probability of matching only the red ball as* 1/70. *I know from my statistics class that the probability of winning is the ratio of the number of successful outcomes to the total number of outcomes. Could you please explain why the probability of matching only the red ball is* 1/70?

Your job is to answer this question, using the probability techniques you have learned in this chapter to justify your answer. In answering the question, assume only one ticket is purchased.

POWERBALL

www.musl.com

Powerball Winners and Prizes

Match	Prize	Approximate Probability
5 white, 1 red	Jackpot	1/120,526,770
5 white	$100,000	1/2,939,677
4 white, 1 red	$5,000	1/502,195
4 white	$100	1/12,249
3 white, 1 red	$100	1/10,685
3 white	$7	1/261
2 white, 1 red	$7	1/697
1 white, 1 red	$4	1/124
1 red	$3	1/70

(Source: Multi-State Lottery Association)

Where Is Powerball Played?
Powerball is played in 26 states, Washington, D.C., and the U.S. Virgin Islands

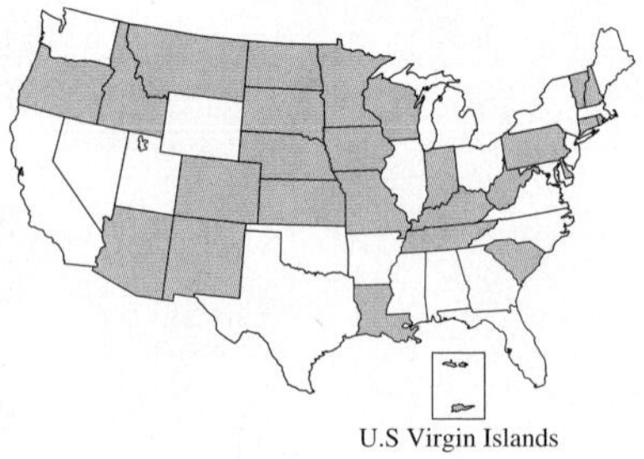

(Source: Multi-State Lottery Association)

Exercises

1. *How Would You Do It?*

(a) How would you investigate the question about the probability of matching only the red ball?

(b) What statistical methods taught in this chapter would you use?

2. *Answering the Question*

Write an explanation that answers the question about the probability of matching only the red ball. Include in your explanation any probability formulas that justify your explanation.

3. *Another Question*

You receive another question asking how the overall probability of winning a prize in the Powerball lottery is determined. The overall probability of winning a prize in the Powerball lottery is 1/36. Write an explanation that answers the question and include any probability formulas that justify your explanation.

TECHNOLOGY

MINITAB **EXCEL** **TI-83**

Simulation: Composing Mozart Variations with Dice

Wolfgang Mozart (1756–1791) composed a wide variety of musical pieces. In his Musical Dice Game, he wrote a Wiener minuet with an almost endless number of variations. Each minuet has 16 bars. In the eighth and sixteenth bars, the player has a choice of two musical phrases. In each of the other 14 bars, the player has a choice of 11 phrases.

To create a minuet, Mozart suggested that the player toss 2 six-sided dice 16 times. For the eighth and sixteenth bars, choose Option 1 if the dice total is odd and Option 2 if it is even. For each of the other 14 bars, subtract 1 from the dice total. The following minuet is the result of the following sequence of numbers.

5	7	1	6	4	10	5	1
6	6	2	4	6	8	8	2

Exercises

1. How many phrases did Mozart write to create the Musical Dice Game minuet? Explain.

2. How many possible variations are there in Mozart's Musical Dice Game minuet? Explain.

3. Use technology to select randomly a number from 1 to 11.
   (a) What is the theoretical probability of each number from 1 to 11 occurring?
   (b) Use this procedure to select 100 integers between 1 and 11. Tally your results and compare them with the probabilities in part (a).

4. What is the probability of randomly selecting options 6, 7, or 8 for the first bar? For all 14 bars? Find each probability using (a) theoretical probability and (b) the results of Exercise 3(b).

5. Use technology to select randomly two numbers from 1, 2, 3, 4, 5, and 6. Find the sum and subtract 1 to obtain a total.
   (a) What is the theoretical probability of each total from 1 to 11?
   (b) Use this procedure to select 100 totals between 1 and 11. Tally your results and compare them with the probabilities in part (a).

6. What is the probability of randomly selecting options 6, 7, or 8 for the first bar? For all 14 bars? Find each probability using (a) theoretical probability and (b) the results of Exercise 5(b).

Extended solutions are given in the *Technology Supplement*.
Technical instruction is provided for MINITAB, Excel, and the TI-83.

CHAPTER 4

Discrete Probability Distributions

4.1 Probability Distributions

4.2 Binomial Distributions

Case Study

4.3 More Discrete Probability Distributions

Uses and Abuses

Real Statistics–Real Decisions

Technology

The National Center for Atmospheric Research (NCAR) is located in Boulder, Colorado. It is part of the complex system of organizations that gathers and analyzes data about weather and other climatic conditions.

Where You've Been

In Chapters 1 through 3, you learned how to collect and describe data and how to find the probability of an event. These skills are used in many different types of careers. For example, data about climatic conditions are used to analyze and forecast the weather throughout the world. On a typical day, 5000 weather stations, 800 to 1100 upper-air balloon stations, 2000 ships, 600 aircraft, several polar-orbiting and geostationary satellites, and a variety of other data-collection devices work together to provide meteorologists with data that are used to forecast the weather. Even with this much data, meteorologists cannot forecast the weather with certainty. Instead, they assign probabilities to certain weather conditions. For instance, a meteorologist might determine that there is a 40% chance of rain (based on the relative frequency of rain under similar weather conditions).

Where You're Going

In Chapter 4, you will learn how to create and use probability distributions. Knowing the shape, central tendency, and variability of a probability distribution will enable you to make decisions in inferential statistics. You are a meteorologist working on a three-day forecast. Assuming that having rain on one day is independent of having rain on another day, you have determined that there is a 40% probability of rain on each of the three days (and 60% probability of no rain). What is the probability that it will rain on 0, 1, 2, or 3 of the days? To answer this, you can create a probability distribution for the possible outcomes.

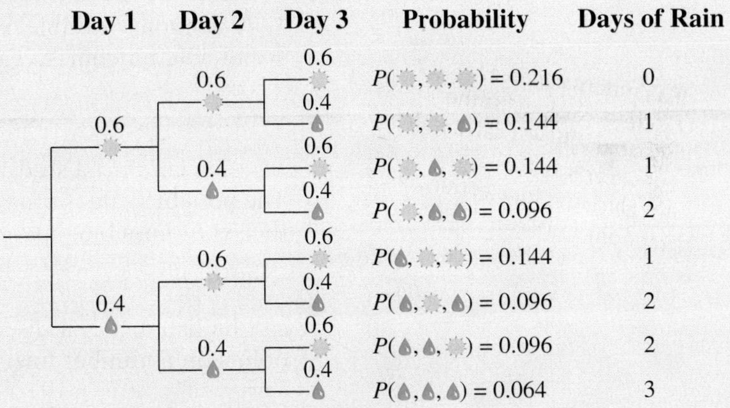

Using the *Addition Rule* with the probabilities in the tree diagram, you can determine the probability of having rain on various numbers of days. You can then use this information to graph a probability distribution.

Probability Distribution		
Days of rain	**Tally**	**Probability**
0	1	0.216
1	3	0.432
2	3	0.288
3	1	0.064

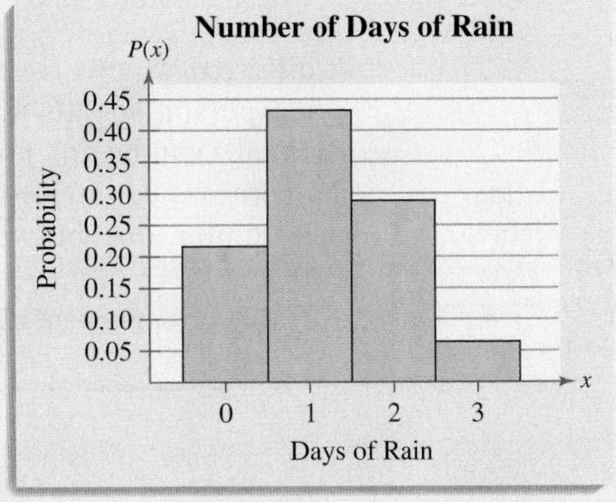

4.1 Probability Distributions

What You Should Learn

- How to distinguish between discrete random variables and continuous random variables
- How to construct a discrete probability distribution and its graph
- How to determine if a distribution is a probability distribution
- How to find the mean, variance, and standard deviation of a discrete probability distribution
- How to find the expected value of a discrete probability distribution

Random Variables • Discrete Probability Distributions • Mean, Variance, and Standard Deviation • Expected Value

Random Variables

The outcome of a probability experiment is often a count or a measure. When this occurs, the outcome is called a random variable.

DEFINITION

A **random variable x** represents a numerical value associated with each outcome of a probability experiment.

The word *random* indicates that x is determined by chance. There are two types of random variables: discrete and continuous.

DEFINITION

A random variable is **discrete** if it has a finite or countable number of possible outcomes that can be listed.

A random variable is **continuous** if it has an uncountable number of possible outcomes, represented by an interval on the number line.

Study Tip

If a random variable is discrete, you can list the possible values it can assume. However, it is impossible to list all values for a continuous random variable.

You conduct a study of the number of calls a salesperson makes in one day. The possible values of the random variable x are 0, 1, 2, 3, 4, and so on. Because the set of possible outcomes

$$\{0, 1, 2, 3, \ldots\}$$

can be listed, x is a discrete random variable. You can represent its values as points on a number line.

Number of Sales Calls (Discrete)

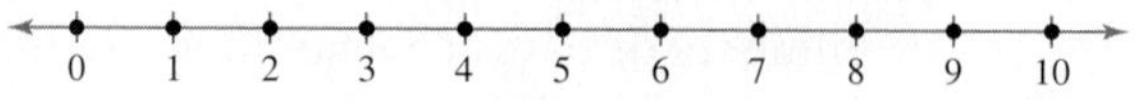

x can have only whole number values: 0, 1, 2, 3, ...

A different way to conduct the study would be to measure the time (in hours) a salesperson spends making calls in one day. Because the time spent making sales calls can be any number from 0 to 24 (including fractions and decimals), x is a continuous random variable. You can represent its values with an interval on a number line, but you cannot list all the possible values.

Hours Spent on Sales Calls (Continuous)

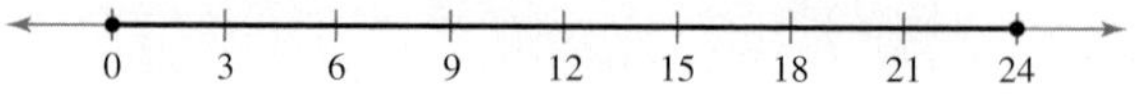

x can have any value between 0 and 24.

Note to Instructor

In more advanced texts, the random variable is denoted X, while x represents particular values of the random variable. To simplify the discussion, we use x to represent both the random variable and its values.

Study Tip

In most practical applications, discrete random variables represent count data, while continuous random variables represent measured data.

Insight

Values of variables such as age, height, and weight are usually rounded to the nearest year, inch, or pound. However, these values represent measured data, and so are continuous random variables.

EXAMPLE 1

Discrete Variables and Continuous Variables

Decide whether the random variable x is discrete or continuous. Explain your reasoning.

1. x represents the number of stocks in the Dow Jones Industrial Average that have share price increases on a given day.
2. x represents the volume of bottled water in a 32-ounce container.

SOLUTION

1. The number of stocks whose share price increases can be counted.

   $$\{0, 1, 2, 3, \ldots, 30\}$$

   So, x is a *discrete* random variable.
2. The amount of water in the container can be any volume between 0 ounces and 32 ounces. So, x is a *continuous* random variable.

Try It Yourself 1

Decide whether the random variable x is discrete or continuous.

1. x represents the length of time it takes to complete a test.
2. x represents the number of home runs hit during a baseball game.

a. Decide if x represents counted data or measured data.
b. Make a conclusion and *explain* your reasoning.

Answer: Page A35

It is important that you can distinguish between discrete and continuous random variables because different statistical techniques are used to analyze each. The remainder of this chapter focuses on discrete random variables and their probability distributions. You will study continuous distributions later.

Discrete Probability Distributions

Each value of a discrete random variable can be assigned a probability. By listing each value of the random variable with its corresponding probability, you are forming a probability distribution.

DEFINITION

A **discrete probability distribution** lists each possible value the random variable can assume, together with its probability. A probability distribution must satisfy the following conditions.

In Words	*In Symbols*
1. The probability of each value of the discrete random variable is between 0 and 1, inclusive.	$0 \leq P(x) \leq 1$
2. The sum of all the probabilities is 1.	$\Sigma P(x) = 1$

Because probabilities represent relative frequencies, a discrete probability distribution can be graphed with a relative frequency histogram.

Note to Instructor

It is important to establish the correspondence between the area under regions of the probability histogram and the probability that the random variable takes on a particular value. Ask students for the area in each bar of the histogram shown.

GUIDELINES

Constructing a Discrete Probability Distribution

Let x be a discrete random variable with possible outcomes $x_1, x_2, \ldots, x_n$.

1. Make a frequency distribution for the possible outcomes.
2. Find the sum of the frequencies.
3. Find the probability of each possible outcome by dividing its frequency by the sum of the frequencies.
4. Check that each probability is between 0 and 1 and that the sum is 1.

Frequency Distribution

Score, x	Frequency, f
1	24
2	33
3	42
4	30
5	21

EXAMPLE 2

Constructing and Graphing a Discrete Probability Distribution

An industrial psychologist administered a personality inventory test for passive-aggressive traits to 150 employees. Individuals were rated on a score from 1 to 5, where 1 was extremely passive and 5 extremely aggressive. A score of 3 indicated neither trait. The results are shown at the left. Construct a probability distribution for the random variable x. Then graph the distribution using a histogram.

SOLUTION Divide the frequency of each score by the total number of individuals in the study to find the probability for each value of the random variable.

$$P(1) = \frac{24}{150} = 0.16 \qquad P(2) = \frac{33}{150} = 0.22 \qquad P(3) = \frac{42}{150} = 0.28$$

$$P(4) = \frac{30}{150} = 0.20 \qquad P(5) = \frac{21}{150} = 0.14$$

The discrete probability distribution is shown in the following table.

x	1	2	3	4	5
$P(x)$	0.16	0.22	0.28	0.20	0.14

Note that $0 \le P(x) \le 1$ and $\Sigma P(x) = 1$.

Passive-Aggressive Traits

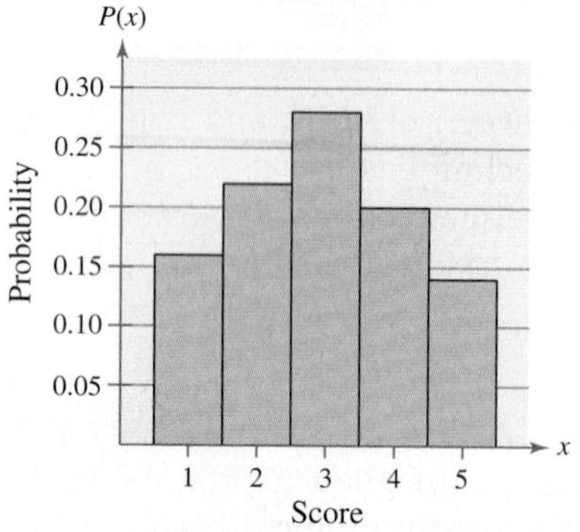

The histogram is shown at the left. Because the width of each bar is one, the area of each bar is equal to the probability of a particular outcome. Also, the probability of an event corresponds to the sum of the areas of the outcomes included in the event. For instance, the probability of the event "having a score of 2 or 3" is equal to the sum of the areas of the second and third bars,

$$(1)(0.22) + (1)(0.28) = 0.22 + 0.28 = 0.50.$$

Interpretation You can see that the distribution is approximately symmetric.

Try It Yourself 2

A company tracks the number of sales new employees make each day during a 100-day probationary period. The results for one new employee are shown at the left. Construct and graph a probability distribution.

Frequency Distribution

Sales per day, x	Number of days, f
0	16
1	19
2	15
3	21
4	9
5	10
6	8
7	2

a. *Find* the probability of each outcome.
b. *Organize* the probabilities in a probability distribution.
c. *Graph* the probability distribution using a histogram.

Answer: Page A35

Probability Distribution

Days of rain, x	Probability, $P(x)$
0	0.216
1	0.432
2	0.288
3	0.064

EXAMPLE 3

Verifying Probability Distributions

Verify that the distribution at the left (see page 171) is a probability distribution.

SOLUTION If the distribution is a probability distribution, then (1) each probability is between 0 and 1, inclusive, and (2) the sum of the probabilities equals 1.

1. Each probability is between 0 and 1.
2. $\Sigma P(x) = 0.216 + 0.432 + 0.288 + 0.064 = 1.$

Because both conditions are met, the distribution is a probability distribution.

Try It Yourself 3

Verify that the distribution you constructed in Try It Yourself 2 is a probability distribution.

a. Verify that the *probability* of each outcome is between 0 and 1.
b. Verify that the *sum* of all the probabilities is 1.
c. Make a *conclusion*.

Answer: Page A35

Picturing the World

In a recent year in the United States, nearly 12 million traffic accidents were reported to the police. A histogram of traffic accidents for various age groups from 16 to 84 is shown. *(Source: U.S. National Highway Traffic Safety Administration)*

U.S. Traffic Accidents by Age

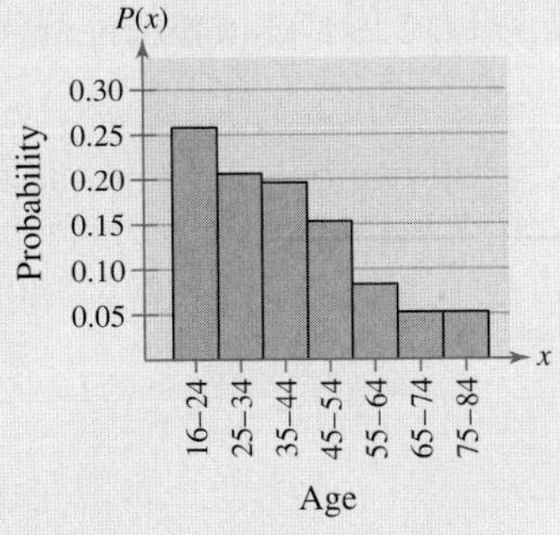

Estimate the probability that a randomly selected person involved in a traffic accident is in the 16 to 34 age group.

EXAMPLE 4

Probability Distributions

Decide whether each distribution is a probability distribution.

1.

x	5	6	7	8
$P(x)$	0.28	0.21	0.43	0.15

2.

x	1	2	3	4
$P(x)$	$\frac{1}{2}$	$\frac{1}{4}$	$\frac{5}{4}$	-1

SOLUTION

1. Each probability is between 0 and 1, but the sum of the probabilities is 1.07, which is greater than 1. So, it is *not* a probability distribution.
2. The sum of the probabilities is equal to 1, but $P(3)$ and $P(4)$ are not between 0 and 1. So, it is *not* a probability distribution. Probabilities can never be negative or greater than 1.

Try It Yourself 4

Decide whether the distribution is a probability distribution. Explain your reasoning.

1.

x	5	6	7	8
$P(x)$	$\frac{1}{16}$	$\frac{5}{8}$	$\frac{1}{4}$	$\frac{3}{16}$

2.

x	1	2	3	4
$P(x)$	0.09	0.36	0.49	0.06

a. Verify that the *probability* of each outcome is between 0 and 1.
b. Verify that the *sum* of all the probabilities is 1.
c. Make a *conclusion*.

Answer: Page A35

Mean, Variance, and Standard Deviation

You can measure the central tendency of a probability distribution with its mean and measure the variability with its variance and standard deviation. The mean of a discrete random variable is defined as follows.

Mean of a Discrete Random Variable

The **mean** of a discrete random variable is given by

$$\mu = \Sigma x P(x).$$

Each value of x is multiplied by its corresponding probability and the products are added.

The mean of the random variable represents the "theoretical average" of a probability experiment and sometimes is not a possible outcome. If the experiment were performed many thousands of times, the mean of all the outcomes would be close to the mean of the random variable.

x	$P(x)$
1	0.16
2	0.22
3	0.28
4	0.20
5	0.14

Note to Instructor

Point out that 2.94 is not a possible value for one person. However, if the scores of the 150 employees were added up and divided by 150, the result would be 2.94.

Study Tip

Notice that the mean in Example 5 is rounded to one decimal place. This rounding was done because the mean of a probability distribution should be rounded to one more decimal place than was used for the random variable x. This *round-off rule* is also used for the variance and standard deviation of a probability distribution.

EXAMPLE 5

Finding the Mean of a Probability Distribution

The probability distribution for the personality inventory test for passive-aggressive traits discussed in Example 2 is given at the left. Find the mean score. What can you conclude?

SOLUTION Use a table to organize your work, as shown below. From the table, you can see that the mean score is approximately 2.9. A score of 3 represents an individual who exhibits neither passive nor aggressive traits. The mean is slightly under 3.

x	$P(x)$	$xP(x)$	
1	0.16	$1(0.16) = 0.16$	
2	0.22	$2(0.22) = 0.44$	
3	0.28	$3(0.28) = 0.84$	
4	0.20	$4(0.20) = 0.80$	
5	0.14	$5(0.14) = 0.70$	
	$\Sigma P(x) = 1$	$\Sigma xP(x) = 2.94$	← Mean

Interpretation You can conclude that the mean personality trait is neither extremely passive nor extremely aggressive, but is slightly closer to passive.

Try It Yourself 5

Find the mean of the probability distribution you constructed in Try It Yourself 2. What can you conclude?

a. *Find the product* of each random outcome and its corresponding probability.
b. *Find the sum* of the products.
c. *What* can you conclude?

Answer: Page A35

Although the mean of the random variable of a probability distribution describes a typical outcome, it gives no information about how the outcomes vary. To study the variation of the outcomes, you can use the variance and standard deviation of the random variable of a probability distribution.

Standard Deviation of a Discrete Random Variable

The **variance** of a discrete random variable is

$$\sigma^2 = \Sigma(x - \mu)^2 P(x).$$

The **standard deviation** is

$$\sigma = \sqrt{\sigma^2}.$$

Study Tip

A shortcut formula for the variance of a probability distribution is

$$\sigma^2 = [\Sigma x^2 P(x)] - \mu^2.$$

EXAMPLE 6

Finding the Variance and Standard Deviation

x	$P(x)$
1	0.16
2	0.22
3	0.28
4	0.20
5	0.14

The probability distribution for the personality inventory test for passive-aggressive traits discussed in Example 2 is given at the left. Find the variance and standard deviation of the probability distribution.

SOLUTION From Example 5, you know that before rounding the mean of the distribution is $\mu = 2.94$. Use a table to organize your work, as shown below.

x	$P(x)$	$x - \mu$	$(x - \mu)^2$	$P(x)(x - \mu)^2$
1	0.16	−1.94	3.764	0.602
2	0.22	−0.94	0.884	0.194
3	0.28	0.06	0.004	0.001
4	0.20	1.06	1.124	0.225
5	0.14	2.06	4.244	0.594
	$\Sigma P(x) = 1$			$\Sigma P(x)(x - \mu)^2 = 1.616$

Variance

So, the variance is

$$\sigma^2 = 1.616 \approx 1.6$$

and the standard deviation is

$$\sigma = \sqrt{\sigma^2} \approx \sqrt{1.616} \approx 1.3.$$

Interpretation Most of the data values differ from the mean by no more than 1.3 points.

Try It Yourself 6

Find the variance and standard deviation for the probability distribution constructed in Try It Yourself 2.

a. *For each value of x, find the square* of the deviation from the mean and multiply that by the corresponding probability of x.
b. *Find the sum* of the products found in part (a) for the variance.
c. *Take the square root* of the variance for the standard deviation.
d. *Interpret* the results.

Answer: Page A36

Expected Value

Note to Instructor

Tell students that expected value plays a role in decision theory. Point out that although probabilities can never be negative, the expected value can be negative.

DEFINITION

The **expected value** of a discrete random variable is equal to the mean of the random variable.

$$\text{Expected Value} = E(x) = \mu = \Sigma xP(x)$$

EXAMPLE 7

Finding an Expected Value

At a raffle, 1500 tickets are sold at $2 each for four prizes of $500, $250, $150, and $75. You buy one ticket. What is the expected value of your gain?

SOLUTION To find the gain for each prize, subtract the price of the ticket from the prize. For instance, your gain for the $500 prize is

$$\$500 - \$2 = \$498$$

and your gain for the $250 prize is

$$\$250 - \$2 = \$248.$$

Write a probability distribution for the possible gains (or outcomes).

Gain, x	$498	$248	$148	$73	−$2
Probability, $P(x)$	$\frac{1}{1500}$	$\frac{1}{1500}$	$\frac{1}{1500}$	$\frac{1}{1500}$	$\frac{1496}{1500}$

Then, using the probability distribution, you can find the expected value.

$$E(x) = \Sigma xP(x)$$

$$= \$498 \cdot \frac{1}{1500} + \$248 \cdot \frac{1}{1500} + \$148 \cdot \frac{1}{1500} + \$73 \cdot \frac{1}{1500} + (-\$2) \cdot \frac{1496}{1500}$$

$$= -\$1.35$$

Interpretation Because the expected value is negative, you can expect to lose an average of $1.35 for each ticket you buy.

Insight

In most applications, an expected value of 0 has a practical interpretation. For instance, in gambling games, an expected value of 0 implies that a game is a fair game (an unlikely occurrence!). In a profit and loss analysis, an expected value of 0 represents the break-even point.

Try It Yourself 7

During a one-year selling period (225 days), a sales representative made between 0 and 9 sales per day, as shown in the table. If this pattern continues, what is the expected value for the number of sales per day?

Number of sales, x	0	1	2	3	4	5	6	7	8	9
Frequency, f (in days)	25	48	60	45	20	10	8	5	3	1

a. *Identify the possible outcomes* and *find the probability* of each.
b. *Find the product* of each outcome and its corresponding probability.
c. *Find the sum* of the products.
d. *Interpret* the results.

Answer: Page A36

4.1 Exercises

Help
MyMathLab
Student Study Pack

Building Basic Skills and Vocabulary

1. What is a random variable? Give an example of a discrete random variable and a continuous random variable. Justify your answer.

2. What is a discrete probability distribution? What are the two conditions that determine a probability distribution?

True or False? In Exercises 3–6, determine whether the statement is true or false. If it is false, rewrite it so that it is a true statement.

3. In most applications, continuous random variables represent counted data, while discrete random variables represent measured data.

4. For a random variable x the word *random* indicates that the value of x is determined by chance.

5. The mean of a random variable represents the "theoretical average" of a probability experiment and sometimes is not a possible outcome.

6. The expected value of a discrete random variable is equal to the standard deviation of the random variable.

Graphical Analysis In Exercises 7–10, decide whether the graph represents a discrete random variable or a continuous random variable. Explain your reasoning.

7. The home attendance for football games at a university

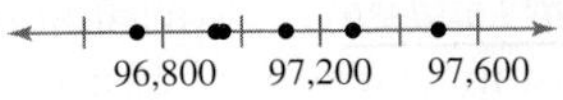

8. The length of time students use a computer each week

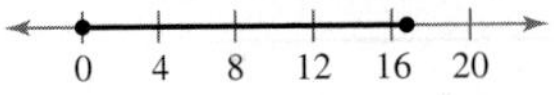

9. The annual vehicle-miles driven in the United States. *(Source: U.S. Federal Highway Administration)*

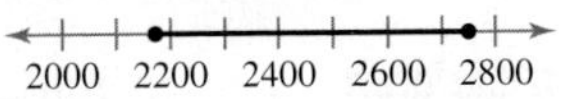

10. The annual traffic fatalities in the United States. *(Source: National Highway Traffic Safety Administration)*

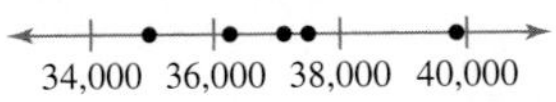

Distinguishing Between Discrete and Continuous Random Variables In Exercises 11–18, decide whether the random variable x is discrete or continuous. Explain your reasoning.

11. x represents the number of highway fatalities in one year in Texas.

12. x represents the length of time it takes to get to work.

13. x represents the volume of blood drawn for a blood test.

14. x represents the number of rainy days in the month of July in Florida.

15. x represents the number of books sold per quarter at a book store.

16. x represents the weight of a chemical compound.

17. x represents the amount of snow (in inches) that fell in Alaska last winter.

18. x represents the number of blackouts in California last summer.

1. A random variable represents a numerical value assigned to an outcome of a probability experiment.
   Examples: Answers will vary.
2. See Selected Answers, page A93.
3. False. In most applications, discrete random variables represent counted data, whereas continuous random variables represent measured data.
4. True 5. True
6. See Selected Answers, page A93.
7. Discrete because home attendance is a random variable that is countable.
8. See Selected Answers, page A93.
9. Continuous because annual vehicle-miles driven is a random variable that cannot be counted.
10. See Selected Answers, page A94.
11. Discrete because number of highway fatalities in one year in Texas is a random variable that is countable.
12. See Selected Answers, page A94.
13. Continuous because volume of blood drawn for a blood test is a random variable that cannot be counted.
14. See Selected Answers, page A94.
15. Discrete because number of books sold per quarter at a book store is a random variable that is countable.
16. See Selected Answers, page A94.
17. See Odd Answers, page A60.
18. See Selected Answers, page A94.

19. (a) 0.35
(b) 0.90
20. (a) 0.45
(b) 0.80
21. 0.22
22. 0.15
23. Yes

Using and Interpreting Concepts

19. *Employee Testing* A company gave psychological tests to prospective employees. The random variable x represents the possible test scores. Use the histogram to find the probability that a person selected at random from the survey's sample had a test score of (a) more than two and (b) less than four.

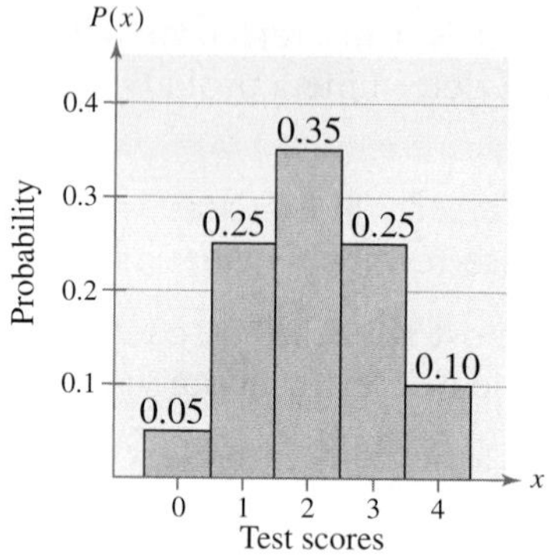

Figure for Exercise 19

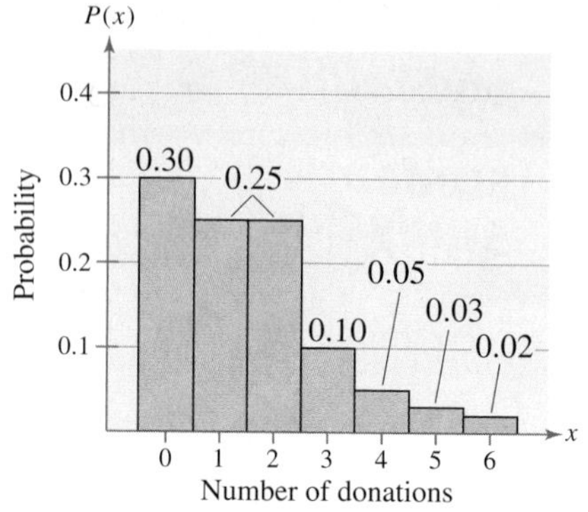

Figure for Exercise 20

20. *Blood Donations* A survey asked a sample of people how many times they donate blood each year. The random variable x represents the number of donations for one year. Use the histogram to find the probability that a person selected at random from the survey's sample donated blood (a) more than once in a year and (b) less than three times in a year.

Determining a Missing Probability In Exercises 21 and 22, determine the probability distribution's missing probability value.

21. *Dependent Children* A sociologist surveyed the households in a small town. The random variable x represents the number of dependent children in the households.

x	0	1	2	3	4
$P(x)$	0.07	0.20	0.38	?	0.13

22. *Dependent Children* The sociologist in Exercise 21 surveyed the households in a neighboring town. The random variable x represents the number of dependent children in the households.

x	0	1	2	3	4	5	6
$P(x)$	0.05	?	0.23	0.21	0.17	0.11	0.08

Identifying Probability Distributions In Exercises 23–26, decide whether the distribution is a probability distribution. If it is not a probability distribution, identify the property (or properties) that are not satisfied.

23. *Psychological Tests* A company gave psychological tests to prospective employees. The random variable x represents the possible test scores.

x	0	1	2	3	4
$P(x)$	0.05	0.25	0.35	0.25	0.10

24. No, $\sum P(x) \approx 1.201$.
25. No, $\sum P(x) = 0.95$ and $P(5) < 0$.
26. Yes
27. See Odd Answers, page A60.
28. See Selected Answers, page A94.
29. See Odd Answers, page A60.
30. See Selected Answers, page A94.

24. *Vehicles Owned* A survey asked a sample of people how many vehicles each owns. The random variable x represents the number of vehicles owned.

x	0	1	2	3
$P(x)$	0.005	0.435	0.555	0.206

25. *Quality Control* A quality inspector checked for imperfections in rolls of fabric. The random variable x represents the number of imperfections found.

x	0	1	2	3	4	5
$P(x)$	$\frac{3}{4}$	$\frac{1}{10}$	$\frac{1}{20}$	$\frac{1}{25}$	$\frac{1}{50}$	$-\frac{1}{100}$

26. *Blood Donations* A survey asked a sample of people how many times they donate blood each year. The random variable x represents the number of donations for one year.

x	0	1	2	3	4	5	6
$P(x)$	0.30	0.25	0.25	0.10	0.05	0.03	0.02

Constructing Probability Distributions In Exercises 27–32, (a) use the frequency distribution to construct a probability distribution, find the (b) mean, (c) variance, and (d) standard deviation of the probability distribution, and (e) interpret the results in the context of the real-life situation.

27. *Dogs* The number of dogs per household in a small town

Dogs	0	1	2	3	4	5
Households	1491	425	168	48	29	14

28. *Cats* The number of cats per household in a small town

Cats	0	1	2	3	4	5
Households	1941	349	203	78	57	40

29. *Computers* The number of computers per household in a small town

Computers	0	1	2	3
Households	300	280	95	20

30. *Accidents* The number of yearly motor vehicle accidents per student

Accidents	0	1	2	3	4	5
Students	260	500	425	305	175	45

31. See Odd Answers, page A60.

32. See Selected Answers, page A94.

33. (a) 5.3
(b) 3.3
(c) 1.8
(d) 5.3
(e) The expected number of questions answered correctly is 5.3 with a standard deviation of 1.8.

34. (a) 1.9
(b) 2.6
(c) 1.6
(d) 1.9
(e) The average number of ATM transactions per customer per day is 1.9 transactions with a standard deviation of 1.6 transactions.

35. (a) 2.1
(b) 1.1
(c) 1.1
(d) 2.1
(e) The average number of hurricanes that hit the U.S. is 2.1 with a standard deviation of 1.1.

36. (a) 1.7
(b) 1.0
(c) 1.0
(d) 1.7
(e) The average car occupancy is 1.7 with a standard deviation of 1.0.

31. *Overtime Hours* The number of overtime hours worked in a week per employee

Overtime hours	0	1	2	3	4	5	6
Employees	6	12	29	57	42	30	16

32. *Extracurricular Activities* The number of school-related extracurricular activities per student

Activities	0	1	2	3	4	5	6	7
Students	19	39	52	57	68	41	27	17

Finding Expected Value In Exercises 33–38, use the probability distribution or histogram to find the (a) mean, (b) variance, (c) standard deviation, and (d) expected value of the probability distribution, and (e) interpret the results.

33. *Quiz* Students in a class take a quiz with eight questions. The number x of questions answered correctly can be approximated by the following probability distribution.

x	0	1	2	3	4	5	6	7	8
$P(x)$	0.02	0.02	0.06	0.06	0.08	0.22	0.30	0.16	0.08

34. *ATM Transactions* A bank recorded the number of ATM transactions made by its customers in one day. The number x of daily ATM transactions per customer can be approximated by the following probability distribution.

x	0	1	2	3	4	5	6
$P(x)$	0.22	0.28	0.20	0.12	0.09	0.07	0.02

35. *Hurricanes* The histogram shows the distribution of hurricanes that have hit the U.S. mainland by category, with 1 the weakest level and 5 the strongest. *(Source: National Oceanic and Atmospheric Administration)*

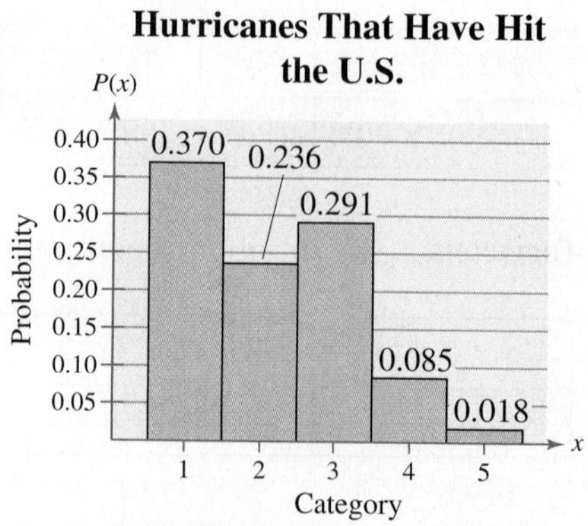

Figure for Exercise 35

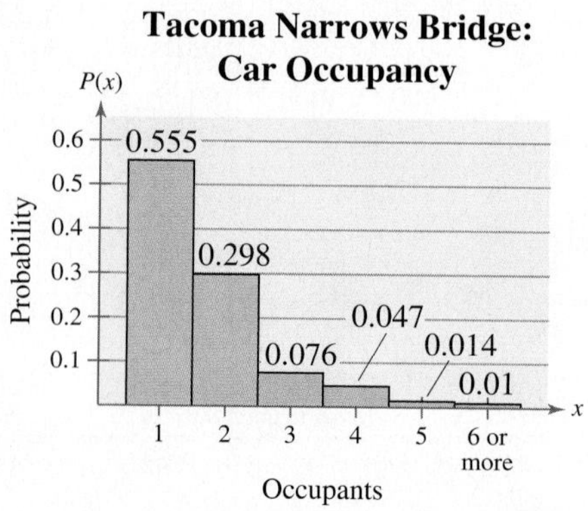

Figure for Exercise 36

36. *Car Occupancy* The histogram shows the distribution of occupants in cars crossing the Tacoma Narrows Bridge in Washington each week. *(Adapted from Washington State Department of Transportation)*

37. (a) 2.6
(b) 1.9
(c) 1.4
(d) 2.6
(e) The average household size is 2.6 with a standard deviation of 1.4.

38. (a) 1.6
(b) 1.9
(c) 1.4
(d) 1.6
(e) The average car occupancy is 1.6 with a standard deviation of 1.4.

39. (a) 0.881
(b) 0.119
(c) 0.112

40. (a) 0.432
(b) 0.569
(c) 0.569

41. A household with three dogs is unusual because the probability of this event is 0.022.

42. A household with no computers is not unusual because the probability of this event is 0.432.

43. −$0.05

44. −$3.21

37. ***Household Size*** The histogram shows the distribution of household sizes in the U.S. for a recent year. *(Adapted from U.S. Census Bureau)*

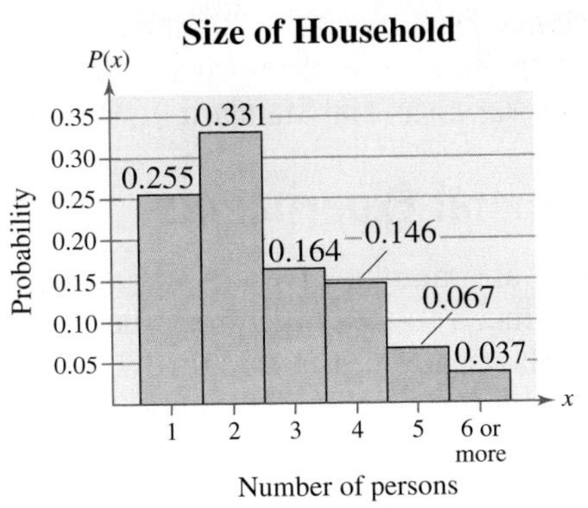

Figure for Exercise 37

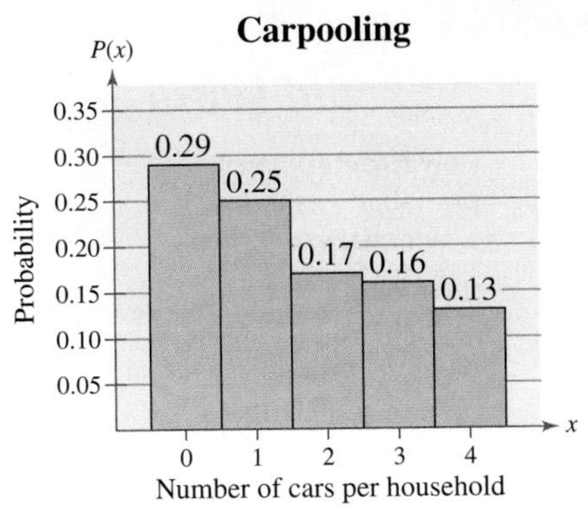

Figure for Exercise 38

38. ***Car Occupancy*** The histogram shows the distribution of carpooling by the number of cars per household. *(Adapted from Federal Highway Administration)*

39. ***Finding Probabilities*** Use the probability distribution you made for Exercise 27 to find the probability of randomly selecting a household that has (a) fewer than two dogs, (b) at least two dogs, and (c) between two and four dogs, inclusive.

40. ***Finding Probabilities*** Use the probability distribution you made for Exercise 29 to find the probability of randomly selecting a household that has (a) no computers, (b) at least one computer, and (c) between one and three computers, inclusive.

Extending Concepts

41. ***Unusual Values*** A person lives in a household with three dogs and claims that having three dogs is not unusual. Use the information in Exercise 27 to determine if this person is correct. Explain your reasoning.

42. ***Unusual Values*** A person lives in a household with no computer and claims that not having a computer is not unusual. Use the information in Exercise 29 to determine if this person is correct. Explain your reasoning.

Games of Chance In Exercises 43 and 44, find the expected net gain to the player for one play of the game. If x is the net gain to a player in a game of chance, then $E(x)$ is usually negative. This value gives the average amount per game the player can expect to lose.

43. In American roulette, the wheel has the 38 numbers

$$00, 0, 1, 2, \ldots, 34, 35, \text{and } 36$$

marked on equally spaced slots. If a player bets $1 on a number and wins, then the player keeps the dollar and receives an additional 35 dollars. Otherwise, the dollar is lost.

44. A charity organization is selling $4 raffle tickets as part of a fund-raising program. The first prize is a boat valued at $3150, and the second prize is a camping tent valued at $450. The rest of the prizes are 15 $25 gift certificates. The number of tickets sold is 5000.

4.2 Binomial Distributions

What You Should Learn

- How to determine if a probability experiment is a binomial experiment
- How to find binomial probabilities using the binomial probability formula
- How to find binomial probabilities using technology, formulas, and a binomial probability table
- How to graph a binomial distribution
- How to find the mean, variance, and standard deviation of a binomial probability distribution

Binomial Experiments • Binomial Probability Formula • Finding Binomial Probabilities • Graphing Binomial Distributions • Mean, Variance, and Standard Deviation

Binomial Experiments

There are many probability experiments for which the results of each trial can be reduced to two outcomes: success and failure. For instance, when a basketball player attempts a free throw, he or she either makes the basket or does not. Probability experiments such as these are called binomial experiments.

DEFINITION

A **binomial experiment** is a probability experiment that satisfies the following conditions.

1. The experiment is repeated for a fixed number of trials, where each trial is independent of the other trials.
2. There are only two possible outcomes of interest for each trial. The outcomes can be classified as a success (S) or as a failure (F).
3. The probability of a success $P(S)$ is the same for each trial.
4. The random variable x counts the number of successful trials.

Note to Instructor

Emphasize that p represents the probability of success on a single trial. Students often have trouble with this concept.

Notation for Binomial Experiments

Symbol	*Description*
n	The number of times a trial is repeated
$p = P(S)$	The probability of success in a single trial
$q = P(F)$	The probability of failure in a single trial $(q = 1 - p)$
x	The random variable represents a count of the number of successes in n trials: $x = 0, 1, 2, 3, \ldots, n.$

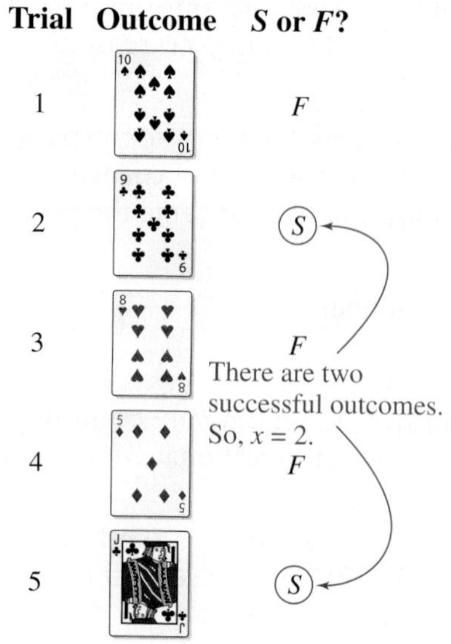

Here is a simple example of a binomial experiment. From a standard deck of cards, you pick a card, note whether it is a club or not, and replace the card. You repeat the experiment five times, so $n = 5$. The outcomes for each trial can be classified in two categories: S = selecting a club and F = selecting another suit. The probabilities of success and failure are

$$p = P(S) = \frac{1}{4} \quad \text{and} \quad q = P(F) = \frac{3}{4}.$$

The random variable x represents the number of clubs selected in the five trials. So, the possible values of the random variable are

0, 1, 2, 3, 4, and 5.

For instance, if $x = 2$, then exactly two of the five cards are clubs and the other three are not clubs. An example of an experiment with $x = 2$ is shown at the left. Note that x is a discrete random variable because its possible values can be listed.

Picturing the World

A recent survey of registered voters in the United States asked whether public school teachers should be required to take drug tests. The respondents' answers were either yes or no. *(Source: Family Research Council)*

Survey question: Should public school teachers be required to take drug tests?

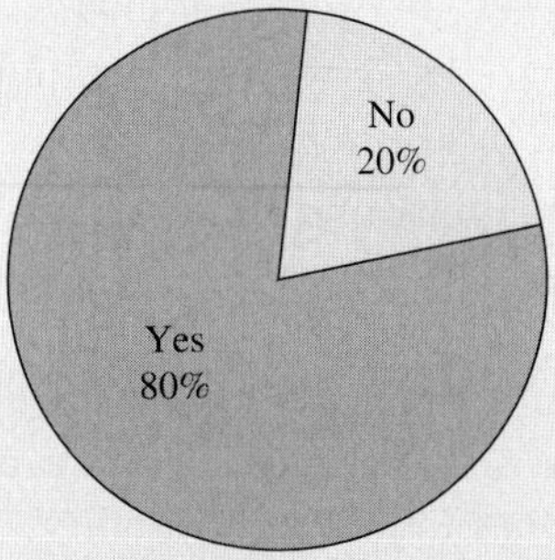

Why is this a binomial experiment? Identify the probability of success, p. Identify the probability of failure, q.

EXAMPLE 1

Binomial Experiments

Decide whether the experiment is a binomial experiment. If it is, specify the values of n, p, and q, and list the possible values of the random variable x. If it is not, explain why.

1. A certain surgical procedure has an 85% chance of success. A doctor performs the procedure on eight patients. The random variable represents the number of successful surgeries.
2. A jar contains five red marbles, nine blue marbles, and six green marbles. You randomly select three marbles from the jar, *without replacement*. The random variable represents the number of red marbles.

SOLUTION

1. The experiment is a binomial experiment because it satisfies the four conditions of a binomial experiment. In the experiment, each surgery represents one trial. There are eight surgeries, and each surgery is independent of the others. There are only two possible outcomes for each surgery—either the surgery is a success or it is a failure. Also, the probability of success for each surgery is 0.85. Finally, the random variable x represents the number of successful surgeries.

   $n = 8$

   $p = 0.85$

   $q = 1 - 0.85 = 0.15$

   $x = 0, 1, 2, 3, 4, 5, 6, 7, 8$

2. The experiment is not a binomial experiment because it does not satisfy all four conditions of a binomial experiment. In the experiment, each marble selection represents one trial, and selecting a red marble is a success. When the first marble is selected, the probability of success is 5/20. However, because the marble is not replaced, the probability of success is no longer 5/20. So, the trials are not independent, and the probability of a success is not the same for each trial.

Try It Yourself 1

Decide whether the following is a binomial experiment. If it is, specify the values of n, p, and q, and list the possible values of the random variable x. If it is not, explain why.

> You take a multiple choice quiz that consists of 10 questions. Each question has four possible answers, only one of which is correct. To complete the quiz, you randomly guess the answer to each question. The random variable represents the number of correct answers.

a. Identify a *trial* of the experiment and what is a "success."
b. Decide if the experiment *satisfies the four conditions* of a binomial experiment.
c. *Make a conclusion* and *identify* n, p, q, and the possible values of x.

Answer: Page A36

Insight

In the binomial probability formula, ${}_nC_x$ determines the number of ways of getting x successes in n trials, regardless of order.

$${}_nC_x = \frac{n!}{(n-x)!\,x!}$$

Study Tip

Recall that $n!$ is read "n factorial" and represents the product of all integers from n to 1. For instance,

$$5! = 5 \cdot 4 \cdot 3 \cdot 2 \cdot 1 = 120.$$

Binomial Probability Formula

There are several ways to find the probability of x successes in n trials of a binomial experiment. One way is to use the binomial probability formula.

Binomial Probability Formula

In a binomial experiment, the probability of exactly x successes in n trials is

$$P(x) = {}_nC_x\, p^x q^{n-x} = \frac{n!}{(n-x)!\,x!}\, p^x q^{n-x}.$$

EXAMPLE 2

Finding Binomial Probabilities

A six-sided die is rolled three times. Find the probability of rolling exactly one 6.

SOLUTION

Method 1: Draw a tree diagram and use the Multiplication Rule.

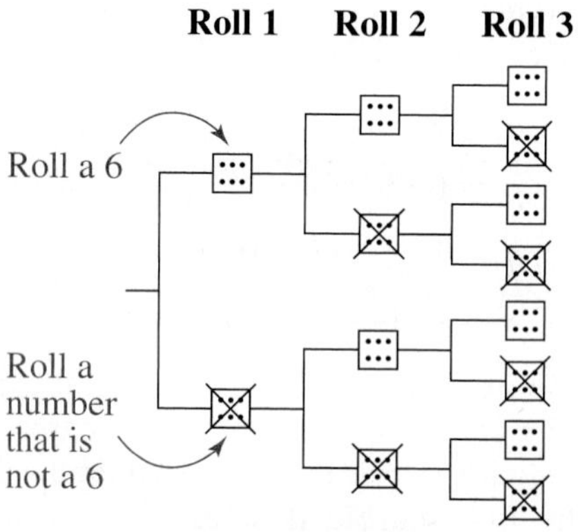

Number of 6s	Probability
3	$\frac{1}{6}\cdot\frac{1}{6}\cdot\frac{1}{6} = \frac{1}{216}$
2	$\frac{1}{6}\cdot\frac{1}{6}\cdot\frac{5}{6} = \frac{5}{216}$
2	$\frac{1}{6}\cdot\frac{5}{6}\cdot\frac{1}{6} = \frac{5}{216}$
1	$\frac{1}{6}\cdot\frac{5}{6}\cdot\frac{5}{6} = \frac{25}{216}$
2	$\frac{5}{6}\cdot\frac{1}{6}\cdot\frac{1}{6} = \frac{5}{216}$
1	$\frac{5}{6}\cdot\frac{1}{6}\cdot\frac{5}{6} = \frac{25}{216}$
1	$\frac{5}{6}\cdot\frac{5}{6}\cdot\frac{1}{6} = \frac{25}{216}$
0	$\frac{5}{6}\cdot\frac{5}{6}\cdot\frac{5}{6} = \frac{125}{216}$

There are three outcomes that have exactly one 6, and each has a probability of $25/216$. So, the probability of rolling exactly one 6 is $3(25/216) \approx 0.347$.

Method 2: Use the binomial probability formula.

In this binomial experiment, rolling a 6 is a success, and rolling any other number is a failure. The values for n, p, q, and x are $n = 3$, $p = \frac{1}{6}$, $q = \frac{5}{6}$, and $x = 1$. The probability of rolling exactly one 6 is

$$P(1) = \frac{3!}{(3-1)!1!}\left(\frac{1}{6}\right)^1\left(\frac{5}{6}\right)^2 = 3\left(\frac{1}{6}\right)\left(\frac{25}{36}\right) = 3\left(\frac{25}{216}\right) = \frac{25}{72} \approx 0.347.$$

Try It Yourself 2

A card is selected from a standard deck and replaced. This experiment is repeated a total of five times. Find the probability of selecting exactly three clubs.

a. *Identify* a trial, a success, and a failure.
b. *Identify* n, p, q, and x.
c. Use the *binomial probability formula*.

Answer: Page A36

By listing the possible values of x with the corresponding probability of each, you can construct a **binomial probability distribution.**

EXAMPLE 3

Constructing a Binomial Distribution

In a survey, workers in the United States were asked to name their expected sources of retirement income. The results are shown in the graph. Seven workers who participated in the survey are randomly selected and asked whether they expect to rely on Social Security for retirement income. Create a binomial probability distribution for the number of workers who respond yes.

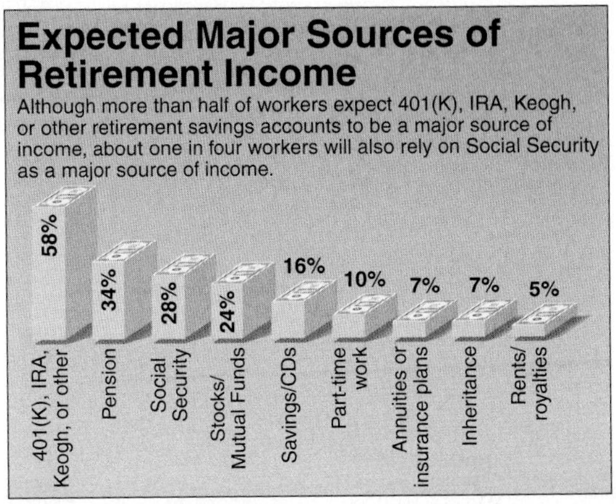

(Source: The Gallup Organization)

SOLUTION From the graph, you can see that 28% of working Americans expect to rely on Social Security for retirement income. So, $p = 0.28$ and $q = 0.72$. Because $n = 7$, the possible values of x are 0, 1, 2, 3, 4, 5, 6, and 7.

$$P(0) = {}_7C_0(0.28)^0(0.72)^7 = 1(0.28)^0(0.72)^7 \approx 0.1003$$
$$P(1) = {}_7C_1(0.28)^1(0.72)^6 = 7(0.28)^1(0.72)^6 \approx 0.2731$$
$$P(2) = {}_7C_2(0.28)^2(0.72)^5 = 21(0.28)^2(0.72)^5 \approx 0.3186$$
$$P(3) = {}_7C_3(0.28)^3(0.72)^4 = 35(0.28)^3(0.72)^4 \approx 0.2065$$
$$P(4) = {}_7C_4(0.28)^4(0.72)^3 = 35(0.28)^4(0.72)^3 \approx 0.0803$$
$$P(5) = {}_7C_5(0.28)^5(0.72)^2 = 21(0.28)^5(0.72)^2 \approx 0.0187$$
$$P(6) = {}_7C_6(0.28)^6(0.72)^1 = 7(0.28)^6(0.72)^1 \approx 0.0024$$
$$P(7) = {}_7C_7(0.28)^7(0.72)^0 = 1(0.28)^7(0.72)^0 \approx 0.0001$$

x	$P(x)$
0	0.1003
1	0.2731
2	0.3186
3	0.2065
4	0.0803
5	0.0187
6	0.0024
7	0.0001
	$\Sigma P(x) = 1$

Notice in the table at the left that all the probabilities are between 0 and 1 and that the sum of the probabilities is 1.

Try It Yourself 3

Seven workers who participated in the survey are randomly selected and asked whether they expect to rely on a pension for retirement income. Create a binomial distribution for the number of retirees who respond yes.

a. *Identify* a trial, a success, and a failure.
b. *Identify* n, p, q, and possible values for x.
c. Use the *binomial probability formula* for each value of x.
d. *Use a table* to show that the properties of a probability distribution are satisfied.

Answer: Page A36

Study Tip

When probabilities are rounded to a fixed number of decimal places, the sum of the probabilities may differ slightly from 1.

Finding Binomial Probabilities

An efficient way to find binomial probabilities is to use a calculator or a computer. For instance, you can find binomial probabilities using MINITAB, Excel, and the TI-83.

EXAMPLE 4

Finding a Binomial Probability Using Technology

The results of a recent survey indicate that 58% of households in the United States own a gas grill. If you randomly select 100 households, what is the probability that exactly 65 households will own a gas grill? Use a technology tool to find the probability. *(Source: Leo Shapiro Research for Weber GrillWatch)*

SOLUTION MINITAB, Excel, and the TI-83 each have features that allow you to find binomial probabilities automatically.

MINITAB

Probability Density Function

Binomial with n = 100 and p = 0.580000

x	P(X=x)
65.00	0.0299

TI-83

```
binompdf(100,.58,65)
.0299216472
```

EXCEL

	A	B	C	D
1	BINOMDIST(65,100,0.58,FALSE)			
2				0.029922

From the displays, you can see that the probability that exactly 65 households will own a gas grill is about 0.03.

Try It Yourself 4

The results of a recent survey indicate that 71% of people in the United States use more than one topping on their hot dogs. If you randomly select 250 people in the United States, what is the probability that exactly 178 of them will use more than one topping? Use a technology tool to find the probability. *(Source: ICR Survey Research Group for Hebrew International)*

a. *Identify* n, p, and x.
b. *Calculate* the binomial probability.
c. *Write* the result as a sentence.

Answer: Page A36

Study Tip

Detailed instructions for using MINITAB, Excel, and the TI-83 are shown in the Technology Guide that accompanies this text. For instance, here are instructions for finding a binomial probability on a TI-83.

[2nd] DISTR

0: binompdf(

Enter the values of n, p, and x separated by commas.

[ENTER]

Note to Instructor

Point out that the illustrated example gives the probability that exactly x successes will occur. Each of these technology tools has a cumulative density function (CDF) that allows an easy computation for the probability of "x or fewer" successes. This is a good time to point out that the CDF adds the areas for the given x-value and all those to its left.

EXAMPLE 5

Finding Binomial Probabilities Using Formulas

A survey indicates that 41% of women in the United States consider reading as their favorite leisure-time activity. You randomly select four U.S. women and ask them if reading is their favorite leisure-time activity. Find the probability that (1) exactly two of them respond yes, (2) at least two of them respond yes, and (3) fewer than two of them respond yes. *(Source: Louis Harris & Associates)*

SOLUTION

1. Using $n = 4$, $p = 0.41$, $q = 0.59$, and $x = 2$, the probability that exactly two women will respond yes is

$$P(2) = {}_4C_2(0.41)^2(0.59)^2 = 6(0.41)^2(0.59)^2 \approx 0.351.$$

2. To find the probability that at least two women will respond yes, you can find the sum of $P(2)$, $P(3)$, and $P(4)$.

$$P(2) = {}_4C_2(0.41)^2(0.59)^2 = 6(0.41)^2(0.59)^2 \approx 0.351094$$
$$P(3) = {}_4C_3(0.41)^3(0.59)^1 = 4(0.41)^3(0.59)^1 \approx 0.162654$$
$$P(4) = {}_4C_4(0.41)^4(0.59)^0 = 1(0.41)^4(0.59)^0 \approx 0.028258$$

So, the probability that at least two will respond yes is

$$P(x \geq 2) = P(2) + P(3) + P(4)$$
$$\approx 0.351094 + 0.162654 + 0.028258$$
$$\approx 0.542.$$

3. To find the probability that fewer than two women will respond yes, you must find the sum of $P(0)$ and $P(1)$.

$$P(0) = {}_4C_0(0.41)^0(0.59)^4 = 1(0.41)^0(0.59)^4 \approx 0.121174$$
$$P(1) = {}_4C_1(0.41)^1(0.59)^3 = 4(0.41)^1(0.59)^3 \approx 0.336822$$

So, the probability that fewer than two will respond yes is

$$P(x < 2) = P(0) + P(1)$$
$$\approx 0.121174 + 0.336822$$
$$\approx 0.458.$$

Study Tip

The complement of "x is at least 2" is "x is less than 2." So, another way to find the probability in part (3) is

$$P(x < 2) = 1 - P(x \geq 2)$$
$$\approx 1 - 0.542$$
$$= 0.458.$$

Try It Yourself 5

A survey indicates that 21% of men in the United States consider fishing as their favorite leisure-time activity. You randomly select five U.S. men and ask them if fishing is their favorite leisure-time activity. Find the probability that (1) exactly two of them respond yes, (2) at least two of them respond yes, and (3) fewer than two of them respond yes. *(Source: Louis Harris & Associates)*

a. Determine the appropriate *values of x* for each situation.
b. Find the *binomial probability* for each value of x. Then find the sum, if necessary.
c. *Write* the result as a sentence.

Answer: Page A36

Note to Instructor

Remind students to decide for each question whether it is easier to calculate a probability as given or to use the complement rule to get their answer.

Finding binomial probabilities with the binomial probability formula can be a tedious process. To make this process easier, you can use a binomial probability table. Table 2 in Appendix B lists the binomial probability for selected values of n and p.

EXAMPLE 6

Finding a Binomial Probability Using a Table

Fifty percent of working adults spend less than 20 minutes each way commuting to their jobs. You randomly select six working adults. What is the probability that exactly three of them spend less than 20 minutes each way commuting to work? Use a table to find the probability. *(Source: Maritz AmeriPoll)*

Note to Instructor

Discuss limitations on the use of the table. It can be used only for particular values of p and particular values of n up to 20. Later in the course students will learn how to calculate probabilities for $n > 20$ using the normal approximation to the binomial.

SOLUTION A portion of Table 2 in Appendix B is shown here. Using the distribution for $n = 6$ and $p = 0.5$, you can find the probability that $x = 3$, as shown by the highlighted areas in the table.

		p												
n	***x***	**.01**	**.05**	**.10**	**.15**	**.20**	**.25**	**.30**	**.35**	**.40**	**.45**	**.50**	**.55**	**.60**
2	0	.980	.902	.810	.723	.640	.563	.490	.423	.360	.303	.250	.203	.160
	1	.020	.095	.180	.255	.320	.375	.420	.455	.480	.495	.500	.495	.480
	2	.000	.002	.010	.023	.040	.063	.090	.123	.160	.203	.250	.303	.360
3	0	.970	.857	.729	.614	.512	.422	.343	.275	.216	.166	.125	.091	.064
	1	.029	.135	.243	.325	.384	.422	.441	.444	.432	.408	.375	.334	.288
	2	.000	.007	.027	.057	.096	.141	.189	.239	.288	.334	.375	.408	.432
	3	.000	.000	.001	.003	.008	.016	.027	.043	.064	.091	.125	.166	.216
6	0	.941	.735	.531	.377	.262	.178	.118	.075	.047	.028	.016	.008	.004
	1	.057	.232	.354	.399	.393	.356	.303	.244	.187	.136	.094	.061	.037
	2	.001	.031	.098	.176	.246	.297	.324	.328	.311	.278	.234	.186	.138
	3	.000	.002	.015	.042	.082	.132	.185	.236	.276	.303	.312	.303	.276
	4	.000	.000	.001	.006	.015	.033	.060	.095	.138	.186	.234	.278	.311
	5	.000	.000	.000	.000	.002	.004	.010	.020	.037	.061	.094	.136	.187
	6	.000	.000	.000	.000	.000	.000	.001	.002	.004	.008	.016	.028	.047

So, the probability that exactly three of the six workers spend less than 20 minutes each way commuting to work is 0.312.

Try It Yourself 6

Thirty percent of all small businesses in the United States have a website. If you randomly select 10 small businesses, what is the probability that exactly four of them have a website? Use a table to find the probability. *(Source: USA TODAY)*

a. *Identify* a trial, a success, and a failure.
b. Identify n, p, and x.
c. *Use Table 2* in Appendix B to find the binomial probability.
d. *Write* the result as a sentence.

Answer: Page A36

Graphing Binomial Distributions

In Section 4.1, you learned how to graph discrete probability distributions. Because a binomial distribution is a discrete probability distribution, you can use the same process.

EXAMPLE 7

Graphing a Binomial Distribution

Sixty-five percent of households in the United States subscribe to cable TV. You randomly select six households and ask each if they subscribe to cable TV. Construct a probability distribution for the random variable x. Then graph the distribution. *(Source: Polk)*

SOLUTION To construct the binomial distribution, find the probability for each value of x. Using $n = 6$, $p = 0.65$, and $q = 0.35$, you can obtain the following.

x	0	1	2	3	4	5	6
$P(x)$	0.002	0.020	0.095	0.235	0.328	0.244	0.075

You can graph the probability distribution using a histogram as shown below.

Subscribing to Cable TV

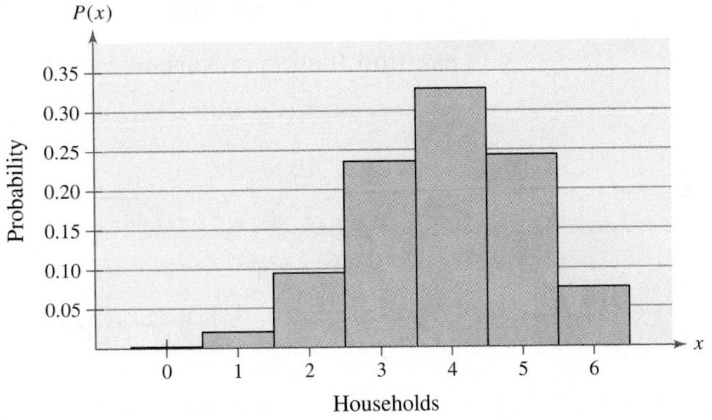

Try It Yourself 7

Fifty-seven percent of households in the United States own a computer. You randomly select six households and ask each if they own a computer. Construct a probability distribution for the random variable x. Then graph the distribution. *(Source: U.S. Department of Commerce)*

a. *Find* the binomial probability for each value of the random variable x.
b. *Organize* the values of x and their corresponding probability in a binomial distribution.
c. Use a histogram to *graph* the binomial distribution. *Answer: Page A36*

Notice in Example 7 that the histogram is skewed left. The graph of a binomial distribution with $p > 0.5$ is skewed left, whereas the graph of a binomial distribution with $p < 0.5$ is skewed right. The graph of a binomial distribution with $p = 0.5$ is symmetric.

Mean, Variance, and Standard Deviation

Although you can use the formulas learned in Section 4.1 for mean, variance, and standard deviation of a discrete probability distribution, the properties of a binomial distribution enable you to use much simpler formulas.

Population Parameters of a Binomial Distribution

Mean: $\mu = np$

Variance: $\sigma^2 = npq$

Standard deviation: $\sigma = \sqrt{npq}$

Note to Instructor

It is useful to have students calculate the mean, variance, and standard deviation for a binomial experiment using the general formulas for discrete probability distributions given in Section 4.1. A quick example (and good review) would be to use the distribution for $n = 4$ and $p = 0.3$.

Point out that the formula $\mu = np$ gives the same result rounded to two decimal places.

x	$P(x)$	$xP(x)$
0	0.240	0
1	0.412	0.412
2	0.265	0.530
3	0.076	0.228
4	0.008	0.032
	$\Sigma = 1.001$	$\Sigma = 1.202$

EXAMPLE 8

Finding and Interpreting Mean, Variance, and Standard Deviation

In Pittsburgh, Pennsylvania, about 56% of the days in a year are cloudy. Find the mean, variance, and standard deviation for the number of cloudy days during the month of June. Interpret the results and determine any unusual values. *(Source: National Climactic Data Center)*

SOLUTION There are 30 days in June. Using

$$n = 30,\ p = 0.56,\ \text{and}\ q = 0.44$$

you can find the mean, variance, and standard deviation as shown below.

$$\mu = np = 30 \cdot 0.56$$
$$= 16.8$$
$$\sigma^2 = npq = 30 \cdot 0.56 \cdot 0.44$$
$$\approx 7.4$$
$$\sigma = \sqrt{npq} = \sqrt{30 \cdot 0.56 \cdot 0.44}$$
$$\approx 2.7$$

Interpretation On average, there are 16.8 cloudy days during the month of June. The standard deviation is about 2.7 days. Values that are more than two standard deviations from the mean are considered unusual. Because $16.8 - 2(2.7) = 11.4$, a June with 11 cloudy days would be unusual. Similarly, because $16.8 + 2(2.7) = 22.2$, a June with 22 cloudy days would also be unusual.

Try It Yourself 8

In San Diego, California, 44% of the days in a year are clear. Find the mean, variance, and standard deviation for the number of clear days during the month of May. Interpret the results and determine any unusual values. *(Source: National Climactic Data Center)*

a. *Identify* a success and the values of n, p, and q.
b. *Find the product* of n and p to calculate the mean.
c. *Find the product* of n, p, and q *for the variance.*
d. *Find the square root* of the variance *for the standard deviation.*
e. *Interpret* the results.

Answer: Page A37

4.2 Exercises

1. (a) $p = 0.50$
   (b) $p = 0.20$
   (c) $p = 0.80$
2. (a) $p = 0.75$
   (b) $p = 0.50$
   (c) $p = 0.25$
3. (a) $n = 12$
   (b) $n = 4$
   (c) $n = 8$
   As n increases, the distribution becomes more symmetric.
4. (a) $n = 10$
   (b) $n = 15$
   (c) $n = 5$
   As n increases, the distribution becomes more symmetric.

Building Basic Skills and Vocabulary

Graphical Analysis In Exercises 1 and 2, match the given probabilities with the correct graph. The histograms each represent binomial distributions. Each distribution has the same number of trials n but different probabilities of success p.

1. $p = 0.20$, $p = 0.50$, $p = 0.80$

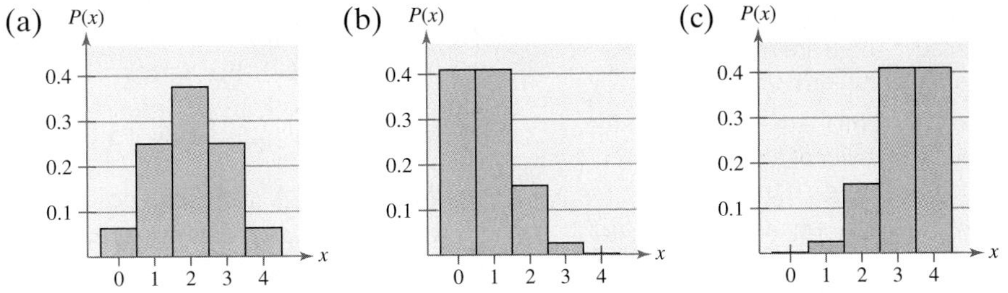

2. $p = 0.25$, $p = 0.50$, $p = 0.75$

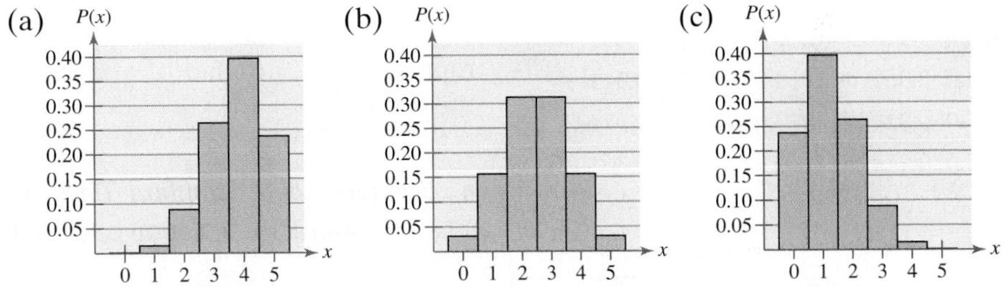

Graphical Analysis In Exercises 3 and 4, match the given values of n with the correct graph. Each histogram shown represents part of a binomial distribution. Each distribution has the same probability of success p but different numbers of trials n. What happens as the value of n increases and the probability of success remains the same?

3. $n = 4$, $n = 8$, $n = 12$

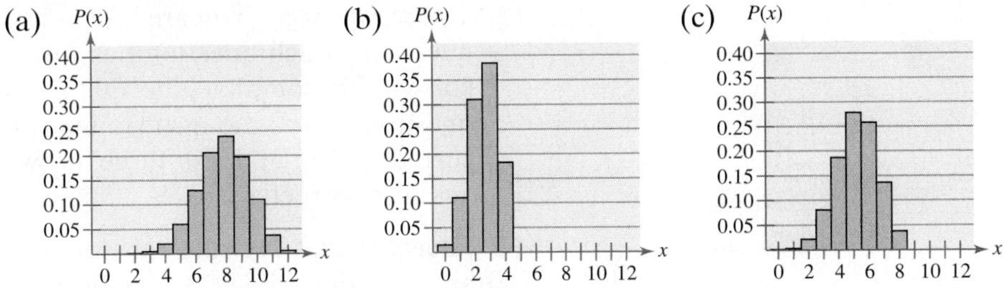

4. $n = 5$, $n = 10$, $n = 15$

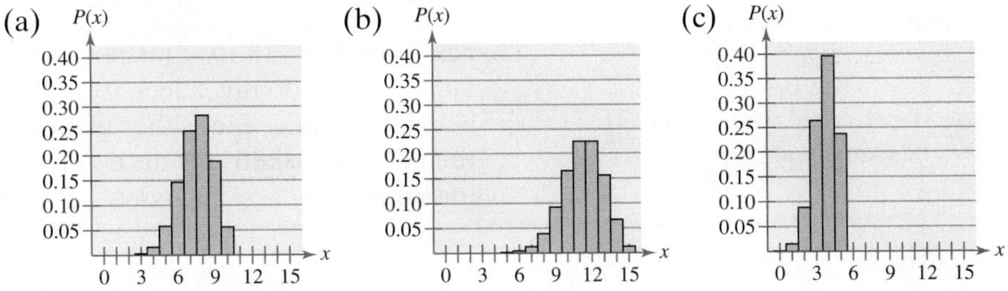

5. Binomial experiment
Success: baby recovers
$n = 5, p = 0.80, q = 0.20,$
$x = 0, 1, 2, \ldots, 5$

6. Binomial experiment
Success: person does not make a purchase
$n = 18, p = 0.74, q = 0.26,$
$x = 0, 1, 2, \ldots, 18$

7. Not a binomial experiment because there are more than two possible outcomes for each trial.

8. Not a binomial experiment because the probability of a success is not the same for each trial.

9. 40; 24; 4.9 **10.** 60.8; 12.16; 3.487

11. 22.08; 18.547; 4.307

12. 215.46; 79.720; 8.929

13. (a) 0.088 (b) 0.104 (c) 0.896

14. (a) 0.318 (b) 0.647 (c) 0.353

15. (a) 0.069 (b) 0.089 (c) 0.911

16. (a) 0.021 (b) 0.026 (c) 0.974

17. (a) 0.213 (b) 0.139 (c) 0.861

18. See Selected Answers, page A94.

19. (a) $n = 6, p = 0.36$

x	$P(x)$
0	0.069
1	0.232
2	0.326
3	0.245
4	0.103
5	0.023
6	0.002

(b)

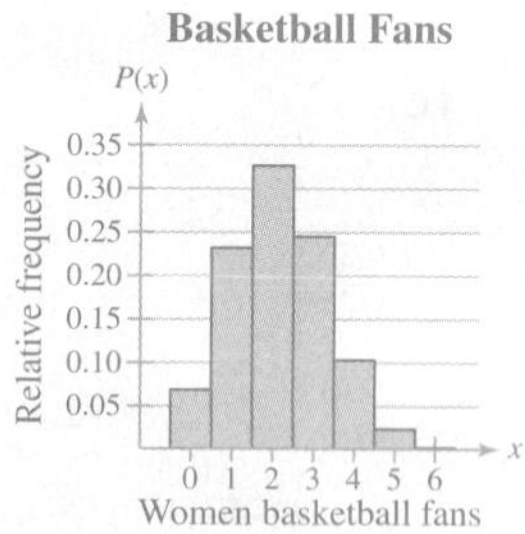

(c) Symmetric

(d) 2.2 (e) 1.4 (f) 1.2

(g) On average 2.2, out of 6, women would consider themselves basketball fans. The standard deviation is 1.2 women.

$x = 0$, 5, or 6 would be uncommon because of their low probabilities.

Identifying and Understanding Binomial Experiments In Exercises 5–8, decide whether the experiment is a binomial experiment. If it is, identify a success, specify the values of n, p, and q, and list the possible values of the random variable x. If it is not a binomial experiment, explain why.

5. ***Cyanosis*** Cyanosis is the condition of having bluish skin due to insufficient oxygen in the blood. About 80% of babies born with cyanosis recover fully. A hospital is caring for five babies born with cyanosis. The random variable represents the number of babies that recover fully. *(Source: The World Book Encyclopedia)*

6. ***Clothing Store Purchases*** From past records, a clothing store finds that 26% of the people who enter the store will make a purchase. During a one-hour period, 18 people enter the store. The random variable represents the number of people who do not make a purchase.

7. ***Political Polls*** A political polling organization calls 1012 people and asks, "Do you approve, disapprove, or have no opinion of the way the president is handling his job?" The random variable represents the number of people who approve of the way the president is handling his job.

8. ***Lottery*** A state lottery randomly chooses six balls numbered from 1 to 40. You choose six numbers and purchase a lottery ticket. The random variable represents the number of matches on your ticket to the numbers drawn in the lottery.

Mean, Variance, and Standard Deviation In Exercises 9–12, find the mean, variance, and standard deviation of the binomial distribution with the given values of n and p.

9. $n = 100, p = 0.4$

10. $n = 76, p = 0.8$

11. $n = 138, p = 0.16$

12. $n = 342, p = 0.63$

Using and Interpreting Concepts

Finding Binomial Probabilities In Exercises 13–18, find the indicated probabilities.

13. ***Answer Guessing*** You are taking a multiple-choice quiz that consists of five questions. Each question has four possible answers, only one of which is correct. To complete the quiz, you randomly guess the answer to each question. Find the probability of guessing (a) exactly three answers correctly, (b) at least three answers correctly, and (c) less than three answers correctly.

14. ***Surgery Success*** A surgical technique is performed on seven patients. You are told there is a 70% chance of success. Find the probability that the surgery is successful for (a) exactly five patients, (b) at least five patients, and (c) less than five patients.

15. ***Basketball Fans*** Fifty-four percent of men consider themselves basketball fans. You randomly select 10 men and ask each if he considers himself a basketball fan. Find the probability that the number who consider themselves basketball fans is (a) exactly eight, (b) at least eight, and (c) less than eight. *(Source: Bruskin-Goldring Research)*

20. See Selected Answers, page A94.

21. (a) $n = 4$
$p = 0.05$

x	$P(x)$
0	0.814506
1	0.171475
2	0.013537
3	0.000475
4	0.000006

(b)

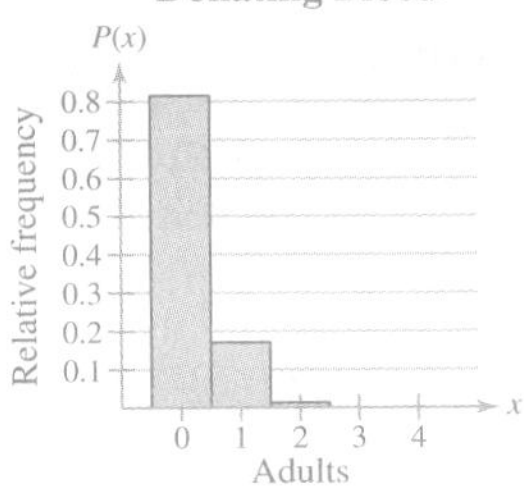

(c) Skewed right

(d) 0.2 (e) 0.2 (f) 0.4

(g) On average 0.2 eligible adults, out of every 4, give blood. The standard deviation is 0.4 adults.

$x = 2, 3,$ or 4 would be uncommon because of their low probabilities.

22. See Selected Answers, page A95.

23. (a) $n = 6, p = 0.61$

x	$P(x)$
0	0.004
1	0.033
2	0.129
3	0.269
4	0.316
5	0.198
6	0.052

(b) 0.129 (c) 0.249

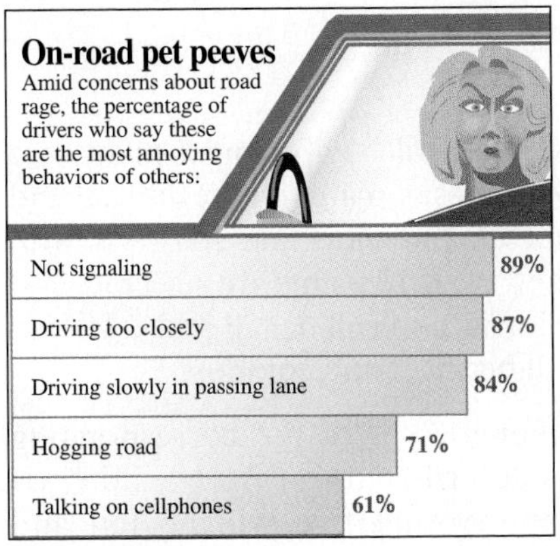

Figure for Exercise 23

16. ***Favorite Cookie*** Ten percent of adults say oatmeal raisin is their favorite cookie. You randomly select 12 adults and ask each to name his or her favorite cookie. Find the probability that the number who say oatmeal raisin is their favorite cookie is (a) exactly four, (b) at least four, and (c) less than four. *(Source: Wearever)*

17. ***Vacation Purpose*** Twenty-one percent of vacationers say the primary purpose of their vacation is outdoor recreation. You randomly select 10 vacationers and ask each to name the primary purpose of his or her vacation. Find the probability that the number who say outdoor recreation is the primary purpose of their vacation is (a) exactly three, (b) more than three, and (c) at most three. *(Source: Travel Industry Association)*

18. ***Honeymoon Financing*** Seventy percent of married couples paid for their honeymoon themselves. You randomly select 20 married couples and ask each if they paid for their honeymoon themselves. Find the probability that the number of couples who say they paid for their honeymoon themselves is (a) exactly one, (b) more than one, and (c) at most one. *(Source: Bride's Magazine)*

Constructing Binomial Distributions In Exercises 19–22, (a) construct a binomial distribution, (b) graph the binomial distribution using a histogram, (c) describe the shape of the histogram, find the (d) mean, (e) variance, and (f) standard deviation of the binomial distribution, and (g) interpret the results in the context of the real-life situation. What values of the random variable x would you consider unusual? Explain your reasoning.

19. ***Women Basketball Fans*** Thirty-six percent of women consider themselves basketball fans. You randomly select six women and ask each if she considers herself a basketball fan. *(Source: Bruskin-Goldring Research)*

20. ***No Trouble Sleeping*** One in four adults says he or she has no trouble sleeping at night. You randomly select five adults and ask each if he or she has no trouble sleeping at night. *(Source: Marist Institute for Public Opinion)*

21. ***Blood Donors*** Five percent of people in the United States eligible to donate blood actually do. You randomly select four eligible blood donors and ask if they donate blood. *(Adapted from American Association of Blood Banks)*

22. ***Blood Types*** Thirty-eight percent of people in the United States have type O^+ blood. You randomly select five Americans and ask them if their blood type is O^+. *(Source: American Association of Blood Banks)*

23. ***Road Rage*** The graph shows the results of a survey of drivers who were asked to name the most annoying habits of other drivers. You randomly select six people who participated in the survey and ask each one of them to name the most annoying habits of other drivers. Let x represent the number who named talking on cell phones as the most annoying habit.

(a) Construct a binomial distribution.

(b) Find the probability that exactly two people will name "talking on cell phones."

(c) Find the probability that at least five people will name "talking on cell phones."

Figure for Exercise 24

24. ***Computer Security*** The graph shows the results of a survey of employees who were asked what types of barriers to effective computer security are in their organizations. You randomly select five employees who participated in the survey and ask each one of them what types of barriers to effective computer security threaten their organizations. Let x represent the number who said that lack of time was a barrier.

(a) Construct a binomial distribution.

(b) Find the probability that exactly two people will say "lack of time."

(c) Find the probability that fewer than four people will say "lack of time."

25. ***Writing*** Find the mean and standard deviation of the binomial distribution in Exercise 23 and interpret the results in the context of the real-life situation. What values of x would you consider unusual? Explain your reasoning.

26. ***Writing*** Find the mean and standard deviation of the binomial distribution in Exercise 24 and interpret the results in the context of the real-life situation. What values of x would you consider unusual? Explain your reasoning.

Extending Concepts

Multinomial Experiments In Exercises 27 and 28, use the following information.

A **multinomial experiment** is a probability experiment that satisfies the following conditions.

1. The experiment is repeated a fixed number of times n where each trial is independent of the other trials.
2. Each trial has k possible mutually exclusive outcomes: $E_1, E_2, E_3, \ldots, E_k$.
3. Each outcome has a fixed probability. Therefore, $P(E_1) = p_1$, $P(E_2) = p_2$, $P(E_3) = p_3, \ldots, P(E_k) = p_k$. The sum of the probabilities for all outcomes is $p_1 + p_2 + p_3 + \cdots + p_k = 1$.
4. x_1 is the number of times E_1 will occur; x_2 is the number of times E_2 will occur; x_3 is the number of times E_3 will occur; and so on.
5. The discrete random variable x counts the number of times $x_1, x_2, x_3, \ldots, x_k$ occurs in n independent trials where $x_1 + x_2 + x_3 + \cdots + x_k = n$. The probability that x will occur is

$$P(x) = \frac{n!}{x_1!x_2!x_3! \cdots x_k!} p_1^{x_1} p_2^{x_2} p_3^{x_3} \ldots p_k^{x_k}.$$

27. ***Genetics*** According to a theory in genetics, if tall and colorful plants are crossed with short and colorless plants, four types of plants will result: tall and colorful, tall and colorless, short and colorful, and short and colorless, with corresponding probabilities of $\frac{9}{16}, \frac{3}{16}, \frac{3}{16}$, and $\frac{1}{16}$. If 10 plants are selected, find the probability that five will be tall and colorful, two will be tall and colorless, two will be short and colorful, and one will be short and colorless.

28. ***Genetics*** Another proposed theory in genetics gives the corresponding probabilities for the four types of plants described as $\frac{5}{16}, \frac{4}{16}, \frac{1}{16}$, and $\frac{6}{16}$. If 10 plants are selected, find the probability that five will be tall and colorful, two will be tall and colorless, two will be short and colorful, and one will be short and colorless.

24. (a) $n = 5, p = 0.41$

x	$P(x)$
0	0.071
1	0.248
2	0.345
3	0.240
4	0.083
5	0.012

(b) 0.345

(c) 0.905

25. 3.7; 1.2

If six drivers are randomly selected, on average 3.7 drivers will name talking on cell phones as the most annoying habit of other drivers.

None of the six randomly selected drivers naming talking on cell phones as the most annoying habit of other drivers is unusual because $P(0) = 0.004$.

26. 2.1; 1.1

If five employees are randomly selected, on average 2.1 employees would claim lack of time was a barrier.

It would be unusual for all 5 employees to claim lack of time was a barrier because $P(5) = 0.012$.

27. 0.033

28. 0.002

Case Study

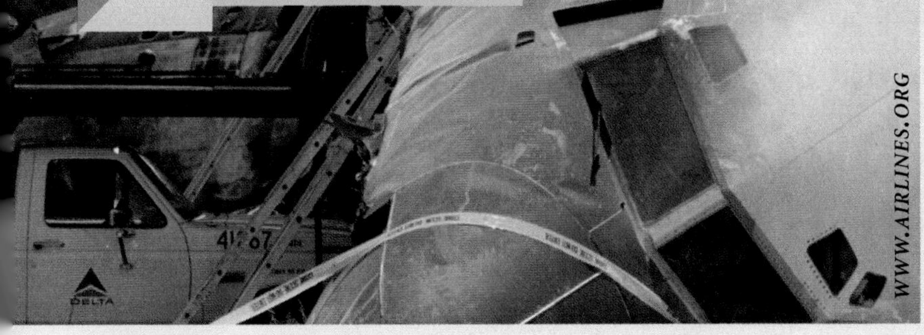

Binomial Distribution of Airplane Accidents

The Air Transport Association of America (ATA) is a support organization for the principal U.S. airlines. Some of the ATA's activities include promoting the air transport industry and conducting industry-wide studies.

The ATA also keeps statistics about commercial airline flights, including those that involve accidents. From 1972 through 2001 for aircraft with 10 or more seats, there were 109 fatal commercial airplane accidents involving U.S. airlines. The distribution of these accidents is shown in the histogram at the right.

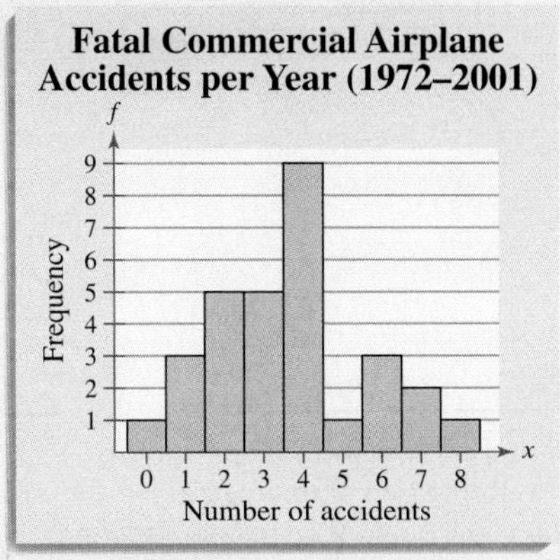

Year	1972	1973	1974	1975	1976	1977	1978	1979	1980	1981	1982	1983	1984	1985	1986
Accidents	7	6	7	2	2	3	5	4	0	4	4	4	1	4	2

Year	1987	1988	1989	1990	1991	1992	1993	1994	1995	1996	1997	1998	1999	2000	2001
Accidents	4	3	8	6	4	4	1	4	2	3	3	1	2	3	6

Exercises

1. In 2001, there were about 9 million commercial flights in the United States. If one is selected at random, what is the probability that it involved a fatal accident?

2. Suppose that the probability of a fatal accident in a given year is 0.0000004. A binomial probability distribution for $n = 9{,}000{,}000$ and $p = 0.0000004$ with $x = 0$ to 10 is shown.

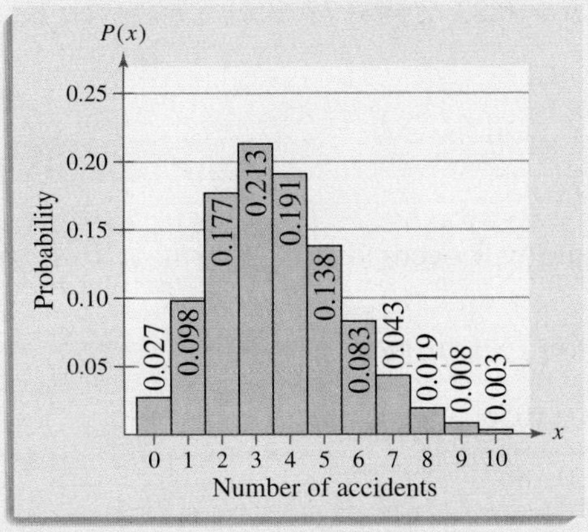

What is the probability that there will be (a) four fatal accidents in a year? (b) 10 fatal accidents? (c) between one and five, inclusive?

3. Construct a binomial distribution for $n = 9{,}000{,}000$ and $p = 0.0000008$ with $x = 0$ to 10. Compare your results with the distribution in Exercise 2.

4. Is a binomial distribution a good model for determining the probability of various numbers of fatal accidents during a year? Explain your reasoning and include a discussion of the four criteria for a binomial experiment.

5. The Federal Aviation Administration says that air flight is so safe that "a person could fly around the clock for over 438 years before being involved in a fatal accident." How can such a statement be justified?

4.3 More Discrete Probability Distributions

What You Should Learn

- How to find probabilities using the geometric distribution
- How to find probabilities using the Poisson distribution

The Geometric Distribution • The Poisson Distribution • Summary of Discrete Probability Distributions

The Geometric Distribution

In this section, you will study two more discrete probability distributions—the geometric distribution and the Poisson distribution.

Many actions in life are repeated until a success occurs. For instance, a CPA candidate might take the CPA exam several times before receiving a passing score, or you might have to dial your Internet connection several times before successfully logging on. Situations such as these can be represented by a geometric distribution.

Note to Instructor

If you choose, this section can be omitted. None of the material included here will be needed in later sections.

DEFINITION

A **geometric distribution** is a discrete probability distribution of a random variable x that satisfies the following conditions.

1. A trial is repeated until a success occurs.
2. The repeated trials are independent of each other.
3. The probability of success p is constant for each trial.

The **probability that the first success will occur on trial number x** is

$$P(x) = p(q)^{x-1}, \text{ where } q = 1 - p.$$

EXAMPLE 1

Finding Probabilities Using the Geometric Distribution

From experience, you know that the probability that you will make a sale on any given telephone call is 0.23. Find the probability that your first sale on any given day will occur on your fourth or fifth sales call.

SOLUTION To find the probability that your first sale will occur on the fourth or fifth call, first find the probability that the sale will occur on the fourth call and the probability that the sale will occur on the fifth call. Then find the sum of the resulting probabilities. Using $p = 0.23$, $q = 0.77$, and $x = 4$, you have

$$P(4) = 0.23 \cdot (0.77)^3 \approx 0.105003.$$

Using $p = 0.23$, $q = 0.77$, and $x = 5$, you have

$$P(5) = 0.23 \cdot (0.77)^4 \approx 0.080852.$$

So, the probability that your first sale will occur on the fourth or fifth sales call is

$$\begin{aligned} P(\text{sale on fourth or fifth call}) &= P(4) + P(5) \\ &\approx 0.105003 + 0.080852 \\ &\approx 0.186. \end{aligned}$$

Study Tip

If the first success occurs on the fourth trial, the outcome is *FFFS* and the probability is

$$P = \overset{F}{(0.77)}\overset{F}{(0.77)}\overset{F}{(0.77)}\overset{S}{(0.23)}.$$

If the first success occurs on the fifth trial, the outcome is *FFFFS* and the probability is

$$P = \overset{F}{(0.77)}\overset{F}{(0.77)}\overset{F}{(0.77)}\overset{F}{(0.77)}\overset{S}{(0.23)}.$$

Note to Instructor

Even though theoretically a success may never occur, the geometric distribution is a discrete probability distribution because the values of x can be listed—1, 2, 3, ... Notice that as x becomes larger, $P(x)$ gets closer to zero. For instance

$$P(50) = 0.23(0.77)^{49} \approx 0.0000006306.$$

The sum of all the probabilities is a geometric series with an infinite number of terms. The sum of this geometric series is $s = a/(1 - r)$, where a is the first term of the series, and r is the common ratio of consecutive terms. In this case $a = p$ and $r = q$, so, $s = p/(1 - q) = p/p = 1$. So, each probability value is between 0 and 1, inclusive, and the sum of all the probabilities is 1.

Try It Yourself 1

Find the probability that your first sale will occur before your fourth sales call.

a. *Use the geometric distribution* to find $P(1)$, $P(2)$, and $P(3)$.
b. *Find the sum* of $P(1)$, $P(2)$, and $P(3)$.
c. *Write* the result as a sentence.

Answer: Page A37

The Poisson Distribution

In a binomial experiment you are interested in finding the probability of a specific number of successes in a given number of trials. Suppose instead that you want to know the probability that a specific number of occurrences takes place within a given unit of time or space. For instance, to determine the probability that an employee will take 15 sick days within a year, you can use the Poisson distribution.

DEFINITION

The **Poisson distribution** is a discrete probability distribution of a random variable x that satisfies the following conditions.

1. The experiment consists of counting the number of times, x, an event occurs in a given interval. The interval can be an interval of time, area, or volume.
2. The probability of the event occurring is the same for each interval.
3. The number of occurrences in one interval is independent of the number of occurrences in other intervals.

The probability of exactly x occurrences in an interval is

$$P(x) = \frac{\mu^x e^{-\mu}}{x!}$$

where e is an irrational number approximately equal to 2.71828 and μ is the mean number of occurrences per interval unit.

EXAMPLE 2

Using the Poisson Distribution

The mean number of accidents per month at a certain intersection is 3. What is the probability that in any given month four accidents will occur at this intersection?

SOLUTION Using $x = 4$ and $\mu = 3$, the probability that four accidents will occur in any given month at the intersection is

$$P(4) = \frac{3^4(2.71828)^{-3}}{4!} \approx 0.168.$$

Try It Yourself 2

What is the probability that more than four accidents will occur in any given month at the intersection?

a. *Use the Poisson distribution* to find $P(0)$, $P(1)$, $P(2)$, $P(3)$, and $P(4)$.
b. *Find the sum* of $P(0)$, $P(1)$, $P(2)$, $P(3)$, and $P(4)$.
c. *Subtract* the sum from 1.
d. *Write* the result as a sentence.

Answer: Page A37

In Example 2 you used a formula to determine a Poisson probability. You can also use a table to find Poisson probabilities. Table 3 in Appendix B lists the Poisson probability for selected values of x and μ. You can use technology tools, such as MINITAB, Excel, and the TI-83, to find Poisson probabilities as well.

Picturing the World

The first successful suspension bridge built in the United States, the Tacoma Narrows Bridge, spans the Tacoma Narrows in Washington State. The average occupancy of vehicles that travel across the bridge is 1.6. The following probability distribution represents the vehicle occupancy on the bridge during a five-day period. *(Source: Washington State Department of Transportation)*

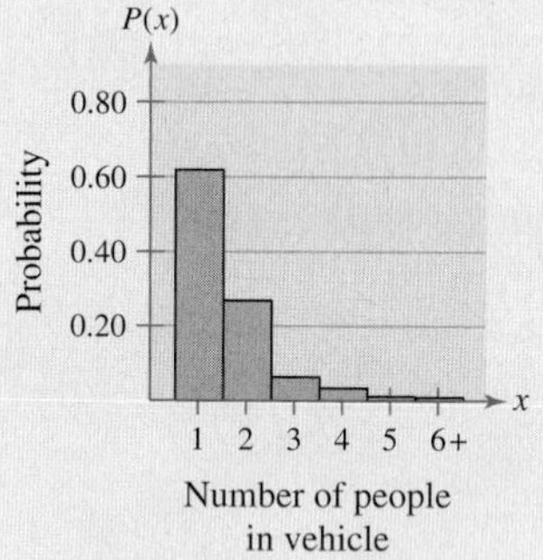

What is the probability that a randomly selected vehicle has two occupants or fewer?

EXAMPLE 3

Finding Poisson Probabilities Using a Table

A population count shows that there is an average of 3.6 rabbits per acre living in a field. Use a table to find the probability that two rabbits are found on any given acre of the field.

SOLUTION A portion of Table 3 in Appendix B is shown here. Using the distribution for $\mu = 3.6$ and $x = 2$, you can find the Poisson probability as shown by the highlighted areas in the table.

	μ						
x	**3.1**	**3.2**	**3.3**	**3.4**	**3.5**	**3.6**	**3.7**
0	.0450	.0408	.0369	.0334	.0302	.0273	.0247
1	.1397	.1304	.1217	.1135	.1057	.0984	.0915
2	.2165	.2087	.2008	.1929	.1850	.1771	.1692
3	.2237	.2226	.2209	.2186	.2158	.2125	.2087
4	.1734	.1781	.1823	.1858	.1888	.1912	.1931
5	.1075	.1140	.1203	.1264	.1322	.1377	.1429
6	.0555	.0608	.0662	.0716	.0771	.0826	.0881
7	.0246	.0278	.0312	.0348	.0385	.0425	.0466
8	.0095	.0111	.0129	.0148	.0169	.0191	.0215
9	.0033	.0040	.0047	.0056	.0066	.0076	.0089
10	.0010	.0013	.0016	.0019	.0023	.0028	.0033

So, the probability that two rabbits are found on any given acre is 0.1771.

Try It Yourself 3

Two thousand brown trout are introduced into a small lake. The lake has a volume of 20,000 cubic meters. Use a table to find the probability that three brown trout are found in any given cubic meter of the lake.

a. *Find the average* number of brown trout per cubic meter.
b. *Identify* μ and x.
c. *Use Table 3* in Appendix B to find the Poisson probability.
d. *Write* the result as a sentence.

Answer: Page A37

Summary of Discrete Probability Distributions

The following table summarizes the discrete probability distributions discussed in this chapter.

Distribution	Summary	Formulas
Binomial Distribution	A binomial experiment satisfies the following conditions. **1.** The experiment is repeated for a fixed number (n) of independent trials. **2.** There are only two possible outcomes for each trial. Each outcome can be classified as a success or as a failure. **3.** The probability of a success must remain constant for each trial. **4.** The random variable x counts the number of successful trials out of the n trials. The parameters of a binomial distribution are n and p.	x = the number of successes in n trials p = probability of success in a single trial q = probability of failure in a single trial $q = 1 - p$ The probability of exactly x successes in n trials is $P(x) = {}_nC_x p^x q^{n-x} = \frac{n!}{(n-x)!x!} p^x q^{n-x}.$
Geometric Distribution	A geometric distribution is a discrete probability distribution of the random variable x that satisfies the following conditions. **1.** A trial is repeated until a success occurs. **2.** The repeated trials are independent of each other. **3.** The probability of success p is constant for each trial. **4.** The random variable x represents the number of the trial in which the first success occurs. The parameter of a geometric distribution is p.	x = the number of the trial in which the first success occurs p = probability of success in a single trial q = probability of failure in a single trial $q = 1 - p$ The probability that the first success occurs on trial number x is $P(x) = p(q)^{x-1}.$
Poisson Distribution	The Poisson distribution is a discrete probability distribution that gives the probability of x occurrences of an event over a specified interval of time, area, or volume. The parameter of a Poisson distribution is μ.	x = the number of occurrences in the given interval μ = the mean number of occurrences in a given time or space unit The probability of exactly x occurrences in an interval is $P(x) = \frac{\mu^x e^{-\mu}}{x!}.$

4.3 Exercises

Help

MyMathLab

Student Study Pack

Building Basic Skills and Vocabulary

In Exercises 1–4, assume the geometric distribution applies. Use the given probability of success p to find the indicated probability.

1. Find $P(3)$ when $p = 0.80$.

2. Find $P(1)$ when $p = 0.25$.

3. Find $P(6)$ when $p = 0.09$.

4. Find $P(4)$ when $p = 0.62$.

In Exercises 5–8, assume the Poisson distribution applies. Use the given mean μ to find the indicated probability.

5. Find $P(4)$ when $\mu = 2$.

6. Find $P(5)$ when $\mu = 6$.

7. Find $P(2)$ when $\mu = 1.5$.

8. Find $P(4)$ when $\mu = 8.1$.

Deciding on a Distribution In Exercises 9–14, decide which probability distribution—binomial, geometric, or Poisson—applies to the question. You do not need to answer the question. Instead, justify your choice.

9. *Pilot's Test* *Given:* The probability that a student passes the written test for a private pilot's license is 0.75. *Question:* What is the probability that a student will fail the test on the first attempt and pass it on the second attempt?

10. *Precipitation* *Given:* In Rapid City, South Dakota, the mean number of days with 0.01 inch or more precipitation for May is 12. *Question:* What is the probability that Rapid City has 18 days with 0.01 inch or more precipitation next May? *(Source: National Climatic Data Center)*

11. *Oil Tankers* *Given:* The mean number of oil tankers at a port city is 8 per day. The port has facilities to handle up to 12 oil tankers in a day. *Question:* What is the probability that too many tankers will arrive on a given day?

12. *Exercise* *Given:* Forty percent of adults in the United States exercise at least 30 minutes a week. In a survey of 120 randomly chosen adults, people were asked, "Do you exercise at least 30 minutes a week?" *Question:* What is the probability that exactly 50 of the people answer yes?

13. *Cheaters* *Given:* Of students ages 16 to 18 with A or B averages who plan to attend college after graduation, 80% cheated to get higher grades. Ten randomly chosen students with A or B averages who plan to attend college after graduation were asked, "Did you cheat to get higher grades?" *Question:* What is the probability that exactly two students answered no? *(Source: The Educational Testing Service)*

14. *No Meat?* *Given:* About 21% of Americans say they could not go one week without eating meat. You select at random 20 Americans. *Question:* What is the probability that the first person who says he or she cannot go one week without eating meat is the fifth person selected? *(Source: Reuters/Zogby)*

1. 0.032
2. 0.25
3. 0.056
4. 0.034
5. 0.09
6. 0.161
7. 0.251
8. 0.054
9. Geometric. You are interested in counting the number of trials until the first success.
10. Poisson. You are interested in counting the number of occurrences that take place within a given unit of time.
11. Poisson. You are interested in counting the number of occurrences that take place within a given unit of space.
12. Binomial. You are interested in counting the number of successes out of *n* trials.
13. Binomial. You are interested in counting the number of successes out of *n* trials.
14. Geometric. You are interested in counting the number of trials until the first success.

15. (a) 0.082
(b) 0.469
(c) 0.531
16. (a) 0.248
(b) 0.789
(c) 0.211
17. (a) 0.009
(b) 0.030
(c) 0.904
18. (a) 0.105
(b) 0.578
(c) 0.317
19. (a) 0.057
(b) 0.958
(c) 0.900
20. (a) 0.195
(b) 0.433
(c) 0.567
21. (a) 0.348
(b) 0.844
(c) 0.156
22. (a) 0.131
(b) 0.627
(c) 0.373

Using and Interpreting Concepts

Using a Geometric Distribution to Find Probabilities In Exercises 15–18, find the indicated probabilities using the geometric distribution. If convenient, use technology to find the probabilities.

15. ***Telephone Sales*** Assume the probability that you will make a sale on any given telephone call is 0.19. Find the probability that you (a) make your first sale on the fifth call, (b) make your first sale on the first, second, or third call, and (c) do not make a sale on the first three calls.

16. ***Free Throws*** Basketball player Shaquille O'Neal makes a free throw shot about 54.1% of the time. Find the probability that (a) the first shot O'Neal makes is the second shot, (b) the first shot O'Neal makes is the first or second shot, and (c) O'Neal does not make two shots. *(Source: CBS SportsLine.com)*

17. ***Auto Parts*** An auto parts seller finds that 1 in every 100 parts sold is defective. Find the probability that (a) the first defective part is the tenth part sold, (b) the first defective part is the first, second, or third part sold, and (c) none of the first 10 parts sold are defective.

18. ***Winning a Prize*** A cereal maker places a game piece in its cereal boxes. The probability of winning a prize in the game is 1 in 4. Find the probability that you (a) win your first prize with your fourth purchase, (b) win your first prize with your first, second, or third purchase, and (c) do not win a prize with your first four purchases.

Using a Poisson Distribution to Find Probabilities In Exercises 19–22, find the indicated probabilities using the Poisson distribution. If convenient, use a Poisson probability table or technology tool to find the probabilities.

19. ***Business Failures*** The mean number of business failures per hour in the United States in a recent year was about 8. Find the probability that (a) exactly four businesses will fail in any given hour, (b) at least four businesses will fail in any given hour, and (c) more than four businesses will fail in any given hour. *(Source: The Wall Street Journal Almanac)*

20. ***Typographical Errors*** A newspaper finds that the mean number of typographical errors per page is 4. Find the probability that (a) exactly three typographical errors will be found on a page, (b) at most three typographical errors will be found on a page, and (c) more than three typographical errors will be found on a page.

21. ***Major Hurricanes*** A major hurricane is a hurricane with wind speeds of 111 miles per hour or greater. During the 20th century, the mean number of major hurricanes to strike the U.S. mainland per year was about 0.7. Find the probability that in a given year (a) exactly one major hurricane will strike the U.S. mainland, (b) at most one major hurricane will strike the U.S. mainland, and (c) more than one major hurricane will strike the U.S. mainland. *(Source: National Hurricane Center)*

22. ***Precipitation*** The mean number of days with 0.01 inch or more precipitation per month for Lewistown, Idaho, is about 8.7. Find the probability that in a given month (a) there are exactly 9 days with 0.01 inch or more precipitation, (b) there are at most 9 days with 0.01 inch or more precipitation, and (c) there are more than 9 days with 0.01 inch or more precipitation. *(Source: National Climatic Data Center)*

23. (a) 0.1453996179
(b) 0.1453688667; The results are approximately the same.
24. (a) 0.629
(b) 0.343
(c) 0.029
25. (a) 1000; 999,000; 999.5
On average you would have to play 1000 times until you won the lottery. The standard deviation is 999.5.
(b) 1000 times
Lose money. On average you would win $500 every 1000 times you play the lottery. So, the net gain would be −$500.
26. (a) 200; 39,800; 199.5
On average 200 records will be examined before finding one that has been miscalculated. The standard deviation is 199.5.
(b) 200
27. (a) $\sigma^2 = 3.9$, $\sigma \approx 2.0$
The standard deviation is 2.0 strokes.
(b) 0.352
28. (a) 4; 2
The standard deviation is 2 inches.
(b) 0.05

Extending Concepts

23. *Approximating the Binomial Distribution* A glass manufacturer finds that 1 in every 1000 glass items produced is warped. (a) Use a binomial distribution to find the probability of finding five defective glass items in a random sample of 6500 glass items. (b) The Poisson distribution can be used to approximate the binomial distribution for large values of n and small values of p. Repeat (a) using a Poisson distribution and compare the results.

24. *Hypergeometric Distribution* Binomial experiments require that any sampling be done with replacement because each trial must be independent of the others. The **hypergeometric distribution** also has two outcomes—success and failure. However, the sampling is done without replacement. Given a population of N items having k successes and $N - k$ failures, the probability of selecting a sample of size n that has x successes and $n - x$ failures is given by

$$P(x) = \frac{({}_kC_x)({}_{N-k}C_{n-x})}{{}_NC_n}.$$

In a shipment of 15 microchips, two are defective and 13 are not defective. A sample of three microchips is chosen at random. Find the probability that (a) all three microchips are not defective, (b) one microchip is defective and two are not defective, and (c) two microchips are defective and one is not defective.

Geometric Distribution: Mean and Variance In Exercises 25 and 26, use the fact that the mean of a geometric distribution is $\mu = 1/p$ and the variance is $\sigma^2 = q/p^2$.

25. *Daily Lottery* A daily number lottery chooses three balls numbered 0 to 9. The probability of winning the lottery is 1/1000. Let x be the number of times you play the lottery before winning the first time. (a) Find the mean, variance, and standard deviation. Interpret the results. (b) How many times would you expect to have to play the lottery before winning? Assume that it costs $1 to play and winners are paid $500. Would you expect to make or lose money playing this lottery? Explain.

26. *Paycheck Errors* A company assumes that 0.5% of the paychecks for a year were calculated incorrectly. The company has 200 employees and examines the payroll records from one month. (a) Find the mean, variance, and standard deviation. Interpret the results. (b) How many employee payroll records would you expect to examine before finding one with an error?

Poisson Distribution: Variance In Exercises 27 and 28, use the fact that the variance of a Poisson distribution is $\sigma^2 = \mu$.

27. *Tiger Woods* At one point in a recent year, the mean number of strokes per hole for golfer Tiger Woods was about 3.9. (a) Find the variance and standard deviation. Interpret the results. (b) How likely is Woods to play an 18-hole round and have more than 72 strokes? *(Source: PGATour.com)*

28. *Snowfall* The mean snowfall in January for Evansville, Indiana, is 4.0 inches. (a) Find the variance and standard deviation. Interpret the results. (b) Find the probability that the snowfall in January for Evansville, Indiana, will exceed 7 inches. *(Source: National Climatic Data Center)*

Uses and Abuses
Statistics in the Real World

Uses

There are countless occurrences of binomial probability distributions in business, science, engineering, and many other fields.

For example, suppose you work for a marketing agency and are in charge of creating a television ad for Brand A toothpaste. The toothpaste manufacturer claims that 40% of toothpaste buyers prefer its brand. To check whether the manufacturer's claim is reasonable, your agency conducts a survey. Of 100 toothpaste buyers selected at random, you find that only 35 (or 35%) prefer Brand A. Could the manufacturer's claim still be true? What if your random sample of 100 found only 25 people (or 25%) who express a preference for Brand A? Would you still be justified in running the advertisement?

Knowing the characteristics of binomial probability distributions will help you answer this type of question. By the time you have completed this course, you will be able make educated decisions about the reasonableness of the manufacturer's claim.

Abuses

Interpreting the "Most Likely" Outcome A common misuse of binomial probability distributions is to think that the "most likely" outcome is the outcome that will occur most of the time. For instance, suppose you randomly choose a committee of four from a large population that is 50% women and 50% men. The most likely composition of the committee is that it will contain two men and two women. Although this is the most likely outcome, the probability that it will occur is only 0.375. There is a 0.5 chance that the committee will contain one man and three women or three men and one woman. So, if either of these outcomes occurs, you should not assume that the selection was unusual or biased.

Exercises

In Exercises 1 and 2, suppose that the manufacturer's claim is true—40% of toothpaste buyers prefer Brand A toothpaste. Use the graph and technology to answer the questions.

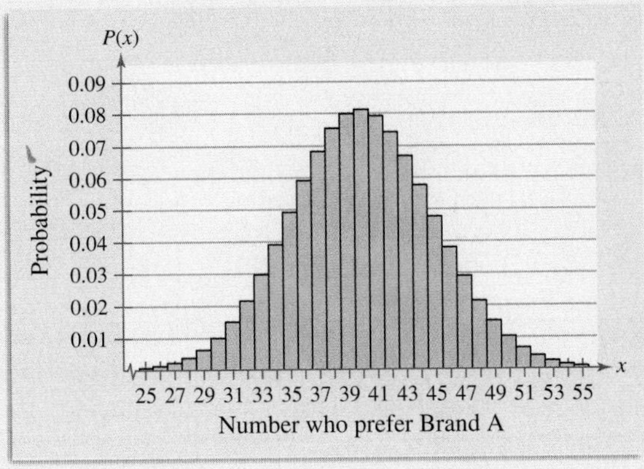

1. ***Interpreting the "Most Likely" Outcome*** In a random sample of 100, what is the most likely outcome? How likely is it?
2. ***Interpreting the "Most Likely" Outcome*** In a random sample of 100, what is the probability that between 35 and 45 people, inclusive, prefer Brand A? Explain your reasoning.
3. Suppose in a random sample of 100, you found 36 who prefer Brand A. Would the manufacturer's claim be believable? Explain your reasoning.
4. Suppose in a random sample of 100, you found 25 who prefer Brand A. Would the manufacturer's claim be believable? Explain your reasoning.

4 Chapter Summary

What did you learn?

Review Exercises

Section 4.1

- How to distinguish between discrete random variables and continuous random variables — *1–4*
- How to determine if a distribution is a probability distribution — *5–10*
- How to construct a discrete probability distribution and its graph and find the mean, variance, and standard deviation of a discrete probability distribution — *11–14*

$$\mu = \Sigma x P(x)$$
$$\sigma^2 = \Sigma(x - \mu)^2 P(x)$$
$$\sigma = \sqrt{\sigma^2}$$

- How to find the expected value of a discrete probability distribution — *15, 16*

Section 4.2

- How to determine if a probability experiment is a binomial experiment — *17, 18*
- How to find binomial probabilities using the binomial probability formula, a binomial probability table, and technology — *19–22*

$$P(x) = {}_nC_x p^x q^{n-x} = \frac{n!}{(n-x)!x!} p^x q^{n-x}$$

- How to construct a binomial distribution and its graph and find the mean, variance, and standard deviation of a binomial probability distribution — *23–26*

$$\mu = np$$
$$\sigma^2 = npq$$
$$\sigma = \sqrt{npq}$$

Section 4.3

- How to find probabilities using the geometric distribution — *27, 28*

$$P(x) = pq^{x-1}$$

- How to find probabilities using the Poisson distribution — *29, 30*

$$P(x) = \frac{\mu^x e^{-\mu}}{x!}$$

4 Review Exercises

1. Discrete
2. Continuous
3. Continuous
4. Discrete
5. No, $\sum P(x) \neq 1$.
6. Yes
7. Yes
8. No, $P(5) > 1$ and $\sum P(x) \neq 1$.
9. Yes
10. No, $\sum P(x) \neq 1$.

Section 4.1

In Exercises 1–4, decide whether the random variable x is discrete or continuous.

1. x represents the number of pumps in use at a gas station.

2. x represents the weight of a truck at a weigh station.

3. x represents the amount of gas pumped at a gas station.

4. x represents the number of defects on a microchip.

In Exercises 5–10, decide whether the distribution is a probability distribution. If it is not, identify the property that is not satisfied.

5. The daily limit for catching bass at a lake is four. The random variable x represents the number of fish caught in a day.

x	0	1	2	3	4
$P(x)$	0.36	0.23	0.08	0.14	0.29

6. The vending machine at a workplace holds 48 bottles of iced tea. The random variable x represents the number of days until the iced tea is gone.

x	1	2	3	4	5	6
$P(x)$	0.05	0.15	0.40	0.25	0.10	0.05

7. A greeting card shop keeps records of customers' buying habits. The random variable x represents the number of cards sold to an individual customer in a shopping visit.

x	1	2	3	4	5	6	7
$P(x)$	0.68	0.14	0.08	0.05	0.02	0.02	0.01

8. The random variable x represents the number of classes in which a student is enrolled in a given semester at a university.

x	1	2	3	4	5	6	7	8
$P(x)$	$\frac{1}{80}$	$\frac{2}{75}$	$\frac{1}{10}$	$\frac{12}{25}$	$\frac{27}{20}$	$\frac{1}{5}$	$\frac{2}{25}$	$\frac{1}{120}$

9. In a survey, Internet users were asked how many email addresses they have. The random variable x represents the number of email addresses.

x	1	2	3
$P(x)$	0.26	0.31	0.43

10. In a survey, adults were asked how many email addresses they have. The random variable x represents the number of email addresses.

x	0	1	2	3
$P(x)$	0.11	0.20	0.28	0.39

In Exercises 11–14,

(a) use the frequency distribution table to construct a probability distribution.

(b) graph the probability distribution using a histogram.

(c) find the mean, variance, and standard deviation of the probability distribution.

11. The number of pages in a section from a sample of statistics texts

Pages	Sections
2	3
3	12
4	72
5	115
6	169
7	120
8	83
9	48
10	22
11	6

12. The number of goals scored by a soccer team during a 32-game season

Goals	Games
0	7
1	8
2	10
3	3
4	3
5	1

13. A survey asked 200 households how many televisions they owned.

Televisions	Households
0	3
1	38
2	83
3	52
4	18
5	5
6	1

14. A television station sells advertising in 15-, 30-, 60-, 90-, and 120-second blocks. The distribution for one 24-hour day is given.

Length (in seconds)	Number
15	76
30	445
60	30
90	3
120	12

In Exercises 15 and 16, find the expected value of the random variable.

15. A person has shares of eight different stocks. The random variable x represents the number of stocks showing a loss on a selected day.

x	0	1	2	3	4	5	6	7	8
$P(x)$	0.02	0.11	0.18	0.32	0.15	0.09	0.05	0.05	0.03

11. (a)

x	f	$P(x)$
2	3	0.005
3	12	0.018
4	72	0.111
5	115	0.177
6	169	0.260
7	120	0.185
8	83	0.128
9	48	0.074
10	22	0.034
11	6	0.009
	$n = 650$	$\Sigma P(x) \approx 1$

(b)

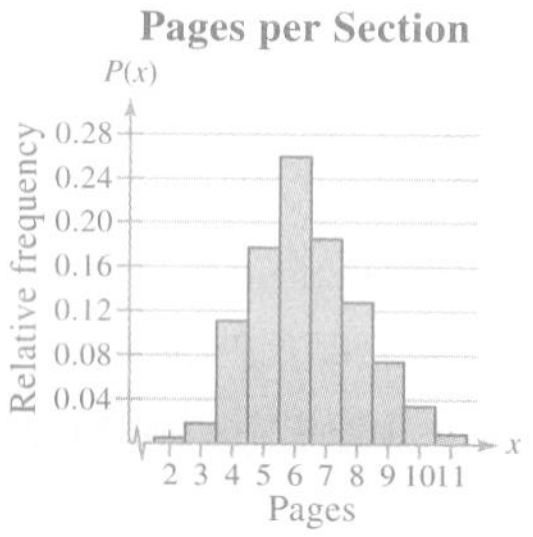

(c) 6.4, 2.9, 1.7

12. See Selected Answers, page A95.

13. (a)

x	f	$P(x)$
0	3	0.015
1	38	0.190
2	83	0.415
3	52	0.260
4	18	0.090
5	5	0.025
6	1	0.005
	$n = 200$	$\Sigma P(x) = 1$

(b)

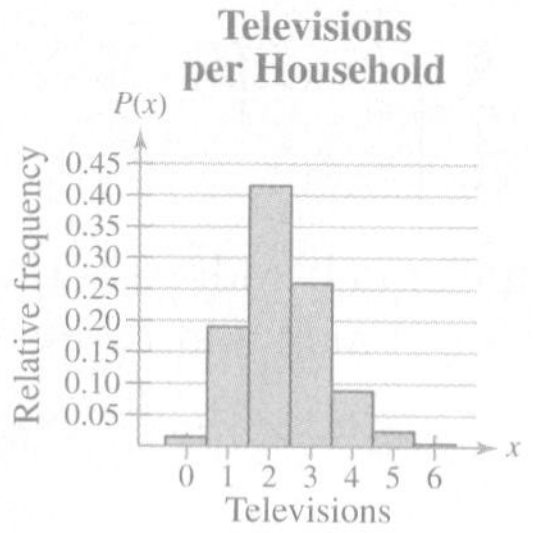

(c) 2.3, 1.1, 1.0

14. (a)

x	f	$P(x)$
15	76	0.134
30	445	0.786
60	30	0.053
90	3	0.005
120	12	0.021
	$n = 566$	$\Sigma P(x) \approx 1$

(b)

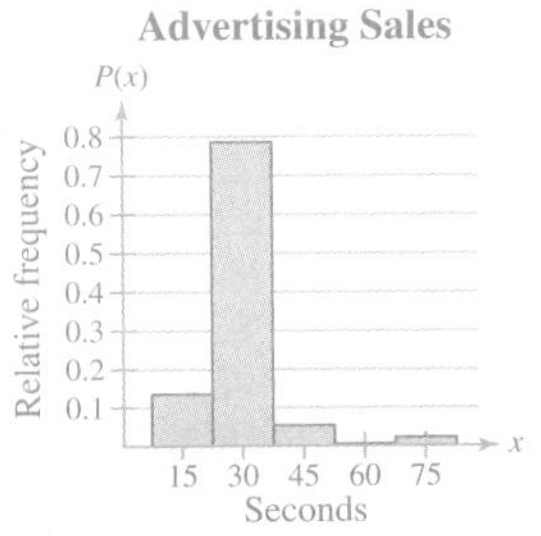

(c) 31.8, 265.5, 16.3

15. 3.4 **16.** 2.5

17. Yes, $n = 12$, $p = 0.24$, $q = 0.76$, $x = 0, 1, \ldots, 12$

18. No, the experiment is not repeated for a fixed number of trials.

19. (a) 0.208 (b) 0.322 (c) 0.114

20. (a) 0.133 (b) 0.205 (c) 0.073

21. (a) 0.294 (b) 0.644 (c) 0.350

22. (a) 0.274 (b) 0.388 (c) 0.114

23. (a)

x	$P(x)$
0	0.007
1	0.059
2	0.201
3	0.342
4	0.291
5	0.099

(b)

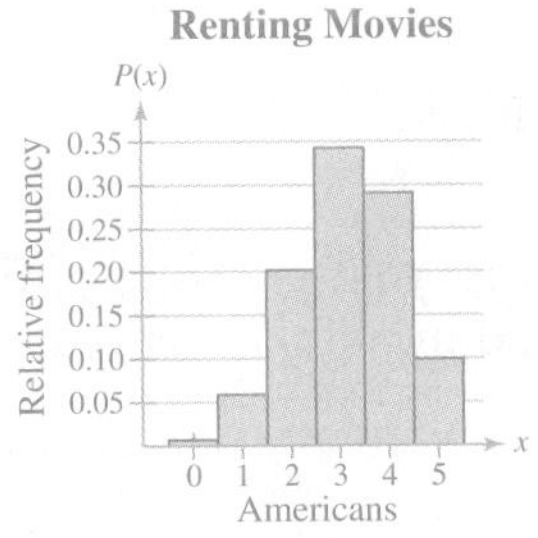

(c) 3.2, 1.2, 1.1

16. A local pub has a chicken wing special on Tuesdays. The pub owners purchase wings in cases of 300. The random variable x represents the number of cases used during the special.

x	1	2	3	4
$P(x)$	$\frac{1}{9}$	$\frac{1}{3}$	$\frac{1}{2}$	$\frac{1}{18}$

Section 4.2

In Exercises 17 and 18, decide whether the experiment is a binomial experiment. If it is not, identify the property that is not satisfied. If it is, list the values of n, p, and q and the values that x can assume.

17. Bags of plain M&M's contain 24% blue candies. One candy is selected from each of 12 bags. The random variable represents the number of blue candies selected. *(Source: Mars, Inc.)*

18. A fair coin is tossed repeatedly until 15 heads are obtained. The random variable x counts the number of tosses.

In Exercises 19–22, find the indicated probabilities.

19. One in four adults is currently on a diet. In a random sample of eight adults, what is the probability that the number currently on a diet is (a) exactly three, (b) at least three, and (c) more than three? *(Source: Wirthlin Worldwide)*

20. One in five people in the United States own individual stocks. In a random sample of 12 people, what is the probability that the number owning individual stocks is (a) exactly four, (b) at least four, and (c) more than four? *(Source: ABC News)*

21. Forty-three percent of adults in the United States receive fewer than five phone calls a day. In a random sample of seven adults, what is the probability that the number receiving fewer than five calls a day is (a) exactly three, (b) at least three, and (c) more than three? *(Source: Wirthlin Worldwide)*

22. In a typical day, 26% of people in the United States with Internet access go online to get news. In a random sample of five people in the United States with Internet access, what is the probability that the number going online to get news is (a) exactly two, (b) at least two, and (c) more than two? *(Source: Horizons Information Systems Consultants)*

In Exercises 23–26,

(a) construct a binomial distribution.

(b) graph the binomial distribution using a histogram.

(c) find the mean, variance, and standard deviation of the binomial distribution.

23. Sixty-three percent of adults in the United States rent videotapes or DVDs at least once a month. Consider a random sample of five Americans who are asked if they rent at least one videotape or DVD a month.

24. (a)

x	$P(x)$
0	0.006
1	0.050
2	0.167
3	0.294
4	0.293
5	0.155
6	0.034

(b)

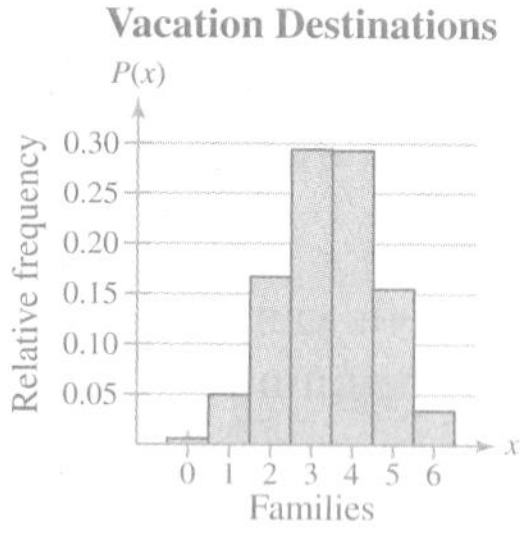

(c) 3.4, 1.5, 1.2

25. (a)

x	$P(x)$
0	0.012
1	0.096
2	0.293
3	0.397
4	0.202

(b)

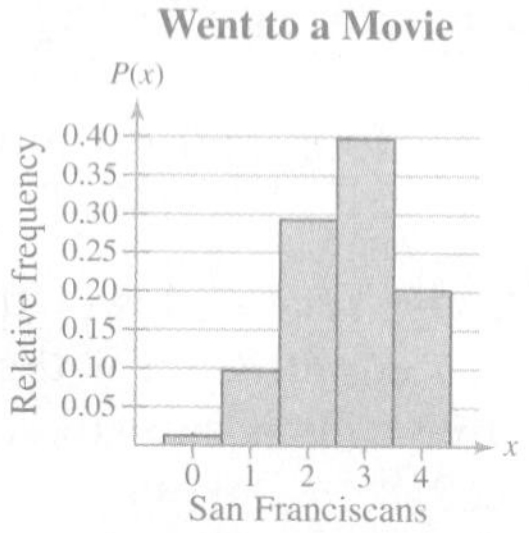

(c) 2.7, 0.9, 0.9

26. See Selected Answers, page A95.

27. (a) 0.096 (b) 0.518 (c) 0.579

28. (a) 0.477 (b) 0.249 (c) 0.726 (d) 0.856

29. (a) 0.810 (b) 0.171 (c) 0.019

30. (a) 0.997 (b) 0.875 (c) 1

24. Fifty-seven percent of families say that their children have an influence on their vacation destinations. Consider a random sample of six families who are asked if their children have an influence on their vacation destinations. *(Source: YP&B/Yankelovich 1997 Travel Monitor)*

25. Over a period of three months, 67% of San Franciscans went out to the movies. Consider a random sample of four San Franciscans who are asked if they went to the movies in those three months. *(Source: Scarborough Research)*

26. In a typical day, 17% of people in the United States with Internet access go online to check the weather. Consider a random sample of five people in the United States with Internet access who are asked if they check the weather when they go online. *(Source: Horizons Information Systems Consultants)*

Section 4.3

In Exercises 27 and 28, find the indicated probabilities using the geometric distribution. If convenient, use technology to find the probabilities.

27. During a promotional contest, a soft drink company places winning caps on one of every six bottles. If you purchase one bottle a day, find the probability that you find your first winning cap (a) on the fourth day, (b) within four days, and (c) sometime after three days.

28. In a recent year, Barry Bonds hit 73 home runs in the 153 games he played. Assume that his home run production stayed at that level the following season. What is the probability that he would hit his first home run (a) on the first game of the season, (b) on the second game of the season, (c) on the first or second game of the season, and (d) within the first three games of the season? *(Source: Baseball-reference.com)*

In Exercises 29 and 30, find the indicated probabilities using the Poisson distribution. If convenient, use a Poisson probability table or technology tool to find the probabilities.

29. During a 36-year period, lightning killed 2768 people in the United States. Assume that this rate holds true today and is constant throughout the year. Find the probability that tomorrow

(a) no one in the United States will be struck and killed by lightning.

(b) one person will be struck and killed.

(c) more than one person will be struck and killed. *(Source: National Oceanic and Atmospheric Administration)*

30. It is estimated that sharks kill 10 people each year worldwide. Find the probability that at least three people are killed by sharks this year

(a) assuming that this rate is true.

(b) if the rate is actually five people a year.

(c) if the rate is actually 15 people a year. *(Source: International Shark Attack File)*

4 Chapter Quiz

Take this quiz as you would take a quiz in class. After you are done, check your work against the answers given in the back of the book.

1. Decide if the random variable, x, is discrete or continuous. Explain your reasoning.

(a) x represents the number of times Yellowstone Park's Old Faithful geyser erupts in one day.

(b) x represents the amount of sugar (in pounds) consumed each day in the United States.

2. The table lists the number of U.S. mainland hurricane strikes (from 1900 to 2000) for various intensities according to the Saffir-Simpson Hurricane Scale. *(Source: National Hurricane Center)*

Intensity	Number of hurricanes
1	61
2	39
3	48
4	14
5	3

(a) Construct a probability distribution of the data.

(b) Graph the discrete probability distribution using a relative frequency histogram.

(c) Find the mean, variance, and standard deviation of the probability distribution and interpret the results.

(d) Find the probability that a hurricane selected at random for further study has an intensity of at least 4.

3. A surgical technique is performed on eight patients. You are told there is an 80% chance of success.

(a) Construct a binomial distribution.

(b) Graph the binomial distribution using a relative frequency histogram.

(c) Find the mean, variance, and standard deviation of the probability distribution and interpret the results.

(d) Find the probability that the surgery is successful for exactly two patients.

(e) Find the probability that the surgery is successful for fewer than two patients.

4. A newspaper finds that the mean number of typographical errors per page is 5. Find the probability that

(a) exactly five typographical errors will be found on a page.

(b) fewer than five typographical errors will be found on a page.

(c) no typographical errors will be found on a page.

---

1. (a) Discrete because the random variable is countable.

(b) Continuous because the random variable has an infinite number of possible outcomes and cannot be counted.

2. (a)

x	f	$P(x)$
1	61	0.370
2	39	0.236
3	48	0.291
4	14	0.085
5	3	0.018
	$n = 165$	$\Sigma P(x) = 1$

(b)

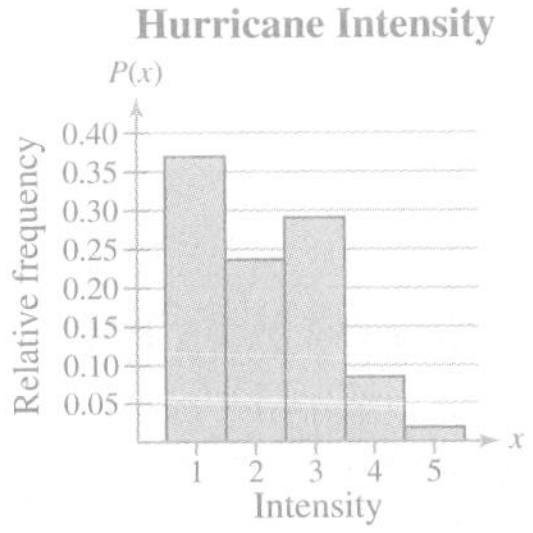

(c) 2.1, 1.1, 1.1

On average, the intensity of a hurricane will be 2.1 with a standard deviation of 1.1.

(d) 0.103

3. (a) $n = 8$

$p = 0.80$

x	$P(x)$
0	0.000003
1	0.000082
2	0.001147
3	0.009175
4	0.045875
5	0.146801
6	0.293601
7	0.335544
8	0.167772

(b)

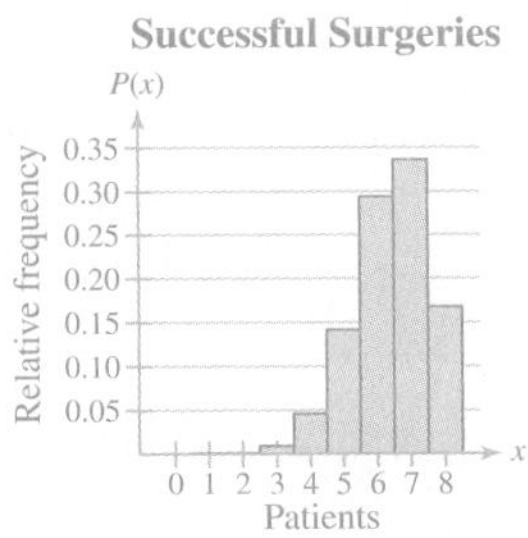

(c) 6.4, 1.3, 1.1

(d) 0.001 (e) 0.000085

4. (a) 0.176 (b) 0.440 (c) 0.007

PUTTING IT ALL TOGETHER

Real Statistics ■ Real Decisions

The Centers for Disease Control and Prevention (CDC) is required by law to publish a report on assisted reproductive technologies (ART). ART includes all fertility treatments in which both the egg and the sperm are used. These procedures generally involve removing eggs from a woman's ovaries, combining them with sperm in the laboratory, and returning them to the woman's body or giving them to another woman.

You are helping to prepare the CDC report and select at random 10 ART cycles for a special review. None of the cycles resulted in a clinical pregnancy. Your manager feels it is impossible to select at random 10 ART cycles that did not result in a clinical pregnancy. Use the information provided at the right and your knowledge of statistics to determine if your manager is correct.

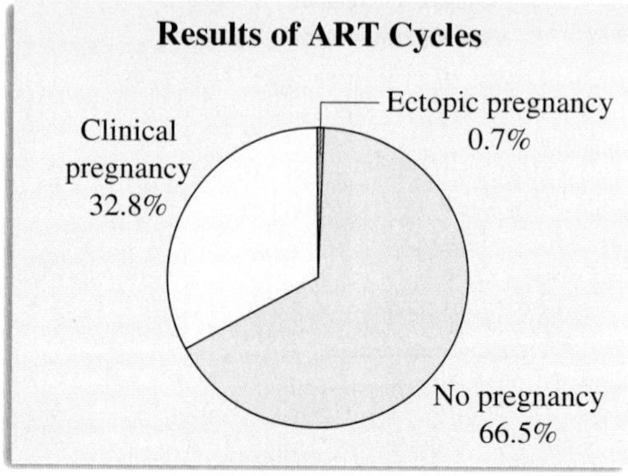

(Source: Centers for Disease Control and Prevention)

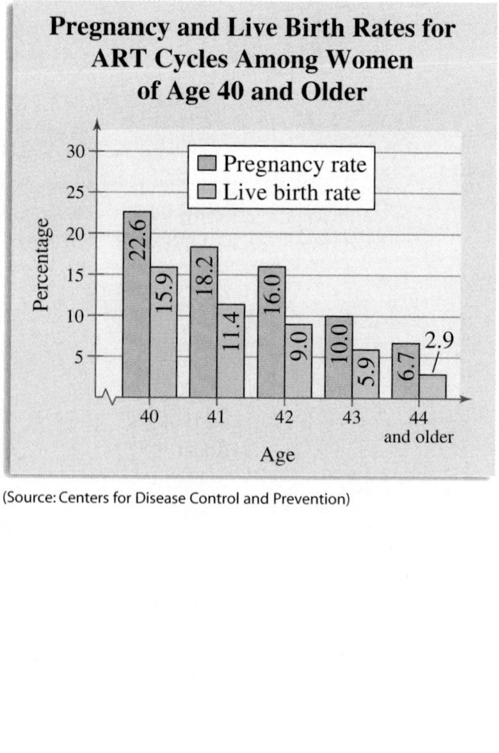

(Source: Centers for Disease Control and Prevention)

Exercises

1. *How Would You Do It?*

(a) How would you determine if your manager's view is correct, that it is impossible to select at random 10 ART cycles that did not result in a clinical pregnancy?

(b) What probability distribution do you think best describes the situation? Do you think the distribution of the number of clinical pregnancies is discrete or continuous? Why?

2. *Answering the Question*

Write an explanation that answers the question, "Is it possible to select at random 10 ART cycles that did not result in a clinical pregnancy?" Include in your explanation the appropriate probability distribution and your calculation of the probability of no clinical pregnancies in 10 ART cycles.

3. *Suspicious Samples?*

Which of the following samples would you consider suspicious if someone told you that the sample was selected at random? Would you believe that the samples were selected at random? Why or why not?

(a) Selecting at random 10 ART cycles among women of age 40, eight of which resulted in clinical pregnancies.

(b) Selecting at random 10 ART cycles among women of age 41, none of which resulted in clinical pregnancies.

TECHNOLOGY | MINITAB | EXCEL | TI-83

Using Poisson Distributions as Queuing Models

Queuing means waiting in line to be served. There are many examples of queuing in everyday life: waiting at a traffic light, waiting in line at a grocery check-out counter, waiting for an elevator, being put on hold for a telephone call, and so on.

Poisson distributions are used to model and predict the number of people (calls, computer programs, vehicles) arriving at the line. In the following exercises, you are asked to use Poisson distributions to analyze the queues at a grocery store check-out counter.

MINITAB

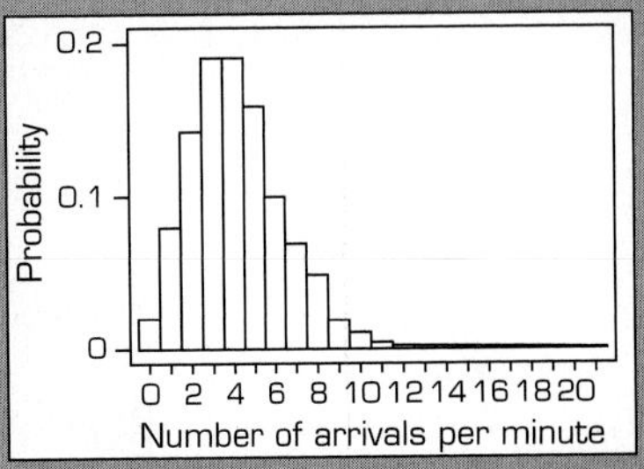

Exercises

In Exercises 1–6, consider a grocery store that can process a total of four customers at its check-out counters each minute.

1. Suppose that the mean number of customers who arrive at the check-out counters each minute is 4. Create a Poisson distribution with $\mu = 4$ for $x = 0$ to 20. Compare your results with the histogram shown at the upper right.

2. MINITAB was used to generate 20 random numbers with a Poisson distribution for $\mu = 4$. Let the random number represent the number of arrivals at the check-out counter each minute for 20 minutes.

3	3	3	3	5	5	6	7	3	6
3	5	6	3	4	6	2	2	4	1

During each of the first four minutes, only three customers arrived. These customers could all be processed, so there were no customers waiting after four minutes.

(a) How many customers were waiting after 5 minutes? 6 minutes? 7 minutes? 8 minutes?

(b) Create a table that shows the number of customers waiting at the end of 1 through 20 minutes.

3. Generate a list of 20 random numbers with a Poisson distribution for $\mu = 4$. Create a table that shows the number of customers waiting at the end of 1 through 20 minutes.

4. Suppose that the mean increases to 5 arrivals per minute. You can still process only four per minute. How many would you expect to be waiting in line after 20 minutes?

5. Simulate the setting in Exercise 4. Do this by generating a list of 20 random numbers with a Poisson distribution for $\mu = 5$. Then create a table that shows the number of customers waiting at the end of 20 minutes.

6. Suppose that the mean number of arrivals per minute is 5. What is the probability that 10 customers will arrive during the first minute?

7. Suppose that the mean number of arrivals per minute is 4.

(a) What is the probability that three, four, or five customers will arrive during the third minute?

(b) What is the probability that more than four customers will arrive during the first minute?

(c) What is the probability that more than four customers will arrive during each of the first four minutes?

Extended solutions are given in the *Technology Supplement.* Technical instruction is provided for MINITAB, Excel, and the TI-83.

CHAPTER 5

Normal Probability Distributions

5.1 Introduction to Normal Distributions and the Standard Normal Distribution

5.2 Normal Distributions: Finding Probabilities

5.3 Normal Distributions: Finding Values

Case Study

5.4 Sampling Distributions and the Central Limit Theorem

5.5 Normal Approximations to Binomial Distributions

Uses and Abuses

Real Statistics–Real Decisions

Technology

In 2000, the National Center for Health Statistics, located in Hyattsville, Maryland, began a 10-year program called *Healthy People 2010* to promote health through changes in people's lifestyles. It is too early to analyze the results of this program, but the results of a similar program that started in 1990, *Healthy People 2000*, are available. During the course of the program, some of the goals were met. For instance, heart disease and stroke death rates were down. Other goals were not met. For instance, although more adults were exercising, a quarter of all adults were still engaged in no physical activity.

Where You've Been

In Chapters 1 through 4, you learned how to collect and describe data, find the probability of an event, and analyze discrete probability distributions. You also learned that if a sample is used to make inferences about a population, then it is critical that the sample not be biased. Suppose, for instance, that you wanted to measure the serum cholesterol levels of adults in the United States. How would you organize the study? When the National Center for Health Statistics performed this study, it used random sampling and then classified the results according to the gender, ethnic background, and age of the participants. One conclusion from the study was that women's cholesterol levels tended to increase throughout their lives, whereas men's increased to age 65, and then decreased.

Where You're Going

In Chapter 5, you will learn how to recognize normal (bell-shaped) distributions and how to use their properties in real-life applications. Suppose that you worked for the U.S. National Center for Health Statistics and were collecting data about various physical traits of people in the United States. Which of the following would you expect to have bell-shaped, symmetric distributions: height, weight, cholesterol level, age, blood pressure, shoe size, reaction times, lung capacity? Of these, all except weight and age have distributions that are approximately normal. For instance, the four graphs below show the height and weight distributions for men and women in the United States aged 20 to 29. Notice that the height distributions are bell shaped, but the weight distributions are skewed right.

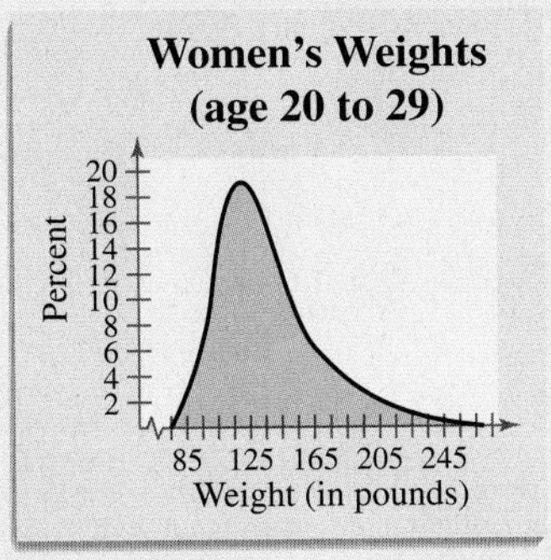

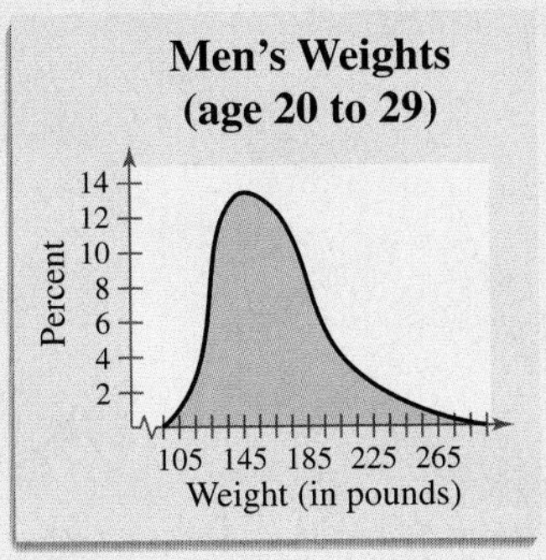

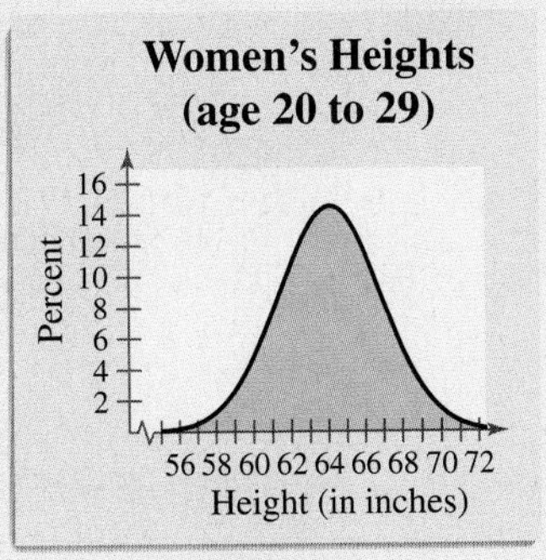

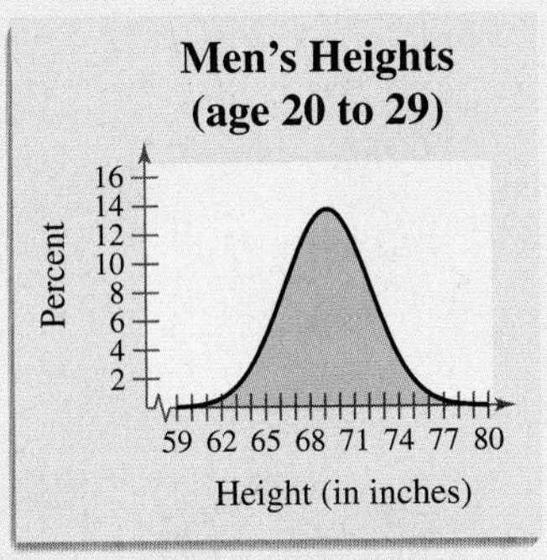

5.1 Introduction to Normal Distributions and the Standard Normal Distribution

What You Should Learn

- How to interpret graphs of normal probability distributions
- How to find areas under the standard normal curve

Properties of a Normal Distribution • The Standard Normal Distribution

Properties of a Normal Distribution

In Section 4.1, you learned that a **continuous random variable** has an infinite number of possible values that can be represented by an interval on the number line. Its probability distribution is called a **continuous probability distribution.** In this chapter, you will study the most important continuous probability distribution in statistics—the normal distribution. Normal distributions can be used to model many sets of measurements in nature, industry, and business. For instance, the systolic blood pressure of humans, the lifetime of television sets, and even housing costs are all normally distributed random variables.

Note to Instructor

Draw several different continuous probability curves. Then point out that the normal (or Gaussian) curve is graphed using the formula shown at the bottom of the page. Have students discuss measures in nature that are normally distributed. Mention that often grades in a statistics class are not normally distributed.

GUIDELINES

Properties of a Normal Distribution

A **normal distribution** is a continuous probability distribution for a random variable x. The graph of a normal distribution is called the **normal curve.** A normal distribution has the following properties.

1. The mean, median, and mode are equal.
2. The normal curve is bell shaped and is symmetric about the mean.
3. The total area under the normal curve is equal to one.
4. The normal curve approaches, but never touches, the x-axis as it extends farther and farther away from the mean.
5. Between $\mu - \sigma$ and $\mu + \sigma$ (in the center of the curve) the graph curves downward. The graph curves upward to the left of $\mu - \sigma$ and to the right of $\mu + \sigma$. The points at which the curve changes from curving upward to curving downward are called *inflection points.*

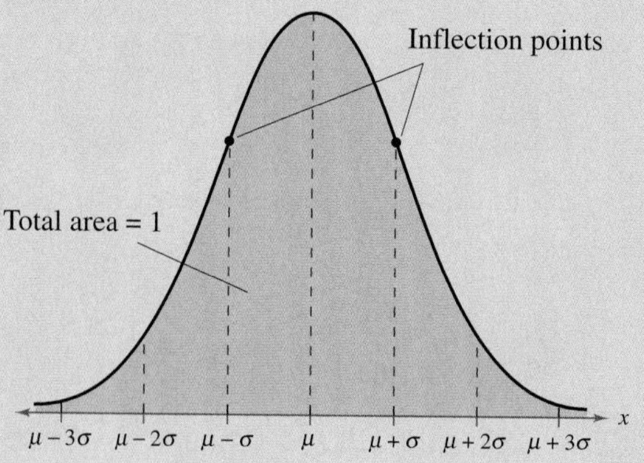

If x is a continuous random variable having a normal distribution with mean μ and standard deviation σ, you can graph a normal curve using the equation

$$y = \frac{1}{\sigma\sqrt{2\pi}}\, e^{-(x-\mu)^2/2\sigma^2}. \qquad e \approx 2.718 \text{ and } \pi \approx 3.14$$

Insight

Because e and π are constants, a normal curve depends completely on two parameters, μ and σ.

A normal distribution can have any mean and any positive standard deviation. These two parameters, μ and σ, completely determine the shape of the normal curve. The mean gives the location of the line of symmetry, and the standard deviation describes how much the data are spread out.

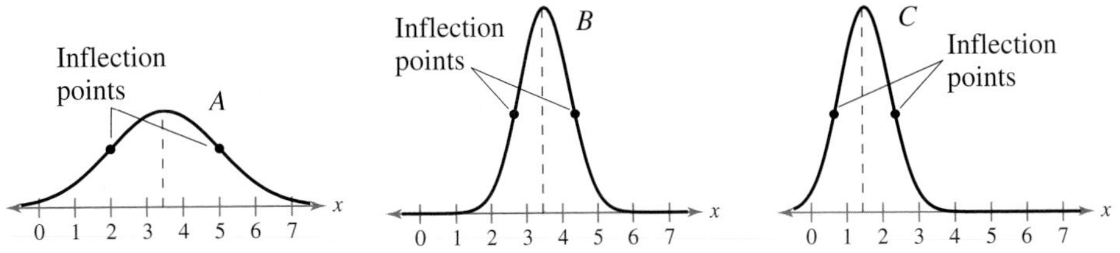

Curve A	Curve B	Curve C
Mean: $\mu = 3.5$	Mean: $\mu = 3.5$	Mean: $\mu = 1.5$
Standard deviation: $\sigma = 1.5$	Standard deviation: $\sigma = 0.7$	Standard deviation: $\sigma = 0.7$

Notice that curve A and curve B above have the same mean, and curve B and curve C have the same standard deviation. The total area under each curve is 1.

EXAMPLE 1

Understanding Mean and Standard Deviation

1. Which normal curve has a greater mean?
2. Which normal curve has a greater standard deviation?

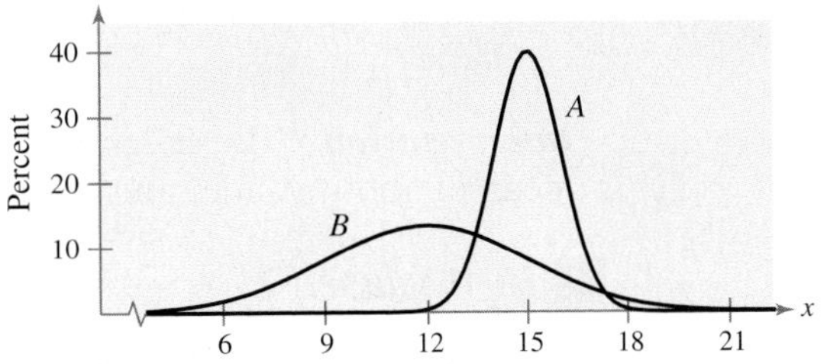

SOLUTION

1. The line of symmetry of curve A occurs at $x = 15$. The line of symmetry of curve B occurs at $x = 12$. So, curve A has a greater mean.
2. Curve B is more spread out than curve A; so, curve B has a greater standard deviation.

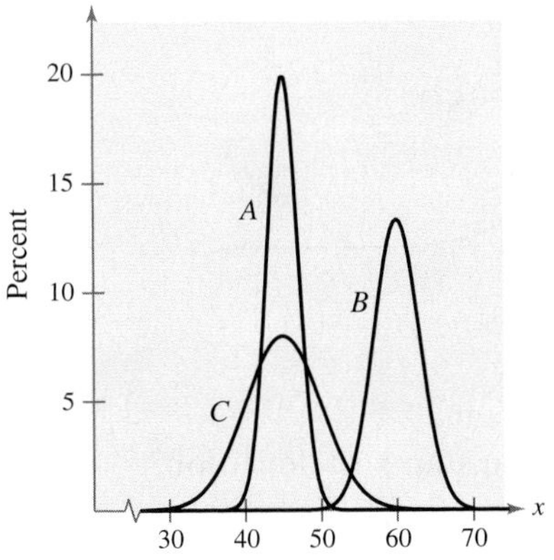

Try It Yourself 1

Consider the normal curves shown at the left. Which normal curve has the greatest mean? Which normal curve has the greatest standard deviation? Justify your answers.

a. Find the location of the *line of symmetry* of each curve. Make a conclusion about which mean is greatest.
b. Determine which normal curve is *more spread out.* Make a conclusion about which standard deviation is greatest.

Answer: Page A37

EXAMPLE 2

Interpreting Graphs of Normal Distributions

The heights (in feet) of fully grown white oak trees are normally distributed. The normal curve shown below represents this distribution. What is the mean height of a fully grown white oak tree? Estimate the standard deviation of this normal distribution.

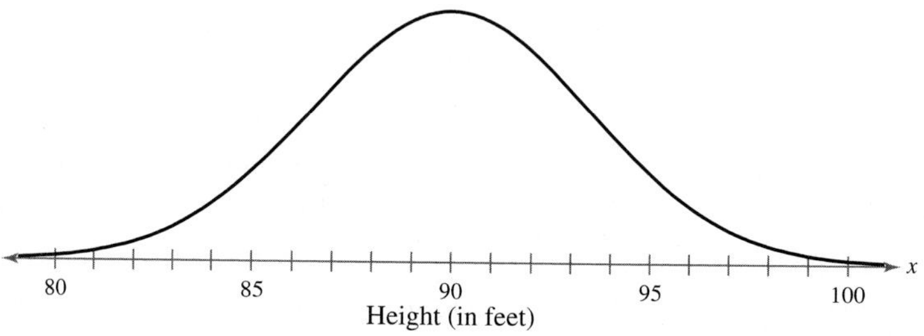

SOLUTION

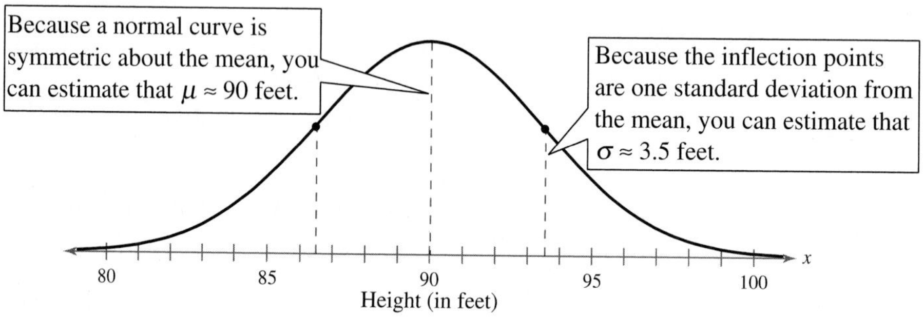

Interpretation The heights of the oak trees are normally distributed with a mean of about 90 feet and a standard deviation of about 3.5 feet.

Try It Yourself 2

The diameters (in feet) of fully grown white oak trees are normally distributed. The normal curve shown below represents this distribution. What is the mean diameter of a fully grown white oak tree? Estimate the standard deviation of this normal distribution.

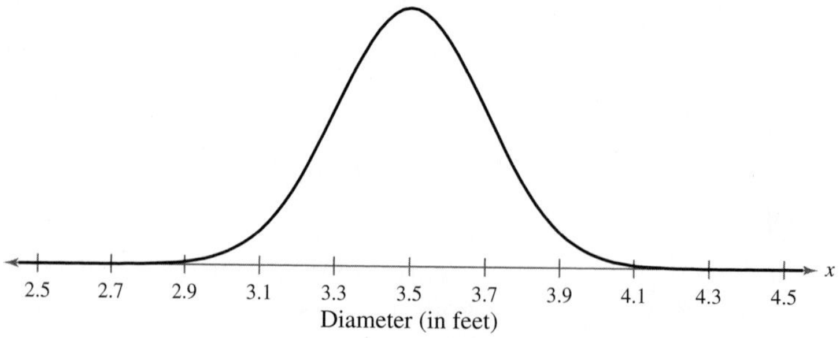

a. Find the *line of symmetry* and identify the mean.
b. Estimate the *inflection points* and identify the standard deviation.

Answer: Page A37

The Standard Normal Distribution

There are infinitely many normal distributions, each with its own mean and standard deviation. The normal distribution with a mean of 0 and a standard deviation of 1 is called **the standard normal distribution.** The horizontal scale of the graph of the standard normal distribution corresponds to z-scores. In Section 2.5, you learned that a z-score is a measure of position that indicates the number of standard deviations a value lies from the mean. Recall that you can transform an x-value to a z-score using the formula

$$z = \frac{\text{Value} - \text{Mean}}{\text{Standard deviation}}$$

$$= \frac{x - \mu}{\sigma}. \qquad \text{Round to the nearest hundredth.}$$

Insight

Because every normal distribution can be transformed to the standard normal distribution, you can use z-scores and the standard normal curve to find areas (and therefore probability) under any normal curve.

DEFINITION

The **standard normal distribution** is a normal distribution with a mean of 0 and a standard deviation of 1.

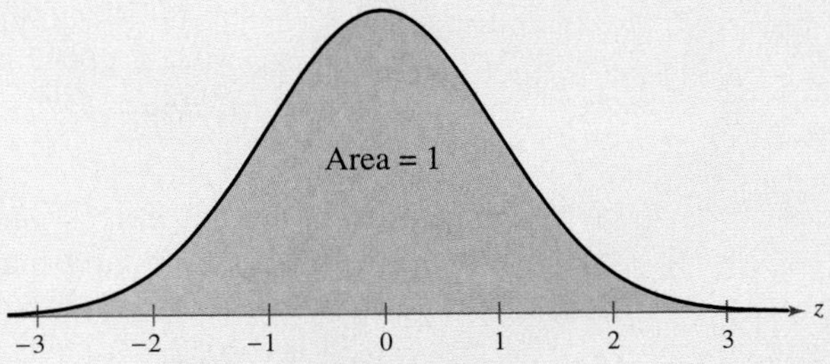

Standard Normal Distribution

Note to Instructor

Mention that the formula for a normal probability density function on page 216 is greatly simplified when $\mu = 0$ and $\sigma = 1$.

$$y = \frac{e^{-x^2/2}}{\sqrt{2\pi}}$$

Study Tip

It is important that you know the difference between x and z. The random variable x is sometimes called a raw score and represents values in a *nonstandard* normal distribution, whereas z represents values in the *standard* normal distribution.

If each data value of a normally distributed random variable x is transformed into a z-score, the result will be the standard normal distribution. When this transformation takes place, the area that falls in the interval under the nonstandard normal curve is the *same* as that under the standard normal curve within the corresponding z-boundaries.

In Section 2.4, you learned to use the Empirical Rule to approximate areas under a normal curve when the values of the random variable x corresponded to $-3, -2, -1, 0, 1, 2,$ or 3 standard deviations from the mean. Now, you will learn to calculate areas corresponding to other x-values. After you use the formula given above to transform an x-value to a z-score, you can use the Standard Normal Table in Appendix B. The table lists the cumulative area under the standard normal curve to the left of z for z-scores from -3.49 to 3.49. As you examine the table, notice the following.

Properties of the Standard Normal Distribution

1. The cumulative area is close to 0 for z-scores close to $z = -3.49$.
2. The cumulative area increases as the z-scores increase.
3. The cumulative area for $z = 0$ is 0.5000.
4. The cumulative area is close to 1 for z-scores close to $z = 3.49$.

Note to Instructor

If you prefer that your students use a 0-to-z table, refer them to Appendix A, where an alternative presentation for this material is given.

EXAMPLE 3

Using the Standard Normal Table

1. Find the cumulative area that corresponds to a z-score of 1.15.
2. Find the cumulative area that corresponds to a z-score of -0.24.

SOLUTION

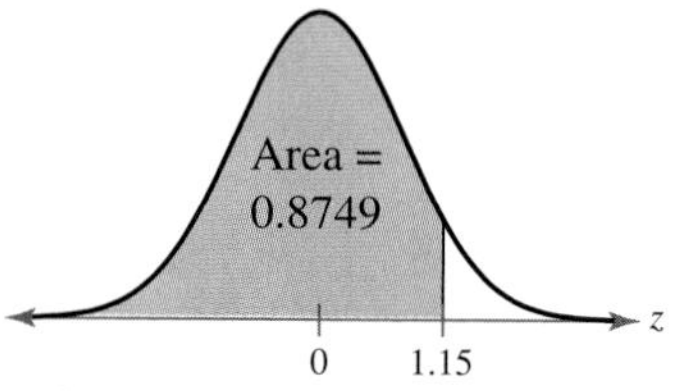

1. Find the area that corresponds to $z = 1.15$ by finding 1.1 in the left column and then moving across the row to the column under 0.05. The number in that row and column is 0.8749. So, the area to the left of $z = 1.15$ is 0.8749.

z	.00	.01	.02	.03	.04	.05	.06
0.0	.5000	.5040	.5080	.5120	.5160	.5199	.5239
0.1	.5398	.5438	.5478	.5517	.5557	.5596	.5636
0.2	.5793	.5832	.5871	.5910	.5948	.5987	.6026
0.9	.8159	.8186	.8212	.8238	.8264	.8289	.8315
1.0	.8413	.8438	.8461	.8485	.8508	.8531	.8554
1.1	.8643	.8665	.8686	.8708	.8729	.8749	.8770
1.2	.8849	.8869	.8888	.8907	.8925	.8944	.8962
1.3	.9032	.9049	.9066	.9082	.9099	.9115	.9131
1.4	.9192	.9207	.9222	.9236	.9251	.9265	.9279

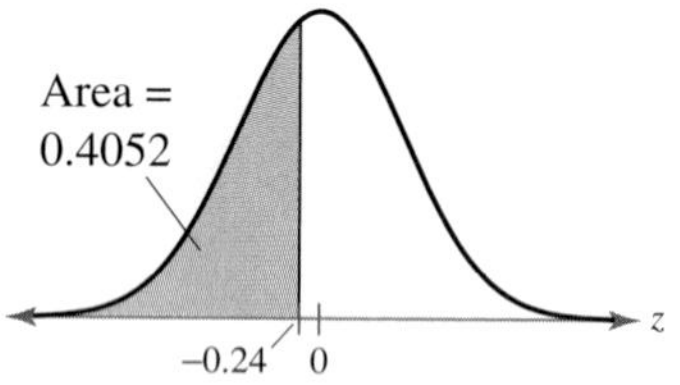

2. Find the area that corresponds to $z = -0.24$ by finding -0.2 in the left column and then moving across the row to the column under 0.04. The number in that row and column is 0.4052. So, the area to the left of $z = -0.24$ is 0.4052.

z	.09	.08	.07	.06	.05	.04	.03
−3.4	.0002	.0003	.0003	.0003	.0003	.0003	.0003
−3.3	.0003	.0004	.0004	.0004	.0004	.0004	.0004
−3.2	.0005	.0005	.0005	.0006	.0006	.0006	.0006
−0.5	.2776	.2810	.2843	.2877	.2912	.2946	.2981
−0.4	.3121	.3156	.3192	.3228	.3264	.3300	.3336
−0.3	.3483	.3520	.3557	.3594	.3632	.3669	.3707
−0.2	.3859	.3897	.3936	.3974	.4013	.4052	.4090
−0.1	.4247	.4286	.4325	.4364	.4404	.4443	.4483
−0.0	.4641	.4681	.4721	.4761	.4801	.4840	.4880

Study Tip

You can use a computer or calculator to find the cumulative area that corresponds to a z-score. For instance, here are instructions for finding the area that corresponds to $z = -0.24$ on a TI-83.

[2nd] [DISTR] [2] −10000,

−.24 [)] [ENTER]

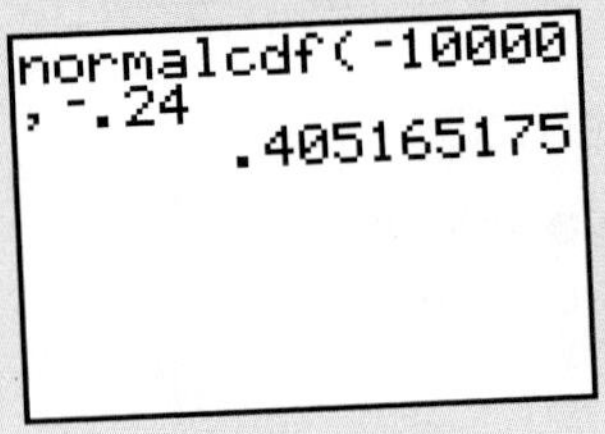

Try It Yourself 3

1. Find the area under the curve to the left of a z-score of -2.19.
2. Find the area under the curve to the left of a z-score of 2.17.

a. Locate the given z-score and *find the area* that corresponds to it in the Standard Normal Table.

Answer: Page A37

When the z-score is not in the table, use the entry closest to it. If the given z-score is exactly midway between two z-scores, then use the area midway between the corresponding areas.

You can use the following guidelines to find various types of areas under the standard normal curve.

Note to Instructor

Students find these three options easy to work with. If you have previously used a 0-to-z table, you will appreciate that students never need be confused as to whether to add 0.5, subtract it from 0.5, or use the table entry to find a required probability.

GUIDELINES

Finding Areas Under the Standard Normal Curve

1. Sketch the standard normal curve and shade the appropriate area under the curve.

2. Find the area by following the directions for each case shown.

a. To find the area to the *left* of z, find the area that corresponds to z in the Standard Normal Table.

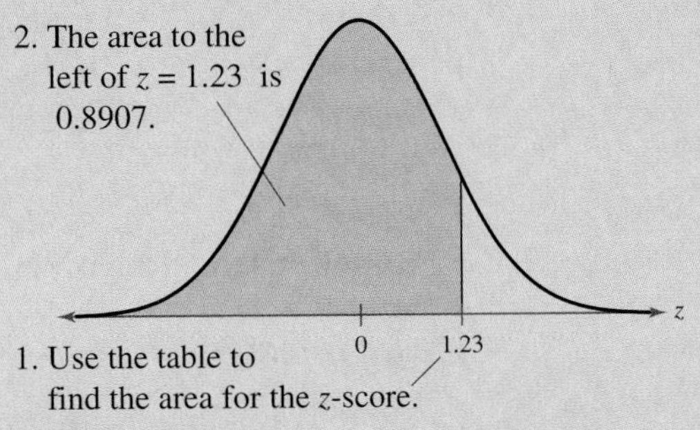

b. To find the area to the *right* of z, use the Standard Normal Table to find the area that corresponds to z. Then subtract the area from 1.

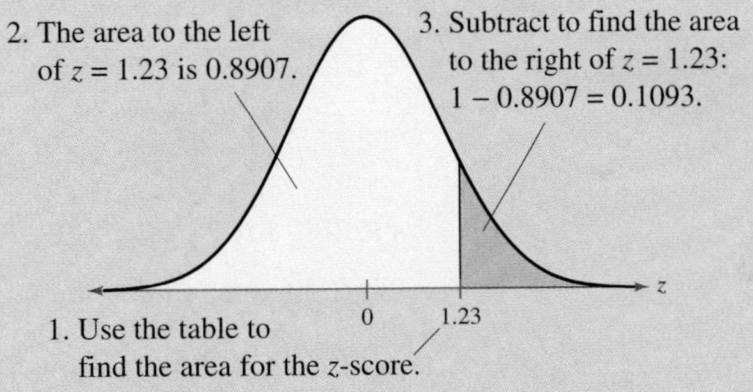

c. To find the area *between* two z-scores, find the area corresponding to each z-score in the Standard Normal Table. Then subtract the smaller area from the larger area.

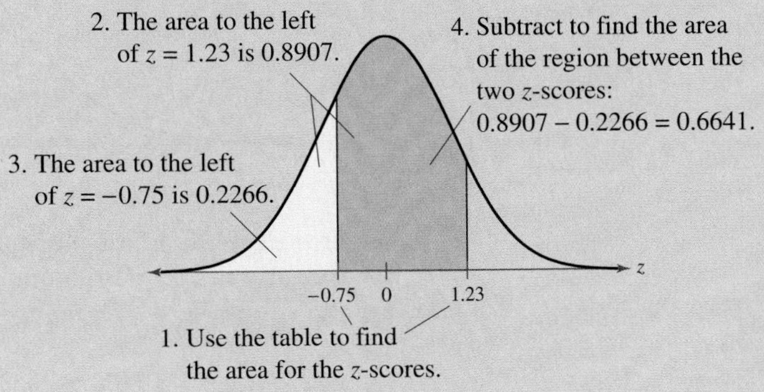

EXAMPLE 4

Finding Area Under the Standard Normal Curve

Find the area under the standard normal curve to the left of $z = -0.99$.

SOLUTION The area under the standard normal curve to the left of $z = -0.99$ is shown.

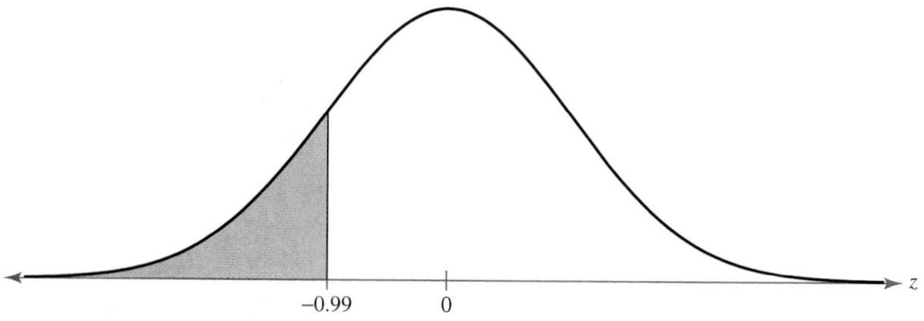

From the Standard Normal Table, this area is equal to 0.1611.

Try It Yourself 4

Find the area under the standard normal curve to the left of $z = 2.13$.

a. *Draw* the standard normal curve and shade the area under the curve and to the left of $z = 2.13$.

b. Use the Standard Normal Table to *find the area* that corresponds to $z = 2.13$.

Answer: Page A37

Insight

Because the normal distribution is a continuous probability distribution, the area under the standard normal curve to the left of a *z*-score gives the probability that *z* is less than that *z*-score. For instance, in Example 4, the area to the left of $z = -0.99$ is 0.1611. So, $P(z < -0.99) = 0.1611$, which is read as "the probability that *z* is less than −0.99 is 0.1611."

EXAMPLE 5

Finding Area Under the Standard Normal Curve

Find the area under the standard normal curve to the right of $z = 1.06$.

SOLUTION The area under the standard normal curve to the right of $z = 1.06$ is shown.

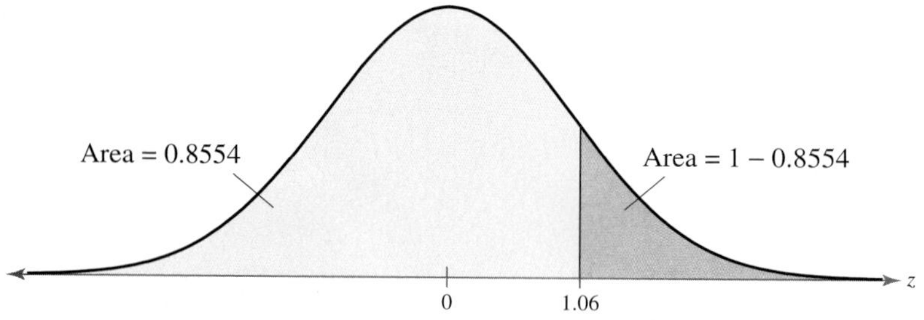

From the Standard Normal Table, the area to the left of $z = 1.06$ is 0.8554. Because the total area under the curve is 1, the area to the right of $z = 1.06$ is

$$\text{Area} = 1 - 0.8554$$
$$= 0.1446.$$

Try It Yourself 5

Find the area under the standard normal curve to the right of $z = -2.16$.

a. *Draw* the standard normal curve and shade the area below the curve and to the right of $z = -2.16$.
b. Use the Standard Normal Table to *find the area* to the left of $z = -2.16$.
c. *Subtract* the area from 1.

Answer: Page A37

EXAMPLE 6

Finding Area Under the Standard Normal Curve

Find the area under the standard normal curve between $z = -1.5$ and $z = 1.25$.

SOLUTION The area under the standard normal curve between $z = -1.5$ and $z = 1.25$ is shown.

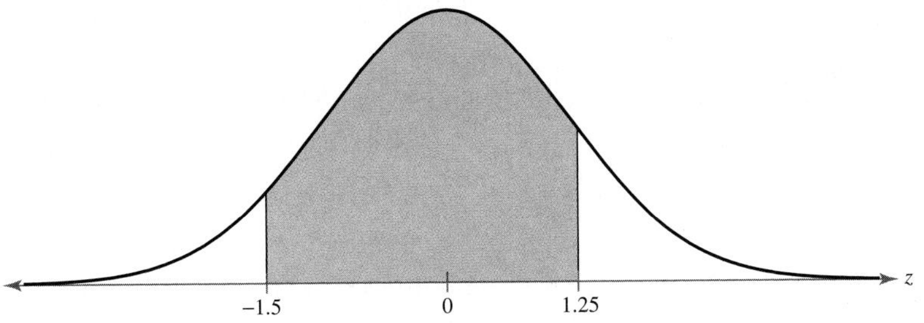

From the Standard Normal Table, the area to the left of $z = 1.25$ is 0.8944 and the area to the left of $z = -1.5$ is 0.0668. So, the area between $z = -1.5$ and $z = 1.25$ is

$$\text{Area} = 0.8944 - 0.0668$$
$$= 0.8276.$$

Interpretation So, 82.76% of the area under the curve falls between $z = -1.5$ and $z = 1.25$.

Try It Yourself 6

Find the area under the standard normal curve between $z = -2.16$ and $z = -1.35$.

a. Use the Standard Normal Table to *find the area* to the left of $z = -1.35$.
b. Use the Standard Normal Table to *find the area* to the left of $z = -2.16$.
c. Subtract the smaller area from the larger area.

Answer: Page A37

Recall in Section 2.5 you learned, using the Empirical Rule, that values lying more than two standard deviations from the mean are considered unusual. Values lying more than three standard deviations from the mean are considered *very* unusual. So if a z-score is greater than 2 or less than -2, it is unusual. If a z-score is greater than 3 or less than -3, it is *very* unusual.

Picturing the World

Each year the Centers for Disease Control and Prevention and the National Center for Health Statistics jointly publish a report summarizing the vital statistics from the previous year. According to one publication, the number of births in a recent year was 4,021,726. The weights of the newborns can be approximated by a normal distribution, as shown by the following graph.

Weights of Newborns

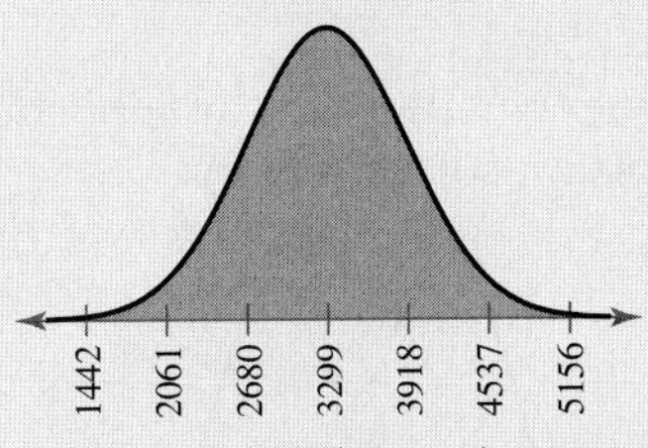

The weights of three newborns are 2000 grams, 3000 grams, and 4000 grams. Find the z-score that corresponds to each weight. Are any of these unusually heavy or light?

5.1 Exercises

1. Answers will vary.
2. 1
3. Answers will vary.
   Similarities: The two curves will have the same line of symmetry.
   Differences: One curve will be more spread out than the other.
4. Answers will vary.
   Similarities: The two curves will have the same shape (i.e., equal standard deviations).
   Differences: The two curves will have different lines of symmetry.
5. $\mu = 0, \sigma = 1$
6. Transform each data value x into a z-score. This is done by subtracting the mean from x and dividing by the standard deviation. In symbols, $z = \dfrac{x - \mu}{\sigma}$.
7. "The" standard normal distribution is used to describe one specific normal distribution ($\mu = 0, \sigma = 1$). "A" normal distribution is used to describe a normal distribution with any mean and standard deviation.
8. (c) is true because a z-score equal to zero indicates that the corresponding x-value is equal to the mean.
9. No, the graph crosses the x-axis.
10. No, the graph is not symmetric.
11. Yes, the graph fulfills the properties of the normal distribution.
12. No, the graph is skewed left.
13. No, the graph is skewed right.
14. No, the graph is not bell shaped.

Building Basic Skills and Vocabulary

1. Find three real-life examples of a continuous variable. Which do you think may be normally distributed? Why?
2. What is the total area under the normal curve?
3. Draw two normal curves that have the same mean but different standard deviations. Describe the similarities and differences.
4. Draw two normal curves that have different means but the same standard deviations. Describe the similarities and differences.
5. What is the mean of the standard normal distribution? What is the standard deviation of the standard normal distribution?
6. Describe how you can transform a nonstandard normal distribution to a standard normal distribution.
7. ***Getting at the Concept*** Why is it correct to say "a" normal distribution and "the" standard normal distribution?
8. ***Getting at the Concept*** If a z-score is zero, which of the following must be true? Explain your reasoning.
   (a) The mean is zero.
   (b) The corresponding x-value is zero.
   (c) The corresponding x-value is equal to the mean.

Graphical Analysis In Exercises 9–14, determine whether the graph could represent a variable with a normal distribution. Explain your reasoning.

9.
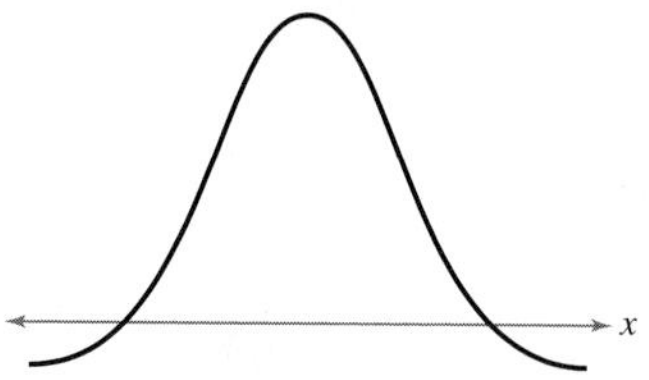

10.
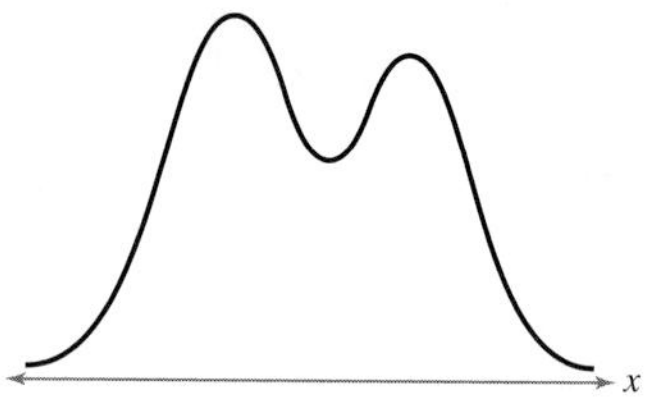

11.
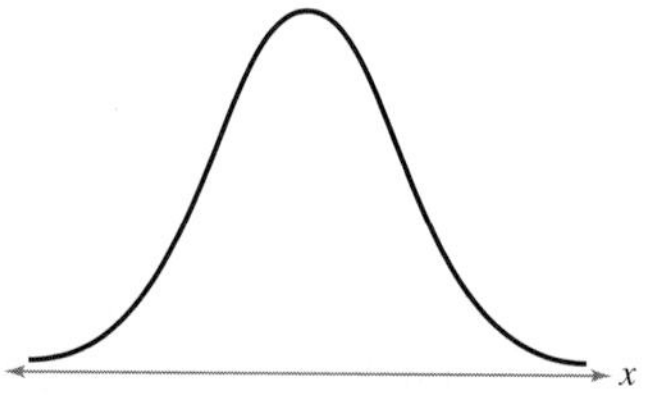

12.
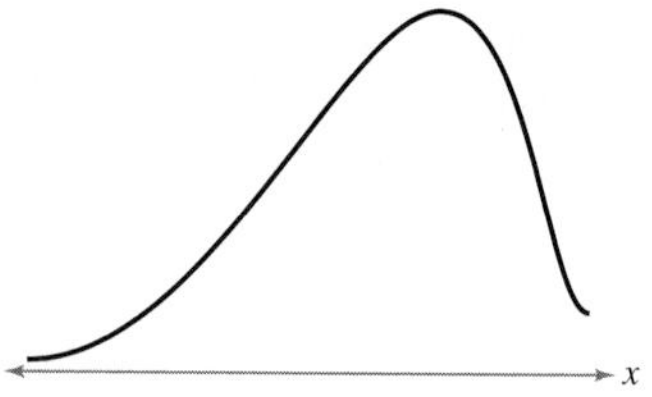

13.
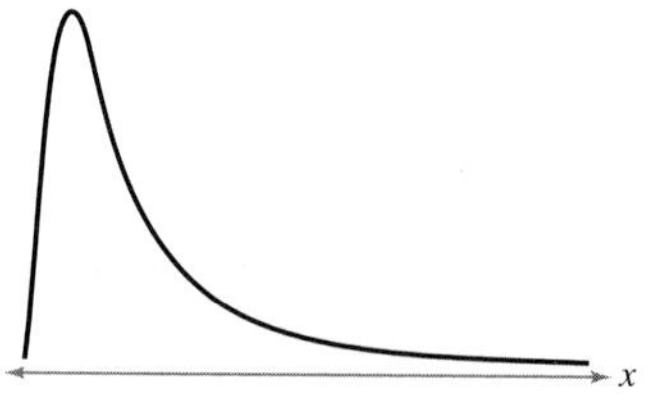

14.
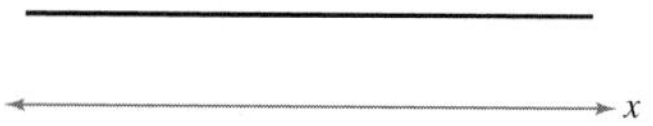

Graphical Analysis In Exercises 15 and 16, determine whether the histogram represents data with a normal distribution. Explain your reasoning.

15.

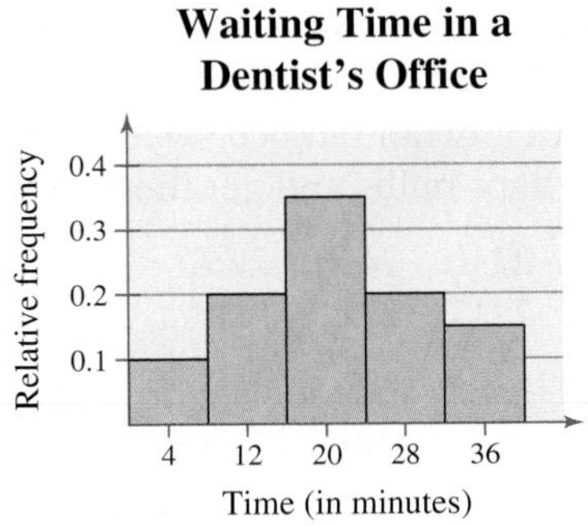

16.

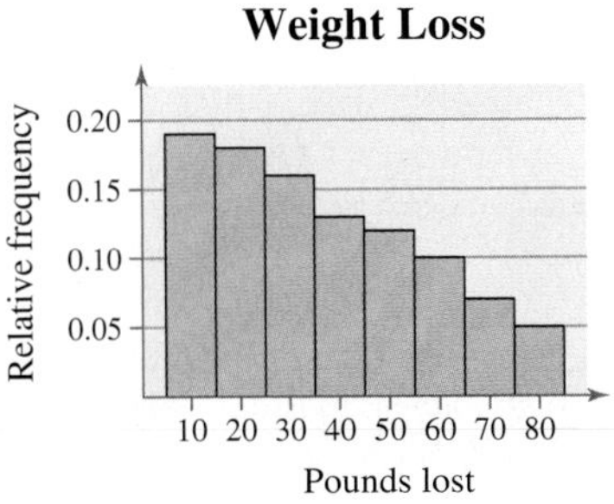

Graphical Analysis In Exercises 17–20, find the area of the indicated region under the standard normal curve.

17.

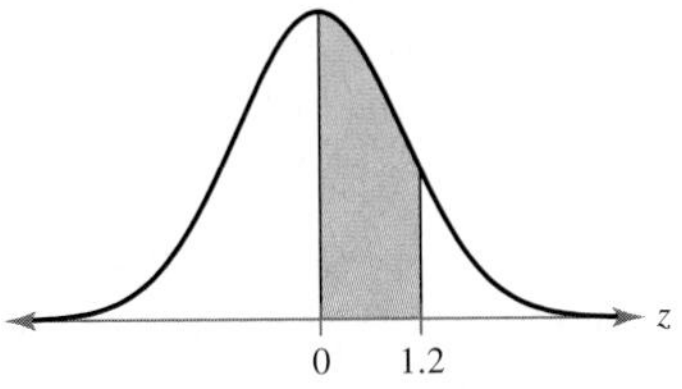

18.

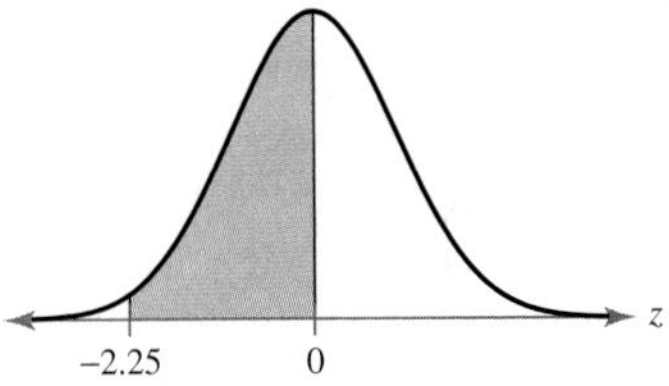

19.

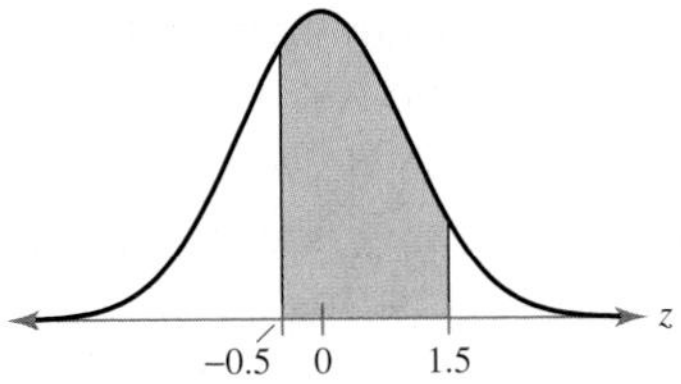

20.

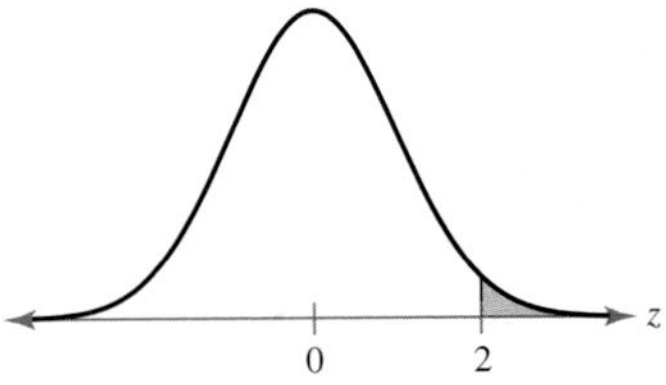

Finding Area In Exercises 21–40, find the indicated area under the standard normal curve.

21. To the left of $z = 1.54$

22. To the left of $z = 0.25$

23. To the left of $z = 1.96$

24. To the left of $z = 1.28$

25. To the right of $z = -0.95$

26. To the right of $z = -1.75$

27. To the right of $z = 1.28$

28. To the right of $z = 2.33$

29. To the left of $z = -2.575$

30. To the left of $z = -3.08$

31. To the right of $z = 1.645$

32. To the right of $z = 2.51$

33. Between $z = 0$ and $z = 1.96$

34. Between $z = 0$ and $z = 3.09$

35. Between $z = -1.53$ and $z = 0$

36. Between $z = -0.51$ and $z = 0$

37. Between $z = -1.96$ and $z = 1.96$

38. Between $z = -2.33$ and $z = 2.33$

39. To the left of $z = -1.28$ or to the right of $z = 1.28$

40. To the left of $z = -1.96$ or to the right of $z = 1.96$

15. It is normal because it is bell shaped and symmetric.
16. It is skewed to the right. So it is not a normal distribution.
17. 0.3849
18. 0.4878
19. 0.6247
20. 0.0228
21. 0.9382
22. 0.5987
23. 0.975
24. 0.8997
25. 0.8289
26. 0.9599
27. 0.1003
28. 0.0099
29. 0.005
30. 0.0010
31. 0.05
32. 0.006
33. 0.475
34. 0.499
35. 0.437
36. 0.195
37. 0.95
38. 0.9802
39. 0.2006
40. 0.05

41. (a)

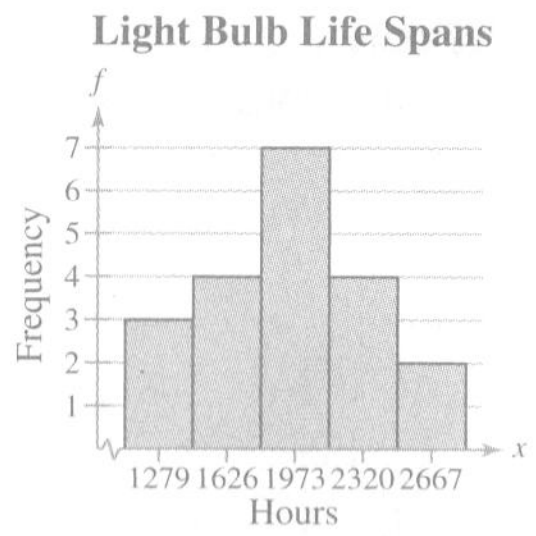

It is reasonable to assume that the life span is normally distributed because the histogram is nearly symmetric and bell shaped.

(b) 1941.35, 432.385

(c) The sample mean of 1941.35 hours is less than the claimed mean, so, on average, the bulbs in the sample lasted for a shorter time. The sample standard deviation of 432 hours is greater than the claimed standard deviation, so the bulbs in the sample had a greater variation in life span than the manufacturer's claim.

42. (a)

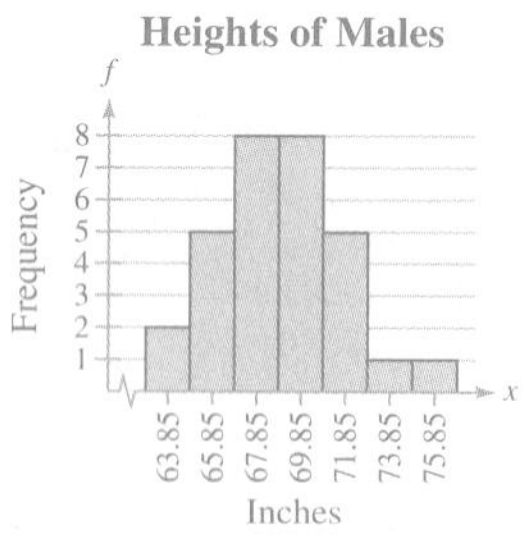

It is reasonable to assume that the heights are normally distributed because the histogram is nearly symmetric and bell shaped.

(b) 68.75, 2.847

(c) The mean of your sample is 0.45 inch less than that of the previous study, so the average height from the sample is less than in the previous study. The standard deviation is about 0.05 inch less than that of the previous study, so the heights are slightly less spread out than in the previous study.

Using and Interpreting Concepts

41. *Manufacturer Claims* DATA You work for a consumer watchdog publication and are testing the advertising claims of a light bulb manufacturer. The manufacturer claims that the life span of the bulb is normally distributed, with a mean of 2000 hours and a standard deviation of 250 hours. You test 20 light bulbs and get the following life spans.

2210, 2406, 2267, 1930, 2005, 2502, 1106, 2140, 1949, 1921,
2217, 2121, 2004, 1397, 1659, 1577, 2840, 1728, 1209, 1639

(a) Draw a frequency histogram to display these data. Use five classes. Is it reasonable to assume that the life span is normally distributed? Why?

(b) Find the mean and standard deviation of your sample.

(c) Compare the mean and standard deviation of your sample with those in the manufacturer's claim. Discuss the differences.

42. *Heights of Men* DATA You are performing a study about the height of 20- to 29-year-old men. A previous study found the height to be normally distributed, with a mean of 69.2 inches and a standard deviation of 2.9 inches. You randomly sample 30 men and find their heights to be as follows. *(Source: National Center for Health Statistics)*

72.1, 71.2, 67.9, 67.3, 69.5, 68.6, 68.8, 69.4, 73.5, 67.1,
69.2, 75.7, 71.1, 69.6, 70.7, 66.9, 71.4, 62.9, 69.2, 64.9,
68.2, 65.2, 69.7, 72.2, 67.5, 66.6, 66.5, 64.2, 65.4, 70.0

(a) Draw a frequency histogram to display these data. Use seven classes with midpoints of 63.85, 65.85, 67.85, 69.85, 71.85, 73.85, and 75.85. Is it reasonable to assume that the heights are normally distributed? Why?

(b) Find the mean and standard deviation of your sample.

(c) Compare the mean and standard deviation of your sample with those in the previous study. Discuss the differences.

Computing and Interpreting z-Scores of Normal Distributions In Exercises 43–46, you are given a normal distribution, the distribution's mean and standard deviation, four values from that distribution, and a graph of the Standard Normal Distribution. (a) Without converting to z-scores, match each value with the letters A, B, C, and D on the given graph of the Standard Normal Distribution. (b) Find the z-score that corresponds to each value and check your answers to part (a). (c) Determine whether any of the values are unusual.

43. *Ball Bearings* Your company manufactures ball bearings. The diameters of the ball bearings are normally distributed, with a mean of 3 inches and a standard deviation of 0.02 inch. The diameters of four ball bearings selected at random are 3.01, 2.97, 2.98, and 3.05.

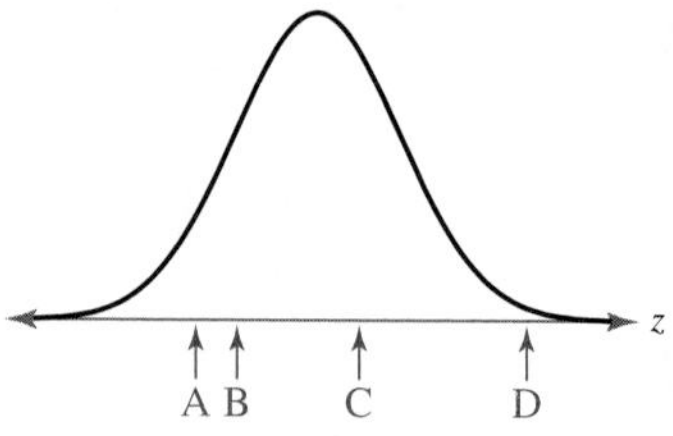

Figure for Exercise 43

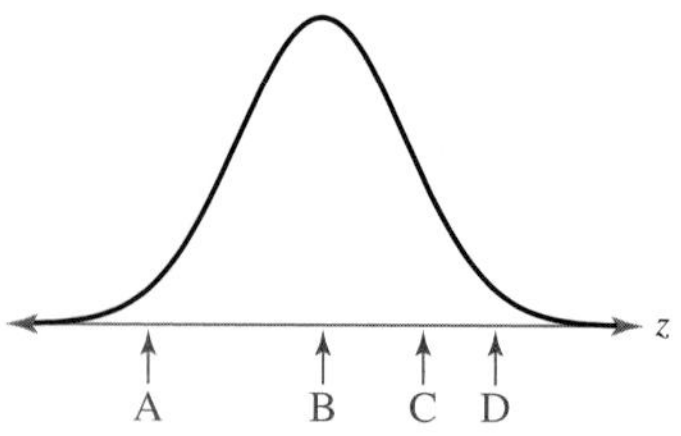

Figure for Exercise 44

43. (a) A = 2.97; B = 2.98; C = 3.01; D = 3.05

(b) 0.5; −1.5; −1; 2.5

(c) $x = 3.05$ is unusual owing to a relatively large z-score (2.5).

44. (a) A = 24,750; B = 30,000; C = 33,000; D = 35,150

(b) 2.06; −2.1; 0; 1.2

(c) $x = 35{,}150$ and $x = 24{,}750$ are unusual owing to their relatively large z-scores (2.06 and −2.1).

45. (a) A = 801; B = 950; C = 1250; D = 1467

(b) −0.36; 1.07; 2.11; −1.08

(c) $x = 1467$ is unusual owing to a relatively large z-score (2.11).

46. (a) A = 14; B = 18; C = 25; D = 32

(b) −0.58; 2.33; −1.42; 0.88

(c) $x = 32$ is unusual owing to a relatively large z-score (2.33).

47. 0.6915

48. 0.1587

49. 0.05

50. 0.8997

51. 0.5328

52. 0.2857

44. ***Tires*** An automobile tire brand has a life expectancy that is normally distributed, with a mean life of 30,000 miles and a standard deviation of 2500 miles. The life spans of four tires selected at random are 35,150 miles, 24,750 miles, 30,000 miles, and 33,000 miles.

45. ***SAT I Scores*** The SAT is an exam used by colleges and universities to evaluate undergraduate applicants. The test scores are normally distributed. In a recent year, the mean test score was 1026 and the standard deviation was 209. The test scores of four students selected at random are 950, 1250, 1467, and 801. *(Source: College Board Online)*

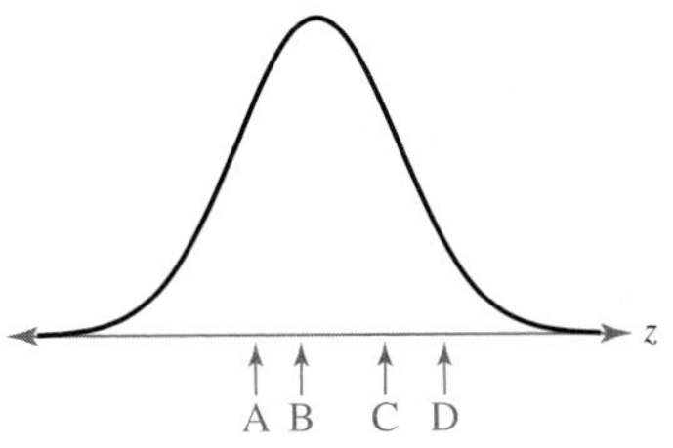

Figure for Exercise 45

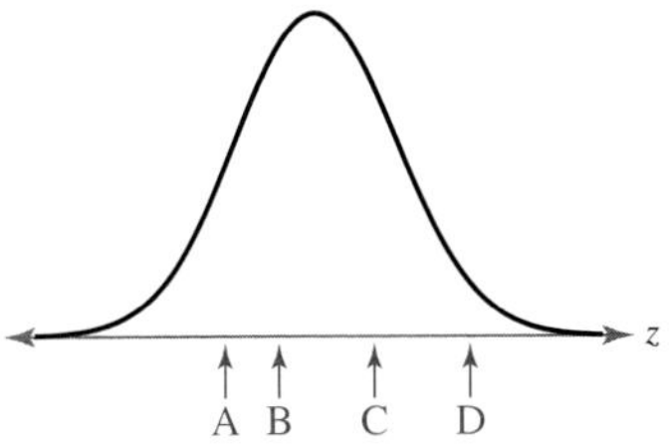

Figure for Exercise 46

46. ***ACT Scores*** The ACT is an exam used by colleges and universities to evaluate undergraduate applicants. The test scores are normally distributed. In a recent year, the mean test score was 20.8 and the standard deviation was 4.8. The test scores of four students selected at random are 18, 32, 14, and 25. *(Source: ACT, Inc.)*

Graphical Analysis In Exercises 47–52, find the probability of z occurring in the indicated region.

47.

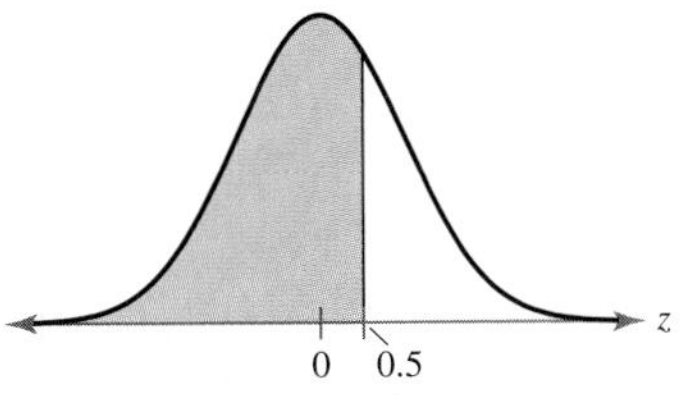

48.

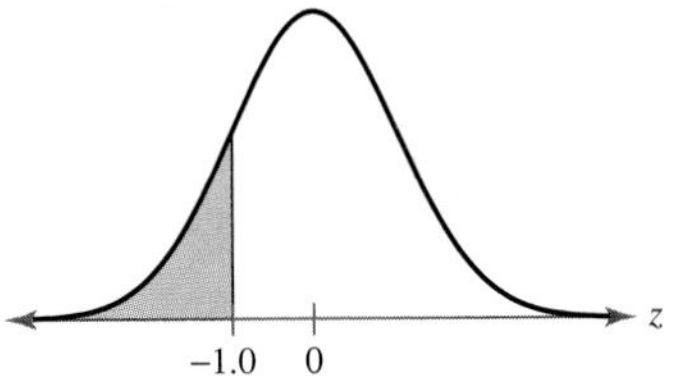

49.

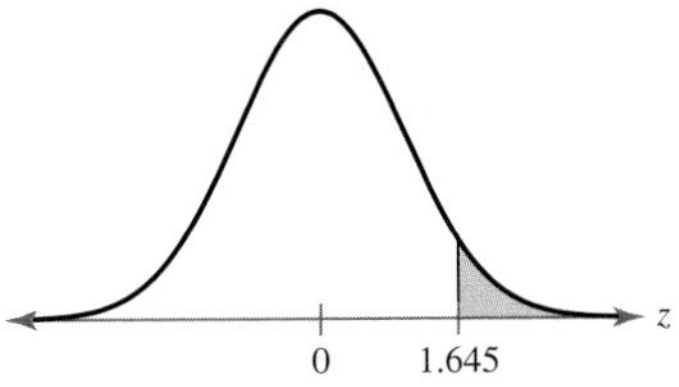

50.

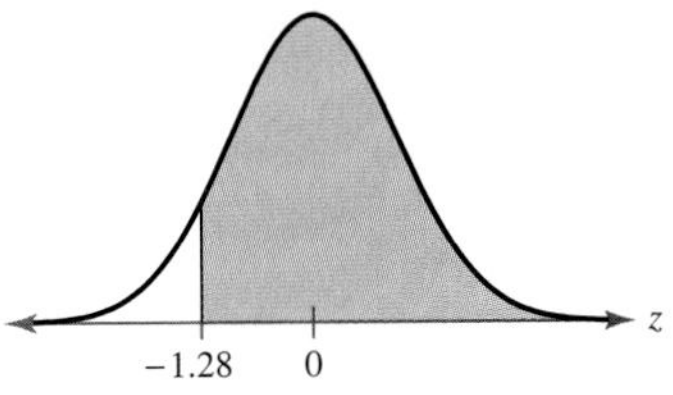

51.

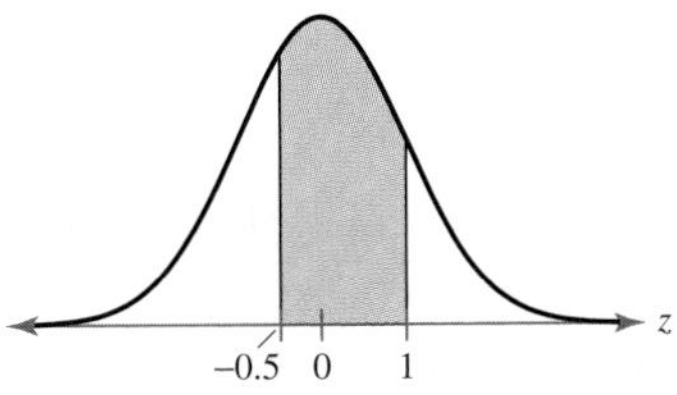

52.

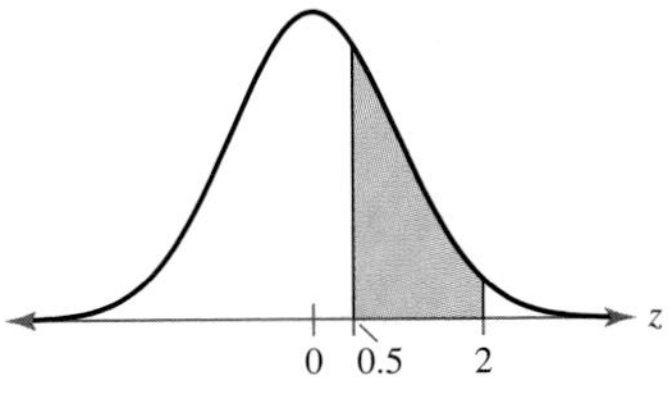

53. 0.9265 **54.** 0.6736
55. 0.9744 **56.** 0.5987
57. 0.3133 **58.** 0.4812
59. 0.901 **60.** 0.95
61. 0.0098 **62.** 0.05

63.

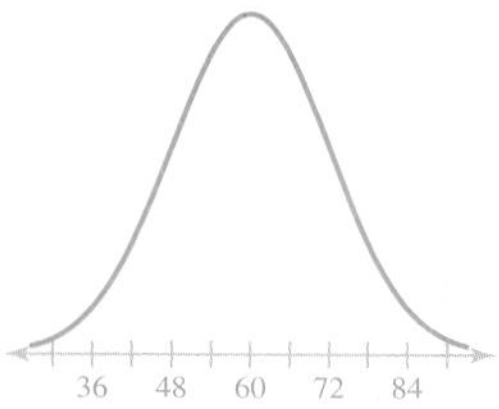

The normal distribution curve is centered at its mean (60) and has 2 points of inflection (48 and 72) representing $\mu \pm \sigma$.

64.

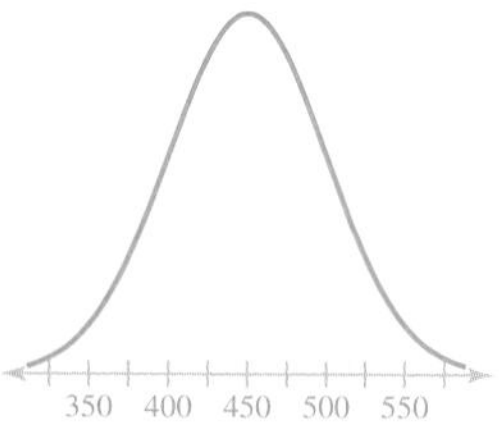

The normal distribution curve is centered at its mean (450) and has 2 points of inflection (400 and 500) representing $\mu \pm \sigma$.

65. (a) Area under curve
$= \text{area of rectangle}$
$= (1)(1)$
$= 1$

(b) 0.25

(c) 0.4

66. (a)

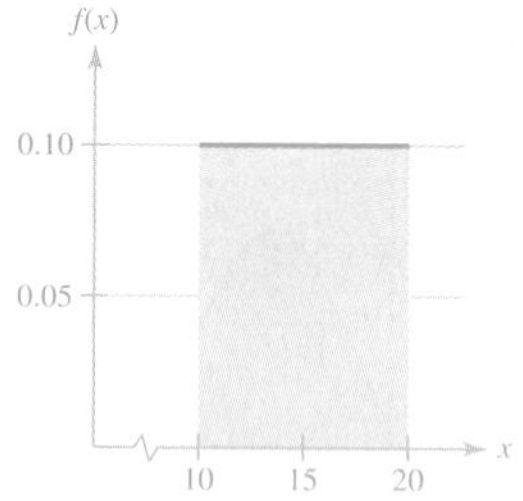

Area under curve
$= \text{area of rectangle}$
$= (20 - 10) \cdot (0.10)$
$= 1$

(b) 0.3

(c) 0.5

Finding Probabilities In Exercises 53–62, find the indicated probability using the standard normal distribution.

53. $P(z < 1.45)$ **54.** $P(z < 0.45)$ **55.** $P(z > -1.95)$

56. $P(z > -0.25)$ **57.** $P(-0.89 < z < 0)$ **58.** $P(-2.08 < z < 0)$

59. $P(-1.65 < z < 1.65)$ **60.** $P(-1.96 < z < 1.96)$

61. $P(z < -2.58 \text{ or } z > 2.58)$ **62.** $P(z < -1.96 \text{ or } z > 1.96)$

Extending Concepts

63. ***Writing*** Draw a normal curve with a mean of 60 and a standard deviation of 12. Describe how you constructed the curve and discuss its features.

64. ***Writing*** Draw a normal curve with a mean of 450 and a standard deviation of 50. Describe how you constructed the curve and discuss its features.

65. ***Uniform Distribution*** Another continuous distribution is the **uniform distribution.** An example is $f(x) = 1$ for $0 \le x \le 1$. The mean of this distribution for this example is 0.5 and the standard deviation is approximately 0.29. The graph of this distribution for this example is a square with the height and width both equal to 1 unit. In general, the density function for a uniform distribution on the interval from $x = a$ to $x = b$ is given by

$$f(x) = \frac{1}{b - a}.$$

The mean is

$$\frac{a + b}{2}$$

and the variance is

$$\frac{(b - a)^2}{12}.$$

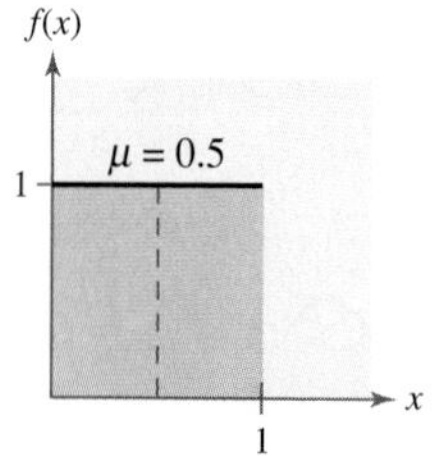

(a) Verify that the area under the curve is 1.

(b) Find the probability that x falls between 0.25 and 0.5.

(c) Find the probability that x falls between 0.3 and 0.7.

66. ***Uniform Distribution*** Consider the uniform density function $f(x) = 0.1$ for $10 \le x \le 20$. The mean of this distribution is 15 and the standard deviation is about 2.89.

(a) Draw a graph of the distribution and show that the area under the curve is 1.

(b) Find the probability that x falls between 12 and 15.

(c) Find the probability that x falls between 13 and 18.

5.2 Normal Distributions: Finding Probabilities

What You Should Learn

- How to find probabilities for normally distributed variables using a table and using technology

Probability and Normal Distributions

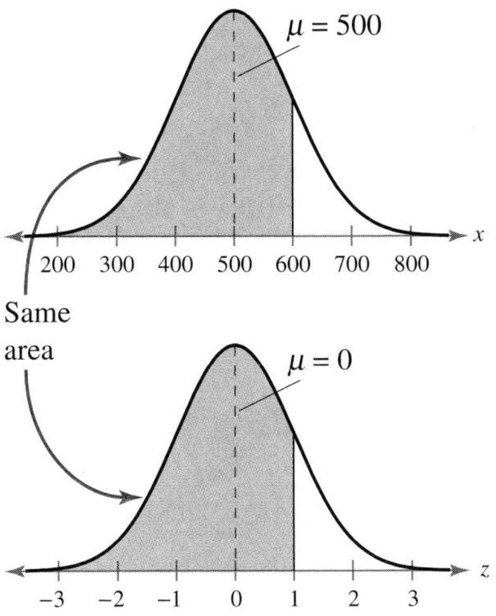

Probability and Normal Distributions

If a random variable x is normally distributed, you can find the probability that x will fall in a given interval by calculating the area under the normal curve for the given interval. To find the area under any normal curve, first convert the upper and lower bounds of the interval to z-scores. Then use the standard normal distribution to find the area. For instance, consider a normal curve with $\mu = 500$ and $\sigma = 100$, as shown at the upper left. The value of x one standard deviation above the mean is $\mu + \sigma = 500 + 100 = 600$. Now consider the standard normal curve shown at the lower left. The value of z one standard deviation above the mean is $\mu + \sigma = 0 + 1 = 1$. Because a z-score of 1 corresponds to an x-value of 600, and areas are not changed with a transformation to a standard normal curve, the shaded areas in the graphs are equal.

EXAMPLE 1

Finding Probabilities for Normal Distributions

A survey indicates that people use their computers an average of 2.4 years before upgrading to a new machine. The standard deviation is 0.5 year. A computer owner is selected at random. Find the probability that he or she will use it for less than 2 years before upgrading. Assume that the variable x is normally distributed.

SOLUTION The graph shows a normal curve with $\mu = 2.4$ and $\sigma = 0.5$ and a shaded area for x less than 2. The z-score that corresponds to 2 years is

$$z = \frac{x - \mu}{\sigma} = \frac{2 - 2.4}{0.5} = -0.80.$$

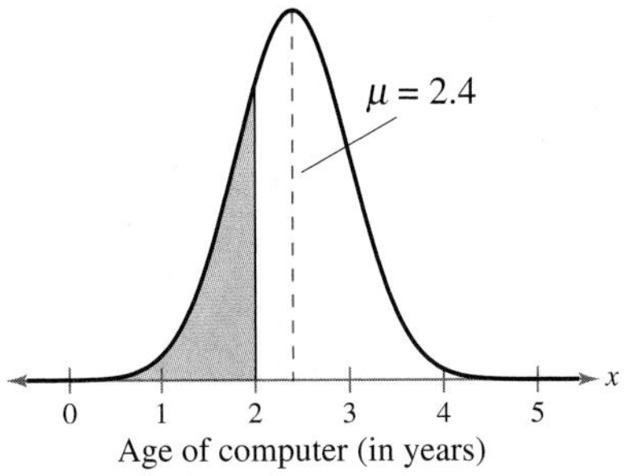

The Standard Normal Table shows that $P(z < -0.8) = 0.2119$. The probability that the computer will be upgraded in less than 2 years is 0.2119. So, 21.19% of new owners will upgrade in less than two years.

Try It Yourself 1

A Ford Focus manual transmission gets an average of 27 miles per gallon (mpg) in city driving with a standard deviation of 1.6 mpg. A Focus is selected at random. What is the probability that it will get more than 31 mpg? Assume that gas mileage is normally distributed. *(Source: U.S. Department of Energy)*

a. *Sketch* a graph.
b. *Find the z-score* that corresponds to 31 miles per gallon.
c. *Find the area* to the right of that z-score.
d. *Write* the result as a sentence.

Answer: Page A37

Study Tip

Another way to write the answer to Example 1 is $P(x < 2) = 0.2119$.

EXAMPLE 2

Finding Probabilities for Normal Distributions

A survey indicates that for each trip to the supermarket, a shopper spends an average of $\mu = 45$ minutes with a standard deviation of $\sigma = 12$ minutes. The length of time spent in the store is normally distributed and is represented by the variable x. A shopper enters the store. (a) Find the probability that the shopper will be in the store for each interval of time listed below. (b) If 200 shoppers enter the store, how many shoppers would you expect to be in the store for each interval of time listed below?

1. Between 24 and 54 minutes 2. More than 39 minutes

SOLUTION

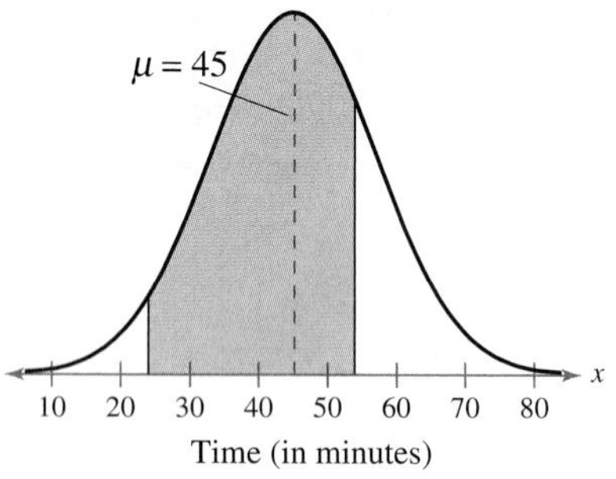

1.(a) The graph at the left shows a normal curve with $\mu = 45$ minutes and $\sigma = 12$ minutes. The area for x between 24 and 54 minutes is shaded. The z-scores that correspond to 24 minutes and to 54 minutes are

$$z_1 = \frac{24 - 45}{12} = -1.75 \quad \text{and} \quad z_2 = \frac{54 - 45}{12} = 0.75.$$

So, the probability that a shopper will be in the store between 24 and 54 minutes is

$$\begin{aligned} P(24 < x < 54) &= P(-1.75 < z < 0.75) \\ &= P(z < 0.75) - P(z < -1.75) \\ &= 0.7734 - 0.0401 = 0.7333. \end{aligned}$$

(b) Another way of interpreting this probability is to say that 73.33% of the shoppers will be in the store between 24 and 54 minutes. If 200 shoppers enter the store, then you would expect $200(0.7333) = 146.66$ (or about 147) shoppers to be in the store between 24 and 54 minutes.

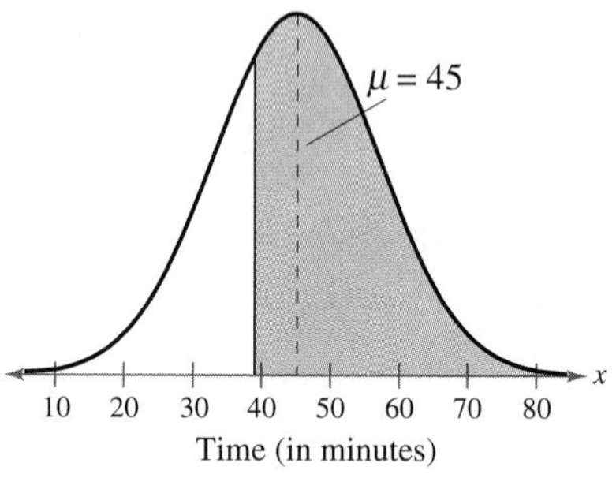

2.(a) The graph at the left shows a normal curve with $\mu = 45$ minutes and $\sigma = 12$ minutes. The area for x greater than 39 minutes is shaded. The z-score that corresponds to 39 minutes is

$$z = \frac{39 - 45}{12} = -0.5.$$

So, the probability that a shopper will be in the store more than 39 minutes is

$$P(x > 39) = P(z > -0.5) = 1 - P(z < -0.5) = 1 - 0.3085 = 0.6915.$$

(b) If 200 shoppers enter the store, then you would expect $200(0.6915) = 138.3$ (or about 138) shoppers to be in the store more than 39 minutes.

Try It Yourself 2

What is the probability that the shopper will be in the supermarket between 33 and 60 minutes?

a. *Sketch* a graph.
b. *Find z-scores* that correspond to 60 minutes and 33 minutes.
c. *Find the cumulative area* for each z-score.
d. *Subtract the smaller area from the larger.*

Answer: Page A37

Another way to find normal probabilities is to use a calculator or a computer. You can find normal probabilities using MINITAB, Excel, and the TI-83.

EXAMPLE 3

Using Technology to Find Normal Probabilities

Assume that cholesterol levels of men in the United States are normally distributed, with a mean of 215 milligrams per deciliter and a standard deviation of 25 milligrams per deciliter. You randomly select a man from the United States. What is the probability that his cholesterol level is less than 175? Use a technology tool to find the probability.

SOLUTION MINITAB, Excel, and the TI-83 each have features that allow you to find normal probabilities without first converting to standard z-scores. For each, you must specify the mean and standard deviation of the population, as well as the x-value(s) that determine the interval.

MINITAB

Cumulative Distribution Function

Normal with mean = 215.000 and standard deviation = 25.0000

x	P(X <= x)
175.0000	0.0548

EXCEL

	A	B	C
1	NORMDIST(175,215,25,TRUE)		
2			0.054799

TI-83

normalcdf(0,175,215,25)
.0547992894

From the displays, you can see that the probability that his cholesterol level is less than 175 is about 0.055, or 5.5%.

Try It Yourself 3

A man from the United States is selected at random. What is the probability that his cholesterol is between 190 and 225? Use a technology tool.

a. *Read the user's guide* for the technology tool you are using.
b. *Enter the appropriate data* to obtain the probability.
c. *Write* the result as a sentence.

Answer: Page A37

Example 3 shows only one of several ways to find normal probabilities using MINITAB, Excel, and the TI-83.

Picturing the World

In baseball, a batting average is the number of hits divided by the number of at-bats. The batting averages of the more than 750 Major League Baseball players in a recent year can be approximated by a normal distribution, as shown in the following graph. The mean of the batting averages is 0.266 and the standard deviation is 0.012.

Major League Baseball

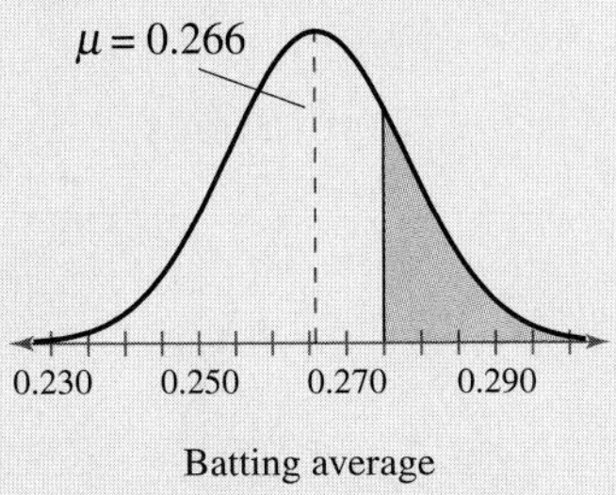

What percent of the players have a batting average of 0.275 or greater? If there are 40 players on a roster, how many would you expect to have a batting average of 0.275 or greater?

5.2 Exercises

1. 0.1151
2. 0.9974
3. 0.1151
4. 0.9861
5. 0.1144
6. 0.5434
7. 0.3022
8. 0.0878
9. 0.2742
10. 0.3462
11. 0.0566
12. 0.4251

Building Basic Skills and Vocabulary

Computing Probabilities In Exercises 1–6, assume the random variable x is normally distributed with mean $\mu = 86$ and standard deviation $\sigma = 5$. Find the indicated probability.

1. $P(x < 80)$

2. $P(x < 100)$

3. $P(x > 92)$

4. $P(x > 75)$

5. $P(70 < x < 80)$

6. $P(85 < x < 95)$

Graphical Analysis In Exercises 7–12, assume a member is selected at random from the population represented by the graph. Find the probability that the member selected at random is from the shaded area of the graph. Assume the variable x is normally distributed.

7. **SAT Verbal Scores**

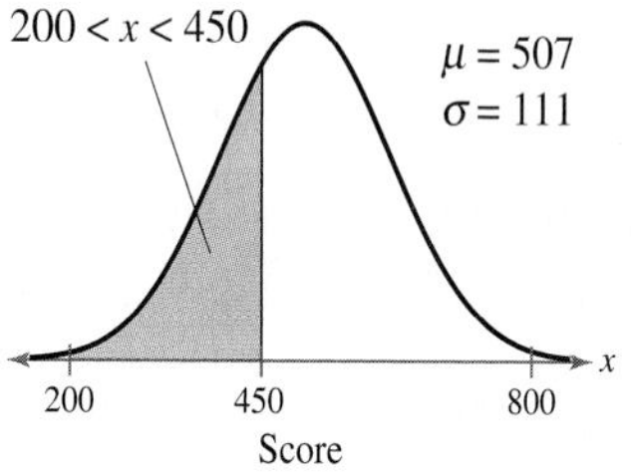

(Source: College Board Online)

8. **SAT Math Scores**

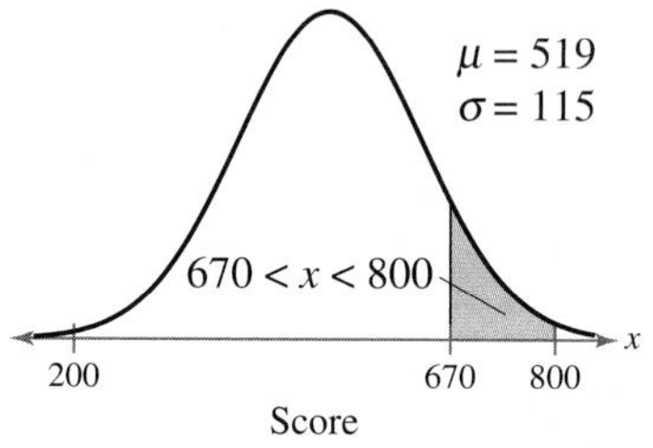

(Source: College Board Online)

9. **U.S. Women Ages 20–34: Total Cholesterol**

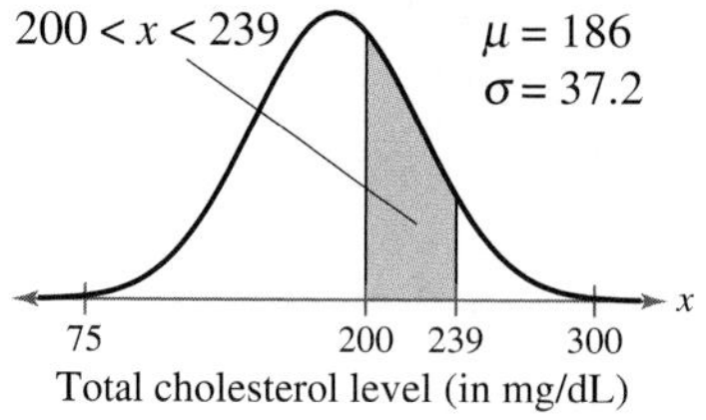

(Adapted from Centers for Disease Control and Prevention)

10. **U.S. Women Ages 55–64: Total Cholesterol**

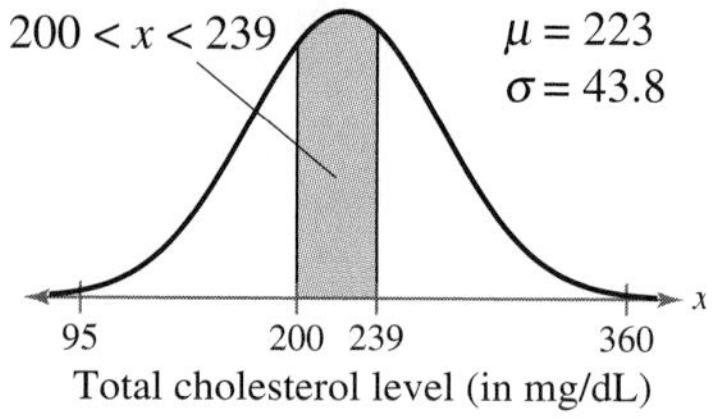

(Adapted from Centers for Disease Control and Prevention)

11. **Chevrolet Blazer: Braking Distance on a Dry Surface**

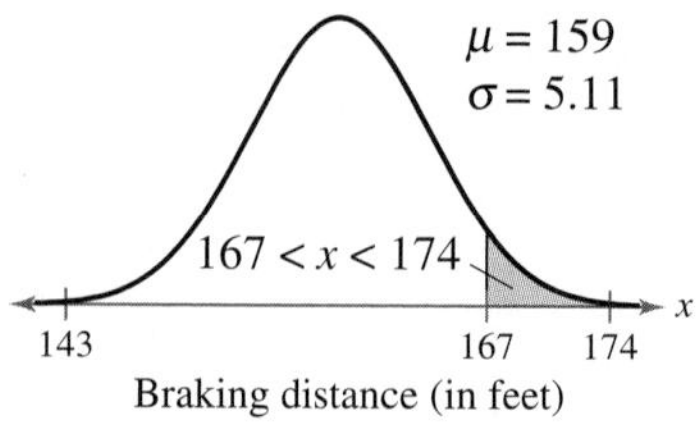

(Source: National Highway Traffic Safety Administration)

12. **Chevrolet Blazer: Braking Distance on a Wet Surface**

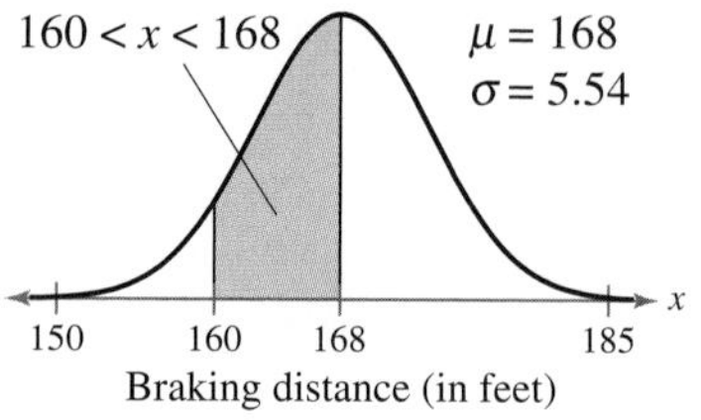

(Source: National Highway Traffic Safety Administration)

13. (a) 0.1357
(b) 0.6983
(c) 0.1660
14. (a) 0.0668
(b) 0.927
(c) 0.0062
15. (a) 0.1711
(b) 0.7018
(c) 0.1271
16. (a) 0.2514
(b) 0.4972
(c) 0.2514
17. (a) 0.0062
(b) 0.9876
(c) 0.0062

Using and Interpreting Concepts

Finding Probabilities In Exercises 13–20, find the indicated probabilities. If convenient, use technology to find the probabilities.

13. ***Heights of Men*** A survey was conducted to measure the height of U.S. men. In the survey, respondents were grouped by age. In the 20–29 age group, the heights were normally distributed, with a mean of 69.2 inches and a standard deviation of 2.9 inches. A study participant is randomly selected. *(Source: U.S. National Center for Health Statistics)*

(a) Find the probability that his height is less than 66 inches.
(b) Find the probability that his height is between 66 and 72 inches.
(c) Find the probability that his height is more than 72 inches.

14. ***Fish Lengths*** The lengths of Atlantic croaker fish are normally distributed, with a mean of 10 inches and a standard deviation of 2 inches. An Atlantic croaker fish is randomly selected. *(Adapted from National Marine Fisheries Service, Fisheries Statistics and Economics Division)*

(a) Find the probability that the length of the fish is less than 7 inches.
(b) Find the probability that the length of the fish is between 7 and 15 inches.
(c) Find the probability that the length of the fish is more than 15 inches.

15. ***ACT Scores*** In a recent year, the ACT scores for high school students with a 3.50 to 4.00 grade point average were normally distributed, with a mean of 24.1 and a standard deviation of 4.3. A student with a 3.50 to 4.00 grade point average who took the ACT during this time is randomly selected. *(Source: ACT, Inc.)*

(a) Find the probability that the student's ACT score is less than 20.
(b) Find the probability that the student's ACT score is between 20 and 29.
(c) Find the probability that the student's ACT score is more than 29.

16. ***Rhesus Monkeys*** The weights of adult male rhesus monkeys are normally distributed, with a mean of 15 pounds and a standard deviation of 3 pounds. A rhesus monkey is randomly selected.

(a) Find the probability that the monkey's weight is less than 13 pounds.
(b) Find the probability that the weight is between 13 and 17 pounds.
(c) Find the probability that the monkey's weight is more than 17 pounds.

17. ***Computer Usage*** A survey was conducted to measure the number of hours per week adults in the United States spend on home computers. In the survey, the number of hours were normally distributed, with a mean of 7 hours and a standard deviation of 1 hour. A survey participant is randomly selected.

(a) Find the probability that the hours spent on the home computer by the participant are less than 4.5 hours per week.
(b) Find the probability that the hours spent on the home computer by the participant are between 4.5 and 9.5 hours per week.
(c) Find the probability that the hours spent on the home computer by the participant are more than 9.5 hours per week.

18. (a) 0.0475
(b) 0.8469
(c) 0.1056
19. (a) 0.0073
(b) 0.806
(c) 0.1867
20. (a) 0.2743
(b) 0.3811
(c) 0.3446
21. (a) 79.95%
(b) 348
22. (a) 43.25%
(b) 363
23. (a) 64.8%
(b) 18

18. *Utility Bills* The monthly utility bills in a city are normally distributed, with a mean of \$100 and a standard deviation of \$12. A utility bill is randomly selected.

(a) Find the probability that the utility bill is less than \$80.
(b) Find the probability that the utility bill is between \$80 and \$115.
(c) Find the probability that the utility bill is more than \$115.

19. *Computer Lab Schedule* The time per week a student uses a lab computer is normally distributed, with a mean of 6.2 hours and a standard deviation of 0.9 hour. A student is randomly selected.

(a) Find the probability that the student uses a lab computer less than 4 hours per week.
(b) Find the probability that the student uses a lab computer between 4 and 7 hours per week.
(c) Find the probability that the student uses a lab computer more than 7 hours per week.

20. *Health Club Schedule* The time per workout an athlete uses a stairclimber is normally distributed, with a mean of 20 minutes and a standard deviation of 5 minutes. An athlete is randomly selected.

(a) Find the probability that the athlete uses a stairclimber for less than 17 minutes.
(b) Find the probability that the athlete uses a stairclimber between 17 and 22 minutes.
(c) Find the probability that the athlete uses a stairclimber for more than 22 minutes.

Using Normal Distributions In Exercises 21–30, answer the questions about the specified normal distribution.

21. *SAT Verbal Scores* Use the normal distribution of SAT verbal scores in Exercise 7 for which the mean is 507 and the standard deviation is 111.

(a) What percent of the SAT verbal scores are less than 600?
(b) If 1000 SAT verbal scores are randomly selected, about how many would you expect to be greater than 550?

22. *SAT Math Scores* Use the normal distribution of SAT math scores in Exercise 8 for which the mean is 519 and the standard deviation is 115.

(a) What percent of the SAT math scores are less than 500?
(b) If 1500 SAT math scores are randomly selected, about how many would you expect to be greater than 600?

23. *Cholesterol* Use the normal distribution of women's total cholesterol levels in Exercise 9 for which the mean is 186 milligrams per deciliter and the standard deviation is 37.2 milligrams per deciliter.

(a) What percent of the women have a total cholesterol level less than 200 milligrams per deciliter of blood?
(b) If 250 U.S. women in the 20–29 age group are randomly selected, about how many would you expect to have a total cholesterol level greater than 240 milligrams per deciliter of blood?

24. (a) 0.6443
(b) 140
25. (a) 30.85%
(b) 31
26. (a) 4.75%
(b) 7
27. (a) 99.87%
(b) 0.798
28. (a) 1.88%
(b) 60
29. 1.5%; It is unusual for a battery to have a life span that is more than 2065 hours because of the relatively large *z*-score (2.17).
30. 5.94%; It is not unusual for a person to consume less than 3.1 pounds of peanuts in a year because the *z*-score is within 2 standard deviations of the mean.

24. *Cholesterol* Use the normal distribution of women's total cholesterol levels in Exercise 10 for which the mean is 223 milligrams per deciliter and the standard deviation is 43.8 milligrams per deciliter.

(a) What percent of the women have a total cholesterol level less than 239 milligrams per deciliter of blood?

(b) If 200 U.S. women in the 50–59 age group are randomly selected, about how many would you expect to have a total cholesterol level greater than 200 milligrams per deciliter of blood?

25. *Fish Lengths* Use the normal distribution of fish lengths in Exercise 14 for which the mean is 10 inches and the standard deviation is 2 inches.

(a) What percent of the fish are longer than 11 inches?

(b) If 200 Atlantic croakers are randomly selected, about how many would you expect to be shorter than 8 inches?

26. *Rhesus Monkeys* Use the normal distribution of monkey weights in Exercise 16 for which the mean is 15 pounds and the standard deviation is 3 pounds.

(a) What percent of the monkeys have a weight that is greater than 20 pounds?

(b) If 50 rhesus monkeys are randomly selected, about how many would you expect to weigh less than 12 pounds?

27. *Computer Usage* Use the normal distribution of computer usage in Exercise 17 for which the mean is 7 hours and the standard deviation is 1 hour.

(a) What percent of the adults spend more than 4 hours per week on a home computer?

(b) If 35 adults in the United States are randomly selected, about how many would you expect to say they spend less than 5 hours per week on a home computer?

28. *Utility Bills* Use the normal distribution of utility bills in Exercise 18 for which the mean is $100 and the standard deviation is $12.

(a) What percent of the utility bills are more than $125?

(b) If 300 utility bills are randomly selected, about how many would you expect to be less than $90?

29. *Battery Life Spans* The life span of a battery is normally distributed, with a mean of 2000 hours and a standard deviation of 30 hours. What percent of batteries have a life span that is more than 2065 hours? Would it be unusual for a battery to have a life span that is more than 2065 hours? Explain your reasoning.

30. *Peanuts* Assume the mean annual consumption of peanuts is normally distributed, with a mean of 5.9 pounds per person and a standard deviation of 1.8 pounds per person. What percent of people annually consume less than 3.1 pounds of peanuts per person? Would it be unusual for a person to consume less than 3.1 pounds of peanuts in a year? Explain your reasoning.

31. Out of control, because there is a point more than 3 standard deviations beyond the mean.
32. Out of control, because two out of three consecutive points lie more than 2 standard deviations from the mean.
33. Out of control, because there are nine consecutive points below the mean, and two out of three consecutive points lie more than 2 standard deviations from the mean.
34. In control, because none of the three warning signals detected a change.

Extending Concepts

Control Charts **Statistical process control (SPC)** is the use of statistics to monitor and improve the quality of a process, such as manufacturing an engine part. In SPC, information about a process is gathered and used to determine if a process is meeting all of the specified requirements. One tool used in SPC is a **control chart.** When individual measurements of a variable x are normally distributed, a control chart can be used to detect processes that are possibly out of statistical control. Three warning signals that a control chart uses to detect a process that may be out of control are as follows:

(1) A point lies beyond three standard deviations of the mean.
(2) There are nine consecutive points that fall on one side of the mean.
(3) At least two of three consecutive points lie more than two standard deviations from the mean.

In Exercises 31–34, a control chart is shown. Each chart has horizontal lines drawn at the mean μ, at $\mu \pm 2\sigma$, and at $\mu \pm 3\sigma$. Determine if the process shown is in control or out of control. Explain.

31. A gear has been designed to have a diameter of 3 inches. The standard deviation of the process is 0.2 inch.

Gears

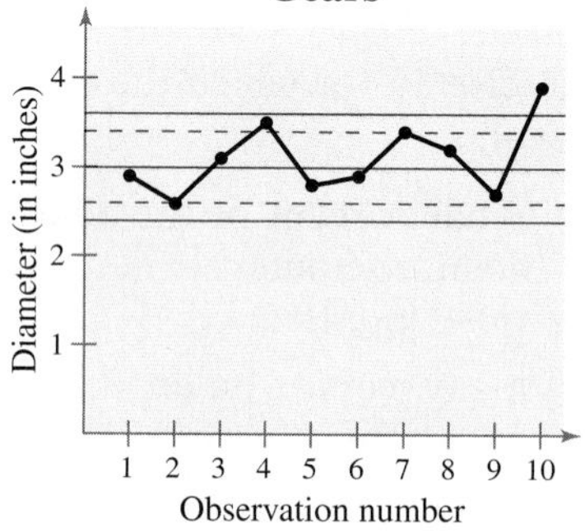

32. A nail has been designed to have a length of 4 inches. The standard deviation of the process is 0.12 inch.

Nails

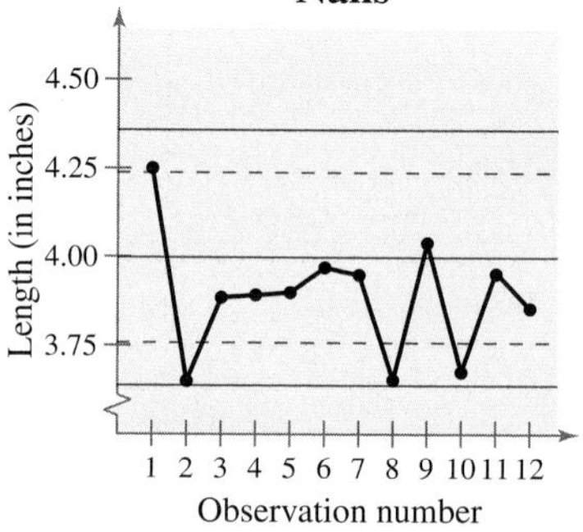

33. A liquid-dispensing machine has been designed to fill bottles with 1 liter of liquid. The standard deviation of the process is 0.1 liter.

Liquid Dispenser

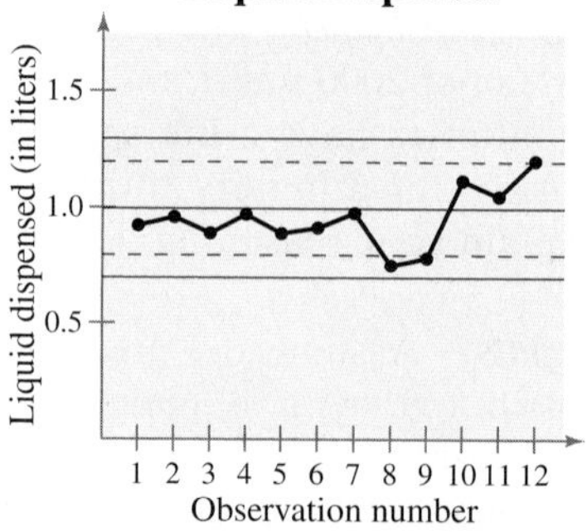

34. An engine part has been designed to have a diameter of 55 millimeters. The standard deviation of the process is 0.001 millimeter.

Engine Part

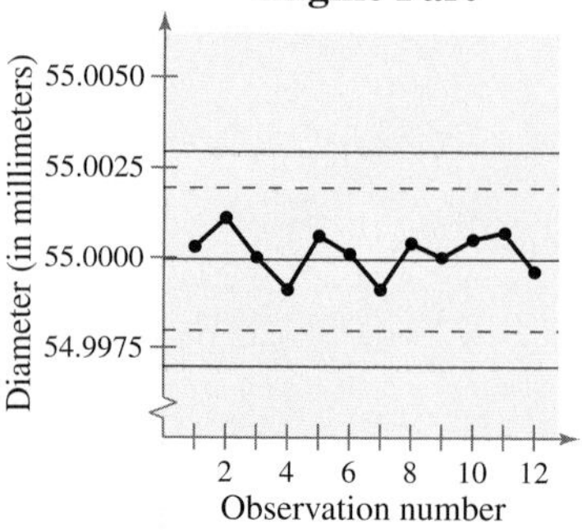

5.3 Normal Distributions: Finding Values

What You Should Learn

- How to find a z-score given the area under the normal curve
- How to transform a z-score to an x-value
- How to find a specific data value of a normal distribution given the probability

Note to Instructor

Have students note that as the z-scores increase, the cumulative areas increase. The CDF is one to one and as such has an inverse function. Discuss how this inverse function (INVCDF) can be used when a cumulative area (percentile) is known and the z-score must be found.

Finding z-Scores • Transforming a z-Score to an x-Value • Finding a Specific Data Value for a Given Probability

Finding z-Scores

In Section 5.2, you were given a normally distributed random variable x and you found the probability that x would fall in a given interval by calculating the area under the normal curve for the given interval.

But what if you are given a probability and want to find a value? For instance, a university might want to know what is the lowest test score a student can have on an entrance exam and still be in the top 10%, or a medical researcher might want to know the cutoff values to select the middle 90% of patients by age. In this section, you will learn how to find a value given an area under a normal curve (or a probability), as shown in the following example.

EXAMPLE 1

Finding a z-Score Given an Area

1. Find the z-score that corresponds to a cumulative area of 0.3632.
2. Find the z-score that has 10.75% of the distribution's area to its right.

SOLUTION

1. Find the z-score that corresponds to an area of 0.3632 by locating 0.3632 in the Standard Normal Table. The values at the beginning of the corresponding row and at the top of the corresponding column give the z-score. For this area, the row value is -0.3 and the column value is 0.05. So, the z-score is -0.35.

z	.09	.08	.07	.06	.05	.04	.03
−3.4	.0002	.0003	.0003	.0003	.0003	.0003	.0003
−0.5	.2776	.2810	.2843	.2877	.2912	.2946	.2981
−0.4	.3121	.3156	.3192	.3228	.3264	.3300	.3336
−0.3	.3483	.3520	.3557	.3594	.3632	.3669	.3707
−0.2	.3859	.3897	.3936	.3974	.4013	.4052	.4090

2. Because the area to the right is 0.1075, the cumulative area is $1 - 0.1075 = 0.8925$. Find the z-score that corresponds to an area of 0.8925 by locating 0.8925 in the Standard Normal Table. For this area, the row value is 1.2 and the column value is 0.04. So, the z-score is 1.24.

z	.00	.01	.02	.03	.04	.05	.06
0.0	.5000	.5040	.5080	.5120	.5160	.5199	.5239
1.0	.8413	.8438	.8461	.8485	.8508	.8531	.8554
1.1	.8643	.8665	.8686	.8708	.8729	.8749	.8770
1.2	.8849	.8869	.8888	.8907	.8925	.8944	.8962
1.3	.9032	.9049	.9066	.9082	.9099	.9115	.9131

Study Tip

You can use a computer or calculator to find the z-score that corresponds to a cumulative area. For instance, here are instructions for finding the z-score that corresponds to an area of 0.3632 on a TI-83.

[2nd] [DISTR] [3] .3632

The calculator will display −.3499183227.

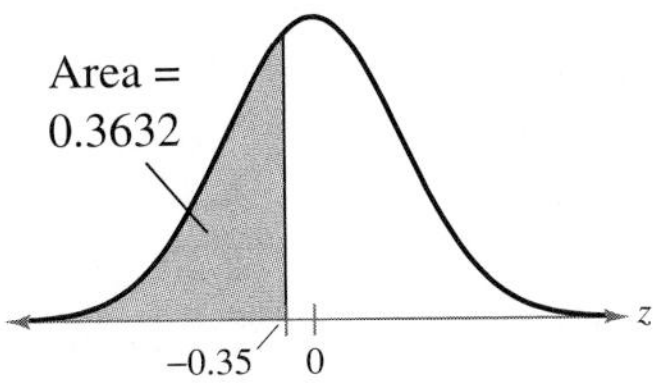

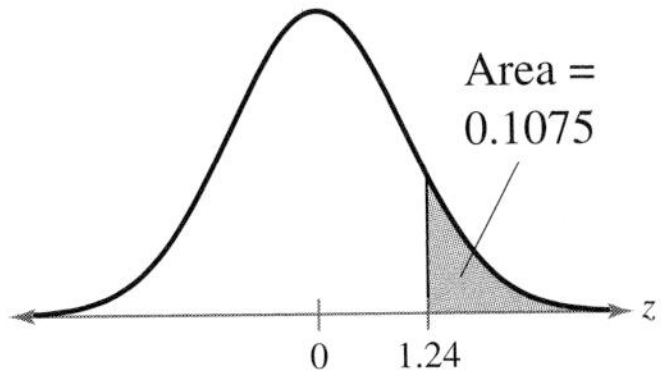

Note to Instructor

If you prefer that your students use a 0-to-z table, refer them to Appendix A where an alternative presentation for this material is given.

Study Tip

In most cases, the given area will not be found in the table, so use the entry closest to it. If the given area is halfway between two area entries, use the z-score halfway between the corresponding z-scores. For instance, in part 1 of Example 2, the z-score between −1.64 and −1.65 is −1.645.

Try It Yourself 1

1. Find the z-score that has 96.16% of the distribution's area to the right.
2. Find the z-score for which 95% of the distribution's area lies between $-z$ and z.

a. *Determine* the cumulative area.
b. *Locate* the area in the Standard Normal Table.
c. *Find the z-score* that corresponds to the area.

Answer: Page A37

In Section 2.5, you learned that percentiles divide a data set into one hundred equal parts. To find a z-score that corresponds to a percentile, you can use the Standard Normal Table. Recall that if a value x represents the 83rd percentile P_{83}, then 83% of the data values are below x and 17% of the data values are above x.

EXAMPLE 2

Finding a z-Score Given a Percentile

Find the z-score that corresponds to each percentile.

1. P_5 2. P_{50} 3. P_{90}

SOLUTION

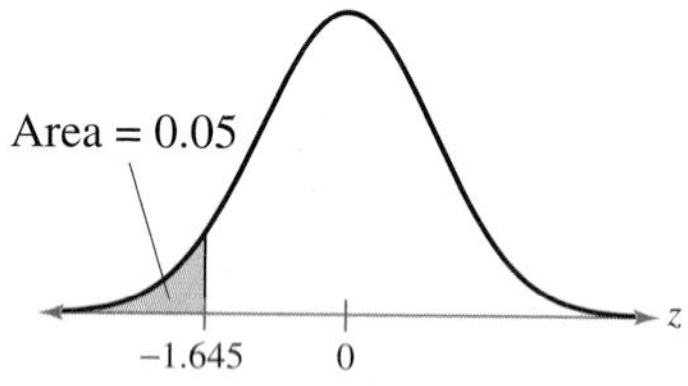

1. To find the z-score that corresponds to P_5, find the z-score that corresponds to an area of 0.05 (see figure) by locating 0.05 in the Standard Normal Table. The areas closest to 0.05 in the table are 0.0495 ($z = -1.65$) and 0.0505 ($z = -1.64$). Because 0.05 is halfway between the two areas in the table, use the z-score that is halfway between −1.64 and −1.65. So, the z-score that corresponds to an area of 0.05 is −1.645.

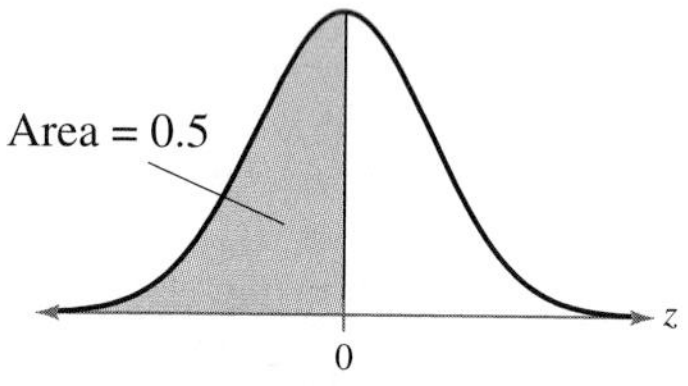

2. To find the z-score that corresponds to P_{50}, find the z-score that corresponds to an area of 0.5 (see figure) by locating 0.5 in the Standard Normal Table. The area closest to 0.5 in the table is 0.5000, so the z-score that corresponds to an area of 0.5 is 0.00.

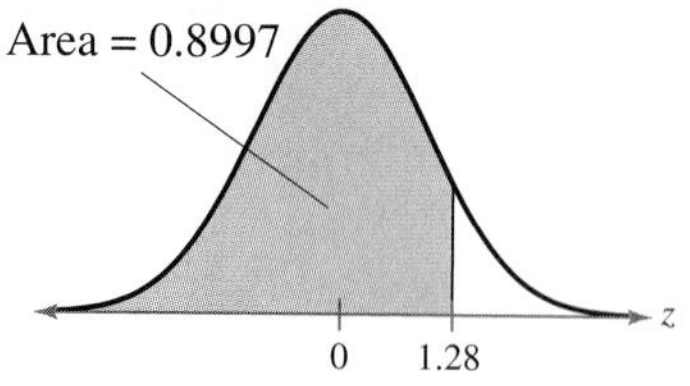

3. To find the z-score that corresponds to P_{90}, find the z-score that corresponds to an area of 0.9 (see figure) by locating 0.9 in the Standard Normal Table. The area closest to 0.9 in the table is 0.8997, so the z-score that corresponds to an area of 0.9 is 1.28.

Try It Yourself 2

Find the z-score that corresponds to each percentile.

1. P_{10} 2. P_{20} 3. P_{99}

a. *Write* the percentile as an area. If necessary, draw a graph of the area to visualize the problem.
b. *Locate* the area in the Standard Normal Table. If the area is not in the table, use the closest area. (See Study Tip above.)
c. *Identify* the z-score that corresponds to the area.

Answer: Page A37

Transforming a z-Score to an x-Value

Recall that to transform an x-value to a z-score, you can use the formula

$$z = \frac{x - \mu}{\sigma}.$$

This formula gives z in terms of x. If you solve this formula for x, you get a new formula that gives x in terms of z.

$z = \dfrac{x - \mu}{\sigma}$	Formula for z in terms of x
$z\sigma = x - \mu$	Multiply each side by σ.
$\mu + z\sigma = x$	Add μ to each side.
$x = \mu + z\sigma$	Interchange sides.

Transforming a z-Score to an x-Value

To transform a standard z-score to a data value x in a given population, use the formula

$$x = \mu + z\sigma.$$

EXAMPLE 3

Finding an x-Value Corresponding to a z-Score

The speeds of vehicles along a stretch of highway are normally distributed, with a mean of 56 miles per hour and a standard deviation of 4 miles per hour. Find the speeds x corresponding to z-scores of 1.96, -2.33, and 0. Interpret your results.

SOLUTION The x-value that corresponds to each standard score is calculated using the formula $x = \mu + z\sigma$.

$z = 1.96$: $\quad x = 56 + 1.96(4) = 63.84$ miles per hour

$z = -2.33$: $\quad x = 56 + (-2.33)(4) = 46.68$ miles per hour

$z = 0$: $\quad x = 56 + 0(4) = 56$ miles per hour

Interpretation You can see that 63.84 miles per hour is above the mean, 46.68 is below the mean, and 56 is equal to the mean.

Try It Yourself 3

The monthly utility bills in a city are normally distributed, with a mean of \$70 and a standard deviation of \$8. Find the x-values that correspond to z-scores of -0.75, 4.29, and -1.82. What can you conclude?

a. *Identify* μ and σ of the nonstandard normal distribution.
b. *Transform* each z-score to an x-value.
c. *Interpret* the results.

Answer: Page A37

Finding a Specific Data Value for a Given Probability

You can also use the normal distribution to find a specific data value (x-value) for a given probability, as shown in Example 4.

Note to Instructor

Mention that to use the cumulative table to find a z-score, you must first express the area given as a cumulative area. It helps to explain these as percentiles. For example, the score in the top 20% represents the 80th percentile. Point out that if students are using a technology tool to find an x-value that corresponds to an area, it is not necessary first to find a z-score.

EXAMPLE 4

Finding a Specific Data Value

Scores for a civil service exam are normally distributed, with a mean of 75 and a standard deviation of 6.5. To be eligible for civil service employment, you must score in the top 5%. What is the lowest score you can earn and still be eligible for employment?

SOLUTION Exam scores in the top 5% correspond to the shaded region shown.

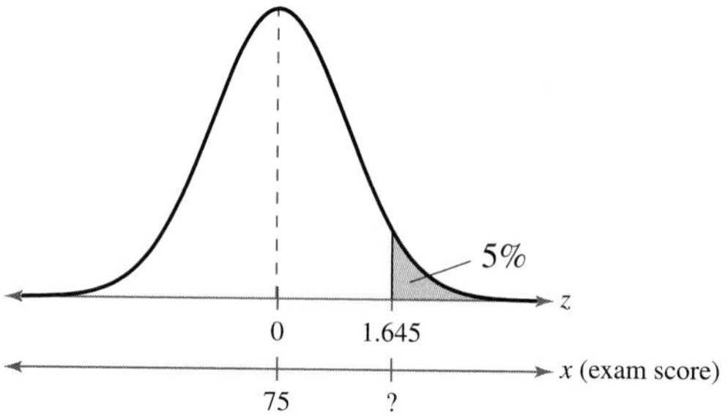

An exam score in the top 5% is any score above the 95th percentile. To find the score that represents the 95th percentile, you must first find the z-score that corresponds to a cumulative area of 0.95. From the Standard Normal Table, you can find that the areas closest to 0.95 are 0.9495 ($z = 1.64$) and 0.9505 ($z = 1.65$). Because 0.95 is halfway between the two areas in the table, use the z-score that is halfway between 1.64 and 1.65. That is, $z = 1.645$. Using the equation $x = \mu + z\sigma$, you have

$$\begin{aligned} x &= \mu + z\sigma \\ &= 75 + 1.645(6.5) \\ &\approx 85.69. \end{aligned}$$

Interpretation The lowest score you can earn and still be eligible for employment is 86.

Try It Yourself 4

The braking distances of a sample of Ford F-150s are normally distributed. On a dry surface, the mean braking distance was 158 feet and the standard deviation was 6.51 feet. What is the longest braking distance on a dry surface one of these Ford F-150s could have and still be in the top 1%? *(Adapted from National Highway Traffic Safety Administration)*

a. *Sketch* a graph.
b. *Find the z-score* that corresponds to the given area.
c. *Find x* using the equation $x = \mu + z\sigma$.
d. *Interpret* the result.

Answer: Page A37

Picturing the World

According to the American Medical Association, the mean number of hours all physicians spend in patient care each week is about 53.2 hours. The hours spent in patient care each week by physicians can be approximated by a normal distribution. Assume the standard deviation is 3 hours.

Hours Physicians Spend in Patient Care

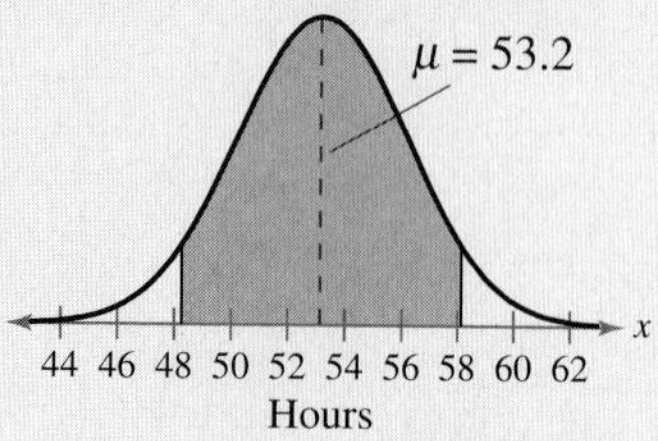

Between what two values does the middle 90% of the data lie?

EXAMPLE 5

Finding a Specific Data Value

In a randomly selected sample of 1169 men ages 35–44, the mean total cholesterol level was 205 milligrams per deciliter with a standard deviation of 39.2 milligrams per deciliter. Assume the total cholesterol levels are normally distributed. Find the highest total cholesterol level a man in this 35–44 age group can have and be in the lowest 1%. *(Adapted from Centers for Disease Control and Prevention)*

SOLUTION

Total cholesterol levels in the lowest 1% correspond to the shaded region shown.

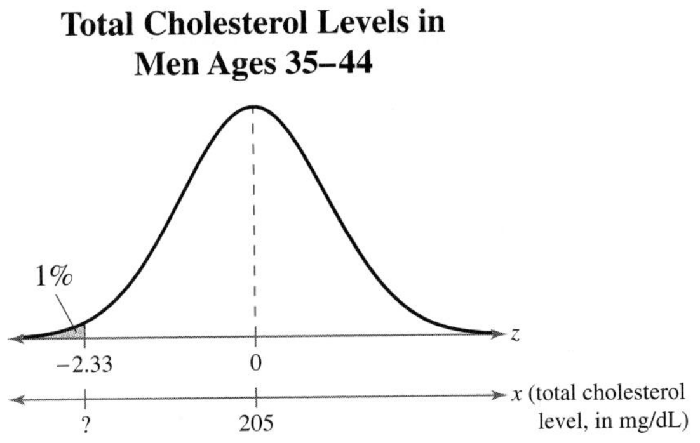

A total cholesterol level in the lowest 1% is any level below the 1st percentile. To find the level that represents the 1st percentile, you must first find the z-score that corresponds to a cumulative area of 0.01. From the Standard Normal Table, you can find that the area closest to 0.01 is 0.0099. So, the z-score that corresponds to an area of 0.01 is $z = -2.33$. Using the equation $x = \mu + z\sigma$, you have

$$\begin{aligned} x &= \mu + z\sigma \\ &= 205 + (-2.33)(39.2) \\ &\approx 113.66. \end{aligned}$$

Interpretation The value that separates the lowest 1% of total cholesterol levels for men in the 35–44 age group from the highest 99% is about 114.

Try It Yourself 5

The length of time employees have worked at a corporation is normally distributed, with a mean of 11.2 years and a standard deviation of 2.1 years. In a company cutback, the lowest 10% in seniority are laid off. What is the maximum length of time an employee could have worked and still be laid off?

a. *Sketch* a graph.
b. *Find the z-score* that corresponds to the given area.
c. *Find x* using the equation $x = \mu + z\sigma$.
d. *Interpret* the result.

Answer: Page A37

5.3 Exercises

Help

MyMathLab

Student Study Pack

1. −2.05
2. −0.81
3. 0.85
4. 0.37
5. −0.16
6. −2.41
7. 2.39
8. 0.73
9. −1.645
10. 1.04
11. 0.995
12. −2.33
13. −2.33
14. −1.04
15. −0.25
16. 0.13
17. 1.175
18. 1.75
19. −0.67
20. 0
21. 0.67
22. −1.28
23. −0.39
24. 0.39
25. −0.38
26. 0.25
27. −0.58
28. 1.99
29. −1.645, 1.645
30. ±1.96
31. 0.325

Building Basic Skills and Vocabulary

In Exercises 1–24, use the Standard Normal Table to find the z-score that corresponds to the given cumulative area or percentile. If the area is not in the table, use the entry closest to the area. If the area is halfway between two entries, use the z-score halfway between the corresponding z-scores.

1. 0.0202 **2.** 0.2090 **3.** 0.8023 **4.** 0.6443

5. 0.4364 **6.** 0.0080 **7.** 0.9916 **8.** 0.7673

9. 0.05 **10.** 0.85 **11.** 0.84 **12.** 0.01

13. P_1 **14.** P_{15} **15.** P_{40} **16.** P_{55}

17. P_{88} **18.** P_{96} **19.** P_{25} **20.** P_{50}

21. P_{75} **22.** P_{10} **23.** P_{35} **24.** P_{65}

Graphical Analysis In Exercises 25–30, find the indicated z-score(s) shown in the graph.

25.

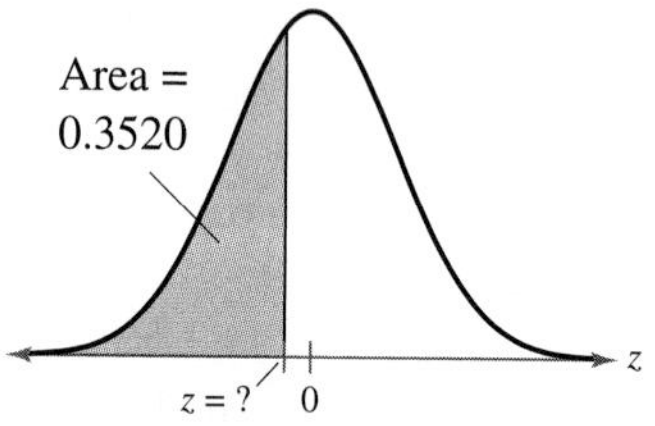

26.

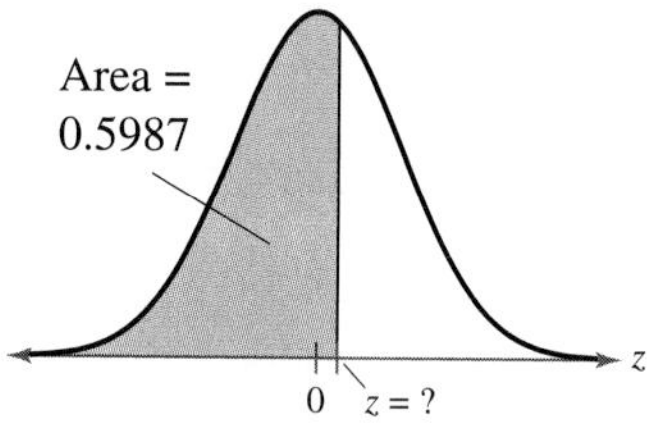

27.

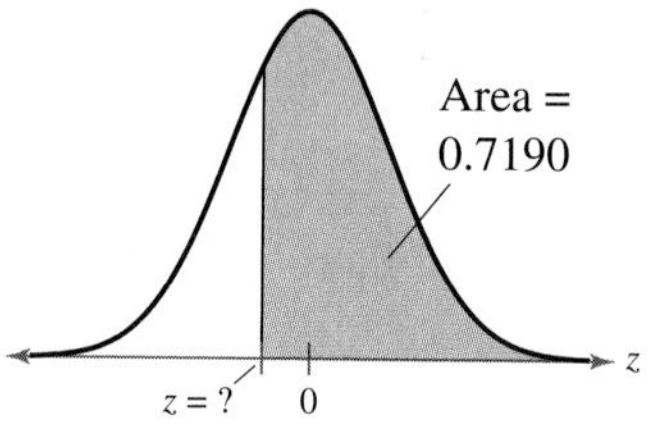

28.

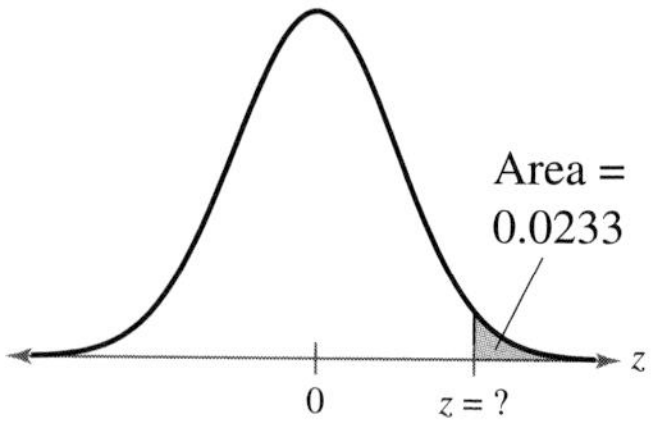

29.

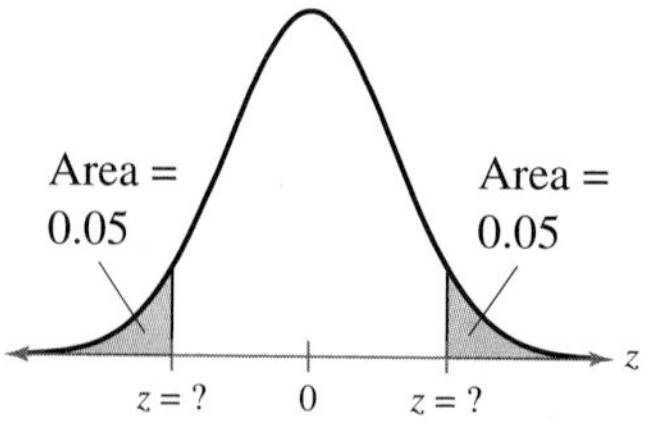

30.

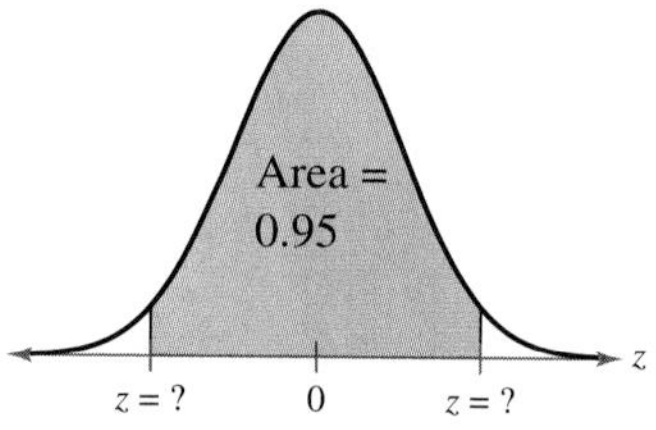

In Exercises 31–38, find the indicated z-score.

31. Find the z-score that has 62.8% of the distribution's area to its left.

32. Find the z-score that has 78.5% of the distribution's area to its left.

33. Find the z-score that has 62.8% of the distribution's area to its right.

34. Find the z-score that has 78.5% of the distribution's area to its right.

35. Find the z-score for which 80% of the distribution's area lies between $-z$ and z.

32. 0.79
33. −0.33
34. −0.79
35. 1.28
36. 2.575
37. ±0.06
38. ±0.15
39. (a) 68.52
(b) 62.14
40. (a) 72.91
(b) 67.24
41. (a) 12.28
(b) 20.08
42. (a) 6.765
(b) 13.725

36. Find the z-score for which 99% of the distribution's area lies between $-z$ and z.

37. Find the z-score for which 5% of the distribution's area lies between $-z$ and z.

38. Find the z-score for which 12% of the distribution's area lies between $-z$ and z.

Using and Interpreting Concepts

Using Normal Distributions In Exercises 39–44, answer the questions about the specified normal distribution.

39. ***Heights of Women*** In a survey of women in the United States (ages 20–29), the mean height was 64 inches with a standard deviation of 2.75 inches. *(Source: National Center for Health Statistics)*

(a) What height represents the 95th percentile?

(b) What height represents the first quartile?

40. ***Heights of Men*** In a survey of men in the United States (ages 20–29), the mean height was 69.2 inches with a standard deviation of 2.9 inches. *(Source: National Center for Health Statistics)*

(a) What height represents the 90th percentile?

(b) What height represents the first quartile?

41. ***Apples*** The annual per capita utilization of apples (in pounds) in the United States can be approximated by a normal distribution, as shown in the graph. *(Adapted from U.S. Department of Agriculture)*

(a) What annual per capita utilization of apples represents the 10th percentile?

(b) What annual per capita utilization of apples represents the third quartile?

Annual U.S. per Capita Apple Utilization

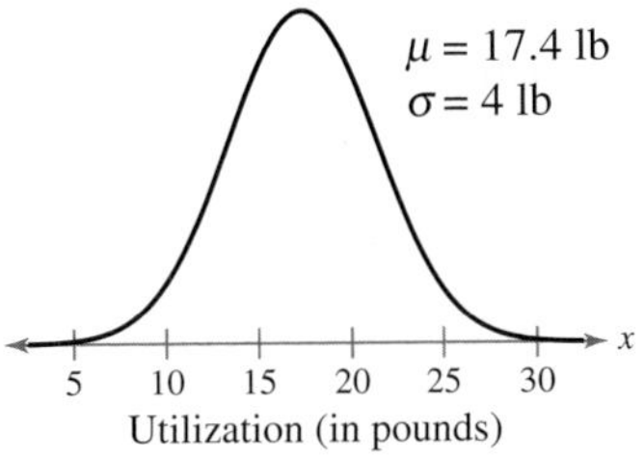

Figure for Exercise 41

Annual U.S. per Capita Orange Utilization

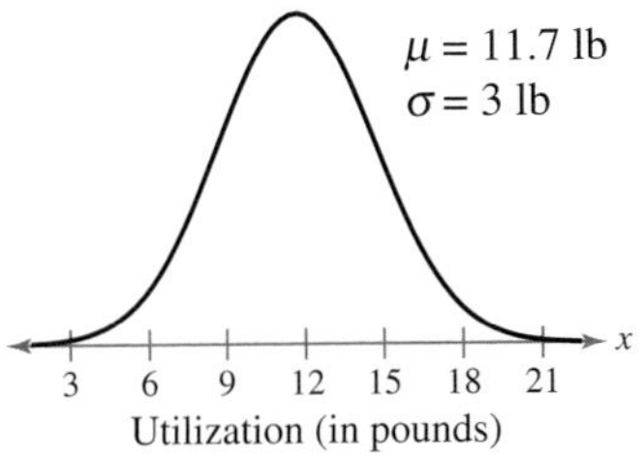

Figure for Exercise 42

42. ***Oranges*** The annual per capita utilization of oranges (in pounds) in the United States can be approximated by a normal distribution, as shown in the graph. *(Adapted from U.S. Department of Agriculture)*

(a) What annual per capita utilization of oranges represents the 5th percentile?

(b) What annual per capita utilization of oranges represents the third quartile?

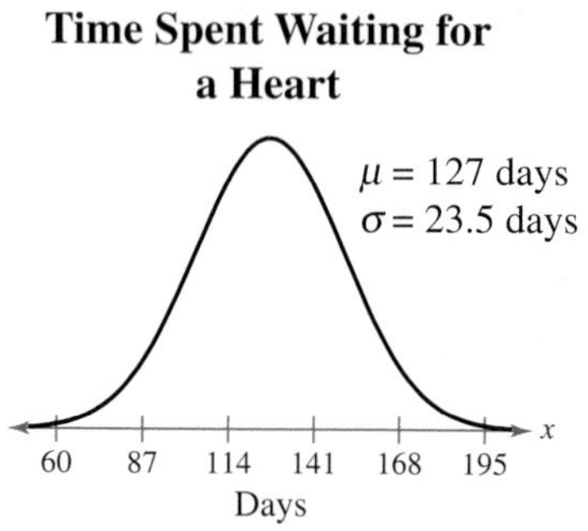

Figure for Exercise 43

43. ***Heart Transplant Waiting Times*** The time spent (in days) waiting for a heart transplant in Ohio and Michigan for patients with type A^+ blood can be approximated by a normal distribution, as shown in the graph. *(Source: Organ Procurement and Transplant Network)*

(a) What is the shortest time spent waiting for a heart that would still place a patient in the top 30% of waiting times?

(b) What is the longest time spent waiting for a heart that would still place a patient in the bottom 10% of waiting times?

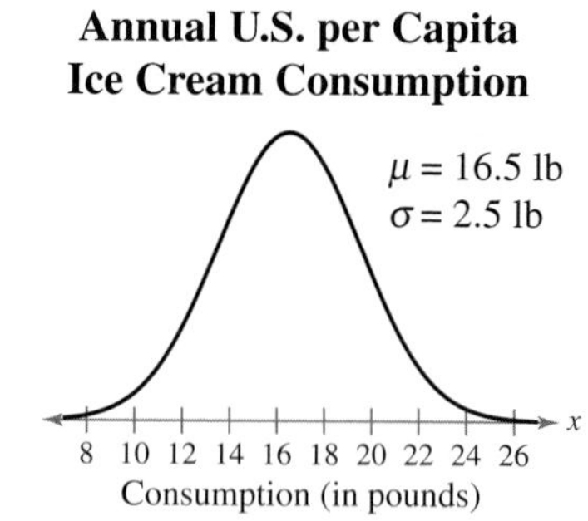

Figure for Exercise 44

44. ***Ice Cream*** The annual per capita consumption of ice cream (in pounds) in the United States can be approximated by a normal distribution, as shown in the graph. *(Adapted from U.S. Department of Agriculture)*

(a) What is the smallest annual per capita consumption of ice cream that can be in the top 25% of consumptions?

(b) What is the largest annual per capita consumption of ice cream that can be in the bottom 15% of consumptions?

45. ***Cereal Boxes*** The weights of the contents of a cereal box are normally distributed with a mean weight of 20 ounces and a standard deviation of 0.07 ounce. Boxes in the lower 5% do not meet the minimum weight requirements and must be repackaged. What is the minimum weight requirement for a cereal box?

46. ***Bags of Cookies*** The weights of bags of cookies are normally distributed with a mean of 15 ounces and a standard deviation of 0.085 ounce. Bags of cookies that have weights in the upper 7.5% are too heavy and must be repackaged. What is the most a bag of cookies can weigh and not need to be repackaged?

43. (a) 139.22
(b) 96.92
44. (a) 18.1875
(b) 13.9125
45. 19.89
46. 15.1224
47. Tires that wear out by 26,800 miles will be replaced free of charge.
48. A = 83.52; B = 76.68; C = 67.32; D = 60.48
49. 7.93

Extending Concepts

47. ***Writing a Guarantee*** You sell a brand of automobile tire that has a life expectancy that is normally distributed, with a mean life of 30,000 miles and a standard deviation of 2500 miles. You want to give a guarantee for free replacement of tires that don't wear well. How should you word your guarantee if you are willing to replace approximately 10% of the tires you sell?

48. ***Statistics Grades*** In a large section of a statistics class, the points for the final exam are normally distributed with a mean of 72 and a standard deviation of 9. Grades are to be assigned according to the following rule.

- The top 10% receive As
- The next 20% receive Bs
- The middle 40% receive Cs
- The next 20% receive Ds
- The bottom 10% receive Fs

Find the lowest score on the final exam that would qualify a student for an A, a B, a C, and a D.

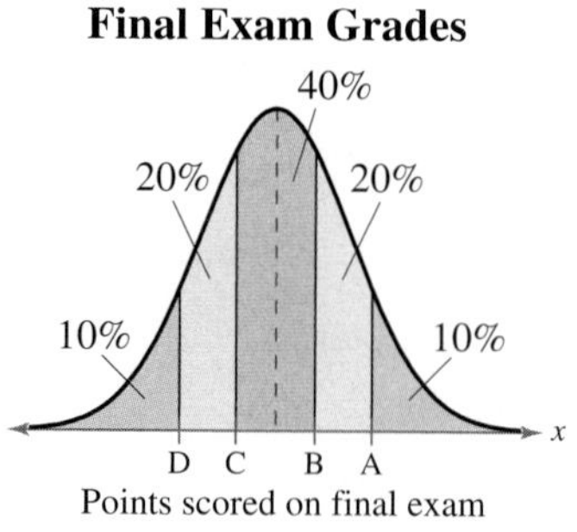

Figure for Exercise 48

49. ***Vending Machine*** A vending machine dispenses coffee into an eight-ounce cup. The amount of coffee dispensed into the cup is normally distributed with a standard deviation of 0.03 ounce. You can allow the cup to overfill 1% of the time. What amount should you set as the mean amount of coffee to be dispensed?

Case Study

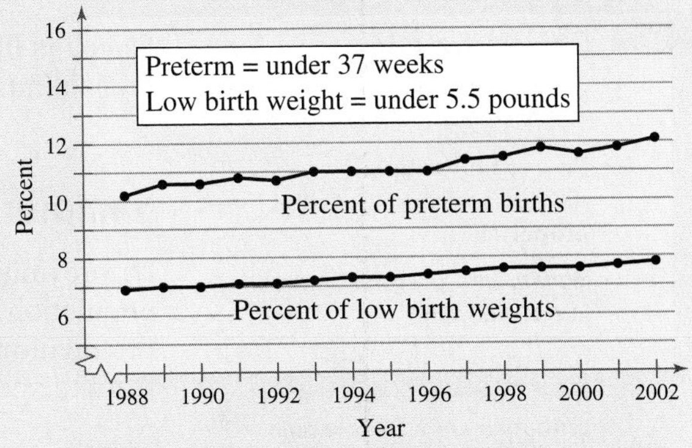

Birth Weights in America

The National Center for Health Statistics (NCHS) keeps records of many health-related aspects of people, including the birth weights of all babies born in the United States.

The birth weight of a baby is related to its gestation period (the time between conception and birth). For a given gestation period, the birth weights can be approximated by a normal distribution. The means and standard deviations of the birth weights for various gestation periods are shown at the right.

One of the many goals of the NCHS is to reduce the percentage of babies born with low birth weights. As you can see from the graph at the upper right, the problem of low birth weights increased from 1988 to 2002.

Gestation period	Mean birth weight	Standard deviation
Under 28 Weeks	1.88 lb	1.19 lb
28 to 31 Weeks	4.07 lb	1.87 lb
32 to 35 Weeks	5.73 lb	1.48 lb
36 Weeks	6.46 lb	1.20 lb
37 to 39 Weeks	7.33 lb	1.09 lb
40 Weeks	7.72 lb	1.05 lb
41 Weeks	7.83 lb	1.08 lb
42 Weeks and over	7.65 lb	1.12 lb

Exercises

1. The distributions of birth weights for three gestation periods are shown. Match the curves with the gestation periods. Explain your reasoning.

(a)

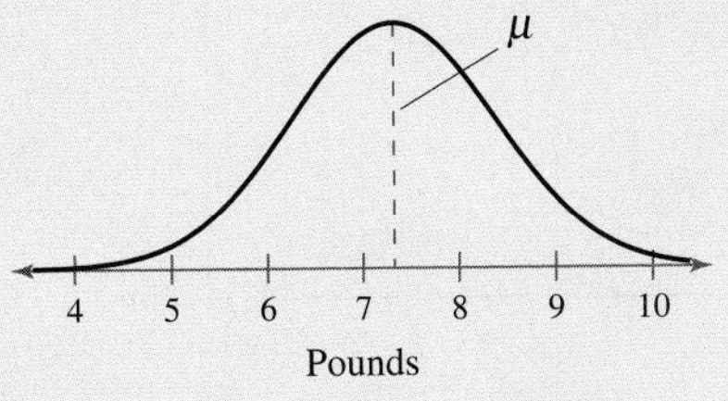

(b)

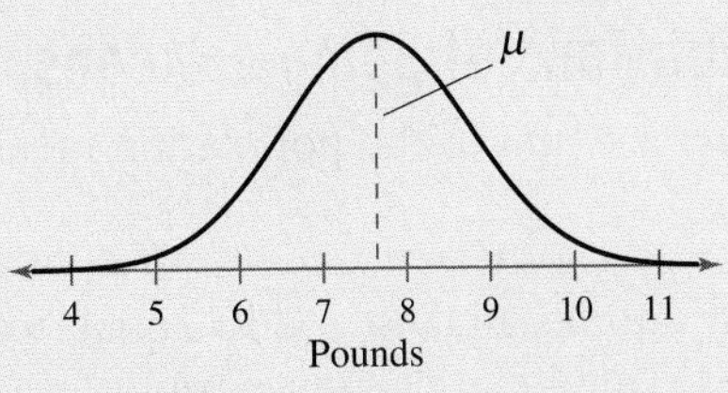

(c)

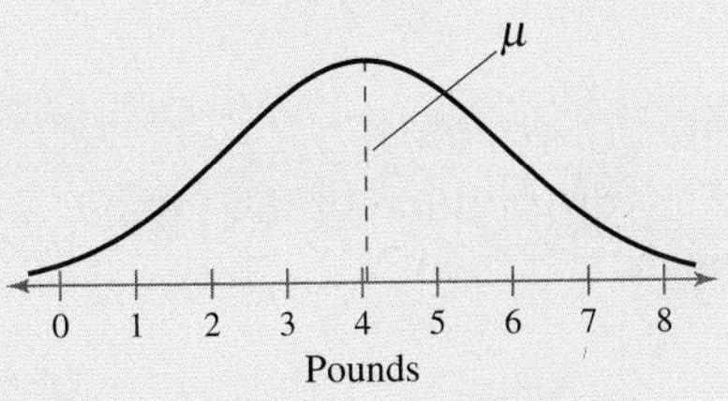

2. What percent of the babies born with each gestation period have a low birth weight (under 5.5 pounds)? Explain your reasoning.

(a) Under 28 weeks (b) 32 to 35 weeks

(c) 37 to 39 weeks (d) 42 weeks and over

3. Describe the weights of the top 10% of the babies born with each gestation period. Explain your reasoning.

(a) 37 to 39 weeks (b) 42 weeks and over

4. For each gestation period, what is the probability that a baby will weigh between 6 and 9 pounds at birth?

(a) 32 to 35 weeks (b) 37 to 39 weeks

(c) 42 weeks and over

5. A birth weight of less than 3.3 pounds is classified by the NCHS as a "very low birth weight." What is the probability that a baby has a very low birth weight for each gestation period?

(a) Under 28 weeks (b) 32 to 35 weeks

(c) 37 to 39 weeks

5.4 Sampling Distributions and the Central Limit Theorem

What You Should Learn

- How to find sampling distributions and verify their properties
- How to interpret the Central Limit Theorem
- How to apply the Central Limit Theorem to find the probability of a sample mean

Sampling Distributions • The Central Limit Theorem • Probability and the Central Limit Theorem

Insight

Sample means can vary from one another and can also vary from the population mean. This type of variation is to be expected and is called *sampling error*.

Note to Instructor

A good exercise that can be used in conjunction with the Venn diagram is to have each student randomly select a place in the random number table and write down the next five digits horizontally. Students can verify that the population of digits $\{0, 1, 2, \ldots, 9\}$ is uniform and has a mean of 4.5 and standard deviation of 2.87. Have each student calculate the mean of his or her sample and write that result on the board. Students can easily see that the sample means vary but are not dispersed as much as the population (range 0 to 9) is. Construct a histogram of the sample means; find the mean of these means and the standard deviation of the means. (With a TI-83, this takes little time even if only one student does the calculations.) Because the population standard deviation is known for this simulation, the results will be approximately normal.

Sampling Distributions

In previous sections, you studied the relationship between the mean of a population and values of a random variable. In this section, you will study the relationship between a population mean and the means of samples taken from the population.

DEFINITION

A **sampling distribution** is the probability distribution of a sample statistic that is formed when samples of size n are repeatedly taken from a population. If the sample statistic is the sample mean, then the distribution is the **sampling distribution of sample means.**

For instance, consider the following Venn diagram. The rectangle represents a large population, and each circle represents a sample of size n. Because the sample entries can differ, the sample means can also differ. The mean of Sample 1 is $\bar{x}_1$; the mean of Sample 2 is $\bar{x}_2$; and so on. The sampling distribution of the sample means for samples of size n for this population consists of $\bar{x}_1$, $\bar{x}_2$, $\bar{x}_3$, and so on. If the samples are drawn with replacement, an infinite number of samples can be drawn from the population.

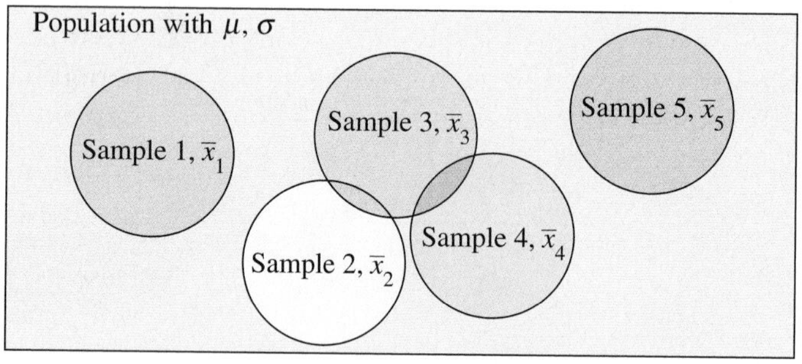

Properties of Sampling Distributions of Sample Means

1. The mean of the sample means $\mu_{\bar{x}}$ is equal to the population mean μ.

$$\mu_{\bar{x}} = \mu$$

2. The standard deviation of the sample means $\sigma_{\bar{x}}$ is equal to the population standard deviation σ divided by the square root of n.

$$\sigma_{\bar{x}} = \frac{\sigma}{\sqrt{n}}$$

The standard deviation of the sampling distribution of the sample means is called the **standard error of the mean.**

Probability Histogram of Population of x

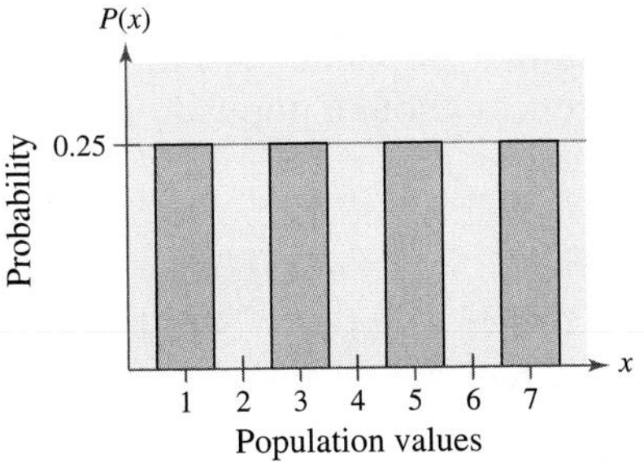

Probability Distribution of Sample Means

$\bar{x}$	f	Probability
1	1	0.0625
2	2	0.1250
3	3	0.1875
4	4	0.2500
5	3	0.1875
6	2	0.1250
7	1	0.0625

Probability Histogram of Sampling Distribution of $\bar{x}$

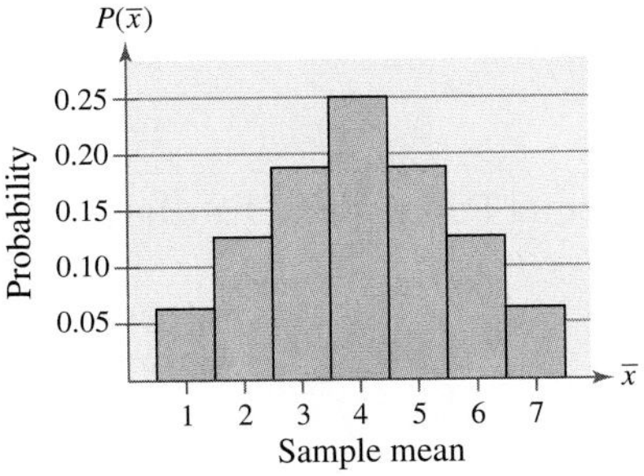

Study Tip

Review Section 4.1 to find the mean and standard deviation of a probability distribution.

EXAMPLE 1

A Sampling Distribution of Sample Means

You write the population values $\{1, 3, 5, 7\}$ on slips of paper and put them in a box. Then you randomly choose two slips of paper, with replacement. List all possible samples of size $n = 2$ and calculate the mean of each. These means form the sampling distribution of the sample means. Find the mean, variance, and standard deviation of the sample means. Compare your results with the mean $\mu = 4$, variance $\sigma^2 = 5$, and standard deviation $\sigma = \sqrt{5} \approx 2.236$ of the population.

SOLUTION List all 16 samples of size 2 from the population and the mean of each sample.

Sample	Sample mean, $\bar{x}$
1, 1	1
1, 3	2
1, 5	3
1, 7	4
3, 1	2
3, 3	3
3, 5	4
3, 7	5

Sample	Sample mean, $\bar{x}$
5, 1	3
5, 3	4
5, 5	5
5, 7	6
7, 1	4
7, 3	5
7, 5	6
7, 7	7

After constructing a probability distribution of the sample means, you can graph the sampling distribution using a probability histogram as shown at the left. Notice that the shape of the histogram is bell shaped and symmetric, similar to a normal curve. The mean, variance, and standard deviation of the 16 sample means are

$$\mu_{\bar{x}} = 4$$

$$(\sigma_{\bar{x}})^2 = \frac{5}{2} = 2.5 \quad \text{and} \quad \sigma_{\bar{x}} = \sqrt{\frac{5}{2}} = \sqrt{2.5} \approx 1.581.$$

These results satisfy the properties of sampling distributions because

$$\mu_{\bar{x}} = \mu = 4 \quad \text{and} \quad \sigma_{\bar{x}} = \frac{\sigma}{\sqrt{n}} = \frac{\sqrt{5}}{\sqrt{2}} \approx \frac{2.236}{\sqrt{2}} \approx 1.581.$$

Try It Yourself 1

List all possible samples of $n = 3$, with replacement, from the population $\{1, 3, 5, 7\}$. Calculate the mean, variance, and standard deviation of the sample means. Compare these values with the corresponding population parameters.

a. *Form* all possible samples of size 3 and find the mean of each.
b. *Make* a probability distribution of the sample means and *find* the mean, variance, and standard deviation.
c. *Compare* the mean, variance, and standard deviation of the sample means with those for the population.

Answer: Page A38

The Central Limit Theorem

The Central Limit Theorem forms the foundation for the inferential branch of statistics. This theorem describes the relationship between the sampling distribution of sample means and the population that the samples are taken from. The Central Limit Theorem is an important tool that provides the information you'll need to use sample statistics to make inferences about a population mean.

Note to Instructor

The sample mean $\bar{x}$ varies from sample to sample and is a random variable. As a random variable, it has a probability distribution, called the sampling distribution of the mean. Mention that other sample statistics, such as s^2, s, and $\hat{p}$, have different sampling distributions that will be studied in the next chapter.

The Central Limit Theorem

1. If samples of size n, where $n \geq 30$, are drawn from any population with a mean μ and a standard deviation σ, then the sampling distribution of sample means approximates a normal distribution. The greater the sample size, the better the approximation.
2. If the population itself is normally distributed, the sampling distribution of sample means is normally distributed for *any* sample size n.

In either case, the sampling distribution of sample means has a mean equal to the population mean.

$$\mu_{\bar{x}} = \mu \qquad \text{Mean}$$

The sampling distribution of sample means has a variance equal to $1/n$ times the variance of the population and a standard deviation equal to the population standard deviation divided by the square root of n.

$$\sigma_{\bar{x}}^2 = \frac{\sigma^2}{n} \qquad \text{Variance}$$

$$\sigma_{\bar{x}} = \frac{\sigma}{\sqrt{n}} \qquad \text{Standard deviation}$$

The standard deviation of the sampling distribution of the sample means, $\sigma_{\bar{x}}$, is also called the **standard error of the mean.**

Insight

The distribution of sample means has the same mean as the population. But its standard deviation is less than the standard deviation of the population. This tells you that the distribution of sample means has the same center as the population, but it is not as spread out.

Moreover, the distribution of sample means becomes less and less spread out (tighter concentration about the mean) as the sample size n increases.

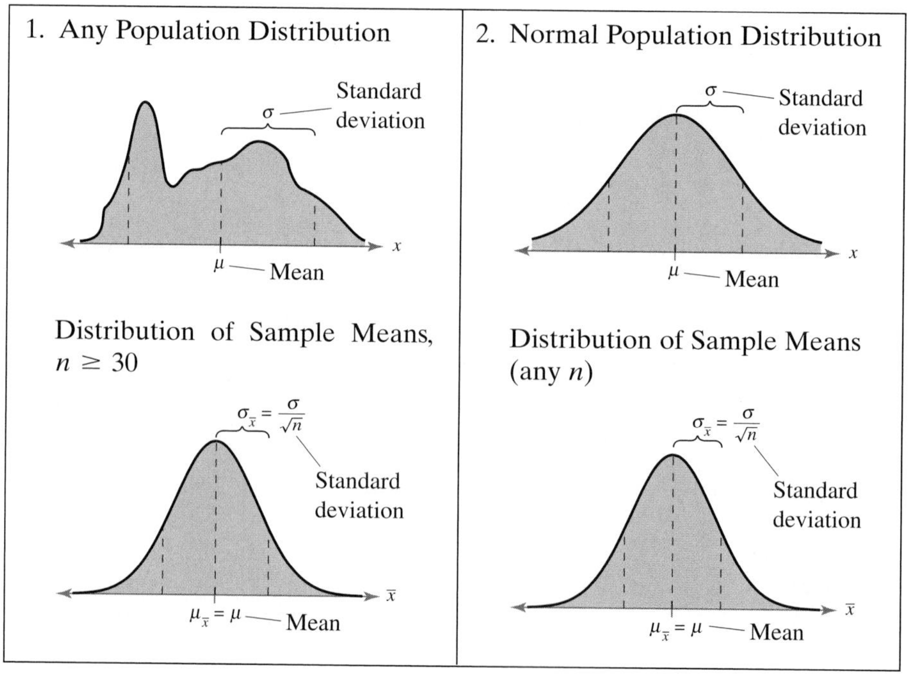

EXAMPLE 2

Interpreting the Central Limit Theorem

Phone bills for residents of a city have a mean of \$64 and a standard deviation of \$9, as shown in the following graph. Random samples of 36 phone bills are drawn from this population, and the mean of each sample is determined. Find the mean and standard error of the mean of the sampling distribution. Then sketch a graph of the sampling distribution of sample means.

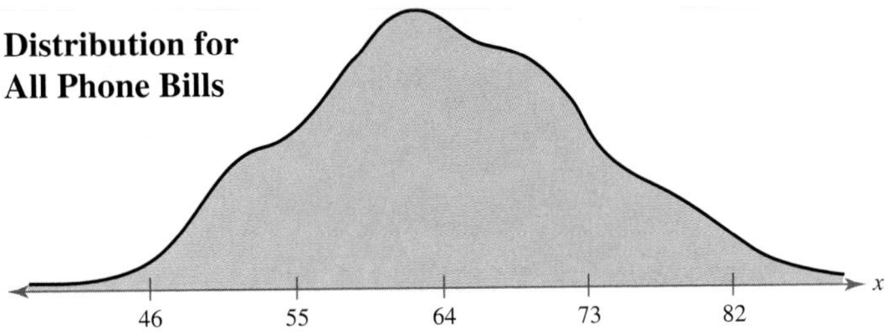

SOLUTION The mean of the sampling distribution is equal to the population mean, and the standard error of the mean is equal to the population standard deviation divided by $\sqrt{n}$. So,

$$\mu_{\bar{x}} = \mu = 64 \quad \text{and} \quad \sigma_{\bar{x}} = \frac{\sigma}{\sqrt{n}} = \frac{9}{\sqrt{36}} = 1.5.$$

Interpretation From the Central Limit Theorem, because the sample size is greater than 30, the sampling distribution can be approximated by a normal distribution with $\mu = \$64$ and $\sigma = \$1.50$, as shown in the graph below.

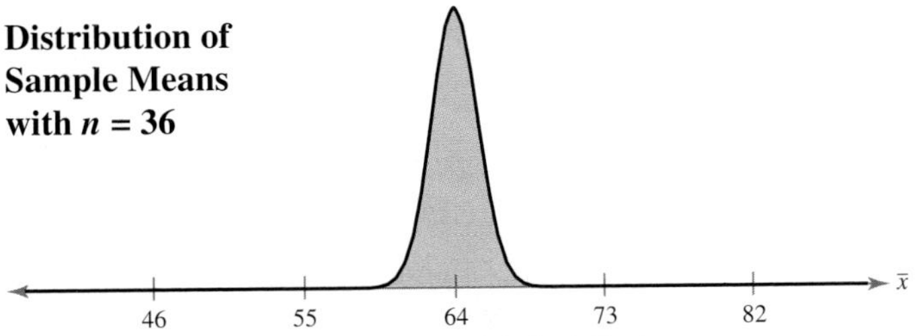

Try It Yourself 2

Suppose random samples of size 100 are drawn from the population in Example 2. Find the mean and standard error of the mean of the sampling distribution. Sketch a graph of the sampling distribution and compare it with the sampling distribution in Example 2.

a. *Find* $\mu_{\bar{x}}$ and $\sigma_{\bar{x}}$.
b. *Identify* the sample size. If $n \geq 30$, *sketch* a normal curve with mean $\mu_{\bar{x}}$ and standard deviation $\sigma_{\bar{x}}$.
c. *Compare* the results with those in Example 2.

Answer: Page A38

Picturing the World

In a recent year, there were more than 5 million parents in the United States who received child support payments. The following histogram shows the distribution of children per custodial parent. The mean number of children was 1.7 and the standard deviation was 0.9. *(Adapted from U.S. Census Bureau)*

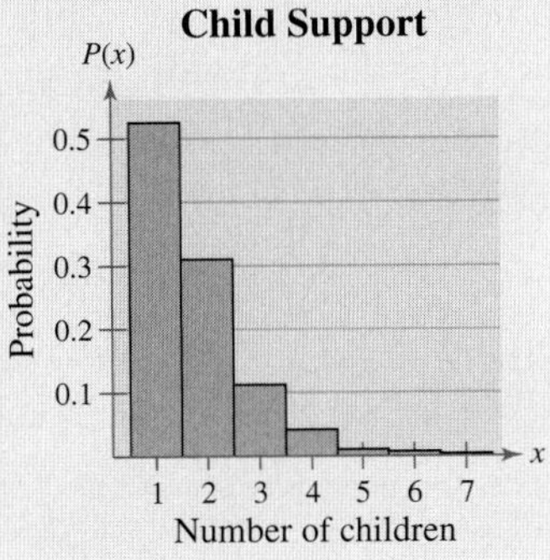

You randomly select 35 parents who receive child support and ask how many children in their custody are receiving child support payments. What is the probability that the mean of the sample is between 1.5 and 1.9 children?

EXAMPLE 3

Interpreting the Central Limit Theorem

The heights of fully grown white oak trees are normally distributed, with a mean of 90 feet and standard deviation of 3.5 feet, as shown in the following graph. Random samples of size 4 are drawn from this population, and the mean of each sample is determined. Find the mean and standard error of the mean of the sampling distribution. Then sketch a graph of the sampling distribution of sample means.

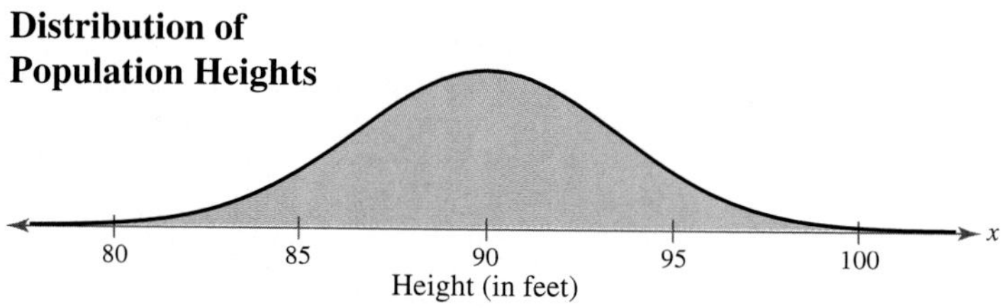

SOLUTION The mean of the sampling distribution is equal to the population mean, and the standard error of the mean is equal to the population standard deviation divided by $\sqrt{n}$. So,

$$\mu_{\bar{x}} = \mu = 90 \text{ feet} \quad \text{and} \quad \sigma_{\bar{x}} = \frac{\sigma}{\sqrt{n}} = \frac{3.5}{\sqrt{4}} = 1.75 \text{ feet.}$$

Interpretation From the Central Limit Theorem, because the population is normally distributed, the sampling distribution of the sample means is also normally distributed, as shown in the graph below.

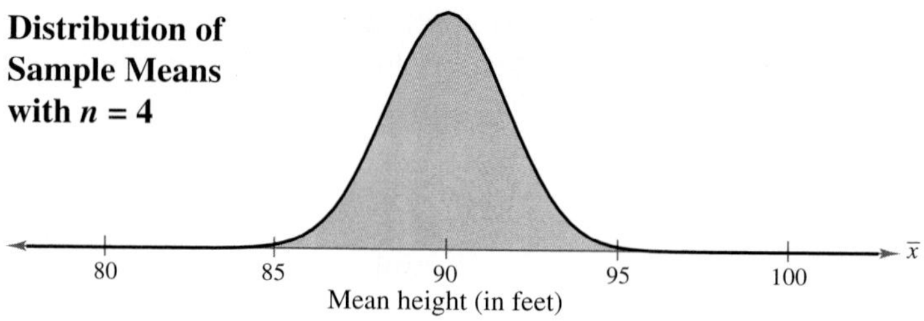

Try It Yourself 3

The diameters of fully grown white oak trees are normally distributed, with a mean of 3.5 feet and a standard deviation of 0.2 foot, as shown in the graph below. Random samples of size 16 are drawn from this population, and the mean of each sample is determined. Find the mean and standard error of the mean of the sampling distribution. Then sketch a graph of the sampling distribution.

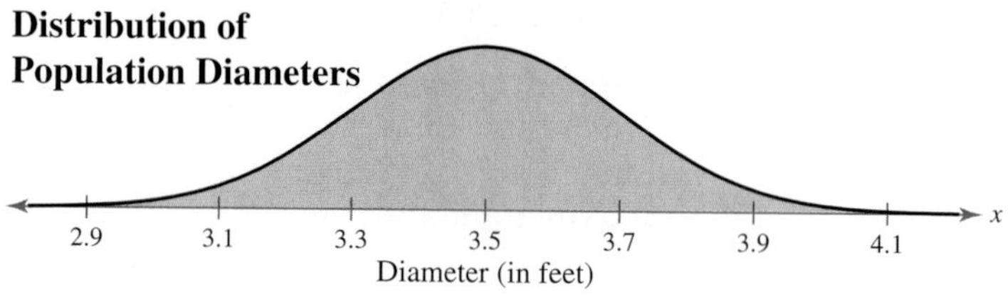

a. *Find* $\mu_{\bar{x}}$ and $\sigma_{\bar{x}}$.
b. *Sketch* a normal curve with mean $\mu_{\bar{x}}$ and standard deviation $\sigma_{\bar{x}}$.

Answer: Page A38

Note to Instructor

For technology users, students need only calculate the standard error before using the normal CDF.

Probability and the Central Limit Theorem

In Section 5.2, you learned how to find the probability that a random variable x will fall in a given interval of population values. In a similar manner, you can find the probability that a sample mean $\overline{x}$ will fall in a given interval of the $\overline{x}$ sampling distribution. To transform $\overline{x}$ to a z-score, you can use the formula

$$z = \frac{\text{Value} - \text{Mean}}{\text{Standard error}} = \frac{\overline{x} - \mu_{\overline{x}}}{\sigma_{\overline{x}}} = \frac{\overline{x} - \mu}{\sigma/\sqrt{n}}.$$

EXAMPLE 4

Finding Probabilities for Sampling Distributions

The graph at the right shows the length of time people spend driving each day. You randomly select 50 drivers ages 15 to 19. What is the probability that the mean time they spend driving each day is between 24.7 and 25.5 minutes? Assume that $\sigma = 1.5$ minutes.

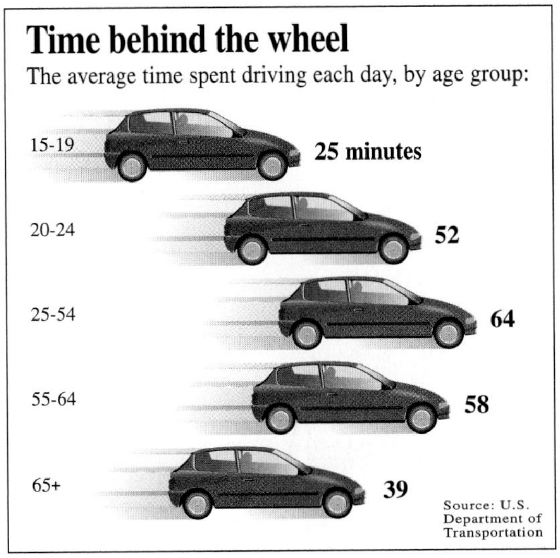

SOLUTION The sample size is greater than 30, so you can use the Central Limit Theorem to conclude that the distribution of sample means is approximately normal with a mean and a standard deviation of

$$\mu_{\overline{x}} = \mu = 25 \text{ minutes} \quad \text{and} \quad \sigma_{\overline{x}} = \frac{\sigma}{\sqrt{n}} = \frac{1.5}{\sqrt{50}} \approx 0.21213 \text{ minute}.$$

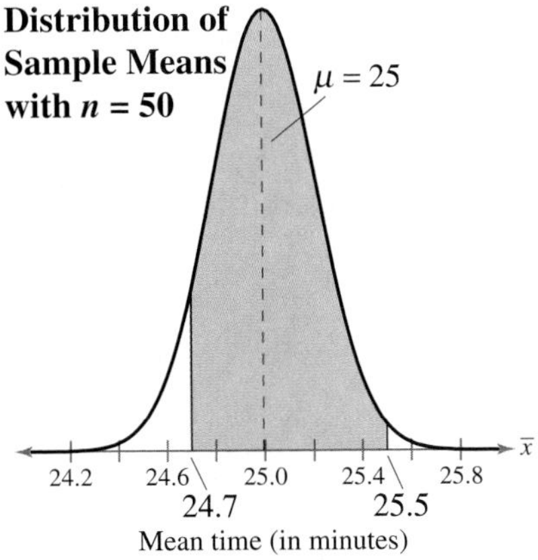

The graph of this distribution is shown at the left with a shaded area between 24.7 and 25.5 minutes. The z-scores that correspond to sample means of 24.7 and 25.5 minutes are

$$z_1 = \frac{24.7 - 25}{1.5/\sqrt{50}} \approx \frac{-0.3}{0.21213} \approx -1.41 \quad \text{and}$$

$$z_2 = \frac{25.5 - 25}{1.5/\sqrt{50}} \approx \frac{0.5}{0.21213} \approx 2.36.$$

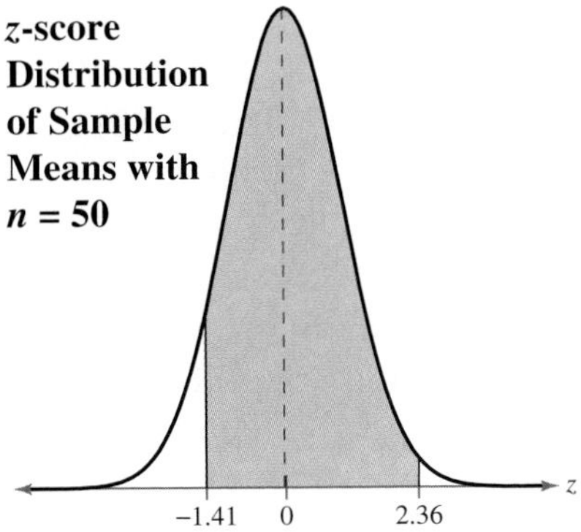

So, the probability that the mean time the 50 people spend driving each day is between 24.7 and 25.5 minutes is

$$\begin{aligned} P(24.7 < \overline{x} < 25.5) &= P(-1.41 < z < 2.36) \\ &= P(z < 2.36) - P(z < -1.41) \\ &= 0.9909 - 0.0793 = 0.9116. \end{aligned}$$

Interpretation Of the samples of 50 drivers ages 15 to 19, 91.16% will have a mean driving time that is between 24.7 and 25.5 minutes, as shown in the graph at the left. This implies that, assuming the value of $\mu = 25$ is correct, only 8.84% of such sample means will lie outside the given interval.

Study Tip

Before you find probabilities for intervals of the sample mean $\overline{x}$, use the Central Limit Theorem to determine the mean and the standard deviation of the sampling distribution of the sample means. That is, calculate $\mu_{\overline{x}}$ and $\sigma_{\overline{x}}$.

Try It Yourself 4

You randomly select 100 drivers ages 15 to 19. What is the probability that the mean time they spend driving each day is between 24.7 and 25.5 minutes? Use $\mu = 25$ and $\sigma = 1.5$ minutes.

a. Use the Central Limit Theorem to *find* $\mu_{\overline{x}}$ and $\sigma_{\overline{x}}$ and *sketch* the sampling distribution of the sample means.
b. *Find the z-scores* that correspond to $\overline{x} = 24.7$ minutes and $\overline{x} = 25.5$ minutes.
c. *Find the cumulative area* that corresponds to each z-score and calculate the probability.

Answer: Page A38

EXAMPLE 5

Finding Probabilities for Sampling Distributions

The mean room and board expense per year at a four-year college is \$5850. You randomly select 9 four-year colleges. What is the probability that the mean room and board is less than \$6180? Assume that the room and board expenses are normally distributed, with a standard deviation of \$1125. *(Source: National Center for Education Statistics)*

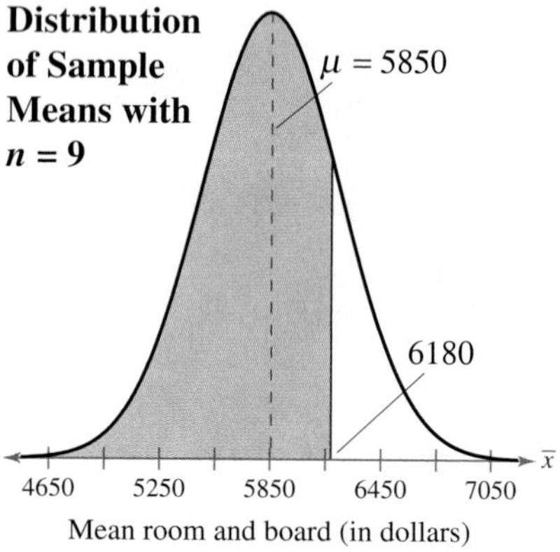

SOLUTION Because the population is normally distributed, you can use the Central Limit Theorem to conclude that the distribution of sample means is normally distributed, with a mean of \$5850 and a standard deviation of \$375.

$$\mu_{\overline{x}} = \mu = 5850 \quad \text{and} \quad \sigma_{\overline{x}} = \frac{\sigma}{\sqrt{n}} = \frac{1125}{\sqrt{9}} = 375$$

The graph of this distribution is shown at the left. The area to the left of \$6180 is shaded. The z-score that corresponds to \$6180 is

$$z = \frac{6180 - 5850}{1125/\sqrt{9}} = \frac{5600}{375} = 0.88.$$

So, the probability that the mean room and board expense is less than \$6180 is

$$P(\overline{x} < 6180) = P(z < 0.88) = 0.8106.$$

Interpretation So, 81.06% of such samples with $n = 9$ will have a mean less than \$6180 and 18.94% of these sample means will lie outside this interval.

Try It Yourself 5

The average sales price of a single-family house in the United States is \$243,756. You randomly select 12 single-family houses. What is the probability that the mean sales price is more than \$200,000? Assume that the sales prices are normally distributed with a standard deviation of \$44,000. *(Source: Federal Housing Finance Board)*

a. Use the Central Limit Theorem to *find* $\mu_{\overline{x}}$ and $\sigma_{\overline{x}}$ and *sketch* the sampling distribution of the sample means.
b. *Find the z-score* that corresponds to $\overline{x} = \$200{,}000$.
c. *Find the cumulative area* that corresponds to the z-score and calculate the probability.

Answer: Page A38

The Central Limit Theorem can also be used to investigate rare occurrences. A rare occurrence is one that occurs with a probability of less than 5%.

EXAMPLE 6

Finding Probabilities for x and $\overline{x}$

A bank auditor claims that credit card balances are normally distributed, with a mean of \$2870 and a standard deviation of \$900.

1. What is the probability that a randomly selected credit card holder has a credit card balance less than \$2500?
2. You randomly select 25 credit card holders. What is the probability that their mean credit card balance is less than \$2500?
3. Compare the probabilities from (1) and (2) and interpret your answer in terms of the auditor's claim.

SOLUTION

1. In this case, you are asked to find the probability associated with a certain value of the random variable x. The z-score that corresponds to $x = \$2500$ is

$$z = \frac{x - \mu}{\sigma} = \frac{2500 - 2870}{900} \approx -0.41.$$

So, the probability that the card holder has a balance less than \$2500 is

$$P(x < 2500) = P(z < -0.41) = 0.3409.$$

2. Here, you are asked to find the probability associated with a sample mean $\overline{x}$. The z-score that corresponds to $\overline{x} = \$2500$ is

$$z = \frac{\overline{x} - \mu_{\overline{x}}}{\sigma_{\overline{x}}} = \frac{\overline{x} - \mu}{\sigma/\sqrt{n}}$$

$$= \frac{2500 - 2870}{900/\sqrt{25}} = \frac{-370}{180} \approx -2.06.$$

So, the probability that the mean credit card balance of the 25 card holders is less than \$2500 is

$$P(\overline{x} < 2500) = P(z < -2.06) = 0.0197.$$

3. ***Interpretation*** Although there is a 34% chance that an individual will have a balance less than \$2500, there is only a 2% chance that the mean of a sample of 25 will have a balance less than \$2500. Because there is only a 2% chance that the mean of a sample of 25 will have a balance less than \$2500, this is a rare occurrence. So, it is possible that the sample is unusual, or it is possible that the auditor's claim that the mean is \$2870 is incorrect.

Study Tip

To find probabilities for individual members of a population with a normally distributed random variable x, use the formula

$$z = \frac{x - \mu}{\sigma}.$$

To find probabilities for the mean $\overline{x}$ of a sample size n, use the formula

$$z = \frac{\overline{x} - \mu_{\overline{x}}}{\sigma_{\overline{x}}}.$$

Try It Yourself 6

A consumer price analyst claims that prices for sound-system receivers are normally distributed, with a mean of \$625 and a standard deviation of \$150. (1) What is the probability that a randomly selected receiver costs less than \$700? (2) You randomly select 10 receivers. What is the probability that their mean cost is less than \$700? (3) Compare these two probabilities.

a. *Find* the z-scores that correspond to x and $\overline{x}$.
b. Use the Standard Normal Table to *find the probability* associated with each z-score.
c. *Compare* the probabilities and *interpret* your answer.

Answer: Page A38

Note to Instructor

You may want to tell students that the second formula can also be used to calculate z-scores for individual values. Consider a sample of $n = 1$ for an individual value.

5.4 Exercises

Building Basic Skills and Vocabulary

In Exercises 1–4, a population has a mean $\mu = 100$ and a standard deviation $\sigma = 15$. Find the mean and standard deviation of a sampling distribution of sample means with the given sample size n.

1. $n = 50$ **2.** $n = 100$

3. $n = 250$ **4.** $n = 1000$

True or False? In Exercises 5–8, determine whether the statement is true or false. If it is false, rewrite it so that it is a true statement.

5. As the size of a sample increases, the mean of the distribution of sample means increases.

6. As the size of a sample increases, the standard deviation of the distribution of sample means increases.

7. The shape of a sampling distribution is normal only if the shape of the population is normal.

8. If the size of a sample is at least 30, you can use z-scores to determine the probability that a sample mean falls in a given interval of the sampling distribution.

Verifying Properties of Sampling Distributions In Exercises 9 and 10, find the mean and standard deviation of the population. List all samples (with replacement) of the given size from that population. Find the mean and standard deviation of the sampling distribution and compare them with the mean and standard deviation of the population.

9. The number of movies that all four people in a family have seen in the past month is 4, 2, 8, and 0. Use a sample size of 3.

10. Four people in a carpool paid the following amounts for textbooks this semester: \$120, \$140, \$180, and \$220. Use a sample size of 2.

Graphical Analysis In Exercises 11 and 12, the graph of a population distribution is shown at the left with its mean and standard deviation. Assume that a sample size of 100 is drawn from each population. Decide which of the graphs labeled (a)–(c) would most closely resemble the sampling distribution of the sample means for each graph. Explain your reasoning.

11. The waiting time (in seconds) at a traffic signal during a red light

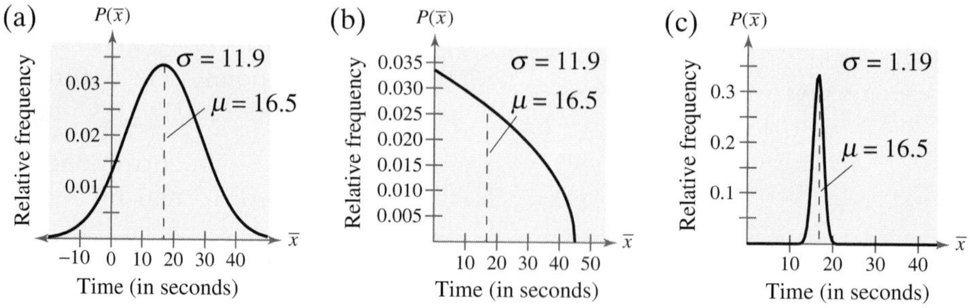

1. 100, 2.12 **2.** 100, 1.5

3. 100, 0.949 **4.** 100, 0.474

5. False. As the size of a sample increases, the mean of the distribution of sample means does not change.

6. False. As the size of the sample increases, the standard deviation of the distribution of sample means decreases.

7. False. The shape of a sampling distribution is normal if either $n \geq 30$ or the shape of the population is normal.

8. True

9. See Odd Answers, page A64.

10. {120 120, 120 140, 120 180, 120 220, 140 120, 140 140, 140 180, 140 220, 180 120, 180 140, 180 180, 180 220, 220 120, 220 140, 220 180, 220 220}

$\mu_{\bar{x}} = 165$, $\sigma_{\bar{x}} \approx 27.157$

$\mu = 165$, $\sigma = 38.406$

11. (c), because $\mu = 16.5$, $\sigma = 1.19$, and the graph approximates a normal curve.

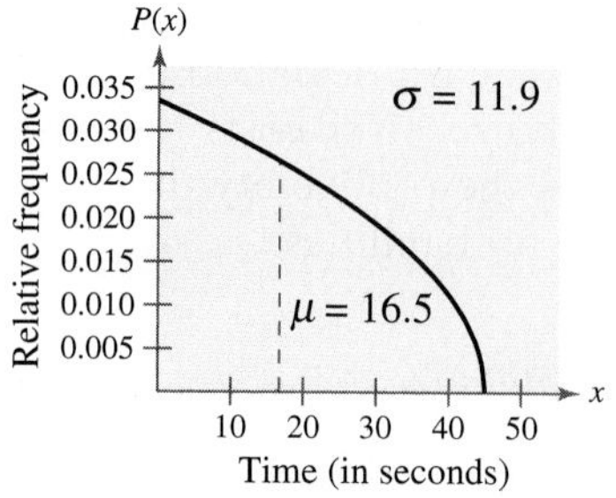

Figure for Exercise 11

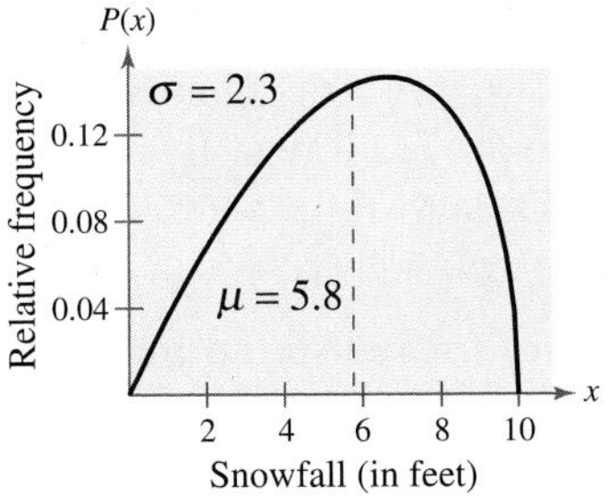

Figure for Exercise 12

12. The annual snowfall (in feet) for a central New York State county

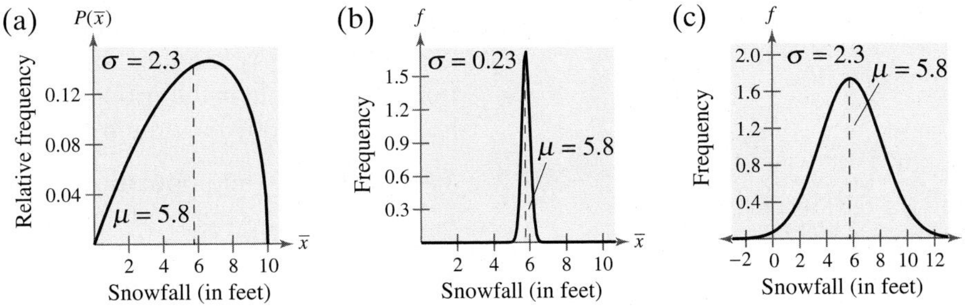

12. See Selected Answers, page A95.

13. 87.5, 1.804

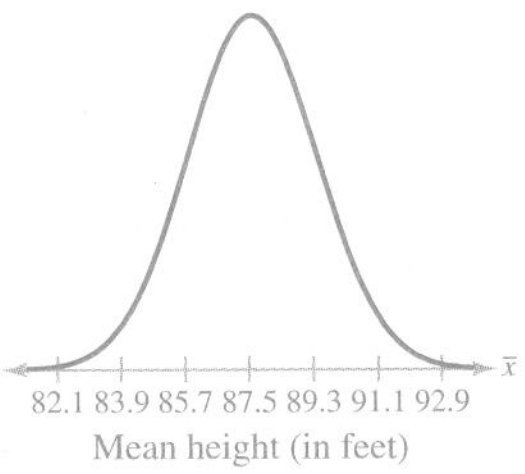

14. See Selected Answers, page A95.

15. 349, 1.26

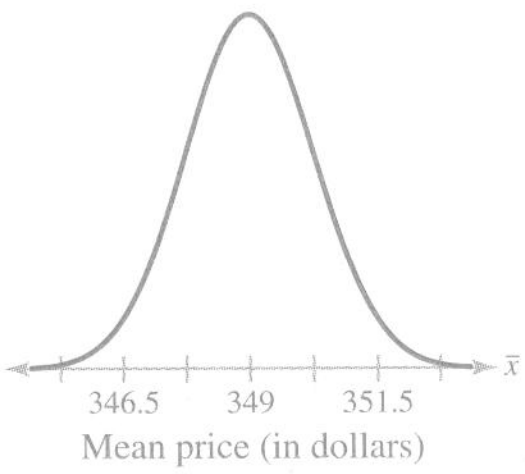

16. See Selected Answers, page A95.

17. 113.5, 8.61

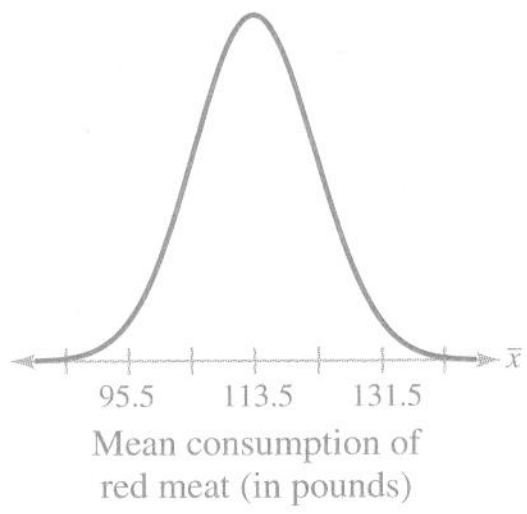

18. See Selected Answers, page A95.

19. See Odd Answers, page A64.

20. See Selected Answers, page A95.

Using and Interpreting Concepts

Using the Central Limit Theorem In Exercises 13–18, use the Central Limit Theorem to find the mean and standard error of the mean of the indicated sampling distribution. Then sketch a graph of the sampling distribution.

13. ***Heights of Trees*** The heights of fully grown sugar maple trees are normally distributed, with a mean of 87.5 feet and a standard deviation of 6.25 feet. Random samples of size 12 are drawn from the population and the mean of each sample is determined.

14. ***Fly Eggs*** The number of eggs a female house fly lays during her lifetime is normally distributed, with a mean of 800 eggs and a standard deviation of 100 eggs. Random samples of size 15 are drawn from this population and the mean of each sample is determined.

15. ***Digital Cameras*** The mean price of digital cameras at an electronics store is \$349, with a standard deviation of \$8. Random samples of size 40 are drawn from this population and the mean of each sample is determined.

16. ***Employees' Ages*** The mean age of employees at a large corporation is 47.2 years, with a standard deviation of 3.6 years. Random samples of size 36 are drawn from this population and the mean of each sample is determined.

17. ***Red Meat Consumed*** The per capita consumption of red meat by people in the United States in a recent year was normally distributed, with a mean of 113.5 pounds and a standard deviation of 38.5 pounds. Random samples of size 20 are drawn from this population and the mean of each sample is determined. *(Adapted from U.S. Department of Agriculture)*

18. ***Soft Drinks*** The per capita consumption of soft drinks by people in the United States in a recent year was normally distributed, with a mean of 49.3 gallons and a standard deviation of 17.1 gallons. Random samples of size 25 are drawn from this population and the mean of each sample is determined. *(Adapted from U.S. Department of Agriculture)*

19. Repeat Exercise 13 for samples of size 24 and 36. What happens to the mean and standard deviation of the distribution of sample means as the size of the sample increases?

20. Repeat Exercise 14 for samples of size 30 and 45. What happens to the mean and to the standard deviation of the distribution of sample means as the size of the sample increases?

21. 0.0019
22. ≈ 0
23. 0.6319
24. 0.2349
25. ≈ 0
26. 0.0162
27. It is more likely to select a sample of 20 women with a mean height less than 70 inches because the sample of 20 has a higher probability.
28. It is more likely to select one man with a height less than 65 inches because the probability is greater.
29. Yes, it is very unlikely that you would have randomly sampled 40 cans with a mean equal to 127.9 ounces.
30. Yes, it is very unlikely that you would have randomly sampled 40 containers with a mean equal to 64.05 ounces.

Finding Probabilities In Exercises 21–26, find the probabilities.

21. *Plumber Salaries* The population mean annual salary for plumbers is $\mu = \$40{,}500$. A random sample of 42 plumbers is drawn from this population. What is the probability that the mean salary of the sample, $\bar{x}$, is less than \$38,000? Assume $\sigma = \$5600$. *(Adapted from Salary.com)*

22. *Nurse Salaries* The population mean annual salary for registered nurses is $\mu = \$45{,}500$. A sample of 35 registered nurses is randomly selected. What is the probability that the mean annual salary of the sample, $\bar{x}$, is less than \$42,000? Assume $\sigma = \$1700$. *(Adapted from Allied Physicians, Inc.)*

23. *Gas Prices: New England* During a certain week the mean price of gasoline in the New England region was $\mu = \$1.689$ per gallon. What is the probability that the mean price $\bar{x}$ for a sample of 32 randomly selected gas stations in that area was between \$1.684 and \$1.699 that week? Assume $\sigma = \$0.045$. *(Adapted from Energy Information Administration)*

24. *Gas Prices: California* During a certain week the mean price of gasoline in California was $\mu = \$2.029$ per gallon. A random sample of 38 gas stations is drawn from this population. What is the probability that $\bar{x}$, the mean price for the sample, was between \$2.034 and \$2.044? Assume $\sigma = \$0.049$. *(Adapted from Energy Information Administration)*

25. *Heights of Women* The mean height of women in the United States (ages 20–29) is $\mu = 64$ inches. A random sample of 60 women in this age group is selected. What is the probability that $\bar{x}$, the mean height for the sample, is greater than 66 inches? Assume $\sigma = 2.75$ inches. *(Source: National Center for Health Statistics)*

26. *Heights of Men* The mean height of men in the United States (ages 20–29) is $\mu = 69.2$ inches. A random sample of 60 men in this age group is selected. What is the probability that $\bar{x}$, the mean height for the sample, is greater than 70 inches? Assume $\sigma = 2.9$ inches. *(Source: National Center for Health Statistics)*

27. *Which Is More Likely?* Assume that the heights given in Exercise 25 are normally distributed. Are you more likely to randomly select one woman with a height less than 70 inches or are you more likely to select a sample of 20 women with a mean height less than 70 inches? Explain.

28. *Which Is More Likely?* Assume that the heights given in Exercise 26 are normally distributed. Are you more likely to randomly select one man with a height less than 65 inches or are you more likely to select a sample of 15 men with a mean height less than 65 inches? Explain.

29. *Make a Decision* A machine used to fill gallon-sized paint cans is regulated so that the amount of paint dispensed has a mean of 128 ounces and a standard deviation of 0.20 ounce. You randomly select 40 cans and carefully measure the contents. The sample mean of the cans is 127.9 ounces. Does the machine need to be reset? Explain your reasoning.

30. *Make a Decision* A machine used to fill half-gallon-sized milk containers is regulated so that the amount of milk dispensed has a mean of 64 ounces and a standard deviation of 0.11 ounce. You randomly select 40 containers and carefully measure the contents. The sample mean of the containers is 64.05 ounces. Does the machine need to be reset? Explain your reasoning.

31. (a) 0.0008
(b) Claim is inaccurate.
(c) No, assuming the manufacturer's claim is true, because 96.25 is within 1 standard deviation of the mean for an individual board.

32. (a) 0.0179
(b) Claim is inaccurate.
(c) No, assuming the manufacturer's claim is true, because 10.21 is within 1 standard deviation of the mean for an individual carton.

33. (a) 0.0002
(b) Claim is inaccurate.
(c) No, assuming the manufacturer's claim is true, because 49,721 is within 1 standard deviation of the mean for an individual tire.

34. (a) 0.0668
(b) Claim is inaccurate.
(c) No, assuming the manufacturer's claim is true, because 37,650 is within 1 standard deviation of the mean for an individual brake pad.

31. *Lumber Cutter* Your lumber company has bought a machine that automatically cuts lumber. The seller of the machine claims that the machine cuts lumber to a mean length of 8 feet (96 inches) with a standard deviation of 0.5 inch. Assume the lengths are normally distributed. You randomly select 40 boards and find that the mean length is 96.25 inches.

(a) Assuming the seller's claim is correct, what is the probability the mean of the sample is 96.25 inches or more?

(b) Using your answer from part (a), what do you think of the seller's claim?

(c) Would it be unusual to have an individual board with a length of 96.25 inches? Why or why not?

32. *Ice Cream Carton Weights* A manufacturer claims that the mean weight of its ice cream cartons is 10 ounces with a standard deviation of 0.5 ounce. Assume the weights are normally distributed. You test 25 cartons and find their mean weight is 10.21 ounces.

(a) Assuming the manufacturer's claim is correct, what is the probability the mean of the sample is 10.21 ounces or more?

(b) Using your answer from part (a), what do you think of the manufacturer's claim?

(c) Would it be unusual to have an individual carton with a weight of 10.21 ounces? Why or why not?

33. *Life of Tires* A manufacturer claims that the life span of its tires is 50,000 miles. You work for a consumer protection agency and you are testing this manufacturer's tires. Assume the life spans of the tires are normally distributed. You select 100 tires at random and test them. The mean life span is 49,721 miles. Assume $\sigma = 800$ miles.

(a) Assuming the manufacturer's claim is correct, what is the probability the mean of the sample is 49,721 miles or less?

(b) Using your answer from part (a), what do you think of the manufacturer's claim?

(c) Would it be unusual to have an individual tire with a life span of 49,721 miles? Why or why not?

34. *Brake Pads* A brake pad manufacturer claims its brake pads will last for 38,000 miles. You work for a consumer protection agency and you are testing this manufacturer's brake pads. Assume the life spans of the brake pads are normally distributed. You randomly select 50 brake pads. In your tests, the mean life of the brake pads is 37,650 miles. Assume $\sigma = 1000$ miles.

(a) Assuming the manufacturer's claim is correct, what is the probability the mean of the sample is 37,650 miles or less?

(b) Using your answer from part (a), what do you think of the manufacturer's claim?

(c) Would it be unusual to have an individual brake pad last for 37,650 miles? Why or why not?

35. Yes, because of the relatively large z-score (2.12).

36. It is very unlikely the machine is calibrated to produce a bolt with a mean of 4 inches.

37. ≈ 0

38. ≈ 1

Extending Concepts

35. ***SAT Scores*** The average math SAT score is 500 with a standard deviation of 100. A particular high school claims that its students have unusually high math SAT scores. A random sample of 50 students from this school was selected, and the mean math SAT score was 530. Is the high school justified in its claim? Explain.

36. ***Machine Calibrations*** A machine in a manufacturing plant is calibrated to produce a bolt that has a mean diameter of 4 inches and a standard deviation of 0.5 inch. An engineer takes a random sample of 100 bolts from this machine and finds the mean diameter is 4.2 inches. What are some possible consequences from these findings?

Finite Correction Factor The formula for the standard error of the mean

$$\sigma_{\bar{x}} = \frac{\sigma}{\sqrt{n}}$$

given in the Central Limit Theorem is based on an assumption that the population has infinitely many members. This is the case whenever sampling is done with replacement (each member is put back after it is selected) because the sampling process could be continued indefinitely. The formula is also valid if the sample size is small in comparison to the population. However, when sampling is done without replacement and the sample size n is more than 5% of the finite population of size N, there is a finite number of possible samples. A **finite correction factor,**

$$\sqrt{\frac{N-n}{N-1}}$$

should be used to adjust the standard error. The sampling distribution of the sample means will be normal with a mean equal to the population mean, and the standard error of the mean will be

$$\sigma_{\bar{x}} = \frac{\sigma}{\sqrt{n}}\sqrt{\frac{N-n}{N-1}}.$$

In Exercises 37 and 38, determine if the finite correction factor should be used. If so, use it in your calculations when you find the probability.

37. ***Gas Prices*** In a sample of 800 gas stations, the mean price for regular gasoline at the pump was \$1.688 per gallon and the standard deviation was \$0.009 per gallon. A random sample of size 55 is drawn from this population. What is the probability that the mean price per gallon is less than \$1.683? *(Adapted from U.S. Department of Energy)*

38. ***Old Faithful*** In a sample of 500 eruptions of the Old Faithful geyser at Yellowstone National Park, the mean duration of the eruptions was 3.32 minutes and the standard deviation was 1.09 minutes. A random sample of size 30 is drawn from this population. What is the probability that the mean duration of eruptions is between 2.5 minutes and 4 minutes? *(Adapted from Yellowstone National Park)*

5.5 Normal Approximations to Binomial Distributions

What You Should Learn

- How to decide when the normal distribution can approximate the binomial distribution
- How to find the correction for continuity
- How to use the normal distribution to approximate binomial probabilities

Approximating a Binomial Distribution • Correction for Continuity • Approximating Binomial Probabilities

Approximating a Binomial Distribution

In Section 4.2, you learned how to find binomial probabilities. For instance, if a surgical procedure has an 85% chance of success and a doctor performs the procedure on 10 patients, it is easy to find the probability of exactly two successful surgeries.

But what if the doctor performs the surgical procedure on 150 patients and you want to find the probability of *fewer than 100* successful surgeries? To do this using the techniques described in Section 4.2, you would have to use the binomial formula 100 times and find the sum of the resulting probabilities. This approach is not practical, of course. A better approach is to use a normal distribution to approximate the binomial distribution.

Normal Approximation to a Binomial Distribution

If $np \geq 5$ and $nq \geq 5$, then the binomial random variable x is approximately normally distributed, with mean

$$\mu = np$$

and standard deviation

$$\sigma = \sqrt{npq}.$$

Study Tip

Properties of a binomial experiment

- n independent trials
- Two possible outcomes: success or failure
- Probability of success is p; probability of failure is $1 - p = q$
- p is constant for each trial

To see why this result is valid, look at the following binomial distributions for $p = 0.25$ and $n = 4$, $n = 10$, $n = 25$, and $n = 50$. Notice that as n increases, the histogram approaches a normal curve.

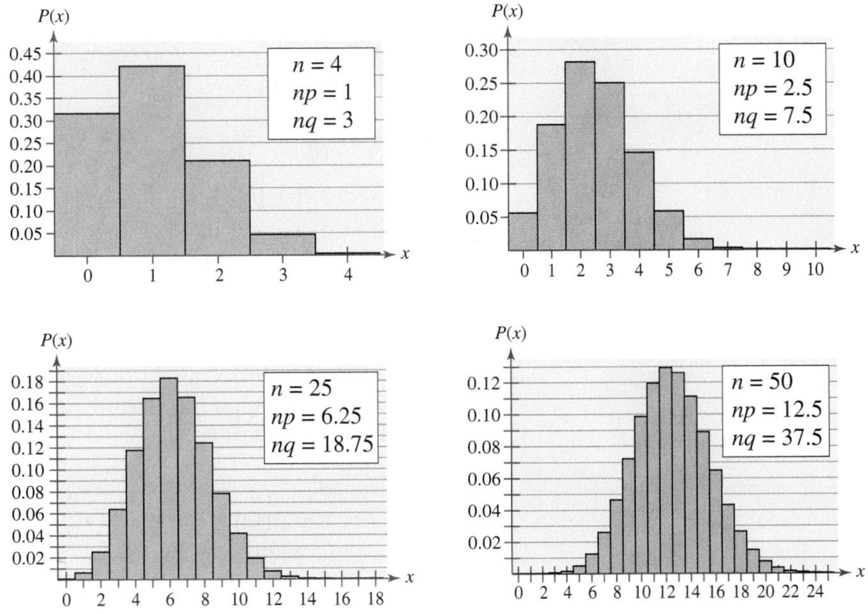

EXAMPLE 1

Approximating the Binomial Distribution

Two binomial experiments are listed. Decide whether you can use the normal distribution to approximate x, the number of people who reply yes. If you can, find the mean and standard deviation. If you cannot, explain why. *(Source: Marist College Institute for Public Opinion)*

1. Thirty-four percent of people in the United States say that they are likely to make a New Year's resolution. You randomly select 15 people in the United States and ask each if he or she is likely to make a New Year's resolution.
2. Six percent of people in the United States who made a New Year's resolution resolved to exercise more. You randomly select 65 people in the United States who made a resolution and ask each if he or she resolved to exercise more.

SOLUTION

1. In this binomial experiment, $n = 15$, $p = 0.34$, and $q = 0.66$. So,

$$np = (15)(0.34) = 5.1$$

and

$$nq = (15)(0.66) = 9.9.$$

Because np and nq are greater than 5, you can use the normal distribution with

$$\mu = 5.10$$

and

$$\sigma = \sqrt{npq} = \sqrt{15 \cdot 0.34 \cdot 0.66} \approx 1.83$$

to approximate the distribution of x.

2. In this binomial experiment, $n = 65$, $p = 0.06$, and $q = 0.94$. So,

$$np = (65)(0.06) = 3.9$$

and

$$nq = (65)(0.94) = 61.1.$$

Because $np < 5$, you cannot use the normal distribution to approximate the distribution of x.

Try It Yourself 1

Consider the following binomial experiment. Decide whether you can use the normal distribution to approximate x, the number of people who reply yes. If you can, find the mean and standard deviation. If you cannot, explain why. *(Source: Marist College Institute for Public Opinion)*

Sixty-one percent of people in the United States who made a New Year's resolution last year kept it. You randomly select 70 people in the United States who made a resolution last year and ask each if he or she kept the resolution.

a. *Identify* n, p, and q.
b. *Find* the products np and nq.
c. *Decide* whether you can use the normal distribution to approximate x.
d. *Find* the mean μ and standard deviation σ, if appropriate. *Answer: Page A38*

Correction for Continuity

The binomial distribution is discrete and can be represented by a probability histogram. To calculate *exact* binomial probabilities, you can use the binomial formula for each value of x and add the results. Geometrically, this corresponds to adding the areas of bars in the probability histogram. Remember that each bar has a width of one unit and x is the midpoint of the interval.

When you use a *continuous* normal distribution to approximate a binomial probability, you need to move 0.5 unit to the left and right of the midpoint to include all possible x-values in the interval. When you do this, you are making a **correction for continuity.**

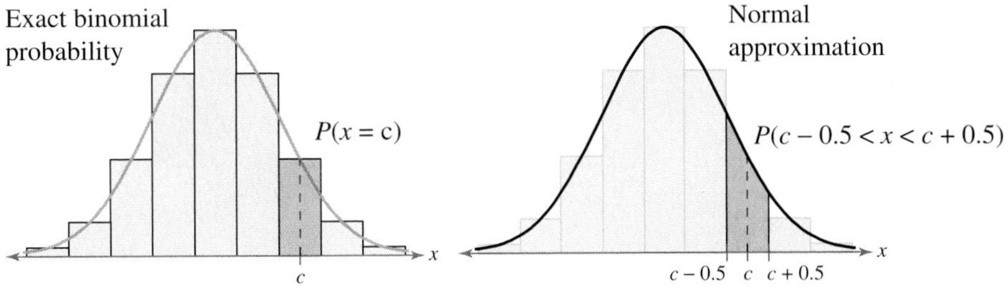

Note to Instructor

For technology users who are not limited to $n = 20$ in the table, many more binomial problems can be calculated without using a normal distribution approximation. However, students should be shown that even technology has limitations. The TI-83 cannot calculate the cumulative binomial probability for $n = 10{,}000$, $p = 0.4$, and $x = 9000$, but that probability can be calculated using a normal approximation. Likewise, depending on the version of MINITAB or Excel you are using, there are memory limitations for the binomial distribution.

EXAMPLE 2

Using a Correction for Continuity

Use a correction for continuity to convert each of the following binomial intervals to a normal distribution interval.

1. The probability of getting between 270 and 310 successes, inclusive
2. The probability of at least 158 successes
3. The probability of getting less than 63 successes

Study Tip

To use a correction for continuity, simply subtract 0.5 from the lowest value and add 0.5 to the highest.

SOLUTION

1. The discrete midpoint values are 270, 271, . . . , 310. The corresponding interval for the continuous normal distribution is

$$269.5 < x < 310.5.$$

2. The discrete midpoint values are 158, 159, 160, ... The corresponding interval for the continuous normal distribution is

$$x > 157.5.$$

3. The discrete midpoint values are . . . , 60, 61, 62. The corresponding interval for the continuous normal distribution is

$$x < 62.5.$$

Try It Yourself 2

Use a correction for continuity to convert each of the following binomial intervals to a normal distribution interval.

1. The probability of getting between 57 and 83 successes, inclusive
2. The probability of getting at most 54 successes

a. List the *midpoint values* for the binomial probability.
b. Use a *correction for continuity* to write the normal distribution interval.

Answer: Page A38

Picturing the World

In a survey of U.S. adults, people were asked if the law should allow doctors to aid dying patients who want to end their lives. The results of the survey are shown in the following pie chart. *(Source: The Harris Poll)*

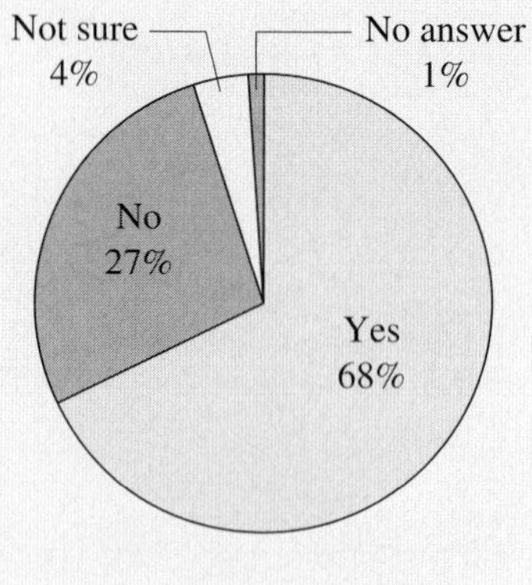

Assume that this Harris Poll is a true indication of the proportion of the population who believe in assisted death for terminally ill patients.

If you sampled 50 adults at random, what is the probability that between 32 and 36, inclusive, would believe in assisted death?

Approximating Binomial Probabilities

GUIDELINES

Using the Normal Distribution to Approximate Binomial Probabilities

In Words	*In Symbols*
1. Verify that the binomial distribution applies.	Specify n, p, and q.
2. Determine if you can use the normal distribution to approximate x, the binomial variable.	Is $np \geq 5$? Is $nq \geq 5$?
3. Find the mean μ and standard deviation σ for the distribution.	$\mu = np$ $\sigma = \sqrt{npq}$
4. Apply the appropriate continuity correction. Shade the corresponding area under the normal curve.	Add or subtract 0.5 from endpoints.
5. Find the corresponding z-score(s).	$z = \dfrac{x - \mu}{\sigma}$
6. Find the probability.	Use the Standard Normal Table.

EXAMPLE 3

Approximating a Binomial Probability

Thirty-four percent of people in the United States say that they are likely to make a New Year's resolution. You randomly select 15 people in the United States and ask each if he or she is likely to make a New Year's resolution. What is the probability that fewer than eight of them respond yes? *(Source: Marist College Institute for Public Opinion)*

SOLUTION From Example 1, you know that you can use a normal distribution with $\mu = 5.1$ and $\sigma \approx 1.83$ to approximate the binomial distribution. Remember to apply the continuity correction for the value of x. In the binomial distribution, the possible midpoint values for "fewer than 8" are

$$\ldots 5, 6, 7.$$

To use the normal distribution, add 0.5 to the right-hand boundary 7 to get $x = 7.5$. The graph at the left shows a normal curve with $\mu = 5.1$ and $\sigma \approx 1.83$ and a shaded area to the left of 7.5. The z-score that corresponds to $x = 7.5$ is

$$z = \frac{7.5 - 5.1}{1.83}$$

$$\approx 1.31.$$

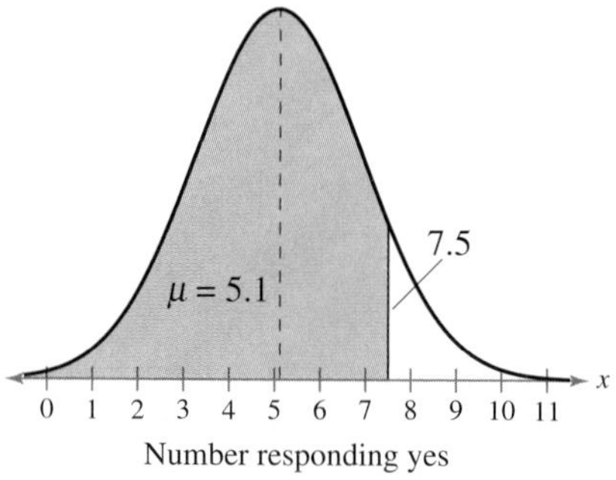

Using the Standard Normal Table,

$$P(z < 1.31) = 0.9049.$$

Interpretation The probability that fewer than eight people respond yes is approximately 0.9049, or about 90%.

Try It Yourself 3

Sixty-one percent of people in the United States who made a New Year's resolution last year kept it. You randomly select 70 people in the United States who made a resolution last year and ask each if he or she kept the resolution. What is the probability that more than 50 respond yes? (See Try It Yourself 1.) *(Source: Marist College Institute for Public Opinion)*

a. *Determine* whether you can use the normal distribution to approximate the binomial variable (see part c of Try It Yourself 1).
b. *Find* the mean μ and the standard deviation σ for the distribution (see part d of Try It Yourself 1).
c. *Apply* the appropriate continuity correction and sketch a graph.
d. *Find* the corresponding z-score.
e. *Use* the Standard Normal Table to find the area to the left of z and calculate the probability.

Answer: Page A38

EXAMPLE 4

Approximating a Binomial Probability

Thirty-eight percent of people in the United States admit that they snoop in other people's medicine cabinets. You randomly select 200 people in the United States and ask each if he or she snoops in other people's medicine cabinets. What is the probability that at least 70 will say yes? *(Source: USA TODAY)*

SOLUTION Because $np = 200 \cdot 0.38 = 76$ and $nq = 200 \cdot 0.62 = 124$, the binomial variable x is approximately normally distributed with

$$\mu = np = 76 \quad \text{and} \quad \sigma = \sqrt{200 \cdot 0.38 \cdot 0.62} \approx 6.86.$$

Using the correction for continuity, you can rewrite the discrete probability $P(x \geq 70)$ as the continuous probability $P(x \geq 69.5)$. The graph shows a normal curve with $\mu = 76$ and $\sigma = 6.86$ and a shaded area to the right of 69.5. The z-score that corresponds to 69.5 is $z = (69.5 - 76)/6.86 \approx -0.95$. So, the probability that at least 70 will say yes is

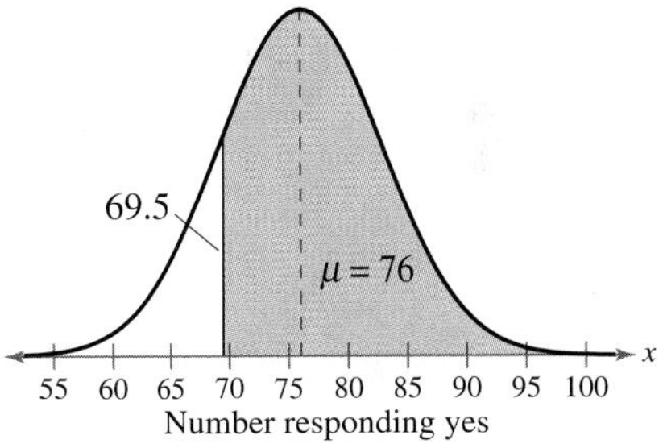

$$\begin{aligned} P(x \geq 69.5) &= P(z \geq -0.95) \\ &= 1 - P(z \leq -0.95) \\ &= 1 - 0.1711 = 0.8289. \end{aligned}$$

Study Tip

In a discrete distribution, there is a difference between $P(x \geq c)$ and $P(x > c)$. This is true because the probability that x is exactly c is not zero. In a continuous distribution, however, there is no difference between $P(x \geq c)$ and $P(x > c)$ because the probability that x is exactly c is zero.

Try It Yourself 4

What is the probability that at most 85 people will say yes?

a. *Determine* whether you can use the normal distribution to approximate the binomial variable (see Example 4).
b. *Find* the mean μ and the standard deviation σ for the distribution.
c. *Apply* a continuity correction to rewrite $P(x \leq 85)$ and sketch a graph.
d. *Find* the corresponding z-score.
e. *Use* the Standard Normal Table to *find the area* to the left of z and calculate the probability.

Answer: Page A38

EXAMPLE 5

Approximating a Binomial Probability

A survey reports that 95% of Internet users use Microsoft Internet Explorer as their browser. You randomly select 200 Internet users and ask each whether he or she uses Microsoft Internet Explorer as his or her browser. What is the probability that exactly 194 will say yes? *(Source: OneStat.com)*

SOLUTION Because $np = 200 \cdot 0.95 = 190$ and $nq = 200 \cdot 0.05 = 10$, the binomial variable x is approximately normally distributed with

$$\mu = np = 190 \quad \text{and} \quad \sigma = \sqrt{200 \cdot 0.95 \cdot 0.05} \approx 3.08.$$

Using the correction for continuity, you can rewrite the discrete probability $P(x = 194)$ as the continuous probability $P(193.5 < x < 194.5)$. The following graph shows a normal curve with $\mu = 190$ and $\sigma = 3.08$ and a shaded area between 193.5 and 194.5.

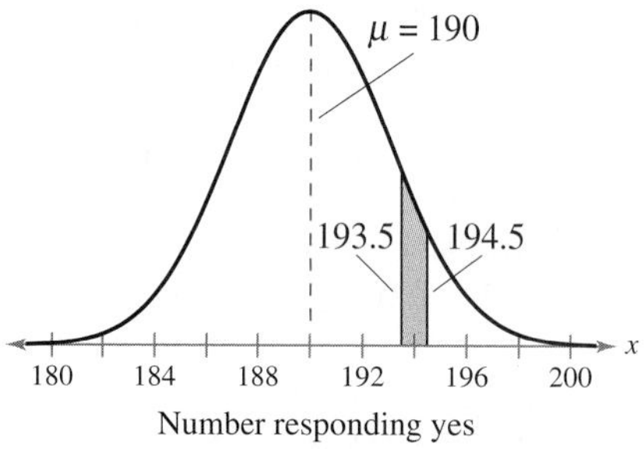

The z-scores that correspond to 193.5 and 194.5 are

$$z_1 = \frac{193.5 - 190}{3.08} \approx 1.14 \quad \text{and} \quad z_2 = \frac{194.5 - 190}{3.08} \approx 1.46.$$

So, the probability that exactly 194 Internet users will say they use Microsoft Internet Explorer is

$$\begin{aligned} P(193.5 < x < 194.5) &= P(1.14 < z < 1.46) \\ &= P(z < 1.46) - P(z < 1.14) \\ &= 0.9279 - 0.8729 \\ &= 0.0550. \end{aligned}$$

Interpretation There is a probability of about 0.06 that exactly 194 of the Internet users will say they use Microsoft Internet Explorer.

Note to Instructor

You may want to have students calculate the probability using the binomial formula from Chapter 4 and compare results.

$P(x = 194)$

$= {}_{200}C_{194}(0.95)^{194}(0.05)^6$

The TI-83 gives 0.061400531.

Try It Yourself 5

What is the probability that exactly 191 people will say yes?

a. *Determine* whether you can use the normal distribution to approximate the binomial variable (see Example 5).
b. *Find* the mean μ and the standard deviation σ for the distribution.
c. *Apply* a continuity correction to rewrite $P(x = 191)$ and sketch a graph.
d. *Find* the corresponding z-scores.
e. *Use* the Standard Normal Table to *find the area* to the left of each z-score and calculate the probability.

Answer: Page A39

5.5 Exercises

Help

MyMathLab

Student Study Pack

Building Basic Skills and Vocabulary

In Exercises 1–4, the sample size n, probability of success p, and probability of failure q are given for a binomial experiment. Decide whether you can use the normal distribution to approximate the random variable x.

1. $n = 20,\ p = 0.80,\ q = 0.20$

2. $n = 12,\ p = 0.60,\ q = 0.40$

3. $n = 15,\ p = 0.65,\ q = 0.35$

4. $n = 18,\ p = 0.85,\ q = 0.15$

Approximating a Binomial Distribution In Exercises 5–8, a binomial experiment is given. Decide whether you can use the normal distribution to approximate the binomial distribution. If you can, find the mean and standard deviation. If you cannot, explain why.

5. ***Credit Card Contract*** A survey of U.S. adults found that 44% read every word of a credit card contract. You ask 10 adults selected at random if he or she reads every word of a credit card contract. *(Source: USA TODAY)*

6. ***Organ Donors*** A survey of U.S. adults found that 63% would want their organs transplanted into a patient who needs them if they were killed in an accident. You randomly select 20 adults and ask each if he or she would want their organs transplanted into a patient who needs them if they were killed in an accident. *(Source: USA TODAY)*

7. ***Prostate Cancer*** In a recent year, the American Cancer Society said that the five-year survival rate for all men diagnosed with prostate cancer was 97%. You randomly select 10 men who were diagnosed with prostate cancer and calculate their five-year survival rate. *(Source: American Cancer Society)*

8. ***Work Weeks*** A survey of workers in the United States found that 8.6% work fewer than 40 hours per week. You randomly select 30 workers in the United States and ask each if he or she works fewer than 40 hours per week.

In Exercises 9–12, match the binomial probability with the correct statement.

Probability	Statement
9. $P(x \geq 45)$	(a) P(there are fewer than 45 successes)
10. $P(x \leq 45)$	(b) P(there are at most 45 successes)
11. $P(x < 45)$	(c) P(there are more than 45 successes)
12. $P(x > 45)$	(d) P(there are at least 45 successes)

In Exercises 13–16, use the correction for continuity and match the binomial probability statement with the corresponding normal distribution statement.

Binomial Probability	Normal Probability
13. $P(x > 89)$	(a) $P(x > 89.5)$
14. $P(x \geq 89)$	(b) $P(x < 88.5)$
15. $P(x \leq 89)$	(c) $P(x \leq 89.5)$
16. $P(x < 89)$	(d) $P(x \geq 88.5)$

1. Cannot use normal distribution.
2. Cannot use normal distribution.
3. Can use normal distribution.
4. Cannot use normal distribution.
5. Cannot use normal distribution because $np < 5$.
6. Can use normal distribution. $\mu = 12.6, \sigma = 2.159$
7. Cannot use normal distribution because $nq < 5$.
8. Cannot use normal distribution because $np < 5$.
9. d
10. b
11. a
12. c
13. a
14. d
15. c
16. b

17. Binomial: 0.549; Normal: 0.5463

18. Binomial: 0.19; Normal: 0.1875

19. Cannot use normal distribution because $np < 5$.

(a) 0.0000199 (b) 0.000023

(c) 0.999977 (d) 0.1635

20. See Selected Answers, page A95.

21. Can use normal distribution.

(a) 0.1174

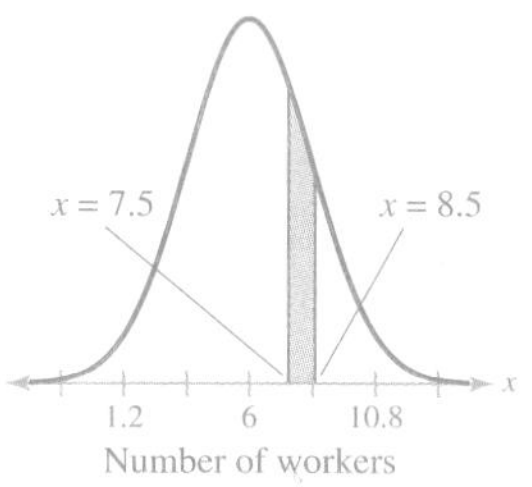

(b) 0.2643

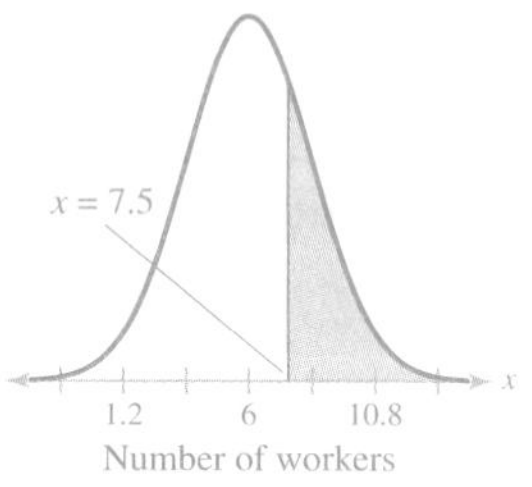

(c) 0.7357

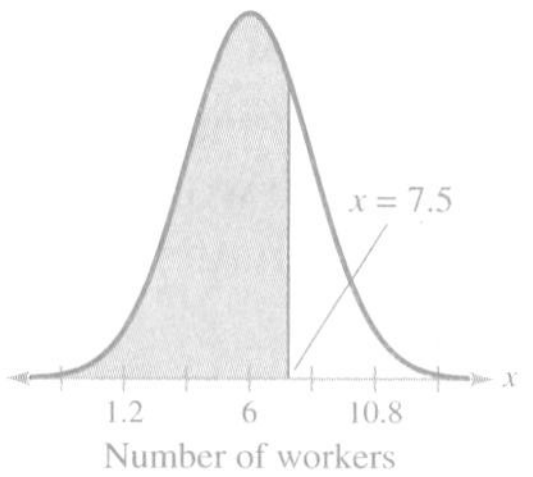

(d) 0.7190

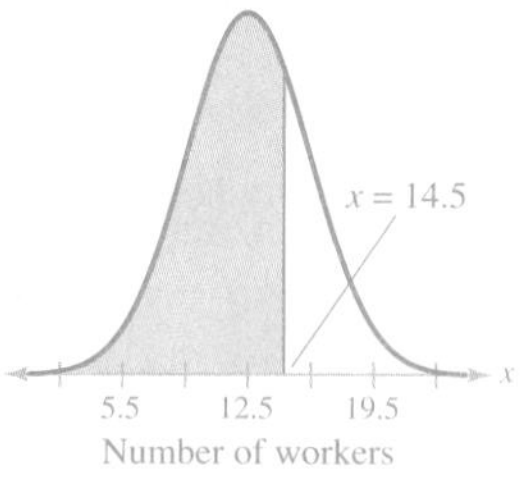

Using and Interpreting Concepts

Graphical Analysis In Exercises 17 and 18, write the binomial probability and the normal probability for the shaded region of the graph. Find the value of each probability and compare the results.

17.

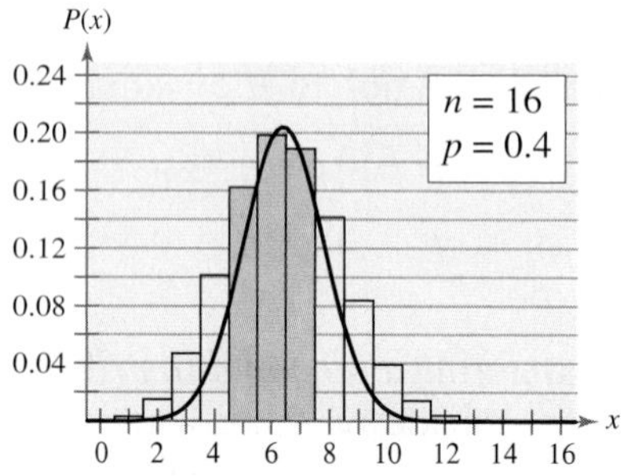

18.

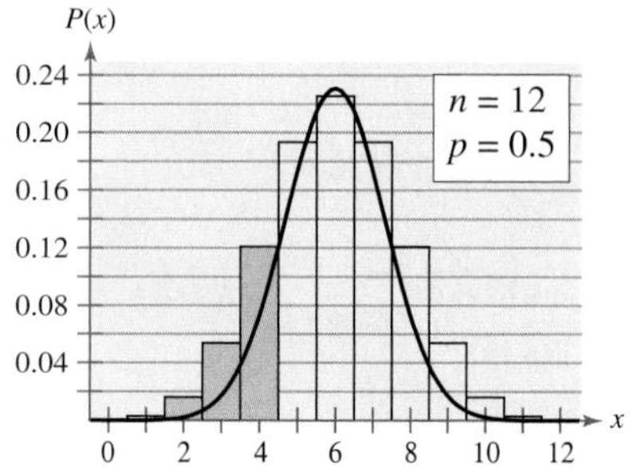

Approximating Binomial Probabilities In Exercises 19–24, decide whether you can use the normal distribution to approximate the binomial distribution. If you can, use the normal distribution to approximate the indicated probabilities and sketch their graphs. If you cannot, explain why and use the binomial distribution to find the indicated probabilities.

19. *Blood Type O^-* Seven percent of people in the United States have type O^- blood. You randomly select 30 people in the United States and ask them if their blood type is O^-. *(Source: American Association of Blood Banks)*

(a) Find the probability that exactly 10 people say they have O^- blood.

(b) Find the probability that at least 10 people say they have O^- blood.

(c) Find the probability that fewer than 10 people say they have O^- blood.

(d) A blood drive would like to get at least five donors with O^- blood. There are 100 donors. What is the probability that there will not be enough O^- blood donors?

20. *Blood Type A^+* Thirty-four percent of people in the United States have type A^+ blood. You randomly select 32 people in the United States and ask them if their blood type is A^+. *(Source: American Association of Blood Banks)*

(a) Find the probability that exactly 12 people say they have A^+ blood.

(b) Find the probability that at least 12 people say they have A^+ blood.

(c) Find the probability that fewer than 12 people say they have A^+ blood.

(d) A blood drive would like to get at least 60 donors with A^+ blood. There are 150 donors. What is the probability that there will not be enough A^+ blood donors?

21. *Public Transportation* Five percent of workers in the United States use public transportation to get to work. You randomly select 120 workers and ask them if they use public transportation to get to work. *(Source: U.S. Census Bureau)*

(a) Find the probability that exactly eight workers will say yes.

(b) Find the probability that at least eight workers will say yes.

(c) Find the probability that fewer than eight workers will say yes.

(d) A transit authority offers discount rates to companies that have at least 15 employees who use public transportation to get to work. There are 250 employees in a company. What is the probability that the company will not get the discount?

22. See Selected Answers, page A95.

23. Can use normal distribution.

(a) 0.0465

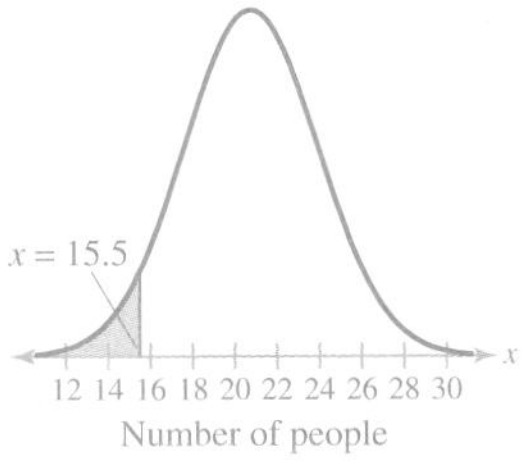

(b) 0.9767

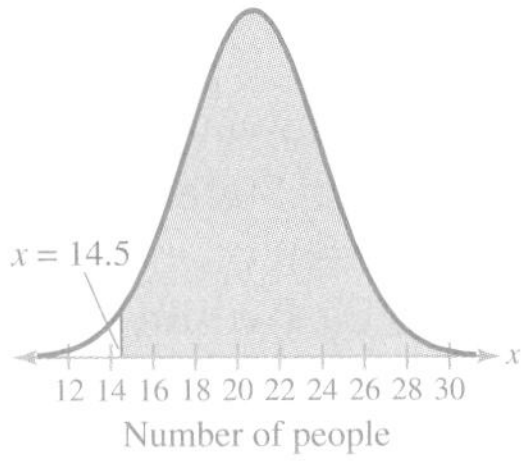

(c) 0.9535

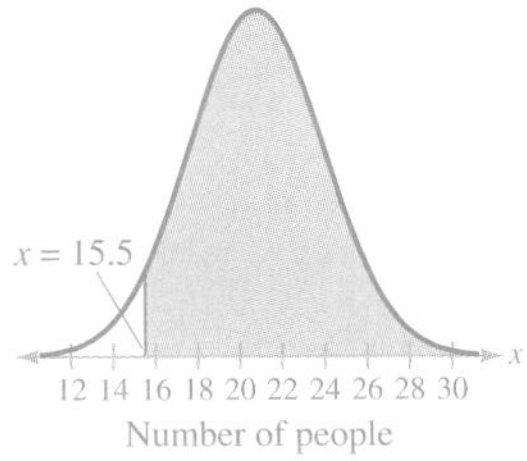

(d) 0.1635

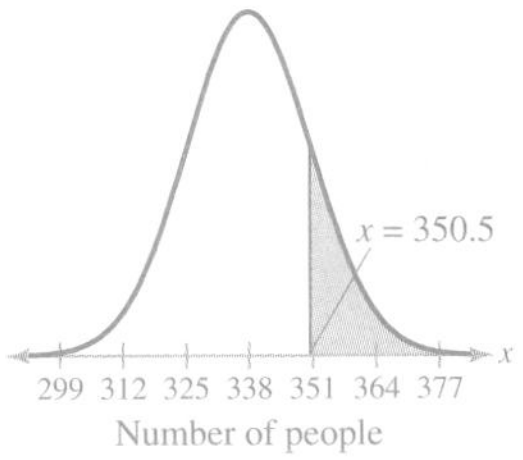

24. Cannot use normal distribution because $np < 5$.

(a) 0.99987

(b) 0.00251

(c) 0.00013

(d) 0.230

25. (a) $np = 6 \geq 5$

$nq = 19 \geq 5$

(b) 0.121

(c) No, because the z-score is within one standard deviation of the mean.

22. *College Graduates* Thirty-one percent of workers in the United States are college graduates. You randomly select 50 workers and ask each if he or she is a college graduate. *(Source: U.S. Bureau of Labor Statistics)*

(a) Find the probability that exactly 14 workers are college graduates.

(b) Find the probability that at least 14 workers are college graduates.

(c) Find the probability that fewer than 14 workers are college graduates.

(d) A committee is looking for 30 working college graduates to volunteer at a career fair. The committee randomly selects 150 workers. What is the probability that there will not be enough college graduates?

23. *Favorite Cookie* Fifty-two percent of adults say chocolate chip is their favorite cookie. You randomly select 40 adults and ask each if chocolate chip is his or her favorite cookie. *(Source: WEAREVER)*

(a) Find the probability that at most 15 people say chocolate chip is their favorite cookie.

(b) Find the probability that at least 15 people say chocolate chip is their favorite cookie.

(c) Find the probability that more than 15 people say chocolate chip is their favorite cookie.

(d) A community bake sale has prepared 350 chocolate chip cookies. The bake sale attracts 650 customers, and they each buy one cookie. What is the probability there will not be enough chocolate chip cookies?

24. *Long Work Weeks* A survey of workers in the United States found that 2.9% work more than 70 hours per week. You randomly select 10 workers in the U.S. and ask each if he or she works more than 70 hours per week.

(a) Find the probability that at most three people say they work more than 70 hours per week.

(b) Find the probability that at least three people say they work more than 70 hours per week.

(c) Find the probability that more than three people say they work more than 70 hours per week.

(d) A large company is concerned about overworked employees who work more than 70 hours per week. The company randomly selects 50 employees. What is the probability there will be no employee working more than 70 hours?

25. *Bigger Home* A survey of homeowners in the United States found that 24% feel their home is too small for their family. You randomly select 25 homeowners and ask them if they feel their home is too small for their family.

(a) Verify that the normal distribution can be used to approximate the binomial distribution.

(b) Find the probability that more than eight homeowners say their home is too small for their family.

(c) Is it unusual for 8 out of 25 homeowners to say their home is too small? Why or why not?

26. (a) $np = 32 \geq 5; nq = 8 \geq 5$
(b) 0.0150
(c) Yes, because the z-score is more than two standard deviations from the mean.
27. Highly unlikely. Answers will vary.
28. Probable. Answers will vary.
29. 0.1020
30. 0.1736

26. *Driving to Work* A survey of workers in the United States found that 80% rely on their own vehicle to get to work. You randomly select 40 workers and ask them if they rely on their own vehicle to get to work.

(a) Verify that the normal distribution can be used to approximate the binomial distribution.

(b) Find the probability that at most 26 workers say they rely on their own vehicle to get to work.

(c) Is it unusual for 26 out of 40 workers to say they rely on their own vehicle to get to work? Why or why not?

Extending Concepts

Getting Physical In Exercises 27 and 28, use the following information. The graph shows the results of a survey of adults in the United States ages 33 to 51 who were asked if they participated in a sport. Seventy percent of adults said they regularly participate in at least one sport, and they gave their favorite sport.

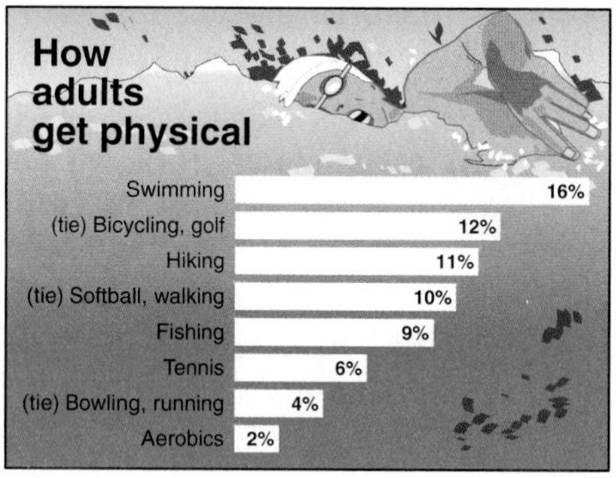

27. You randomly select 250 people in the United States ages 33 to 51 and ask each if he or she regularly participates in at least one sport. You find that 60% say no. How likely is this result? Do you think the sample is a good one? Explain your reasoning.

28. You randomly select 300 people in the United States ages 33 to 51 and ask each if he or she regularly participates in at least one sport. Of the 200 who say yes, 9% say they participate in hiking. How likely is this result? Is the sample a good one? Explain your reasoning.

Testing a Drug In Exercises 29 and 30, use the following information. A drug manufacturer claims that a drug cures a rare skin disease 75% of the time. The claim is checked by testing the drug on 100 patients. If at least 70 patients are cured, the claim will be accepted.

29. Find the probability that the claim will be rejected assuming that the manufacturer's claim is true.

30. Find the probability that the claim will be accepted assuming that the actual probability that the drug cures the skin disease is 65%.

Uses and Abuses

Statistics in the Real World

Uses

Normal Distributions Normal distributions can be used to describe many real-life situations and are widely used in the fields of science, business, and psychology. They are the most important probability distributions in statistics and can be used to approximate other distributions, such as discrete binomial distributions.

The most incredible application of the normal distribution lies in the Central Limit Theorem. This theorem states that no matter what type of distribution a population may have, as long as the sample size is at least 30, the distribution of sample means will be normal. If the population is itself normal, then the distribution of sample means will be normal no matter how small the sample is.

The normal distribution is essential to sampling theory. Sampling theory forms the basis of statistical inference, which you will begin to study in the next chapter.

Abuses

Confusing Likelihood with Certainty A common abuse of normal probability distributions is to confuse the concept of likelihood with the concept of certainty. For instance, if you randomly select a member from a population that is normally distributed, you know the probability is approximately 95% that you will obtain a value that lies within two standard deviations of the mean. This *does not* imply, however, that you cannot get an unusual result. In fact, 5% of the time you should expect to get a value that is more than two standard deviations from the mean.

Suppose a population is normally distributed with a mean of 100 and standard deviation of 15. It would not be unusual for an individual value taken from this population to be 112 or more. It *would* be, however, highly unusual to obtain a sample mean of 112 or more from a sample with 100 members.

Exercises

1. ***Confusing Likelihood with Certainty*** You are randomly selecting 100 people from a population that is normally distributed. Are you certain to get exactly 95 people who lie within two standard deviations of the mean? Explain your reasoning.

2. ***Confusing Likelihood with Certainty*** You are randomly selecting 10 people from a large population that is normally distributed. Which of the following is more likely? Explain your reasoning.

a. All 10 lie within 2 standard deviations of the mean.

b. At least one person does not lie within 2 standard deviations of the mean.

5 Chapter Summary

What did you learn?

Review Exercises

Section 5.1

- How to interpret graphs of normal probability distributions — *1, 2*
- How to find and interpret z-scores — *3, 4*

$$z = \frac{x - \mu}{\sigma}$$

- How to find areas under the standard normal curve — *5–16*

Section 5.2

- How to find probabilities for normally distributed variables — *17–24*

Section 5.3

- How to find a z-score given the area under the normal curve — *25–30*
- How to transform a z-score to an x-value — *31, 32*

$$x = \mu + z\sigma$$

- How to find a specific data value of a normal distribution given the probability — *33–36*

Section 5.4

- How to find sampling distributions and verify their properties — *37, 38*
- How to interpret the Central Limit Theorem — *39, 40*

$$\mu_{\bar{x}} = \mu, \sigma_{\bar{x}} = \frac{\sigma}{\sqrt{n}}$$

- How to apply the Central Limit Theorem to find the probability of a sample mean — *41–46*

Section 5.5

- How to decide when the normal distribution can approximate the binomial distribution — *47, 48*

$$\mu = np, \sigma = \sqrt{npq}$$

- How to find the correction for continuity — *49–52*
- How to use the normal distribution to approximate binomial probabilities — *53, 54*

5 Review Exercises

1. $\mu = 15, \sigma = 3$
2. $\mu = -3, \sigma = 5$
3. -2.25; 0.5; 2; 3.5
4. 1.32 and 1.78 are unusual.
5. 0.2005
6. 0.9946
7. 0.3936
8. 0.8962
9. 0.0465
10. 0.7967
11. 0.4495
12. 0.2224
13. 0.3519
14. 0.95
15. 0.1336
16. 0.5905
17. 0.8997
18. 0.7704
19. 0.9236
20. 0.3364
21. 0.0124
22. 0.5465

Section 5.1

In Exercises 1 and 2, use the graph to estimate μ and σ.

1.

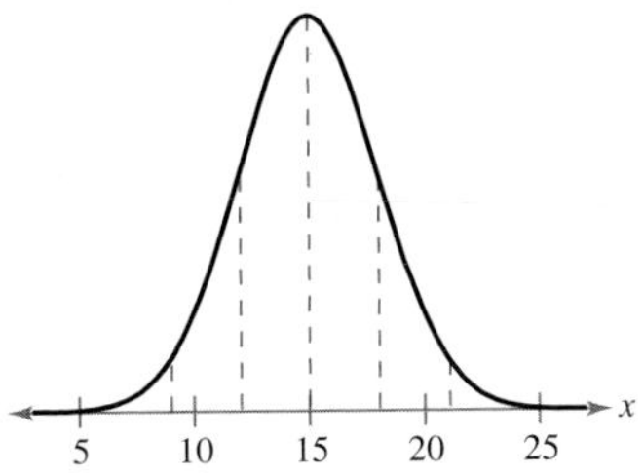

2.

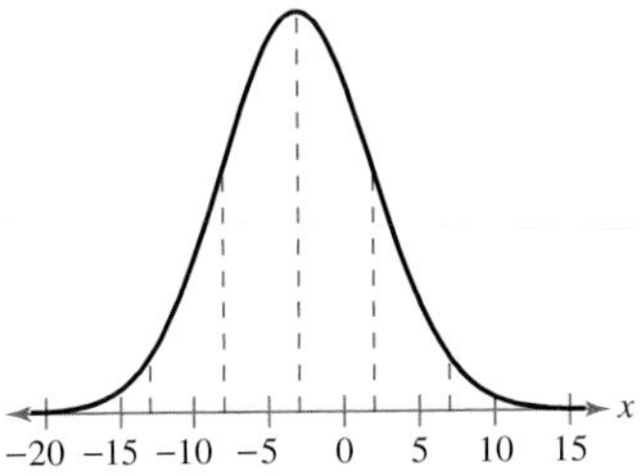

In Exercises 3 and 4, use the following information and standard scores to investigate observations about a normal population. A batch of 2500 resistors is normally distributed, with a mean resistance of 1.5 ohms and a standard deviation of 0.08 ohm. Four resistors are randomly selected and tested. Their resistances were measured at 1.32, 1.54, 1.66, and 1.78 ohms.

3. How many standard deviations from the mean are these observations?

4. Are there any unusual observations?

In Exercises 5–16, use the Standard Normal Table to find the indicated area under the standard normal curve.

5. To the left of $z = -0.84$

6. To the left of $z = 2.55$

7. To the left of $z = -0.27$

8. To the left of $z = 1.26$

9. To the right of $z = 1.68$

10. To the right of $z = -0.83$

11. Between $z = -1.64$ and the mean

12. Between $z = -1.22$ and $z = -0.43$

13. Between $z = 0.15$ and $z = 1.35$

14. Between $z = -1.96$ and $z = 1.96$

15. To the left of $z = -1.5$ and to the right of $z = 1.5$

16. To the left of $z = 0.12$ and to the right of $z = 1.72$

Section 5.2

In Exercises 17–22, find the indicated probabilities.

17. $P(z < 1.28)$

18. $P(z > -0.74)$

19. $P(-2.15 < z < 1.55)$

20. $P(0.42 < z < 3.15)$

21. $P(z < -2.50 \text{ or } z > 2.50)$

22. $P(z < 0 \text{ or } z > 1.68)$

23. (a) 0.3156
(b) 0.3099
(c) 0.3446
24. (a) 0.9544
(b) 0.3420
(c) 0.0026
25. −0.07
26. −1.28
27. 1.13
28. −2.05
29. 1.04
30. −0.84
31. 43.9 meters
32. 45.7 meters
33. 45.9 meters
34. 45.435 meters
35. 45.74 meters
36. 44.28 meters

In Exercises 23 and 24, find the indicated probabilities.

23. A study found that the mean migration distance of the green turtle was 2200 kilometers and the standard deviation was 625 kilometers. Assuming that the distances are normally distributed, find the probability that a randomly selected green turtle migrates a distance of

(a) less than 1900 kilometers.

(b) between 2000 kilometers and 2500 kilometers.

(c) greater than 2450 kilometers.

(Adapted from Dorling Kindersley Visual Encyclopedia)

24. The world's smallest mammal is the Kitti's hog-nosed bat, with a mean weight of 1.5 grams and a standard deviation of 0.25 gram. Assuming that the weights are normally distributed, find the probability of randomly selecting a bat that weighs

(a) between 1.0 gram and 2.0 grams.

(b) between 1.6 grams and 2.2 grams.

(c) more than 2.2 grams.

(Adapted from Dorling Kindersley Visual Encyclopedia)

Section 5.3

In Exercises 25–30, use the Standard Normal Table to find the *z*-score that corresponds to the given cumulative area or percentile. If the area is not in the table, use the entry closest to the area.

25. 0.4721 **26.** 0.1 **27.** 0.8708

28. P_2 **29.** P_{85} **30.** P_{20}

In Exercises 31–36, use the following information. On a dry surface, the braking distance (in meters) of a Pontiac Grand AM SE can be approximated by a normal distribution, as shown in the graph. *(Source: National Highway Traffic Safety Administration)*

Braking Distance of a Pontiac Grand Am SE

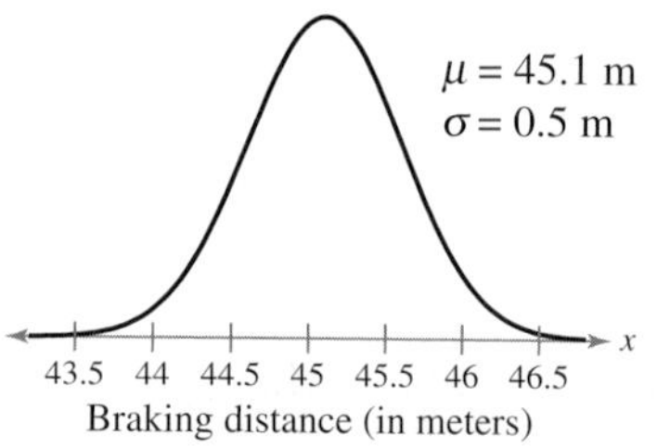

31. Find the braking distance of a Pontiac Grand AM SE that corresponds to $z = -2.4$.

32. Find the braking distance of a Pontiac Grand AM SE that corresponds to $z = 1.2$.

33. What braking distance of a Pontiac Grand AM SE represents the 95th percentile?

34. What braking distance of a Pontiac Grand AM SE represents the third quartile?

35. What is the shortest braking distance of a Pontiac Grand AM SE that can be in the top 10% of braking distances?

36. What is the longest braking distance of a Pontiac Grand AM SE that can be in the bottom 5% of braking distances?

37. See Odd Answers, page A65.

38. $\{00, 01, 02, 03, 10, 11, 12, 13, 20, 21, 22, 23, 30, 31, 32, 33\}$

1.5, 1.118; 1.5, 0.791

39. 152.7, 8.7

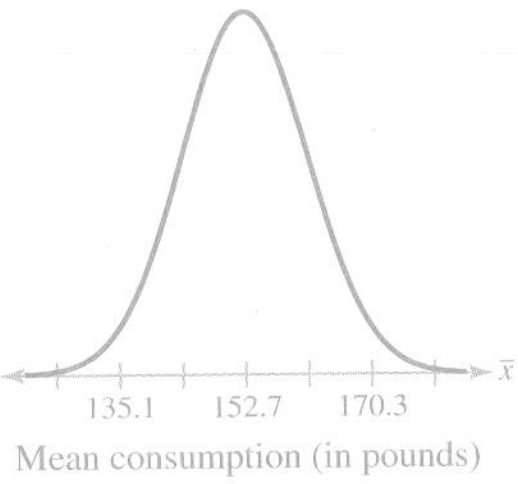

40. 226.6, 10.768

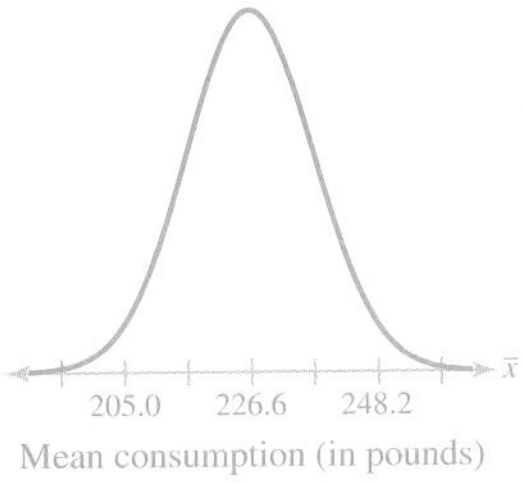

41. (a) 0.0485

(b) 0.8180

(c) 0.0823

(a) and (c) are smaller, (b) is larger. This is to be expected because the standard error of the sample means is smaller.

42. (a) ≈ 1

(b) 0.1446

(c) ≈ 0

(a) is larger and (b) and (c) are smaller.

43. (a) ≈ 0 (b) ≈ 0

44. (a) 0.9918

(b) 0.9998

Section 5.4

In Exercises 37 and 38, use the given population to find the sampling distribution of the sample means for the indicated sample sizes. Find the mean and standard deviation of the population and the mean and standard deviation of the sampling distribution. Compare the values.

37. A corporation has five executives. The number of minutes each exercises a week is reported as 40, 200, 80, 0, and 600. Draw three executives' names from this population, with replacement, and form a sampling distribution of the sample mean of the minutes they exercise.

38. There are four residents sharing a house. The number of times each washes his or her car each month is 1, 2, 0, and 3. Draw two names from this population, with replacement, and form a sampling distribution for the sample mean of the number of times their cars are washed each month.

In Exercises 39 and 40, use the Central Limit Theorem to find the mean and standard error of the mean of the indicated sampling distribution. Then sketch a graph of the sampling distribution.

39. The consumption of processed fruits by people in the United States in a recent year was normally distributed, with a mean of 152.7 pounds and a standard deviation of 51.6 pounds. Random samples of size 35 are drawn from this population. *(Adapted from U.S. Department of Agriculture)*

40. The consumption of processed vegetables by people in the United States in a recent year was normally distributed, with a mean of 226.6 pounds and a standard deviation of 68.1 pounds. Random samples of size 40 are drawn from this population. *(Adapted from U.S. Department of Agriculture)*

In Exercises 41–46, find the probabilities for the sampling distributions.

41. Refer to Exercise 23. A sample of 12 green turtles is randomly selected. Find the probability that the sample mean of the distance migrated is (a) less than 1900 kilometers, (b) between 2000 kilometers and 2500 kilometers, and (c) greater than 2450 kilometers. Compare your answers with those in Exercise 23.

42. Refer to Exercise 24. A sample of seven Kitti's hog-nosed bats is randomly selected. Find the probability that the sample mean is (a) between 1.0 gram and 2.0 grams, (b) between 1.6 grams and 2.2 grams, and (c) more than 2.2 grams. Compare your answers with those in Exercise 24.

43. The mean annual salary for chauffeurs is \$24,700. A random sample of size 45 is drawn from this population. What is the probability that the mean annual salary is (a) less than \$23,700 and (b) more than \$26,200? Assume $\sigma = \$1500$. *(Source: Salary.com)*

44. The mean value of land and buildings per acre for farms is \$1300. A random sample of size 36 is drawn. What is the probability that the mean value of land and buildings per acre is (a) less than \$1400 and (b) more than \$1150? Assume $\sigma = \$250$.

45. 0.0019

46. 0.0006

47. Cannot use normal distribution because $nq < 5$.

48. Can use normal distribution. $\mu = 8.85, \sigma = 1.905$

49. $P(x > 24.5)$

50. $P(x < 36.5)$

51. $P(44.5 < x < 45.5)$

52. $P(49.5 < x < 50.5)$

53. Can use normal distribution.

0.0032

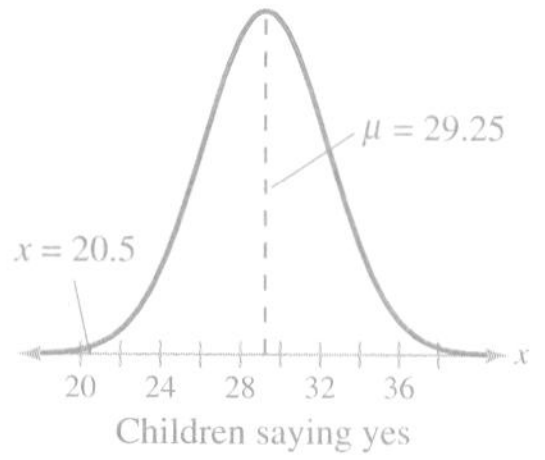

54. Cannot use normal distribution because $np < 5$.

0.171

45. The mean price of houses in a city is \$1.5 million with a standard deviation of \$500,000. The house prices are normally distributed. You randomly select 15 houses in this city. What is the probability that the mean price will be less than \$1.125 million?

46. Mean rent in a city is \$5000 per month with a standard deviation of \$300. The rents are normally distributed. You randomly select 15 apartments in this city. What is the probability that the mean price will be more than \$5250?

Section 5.5

In Exercises 47 and 48, a binomial experiment is given. Decide whether you can use the normal distribution to approximate the binomial distribution. If you can, find the mean and standard deviation. If you cannot, explain why.

47. In a recent year, the American Cancer Society predicted that the five-year survival rate for new cases of kidney cancer would be 90%. You randomly select 12 men who were new kidney cancer cases this year and calculate their five-year survival rate. *(Source: American Cancer Society)*

48. A survey indicates that 59% of men purchased perfume in the past year. You randomly select 15 men and ask them if they have purchased perfume in the past year. *(Source: USA TODAY)*

In Exercises 49–52, write the binomial probability as a normal probability using the continuity correction.

	Binomial Probability	Normal Probability
49.	$P(x \geq 25)$	$P(x > ?)$
50.	$P(x \leq 36)$	$P(x < ?)$
51.	$P(x = 45)$	$P(? < x < ?)$
52.	$P(x = 50)$	$P(? < x < ?)$

In Exercises 53 and 54, decide whether you can use the normal distribution to approximate the binomial distribution. If you can, use the normal distribution to approximate the indicated probabilities and sketch their graphs. If you cannot, explain why and use the binomial distribution to find the indicated probabilities.

53. Sixty-five percent of children ages 12 to 17 keep at least part of their savings in a savings account. You randomly select 45 children and ask each if he or she keeps at least part of his or her savings in a savings account. Find the probability that at most 20 children will say yes. *(Source: International Communications Research for Merrill Lynch)*

54. Thirty-three percent of adults graded public schools as excellent or good at preparing students for college. You randomly select 12 adults and ask them if they think public schools are excellent or good at preparing students for college. Find the probability that more than five adults will say yes. *(Source: Marist Institute for Public Opinion)*

5 Chapter Quiz

1. (a) 0.9821
   (b) 0.9994
   (c) 0.9802
   (d) 0.8135
2. (a) 0.9198
   (b) 0.1940
   (c) 0.0456
3. 0.1611
4. 0.5739
5. 81.59%
6. 1417.6
7. 337.588
8. 257.952
9. ≈ 0
10. More likely to select one student with a test score greater than 300 because the standard error of the mean is less than the standard deviation.
11. Can use normal distribution.
    $\mu = 16.32, \sigma \approx 2.285$
12. 0.3594

Take this quiz as you would take a quiz in class. After you are done, check your work against the answers given in the back of the book.

1. Find each standard normal probability.

(a) $P(z > -2.10)$
(b) $P(z < 3.22)$
(c) $P(-2.33 < z < 2.33)$
(d) $P(z < -1.75 \text{ or } z > -0.75)$

2. Find each normal probability for the given parameters.

(a) $\mu = 5.5$, $\sigma = 0.08$, $P(5.36 < x < 5.64)$
(b) $\mu = -8.2$, $\sigma = 7.84$, $P(-5.00 < x < 0)$
(c) $\mu = 18.5$, $\sigma = 9.25$, $P(x < 0 \text{ or } x > 37)$

In Exercises 3–10, use the following information. In a recent year, grade 8 Washington State public school students taking a mathematics assessment test had a mean score of 281 with a standard deviation of 34.4. Possible test scores could range from 0 to 500. Assume that the scores are normally distributed. *(Source: National Center for Educational Statistics)*

3. Find the probability that a student had a score higher than 315.

4. Find the probability that a student had a score between 250 and 305.

5. What percent of the students had a test score that is greater than 250?

6. If 2000 students are randomly selected, how many would be expected to have a test score that is less than 300?

7. What is the lowest score that would still place a student in the top 5% of the scores?

8. What is the highest score that would still place a student in the bottom 25% of the scores?

9. A random sample of 60 students is drawn from this population. What is the probability that the mean test score is greater than 300?

10. Are you more likely to randomly select one student with a test score greater than 300 or are you more likely to select a sample of 15 students with a mean test score greater than 300? Explain.

In Exercises 11 and 12, use the following information. In a survey of adults, 68% thought that DNA tests for identifying an individual were very reliable. You randomly select 24 adults and ask each if he or she thinks DNA tests for identifying an individual are very reliable. *(Source: CBS News)*

11. Decide whether you can use the normal distribution to approximate the binomial distribution. If you can, find the mean and standard deviation. If you cannot, explain why.

12. Find the probability that at most 15 people say DNA tests for identifying an individual are very reliable.

PUTTING IT ALL TOGETHER

Real Statistics ■ Real Decisions

You work for a manufacturing company as a statistical process analyst. Your job is to analyze processes and make sure they are in statistical control. In one process, a machine cuts wood boards to a thickness of 25 millimeters with an acceptable margin of error of ± 0.6 millimeter. (Assume this process can be approximated by a normal distribution.) So, the acceptable range of thicknesses for the boards is 24.4 millimeters to 25.6 millimeters, inclusive.

Because of machine vibrations and other factors, the setting of the wood-cutting machine "shifts" from 25 millimeters. To check that the machine is cutting the boards to the correct thickness, you select at random three samples of four boards and find the mean thickness (in millimeters) of each sample. A coworker asks you why you take three samples of size 4 and find the mean instead of randomly choosing and measuring 12 boards individually to check the machine's settings. (*Note:* Both samples are chosen without replacement.)

Exercises

1. *Sampling Individuals*

You select one board and measure its thickness. Assume the machine shifts and is cutting boards with a mean thickness of 25.4 millimeters and a standard deviation of 0.2 millimeter.

(a) What is the probability that you select a board that is *not* outside the acceptable range (in other words, you do not detect that the machine has shifted)? (See figure.)

(b) You randomly select 12 boards. What is the probability that you select at least one board that is *not* outside the acceptable range?

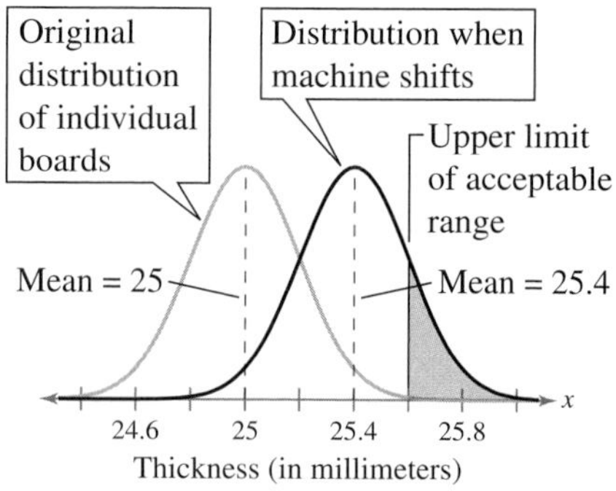

Figure for Exercise 1

2. *Sampling Groups of Four*

You select four boards and find their mean thickness. Assume the machine shifts and is cutting boards with a mean thickness of 25.4 millimeters and a standard deviation of 0.2 millimeter.

(a) What is the probability that you select a sample of four boards that has a mean that is *not* outside the acceptable range? (See figure.)

(b) You randomly select three samples of four boards. What is the probability that you select at least one sample of four boards that has a mean that is *not* outside the acceptable range?

(c) What is more sensitive to change—an individual measure or the mean?

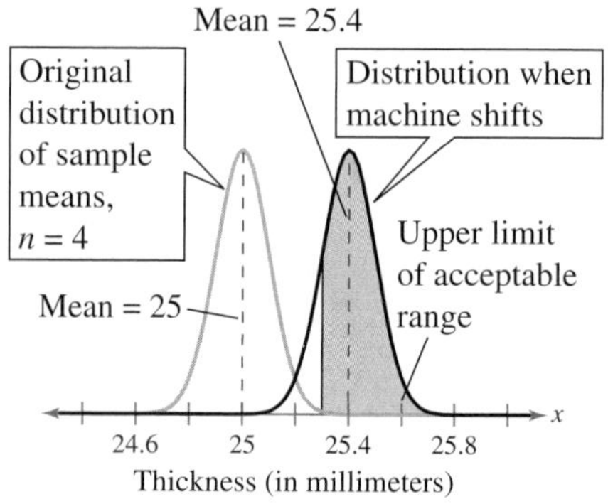

Figure for Exercise 2

3. *Writing an Explanation*

Write a paragraph to your coworker explaining why you take three samples of size 4 and find the mean of each sample instead of randomly choosing and measuring 12 boards individually to check the machine's settings.

TECHNOLOGY

MINITAB EXCEL TI-83

U.S. Census Bureau

www.census.gov

Age Distribution in the United States

One of the jobs of the U.S. Census Bureau is to keep track of the age distribution in the country. The age distribution in 2003 is shown below.

Age Distribution in the U.S.

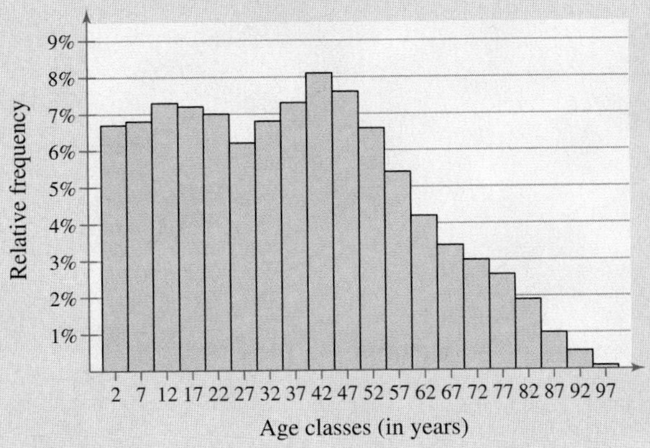

Class boundaries	Class midpoint	Relative frequency
0–4	2	6.7%
5–9	7	6.8%
10–14	12	7.4%
15–19	17	7.2%
20–24	22	7.0%
25–29	27	6.2%
30–34	32	6.8%
35–39	37	7.3%
40–44	42	8.1%
45–49	47	7.6%
50–54	52	6.6%
55–59	57	5.5%
60–64	62	4.2%
65–69	67	3.4%
70–74	72	3.0%
75–79	77	2.6%
80–84	82	1.9%
85–89	87	1.0%
90–94	92	0.5%
95–99	97	0.2%

Exercises

We used a technology tool to select random samples with $n = 40$ from the age distribution of the United States. The means of the 36 samples were as follows.

DATA

28.14, 31.56, 36.86, 32.37, 36.12, 39.53, 36.19, 39.02, 35.62, 36.30, 34.38, 32.98, 36.41, 30.24, 34.19, 44.72, 38.84, 42.87, 38.90, 34.71, 34.13, 38.25, 38.04, 34.07, 39.74, 40.91, 42.63, 35.29, 35.91, 34.36, 36.51, 36.47, 32.88, 37.33, 31.27, 35.80

1. Enter the age distribution of the United States into a technology tool. Use the tool to find the mean age in the United States.
2. Enter the set of sample means into a technology tool. Find the mean of the set of sample means. How does it compare with the mean age in the United States? Does this agree with the result predicted by the Central Limit Theorem?
3. Are the ages of people in the United States normally distributed? Explain your reasoning.
4. Sketch a relative frequency histogram for the 36 sample means. Use nine classes. Is the histogram approximately bell shaped and symmetric? Does this agree with the result predicted by the Central Limit Theorem?
5. Use a technology tool to find the standard deviation of the ages of people in the United States.
6. Use a technology tool to find the standard deviation of the set of 36 sample means. How does it compare with the standard deviation of the ages? Does this agree with the result predicted by the Central Limit Theorem?

Extended solutions are given in the *Technology Supplement.* Technical instruction is provided for MINITAB, Excel, and the TI-83.

CHAPTER 6

Confidence Intervals

6.1 Confidence Intervals for the Mean (Large Samples)

Case Study

6.2 Confidence Intervals for the Mean (Small Samples)

6.3 Confidence Intervals for Population Proportions

6.4 Confidence Intervals for Variance and Standard Deviation

Uses and Abuses

Real Statistics–Real Decisions

Technology

The Energy Policy Conservation Act (EPCA) was enacted into law in 1975. This act established the Corporate Average Fuel Economy (CAFE) standards for passenger cars and light trucks. These standards require that the fuel economy rating for a manufacturer's entire line of passenger cars must average at least 27.5 miles per gallon, and the fuel economy rating for a manufacturer's entire line of light trucks must average at least 20.7 miles per gallon.

Where You've Been

In Chapters 1 through 5, you studied descriptive statistics (how to collect and describe data) and probability (how to find probabilities and analyze discrete and continuous probability distributions). For instance, automobile manufacturers use descriptive statistics to analyze the data collected during vehicle tests conducted in their laboratories.

To meet the CAFE requirements, automobile manufacturers test the mean fuel economy rating of preproduction prototypes of their new vehicles and submit their test results to the Environmental Protection Agency (EPA). The EPA retests the mean fuel economy rating of a sample of the new vehicles, about 10% of the tested vehicles, to confirm the manufacturer's results. In a recent year, the EPA tested a sample of 73 cars from an automobile manufacturer's line of passenger cars. The mean fuel economy rating was 28.8 miles per gallon.

Where You're Going

In this chapter, you will begin your study of inferential statistics—the second major branch of statistics. For instance, from the mean of the sample of a manufacturer's line of passenger cars, the EPA can estimate the mean fuel economy rating to be 28.8 miles per gallon for the manufacturer's *entire* line of passenger cars. Because this estimate consists of a single number represented by a point on a number line, it is called a point estimate. The problem with using a point estimate is that it is rarely equal to the exact parameter (mean, standard deviation, or proportion) of the population.

In this chapter, you will learn how to make a more meaningful estimate by specifying an interval of values on a number line, together with a statement of how confident you are that your interval contains the population parameter. Suppose the EPA wanted to be 90% confident of its estimate for the mean fuel economy rating for the manufacturer's entire line of passenger cars. Here is an overview of how to construct an interval estimate.

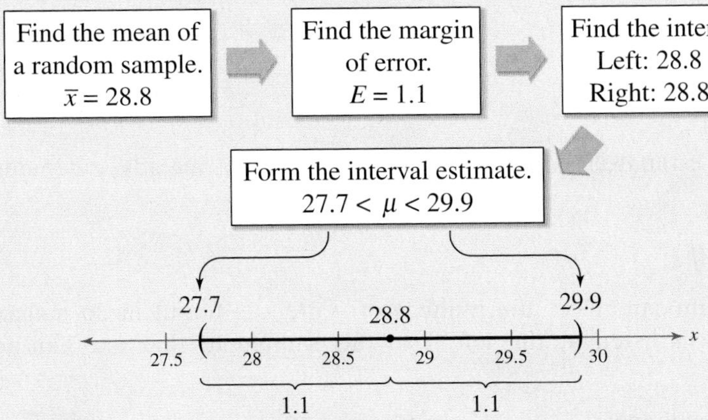

So, the EPA can be 90% confident that the mean fuel economy rating for the manufacturer's entire line of passenger cars is between 27.7 and 29.9 miles per gallon.

6.1 Confidence Intervals for the Mean (Large Samples)

What You Should Learn

- How to find a point estimate and a margin of error
- How to construct and interpret confidence intervals for the population mean
- How to determine the minimum sample size required when estimating μ

Estimating Population Parameters • Confidence Intervals for the Population Mean • Sample Size

Estimating Population Parameters

In this chapter, you will learn an important technique of statistical inference—to use sample statistics to estimate the value of an unknown population parameter. In this section, you will learn how to use sample statistics to make an estimate of the population parameter μ when the sample size is at least 30 or when the population is normally distributed and the standard deviation σ is known. To make such an inference, begin by finding a point estimate.

DEFINITION

A **point estimate** is a single value estimate for a population parameter. The most unbiased point estimate of the population mean μ is the sample mean $\bar{x}$.

Insight

A statistic is unbiased if it does not overestimate or underestimate the parameter. Because the mean of all possible sample means of the same size equals the population mean, $\bar{x}$ is an unbiased estimator of μ.

EXAMPLE 1

Finding a Point Estimate

Market researchers use the number of sentences per advertisement as a measure of readability for magazine advertisements. The following represents a random sample of the number of sentences found in 54 advertisements. Find a point estimate of the population mean μ. *(Source: Journal of Advertising Research)*

9	20	18	16	9	16	16	9	11	13	22	16	5	18	6	6	5	12
25	17	23	7	10	9	10	10	5	11	18	18	9	9	17	13	11	7
14	6	11	12	11	15	6	12	14	11	4	9	18	12	12	17	11	20

SOLUTION The sample mean of the data is

$$\bar{x} = \frac{\Sigma x}{n}$$

$$= \frac{671}{54}$$

$$\approx 12.4.$$

So, your point estimate for the mean length of all magazine advertisements is 12.4 sentences.

Try It Yourself 1

Another random sample of the number of sentences found in 30 magazine advertisements is listed at the left. Use this sample to find another point estimate for μ.

a. *Find* the sample mean.
b. *Estimate* the mean sentence length of the population.

Answer: Page A39

Sample Data

Number of Sentences					
16	9	14	11	17	12
99	18	13	12	5	9
17	6	11	17	18	20
6	14	7	11	12	12
5	11	18	6	4	13

In Example 1, the probability that the population mean is exactly 12.4 is virtually zero. So, instead of estimating μ to be exactly 12.4 using a *point* estimate, you can estimate that μ lies in an *interval*. This is called *making an interval estimate.*

DEFINITION

An **interval estimate** is an interval, or range of values, used to estimate a population parameter.

Although you can assume that the point estimate in Example 1 is not equal to the actual population mean, it is probably close to it. To form an interval estimate, use the point estimate as the center of the interval, then add and subtract a margin of error. For instance, if the margin of error is 2.1, then an interval estimate would be given by 12.4 ± 2.1 or $10.3 < \mu < 14.5$. The point estimate and interval estimate are as follows.

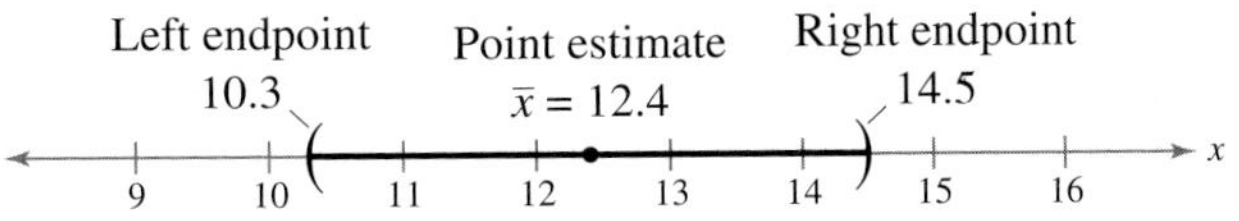

Interval estimate

Study Tip

In this course, you will usually use 90%, 95%, and 99% levels of confidence. The following z-scores correspond to these levels of confidence.

Level of Confidence	z_c
90%	1.645
95%	1.96
99%	2.575

Before finding an interval estimate, you should first determine how confident you need to be that your interval estimate contains the population mean μ.

DEFINITION

The **level of confidence** ***c*** is the probability that the interval estimate contains the population parameter.

You know from the Central Limit Theorem that when $n \geq 30$, the sampling distribution of sample means is a normal distribution. The level of confidence c is the area under the standard normal curve between the **critical values,** $-z_c$ and z_c. You can see from the graph that c is the percent of the area under the normal curve between $-z_c$ and z_c. The area remaining is $1 - c$, so the area in each tail is $\frac{1}{2}(1 - c)$. For instance, if $c = 90\%$, then 5% of the area lies to the left of $-z_c = -1.645$ and 5% lies to the right of $z_c = 1.645$.

Note to Instructor

Work through the details of finding z_c with different values of c. Point out to students that there is a natural variation from one sample to another, but c is the percent of these sample means that will be between $-z_c$ and z_c.

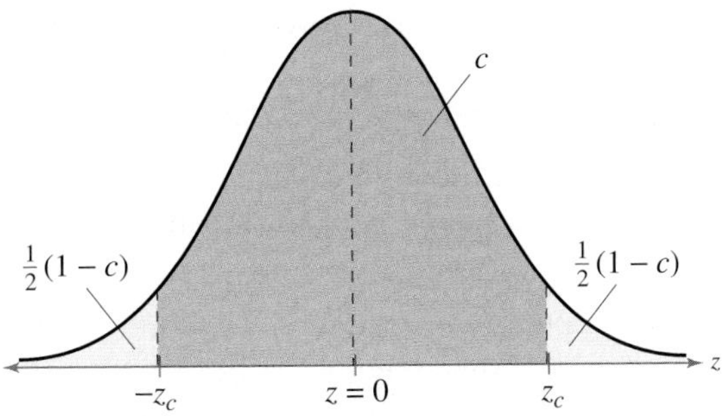

The difference between the point estimate and the actual parameter value is called the **sampling error.** When μ is estimated, the sampling error is the difference $\bar{x} - \mu$. In most cases, of course, μ is unknown, and $\bar{x}$ varies from sample to sample. However, you can calculate a maximum value for the error if you know the level of confidence and the sampling distribution.

DEFINITION

Given a level of confidence c, the **margin of error** (sometimes also called the maximum error of estimate or error tolerance) $\boldsymbol{E}$ is the greatest possible distance between the point estimate and the value of the parameter it is estimating.

$$E = z_c \sigma_{\bar{x}} = z_c \frac{\sigma}{\sqrt{n}}$$

When $n \geq 30$, the sample standard deviation s can be used in place of σ.

EXAMPLE 2

Finding the Margin of Error

Use the data in Example 1 and a 95% confidence level to find the margin of error for the mean number of sentences in all magazine advertisements.

SOLUTION The z-score that corresponds to a 95% confidence level is 1.96. This implies that 95% of the area under the standard normal curve falls within 1.96 standard deviations of the mean. (You can approximate the distribution of the sample means with a normal curve by the Central Limit Theorem, because $n = 54 \geq 30$.) You don't know the population standard deviation σ. But because $n \geq 30$, you can use s in place of σ.

$$s = \sqrt{\frac{\Sigma(x - \bar{x})^2}{n - 1}} = \sqrt{\frac{1333.24}{53}} \approx 5.0$$

Using the values $z_c = 1.96$, $\sigma \approx s \approx 5.0$, and $n = 54$,

$$E = z_c \frac{\sigma}{\sqrt{n}} \approx 1.96 \cdot \frac{5.0}{\sqrt{54}} \approx 1.3.$$

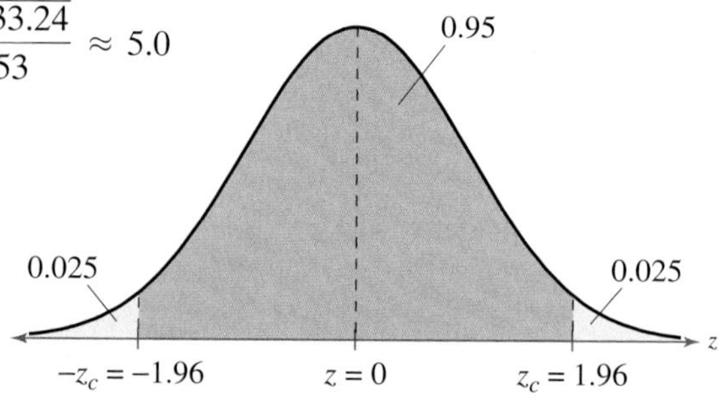

Interpretation You are 95% confident that the margin of error for the population mean is about 1.3 sentences.

Try It Yourself 2

Use the data given in Try It Yourself 1 and a 95% confidence level to find the margin of error for the mean number of sentences in a magazine advertisement.

a. *Identify* z_c, n, and s.
b. *Find* E using z_c, $\sigma \approx s$, and n.
c. *State* the margin of error.

Answer: Page A39

Picturing the World

Many investors choose mutual funds as a way to invest in the stock market. The mean annual rate of return for mutual funds in a recent year was estimated by taking a random sample of 32 mutual funds. The mean annual rate of return for the sample was 37.18%, with a standard deviation of 12.67%. *(Source: CBS Marketwatch)*

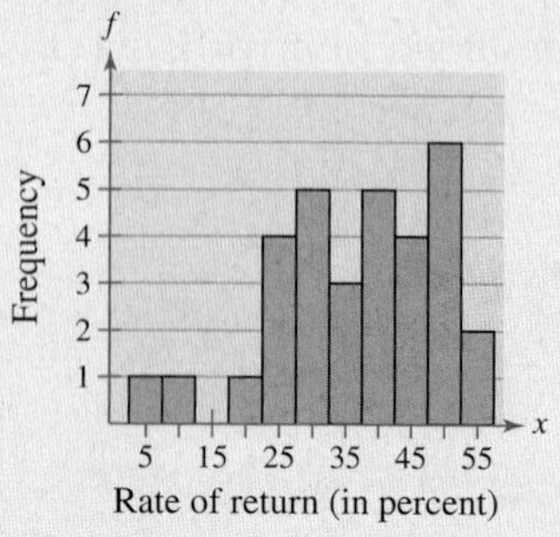

For a 95% confidence interval, what would be the margin of error for the population mean rate of return?

Study Tip

Remember that you can calculate the sample standard deviation s using the formula

$$s = \sqrt{\frac{\Sigma(x - \bar{x})^2}{n - 1}}$$

or the shortcut formula

$$s = \sqrt{\frac{\Sigma x^2 - (\Sigma x)^2/n}{n - 1}}.$$

Confidence Intervals for the Population Mean

Using a point estimate and a margin of error, you can construct an interval estimate of a population parameter such as μ. This interval estimate is called a confidence interval.

> **Study Tip**
>
> When you compute a confidence interval for a population mean, the general *round-off rule* is to round off to the same number of decimal places given for the sample mean.

DEFINITION

A ***c*-confidence interval** for the population mean μ is

$$\bar{x} - E < \mu < \bar{x} + E.$$

The probability that the confidence interval contains μ is c.

GUIDELINES

Finding a Confidence Interval for a Population Mean ($n \geq 30$ or σ known with a normally distributed population)

In Words	*In Symbols*
1. Find the sample statistics n and $\bar{x}$.	$\bar{x} = \dfrac{\Sigma x}{n}$
2. Specify σ, if known. Otherwise, if $n \geq 30$, find the sample standard deviation s and use it as an estimate for σ.	$s = \sqrt{\dfrac{\Sigma(x - \bar{x})^2}{n - 1}}$
3. Find the critical value z_c that corresponds to the given level of confidence.	Use the Standard Normal Table.
4. Find the margin of error E.	$E = z_c \dfrac{\sigma}{\sqrt{n}}$
5. Find the left and right endpoints and form the confidence interval.	Left endpoint: $\bar{x} - E$ Right endpoint: $\bar{x} + E$ Interval: $\bar{x} - E < \mu < \bar{x} + E$

> **Study Tip**
>
> The left and right endpoints of a confidence interval are $\bar{x} - E$ and $\bar{x} + E$, respectively. Other ways to represent a confidence interval are $(\bar{x} - E, \bar{x} + E)$ and $\bar{x} \pm E$. For instance, in Example 3, you could write the confidence interval as $(11.1, 13.7)$ or 12.4 ± 1.3.

EXAMPLE 3

Constructing a Confidence Interval

See MINITAB steps on page 328.

Construct a 95% confidence interval for the mean number of sentences in all magazine advertisements.

SOLUTION In Examples 1 and 2, you found that $\bar{x} = 12.4$ and $E = 1.3$. The confidence interval is as follows.

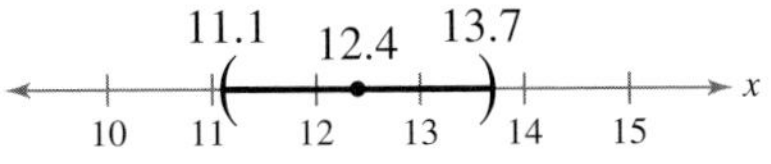

Left Endpoint	Right Endpoint
$\bar{x} - E = 12.4 - 1.3 = 11.1$	$\bar{x} + E = 12.4 + 1.3 = 13.7$

$$11.1 < \mu < 13.7$$

Interpretation With 95% confidence, you can say that the population mean number of sentences is between 11.1 and 13.7.

Try It Yourself 3

Use the data given in Try It Yourself 1 to construct a 95% confidence interval for the mean number of sentences in all magazine advertisements. Compare your result with the interval found in Example 3.

a. *Find* $\bar{x}$ and E.
b. *Find* the left and right endpoints of the confidence interval.
c. *State* the 95% confidence interval and compare it with Example 3.

Answer: Page A39

Insight

The width of a confidence interval is $2E$. Examine the formula for E to see why a larger sample size tends to give you a narrower confidence interval for the same level of confidence.

EXAMPLE 4

Constructing a Confidence Interval Using Technology

Use a computer or graphing calculator to construct a 99% confidence interval for the mean number of sentences in all magazine advertisements, using the sample in Example 1.

SOLUTION To use a technology tool to solve the problem, enter the data and find that the sample standard deviation is $s \approx 5.0$. Then, use the confidence interval command to calculate the confidence interval (*ZInterval* for TI-83, *1-Sample Z* for MINITAB). The displays should look like the ones shown here.

MINITAB

Z Confidence Intervals

The assumed sigma = 5

Variable	N	Mean	StDev	SE Mean	99.0 % CI
C1	54	12.426	5.015	0.680	(10.673, 14.179)

TI-83

ZInterval
(10.673, 14.179)
x̄=12.42592593
Sx=5.015454801
n=54

So, a 99% confidence interval for μ is $(10.7, 14.2)$.

Interpretation With 99% confidence, you can say that the population mean number of sentences is between 10.7 and 14.2.

Try It Yourself 4

Use the sample data in Example 1 and a computer or graphing calculator to construct 75% and 85% confidence intervals for the mean number of sentences in all magazine advertisements. How does the width of the confidence interval change as the level of confidence increases?

a. *Enter* the data.
b. *Use* the appropriate command to construct each confidence interval.
c. *Compare* the widths of the confidence intervals for $c = 0.75, 0.85$, and 0.99.

Answer: Page A39

In Example 4 and Try It Yourself 4, the same sample data were used to construct confidence intervals with different levels of confidence. Notice that as the level of confidence increases, the width of the confidence interval also increases. In other words, when the same sample data are used, *the greater the level of confidence, the wider the interval.*

If the population is normally distributed and the population standard deviation σ is known, you may use the normal sampling distribution for any sample size, as shown in Example 5.

EXAMPLE 5

Constructing a Confidence Interval, σ Known

See TI-83 steps on page 329.

A college admissions director wishes to estimate the mean age of all students currently enrolled. In a random sample of 20 students, the mean age is found to be 22.9 years. From past studies, the standard deviation is known to be 1.5 years, and the population is normally distributed. Construct a 90% confidence interval of the population mean age.

SOLUTION Using $n = 20$, $\bar{x} = 22.9$, $\sigma = 1.5$, and $z_c = 1.645$, the margin of error at the 90% confidence interval is

$$E = z_c \frac{\sigma}{\sqrt{n}} = 1.645 \cdot \frac{1.5}{\sqrt{20}} \approx 0.6.$$

The 90% confidence interval can be written as $\bar{x} \pm E = 22.9 \pm 0.6$ or as follows.

Left Endpoint	Right Endpoint
$\bar{x} - E = 22.9 - 0.6 = 22.3$	$\bar{x} + E = 22.9 + 0.6 = 23.5$

$$22.3 < \mu < 23.5$$

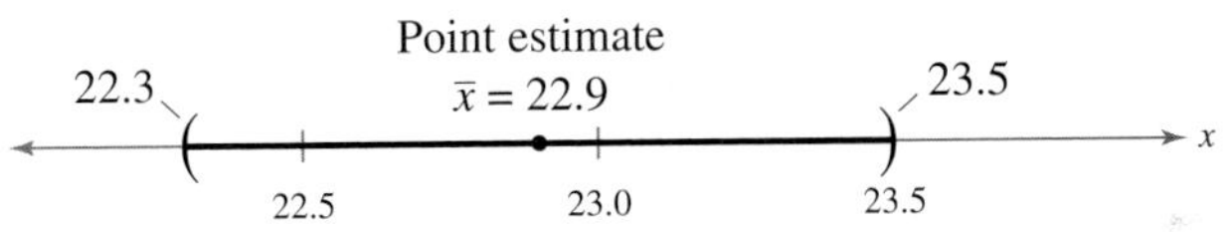

Interpretation With 90% confidence, you can say that the mean age of all the students is between 22.3 and 23.5 years.

Try It Yourself 5

Construct a 90% confidence interval of the population mean age for the college students in Example 5 if the sample size is increased to 30 students. Compare your answer with Example 5.

a. *Identify* n, $\bar{x}$, σ, and z_c, and *find* E.
b. *Find* the left and right endpoints of the confidence interval.
c. *Specify* the 90% confidence interval and compare your answer with Example 5.

Answer: Page A39

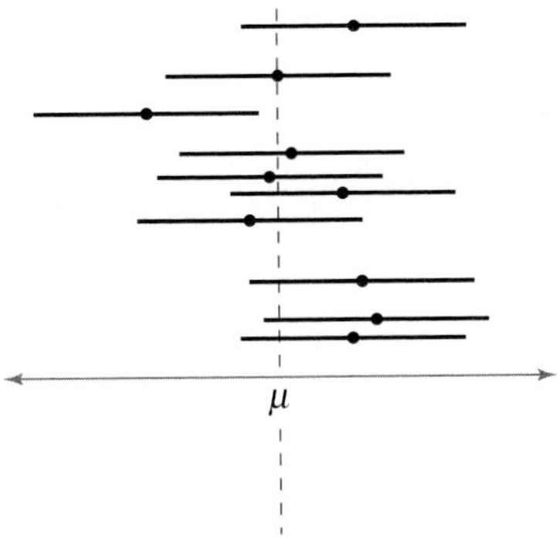

The horizontal segments represent 90% confidence intervals for different samples of the same size. In the long run, 9 of every 10 such intervals will contain μ.

After constructing a confidence interval, it is important that you interpret the results correctly. Consider the 90% confidence interval constructed in Example 5. Because μ already exists, it is either in the interval or not. It is *not* correct to say "There is a 90% probability that the actual mean is in the interval (22.3, 23.5)." The correct way to interpret your confidence interval is "There is a 90% probability that the confidence interval you described contains μ." This also means, of course, that there is a 10% probability that your confidence interval will not contain μ.

Sample Size

For the same sample statistics, as the level of confidence increases, the confidence interval widens. As the confidence interval widens, the precision of the estimate decreases. One way to improve the precision of an estimate without decreasing the level of confidence is to increase the sample size. But how large a sample size is needed to guarantee a certain level of confidence for a given margin of error?

Insight

Using the formula for the margin of error E, you can derive n as shown below.

$$E = z_c \frac{\sigma}{\sqrt{n}}$$

$$E\sqrt{n} = z_c\sigma$$

$$\sqrt{n} = \frac{z_c\sigma}{E}$$

$$n = \left(\frac{z_c\sigma}{E}\right)^2$$

Find a Minimum Sample Size to Estimate μ

Given a c-confidence level and a margin of error E, the minimum sample size n needed to estimate μ the population mean is

$$n = \left(\frac{z_c\sigma}{E}\right)^2.$$

If σ is unknown, you can estimate it using s, provided you have a preliminary sample with at least 30 members.

EXAMPLE 6

Determining a Minimum Sample Size

You want to estimate the mean number of sentences in a magazine advertisement. How many magazine advertisements must be included in the sample if you want to be 95% confident that the sample mean is within one sentence of the population mean?

SOLUTION Using $c = 0.95$, $z_c = 1.96$, $\sigma \approx s \approx 5.0$ (from Example 2), and $E = 1$, you can solve for the minimum sample size n.

$$n = \left(\frac{z_c\sigma}{E}\right)^2$$

$$\approx \left(\frac{1.96 \cdot 5.0}{1}\right)^2$$

$$= 96.04$$

When necessary, round up to obtain a whole number. So, you should include at least 97 magazine advertisements in your sample. (You already have 54, so you need 43 more.) Note that 97 is the *minimum* number of magazines advertisements to include in the sample. You could include more, if desired.

Try It Yourself 6

How many magazine advertisements must be included in the sample if you want to be 95% confident that the sample mean is within two sentences of the population mean? Compare your answer with Example 6.

a. *Identify* z_c, E, and s.
b. *Use* z_c, E, and $\sigma \approx s$ to find the minimum sample size n.
c. *State* how many magazine advertisements must be included in the sample and compare your answer with Example 6.

Answer: Page A39

Note to Instructor

Students often have trouble identifying the value of E. Remind them that E represents the now given error tolerance (or margin of error). Also point out that the precision is not related to the size of the population—only to the size of the sample.

6.1 Exercises

Building Basic Skills and Vocabulary

1. When estimating a population mean, are you more likely to be correct if you use a point estimate or an interval estimate? Explain your reasoning.

2. Which statistic is the best unbiased estimator for μ?

(a) s (b) $\bar{x}$ (c) the median (d) the mode

3. Given the same sample statistics, which level of confidence would produce the widest confidence interval? Explain your reasoning.

(a) 90% (b) 95% (c) 98% (d) 99%

4. What is the effect on the width of the confidence interval when the sample size is increased? Explain your reasoning.

(a) The width increases (b) The width decreases (c) No effect

In Exercises 5–8, find the critical value z_c necessary to form a confidence interval at the given level of confidence.

5. $c = 0.80$ **6.** $c = 0.85$ **7.** $c = 0.75$ **8.** $c = 0.97$

Graphical Analysis In Exercises 9–14, use the values on the number line to find the sampling error.

9.
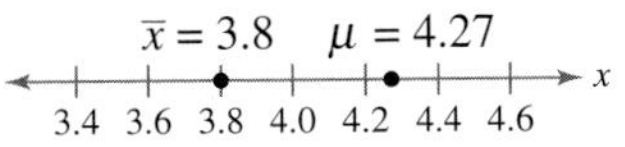

10.
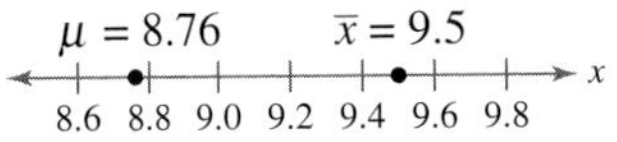

11.
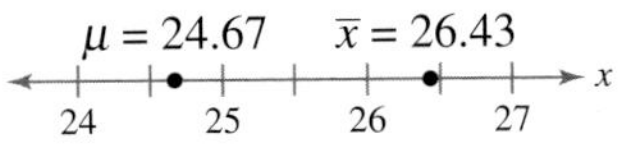

12.
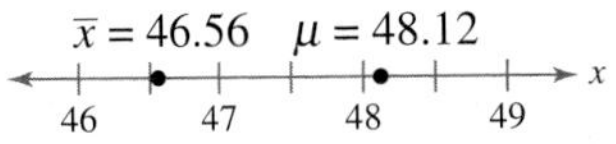

13.
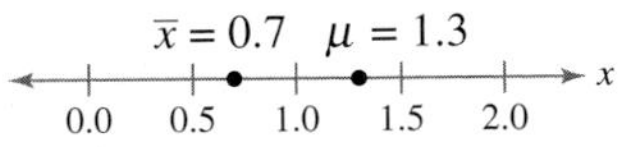

14.
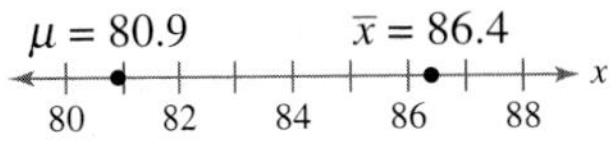

In Exercises 15–18, find the margin of error for the given values of c, s, and n.

15. $c = 0.90, s = 2.5, n = 36$

16. $c = 0.95, s = 3.0, n = 60$

17. $c = 0.65, s = 1.5, n = 50$

18. $c = 0.975, s = 3.4, n = 100$

Matching In Exercises 19–22, match the level of confidence c with its representation on the number line, given $\bar{x} = 57.2$, $s = 7.1$, and $n = 50$.

19. $c = 0.88$ **20.** $c = 0.90$ **21.** $c = 0.95$ **22.** $c = 0.98$

(a)
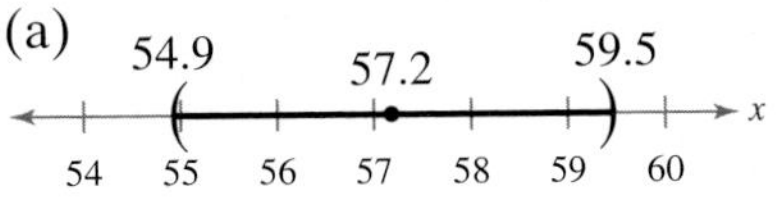

(b)
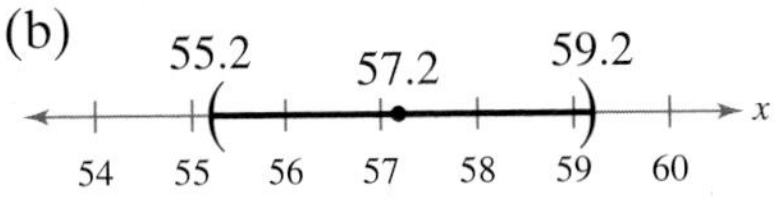

(c)
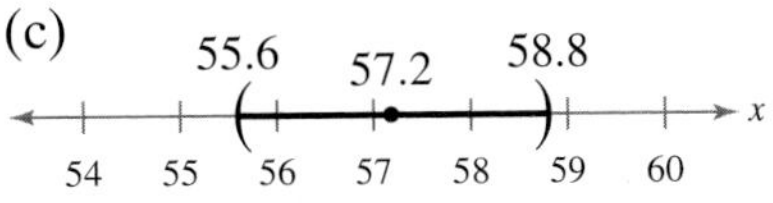

(d)
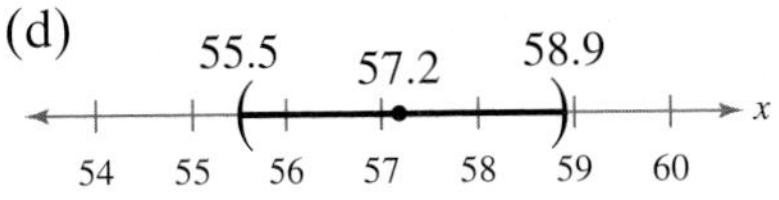

Help

MyMathLab

Student Study Pack

1. You are more likely to be correct using an interval estimate because it is unlikely that a point estimate will equal the population mean exactly.
2. b
3. d; As the level of confidence increases, z_c increases therefore causing wider intervals.
4. b; As n increases, E decreases because $\sqrt{n}$ is in the denominator of the formula for E. So, the intervals become narrower.
5. 1.28
6. 1.44
7. 1.15
8. 2.17
9. 0.47
10. 0.74
11. 1.76
12. 1.56
13. 0.60
14. 5.5
15. 0.685
16. 0.759
17. 0.197
18. 0.762
19. c
20. d
21. b
22. a

23. (14.8, 15.6)
24. (31.22, 31.56)
25. (4.18, 4.36)
26. (13.1, 13.9)
27. 126
28. 25
29. 7
30. 139
31. 0.7, 2.8
32. 18.45, 62.52
33. (614.41, 647.39); (611.25, 650.55) The 95% CI is wider.
34. (18.66, 19.96); (18.54, 20.08) The 95% CI is wider.
35. (25.7, 27.9); (25.5, 28.1) The 95% CI is wider.
36. (21, 25); (21, 25) The 95% CI is wider.
37. (94.58, 105.42)

In Exercises 23–26, construct the indicated confidence interval for the population mean μ.

23. $c = 0.90, \bar{x} = 15.2, s = 2.0, n = 60$

24. $c = 0.95, \bar{x} = 31.39, s = 0.8, n = 82$

25. $c = 0.95, \bar{x} = 4.27, s = 0.3, n = 42$

26. $c = 0.99, \bar{x} = 13.5, s = 1.5, n = 100$

In Exercises 27–30, find the minimum sample size n needed to estimate μ for the given values of c, s, and E.

27. $c = 0.90, s = 6.8, E = 1$

28. $c = 0.95, s = 2.5, E = 1$

29. $c = 0.80, s = 4.1, E = 2$

30. $c = 0.98, s = 10.1, E = 2$

Using and Interpreting Concepts

Finding the Margin of Error In Exercises 31 and 32, use the given confidence interval to find the estimated margin of error. Then find the sample mean.

31. *Seedlings* A biologist reports a confidence interval of (2.1, 3.5) when estimating the mean height (in centimeters) of a sample of seedlings.

32. *Book Prices* A store manager reports a confidence interval of (44.07, 80.97) when estimating the mean price (in dollars) of a sample of textbooks.

Constructing Confidence Intervals In Exercises 33–36, you are given the sample mean and the sample standard deviation. Use this information to construct the 90% and 95% confidence intervals for the population mean. Which interval is wider?

33. *Gas Grills* A random sample of 32 gas grills has a mean price of \$630.90 and a standard deviation of \$56.70.

34. *Stock Prices* From a random sample of 36 days in a recent year, the closing stock prices for Hasbro had a mean of \$19.31 and a standard deviation of \$2.37. *(Adapted from Interactive Data Corp.)*

35. *Wheat Yields* A random sample of 156 fields of durum wheat has a mean yield of 26.8 bushels per acre and standard deviation of 8.0 bushels per acre. *(Source: Wheat Quality Council)*

36. *Sodium Chloride Concentration* In 36 randomly selected seawater samples, the mean sodium chloride concentration was 23 cubic centimeters per cubic meter and the standard deviation was 6.7 cubic centimeters per cubic meter. *(Adapted from Dorling Kindersley Visual Encyclopedia)*

37. *Repair Costs: Washing Machines* You work for a consumer advocate agency and want to find the mean repair cost of a washing machine. As part of your study, you randomly select 40 repair costs and find the mean to be \$100.00. The sample standard deviation is \$17.50. Construct a 95% confidence interval for the population mean repair cost. *(Adapted from Consumer Reports)*

38. (70, 80)

39. (96.17, 103.84)

$n = 40$ CI is wider because a smaller sample is taken, giving less information about the population.

40. (72, 78)

$n = 50$ CI is wider because a smaller sample is taken giving less information about the population.

41. (9.719, 11.185)

42. (0.18, 0.22)

43. (8.687, 12.217)

$s = 5.130$ CI is wider because of the increased variability within the population.

44. (0.17, 0.23)

$s = 0.1$ CI is wider because of the increased variability within the population.

45. (a) An increase in the level of confidence will widen the confidence interval.

(b) An increase in the sample size will narrow the confidence interval.

(c) An increase in the standard deviation will widen the confidence interval.

46. Answers will vary.

47. (8.4, 9.7); (8.1, 10.1)

The 99% CI is wider.

48. (1.5, 1.9); (1.4, 2.0)

The 99% CI is wider.

38. ***Repair Costs: VCRs*** In a random sample of 50 VCRs, the mean repair cost was \$75 and the standard deviation was \$12.50. Construct a 99% confidence interval for the population mean repair cost. *(Adapted from Consumer Reports)*

39. Repeat Exercise 37, changing the sample size to $n = 80$. Which confidence interval is wider? Explain.

40. Repeat Exercise 38, changing the sample size to $n = 90$. Which confidence interval is wider? Explain.

41. ***Heights of Beech Trees*** A random sample of 56 American beech trees has a mean height of 10.452 meters and a standard deviation of 2.130 meters. Construct a 99% confidence interval for the population mean height.

42. ***Diameters of Beech Trees*** In a random sample of 56 American beech trees, the mean diameter was 0.20 meter and the standard deviation was 0.06 meter. Construct a 95% confidence interval for the population mean diameter.

43. Repeat Exercise 41, using a standard deviation of $s = 5.130$ meters. Which confidence interval is wider? Explain.

44. Repeat Exercise 42, using a standard deviation of $s = 0.1$ meter. Which confidence interval is wider? Explain.

45. If all other quantities remain the same, how does the indicated change affect the width of a confidence interval?

(a) Increase in the level of confidence

(b) Increase in the sample size

(c) Increase in the standard deviation

46. Describe how you would form a 90% confidence interval to estimate the population mean age for students at your school.

Constructing Confidence Intervals In Exercises 47 and 48, you are given the sample mean and the population standard deviation. Use this information to construct the 90% and 99% confidence intervals for the population mean. Which interval is wider?

47. ***Newspaper Reading Times*** A publisher wants to estimate the mean length of time (in minutes) all adults spend reading newspapers. To determine this estimate, the publisher takes a random sample of 15 people and obtains the following results.

11, 9, 8, 10, 10, 9, 7, 11, 11, 7, 6, 9, 10, 8, 10

From past studies, the publisher assumes σ is 1.5 minutes and that the population of times is normally distributed.

48. ***Computer Usage*** A computer company wants to estimate the mean number of hours per day all adults use computers at home. In a random sample of 21 adults, the mean length of time a computer was used at home was 1.7 hours. From past studies, the company assumes σ is 0.6 hour and that the population of times is normally distributed. *(Adapted from National Association for Sport & Physical Education)*

49. 89

50. 4

51. (a) 121

(b) 208

99% CI requires larger sample because more information is needed from the population to be 99% confident.

52. (a) 4

(b) 10

99% CI requires larger sample because more information is needed from the population to be 99% confident.

53. (a) 32

(b) 87

$E = 0.15$ requires a larger sample size. As the error size decreases, a larger sample must be taken to obtain enough information from the population to ensure desired accuracy.

49. Determine the minimum required sample size if you want to be 95% confident that the sample mean is within one unit of the population mean given $\sigma = 4.8$. Assume the population is normally distributed.

50. Determine the minimum required sample size if you want to be 99% confident that the sample mean is within two units of the population mean given $\sigma = 1.4$. Assume the population is normally distributed.

51. ***Cholesterol Contents of Cheese*** A cheese processing company wants to estimate the mean cholesterol content of all one-ounce servings of cheese. The estimate must be within 0.5 milligram of the population mean.

(a) Determine the minimum required sample size to construct a 95% confidence interval for the population mean. Assume the population standard deviation is 2.8 milligrams.

(b) Repeat part (a) using a 99% confidence interval.

(c) Which level of confidence requires a larger sample size? Explain.

52. ***Ages of College Students*** An admissions director wants to estimate the mean age of all students enrolled at a college. The estimate must be within 1 year of the population mean. Assume the population of ages is normally distributed.

(a) Determine the minimum required sample size to construct a 90% confidence interval for the population mean. Assume the population standard deviation is 1.2 years.

(b) Repeat part (a) using a 99% confidence interval.

(c) Which level of confidence requires a larger sample size? Explain.

53. ***Paint Can Volumes*** A paint manufacturer uses a machine to fill gallon cans with paint (see figure).

Error tolerance = 0.25 oz

(a) The manufacturer wants to estimate the mean volume of paint the machine is putting in the cans within 0.25 ounce. Determine the minimum sample size required to construct a 90% confidence interval for the population mean. Assume the population standard deviation is 0.85 ounce.

(b) Repeat part (a) using an error tolerance of 0.15 ounce. Which error tolerance requires a larger sample size? Explain.

54. (a) 35

(b) 9

$E = 1$ requires a larger sample size. As the error size decreases, a larger sample must be taken to obtain enough information from the population to ensure desired accuracy.

55. (a) 16

(b) 62

$E = 0.0625$ requires a larger sample size. As the error size decreases, a larger sample must be taken to obtain enough information from the population to ensure desired accuracy.

56. (a) 44

(b) 174

$E = 0.03125$ requires a larger sample size. As the error size decreases, a larger sample must be taken to obtain enough information from the population to ensure desired accuracy.

57. (a) 42

(b) 60

$\sigma = 0.30$ requires a larger sample size. Owing to the increased variability in the population, a larger sample size is needed to ensure the desired accuracy.

54. *Water Dispensing Machine* A beverage company uses a machine to fill one-liter bottles with water (see figure). Assume that the population of volumes is normally distributed.

Error tolerance = 1 mL

(a) The company wants to estimate the mean volume of water the machine is putting in the bottles within 1 milliliter. Determine the minimum sample size required to construct a 95% confidence interval for the population mean. Assume the population standard deviation is 3 milliliters.

(b) Repeat part (a) using an error tolerance of 2 milliliters. Which error tolerance requires a larger sample size? Explain.

55. *Plastic Sheet Cutting* A machine cuts plastic into sheets that are 50 feet (600 inches) long. Assume that the population of lengths is normally distributed.

(a) The company wants to estimate the mean length the machine is cutting the plastic within 0.125 inch. Determine the minimum sample size required to construct a 95% confidence interval for the population mean. Assume the population standard deviation is 0.25 inch.

(b) Repeat part (a) using an error tolerance of 0.0625 inch. Which error tolerance requires a larger sample size? Explain.

56. *Paint Sprayer* A company uses an automated sprayer to apply paint to metal. The company sets the sprayer to apply the paint one-inch thick.

(a) The company wants to estimate the mean thickness the sprayer is applying the paint within 0.0625 inch. Determine the minimum sample size required to construct a 90% confidence interval for the population mean. Assume the population standard deviation is 0.25 inch.

(b) Repeat part (a) using an error tolerance of 0.03125 inch. Which error tolerance requires a larger sample size? Explain.

57. *Soccer Balls* A soccer ball manufacturer wants to estimate the mean circumference of soccer balls within 0.1 inch.

(a) Determine the minimum sample size required to construct a 99% confidence interval for the population mean. Assume the population standard deviation is 0.25 inch.

(b) Repeat part (a) using a standard deviation of 0.3 inch. Which standard deviation requires a larger sample size? Explain.

58. (a) 12

(b) 3

$\sigma = 0.20$ requires a larger sample size. Owing to the decreased variability in the population, a smaller sample size is needed to ensure the desired accuracy.

59. (a) An increase in the level of confidence will increase the minimum sample size required.

(b) An increase (larger E) in the error tolerance will decrease the minimum sample size required.

(c) An increase in the population standard deviation will increase the minimum sample size required.

60. A 99% CI may not be practical to use in all situations. It may produce a CI so wide that it has no practical application.

61. (303.5, 311.3)

62. (99.6, 104.0)

63. (13.680, 15.200)

58. ***Mini-Soccer Balls*** A soccer ball manufacturer wants to estimate the mean circumference of mini-soccer balls within 0.15 inch. Assume that the population of circumferences is normally distributed.

(a) Determine the minimum sample size required to construct a 99% confidence interval for the population mean. Assume the population standard deviation is 0.20 inch.

(b) Repeat part (a) using a standard deviation of 0.10 inch. Which standard deviation requires a larger sample size? Explain.

59. If all other quantities remain the same, how does the indicated change affect the minimum sample size requirement?

(a) Increase in the level of confidence

(b) Increase in the error tolerance

(c) Increase in the standard deviation

60. When estimating the population mean, why not construct a 99% confidence interval every time?

Using Technology In Exercises 61–64, you are given a data sample. Use a computer or graphing calculator to construct a 95% confidence interval for the population mean. Interpret your answer.

DATA

61. ***Airfare*** A random sample of airfare prices (in dollars) for a one-way ticket from New York to Houston *(Adapted from Newsweek)*

Key: 29|2 = 292

28	8
29	0 2 3 3 5 6 8
30	0 2 3 4 5 5 5 5 6 6 7 7
31	3 4 4 6
32	0 0 1 1 2 2 6 7

DATA

62. ***Airfare*** A random sample of airfare prices (in dollars) for a one-way ticket from Atlanta to Chicago *(Adapted from Newsweek)*

Key: 8|7 = 87

8	7 7
9	0 4 4
9	5 6 8 8 8 9
10	0 1 1 1 1 2 3 3 3 4 4 4
10	5 5 5 5 6 7 7 8 9
11	1 4
11	7

DATA

63. ***Annual Precipitation*** A random sample of the annual precipitation (in inches) for Anchorage, Alaska *(Source: Alaska Climate Research Center)*

14.62	14.66	12.76	16.10	13.09
12.87	14.93	19.48	12.49	15.44
13.67	11.49	15.89	15.17	14.63
13.42	11.81	15.46	12.25	14.97
14.20	16.73	17.64	14.75	14.51
16.89	17.06	13.06	8.61	14.54

64. (16.664, 19.611)

65. (a) 0.707
(b) 0.949
(c) 0.962
(d) 0.975
(e) The finite population correction factor approaches 1 as the sample size decreases while the population size remains the same.

66. (a) 0.711
(b) 0.937
(c) 0.964
(d) 0.988
(e) The finite population correction factor approaches 1 as the population size increases while the sample size remains the same.

67. $E = \dfrac{z_c \sigma}{\sqrt{n}} \Rightarrow E\sqrt{n} = z_c \sigma$

$\Rightarrow \sqrt{n} = \dfrac{z_c \sigma}{E}$

$\Rightarrow n = \left(\dfrac{z_c \sigma}{E}\right)^2$

DATA

64. ***Annual Precipitation*** A random sample of the annual precipitation (in inches) for Nome, Alaska *(Source: Alaska Climate Research Center)*

16.69	11.17	17.47	18.31	13.41	19.04	29.49	19.27	25.61
20.66	19.91	15.24	19.78	22.15	24.25	15.54	21.66	20.80
14.30	13.56	18.74	7.42	14.41	20.09	12.59	14.97	18.06
22.06	16.96	19.25	19.18	20.22	16.27			

Extending Concepts

Finite Population Correction Factor In Exercises 65 and 66, use the following information.

In this section, you studied the formation of a confidence interval to estimate a population mean when the population is large or infinite. When a population is finite, the formula that determines the standard error of the mean $\sigma_{\bar{x}}$ needs to be adjusted. If N is the size of the population and n is the size of the sample (*where* $n \geq 0.05\,N$), the standard error of the mean is

$$\sigma_{\bar{x}} = \frac{\sigma}{\sqrt{n}}\sqrt{\frac{N-n}{N-1}}.$$

The expression $\sqrt{(N-n)/(N-1)}$ is called the **finite population correction factor.** The margin of error is

$$E = z_c \frac{\sigma}{\sqrt{n}}\sqrt{\frac{N-n}{N-1}}.$$

65. Determine the finite population correction factor for each of the following.

(a) $N = 1000$ and $n = 500$
(b) $N = 1000$ and $n = 100$
(c) $N = 1000$ and $n = 75$
(d) $N = 1000$ and $n = 50$
(e) What happens to the finite population correction factor as the sample size n decreases but the population size N remains the same?

66. Determine the finite population correction factor for each of the following.

(a) $N = 100$ and $n = 50$
(b) $N = 400$ and $n = 50$
(c) $N = 700$ and $n = 50$
(d) $N = 2000$ and $n = 50$
(e) What happens to the finite population correction factor as the population size N increases but the sample size n remains the same?

67. ***Sample Size*** The equation for determining the sample size

$$n = \left(\frac{z_c \sigma}{E}\right)^2$$

can be obtained by solving the equation for the margin of error

$$E = \frac{z_c \sigma}{\sqrt{n}}$$

for n. Show that this is true and justify each step.

Case Study

Shoulder Heights of Appalachian Black Bears

The Appalachian Bear Center (ABC) is a not-for-profit organization located near the Great Smoky Mountains National Park. ABC's programs include the rehabilitation of orphaned and injured black bears, as well as research and education about Appalachian black bears. ABC provides the most natural environment possible for rehabilitating black bears before their release back into the wild.

Katie Settlage performed a study during the summer of 2002 to learn more about the Appalachian black bear population in the Great Smoky Mountains National Park. She and a team of researchers found 68 black bears in the park and took measurements such as paw size, weight, and shoulder height. The stem-and-leaf plots below show the shoulder heights (in centimeters) of the 40 male and 28 female bears from the study.

Shoulder Heights (in cm) of Male Bears

Key: $4|9 = 49$

4	9
5	7
6	8 9
7	1 1 2 2 2 2 2 3 3 3 4 4 5 5 6 6 7 8 8
8	1 2 3 4 4 5 6 7 9
9	0 0 3 6 7 9
10	2
11	4

Shoulder Heights (in cm) of Female Bears

Key: $5|0 = 50$

5	0
6	7 8
7	1 2 3 3 3 3 3 4 4 5 5 5 5 5 5 6 9
8	1 2 2 3 3 4 5
9	3

Exercises

1. Use the sample to find a point estimate for the mean shoulder height of

(a) male bears.

(b) female bears.

2. Find the standard deviation of the sample of shoulder heights for the

(a) male bears.

(b) female bears.

3. Use the sample to construct a 95% confidence interval for the mean shoulder height of

(a) male bears.

(b) female bears.

4. Use the sample to construct a 95% confidence interval for the mean shoulder height of all bears in the study. How do your results differ from those in Exercise 3? Explain.

6.2 Confidence Intervals for the Mean (Small Samples)

What You Should Learn

- How to interpret the t-distribution and use a t-distribution table
- How to construct confidence intervals when $n < 30$, the population is normally distributed, and σ is unknown

The t-Distribution • Confidence Intervals and t-Distributions

The t-Distribution

In many real-life situations, the population standard deviation is unknown. Moreover, because of various constraints such as time and cost, it is often not practical to collect samples of size 30 or more. So, how can you construct a confidence interval for a population mean given such circumstances? If the random variable is normally distributed (or approximately normally distributed), you can use a t-distribution.

DEFINITION

If the distribution of a random variable x is approximately normal, then

$$t = \frac{\bar{x} - \mu}{\frac{s}{\sqrt{n}}}$$

follows a ***t*-distribution.**

Critical values of t are denoted by t_c. Several properties of the t-distribution are as follows.

1. The t-distribution is bell shaped and symmetric about the mean.
2. The t-distribution is a family of curves, each determined by a parameter called the degrees of freedom. The **degrees of freedom** are the number of free choices left after a sample statistic such as $\bar{x}$ is calculated. When you use a t-distribution to estimate a population mean, the degrees of freedom are equal to one less than the sample size.

   $$\text{d.f.} = n - 1 \qquad \text{Degrees of freedom}$$

3. The total area under a t-curve is 1 or 100%.
4. The mean, median, and mode of the t-distribution are equal to zero.
5. As the degrees of freedom increase, the t-distribution approaches the normal distribution. After 30 d.f. the t-distribution is very close to the standard normal z-distribution.

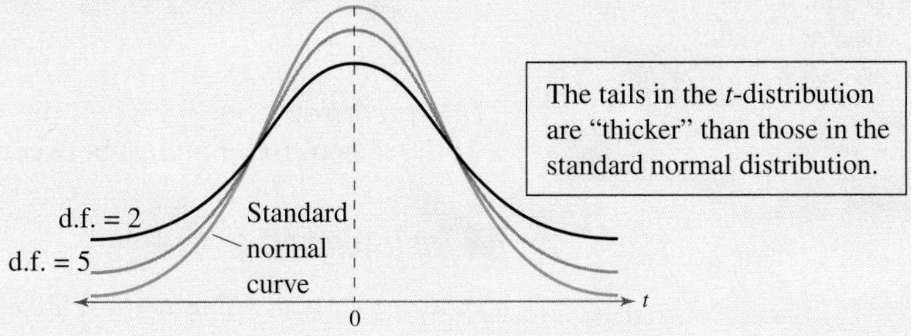

Historical Reference

William S. Gosset (1876–1937) developed the t-distribution while employed by the Guinness Brewing Company in Dublin, Ireland. Gosset published his findings using the pseudonym Student. The t-distribution is sometimes referred to as Student's t-distribution. (See page 29 for others who were important in the history of statistics.)

Table 5 of Appendix B lists critical values of t for selected confidence intervals and degrees of freedom.

EXAMPLE 1

Finding Critical Values of t

Find the critical value t_c for a 95% confidence when the sample size is 15.

SOLUTION Because $n = 15$, the degrees of freedom are

$$\text{d.f.} = n - 1 = 15 - 1 = 14.$$

A portion of Table 5 is shown. Using d.f. $= 14$ and $c = 0.95$, you can find the critical value t_c as shown by the highlighted areas in the table.

	Level of confidence, c	**0.50**	**0.80**	**0.90**	**0.95**	**0.98**
	One tail, α	**0.25**	**0.10**	**0.05**	**0.025**	**0.01**
d.f.	**Two tails, α**	**0.50**	**0.20**	**0.10**	**0.05**	**0.02**
1		1.000	3.078	6.314	12.706	31.821
2		.816	1.886	2.920	4.303	6.965
3		.765	1.638	2.353	3.182	4.541
4		.741	1.533	2.132	2.776	3.747
5		.727	1.476	2.015	2.571	3.365
11		.697	1.363	1.796	2.201	2.718
12		.695	1.356	1.782	2.179	2.681
13		.694	1.350	1.771	2.160	2.650
14		.692	1.345	1.761	2.145	2.624
15		.691	1.341	1.753	2.131	2.602
16		.690	1.337	1.746	2.120	2.583

From the table, you can see that $t_c = 2.145$. The graph shows the t-distribution for 14 degrees of freedom, $c = 0.95$, and $t_c = 2.145$.

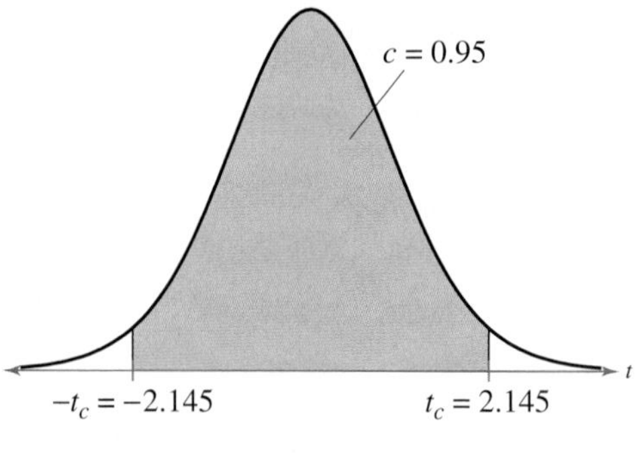

Interpretation So, 95% of the area under the t-distribution curve with 14 degrees of freedom lies between $t = \pm 2.145$.

Try It Yourself 1

Find the critical value t_c for a 90% confidence when the sample size is 22.

a. *Identify* the degrees of freedom.
b. *Identify* the level of confidence c.
c. *Use* Table 5 of Appendix B to find t_c.

Answer: Page A39

Study Tip

Unlike the z-table, critical values for a specific confidence interval can be found in the column headed by c in the appropriate d.f. row. (The symbol α will be explained in Chapter 7.)

Insight

For 30 or more degrees of freedom, the critical values for the t-distribution are close to the corresponding critical values for the normal distribution. Moreover, the values in the last row of the table marked ∞ d.f. correspond *exactly* to the normal distribution values.

Confidence Intervals and t-Distributions

Constructing a confidence interval using the t-distribution is similar to constructing a confidence interval using the normal distribution—both use a point estimate $\bar{x}$ and a margin of error E.

GUIDELINES

Constructing a Confidence Interval for the Mean: t-Distribution

In Words	*In Symbols*
1. Identify the sample statistics n, $\bar{x}$, and s.	$\bar{x} = \dfrac{\Sigma x}{n}$, $s = \sqrt{\dfrac{\Sigma(x - \bar{x})^2}{n - 1}}$
2. Identify the degrees of freedom, the level of confidence c, and the critical value t_c.	d.f. $= n - 1$
3. Find the margin of error E.	$E = t_c \dfrac{s}{\sqrt{n}}$
4. Find the left and right endpoints and form the confidence interval.	Left endpoint: $\bar{x} - E$ Right endpoint: $\bar{x} + E$ Interval: $\bar{x} - E < \mu < \bar{x} + E$

EXAMPLE 2

Constructing a Confidence Interval

See MINITAB steps on page 328.

Note to Instructor

Illustrate the concept of degrees of freedom with the following example. Suppose the number of chairs in a classroom equals the number of students: 25 chairs and 25 students. Each of the first 24 students to enter the classroom has a choice as to which chair he or she will sit in. There is no freedom of choice, however, for the 25th student who enters the room.

You randomly select 16 restaurants and measure the temperature of the coffee sold at each. The sample mean temperature is 162.0°F with a sample standard deviation of 10.0°F. Find the 95% confidence interval for the mean temperature. Assume the temperatures are approximately normally distributed.

SOLUTION Because the sample size is less than 30, σ is unknown, and the temperatures are approximately normally distributed, you can use the t-distribution. Using $n = 16$, $\bar{x} = 162.0$, $s = 10.0$, $c = 0.95$, and d.f. $= 15$, you can use Table 5 to find that $t_c = 2.131$. The margin of error at the 95% confidence interval is

$$E = t_c \frac{s}{\sqrt{n}} = 2.131 \cdot \frac{10}{\sqrt{16}} \approx 5.3.$$

The confidence interval is as follows.

Left Endpoint	Right Endpoint
$\bar{x} - E = 162 - 5.3 = 156.7$	$\bar{x} + E = 162 + 5.3 = 167.3$

$$156.7 < \mu < 167.3$$

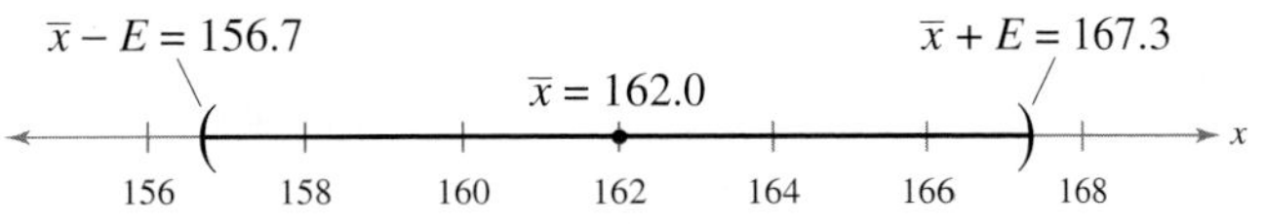

Interpretation With 95% confidence, you can say that the mean temperature of coffee sold is between 156.7°F and 167.3°F.

Try It Yourself 2

Find the 90% and 99% confidence intervals for the mean temperature.

a. *Find* t_c and E for each level of confidence.
b. *Use* $\overline{x}$ and E to find the left and right endpoints.
c. *State* the 90% and 99% confidence intervals for the mean temperature.

Answer: Page A39

EXAMPLE 3

See TI-83 steps on page 329.

Constructing a Confidence Interval

You randomly select 20 mortgage institutions and determine the current mortgage interest rate at each. The sample mean rate is 6.93% with a sample standard deviation of 0.42%. Find the 99% confidence interval for the population mean mortgage interest rate. Assume the interest rates are approximately normally distributed.

SOLUTION Because the sample size is less than 30, σ is unknown, and the interest rates are approximately normally distributed, you can use the t-distribution. Using $n = 20$, $\overline{x} = 6.93$, $s = 0.42$, $c = 0.99$, and d.f. $= 19$, you can use Table 5 to find that $t_c = 2.861$. The margin of error at the 99% confidence interval is

$$E = t_c\frac{s}{\sqrt{n}}$$

$$= 2.861 \cdot \frac{0.42}{\sqrt{20}}$$

$$\approx 0.27.$$

The confidence interval is as follows.

Left Endpoint	Right Endpoint
$\overline{x} - E = 6.93 - 0.27 = 6.66$	$\overline{x} + E = 6.93 + 0.27 = 7.20$

$$6.66 < \mu < 7.20$$

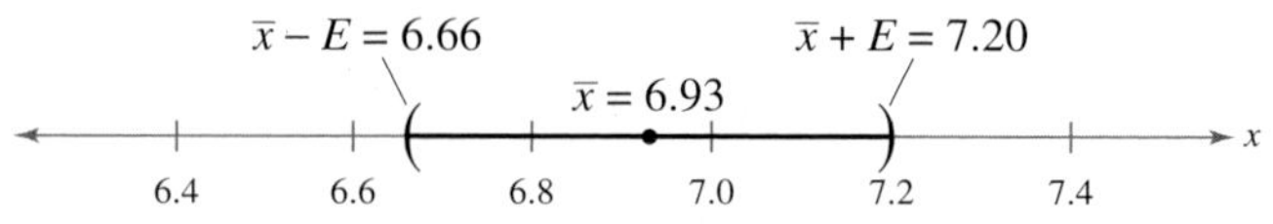

Interpretation With 99% confidence, you can say that the population mean mortgage interest rate is between 6.66% and 7.20%.

Try It Yourself 3

Find the 90% and 95% confidence intervals for the population mean mortgage interest rate. Compare the widths of the intervals.

a. *Find* t_c and E for each level of confidence.
b. *Use* $\overline{x}$ and E to find the left and right endpoints.
c. *State* the 90% and 95% confidence intervals for the population mean mortgage interest rate and compare their widths.

Answer: Page A39

Picturing the World

Two footballs, one filled with air and the other filled with helium, were kicked on a windless day at Ohio State University. The footballs were alternated with each kick. After 10 practice kicks, each football was kicked 29 more times. The distances (in yards) are listed. *(Source: The Columbus Dispatch)*

Air Filled

1	9
2	0 0 2 2 2
2	5 5 5 5 6 6
2	7 7 7 8 8 8 8 8 9 9 9
3	1 1 1 2
3	3 4

Key: 1|9 = 19

Helium Filled

1	1 2
1	4
1	
2	2
2	3 4 6 6 6
2	7 8 8 8 9 9 9 9
3	0 0 0 0 1 1 2 2
3	3 4 5
3	9

Key: 1|1 = 11

Assume that the distances are normally distributed for each football. Apply the flowchart at the right to each sample. Find a 95% confidence interval for the mean distance each football traveled. Do the confidence intervals overlap? What does this result tell you?

The flowchart describes when to use the normal distribution to construct a confidence interval for the population mean and when to use a t-distribution.

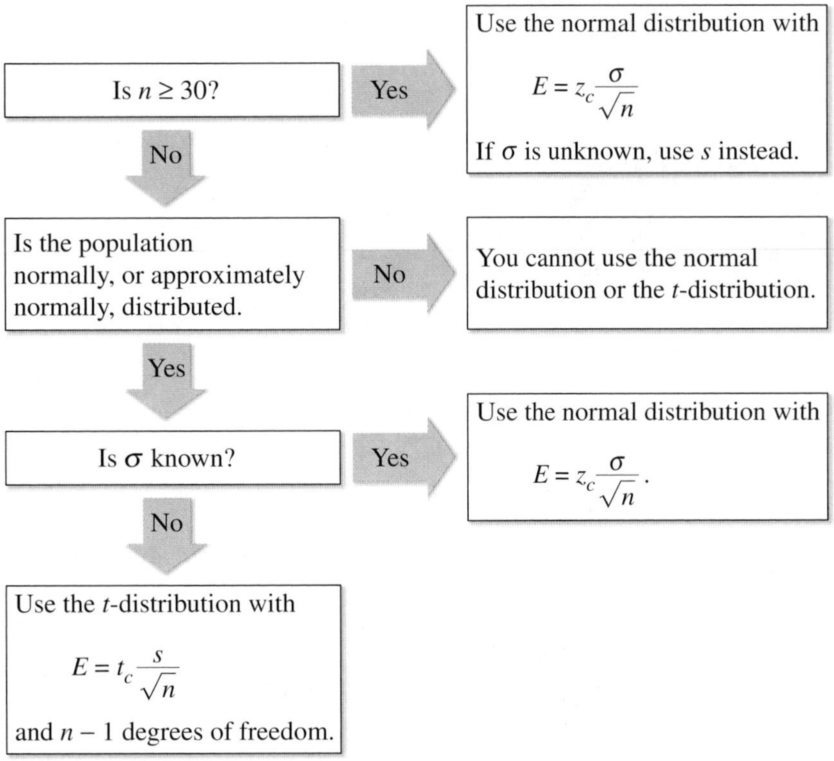

EXAMPLE 4

Choosing the Normal or t-Distribution

You randomly select 25 newly constructed houses. The sample mean construction cost is \$181,000 and the population standard deviation is \$28,000. Assuming construction costs are normally distributed, should you use the normal distribution, the t-distribution, or neither to construct a 95% confidence interval for the population mean construction cost? Explain your reasoning.

SOLUTION Because the population is normally distributed and the population standard deviation is known, you should use the normal distribution.

Try It Yourself 4

You randomly select 18 adult male athletes and measure the resting heart rate of each. The sample mean heart rate is 64 beats per minute with a sample standard deviation of 2.5 beats per minute. Assuming the heart rates are normally distributed, should you use the normal distribution, the t-distribution, or neither to construct a 90% confidence interval for the mean heart rate? Explain your reasoning.

a. *Use* the flowchart to determine which distribution you should use to construct the 90% confidence interval for the mean heart rate.

Answer: Page A39

6.2 Exercises

Building Basic Skills and Vocabulary

In Exercises 1–4, find the critical value t_c for the given confidence level c and sample size n.

1. $c = 0.90,\ n = 10$

2. $c = 0.95,\ n = 12$

3. $c = 0.99,\ n = 16$

4. $c = 0.98,\ n = 20$

In Exercises 5 and 6, you incorrectly used the normal distribution to find the margin of error for the given values of c, s, and n.

(a) Find the value of E using the normal distribution.

(b) Find the correct value using a t-distribution. Compare the results.

5. $c = 0.95,\ s = 5,\ n = 16$

6. $c = 0.99,\ s = 3,\ n = 6$

In Exercises 7–10, construct the indicated confidence interval for the population mean μ using (a) a t-distribution. (b) If you had incorrectly used a normal distribution, which interval would be wider?

7. $c = 0.90,\ \bar{x} = 12.5,\ s = 2.0,\ n = 6$

8. $c = 0.95,\ \bar{x} = 13.4,\ s = 0.85,\ n = 8$

9. $c = 0.98,\ \bar{x} = 4.3,\ s = 0.34,\ n = 14$

10. $c = 0.99,\ \bar{x} = 14,\ s = 2.0,\ n = 10$

Using and Interpreting Concepts

Constructing Confidence Intervals In Exercises 11 and 12, you are given the sample mean and the sample standard deviation. Assume the variable is normally distributed and use a t-distribution to construct a 95% confidence interval for the population mean μ. What is the margin of error of μ?

11. ***Repair Costs: Microwaves*** In a random sample of five microwave ovens, the mean repair cost was \$75.00 and the standard deviation was \$12.50. *(Adapted from Consumer Reports)*

12. ***Repair Costs: Computers*** In a random sample of seven computers, the mean repair cost was \$100.00 and the standard deviation was \$42.50. *(Adapted from Consumer Reports)*

13. You did some research on repair costs of microwave ovens and found that the standard deviation is $\sigma = \$15$. Repeat Exercise 11, using a normal distribution with the appropriate calculations for a standard deviation that is known. Compare the results.

14. You did some research on repair costs of computers and found that the standard deviation is $\sigma = \$50$. Repeat Exercise 12, using a normal distribution with the appropriate calculations for a standard deviation that is known. Compare the results.

Help
MyMathLab
Student Study Pack

1. 1.833
2. 2.201
3. 2.947
4. 2.539
5. (a) 2.450
   (b) 2.664
6. (a) 3.154
   (b) 4.938
7. (a) (10.9, 14.1)
   (b) t-CI is wider.
8. (a) (12.7, 14.1)
   (b) t-CI is wider.
9. (a) (4.1, 4.5)
   (b) t-CI is wider.
10. (a) (12, 16)
    (b) t-CI is wider.
11. (59.48, 90.52); 15.518
12. (60.69, 139.31); 39.307
13. (61.85, 88.15); 13.148
    The t-CI is wider.
14. (62.96, 137.04); 37.041
    The t-CI is wider.

15. (a) (3.8, 5.2)
(b) (4.4, 4.6); t-CI is wider.
16. (a) (1.2, 1.6)
(b) (1.38, 1.42)
17. (a) 4174.75
(b) 100.341
(c) (4071.63, 4277.87)
18. (a) 4418.817
(b) 389.294
(c) (4105.439, 4732.195)
19. (a) 909.1
(b) 305.3
(c) (635.4, 1182.8)
20. (a) 2.35
(b) 1.03
(c) (1.56, 3.14)

Constructing Confidence Intervals In Exercises 15 and 16, you are given the sample mean and the sample standard deviation. Assume the variable is normally distributed and use a normal distribution or a t-distribution to construct a 90% confidence interval for the population mean μ.

15. ***Waste Generated*** (a) In a random sample of 10 adults from the United States, the mean waste generated per person per day was 4.5 pounds and the standard deviation was 1.2 pounds. (b) Repeat part (a), assuming the same statistics came from a sample size of 500. Compare the results. *(Adapted from U.S. Environmental Protection Agency)*

16. ***Waste Recycled*** (a) In a random sample of 12 adults from the United States, the mean waste recycled per person per day was 1.4 pounds and the standard deviation was 0.3 pound. (b) Repeat part (a), assuming the same statistics came from a sample size of 600. Compare the results. *(Adapted from U.S. Environmental Protection Agency)*

Constructing Confidence Intervals In Exercises 17–20, a data set is given. For each data set, (a) find the sample mean, (b) find the sample standard deviation, and (c) construct a 99% confidence interval for the population mean μ. Assume the population of each data set is normally distributed.

17. ***Biology*** The monthly incomes for 10 randomly selected people, each with a bachelor's degree in biology *(Adapted from U.S. Bureau of Labor Statistics)*

4148.51	3978.27	4093.63	4091.95	4282.18
4223.64	4276.50	4207.41	4285.69	4159.72

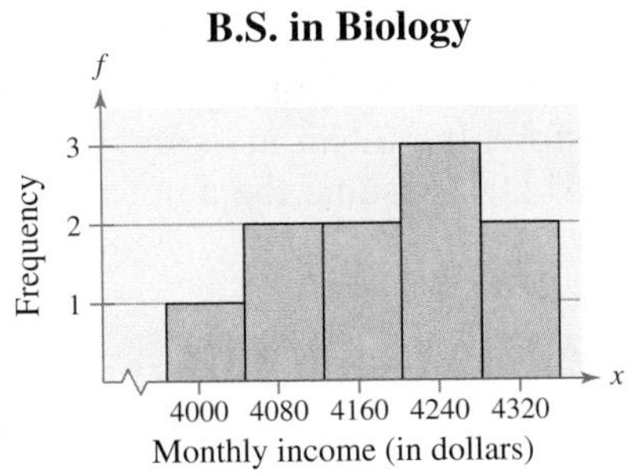

Figure for Exercise 17

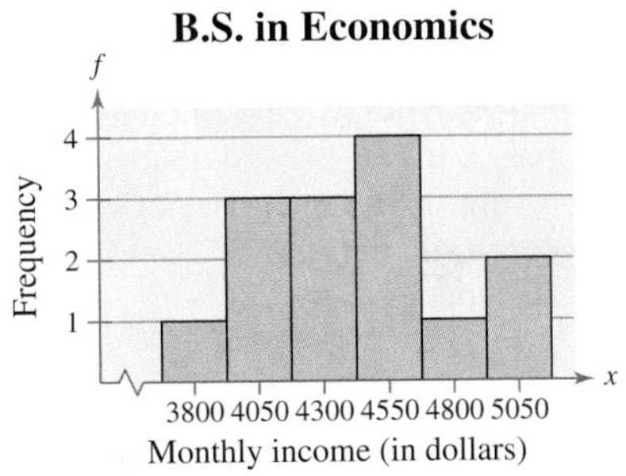

Figure for Exercise 18

18. ***Economics*** The monthly incomes for 14 randomly selected people, each with a bachelor's degree in economics *(Adapted from U.S. Bureau of Labor Statistics)*

4450.66	4596.73	4366.66	4455.40	4151.70	3727.08	4283.76
4527.64	4407.34	5036.64	5083.73	3946.47	4023.41	4806.22

19. ***SAT Scores*** The SAT scores for 12 randomly selected high school seniors

1424	1223	987	692	947	723
837	721	747	540	623	1445

20. ***GPA*** The grade point averages (GPA) for 15 randomly selected college students

2.3	3.3	2.6	1.8	0.2	3.1	4.0	0.7
2.3	2.0	3.1	3.4	1.3	2.6	2.6	

Choosing a Distribution In Exercises 21–26, use a normal distribution or a t-distribution to construct a 95% confidence interval for the population mean. Justify your decision. If neither distribution can be used, explain why not.

21. ***Lengths of Bolts*** In a random sample of 70 bolts, the mean length was 1.25 inches and the standard deviation was 0.01 inch.

22. ***Prices of Toasters*** You took a random sample of 12 two-slice toasters and found the mean price was \$61.12 and the standard deviation was \$24.62. Assume the prices are normally distributed. *(Source: Cooking.com)*

DATA **23.** ***Sports Cars: Miles per Gallon*** You take a random survey of 25 sports cars and record the miles per gallon for each. The data are listed below. Assume the miles per gallon are normally distributed.

24	24	27	20	26	23	18	29	24
22	22	27	26	20	28	30	23	24
19	22	24	26	23	24	25		

DATA **24.** ***ACT Scores*** In a recent year, the standard deviation of ACT scores for all students was 4.8. You take a random survey of 20 students and determine the ACT score for each. The scores are listed below. Assume the test scores are normally distributed. *(Source: ACT, Inc.)*

26	22	23	12	19	25	23	21	25	10
17	26	23	24	20	14	21	23	20	22

25. ***Hospital Waiting Times*** In a random sample of 19 patients at a hospital's minor emergency department, the mean waiting time (in minutes) before seeing a medical professional was 23 minutes and the standard deviation was 11 minutes. Assume the waiting times are not normally distributed.

26. ***Grocery Store Spending*** In a random sample of 17 shoppers at a grocery store, the mean amount spent was \$28.13 and the standard deviation was \$12.05. Assume the amounts spent are normally distributed.

Extending Concepts

27. ***Tennis Ball Manufacturing*** A company manufactures tennis balls. When its tennis balls are dropped onto a concrete surface from a height of 100 inches, the company wants the mean height the balls bounce upward to be 55.5 inches. This average is maintained by periodically testing random samples of 25 tennis balls. If the t-value falls between $-t_{0.99}$ and $t_{0.99}$, the company will be satisfied that it is manufacturing acceptable tennis balls. A sample of 25 balls is randomly selected and tested. The mean bounce height of the sample is 56.0 inches and the standard deviation is 0.25 inch. Is the company making acceptable tennis balls? Explain your reasoning.

28. ***Light Bulb Manufacturing*** A company manufactures light bulbs. The company wants the bulbs to have a mean life span of 1000 hours. This average is maintained by periodically testing random samples of 16 light bulbs. If the t-value falls between $-t_{0.99}$ and $t_{0.99}$, the company will be satisfied that it is manufacturing acceptable light bulbs. A sample of 16 light bulbs is randomly selected and tested. The mean life span of the sample is 1015 hours and the standard deviation is 25 hours. Is the company making acceptable light bulbs? Explain your reasoning.

21. Use normal distribution because $n \geq 30$.

$(1.248, 1.252)$

22. Use the t-distribution, because $n < 30$, the prices are normally distributed, and σ is unknown.

$(45.48, 76.76)$

23. Use t-distribution because $n < 30$, the miles per gallon are normally distributed, and σ is unknown.

$(23, 25)$

24. Use normal distribution, because σ is known and the scores are normally distributed.

$(18.7, 22.9)$

25. Cannot use normal or t-distribution because $n < 30$ and the times are not normally distributed.

26. Use t-distribution, because $n < 30$, the amounts are normally distributed, and σ is unknown.

$(21.93, 34.33)$

27. $n = 25, \bar{x} = 56.0, s = 0.25$

$\pm t_{0.99} \Rightarrow 99\%$ t-CI

$$\bar{x} \pm t_c \frac{s}{\sqrt{n}} = 56.0 \pm 2.797 \frac{0.25}{\sqrt{25}}$$

$$\approx (55.9, 56.1)$$

They are not making good tennis balls because desired bounce height of 55.5 inches is not contained between 55.9 and 56.1 inches.

28. $n = 16, \bar{x} = 1015, s = 25$

$\pm t_{0.99} \Rightarrow 99\%$ t-CI

$$\bar{x} \pm t_c \frac{s}{\sqrt{n}} = 1015 \pm 2.947 \frac{25}{\sqrt{16}}$$

$$\approx (997, 1033)$$

They are making good light bulbs because the desired bulb life of 1000 hours is contained between 997 and 1033 hours.

6.3 Confidence Intervals for Population Proportions

What You Should Learn

- How to find a point estimate for the population proportion
- How to construct a confidence interval for a population proportion
- How to determine the minimum sample size required when estimating a population proportion

Point Estimate for the Population Proportion p • Confidence Intervals for a Population Proportion p • Increasing Sample Size to Increase Precision

Point Estimate for the Population Proportion p

Recall from Section 4.2, that the probability of success in a single trial of a binomial experiment is p. This probability is a population **proportion.** In this section, you will learn how to estimate a population proportion p using a confidence interval. As with confidence intervals for μ, you will start with a point estimate.

DEFINITION

The point estimate for p, the population proportion of successes, is given by the proportion of successes in a sample and is denoted by

$$\hat{p} = \frac{x}{n}$$

where x is the number of successes in the sample and n is the number of in the sample. The point estimate for the proportion of failures is $\hat{q} = 1 - \hat{p}$. The symbols $\hat{p}$ and $\hat{q}$ are read as "p hat" and "q hat."

Insight

In the first two sections, estimates were made for quantitative data. In this section, sample proportions are used to make estimates for qualitative data.

EXAMPLE 1

Finding a Point Estimate for p

In a survey of 1011 U.S. adults, 293 said that their favorite sport to watch is football. Find a point estimate for the population proportion of U.S. adults who say their favorite sport to watch is football. *(Source: The Harris Poll)*

SOLUTION Using $n = 1011$ and $x = 293$,

$$\begin{aligned}\hat{p} &= \frac{x}{n} \\ &= \frac{293}{1011} \\ &\approx 0.290 \\ &= 29.0\%.\end{aligned}$$

Try It Yourself 1

In a survey of 3859 adults from the U.S., 1177 said that of all the presidents from the 20th century, Franklin D. Roosevelt was the greatest. Find a point estimate for the population proportion of adults who say Franklin D. Roosevelt was the greatest president of the 20th century. *(Source: The Harris Poll)*

a. *Identify* x and n.
b. Use x and n to *find* $\hat{p}$.

Answer: Page A39

Picturing the World

A Gallup poll surveyed 1016 households in the United States about their pets. Of those surveyed, 599 said they had at least one dog or cat as a pet.

Find a 90% confidence interval for the proportion of households in the United States that own at least one dog or cat.

Confidence Intervals for a Population Proportion p

Constructing a confidence interval for a population proportion p is similar to constructing a confidence interval for a population mean. You start with a point estimate and calculate a margin of error.

DEFINITION

A ***c*-confidence interval** for the population proportion p is

$$\hat{p} - E < p < \hat{p} + E$$

where

$$E = z_c \sqrt{\frac{\hat{p}\hat{q}}{n}}.$$

The probability that the confidence interval contains p is c.

In Section 5.5, you learned that a binomial distribution can be approximated by the normal distribution if $np \geq 5$ and $nq \geq 5$. When $n\hat{p} \geq 5$ and $n\hat{q} \geq 5$, the sampling distribution for $\hat{p}$ is approximately normal with a mean of

$$\mu_{\hat{p}} = p$$

and a standard error of

$$\sigma_{\hat{p}} = \sqrt{\frac{pq}{n}}.$$

GUIDELINES

Constructing a Confidence Interval for a Population Proportion

In Words	*In Symbols*
1. Identify the sample statistics n and x.	
2. Find the point estimate $\hat{p}$.	$\hat{p} = \frac{x}{n}$
3. Verify that the sampling distribution of $\hat{p}$ can be approximated by the normal distribution.	$n\hat{p} \geq 5,\ n\hat{q} \geq 5$
4. Find the critical value z_c that corresponds to the given level of confidence c.	Use the Standard Normal Table.
5. Find the margin of error E.	$E = z_c \sqrt{\frac{\hat{p}\hat{q}}{n}}$
6. Find the left and right endpoints and form the confidence interval.	Left endpoint: $\hat{p} - E$ Right endpoint: $\hat{p} + E$ Interval: $\hat{p} - E < p < \hat{p} + E$

MINITAB and TI-83 steps are shown on pages 328 and 329.

Note to Instructor

Point out that the value for E is calculated by multiplying the z-score (the number of standard deviations from the mean) times the standard error $\hat{p}$. The z-scores are found the same way they were found in Section 6.1.

Study Tip

Notice in Example 2 that the confidence interval for the proportion p was rounded to three decimal places. This *round-off rule* will be used throughout the text.

EXAMPLE 2

Constructing a Confidence Interval for p

Construct a 95% confidence interval for the proportion of adults in the United States who say that their favorite sport to watch is football.

SOLUTION From Example 1, $\hat{p} \approx 0.29$. So,

$$\hat{q} = 1 - 0.29 = 0.71.$$

Using $n = 1011$, you can verify that the sampling distribution of $\hat{p}$ can be approximated by the normal distribution.

$$n\hat{p} \approx 1011 \cdot 0.29 \approx 293 > 5$$

and

$$n\hat{q} \approx 1011 \cdot 0.71 \approx 718 > 5$$

Using $z_c = 1.96$, the margin of error is

$$E = z_c\sqrt{\frac{\hat{p}\hat{q}}{n}} \approx 1.96\sqrt{\frac{(0.29)(0.71)}{1011}} \approx 0.028.$$

The 95% confidence interval is as follows.

Left Endpoint

$$\hat{p} - E = 0.29 - 0.028 = 0.262$$

Right Endpoint

$$\hat{p} + E = 0.29 + 0.028 = 0.318$$

$$0.262 < p < 0.318$$

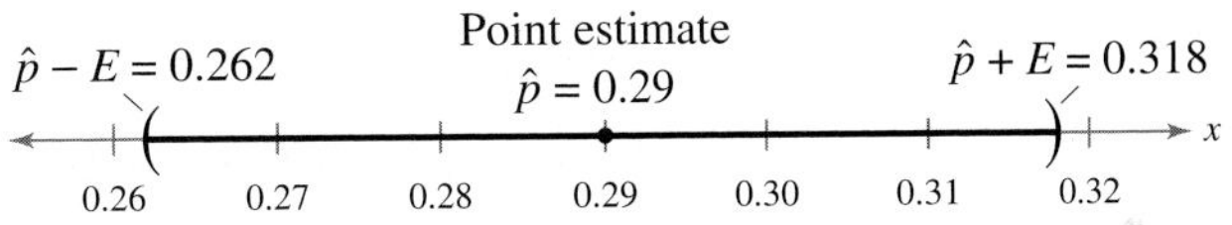

Interpretation With 95% confidence, you can say that the proportion of adults who say football is their favorite sport is between 26.2% and 31.8%.

Try It Yourself 2

Use the data given in Try It Yourself 1 to construct a 90% confidence interval for the proportion of adults who say that of all the presidents from the 20th century, Franklin D. Roosevelt was the greatest.

a. *Find* $\hat{p}$ and $\hat{q}$.
b. *Verify* that the sampling distribution of $\hat{p}$ can be approximated by the normal distribution.
c. *Find* z_c and E.
d. *Use* $\hat{p}$ and E to find the left and right endpoints.
e. *Specify* the 90% confidence interval for the proportion of adults who say that Franklin D. Roosevelt was the greatest president of the 20th century.

Answer: Page A39

The confidence level of 95% used in Example 2 is typical of opinion polls. The result, however, is usually not stated as a confidence interval. Instead, the result of Example 2 would be stated as "29% with a margin of error of ±2.8%."

EXAMPLE 3

Constructing a Confidence Interval for p

The graph shown at the right is from a survey of 900 U.S. adults. Construct a 99% confidence interval for the proportion of adults who think that teenagers are the more dangerous drivers. *(Adapted from The Gallup Organization)*

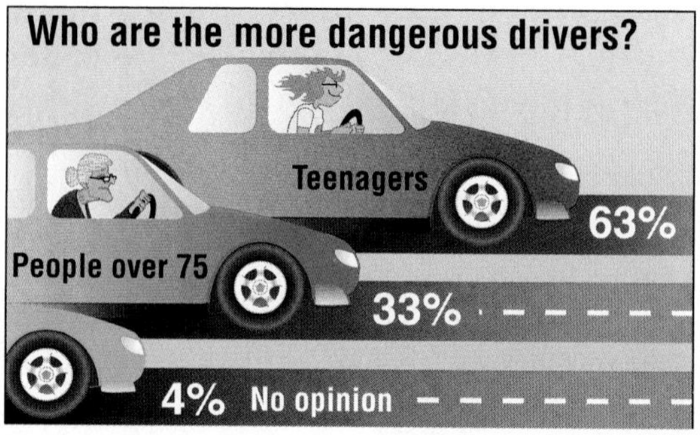

Insight

In Example 3, note that $np \geq 5$ and $nq \geq 5$. So, the sampling distribution of $\hat{p}$ is approximately normal.

SOLUTION From the graph, $\hat{p} = 0.63$. So,

$$\hat{q} = 1 - 0.63 = 0.37.$$

Using these values and the values $n = 900$ and $z_c = 2.575$, the margin of error is

$$E = z_c\sqrt{\frac{\hat{p}\hat{q}}{n}}$$

$$\approx 2.575\sqrt{\frac{(0.63)(0.37)}{900}}$$ Use Table 4 in Appendix B to estimate that z_c is halfway between 2.57 and 2.58.

$$\approx 0.041.$$

The 99% confidence interval is as follows.

Left Endpoint	Right Endpoint
$\hat{p} - E = 0.63 - 0.041 = 0.589$	$\hat{p} + E = 0.63 + 0.041 = 0.671$

$$0.589 < p < 0.671$$

Interpretation With 99% confidence, you can say that the proportion of adults who think that teenagers are the more dangerous drivers is between 58.9% and 67.1%.

Try It Yourself 3

Use the survey information in Example 3 to construct a 99% confidence interval for the proportion of adults who think that people over 75 are the more dangerous drivers.

a. *Identify* n and $\hat{p}$.
b. *Use* $\hat{p}$ to find $\hat{q}$.
c. *Verify* that the sampling distribution of $\hat{p}$ is approximately normal.
d. *Identify* the critical value z_c that corresponds to the given level of confidence.
e. *Find* the left and right endpoints of the confidence interval.
f. *Specify* the 99% confidence interval for the proportion of adults who think that people over 75 are the more dangerous drivers.

Answer: Page A39

Increasing Sample Size to Increase Precision

One way to increase the precision of the confidence interval without decreasing the level of confidence is to increase the sample size.

Finding a Minimum Sample Size to Estimate *p*

Given a *c*-confidence level and a margin or error *E*, the minimum sample size *n* needed to estimate *p* is

$$n = \hat{p}\hat{q}\left(\frac{z_c}{E}\right)^2.$$

This formula assumes that you have a preliminary estimate for $\hat{p}$ and $\hat{q}$. If not, use $\hat{p} = 0.5$ and $\hat{q} = 0.5$.

Insight

The reason for using 0.5 as values for $\hat{p}$ and $\hat{q}$ when no preliminary estimate is available is that these values yield a maximum value for the product $\hat{p}\hat{q} = \hat{p}(1 - \hat{p})$. In other words, if you don't estimate the values of $\hat{p}$ and $\hat{q}$, you must pay the penalty of using a larger sample.

EXAMPLE 4

Determining a Minimum Sample Size

You are running a political campaign and wish to estimate, with 95% confidence, the proportion of registered voters who will vote for your candidate. Your estimate must be accurate within 3% of the true population. Find the minimum sample size needed if (1) no preliminary estimate is available and (2) a preliminary estimate gives $\hat{p} = 0.31$.

SOLUTION

1. Because you do not have a preliminary estimate for $\hat{p}$, use $\hat{p} = 0.5$ and $\hat{q} = 0.5$. Using $z_c = 1.96$ and $E = 0.03$, you can solve for n.

$$n = \hat{p}\hat{q}\left(\frac{z_c}{E}\right)^2 = (0.5)(0.5)\left(\frac{1.96}{0.03}\right)^2 \approx 1067.11$$

Because n is a decimal, round up to the nearest whole number. So, at least 1068 registered voters should be included in the sample.

2. You have a preliminary estimate of $\hat{p} = 0.31$. So, $\hat{q} = 0.69$. Using $z_c = 1.96$ and $E = 0.03$, you can solve for n.

$$n = \hat{p}\hat{q}\left(\frac{z_c}{E}\right)^2 = (0.31)(0.69)\left(\frac{1.96}{0.03}\right)^2 \approx 913.02$$

Because n is a decimal, round up to the nearest whole number. So, at least 914 registered voters should be included in the sample.

Try It Yourself 4

You wish to estimate, with 90% confidence and within 2% of the true population, the proportion of adults age 18 to 29 who have high blood pressure. Find the minimum sample size needed if (1) no preliminary estimate is available and (2) a previous survey found that 4% of adults in this age group had high blood pressure. *(Source: Centers for Disease Control and Prevention)*

a. *Identify* $\hat{p}$, $\hat{q}$, z_c, and E. If $\hat{p}$ is unknown, use 0.5.
b. Use $\hat{p}$, $\hat{q}$, z_c, and E to *find* the minimum sample size n.
c. *Determine* how many adults should be included in the sample.

Answer: Page A39

6.3 Exercises

Help

MyMathLab

Student Study Pack

1. False. To estimate the value of p, the population proportion of successes, use the point estimate $\hat{p} = x/n$.
2. True
3. 0.119, 0.881
4. 0.500, 0.500
5. 0.64, 0.36
6. 0.398, 0.602
7. 0.12, 0.88
8. 0.150, 0.850
9. 0.694, 0.306
10. 0.660, 0.340
11. 0.451, 0.549
12. 0.470, 0.530
13. (0.45, 0.51)
14. (0.458, 0.562)

Building Basic Skills and Vocabulary

True or False? In Exercises 1 and 2, determine whether the statement is true or false. If it is false, rewrite it so that it is a true statement.

1. To estimate the value of p, the population proportion of successes, use the point estimate x.

2. The point estimate for the proportion of failures is $1 - \hat{p}$.

Finding $\hat{p}$ and $\hat{q}$ In Exercises 3–12, let p be the population proportion for the given condition. Find point estimates for p and q.

3. ***U.S. Food Safe?*** In a survey of 1024 adults from the United States, 122 said they were not confident that the food they eat in the United States is safe. *(Source: ABC News)*

4. ***Unions*** In a survey of 1001 U.S. adults, 501 believed the role of unions in protecting workers' rights is as important today as it was in years past. *(Source: Wirthlin Worldwide)*

5. ***Obese or Overweight?*** A study of 4115 U.S. adults found that 2634 were obese or overweight. *(Source: National Health and Nutrition Examination Survey)*

6. ***Accidents and Alcohol*** A study of 2008 traffic fatalities found that 800 of the fatalities were alcohol related. *(Adapted from U.S. National Highway Traffic Safety Administration)*

7. ***Planning for the Future*** Of 848 children surveyed, 102 plan to join the armed forces in the future. *(Source: National Geographic Kids)*

8. ***New Year's Resolutions*** Of 1004 adults surveyed, 151 intend to get a better job as a New Year's resolution. *(Source: Marist College Institute for Public Opinion)*

9. ***Quitting Smoking?*** In a survey of 284 smokers, 197 smokers reported that they wanted to quit smoking. *(Adapted from American Lung Association)*

10. ***Happy at Work?*** In a survey of 1003 U.S. adults, 662 would be happy spending the rest of their career with their current employer. *(Source: Maritz Poll)*

11. ***Home Repairs*** In a survey of 586 homeowners, 264 said they have concerns about contractors doing repairs in their homes. *(Adapted from Opinion Research for Scott Rags)*

12. ***Stressful Travel*** In a survey of 3224 U.S. adults, 1515 said flying is the most stressful form of travel. *(Adapted from Travelocity)*

Using and Interpreting Concepts

13. ***Election Poll*** An election poll reported that a candidate had an approval rating of 48% with a margin of error E of 3%. Construct a confidence interval for the proportion of adults who approve of the candidate.

14. ***Direct Mail*** A survey shows that 51% of adults prefer to receive advertisements through the mail. The margin of error E is 5.2%. Construct a confidence interval for the proportion of adults who prefer to receive advertisements through the mail. *(Source: PRIMEDIA Business Magazines & Media, Inc.)*

15. (0.099, 0.139); (0.093, 0.145)
The 99% CI is wider.
16. (0.469, 0.531); (0.459, 0.541)
The 99% CI is wider.
17. (0.625, 0.655); (0.621, 0.659)
The 99% CI is wider.
18. (0.377, 0.419); (0.370, 0.426)
The 99% CI is wider.
19. (0.098, 0.142); (0.091, 0.149)
The 99% CI is wider.
20. (0.128, 0.172); (0.121, 0.179)
The 99% CI is wider.
21. (a) 1068
(b) 822
(c) Having an estimate of the proportion reduces the minimum sample size needed.
22. (a) 849
(b) 542
(c) Having an estimate of the proportion reduces the minimum sample size needed.
23. (a) 1690
(b) 1267
(c) Having an estimate of the proportion reduces the minimum sample size needed.
24. (a) 961
(b) 592
(c) Having an estimate of the proportion reduces the minimum sample size needed.

Constructing a Confidence Interval In Exercises 15–20, construct the 95% and 99% confidence intervals for the population proportion p using the indicated sample statistics. Which interval is wider?

15. Use the statistics in Exercise 3.

16. Use the statistics in Exercise 4.

17. Use the statistics in Exercise 5.

18. Use the statistics in Exercise 6.

19. Use the statistics in Exercise 7.

20. Use the statistics in Exercise 8.

21. ***Travel Plans*** You are a travel agent and wish to estimate, with 95% confidence, the proportion of vacationers who plan to travel outside the United States in the next 12 months. Your estimate must be accurate within 3% of the true proportion.

(a) No preliminary estimate is available. Find the minimum sample size needed.

(b) Find the minimum sample size needed, using a prior study that found that 26% of the respondents said they planned to travel outside the United States in the next 12 months. *(Source: Wirthlin Worldwide)*

(c) Compare the results from parts (a) and (b).

22. ***Online Service Usage*** You are a travel agent and wish to estimate, with 98% confidence, the proportion of vacationers who use an online service or the Internet to make reservations for lodging. Your estimate must be accurate within 4% of the population proportion.

(a) No preliminary estimate is available. Find the minimum sample size needed.

(b) Find the minimum sample size needed, using a prior study that found that 20% of the respondents said they used an online service or the Internet to make reservations for lodging. *(Source: Travel Industry Association of America)*

(c) Compare the results from parts (a) and (b).

23. ***Camcorder Repairs*** You wish to estimate, with 96% confidence, the proportion of camcorders that need repairs or have problems by the time the product is five years old. Your estimate must be accurate within 2.5% of the true proportion.

(a) No preliminary estimate is available. Find the minimum sample size needed.

(b) Find the minimum sample size needed, using a prior study that found that 25% of camcorders needed repairs or had problems by the time the product was five years old. *(Source: Consumer Reports)*

(c) Compare the results from parts (a) and (b).

24. ***Computer Repairs*** You wish to estimate, with 97% confidence and within 3.5% of the true population, the proportion of computers that need repairs or have problems by the time the product is three years old.

(a) No preliminary estimate is available. Find the minimum sample size needed.

(b) Find the minimum sample size needed, using a prior study that found that 19% of computers needed repairs or had problems by the time the product was three years old. *(Source: Consumer Reports)*

(c) Compare the results from parts (a) and (b).

25. (a) (0.554, 0.666)
(b) (0.383, 0.497)
It is unlikely that the two proportions are equal because the confidence intervals estimating the proportions do not overlap.

26. (a) (0.482, 0.618)
(b) (0.333, 0.467)
It is unlikely that the two proportions are equal because the confidence intervals estimating the proportions do not overlap.

Constructing Confidence Intervals In Exercises 25 and 26, use the following information. The graph shows the results of a survey in which 500 men, 500 women, 350 people who use microwaves often, and 350 people who rarely use microwaves were asked if they favored irradiation of red meat to kill disease microbes. *(Adapted from Peter D. Hart Research Associates for Grocery Manufacturers of America)*

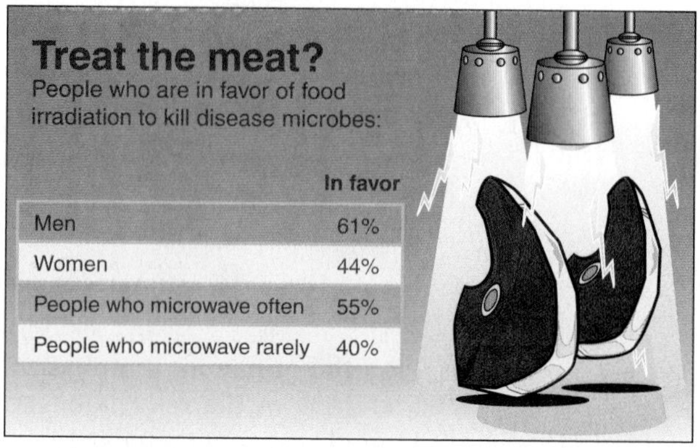

25. ***Men and Women*** Construct a 99% confidence interval for

(a) the proportion of men who favor irradiation of red meat.

(b) the proportion of women who favor irradiation of red meat. Is it possible that these two proportions are equal? Explain your reasoning.

26. ***Frequent Microwavers and Infrequent Microwavers*** Construct a 99% confidence interval for

(a) the proportion of frequent microwave users who favor irradiation of red meat.

(b) the proportion of infrequent microwave users who favor irradiation of red meat. Is it possible that these two proportions are equal? Explain your reasoning.

Constructing Confidence Intervals In Exercises 27 and 28, use the following information. The graph shows the results of a survey in which 600 adults from Generation X, 600 adults from the baby boomer generation, and 429 senior citizens were asked if they use an investment professional. *(Adapted from New York Life Investment Management)*

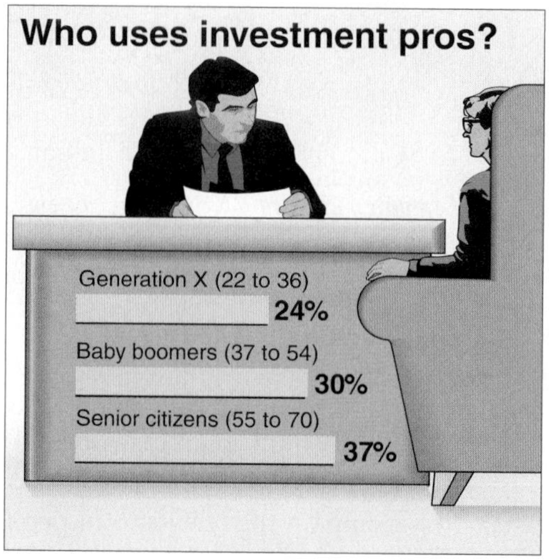

27. (a) (0.195, 0.285)
(b) (0.252, 0.348)
(c) (0.31, 0.43)

28. (a) Yes, the confidence intervals overlap.
(b) Yes, the confidence intervals overlap.
(c) No, the confidence intervals do not overlap.

29. (30.4%, 32.4%) is approximately a 95.2% CI.

30. (24%, 30%) is approximately a 96.8% CI.

31. If $n\hat{p} < 5$ or $n\hat{q} < 5$, the sampling distribution of $\hat{p}$ may not be normally distributed, therefore preventing the use of z_c when calculating the confidence interval.

32. $E = z_c\sqrt{\frac{\hat{p}\hat{q}}{n}} \Rightarrow \frac{E}{z_c} = \sqrt{\frac{\hat{p}\hat{q}}{n}} \Rightarrow$

$\left(\frac{E}{z_c}\right)^2 = \frac{\hat{p}\hat{q}}{n} \Rightarrow n = \hat{p}\hat{q}\left(\frac{z_c}{E}\right)^2$

33.

$\hat{p}$	$\hat{q} = 1 - \hat{p}$	$\hat{p}\hat{q}$
0.0	1.0	0.00
0.1	0.9	0.09
0.2	0.8	0.16
0.3	0.7	0.21
0.4	0.6	0.24
0.5	0.5	0.25
0.6	0.4	0.24
0.7	0.3	0.21
0.8	0.2	0.16
0.9	0.1	0.09
1.0	0.0	0.00

$\hat{p}$	$\hat{q} = 1 - \hat{p}$	$\hat{p}\hat{q}$
0.45	0.55	0.2475
0.46	0.54	0.2484
0.47	0.53	0.2491
0.48	0.52	0.2496
0.49	0.51	0.2499
0.50	0.50	0.2500
0.51	0.49	0.2499
0.52	0.48	0.2496
0.53	0.47	0.2491
0.54	0.46	0.2484
0.55	0.45	0.2475

$\hat{p} = 0.5$ gives the maximum value of $\hat{p}\hat{q}$.

27. ***Investment Professionals*** Construct a 99% confidence interval for

(a) the proportion of adults from Generation X who use an investment professional.

(b) the proportion of adults from the baby boomer generation who use an investment professional.

(c) the proportion of adults who are senior citizens who use an investment professional.

28. ***Investment Professionals*** Determine whether the following proportions are equal and explain your reasoning.

(a) The proportion of adults from Exercise 27(a) and the proportion of adults from Exercise 27(b).

(b) The proportion of adults from Exercise 27(b) and the proportion of adults from Exercise 27(c).

(c) The proportion of adults from Exercise 27(a) and the proportion of adults from Exercise 27(c).

Extending Concepts

Newspaper Surveys In Exercises 29 and 30, translate the newspaper excerpt into a confidence interval for p. Approximate the value of the level of confidence, c.

29. In a survey of 8451 U.S. adults, 31.4% said they were taking vitamin E as a supplement. The survey's margin of error is plus or minus 1%. *(Source: Decision Analyst, Inc.)*

30. In a survey of 1001 U.S. adults, 27% said they had smoked a cigarette in the past week. The survey's margin of error is plus or minus 3%. *(Source: The Gallup Organization)*

31. ***Why Check It?*** Why is it necessary to check that $n\hat{p} \geq 5$ and $n\hat{q} \geq 5$?

32. ***Sample Size*** The equation for determining the sample size $n = \hat{p}\hat{q}[(z_c)/E]^2$ can be obtained by solving the equation for the margin of error $E = z_c\sqrt{(\hat{p}\hat{q})/n}$ for n. Show that this is true and justify each step.

33. ***Maximum Value of $\hat{p}\hat{q}$*** Complete the tables for different values of $\hat{p}$ and $\hat{q} = 1 - \hat{p}$. From the table, which value of $\hat{p}$ appears to give the maximum value of the product $\hat{p}\hat{q}$?

$\hat{p}$	$\hat{q} = 1 - \hat{p}$	$\hat{p}\hat{q}$
0.0	1.0	0.00
0.1	0.9	0.09
0.2	0.8	
0.3		
0.4		
0.5		
0.6		
0.7		
0.8		
0.9		
1.0		

$\hat{p}$	$\hat{q} = 1 - \hat{p}$	$\hat{p}\hat{q}$
0.45		
0.46		
0.47		
0.48		
0.49		
0.50		
0.51		
0.52		
0.53		
0.54		
0.55		

6.4 Confidence Intervals for Variance and Standard Deviation

What You Should Learn

- How to interpret the chi-square distribution and use a chi-square distribution table
- How to use the chi-square distribution to construct a confidence interval for the variance and standard deviation

The Chi-Square Distribution • Confidence Intervals for σ^2 and σ

The Chi-Square Distribution

In manufacturing, it is necessary to control the amount that a process varies. For instance, an automobile part manufacturer must produce thousands of parts to be used in the manufacturing process. It is important that the parts vary little or not at all. How can you measure, and consequently control, the amount of variation in the parts? You can start with a point estimate.

DEFINITION

The **point estimate for σ^2** is s^2 and the **point estimate for σ** is s. s^2 is the most unbiased estimate for σ^2.

Study Tip

The Greek letter χ is pronounced "ki," which rhymes with the more familiar Greek letter π.

You can use a *chi-square distribution* to construct a confidence interval for the variance and standard deviation.

DEFINITION

If the random variable x has a normal distribution, then the distribution of

$$\chi^2 = \frac{(n-1)s^2}{\sigma^2}$$

forms a **chi-square distribution** for samples of any size $n > 1$. Four properties of the chi-square distribution are as follows.

1. All chi-square values χ^2 are greater than or equal to zero.
2. The chi-square distribution is a family of curves, each determined by the degrees of freedom. To form a confidence interval for σ^2, use the χ^2-distribution with degrees of freedom equal to one less than the sample size.

   d.f. $= n - 1$ Degrees of freedom
3. The area under each curve of the chi-square distribution equals one.
4. Chi-square distributions are positively skewed.

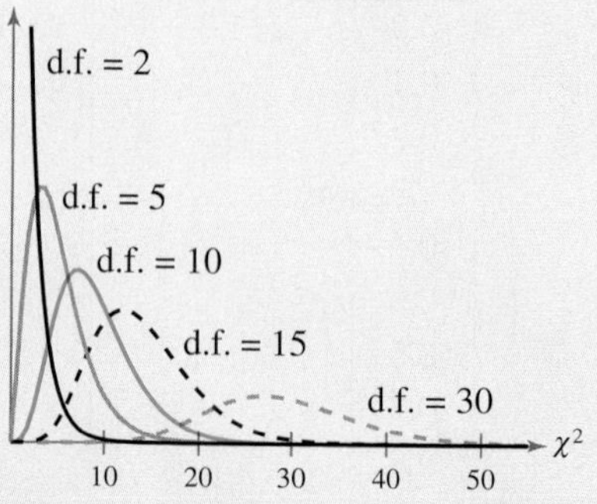

Chi-square distributions

Note to Instructor

If you are short of time, this section can be omitted. Or, if you prefer, it can be covered with Chapter 10 when additional chi-square applications are presented.

Study Tip

For chi-square critical values with a *c*-confidence level, the following values are what you look up in Table 6 in Appendix B.

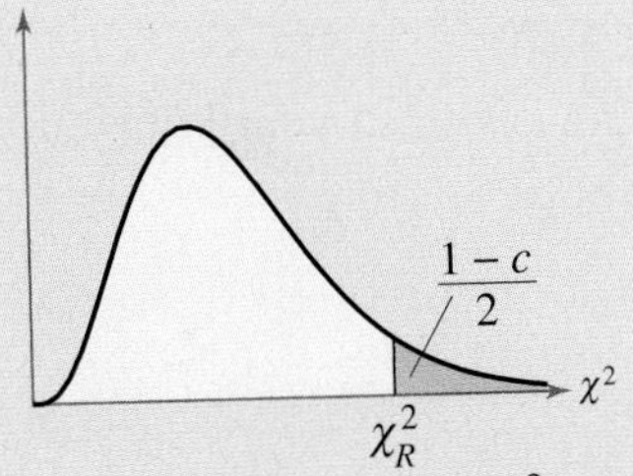

Area to the right of χ_R^2

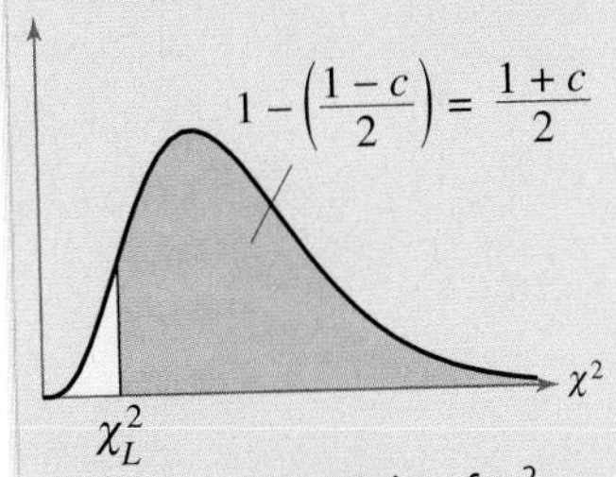

Area to the right of χ_L^2

The result is that you can conclude that the area between the left and right critical values is *c*.

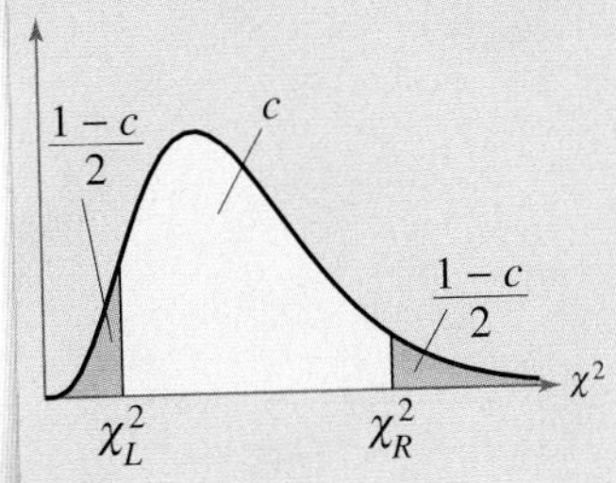

There are two critical values for each level of confidence. The value χ_R^2 represents the right-tail critical value and χ_L^2 represents the left-tail critical value. Table 6 in Appendix B lists critical values of χ^2 for various degrees of freedom and areas. Each area in the table represents the region under the chi-square curve to the *right* of the critical value.

EXAMPLE 1

Finding Critical Values for χ^2

Find the critical values χ_R^2 and χ_L^2 for a 90% confidence interval when the sample size is 20.

SOLUTION Because the sample size is 20, there are

d.f. $= n - 1 = 20 - 1 = 19$ degrees of freedom.

The areas to the right of χ_R^2 and χ_L^2 are

$$\text{Area to right of } \chi_R^2 = \frac{1-c}{2} = \frac{1-0.90}{2} = 0.05$$

and

$$\text{Area to right of } \chi_L^2 = \frac{1+c}{2} = \frac{1+0.90}{2} = 0.95.$$

Part of Table 6 is shown. Using d.f. = 19 and the areas 0.95 and 0.05, you can find the critical values, as shown by the highlighted areas in the table.

Degrees of freedom	α						
	0.995	0.99	0.975	0.95	0.90	0.10	0.05
1	—	—	0.001	0.004	0.016	2.706	3.841
2	0.010	0.020	0.051	0.103	0.211	4.605	5.991
3	0.072	0.115	0.216	0.352	0.584	6.251	7.815
15	4.601	5.229	6.262	7.261	8.547	22.307	24.996
16	5.142	5.812	6.908	7.962	9.312	23.542	26.296
17	5.697	6.408	7.564	8.672	10.085	24.769	27.587
18	6.265	7.015	8.231	9.390	10.865	25.989	28.869
19	6.844	7.633	8.907	10.117	11.651	27.204	30.144
20	7.434	8.260	9.591	10.851	12.443	28.412	31.410

χ_L^2 χ_R^2

From the table, you can see that $\chi_R^2 = 30.144$ and $\chi_L^2 = 10.117$.

Interpretation So, 90% of the area under the curve lies between 10.117 and 30.144.

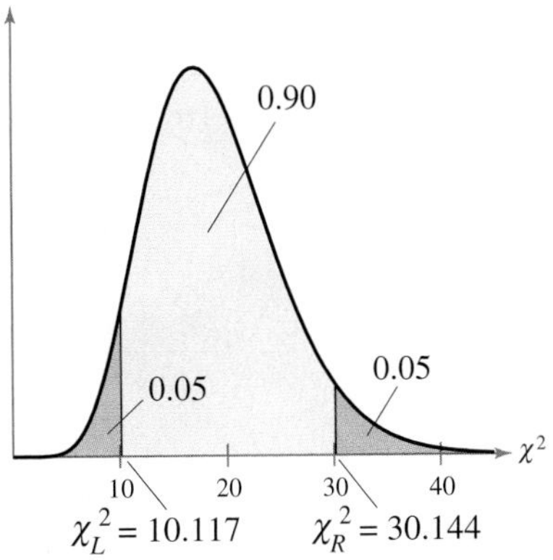

Try It Yourself 1

Find the critical values χ_R^2 and χ_L^2 for a 95% confidence interval when the sample size is 25.

a. *Identify* the degrees of freedom and the level of confidence.
b. *Find* the area to the right of χ_R^2 and χ_L^2.
c. *Use* Table 6 of Appendix B to find χ_R^2 and χ_L^2.

Answer: Page A39

Confidence Intervals for σ^2 and σ

You can use the critical values χ_R^2 and χ_L^2 to construct confidence intervals for a population variance and standard deviation. As you would expect, the best point estimate for the variance is s^2 and the best point estimate for the standard deviation is s.

DEFINITION

A c-confidence interval for a population variance and standard deviation is as follows.

Confidence Interval for σ^2:

$$\frac{(n-1)s^2}{\chi_R^2} < \sigma^2 < \frac{(n-1)s^2}{\chi_L^2}$$

Confidence Interval for σ:

$$\sqrt{\frac{(n-1)s^2}{\chi_R^2}} < \sigma < \sqrt{\frac{(n-1)s^2}{\chi_L^2}}$$

The probability that the confidence intervals contain σ^2 or σ is c.

GUIDELINES

Constructing a Confidence Interval for a Variance and Standard Deviation

In Words	*In Symbols*
1. Verify that the population has a normal distribution.	
2. Identify the sample statistic n and the degrees of freedom.	d.f. $= n - 1$
3. Find the point estimate s^2.	$s^2 = \dfrac{\Sigma(x-\bar{x})^2}{n-1}$
4. Find the critical values χ_R^2 and χ_L^2 that correspond to the given level of confidence c.	Use Table 6 in Appendix B.
5. Find the left and right endpoints and form the confidence interval for the population variance.	Left Endpoint, Right Endpoint: $\dfrac{(n-1)s^2}{\chi_R^2} < \sigma^2 < \dfrac{(n-1)s^2}{\chi_L^2}$
6. Find the confidence interval for the population standard deviation by taking the square root of each endpoint.	$\sqrt{\dfrac{(n-1)s^2}{\chi_R^2}} < \sigma < \sqrt{\dfrac{(n-1)s^2}{\chi_L^2}}$

Picturing the World

The gray whale has the longest annual migration distance of any mammal. Gray whales leave from Baja, California and western Mexico in the spring, migrating to the Bering and Chukchi seas for the summer months. Tracking a sample of 50 whales for a year provided a mean migration distance of 11,064 miles with a standard deviation of 860 miles. *(Source: National Marine Fisheries Services)*

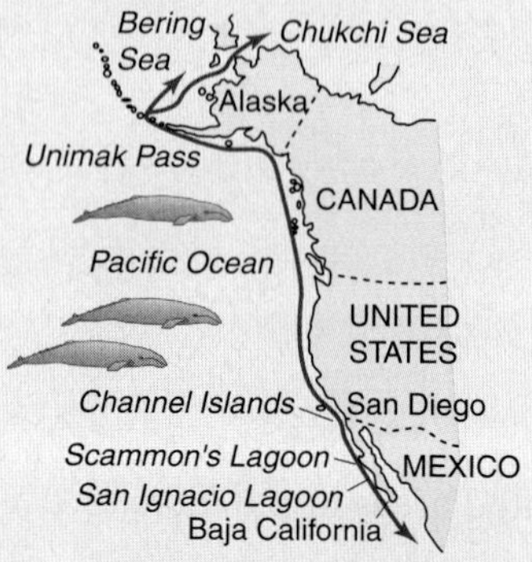

Construct a 90% confidence interval for the standard deviation for the migration distance of gray whales. Assume that the population of migration distances has a normal distribution.

EXAMPLE 2

Constructing a Confidence Interval

You randomly select and weigh 30 samples of an allergy medicine. The sample standard deviation is 1.20 milligrams. Assuming the weights are normally distributed, construct 99% confidence intervals for the population variance and standard deviation.

SOLUTION The areas to the right of χ_R^2 and χ_L^2 are

$$\text{Area to right of } \chi_R^2 = \frac{1-c}{2} = \frac{1-0.99}{2} = 0.005$$

and

$$\text{Area to right of } \chi_L^2 = \frac{1+c}{2} = \frac{1+0.99}{2} = 0.995.$$

Using the values $n = 30$, d.f. $= 29$, and $c = 0.99$, the critical values χ_R^2 and χ_L^2 are

$$\chi_R^2 = 52.336 \quad \text{and} \quad \chi_L^2 = 13.121.$$

Using these critical values and $s = 1.20$, the confidence interval for σ^2 is as follows.

Left Endpoint

$$\frac{(n-1)s^2}{\chi_R^2} = \frac{(30-1)(1.20)^2}{52.336} \approx 0.80$$

Right Endpoint

$$\frac{(n-1)s^2}{\chi_L^2} = \frac{(30-1)(1.20)^2}{13.121} \approx 3.18$$

$$0.80 < \sigma^2 < 3.18$$

The confidence interval for σ is

$$\sqrt{0.80} < \sigma < \sqrt{3.18}$$

$$0.89 < \sigma < 1.78.$$

Interpretation With 99% confidence, you can say that the population variance is between 0.80 and 3.18. The population standard deviation is between 0.89 and 1.78 milligrams.

Try It Yourself 2

Find the 90% and 95% confidence intervals for the population variance and standard deviation of the medicine weights.

a. *Find* the critical values χ_R^2 and χ_L^2 for each confidence interval.
b. *Use* n, s, χ_R^2, and χ_L^2 to find the left and right endpoints for each confidence interval for the variance.
c. *Find* the square roots of the endpoints of each confidence interval.
d. *Specify* the 90% and 95% confidence intervals for the population variance and standard deviation.

Answer: Page A39

Note to Instructor

Point out that the left endpoint requires using χ_R^2 and the right endpoint requires using χ_L^2. This is true because $\chi_R^2 > \chi_L^2$ and dividing the same numerator by a larger value will produce a smaller quotient.

Study Tip

When a confidence interval for a population variance or standard deviation is computed, the general *round-off rule* is to round off to the same number of decimal places given for the sample variance or standard deviation.

6.4 Exercises

Help

MyMathLab

Student Study Pack

Building Basic Skills and Vocabulary

In Exercises 1–6, find the critical values χ_R^2 and χ_L^2 for the given confidence level c and sample size n.

1. $c = 0.90,\ n = 10$

2. $c = 0.99,\ n = 13$

3. $c = 0.95,\ n = 22$

4. $c = 0.98,\ n = 26$

5. $c = 0.99,\ n = 30$

6. $c = 0.80,\ n = 29$

Using and Interpreting Concepts

Constructing Confidence Intervals In Exercises 7–16, assume each sample is taken from a normally distributed population and construct the indicated confidence intervals for (a) the population variance σ^2 and (b) the population standard deviation σ.

7. *Vitamins* To analyze the variation of vitamin supplement tablets, you randomly select and weigh 14 tablets. The results (in milligrams) are shown. Use a 90% level of confidence.

500.000	499.995	500.010	499.997	500.015
499.988	500.000	499.996	500.020	500.002
499.998	499.996	500.003	500.000	

8. *Cough Syrup* You randomly select and measure the contents of 15 bottles of cough syrup. The results (in fluid ounces) are shown. Use a 90% level of confidence.

4.211	4.246	4.269	4.241	4.260
4.293	4.189	4.248	4.220	4.239
4.253	4.209	4.300	4.256	4.290

9. *Car Batteries* The number of hours of reserve capacity of 18 randomly selected automotive batteries is shown. Use a 99% level of confidence. *(Adapted from Consumer Reports)*

1.70	1.60	1.94	1.58	1.74	1.60
1.86	1.72	1.38	1.46	1.64	1.49
1.55	1.70	1.75	0.88	1.77	2.07

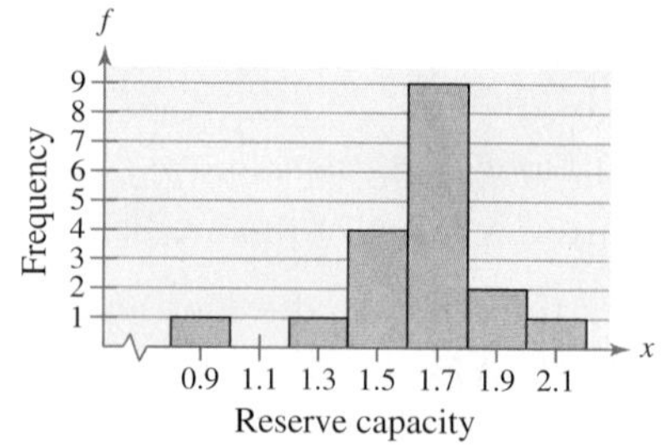

1. 16.919, 3.325
2. 28.299, 3.074
3. 35.479, 10.283
4. 44.314, 11.524
5. 52.336, 13.121
6. 37.916, 18.939
7. (a) (0.0000413, 0.000157)
   (b) (0.00643, 0.0125)
8. (a) (0.000609, 0.00220)
   (b) (0.0247, 0.0469)
9. (a) (0.031, 0.191)
   (b) (0.175, 0.437)

10. (a) (0.00467, 0.0195)
(b) (0.0683, 0.140)

11. (a) (4.34, 44.64)
(b) (2.08, 6.68)

12. (a) (759, 3171)
(b) (28, 56)

13. (a) (360, 1830)
(b) (19, 43)

14. (a) (22.4, 98.1)
(b) (4.7, 9.9)

10. *Bolts* You randomly select and measure the lengths of 17 bolts. The results (in inches) are shown. Use a 95% level of confidence.

1.286	1.138	1.240	1.132	1.381	1.137
1.300	1.167	1.240	1.401	1.241	1.171
1.217	1.360	1.302	1.331	1.383	

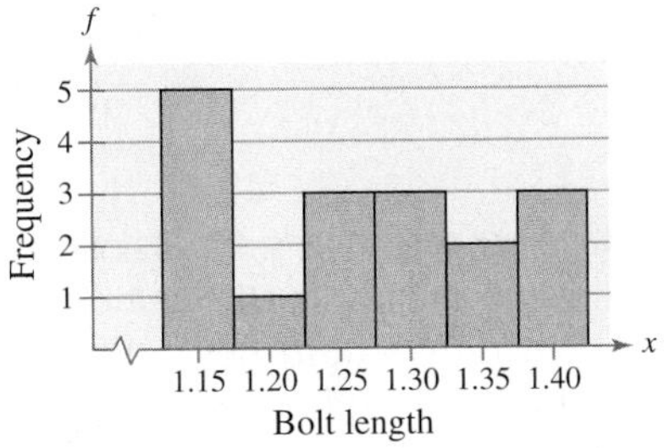

11. *Lawn Mower* A lawn mower manufacturer is trying to determine the standard deviation of the life of one of its lawn mower models. To do this, it randomly selects 12 lawn mowers that were sold several years ago and finds that the sample standard deviation is 3.25 years. Use a 99% level of confidence. *(Adapted from Consumer Reports)*

12. *DVD Players* A magazine includes a report on the prices of DVD players. The article states that 17 randomly selected DVD players had a standard deviation of \$37. Use a 95% level of confidence. *(Adapted from Consumer Reports)*

13. *Hotels* As part of your vacation planning, you randomly contact 10 hotels in your destination area and record the room rate of each. The results are shown in the stem-and-leaf plot. Use a 90% level of confidence.

6	0 3
7	
8	3
9	0
10	2 8
11	3 8
12	2
13	
14	1

Key: 8|3 = 83

14. *Pulse Rates* The pulse rates of a random sample of 16 adults are shown in the stem-and-leaf plot. Use a 95% level of confidence.

6	2
6	5 8 9
7	0 1 4 4
7	6 6 7 9 9
8	0 0
8	7

Key: 6|2 = 62

15. (a) (128, 492)
 (b) (11, 22)
16. (a) (8,831,450, 21,224,305)
 (b) (2972, 4607)
17. (a) (6622, 24,422)
 (b) (81, 156)
18. (a) (20.5, 150.8)
 (b) (4.5, 12.3)
19. (a) (0.08, 0.61)
 (b) (0.28, 0.78)
20. (a) (25,240,874, 80,295,203)
 (b) (5024, 8961)
21. Yes, because the confidence interval is below 0.015.
22. No, because the majority of the confidence interval is above 0.025.

15. *Water Quality* As part of a water quality survey, you test the water hardness in several randomly selected streams. The results are shown in the figure. Use a 95% level of confidence.

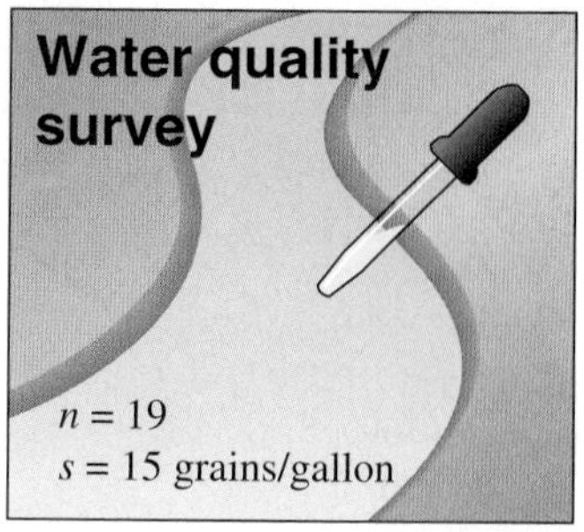

16. *Website Costs* As part of a survey, you ask a random sample of business owners how much they would be willing to pay for a website for their company. The results are shown in the figure. Use a 90% level of confidence.

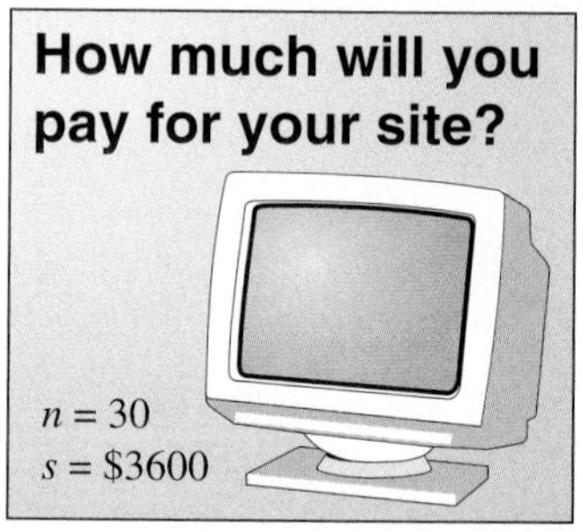

17. *Monthly Income* The monthly incomes of 20 randomly selected individuals who have recently graduated with a bachelor's degree in social science have a sample standard deviation of \$107. Use a 95% level of confidence. *(Adapted from U.S. Census Bureau)*

18. *Sodium Chloride Concentration* The sodium chloride concentrations of 13 randomly selected seawater samples have a standard deviation of 6.7 cubic centimeters per cubic meter. Use a 98% level of confidence. *(Adapted from Dorling Kindersley Visual Encyclopedia)*

19. *Cereal Boxes* The weights of a random sample of 15 cereal boxes have a sample standard deviation of 0.42 ounce. Use a 99% level of confidence.

20. *SUVs* The prices of a random sample of 18 new SUVs have a sample standard deviation of \$6400. Use a 90% level of confidence.

Extending Concepts

21. *Vitamin Tablet Weights* You are analyzing the sample of vitamin supplement tablets in Exercise 7. The population standard deviation of the tablets' weights should be less than 0.015 milligram. Does the confidence interval you constructed for σ suggest that the variation in the tablets' weights is at an acceptable level? Explain your reasoning.

22. *Cough Syrup Bottle Contents* You are analyzing the sample of cough syrup bottles in Exercise 8. The population standard deviation of the bottles' contents should be less than 0.025 fluid ounce. Does the confidence interval you constructed for σ suggest that the variation in the bottles' contents is at an acceptable level? Explain your reasoning.

Uses and Abuses

Statistics in the Real World

Uses

By now, you know that complete information about population parameters is often not available. The techniques of this chapter can be used to make interval estimates of these parameters so that you can make informed decisions.

From what you learned in this chapter, you know that point estimates of population parameters are rarely exact. Remembering this can help you make good decisions in your career and in everyday life. For instance, suppose the results of a survey tell you that 52% of the population plans to vote for a certain item on a ballot. You know that this is only a point estimate of the actual proportion that will vote for the item. If the interval estimate is $0.49 < p < 0.55$, then you know this means it is possible that the item will not receive a majority vote.

Abuses

Unrepresentative Samples There are many ways that surveys can result in incorrect predictions. When you read the results of a survey, remember to question the sample size, the sampling technique, and the questions asked. For example, suppose you want to know how a group of people will vote in an election. From the diagram at the right, you can see that even if your sample is large enough, it may not consist of actual voters.

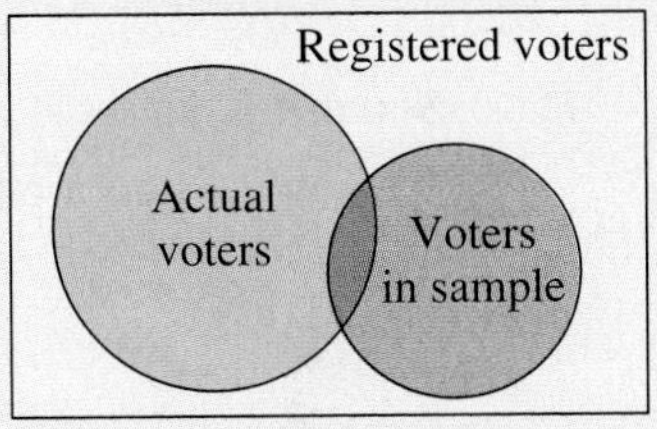

Biased Survey Questions In surveys, it is also important to analyze the wording of the questions. For example, a question about rezoning might be presented as: "Knowing that rezoning will result in more businesses contributing to school taxes, would you support the rezoning?"

Exercises

1. ***Unrepresentative Samples*** Find an example of a survey that is reported in a newspaper or magazine. Describe different ways that the sample could have been unrepresentative of the population.

2. ***Biased Survey Questions*** Find an example of a survey that is reported in a newspaper or magazine. Describe different ways that the survey questions could have been biased.

6 Chapter Summary

What did you learn?

Review Exercises

Section 6.1

- How to find a point estimate and a margin of error — *1, 2*

$$E = z_c \frac{\sigma}{\sqrt{n}}$$

- How to construct and interpret confidence intervals for the population mean — *3, 4*

$$\bar{x} - E < \mu < \bar{x} + E$$

- How to determine the minimum sample size required when estimating μ — *5–8*

$$n = \left(\frac{z_c \sigma}{E}\right)^2$$

Section 6.2

- How to interpret the t-distribution and use a t-distribution table — *9–12*

$$t = (\bar{x} - \mu)/(s/\sqrt{n})$$

- How to construct confidence intervals when $n < 30$ and σ is unknown — *13–22*

$$\bar{x} - E < \mu < \bar{x} + E, \quad E = t_c \frac{s}{\sqrt{n}}$$

Section 6.3

- How to find a point estimate for the population proportion — *23–30*

$$\hat{p} = \frac{x}{n}$$

- How to construct a confidence interval for a population proportion — *31–38*

$$\hat{p} - E < p < \hat{p} + E, \quad E = z_c \sqrt{\frac{\hat{p}\hat{q}}{n}}$$

- How to determine the minimum sample size required when estimating a population proportion — *39, 40*

$$n = \hat{p}\hat{q}\left(\frac{z_c}{E}\right)^2$$

Section 6.4

- How to interpret the chi-square distribution and use a chi-square distribution table — *41–44*

$$\chi^2 = \frac{(n-1)s^2}{\sigma^2}$$

- How to use the chi-square distribution to construct a confidence interval for the variance and standard deviation — *45–48*

$$\frac{(n-1)s^2}{\chi_R^2} < \sigma^2 < \frac{(n-1)s^2}{\chi_L^2}, \quad \sqrt{\frac{(n-1)s^2}{\chi_R^2}} < \sigma < \sqrt{\frac{(n-1)s^2}{\chi_L^2}}$$

6 Review Exercises

1. (a) 103.5
   (b) 9.016
2. (a) 9.5
   (b) 2.064
3. (10.2, 10.4)
4. (0.0922, 0.0928)
5. 47
6. 1992
7. 49
8. 1091

Section 6.1

In Exercises 1 and 2, find (a) the point estimate of the population mean and (b) the margin of error for a 90% confidence interval.

DATA **1.** Waking times of 40 people who start work at 8:00 A.M. (in minutes past 5:00 A.M.)

135	145	95	140	135	95	110	50
90	165	110	125	80	125	130	110
25	75	65	100	60	125	115	135
95	90	140	40	75	50	130	85
100	160	135	45	135	115	75	130

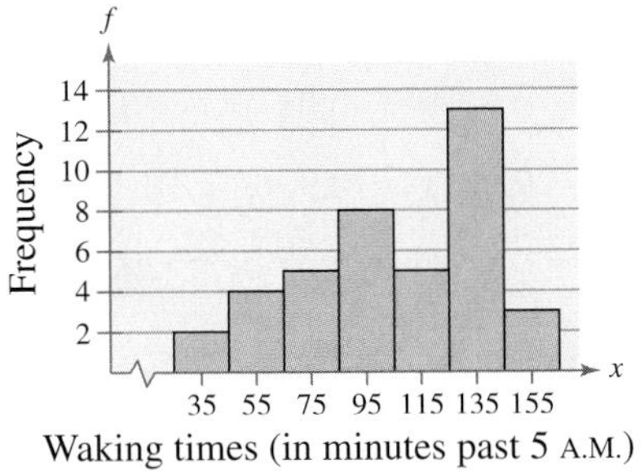

Figure for Exercise 1

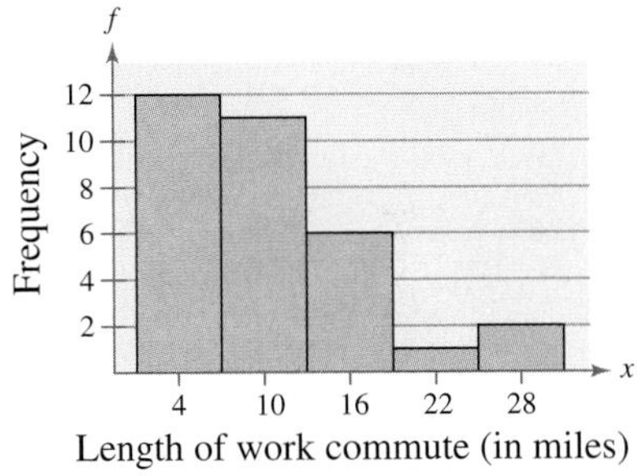

Figure for Exercise 2

DATA **2.** Length of work commute of 32 people (in miles)

12	9	7	2	8	7	3	27
21	10	13	3	7	2	30	7
6	13	6	14	4	1	10	3
13	6	2	9	2	12	16	18

In Exercises 3 and 4, construct the indicated confidence interval for the population mean μ.

3. $c = 0.95, \bar{x} = 10.3, s = 0.277, n = 100$

4. $c = 0.90, \bar{x} = 0.0925, s = 0.0013, n = 45$

In Exercises 5–8, determine the minimum sample size to estimate μ.

5. Use the results from Exercise 1. Determine the minimum survey size that is necessary to be 95% confident that the sample mean waking time is within 10 minutes of the actual mean waking time.

6. Use the results from Exercise 1. Now suppose you want 99% confidence with a maximum error of 2 minutes. How many people would you need to survey?

7. Use the results from Exercise 2. Determine the minimum survey size that is necessary to be 95% confident that the sample mean length of work commute is within 2 miles of the actual mean length of work commute.

8. Use the results from Exercise 2. Now suppose you want 98% confidence with a maximum error of 0.5 mile. How many people would you need to survey?

9. 1.415
10. 1.323
11. 2.624
12. 2.756
13. 10.255
14. 0.021
15. 0.706
16. 10.556
17. (42.5, 63.1)
18. (3.479, 3.521)
19. (6.1, 7.5)
20. (14.6, 35.8)
21. (74, 86)
22. (69, 91)
23. 0.65, 0.35
24. 0.290, 0.710

Section 6.2

In Exercises 9–12, find the critical value t_c for the given confidence level c and sample size n.

9. $c = 0.80, n = 8$

10. $c = 0.80, n = 22$

11. $c = 0.98, n = 15$

12. $c = 0.99, n = 30$

In Exercises 13–16, find the margin of error for μ.

13. $c = 0.90, s = 23.4, n = 16, \bar{x} = 52.8$

14. $c = 0.95, s = 0.05, n = 25, \bar{x} = 3.5$

15. $c = 0.98, s = 0.9, n = 12, \bar{x} = 6.8$

16. $c = 0.99, s = 16.5, n = 20, \bar{x} = 25.2$

17. Construct the confidence interval for μ using the statistics in Exercise 13.

18. Construct the confidence interval for μ using the statistics in Exercise 14.

19. Construct the confidence interval for μ using the statistics in Exercise 15.

20. Construct the confidence interval for μ using the statistics in Exercise 16.

21. In a random sample of 15 CD players brought in for repair, the average repair cost was $80 and the standard deviation was $14. Construct a 90% confidence interval for μ. Assume the repair costs are normally distributed. *(Adapted from Consumer Reports)*

22. Repeat Exercise 21 using a 99% confidence interval.

Section 6.3

In Exercises 23–30, let p be the proportion of the population who respond yes. Use the given information to find $\hat{p}$ and $\hat{q}$.

23. In a survey of 2120 college students, 1378 have credit cards. *(Source: The Harris Poll)*

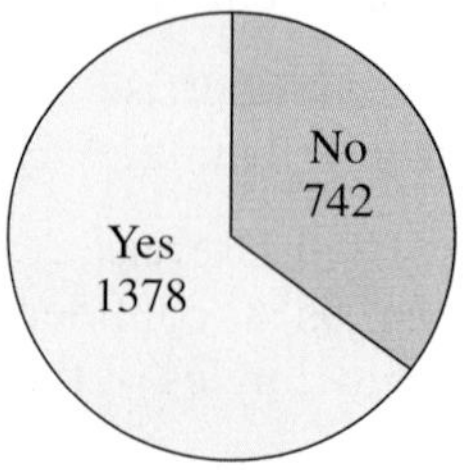

24. In a survey of 3462 U.S. adults, 1004 felt that education should be the highest priority for future economic growth. *(Source: The Harris Poll)*

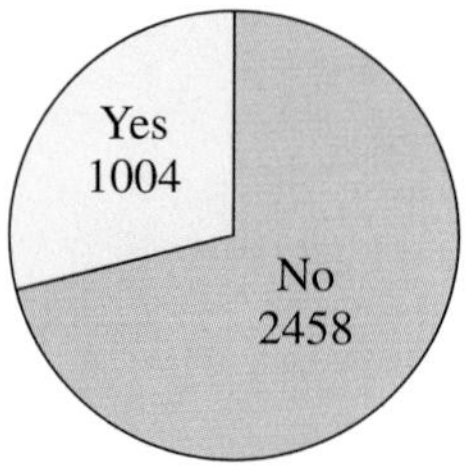

25. 0.591, 0.409
26. 0.125, 0.875
27. 0.18, 0.82
28. 0.698, 0.302
29. 0.4, 0.6
30. 0.520, 0.480
31. (0.630, 0.670)
32. (0.270, 0.310)
33. (0.566, 0.616)
34. (0.101, 0.149)

25. In a survey of 1011 adults, 597 said that they follow professional football. *(Source: The Harris Poll)*

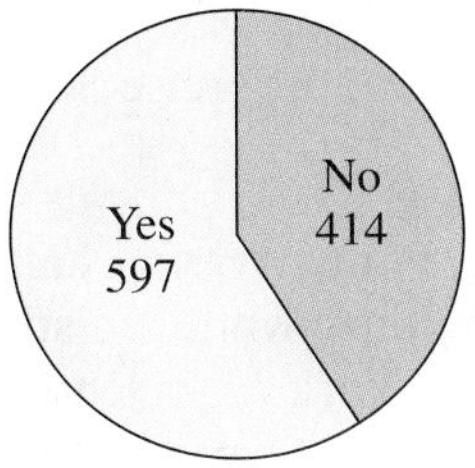

26. In a survey of 1000 U.S. adults, 125 are making the minimum payment(s) on their credit card(s). *(Source: Cambridge Consumer Credit Index)*

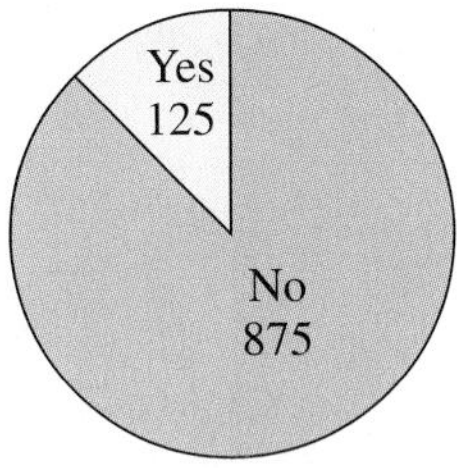

27. In a survey of 644 U.S. teens, 116 said they think their business leaders are ethical. *(Source: USA TODAY)*

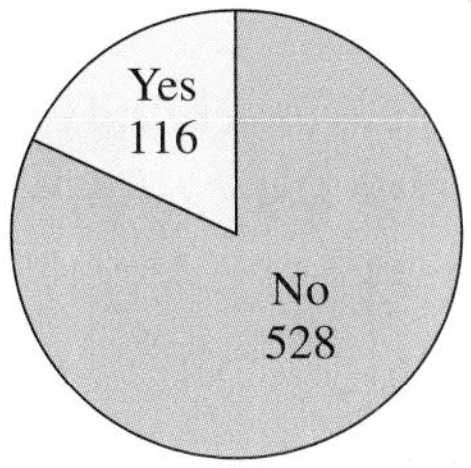

28. In a survey of 957 U.S. adults, 668 said they approve of labor unions. *(Adapted from The Gallup Organization)*

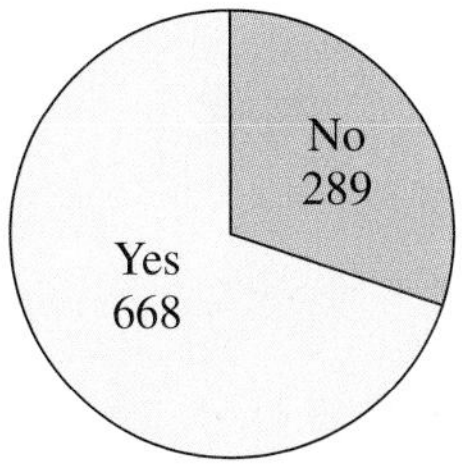

29. In a survey of 1371 U.S. adult Internet users, 548 said that they have high-speed Internet access in their homes. *(Source: Pew Internet and American Life Project)*

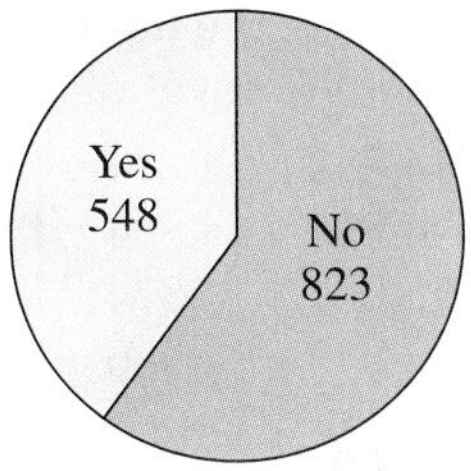

30. In a survey of 2365 U.S. adults, 1230 said they worry most about missing deductions when filing their taxes. *(Adapted from USA TODAY)*

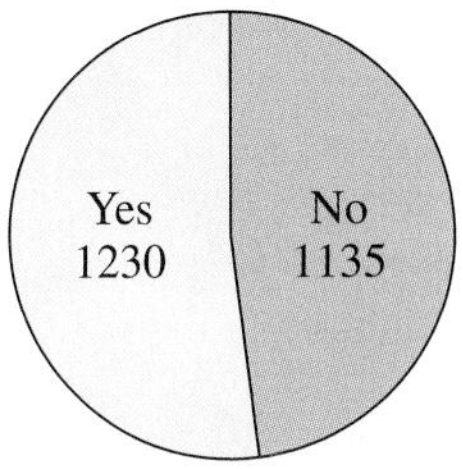

In Exercises 31–38, construct the indicated confidence interval for the population proportion p.

31. Use the sample in Exercise 23 with $c = 0.95$.

32. Use the sample in Exercise 24 with $c = 0.99$.

33. Use the sample in Exercise 25 with $c = 0.90$.

34. Use the sample in Exercise 26 with $c = 0.98$.

35. (0.141, 0.219)
36. (0.674, 0.722)
37. (0.374, 0.426)
38. (0.496, 0.544)
39. 273
40. 1879; the sample size is larger.
41. 23.337, 4.404
42. 42.980, 10.856
43. 14.067, 2.167
44. 23.589, 1.735
45. (0.0029, 0.0127); (0.0539, 0.1127)
46. (0.002, 0.017); (0.045, 0.130)
47. (0.82, 2.20); (0.91, 1.48)
48. (0.76, 2.47); (0.87, 1.57)

35. Use the sample in Exercise 27 with $c = 0.99$.

36. Use the sample in Exercise 28 with $c = 0.90$.

37. Use the sample in Exercise 29 with $c = 0.95$.

38. Use the sample in Exercise 30 with $c = 0.98$.

39. A study of Pennsylvania high school wrestlers found that 23% had sustained an injury while wrestling that year. You want to conduct a similar study on New York high school wrestlers, and you wish to estimate, with 95% confidence, the proportion of injured wrestlers within 5%. Use the prior study's proportion to find the minimum sample size required. *(Source: The Pennsylvania Athletic Trainers' Society, Inc.)*

40. Repeat Exercise 39, using a 99% confidence level and a margin of error of 2.5%. How does this sample size compare with your answer from Exercise 39?

Section 6.4

In Exercises 41–44, find the critical values χ_R^2 and χ_L^2 needed to estimate σ^2 for the given confidence level c and sample size n.

41. $c = 0.95, n = 13$

42. $c = 0.98, n = 25$

43. $c = 0.90, n = 8$

44. $c = 0.99, n = 10$

In Exercises 45–48, construct the indicated confidence intervals for σ^2 and σ. Assume the samples are each taken from a normally distributed population.

45. A random sample of the liquid content (in fluid ounces) of 16 beverage cans is shown. Use a 95% level of confidence.

14.816	14.863	14.814	14.998
14.965	14.824	14.884	14.838
14.916	15.021	14.874	14.856
14.860	14.772	14.980	14.919

46. Repeat Exercise 45 using a 99% level of confidence.

DATA **47.** A random sample of the sodium content (in milligrams) of 24 boxes of corn flakes cereal is shown. Use a 90% level of confidence.

290.8	292.5	288.0	291.9	290.3
291.4	290.0	290.3	289.9	289.3
291.2	290.6	289.8	289.7	291.1
292.9	288.7	290.9	289.9	290.6
291.4	289.7	290.0	291.3	

48. Repeat Exercise 47 using a 95% level of confidence.

6 Chapter Quiz

Take this quiz as you would take a quiz in class. After you are done, check your work against the answers given in the back of the book.

DATA **1.** The following data set represents the repair costs (in dollars) for a random sample of 30 dishwashers. *(Adapted from Consumer Reports)*

41.82	52.81	57.80	68.16	73.48	78.88	88.13	88.79
90.07	90.35	91.68	91.72	93.01	95.21	95.34	96.50
100.05	101.32	103.59	104.19	105.62	111.32	117.14	118.42
118.77	119.01	120.70	140.52	141.84	147.06		

(a) Find the point estimate of the population mean.

(b) Find the margin of error for a 95% level of confidence.

(c) Construct a 95% confidence interval for the population mean and interpret the results.

2. You want to estimate the mean repair cost for dishwashers. The estimate must be within \$10 of the population mean. Determine the required sample size to construct a 99% confidence interval for the population mean. Assume the population standard deviation is \$22.50. *(Adapted from Consumer Reports)*

3. The following data set represents the time (in minutes) for a random sample of phone calls made by employees at a company.

7.5 2.0 12.1 8.8 9.4 7.3 1.9 2.8 7.0 7.3

(a) Find the sample mean.

(b) Find the sample standard deviation.

(c) Use the t-distribution to construct a 90% confidence interval for the population mean and interpret the results. Assume the population of the data set is normally distributed.

(d) Repeat part (c), assuming $\sigma = 3.5$ minutes. Compare the results.

4. In a random sample of eight people with advanced degrees in biology, the mean monthly income was \$4744 and the standard deviation was \$580. Assume the monthly incomes are normally distributed and use a t-distribution to construct a 95% confidence interval for the population mean monthly income for people with advanced degrees in biology. *(Adapted from U.S. Bureau of Labor Statistics)*

5. In a survey of 1037 adults from the United States age 65 and over, 643 were concerned about getting the flu. *(Source: Harvard School for Public Health)*

(a) Find a point estimate for the population proportion p of those concerned about getting the flu.

(b) Construct a 90% confidence interval for the population proportion.

(c) Find the minimum sample size needed to estimate the population proportion at the 99% confidence level in order to ensure that the estimate is accurate within 4% of the population proportion.

6. Refer to the data set in Exercise 1. Assume the population of dishwasher repair costs is normally distributed.

(a) Construct a 95% confidence interval for the population variance.

(b) Construct a 95% confidence interval for the population standard deviation.

1. (a) 98.110
(b) 8.847
(c) (89.263, 106.957)
You are 95% confident that the mean repair cost is between \$89.26 and \$106.96.

2. 34

3. (a) 6.61
(b) 3.38
(c) (4.65, 8.57)
(d) (4.79, 8.43)
The t-CI is wider because less information is available.

4. (4259, 5229)

5. (a) 0.620
(b) (0.595, 0.645)
(c) 977

6. (a) (387.650, 1104.514)
(b) (19.689, 33.234)

PUTTING IT ALL TOGETHER

Real Statistics ■ Real Decisions

United States Environmental Protection Agency
www.epa.gov

As part of the U.S. Environmental Protection Agency's (EPA) efforts to "protect human health and safeguard the natural environment," the EPA conducts the Urban Air Toxics Monitoring Program (UATMP). The program has gathered thousands of air samples and analyzed them for concentrations of over 50 different organic compounds, such as acetylene (used for cutting and welding metals and in the production of neoprene rubber, plastics, and resins). The results from UATMP are used to gain insight into the effects of air pollution and if efforts to clean up the air are working.

For instance, using air samples from a major city, the EPA can analyze the results and estimate the mean concentration of acetylene in the air using a 95% confidence interval. They can then compare the interval with previous years' results to see if there are any trends and if there has been a significant change in the amount of acetylene in the air.

You work for the EPA and are asked to interpret the results shown in the graph at the right. The graph shows the point estimate for the population mean concentration and the 95% confidence interval for μ for acetylene over a three-year period. The data are based on air samples taken at one major city.

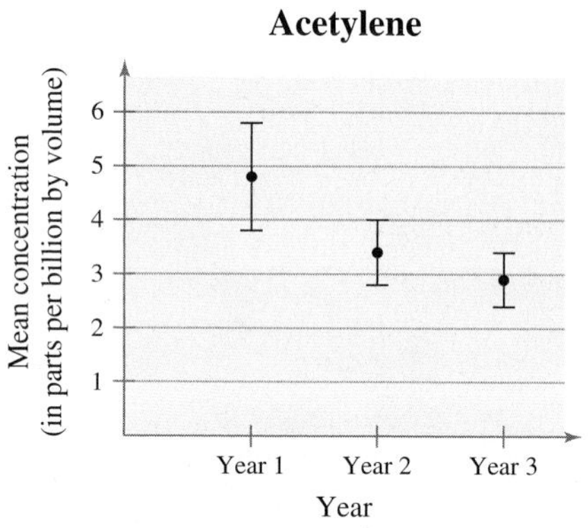

Exercises

1. *Interpreting the Results*

Consider the graph of the mean concentration levels of acetylene. For the following years, decide if there has been a change in the mean concentration level of acetylene. Explain your reasoning.

(a) From Year 1 to Year 2

(b) From Year 2 to Year 3

(c) From Year 1 to Year 3

2. *What Can You Conclude?*

Using the results of Exercise 1, what can you conclude about the efforts to reduce the concentration of acetylene in the air?

3. *How Do You Think They Did It?*

How do you think the EPA constructed the 95% confidence interval for the population mean concentration of the organic compounds in the air? Do the following to answer the question (you do not need to make any calculations).

(a) What sampling distribution do you think they used? Why?

(b) Do you think they used the population standard deviation in calculating the margin of error? Why or why not? If not, what could they have used?

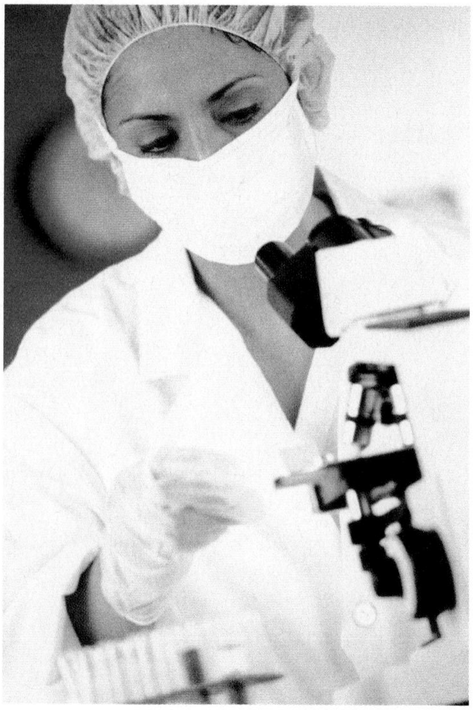

TECHNOLOGY MINITAB EXCEL TI-83

THE GALLUP ORGANIZATION

Most Admired Polls

WWW.GALLUP.COM

Since 1948, the Gallup Organization has conducted a "most admired" poll. The methodology for the 2003 poll is described at the right.

Survey Question

What man that you have heard or read about, living today in any part of the world, do you admire most? And who is your second choice?*

*Survey respondents are asked an identical question about most admired woman.

"The results are based on telephone interviews with a randomly selected national sample of 1004 adults, 18 and over, conducted December 5–7, 2003. For results based on samples of this size, one can say with 95 percent confidence that the error attributable to sampling and other random effects could be plus or minus 3 percentage points. In addition to sampling error, question wording and practical difficulties in conducting surveys can introduce error or bias into the findings of public opinion polls."

Exercises

1. In 2003, 40 people named Pope John Paul II as their most admired man. Use a technology tool to find a 95% confidence interval for the proportion that would have chosen the pope.

2. Does the confidence interval you obtained in Exercise 1 agree with the statement issued by the Gallup Organization that the proportion is 4% plus or minus 3%? Explain.

3. In 2003, the most named woman was Hillary Clinton. The second named woman was Oprah Winfrey, who was named by 7% of the people in the sample. Use a technology tool to find a 95% confidence interval for the proportion of the population that would have chosen Oprah Winfrey.

4. Use a technology tool to simulate a most admired poll. Assume that the actual population proportion who most admire Oprah Winfrey is 10%. Run the simulation several times using $n = 1004$.

(a) What was the least value you obtained for $\hat{p}$?

(b) What was the greatest value you obtained for $\hat{p}$?

MINITAB

Generate 200 rows of data

Store in column(s): C1

Number of trials: 1004

Probability of success: .1

5. Is it possible that the actual proportion of the population that most admired Oprah Winfrey was 10% or greater? Explain your reasoning.

Extended solutions are given in the *Technology Supplement.* Technical instruction is provided for MINITAB, Excel, and the TI-83.

6 Using Technology to Construct Confidence Intervals

Here are some MINITAB and TI-83 printouts for some examples in this chapter. To duplicate the MINITAB results, you need the original data. For the TI-83, you can simply enter the descriptive statistics. Answers may be slightly different because of rounding.

Display Descriptive Statistics...
Store Descriptive Statistics...
1-Sample Z...
1-Sample t...
2-Sample t...
Paired t...
1 Proportion...
2 Proportions...

(See Example 3, page 283.)

9	20	18	16	9	16	16	9	11	13	22	16	5	18	6	6	5	12
25	17	23	7	10	9	10	10	5	11	18	18	9	9	17	13	11	7
14	6	11	12	11	15	6	12	14	11	4	9	18	12	12	17	11	20

MINITAB

Z Confidence Intervals

The assumed sigma = 5

Variable	N	Mean	StDev	SE Mean	95.0 % CI
C1	54	12.426	5.015	0.680	(11.092, 13.760)

Display Descriptive Statistics...
Store Descriptive Statistics...
1-Sample Z...
1-Sample t...
2-Sample t...
Paired t...
1 Proportion...
2 Proportions...

(See Example 2, page 297.)

159°F	173°F	162°F	151°F	173°F	162°F	148°F	172°F
167°F	170°F	151°F	153°F	172°F	143°F	166°F	170°F

MINITAB

T Confidence Intervals

Variable	N	Mean	StDev	SE Mean	95.0 % CI
C3	16	162.00	10.00	2.50	(156.67, 167.33)

(See Example 2, page 305.)

Display Descriptive Statistics...
Store Descriptive Statistics...
1-Sample Z...
1-Sample t...
2-Sample t...
Paired t...
1 Proportion...
2 Proportions...

MINITAB

Test and Confidence Interval for One Proportion

Test of p = 0.29 vs p not = 0.29

Sample	X	N	Sample p	95.0 % CI	Z-Value	P-Value
1	293	1011	0.289812	(0.261847, 0.317777)	-0.01	0.989

(See Example 5, page 285.)

TI-83

EDIT CALC **TESTS**
1: Z–Test...
2: T–Test...
3: 2–SampZTest...
4: 2–SampTTest...
5: 1–PropZTest...
6: 2–PropZTest...
7↓ ZInterval...

↓

TI-83

ZInterval
Inpt: Data **Stats**
σ: 1.5
$\bar{x}$: 22.9
n: 20
C–Level: .9
Calculate

↓

TI-83

ZInterval
(22.348, 23.452)
$\bar{x}$= 22.9
n= 20

(See Example 3, page 298.)

TI-83

EDIT CALC **TESTS**
2↑ T–Test...
3: 2–SampZTest...
4: 2–SampTTest...
5: 1–PropZTest...
6: 2–PropZTest...
7: ZInterval...
8↓ TInterval...

↓

TI-83

TInterval
Inpt: Data **Stats**
$\bar{x}$: 6.93
Sx: .42
n: 20
C–Level: .99
Calculate

↓

TI-83

TInterval
(6.6613, 7.1987)
$\bar{x}$= 6.93
Sx= .42
n= 20

(See Example 2, page 305.)

TI-83

EDIT CALC **TESTS**
5↑ 1–PropZTest...
6: 2–PropZTest...
7: ZInterval...
8: TInterval...
9: 2–SampZInt...
0: 2–SampTInt...
A↓ 1–PropZInt...

↓

TI-83

1–PropZInt
x: 293
n: 1011
C–Level: .95
Calculate

↓

TI-83

1–PropZInt
(.26185, .31778)
$\hat{p}$= .2898120673
n= 1011

CHAPTER 7

Hypothesis Testing with One Sample

7.1 Introduction to Hypothesis Testing

7.2 Hypothesis Testing for the Mean (Large Samples)

Case Study

7.3 Hypothesis Testing for the Mean (Small Samples)

7.4 Hypothesis Testing for Proportions

7.5 Hypothesis Testing for Variance and Standard Deviation

Uses and Abuses

Real Statistics–Real Decisions

Technology

Computer software is protected by federal copyright laws. Each year, software companies lose billions of dollars because of pirated software. Federal criminal penalties for software piracy can include fines up to $250,000 and jail terms of up to five years.

Where You've Been

In Chapter 6, you began your study of inferential statistics. There, you learned how to form a confidence interval estimate about a population parameter, such as the proportion of people in the United States who agree with a certain statement. For instance, in a nationwide poll conducted by *Harris Interactive* on behalf of the Business Software Alliance (BSA), U.S. students age 8 to 18 years were asked several questions about their attitudes toward copyright law and Internet behavior. Here are some of the results.

Survey Question	Number Surveyed	Number Who Said Yes
Is software protected by copyright law?	1183	1017
Did you learn about the laws against downloading files without permission from your teachers?	1017	183
Have you ever downloaded software from the Internet without paying for it?	1183	260
Did you download the software from home?	260	143

Where You're Going

In this chapter, you will continue your study of inferential statistics. But now, instead of making an estimate about a population parameter, you will learn how to test a claim about a parameter.

For instance, suppose that you work for *Harris Interactive* and are asked to test a claim that the proportion of U.S. students age 8 to 18 who are aware that software is protected by copyright law is $p = 0.80$. To test the claim, you take a random sample of $n = 1183$ students and find that 1017 of them are aware that software is protected by copyright law. Your sample statistic is $\hat{p} = 0.860$.

Is your sample statistic different enough from the claim ($p = 0.80$) to decide that the claim is false? The answer lies in the sampling distribution of sample proportions taken from a population in which $p = 0.80$. The graph below shows that your sample statistic is over 5 standard errors from the claim value. If the claim is true, the probability of the sample statistic's being 5 standard errors or more from the claimed value is extremely small. Something is wrong! If your sample was truly random, then you can conclude that the actual proportion of the student population is not 0.80. In other words, you tested the original claim (hypothesis), and you decided to reject it.

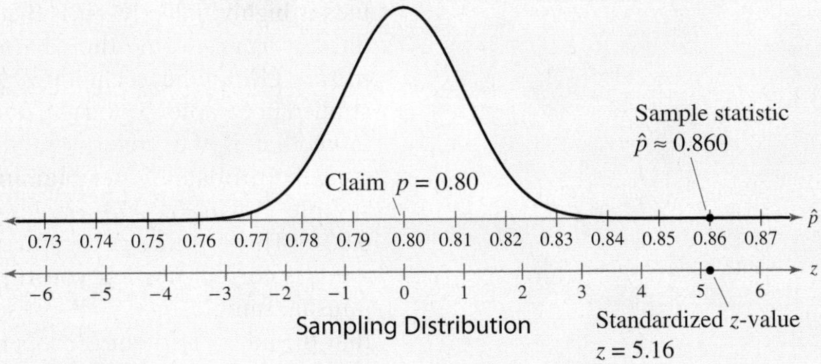

7.1 Introduction to Hypothesis Testing

What You Should Learn

- A practical introduction to hypothesis tests
- How to state a null hypothesis and an alternative hypothesis
- How to identify type I and type II errors and interpret the level of significance
- How to know whether to use a one-tailed or two-tailed statistical test and finding a *P*-value
- How to make and interpret a decision based on the results of a statistical test
- How to write a claim for a hypothesis test

Hypothesis Tests • Stating a Hypothesis • Types of Errors and Level of Significance • Statistical Tests and *P*-values • Making a Decision and Interpreting the Decision • Strategies for Hypothesis Testing

Hypothesis Tests

Throughout the remainder of this course, you will study an important technique in inferential statistics called hypothesis testing. A **hypothesis test** is a process that uses sample statistics to test a claim about the value of a population parameter. Researchers in fields such as medicine, psychology, and business rely on hypothesis testing to make informed decisions about new medicines, treatments, and marketing strategies.

For instance, suppose a cellular phone company advertises that the mean talk time of a new cellular phone battery is 180 minutes. If you suspect that the mean talk time is not 180 minutes, how could you show that the advertisement is false?

Obviously, you can't test *all* the batteries, but you can still make a reasonable decision about the mean talk time by taking a random sample from the population of batteries and measuring the talk time of each. If the sample mean differs enough from the advertisement's mean, you can decide that the advertisement is wrong.

For instance, to test that the mean talk time of all cellular phone batteries of this type is $\mu = 180$ minutes, you could take a random sample of $n = 100$ batteries and measure the talk time of each. Suppose you obtain a sample mean of $\bar{x} = 174$ minutes with a sample standard deviation of $s = 20$ minutes. Does this indicate that the company's advertisement is false?

To decide, you do something unusual—*you assume the advertisement is correct!* That is, you assume that $\mu = 180$. Then, you examine the sampling distribution of sample means (with $n = 100$) taken from a population in which $\mu = 180$ and $\sigma = 20$. From the Central Limit Theorem, you know this sampling distribution is normal with a mean of 180 and standard error of

$$\frac{20}{\sqrt{100}} = 2.$$

Sampling Distribution of $\bar{x}$

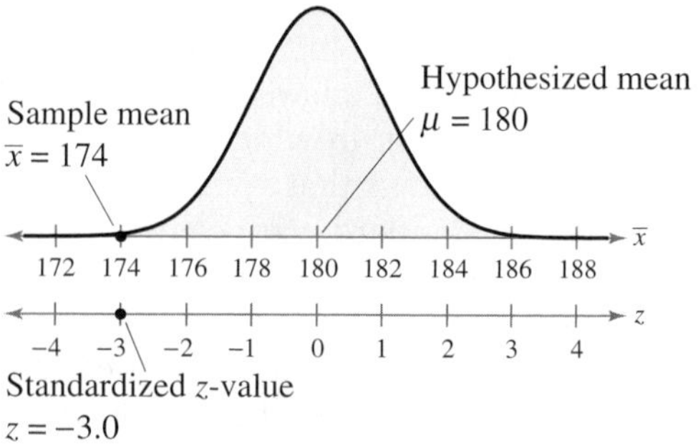

In the graph at the right, notice that your sample mean of $\bar{x} = 174$ minutes is highly unlikely—it is 3 standard errors from the claimed mean! Using the techniques you studied in Chapter 5, you can determine that if the advertisement is true, the probability of obtaining a sample mean of 174 or less is 0.0013. This is a rare occurrence! Your assumption that the company's advertisement is correct has led you to an improbable result. So, either you had a very unusual sample, or the advertisement is probably false. The logical conclusion is that the advertisement is probably false.

Insight

As you study this chapter, don't get confused regarding concepts of certainty and importance. For instance, even if you were very certain that the mean talk time of a type of cellular phone battery is not 180 minutes, the actual mean talk time might be very close to this value and the difference might not be important.

Note to Instructor

Some texts state the null hypothesis using the strict equality symbol. We use the symbol that is complementary to the alternative hypothesis.

Insight

The term *null hypothesis* was introduced by Ronald Fisher (see page 29). If the statement in the null hypothesis is not true, then the alternative hypothesis must be true.

Stating a Hypothesis

A verbal statement, or claim, about a population parameter is called a **statistical hypothesis.** To test a population parameter, you should carefully state a pair of hypotheses—one that represents the claim and the other, its complement. When one of these hypotheses is false, the other must be true. Either hypothesis—the *null hypothesis* or the *alternative hypothesis*—may represent the original claim.

DEFINITION

1. A **null hypothesis** $\boldsymbol{H_0}$ is a statistical hypothesis that contains a statement of equality, such as $\leq$, $=$, or $\geq$.
2. The **alternative hypothesis** $\boldsymbol{H_a}$ is the complement of the null hypothesis. It is a statement that must be true if H_0 is false and it contains a statement of inequality, such as $>$, $\neq$, or $<$.

H_0 is read as "H subzero" or "H naught" and H_a is read as "H sub-a."

To write the null and alternative hypotheses, translate the claim made about the population parameter from a verbal statement to a mathematical statement. Then, write its complement. For instance, if the claim value is k and the population parameter is μ, then some possible pairs of null and alternative hypotheses are

$$\begin{cases} H_0: \mu \leq k \\ H_a: \mu > k \end{cases} \qquad \begin{cases} H_0: \mu \geq k \\ H_a: \mu < k \end{cases} \qquad \begin{cases} H_0: \mu = k \\ H_a: \mu \neq k \end{cases}.$$

Regardless of which of the three pairs of hypotheses you use, you always assume $\mu = k$ and examine the sampling distribution on the basis of this assumption.

The following table shows the relationship between possible verbal statements about the parameter μ and the corresponding null and alternative hypotheses. Similar statements can be made to test other population parameters, such as p, σ, or σ^2.

Verbal Statement H_0 *The mean is . . .*	Mathematical Statements	Verbal Statement H_a *The mean is . . .*
. . . greater than or equal to k. *. . . at least k.* *. . . not less than k.*	$\begin{cases} H_0: \mu \geq k \\ H_a: \mu < k \end{cases}$	*. . . less than k.* *. . . below k.* *. . . fewer than k.*
. . . less than or equal to k. *. . . at most k.* *. . . not more than k.*	$\begin{cases} H_0: \mu \leq k \\ H_a: \mu > k \end{cases}$	*. . . greater than k.* *. . . above k.* *. . . more than k.*
. . . equal to k. *. . . k.* *. . . exactly k.*	$\begin{cases} H_0: \mu = k \\ H_a: \mu \neq k \end{cases}$	*. . . not equal to k.* *. . . different from k.* *. . . not k.*

Picturing the World

In a study performed at the Cleveland Clinic, a sample of 50 randomly chosen adults who developed symptoms of the common cold were given five zinc gluconate throat lozenges per day. The sample mean duration of nasal congestion was four days. So, the researchers claimed that the mean duration of nasal congestion of all adults who take zinc lozenges after developing cold symptoms is four days. *(Adapted from Annals of Internal Medicine)*

Determine a null hypothesis and alternative hypothesis for this claim.

Note to Instructor

Begin with a hypothesis statement and ask students to state its logical complement. Some students will have difficulty with the fact that the complement of $\mu \neq k$ is $\mu = k$. Discuss the role of a double negative in English. The important point is that if you conclude that H_0 is false, then you are also concluding that H_a is true.

EXAMPLE 1

Stating the Null and Alternative Hypotheses

Write the claim as a mathematical sentence. State the null and alternative hypotheses, and identify which represents the claim.

1. A university publicizes that the proportion of its students who graduate in 4 years is 82%.
2. A water faucet manufacturer announces that the mean flow rate of a certain type of faucet is less than 2.5 gallons per minute.
3. A cereal company advertises that the mean weight of the contents of its 20-ounce size cereal boxes is more than 20 ounces.

SOLUTION

1. The claim "the proportion . . . is 82%" can be written as $p = 0.82$. Its complement is $p \neq 0.82$. Because $p = 0.82$ contains the statement of equality, it becomes the null hypothesis. In this case, the null hypothesis represents the claim.

   H_0: $p = 0.82$ (Claim)

   H_a: $p \neq 0.82$

2. The claim "the mean . . . is less than 2.5 gallons per minute" can be written as $\mu < 2.5$. Its complement is $\mu \geq 2.5$. Because $\mu \geq 2.5$ contains the statement of equality, it becomes the null hypothesis. In this case, the alternative hypothesis represents the claim.

   H_0: $\mu \geq 2.5$ gallons per minute

   H_a: $\mu < 2.5$ gallons per minute (Claim)

3. The claim "the mean . . . is more than 20 ounces" can be written as $\mu > 20$. Its complement is $\mu \leq 20$. Because $\mu \leq 20$ contains the statement of equality, it becomes the null hypothesis. In this case, the alternative hypothesis represents the claim.

   H_0: $\mu \leq 20$ ounces

   H_a: $\mu > 20$ ounces (Claim)

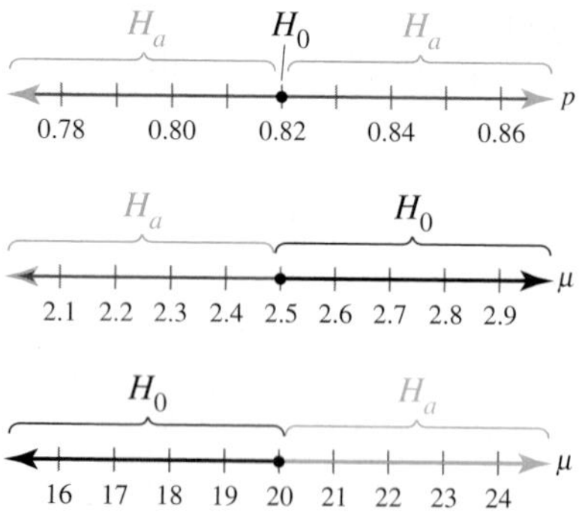

In each of these graphs, notice that each point on the number line is in H_0 or H_a, but no point is in both.

Try It Yourself 1

Write the claim as a mathematical sentence. State the null and alternative hypotheses, and identify which represents the claim.

1. A consumer analyst reports that the mean life of a certain type of automobile battery is not 74 months.
2. A television manufacturer publishes that the variance of the life of a certain type of television is less than or equal to 3.5.
3. A radio station publicizes that its proportion of the local listening audience is greater than 39%.

a. *Identify* the verbal claim and *write* it as a mathematical statement.
b. *Write* the complement of the claim.
c. *Identify* the null and alternative hypotheses and *determine* which one represents the claim.

Answer: Page A39

Types of Errors and Level of Significance

No matter which hypothesis represents the claim, you always begin a hypothesis test by assuming that the equality condition in the null hypothesis is true. So, when you perform a hypothesis test, you make one of two decisions:

1. reject the null hypothesis or
2. fail to reject the null hypothesis.

Because your decision is based on a sample rather than the entire population, there is always the possibility you will make the wrong decision.

For instance, suppose you claim that a certain coin is not fair. To test your claim, you flip the coin 100 times and get 49 heads and 51 tails. You would probably agree that you do not have enough evidence to support your claim. Even so, it is possible that the coin is actually not fair and you had an unusual sample.

But what if you flip the coin 100 times and get 21 heads and 79 tails? It would be a rare occurrence to get only 21 heads out of 100 tosses with a fair coin. So, you probably have sufficient evidence to support your claim that the coin is not fair. However, you can't be 100% sure. It is possible that the coin is fair and you had an unusual sample.

If p represents the proportion of heads, the claim that "the coin is not fair" can be written as the mathematical statement $p \neq 0.5$. Its complement, "the coin is fair," is written as $p = 0.5$. So, your null hypothesis and alternative hypothesis are

$$H_0\text{: } p = 0.5$$

and

$$H_a\text{: } p \neq 0.5. \text{ (Claim)}$$

Remember, the only way to be absolutely certain of whether H_0 is true or false is to test the entire population. Because your decision—to reject H_0 or to fail to reject H_0—is based on a sample, you must accept the fact that your decision might be incorrect. You might reject a null hypothesis when it is actually true. Or, you might fail to reject a null hypothesis when it is actually false.

DEFINITION

A **type I error** occurs if the null hypothesis is rejected when it is true.

A **type II error** occurs if the null hypothesis is not rejected when it is false.

The following table shows the four possible outcomes of a hypothesis test.

	Truth of H_0	
Decision	H_0 is true.	H_0 is false.
Do not reject H_0.	Correct decision	Type II error
Reject H_0.	Type I error	Correct decision

	Truth About Defendant	
Verdict	Innocent	Guilty
Not guilty	Justice	Type II error
Guilty	Type I error	Justice

Hypothesis testing is sometimes compared to the legal system used in the United States. Under this system, the following steps are used.

1. A carefully worded accusation is written.
2. The defendant is assumed innocent (H_0) until proven guilty. The burden of proof lies with the prosecution. If the evidence is not strong enough, there is no conviction. A "not guilty" verdict does not prove that a defendant is innocent.
3. The evidence needs to be conclusive beyond a reasonable doubt. The system assumes that more harm is done by convicting the innocent (type I error) than by not convicting the guilty (type II error).

EXAMPLE 2

Identifying Type I and Type II Errors

The USDA limit for salmonella contamination for chicken is 20%. A meat inspector reports that the chicken produced by a company exceeds the USDA limit. You perform a hypothesis test to determine whether the meat inspector's claim is true. When will a type I or type II error occur? Which is more serious? *(Source: United States Department of Agriculture)*

SOLUTION Let p represent the proportion of the chicken that is contaminated. The meat inspector's claim is "more than 20% is contaminated." You can write the null and alternative hypotheses as follows.

H_0: $p \leq 0.2$ The proportion is less than or equal to 20%.

H_a: $p > 0.2$ (Claim) The proportion is greater than 20%.

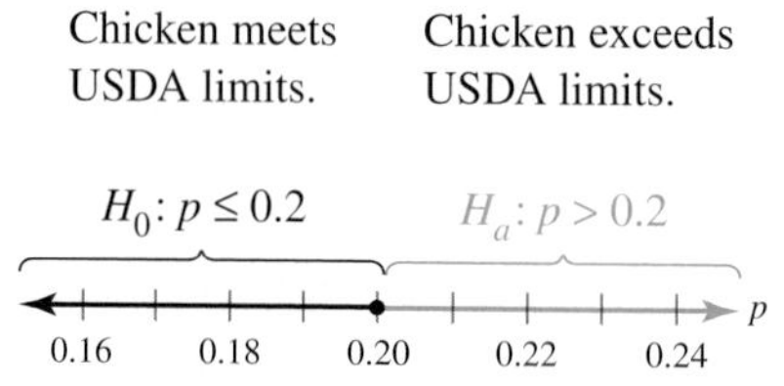

A type I error will occur if the actual proportion of contaminated chicken is less than or equal to 0.2, but you decide to reject H_0. A type II error will occur if the actual proportion of contaminated chicken is greater than 0.2, but you do not reject H_0. With a type I error, you might create a health scare and hurt the sales of chicken producers who were actually meeting the USDA limits. With a type II error, you could be allowing chicken that exceeded the USDA contamination limit to be sold to consumers. A type II error could result in sickness or even death.

Try It Yourself 2

A company specializing in parachute assembly states that its main parachute failure rate is not more than 1%. You perform a hypothesis test to determine whether the company's claim is false. When will a type I or type II error occur? Which is more serious?

a. *State* the null and alternative hypotheses.
b. *Write* the possible type I and type II errors.
c. *Determine* which error is more serious.

Answer: Page A39

Because there is variation from sample to sample, there is always a possibility that you will reject a null hypothesis when it is actually true. You can decrease the probability of doing so by lowering the *level of significance*.

Insight

When you decrease α (the maximum allowable probability of making a type I error), you are likely to be increasing β.

DEFINITION

In a hypothesis test, the **level of significance** is your maximum allowable probability of making a type I error. It is denoted by α, the lowercase Greek letter alpha.

The probability of a type II error is denoted by β, the lowercase Greek letter beta.

By setting the level of significance at a small value, you are saying that you want the probability of rejecting a true null hypothesis to be small. Three commonly used levels of significance are $\alpha = 0.10$, $\alpha = 0.05$, and $\alpha = 0.01$.

Statistical Tests and *P*-values

Note to Instructor

You can use an example of "false positive" and "false negative" results for a medical test (say diabetes) to discuss type I and type II errors. You might also want to point out that computation of β is beyond the scope of an introductory statistics text.

After stating the null and alternative hypotheses and specifying the level of significance, the next step in a hypothesis test is to obtain a random sample from the population and calculate sample statistics such as the mean and the standard deviation. The statistic that is compared with the parameter in the null hypothesis is called the **test statistic.** The type of test used and the sampling distribution are based on the test statistic.

In this chapter, you will learn about several one-sample statistical tests. The following table shows the relationships between population parameters and their corresponding test statistics and standardized test statistics.

Population parameter	Test statistic	Standardized test statistic
μ	$\bar{x}$	z (Section 7.2, $n \geq 30$)
		t (Section 7.3, $n < 30$)
p	$\hat{p}$	z (Section 7.4)
σ^2	s^2	χ^2 (Section 7.5)

One way to decide whether to reject the null hypothesis is to determine whether the probability of obtaining the standardized test statistic (or one that is more extreme) is less than the level of significance.

DEFINITION

If the null hypothesis is true, a ***P*-value** (or **probability value**) of a hypothesis test is the probability of obtaining a sample statistic with a value as extreme or more extreme than the one determined from the sample data.

The P-value of a hypothesis test depends on the nature of the test. There are three types of hypothesis tests—a left-, right-, or two-tailed test. The type of test depends on the region of the sampling distribution that favors a rejection of H_0. This region is indicated by the alternative hypothesis.

DEFINITION

1. If the alternative hypothesis H_a contains the less-than inequality symbol ($<$), the hypothesis test is a **left-tailed test.**

H_0: $\mu \geq k$
H_a: $\mu < k$

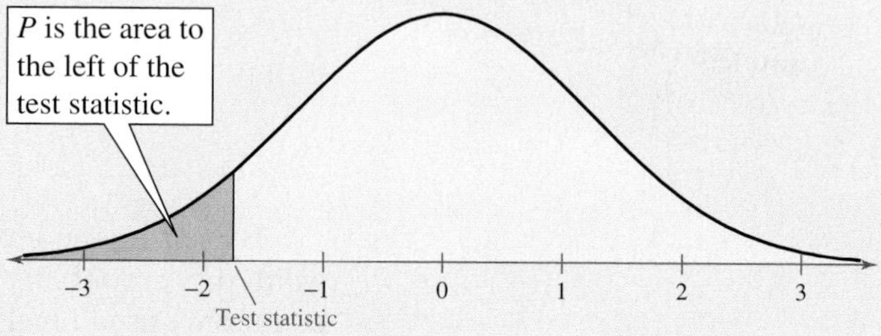

Left-Tailed Test

2. If the alternative hypothesis H_a contains the greater-than inequality symbol ($>$), the hypothesis test is a **right-tailed test.**

H_0: $\mu \leq k$
H_a: $\mu > k$

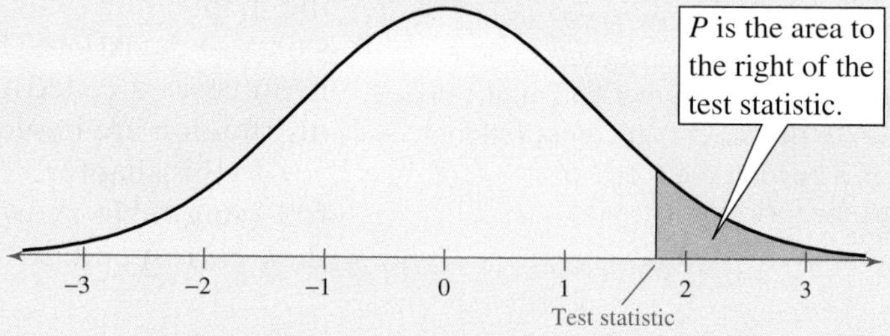

Right-Tailed Test

3. If the alternative hypothesis H_a contains the not-equal-to symbol ($\neq$), the hypothesis test is a **two-tailed test.** In a two-tailed test, each tail has an area of $\frac{1}{2}P$.

H_0: $\mu = k$
H_a: $\mu \neq k$

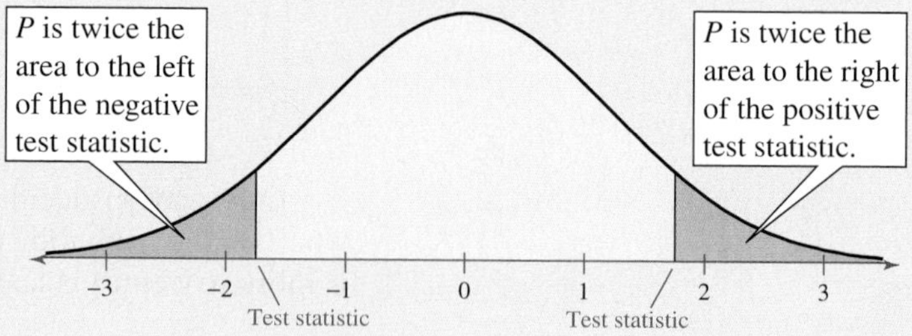

Two-Tailed Test

Study Tip

The third type of test is called a two-tailed test because evidence that would support the alternative hypothesis could lie in either tail of the sampling distribution.

The smaller the P-value of the test, the more evidence there is to reject the null hypothesis. A very small P-value indicates a rare occurrence. Remember, however, that even a very low P-value does not constitute proof that the null hypothesis is false.

EXAMPLE 3

Identifying the Nature of a Hypothesis Test

For each claim, state H_0 and H_a in words and in symbols. Then determine whether the hypothesis test is a left-tailed test, right-tailed test, or two-tailed test. Sketch a normal sampling distribution and shade the area for the P-value.

1. A university publicizes that the proportion of its students who graduate in 4 years is 82%.
2. A water faucet manufacturer announces that the mean flow rate of a certain type of faucet is less than 2.5 gallons per minute (gpm).
3. A cereal company advertises that the mean weight of the contents of its 20-ounce size cereal boxes is more than 20 ounces.

SOLUTION

In Symbols	***In Words***
1. H_0: $p = 0.82$	The proportion of students who graduate in 4 years is 82%.
H_a: $p \neq 0.82$	The proportion of students who graduate in 4 years is not 82%.

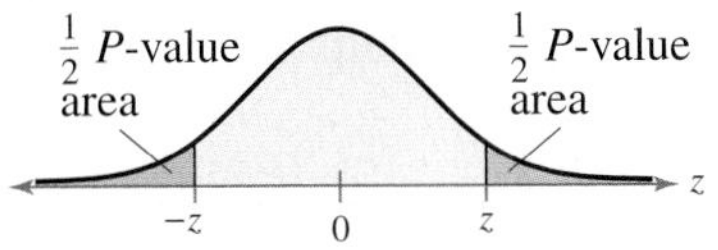

Because H_a contains the $\neq$ symbol, the test is a two-tailed hypothesis test. The graph of the normal sampling distribution at the left shows the shaded area for the P-value.

2. H_0: $\mu \geq 2.5$ gpm	The mean flow rate of a certain type of faucet is greater than or equal to 2.5 gallons per minute.
H_a: $\mu < 2.5$ gpm	The mean flow rate of a certain type of faucet is less than 2.5 gallons per minute.

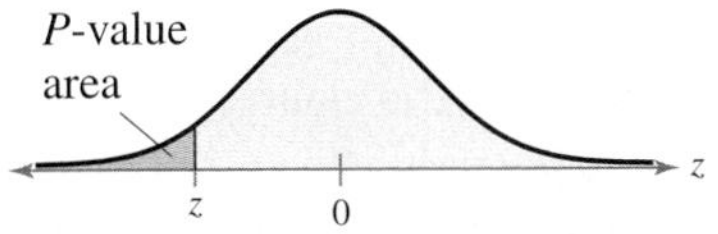

Because H_a contains the $<$ symbol, the test is a left-tailed hypothesis test. The graph of the normal sampling distribution at the left shows the shaded area for the P-value.

3. H_0: $\mu \leq 20$ oz	The mean weight of the contents of the cereal boxes is less than or equal to 20 ounces.
H_a: $\mu > 20$ oz	The mean weight of the contents of the cereal boxes is greater than 20 ounces.

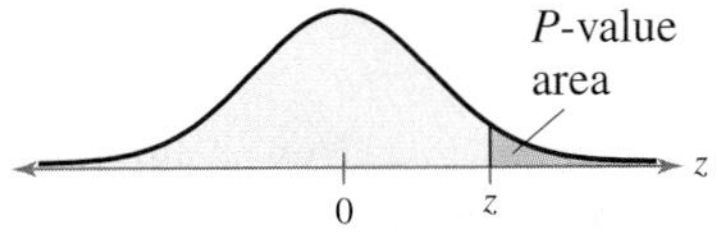

Because H_a contains the $>$ symbol, the test is a right-tailed hypothesis test. The graph of the normal sampling distribution at the left shows the shaded area for the P-value.

Try It Yourself 3

For each claim, determine whether the hypothesis test is a left-, right-, or two-tailed test. Sketch a normal sampling distribution and shade the area for the P-value.

1. A consumer analyst reports that the mean life of a certain type of automobile battery is 74 months.
2. A radio station publicizes that its proportion of the local listening audience is greater than 39%.

a. *Write* H_0 and H_a.
b. *Determine* whether the test is left-tailed, right-tailed, or two-tailed.
c. *Sketch* the sampling distribution and shade the area for the P-value.

Answer: Page A40

Making a Decision and Interpreting the Decision

To conclude a hypothesis test, you make a decision and interpret that decision. There are only two possible outcomes to a hypothesis test: (1) reject the null hypothesis, and (2) fail to reject the null hypothesis.

Insight

In this chapter, you will learn that there are two basic types of **decision rules** for deciding to reject H_0 or fail to reject H_0. The decision rule described on this page is based on *P*-values. The second basic type of decision rule is based on rejection regions. When the standardized test statistic falls in the rejection region, the observed probability (*P*-value) of a type I error is less than α. You will learn more about rejection regions in the next section.

Decision Rule Based on *P*-value

To use a *P*-value to make a conclusion in a hypothesis test, compare the *P*-value with α.

1. If $P \leq \alpha$, then reject H_0.

2. If $P > \alpha$, then fail to reject H_0.

Failing to reject the null hypothesis does not mean that you have accepted the null hypothesis as true. It simply means that there is not enough evidence to reject the null hypothesis. If you want to support a claim, state it so that it becomes the alternative hypothesis. If you want to disprove a claim, state it so that it becomes the null hypothesis. The following table will help you interpret your decision.

	Claim	
Decision	Claim is H_0.	Claim is H_a.
Reject H_0.	There is enough evidence to reject the claim.	There is enough evidence to support the claim.
Fail to reject H_0.	There is not enough evidence to reject the claim.	There is not enough evidence to support the claim.

EXAMPLE 4

Interpreting a Decision

You perform a hypothesis test for each of the following claims. How should you interpret your decision if you reject H_0? If you fail to reject H_0?

1. H_0 (Claim): A university publicizes that the proportion of its students who graduate in 4 years is 82%.
2. H_a (Claim): A government safety administration reports that the mean stopping distance (on a dry surface) for a Chevrolet Malibu is less than 148 feet.

SOLUTION

1. If you reject H_0, then you should conclude "there is sufficient evidence to indicate that the university's claim is false." If you fail to reject H_0, then you should conclude "there is insufficient evidence to indicate that the university's claim (of a four-year graduation rate of 82%) is false."
2. The null hypothesis is "the mean stopping distance ... is greater than or equal to 148 feet." If you reject H_0, then you should conclude "there is enough evidence to support the government safety administration's claim (the stopping distance for a Chevrolet Malibu is less than 148 feet)." If you fail to reject H_0, then you should conclude "there is not enough evidence to support the government safety administration's claim that the stopping distance for a Chevrolet Malibu is less than 148 feet."

Try It Yourself 4

You perform a hypothesis test for the following claim. How should you interpret your decision if you reject H_0? If you fail to reject H_0?

H_a (Claim): A radio station publicizes that its proportion of the local listening audience is greater than 39%.

a. Interpret your decision if you reject the null hypothesis.
b. Interpret your decision if you do not reject the null hypothesis.

Answer: Page A40

The general steps for a hypothesis test are summarized below.

Steps for Hypothesis Testing

1. State the claim mathematically and verbally. Identify the null and alternative hypotheses.

H_0: ? $\quad H_a$: ?

2. Specify the level of significance.

$\alpha =$?

3. Determine the standardized sampling distribution and sketch its graph.

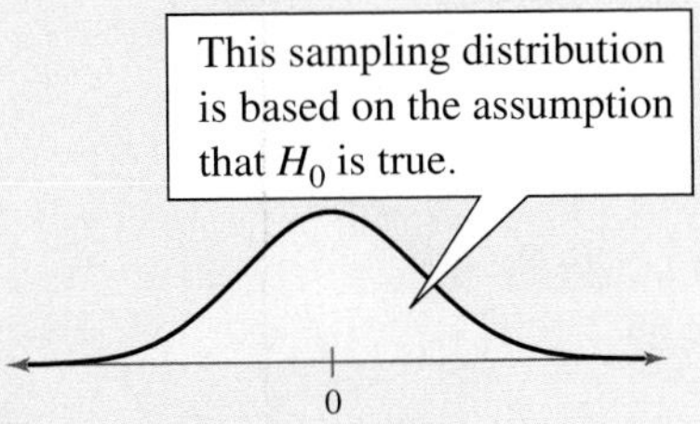

4. Calculate the test statistic and its standardized value. Add it to your sketch.

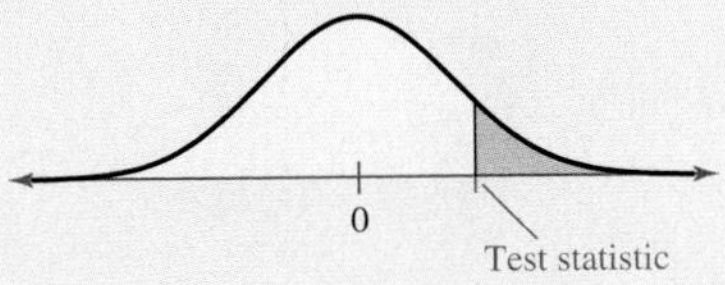

5. Find the P-value.

6. Use the following decision rule.

Is the P-value less than or equal to the level of significance?

No → Fail to reject H_0.

Yes → Reject H_0.

7. Write a statement to interpret the decision in the context of the original claim.

In the steps above, the graphs show a right-tailed test. However, the same basic steps also apply to left-tailed and two-tailed tests.

Strategies for Hypothesis Testing

In a courtroom, the strategy used by an attorney depends on whether the attorney is representing the defense or the prosecution. In a similar way, the strategy that you will use in hypothesis testing should depend on whether you are trying to support or reject a claim. Remember that you cannot use a hypothesis test to support your claim if your claim is the null hypothesis. So, as a researcher, if you want a conclusion that supports your claim, word your claim so it is the alternative hypothesis. If you want to reject a claim, word it so it is the null hypothesis.

EXAMPLE 5

Writing the Hypotheses

A medical research team is investigating the benefits of a new surgical treatment. One of the claims is that the mean recovery time for patients after the new treatment is about 96 hours. How would you write the null and alternative hypotheses if (1) you are on the research team and want to support the claim? (2) you are on an opposing team and want to reject the claim?

SOLUTION

1. To answer this question, first think about the context of the claim. Because you are investigating the benefits of a new treatment, your concern is that the recovery time for patients after the new treatment will be less than 96 hours. Because you want to support this claim, make it the alternative hypothesis. So, H_a: $\mu < 96$ hours. Its complement, $\mu \geq 96$ hours, would be the null hypothesis.

   H_0: $\mu \geq 96$

   H_a: $\mu < 96$ (Claim)

2. First think about the context of the claim. As an opposing researcher, you do not want the recovery time to be 96 hours or less. Because you want to reject this claim, make it the null hypothesis. So, H_0: $\mu \leq 96$ hours. Its complement, $\mu > 96$ hours, would be the alternative hypothesis.

   H_0: $\mu \leq 96$ (Claim)

   H_a: $\mu > 96$

Try It Yourself 5

1. You represent a chemical company that is being sued for paint damage to automobiles. You want to support the claim that the mean repair cost per automobile is about \$650. How would you write the null and alternative hypotheses?
2. You are on a research team that is investigating the mean temperature of adult humans (see page 363). The commonly accepted claim is that the mean temperature is about 98.6°F. You want to show that this claim is false. How would you write the null and alternative hypotheses?

a. *Think* about the context of the claim.
b. *Determine* whether you want to support or reject the claim.
c. *Write* the null and alternative hypotheses.

Answer: Page A40

7.1 Exercises

Help
MyMathLab
Student Study Pack

Building Basic Skills and Vocabulary

1. What are the two types of hypotheses used in a hypothesis test? How are they related?

2. Describe the two types of error possible in a hypothesis test decision.

True or False? In Exercises 3–8, determine whether the statement is true of false. If it is false, rewrite it as a true statement.

3. In a hypothesis test, you assume the alternative hypothesis is true.

4. A statistical hypothesis is a statement about a sample.

5. If you decide to reject the null hypothesis, you can support the alternative hypothesis.

6. The level of significance is the maximum probability you allow for rejecting a null hypothesis when it is actually true.

7. A large P-value in a test will favor a rejection of the null hypothesis.

8. If you want to support a claim, write it as your null hypothesis.

Stating Hypotheses In Exercises 9–14, use the given statement to represent a claim. Write its complement and state which is H_0 and which is H_a.

9. $\mu \leq 645$

10. $\mu < 128$

11. $\sigma \neq 5$

12. $\sigma^2 \geq 1.2$

13. $p < 0.45$

14. $p = 0.21$

Graphical Analysis In Exercises 15–18, match the alternative hypothesis with its graph. Then state the null hypothesis and sketch its graph.

15. H_a: $\mu > 3$

16. H_a: $\mu < 3$

17. H_a: $\mu \neq 3$

18. H_a: $\mu > 2$

(a) 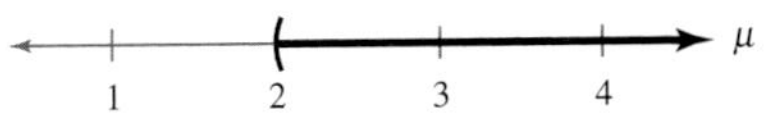

(b) 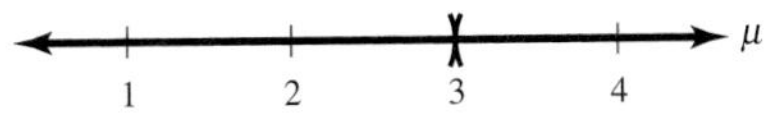

(c) 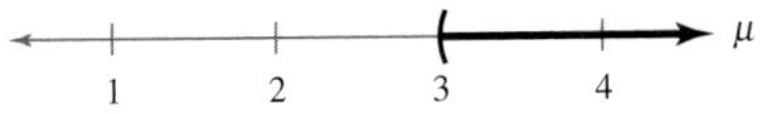

(d) 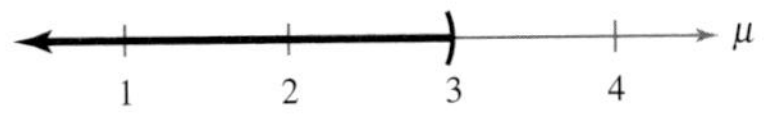

Identifying Tests In Exercises 19–22, determine whether the hypothesis test with the given null and alternative hypotheses is left-tailed, right-tailed, or two-tailed.

19. H_0: $\mu \leq 8.0$
H_a: $\mu > 8.0$

20. H_0: $\sigma \geq 5.2$
H_a: $\sigma < 5.2$

21. H_0: $\sigma^2 = 142$
H_a: $\sigma^2 \neq 142$

22. H_0: $p = 0.25$
H_a: $p \neq 0.25$

---

1. The two types of hypotheses used in a hypothesis test are the null hypothesis and the alternative hypothesis.

   The alternative hypothesis is the complement of the null hypothesis.

2. A type I error occurs if the null hypothesis is rejected when it is true.

   A type II error occurs if the null hypothesis is not rejected when it is false.

3. False. In a hypothesis test, you assume the null hypothesis is true.

4. False. A statistical hypothesis is a statement about a population.

5. True 6. True

7. False. A small P-value in a test will favor a rejection of the null hypothesis.

8. See Selected Answers, page A96.

9. H_0: $\mu \leq 645$; H_a: $\mu > 645$

10. H_0: $\mu \geq 128$; H_a: $\mu < 128$

11. H_0: $\sigma = 5$; H_a: $\sigma \neq 5$

12. H_0: $\sigma^2 \geq 1.2$; H_a: $\sigma^2 < 1.2$

13. H_0: $p \geq 0.45$: H_a: $p < 0.45$

14. H_0: $p = 0.21$; H_a: $p \neq 0.21$

15. c; H_0: $\mu \leq 3$ 

16. d; H_0: $\mu \geq 3$ 

17. b; H_0: $\mu = 3$ 

18. a; H_0: $\mu \leq 2$ 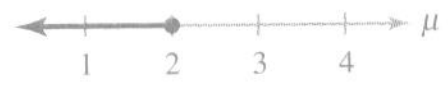

19. Right-tailed 20. Left-tailed

21. Two-tailed 22. Two-tailed

Using and Interpreting Concepts

Stating the Hypotheses In Exercises 23–28, state the claim mathematically. Write the null and alternative hypotheses. Identify which is the claim.

23. ***Light Bulbs*** A light bulb manufacturer claims that the mean life of a certain type of light bulb is more than 750 hours.

24. ***Shipping Errors*** As stated by a company's shipping department, the number of shipping errors per million shipments has a standard deviation that is less than 3.

25. ***Base Price of a Car*** The standard deviation of the base price of a certain type of car is no more than $1220.

26. ***College Students*** A research organization reports that 33% of the residents in Ann Arbor, Michigan are college students. *(Source: U.S. Census Bureau)*

27. ***Seat Belts*** The results of a recent study show that the proportion of people in the western United States who use seat belts when riding in a car or truck is 84%. *(Source: National Center for Statistics and Analysis)*

28. ***Survival Times*** A study claims that the mean survival time for certain cancer patients treated immediately with chemotherapy and radiation is 24 months.

Identifying Errors In Exercises 29–34, write sentences describing type I and type II errors for a hypothesis test of the indicated claim.

29. ***Repeat Buyers*** A car dealer claims that at least 24% of its new customers will return to buy their next car.

30. ***Literary Skills*** A study claims that the proportion of adults in the United States with rudimentary literary skills is 21%. *(Source: U.S. Department of Education)*

31. ***Chess*** A local chess club claims that the length of time to play a game has a standard deviation of more than 23 minutes.

32. ***ER Visits*** A hospital spokesperson states that 2% of emergency room visits by college undergraduates are for alcohol-related health problems.

33. ***Computers*** According to a recent survey, 88% of college students own a computer. *(Source: Harris Poll)*

34. ***Wristwatch Batteries*** A battery manufacturer guarantees that the standard deviation of the life of its wristwatch batteries is less than 5 months.

Identifying Tests In Exercises 35–40, determine whether the hypothesis test for each claim is left-tailed, right-tailed, or two-tailed. Explain your reasoning.

35. ***Security Alarms*** At least 14% of all homeowners have a home security alarm.

36. ***Clocks*** A manufacturer of grandfather clocks claims that the mean time its clocks lose is no more than 0.02 second per day.

37. ***Lung Cancer*** A government report claims the proportion of lung cancer cases that are due to smoking is 90%.

38. ***Tires*** The mean life of a certain tire is no less than 50,000 miles. *(Adapted from Goodyear)*

23. $\mu > 750$
H_0: $\mu \le 750$; H_a: $\mu > 750$ (claim)

24. $\sigma < 3$
H_0: $\sigma \ge 3$; H_a: $\sigma < 3$ (claim)

25. $\sigma \le 1220$
H_0: $\sigma \le 1220$ (claim)
H_a: $\sigma > 1220$

26. $p = 0.33$
H_0: $p = 0.33$ (claim); H_a: $p \ne 0.33$

27. $p = 0.84$
H_0: $p = 0.84$ (claim); H_a: $p \ne 0.84$

28. $\mu = 24$
H_0: $\mu = 24$ (claim); H_a: $\mu \ne 24$

29. Type I: Rejecting H_0: $p \ge 0.24$ when actually $p \ge 0.24$
Type II: Not rejecting H_0: $p \ge 0.24$ when actually $p < 0.24$

30. Type I: Rejecting H_0: $p = 0.21$ when actually $p = 0.21$
Type II: Not rejecting H_0: $p = 0.21$ when actually $p \ne 0.21$

31. Type I: Rejecting H_0: $\sigma \le 23$ when actually $\sigma \le 23$
Type II: Not rejecting H_0: $\sigma \le 23$ when actually $\sigma > 23$

32. Type I: Rejecting H_0: $p = 0.02$ when actually $p = 0.02$
Type II: Not rejecting H_0: $p = 0.02$ when actually $p \ne 0.02$

33. Type I: Rejecting H_0: $p = 0.88$ when actually $p = 0.88$
Type II: Not rejecting H_0: $p = 0.88$ when actually $p \ne 0.88$

34. Type I: Rejecting H_0: $\sigma \ge 5$ when actually $\sigma \ge 5$
Type II: Not rejecting H_0: $\sigma \ge 5$ when actually $\sigma < 5$

35. Left-tailed because the alternative hypothesis contains $<$.

36. Right-tailed because the alternative hypothesis contains $>$.

37. Two-tailed because the alternative hypothesis contains $\ne$.

38. Left-tailed because the alternative hypothesis contains $<$.

39. Two-tailed because the alternative hypothesis contains $\neq$.
40. Right-tailed because the alternative hypothesis contains $>$.
41. (a) There is enough evidence to support the company's claim.
(b) There is not enough evidence to support the company's claim.
42. (a) There is enough evidence to reject the Postal Service's claim.
(b) There is not enough evidence to reject the Postal Service's claim.
43. (a) There is enough evidence to support the Department of Labor's claim.
(b) There is not enough evidence to decide that the Department of Labor's claim is true.
44. (a) There is enough evidence to reject the manufacturer's claim.
(b) There is not enough evidence to reject the manufacturer's claim.
45. (a) There is enough evidence to reject the manufacturer's claim.
(b) There is not enough evidence to reject the manufacturer's claim.
46. (a) There is enough evidence to reject the soft-drink maker's claim.
(b) There is not enough evidence to reject the soft-drink maker's claim.
47. H_0: $\mu = 10$; H_a: $\mu \neq 10$
48. H_0: $\mu = 21$; H_a: $\mu \neq 21$
49. (a) H_0: $\mu \leq 15$; H_a: $\mu > 15$
(b) H_0: $\mu \geq 15$; H_a: $\mu < 15$
50. (a) H_0: $\mu \leq 28$
(b) H_0: $\mu \geq 28$

39. *Return Rate* A financial analyst claims that the return rate of a 15-year U.S. bond has a standard deviation of 5.3%.

40. *Dreams* A research institute claims the mean length of most dreams is greater than 10 minutes. *(Adapted from The Lucidity Institute)*

Interpreting a Decision In Exercises 41–46, consider each claim. If a hypothesis test is performed, how should you interpret a decision that

(a) rejects the null hypothesis?

(b) fails to reject the null hypothesis?

41. *Pictures Developed* A film company claims that the mean number of pictures developed for a standard roll of film with 24 exposures is more than 22.

42. *Shipment Weights* A government worker claims that the standard deviation of the mean weight of all U.S. Postal Service shipments is 0.40 pound.

43. *Hourly Wages* The U.S. Department of Labor claims the proportion of hourly workers earning over $10.40 per hour is greater than 75%. *(Adapted from U.S. Department of Labor)*

44. *Gas Mileage* An automotive manufacturer claims the standard deviation for the gas mileage of its models is 3.9 miles per gallon.

45. *Car Prices* The mean price of a new compact car is $22,035.

46. *Calories* A soft-drink maker claims the mean calorie content of its beverages is 26 calories per serving.

47. *Writing a Claim: Medical Research* Your medical research team is investigating the proper dose of a certain heart medication. The medicine manufacturer thinks that the mean dose should be 10 milligrams. You support this claim. How would you write the null and alternative hypotheses?

48. *Writing a Claim: Taxicab Company* A taxicab company claims that the mean travel time between two destinations is about 21 minutes. You work for the bus company and want to reject this claim. How would you write the null and alternative hypotheses?

49. *Writing Hypotheses: Refrigerator Manufacturer* A refrigerator manufacturer claims that the mean life of its refrigerators is about 15 years. You are asked to perform a hypothesis test to test this claim. How would you write the null and alternative hypothesis if

(a) you represent the manufacturer and want to support the claim?

(b) you represent a consumer group and want to reject the claim?

50. *Writing Hypotheses: Internet Provider* An Internet provider is trying to gain advertising deals and claims that the mean time a customer spends online per day is about 28 minutes. You are asked to test this claim. How would you write the null hypothesis if

(a) you represent the Internet provider and want to support the claim?

(b) you represent a competing advertiser and want to reject the claim?

51. If you decrease α, you are decreasing the probability that you will reject H_0. Therefore, you are increasing the probability of failing to reject H_0. This could increase β, the probability of failing to reject H_0 when H_0 is false.

52. There are no z-values that correspond to $\alpha = 0$. If $\alpha = 0$, the null hypothesis cannot be rejected and the hypothesis test is useless.

53. (a) Reject H_0 because the confidence interval is located below 70.
(b) Do not reject H_0 because the confidence interval includes values larger than 70.
(c) Do not reject H_0 because the confidence interval includes values larger than 70.

54. (a) Reject H_0 because the confidence interval is located above 54.
(b) Do not reject H_0 because the confidence interval includes values below 54.
(c) Do not reject H_0 because the confidence interval includes values below 54.

55. (a) Reject H_0 because the confidence interval is located above 0.20.
(b) Do not reject H_0 because the confidence interval includes values less than 0.20.
(c) Do not reject H_0 because the confidence interval includes values less than 0.20.

56. (a) Reject H_0 because the confidence interval is located below 0.73.
(b) Do not reject H_0 because the confidence interval includes values above 0.73.
(c) Do not reject H_0 because the confidence interval includes values above 0.73.

Extending Concepts

51. *Getting at the Concept* Why can decreasing the probability of a type I error cause an increase in the probability of a type II error?

52. *Getting at the Concept* Why not use a level of significance of $\alpha = 0$?

Graphical Analysis In Exercises 53–56, you are given a null hypothesis and three confidence intervals that represent three samplings. Decide whether each confidence interval indicates that you should reject H_0. Explain your reasoning.

53. H_0: $\mu \geq 70$

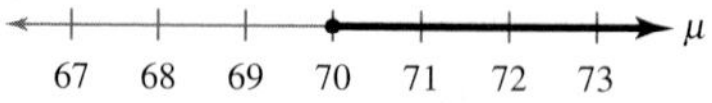

(a) $67 < \mu < 69$

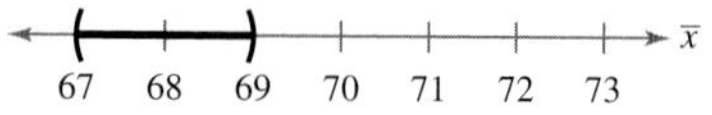

(b) $67 < \mu < 71$

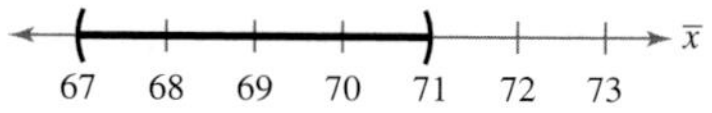

(c) $69.5 < \mu < 72.5$

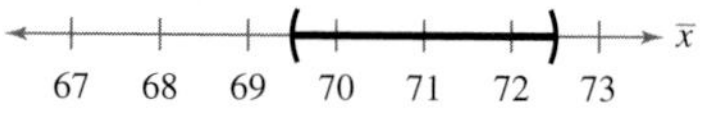

54. H_0: $\mu \leq 54$

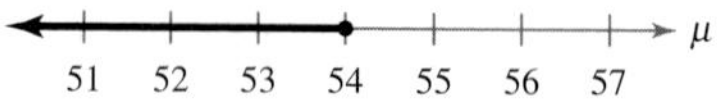

(a) $54.5 < \mu < 55.5$

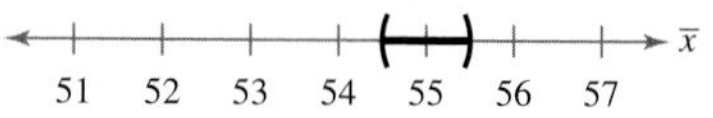

(b) $51.5 < \mu < 54.5$

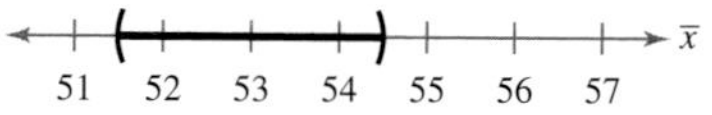

(c) $53.5 < \mu < 56.5$

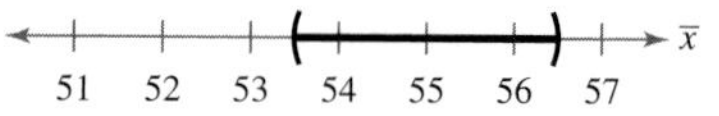

55. H_0: $p \leq 0.20$

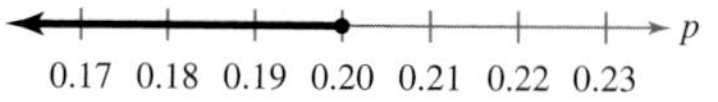

(a) $0.21 < p < 0.23$

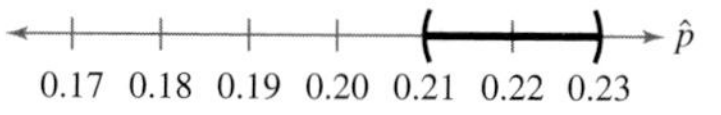

(b) $0.19 < p < 0.23$

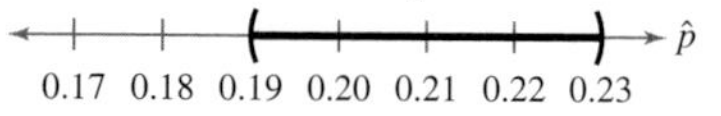

(c) $0.175 < p < 0.205$

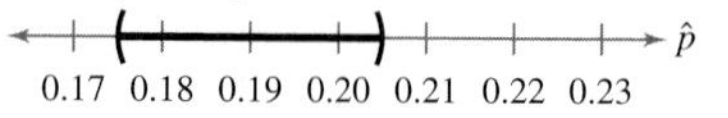

56. H_0: $p \geq 0.73$

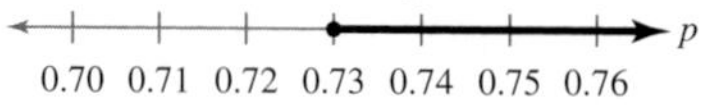

(a) $0.715 < p < 0.725$

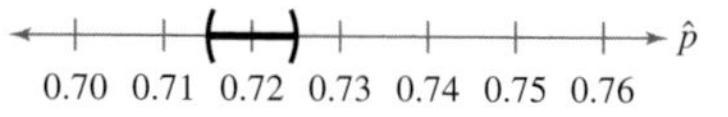

(b) $0.73 < p < 0.75$

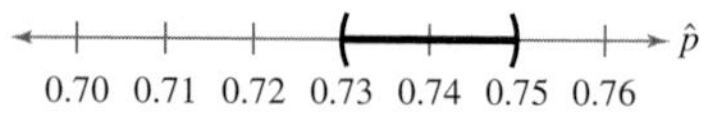

(c) $0.695 < p < 0.745$

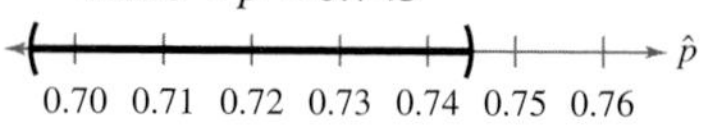

7.2 Hypothesis Testing for the Mean (Large Samples)

Using P-values to Make Decisions • Using P-values for a z-Test • Rejection Regions and Critical Values • Using Rejection Regions for a z-Test

What You Should Learn

- How to find P-values and use them to test a mean μ
- How to use P-values for a z-test
- How to find critical values and rejection regions in a normal distribution
- How to use rejection regions for a z-test

Using P-values to Make Decisions

In Chapter 5, you learned that when the sample size is at least 30, the sampling distribution for $\overline{x}$ (the sample mean) is normal. In Section 7.1, you learned that a way to reach a conclusion in a hypothesis test is to use a P-value for the sample statistic, such as $\overline{x}$. Recall that when you assume the null hypothesis is true, a P-value (or probability value) of a hypothesis test is the probability of obtaining a sample statistic with a value as extreme or more extreme than the one determined from the sample data. The decision rule for a hypothesis test based on a P-value is as follows.

Note to Instructor

If a P-value is less than 0.01, the null hypothesis will be rejected at the common levels of $\alpha = 0.01$, $\alpha = 0.05$, and $\alpha = 0.10$. If the P-value is greater than 0.10, then you would fail to reject H_0 for these common levels. Make sure students know that the same conclusion will be reached regardless of whether they use the critical value method or the P-value method.

Decision Rule Based on P-value

To use a P-value to make a conclusion in a hypothesis test, compare the P-value with α.

1. If $P \le \alpha$, then reject H_0.

2. If $P > \alpha$, then fail to reject H_0.

EXAMPLE 1

Interpreting a P-value

The P-value for a hypothesis test is $P = 0.0237$. What is your decision if the level of significance is (1) $\alpha = 0.05$ and (2) $\alpha = 0.01$?

SOLUTION

1. Because $0.0237 < 0.05$, you should reject the null hypothesis.
2. Because $0.0237 > 0.01$, you should fail to reject the null hypothesis.

Try It Yourself 1

The P-value for a hypothesis test is $P = 0.0347$. What is your decision if the level of significance is (1) $\alpha = 0.01$ and (2) $\alpha = 0.05$?

a. *Compare* the P-value with the level of significance.
b. *Make* your decision.

Answer: Page A40

Insight

The lower the P-value, the more evidence there is in favor of rejecting H_0. The P-value gives you the lowest level of significance for which the sample statistic allows you to reject the null hypothesis. In Example 1, you would reject H_0 at any level of significance greater than or equal to 0.0237.

Finding the P-value for a Hypothesis Test

After determining the hypothesis test's standardized test statistic and the test statistic's corresponding area, do one of the following to find the P-value.

a. For a left-tailed test, $P =$ (Area in left tail).

b. For a right-tailed test, $P =$ (Area in right tail).

c. For a two-tailed test, $P = 2$(Area in tail of test statistic).

EXAMPLE 2

Finding a P-value for a Left-Tailed Test

Find the P-value for a left-tailed hypothesis test with a test statistic of $z = -2.23$. Decide whether to reject H_0 if the level of significance is $\alpha = 0.01$.

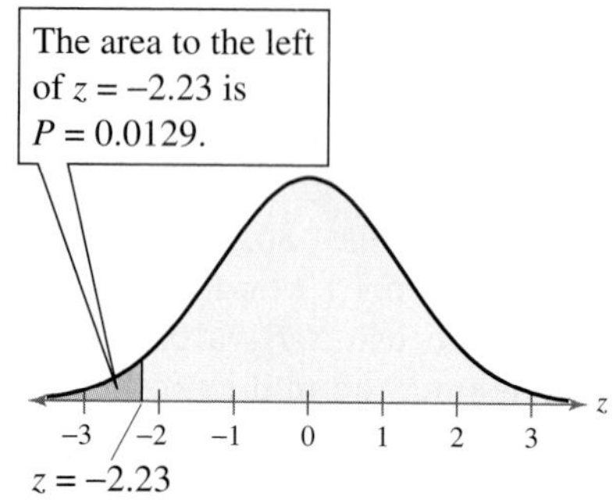

Left-Tailed Test

SOLUTION

The graph shows a standard normal curve with a shaded area to the left of $z = -2.23$. For a left-tailed test,

$$P = (\text{Area in left tail}).$$

From Table 4 in Appendix B, the area corresponding to $z = -2.23$ is 0.0129, which is the area in the left tail. So, the P-value for a left-tailed hypothesis test with a test statistic of $z = -2.23$ is $P = 0.0129$. Because $0.0129 > 0.01$, you should fail to reject H_0.

Try It Yourself 2

Find the P-value for a left-tailed hypothesis test with a test statistic of $z = -1.62$. Decide whether to reject H_0 if the level of significance is $\alpha = 0.05$.

a. *Use* Table 4 in Appendix B to locate the area that corresponds to $z = -1.62$.
b. *Calculate* the P-value for a left-tailed test, the area in the left tail.
c. *Compare* the P-value with α and *decide* whether to reject H_0.

Answer: Page A40

EXAMPLE 3

Finding a P-value for a Two-Tailed Test

Find the P-value for a two-tailed hypothesis test with a test statistic of $z = 2.14$. Decide whether to reject H_0 if the level of significance is $\alpha = 0.05$.

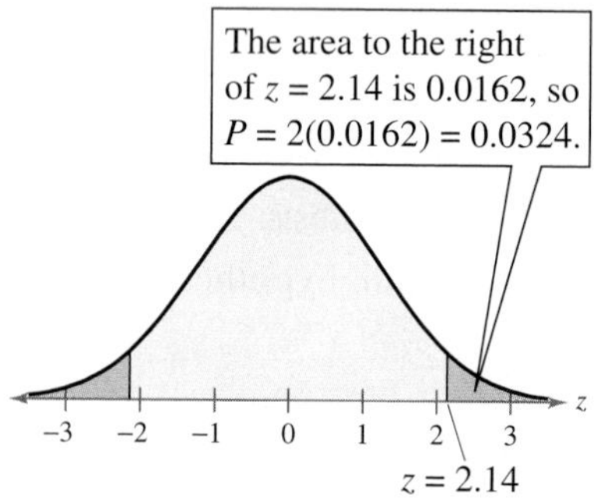

Two-Tailed Test

SOLUTION

The graph shows a standard normal curve with shaded areas to the left of $z = -2.14$ and to the right of $z = 2.14$. For a two-tailed test,

$$P = 2(\text{Area in tail of test statistic}).$$

From Table 4, the area corresponding to $z = 2.14$ is 0.9838. The area in the right tail is $1 - 0.9838 = 0.0162$. So, the P-value for a two-tailed hypothesis test with a test statistic of $z = 2.14$ is $P = 2(0.0162) = 0.0324$. Because $0.0324 < 0.05$, you should reject H_0.

Try It Yourself 3

Find the P-value for a two-tailed hypothesis test with a test statistic of $z = 2.31$. Decide whether to reject H_0 if the level of significance is $\alpha = 0.01$.

a. *Use* Table 4 to locate the area that corresponds to $z = 2.31$.
b. *Calculate* the P-value for a two-tailed test, twice the area in the tail of the test statistic.
c. *Compare* the P-value with α and *decide* whether to reject H_0.

Answer: Page A40

Using *P*-values for a *z*-Test

The z-test for the mean is used in populations for which the sampling distribution of sample means is normal. To use the z-test, you need to find the standardized value for your test statistic $\overline{x}$.

$$z = \frac{(\text{Sample mean}) - (\text{Hypothesized mean})}{\text{Standard error}}$$

z-Test for a Mean μ

The ***z*-test for a mean** is a statistical test for a population mean. The z-test can be used when the population is normal and σ is known, or for any population when the sample size n is at least 30. The **test statistic** is the sample mean $\overline{x}$ and the **standardized test statistic** is z.

$$z = \frac{\overline{x} - \mu}{\sigma/\sqrt{n}}$$

Recall that $\frac{\sigma}{\sqrt{n}}$ = standard error = $\sigma_{\overline{x}}$.

When $n \geq 30$, you can use the sample standard deviation s in place of σ.

Study Tip

With all hypothesis tests, it is helpful to sketch the sampling distribution. Your sketch should include the standardized test statistic.

Insight

When the sample size is at least 30, you know the following about the sampling distribution of sample means.

(1) The shape is normal.

(2) The mean is the hypothesized mean.

(3) The standard error is $s/\sqrt{n}$, where s can be used in place of σ.

GUIDELINES

Using *P*-values for a *z*-Test for Mean μ

In Words	*In Symbols*
1. State the claim mathematically and verbally. Identify the null and alternative hypotheses.	State H_0 and H_a.
2. Specify the level of significance.	Identify α.
3. Determine the standardized test statistic.	$z = \frac{\overline{x} - \mu}{\sigma/\sqrt{n}}$ or if $n \geq 30$, use $\sigma \approx s$.
4. Find the area that corresponds to z.	Use Table 4 in Appendix B.

5. Find the P-value.

a. For a left-tailed test, P = (Area in left tail).

b. For a right-tailed test, P = (Area in right tail).

c. For a two-tailed test, P = 2(Area in tail of test statistic).

6. Make a decision to reject or fail to reject the null hypothesis.	Reject H_0 if P-value is less than or equal to α. Otherwise, fail to reject H_0.
7. Interpret the decision in the context of the original claim.	

Note to Instructor

We use the same format for all hypothesis testing throughout the text. Using the same format makes it easier for students to understand the logic of the test. Emphasize that the sampling distribution, and consequently the logic of the test, is based on the assumption that the equality condition of the null hypothesis is true.

EXAMPLE 4

Hypothesis Testing Using P-values

In an advertisement, a pizza shop claims that its mean delivery time is less than 30 minutes. A random selection of 36 delivery times has a sample mean of 28.5 minutes and a standard deviation of 3.5 minutes. Is there enough evidence to support the claim at $\alpha = 0.01$? Use a P-value.

SOLUTION The claim is "the mean delivery time is less than 30 minutes." So, the null and alternative hypotheses are

H_0: $\mu \geq 30$ minutes and H_a: $\mu < 30$ minutes. (Claim)

The level of significance is $\alpha = 0.01$. The standardized test statistic is

$$z = \frac{\bar{x} - \mu}{\sigma/\sqrt{n}}$$ Because $n \geq 30$, use the z-test.

$$= \frac{28.5 - 30}{3.5/\sqrt{36}}$$ Because $n \geq 30$, use $\sigma \approx s = 3.5$. Assume $\mu = 30$.

$$\approx -2.57.$$

In Table 4 of Appendix B, the area corresponding to $z = -2.57$ is 0.0051. Because this test is a left-tailed test, the P-value is equal to the area to the left of $z = -2.57$. So, $P = 0.0051$. Because the P-value is less than $\alpha = 0.01$, you should decide to reject the null hypothesis.

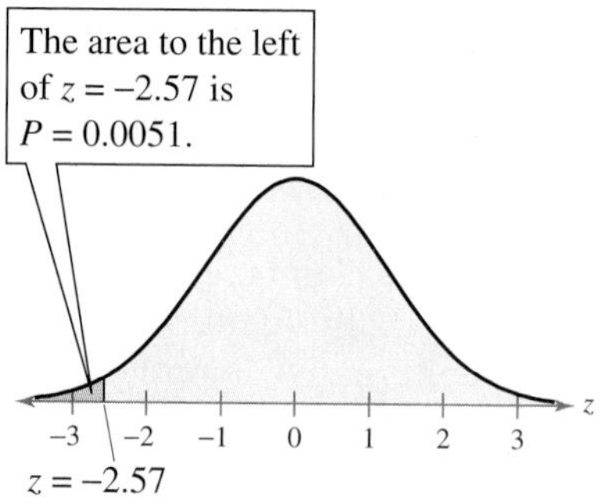

Left-Tailed Test

Interpretation At the 1% level of significance, you have sufficient evidence to conclude that the mean delivery time is less than 30 minutes.

Try It Yourself 4

Home owners claim that the mean speed of automobiles traveling on their street is greater than the speed limit of 35 miles per hour. A random sample of 100 automobiles has a mean speed of 36 miles per hour and a standard deviation of 4 miles per hour. Is there enough evidence to support the claim at $\alpha = 0.05$? Use a P-value.

a. *Identify* the claim. Then *state* the null and alternative hypotheses.
b. *Identify* the level of significance.
c. *Find* the standardized test statistic z.
d. *Find* the P-value.
e. *Decide* whether to reject the null hypothesis.
f. *Interpret* the decision in the context of the original claim. *Answer: Page A40*

EXAMPLE 5

See MINITAB steps on page 400.

Hypothesis Testing Using P-values

You think that the average franchise investment information shown in the graph is incorrect, so you randomly select 30 franchises and determine the necessary investment for each. The sample mean investment is $135,000 with a standard deviation of $30,000. Is there enough evidence to support your claim at $\alpha = 0.05$? Use a P-value.

Franchise Investment

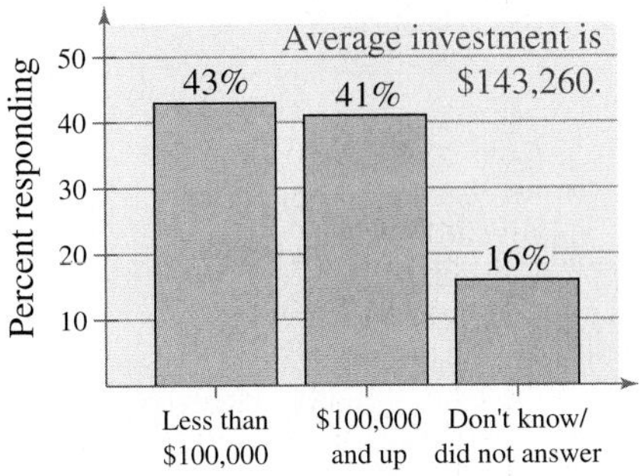

SOLUTION The claim is "the mean is different from $143,260." So, the null and alternative hypotheses are

$$H_0\colon \mu = \$143{,}260 \quad \text{and} \quad H_a\colon \mu \neq \$143{,}260. \text{ (Claim)}$$

The level of significance is $\alpha = 0.05$. The standardized test statistic is

$$z = \frac{\bar{x} - \mu}{\sigma/\sqrt{n}} \qquad \text{Because } n \geq 30 \text{, use the } z\text{-test.}$$

$$= \frac{135{,}000 - 143{,}260}{30{,}000/\sqrt{30}} \qquad \text{Because } n \geq 30 \text{, use } \sigma \approx s = 30{,}000. \text{ Assume } \mu = 143{,}260.$$

$$\approx -1.51.$$

In Table 4, the area corresponding to $z = -1.51$ is 0.0655. Because the test is a two-tailed test, the P-value is equal to twice the area to the left of $z = -1.51$. So,

$$P = 2(0.0655) = 0.1310.$$

Because the P-value is greater than α, you should fail to reject the null hypothesis.

Interpretation There is not enough evidence at the 5% level of significance to conclude that the mean franchise investment is different from $143,260.

The area to the left of $z = -1.51$ is 0.0655, so $P = 2(0.0655) = 0.1310$.

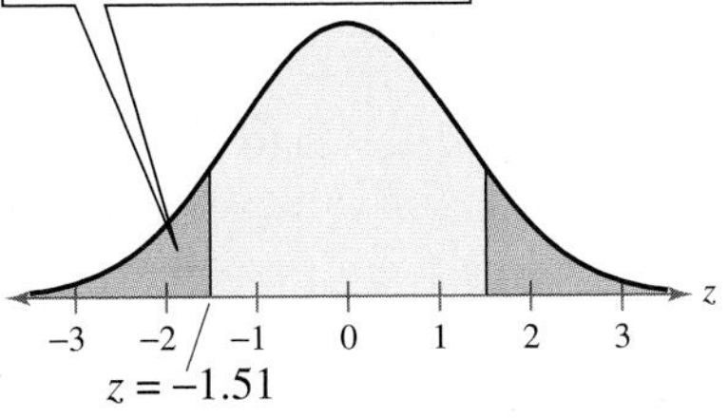

Two-Tailed Test

Try It Yourself 5

One of your distributors reports an average of 150 sales per day for the distributorship. You suspect that this average is not accurate, so you randomly select 35 days and determine the number of sales each day. The sample mean is 143 daily sales with a standard deviation of 15 sales. At $\alpha = 0.01$, is there enough evidence to doubt the distributor's reported average? Use a P-value.

a. *Identify* the claim. Then *state* the null and alternative hypotheses.
b. *Identify* the level of significance.
c. *Find* the standardized test statistic z.
d. *Find* the P-value.
e. *Decide* whether to reject the null hypothesis.
f. *Interpret* the decision in the context of the original claim.

Answer: Page A40

EXAMPLE 6

Using a Technology Tool to Find a P-value

What decision should you make for the following MINITAB printout, using a level of significance of $\alpha = 0.05$?

MINITAB

Test of mu = 6.200 vs mu not = 6.2000
The assumed sigma = 0.470

Variable	N	Mean	StDev	SE Mean	Z	P
Sample	53	6.0666	0.4146	0.0646	-2.07	0.039

Study Tip

For this test, a sample of 53 values was entered into a column called "Sample." The hypotheses are

H_0: $\mu = 6.2$

and

H_a: $\mu \neq 6.2$.

SOLUTION The *P*-value for this test is given as 0.039. Because the *P*-value is less than 0.05, reject the null hypothesis.

Try It Yourself 6

For the MINITAB hypothesis test shown in Example 6, make a decision at the $\alpha = 0.01$ level of significance.

a. *Compare* the *P*-value with the level of significance.
b. *Make* your decision.

Answer: Page A40

Rejection Regions and Critical Values

Another method to decide whether to reject the null hypothesis is to determine whether the standardized test statistic falls within a range of values called the rejection region of the sampling distribution.

Insight

If the test statistic falls in a rejection region, it would be considered a rare occurrence.

DEFINITION

A **rejection region** (or **critical region**) of the sampling distribution is the range of values for which the null hypothesis is not probable. If a test statistic falls in this region, the null hypothesis is rejected. A **critical value** z_0 separates the rejection region from the nonrejection region.

GUIDELINES

Finding Critical Values in a Normal Distribution

1. Specify the level of significance α.
2. Decide whether the test is left-tailed, right-tailed, or two-tailed.
3. Find the critical value(s) z_0. If the hypothesis test is
   a. *left-tailed*, find the z-score that corresponds to an area of α.
   b. *right-tailed*, find the z-score that corresponds to an area of $1 - \alpha$.
   c. *two-tailed*, find the z-scores that correspond to $\frac{1}{2}\alpha$ and $1 - \frac{1}{2}\alpha$.
4. Sketch the standard normal distribution. Draw a vertical line at each critical value and shade the rejection region(s).

EXAMPLE 7

Finding a Critical Value for a Left-Tailed Test

Find the critical value and rejection region for a left-tailed test with $\alpha = 0.01$.

SOLUTION The graph at the right shows a standard normal curve with a shaded area of 0.01 in the left tail. In Table 4, the z-score that corresponds to an area of 0.01 is -2.33. So, the critical value is $z_0 = -2.33$. The rejection region is to the left of this critical value.

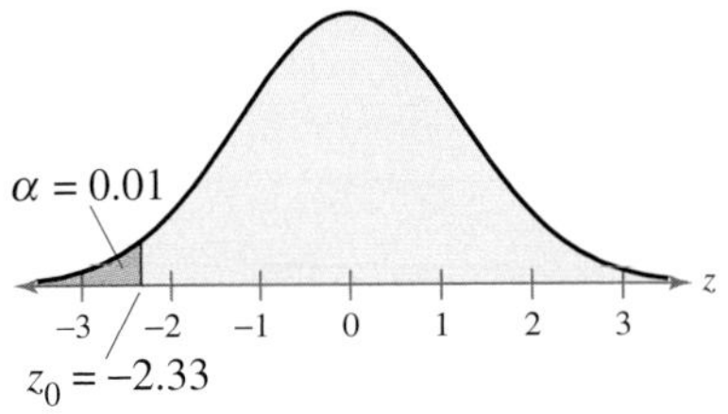

1% Level of Significance

Try It Yourself 7

Find the critical value and rejection region for a left-tailed test with $\alpha = 0.10$.

a. *Draw* a graph of the standard normal curve with an area of α in the left tail.
b. *Use* Table 4 to locate the area that is closest to α.
c. *Find* the z-score that corresponds to this area.
d. *Identify* the rejection region.

Answer: Page A40

If you cannot find the exact area in Table 4, use the area that is closest. For instance, in Example 7, the area closest to 0.01 is 0.0099.

EXAMPLE 8

Finding Critical Values for a Two-Tailed Test

Find the critical values and rejection regions for a two-tailed test with $\alpha = 0.05$.

SOLUTION The graph at the right shows a standard normal curve with shaded areas of $\frac{1}{2}\alpha = 0.025$ in each tail. The area to the left of $-z_0$ is $\frac{1}{2}\alpha = 0.025$, and the area to the left of z_0 is $1 - \frac{1}{2}\alpha = 0.975$. In Table 4, the z-scores that correspond to the areas 0.025 and 0.975 are -1.96 and 1.96, respectively. So, the critical values are $-z_0 = -1.96$ and $z_0 = 1.96$. The rejection regions are to the left of $-z_0$ and to the right of z_0.

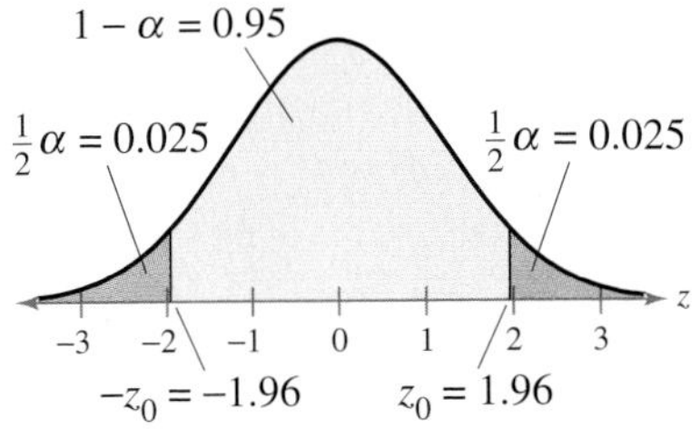

5% Level of Significance

Try It Yourself 8

Find the critical values and rejection regions for a two-tailed test with $\alpha = 0.08$.

a. *Draw* a graph of the standard normal curve with an area of $\frac{1}{2}\alpha$ in each tail.
b. *Use* Table 4 to locate the areas that are closest to $\frac{1}{2}\alpha$ and $1 - \frac{1}{2}\alpha$.
c. *Find* the z-scores that correspond to these areas.
d. *Identify* the rejection regions.

Answer: Page A40

Study Tip

Notice in Example 8 that the critical values are opposites. This is always true for two-tailed z-tests.

The table lists the critical values for commonly used levels of significance.

Alpha	Tail	z
0.10	Left	−1.28
	Right	1.28
	Two	±1.645
0.05	Left	−1.645
	Right	1.645
	Two	±1.96
0.01	Left	−2.33
	Right	2.33
	Two	±2.575

Using Rejection Regions for a z-Test

To conclude a hypothesis test using rejection region(s), you make a decision and interpret the decision as follows.

Decision Rule Based on Rejection Region

To use a rejection region to conduct a hypothesis test, calculate the standardized test statistic z. If the standardized test statistic

1. is in the rejection region, then reject H_0.

2. is *not* in the rejection region, then fail to reject H_0.

Left-Tailed Test

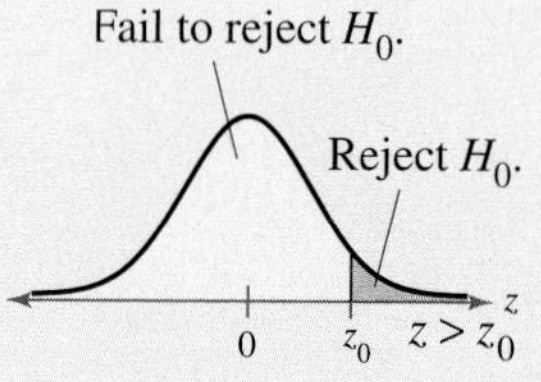

Right-Tailed Test

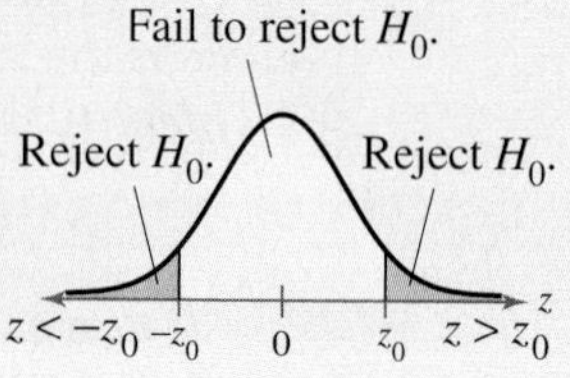

Two-Tailed Test

Failing to reject the null hypothesis does not mean that you have accepted the null hypothesis as true. It simply means that there is not enough evidence to reject the null hypothesis.

GUIDELINES

Using Rejection Regions for a z-Test for a Mean μ

In Words	*In Symbols*
1. State the claim mathematically and verbally. Identify the null and alternative hypotheses.	State H_0 and H_a.
2. Specify the level of significance.	Identify α.
3. Sketch the sampling distribution.	
4. Determine the critical value(s).	Use Table 4 in Appendix B.
5. Determine the rejection region(s).	
6. Find the standardized test statistic.	$z = \dfrac{\bar{x} - \mu}{\sigma/\sqrt{n}}$ or if $n \geq 30$ use $\sigma \approx s$.
7. Make a decision to reject or fail to reject the null hypothesis.	If z is in the rejection region, reject H_0. Otherwise, fail to reject H_0.
8. Interpret the decision in the context of the original claim.	

EXAMPLE 9

See TI-83 steps on page 401.

Testing μ with a Large Sample

Employees in a large accounting firm claim that the mean salary of the firm's accountants is less than that of its competitor's, which is \$45,000. A random sample of 30 of the firm's accountants has a mean salary of \$43,500 with a standard deviation of \$5200. At $\alpha = 0.05$, test the employees' claim.

SOLUTION The claim is "the mean salary is less than \$45,000." So, the null and alternative hypotheses can be written as

$$H_0\text{: } \mu \geq \$45{,}000 \quad \text{and} \quad H_a\text{: } \mu < \$45{,}000. \text{ (Claim)}$$

Because the test is a left-tailed test and the level of significance is $\alpha = 0.05$, the critical value is $z_0 = -1.645$ and the rejection region is $z < -1.645$. The standardized test statistic is

$$z = \frac{\overline{x} - \mu}{\sigma/\sqrt{n}} \qquad \text{Because } n \geq 30\text{, use the } z\text{-test.}$$

$$= \frac{43{,}500 - 45{,}000}{5200/\sqrt{30}} \qquad \text{Because } n \geq 30\text{, use } \sigma \approx s = 5200. \text{ Assume } \mu = 45{,}000.$$

$$\approx -1.58.$$

The graph shows the location of the rejection region and the standardized test statistic z. Because z is not in the rejection region, you fail to reject the null hypothesis.

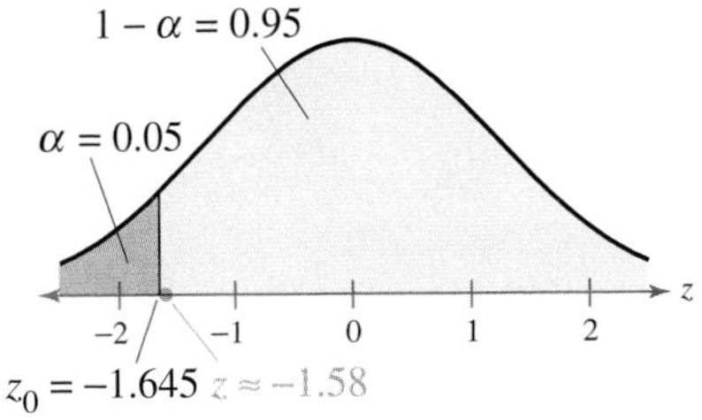

5% Level of Significance

Interpretation There is not enough evidence at the 5% level of significance to support the employees' claim that the mean salary is less than \$45,000.

Be sure you understand the decision made in this example. Even though your sample has a mean of \$43,500, you cannot (at a 5% level of significance) support the claim that the mean of all the accountants' salaries is less than \$45,000. The difference between your test statistic and the hypothesized mean is probably due to sampling error.

Try It Yourself 9

The CEO of the firm claims that the mean work day of the firm's accountants is less than 8.5 hours. A random sample of 35 of the firm's accountants has a mean work day of 8.2 hours with a standard deviation of 0.5 hour. At $\alpha = 0.01$, test the CEO's claim.

a. *Identify* the claim and state H_0 and H_a.
b. *Identify* the level of significance α.
c. *Find* the critical value z_0 and identify the rejection region.
d. *Find* the standardized test statistic z.
e. Sketch a graph. *Decide* whether to reject the null hypothesis.
f. *Interpret* the decision in the context of the original claim.

Answer: Page A40

Picturing the World

Each year, the Environmental Protection Agency (EPA) publishes reports of gas mileage for all makes and models of passenger vehicles. In a recent year, the compact car with an automatic transmission that posted the best mileage was the Honda Civic Hybrid. It had a mean mileage of 47 miles per gallon (city) and 48 miles per gallon (highway). Suppose that Honda believes a Civic Hybrid exceeds 48 miles per gallon on the highway. To support its claim, it tests 34 cars on highway driving and obtains a sample mean of 49.2 miles per gallon with a standard deviation of 2.3 miles per gallon. *(Source: EPA)*

Is the evidence strong enough to prove that the Civic's highway miles per gallon exceeds the EPA estimate? Use a z-test with $\alpha = 0.01$.

EXAMPLE 10

Testing μ with a Large Sample

The U.S. Department of Agriculture reports that the mean cost of raising a child from birth to age 2 in a rural area is \$9405. You believe this value is incorrect, so you select a random sample of 900 children (age 2) and find that the mean cost is \$9290 with a standard deviation of \$1540. At $\alpha = 0.05$, is there enough evidence to conclude that the mean cost is different from \$9405? *(Adapted from U.S. Department of Agriculture Center for Nutrition Policy and Promotion)*

SOLUTION You want to support the claim that "the mean cost is different from \$9405." So, the null and alternative hypotheses are

H_0: $\mu = \$9405$

and

H_a: $\mu \neq \$9405$. (Claim)

Because the test is a two-tailed test and the level of significance is $\alpha = 0.05$, the critical values are $-z_0 = -1.96$ and $z_0 = 1.96$. The rejection regions are $z < -1.96$ and $z > 1.96$. The standardized test statistic is

$$z = \frac{\bar{x} - \mu}{\sigma/\sqrt{n}}$$ Because $n \geq 30$, use the z-test.

$$= \frac{9290 - 9405}{1540/\sqrt{900}}$$ Because $n \geq 30$, use $\sigma \approx s = 1540$. Assume $\mu = 9450$.

$$\approx -2.24.$$

The graph shows the location of the rejection regions and the standardized test statistic z. Because z is in the rejection region, you should reject the null hypothesis.

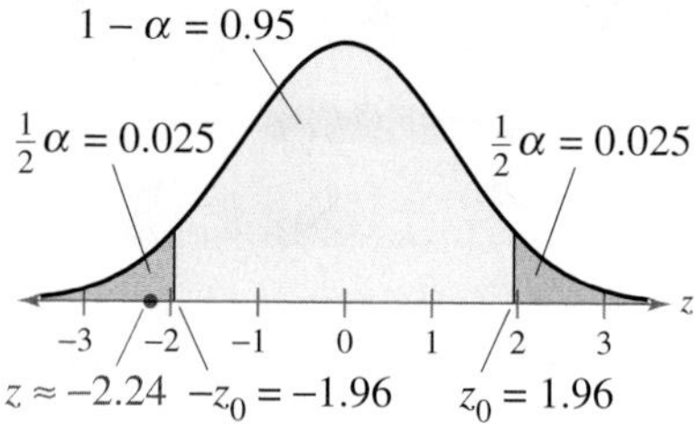

5% Level of Significance

Interpretation You have enough evidence to conclude that the mean cost of raising a child from birth to age 2 in a rural area is significantly different from \$9405 at the 5% level of significance.

Try It Yourself 10

Using the information and results of Example 10, determine whether there is enough evidence to support the claim that the mean cost of raising a child from birth to age 2 in a rural area is different from \$9405. Use $\alpha = 0.01$.

a. *Identify* the level of significance α.
b. *Find* the critical values $\pm z_0$ and identify the rejection regions.
c. Sketch a graph. *Decide* whether to reject the null hypothesis.
d. *Interpret* the decision in the context of the original claim.

Answer: Page A40

7.2 Exercises

Help

MyMathLab

Student Study Pack

1. $P = 0.1151$; fail to reject H_0.
2. $P = 0.0455$; reject H_0.
3. $P = 0.0096$; reject H_0.
4. $P = 0.1093$; fail to reject H_0.
5. $P = 0.1188$; fail to reject H_0.
6. $P = 0.0214$; fail to reject H_0.
7. c
8. d
9. e
10. f
11. b
12. a
13. (a) Fail to reject H_0.
    (b) Reject H_0.

Building Basic Skills and Vocabulary

In Exercises 1–6, find the P-value for the indicated hypothesis test with the given standardized test statistic z. Decide whether to reject H_0 for the given level of significance α.

1. Left-tailed test, $z = -1.20$, $\alpha = 0.10$

2. Left-tailed test, $z = -1.69$, $\alpha = 0.05$

3. Right-tailed test, $z = 2.34$, $\alpha = 0.01$

4. Right-tailed test, $z = 1.23$, $\alpha = 0.10$

5. Two-tailed test, $z = -1.56$, $\alpha = 0.05$

6. Two-tailed test, $z = 2.30$, $\alpha = 0.01$

Graphical Analysis In Exercises 7–12, match each P-value with the graph that displays its area. The graphs are labeled (a)–(f).

7. $P = 0.0089$

8. $P = 0.0132$

9. $P = 0.3050$

10. $P = 0.0688$

11. $P = 0.0233$

12. $P = 0.0287$

(a)

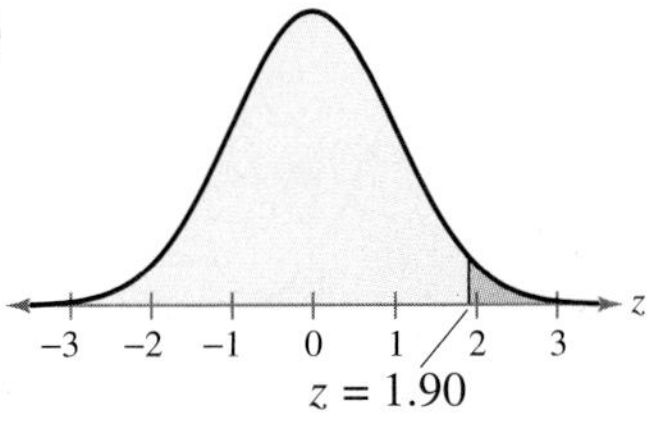

(b)

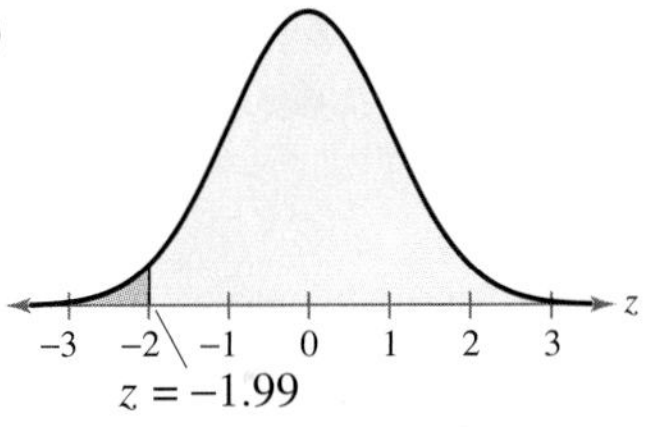

(c)

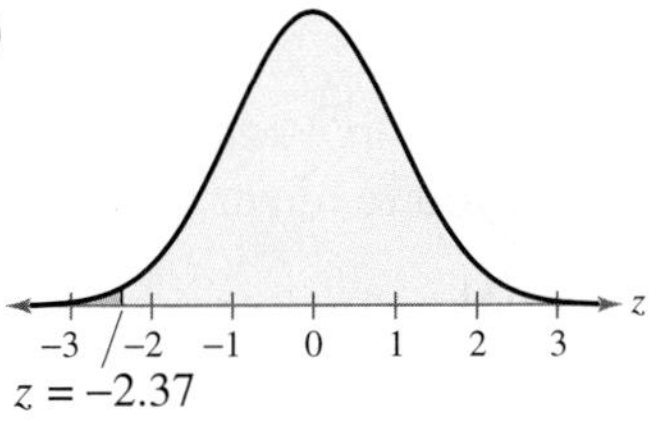

(d)

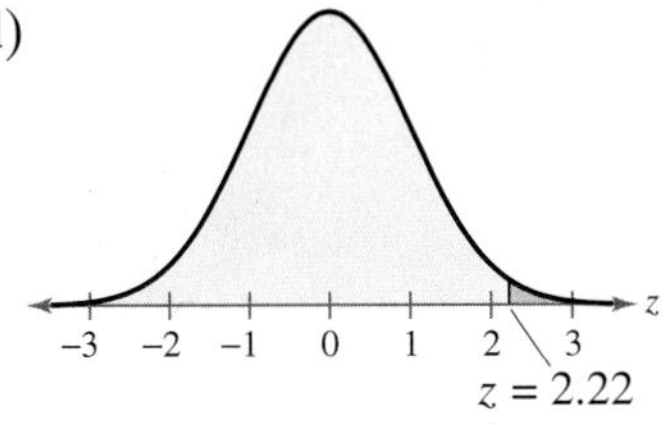

(e)

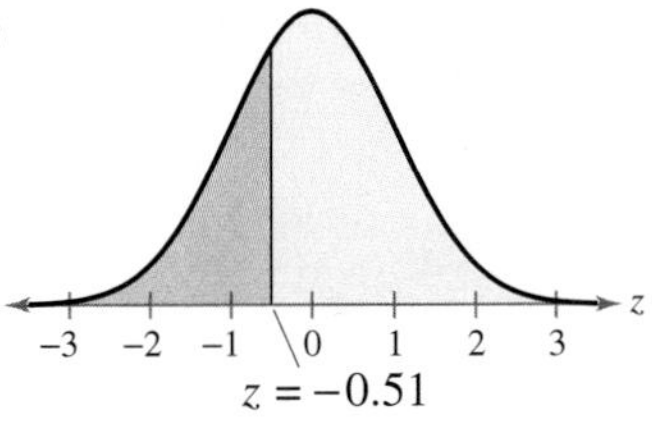

(f)

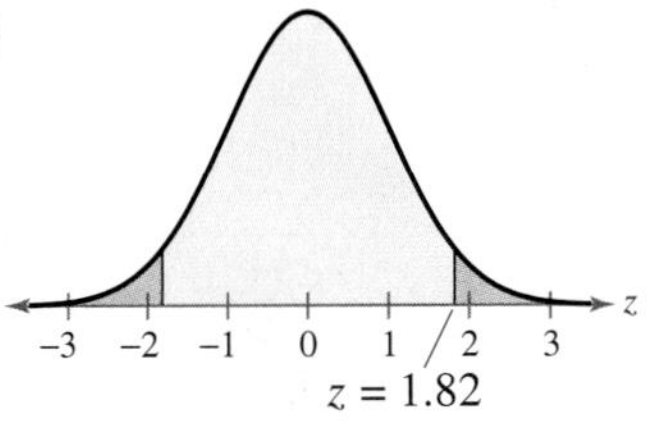

13. Given H_0: $\mu = 100$, H_a: $\mu \neq 100$, and $P = 0.0461$.

(a) Do you reject or fail to reject H_0 at the 0.01 level of significance?

(b) Do you reject or fail to reject H_0 at the 0.05 level of significance?

14. (a) Fail to reject H_0.
(b) Fail to reject H_0.
15. 1.645
16. 1.41
17. −1.88
18. −1.34
19. ±2.33
20. ±1.645
21. Right-tailed ($\alpha = 0.01$)
22. Two-tailed ($\alpha = 0.05$)
23. Two-tailed ($\alpha = 0.10$)
24. Left-tailed ($\alpha = 0.05$)
25. (a) Fail to reject H_0 because $-1.645 < z < 1.645$.
(b) Reject H_0 because $z > 1.645$.
(c) Fail to reject H_0 because $-1.645 < z < 1.645$.
(d) Reject H_0 because $z < -1.645$.
26. (a) Reject H_0 because $z > 1.96$.
(b) Fail to reject H_0 because $-1.96 < z < 1.96$.
(c) Fail to reject H_0 because $-1.96 < z < 1.96$.
(d) Reject H_0 because $z < -1.96$.

14. Given H_0: $\mu \geq 8.5$, H_a: $\mu < 8.5$, and $P = 0.0691$.

(a) Do you reject or fail to reject H_0 at the 0.01 level of significance?

(b) Do you reject or fail to reject H_0 at the 0.05 level of significance?

Finding Critical Values In Exercises 15–20, find the critical value(s) for the indicated type of test and level of significance α.

15. Right-tailed test, $\alpha = 0.05$

16. Right-tailed test, $\alpha = 0.08$

17. Left-tailed test, $\alpha = 0.03$

18. Left-tailed test, $\alpha = 0.09$

19. Two-tailed test, $\alpha = 0.02$

20. Two-tailed test, $\alpha = 0.10$

Graphical Analysis In Exercises 21–24,

(a) state whether the graph shows a left-tailed, right-tailed, or two-tailed test.

(b) state whether $\alpha = 0.01$, 0.05, or 0.10.

21.

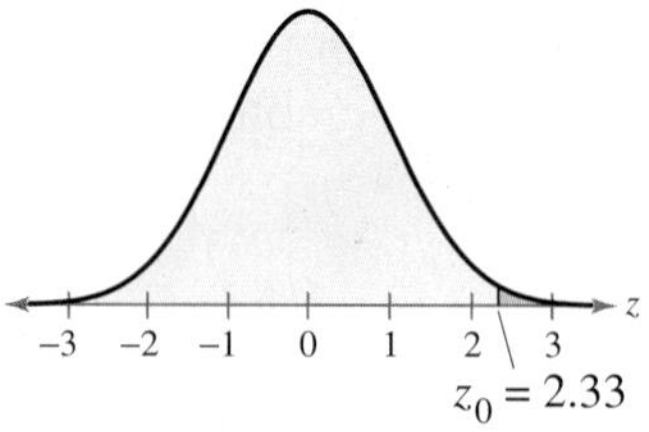

22.

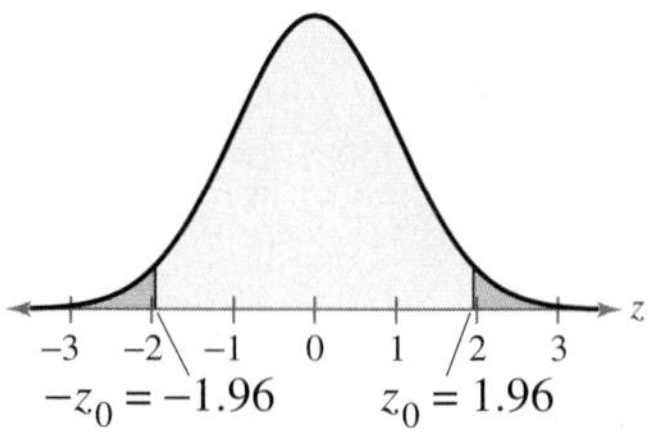

23.

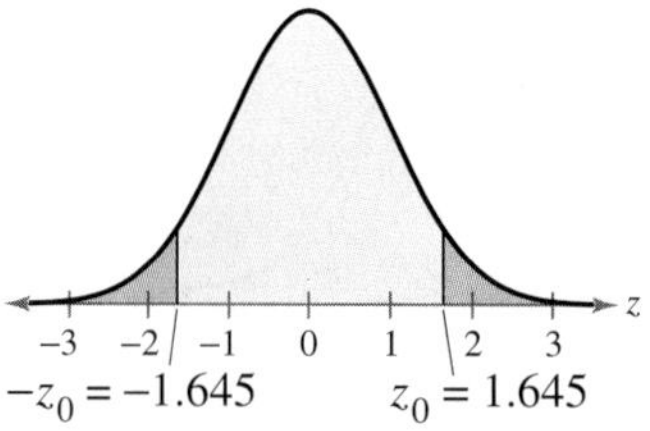

24.

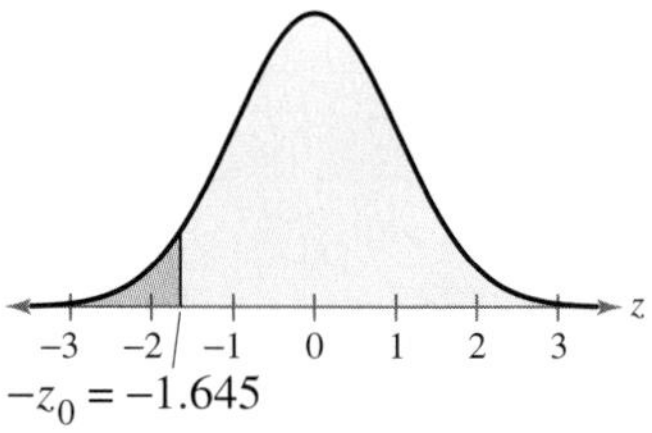

Graphical Analysis In Exercises 25–28, state whether each standardized test statistic z allows you to reject the null hypothesis. Explain your reasoning.

25. (a) $z = 1.631$
(b) $z = 1.723$
(c) $z = -1.464$
(d) $z = -1.655$

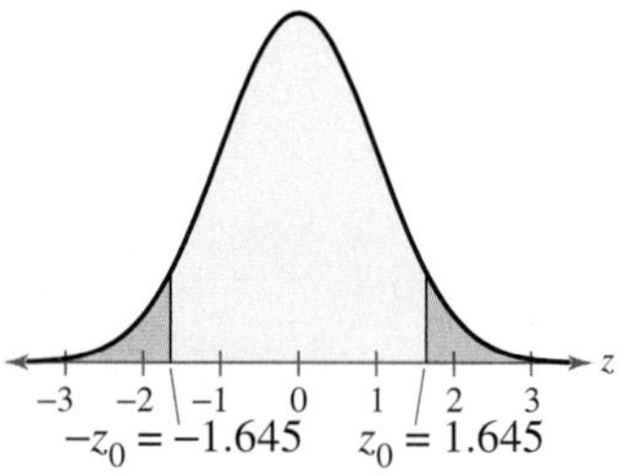

26. (a) $z = 1.98$
(b) $z = -1.89$
(c) $z = 1.65$
(d) $z = -1.99$

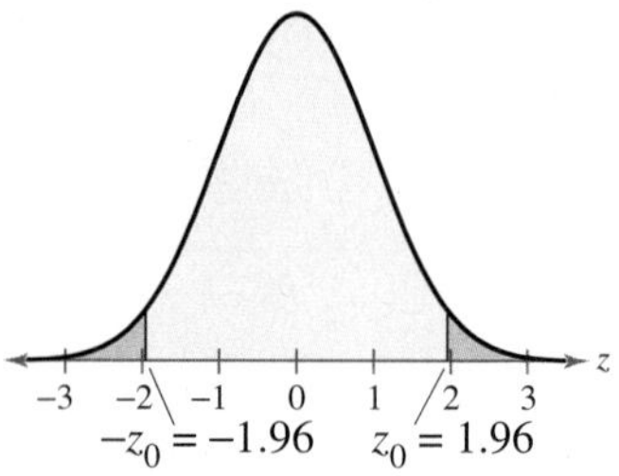

27. (a) Fail to reject H_0 because $z < 1.285$.
(b) Fail to reject H_0 because $z < 1.285$.
(c) Fail to reject H_0 because $z < 1.285$.
(d) Reject H_0 because $z > 1.285$.

28. (a) Fail to reject H_0 because $-2.575 < z < 2.575$.
(b) Reject H_0 because $z < -2.575$.
(c) Reject H_0 because $z > 2.575$.
(d) Fail to reject H_0 because $-2.575 < z < 2.575$.

29. Reject H_0. There is not enough evidence at the 5% level to support the claim.

30. Fail to reject H_0. There is not enough evidence at the 5% level to support the claim.

31. Reject H_0. There is enough evidence at the 1% level to support the claim.

32. Reject H_0. There is enough evidence at the 1% level to reject the claim.

33. (a) H_0: $\mu \leq 260$
H_a: $\mu > 260$ (claim)
(b) 0.838; 0.7995
(c) 0.2005
(d) Fail to reject H_0.
(e) There is insufficient evidence at the 4% level of significance to support the claim that the mean score for Illinois' eighth graders is more than 260.

34. (a) H_0: $\mu \leq 1.5$
H_a: $\mu > 1.5$ (claim)
(b) 1.105; 0.8654
(c) 0.1335
(d) Fail to reject H_0.
(e) There is insufficient evidence at the 10% level of significance to support the claim that the mean reserve capacity of the battery is greater than 1.5 hours.

27. (a) $z = -1.301$
(b) $z = 1.203$
(c) $z = 1.280$
(d) $z = 1.286$

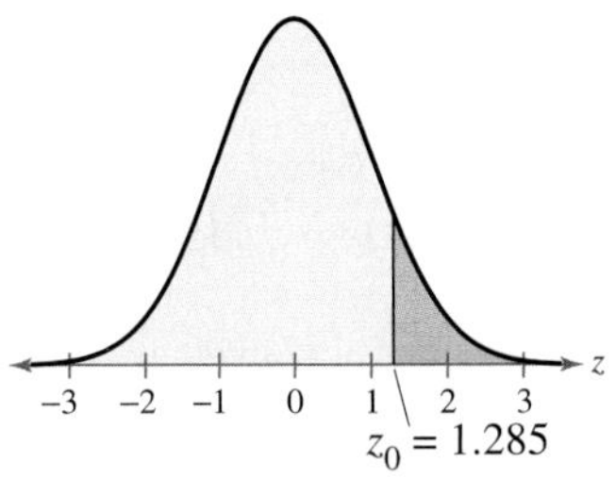

28. (a) $z = 2.557$
(b) $z = -2.755$
(c) $z = 2.585$
(d) $z = -2.475$

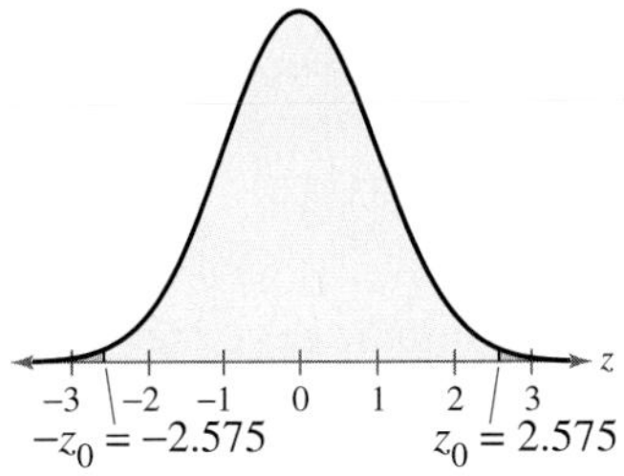

In Exercises 29–32, test the claim about the population mean μ at the given level of significance using the given sample statistics.

29. Claim: $\mu = 40$; $\alpha = 0.05$. Sample statistics: $\bar{x} = 39.2$, $s = 3.23$, $n = 75$

30. Claim: $\mu > 1030$; $\alpha = 0.05$. Sample statistics: $\bar{x} = 1035$, $s = 23$, $n = 50$

31. Claim: $\mu \neq 6000$; $\alpha = 0.01$. Sample statistics: $\bar{x} = 5800$, $s = 350$, $n = 35$

32. Claim: $\mu \leq 22{,}500$; $\alpha = 0.01$. Sample statistics: $\bar{x} = 23{,}250$, $s = 1200$, $n = 45$

Using and Interpreting Concepts

Testing Claims Using P-values In Exercises 33–38,

(a) write the claim mathematically and identify H_0 and H_a.
(b) find the standardized test statistic z and its corresponding area.
(c) find the P-value.
(d) decide whether to reject or fail to reject the null hypothesis.
(e) interpret the decision in the context of the original claim.

33. ***Mathematics Assessment Tests*** In Illinois, a random sample of 85 eighth grade students has a mean score of 265 with a standard deviation of 55 on a national mathematics assessment test. This test result prompts a state school administrator to declare that the mean score for the state's eighth graders on the examination is more than 260. At $\alpha = 0.04$, is there enough evidence to support the administrator's claim? *(Adapted from National Center for Education Statistics)*

34. ***Automotive Battery Reserves*** An automotive battery manufacturer guarantees that the mean reserve capacity of a certain battery is greater than 1.5 hours. To test this claim, you randomly select a sample of 50 batteries and find the mean reserve capacity to be 1.55 hours with a standard deviation of 0.32 hour. At $\alpha = 0.10$, do you have enough evidence to support the manufacturer's claim?

35. (a) H_0: $\mu \leq 7$; H_a: $\mu > 7$ (claim)
(b) 2.996; 0.9987
(c) 0.0013 (d) Reject H_0.
(e) There is sufficient evidence at the 7% level to support the claim that the mean consumption of tea by a person in the United States is more than 7 gallons per year.

36. (a) H_0: $\mu = 3.4$ (claim) H_a: $\mu \neq 3.4$
(b) -1.371; 0.0853
(c) 0.1706 (d) Fail to reject H_0.
(e) There is insufficient evidence at the 8% level to reject the claim that the mean tuna consumption by a person in the United States is 3.4 pounds per year.

37. (a) H_0: $\mu = 15$ (claim); H_a: $\mu \neq 15$
(b) -0.219; 0.4129
(c) 0.8258 (d) Fail to reject H_0.
(e) There is insufficient evidence at the 5% level to reject the claim that the mean time it takes smokers to quit smoking permanently is 15 years.

38. (a) H_0: $\mu \leq \$90{,}900$ (claim) H_a: $\mu > \$90{,}900$
(b) 0.238; 0.5948
(c) 0.5948 (d) Fail to reject H_0.
(e) There is insufficient evidence at the 3% level to reject the claim that the mean annual salary for engineering managers in Alabama is at least \$90,900.

39. (a) H_0: $\mu = 40$ (claim); H_a: $\mu \neq 40$
(b) $-z_0 = -2.575$ and $z_0 = 2.575$
Rejection regions: $z < -2.575$, $z > 2.575$
(c) -0.584 (d) Fail to reject H_0.
(e) There is insufficient evidence at the 1% level to reject the claim that the mean caffeine content per one 12-ounce bottle of cola is 40 milligrams.

40. (a) H_0: $\mu = 80$ (claim); H_a: $\mu \neq 80$
(b) $-z_0 = -1.96$ and $z_0 = 1.96$
Rejection regions: $z < -1.96$, $z > 1.96$
(c) 0.555 (d) Fail to reject H_0.

35. *Tea Drinkers* A tea drinker's society estimates that the mean consumption of tea by a person in the United States is more than 7 gallons per year. In a sample of 100 people, you find that the mean consumption of tea is 7.8 gallons per year with a standard deviation of 2.67 gallons. At $\alpha = 0.07$, can you support the society's claim? *(Adapted from U.S. Department of Agriculture)*

36. *Tuna Consumption* A nutritionist claims that the mean tuna consumption by a person in the United States is 3.4 pounds per year. A sample of 60 people in the United States shows that the mean tuna consumption by a person is 3.2 pounds per year with a standard deviation of 1.13 pounds. At $\alpha = 0.08$, can you reject the nutritionist's claim? *(Adapted from U.S. Department of Agriculture)*

DATA **37. *Quitting Smoking*** The number of years it took a random sample of 32 former smokers to quit permanently is listed. At $\alpha = 0.05$, is there enough evidence to reject the claim that the mean time it takes smokers to quit smoking permanently is 15 years? *(Adapted from The Gallup Organization)*

15.7	13.2	22.6	13.0	10.7	18.1	14.7	7.0	17.3	7.5	21.8
12.3	19.8	13.8	16.0	15.5	13.1	20.7	15.5	9.8	11.9	16.9
7.0	19.3	13.2	14.6	20.9	15.4	13.3	11.6	10.9	21.6	

DATA **38. *Salaries*** An Alabama politician claims that the mean annual salary for engineering managers in Alabama is more than the national mean, \$90,900. The annual salaries (in dollars) for a random sample of 34 engineering managers in Alabama are listed. At $\alpha = 0.03$, is there enough evidence to support the politician's claim? *(Adapted from America's Career InfoNet)*

87,512	89,510	82,639	98,897	104,877	87,592
103,632	105,202	93,672	104,878	101,508	88,302
89,231	105,060	95,974	76,955	101,396	69,838
87,728	98,314	104,349	93,493	74,918	89,736
68,060	96,777	82,723	92,944	86,114	97,062
85,975	84,956	95,905	79,042		

Testing Claims In Exercises 39–46, (a) write the claim mathematically and identify H_0 and H_a, (b) find the critical values and identify the rejection regions, (c) find the standardized test statistic, (d) decide whether to reject or fail to reject the null hypothesis, and (e) interpret the decision in the context of the original claim.

39. *Caffeine Content in Colas* A company that makes cola drinks states that the mean caffeine content per one 12-ounce bottle of cola is 40 milligrams. You work as a quality control manager and are asked to test this claim. During your tests, you find that a random sample of thirty 12-ounce bottles of cola has a mean caffeine content of 39.2 milligrams with a standard deviation of 7.5 milligrams. At $\alpha = 0.01$, can you reject the company's claim? *(Adapted from Reader's Digest Eating for Good Health)*

40. *Caffeine Content* A coffee shop claims that its fresh-brewed drinks have a mean caffeine content of 80 milligrams per 5 ounces. You work for a city health agency and are asked to test this claim. You find that a random sample of 42 five-ounce servings has a mean caffeine content of 83 milligrams and a standard deviation of 35 milligrams. At $\alpha = 0.05$, do you have enough evidence to reject the shop's claim? *(Adapted from Reader's Digest Eating for Good Health)*

(e) There is insufficient evidence at the 5% level to reject the claim that the mean caffeine content is 80 milligrams per 5 ounces.

41. (a) H_0: $\mu \geq 750$ (claim)
H_a: $\mu < 750$

(b) $z_0 = -2.05$
Rejection region: $z < -2.05$

(c) -0.500 (d) Fail to reject H_0.

(e) There is insufficient evidence at the 2% level to reject the claim that the mean life of the bulb is at least 750 hours.

42. (a) H_0: $\mu \leq 230$ (claim)
H_a: $\mu > 230$

(b) $z_0 = 1.75$
Rejection region: $z > 1.75$

(c) 1.442 (d) Fail to reject H_0.

(e) There is insufficient evidence at the 4% level to reject the claim that the mean sodium content per serving of cereal is no more than 230 milligrams.

43. (a) H_0: $\mu \leq 28$
H_a: $\mu > 28$ (claim)

(b) $z_0 = 1.55$
Rejection region: $z > 1.55$

(c) 1.318 (d) Fail to reject H_0.

(e) There is insufficient evidence at the 6% level to support the claim that the mean nitrogen dioxide level in West London is greater than 28 parts per billion.

44. (a) H_0: $\mu \geq 10{,}000$ (claim)
H_a: $\mu < 10{,}000$

(b) $z_0 = -1.34$
Rejection region: $z < -1.34$

(c) -1.377 (d) Reject H_0.

(e) There is sufficient evidence at the 9% level to reject the claim that the mean life of fluorescent lamps is at least 10,000 hours.

45. (a) H_0: $\mu \geq 10$ (claim); H_a: $\mu < 10$

(b) $z_0 = -1.88$
Rejection region: $z < -1.88$

(c) -0.510 (d) Fail to reject H_0.

(e) There is insufficient evidence at the 3% level to reject the claim that the mean weight loss after one month is at least 10 pounds.

41. ***Light Bulbs*** A light bulb manufacturer guarantees that the mean life of a certain type of light bulb is at least 750 hours. A random sample of 36 light bulbs has a mean life of 745 hours with a standard deviation of 60 hours. Do you have enough evidence to reject the manufacturer's claim? Use $\alpha = 0.02$.

42. ***Sodium Content in Cereal*** In your work for a national health organization, you are asked to monitor the amount of sodium in a certain brand of cereal. You find that a random sample of 52 cereal servings has a mean sodium content of 232 milligrams with a standard deviation of 10 milligrams. At $\alpha = 0.04$, can you conclude that the mean sodium content per serving of cereal is less than than 230 milligrams?

DATA

43. ***Nitrogen Dioxide Levels*** A scientist estimates that the mean nitrogen dioxide level in West London is greater than 28 parts per billion. You want to test this estimate. To do so, you determine the nitrogen dioxide levels for 36 randomly selected days. The results (in parts per billion) are listed below. At $\alpha = 0.06$, can you support the scientist's estimate? *(Adapted from National Environmental Technology Centre)*

27	29	53	31	16	47	22	17	13	46	99	15
20	17	28	10	14	9	35	29	32	67	24	31
43	29	12	39	65	94	12	27	13	16	40	62

DATA

44. ***Fluorescent Lamps*** A fluorescent lamp manufacturer guarantees that the mean life of a certain type of lamp is at least 10,000 hours. You want to test this guarantee. To do so, you record the life of a random sample of 32 fluorescent lamps. The results (in hours) are shown below. At $\alpha = 0.09$, do you have enough evidence to reject the manufacturer's claim?

8,800	9,155	13,001	10,250	10,002	11,413	8,234	10,402
10,016	8,015	6,110	11,005	11,555	9,254	6,991	12,006
10,420	8,302	8,151	10,980	10,186	10,003	8,814	11,445
6,277	8,632	7,265	10,584	9,397	11,987	7,556	10,380

DATA

45. ***Weight Loss*** A weight loss program claims that program participants have a mean weight loss of at least 10 pounds after one month. You work for a medical association and are asked to test this claim. A random sample of 30 program participants and their weight losses (in pounds) after one month is listed in the stem-and-leaf plot below. At $\alpha = 0.03$, do you have enough evidence to reject the program's claim?

Weight Loss (in pounds) after One Month

Key: 5|7 = 5.7

5	7 7
6	6 7
7	0 1 9
8	2 2 7 9
9	0 3 5 6 8
10	2 5 6 6
11	1 2 5 7 8
12	0 7 8
13	8
14	
15	0

46. (a) H_0: $\mu \leq 10$
H_a: $\mu > 10$ (claim)

(b) $z_0 = 1.48$
Rejection region: $z > 1.48$

(c) 1.393

(d) Fail to reject H_0.

(e) There is insufficient evidence at the 7% level to support the claim that the mean life span of a microwave oven for that company is more than 10 years.

47. Fail to reject H_0 because the standardized test statistic $z = -1.96$ is greater than the critical value $z_0 = -2.33$.

48. Fail to reject H_0 because the standardized test statistic $z = 1.548$ is less than the critical value $z_0 = 1.645$.

49. b, c, d

50. a, b, d

51. In the classical z-test, the test statistic is compared with critical values. The z-test using a P-value compares the P-value with the level of significance α.

DATA

46. *Microwave Ovens* A company that manufactures appliances claims that the mean life span of its microwave ovens is more than 10 years. You want to test this claim. A random sample of 45 microwave ovens and their life spans (in years) is listed in the stem-and-leaf plot below. At $\alpha = 0.07$, can you support the company's claim?

Mean Life Span (in years)

Key: $4|2 = 4.2$

4	2 5
5	
6	
7	1 2 5
8	3 5 6 6 7
9	0 0 3 4 4 6 7 8
10	2 3 4 5 5 7 7 7 8 9
11	2 4 7 8 9
12	6 8 8 8 9 9
13	1 6 7 8
14	2 9

Extending Concepts

47. *Electric Usage* You believe the mean annual kilowatt usage of U.S. residential customers is less than 10,800. You do some research and find that a random sample of 30 residential customers has a mean kilowatt usage of 10,700 with a standard deviation of 280. You conduct a statistical experiment where H_0: $\mu \geq 10{,}800$ and H_a: $\mu < 10{,}800$. At $\alpha = 0.01$, explain why you cannot reject H_0. *(Adapted from Edison Electric Institute)*

48. *Vehicle Miles of Travel* You believe the mean vehicle miles of travel (VMT) per U.S. household is greater than 22,000 miles. You do some research and find that a random sample of 36 U.S. households has a mean VMT of 22,200 miles with a standard deviation of 775 miles. You conduct a statistical experiment where H_0: $\mu \leq 22{,}000$ and H_a: $\mu > 22{,}000$. At $\alpha = 0.05$, explain why you cannot reject H_0. *(Adapted from U.S. Federal Highway Administration)*

49. *Using Different Values of α and n* In Exercise 47, you believe that H_0 is not valid. Which of the following allows you to reject H_0?

(a) Use the same values but increase α from 0.01 to 0.02.

(b) Use the same values but increase α from 0.01 to 0.03.

(c) Use the same values but increase n from 30 to 50.

(d) Use the same values but increase n from 30 to 100.

50. *Using Different Values of α and n* In Exercise 48, you believe that H_0 is not valid. Which of the following allows you to reject H_0?

(a) Use the same values but increase α from 0.05 to 0.06.

(b) Use the same values but increase α from 0.05 to 0.07.

(c) Use the same values but increase n from 36 to 40.

(d) Use the same values but increase n from 36 to 80.

51. *Writing* Explain the difference between the classical z-test for μ and the z-test for μ using a P-value.

Case Study

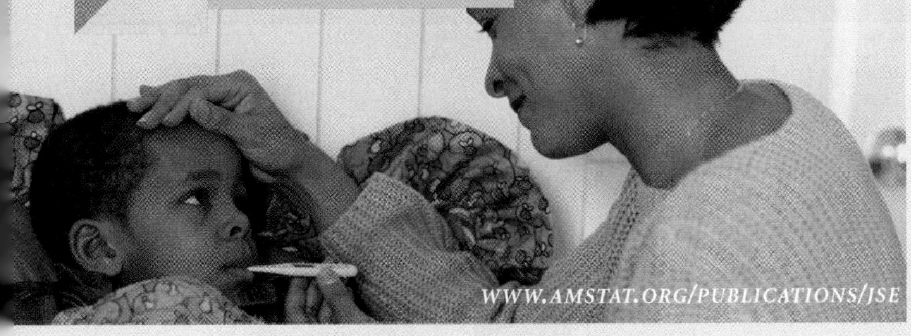

Human Body Temperature: What's Normal?

In an article in the *Journal of Statistics Education* (vol. 4, no. 2), Allen Shoemaker describes a study that was reported in the *Journal of the American Medical Association (JAMA)*.* It is generally accepted that the mean body temperature of adult humans is 98.6°F. In his article, Shoemaker uses the data from the JAMA article to test this hypothesis. Here is a summary of his test.

Claim: The body temperature of adults is 98.6°F.

H_0: $\mu = 98.6$°F (Claim) H_a: $\mu \neq 98.6$°F

Sample Size: $n = 130$

Population: Adult human temperatures (Fahrenheit)

Distribution: Approximately normal

Test Statistics: $\bar{x} = 98.25$, $s = 0.73$

* Data for the JAMA article were collected from healthy men and women, aged 18 to 40, at the University of Maryland Center for Vaccine Development, Baltimore.

Men's Temperatures (in degrees Fahrenheit)

Stem	Leaves
96	3
96	7 9
97	0 1 1 1 2 3 4 4 4 4
97	5 5 6 6 6 7 8 8 8 9 9
98	0 0 0 0 0 0 1 1 2 2 2 2 3 3 4 4 4 4
98	5 5 6 6 6 6 6 6 7 7 8 8 8 9
99	0 0 0 1 2 3 4
99	5
100	
100	

Key: 96|3 = 96.3

Women's Temperatures (in degrees Fahrenheit)

Stem	Leaves
96	4
96	7 8
97	2 2 4
97	6 7 7 8 8 8 9 9 9
98	0 0 0 0 0 1 2 2 2 2 2 2 3 3 3 4 4 4 4 4
98	5 6 6 6 6 7 7 7 7 7 7 8 8 8 8 8 8 9
99	0 0 1 1 2 2 3 4
99	9
100	0
100	8

Key: 96|4 = 96.4

Exercises

1. Complete the hypothesis test for all adults (men and women) by performing the following steps. Use a level of significance of $\alpha = 0.05$.
   (a) Sketch the sampling distribution.
   (b) Determine the critical values and add them to your sketch.
   (c) Determine the rejection regions and shade them in your sketch.
   (d) Find the standardized test statistic. Add it to your sketch.
   (e) Make a decision to reject or fail to reject the null hypothesis.
   (f) Interpret the decision in the context of the original claim.
2. If you lower the level of significance to $\alpha = 0.01$, does your decision change? Explain your reasoning.
3. Test the hypothesis that the mean temperature of men is 98.6°F. What can you conclude at a level of significance of $\alpha = 0.01$?
4. Test the hypothesis that the mean temperature of women is 98.6°F. What can you conclude at a level of significance of $\alpha = 0.01$?
5. Use the sample of 130 temperatures to form a 99% confidence interval for the mean body temperature of adult humans.
6. The conventional "normal" body temperature was established by Carl Wunderlich over 100 years ago. What, in Wunderlich's sampling procedure, do you think might have led him to an incorrect conclusion?

7.3 Hypothesis Testing for the Mean (Small Samples)

What You Should Learn

- How to find critical values in a *t*-distribution
- How to use the *t*-test to test a mean μ
- How to use technology to find *P*-values and use them with a *t*-test to test a mean μ

Critical Values in a *t*-Distribution • The *t*-Test for a Mean μ ($n < 30$, σ unknown) • Using *P*-values with *t*-Tests

Critical Values in a *t*-Distribution

In Section 7.2, you learned how to perform a hypothesis test for a population mean when the sample size was at least 30. In real life, it is often not practical to collect samples of size 30 or more. However, if the population has a normal, or nearly normal, distribution, you can still test the population mean μ. To do so, you can use the *t*-sampling distribution with $n - 1$ degrees of freedom.

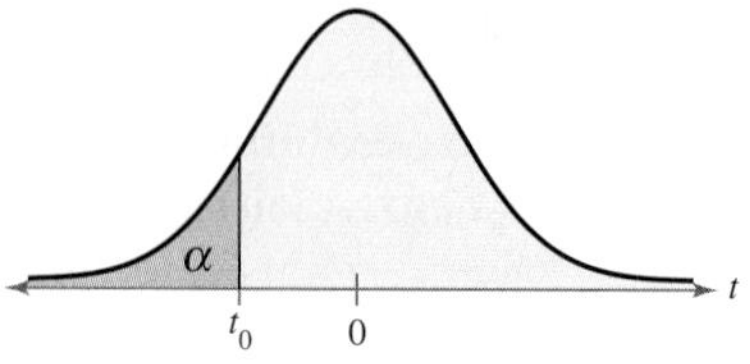

Left-Tailed Test

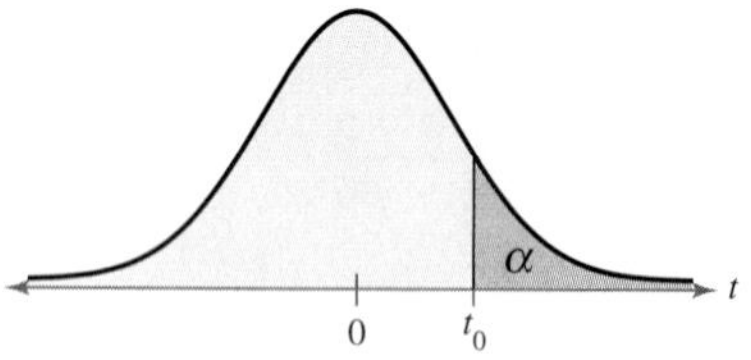

Right-Tailed Test

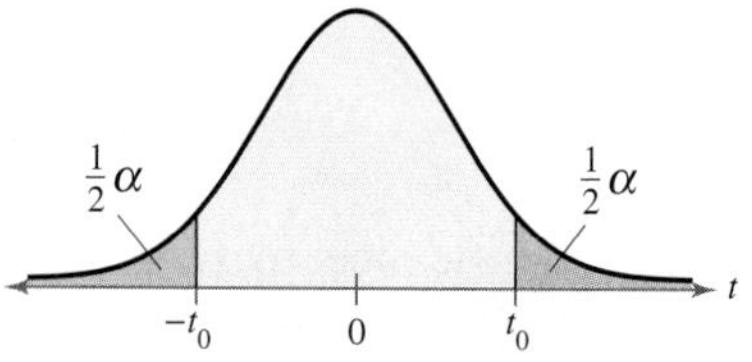

Two-Tailed Test

GUIDELINES

Finding Critical Values in a *t*-Distribution

1. Identify the level of significance α.
2. Identify the degrees of freedom d.f. $= n - 1$.
3. Find the critical value(s) using Table 5 in Appendix B in the row with $n - 1$ degrees of freedom. If the hypothesis test is
   - **a.** *left-tailed,* use "One Tail, α" column with a negative sign.
   - **b.** *right-tailed,* use "One Tail, α" column with a positive sign.
   - **c.** *two-tailed,* use "Two Tails, α" column with a negative and a positive sign.

EXAMPLE 1

Finding Critical Values for t

Find the critical value t_0 for a left-tailed test given $\alpha = 0.05$ and $n = 21$.

SOLUTION The degrees of freedom are

$$\text{d.f.} = n - 1 = 21 - 1 = 20.$$

To find the critical value, use Table 5 in Appendix B with d.f. $= 20$ and 0.05 in the "One Tail, α" column. Because the test is a left-tailed test, the critical value is negative. So,

$$t_0 = -1.725.$$

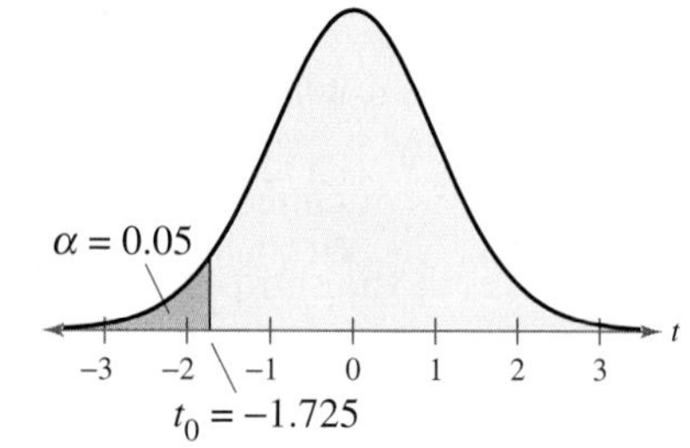

Try It Yourself 1

Find the critical value t_0 for a left-tailed test with $\alpha = 0.01$ and $n = 14$.

a. *Find* the *t*-value in Table 5 in Appendix B. Use d.f. $= 13$ and $\alpha = 0.01$ in the "One Tail, α" column.
b. *Use* a negative sign.

Answer: Page A40

Note to Instructor

A thoughtful student might ask what should be done if the sample size is small, the standard deviation is not known, and you cannot assume that the population is normally distributed. Chapter 11 will cover this case (see nonparametric tests). You can cover these tests immediately after this section if desirable.

EXAMPLE 2

Finding Critical Values for t

Find the critical value t_0 for a right-tailed test with $\alpha = 0.01$ and $n = 17$.

SOLUTION The degrees of freedom are

$$\text{d.f.} = n - 1 = 17 - 1 = 16.$$

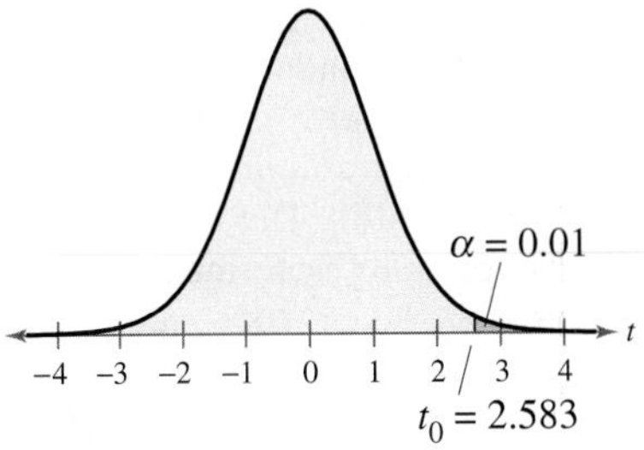

To find the critical value, use Table 5 with d.f. = 16 and $\alpha = 0.01$ in the "One tail, α" column. Because the test is right tailed, the critical value is positive. So,

$$t_0 = 2.583.$$

Try It Yourself 2

Find the critical value t_0 for a right-tailed test with $\alpha = 0.05$ and $n = 9$.

a. *Find* the t-value in Table 5 using d.f. = 8 and $\alpha = 0.05$ in the "One tail, α" column.

b. *Use* a positive sign.

Answer: Page A40

EXAMPLE 3

Finding Critical Values for t

Find the critical values t_0 and $-t_0$ for a two-tailed test with $\alpha = 0.05$ and $n = 26$.

SOLUTION The degrees of freedom are

$$\text{d.f.} = n - 1 = 26 - 1 = 25.$$

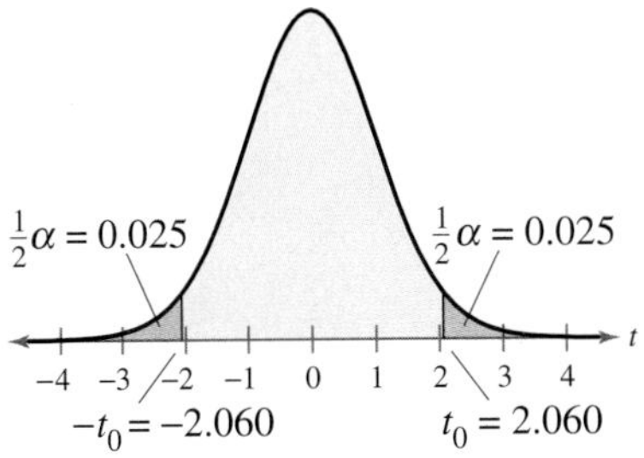

To find the critical value, use Table 5 with d.f. = 25 and $\alpha = 0.05$ in the "Two tail, α" column. Because the test is two-tailed, one critical value is negative and one is positive. So,

$$-t_0 = -2.060 \text{ and } t_0 = 2.060.$$

Try It Yourself 3

Find the critical values $\pm t_0$ for a two-tailed test with $\alpha = 0.01$ and $n = 16$.

a. *Find* the t-value in Table 5 using d.f. = 15 and $\alpha = 0.01$ in the "Two tail, α" column.

b. *Use* a negative and a positive sign.

Answer: Page A40

The t-Test for a Mean μ ($n < 30$, σ unknown)

Picturing the World

On the basis of a t-test, a decision was made whether to send truckloads of waste contaminated with cadmium to a sanitary landfill or a hazardous waste landfill. The trucks were sampled to determine if the mean level of cadmium exceeded the allowable amount of 1 milligram per liter for a sanitary landfill. In the study, the null hypothesis was $\mu \leq 1$. *(Source: Pacific Northwest National Laboratory)*

	H_0 True	H_0 False
Fail to reject H_0.		
Reject H_0.		

Describe the possible type I and type II errors of this situation.

To test a claim about a mean μ using a small sample ($n < 30$) from a normal, or nearly normal, distribution when σ is unknown, you can use a t-sampling distribution.

$$t = \frac{\text{(Sample mean)} - \text{(Hypothesized mean)}}{\text{Standard error}}$$

t-Test for a Mean μ

The ***t*-test for a mean** is a statistical test for a population mean. The t-test can be used when the population is normal or nearly normal, σ is unknown, and $n < 30$. The **test statistic** is $\bar{x}$ and the **standardized test statistic** is t.

$$t = \frac{\bar{x} - \mu}{s/\sqrt{n}}$$

The degrees of freedom are

$$\text{d.f.} = n - 1.$$

GUIDELINES

Using the t-Test for a Mean μ (Small Sample)

In Words	*In Symbols*
1. State the claim mathematically and verbally. Identify the null and alternative hypotheses.	State H_0 and H_a.
2. Specify the level of significance.	Identify α.
3. Identify the degrees of freedom and sketch the sampling distribution.	$\text{d.f.} = n - 1$
4. Determine any critical values.	Use Table 5 in Appendix B.
5. Determine any rejection regions.	
6. Find the standardized test statistic.	$t = \dfrac{\bar{x} - \mu}{s/\sqrt{n}}$
7. Make a decision to reject or fail to reject the null hypothesis.	If t is in the rejection region, reject H_0. Otherwise, fail to reject H_0.
8. Interpret the decision in the context of the original claim.	

Remember that when you make a decision, the possibility of a type I or a type II error exists.

If you prefer using P-values, turn to page 369 to learn how to use P-values for a t-test for a mean μ (small sample).

EXAMPLE 4

See MINITAB steps on page 400.

Testing μ with a Small Sample

A used car dealer says that the mean price of a 2002 Ford F-150 Super Cab is at least \$18,800. You suspect this claim is incorrect and find that a random sample of 14 similar vehicles has a mean price of \$18,000 and a standard deviation of \$1250. Is there enough evidence to reject the dealer's claim at $\alpha = 0.05$? Assume the population is normally distributed. *(Adapted from Kelley Blue Book)*

SOLUTION The claim is "the mean price is at least \$18,800." So, the null and alternative hypotheses are

$$H_0: \mu \geq \$18{,}800 \text{ (Claim)}$$

and

$$H_a: \mu < \$18{,}800.$$

The test is a left-tailed test, the level of significance is $\alpha = 0.05$, and there are d.f. $= 14 - 1 = 13$ degrees of freedom. So, the critical value is $t_0 = -1.771$. The rejection region is $t < -1.771$. The standardized test statistic is

$$t = \frac{\bar{x} - \mu}{s/\sqrt{n}} \qquad \text{Because } n < 30, \text{ use the } t\text{-test.}$$

$$= \frac{18{,}000 - 18{,}800}{1250/\sqrt{14}} \qquad \text{Assume } \mu = 18{,}800.$$

$$\approx -2.39.$$

The graph shows the location of the rejection region and the standardized test statistic t. Because t is in the rejection region, you should decide to reject the null hypothesis.

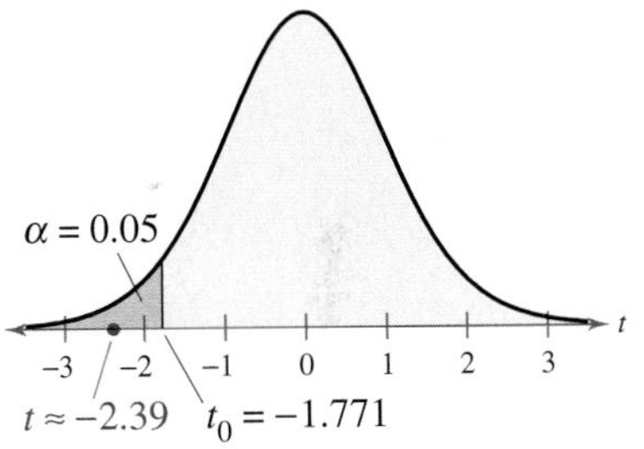

Interpretation There is enough evidence at the 5% level of significance to reject the claim that the mean price of a 2002 Ford F-150 Super Cab is at least \$18,800.

Try It Yourself 4

An insurance agent says that the mean cost of insuring a 2002 Ford F-150 Super Cab is at least \$1240. A random sample of nine similar insurance quotes has a mean cost of \$1190 and a standard deviation of \$62. Is there enough evidence to reject the agent's claim at $\alpha = 0.01$? Assume the population is normally distributed.

a. *Identify* the claim and state H_0 and H_a.
b. *Identify* the level of significance α and the degrees of freedom d.f.
c. *Find* the critical value t_0 and identify the rejection region.
d. *Use* the t-test to find the standardized test statistic t.
e. Sketch a graph. *Decide* whether to reject the null hypothesis.
f. *Interpret* the decision in the context of the original claim.

Answer: Page A40

EXAMPLE 5

See TI-83 steps on page 401.

Testing μ with a Small Sample

An industrial company claims that the mean pH level of the water in a nearby river is 6.8. You randomly select 19 water samples and measure the pH of each. The sample mean and standard deviation are 6.7 and 0.24, respectively. Is there enough evidence to reject the company's claim at $\alpha = 0.05$? Assume the population is normally distributed.

SOLUTION The claim is "the mean pH level is 6.8." So, the null and alternative hypotheses are

$$H_0: \mu = 6.8 \text{ (Claim)}$$

and

$$H_a: \mu \neq 6.8.$$

The test is a two-tailed test, the level of significance is $\alpha = 0.05$, and there are d.f. $= 19 - 1 = 18$ degrees of freedom. So, the critical values are $-t_0 = -2.101$ and $t_0 = 2.101$. The rejection regions are $t < -2.101$ and $t > 2.101$. The standardized test statistic is

$$t = \frac{\bar{x} - \mu}{s/\sqrt{n}} \qquad \text{Because } n < 30\text{, use the } t\text{-test.}$$

$$= \frac{6.7 - 6.8}{0.24/\sqrt{19}} \qquad \text{Assume } \mu = 6.8.$$

$$\approx -1.82.$$

The graph shows the location of the rejection region and the standardized test statistic t. Because t is not in the rejection region, you should decide not to reject the null hypothesis.

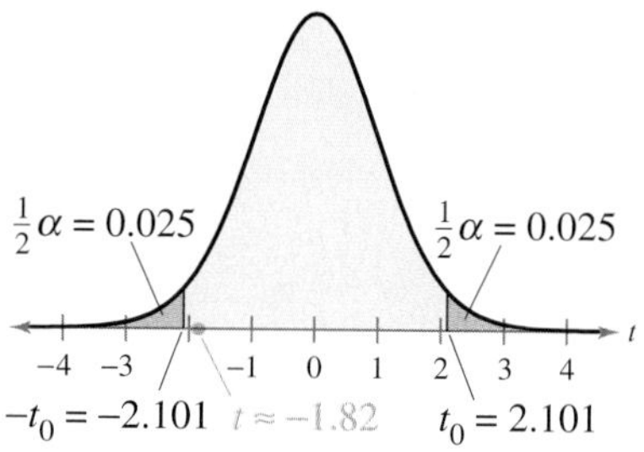

5% Level of Significance

Interpretation There is not enough evidence at the 5% level of significance to reject the claim that the mean pH is 6.8.

Try It Yourself 5

The company also claims that the mean conductivity of the river is 1890 milligrams per liter. The conductivity of a water sample is a measure of the total dissolved solids in the sample. You randomly select 19 water samples and measure the conductivity of each. The sample mean and standard deviation are 2500 milligrams per liter and 700 milligrams per liter, respectively. Is there enough evidence to reject the company's claim at $\alpha = 0.01$? Assume the population is normally distributed.

a. *Identify* the claim and state H_0 and H_a.
b. *Identify* the level of significance α and the degrees of freedom d.f.
c. *Find* the critical values $\pm t_0$ and identify the rejection regions.
d. *Use* the t-test to find the standardized test statistic t.
e. Sketch a graph. *Decide* whether to reject the null hypothesis.
f. *Interpret* the decision in the context of the original claim.

Answer: Page A40

Using *P*-values with *t*-Tests

Suppose you wanted to find a *P*-value given $t = 1.98$, 15 degrees of freedom, and a right-tailed test. Using Table 5 in Appendix B, you can determine that *P* falls between $\alpha = 0.025$ and $\alpha = 0.05$, but you cannot determine an exact value for *P*. In such cases, you can use technology to perform a hypothesis test and find exact *P*-values.

EXAMPLE 6

Using P-values with a t-Test

The American Automobile Association claims that the mean daily meal cost for a family of four traveling on vacation in California is \$132. A random sample of 11 such families has a mean daily meal cost of \$141 with a standard deviation of \$20. Is there enough evidence to reject the claim at $\alpha = 0.10$? Assume the population is normally distributed. *(Adapted from American Automobile Association)*

SOLUTION The TI-83 display at the far left shows how to set up the hypothesis test. The two displays on the right show the possible results, depending on whether you select "Calculate" or "Draw."

TI-83

```
T-Test
Inpt:Data Stats
μ0:132
x̄:141
Sx:20
n:11
μ: ≠μ0  <μ0  >μ0
Calculate Draw
```

TI-83

```
T-Test
μ≠ 132
t=1.492481156
p=.1664335286
x̄=141
Sx=20
n=11
```

TI-83

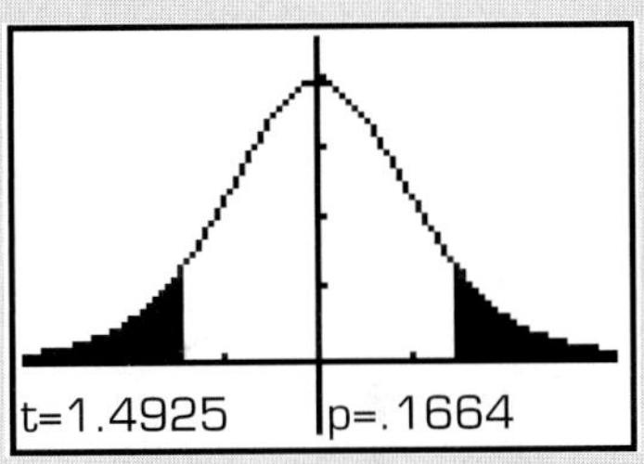

From the displays, you can see that $P \approx 0.1664$. Because $P > 0.10$, there is not enough evidence to reject the claim at the 10% level of significance. In other words, you should not reject the null hypothesis.

> **Study Tip**
>
> Note that the TI-83 display on the far right in Example 6 also displays the standardized test statistic $t \approx 1.4925$.
>
> If you are using the TI-83 and have the raw data from a sample, note that you can enter it into a list and use the *Data* input option instead of the *Stats* input option. To use MINITAB to perform a one-sample *t*-test, you must have the raw data.

Try It Yourself 6

The American Automobile Association claims that the mean nightly lodging rate for a family of four traveling on vacation in California is at least \$136. A random sample of six such families has a mean nightly lodging rate of \$126 with a standard deviation of \$12. Is there enough evidence to reject the claim at $\alpha = 0.05$? Assume the population is normally distributed. *(Adapted from American Automobile Association)*

a. *Use* a TI-83 to find the *P*-value.
b. *Compare* the *P*-value with the level of significance α.
c. *Make* a decision.
d. *Interpret* the decision in the context of the original claim.

Answer: Page A41

7.3 Exercises

Building Basic Skills and Vocabulary

1. Explain how to find critical values for a t sampling distribution.
2. Explain how to use a t-test to test a hypothesized mean μ given a small sample $(n < 30)$. What assumption about the population is necessary?

In Exercises 3–14, find the critical value(s) for the indicated t-test, level of significance α, and sample size n.

3. Right-tailed test, $\alpha = 0.05$, $n = 23$
4. Right-tailed test, $\alpha = 0.01$, $n = 11$
5. Left-tailed test, $\alpha = 0.025$, $n = 19$
6. Left-tailed test, $\alpha = 0.05$, $n = 14$
7. Two-tailed test, $\alpha = 0.01$, $n = 27$
8. Two-tailed test, $\alpha = 0.05$, $n = 10$
9. Right-tailed test, $\alpha = 0.10$, $n = 20$
10. Right-tailed test, $\alpha = 0.05$, $n = 15$
11. Left-tailed test, $\alpha = 0.01$, $n = 28$
12. Left-tailed test, $\alpha = 0.005$, $n = 12$
13. Two-tailed test, $\alpha = 0.02$, $n = 5$
14. Two-tailed test, $\alpha = 0.10$, $n = 22$

Graphical Analysis In Exercises 15–18, state whether the standardized test statistic t indicates that you should reject the null hypothesis. Explain.

15. (a) $t = 2.091$
    (b) $t = 0$
    (c) $t = -1.08$
    (d) $t = -2.096$

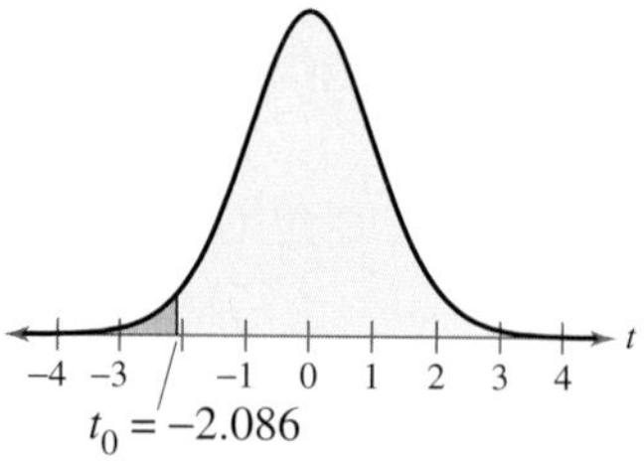

16. (a) $t = 1.308$
    (b) $t = -1.389$
    (c) $t = 1.650$
    (d) $t = -0.998$

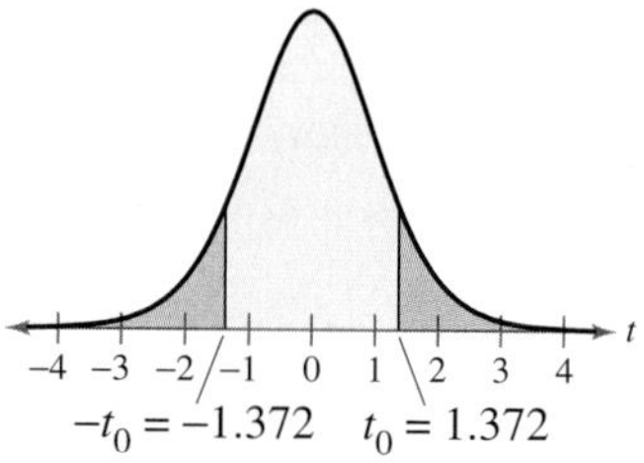

17. (a) $t = -2.502$
    (b) $t = 2.203$
    (c) $t = 2.680$
    (d) $t = -2.703$

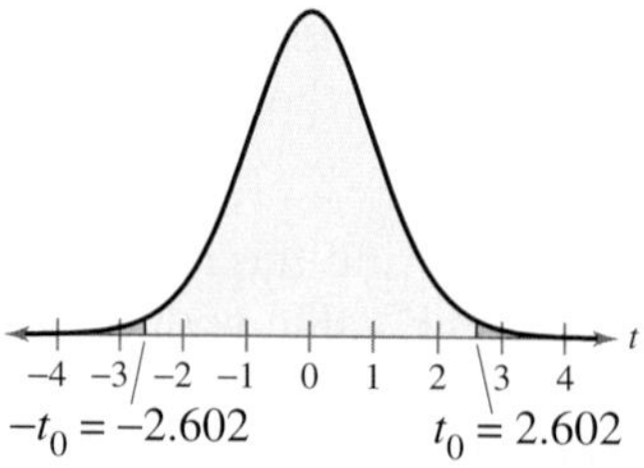

18. (a) $t = 1.705$
    (b) $t = -1.755$
    (c) $t = -1.585$
    (d) $t = 1.745$

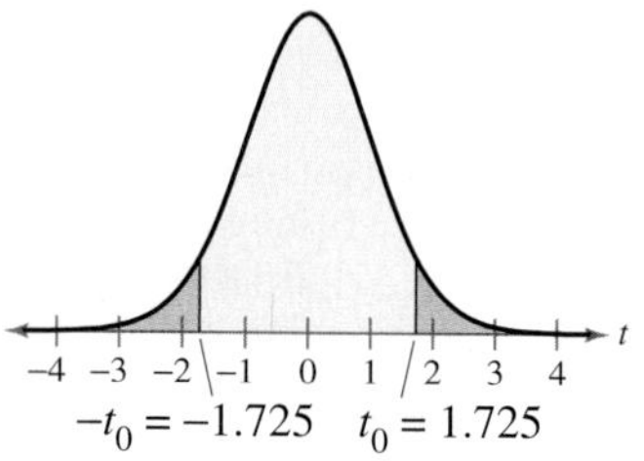

1. Identify the level of significance α and the degrees of freedom, d.f. $= n - 1$. Find the critical value(s) using the t-distribution table in the row with $n - 1$ d.f. If the hypothesis test is
   (1) left-tailed, use "One Tail, α" column with a negative sign.
   (2) right-tailed, use "One Tail, α" column with a positive sign.
   (3) two-tailed, use "Two Tail, α" column with a negative and a positive sign.
2. See Selected Answers, page A96.
3. 1.717 **4.** 2.764 **5.** −2.101
6. −1.771 **7.** ±2.779
8. ±2.262 **9.** 1.328 **10.** 1.761
11. −2.473 **12.** −3.106
13. ±3.747 **14.** ±1.721
15. (a) Fail to reject H_0 because $t > -2.086$.
    (b) Fail to reject H_0 because $t > -2.086$.
    (c) Fail to reject H_0 because $t > -2.086$.
    (d) Reject H_0 because $t < -2.086$.
16. See Selected Answers, page A96.
17. (a) Fail to reject H_0 because $-2.602 < t < 2.602$.
    (b) Fail to reject H_0 because $-2.602 < t < 2.602$.
    (c) Reject H_0 because $t > 2.602$.
    (d) Reject H_0 because $t < -2.602$.
18. See Selected Answers, page A96.

In Exercises 19–22, use a t-test to test the claim about the population mean μ at the given level of significance α using the given sample statistics. For each claim, assume the population is normally distributed.

19. Claim: $\mu = 15$; $\alpha = 0.01$.
Sample statistics: $\bar{x} = 13.9$, $s = 3.23$, $n = 6$

20. Claim: $\mu > 25$; $\alpha = 0.05$.
Sample statistics: $\bar{x} = 26.2$, $s = 2.32$, $n = 17$

21. Claim: $\mu \geq 8000$; $\alpha = 0.01$.
Sample statistics: $\bar{x} = 7700$, $s = 450$, $n = 25$

22. Claim: $\mu \neq 52{,}200$; $\alpha = 0.05$.
Sample statistics: $\bar{x} = 53{,}220$, $s = 1200$, $n = 4$

Using and Interpreting Concepts

Testing Claims In Exercises 23–28,

(a) write the claim mathematically and identify H_0 and H_a.

(b) find the critical value(s) and identify the rejection region(s).

(c) find the standardized test statistic.

(d) decide whether to reject or fail to reject the null hypothesis.

(e) interpret the decision in the context of the original claim.

For each claim, assume the population is normally distributed.

23. ***Microwave Repair Costs*** A microwave oven repairer says that the mean repair cost for damaged microwave ovens is less than \$100. You work for the repairer and want to test this claim. You find that a random sample of five microwave ovens has a mean repair cost of \$75 and a standard deviation of \$12.50. At $\alpha = 0.01$, do you have enough evidence to support the repairer's claim? *(Adapted from Consumer Reports)*

24. ***Computer Repair Costs*** A computer repairer believes that the mean repair cost for damaged computers is more than \$95. To test this claim, you determine the repair costs for seven randomly selected computers and find that the mean repair cost is \$100 per computer with a standard deviation of \$42.50. At $\alpha = 0.01$, do you have enough evidence to support the repairer's claim? *(Adapted from Consumer Reports)*

25. ***Waste Recycled*** An environmentalist estimates that the mean waste recycled by adults in the United States is more than 1 pound per person per day. You want to test this claim. You find that the mean waste recycled per person per day for a random sample of 12 adults in the United States is 1.2 pounds and the standard deviation is 0.3 pound. At $\alpha = 0.05$, can you support the claim? *(Adapted from Franklin Association, Ltd.)*

26. ***Waste Generated*** As part of your work for an environmental awareness group, you want to test a claim that the mean waste generated by adults in the United States is more than 4 pounds per day. In a random sample of 10 adults in the United States, you find that the mean waste generated per person per day is 4.3 pounds with a standard deviation of 1.2 pounds. At $\alpha = 0.05$, can you support the claim? *(Adapted from Franklin Association, Ltd.)*

19. Fail to reject H_0. There is not enough evidence at the 1% level to reject the claim.

20. Reject H_0. There is enough evidence at the 5% level to support the claim.

21. Reject H_0. There is enough evidence at the 1% level to reject the claim.

22. Fail to reject H_0. There is not enough evidence at the 5% level to support the claim.

23. (a) H_0: $\mu \geq 100$
H_a: $\mu < 100$ (claim)
(b) $t_0 = -3.747$
Rejection region: $t < -3.747$
(c) -4.472 (d) Reject H_0.
(e) There is sufficient evidence at the 1% level to support the claim that the mean repair cost for damaged microwave ovens is less than \$100.

24. (a) H_0: $\mu \leq \$95$
H_a: $\mu > \$95$ (claim)
(b) $t_0 = 3.143$
Rejection region: $t > 3.143$
(c) 0.311 (d) Fail to reject H_0.
(e) There is insufficient evidence at the 1% level to support the claim that the mean repair cost for damaged computers is more than \$95.

25. (a) H_0: $\mu \leq 1$; H_a: $\mu > 1$ (claim)
(b) $t_0 = 1.796$
Rejection region: $t > 1.796$
(c) 2.309 (d) Reject H_0.
(e) There is sufficient evidence at the 5% level to support the claim that the mean waste recycled by adults in the United States is more than 1 pound per person per day.

26. (a) H_0: $\mu \leq 4$; H_a: $\mu > 4$ (claim)
(b) $t_0 = 1.833$
Rejection region: $t > 1.833$
(c) 0.791 (d) Fail to reject H_0.
(e) There is insufficient evidence at the 5% level to support the claim that the mean waste generated by adults in the United States is more than 4 pounds per day.

27. (a) H_0: $\mu = \$22{,}300$ (claim)
H_a: $\mu \neq \$22{,}300$
(b) $-t_0 = -2.262$ and $t_0 = 2.262$
Rejection regions: $t < -2.262$, $t > 2.262$
(c) -0.572
(d) Fail to reject H_0.
(e) There is insufficient evidence at the 5% level to reject the claim that the mean salary for full-time male workers over age 25 without a high school diploma is \$22,300.

28. (a) H_0: $\mu = \$17{,}100$ (claim)
H_a: $\mu \neq \$17{,}100$
(b) $-t_0 = -2.201$ and $t_0 = 2.201$
Rejection regions: $t < -2.201$, $t > 2.201$
(c) 0.948
(d) Fail to reject H_0.
(e) There is insufficient evidence at the 5% level to reject the claim that the mean annual pay for full-time female workers over age 25 without a high school diploma is \$17,100.

29. (a) H_0: $\mu \geq 3.0$
H_a: $\mu < 3.0$ (claim)
(b) 0.130
(c) Fail to reject H_0.
(d) There is insufficient evidence at the 5% level to support the claim that teenage males drink fewer than three 12-ounce servings of soda per day.

30. (a) H_0: $\mu \leq \$580$
H_a: $\mu > \$580$ (claim)
(b) 0.038
(c) Reject H_0.
(d) There is sufficient evidence at the 5% level to support the claim that teachers spend a mean of more than \$580 of their own money on school supplies in a year.

31. (a) H_0: $\mu \geq 32$; H_a: $\mu < 32$ (claim)
(b) 0.034
(c) Fail to reject H_0.
(d) There is insufficient evidence at the 1% level to support the claim that the mean class size for full-time faculty is fewer than 32 students.

27. *Annual Pay* An employment information service claims the mean annual pay for full-time male workers over age 25 and without a high school diploma is \$22,300. The annual pay for a random sample of 10 full-time male workers without a high school diploma is listed. At $\alpha = 0.05$, test the claim that the mean salary is \$22,300. *(Adapted from U.S. Census Bureau)*

20,654 21,138 22,355 21,395 22,975
16,912 19,156 23,193 24,180 26,285

28. *Annual Pay* An employment information service claims the mean annual pay for full-time female workers over age 25 and without a high school diploma is \$17,100. The annual pay for a random sample of 12 full-time female workers without a high school diploma is listed. At $\alpha = 0.05$, test the claim that the mean salary is \$17,100. *(Adapted from U.S. Census Bureau)*

16,009 16,790 17,328 18,161 16,631 21,028
16,114 17,176 17,503 19,764 15,316 18,801

Testing Claims Using P-values In Exercises 29–34, (a) write the claim mathematically and identify H_0 and H_a, (b) use technology to find the *P*-value, (c) decide whether to reject or fail to reject the null hypothesis, and (d) interpret the decision in the context of the original claim. Assume the population is normally distributed.

DATA **29. *Soda Consumption*** For your study on the food consumption habits of teenage males, you randomly select 20 teenage males and ask each how many 12-ounce servings of soda he drinks each day. The results are listed below. At $\alpha = 0.05$, is there enough evidence to support the claim that teenage males drink fewer than three 12-ounce servings of soda per day? *(Adapted from Center for Science in the Public Interest)*

3.3 2.1 2.5 2.1 3.4 3.3 4.4 3.4 2.5 3.2
2.5 3.8 2.0 2.9 1.9 1.3 1.9 2.8 4.2 2.2

DATA **30. *School Supplies*** A company that manufactures school supplies says that teachers spend a mean of more than \$580 of their own money on school supplies in a year. A random sample of the amounts (in dollars) that 24 teachers spent on school supplies in a recent year is listed below. At $\alpha = 0.05$, is there enough evidence to support the company's claim? *(Adapted from Washington Post Company)*

523 581 476 600 721 622 575 320 888 875 910 643
596 587 398 622 585 700 590 710 729 857 630 520

31. *Class Size* You receive a brochure from a large university. The brochure indicates that the mean class size for full-time faculty is fewer than 32 students. You want to test this claim. You randomly select 18 classes taught by full-time faculty and determine the class size of each. The results are listed below. At $\alpha = 0.01$, can you support the university's claim? *(Adapted from National Center for Education Statistics)*

35 28 29 33 32 40 26 25 29 28 30 36 33 29 27 30 28 25

32. *Faculty Classroom Hours* The dean of a university estimates that the mean number of classroom hours per week for full-time faculty is 11.0. As a member of the student council, you want to test this claim. A random sample of the number of classroom hours for full-time faculty for one week is listed below. At $\alpha = 0.01$, can you reject the dean's claim? *(Adapted from National Center for Education Statistics)*

11.8 8.6 12.6 7.9 6.4 10.4 13.6 9.1

33. *Eating Out* A restaurant association says the typical household in the United States spends a mean of \$2116 per year on food away from home. You are a consumer reporter for a national publication and want to test this claim. You randomly select 12 U.S. households and find out how much each spent on food away from home per year. Can you reject the restaurant association's claim at $\alpha = 0.02$? *(Adapted from the National Restaurant Association)*

3474 1262 853 1988 2778 2283 2797 1916 3415 2645 2726 2948

34. *Lodging Costs* A travel association says the daily lodging costs for a family in the United States is \$132. You work for a tourist publication and want to test this claim. You randomly select 10 U.S. families and find out how much each spent on lodging for one overnight trip. At $\alpha = 0.02$, can you reject the travel association's claim? *(Adapted from the American Automobile Association)*

164 137 142 155 119 104 74 204 148 181

Extending Concepts

35. *Credit Card Balances* To test the claim that the mean credit card balance among families that have a balance is greater than \$2294, you do some research and find that a random sample of six cardholders has a mean credit balance of \$2494 with a standard deviation of \$325. You conduct a statistical experiment where H_0: $\mu \leq \$2294$ and H_a: $\mu > \$2294$. At $\alpha = 0.01$, explain why you cannot reject H_0. Assume the population is normally distributed. *(Adapted from American City Business Journals Inc.)*

36. *Using Different Values of α and n* In Exercise 35, you believe that H_0 is not valid. Which of the following allows you to reject H_0? (a) Use the same values but increase α from 0.01 to 0.05. (b) Use the same values but increase α from 0.01 to 0.10. (c) Use the same values but increase n from 6 to 12. (d) Use the same values but increase n from 6 to 24.

Deciding on a Distribution In Exercises 37 and 38, decide whether you should use a normal sampling distribution or a *t*-sampling distribution to perform the hypothesis test. Justify your decision. Then use the distribution to test the claim. Write a short paragraph about the results of the test and what you can conclude about the claim.

37. *Gas Mileage* A car company says that the mean gas mileage for its luxury sedan is at least 21 miles per gallon (mpg). You believe the claim is incorrect and find that a random sample of five cars has a mean gas mileage of 19 mpg and a standard deviation of 4 mpg. Assume the gas mileage of all of the company's luxury sedans is normally distributed. At $\alpha = 0.05$, test the company's claim. *(Adapted from Consumer Reports)*

38. *Faculty Time* An administrator of a state university system says that the mean time full-time faculty spend with students outside of class is 337 hours per week. (The hours were counted for each student, so a professor's spending five hours with a group of four students would count as 20 hours.) The mean time a random sample of 50 full-time faculty members spent with students outside of class was 332 hours per week and the standard deviation was 10 hours per week. At $\alpha = 0.01$, test the administrator's claim. *(Adapted from National Center for Education Statistics)*

32. (a) H_0: $\mu = 11.0$ (claim)
H_a: $\mu \neq 11.0$
(b) 0.316
(c) Fail to reject H_0.
(d) There is insufficient evidence at the 1% level to reject the claim that the mean number of classroom hours per week for full-time faculty is 11.0.

33. (a) H_0: $\mu = \$2116$ (claim)
H_a: $\mu \neq \$2116$
(b) 0.210
(c) Fail to reject H_0.
(d) There is sufficient evidence at the 2% level to reject the claim that the typical household in the United States spends a mean amount of \$2116 per year on food away from home.

34. (a) H_0: $\mu = \$132$ (claim)
H_a: $\mu \neq \$132$
(b) 0.386
(c) Fail to reject H_0.
(d) There is insufficient evidence at the 2% level to reject the claim that the daily lodging costs for a family in the United States is \$132.

35. Because the *P*-value $= 0.096 > 0.01$, fail to reject H_0.

36. b, d

37. Use the *t*-distribution because the population is normal, $n < 30$, and σ is unknown.

Fail to reject H_0. There is insufficient evidence at the 5% level to reject the claim that the mean gas mileage for the luxury sedan is at least 21 miles per gallon.

38. Use the *z*-distribution because σ is unknown and $n \geq 30$.

Reject H_0. There is sufficient evidence at the 1% level to reject the claim that the mean time full-time faculty spend with students outside of class is 337 hours per week.

7.4 Hypothesis Testing for Proportions

What You Should Learn

- How to use the z-test to test a population proportion p

Hypothesis Test for Proportions

Hypothesis Test for Proportions

In Sections 7.2 and 7.3, you learned how to perform a hypothesis test for a population mean. In this section, you will learn how to test a population proportion p.

Hypothesis tests for proportions occur when a politician wants to know the proportion of his or her constituents who favor a certain bill or when a quality assurance engineer tests the proportion of parts that are defective.

If $np \geq 5$ and $nq \geq 5$ for a binomial distribution, then the sampling distribution for $\hat{p}$ is normal with

$$\mu_{\hat{p}} = p \quad \text{and} \quad \sigma_{\hat{p}} = \sqrt{pq/n}.$$

z-Test for a Proportion p

The **z-test for a proportion** is a statistical test for a population proportion p. The z-test can be used when a binomial distribution is given such that $np \geq 5$ and $nq \geq 5$. The **test statistic** is the sample proportion $\hat{p}$ and the **standardized test statistic** is z.

$$z = \frac{\hat{p} - \mu_{\hat{p}}}{\sigma_{\hat{p}}} = \frac{\hat{p} - p}{\sqrt{pq/n}}$$

GUIDELINES

Using a z-Test for a Proportion p

Verify that $np \geq 5$ and $nq \geq 5$.

In Words	*In Symbols*
1. State the claim mathematically and verbally. Identify the null and alternative hypotheses.	State H_0 and H_a.
2. Specify the level of significance.	Identify α.
3. Sketch the sampling distribution.	
4. Determine any critical values.	Use Table 4 in Appendix B.
5. Determine any rejection regions.	
6. Find the standardized test statistic.	$z = \frac{\hat{p} - p}{\sqrt{pq/n}}$
7. Make a decision to reject or fail to reject the null hypothesis.	If z is in the rejection region, reject H_0. Otherwise, fail to reject H_0.
8. Interpret the decision in the context of the original claim.	

Insight

A hypothesis test for a proportion p can also be performed using P-values. Use the guidelines on page 349 for using P-values for a z-test for a mean μ, but in Step 3 find the standardized test statistic by using the formula

$$z = \frac{\hat{p} - p}{\sqrt{pq/n}}.$$

The other steps in the test are the same.

EXAMPLE 1

See TI-83 steps on page 401.

Hypothesis Test for a Proportion

A medical researcher claims that less than 20% of adults in the United States are allergic to a medication. In a random sample of 100 adults, 15% say they have such an allergy. At $\alpha = 0.01$, is there enough evidence to support the reseacher's claim? *(Adapted from Marist College Institute for Public Opinion)*

SOLUTION The products $np = 100(0.20) = 20$ and $nq = 100(0.80) = 80$ are both greater than 5. So, you can use a z-test. The claim is "less than 20% are allergic to a medication." So, the null and alternative hypotheses are

$$H_0\colon p \geq 0.2 \quad \text{and} \quad H_a\colon p < 0.2. \text{ (Claim)}$$

Because the test is a left-tailed test and the level of significance is $\alpha = 0.01$, the critical value is $z_0 = -2.33$ and the rejection region is $z < -2.33$. The standardized test statistic is

$$z = \frac{\hat{p} - p}{\sqrt{pq/n}} \qquad \text{Because } np \geq 5 \text{ and } nq \geq 5, \text{ you can use the } z\text{-test.}$$

$$= \frac{0.15 - 0.2}{\sqrt{(0.2)(0.8)/100}} \qquad \text{Assume } p = 0.2.$$

$$= -1.25.$$

The graph shows the location of the rejection region and the standardized test statistic z. Because z is not in the rejection region, you should decide not to reject the null hypothesis.

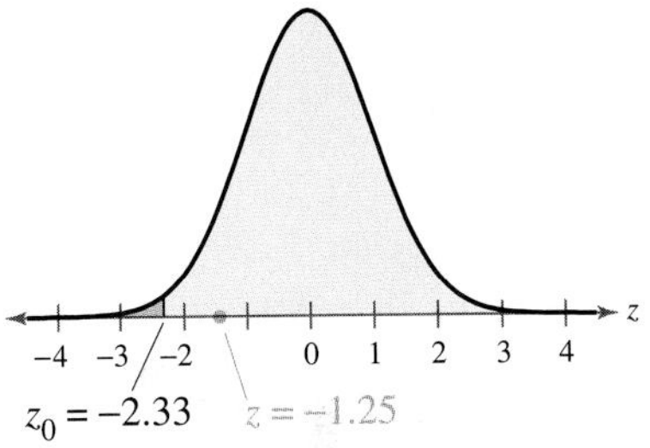

Interpretation There is not enough evidence to support the claim that less than 20% of adults in the United States are allergic to the medication.

Study Tip

Remember that when you fail to reject H_0, a type II error is possible. For instance, in Example 1 the null hypothesis, $p \geq 0.2$, may be false.

Try It Yourself 1

A researcher claims that less than 30% of adults in the United States are allergic to trees, weeds, flowers, and grasses. In a random sample of 86 adults, 20% say they have such an allergy. At $\alpha = 0.05$, is there enough evidence to support the researcher's claim? *(Adapted from Marist College Institute for Public Opinion)*

a. *Verify* that $np \geq 5$ and $nq \geq 5$.
b. *Identify* the claim and state H_0 and H_a.
c. *Identify* the level of significance α.
d. *Find* the critical value z_0 and *identify* the rejection region.
e. *Use* the z-test to find the standardized test statistic z.
f. *Decide* whether to reject the null hypothesis. Use a graph if necessary.
g. *Interpret* the decision in the context of the original claim. *Answer: Page A41*

To use a P-value to perform the hypothesis test in Example 1, use Table 4 to find the area corresponding to $z = -1.25$. The area is 0.1056. Because this is a left-tailed test, the P-value is equal to the area to the left of $z = -1.25$. So, $P = 0.1056$. Because the P-value is greater than $\alpha = 0.01$, you should fail to reject the null hypothesis. Note this is the same result obtained in Example 1.

Picturing the World

A recent study claimed that 53.7% of three-year-olds in the United States were given over-the-counter drugs (such as Tylenol) in a period of 30 days. To test this claim, you conduct a random telephone survey of 300 parents of three-year-olds. In the survey, you find that 166 of the three-year-olds were given an over-the-counter drug during the past 30 days. *(Source: Journal of the American Medical Association)*

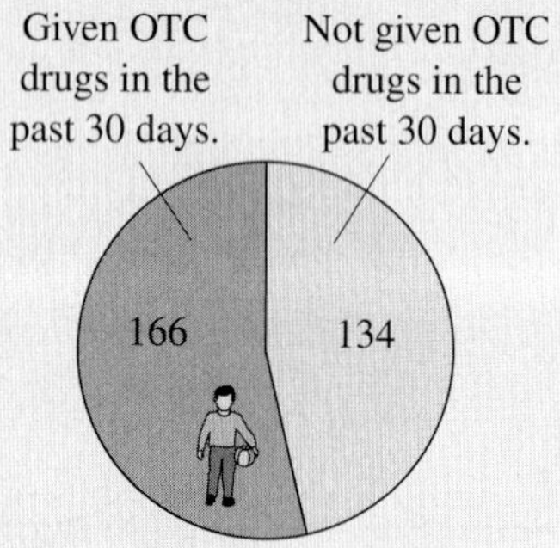

At $\alpha = 0.05$, is there enough evidence to reject the claim?

EXAMPLE 2

Hypothesis Test for a Proportion

See MINITAB steps on page 400.

Harper's Index claims that 23% of people in the United States are in favor of outlawing cigarettes. You decide to test this claim and ask a random sample of 200 people in the United States whether they are in favor of outlawing cigarettes. Of the 200 people, 27% are in favor. At $\alpha = 0.05$, is there enough evidence to reject the claim?

SOLUTION The products $np = 200(0.23) = 46$ and $nq = 200(0.77) = 154$ are both greater than 5. So, you can use a z-test. The claim is "23% of people in the United States are in favor of outlawing cigarettes." So, the null and alternative hypotheses are

$$H_0\text{: } p = 0.23 \text{ (Claim)} \quad \text{and} \quad H_a\text{: } p \neq 0.23.$$

Because the test is a two-tailed test and the level of significance is $\alpha = 0.05$, the critical values are $-z_0 = -1.96$ and $z_0 = 1.96$. The rejection regions are $z < -1.96$ and $z > 1.96$. The standardized test statistic is

$$z = \frac{\hat{p} - p}{\sqrt{pq/n}}$$ Because $np \geq 5$ and $nq \geq 5$, you can use the z-test.

$$= \frac{0.27 - 0.23}{\sqrt{(0.23)(0.77)/200}}$$ Assume $p = 0.23$.

$$\approx 1.34.$$

The graph shows the location of the rejection regions and the standardized test statistic z. Because z is not in the rejection region, you should fail to reject the null hypothesis.

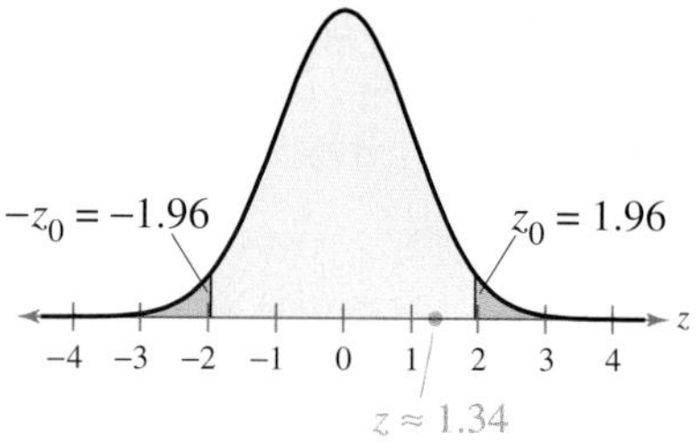

Interpretation At the 5% level of significance, there is not enough evidence to reject the claim that 23% of people in the United States are in favor of outlawing cigarettes.

Try It Yourself 2

USA TODAY reports that 5% of U.S. adults have seen an extraterrestrial being. You decide to test this claim and ask a random sample of 250 U.S. adults whether they have ever seen an extraterrestrial being. Of those surveyed, 8% reply yes. At $\alpha = 0.01$, is there enough evidence to reject the claim?

a. *Verify* that $np \geq 5$ and $nq \geq 5$.
b. *Identify* the claim and state H_0 and H_a.
c. *Identify* the level of significance α.
d. *Find* the critical values $-z_0$ and z_0 and *identify* the rejection regions.
e. *Use* the z-test to find the standardized test statistic z.
f. *Decide* whether to reject the null hypothesis. Use a graph if necessary.
g. *Interpret* the decision in the context of the original claim.

Answer: Page A41

EXAMPLE 3

Hypothesis Test for a Proportion

The Pew Research Center claims that more than 55% of U.S. adults regularly watch a network news broadcast. You decide to test this claim and ask a random sample of 425 adults in the United States whether they regularly watch a network news broadcast. Of the 425 adults, 255 respond yes. At $\alpha = 0.05$, is there enough evidence to support the claim?

SOLUTION The products $np = 425(0.55) \approx 234$ and $nq = 425(0.45) \approx 191$ are both greater than 5. So, you can use a z-test. The claim is "more than 55% of U.S. adults watch a network news broadcast." So, the null and alternative hypotheses are

H_0: $p \leq 0.55$ and H_a: $p > 0.55$. (Claim)

Because the test is a right-tailed test and the level of significance is $\alpha = 0.05$, the critical value is $z_0 = 1.645$ and the rejection region is $z > 1.645$. The standardized test statistic is

$$z = \frac{\hat{p} - p}{\sqrt{pq/n}}$$ Because $np \geq 5$ and $nq \geq 5$, you can use the z-test.

$$= \frac{(x/n) - p}{\sqrt{pq/n}}$$ $\hat{p} = x/n$.

$$= \frac{(255/425) - 0.55}{\sqrt{(0.55)(0.45)/425}}$$ Assume $p = 0.55$.

$$\approx 2.07.$$

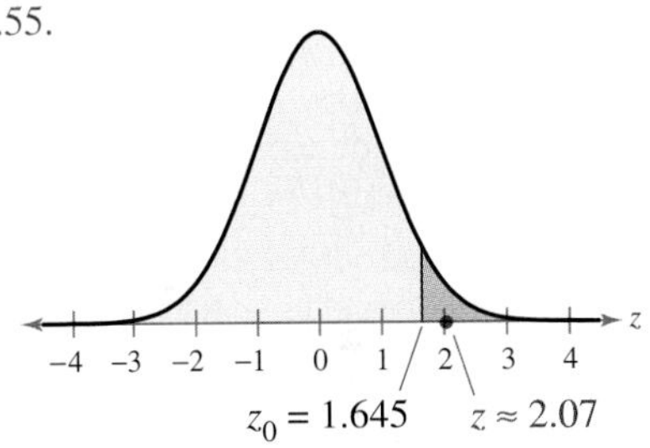

The graph shows the location of the rejection region and the standardized test statistic z. Because z is in the rejection region, you should decide to reject the null hypothesis.

Interpretation There is enough evidence at the 5% level of significance to support the claim that more than 55% of U.S. adults regularly watch a network news broadcast.

Try It Yourself 3

The Pew Research Center claims that more than 38% of U.S. adults regularly watch a cable news broadcast. You decide to test this claim and ask a random sample of 75 adults whether they regularly watch a cable news broadcast. Of the 75 adults, 33 respond yes. At $\alpha = 0.01$, is there enough evidence to support the claim?

a. *Verify* that $np \geq 5$ and $nq \geq 5$.
b. *Identify* the claim and state H_0 and H_a.
c. *Identify* the level of significance α.
d. *Find* the critical value z_0 and *identify* the rejection region.
e. *Use* the z-test to find the standardized test statistic z.
f. *Decide* whether to reject the null hypothesis. Use a graph if necessary.
g. *Interpret* the decision in the context of the original claim.

Answer: Page A41

7.4 Exercises

Help

MyMathLab

Student Study Pack

Building Basic Skills and Vocabulary

1. Explain how to test a population proportion p.

2. Explain how to decide when a normal distribution can be used to approximate a binomial distribution.

In Exercises 3–8, decide whether the normal sampling distribution can be used. If it can be used, test the claim about the population proportion p at the given level of significance α using the given sample statistics.

3. Claim: $p \neq 0.25$; $\alpha = 0.05$. Sample statistics: $\hat{p} = 0.239$, $n = 105$

4. Claim: $p \leq 0.30$; $\alpha = 0.05$. Sample statistics: $\hat{p} = 0.35$, $n = 500$

5. Claim: $p < 0.12$; $\alpha = 0.01$. Sample statistics: $\hat{p} = 0.10$, $n = 20$

6. Claim: $p > 0.125$; $\alpha = 0.01$. Sample statistics: $\hat{p} = 0.2325$, $n = 45$

7. Claim: $p \geq 0.48$; $\alpha = 0.10$. Sample statistics: $\hat{p} = 0.40$, $n = 70$

8. Claim: $p = 0.80$; $\alpha = 0.10$. Sample statistics: $\hat{p} = 0.875$, $n = 16$

Using and Interpreting Concepts

Testing Claims In Exercises 9–14, (a) write the claim mathematically and identify H_0 and H_a, (b) find the critical value(s) and identify the rejection region(s), (c) find the standardized test statistic, (d) decide whether to reject or fail to reject the null hypothesis, and (e) interpret the decision in the context of the original claim.

9. ***Smokers*** A medical researcher says that at least 23% of U.S. adults are smokers. In a random sample of 200 U.S. adults, 22.5% say that they are smokers. At $\alpha = 0.01$, do you have enough evidence to reject the researcher's claim? *(Adapted from U.S. National Center for Health Statistics)*

10. ***Do You Eat Breakfast?*** A medical researcher estimates that no more than 55% of U.S. adults eat breakfast every day. In a random sample of 250 U.S. adults, 56.4% say they eat breakfast every day. At $\alpha = 0.01$, is there enough evidence to reject the researcher's claim? *(Adapted from U.S. National Center for Health Statistics)*

11. ***Environmentally Conscious Consumers*** You are employed by an environmental conservation agency that recently claimed more than 30% of U.S. consumers have stopped buying a certain product because the manufacturing of the product pollutes the environment. You want to test this claim. To do so, you randomly select 1050 U.S. consumers and find that 32% have stopped buying this product because of pollution concerns. At $\alpha = 0.03$, can you support the claim? *(Adapted from Wirthlin Worldwide)*

12. ***Genetically Modified Foods*** An environmentalist claims that more than 58% of British consumers want supermarkets to stop selling genetically modified foods. You want to test this claim. You find that in a random sample of 100 British consumers, 61% say that they want supermarkets to stop selling genetically modified foods. At $\alpha = 0.10$, can you support the environmentalist's claim? *(Adapted from Friends of the Earth)*

1. Verify that $np \geq 5$ and $nq \geq 5$. State H_0 and H_a. Specify the level of significance α. Determine the critical value(s) and rejection region(s). Find the standardized test statistic. Make a decision and interpret in the context of the original claim.
2. If $np \geq 5$ and $nq \geq 5$, the normal distribution can be used.
3. Use normal distribution.
   Fail to reject H_0. There is not enough evidence at the 5% level to support the claim.
4. Reject H_0. There is enough evidence at the 5% level to reject the claim.
5. Cannot use normal distribution.
6. Use normal distribution.
   Fail to reject H_0. There is not enough evidence at the 1% level to support the claim.
7. Use normal distribution.
   Reject H_0. There is enough evidence at the 10% level to reject the claim.
8. Cannot use normal distribution.
9. (a) H_0: $p \geq 0.23$ (claim)
   H_a: $p < 0.23$
   (b) $z_0 = -2.33$
   Rejection region: $z < -2.33$
   (c) -0.168
   (d) Fail to reject H_0.
   (e) There is insufficient evidence at the 1% level to reject the claim that at least 23% of U.S. adults are smokers.

10. (a) H_0: $p \leq 0.55$ (claim)
H_a: $p > 0.55$
(b) $z_0 = 2.33$
Rejection region: $z > 2.33$
(c) 0.445
(d) Fail to reject H_0.
(e) There is insufficient evidence at the 1% level to reject the claim that no more than 55% of U.S. adults eat breakfast every day.

11. (a) H_0: $p \leq 0.30$
H_a: $p > 0.30$ (claim)
(b) $z_0 = 1.88$
Rejection region: $z > 1.88$
(c) 1.414
(d) Fail to reject H_0.
(e) There is insufficient evidence at the 3% level to support the claim that more than 30% of U.S. consumers have stopped buying the product because the manufacturing of the product pollutes the environment.

12. See Selected Answers, page A96.

13. (a) H_0: $p = 0.44$ (claim)
H_a: $p \neq 0.44$
(b) $-z_0 = -2.33$ and $z_0 = 2.33$
Rejection regions: $z < -2.33$, $z > 2.33$
(c) -2.537
(d) Reject H_0.
(e) There is sufficient evidence at the 2% level to reject the claim that 44% of home buyers found their real estate agent through a friend.

14. See Selected Answers, page A96.

15. Fail to reject H_0. There is not enough evidence at the 5% level to reject the claim that at least 52% of the adults are more likely to buy a product when there are free samples.

16. The company should continue the use of giveaways because there is not enough evidence to say that less than 52% of the adults would be more likely to buy a product when there are free samples.

17. Reject H_0.
The results are the same.

18. See Selected Answers, page A96.

13. ***Finding a Real Estate Agent*** In your work for a real estate company, you find that in a sample of 1762 home buyers, 722 found their real estate agent through a friend. At $\alpha = 0.02$, can you reject the claim that 44% of home buyers find their real estate agent through a friend? *(Adapted from USA TODAY)*

14. ***Do You Volunteer?*** A government association claims that 44% of adults in the United States do volunteer work. You work for a volunteer organization and are asked to test this claim. You find that in a random sample of 1165 adults, 556 do volunteer work. At $\alpha = 0.05$, do you have enough evidence to reject the association's claim? *(Adapted from U.S. Census Bureau)*

Free Samples In Exercises 15 and 16, use the graph, which shows what adults think about the effectiveness of free samples.

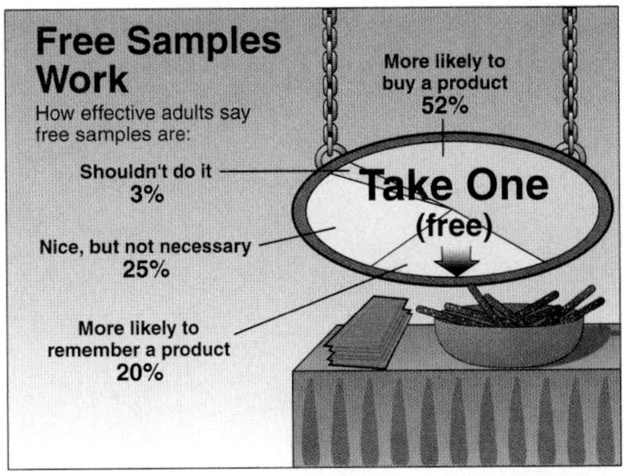

15. ***Do Free Samples Work?*** You interview a random sample of 50 adults. The results of the survey show that 48% of the adults said they were more likely to buy a product when there are free samples. At $\alpha = 0.05$, can you reject the claim that at least 52% of the adults are more likely to buy a product when there are free samples?

16. ***Should Free Samples Be Used?*** Use your conclusion from Exercise 15 to write a paragraph on the use of free samples. Do you think a company should use free samples to get people to buy a product? Explain.

Extending Concepts

Alternative Formula In Exercises 17 and 18, use the following information. When you know the number of successes x, the sample size n, and probability p, it can be easier to use the formula

$$z = \frac{x - np}{\sqrt{npq}}$$

to find the standardized test statistic when using a z-test for a proportion p.

17. Rework Exercise 13 using the alternative formula and compare the results.

18. The alternative formula is derived from the formula

$$z = \frac{\hat{p} - p}{\sqrt{pq/n}} = \frac{(x/n) - p}{\sqrt{pq/n}}.$$

Use this formula to derive the alternative formula. Justify each step.

7.5 Hypothesis Testing for Variance and Standard Deviation

What You Should Learn

- How to find critical values for a χ^2-test
- How to use the χ^2-test to test a variance or a standard deviation

Critical Values for a χ^2-Test • The Chi-Square Test

Critical Values for a χ^2-Test

In real life, it is often important to produce consistent predictable results. For instance, consider a company that manufactures golf balls. The manufacturer must produce millions of golf balls, each having the same size and the same weight. There is a very low tolerance for variation. If the population is normal, you can test the variance and standard deviation of the process using the chi-square distribution with $n - 1$ degrees of freedom.

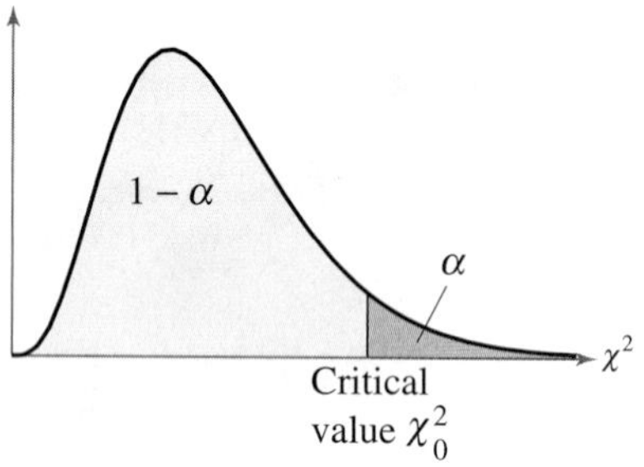

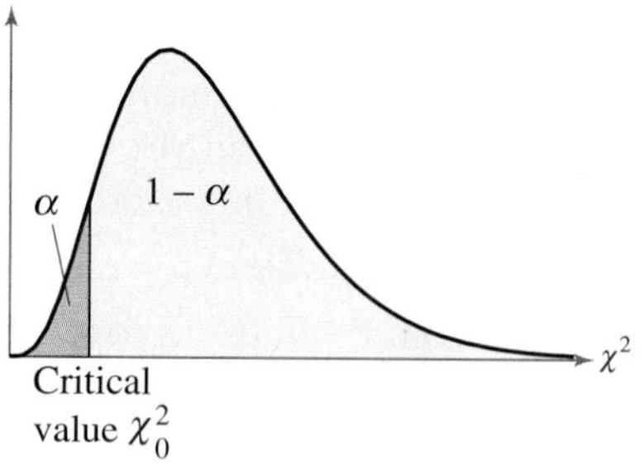

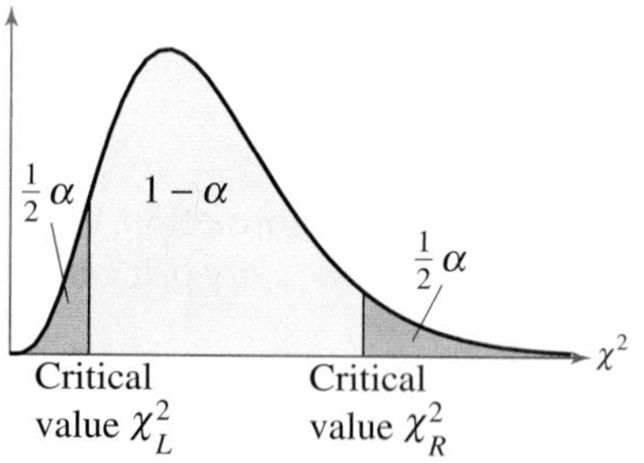

GUIDELINES

Finding Critical Values for the χ^2-Test

1. Specify the level of significance α.
2. Determine the degrees of freedom d.f. $= n - 1$.
3. The critical values for the χ^2-distribution are found in Table 6 of Appendix B. To find the critical value(s) for a
   a. *right-tailed test,* use the value that corresponds to d.f. and α.
   b. *left-tailed test,* use the value that corresponds to d.f. and $1 - \alpha$.
   c. *two-tailed test,* use the values that correspond to d.f. and $\frac{1}{2}\alpha$ and d.f. and $1 - \frac{1}{2}\alpha$.

EXAMPLE 1

Finding Critical Values for χ^2

Find the critical χ^2-value for a right-tailed test when $n = 26$ and $\alpha = 0.10$.

SOLUTION The degrees of freedom are

$$\text{d.f.} = n - 1 = 26 - 1 = 25.$$

The graph at the right shows a χ^2-distribution with 25 degrees of freedom and a shaded area of $\alpha = 0.10$ in the right tail. In Table 6 in Appendix B with d.f. $= 25$ and $\alpha = 0.10$, the critical value is

$$\chi_0^2 = 34.382.$$

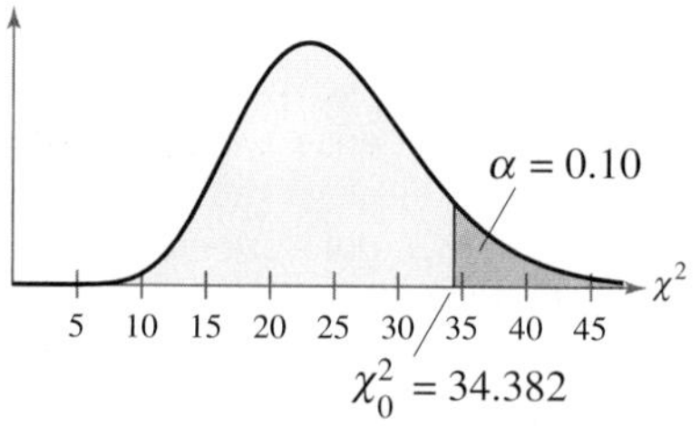

Try It Yourself 1

Find the critical χ^2-value for a right-tailed test when $n = 18$ and $\alpha = 0.01$.

a. *Find* the value using Table 6 in Appendix B with d.f. $= n - 1$ and the area α.

Answer: Page A41

Note to Instructor

This section can be omitted or covered later (with Chapter 10) without loss of continuity.

EXAMPLE 2

Finding Critical Values for χ^2

Find the critical χ^2-value for a left-tailed test when $n = 11$ and $\alpha = 0.01$.

SOLUTION The degrees of freedom are d.f. $= n - 1 = 11 - 1 = 10$. The graph at the right shows a χ^2-distribution with 10 degrees of freedom and a shaded area of $\alpha = 0.01$ in the left tail. The area to the right of the critical value is

$$1 - \alpha = 1 - 0.01 = 0.99.$$

In Table 6 with d.f. $= 10$ and the area $1 - \alpha = 0.99$, the critical value is $\chi_0^2 = 2.558$.

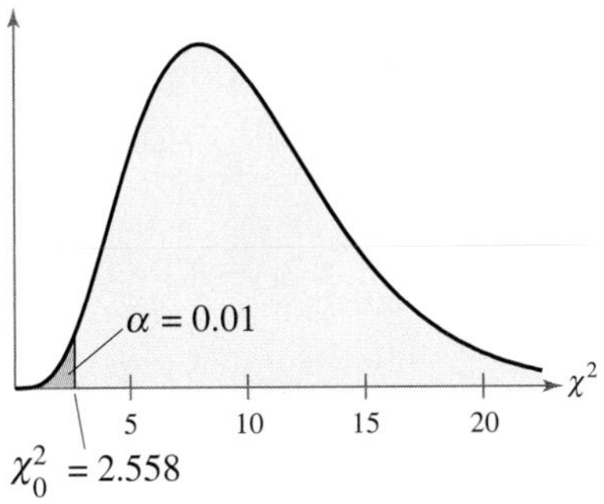

Try It Yourself 2

Find the critical χ^2-value for a left-tailed test when $n = 30$ and $\alpha = 0.05$.

a. *Find* the value using Table 6 with d.f. $= n - 1$ and the area $1 - \alpha$.

Answer: Page A41

EXAMPLE 3

Finding Critical Values for χ^2

Find the critical χ^2-values for a two-tailed test when $n = 13$ and $\alpha = 0.01$.

SOLUTION The degrees of freedom are d.f. $= n - 1 = 13 - 1 = 12$. The graph at the right shows a χ^2-distribution with 12 degrees of freedom and a shaded area of $\frac{1}{2}\alpha = 0.005$ in each tail. The areas to the right of the critical values are

$$\tfrac{1}{2}\alpha = 0.005$$

and

$$1 - \tfrac{1}{2}\alpha = 0.995.$$

In Table 6 with d.f. $= 12$ and the areas 0.005 and 0.995, the critical values are $\chi_L^2 = 3.074$ and $\chi_R^2 = 28.299$.

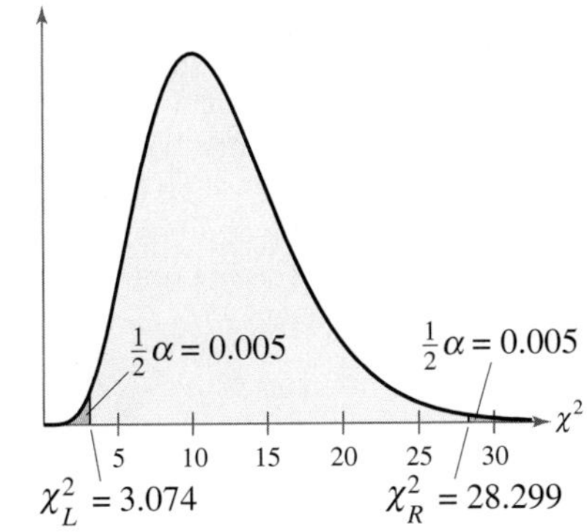

Try It Yourself 3

Find the critical χ^2-values for a two-tailed test when $n = 19$ and $\alpha = 0.05$.

a. *Find* the first critical value χ_R^2 using Table 6 with d.f. $= n - 1$ and the area $\frac{1}{2}\alpha$.

b. *Find* the second critical value χ_L^2 using Table 6 with d.f. $= n - 1$ and the area $1 - \frac{1}{2}\alpha$.

Answer: Page A41

Note that because chi-square distributions are not symmetric (like normal or t-distributions), in a two-tailed test the two critical values are not opposites. Each critical value must be calculated separately.

The Chi-Square Test

To test a variance σ^2 or a standard deviation σ of a population that is normally distributed, you can use the χ^2-test. The χ^2-test for a variance or standard deviation is not as robust as the tests for the population mean μ or the population proportion p. So, it is essential in performing a χ^2-test for a variance or standard deviation that the population be normally distributed. The results can be misleading if the population is not normal.

χ^2-Test for a Variance σ^2 or Standard Deviation σ

The **χ^2-test for a variance or standard deviation** is a statistical test for a population variance or standard deviation. The χ^2-test can be used when the population is normal. The **test statistic** is s^2 and the **standardized test statistic**

$$\chi^2 = \frac{(n-1)s^2}{\sigma^2}$$

follows a chi-square distribution with degrees of freedom

$$\text{d.f.} = n - 1.$$

GUIDELINES

Using the χ^2-Test for a Variance or Standard Deviation

In Words	*In Symbols*
1. State the claim mathematically and verbally. Identify the null and alternative hypotheses.	State H_0 and H_a.
2. Specify the level of significance.	Identify α.
3. Determine the degrees of freedom and sketch the sampling distribution.	$\text{d.f.} = n - 1$
4. Determine any critical values.	Use Table 6 in Appendix B.
5. Determine any rejection regions.	
6. Find the standardized test statistic.	$\chi^2 = \frac{(n-1)s^2}{\sigma^2}$
7. Make a decision to reject or fail to reject the null hypothesis.	If χ^2 is in the rejection region, reject H_0. Otherwise, fail to reject H_0.
8. Interpret the decision in the context of the original claim.	

Note to Instructor

Review the properties of chi-square distributions. Tell students that this family of distributions will be used in later chapters, but the degrees of freedom for those tests are not necessarily $n - 1$.

Note to Instructor

Although other methods of testing hypotheses have also required that data come from normally distributed populations, the tests for variances and standard deviations can be misleading if the populations are not normal. Explain that the condition for a normal distribution is more important for tests of variances or standard deviations.

Picturing the World

A community center claims that the chlorine level in its pool has a standard deviation of 0.46 parts per million (ppm). A sampling of the pool's chlorine levels at 25 random times during a month yields a standard deviation of 0.61 ppm. *(Adapted from American Pool Supply)*

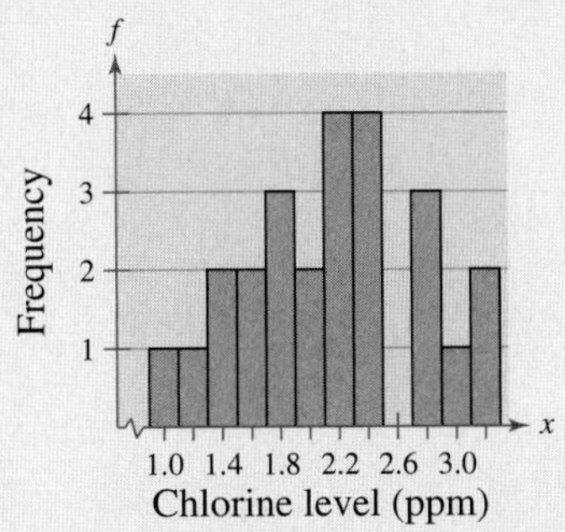

At $\alpha = 0.05$ is there enough evidence to reject the claim?

EXAMPLE 4

Using a Hypothesis Test for the Population Variance

A dairy processing company claims that the variance of the amount of fat in the whole milk processed by the company is no more than 0.25. You suspect this is wrong and find that a random sample of 41 milk containers has a variance of 0.27. At $\alpha = 0.05$, is there enough evidence to reject the company's claim? Assume the population is normally distributed.

SOLUTION The claim is "the variance is no more than 0.25." So, the null and alternative hypotheses are

$$H_0\colon \sigma^2 \le 0.25 \text{ (Claim)} \quad \text{and} \quad H_a\colon \sigma^2 > 0.25.$$

The test is a right-tailed test, the level of significance is $\alpha = 0.05$, and there are d.f. $= 41 - 1 = 40$ degrees of freedom. So, the critical value is

$$\chi_0^2 = 55.758.$$

The rejection region is $\chi^2 > 55.758$. The standardized test statistic is

$$\chi^2 = \frac{(n-1)s^2}{\sigma^2} \qquad \text{Use the chi-square test.}$$

$$= \frac{(41-1)(0.27)}{0.25} \qquad \text{Assume } \sigma^2 = 0.25.$$

$$= 43.2.$$

The graph shows the location of the rejection region and the standardized test statistic χ^2. Because χ^2 is not in the rejection region, you should decide not to reject the null hypothesis.

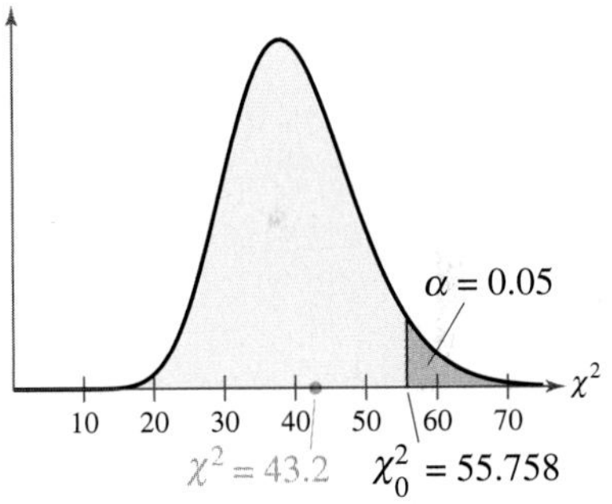

Interpretation You don't have enough evidence at the 5% level of significance to reject the company's claim that the variance of the amount of fat in the whole milk is no more than 0.25.

Try It Yourself 4

A bottling company claims that the variance of the amount of sports drink in a 12-ounce bottle is no more than 0.40. A random sample of 31 bottles has a variance of 0.75. At $\alpha = 0.01$, is there enough evidence to reject the company's claim? Assume the population is normally distributed.

a. *Identify* the claim and state H_0 and H_a.
b. *Identify* the level of significance α and the degrees of freedom d.f.
c. *Find* the critical value and *identify* the rejection region.
d. *Use* the χ^2-test to find the standardized test statistic χ^2.
e. *Decide* whether to reject the null hypothesis. Use a graph if necessary.
f. *Interpret* the decision in the context of the original claim.

Answer: Page A41

EXAMPLE 5

Using a Hypothesis Test for the Standard Deviation

A restaurant claims that the standard deviation in the length of serving times is less than 2.9 minutes. A random sample of 23 serving times has a standard deviation of 2.1 minutes. At $\alpha = 0.10$, is there enough evidence to support the restaurant's claim? Assume the population is normally distributed.

Study Tip

Although you are testing a standard deviation in Example 5, the χ^2-statistic requires variances. Don't forget to square the given standard deviations to calculate these variances.

SOLUTION The claim is "the standard deviation is less than 2.9 minutes." So, the null and alternative hypotheses are

H_0: $\sigma \geq 2.9$ minutes and H_a: $\sigma < 2.9$ minutes. (Claim)

The test is a left-tailed test, the level of significance is $\alpha = 0.10$, and there are

$$\text{d.f.} = 23 - 1 = 22$$

degrees of freedom. So, the critical value is

$$\chi_0^2 = 14.042.$$

The rejection region is $\chi^2 < 14.042$. The standardized test statistic is

$$\chi^2 = \frac{(n-1)s^2}{\sigma^2} \quad \text{Use the chi-square test.}$$

$$= \frac{(23-1)(2.1)^2}{2.9^2} \quad \text{Assume } \sigma = 2.9.$$

$$\approx 11.54.$$

The graph shows the location of the rejection region and the standardized test statistic χ^2. Because χ^2 is in the rejection region, you should decide to reject the null hypothesis.

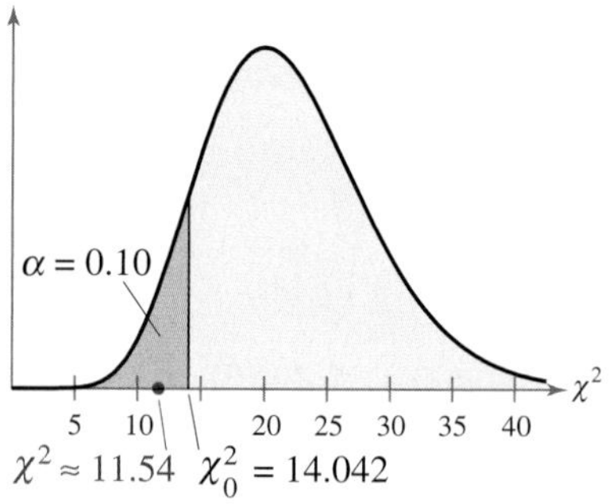

Interpretation There is enough evidence at the 10% level of significance to support the claim that the standard deviation for the length of serving times is less than 2.9 minutes.

Try It Yourself 5

A police chief claims that the standard deviation in the length of response times is less than 3.7 minutes. A random sample of nine response times has a standard deviation of 3.0 minutes. At $\alpha = 0.05$, is there enough evidence to support the police chief's claim? Assume the population is normally distributed.

a. *Identify* the claim and state H_0 and H_a.
b. *Identify* the level of significance α and the degrees of freedom d.f.
c. *Find* the critical value and *identify* the rejection region.
d. *Use* the χ^2-test to find the standardized test statistic χ^2.
e. *Decide* whether to reject the null hypothesis. Use a graph if necessary.
f. *Interpret* the decision in the context of the original claim.

Answer: Page A41

EXAMPLE 6

Using a Hypothesis Test for the Population Variance

A sporting goods manufacturer claims that the variance of the strength in a certain fishing line is 15.9. A random sample of 15 fishing line spools has a variance of 21.8. At $\alpha = 0.05$, is there enough evidence to reject the manufacturer's claim? Assume the population is normally distributed.

SOLUTION The claim is "the variance is 15.9." So, the null and alternative hypotheses are

$$H_0\text{: } \sigma^2 = 15.9 \text{ (Claim)} \quad \text{and} \quad H_a\text{: } \sigma^2 \neq 15.9.$$

The test is a two-tailed test, the level of significance is $\alpha = 0.05$, and there are

$$\text{d.f.} = 15 - 1 = 14$$

degrees of freedom. So, the critical values are $\chi_L^2 = 5.629$ and $\chi_R^2 = 26.119$. The rejection regions are $\chi^2 < 5.629$ and $\chi^2 > 26.119$. The standardized test statistic is

$$\chi^2 = \frac{(n-1)s^2}{\sigma^2} \qquad \text{Use the chi-square test.}$$

$$= \frac{(15-1)(21.8)}{15.9} \qquad \text{Assume } \sigma^2 = 15.9.$$

$$\approx 19.19.$$

The graph shows the location of the rejection regions and the standardized test statistic χ^2. Because χ^2 is not in the rejection regions, you should decide not to reject the null hypothesis.

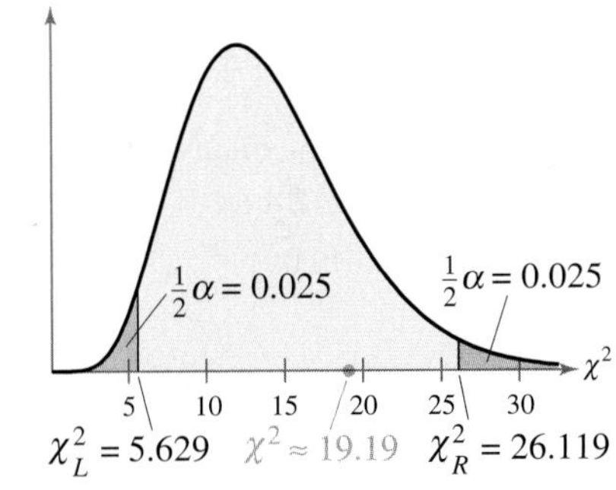

Interpretation At the 5% level of significance, there is not enough evidence to reject the claim that the variance in the strength of the fishing line is 15.9.

Try It Yourself 6

A tire manufacturer claims that the variance of the diameters in a certain tire model is 8.6. A random sample of 10 tires has a variance of 4.3. At $\alpha = 0.01$, is there enough evidence to reject the manufacturer's claim? Assume the population is normally distributed.

a. *Identify* the claim and state H_0 and H_a.
b. *Identify* the level of significance α and the degrees of freedom d.f.
c. *Find* the critical values and *identify* the rejection regions.
d. *Use* the χ^2-test to find the standardized test statistic χ^2.
e. *Decide* whether to reject the null hypothesis. Use a graph if necessary.
f. *Interpret* the decision in the context of the original claim.

Answer: Page A41

7.5 Exercises

Building Basic Skills and Vocabulary

1. Explain how to find critical values in a χ^2 sampling distribution.

2. Explain how to test a population variance or a population standard deviation.

In Exercises 3–8, find the critical value(s) for the indicated test for a population variance, sample size n, and level of significance α.

3. Right-tailed test, $n = 27, \alpha = 0.05$

4. Right-tailed test, $n = 10, \alpha = 0.10$

5. Left-tailed test, $n = 7, \alpha = 0.01$

6. Left-tailed test, $n = 24, \alpha = 0.05$

7. Two-tailed test, $n = 16, \alpha = 0.10$

8. Two-tailed test, $n = 29, \alpha = 0.01$

Graphical Analysis In Exercises 9–12, state whether the standardized test statistic χ^2 allows you to reject the null hypothesis.

9. (a) $\chi^2 = 2.091$
(b) $\chi^2 = 0$
(c) $\chi^2 = 1.086$
(d) $\chi^2 = 6.3471$

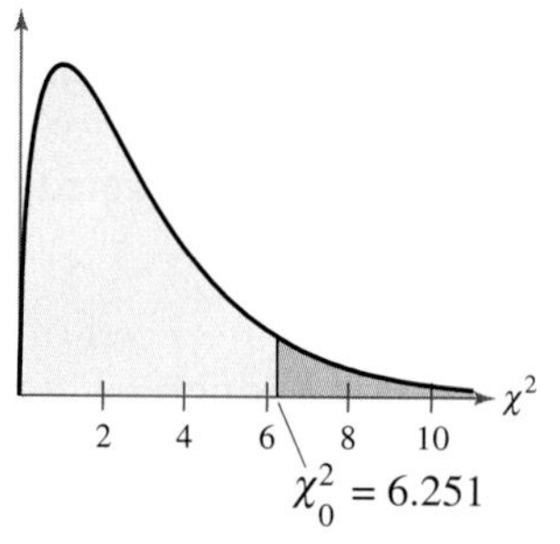

10. (a) $\chi^2 = 0.771$
(b) $\chi^2 = 9.486$
(c) $\chi^2 = 0.701$
(d) $\chi^2 = 9.508$

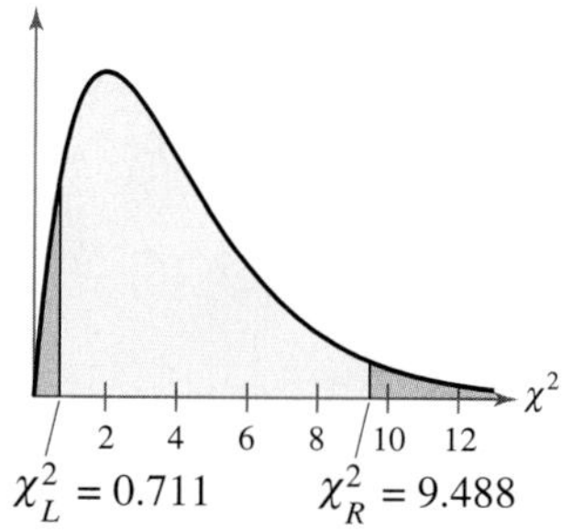

11. (a) $\chi^2 = 22.302$
(b) $\chi^2 = 23.309$
(c) $\chi^2 = 8.457$
(d) $\chi^2 = 8.577$

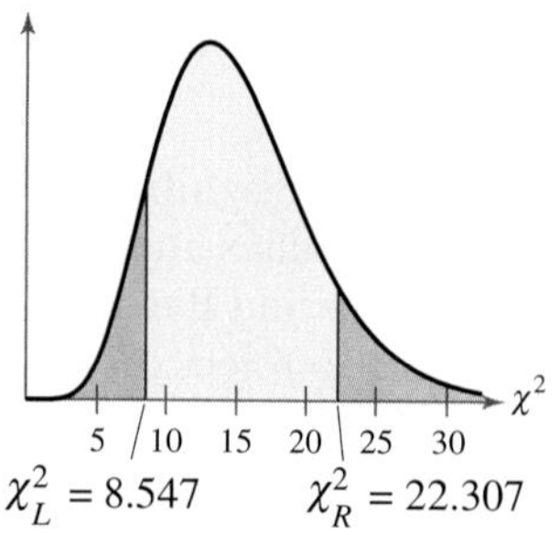

12. (a) $\chi^2 = 10.065$
(b) $\chi^2 = 10.075$
(c) $\chi^2 = 10.585$
(d) $\chi^2 = 10.745$

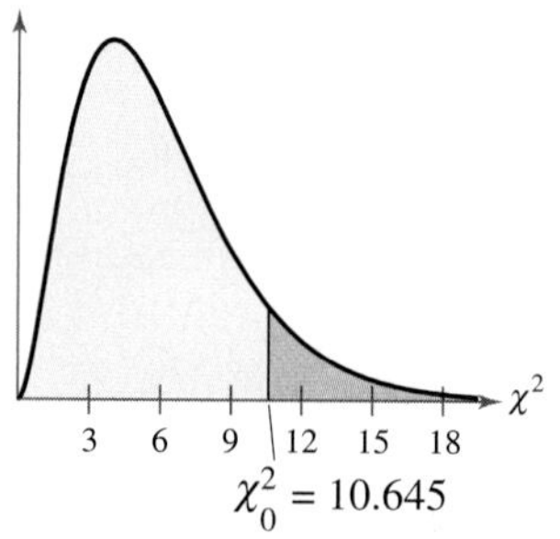

1. Specify the level of significance α. Determine the degrees of freedom. Determine the critical values using the χ^2 distribution. If (a) right-tailed test, use the value that corresponds to d.f. and α; (b) left-tailed test, use the value that corresponds to d.f. and $1 - \alpha$; (c) two-tailed test, use the value that corresponds to d.f. and $\frac{1}{2}\alpha$ and $1 - \frac{1}{2}\alpha$.
2. State H_0 and H_a. Specify the level of significance. Determine the degrees of freedom. Determine the critical value(s) and rejection region(s). Find the standardized test statistic. Make a decision and interpret in the context of the original claim.
3. 38.885 **4.** 14.684
5. 0.872 **6.** 13.091
7. $\chi_L^2 = 7.261, \chi_R^2 = 24.996$
8. $\chi_L^2 = 12.461, \chi_R^2 = 50.993$
9. (a) Fail to reject H_0.
(b) Fail to reject H_0.
(c) Fail to reject H_0.
(d) Reject H_0.
10. (a) Fail to reject H_0.
(b) Fail to reject H_0.
(c) Reject H_0.
(d) Reject H_0.
11. (a) Fail to reject H_0.
(b) Reject H_0.
(c) Reject H_0.
(d) Fail to reject H_0.
12. (a) Fail to reject H_0.
(b) Fail to reject H_0.
(c) Fail to reject H_0.
(d) Reject H_0.

In Exercises 13 and 14, use a χ^2-test to test the claim about the population variance σ^2 or standard deviation σ at the given level of significance α using the given sample statistics. Assume the population is normally distributed.

13. Claim: $\sigma^2 = 0.52$; $\alpha = 0.05$. Sample statistics: $s^2 = 0.508$, $n = 18$

14. Claim: $\sigma < 40$; $\alpha = 0.01$. Sample statistics: $s = 40.8$, $n = 12$

Using and Interpreting Concepts

Testing Claims In Exercises 15–24, (a) write the claim mathematically and identify H_0 and H_a, (b) find the critical value(s) and identify the rejection region(s), (c) use the χ^2-test to find the standardized test statistic, (d) decide whether to reject or fail to reject the null hypothesis, and (e) interpret the decision in the context of the original claim. For each claim, assume the population is normally distributed.

15. ***Life of Appliances*** A large appliance company estimates that the variance of the life of its appliances is 3. You work for a consumer advocacy group and are asked to test this claim. You find that a random sample of the lives of 27 of the company's appliances has a variance of 2.8. At $\alpha = 0.05$, do you have enough evidence to reject the company's claim? *(Adapted from Consumer Reports)*

16. ***Luxury Sedan Gas Mileage*** An automotive manufacturer believes that the variance of the gas mileage for its luxury sedans is 5. You work for an energy conservation agency and want to test this claim. You find that a random sample of the miles per gallon of 23 of the manufacturer's sedans has a variance of 4.5. At $\alpha = 0.05$, do you have enough evidence to reject the manufacturer's claim? *(Adapted from Consumer Reports)*

17. ***Physical Science Assessment Tests*** On a physical science assessment test, the scores of a random sample of 22 eighth grade students have a standard deviation of 27.7. This result prompts a test administrator to claim that the standard deviation for eighth graders on the examination is less than 29. At $\alpha = 0.10$, is there enough evidence to support the administrator's claim? *(Adapted from National Center for Educational Statistics)*

18. ***Life Science Assessment Tests*** A state school administrator says that the standard deviation of test scores for eighth grade students who took a life science assessment test is less than 30. You work for the administrator and are asked to test this claim. To do so, you randomly select 10 tests and find that the tests have a standard deviation of 28.8. At $\alpha = 0.01$, is there enough evidence to support the administrator's claim? *(Adapted from National Center for Educational Statistics)*

19. ***Hospital Waiting Times*** A hospital spokesperson claims that the standard deviation of the waiting times experienced by patients in its minor emergency department is no more than 0.5 minute. You doubt the validity of this claim. If a random sample of 25 waiting times has a standard deviation of 0.7 minute, can you reject the spokesperson's claim? Use $\alpha = 0.10$.

20. ***Length of Stay*** A doctor says the standard deviation of the lengths of stay for patients involved in a crash in which the vehicle struck a tree is 6.14 days. A random sample of 20 lengths of stay for patients involved in this type of crash has a standard deviation of 6.5 days. At $\alpha = 0.05$, can you reject the doctor's claim? *(Adapted from National Highway Traffic Safety Administration)*

13. Fail to reject H_0. There is not enough evidence at the 5% level to reject the claim.

14. Fail to reject H_0. There is not enough evidence at the 1% level to support the claim.

15. (a) H_0: $\sigma^2 = 3$ (claim); H_a: $\sigma^2 \neq 3$

(b) $\chi^2_L = 13.844$, $\chi^2_R = 41.923$

Rejection regions: $\chi^2 > 41.923$, $\chi^2 < 13.844$

(c) 24.267

(d) Fail to reject H_0.

(e) There is insufficient evidence at the 5% level of significance to reject the claim that the variance of the life of the appliances is 3.

16. (a) H_0: $\sigma^2 = 5$ (claim); H_a: $\sigma^2 \neq 5$

(b) $\chi^2_L = 10.982$, $\chi^2_R = 36.781$

Rejection regions: $\chi^2 > 36.781$, $\chi^2 < 10.982$

(c) 19.800

(d) Fail to reject H_0.

(e) There is insufficient evidence at the 5% level to reject the claim that the variance of the gas mileage is 5.

17. (a) H_0: $\sigma \geq 29$; H_a: $\sigma < 29$ (claim)

(b) $\chi^2_0 = 13.240$

Rejection region: $\chi^2 < 13.240$

(c) 19.159

(d) Fail to reject H_0.

(e) There is insufficient evidence at the 10% level of significance to support the claim that the standard deviation for eighth graders on the examination is less than 29.

18. See Selected Answers, page A96.

19. (a) H_0: $\sigma \leq 0.5$ (claim)
H_a: $\sigma > 0.5$

(b) $\chi^2_0 = 33.196$

Rejection region: $\chi^2 > 33.196$

(c) 47.040

(d) Reject H_0.

(e) There is sufficient evidence at the 10% level to reject the claim that the standard deviation of the waiting times is no more than 0.5 minute.

20. See Selected Answers, page A96.

21. (a) H_0: $\sigma \geq \$18{,}000$
H_a: $\sigma < \$18{,}000$ (claim)
(b) $\chi_0^2 = 18.114$
Rejection region: $\chi^2 < 18.114$
(c) 28.521
(d) Fail to reject H_0.
(e) There is insufficient evidence at the 10% level to support the claim that the standard deviation of the total charge for patients involved in a crash in which the vehicle struck a wall is less than \$18,000.

22. (a) H_0: $\sigma \leq \$25$ (claim)
H_a: $\sigma > \$25$
(b) $\chi_0^2 = 26.217$
Rejection region: $\chi^2 > 26.217$
(c) 14.520
(d) Fail to reject H_0.
(e) There is insufficient evidence at the 1% level to reject the claim that the standard deviation of the room rates of hotels in the city is no more than \$25.

23. (a) H_0: $\sigma \leq 20{,}000$
H_a: $\sigma > 20{,}000$ (claim)
(b) $\chi_0^2 = 24.996$
Rejection region: $\chi^2 > 24.996$
(c) 16.011
(d) Fail to reject H_0.
(e) There is insufficient evidence at the 5% level to support the claim that the standard deviation of the annual salaries for actuaries is more than \$20,000.

24. (a) H_0: $\sigma \geq \$14{,}500$ (claim)
H_a: $\sigma < \$14{,}500$
(b) $\chi_0^2 = 10.085$
Rejection region: $\chi^2 < 10.085$
(c) 15.985
(d) Fail to reject H_0.
(e) There is insufficient evidence at the 10% level to reject the claim that the standard deviation of the annual salaries for public relations managers is at least \$14,500.

25. P-value = 0.615; Fail to reject H_0.
26. P-value = 0.269; Fail to reject H_0.
27. P-value = 0.381; Fail to reject H_0.
28. P-value = 0.475; Fail to reject H_0.

21. *Total Charge* An insurance agent says the standard deviation of the total charge for patients involved in a crash in which the vehicle struck a wall is less than \$18,000. A random sample of 28 total charges for patients involved in this type of crash has a standard deviation of \$18,500. At $\alpha = 0.10$, can you support the agent's claim? *(Adapted from National Highway Traffic Safety Administration)*

22. *Hotel Room Rates* A travel agency estimates that the standard deviation of the room rates of hotels in a certain city is no more than \$25. You work for a consumer advocacy group and are asked to test this claim. You find that a random sample of 13 hotels has a standard deviation of \$27.50. At $\alpha = 0.01$, do you have enough evidence to reject the agency's claim? *(Adapted from Smith Travel Research)*

23. *Salaries* The annual salaries of 16 randomly chosen actuaries are listed below. At $\alpha = 0.05$, can you conclude that the standard deviation of the annual salaries is greater than \$20,000? *(Adapted from America's Career InfoNet)*

55,018	44,952	95,517	55,553	51,900	86,384	58,862	47,615
71,104	106,710	107,875	78,245	49,945	63,582	75,252	82,522

24. *Salaries* An employment information service says that the standard deviation of the annual salaries for public relations managers is at least \$14,500. The annual salaries for 18 randomly chosen public relations managers are listed. At $\alpha = 0.10$, can you reject the claim? *(Adapted from America's Career InfoNet)*

47,517	60,217	39,177	61,744	79,422	70,770
60,549	60,263	72,939	72,372	75,014	59,164
44,811	65,413	61,310	90,433	44,185	41,805

Extending Concepts

***P*-values** You can calculate the P-value for a χ^2-test using technology. After calculating the χ^2-test value, you can use the cumulative density function (CDF) to calculate the area under the curve. From Example 4 on page 383, $\chi^2 = 43.2$. Using a TI-83 (choose 7 from the DISTR menu), enter 0 for the lower bound, 43.2 for the upper bound, and 40 for the degrees of freedom as shown.

TI-83

```
χ² cdf (0, 43.2, 40)
              .6637768667
```

The P-value is approximately $1 - 0.664 = 0.336$. Because $P > \alpha = 0.05$, the conclusion is to fail to reject H_0.

25. Use the P-value method to perform the hypothesis test for Exercise 21.

26. Use the P-value method to perform the hypothesis test for Exercise 22.

27. Use the P-value method to perform the hypothesis test for Exercise 23.

28. Use the P-value method to perform the hypothesis test for Exercise 24.

Uses and Abuses

Statistics in the Real World

Uses

Hypothesis Testing Hypothesis testing is important in many different fields because it gives a scientific procedure for assessing the validity of a claim about a population. Some of the concepts in hypothesis testing are intuitive, but some are not. For instance, in this chapter you learned that a test statistic does not have to differ much from a hypothesized population parameter to conclude that the null hypothesis should be rejected.

Abuses

Not Using a Random Sample The entire theory of hypothesis testing is based on the fact that the sample is randomly selected. If the sample is not random, then you cannot use it to infer anything about a population parameter.

Attempting to Prove the Null Hypothesis Another abuse is to assume that you have proven the null hypothesis when the P-value is greater than the level of significance (or the test statistic does not fall in the rejection region). Remember that hypothesis testing can never prove that the null hypothesis is true—only that there is not enough evidence to reject it. This is similar to a situation in which a defendant is found not guilty in a court of law. The defendant is not declared to be innocent—only that there was not enough evidence to convict him or her.

Making Type I or Type II Errors Remember that a type I error is rejecting a null hypothesis that is true and a type II error is failing to reject a null hypothesis that is false. You can decrease the chance of making both types of errors by increasing the sample size.

Exercises

In Exercises 1–4, assume that you work in a market research department. You are asked to write a report about the claim that 47% of all new car buyers prefer a four-door car.

1. ***Not Using a Random Sample*** How could you choose a random sample to test this hypothesis?
2. ***Attempting to Prove the Null Hypothesis*** What is the null hypothesis in this situation? Describe how your report could be incorrect by trying to prove the null hypothesis.
3. ***Making a Type I Error*** Describe how your report could make a type I error.
4. ***Making a Type II Error*** Describe how your report could make a type II error.

What Type of Car Do You Prefer?

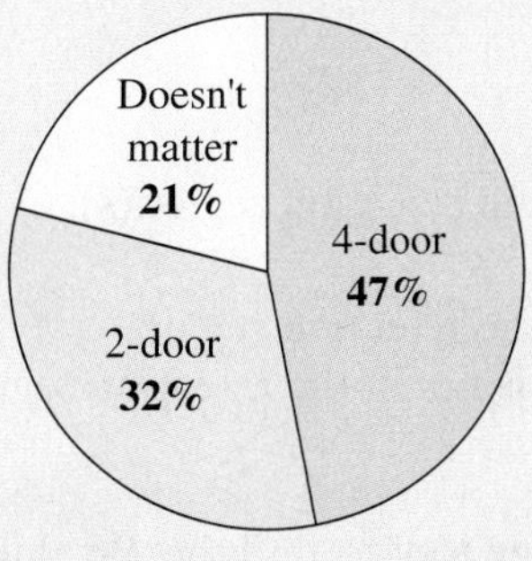

7 A Summary of Hypothesis Testing

Insight

Large sample sizes will usually increase the cost and effort of testing a hypothesis, but they also tend to make your decision more reliable.

With hypothesis testing, perhaps more than any other area of statistics, it can be difficult to see the forest for all the trees. To help you see the forest—the overall picture—a summary of what you studied in this chapter is provided.

Writing the Hypotheses

- You are given a claim about a population parameter μ, p, σ^2, or σ.
- Rewrite the claim and its complement using $\underbrace{\leq, \geq, =}_{H_0}$ and $\underbrace{>, <, \neq}_{H_a}$.
- Identify the claim. Is it H_0 or H_a?

Specifying a Level of Significance

- Specify α, the maximum acceptable probability of rejecting a valid H_0 (a type I error).

Specifying the Sample Size

- Specify your sample size n.

Choosing the Test

■ Any population ■ Normally distributed population

- **Mean:** H_0 describes a hypothesized population mean μ.
  - ■ Use a ***z*-test** for *any* population if $n \geq 30$.
  - ■ Use a ***z*-test** if the population is normal and σ is known for any n.
  - ■ Use a ***t*-test** if the population is normal and $n < 30$, but σ is unknown.
- **Proportion:** H_0 describes a hypothesized population proportion p.
  - ■ Use a ***z*-test** for *any* population if $np \geq 5$ and $nq \geq 5$.
- **Variance or Standard Deviation:** H_0 describes a hypothesized population variance σ^2 or standard deviation σ.
  - ■ Use a **χ^2-test** if the population is normal.

Sketching the Sampling Distribution

- Use H_a to decide if the test is left-tailed, right-tailed, or two-tailed.

Finding the Standardized Test Statistic

- Take a random sample of size n from the population.
- Compute the test statistic $\bar{x}$, $\hat{p}$, or s^2.
- Find the standardized test statistic z, t, or χ^2.

Making a Decision

Option 1. Decision based on rejection region

- Use α to find the critical value(s) z_0, t_0, or χ_0^2 and rejection region(s).
- **Decision Rule:**
  Reject H_0 if the standardized test statistic is in the rejection region.
  Fail to reject H_0 if the standardized test statistic is not in the rejection region.

Option 2. Decision based on P-value

- Use the standardized test statistic or a technology tool to find the P-value.
- **Decision Rule:**
  Reject H_0 if $P \leq \alpha$.
  Fail to reject H_0 if $P > \alpha$.

Study Tip

If your standardized test statistic is z or t, remember that these values measure standard deviations from the mean. Values that are outside of ± 3 indicate that H_0 is very unlikely. Values that are outside of ± 5 indicate that H_0 is almost impossible.

z-Test for a Hypothesized Mean μ *(Section 7.2)*

Test Statistic: $\bar{x}$

Critical value: z_0 (Use Table 4.)

If $n \geq 30$, s can be used in place of σ.

Sampling distribution of sample means is a normal distribution.

Standardized Test Statistic: z

$$z = \frac{\bar{x} - \mu}{\sigma/\sqrt{n}}$$

Sample mean ($\bar{x}$); Hypothesized mean (μ); Population standard deviation (σ); Sample size (n)

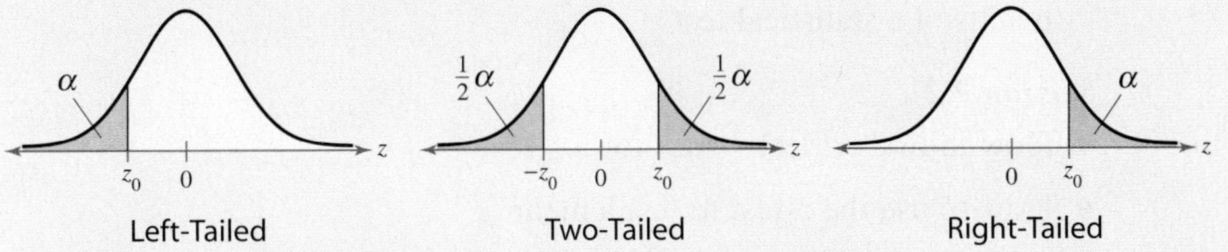

z-Test for a Hypothesized Proportion p *(Section 7.4)*

Test Statistic: $\hat{p}$

Critical value: z_0 (Use Table 4.)

Sampling distribution of sample proportions is a normal distribution.

Standardized Test Statistic: z

$$z = \frac{\hat{p} - p}{\sqrt{pq/n}}$$

Sample proportion ($\hat{p}$); Hypothesized proportion (p); $q = 1 - p$; Sample size (n)

t-Test for a Hypothesized Mean μ *(Section 7.3)*

Test Statistic: $\bar{x}$

Critical value: t_0 (Use Table 5.)

Sampling distribution of sample means is approximated by a t-distribution with d.f. $= n - 1$.

Standardized Test Statistic: t

$$t = \frac{\bar{x} - \mu}{s/\sqrt{n}}$$

Sample mean ($\bar{x}$); Hypothesized mean (μ); Sample standard deviation (s); Sample size (n)

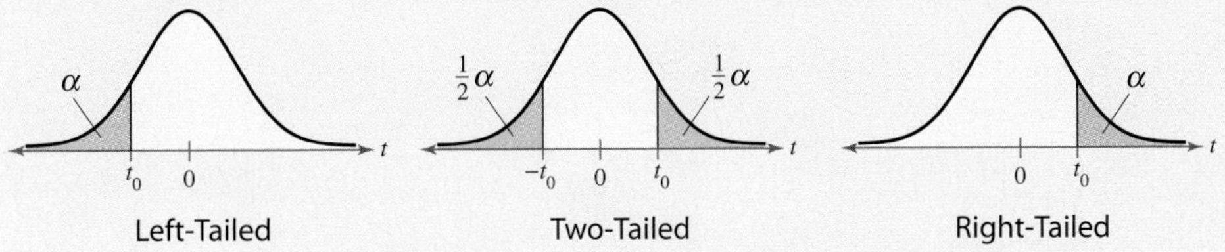

χ^2-Test for a Hypothesized Variance σ^2 or Standard Deviation σ *(Section 7.5)*

Test Statistic: s^2

Critical value: χ^2_0 (Use Table 6.)

Sampling distribution is approximated by a chi-square distribution with d.f. $= n - 1$.

Standardized Test Statistic: χ^2

$$\chi^2 = \frac{(n - 1)s^2}{\sigma^2}$$

Sample size (n); Sample variance (s^2); Hypothesized variance (σ^2)

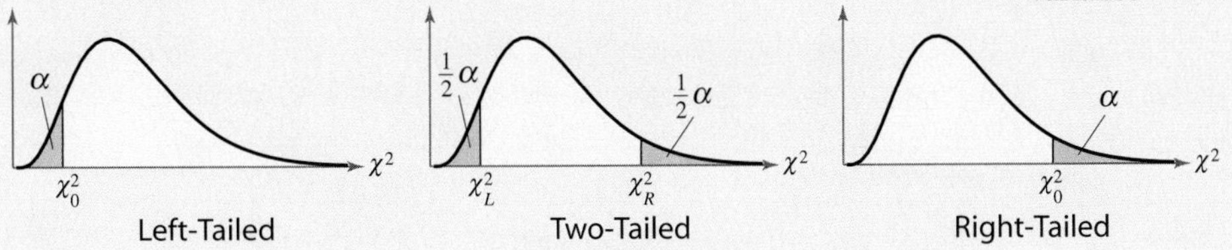

7 Chapter Summary

What did you learn?	Review Exercises
Section 7.1	
◆ How to state a null hypothesis and an alternative hypothesis	*1–6*
◆ How to interpret the level of significance of a hypothesis test and identify type I and type II errors, how to know whether to use a one-tailed or a two-tailed statistical test, and how to interpret a decision based on the results of a statistical test	*7–10*
Section 7.2	
◆ How to find critical values for a z-test	*11–14*
◆ How to use the z-test to test a mean μ	*15–18, 21, 22*
◆ How to find P-values and use them to test a mean μ	*19–22*
Section 7.3	
◆ How to find critical values in a t-distribution	*23–26*
◆ How to use the t-test to test a mean μ	*27–34*
◆ How to use technology to find P-values and use them with a t-test to test a mean μ	*35, 36*
Section 7.4	
◆ How to use the z-test to test a population proportion p	*37–46*
Section 7.5	
◆ How to find critical values for a χ^2-test	*47–50*
◆ How to use the χ^2-test to test a variance or a standard deviation	*51–56*

7 Review Exercises

1. H_0: $\mu \le 1593$ (claim)
   H_a: $\mu > 1593$
2. H_0: $\mu = 35$ (claim); H_a: $\mu \ne 35$
3. H_0: $p \ge 0.205$
   H_a: $p < 0.205$ (claim)
4. H_0: $\mu = 150{,}020$
   H_a: $\mu \ne 150{,}020$ (claim)
5. H_0: $\sigma \le 4.5$; H_a: $\sigma > 4.5$ (claim)
6. H_0: $p \ge 0.70$ (claim); H_a: $p < 0.70$
7. (a) H_0: $p = 0.73$ (claim)
   H_a: $p \ne 0.73$
   (b) Type I error will occur if H_0 is rejected when the actual proportion of American adults who use nonprescription pain relievers is 0.73.
   Type II error will occur if H_0 is not rejected when the actual proportion of American adults who use nonprescription pain relievers is not 0.73.
   (c) Two-tailed because the alternative hypothesis contains $\ne$.
   (d) There is enough evidence to reject the claim.
   (e) There is not enough evidence to reject the claim.
8. See Selected Answers, page A96.
9. See Odd Answers, page A71.
10. See Selected Answers, page A96.
11. -2.05 12. ± 2.81
13. 1.96 14. ± 1.75
15. Reject H_0. There is enough evidence at the 5% level to reject the claim.
16. Reject H_0. There is enough evidence at the 5% level to support the claim.
17. Fail to reject H_0. There is not enough evidence at the 1% level to support the claim.
18. Fail to reject H_0. There is not enough evidence at the 5% level to reject the claim.

Section 7.1

In Exercises 1–6, use the given claim to state a null and an alternative hypothesis. Identify which hypothesis represents the claim.

1. Claim: $\mu \le 1593$

2. Claim: $\mu = 35$

3. Claim: $p < 0.205$

4. Claim: $\mu \ne 150{,}020$

5. Claim: $\sigma > 4.5$

6. Claim: $p \ge 0.70$

In Exercises 7–10, do the following.

(a) State the null and alternative hypotheses.

(b) Determine when a type I or type II error occurs for a hypothesis test of the claim.

(c) Determine whether the hypothesis test is a left-tailed test, a right-tailed test, or a two-tailed test. Explain your reasoning.

(d) How should you interpret a decision that rejects the null hypothesis?

(e) How should you interpret a decision that fails to reject the null hypothesis?

7. An organization believes that the proportion of U.S. adults who use nonprescription pain relievers is 73%. *(Source: CBS News)*

8. A tire manufacturer guarantees that the mean life of a certain type of tire is at least 30,000 miles.

9. A soup maker says that the standard deviation of the sodium content in one serving of a certain soup is no more than 50 milligrams. *(Adapted from Consumer Reports)*

10. A cereal maker claims that the mean number of fat calories in one serving of its cereal is less than 20.

Section 7.2

In Exercises 11–14, find the critical value(s) for the indicated *z*-test and level of significance α.

11. Left-tailed test, $\alpha = 0.02$

12. Two-tailed test, $\alpha = 0.005$

13. Right-tailed test, $\alpha = 0.025$

14. Two-tailed test, $\alpha = 0.08$

In Exercises 15–18, use a *z*-test to test the claim about the population mean μ at the given level of significance α using the given sample statistics.

15. Claim: $\mu \le 45$; $\alpha = 0.05$. Sample statistics: $\bar{x} = 47.2$, $s = 6.7$, $n = 42$

16. Claim: $\mu \ne 0$; $\alpha = 0.05$. Sample statistics: $\bar{x} = -0.69$, $s = 2.62$, $n = 60$

17. Claim: $\mu < 5.500$; $\alpha = 0.01$. Sample statistics: $\bar{x} = 5.497$, $s = 0.011$, $n = 36$

18. Claim: $\mu = 7450$; $\alpha = 0.05$. Sample statistics: $\bar{x} = 7512$, $s = 243$, $n = 57$

In Exercises 19 and 20, use a P-value to test the claim about the population mean μ using the given sample statistics. State your decision for $\alpha = 0.10$, $\alpha = 0.05$, and $\alpha = 0.01$ levels of significance.

19. Claim: $\mu \le 0.05$; Sample statistics: $\bar{x} = 0.057$, $s = 0.018$, $n = 32$

20. Claim: $\mu \ne 230$; Sample statistics: $\bar{x} = 216.5$, $s = 17.3$, $n = 48$

In Exercises 21 and 22, test the claim about the population mean μ using rejection region(s) or a P-value.

21. A tourist agency in Florida claims the mean daily cost of meals and lodging for a family of four traveling in Florida is \$284. You work for a consumer protection advocate and want to test this claim. In a random sample of 50 families of four traveling in Florida, the mean daily cost of meals and lodging is \$292 and the standard deviation is \$25. At $\alpha = 0.05$, do you have enough evidence to reject the agency's claim? *(Adapted from American Automobile Association)*

22. A tourist agency in Hawaii claims the mean daily cost of meals and lodging for a family of four traveling in Hawaii is at most \$533. In an annual survey performed by the consumer protection agency you work for, a random sample of 45 families of four traveling in Hawaii had a mean daily cost of meals and lodging of \$542 with a standard deviation of \$40. At $\alpha = 0.05$, do you have enough evidence to reject the tourist agency's claim? *(Adapted from American Automobile Association)*

Section 7.3

In Exercises 23–26, find the critical value(s) for the indicated t-test, level of significance α, and sample size N.

23. Two-tailed test, $\alpha = 0.05$, $n = 20$

24. Right-tailed test, $\alpha = 0.01$, $n = 8$

25. Left-tailed test, $\alpha = 0.10$, $n = 15$

26. Two-tailed test, $\alpha = 0.05$, $n = 12$

In Exercises 27–32, use a t-test to test the claim about the population mean μ at the given level of significance α using the given sample statistics. For each claim, assume the population is normally distributed.

27. Claim: $\mu \ne 95$; $\alpha = 0.05$. Sample statistics: $\bar{x} = 94.1$, $s = 1.53$, $n = 12$

28. Claim: $\mu > 12{,}700$; $\alpha = 0.05$.
Sample statistics: $\bar{x} = 12{,}804$, $s = 248$, $n = 21$

29. Claim: $\mu \ge 0$; $\alpha = 0.10$. Sample statistics: $\bar{x} = -0.45$, $s = 1.38$, $n = 16$

30. Claim: $\mu = 4.20$; $\alpha = 0.02$. Sample statistics: $\bar{x} = 4.41$, $s = 0.26$, $n = 9$

31. Claim: $\mu \le 48$; $\alpha = 0.01$. Sample statistics: $\bar{x} = 52$, $s = 2.5$, $n = 7$

32. Claim: $\mu < 850$; $\alpha = 0.025$. Sample statistics: $\bar{x} = 875$, $s = 25$, $n = 14$

19. H_0: $\mu \le 0.05$ (claim); H_a: $\mu > 0.05$
$z \approx 2.200$; P-value $= 0.0139$
$\alpha = 0.10 \Rightarrow$ Reject H_0.
$\alpha = 0.05 \Rightarrow$ Reject H_0.
$\alpha = 0.01 \Rightarrow$ Fail to reject H_0.

20. H_0: $\mu = 230$; H_a: $\mu \ne 230$ (claim)
$z \approx -5.406$; P-value ≈ 0
$\alpha = 0.10 \Rightarrow$ Reject H_0.
$\alpha = 0.05 \Rightarrow$ Reject H_0.
$\alpha = 0.01 \Rightarrow$ Reject H_0.

21. Reject H_0. There is sufficient evidence to reject the claim that mean daily cost of meals and lodging is \$284.

22. Fail to reject H_0. There is not enough evidence to reject the claim.

23. ± 2.093

24. $t_0 = 2.998$

25. -1.345

26. $t_0 = \pm 2.201$

27. Fail to reject H_0. There is not enough evidence at the 5% level to support the claim.

28. Reject H_0. There is enough evidence at the 5% level to support the claim.

29. Fail to reject H_0. There is not enough evidence at the 10% level to reject the claim.

30. Fail to reject H_0. There is not enough evidence at the 2% level to reject the claim.

31. Reject H_0. There is enough evidence at the 1% level to reject the claim.

32. Fail to reject H_0. There is not enough evidence at the 2.5% level to support the claim.

33. Fail to reject H_0. There is not enough evidence at the 10% level to reject the claim that the average monthly cost of joining a health club is \$25.

34. Reject H_0. There is enough evidence at the 10% level to reject the claim that the restaurant's hamburgers have no more than 10 grams of fat.

35. Fail to reject H_0. There is not enough evidence at the 1% level to reject the claim that the mean number of hours of sleep high school students get each night is at least 6 hours.

36. Fail to reject H_0. There is not enough evidence at the 5% level to support the claim that the mean number of classroom hours per week for full-time faculty is more than 9 hours per week.

37. Fail to reject H_0. There is not enough evidence at the 5% level to reject the claim.

38. Reject H_0. There is enough evidence at the 1% level to support the claim.

39. Because $np = 3.6$ is less than 5, the normal distribution cannot be used to approximate the binomial distribution.

40. Reject H_0. There is enough evidence at the 10% level to reject the claim.

41. Because $np = 1.2$ is less than 5, the normal distribution cannot be used to approximate the binomial distribution.

42. Fail to reject H_0. There is not enough evidence at the 1% level to support the claim.

43. Fail to reject H_0. There is not enough evidence at the 1% level to support the claim.

44. Fail to reject H_0. There is not enough evidence at the 10% level to reject the claim.

In Exercises 33 and 34, use a *t*-test to investigate the claim. For each claim, assume each population is normally distributed.

33. A fitness magazine advertises that the average monthly cost of joining a health club is \$25. You work for a consumer advocacy group and are asked to test this claim. You find that a random sample of 18 clubs has a mean monthly cost of \$26.25 and a standard deviation of \$3.23. At $\alpha = 0.10$, do you have enough evidence to reject the advertisement's claim?

34. A certain restaurant claims that its hamburgers have no more than 10 grams of fat. You work for a nutritional health agency and are asked to test this claim. You find that a random sample of nine hamburgers has a mean fat content of 13.5 grams and a standard deviation of 5.8 grams. At $\alpha = 0.10$, do you have enough evidence to reject the restaurant's claim?

In Exercises 35 and 36, use a *t*-statistic and its *P*-value to test the claim about the population mean μ using the given sample statistics. Assume the population is normally distributed.

DATA **35.** A medical association says that the mean number of hours of sleep high school students get each night is at least 6. The number of hours of sleep a random sample of 20 high school students got in one night is listed. At $\alpha = 0.01$, can you reject the association's claim? *(Adapted from the Harris Poll)*

5.2	6.3	4.2	4.1	8.2	3.5	6.4	9.1	6.2	5.8
7.4	5.0	4.0	5.3	4.2	3.1	7.9	8.2	5.3	5.9

36. A large university says the mean number of classroom hours per week for full-time faculty is more than 9. A random sample of the number of classroom hours for full-time faculty for one week is listed. At $\alpha = 0.05$, test the university's claim. *(Adapted from National Center for Education Statistics)*

10.7 9.8 11.6 9.7 7.6 11.3 14.1 8.1 11.5 8.5 6.9

Section 7.4

In Exercises 37–44, decide whether the normal sampling distribution can be used to approximate the binomial distribution. If it can, use the *z*-test to test the claim about the population proportion *p* at the given level of significance α using the given sample statistics.

37. Claim: $p = 0.15$; $\alpha = 0.05$. Sample statistics: $\hat{p} = 0.09$, $n = 40$

38. Claim: $p < 0.70$; $\alpha = 0.01$. Sample statistics: $\hat{p} = 0.50$, $n = 68$

39. Claim: $p < 0.08$; $\alpha = 0.05$. Sample statistics: $\hat{p} = 0.03$, $n = 45$

40. Claim: $p = 0.50$; $\alpha = 0.10$. Sample statistics: $\hat{p} = 0.71$, $n = 129$

41. Claim: $p \geq 0.04$; $\alpha = 0.10$. Sample statistics: $\hat{p} = 0.03$, $n = 30$

42. Claim: $p \neq 0.34$; $\alpha = 0.01$. Sample statistics: $\hat{p} = 0.29$, $n = 60$

43. Claim: $p \neq 0.20$; $\alpha = 0.01$. Sample statistics: $\hat{p} = 0.23$, $n = 56$

44. Claim: $p \leq 0.80$; $\alpha = 0.10$. Sample statistics: $\hat{p} = 0.85$, $n = 43$

45. Reject H_0. There is enough evidence at the 10% level to support the claim that over 75% of people in the United States prepare and file their taxes before April 15th.

46. Fail to reject H_0. There is not enough evidence at the 5% level to reject the claim that the rate of false positives for a Western blot assay is 2%.

47. 30.144

48. $\chi^2_L = 3.565$; $\chi^2_R = 29.819$

49. 33.196

50. 1.145

51. Fail to reject H_0. There is not enough evidence at the 10% level to support the claim.

52. Fail to reject H_0. There is not enough evidence at the 2.5% level to reject the claim.

53. Fail to reject H_0. There is not enough evidence at the 5% level to reject the claim.

54. Fail to reject H_0. There is not enough evidence at the 1% level to support the claim.

55. Reject H_0. There is enough evidence at the 0.5% level to reject the claim that the variance is at most 0.01.

56. Fail to reject H_0. There is not enough evidence at the 1% level to reject the claim that the standard deviation is no more than 0.0025 liter.

In Exercises 45 and 46, test the claim about the population proportion p.

45. A polling agency says that over 75% of people in the United States prepare and file their income taxes before April 15th. In a random survey of 1036 people in the United States, 818 said they prepare and file their income taxes before April 15th. Test the agency's claim at the $\alpha = 0.10$ level. What can you conclude? *(Adapted from The Gallup Poll)*

46. The Western blot assay is a blood test for the presence of HIV. It has been found that this test sometimes gives false positive results for HIV; specifically, when it does not find a certain antibody called p-31 antibody. A medical researcher claims that the rate of false positives in this case is 2%. A recent study of 121 randomly selected U.S. blood donors who tested positive for HIV but did not have p-31 antibodies in their blood found that 4 were actually HIV negative. Test the researcher's claim of a 2% false positive rate at the $\alpha = 0.05$ level. What can you conclude? *(Adapted from Centers for Disease Control and Prevention)*

Section 7.5

In Exercises 47–50, find the critical value(s) for the indicated χ^2-test for a population variance, sample size n, and level of significance α.

47. Right-tailed test, $n = 20$, $\alpha = 0.05$

48. Two-tailed test, $n = 14$, $\alpha = 0.01$

49. Right-tailed test, $n = 25$, $\alpha = 0.10$

50. Left-tailed test, $n = 6$, $\alpha = 0.05$

In Exercises 51–54, use a χ^2-test to test the claim about the population variance σ^2 or standard deviation σ at the given level of significance α and using the given sample statistics. Assume the population is normally distributed.

51. Claim: $\sigma^2 > 2$; $\alpha = 0.10$. Sample statistics: $s^2 = 2.38$, $n = 18$

52. Claim: $\sigma^2 \leq 60$; $\alpha = 0.025$. Sample statistics: $s^2 = 72.7$, $n = 15$

53. Claim: $\sigma = 1.25$; $\alpha = 0.05$. Sample statistics: $s = 1.03$, $n = 6$

54. Claim: $\sigma \neq 0.035$; $\alpha = 0.01$. Sample statistics: $s = 0.026$, $n = 16$

In Exercises 55 and 56, test the claim about the population variance or standard deviation. For each claim, assume the population is normally distributed.

55. A bolt manufacturer makes a type of bolt to be used in airtight containers. The manufacturer needs to be sure that all of its bolts are very similar in width, so it sets an upper tolerance limit for the variance of bolt width at 0.01. A random sample of 28 bolts yields a variance of 0.064 for bolt width. Test the manufacturer's claim that the variance is at most 0.01 at the $\alpha = 0.005$ level. What can you conclude?

56. A bottler needs to be sure that its liquid dispensers are set properly. The standard deviation of liquid dispensed must be no more than 0.0025 liter. A random sample of 14 bottles has a standard deviation of 0.0031 liter. Test the bottler's claim that the standard deviation is no more than 0.0025 liter at the $\alpha = 0.01$ level. What can you conclude?

7 Chapter Quiz

Take this quiz as you would take a quiz in class. After you are done, check your work against the answers given in the back of the book.

For this quiz, do the following.

(a) Write the claim mathematically. Identify H_0 and H_a.

(b) Determine whether the hypothesis test is a one-tailed or a two-tailed test and whether to use a *z*-test, a *t*-test, or a χ^2-test. Explain your reasoning.

(c) If necessary, find the critical value(s) and identify the rejection region(s).

(d) Find the appropriate test statistic. If necessary, find the *P*-value.

(e) Decide whether to reject or fail to reject the null hypothesis. Then interpret the decision in the context of the original claim.

1. A citrus grower's association believes that the mean utilization of fresh citrus fruits by people in the United States is at least 94 pounds per year. A random sample of 103 people in the United States has a mean utilization of fresh citrus fruits of 93.5 pounds per year and a standard deviation of 30 pounds. At $\alpha = 0.02$, can you reject the association's claim that the mean utilization of fresh citrus fruits by people in the United States is at least 94 pounds per year? *(Adapted from U.S. Department of Agriculture)*

2. An auto maker estimates that the mean gas mileage of its luxury sedan is at least 25 miles per gallon. A random sample of eight such cars had a mean of 23 miles per gallon and a standard deviation of 5 miles per gallon. At $\alpha = 0.05$, can you reject the auto maker's claim that the mean gas mileage of its luxury sedan is at least 25 miles per gallon? Assume the population is normally distributed. *(Adapted from Consumer Reports)*

3. A maker of microwave ovens advertises that no more than 10% of its microwaves need repair during the first five years of use. In a random sample of 57 microwaves five years old, 13% needed repairs. At $\alpha = 0.04$, can you reject the maker's claim that no more than 10% of its microwaves need repair during the first five years of use? *(Adapted from Consumer Reports)*

4. A state school administrator says that the standard deviation of SAT verbal test scores is 105. A random sample of 14 SAT verbal test scores has a standard deviation of 113. At $\alpha = 0.01$, test the administrator's claim. What can you conclude? Assume the population is normally distributed. *(Adapted from National Center for Educational Statistics)*

5. An employment information service reports that the mean annual salary for full-time male workers age 25 to 34 with a bachelor's degree is $59,482. You doubt the validity of this claim. If a random sample of 12 full-time male workers age 25 to 34 with a bachelor's degree has a mean annual salary of $58,581 with a standard deviation of $6500, can you reject the service's claim? Use a *P*-value and $\alpha = 0.05$. Assume the population is normally distributed. *(Adapted from U.S. Census Bureau)*

6. A tourist agency in Massachusetts claims the mean daily cost of meals and lodging for a family of four traveling in the state is $276. You work for a consumer protection advocate and want to test this claim. In a random sample of 35 families of four traveling in Massachusetts, the mean daily cost of meals and lodging is $285 and the standard deviation is $30. Do you have enough evidence to reject the agency's claim? Use a *P*-value and $\alpha = 0.05$. *(Adapted from American Automobile Association)*

---

1. (a) H_0: $\mu \geq 94$ (claim); H_a: $\mu < 94$

(b) Left-tailed because the alternative hypothesis contains $<$; *z*-test because $n \geq 30$.

(c) $z_0 = -2.05$
Rejection region: $z < -2.05$

(d) -0.169

(e) Fail to reject H_0. There is insufficient evidence at the 2% level to reject the claim that the mean consumption of fresh citrus fruits by people in the United States is at least 94 pounds per year.

2. (a) H_0: $\mu \geq 25$ (claim); H_a: $\mu < 25$

(b) Left-tailed because the alternative hypothesis contains $<$; *t*-test because the population is normal, $n < 30$, and σ is unknown.

(c) $t_0 = -1.895$
Rejection region: $t < -1.895$

(d) -1.131

(e) Fail to reject H_0. There is insufficient evidence at the 5% level to reject the claim that the mean gas mileage is at least 25 miles per gallon.

3. (a) H_0: $p \leq 0.10$ (claim)
H_a: $p > 0.10$

(b) Right-tailed because the alternative hypothesis contains $>$; *z*-test because np, $nq \geq 5$.

(c) $z_0 = 1.75$
Rejection region: $z > 1.75$

(d) 0.755

(e) Fail to reject H_0. There is insufficient evidence at the 4% level to reject the claim that no more than 10% of the microwaves need repair during the first five years of use.

4. See Odd Answers, page A72.

5. See Odd Answers, page A72.

6. See Odd Answers, page A72.

PUTTING IT ALL TOGETHER

Real Statistics ▪ Real Decisions

You work for the public relations department of the Social Security Administration. In an effort to design better advertising campaigns, your department decides to conduct a survey to find out the opinions people in the United States have about the Social Security system. One of the questions asked and the results of each response and the respondent's age are shown in the table.

Your department believes that less than 40% of people in the United States think the Social Security system will have the money available to provide the benefits they expect for their retirements. Also, your department believes that the mean age of people in the United States who would say yes to this question is 60 years or older. As the department's research analyst, you must work with the data and determine if these claims can be supported or rejected.

Exercises

1. *How Would You Do It?*

(a) What sampling technique would you use to select the sample for the study? Why? What sampling technique would you use if you wanted to select samples from four age groups: 18–29, 30–49, 50–64, and 65 and over?

(b) Which technique in part (a) will give you a sample that is representative of the population?

(c) Identify possible flaws or biases in your study.

2. *Testing a Proportion*

Test the claim that less than 40% of people in the United States think the Social Security system will have the money available to provide the benefits they expect for their retirements. Use $\alpha = 0.10$. Write a paragraph that interprets the test's decision. Does the decision support your department's claim?

3. *Testing a Mean*

Test the claim that the mean age of people in the United States who would say yes to the survey question shown in the table is 60 years or older. Use $\alpha = 0.10$ and assume that the population is normally distributed. Write a paragraph that interprets the test's decision. Is there enough evidence to reject your department's claim?

4. *Your Conclusions*

On the basis of your analysis of the responses to this survey question, what would you tell your department?

SOCIAL SECURITY ADMINISTRATION USA

www.socialsecurity.gov

Do you think the Social Security system will have the money available to provide the benefits you expect for your retirement?

Age	Response
83	Yes
37	No
49	No
27	Yes
27	No
59	Yes
44	Yes
46	No
25	No
30	No
72	Yes
66	No
47	No
47	No
50	Yes
76	Yes
51	Yes
58	No
79	Yes
69	Yes
66	No
35	Yes
47	No
29	Yes
76	No
31	Yes
40	No
61	Yes
30	Yes
71	No
64	No
18	No

(Adapted from Newsweek)

TECHNOLOGY

MINITAB | EXCEL | TI-83

The Case of the Vanishing Women

53% ➡ 29% ➡ 9% ➡ 0%

From 1966 to 1968, Dr. Benjamin Spock and others were tried for conspiracy to violate the Selective Service Act by encouraging resistance to the Vietnam War. By a series of three selections, there ended up being no women on the jury. In 1969, Hans Zeisel wrote an article in *The University of Chicago Law Review* using statistics and hypothesis testing to argue that the jury selection was biased against Dr. Spock. Dr. Spock was a well-known pediatrician and author of books about raising children. Millions of mothers had read his books and followed his advice. Zeisel argued that, by keeping women off the jury, the court prejudiced the verdict.

The jury selection process for Dr. Spock's trial is shown at the right.

Stage 1. The clerk of the Federal District Court selected 350 people "at random" from the Boston City Directory. The directory contained several hundred names, 53% of whom were women. However, only 102 of the 350 people selected were women.

Stage 2. The trial judge, Judge Ford, selected 100 people "at random" from the 350 people. This group was called a venire and it contained only nine women.

Stage 3. The court clerk assigned numbers to the members of the venire and, one by one, they were interrogated by the attorneys for the prosecution and defense until 12 members of the jury were chosen. At this stage, only one potential female juror was questioned, and she was eliminated by the prosecutor under his quota of peremptory challenges (for which he did not have to give a reason).

Exercises

1. The MINITAB display below shows a hypothesis test for a claim that the proportion of women in the city directory is $p = 0.53$. In the test, $n = 350$ and $\hat{p} = 0.2914$. Should you reject the claim? What is the level of significance? Explain.

2. In Exercise 1, you rejected the claim that $p = 0.53$. But this claim was true. What type of error is this?

3. If you reject a true claim with a level of significance that is virtually zero, what can you infer about the randomness of your sampling process?

4. Describe a hypothesis test for Judge Ford's "random" selection of the venire. Use a claim of

$$p = \frac{102}{350} \approx 0.2914.$$

(a) Write the null and alternative hypotheses.

(b) Use a technology tool to perform the test.

(c) Make a decision.

(d) Interpret the decision in the context of the original claim. Could Judge Ford's selection of 100 venire members have been random?

MINITAB

Test and Confidence Interval for One Proportion

Test of p = 0.53 vs p not = 0.53

Sample	X	N	Sample p	99.0 % CI	Z-Value	P-Value
1	102	350	0.291429	(0.228862, 0.353995)	-8.94	0.000

Extended solutions are given in the *Technology Supplement*.
Technical instruction is provided for MINITAB, Excel, and the TI-83.

7 Using Technology to Perform Hypothesis Tests

Here are some MINITAB and TI-83 printouts for some of the examples in this chapter. To duplicate the MINITAB results, you need the original data. For the TI-83, you can simply enter the descriptive statistics.

(See Example 5, page 351.)

Data for Investments for 30 Franchises

\$70,700	\$69,400	\$90,600	\$85,500	\$97,100	\$100,800
\$114,700	\$119,600	\$123,600	\$127,200	\$131,200	\$132,000
\$134,400	\$136,700	\$138,500	\$140,900	\$143,300	\$143,900
\$151,100	\$151,700	\$157,200	\$159,800	\$163,000	\$163,500
\$169,300	\$167,400	\$168,800	\$168,800	\$159,100	\$170,200

Display Descriptive Statistics...
Store Descriptive Statistics...
1-Sample Z...
1-Sample t...
2-Sample t...
Paired t...
1 Proportion...
2 Proportions...

MINITAB

Z-Test of the Mean

Test of mu = 143260 vs mu not = 143260
The assumed sigma = 30000

Variable	N	Mean	StDev	SE Mean	Z	P
C1	30	135000	30000	5477	−1.51	0.13

(See Example 4, page 367.)

Data for prices of 14 Ford F-150 Super Cab Pickups

\$16,800	\$16,090	\$16,100	\$16,330	\$17,700
\$17,860	\$18,155	\$18,235	\$18,450	\$18,600
\$19,400	\$19,400	\$19,370	\$19,510	

Display Descriptive Statistics...
Store Descriptive Statistics...
1-Sample Z...
1-Sample t...
2-Sample t...
Paired t...
1 Proportion...
2 Proportions...

MINITAB

T-Test of the Mean

Test of mu = 18800 vs mu < 18800

Variable	N	Mean	StDev	SE Mean	T	P
C1	14	18000	1250	334	−2.39	0.016

(See Example 2, page 376.)

Display Descriptive Statistics...
Store Descriptive Statistics...
1-Sample Z...
1-Sample t...
2-Sample t...
Paired t...
1 Proportion...
2 Proportions...

MINITAB

Test and Confidence Interval for One Proportion

Test of p = 0.23 vs p not = 0.23

Sample	X	N	Sample p	95.0 % CI	Z-Value	P-Value
1	54	200	0.270000	(0.208471, 0.331529)	1.34	0.179

(See Example 9, page 355.)

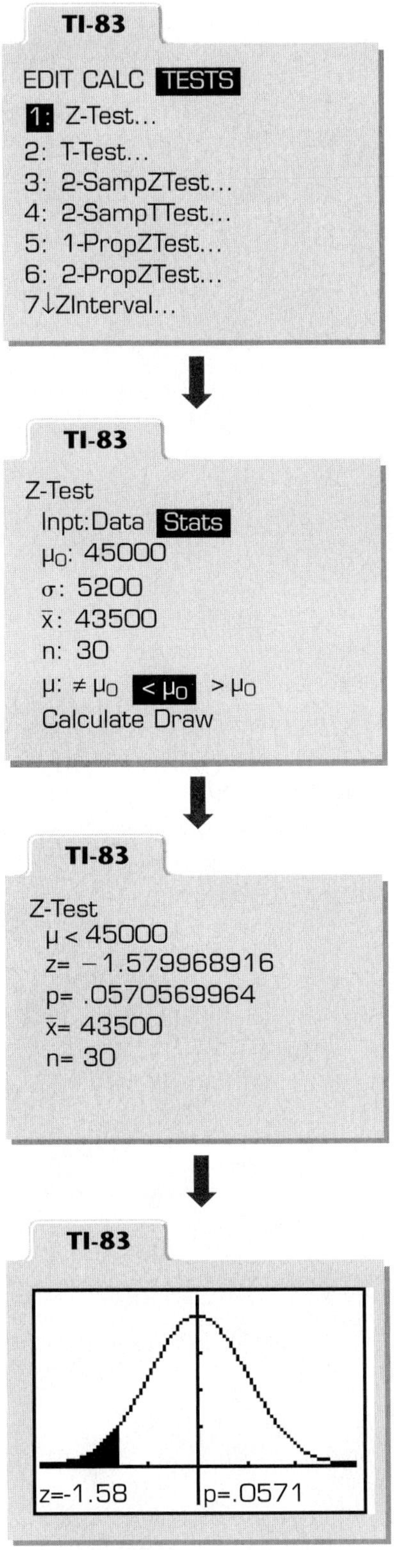

(See Example 5, page 368.)

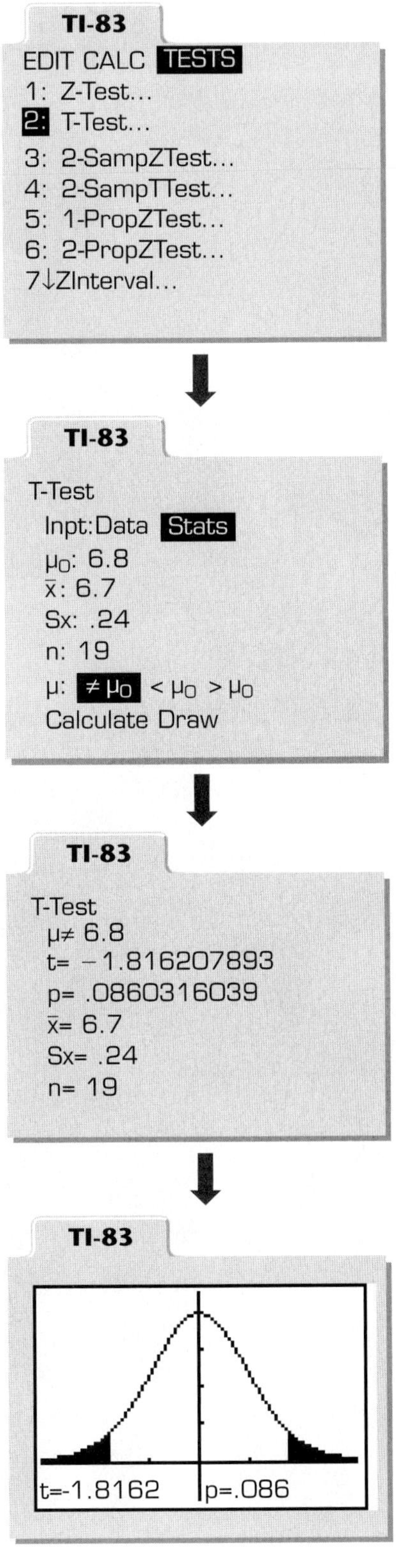

(See Example 1, page 375.)

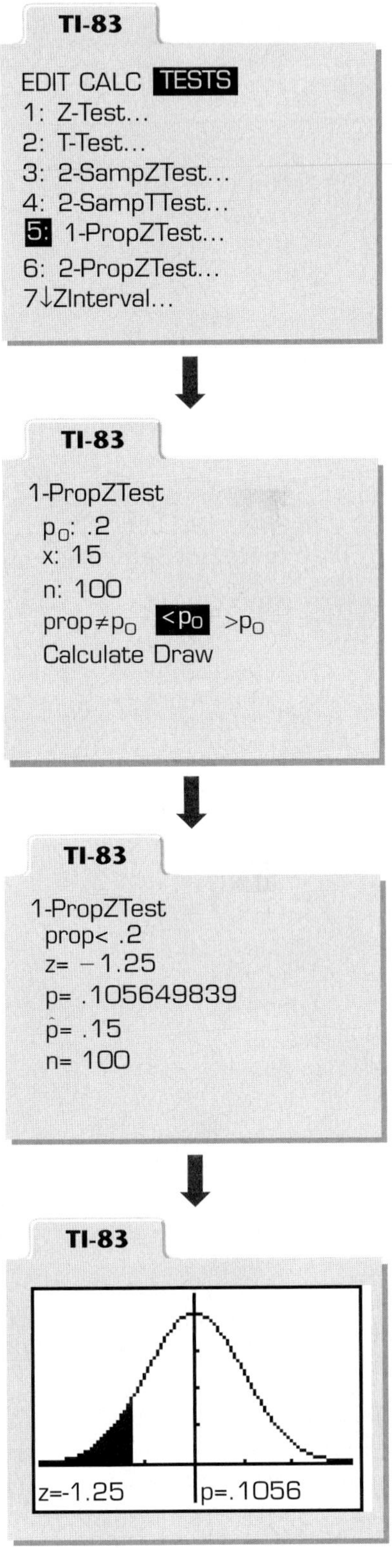

CHAPTER 8

Hypothesis Testing with Two Samples

8.1 Testing the Difference Between Means (Large Independent Samples)

Case Study

8.2 Testing the Difference Between Means (Small Independent Samples)

8.3 Testing the Difference Between Means (Dependent Samples)

8.4 Testing the Difference Between Proportions

Uses and Abuses

Real Statistics–Real Decisions

Technology

Every day in the United States, approximately 4400 students age 12–17 years try their first cigarette. About one-third of these students can expect to die from a smoking-related disease. Each year, the American Legacy Foundation and the Centers for Disease Control and Prevention publish a report, the *National Youth Tobacco Survey* (*NYTS*), to provide information on the most widely used tobacco products among U.S. students.

Where You've Been

In Chapter 6, you were introduced to inferential statistics and you learned how to form confidence intervals to estimate a parameter. Then, in Chapter 7, you learned how to test a claim about a population parameter, basing your decision on sample statistics and their distributions.

The *National Youth Tobacco Survey* (*NYTS*) is a study conducted by the American Legacy Foundation and the Centers for Disease Control and Prevention to provide information about student use of tobacco products. As part of this study in a recent year, a random sample of 6771 U.S. male high school students was surveyed. The following proportions were found.

Male High School Students ($n = 6771$)

Characteristic	Frequency	Proportion
Smoke cigarettes (at least one in the last 30 days)	1666	0.246
Smoke cigars (at least one in the last 30 days)	1144	0.169
Use smokeless tobacco (at least once in the last 30 days)	731	0.108

Where You're Going

In this chapter, you will continue your study of inferential statistics and hypothesis testing. Now, however, instead of testing a hypothesis about a single population, you will learn how to test a hypothesis that compares two populations.

For instance, in the *NYTS* study a second random sample of 6767 U.S. high school students was surveyed. These students were similar to the male high school student sample in terms of race, age, type of school attended, and average class size. The second group, however, consisted of female high school students. Here are the study's findings for this second group.

Female High School Students ($n = 6767$)

Characteristic	Frequency	Proportion
Smoke cigarettes (at least one in the last 30 days)	1435	0.212
Smoke cigars (at least one in the last 30 days)	420	0.062
Use smokeless tobacco (at least once in the last 30 days)	95	0.014

From these two samples, can you conclude that there is a significantly greater proportion of high school students who smoke cigarettes, smoke cigars, or use smokeless tobacco among males than among females? Or, might the differences in the proportions be due to chance?

In this chapter, you will learn that you can answer these questions by testing the hypothesis that the two proportions are equal. For the proportions of students who use smokeless tobacco, for instance, the *P*-value for the hypothesis that $p_1 = p_2$ is about 0.0000. So, it is almost impossible that the two groups experienced the same proportion that use smokeless tobacco.

8.1 Testing the Difference Between Means (Large Independent Samples)

What You Should Learn

- An introduction to two-sample hypothesis testing for the difference between two population parameters
- How to perform a two-sample *z*-test for the difference between two means μ_1 and μ_2 using large independent samples

An Overview of Two-Sample Hypothesis Testing • Two-Sample *z*-Test for the Difference Between Means

An Overview of Two-Sample Hypothesis Testing

In Chapter 7, you studied methods for testing a claim about the value of a population parameter. In this chapter, you will learn how to test a claim comparing parameters from two populations.

For instance, suppose you are developing a marketing plan for an Internet service provider and want to determine whether there is a difference in the amount of time male and female college students spend online each day. The only way you can conclude with certainty that there is a difference is to take a census of all college students, calculate the mean daily times male students and female students spend online, and find the difference. Of course, it is not practical to take such a census. However, you can still determine with some degree of certainty whether such a difference exists.

You can begin by assuming that there is no difference in the mean times of the two populations. That is, $\mu_1 - \mu_2 = 0$. Then by taking a random sample from each population, and using the resulting two-sample test statistic $\bar{x}_1 - \bar{x}_2$, you can perform a two-sample hypothesis test. Suppose you obtain the following results:

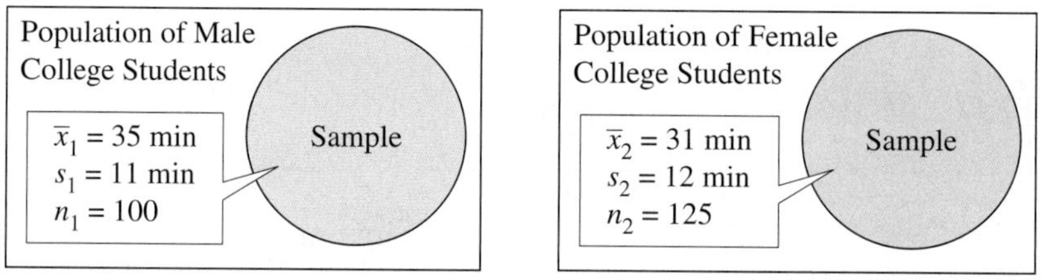

The graph below shows the sampling distribution of $\bar{x}_1 - \bar{x}_2$ for many similar samples taken from each population, under the assumption that $\mu_1 - \mu_2 = 0$. From the graph, you can see that it would be quite unlikely to obtain sample means that differ by 4 minutes if the actual difference were 0. The difference of the sample means would be more than 2.5 standard errors from the hypothesized difference of 0! So, you can conclude that there is a significant difference in the amount of time male college students and female college students spend online each day.

Sampling Distribution

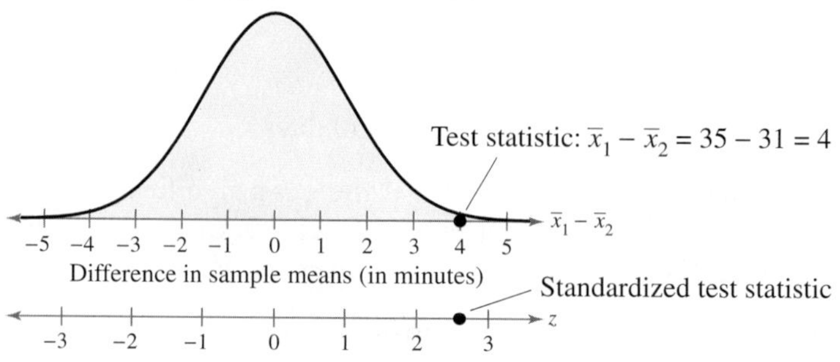

Note to Instructor

In this chapter, students will primarily use hypothesis tests for which the null hypothesis states that there is no difference in the population parameters. The opportunity to conduct tests for differences other than zero, however, is discussed in the Extending Concepts exercises.

It is important to remember that when you perform a two-sample hypothesis test, you are testing a claim concerning the difference between the parameters in two populations, not the values of the parameters themselves.

DEFINITION

For a two-sample hypothesis test,

1. the **null hypothesis H_0** is a statistical hypothesis that usually states there is no difference between the parameters of two populations. The null hypothesis always contains the symbol $\leq$, $=$, or $\geq$.
2. the **alternative hypothesis H_a** is a statistical hypothesis that is true when H_0 is false. The alternative hypothesis contains the symbol $>$, $\neq$, or $<$.

Study Tip

You can also write the null and alternative hypotheses as follows.

$$\begin{cases} H_0\colon \mu_1 - \mu_2 = 0 \\ H_a\colon \mu_1 - \mu_2 \neq 0 \end{cases}$$

$$\begin{cases} H_0\colon \mu_1 - \mu_2 \leq 0 \\ H_a\colon \mu_1 - \mu_2 > 0 \end{cases}$$

$$\begin{cases} H_0\colon \mu_1 - \mu_2 \geq 0 \\ H_a\colon \mu_1 - \mu_2 < 0 \end{cases}$$

To write a null and an alternative hypothesis for a two-sample hypothesis test, translate the claim made about the population parameters from a verbal statement to a mathematical statement. Then, write its complementary statement. For instance, if the claim is about two population parameters μ_1 and μ_2, then some possible pairs of null and alternative hypotheses are

$$\begin{cases} H_0\colon \mu_1 = \mu_2 \\ H_a\colon \mu_1 \neq \mu_2 \end{cases}, \quad \begin{cases} H_0\colon \mu_1 \leq \mu_2 \\ H_a\colon \mu_1 > \mu_2 \end{cases}, \quad \text{and} \quad \begin{cases} H_0\colon \mu_1 \geq \mu_2 \\ H_a\colon \mu_1 < \mu_2 \end{cases}.$$

Regardless of which hypotheses you use, you always assume $\mu_1 = \mu_2$.

Two-Sample z-Test for the Difference Between Means

In the remainder of this section, you will learn how to perform a z-test for the difference between two population means μ_1 and μ_2. Three conditions are necessary to perform such a test.

1. The samples must be randomly selected.
2. The samples must be independent. Two samples are **independent** if the sample selected from one population is not related to the sample selected from the second population. (You will study the distinction between independent samples and dependent samples in Section 8.3.)
3. Each sample size must be at least 30 or, if not, each population must have a normal distribution with a known standard deviation.

If these requirements are met, then the sampling distribution for $\bar{x}_1 - \bar{x}_2$ (the difference of the sample means) is a normal distribution with mean and standard error of

$$\mu_{\bar{x}_1 - \bar{x}_2} = \mu_{\bar{x}_1} - \mu_{\bar{x}_2} = \mu_1 - \mu_2$$

and

$$\sigma_{\bar{x}_1 - \bar{x}_2} = \sqrt{\sigma_{\bar{x}_1}^2 + \sigma_{\bar{x}_2}^2} = \sqrt{\frac{\sigma_1^2}{n_1} + \frac{\sigma_2^2}{n_2}}.$$

Sampling Distribution for $\bar{x}_1 - \bar{x}_2$

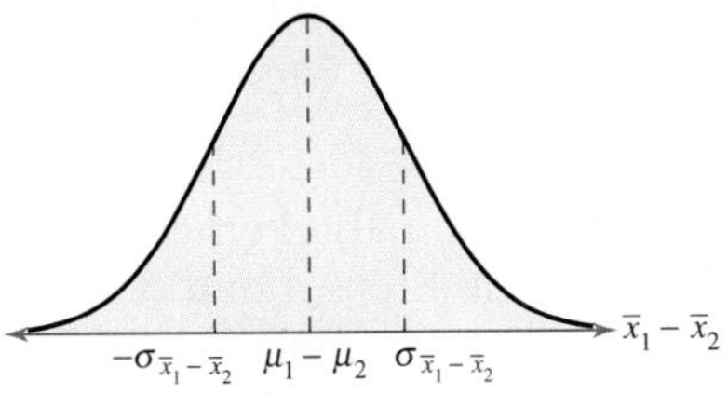

Notice that the variance of the sampling distribution $\sigma_{\bar{x}_1 - \bar{x}_2}^2$ is the sum of the variances of the individual sampling distributions for $\bar{x}_1$ and $\bar{x}_2$.

Note to Instructor

Point out the similarities in conducting two-sample hypothesis tests with one-sample tests presented in Chapter 7. Also, point out that *P*-values can be used in this chapter in much the same way they were used in Chapter 7.

Because the sampling distribution for $\bar{x}_1 - \bar{x}_2$ is a normal distribution, you can use the z-test to test the difference between two population means μ_1 and μ_2. Notice that the standardized test statistic takes the form of

$$z = \frac{(\text{Observed difference}) - (\text{Hypothesized difference})}{(\text{Standard error})}.$$

Two-Sample z-Test for the Difference Between Means

A **two-sample** z**-test** can be used to test the difference between two population means μ_1 and μ_2 when a large sample (at least 30) is randomly selected from each population and the samples are *independent*. The **test statistic** is $\bar{x}_1 - \bar{x}_2$, and the **standardized test statistic** is

$$z = \frac{(\bar{x}_1 - \bar{x}_2) - (\mu_1 - \mu_2)}{\sigma_{\bar{x}_1 - \bar{x}_2}} \quad \text{where} \quad \sigma_{\bar{x}_1 - \bar{x}_2} = \sqrt{\frac{\sigma_1^2}{n_1} + \frac{\sigma_2^2}{n_2}}.$$

When the samples are large, you can use s_1 and s_2 in place of σ_1 and σ_2. If the samples are not large, you can still use a two-sample z-test, provided the populations are normally distributed and the population standard deviations are known.

If the null hypothesis states $\mu_1 = \mu_2$, $\mu_1 \leq \mu_2$, or $\mu_1 \geq \mu_2$, then $\mu_1 = \mu_2$ is assumed and the expression $\mu_1 - \mu_2$ is equal to 0 in the preceding test.

GUIDELINES

Using a Two-Sample z-Test for the Difference Between Means (Large Independent Samples)

In Words	*In Symbols*
1. State the claim mathematically. Identify the null and alternative hypotheses.	State H_0 and H_a.
2. Specify the level of significance.	Identify α.
3. Sketch the sampling distribution.	
4. Determine the critical value(s).	Use Table 4 in Appendix B.
5. Determine the rejection region(s).	
6. Find the standardized test statistic.	$z = \frac{(\bar{x}_1 - \bar{x}_2) - (\mu_1 - \mu_2)}{\sigma_{\bar{x}_1 - \bar{x}_2}}$
7. Make a decision to reject or fail to reject the null hypothesis.	If z is in the rejection region, reject H_0. Otherwise, fail to reject H_0.
8. Interpret the decision in the context of the original claim.	

A hypothesis test for the difference between means can also be performed using P-values. Use the same guidelines above, skipping Steps 4 and 5. After finding the standardized test statistic, use the Standard Normal Table to calculate the P-value. Then make a decision to reject or fail to reject the null hypothesis. If P is less than or equal to α, reject H_0. Otherwise, fail to reject H_0.

Picturing the World

There are about 21,400 civilian federal employees working in Michigan and about 24,400 in Massachusetts. In a survey, 200 federal employees in each state were asked to report their salary. The results were as follows. *(Adapted from United States Office of Personnel Management)*

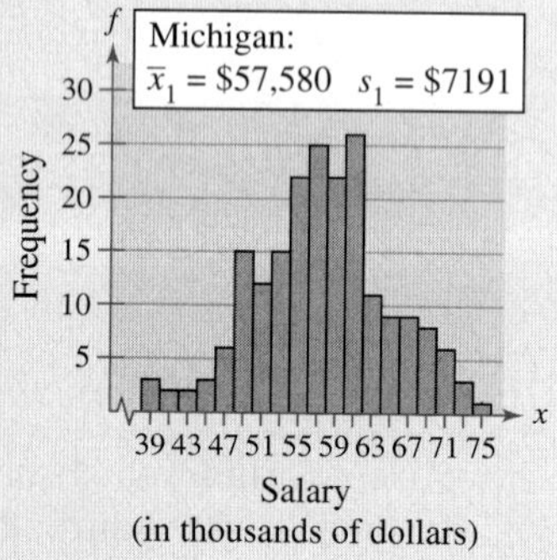

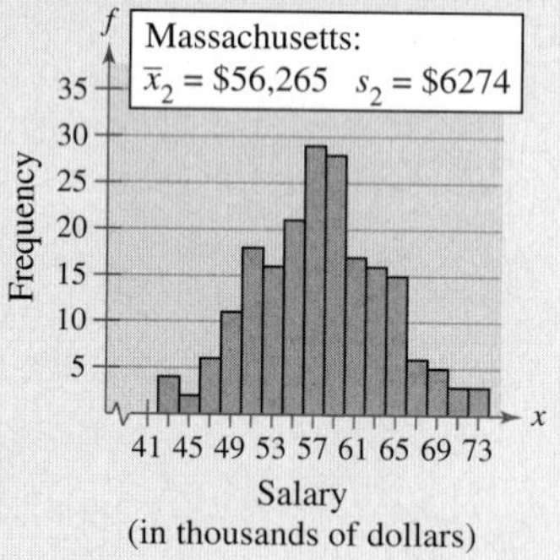

Is there enough evidence to conclude that there is a difference in the mean incomes of civilian federal employees in Michigan and Massachusetts using $\alpha = 0.05$?

EXAMPLE 1

A Two-Sample z-Test for the Difference Between Means

See TI-83 steps on page 455.

Sample Statistics for Household Incomes

Visa Gold	Gold MasterCard
$\bar{x}_1 = \$60{,}900$	$\bar{x}_2 = \$64{,}300$
$s_1 = \$12{,}000$	$s_2 = \$15{,}000$
$n_1 = 100$	$n_2 = 100$

An advertising executive claims that there is a difference in the mean household income for credit card holders of Visa Gold and of Gold MasterCard. The results of a random survey of 100 customers from each group are shown at the left. The two samples are independent. Do the results support the executive's claim? Use $\alpha = 0.05$. *(Adapted from Claritas Inc.)*

SOLUTION You want to test the claim that there is a difference in the mean household incomes for Visa Gold and Gold MasterCard credit card holders. So, the null and alternative hypotheses are

$$H_0\colon \mu_1 = \mu_2 \quad \text{and} \quad H_a\colon \mu_1 \neq \mu_2. \text{ (Claim)}$$

Because the test is a two-tailed test and the level of significance is $\alpha = 0.05$, the critical values are $-z_0 = -1.96$ and $z_0 = 1.96$. The rejection regions are $z < -1.96$ and $z > 1.96$. Because both samples are large, s_1 and s_2 are used in place of σ_1 and σ_2 to calculate the standard error.

$$\sigma_{\bar{x}_1-\bar{x}_2} = \sqrt{\frac{s_1^2}{n_1} + \frac{s_2^2}{n_2}} = \sqrt{\frac{12{,}000^2}{100} + \frac{15{,}000^2}{100}} \approx 1921$$

The standardized test statistic is

$$z = \frac{(\bar{x}_1 - \bar{x}_2) - (\mu_1 - \mu_2)}{\sigma_{\bar{x}_1-\bar{x}_2}} \qquad \text{Use the } z\text{-test.}$$

$$\approx \frac{(60{,}900 - 64{,}300) - 0}{1921} \qquad \text{Assume } \mu_1 = \mu_2, \text{ so } \mu_1 - \mu_2 = 0.$$

$$\approx -1.77.$$

Study Tip

In Example 1, you can also use a *P*-value to perform the hypothesis test. For instance, the test is a two-tailed test, so the *P*-value is equal to twice the area to the left of $z = -1.77$, or

$$2(0.0384) = 0.0768.$$

Because $0.0768 > 0.05$, you should fail to reject H_0.

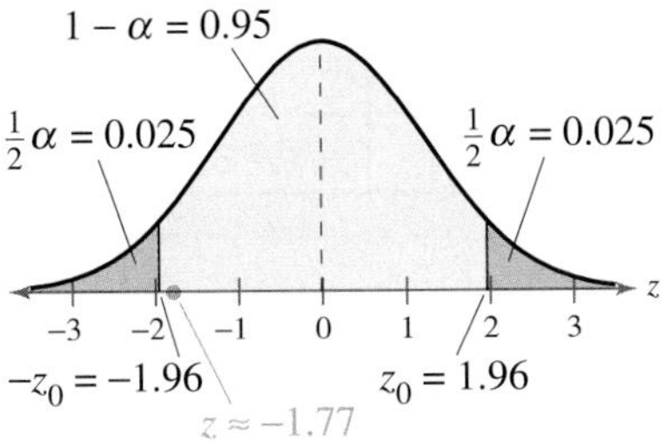

The graph at the left shows the location of the rejection regions and the standardized test statistic z. Because z is not in the rejection region, you should fail to reject the null hypothesis.

Interpretation At the 5% level, there is not enough evidence to support the executive's claim that there is a difference in the mean household incomes.

Try It Yourself 1

A survey indicates that the mean per capita credit card charge for residents of New Hampshire and New York is \$3900 and \$3500 per year, respectively. The survey included a randomly selected sample of size 50 from each state, and sample standard deviations are \$900 (NH) and \$500 (NY). The two samples are independent. At $\alpha = 0.01$, is there enough evidence to conclude that there is a difference in the mean credit card charges? *(Adapted from Card Management Information Services and the U.S. Census Bureau)*

a. *Identify* the claim and state H_0 and H_a.
b. *Specify* the level of significance α.
c. *Find* the critical values and *identify* the rejection regions.
d. *Use* the z-test to find the standardized test statistic z.
e. *Decide* whether to reject the null hypothesis. Use a graph if necessary.
f. *Interpret* the decision in the context of the original claim.

Answer: Page A41

EXAMPLE 2

Using Technology to Perform a Two-Sample z-Test

Sample Statistics for Daily Cost of Meals and Lodging for a Family of Four

Texas	Washington
$\bar{x}_1 = \$208$	$\bar{x}_2 = \$218$
$s_1 = \$15$	$s_2 = \$28$
$n_1 = 50$	$n_2 = 35$

The American Automobile Association claims that the average daily cost for meals and lodging for vacationing in Texas is less than the same average costs for vacationing in Washington State. The table at the left shows the results of a random survey of vacationers in each state. The two samples are independent. At $\alpha = 0.01$, is there enough evidence to support the claim? (H_0: $\mu_1 \geq \mu_2$ and H_a: $\mu_1 < \mu_2$ (claim))

SOLUTION The top two displays show how to set up the hypothesis test using a TI-83. The remaining displays show the possible results, depending on whether you select *Calculate* or *Draw*.

TI-83

```
2-SampZTest
 Inpt:Data Stats
 σ1:15
 σ2:28
 x̄1:208
 n1:50
 x̄2:218
↓n2:35
```

TI-83

```
2-SampZTest
↑σ2:28
 x̄1:208
 n1:50
 x̄2:218
 n2:35
 μ1:≠μ2 <μ2 >μ2
 Calculate Draw
```

TI-83

```
2-SampZTest
 μ1<μ2
 z=-1.928074718
 p=.0269228477
 x̄1:208
 x̄2:218
↓n1:50
```

TI-83

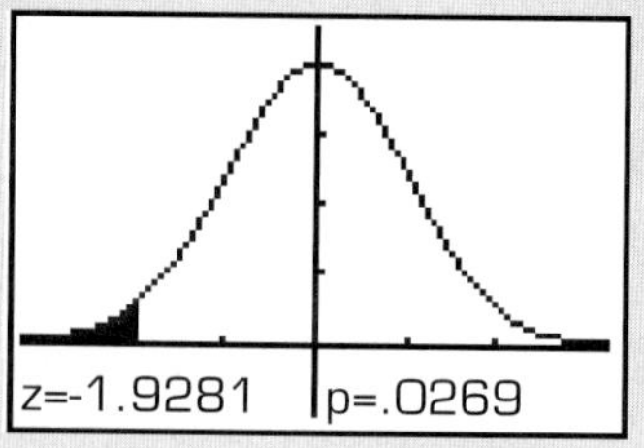

Study Tip

Note that the TI-83 displays $P \approx 0.0269$. Because $P > \alpha$, you should fail to reject the null hypothesis.

To use MINITAB to perform a two-sample *z*-test, you must use raw data.

Note to Instructor

Students might be confused about designating Texas as population 1 and Washington as population 2. Explain that the final result of the test would be identical if the subscripts were reversed. Show that reversing the subscripts would produce a right-tailed test for which the test statistic is 1.93, the critical value is $z = 2.33$, and the rejection region is $z > 2.33$.

Because the test is a left-tailed test and $\alpha = 0.01$, the rejection region is $z < -2.33$. The standardized test statistic $z \approx -1.93$ is not in the rejection region, so you should fail to reject the null hypothesis.

Interpretation At the 1% level, there is not enough evidence to support the American Automobile Association's claim.

Try It Yourself 2

Sample Statistics for Daily Cost of Meals and Lodging for a Family of Four

Florida	Delaware
$\bar{x}_1 = \$252$	$\bar{x}_2 = \$242$
$s_1 = \$22$	$s_2 = \$18$
$n_1 = 150$	$n_2 = 200$

The American Automobile Association claims that the average daily meal and lodging costs for vacationing in Florida are greater than the same average costs for vacationing in Delaware. The table at the left shows the results of a random survey of vacationers in each state. The two samples are independent. At $\alpha = 0.05$, is there enough evidence to support the claim?

a. *Use* a TI-83 to find the test statistic or the *P*-value.
b. *Make* a decision.

Answer: Page A41

8.1 Exercises

Help

MyMathLab

Student Study Pack

Building Basic Skills and Vocabulary

1. Explain how to perform a two-sample z-test for the difference between the means of two populations using large independent samples.

2. What two conditions are necessary in order to use the z-test to test the difference between two population means?

In Exercises 3–6, (a) find the test statistic, (b) find the standardized test statistic, (c) decide whether the standardized test statistic is in the rejection region, and (d) decide whether you should reject or fail to reject the null hypothesis. The samples are random and independent.

3. Claim: $\mu_1 = \mu_2$, $\alpha = 0.05$. Sample statistics: $\bar{x}_1 = 16$, $s_1 = 1.1$, $n_1 = 50$ and $\bar{x}_2 = 14$, $s_2 = 1.5$, $n_2 = 50$

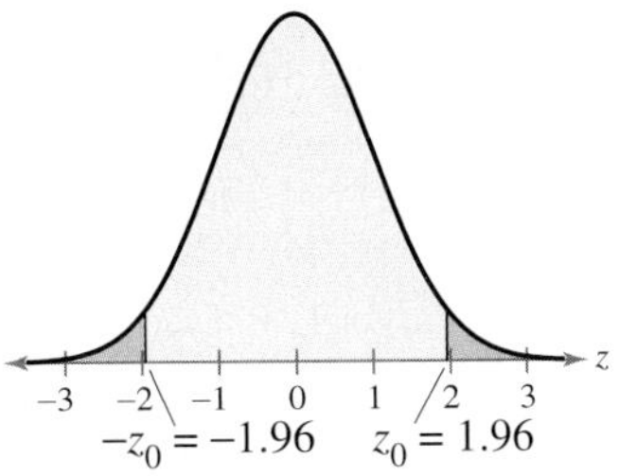

Figure for Exercise 3

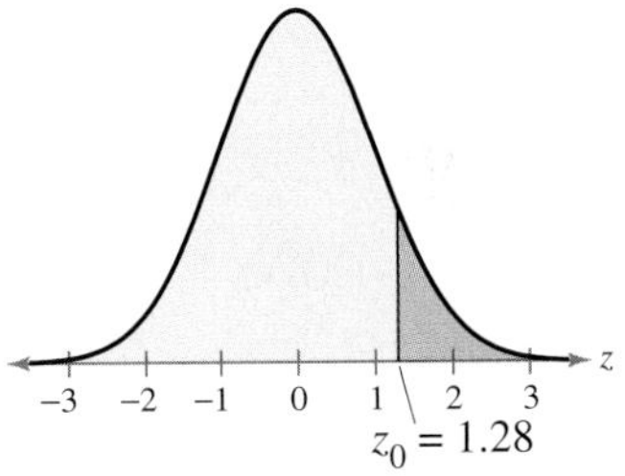

Figure for Exercise 4

4. Claim: $\mu_1 > \mu_2$, $\alpha = 0.10$. Sample statistics: $\bar{x}_1 = 500$, $s_1 = 30$, $n_1 = 100$ and $\bar{x}_2 = 510$, $s_2 = 15$, $n_2 = 75$

5. Claim: $\mu_1 < \mu_2$, $\alpha = 0.01$. Sample statistics: $\bar{x}_1 = 1225$, $s_1 = 75$, $n_1 = 35$ and $\bar{x}_2 = 1195$, $s_2 = 105$, $n_2 = 105$

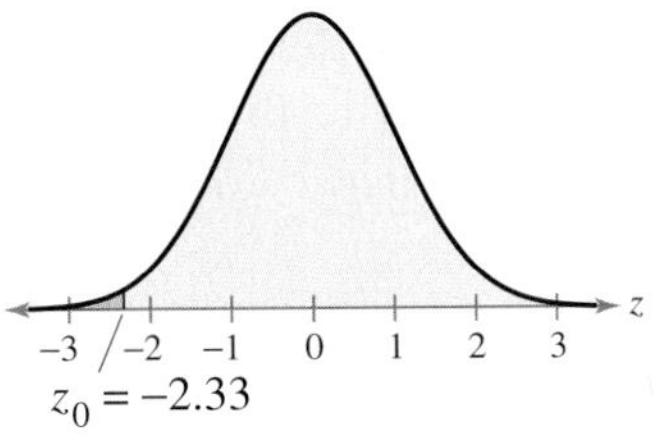

Figure for Exercise 5

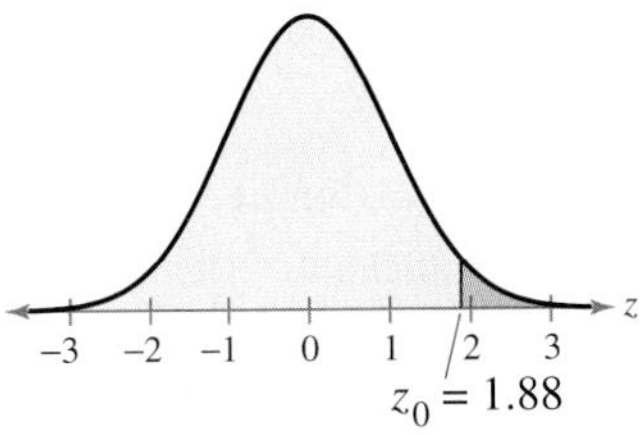

Figure for Exercise 6

6. Claim: $\mu_1 \leq \mu_2$, $\alpha = 0.03$. Sample statistics: $\bar{x}_1 = 5004$, $s_1 = 136$, $n_1 = 144$ and $\bar{x}_2 = 4895$, $s_2 = 215$, $n_2 = 156$

In Exercises 7 and 8, use the given sample statistics to test the claim about the difference between two population means μ_1 and μ_2 at the given level of significance α.

7. Claim: $\mu_1 > \mu_2$, $\alpha = 0.01$. Statistics: $\bar{x}_1 = 5.2$, $s_1 = 0.2$, $n_1 = 45$ and $\bar{x}_2 = 5.5$, $s_2 = 0.3$, $n_2 = 37$

8. Claim: $\mu_1 \neq \mu_2$, $\alpha = 0.05$. Statistics: $\bar{x}_1 = 52$, $s_1 = 2.5$, $n_1 = 70$ and $\bar{x}_2 = 45$, $s_2 = 5.5$, $n_2 = 60$

1. State the hypotheses and identify the claim. Specify the level of significance. Find the critical value(s) and identify the rejection region(s). Find the standardized test statistic. Make a decision and interpret in the context of the claim.

2. (1) The samples must be independent. (2) $\{n_1 \geq 30$ and $n_2 \geq 30\}$ or {each population must have a normal distribution with a known standard deviation}

3. (a) 2 (b) 7.603
(c) z is in the rejection region.
(d) Reject H_0. There is enough evidence at the 5% level to reject the claim.

4. (a) −10 (b) −2.887
(c) z is not in the rejection region.
(d) Fail to reject H_0. There is not enough evidence at the 10% level to support the claim.

5. (a) 30 (b) 1.84
(c) z is not in the rejection region.
(d) Fail to reject H_0. There is not enough evidence at the 1% level to support the claim.

6. (a) 109 (b) 5.289
(c) z is in the rejection region.
(d) Reject H_0. There is enough evidence at the 3% level to reject the claim.

7. Fail to reject H_0. There is enough evidence at the 1% level to support the claim.

8. Reject H_0. There is enough evidence at the 5% level to support the claim.

9. (a) H_0: $\mu_1 = \mu_2$
H_a: $\mu_1 \neq \mu_2$ (claim)
(b) $-z_0 = -1.645$ and $z_0 = 1.645$
Rejection regions: $z < -1.645$, $z > 1.645$
(c) -2.786 (d) Reject H_0.
(e) There is sufficient evidence at the 10% level to support the claim that the mean braking distance is different for the two types of tires.

10. (a) H_0: $\mu_1 \leq \mu_2$
H_a: $\mu_1 > \mu_2$ (claim)
(b) $z_0 = 1.28$
Rejection region: $z > 1.28$
(c) 3.919 (d) Reject H_0.
(e) There is sufficient evidence at the 10% level to support the claim that the mean braking distance for Type C is greater than for Type D.

11. (a) H_0: $\mu_1 \geq \mu_2$
H_a: $\mu_1 < \mu_2$ (claim)
(b) $z_0 = -2.33$
Rejection region: $z < -2.33$
(c) -1.536 (d) Fail to reject H_0.
(e) There is insufficient evidence at the 1% level to conclude that the repair costs for Model A are lower than for Model B.

12. (a) H_0: $\mu_1 = \mu_2$ (claim)
H_a: $\mu_1 \neq \mu_2$
(b) $-z_0 = -2.575$ and $z_0 = 2.575$
Rejection regions: $z < -2.575$, $z > 2.575$
(c) -3.165 (d) Reject H_0.
(e) There is sufficient evidence at the 1% level to reject the claim that the mean repair costs for Model A and Model B are the same.

13. (a) H_0: $\mu_1 = \mu_2$ (claim)
H_a: $\mu_1 \neq \mu_2$
(b) $-z_0 = -2.575$ and $z_0 = 2.575$
Rejection regions: $z < -2.575$, $z > 2.575$
(c) 0.202 (d) Fail to reject H_0.
(e) There is insufficient evidence at the 1% level to reject the claim that male and female high school students have equal ACT scores.

Using and Interpreting Concepts

Testing the Difference Between Two Means In Exercises 9–22, (a) identify the claim and state H_0 and H_a, (b) use Table 4 in Appendix B to find the critical value(s) and identify the rejection region(s), (c) find the standardized test statistic z, (d) decide whether to reject or fail to reject the null hypothesis, and (e) interpret the decision in the context of the original claim. If convenient, use technology to solve the problem. In each exercise, assume the samples are randomly selected and that the samples are independent.

9. ***Braking Distances*** A safety engineer records the braking distances of two types of tires. Each randomly selected sample has 35 tires. The results of the tests are shown in the figure. At $\alpha = 0.10$, can the engineer support the claim that the mean braking distance is different for the two types of tires? *(Adapted from Consumer Reports)*

Figure for Exercise 9

Figure for Exercise 10

10. ***Braking Distances*** To compare the braking distances for two types of tires, a safety engineer conducts 50 braking tests for each of the two types of tires. The results of the tests are shown in the figure. At $\alpha = 0.10$, can the engineer support the claim that the mean braking distance for Type C is greater than the mean braking distance for Type D? *(Adapted from Consumer Reports)*

11. ***Repair Costs: Microwave Ovens*** You want to buy a microwave oven and will choose Model A if its repair costs are lower than Model B's. You research the repair costs of 47 Model A ovens and 55 Model B ovens. The results of your research are shown in the figure. At $\alpha = 0.01$, would you buy Model A? *(Adapted from Consumer Reports)*

Figure for Exercise 11

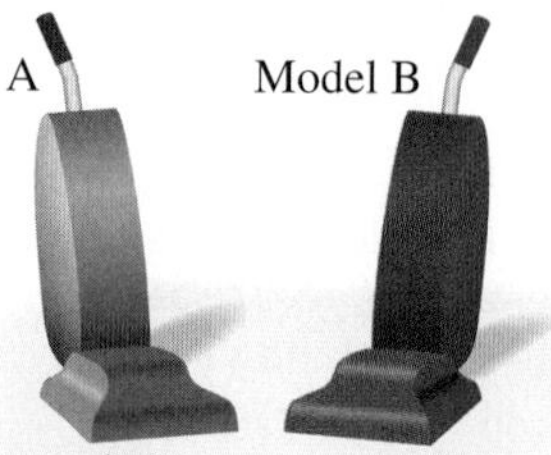

Figure for Exercise 12

12. ***Repair Costs: Vacuum Cleaners*** You want to buy a vacuum cleaner, and a salesperson tells you the repair costs for Model A and Model B are equal. You research the repair costs of 34 Model A vacuum cleaners and 46 Model B vacuum cleaners. The results of your research are shown in the figure. At $\alpha = 0.01$, can you reject the salesperson's claim? *(Adapted from Consumer Reports)*

14. (a) H_0: $\mu_1 \le \mu_2$
H_a: $\mu_1 > \mu_2$ (claim)

(b) $z_0 = 1.28$
Rejection region: $z > 1.28$

(c) 2.161

(d) Reject H_0.

(e) There is sufficient evidence at the 10% level to support the claim that the ACT scores are higher for high school students in a college prep program.

15. (a) H_0: $\mu_1 = \mu_2$ (claim)
H_a: $\mu_1 \ne \mu_2$

(b) $-z_0 = -1.645$ and $z_0 = 1.645$
Rejection regions: $z < -1.645$, $z > 1.645$

(c) -0.606

(d) Fail to reject H_0.

(e) There is insufficient evidence at the 10% level to reject the claim that the lodging cost for a family traveling in California is the same as in Florida.

16. See Selected Answers, page A96.

17. (a) H_0: $\mu_1 = \mu_2$ (claim)
H_a: $\mu_1 \ne \mu_2$

(b) $-z_0 = -1.645$ and $z_0 = 1.645$
Rejection regions: $z < -1.645$, $z > 1.645$

(c) 0.645

(d) Fail to reject H_0.

(e) There is insufficient evidence at the 10% level to reject the claim that the lodging cost for a family traveling in California is the same as in Florida.

The new samples do not lead to a different conclusion.

18. See Selected Answers, page A97.

19. (a) H_0: $\mu_1 \le \mu_2$
H_a: $\mu_1 > \mu_2$ (claim)

(b) $z_0 = 1.96$
Rejection region: $z > 1.96$

(c) 4.988 (d) Reject H_0.

(e) At the 2.5% level of significance, there is sufficient evidence to support the claim that children spent more time watching television in 1981 than children do today.

13. *ACT Scores* The mean ACT score for 43 male high school students is 21.0 and the standard deviation is 5.0. The mean ACT score for 56 female high school students is 20.8 and the standard deviation is 4.7. At $\alpha = 0.01$, can you support the claim that male and female high school students have equal ACT scores? *(Adapted from ACT Inc.)*

14. *ACT Scores* A guidance counselor claims high school students in a college prep program have higher ACT scores than those in a general program. The mean ACT score for 49 high school students who are in a college prep program is 22.1 and the standard deviation is 4.8. The mean ACT score for 44 high school students who are in a general program is 19.8 and the standard deviation is 5.4. At $\alpha = 0.10$, can you support the guidance counselor's claim? *(Adapted from ACT Inc.)*

15. *Lodging Costs* A travel association says the mean daily lodging cost for a family traveling in California is the same as in Florida. The mean daily lodging cost for 35 families traveling in California is \$136 and the standard deviation is \$25. The mean daily lodging cost for 35 families traveling in Florida is \$140 and the standard deviation is \$30. At $\alpha = 0.10$, is there enough evidence to reject the travel association's claim? *(Adapted from the American Automobile Association)*

16. *Money Spent Eating Out* A restaurant association says that households in the United States headed by people under the age of 25 spend less on food away from home than households headed by people ages 55–64. The mean amount spent by 30 households headed by people under the age of 25 is \$1526 and the standard deviation is \$225. The mean amount spent by 30 households headed by people ages 55–64 is \$2136 and the standard deviation is \$350. Can you support the restaurant association's claim at $\alpha = 0.05$? *(Adapted from the National Restaurant Association)*

17. *Lodging Costs* Refer to Exercise 15. Two more samples are taken, one from California and one from Florida. For 50 families traveling in California, $\bar{x}_1 = \$146$ and $s_1 = \$30$. For 50 families traveling in Florida, $\bar{x}_2 = \$142$ and $s_2 = \$32$. Use $\alpha = 0.10$. Do the new samples lead to a different conclusion?

18. *Money Spent Eating Out* Refer to Exercise 16. Two more samples are taken, one from each age group. For 40 households headed by people under the age of 25, $\bar{x}_1 = \$1600$ and $s_1 = \$230$. For 40 households headed by people ages 55–64, $\bar{x}_2 = \$2040$ and $s_2 = \$380$. Use $\alpha = 0.05$. Do the new samples lead to a different conclusion?

DATA

19. *Watching More TV?* A sociologist claims that children ages 3–12 spent more time watching television in 1981 than children ages 3–12 do today. A study was conducted in 1981 to find the time that children ages 3–12 watched television on weekdays. The results (in hours per weekday) are shown below.

2.0 2.5 2.1 2.3 2.1 1.6 2.6 2.1 2.1 2.4 2.1 2.1 1.5
1.7 2.1 2.3 2.5 3.3 2.2 2.9 1.5 1.9 2.4 2.2 1.2 3.0
1.0 2.1 1.9 2.2

Recently, a similar study was conducted. The results are shown below.

1.9 1.8 0.9 1.6 2.0 1.7 1.1 1.1 1.6 2.0 1.4 1.5 1.7
1.6 1.6 1.7 1.2 2.0 2.2 1.6 1.5 2.0 1.6 1.8 1.7 1.3
1.1 1.4 1.2 2.0

At $\alpha = 0.025$, can you support the sociologist's claim? *(Adapted from University of Michigan's Institute for Social Research)*

20. (a) H_0: $\mu_1 \geq \mu_2$
H_a: $\mu_1 < \mu_2$ (claim)
(b) $z_0 = -1.88$
Rejection region: $z < -1.88$
(c) -6.388
(d) Reject H_0.
(e) There is sufficient evidence at the 3% level to support the sociologist's claim that middle school boys spent less time studying in 1981 than middle school boys do today.

21. (a) H_0: $\mu_1 = \mu_2$ (claim)
H_a: $\mu_1 \neq \mu_2$
(b) $-z_0 = -2.575$ and $z_0 = 2.575$
Rejection regions: $z < -2.575$, $z > 2.575$
(c) 66.172
(d) Reject H_0.
(e) At the 1% level of significance, there is sufficient evidence to reject the claim that there is no difference in mean washer diameters.

22. (a) H_0: $\mu_1 = \mu_2$ (claim)
H_a: $\mu_1 \neq \mu_2$
(b) $-z_0 = -2.05$ and $z_0 = 2.05$
Rejection regions: $z < -2.05$, $z > 2.05$
(c) -69.346
(d) Reject H_0.
(e) There is sufficient evidence at the 4% level to reject the engineer's claim that there is no difference in mean nut diameters.

23. They are equivalent through algebraic manipulation of the equation.

$\mu_1 = \mu_2 \Rightarrow \mu_1 - \mu_2 = 0$

24. They are equivalent through algebraic manipulation of the equation.

$\mu_1 \geq \mu_2 \Rightarrow \mu_1 - \mu_2 \geq 0$

DATA **20.** ***Spending More Time Studying?*** A sociologist thinks middle school boys spent less time studying in 1981 than middle school boys do today. A study was conducted in 1981 to find the time that middle school boys spent studying on weekdays. The results (in minutes per weekday) are shown below.

13.9	17.3	10.0	21.2	12.8	13.7	15.0	11.9	17.3	12.1	13.5
17.9	19.1	18.2	14.8	19.1	6.2	10.6	10.7	23.1	15.7	14.1
10.7	17.2	18.6	16.3	17.8	16.2	15.8	7.5	9.7	3.7	5.3
18.8	8.9									

Recently, a similar study was conducted. The results are shown below.

17.5	27.9	14.1	19.9	20.9	19.6	15.9	21.8	23.0	20.6	20.5
31.0	24.0	14.4	22.5	24.6	12.0	18.6	22.8	27.6	20.1	18.5
25.5	22.5	20.0	27.8	13.2	18.4	21.6	23.0	24.5	28.0	17.4
15.2	25.0									

At $\alpha = 0.03$, can you support the sociologist's claim? *(Adapted from University of Michigan's Institute for Social Research)*

DATA **21.** ***Washer Diameters*** A production engineer claims that there is no difference in the mean washer diameter manufactured by two different methods. The first method produces washers with the following diameters (in inches).

0.861	0.864	0.882	0.887	0.858	0.879	0.887	0.876	0.870
0.894	0.884	0.882	0.869	0.859	0.887	0.875	0.863	0.887
0.882	0.862	0.906	0.880	0.877	0.864	0.873	0.860	0.866
0.869	0.877	0.863	0.875	0.883	0.872	0.879	0.861	

The second method produces washers with these diameters (in inches).

0.705	0.703	0.715	0.711	0.690	0.720	0.702	0.686	0.704
0.712	0.718	0.695	0.708	0.695	0.699	0.715	0.691	0.696
0.680	0.703	0.697	0.694	0.714	0.694	0.672	0.688	0.700
0.715	0.709	0.698	0.696	0.700	0.706	0.695	0.715	

At $\alpha = 0.01$, can you reject the production engineer's claim?

DATA **22.** ***Nut Diameters*** A production engineer claims that there is no difference in the mean nut diameter manufactured by two different methods. The first method produces nuts with the following diameters (in centimeters).

3.330	3.337	3.329	3.354	3.325	3.343	3.333	3.347	3.332	3.358
3.353	3.335	3.341	3.331	3.327	3.326	3.337	3.336	3.323	3.347
3.329	3.345	3.329	3.338	3.353	3.339	3.338	3.338	3.350	3.320
3.364	3.340	3.348	3.339	3.336	3.321	3.316	3.352	3.320	3.336

The second method produces nuts with these diameters (in centimeters).

3.513	3.490	3.498	3.504	3.483	3.512	3.494	3.514	3.495	3.489
3.493	3.499	3.497	3.495	3.496	3.485	3.506	3.517	3.484	3.498
3.522	3.505	3.501	3.491	3.500	3.499	3.475	3.486	3.501	3.496
3.504	3.513	3.511	3.501	3.487	3.508	3.515	3.505	3.496	3.505

At $\alpha = 0.04$, can you reject the production engineer's claim?

23. ***Getting at the Concept*** Explain why the null hypothesis $H_0: \mu_1 = \mu_2$ is equivalent to the null hypothesis $H_0: \mu_1 - \mu_2 = 0$.

24. ***Getting at the Concept*** Explain why the null hypothesis $H_0: \mu_1 \geq \mu_2$ is equivalent to the null hypothesis $H_0: \mu_1 - \mu_2 \geq 0$.

Extending Concepts

Testing a Difference Other Than Zero Sometimes a researcher is interested in testing a difference in means other than zero. For instance, you may want to determine if children today spend an average of nine hours a week more in day care (or preschool) than children did 20 years ago. In Exercises 25–28, you will test the difference between two means using a null hypothesis of $H_0\colon \mu_1 - \mu_2 = k$, $H_0\colon \mu_1 - \mu_2 \geq k$, or $H_0\colon \mu_1 - \mu_2 \leq k$. The formula for the z-test is still

$$z = \frac{(\bar{x}_1 - \bar{x}_2) - (\mu_1 - \mu_2)}{\sigma_{\bar{x}_1 - \bar{x}_2}} \quad \text{where} \quad \sigma_{\bar{x}_1 - \bar{x}_2} = \sqrt{\frac{\sigma_1^2}{n_1} + \frac{\sigma_2^2}{n_2}}.$$

25. ***Time in Day Care or Preschool*** In 1981, a study of 70 randomly selected children (under 3 years old) found that the mean length of time spent in day care or preschool per week was 11.5 hours with a standard deviation of 3.8 hours. A recent study of 65 randomly selected children (under 3 years old) found that the mean length of time spent in day care or preschool per week was 20 hours and the standard deviation was 6.7 hours. At $\alpha = 0.01$, test the claim that children spend 9 hours a week more in day care or preschool today than they did in 1981. *(Adapted from University of Michigan's Institute for Social Research)*

26. ***Time Watching TV*** A recent study of 48 randomly selected children (ages 6–8) found that the mean length of time spent watching television each week was 12.63 hours and the standard deviation was 4.21 hours. The mean time 56 randomly selected children (ages 9–12) watched television each week was 13.60 hours and the standard deviation was 4.53 hours. At $\alpha = 0.05$, test the claim that the mean time per week children ages 6–8 watch television is 1 hour less than that of children ages 9–12. *(Adapted from University of Michigan's Institute for Social Research)*

27. ***Statistician Salaries*** Is the difference between the mean annual salaries of statisticians in Massachusetts and New York more than \$6000? To decide, you select a random sample of statisticians from each state. The results of each survey are shown in the figure. At $\alpha = 0.10$, what should you conclude? *(Adapted from America's Career InfoNet)*

Statisticians in Massachusetts

$\bar{x}_1 = \$55{,}900$
$s_1 = \$8875$
$n_1 = 45$

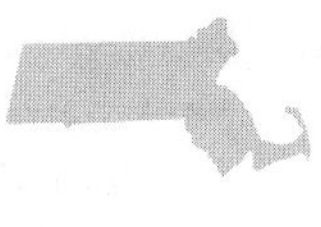

Statisticians in New York

$\bar{x}_2 = \$54{,}200$
$s_2 = \$9175$
$n_2 = 37$

28. ***Photographer and Computer Programmer Salaries*** At $\alpha = 0.05$, test the claim that the difference between the mean salary for computer programmers and the mean salary for photographers in Wisconsin is less than \$30,000. The results of a survey of randomly selected computer programmers and photographers in Wisconsin are shown in the figure. *(Adapted from America's Career InfoNet)*

Computer Programmers in Wisconsin

$\bar{x}_1 = \$52{,}300$
$s_1 = \$8250$
$n_1 = 31$

Photographers in Wisconsin

$\bar{x}_2 = \$25{,}300$
$s_2 = \$3200$
$n_2 = 33$

25. $H_0\colon \mu_1 - \mu_2 = -9$ (claim)
$H_a\colon \mu_1 - \mu_2 \neq -9$
Fail to reject H_0. There is not enough evidence at the 1% level to reject the claim that children spend 9 hours a week more in day care or preschool.

26. $H_0\colon \mu_1 - \mu_2 = -1$ (claim)
$H_a\colon \mu_1 - \mu_2 \neq -1$
Fail to reject H_0. There is not enough evidence at the 5% level to reject the claim that the mean time per week children ages 6–8 watch television is 1 hour less than that of children ages 9–12.

27. $H_0\colon \mu_1 - \mu_2 \leq 6000$
$H_a\colon \mu_1 - \mu_2 > 6000$ (claim)
Fail to reject H_0. There is not enough evidence at the 10% level to support the claim that the difference in mean annual salaries of statisticians is more than \$6000.

28. $H_0\colon \mu_1 - \mu_2 \geq 30{,}000$
$H_a\colon \mu_1 - \mu_2 < 30{,}000$ (claim)
Reject H_0. There is sufficient evidence at the 5% level to support the claim that the difference in mean annual salaries is less than \$30,000.

Constructing Confidence Intervals for $\mu_1 - \mu_2$ You can construct a confidence interval for the difference in two population means $\mu_1 - \mu_2$ by using the following if $n_1 \geq 30$ and $n_2 \geq 30$, or both populations are normally distributed. Also, the samples must be randomly selected and independent.

$$(\bar{x}_1 - \bar{x}_2) - z_c\sqrt{\frac{\sigma_1^2}{n_1} + \frac{\sigma_2^2}{n_2}} < \mu_1 - \mu_2 < (\bar{x}_1 - \bar{x}_2) + z_c\sqrt{\frac{\sigma_1^2}{n_1} + \frac{\sigma_2^2}{n_2}}$$

In Exercises 29 and 30, construct the indicated confidence interval for $\mu_1 - \mu_2$.

29. ***Herbal Supplement and Weight Loss*** A study was conducted to see if an herbal supplement helped to reduce weight. After 12 weeks, 42 subjects using the herbal supplement and a high-fiber, low-calorie diet had a mean weight loss of 3.2 kilograms and a standard deviation of 3.3 kilograms. During the same time period, 42 subjects using a placebo and a high-fiber, low-calorie diet had a mean weight loss of 4.1 kilograms and a standard deviation of 3.9 kilograms. Construct a 95% confidence interval for $\mu_1 - \mu_2$, where μ_1 is the mean weight loss for the group using the herbal supplement and μ_2 is the mean weight loss for the group using the placebo. *(Source: The Journal of the American Medical Association)*

30. ***Comparing Cancer Drugs*** Two groups of patients with colorectal cancer are treated with a different drug. Group A's 140 patients are treated using the drug Irinotecan and Group B's 127 patients are treated using the drug Fluorouracil. The mean number of months that Group A reported no cancer-related pain was 10.3 and the standard deviation was 1.2. The mean number of months that Group B reported no cancer-related pain was 8.5 and the standard deviation was 1.5. Construct a 95% confidence interval for $\mu_1 - \mu_2$, where μ_1 is the mean number of months that Group A reported no cancer-related pain and μ_2 is the mean number of months that Group B reported no cancer-related pain. *(Adapted from The Lancet)*

31. ***Make a Decision*** Refer to the study in Exercise 29. At $\alpha = 0.05$, test the claim that the mean weight loss for the group using the herbal supplement is greater than the mean weight loss for the group using the placebo. Would you recommend using the herbal supplement with a high-fiber, low-calorie diet to lose weight? Explain your reasoning.

32. ***Make a Decision*** Refer to the study in Exercise 30. At $\alpha = 0.05$, test the claim that the mean number of months of cancer-related pain relief with Irinotecan is greater than the mean number of months of cancer-related pain relief obtained with Fluorouracil. Would you recommend using Irinotecan over Fluorouracil to relieve cancer-related pain? Explain your reasoning.

33. ***Getting at the Concept*** Compare the confidence interval you constructed in Exercise 29 with the hypothesis test result in Exercise 31. Explain why you would fail to reject the null hypothesis if the confidence interval contains 0.

34. ***Getting at the Concept*** Compare the confidence interval you constructed in Exercise 30 with the hypothesis test result in Exercise 32. Explain why you would reject the null hypothesis if the confidence interval contains only positive numbers.

---

29. $-2.45 < \mu_1 - \mu_2 < 0.65$

30. $1.47 < \mu_1 - \mu_2 < 2.13$

31. There is not enough evidence at the 5% level to support the claim. You should not recommend using the herbal supplement with a high-fiber, low-calorie diet to lose weight because there was not a significant difference between the weight loss it produced and that of the placebo.

32. There is enough evidence at the 5% level to support the claim. You should recommend using Irinotecan over Fluorouracil because the average number of months with no reported cancer-related pain was significantly higher.

33. The 95% CI for $\mu_1 - \mu_2$ in Exercise 29 contained values less than or equal to zero and, as found in Exercise 31, there was not enough evidence at the 5% level of significance to support the claim.

    If zero is contained in the CI for $\mu_1 - \mu_2$, you fail to reject H_0 because the null hypothesis states that $\mu_1 - \mu_2$ is less than or equal to zero.

34. The 95% CI for $\mu_1 - \mu_2$ in Exercise 30 contained only values greater than zero and, as found in Exercise 32, there was sufficient evidence at the 5% level of significance to support the claim.

    If the CI for $\mu_1 - \mu_2$ contains only positive numbers, you reject H_0 because the null hypothesis states that $\mu_1 - \mu_2$ is less than or equal to zero, i.e., $\mu_1 - \mu_2$ is not positive.

Case Study

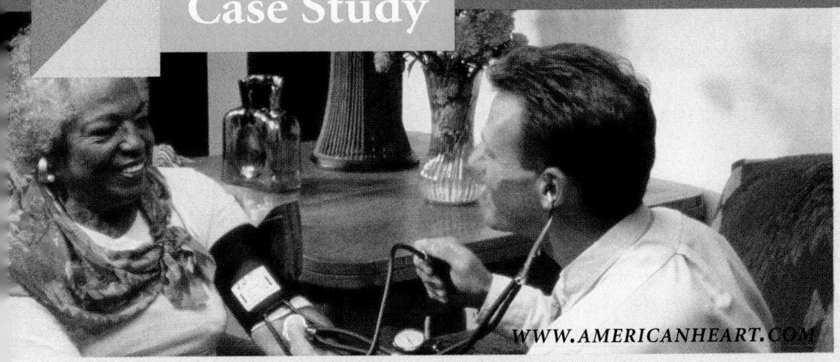
WWW.AMERICANHEART.COM

DASH Diet and Blood Pressure

A study in *The Journal of the American Medical Association* describes the effects of comprehensive lifestyle modification on blood pressure control. In the study, 810 men and women were randomly divided into three groups. The first group, the "advice only" group, received only advice and printed educational materials about the effects of weight, sodium intake, physical activity, and diet on blood pressure. The second group, the "established" group, participated in a traditional diet and exercise program. The third group, the "established plus DASH" group, participated in a specific diet and exercise program, called the Dietary Approaches to Stop Hypertension (DASH) program, which emphasizes the consumption of fruits, vegetables, and low-fat dairy products.

Before the study, each group had a mean systolic blood pressure of about 135 mm Hg and a mean diastolic blood pressure of about 85 mm Hg. The mean systolic and diastolic blood pressures of each group after the six-month study are shown in the table at the right.

	Established plus DASH $n_1 = 269$	Established $n_1 = 268$	Advice only $n_1 = 273$
Systolic blood pressure	$\bar{x}_1 = 123.9$ $s_1 = 9.9$	$\bar{x}_2 = 124.5$ $s_2 = 10.1$	$\bar{x}_3 = 128.4$ $s_3 = 9.2$
Diastolic blood pressure	$\bar{x}_1 = 78.6$ $s_1 = 6.8$	$\bar{x}_2 = 79.5$ $s_2 = 6.7$	$\bar{x}_3 = 81.2$ $s_3 = 6.3$

Blood pressure is measured in millimeters of mercury (mm Hg).

Exercises

In Exercises 1–4, perform a two-sample z-test to determine whether the mean blood pressures of the two indicated groups are different. For each exercise, write your conclusion as a sentence. Use $\alpha = 0.05$.

1. Test the systolic blood pressures of people in the established plus DASH group against those in the established group.

2. Test the systolic blood pressures of people in the established plus DASH group against those in the advice only group.

3. Test the diastolic blood pressures of people in the established plus DASH group against those in the established group.

4. Test the diastolic blood pressures of people in the established plus DASH group against those in the advice only group.

5. Suppose that in another study, 100 people were given a blood pressure-reducing medication. The mean systolic blood pressure before the study was 145 mm Hg and the mean diastolic blood pressure was 92 mm Hg. After the study, the mean blood pressures were 122 mm Hg and 83 mm Hg.

(a) Test the mean systolic blood pressure of this group with the established plus DASH group described previously. Use $\alpha = 0.01$ and assume $s = 9$ mm Hg for the medication group.

(b) Test the mean diastolic blood pressure of this group with the established plus DASH group described previously. Use $\alpha = 0.01$ and assume $s = 6$ mm Hg for the medication group.

6. In which comparisons in Exercises 1–5 did you find a difference in blood pressures? Write a summary of your findings.

8.2 Testing the Difference Between Means (Small Independent Samples)

What You Should Learn

- How to perform a *t*-test for the difference between two population means μ_1 and μ_2 using small independent samples

The Two-Sample *t*-Test for the Difference Between Means

The Two-Sample *t*-Test for the Difference Between Means

As you have learned, in real life, it is often not practical to collect samples of size 30 or more from each of two populations. However, if both populations have a normal distribution, you can still test the difference between their means. In this section, you will learn how to use a *t*-test to test the difference between two population means μ_1 and μ_2 using a sample from each population. The following conditions are necessary to use a *t*-test for small independent samples.

1. The samples must be randomly selected.
2. The samples must be independent. Recall that two samples are *independent* if the sample selected from one population is not related to the sample selected from the second population.
3. Each population must have a normal distribution.

When these conditions are met, the sampling distribution for the difference between the sample means $\bar{x}_1 - \bar{x}_2$ is approximated by a *t*-distribution with mean $\mu_1 - \mu_2$. So, you can use a two-sample *t*-test to test the difference between the two population means μ_1 and μ_2. The standard error and the degrees of freedom of the sampling distribution depend on whether the population variances σ_1^2 and σ_2^2 are equal.

Study Tip

You will learn to test for differences in variances in two populations in Chapter 10. In this chapter, each example and exercise will state whether the variances are equal.

Two-Sample *t*-Test for the Difference Between Means

A **two-sample *t*-test** is used to test the difference between two population means μ_1 and μ_2 when a sample is randomly selected from each population. Performing this test requires each population to be normally distributed, and the samples should be independent. The standardized test statistic is

$$t = \frac{(\bar{x}_1 - \bar{x}_2) - (\mu_1 - \mu_2)}{\sigma_{\bar{x}_1 - \bar{x}_2}}.$$

If the population variances are equal, then information from the two samples is combined to calculate a **pooled estimate of the standard deviation $\hat{\sigma}$.**

$$\hat{\sigma} = \sqrt{\frac{(n_1 - 1)s_1^2 + (n_2 - 1)s_2^2}{n_1 + n_2 - 2}}$$

The standard error for the sampling distribution of $\bar{x}_1 - x_2$ is

$$\sigma_{\bar{x}_1 - \bar{x}_2} = \hat{\sigma} \cdot \sqrt{\frac{1}{n_1} + \frac{1}{n_2}} \qquad \text{Variances equal}$$

and d.f. $= n_1 + n_2 - 2$.

If the population variances are not equal, then the standard error is

$$\sigma_{\bar{x}_1 - \bar{x}_2} = \sqrt{\frac{s_1^2}{n_1} + \frac{s_2^2}{n_2}} \qquad \text{Variances not equal}$$

and d.f. = smaller of $n_1 - 1$ or $n_2 - 1$.

Note to Instructor

If you choose, you can cover tests for equal variances (Section 10.3) before doing these *t*-tests. In case you cover Section 10.3 later or do not have time to cover it at all, students will be informed whether to assume equal variances in each example, Try-It, and exercise in this section. In each instance, we have run an *F*-test at the level of significance stated in the problem and have reported the outcome of the test.

Picturing the World

A study published by the American Psychological Association in the journal *Neuropsychology* reported that children with musical training showed better verbal memory than children with no musical training. The study also showed that the longer the musical training, the better the verbal memory. Suppose you tried to duplicate the results as follows. A verbal memory test with a possible 100 points was administered to 90 children. Half had musical training, while the other half acted as the control group and had no musical training. The 45 children with training had an average score of 83.12 with a standard deviation of 5.7. The 45 students in the control group had an average score of 79.9 with a standard deviation of 6.2.

At $\alpha = 0.05$, is there enough evidence to support the claim that musical training increases verbal memory test scores? Assume the populations are normally distributed and the population variances are equal.

The requirements for the z-test described in Section 8.1 and the t-test described in this section are shown in the flowchart below.

Two-Sample Tests for Independent Samples

Are both sample sizes at least 30? — Yes → Use the z-test.

No ↓

Are both populations normal? — No → You cannot use the z-test or the t-test.

Yes ↓

Are both population standard deviations known? — Yes ↓ Use the z-test.

No → Are the population variances equal?

Yes → Use the t-test with $\sigma_{\bar{x}_1 - \bar{x}_2} = \hat{\sigma}\sqrt{\frac{1}{n_1} + \frac{1}{n_2}}$ and d.f. $= n_1 + n_2 - 2$.

No ↓ Use the t-test with $\sigma_{\bar{x}_1 - \bar{x}_2} = \sqrt{\frac{s_1^2}{n_1} + \frac{s_2^2}{n_2}}$ and d.f. = smaller of $n_1 - 1$ or $n_2 - 1$.

GUIDELINES

Using a Two-Sample t-Test for the Difference Between Means (Small Independent Samples)

In Words	*In Symbols*
1. State the claim. Identify the null and alternative hypotheses.	State H_0 and H_a.
2. Specify the level of significance.	Identify α.
3. Identify the degrees of freedom and sketch the sampling distribution.	d.f. $= n_1 + n_2 - 2$ or d.f. $=$ smaller of $n_1 - 1$ or $n_2 - 1$
4. Determine the critical value(s).	Use Table 5 in Appendix B.
5. Determine the rejection region(s).	
6. Find the standardized test statistic.	$t = \frac{(\bar{x}_1 - \bar{x}_2) - (\mu_1 - \mu_2)}{\sigma_{\bar{x}_1 - \bar{x}_2}}$
7. Make a decision to reject or fail to reject the null hypothesis.	If t is in the rejection region, reject H_0. Otherwise, fail to reject H_0.
8. Interpret the decision in the context of the original claim.	

EXAMPLE 1

A Two-Sample t-Test for the Difference Between Means

See MINITAB steps on page 454.

Sample Statistics for Stopping Distances under Winter Conditions

Winterfire	Alpin
$\bar{x}_1 = 51$ ft	$\bar{x}_2 = 55$ ft
$s_1 = 8$ ft	$s_2 = 3$ ft
$n_1 = 10$	$n_2 = 12$

Consumer Reports tested the stopping distances (in feet) of 10 Firestone Winterfire tires and 12 Michelin XM+S Alpin tires when traveling on ice at 15 miles per hour. The results are shown at the left. Can you conclude that there is a difference in the mean stopping distances of the two types of tires? Use $\alpha = 0.01$. Assume the populations are normally distributed and the population variances are not equal.

SOLUTION You want to test whether the mean stopping distances are different. So, the null and alternative hypotheses are

$$H_0\colon \mu_1 = \mu_2 \quad \text{and} \quad H_a\colon \mu_1 \neq \mu_2. \text{ (Claim)}$$

Because the variances are not equal and the smaller sample size is 10, use d.f. = 10 − 1 = 9. Because the test is a two-tailed test with d.f. = 9 and $\alpha = 0.01$, the critical values are $-t_0 = -3.250$ and $t_0 = 3.250$. The rejection regions are $t < -3.250$ and $t > 3.250$. The standard error is

$$\sigma_{\bar{x}_1-\bar{x}_2} = \sqrt{\frac{s_1^2}{n_1} + \frac{s_2^2}{n_2}} = \sqrt{\frac{8^2}{10} + \frac{3^2}{12}} \approx 2.674.$$

The standardized test statistic is

$$t = \frac{(\bar{x}_1 - \bar{x}_2) - (\mu_1 - \mu_2)}{\sigma_{\bar{x}_1-\bar{x}_2}} \quad \text{Use the } t\text{-test.}$$

$$\approx \frac{(51 - 55) - 0}{2.674} \quad \text{Assume } \mu_1 = \mu_2 \text{, so } \mu_1 - \mu_2 = 0.$$

$$\approx -1.496.$$

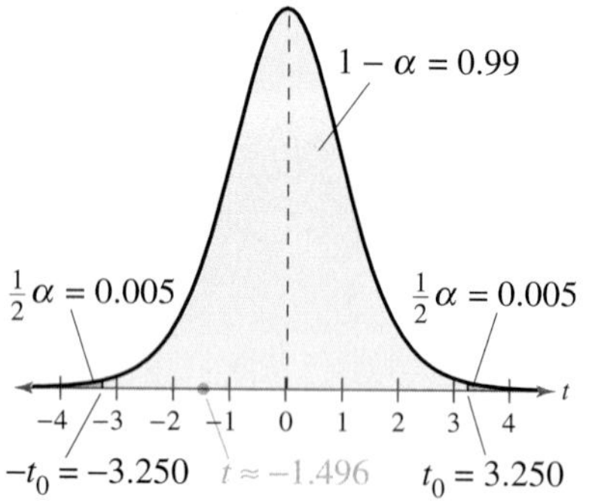

The graph at the left shows the location of the rejection regions and the standardized test statistic t. Because t is not in a rejection region, you should fail to reject the null hypothesis.

Interpretation At the 1% level, there is not enough evidence to conclude that the mean stopping distances of the tires are different.

Try It Yourself 1

Sample Statistics for Stopping Distances on Wet Pavement

Winterfire	Alpin
$\bar{x}_1 = 102$ ft	$\bar{x}_2 = 94$ ft
$s_1 = 10$ ft	$s_2 = 4$ ft
$n_1 = 10$	$n_2 = 12$

Consumer Reports tested the stopping distances (in feet) of the same tires from Example 1 when traveling on wet pavement at 40 miles per hour. The results are shown at the left. Can you conclude that the mean stopping distances are different? Use $\alpha = 0.05$. Assume the populations are normally distributed and the population variances are not equal.

a. *Identify* the claim and state H_0 and H_a.
b. *Specify* the level of significance α.
c. *Determine* the degrees of freedom.
d. *Find* the critical values and *identify* the rejection regions.
e. *Use* the t-test to find the standardized test statistic t.
f. *Decide* whether to reject the null hypothesis. Use a graph if necessary.
g. *Interpret* the decision in the context of the original claim.

Answer: Page A41

EXAMPLE 2

A Two-Sample t-Test for the Difference Between Means

See TI-83 steps on page 455.

Sample Statistics for Calling Range

Manufacturer	Competition
$\bar{x}_1 = 1275$ ft	$\bar{x}_2 = 1250$ ft
$s_1 = 45$ ft	$s_2 = 30$ ft
$n_1 = 14$	$n_2 = 16$

A manufacturer claims that the calling range (in feet) of its 2.4-GHz cordless telephone is greater than that of its leading competitor. You perform a study using 14 randomly selected phones from the manufacturer and 16 randomly selected similar phones from its competitor. The results are shown at the left. At $\alpha = 0.05$, can you support the manufacturer's claim? Assume the populations are normally distributed and the population variances are equal.

SOLUTION The claim is "the mean range of our cordless phone is greater than the mean range of yours." So, the null and alternative hypotheses are

$$H_0: \mu_1 \leq \mu_2 \quad \text{and} \quad H_a: \mu_1 > \mu_2. \text{ (Claim)}$$

Because the variances are equal, d.f. $= n_1 + n_2 - 2 = 14 + 16 - 2 = 28$. Because the test is a right-tailed test, d.f. $= 28$, and $\alpha = 0.05$, the critical value is $t_0 = 1.701$. The rejection region is $t > 1.701$. The standard error is

$$\sigma_{\bar{x}_1-\bar{x}_2} = \sqrt{\frac{(n_1-1)s_1^2 + (n_2-1)s_2^2}{n_1+n_2-2}} \cdot \sqrt{\frac{1}{n_1} + \frac{1}{n_2}}$$

$$= \sqrt{\frac{(13)(45^2) + (15)(30^2)}{14+16-2}} \cdot \sqrt{\frac{1}{14} + \frac{1}{16}} \approx 13.802.$$

The standardized test statistic is

$$t = \frac{(\bar{x}_1 - \bar{x}_2) - (\mu_1 - \mu_2)}{\sigma_{\bar{x}_1-\bar{x}_2}} \qquad \text{Use the } t\text{-test.}$$

$$\approx \frac{(1275 - 1250) - 0}{13.802} \qquad \text{Assume } \mu_1 = \mu_2, \text{ so } \mu_1 - \mu_2 = 0.$$

$$\approx 1.811.$$

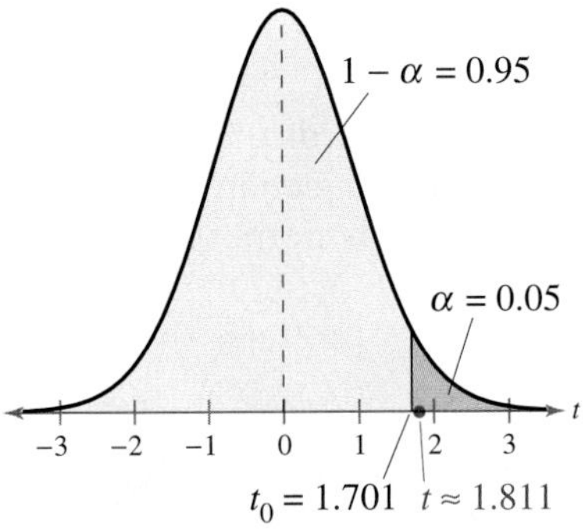

The graph at the left shows the location of the rejection region and the standardized test statistic t. Because t is in the rejection region, you should decide to reject the null hypothesis.

Interpretation At the 5% level, there is enough evidence to support the manufacturer's claim that its phone has a greater calling range than its competitor's.

Try It Yourself 2

Sample Statistics for Watt Usage

Manufacturer	Competition
$\bar{x}_1 = 73$	$\bar{x}_2 = 74$
$s_1 = 2.4$	$s_2 = 3.2$
$n_1 = 12$	$n_2 = 15$

A manufacturer claims that the watt usage of its 17-inch flat panel monitors is less than that of its leading competitor. You perform a study and obtain the results shown at the left. At $\alpha = 0.10$, is there enough evidence to support the manufacturer's claim? Assume the populations are normally distributed and the population variances are equal.

a. *Identify* the claim and state H_0 and H_a.
b. *Specify* the level of significance α.
c. *Determine* the degrees of freedom.
d. *Find* the critical value and *identify* the rejection region.
e. *Use* the t-test to find the standardized test statistic t.
f. *Decide* whether to reject the null hypothesis. Use a graph if necessary.
g. *Interpret* the decision in the context of the original claim. *Answer: Page A41*

8.2 Exercises

Building Basic Skills and Vocabulary

1. Explain how to perform a two-sample t-test for the difference between the means of two populations.

2. What conditions are necessary in order to use a t-test to test the difference between two population means?

In Exercises 3–10, use Table 5 in Appendix B to find the critical value(s) for the indicated alternative hypothesis, level of significance α, and sample sizes n_1 and n_2. Assume that the samples are independent, normal, and random and that the population variances are (a) equal and (b) not equal.

3. H_a: $\mu_1 \neq \mu_2$, $\alpha = 0.10$, $n_1 = 10$, $n_2 = 12$

4. H_a: $\mu_1 > \mu_2$, $\alpha = 0.01$, $n_1 = 12$, $n_2 = 15$

5. H_a: $\mu_1 < \mu_2$, $\alpha = 0.025$, $n_1 = 15$, $n_2 = 9$

6. H_a: $\mu_1 \neq \mu_2$, $\alpha = 0.05$, $n_1 = 19$, $n_2 = 22$

7. H_a: $\mu_1 > \mu_2$, $\alpha = 0.05$, $n_1 = 13$, $n_2 = 8$

8. H_a: $\mu_1 < \mu_2$, $\alpha = 0.10$, $n_1 = 7$, $n_2 = 2$

9. H_a: $\mu_1 \neq \mu_2$, $\alpha = 0.01$, $n_1 = 12$, $n_2 = 17$

10. H_a: $\mu_1 > \mu_2$, $\alpha = 0.005$, $n_1 = 5$, $n_2 = 11$

In Exercises 11–14, (a) find the test statistic, (b) find the standardized test statistic, (c) decide whether the standardized test statistic is in the rejection region, and (d) decide whether you should reject or fail to reject the null hypothesis.

11. H_0: $\mu_1 = \mu_2$, $\alpha = 0.01$
Sample statistics: $\bar{x}_1 = 33.7$, $s_1 = 3.5$, $n_1 = 10$ and $\bar{x}_2 = 35.5$, $s_2 = 2.2$, $n_2 = 7$
Assume $\sigma_1^2 = \sigma_2^2$.

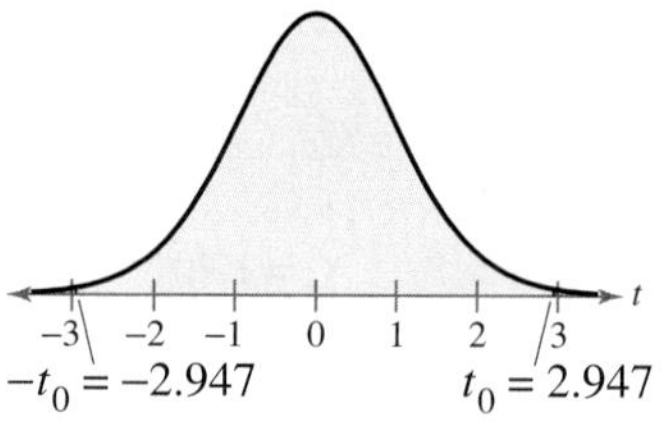

12. H_0: $\mu_1 \geq \mu_2$, $\alpha = 0.10$
Sample statistics: $\bar{x}_1 = 0.515$, $s_1 = 0.305$, $n_1 = 11$ and $\bar{x}_2 = 0.475$, $s_2 = 0.215$, $n_2 = 9$
Assume $\sigma_1^2 = \sigma_2^2$.

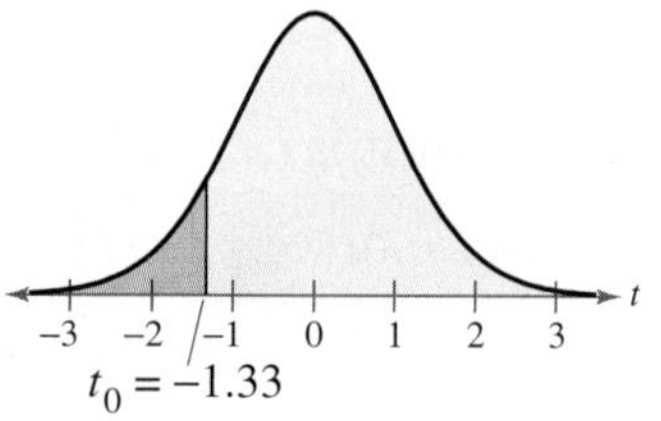

13. H_0: $\mu_1 \leq \mu_2$, $\alpha = 0.05$
Sample statistics: $\bar{x}_1 = 2250$, $s_1 = 175$, $n_1 = 13$ and $\bar{x}_2 = 2305$, $s_2 = 52$, $n_2 = 10$
Assume $\sigma_1^2 \neq \sigma_2^2$.

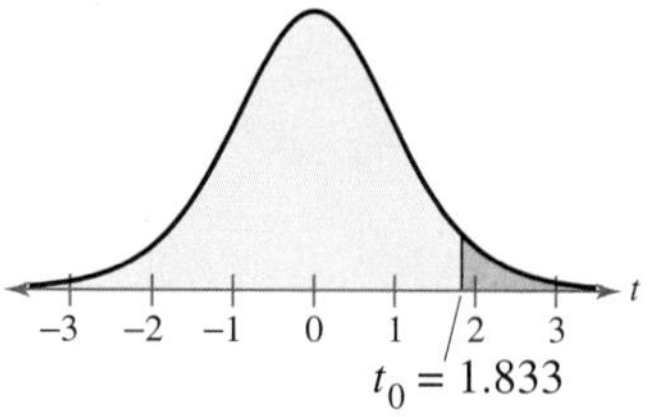

Help

MyMathLab

Student Study Pack

1. State hypotheses and identify the claim. Specify the level of significance. Determine the degrees of freedom. Find the critical value(s) and identify the rejection region(s). Find the standardized test statistic. Make a decision and interpret in the context of the original claim.

2. (1) The samples must be independent.
(2) Each population must have a normal distribution.
(3) The size of at least one of the samples is less than 30.

3. (a) $t_0 = \pm 1.725$
(b) $t_0 = \pm 1.833$

4. (a) $t_0 = 2.485$
(b) $t_0 = 2.718$

5. (a) $t_0 = -2.074$
(b) $t_0 = -2.306$

6. (a) $t_0 = \pm 1.96$
(b) $t_0 = \pm 2.101$

7. (a) $t_0 = 1.729$
(b) $t_0 = 1.895$

8. (a) $t_0 = -1.415$
(b) $t_0 = -3.078$

9. (a) $t_0 = \pm 2.771$
(b) $t_0 = \pm 3.106$

10. (a) $t_0 = 2.977$
(b) $t_0 = 4.604$

11. (a) -1.8 (b) -1.199
(c) t is not in the rejection region.
(d) Fail to reject H_0. There is not enough evidence at the 1% level to reject the claim.

12. (a) 0.04 (b) 0.331

(c) t is not in the rejection region.

(d) Fail to reject H_0. There is not enough at the 10% level evidence to reject the claim.

13. There is no need to run the test because this is a right-tailed test and the test statistic is negative. It is obvious that the standardized test statistic will also be negative and fall outside of the rejection region. So, the decision is to fail to reject H_0.

14. (a) -5 (b) -4.025

(c) t is not in the rejection region.

(d) Fail to reject H_0. There is not enough at the 1% level evidence to reject the claim.

15. (a) H_0: $\mu_1 = \mu_2$ (claim)
H_a: $\mu_1 \neq \mu_2$

(b) $-t_0 = -1.706$ and $t_0 = 1.706$

Rejection regions: $t < -1.706$, $t > 1.706$

(c) 0.876 (d) Fail to reject H_0.

(e) There is not enough evidence at the 10% level to reject the claim that the mean footwell intrusions for small and midsize cars are equal.

16. (a) H_0: $\mu_1 = \mu_2$ (claim)
H_a: $\mu_1 \neq \mu_2$

(b) $-t_0 = -2.201$ and $t_0 = 2.201$

Rejection regions: $t < -2.201$, $t > 2.201$

(c) 2.145 (d) Fail to reject H_0.

(e) There is not enough evidence at the 5% level to reject the claim that the mean footwell intrusion for small pickups and utility vehicles are equal.

17. (a) H_0: $\mu_1 \geq \mu_2$
H_a: $\mu_1 < \mu_2$ (claim)

(b) $t_0 = -1.350$

Rejection region: $t < -1.350$

(c) -4.038 (d) Reject H_0.

(e) There is enough evidence at the 10% level to support the claim that mean bumper repair cost is less for small cars than it is for midsize cars.

18. See Selected Answers, page A97.

14. H_0: $\mu_1 \leq \mu_2$, $\alpha = 0.01$
Sample statistics: $\overline{x}_1 = 45$, $s_1 = 4.8$,
$n_1 = 16$ and $\overline{x}_2 = 50$, $s_2 = 1.2$, $n_2 = 14$
Assume $\sigma_1^2 \neq \sigma_2^2$.

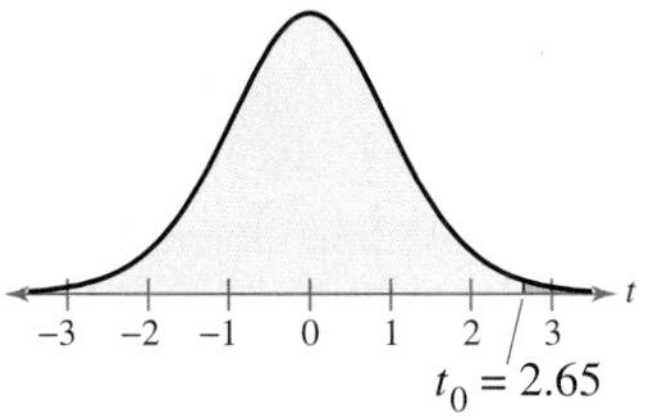

Using and Interpreting Concepts

Testing the Difference Between Two Means In Exercises 15–24, (a) identify the claim and state H_0 and H_a, (b) find the critical value(s) and identify the rejection region(s), (c) find the standardized test statistic, (d) decide whether to reject or fail to reject the null hypothesis, and (e) interpret the decision in the context of the original claim. Assume the populations have approximately normal distributions.

15. ***Footwell Intrusion*** An insurance actuary claims that the mean footwell intrusions for small and midsize cars are equal. Crash tests at 40 miles per hour were performed on 13 randomly selected small cars and 15 randomly selected midsize cars. The amount that the footwell intruded on the driver's legs was measured. The mean footwell intrusion for the small cars was 15.1 centimeters and the standard deviation was 7.55 centimeters. The mean footwell intrusion for the midsize cars was 12.8 centimeters and the standard deviation was 6.35 centimeters. At $\alpha = 0.10$, can you reject the insurance actuary's claim? Assume the population variances are equal. *(Adapted from Insurance Institute for Highway Safety)*

16. ***Footwell Intrusion*** The footwell intrusions for small pickups and utility vehicles in several crash tests at 40 miles per hour were compared. For the five randomly selected pickups that were crashed, the mean footwell intrusion was 20.1 centimeters with a standard deviation of 5.03 centimeters. For the eight randomly selected utility vehicles that were crashed, the mean footwell intrusion was 12.0 centimeters with a standard deviation of 7.38 centimeters. At $\alpha = 0.05$, can you reject the claim that the mean footwell intrusions for these vehicle types are equal? Assume the population variances are equal. *(Adapted from Insurance Institute for Highway Safety)*

17. ***Repair Costs: Bumpers*** In crash tests at 5 miles per hour, the mean bumper repair cost for 14 randomly selected small cars is \$473 with a standard deviation of \$190. In similar tests of 23 randomly selected midsize cars, the mean bumper repair cost is \$741 with a standard deviation of \$205. At $\alpha = 0.10$, can you conclude that the mean bumper repair cost is less for small cars than it is for midsize cars? Assume that the population variances are not equal. *(Adapted from Insurance Institute for Highway Safety)*

18. ***Repair Costs: Bumpers*** Crash tests at 5 miles per hour were performed on five randomly selected small pickups and eight randomly selected small utility vehicles. For the small pickups, the mean bumper repair cost was \$1090 and the standard deviation was \$403. For the small utility vehicles, the mean bumper repair cost was \$485 and the standard deviation was \$382. At $\alpha = 0.10$, can you conclude that the mean bumper repair cost is greater for small pickups than for small utility vehicles? Assume the population variances are equal. *(Adapted from Insurance Institute for Highway Safety)*

19. (a) H_0: $\mu_1 \leq \mu_2$
H_a: $\mu_1 > \mu_2$ (claim)

(b) $t_0 = 1.345$

Rejection region: $t > 1.345$

(c) 1.440 (d) Reject H_0.

(e) There is enough evidence at the 10% level to support the claim that the mean annual wage is greater in Allegheny County than it is in Erie County.

20. (a) H_0: $\mu_1 = \mu_2$
H_a: $\mu_1 \neq \mu_2$ (claim)

(b) $-t_0 = -2.576$ and $t_0 = 2.576$

Rejection regions: $t < -2.576$, $t > 2.576$

(c) 0.273 (d) Fail to reject H_0.

(e) There is not enough evidence at the 1% level to support the claim that the mean annual incomes in Delaware and Greene Counties are not the same.

21. (a) H_0: $\mu_1 = \mu_2$
H_a: $\mu_1 \neq \mu_2$ (claim)

(b) $-t_0 = -2.831$ and $t_0 = 2.831$

Rejection regions: $t < -2.831$, $t > 2.831$

(c) -6.410 (d) Reject H_0.

(e) There is enough evidence at the 1% level to support the claim that the new treatment makes a difference in the tensile strength of steel bars.

22. (a) H_0: $\mu_1 \leq \mu_2$
H_a: $\mu_1 > \mu_2$ (claim)

(b) $t_0 = 1.350$

Rejection region: $t > 1.350$

(c) 3.429 (d) Reject H_0.

(e) There is enough evidence at the 10% level to support the claim and to recommend using the experimental method.

19. ***Annual Wages*** A personnel director claims that the mean annual wage is greater in Allegheny County than it is in Erie County. In Allegheny County, a random sample of 19 residents has a mean annual income of \$41,300 and a standard deviation of \$8600. In Erie County, a random sample of 15 residents has a mean annual income of \$37,800 and a standard deviation of \$5500. At $\alpha = 0.10$, can you support the personnel director's claim? Assume the population variances are not equal. *(Adapted from U.S. Census Bureau, Housing and Household Economics Statistics Division)*

20. ***Annual Wages*** A random sample of 17 residents of Delaware County in Pennsylvania has a mean annual income of \$35,800 and a standard deviation of \$7800. In Greene County in Pennsylvania, a random sample of 18 residents has a mean annual income of \$35,100 and a standard deviation of \$7375. Test the claim at $\alpha = 0.01$ that the mean annual incomes in Delaware and Greene Counties are not the same. Assume the population variances are equal. *Hint:* For $n > 29$, use the last row (∞) in the t-distribution table. *(Adapted from U.S. Census Bureau, Housing and Household Economics Statistics Division)*

DATA **21.** ***Tensile Strength*** The tensile strength of a metal is a measure of its ability to resist tearing when it is pulled lengthwise. A new experimental type of treatment produced steel bars with the following tensile strengths (in newtons per square millimeter).

Experimental Method:

363	355	305	350	340
373	311	348	338	320

The old method produced steel bars with the following tensile strengths (in newtons per square millimeter).

Old Method:

362	382	368	398	381	391	400
410	396	411	385	385	395	

At $\alpha = 0.01$, does the new treatment make a difference in the tensile strength of steel bars? Assume the population variances are equal.

DATA **22.** ***Tensile Strength*** An engineer wants to compare the tensile strengths of steel bars that are produced using a conventional method and an experimental method. (The tensile strength of a metal is a measure of its ability to resist tearing when pulled lengthwise.) To do so, the engineer randomly selects steel bars that are manufactured using each method and records the following tensile strengths (in newtons per square millimeter).

Experimental Method:

395	389	421	394	407	411	389	402	422
416	402	408	400	386	411	405	389	

Conventional Method:

362	352	380	382	413	384	400
378	419	379	384	388	372	383

At $\alpha = 0.10$, can the engineer claim that the experimental method produces steel with greater mean tensile strength? Should the engineer recommend using the experimental method? Assume the population variances are not equal.

23. (a) H_0: $\mu_1 \geq \mu_2$
H_a: $\mu_1 < \mu_2$ (claim)
(b) $t_0 = -1.282$
Rejection region: $t < -1.282$
(c) -4.295 (d) Reject H_0.
(e) There is enough evidence at the 10% level to support the claim and to recommend changing to the new method.

24. (a) H_0: $\mu_1 \geq \mu_2$
H_a: $\mu_1 < \mu_2$ (claim)
(b) $t_0 = -1.645$
Rejection region: $t < -1.645$
(c) -1.721 (d) Reject H_0.
(e) There is enough evidence at the 5% level to support the claim that the mean science test score is lower for students taught using the lab method than it is for students taught using the software.

DATA **23. *Teaching Methods*** A new method of teaching reading is being tested on third grade students. A group of randomly selected students is taught using the new curriculum. A control group of randomly selected students is taught using the old curriculum. The reading test scores of the two groups are shown in the stem-and-leaf plot.

Old Curriculum		New Curriculum
9	3	
9 9	4	3
9 8 8 4 3 3 2 1	5	2 4
7 6 4 2 2 1 0 0	6	0 1 1 4 7 7 7 7 8 9 9
	7	0 1 1 2 3 3 4 9
	8	2 4

Key:
9|4 = 49 (old curriculum)
4|3 = 43 (new curriculum)

At $\alpha = 0.10$, is there enough evidence to conclude that the new method of teaching reading produces higher reading test scores than the old method? Would you recommend changing to the new method? Assume the population variances are equal.

DATA **24. *Teaching Methods*** Two teaching methods and their effects on science test scores are being reviewed. A randomly selected group of students is taught in traditional lab sessions. A second randomly selected group of students is taught using interactive simulation software. The science test scores of the two groups are shown in the stem-and-leaf plot.

Traditional Lab		Interactive Simulation Software
4	6	
9 9 8 8 7 6 6 3 2 1 0	7	0 4 5 5 7 7 8
9 8 5 1 1 1 0 0	8	0 0 3 4 7 8 8 9 9
2 0	9	1 3 9

Key:
0|9 = 90 (traditional)
9|1 = 91 (interactive)

At $\alpha = 0.05$, can you support the claim that the mean science test score is lower for students taught using the lab method than it is for students taught using the interactive simulation software? Assume the population variances are equal.

Extending Concepts

Constructing Confidence Intervals for $\mu_1 - \mu_2$ If the sampling distribution for $\bar{x}_1 - \bar{x}_2$ is approximated by a t-distribution and the populations have equal variances, you can construct a confidence interval for $\mu_1 - \mu_2$ by using the following.

$$(\bar{x}_1 - \bar{x}_2) - t_c\hat{\sigma}\sqrt{\frac{1}{n_1} + \frac{1}{n_2}} < \mu_1 - \mu_2 < (\bar{x}_1 - \bar{x}_2) + t_c\hat{\sigma}\sqrt{\frac{1}{n_1} + \frac{1}{n_2}}$$

where $\hat{\sigma} = \sqrt{\dfrac{(n_1 - 1)s_1^2 + (n_2 - 1)s_2^2}{n_1 + n_2 - 2}}$

and

$\text{d.f.} = n_1 + n_2 - 2$

In Exercises 25 and 26, construct a confidence interval for $\mu_1 - \mu_2$. Assume the populations are approximately normal with equal variances.

25. $4.336 < \mu_1 - \mu_2 < 15.664$
26. $0.425 < \mu_1 - \mu_2 < 3.575$
27. $-11.580 < \mu_1 - \mu_2 < -8.420$
28. $1.074 < \mu_1 - \mu_2 < 2.926$

25. *Calories* In a study of various fast foods, you find that the mean calorie content of 15 grilled chicken sandwiches from Arby's is $\overline{x}_1 = 410$ calories with a standard deviation of $s_1 = 6.2$ calories. You also find that the mean calorie content of 12 similar chicken sandwiches from McDonald's is $\overline{x}_2 = 400$ calories with a standard deviation of $s_2 = 8.1$ calories. Construct a 95% confidence interval for the difference in mean calorie content of grilled chicken sandwiches. *(Adapted from Arby's Inc. and McDonald's Corporation)*

26. *Protein* A nutritionist wants to compare the mean protein content of grilled chicken sandwiches from Arby's and McDonald's. To do so, you randomly select several grilled chicken sandwiches from each restaurant and measure the protein content (in grams) of each. The results are listed below. Construct a 95% confidence interval for the difference in mean protein content of grilled chicken sandwiches. *(Adapted from Arby's Inc. and McDonald's Corporation)*

Restaurant	Mean protein content	Standard deviation	Sample size
Arby's	$\overline{x}_1 = 29$ grams	$s_1 = 2.1$ grams	$n_1 = 15$
McDonald's	$\overline{x}_2 = 27$ grams	$s_2 = 1.8$ grams	$n_2 = 12$

Constructing Confidence Intervals for $\mu_1 - \mu_2$ If the sampling distribution for $\overline{x}_1 - \overline{x}_2$ is approximated by a *t*-distribution and the population variances are not equal, you can construct a confidence interval for $\mu_1 - \mu_2$ by using the following.

$$(\overline{x}_1 - \overline{x}_2) - t_c\sqrt{\frac{s_1^2}{n_1} + \frac{s_2^2}{n_2}} < u_1 - u_2 < (\overline{x}_1 - \overline{x}_2) + t_c\sqrt{\frac{s_1^2}{n_1} + \frac{s_2^2}{n_2}}$$

and d.f. is the smaller of $n_1 - 1$ or $n_2 - 1$.

In Exercises 27 and 28, construct the indicated confidence interval for $\mu_1 - \mu_2$. Assume the populations are approximately normal with unequal variances.

27. *Cholesterol* To compare the mean cholesterol content (in milligrams) of grilled chicken sandwiches from Arby's and McDonald's, you randomly select several sandwiches from each restaurant and measure their cholesterol contents. The results are listed below. Construct an 80% confidence interval for the difference in mean cholesterol content of grilled chicken sandwiches. *(Adapted from Arby's Inc. and McDonald's Corporation)*

Restaurant	Mean cholesterol content	Standard deviation	Sample size
Arby's	$\overline{x}_1 = 60$ mg	$s_1 = 3.59$ mg	$n_1 = 15$
McDonald's	$\overline{x}_2 = 70$ mg	$s_2 = 2.41$ mg	$n_2 = 12$

28. *Carbohydrates* A study of fast food finds that the mean carbohydrate content of 20 grilled chicken sandwiches from Arby's is $\overline{x}_1 = 39$ grams with a standard deviation of $s_1 = 2.42$ grams. The study also finds that the mean carbohydrate content of 15 grilled chicken sandwiches from McDonald's is $\overline{x}_2 = 37$ grams with a standard deviation of $s_2 = 1.65$ grams. Construct an 80% confidence interval for the difference in mean carbohydrate content. *(Adapted from Arby's Inc. and McDonald's Corporation)*

8.3 Testing the Difference Between Means (Dependent Samples)

What You Should Learn

- How to decide whether two samples are independent or dependent
- How to perform a *t*-test to test the mean of the differences for a population of paired data

Independent and Dependent Samples • The *t*-Test for the Difference Between Means

Independent and Dependent Samples

In Sections 8.1 and 8.2, you studied two-sample hypothesis tests in which the samples were independent. In this section, you will learn how to perform two-sample hypothesis tests using dependent samples.

DEFINITION

Two samples are **independent** if the sample selected from one population is not related to the sample selected from the second population. Two samples are **dependent** if each member of one sample corresponds to a member of the other sample. Dependent samples are also called **paired samples** or **matched samples.**

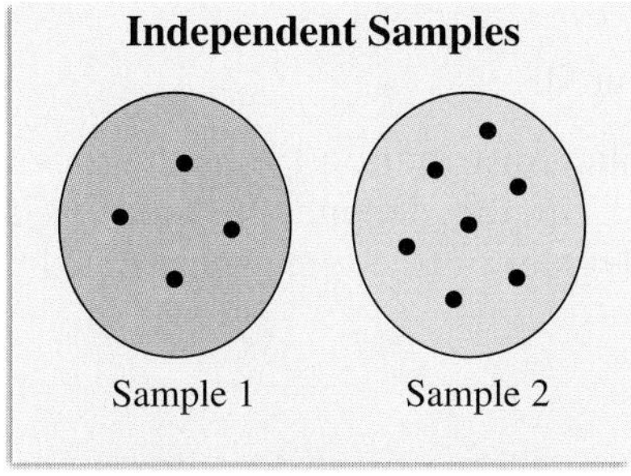

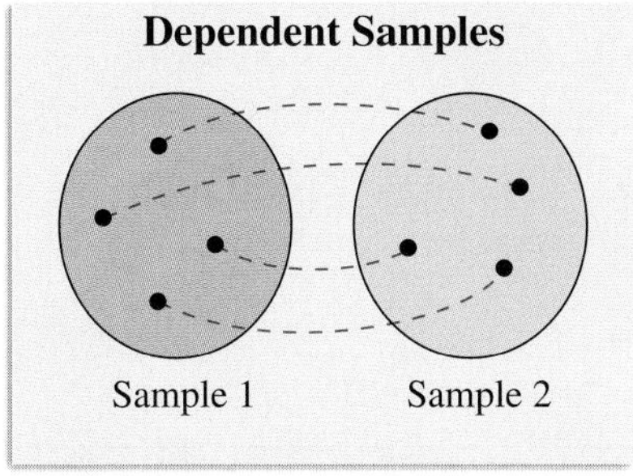

EXAMPLE 1

Independent and Dependent Samples

Classify each pair of samples as independent or dependent and justify your answer.

1. Sample 1: Resting heart rates of 35 individuals before drinking coffee
   Sample 2: Resting heart rates of the same individuals after drinking two cups of coffee
2. Sample 1: Test scores for 35 statistics students
   Sample 2: Test scores for 42 biology students who do not study statistics

SOLUTION

1. These samples are dependent. Because the resting heart rates of the same individuals are taken, the samples are related. The samples can be paired with respect to each individual.
2. These samples are independent. It is not possible to form a pairing between the members of samples; the sample sizes are different, and the data represent test scores for different individuals.

Insight

Dependent samples often involve identical twins, before and after results for the same person or object, or results of individuals matched for specific characteristics.

Try It Yourself 1

Classify each pair of samples as independent or dependent.

1. Sample 1: Heights of 27 adult females
   Sample 2: Heights of 27 adult males
2. Sample 1: Midterm exam scores of 14 chemistry students
   Sample 2: Final exam scores of the same 14 chemistry students

a. Determine whether the samples are related.

Answer: Page A41

The t-Test for the Difference Between Means

In Sections 8.1 and 8.2, you performed two-sample hypothesis tests with independent samples using the test statistic $\overline{x}_1 - \overline{x}_2$ (the difference in the means of the two samples). To perform a two-sample hypothesis test with dependent samples, you will use a different technique. You will first find the difference d for each data pair:

$$d = x_1 - x_2.$$ Difference between entries for a data pair

The test statistic is the mean $\overline{d}$ of these differences

$$\overline{d} = \frac{\Sigma d}{n}.$$ Mean of the differences between paired data entries in the dependent samples

The following conditions are required to conduct the test.

1. The samples must be randomly selected.
2. The samples must be dependent (paired).
3. Both populations must be normally distributed.

If these requirements are met, then the sampling distribution for $\overline{d}$, the mean of the differences of the paired data entries in the dependent samples, is approximated by a t-distribution with $n - 1$ degrees of freedom, where n is the number of data pairs.

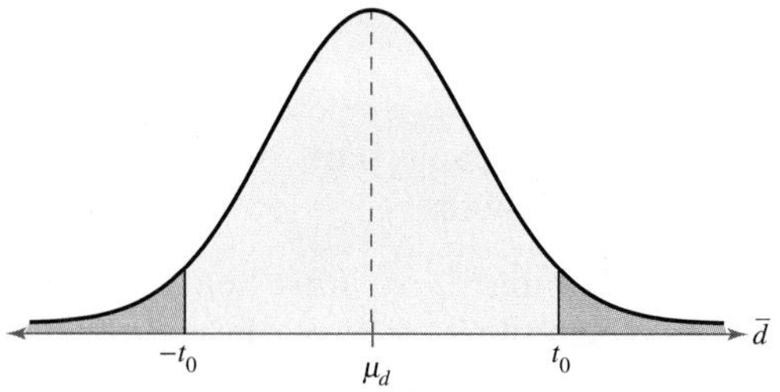

The following symbols are used for the t-test for μ_d.

Symbol	Description
n	The number of pairs of data
d	The difference between entries for a data pair, $d = x_1 - x_2$
μ_d	The hypothesized mean of the differences of paired data in the population
$\overline{d}$	The mean of the differences between the paired data entries in the dependent samples $\overline{d} = \frac{\Sigma d}{n}$
s_d	The standard deviation of the differences between the paired data entries in the dependent samples $s_d = \sqrt{\frac{n(\Sigma d^2) - (\Sigma d)^2}{n(n-1)}}$

Although formulas are given for the mean and standard deviation of differences, you should use a technology tool to calculate these statistics.

Picturing the World

The manufacturer of an appetite suppressant claims that when its product is taken while following a low-fat diet with regular exercise for 4 months, the average weight loss is 20 pounds. *(Adapted from NetHealth, Inc.)* To test this claim, you studied 12 dieters for 4 months. The dieters followed a low-fat diet with regular exercise all 4 months. The results are shown in the following table.

Weights (in pounds) of 12 dieters

	Original weight	4th month
1	185	168
2	194	177
3	213	196
4	198	180
5	244	229
6	162	144
7	211	197
8	273	252
9	178	161
10	192	178
11	181	161
12	209	193

Does your study provide sufficient evidence to reject the manufacturer's claim at a level of significance of $\alpha = 0.10$? Assume the weights are normally distributed.

When you use a t-distribution to approximate the sampling distribution for $\bar{d}$, you can use a t-test to test a claim about the mean of the differences for a population of paired data.

t-Test for the Difference Between Means

A t-test can be used to test the difference of two population means when a sample is randomly selected from each population. The requirements for performing the test are that each population must be normal and each member of the first sample must be paired with a member of the second sample. The **test statistic** is

$$\bar{d} = \frac{\Sigma d}{n}$$

and the **standardized test statistic** is

$$t = \frac{\bar{d} - \mu_d}{s_d/\sqrt{n}}.$$

The degrees of freedom are

$$\text{d.f.} = n - 1.$$

GUIDELINES

Using the t-Test for the Difference Between Means (Dependent Samples)

In Words	*In Symbols*
1. State the claim mathematically. Identify the null and alternative hypotheses.	State H_0 and H_a.
2. Specify the level of significance.	Identify α.
3. Identify the degrees of freedom and sketch the sampling distribution.	d.f. $= n - 1$
4. Determine the critical value(s).	Use Table 5 in Appendix B.
5. Determine the rejection region(s).	
6. Calculate $\bar{d}$ and s_d. Use a table.	$\bar{d} = \frac{\Sigma d}{n}$ $s_d = \sqrt{\frac{n(\Sigma d^2) - (\Sigma d)^2}{n(n-1)}}$
7. Find the standardized test statistic.	$t = \frac{\bar{d} - \mu_d}{s_d/\sqrt{n}}$
8. Make a decision to reject or fail to reject the null hypothesis.	If t is in the rejection region, reject H_0. Otherwise, fail to reject H_0.
9. Interpret the decision in the context of the original claim.	

Study Tip

If $n > 29$, use the last row (∞) in the t-distribution table.

EXAMPLE 2

The t-Test for the Difference Between Means

See MINITAB steps on page 454.

A golf club manufacturer claims that golfers can lower their scores by using the manufacturer's newly designed golf clubs. Eight golfers are randomly selected, and each is asked to give his or her most recent score. After using the new clubs for one month, the golfers are again asked to give their most recent score. The scores for each golfer are shown in the table. Assuming the golf scores are normally distributed, is there enough evidence to support the manufacturer's claim at $\alpha = 0.10$?

Golfer	1	2	3	4	5	6	7	8
Score (old design)	89	84	96	82	74	92	85	91
Score (new design)	83	83	92	84	76	91	80	91

SOLUTION The claim is that "golfers can lower their scores." In other words, the manufacturer claims that the score using the old clubs will be greater than the score using the new clubs. Each difference is given by

$$d = (\text{old score}) - (\text{new score}).$$

The null and alternative hypotheses are

$$H_0: \mu_d \leq 0 \quad \text{and} \quad H_a: \mu_d > 0. \text{ (Claim)}$$

Because the test is a right-tailed test, $\alpha = 0.10$, and d.f. $= 8 - 1 = 7$, the critical value is $t_0 = 1.415$. The rejection region is $t > 1.415$. Using the table at the left, you can calculate $\bar{d}$ and s_d as follows.

Old	New	d	d^2
89	83	6	36
84	83	1	1
96	92	4	16
82	84	−2	4
74	76	−2	4
92	91	1	1
85	80	5	25
91	91	0	0
		$\Sigma = 13$	$\Sigma = 87$

$$\bar{d} = \frac{\Sigma d}{n} = \frac{13}{8} = 1.625$$

$$s_d = \sqrt{\frac{n(\Sigma d^2) - (\Sigma d)^2}{n(n-1)}} = \sqrt{\frac{8(87) - 13^2}{8(8-1)}} \approx 3.07$$

The standardized test statistic is

$$t = \frac{\bar{d} - \mu_d}{s_d/\sqrt{n}} \quad \text{Use the } t\text{-test.}$$

$$\approx \frac{1.625 - 0}{3.07/\sqrt{8}} \quad \text{Assume } \mu_d = 0.$$

$$\approx 1.497.$$

The graph at the right shows the location of the rejection region and the standardized test statistic t. Because t is in the rejection region, you should decide to reject the null hypothesis. There is enough evidence to support the golf club manufacturer's claim at the 10% level.

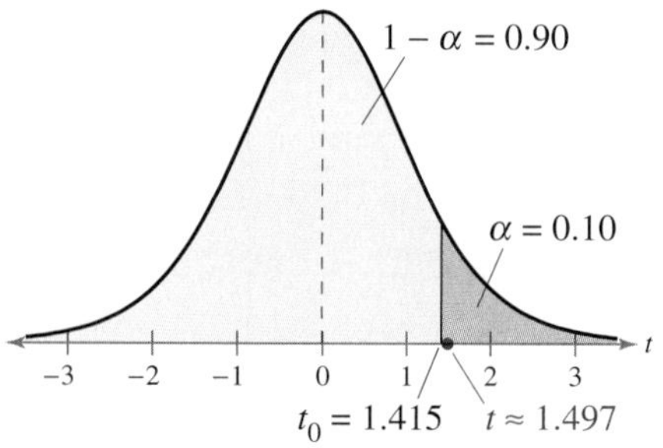

Interpretation The results of this test indicate that after the golfers used the new clubs, their scores were significantly lower.

Study Tip

You can also use technology and a P-value to perform a hypothesis test for the difference between means. For instance, in Example 2, you can enter the data in MINITAB (as shown on page 454) and find $P \approx 0.089$. Because $P < \alpha$, you should decide to reject the null hypothesis.

Insight

Note that in Example 2 it is possible that the score improved because of other reasons. Many advertisements misuse statistical results by implying a cause and effect that was not substantiated by testing.

Try It Yourself 2

Before	After
72	73
81	80
76	79
74	76
75	76
80	80
68	74
75	77
78	75
76	74
74	76
77	78

A physician claims that an experimental medication increases an individual's heart rate. Twelve test subjects are randomly selected, and the heart rate of each is measured. The subjects are then injected with the medication and, after 1 hour, the heart rate of each is measured again. The results are listed at the left. Assuming the heart rates are normally distributed, is there enough evidence to support the physician's claim at $\alpha = 0.05$?

a. *Identify* the claim and state H_0 and H_a.
b. *Specify* the level of significance α and the degrees of freedom d.f.
c. *Find* the critical value t_0 and *identify* the rejection region.
d. Calculate $\bar{d}$ and s_d.
e. *Use* the t-test to find the standardized test statistic t.
f. *Decide* whether to reject the null hypothesis. Use a graph if necessary.
g. *Interpret* the decision in the context of the original claim.

Answer: Page A41

EXAMPLE 3

The t-Test for the Difference Between Means

A state legislator wants to determine whether her performance rating (0–100) has changed from last year to this year. The following table shows the legislator's performance rating from the same 16 randomly selected voters for last year and this year. At $\alpha = 0.01$, is there enough evidence to conclude that the legislator's performance rating has changed? Assume the performance ratings are normally distributed.

Voter	1	2	3	4	5	6	7	8
Rating (last year)	60	54	78	84	91	25	50	65
Rating (this year)	56	48	70	60	85	40	40	55

Voter	9	10	11	12	13	14	15	16
Rating (last year)	68	81	75	45	62	79	58	63
Rating (this year)	80	75	78	50	50	85	53	60

Study Tip

If you prefer to use a technology tool for this type of test, enter the data in two columns and form a third column in which you calculate the difference for each pair. You can now perform a one-sample t-test on the difference column as shown in Chapter 7.

SOLUTION If there is a change in the legislator's rating, there will be a difference between "this year's" ratings and "last year's" ratings. Because the legislator wants to see if there is a difference, the null and alternative hypotheses are

$$H_0\text{: } \mu_d = 0$$

and

$$H_a\text{: } \mu_d \neq 0. \text{ (Claim)}$$

Because the test is a two-tailed test, $\alpha = 0.01$, and d.f. $= 16 - 1 = 15$, the critical values are $-t_0 = -2.947$ and $t_0 = 2.947$. The rejection regions are $t < -2.947$ and $t > 2.947$.

Before	After	d	d^2
60	56	4	16
54	48	6	36
78	70	8	64
84	60	24	576
91	85	6	36
25	40	−15	225
50	40	10	100
65	55	10	100
68	80	−12	144
81	75	6	36
75	78	−3	9
45	50	−5	25
62	50	12	144
79	85	−6	36
58	53	5	25
63	60	3	9
		$\Sigma = 53$	$\Sigma = 1581$

Using the table at the left, you can calculate $\bar{d}$ and s_d as shown below.

$$\bar{d} = \frac{\Sigma d}{n} = \frac{53}{16} = 3.3125$$

$$s_d = \sqrt{\frac{n(\Sigma d^2) - (\Sigma d)^2}{n(n-1)}} = \sqrt{\frac{16(1581) - 53^2}{16(16-1)}} \approx 9.68$$

The standardized test statistic is

$$t = \frac{\bar{d} - \mu_d}{s_d/\sqrt{n}} \qquad \text{Use the } t\text{-test.}$$

$$\approx \frac{3.3125 - 0}{9.68/\sqrt{16}} \qquad \text{Assume } \mu_d = 0.$$

$$\approx 1.369.$$

The graph at the right shows the location of the rejection region and the standardized test statistic t. Because t is not in the rejection region, you should fail to reject the null hypothesis at the 1% level.

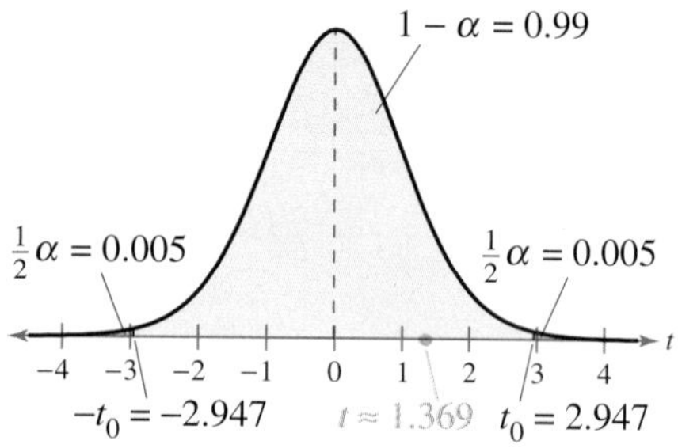

Interpretation There is not enough evidence to conclude that the legislator's performance rating has changed.

Try It Yourself 3

A medical researcher wants to determine whether a drug changes the body's temperature. Seven test subjects are randomly selected, and the body temperature (in degrees Fahrenheit) of each is measured. The subjects are then given the drug and, after 20 minutes, the body temperature of each is measured again. The results are listed below. At $\alpha = 0.05$, is there enough evidence to conclude that the drug changes the body's temperature? Assume the body temperatures are normally distributed.

Subject	1	2	3	4	5	6	7
Initial temperature	101.8	98.5	98.1	99.4	98.9	100.2	97.9
Second temperature	99.2	98.4	98.2	99	98.6	99.7	97.8

a. *Identify* the claim and state H_0 and H_a.
b. *Specify* the level of significance α and the degrees of freedom d.f.
c. *Find* the critical values and the rejection regions.
d. Calculate $\bar{d}$ and s_d.
e. *Use* the t-test to find the standardized test statistic t.
f. *Decide* whether to reject the null hypothesis. Use a graph if necessary.
g. *Interpret* the decision in the context of the original claim.

Answer: Page A41

8.3 Exercises

Help

MyMathLab

Student Study Pack

Building Basic Skills and Vocabulary

1. What is the difference between two samples that are dependent and two samples that are independent? Give an example of two dependent samples and two independent samples.

2. What conditions are necessary in order to use the dependent samples t-test for the mean of the difference of two populations?

In Exercises 3–10, classify the two given samples as independent or dependent. Explain your reasoning.

3. Sample 1: The SAT scores for 35 high school students who did not take an SAT preparation course
Sample 2: The SAT scores for 40 high school students who did take an SAT preparation course

4. Sample 1: The SAT scores for 44 high school students
Sample 2: The SAT scores for the same 44 high school students after taking an SAT preparation course

5. Sample 1: The weights of 51 adults
Sample 2: The weights of the same 51 adults after participating in a diet and exercise program for one month

6. Sample 1: The weights of 40 females
Sample 2: The weights of 40 males

7. Sample 1: The average speed of 23 powerboats using an old hull design
Sample 2: The average speed of 14 powerboats using a new hull design

8. Sample 1: The fuel mileage of 10 cars
Sample 2: The fuel mileage of the same 10 cars using a fuel additive

9. The table shows the braking distances (in feet) for each of four different sets of tires with the car's anti-lock braking system (ABS) on and with ABS off. The tests were done on ice with cars traveling at 15 miles per hour. *(Source: Consumer Reports)*

Tire Set	1	2	3	4
Braking distance with ABS	42	55	43	61
Braking distance without ABS	58	67	59	75

10. The table shows the heart rates (in beats per minute) of five people before exercising and after.

Person	1	2	3	4	5
Heart rate before exercising	65	72	85	78	93
Heart rate after exercising	127	135	140	136	150

1. Two samples are dependent if each member of one sample corresponds to a member of the other sample. Example: The weights of 22 people before starting an exercise program and the weights of the same 22 people 6 weeks after starting the exercise program.

   Two samples are independent if the sample selected from one population is not related to the sample selected from the second population. Example: The weights of 25 cats and the weights of 20 dogs.
2. (1) Each sample must be randomly selected from a normal population. (2) Each member of the first sample must be paired with a member of the second sample.
3. Independent because the scores are from different students.
4. Dependent because the sample can be paired with respect to each individual.
5. Dependent because the same adults were sampled.
6. Independent because the data represent weights from different individuals.
7. Independent because different boats were sampled.
8. Dependent because the samples can be paired with respect to each car.
9. Dependent because the same tire sets were sampled.
10. Dependent because the samples can be paired with respect to each individual.

11. Left-tailed test; Fail to reject H_0.
12. Two-tailed test; Reject H_0.
13. Right-tailed test; Reject H_0.
14. Right-tailed test; Reject H_0.
15. Left-tailed test; Reject H_0.
16. Two-tailed test; Reject H_0.
17. (a) H_0: $\mu_d \geq 0$
H_a: $\mu_d < 0$ (claim)
(b) $t_0 = -2.650$
Rejection region: $t < -2.650$
(c) $\bar{d} \approx -33.714$; $s_d \approx 42.034$
(d) -3.001
(e) Reject H_0.
(f) There is enough evidence at the 1% level to support the claim that the second SAT scores are improved.
18. (a) H_0: $\mu_d \geq 0$
H_a: $\mu_d < 0$ (claim)
(b) $t_0 = -2.821$
Rejection region: $t < -2.821$
(c) $\bar{d} \approx -59.9$; $s_d \approx 26.831$
(d) -7.06
(e) Reject H_0.
(f) There is enough evidence at the 1% level to support the claim that the SAT prep course improves verbal SAT scores.

In Exercises 11–16, test the claim about the mean of the difference of two populations. Use a t-test for dependent, random samples at the given level of significance with the given statistics. Is the test right tailed, left tailed, or two tailed? Assume the populations are normally distributed.

11. Claim: $\mu_d < 0, \alpha = 0.05$. Statistics: $\bar{d} = 10, s_d = 1.5, n = 10$

12. Claim: $\mu_d = 0, \alpha = 0.01$. Statistics: $\bar{d} = 3.2, s_d = 0.25, n = 8$

13. Claim: $\mu_d \leq 0, \alpha = 0.10$. Statistics: $\bar{d} = 6.1, s_d = 0.36, n = 16$

14. Claim: $\mu_d > 0, \alpha = 0.05$. Statistics: $\bar{d} = 5.3, s_d = 0.41, n = 28$

15. Claim: $\mu_d \geq 0, \alpha = 0.01$. Statistics: $\bar{d} = -2.3, s_d = 1.2, n = 15$

16. Claim: $\mu_d \neq 0, \alpha = 0.10$. Statistics: $\bar{d} = 4.2, s_d = 0.8, n = 20$

Using and Interpreting Concepts

Testing the Difference Between Two Means In Exercises 17–28, (a) identify the claim and state H_0 and H_a, (b) find the critical value(s) and identify the rejection region(s), (c) calculate $\bar{d}$ and s_d, (d) use the t-test to find the standardized test statistic t, (e) decide whether to reject or fail to reject the null hypothesis, and (f) interpret the decision in the context of the original claim. For each randomly selected sample, assume the distribution of the population is normal.

DATA **17.** ***SAT Scores*** The table shows the scores for 14 students the first two times they took the verbal SAT. At $\alpha = 0.01$, is there enough evidence to conclude that the students' verbal SAT scores improved the second time they took the verbal SAT?

Student	1	2	3	4	5	6	7
Score on first SAT	445	510	429	452	629	433	551
Score on second SAT	446	571	517	478	610	453	516

Student	8	9	10	11	12	13	14
Score on first SAT	358	477	325	513	636	571	442
Score on second SAT	478	532	399	531	648	603	461

DATA **18.** ***SAT Scores*** An SAT prep course claims to improve the test scores of students. The table shows the scores for 10 students the first two times they took the verbal SAT. Before taking the SAT for the second time, each student took a course to try to improve his or her verbal SAT scores. Test the claim at $\alpha = 0.01$.

Student	1	2	3	4	5	6	7
Score on first SAT	308	456	352	433	306	471	422
Score on second SAT	400	524	409	491	348	583	451

Student	8	9	10
Score on first SAT	370	320	418
Score on second SAT	408	391	450

19. (a) H_0: $\mu_d \geq 0$
H_a: $\mu_d < 0$ (claim)
(b) $t_0 = -1.415$
Rejection region: $t < -1.415$
(c) $\bar{d} \approx -1.125$; $s_d \approx 0.871$
(d) -3.653
(e) Reject H_0.
(f) There is enough evidence at the 10% level to support the claim that the fuel additive improved gas mileage.

20. (a) H_0: $\mu_d \geq 0$; H_a: $\mu_d < 0$ (claim)
(b) $t_0 = -1.397$
Rejection region: $t < -1.397$
(c) $\bar{d} \approx -1.611$; $s_d \approx 0.770$
(d) -6.277
(e) Reject H_0.
(f) There is enough evidence at the 10% level to support the claim that the fuel additive improved gas mileage.

21. (a) H_0: $\mu_d \leq 0$; H_a: $\mu_d > 0$ (claim)
(b) $t_0 = 1.363$
Rejection region: $t > 1.363$
(c) $\bar{d} = 3.75$; $s_d \approx 7.841$
(d) 1.657
(e) Reject H_0.
(f) There is enough evidence at the 10% level to support the claim that the exercise program helps participants lose weight.

19. ***Gas Mileage*** The table shows the gas mileage (in miles per gallon) of eight cars with and without using a fuel additive. At $\alpha = 0.10$, is there enough evidence to conclude that the fuel additive improved gas mileage?

Car	1	2	3	4	5	6	7	8
Gas mileage without additive	35.8	37.7	39.4	36.8	36.6	33.7	38.4	37.3
Gas mileage with fuel additive	36.2	39.8	40.1	39.3	36.9	34.5	38.8	39.1

20. ***Gas Mileage*** To test whether a fuel additive improves gas mileage, investigators measured the gas mileage (in miles per gallon) of nine cars with and without the fuel additive. The results are shown below. At $\alpha = 0.10$, can you conclude that the fuel additive improved gas mileage?

Car	1	2	3	4	5	6	7	8	9
Gas mileage without additive	34.5	36.7	34.4	39.8	33.6	35.4	38.4	35.3	37.9
Gas mileage with fuel additive	36.4	38.8	36.1	40.1	34.7	38.3	40.2	37.2	38.7

DATA **21.** ***Losing Weight*** A nutritionist claims that a particular exercise program will help participants lose weight after one month. The table shows the weights of 12 adults before participating in the exercise program and one month after participating in the exercise program. At $\alpha = 0.10$, can you conclude that the exercise program helps participants lose weight?

Participant	1	2	3	4	5	6	7	8	9	10	11	12
Weight before exercise program	157	185	120	212	230	165	207	251	196	140	137	172
Weight after exercise program	150	181	121	206	215	169	210	232	188	138	145	172

22. (a) H_0: $\mu_d \leq 0$; H_a: $\mu_d > 0$ (claim)
(b) $t_0 = 1.350$
Rejection region: $t > 1.350$
(c) $\bar{d} \approx 1.357$; $s_d \approx 3.973$
(d) 1.278
(e) Fail to reject H_0.
(f) There is not enough evidence at the 10% level to support the claim that the dieting program helped adults lose weight.

23. (a) H_0: $\mu_d \leq 0$; H_a: $\mu_d > 0$ (claim)
(b) $t_0 = 2.764$
Rejection region: $t > 2.764$
(c) $\bar{d} \approx 1.255$; $s_d \approx 0.441$
(d) 9.438
(e) Reject H_0.
(f) There is enough evidence at the 1% level to support the claim that soft tissue therapy and spinal manipulation help reduce the length of time patients suffer from headaches.

24. (a) H_0: $\mu_d \leq 0$; H_a: $\mu_d > 0$ (claim)
(b) $t_0 = 2.681$
Rejection region: $t > 2.681$
(c) $\bar{d} \approx 1.469$; $s_d \approx 0.494$
(d) 10.722
(e) Reject H_0.
(f) There is enough evidence at the 1% level to support the claim that the new drug helps to reduce the number of daily headache hours.

DATA **22. *Losing Weight*** The table shows the weights of 14 adults before a dieting program and two weeks after the dieting program. At $\alpha = 0.10$, is there enough evidence to conclude that the program helped adults lose weight?

Participant	1	2	3	4	5	6	7
Weight before dieting program	194	234	265	188	170	212	139
Weight after dieting program	190	235	255	187	175	209	139

Participant	8	9	10	11	12	13	14
Weight before exercise program	280	291	210	190	155	166	198
Weight after exercise program	277	285	212	194	152	167	196

DATA **23. *Headaches*** A physical therapist suggests that soft tissue therapy and spinal manipulation help to reduce the length of time patients suffer from headaches. The table shows the number of hours per day a sample of 11 patients suffered from headaches, before and after seven weeks of receiving soft tissue therapy and spinal manipulation. At $\alpha = 0.01$, is there enough evidence to support the therapist's claim? *(Adapted from The Journal of the American Medical Association)*

Patient	1	2	3	4	5	6
Daily headache hours (before)	2.8	2.4	2.8	2.6	2.7	2.9
Daily headache hours (after)	1.6	1.3	1.6	1.4	1.5	1.6

Patient	7	8	9	10	11
Daily headache hours (before)	3.2	2.9	4.1	1.6	2.5
Daily headache hours (after)	1.7	1.6	1.8	1.2	1.4

DATA **24. *Headaches*** The table shows the number of hours per day a sample of 13 patients suffered from headaches before and after seven weeks of receiving soft tissue therapy and laser treatment. At $\alpha = 0.01$, test the claim that the therapy and treatment help to reduce the number of daily headache hours. *(Adapted from The Journal of the American Medical Association)*

Patient	1	2	3	4	5	6	7
Daily headache hours (before)	4.1	3.9	3.8	4.5	2.4	3.6	3.4
Daily headache hours (after)	2.2	2.8	2.5	2.6	1.9	1.8	2.0

Patient	8	9	10	11	12	13
Daily headache hours (before)	3.4	3.5	2.7	3.7	4.4	3.2
Daily headache hours (after)	1.6	1.5	2.1	1.8	3.0	1.7

25. (a) H_0: $\mu_d \leq 0$; H_a: $\mu_d > 0$ (claim)
(b) $t_0 = 1.895$
Rejection region: $t > 1.895$
(c) $\bar{d} = 14.75$; $s_d \approx 6.861$
(d) 6.082
(e) Reject H_0.
(f) There is enough evidence at the 5% level to support the claim that the new drug reduces systolic blood pressure.

26. (a) H_0: $\mu_d \leq 0$; H_a: $\mu_d > 0$ (claim)
(b) $t_0 = 1.860$
Rejection region: $t > 1.860$
(c) $\bar{d} \approx 6.333$; $s_d \approx 5.362$
(d) 3.544
(e) Reject H_0.
(f) There is enough evidence at the 5% level to support the claim that the new drug reduces diastolic blood pressure.

27. (a) H_0: $\mu_d = 0$; H_a: $\mu_d \neq 0$ (claim)
(b) $-t_0 = -2.365$ and $t_0 = 2.365$
Rejection regions: $t < -2.365$, $t > 2.365$
(c) $\bar{d} = -1$; $s_d \approx 1.309$
(d) -2.159
(e) Fail to reject H_0.
(f) There is not enough evidence at the 5% level to support the claim that the product ratings have changed from last year to this year.

28. (a) H_0: $\mu_d = 0$; H_a: $\mu_d \neq 0$ (claim)
(b) $-t_0 = -2.571$ and $t_0 = 2.571$
Rejection regions: $t < -2.571$, $t > 2.571$
(c) $\bar{d} \approx -7.167$; $s_d \approx 10.245$
(d) -1.715
(e) Fail to reject H_0.
(f) There is not enough evidence at the 5% level to support the claim that the plant's performance ratings have changed.

25. *Blood Pressure* A pharmaceutical company guarantees that its new drug reduces systolic blood pressure. The table shows the systolic blood pressure (in millimeters of mercury) for eight patients before taking the new drug and two hours after taking the drug. At $\alpha = 0.05$, can you conclude that the new drug reduces systolic blood pressure?

Patient	1	2	3	4	5	6	7	8
Systolic blood pressure (before)	201	171	186	162	165	167	175	148
Systolic blood pressure (after)	192	165	167	155	148	144	152	134

26. *Blood Pressure* The diastolic blood pressure (in millimeters of mercury) for nine patients before taking a new drug and two hours after taking the drug is shown in the table below. At $\alpha = 0.05$, can you conclude that the new drug reduces diastolic blood pressure?

Patient	1	2	3	4	5
Diastolic blood pressure (before)	103	122	106	112	125
Diastolic blood pressure (after)	98	121	107	105	108

Patient	6	7	8	9
Diastolic blood pressure (before)	97	107	118	112
Diastolic blood pressure (after)	89	102	114	101

27. *Product Ratings* A company wants to determine whether its consumer product ratings (0–10) have changed from last year to this year. The table shows the company's product ratings from the same eight consumers for last year and this year. At $\alpha = 0.05$, is there enough evidence to conclude that the product ratings have changed?

Consumer	1	2	3	4	5	6	7	8
Rating (last year)	5	7	2	3	9	10	8	7
Rating (this year)	5	9	4	6	9	9	9	8

28. *Performance Ratings* A manufacturing plant wants to determine whether its manager performance ratings (0–100) have changed from last month to this month. The table shows the plant's performance ratings from the same six managers for last month and this month. At $\alpha = 0.05$, is there enough evidence to conclude that the plant's performance ratings have changed?

Manager	1	2	3	4	5	6
Rating (last month)	85	96	70	76	81	78
Rating (this month)	88	85	89	86	92	89

29. $-1.763 < \mu_d < -1.287$
30. $-0.827 < \mu_d < -0.045$

Extending Concepts

Constructing Confidence Intervals for μ_d To construct a confidence interval for μ_d, use the following inequality.

$$\bar{d} - t_c\frac{s_d}{\sqrt{n}} < \mu_d < \bar{d} + t_c\frac{s_d}{\sqrt{n}}$$

In Exercises 29 and 30, construct the indicated confidence interval for μ_d. Assume the populations are normally distributed.

DATA **29.** ***Drug Testing*** A sleep disorder specialist wants to test the effectiveness of a new drug that is reported to increase the number of hours of sleep patients get during the night. To do so, the specialist randomly selects 16 patients and records the number of hours of sleep each gets with and without the new drug. The results of the two-night study are listed below. Construct a 90% confidence interval for μ_d.

Patient	1	2	3	4	5	6	7	8	9	10	11	12	13	14	15	16
Hours of sleep without the drug	1.8	2.0	3.4	3.5	3.7	3.8	3.9	3.9	4.0	4.9	5.1	5.2	5.0	4.5	4.2	4.7
Hours of sleep using the new drug	3.0	3.6	4.0	4.4	4.5	5.2	5.5	5.7	6.2	6.3	6.6	7.8	7.2	6.5	5.6	5.9

DATA **30.** ***Herbal Medicine Testing*** An herbal medicine is tested on 14 randomly selected patients with sleeping disorders. The table shows the hours of sleep patients got during one night without using the herbal medicine and the hours of sleep the patients got on another night after the herbal medicine was administered. Construct a 95% confidence interval for μ_d.

Patient	1	2	3	4	5	6	7	8	9	10	11	12	13	14
Hours of sleep without medicine	1.0	1.4	3.4	3.7	5.1	5.1	5.2	5.3	5.5	5.8	4.2	4.8	2.9	4.5
Hours of sleep using the herbal medicine	2.9	3.3	3.5	4.4	5.0	5.0	5.2	5.3	6.0	6.5	4.4	4.7	3.1	4.7

8.4 Testing the Difference Between Proportions

What You Should Learn

- How to perform a z-test for the difference between two population proportions p_1 and p_2

Two-Sample z-Test for the Difference Between Proportions

Two-Sample z-Test for the Difference Between Proportions

In this section, you will learn how to use a z-test to test the difference between two population proportions p_1 and p_2 using a sample proportion from each population. For instance, suppose you want to determine whether the proportion of female college students who earn a bachelor's degree in four years is different from the proportion of male college students who earn a bachelor's degree in four years. The following conditions are necessary to use a z-test to test such a difference.

1. The samples must be randomly selected.
2. The samples must be independent.
3. The samples must be large enough to use a normal sampling distribution. That is,

$$n_1 p_1 \geq 5,$$
$$n_1 q_1 \geq 5,$$
$$n_2 p_2 \geq 5, \quad \text{and}$$
$$n_2 q_2 \geq 5.$$

If these conditions are met, then the sampling distribution for $\hat{p}_1 - \hat{p}_2$, the difference between the sample proportions, is a normal distribution with mean

$$\mu_{\hat{p}_1-\hat{p}_2} = p_1 - p_2$$

and standard error

$$\sigma_{\hat{p}_1-\hat{p}_2} = \sqrt{\frac{p_1 q_1}{n_1} + \frac{p_2 q_2}{n_2}}.$$

Notice that you need to know the population proportions to calculate the standard error. Because a hypothesis test for $p_1 - p_2$ is based on the assumption that $p_1 = p_2$, you can calculate a weighted estimate of p_1 and p_2 using

$$\bar{p} = \frac{x_1 + x_2}{n_1 + n_2}, \text{ where } x_1 = n_1 \hat{p}_1 \text{ and } x_2 = n_2 \hat{p}_2.$$

With the weighted estimate $\bar{p}$, the standard error of the sampling distribution for $\hat{p}_1 - \hat{p}_2$ is

$$\sigma_{\hat{p}_1-\hat{p}_2} = \sqrt{\bar{p}\,\bar{q}\left(\frac{1}{n_1} + \frac{1}{n_2}\right)}, \text{ where } \bar{q} = 1 - \bar{p}.$$

Also observe that you need to know the population proportions in verifying that the samples are large enough to be approximated by the normal distribution. But when determining whether the z-test can be used for the difference between proportions for a binomial experiment, you should use $\bar{p}$ in place of p_1 and p_2 and use $\bar{q}$ in place of q_1 and q_2.

Study Tip

The following symbols are used in the z-test for $p_1 - p_2$. See Sections 4.2 and 5.5 to review the binomial distribution.

Symbol	Description
p_1, p_2	Population proportions
x_1, x_2	Number of successes in each sample
n_1, n_2	Size of each sample
$\hat{p}_1, \hat{p}_2$	Sample proportions of successes
$\bar{p}$	Weighted estimate for p_1 and p_2

Picturing the World

A study found a difference in the proportion of adult men and women who use alternative medicines, such as herbal medicines, massage, spiritual healing, and acupuncture. In the study, 400 randomly selected women and 450 randomly selected men were asked if they had used at least one alternative medicine therapy during the past year. The results are shown below. *(Adapted from The Journal of the American Medical Association)*

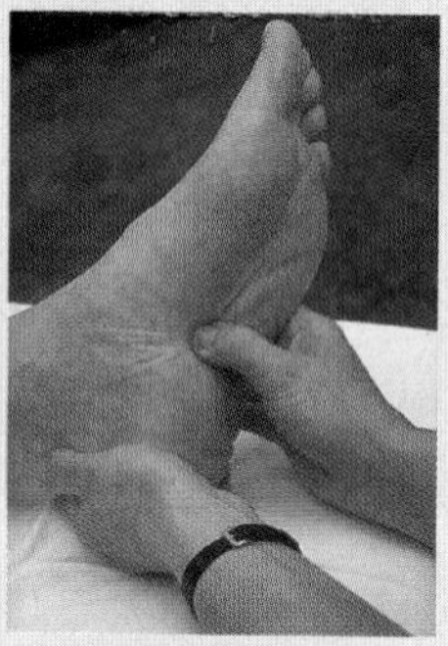

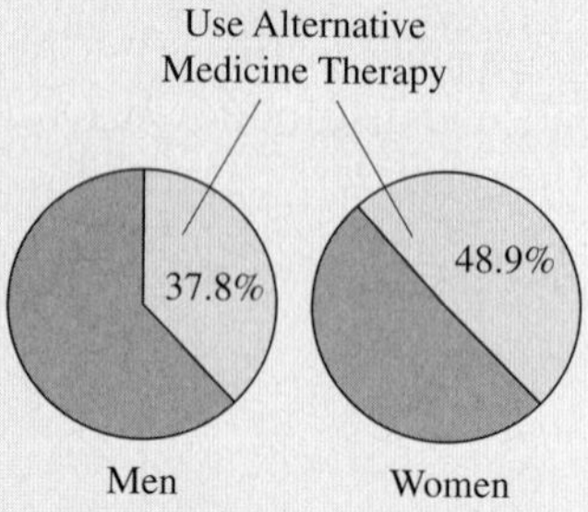

At $\alpha = 0.05$, is there enough evidence to support the claim that there is a significant difference in use of alternative medicines between men and women?

If the sampling distribution for $\hat{p}_1 - \hat{p}_2$ is normal, you can use a two-sample z-test to test the difference between two population proportions p_1 and p_2.

Two-Sample z-Test for the Difference Between Proportions

A two-sample z-test is used to test the difference between two population proportions p_1 and p_2 when a sample is randomly selected from each population. The **test statistic** is

$$\hat{p}_1 - \hat{p}_2$$

and the **standardized test statistic** is

$$z = \frac{(\hat{p}_1 - \hat{p}_2) - (p_1 - p_2)}{\sqrt{\overline{p}\,\overline{q}\left(\dfrac{1}{n_1} + \dfrac{1}{n_2}\right)}}$$

where

$$\overline{p} = \frac{x_1 + x_2}{n_1 + n_2} \quad \text{and} \quad \overline{q} = 1 - \overline{p}.$$

Note: $n_1\overline{p}$, $n_1\overline{q}$, $n_2\overline{p}$, and $n_2\overline{q}$ must be at least 5.

If the null hypothesis states $p_1 = p_2$, $p_1 \leq p_2$, or $p_1 \geq p_2$, then $p_1 = p_2$ is assumed and the expression $p_1 - p_2$ is equal to 0 in the preceding test.

GUIDELINES

Using a Two-Sample z-Test for the Difference Between Proportions

In Words	*In Symbols*
1. State the claim. Identify the null and alternative hypotheses.	State H_0 and H_a.
2. Specify the level of significance.	Identify α.
3. Determine the critical value(s).	Use Table 4 in Appendix B.
4. Determine the rejection region(s).	
5. Find the weighted estimate of p_1 and p_2.	$\overline{p} = \dfrac{x_1 + x_2}{n_1 + n_2}$
6. Find the standardized test statistic.	$z = \dfrac{(\hat{p}_1 - \hat{p}_2) - (p_1 - p_2)}{\sqrt{\overline{p}\,\overline{q}\left(\frac{1}{n_1} + \frac{1}{n_2}\right)}}$
7. Make a decision to reject or fail to reject the null hypothesis.	If z is in the rejection region, reject H_0. Otherwise, fail to reject H_0.
8. Interpret the decision in the context of the original claim.	

To use P-values in a test for the difference in proportions, use the same guidelines as above, skipping Steps 3 and 4. After finding the standardized test statistic, use the Standard Normal Table to calculate the P-value. Then make a decision to reject or fail to reject the null hypothesis. If P is less than or equal to α, reject H_0. Otherwise fail to reject H_0.

See TI-83 steps on page 455.

EXAMPLE 1

A Two-Sample z-Test for the Difference Between Proportions

In a study of 200 randomly selected adult female and 250 randomly selected adult male Internet users, 30% of the females and 38% of the males said that they plan to shop online at least once during the next month. At $\alpha = 0.10$, test the claim that there is a difference in the proportion of female and the proportion of male Internet users who plan to shop online.

Sample Statistics for Internet Users

Female	Male
$n_1 = 200$	$n_2 = 250$
$\hat{p}_1 = 0.30$	$\hat{p}_2 = 0.38$
$n_1\hat{p}_1 = 60$	$n_2\hat{p}_2 = 95$

SOLUTION You want to determine whether there is a difference in the proportions. So, the null and alternative hypotheses are

$$H_0\text{: } p_1 = p_2 \quad \text{and} \quad H_a\text{: } p_1 \neq p_2. \text{ (Claim)}$$

Because the test is two tailed and the level of significance is $\alpha = 0.10$, the critical values are $-z_0 = -1.645$ and $z_0 = 1.645$. The rejection regions are $z < -1.645$ and $z > 1.645$. The weighted estimate of the population proportions is

$$\overline{p} = \frac{x_1 + x_2}{n_1 + n_2} = \frac{60 + 95}{200 + 250} = \frac{155}{450} \approx 0.344$$

and

$$\overline{q} = 1 - \overline{p} = 1 - 0.344 = 0.656.$$

Study Tip

To find x_1 and x_2, use
$x_1 = n_1\hat{p}_1$ and
$x_2 = n_2\hat{p}_2$.

Because 200(0.344), 200(0.656), 250(0.344), and 250(0.656) are at least 5, you can use a two-sample z-test. The standardized test statistic is

$$z = \frac{(\hat{p}_1 - \hat{p}_2) - (p_1 - p_2)}{\sqrt{\overline{p}\,\overline{q}\left(\frac{1}{n_1} + \frac{1}{n_2}\right)}} \approx \frac{(0.30 - 0.38) - 0}{\sqrt{(0.344)(0.656)\left(\frac{1}{200} + \frac{1}{250}\right)}} \approx -1.78.$$

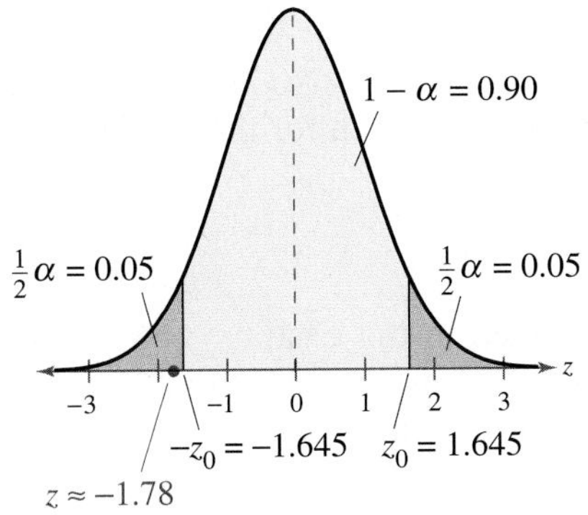

The graph at the left shows the location of the rejection regions and the standardized test statistic. Because z is in the rejection region, you should decide to reject the null hypothesis.

Interpretation You have enough evidence at the 10% level to conclude that there is a difference in the proportion of female and the proportion of male Internet users who plan to shop online.

Try It Yourself 1

Consider the results of the *NYTS* study discussed in the Chapter Opener. At $\alpha = 0.05$, is there a difference between the proportion of male high school students who smoke cigarettes and the proportion of female high school students who smoke cigarettes?

a. *Identify* the claim and state H_0 and H_a.
b. *Specify* the level of significance α.
c. *Find* the critical values and *identify* the rejection regions.
d. *Find* $\overline{p}$ and $\overline{q}$.
e. *Verify* that $n_1\overline{p}$, $n_1\overline{q}$, $n_2\overline{p}$, and $n_2\overline{q}$ are at least 5.
f. *Use* the two-sample z-test to find the standardized test statistic z.
g. *Decide* whether to reject the null hypothesis.
h. *Interpret* the decision in the context of the original claim.

Answer: Page A42

Sample Statistics for Cholesterol-Reducing Medication

Received medication	Received placebo
$n_1 = 4700$	$n_2 = 4300$
$x_1 = 301$	$x_2 = 357$
$\hat{p}_1 = 0.064$	$\hat{p}_2 = 0.083$

Study Tip

To find $\hat{p}_1$ and $\hat{p}_2$, use

$$\hat{p}_1 = \frac{x_1}{n_1} \text{ and}$$

$$\hat{p}_2 = \frac{x_2}{n_2}.$$

EXAMPLE 2

A Two-Sample z-Test for the Difference Between Proportions

A medical research team conducted a study to test the effect of a cholesterol-reducing medication. At the end of the study, the researchers found that of the 4700 subjects who took the medication, 301 died of heart disease. Of the 4300 subjects who took a placebo, 357 died of heart disease. At $\alpha = 0.01$, can you conclude that the death rate is lower for those who took the medication than for those who took the placebo? *(Source: New England Journal of Medicine)*

SOLUTION You want to determine whether the death rate is lower for those who took the medication than for those who took the placebo. So, the null and alternative hypotheses are

$$H_0\text{: } p_1 \geq p_2 \quad \text{and} \quad H_a\text{: } p_1 < p_2. \text{ (Claim)}$$

Because the test is left tailed and the level of significance is $\alpha = 0.01$, the critical value is $z_0 = -2.33$. The rejection region is $z < -2.33$. The weighted estimate of p_1 and p_2 is

$$\bar{p} = \frac{x_1 + x_2}{n_1 + n_2} = \frac{301 + 357}{4700 + 4300} = \frac{658}{9000} \approx 0.073$$

and $\bar{q} = 1 - \bar{p} = 1 - 0.073 = 0.927$. Because $4700(0.073)$, $4700(0.927)$, $4300(0.073)$, and $4300(0.927)$ are at least 5, you can use a two-sample z-test.

$$z = \frac{(\hat{p}_1 - \hat{p}_2) - (p_1 - p_2)}{\sqrt{\bar{p}\bar{q}\left(\frac{1}{n_1} + \frac{1}{n_2}\right)}} \approx \frac{(0.064 - 0.083) - 0}{\sqrt{(0.073)(0.927)\left(\frac{1}{4700} + \frac{1}{4300}\right)}} \approx -3.46$$

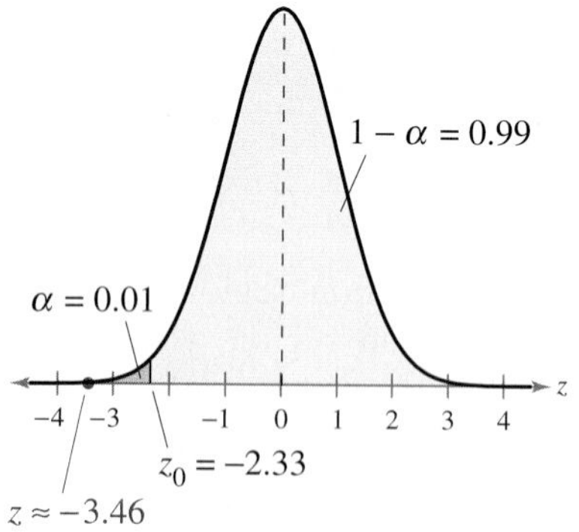

The graph at the left shows the location of the rejection region and the standardized test statistic. Because z is in the rejection region, you should decide to reject the null hypothesis.

Interpretation At the 1% level, there is enough evidence to conclude that the death rate is lower for those who took the medication than for those who took the placebo.

Try It Yourself 2

Consider the results of the *NYTS* study discussed in the Chapter Opener. At $\alpha = 0.05$, can you conclude that the proportion of male high school students who smoke cigars is greater than the proportion of female high school students who smoke cigars?

a. *Identify* the claim and state H_0 and H_a.
b. *Specify* the level of significance α.
c. *Find* the critical value and *identify* the rejection region.
d. *Find* $\bar{p}$ and $\bar{q}$.
e. *Verify* that $n_1\bar{p}$, $n_1\bar{q}$, $n_2\bar{p}$, and $n_2\bar{q}$ are at least 5.
f. *Use* the two-sample z-test to find the standardized test statistic z.
g. *Decide* whether to reject the null hypothesis.
h. *Interpret* the decision in the context of the original claim.

Answer: Page A42

8.4 Exercises

Building Basic Skills and Vocabulary

1. Explain how to perform a two-sample z-test for the difference between two population proportions.

2. What conditions are necessary in order to use the z-test to test the difference between two population proportions?

In Exercises 3–6, test the claim about the difference between two population proportions p_1 and p_2 for the given level of significance α and the given sample statistics. Is the test right tailed, left tailed, or two tailed? Assume the sample statistics are from independent random samples.

3. Claim: $p_1 \neq p_2$, $\alpha = 0.01$.
Sample statistics: $x_1 = 35$, $n_1 = 70$ and $x_2 = 36$, $n_2 = 60$

4. Claim: $p_1 < p_2$, $\alpha = 0.05$.
Sample statistics: $x_1 = 471$, $n_1 = 785$ and $x_2 = 372$, $n_2 = 465$

5. Claim: $p_1 \leq p_2$, $\alpha = 0.10$.
Sample statistics: $x_1 = 344$, $n_1 = 860$ and $x_2 = 304$, $n_2 = 800$

6. Claim: $p_1 = p_2$, $\alpha = 0.05$.
Sample statistics: $x_1 = 29$, $n_1 = 45$ and $x_2 = 25$, $n_2 = 30$

Using and Interpreting Concepts

Testing the Difference Between Two Proportions In Exercises 7–16, (a) identify the claim and state H_0 and H_a, (b) find the critical value(s) and identify the rejection region(s), (c) find the standardized test statistic, (d) decide whether to reject or fail to reject the null hypothesis, and (e) interpret the decision in the context of the original claim. Assume the random samples are independent. If convenient, use technology to solve the problem.

7. ***Alternative Medicine Use*** In a 1991 study of 1539 adults, 520 said they had used alternative medicines (for example, folk remedies and homeopathy) in the previous year. In a more recent study of 2055 adults, 865 said they had used alternative medicines in the previous year. At $\alpha = 0.05$, can you reject the claim that the proportion of adults using alternative medicines has not changed since 1991? *(Source: The Journal of the American Medical Association)*

Have You Used Alternative Medicines in the Past Year?

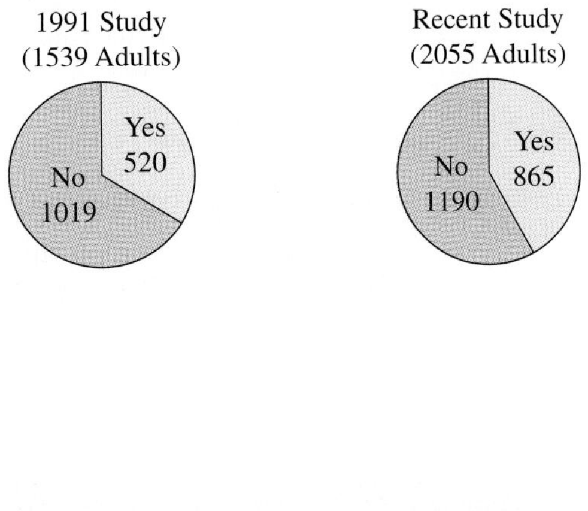

1. State the hypotheses and identify the claim. Specify the level of significance. Find the critical value(s) and rejection region(s). Find $\bar{p}$ and $\bar{q}$. Find the standardized test statistic. Make a decision and interpret in the context of the claim.
2. (1) The sample must be independent.
   (2) $n_1p_1 \geq 5$, $n_1q_1 \geq 5$, $n_2p_2 \geq 5$, and $n_2q_2 \geq 5$
3. Two-tailed test; Fail to reject H_0.
4. Left-tailed test; Reject H_0.
5. Right-tailed test; Fail to reject H_0.
6. Two-tailed test; Fail to reject H_0.
7. (a) H_0: $p_1 = p_2$ (claim); H_a: $p_1 \neq p_2$
   (b) $-z_0 = -1.96$ and $z_0 = 1.96$
   Rejection regions: $z < -1.96$, $z > 1.96$
   (c) -5.060
   (d) Reject H_0.
   (e) There is sufficient evidence at the 5% level to reject the claim that the proportion of adults using alternative medicines has not changed since 1991.

8. (a) H_0: $p_1 = p_2$ (claim)
H_a: $p_1 \neq p_2$
(b) $-z_0 = -1.645$ and $z_0 = 1.645$
Rejection regions: $z < -1.645$, $z > 1.645$
(c) -2.866
(d) Reject H_0.
(e) There is sufficient evidence at the 10% level to reject the claim that the proportions of patients suffering new bouts of depression are the same for the two groups.

9. (a) H_0: $p_1 = p_2$ (claim)
H_a: $p_1 \neq p_2$
(b) $-z_0 = -1.645$ and $z_0 = 1.645$
Rejection regions: $z < -1.645$, $z > 1.645$
(c) 4.351
(d) Reject H_0.
(e) There is sufficient evidence at the 10% level to reject the claim that the proportions of male and female senior citizens who eat the daily recommended number of servings of vegetables are the same.

10. (a) H_0: $p_1 \geq p_2$; H_a: $p_1 < p_2$ (claim)
(b) $z_0 = -2.33$
Rejection region: $z < -2.33$
(c) -1.563
(d) Fail to reject H_0.
(e) There is insufficient evidence at the 1% level to support the claim that the proportion of senior citizens who said they eat the daily recommended number of servings of fruit is lower for males than for females.

11. (a) H_0: $p_1 \geq p_2$
H_a: $p_1 < p_2$ (claim)
(b) $z_0 = -2.33$
Rejection region: $z < -2.33$
(c) -1.596
(d) Fail to reject H_0.
(e) There is insufficient evidence at the 1% level to support the claim that the proportion of adults who are smokers is lower in Alabama than in Missouri.

12. See Selected Answers, page A97.

8. *Antidepressants* A medical research team studied patients using an antidepressant drug to recover from chronic depression. Of 77 patients who used the drug for two years, five suffered a new bout of depression. Of 84 patients who used the drug for seven months, 19 suffered a new bout of depression. At $\alpha = 0.10$, can you reject the claim that the proportions of patients suffering new bouts of depression are the same for the two groups? *(Adapted from The Journal of the American Medical Association)*

Have You Suffered a New Bout of Depression?

Used Antidepressant for 2 Years (77 Patients) | Used Antidepressant for 7 Months (84 Patients)

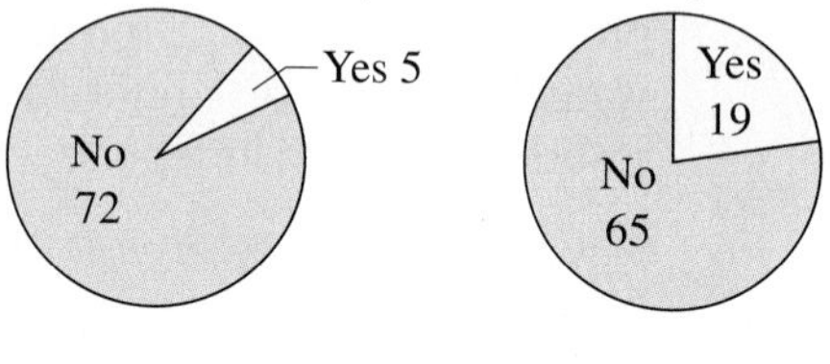

9. *Recommended Servings of Vegetables* In a survey of 5240 male senior citizens, 2201 said they eat the daily recommended number of servings of vegetables. In a survey of 6180 female senior citizens, 2348 said they eat the daily recommended number of servings of vegetables. At $\alpha = 0.10$, can you reject the claim that the proportions of senior citizens who said they eat the daily recommended number of servings of vegetables are the same for the two groups? *(Adapted from USDA Economic Research Service)*

10. *Recommended Servings of Fruit* In a survey of 1245 male senior citizens, 361 said they eat the daily recommended number of servings of fruit. In a survey of 1065 female senior citizens, 341 said they eat the daily recommended number of servings of fruit. At $\alpha = 0.01$, can you support the claim that the proportion of senior citizens who said they eat the daily recommended number of servings of fruit is lower for males than for females? *(Adapted from USDA Economic Research Service)*

11. *Smokers in Alabama and Missouri* A state-by-state survey found that the proportions of adults who are smokers in Alabama and Missouri were 24.4% and 26.6%, respectively. (Suppose the number of respondents from each state was 2000.) At $\alpha = 0.01$, can you support the claim that the proportion of adults who are smokers is lower in Alabama than in Missouri? *(Adapted from The Centers for Disease Control and Prevention)*

12. *Smokers in California and Oregon* In a random survey of 1500 adults in California and 1000 adults in Oregon, you find that the percentages of smokers are 16.4% and 22.4%, respectively. At $\alpha = 0.05$, is there enough evidence to conclude that the proportion of adults who are smokers is lower in California than in Oregon? *(Adapted from The Centers for Disease Control and Prevention)*

13. *Smokers: Male and Female* In a survey of 5800 twelfth grade males, 1589 said they had smoked in the last 30 days. In a survey of 6600 twelfth grade females, 1683 said they had smoked in the last 30 days. At $\alpha = 0.01$, can you support the claim that the proportion of twelfth grade males who said they had smoked in the last 30 days is less than the proportion of twelfth grade females? *(Source: Monitoring the Future Study, University of Michigan)*

13. (a) H_0: $p_1 \geq p_2$
H_a: $p_1 < p_2$ (claim)
(b) $z_0 = -2.33$
Rejection region: $z < -2.33$
(c) 2.395 (d) Fail to reject H_0.
(e) There is insufficient evidence at the 1% level to support the claim that the proportion of twelfth grade males who said they had smoked in the last 30 days is less than the proportion of twelfth grade females.

14. (a) H_0: $p_1 = p_2$ (claim)
H_a: $p_1 \neq p_2$
(b) $-z_0 = -1.645$ and $z_0 = 1.645$
Rejection regions: $z < -1.645$, $z > 1.645$
(c) 15.145 (d) Reject H_0.
(e) There is sufficient evidence at the 10% level to reject the claim that the proportion of college students who said they had smoked in the last 30 days has not changed.

15. (a) H_0: $p_1 = p_2$ (claim)
H_a: $p_1 \neq p_2$
(b) $-z_0 = -1.96$ and $z_0 = 1.96$
Rejection regions: $z < -1.96$, $z > 1.96$
(c) 26.135 (d) Reject H_0.
(e) There is sufficient evidence at the 5% level to reject the claim that the proportions of adults who have chronic backaches are the same for the two groups.

16. See Selected Answers, page A97.

17. There is insufficient evidence at the 1% level to support the claim that the proportions of teens who watch less than 1 hour of TV per week is greater for girls than for boys.

18. There is insufficient evidence at the 1% level to support the representative's belief that the proportion of teens who watch 10 to 20 hours of TV per week is greater for boys than for girls.

19. There is sufficient evidence at the 5% level to reject the organization's claim that the proportions of teens who watch 1 to 5 hours of TV per week is the same for boys and for girls.

20. See Selected Answers, page A97.

14. *Smokers: Then and Now* In a survey of 15,200 college students, 5335 said they had smoked in the last 30 days. In another survey of 12,900 college students taken four years later, 3444 said they had smoked in the last 30 days. At $\alpha = 0.10$, can you reject the claim that the proportion of college students who said they had smoked in the last 30 days has not changed? *(Adapted from the Monitoring the Future Study, University of Michigan)*

15. *Anxiety and Backaches* A pharmaceutical marketing group studied the ailments of anxiety sufferers. The group found that of 9600 adults who suffer from anxiety, 2496 have chronic backaches. Of 11,500 adults who do not suffer from anxiety, 1380 have chronic backaches. At $\alpha = 0.05$, can you reject the claim that the proportions of adults who have chronic backaches are the same for the two groups? *(Adapted from Mars OTC/DTC Survey)*

16. *Diabetes and High Cholesterol* A medical group studied the ailments of adults with diabetes. Of 8750 adults who are diabetic, 2100 have high cholesterol. Of 12,350 adults who are not diabetic, 1482 have high cholesterol. At $\alpha = 0.05$, can you reject the claim that the proportions of adults who have high cholesterol are the same for the two groups? *(Adapted from Mars OTC/DTC Survey)*

Teen TV Time In Exercises 17–20, refer to the figure. Assume the survey included 700 boys and 500 girls, and assume the samples are random and independent.

17. *Less Than One Hour of TV* At $\alpha = 0.01$, can you support the claim that the proportion of teens who watch less than 1 hour of TV per week is greater for girls than for boys?

18. *Ten to Twenty Hours of TV* A television marketing representative believes that the proportion of teens who watch 10 to 20 hours of TV per week is greater for boys than for girls. If the representative's belief is based on the results of the survey, is there enough evidence to support the representative's belief? Use $\alpha = 0.01$.

19. *One to Five Hours of TV* On the basis of the survey, an educational organization reports that the proportion of teens who watch 1 to 5 hours of TV per week is the same for boys and for girls. Using $\alpha = 0.05$, can you reject the organization's claim?

20. *Five to Ten Hours of TV* At $\alpha = 0.05$, can you reject the claim that the proportion of teens who watch 5 to 10 hours of TV per week is the same for boys and for girls?

21. There is sufficient evidence at the 5% level to support the claim that the proportions of men age 18 to 24 living in parents' homes was more in 1992 than in 2002.
22. There is insufficient evidence at the 5% level to support the claim that the proportion of women age 18 to 24 living in parents' homes was more in 1992 than in 2002.
23. There is sufficient evidence at the 1% level to reject the claim that the proportions of 18- to 24-year-olds living in parents' homes in 1992 was the same for men and women.
24. There is sufficient evidence at the 1% level to reject the claim that the proportion of 18- to 24-year-olds living in parents' homes in 2002 was the same for men and women.
25. $0.0312 < p_1 - p_2 < 0.0328$
26. $-0.039 < p_1 - p_2 < -0.037$

Moving Out In Exercises 21–24, refer to the figure. Assume the survey included 580 men and 500 women in 1992 and 500 men and 425 women in 2002, and assume the samples are random and independent.

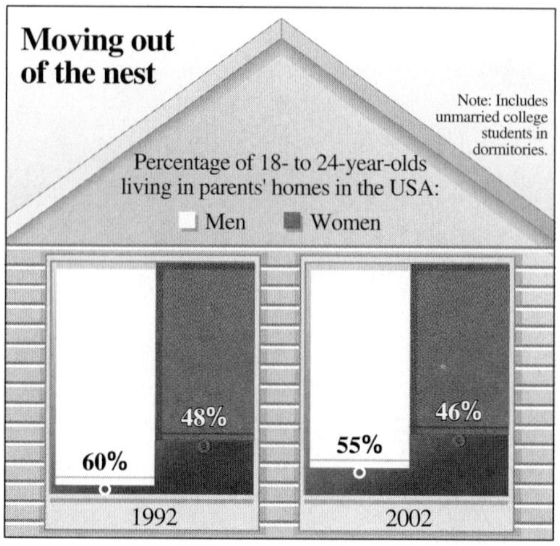

21. ***Men: Then and Now*** At $\alpha = 0.05$, can you support the claim that the proportion of men age 18 to 24 living in parents' homes was more in 1992 than in 2002?

22. ***Women: Then and Now*** At $\alpha = 0.05$, can you support the claim that the proportion of women age 18 to 24 living in parents' homes was more in 1992 than in 2002?

23. ***Then: Men and Women*** At $\alpha = 0.01$, can you reject the claim that the proportion of 18- to 24-year-olds living in parents' homes in 1992 was the same for men and women?

24. ***Now: Men and Women*** At $\alpha = 0.01$, can you reject the claim that the proportion of 18- to 24-year-olds living in parents' homes in 2002 was the same for men and women?

Extending Concepts

Constructing Confidence Intervals for $p_1 - p_2$ You can construct a confidence interval for the difference of two population means $p_1 - p_2$ by using the following inequality.

$$(\hat{p}_1 - \hat{p}_2) - z_c\sqrt{\frac{\hat{p}_1\hat{q}_1}{n_1} + \frac{\hat{p}_2\hat{q}_2}{n_2}} < p_1 - p_2 < (\hat{p}_1 - \hat{p}_2) + z_c\sqrt{\frac{\hat{p}_1\hat{q}_1}{n_1} + \frac{\hat{p}_2\hat{q}_2}{n_2}}$$

In Exercises 25 and 26, construct the indicated confidence interval for $p_1 - p_2$. Assume the samples are random and independent.

25. ***Students Planning to Study Engineering*** Several years ago, a survey of 977,000 students taking the SAT revealed that 11.7% of the students were planning to study engineering in college. In a recent survey of 1,085,000 students taking the SAT, 8.5% of the students were planning to study engineering. Construct a 95% confidence interval for the difference in proportions $p_1 - p_2$. *(Source: College Entrance Examination Board)*

26. ***Students Planning to Study Social Science*** In a certain year, the percent of students taking the SAT who said they intended to study social science in college was 7.5%. Eleven years later, 11.3% said they intended to study social science in college. Construct a 95% confidence interval for the difference in proportions $p_1 - p_2$. Assume 997,000 students were surveyed the first year and 1,085,000 students were surveyed 11 years later. *(Adapted from College Entrance Examination Board)*

Uses and Abuses

Statistics in the Real World

Uses

Hypothesis Testing with Two Samples Hypothesis testing enables you to decide whether differences in samples indicate actual differences in populations or are merely due to sampling error. Here are some of the kinds of questions that two-sample hypothesis testing can help you analyze. Is one medication more effective than another? Are boys more likely to have attention deficit disorder than girls? Does a new marketing method produce better results than the previous marketing method?

Abuses

Using Nonrepresentative Samples In comparisons of data collected from two different samples, care should be taken to ensure that the samples are similar in all characteristics except the variable being tested. For instance, suppose you are examining a claim that a new arthritis medication lessens joint pain.

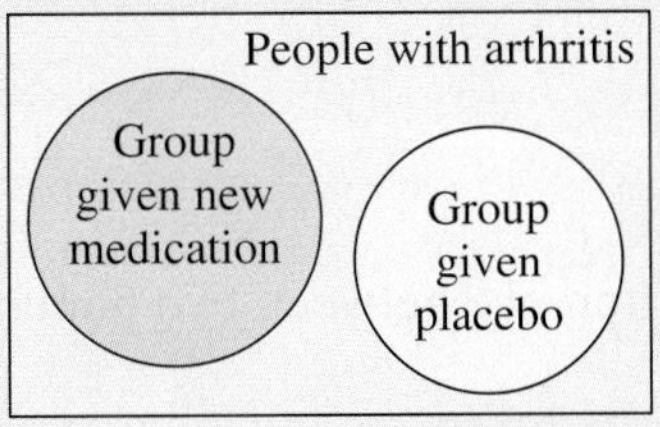

If the group that is given a medication is over 60 years old and the group given the placebo is under 40, variables other than the medication might affect the outcome of the study. When you look for other abuses in a study, consider how the claim in the study was determined. What were the sample sizes? Were the samples random? Were they independent? Was the sampling conducted by an unbiased researcher?

Exercises

1. ***Using Nonrepresentative Samples*** Assume that you work for the Food and Drug Administration (*www.fda.gov*). A pharmaceutical company has applied for approval to market a new arthritis mediation. The research involved a test group that was given the medication and another test group that was given a placebo. Describe some ways that the test groups might not have been representative of the entire population of people with arthritis.

2. Medical research often involves blind and double-blind testing. Explain what these two terms mean. If necessary, use your library or another research resource to help you answer this question.

8 Chapter Summary

What did you learn?

Review Exercises

Section 8.1

- How to perform a two-sample z-test for the difference between two means μ_1 and μ_2 using large independent samples — *1–6*

$$z = \frac{(\bar{x}_1 - \bar{x}_2) - (\mu_1 - \mu_2)}{\sigma_{\bar{x}_1 - \bar{x}_2}}$$

Section 8.2

- How to perform a t-test for the difference between two population means μ_1 and μ_2 using small independent samples — *7–14*

$$t = \frac{(\bar{x}_1 - \bar{x}_2) - (\mu_1 - \mu_2)}{\sigma_{\bar{x}_1 - \bar{x}_2}}$$

Section 8.3

- How to decide whether two samples are independent or dependent — *15, 16*
- How to perform a t-test to test the mean of the differences for a population of paired data — *17–22*

$$t = \frac{\bar{d} - \mu_d}{s_d / \sqrt{n}}$$

Section 8.4

- How to perform a z-test for the difference between two population proportions p_1 and p_2 — *23–28*

$$z = \frac{(\hat{p}_1 - \hat{p}_2) - (p_1 - p_2)}{\sqrt{\bar{p}\bar{q}\left(\frac{1}{n_1} + \frac{1}{n_2}\right)}}$$

Two-Sample Hypothesis Testing for Population Means

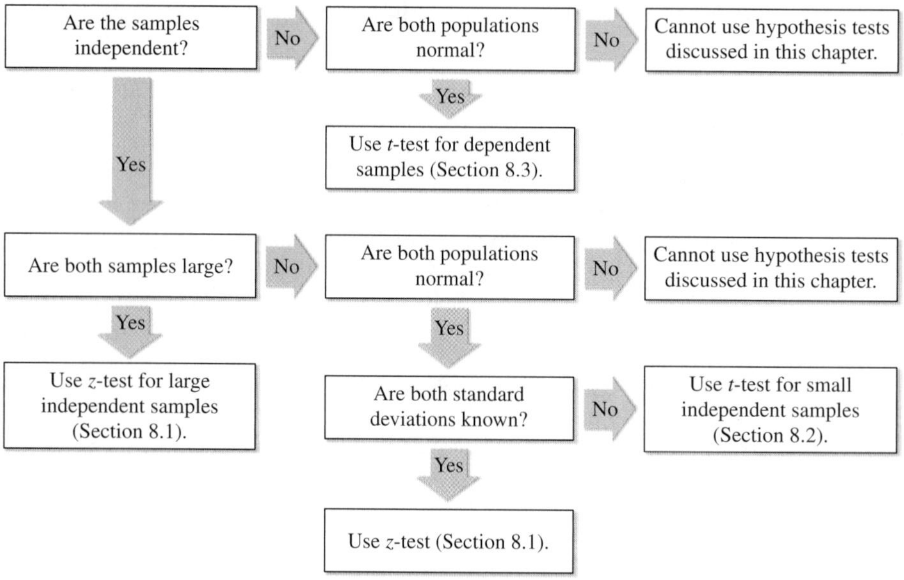

8 Review Exercises

1. Reject H_0. There is enough evidence at the 5% level to reject the claim.
2. Reject H_0. There is enough evidence at the 1% level to reject the claim.
3. Reject H_0. There is enough evidence at the 10% level to support the claim.
4. Reject H_0. There is enough evidence at the 5% level to support the claim.
5. (a) H_0: $\mu_1 \leq \mu_2$
   H_a: $\mu_1 > \mu_2$ (claim)
   (b) $z_0 = 1.645$
   Rejection region: $z > 1.645$
   (c) 1.750
   (d) Reject H_0.
   (e) There is sufficient evidence at the 5% level to conclude that the McDonald's sandwich has fewer calories than the Long John Silver's sandwich.
6. (a) H_0: $\mu_1 = \mu_2$
   H_a: $\mu_1 \neq \mu_2$ (claim)
   (b) $-z_0 = -1.645$ and $z_0 = 1.645$
   Rejection regions: $z < -1.645$, $z > 1.645$
   (c) -2.698
   (d) Reject H_0.
   (e) There is sufficient evidence at the 10% level to support the claim that the caloric contents of the two types of french fries are different.
7. Fail to reject H_0. There is not enough evidence at the 5% level to reject the claim.
8. Fail to reject H_0. There is not enough evidence at the 10% level to reject the claim.
9. Fail to reject H_0. There is not enough evidence at the 10% level to reject the claim.

Section 8.1

In Exercises 1–4, use the given sample statistics (independent samples) to test the claim about the difference between two population means μ_1 and μ_2 at the given level of significance α.

1. Claim: $\mu_1 \geq \mu_2$, $\alpha = 0.05$. Sample statistics: $\bar{x}_1 = 1.28$, $s_1 = 0.28$, $n_1 = 76$ and $\bar{x}_2 = 1.36$, $s_2 = 0.23$, $n_2 = 65$

2. Claim: $\mu_1 = \mu_2$, $\alpha = 0.01$. Sample statistics: $\bar{x}_1 = 5595$, $s_1 = 52$, $n_1 = 156$ and $\bar{x}_2 = 5575$, $s_2 = 68$, $n_2 = 216$

3. Claim: $\mu_1 < \mu_2$, $\alpha = 0.10$. Sample statistics: $\bar{x}_1 = 0.28$, $s_1 = 0.11$, $n_1 = 41$ and $\bar{x}_2 = 0.33$, $s_2 = 0.10$, $n_2 = 34$

4. Claim: $\mu_1 \neq \mu_2$, $\alpha = 0.05$. Sample statistics: $\bar{x}_1 = 82$, $s_1 = 11$, $n_1 = 410$ and $\bar{x}_2 = 90$, $s_2 = 10$, $n_2 = 340$

In Exercises 5 and 6, (a) identify the claim and state H_0 and H_a, (b) find the critical value(s) and identify the rejection region(s), (c) find the standardized test statistic z, (d) decide whether to reject or fail to reject the null hypothesis, and (e) interpret the decision in the context of the original claim.

5. In a fast food study, a nutritionist finds that the mean calorie content of 36 randomly selected Long John Silver's fish sandwiches is 430 calories with a standard deviation of 43 calories. The mean calorie content of 41 randomly selected McDonald's fish sandwiches is 410 calories with a standard deviation of 57 calories. At $\alpha = 0.05$, is there enough evidence for the nutritionist to conclude that the McDonald's sandwich has fewer calories than the Long John Silver's sandwich? *(Adapted from DietFacts)*

6. A study of fast food nutrition compared the caloric content of french fries. Thirty-eight randomly selected servings of Burger King medium french fries had a mean of 360 calories and a standard deviation of 50 calories, and 35 randomly selected servings of Wendy's medium french fries had a mean of 390 calories and a standard deviation of 45 calories. At $\alpha = 0.10$, can you support the claim that the caloric contents of the two types of french fries are different? *(Adapted from Burger King Brands, Inc. and DietFacts)*

Section 8.2

In Exercises 7–12, use the given sample statistics to test the claim about the difference between two population means μ_1 and μ_2 at the given level of significance α. Assume the samples are random and independent and that the populations are approximately normally distributed.

7. Claim: $\mu_1 = \mu_2$, $\alpha = 0.05$. Sample statistics: $\bar{x}_1 = 250$, $s_1 = 26$, $n_1 = 21$ and $\bar{x}_2 = 240$, $s_2 = 22$, $n_2 = 12$. Assume equal variances.

8. Claim: $\mu_1 = \mu_2$, $\alpha = 0.10$. Sample statistics: $\bar{x}_1 = 0.015$, $s_1 = 0.011$, $n_1 = 8$ and $\bar{x}_2 = 0.019$, $s_2 = 0.004$, $n_2 = 6$. Assume variances are not equal.

9. Claim: $\mu_1 \leq \mu_2$, $\alpha = 0.05$. Sample statistics: $\bar{x}_1 = 183.5$, $s_1 = 1.3$, $n_1 = 25$ and $\bar{x}_2 = 184.7$, $s_2 = 3.9$, $n_2 = 25$. Assume variances are not equal.

10. Fail to reject H_0. There is not enough evidence at the 1% level to reject the claim.

11. Reject H_0. There is enough evidence at the 1% level to support the claim.

12. Fail to reject H_0. There is not enough evidence at the 10% level to reject the claim.

13. (a) H_0: $\mu_1 \leq \mu_2$
H_a: $\mu_1 > \mu_2$ (claim)
(b) $t_0 = 1.645$
Rejection region: $t > 1.645$
(c) 2.266
(d) Reject H_0.
(e) There is sufficient evidence at the 5% level to support the claim that third graders taught with the directed reading activities scored higher than those taught without the activities.

14. (a) H_0: $\mu_1 = \mu_2$
H_a: $\mu_1 \neq \mu_2$ (claim)
(b) $-t_0 = -2.086$ and $t_0 = 2.086$
Rejection regions: $t < -2.086$, $t > 2.086$
(c) 1.600
(d) Fail to reject H_0.
(e) There is not enough evidence at the 5% level to support the agent's claim that there is no difference between the mean household incomes of the two neighborhoods.

15. Independent because the two samples of laboratory mice are different groups.

10. Claim: $\mu_1 \geq \mu_2$, $\alpha = 0.01$. Sample statistics: $\bar{x}_1 = 25.6$, $s_1 = 8.25$, $n_1 = 15$ and $\bar{x}_2 = 22.4$, $s_2 = 7.85$, $n_2 = 34$. Assume equal variances.

11. Claim: $\mu_1 \neq \mu_2$, $\alpha = 0.01$. Sample statistics: $\bar{x}_1 = 61$, $s_1 = 3.3$, $n_1 = 5$ and $\bar{x}_2 = 55$, $s_2 = 1.2$, $n_2 = 7$. Assume equal variances.

12. Claim: $\mu_1 \geq \mu_2$, $\alpha = 0.10$. Sample statistics: $\bar{x}_1 = 520$, $s_1 = 25$, $n_1 = 7$ and $\bar{x}_2 = 500$, $s_2 = 55$, $n_2 = 6$. Assume variances are not equal.

In Exercises 13 and 14, (a) identify the claim and state H_0 and H_a, (b) find the critical value(s) and identify the rejection region(s), (c) find the standardized test statistic, (d) decide whether to reject or fail to reject the null hypothesis, and (e) interpret the decision in the context of the original claim.

DATA

13. A study of methods for teaching reading in the third grade was conducted. One classroom of 21 students participated in directed reading activities for eight weeks. Another classroom of 23 students followed the same curriculum without the activities. Students in both classrooms then took the same reading test.

The following are reading scores for the first (treated) classroom.

24	43	58	71	43	49	61	44	67	49	53	
56	59	52	62	54	57	33	46	43	57		

The following are reading scores for the second (control) classroom.

42	43	55	26	62	37	33	41	19	54	20	85
46	10	17	60	53	42	37	42	55	28	48	

Diagnostics suggest that the data are sampled from a normal population and the population variances are equal. Test the claim that third graders taught with the directed reading activities scored higher than those taught without the activities. Use $\alpha = 0.05$. *(Source: StatLib/Schmitt, Maribeth C., The Effects of an Elaborated Directed Reading Activity on the Metacomprehension Skills of Third Graders)*

14. A real estate agent claims that there is no difference between the mean household incomes of two neighborhoods. The mean income of 12 randomly selected households from the first neighborhood was \$18,250 with a standard deviation of \$1200. In the second neighborhood, 10 randomly selected households had a mean income of \$17,500 with a standard deviation of \$950. Assume normal distributions and equal population variances. Test the claim at $\alpha = 0.05$.

Section 8.3

In Exercises 15 and 16, classify the two given samples as independent or dependent. Explain your reasoning.

15. Sample 1: Maze completion times for 14 standard laboratory mice
Sample 2: Maze completion times for 14 laboratory mice bred for higher metabolic rate

16. Dependent because the same mice were used for both experiments.
17. Two-tailed test; Reject H_0.
18. Left-tailed test; Fail to reject H_0.
19. Right-tailed test; Reject H_0.
20. Two-tailed test; Reject H_0.
21. (a) H_0: $\mu_d \leq 0$; H_a: $\mu_d > 0$ (claim)
(b) $t_0 = 1.383$
Rejection region: $t > 1.383$
(c) $\bar{d} = 5$; $s_d \approx 8.743$
(d) 1.808
(e) Reject H_0.
(f) There is enough evidence at the 10% level to support the claim that the men's diastolic blood pressure decreased.
22. (a) H_0: $\mu_d \leq 0$; H_a: $\mu_d > 0$ (claim)
(b) $t_0 = 1.372$
Rejection region: $t > 1.372$
(c) $\bar{d} \approx -0.636$; $s_d \approx 5.870$
(d) -0.359
(e) Fail to reject H_0.
(f) There is not enough evidence at the 10% level to support the claim that systolic blood pressure was lowered after the herbal supplement was taken.

16. Sample 1: Maze completion times for 43 mice
Sample 2: Maze completion times for those 43 mice after two weeks of maze practice

In Exercises 17–20, using a test for paired data, test the claim about the mean of the difference of the two populations at the given level of significance α using the given statistics. Is the test right tailed, left tailed, or two tailed? Assume the populations are normally distributed.

17. Claim: $\mu_d = 0, \alpha = 0.05$. Statistics: $\bar{d} = 10, s_d = 12.4, n = 100$

18. Claim: $\mu_d < 0, \alpha = 0.01$. Statistics: $\bar{d} = 3.2, s_d = 1.38, n = 25$

19. Claim: $\mu_d \leq 6, \alpha = 0.10$. Statistics: $\bar{d} = 10.3, s_d = 1.24, n = 33$

20. Claim: $\mu_d \neq 15, \alpha = 0.05$. Statistics: $\bar{d} = 17.5, s_d = 4.05, n = 37$

In Exercises 21 and 22, (a) identify the claim and state H_0 and H_a, (b) find the critical value(s) and identify the rejection region(s), (c) calculate $\bar{d}$ and s_d, (d) use the t-test to find the standardized test statistic, (e) decide whether to reject or fail to reject the null hypothesis, and (f) interpret the decision in the context of the original claim. For each sample, assume the distribution of the population is normal.

DATA **21.** A medical researcher wants to test the effects of calcium supplements on men's blood pressure. In part of the study, 10 randomly selected men are given a calcium supplement for 12 weeks. The researcher measures the men's diastolic blood pressure at the beginning and at the end of the 12-week study and records the results shown below. At $\alpha = 0.10$, can the researcher claim that the men's diastolic blood pressure decreased? *(Source: The Journal of the American Medical Association)*

Patient	1	2	3	4	5	6	7	8	9	10
Before	107	110	123	129	112	111	107	112	136	102
After	100	114	105	112	115	116	106	102	125	104

DATA **22.** In a study testing the effects of an herbal supplement on blood pressure in men, 11 randomly selected men were given an herbal supplement for 15 weeks. The following measurements are for each subject's diastolic blood pressure taken before and after the 15-week treatment period. At $\alpha = 0.10$, can you support the claim that systolic blood pressure was lowered? *(Adapted from The Journal of the American Medical Association)*

Patient	1	2	3	4	5	6	7	8	9	10	11
Before	123	109	112	102	98	114	119	112	110	117	130
After	124	97	113	105	95	119	114	114	121	118	133

23. Two-tailed test; Fail to reject H_0.
24. Right-tailed test; Fail to reject H_0.
25. Right-tailed test; Fail to reject H_0.
26. Left-tailed test; Fail to reject H_0.
27. (a) H_0: $p_1 = p_2$ (claim)
H_a: $p_1 \neq p_2$
(b) $-z_0 = -1.645$ and $z_0 = 1.645$
Rejection regions: $z < -1.645$, $z > 1.645$
(c) -0.718
(d) Fail to reject H_0.
(e) There is not enough evidence at the 10% level to reject the claim that the proportions of Canadians with a college degree were the same for the two years.
28. (a) H_0: $p_1 \leq p_2$
H_a: $p_1 > p_2$ (claim)
(b) $z_0 = 1.645$
Rejection region: $z > 1.645$
(c) 51.458 (d) Reject H_0.
(e) There is enough evidence at the 5% level to support the claim that the proportion of females with intended education majors was higher than the proportion of females with intended biological science majors.

Section 8.4

In Exercises 23–26, test the claim about the difference between two population proportions p_1 and p_2 at the given level of significance α using the given sample statistics. Is the test right tailed, left tailed, or two tailed? Assume the sample statistics are from independent samples.

23. Claim: $p_1 = p_2$, $\alpha = 0.05$. Sample statistics: $x_1 = 375$, $n_1 = 720$ and $x_2 = 365$, $n_2 = 660$

24. Claim: $p_1 \leq p_2$, $\alpha = 0.01$. Sample statistics: $x_1 = 15$, $n_1 = 100$ and $x_2 = 42$, $n_2 = 200$

25. Claim: $p_1 > p_2$, $\alpha = 0.10$. Sample statistics: $x_1 = 227$, $n_1 = 556$ and $x_2 = 198$, $n_2 = 420$

26. Claim: $p_1 < p_2$, $\alpha = 0.05$. Sample statistics: $x_1 = 86$, $n_1 = 900$ and $x_2 = 107$, $n_2 = 1200$

In Exercises 27 and 28, (a) identify the claim and state H_0 and H_a, (b) find the critical value(s) and identify the rejection region(s), (c) find the standardized test statistic, (d) decide whether to reject or fail to reject the null hypothesis, and (e) interpret the decision in the context of the original claim. Assume the samples are independent.

27. In a random sample of 200 Canadians eight years ago, 26 had college degrees. In a recent year, in a random sample of 300 Canadians, 46 had a college degree. At $\alpha = 0.10$, can you reject the claim that the proportions of Canadians with a college degree were the same for the two years? *(Adapted from Statistics Canada)*

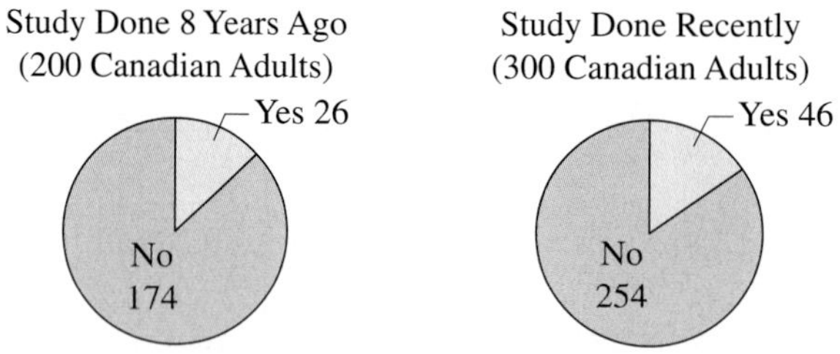

28. In a recent year, of 79,160 students taking the SAT I who were planning to major in education, 61,745 were female. That year, of the 50,829 students planning to major in biological sciences, 33,039 were female. At $\alpha = 0.05$, test the claim that the proportion of females with intended education majors was higher than the proportion of females with intended biological sciences majors. *(Source: College Board)*

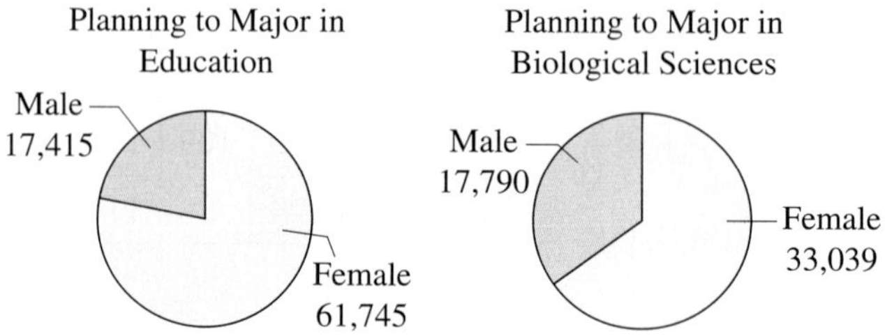

8 Chapter Quiz

Take this quiz as you would take a quiz in class. After you are done, check your work against the answers given in the back of the book.

For this quiz, do the following.

(a) Write the claim mathematically and identify H_0 and H_a.

(b) Determine whether the hypothesis test is a one-tailed test or a two-tailed test and whether to use a z-test or a t-test. Explain your reasoning.

(c) Find the critical value(s) and identify the rejection region(s).

(d) Use the appropriate test to find the appropriate standardized test statistic.

(e) Decide whether to reject or fail to reject the null hypothesis.

(f) Interpret the decision in the context of the original claim.

1. The mean score on a science assessment for 49 randomly selected male high school students was 300.4 and the standard deviation was 1.6. The mean score on the same test for 50 randomly selected female high school students was 290.6 and the standard deviation was 1.5. At $\alpha = 0.05$, can you support the claim that the mean score on the science assessment for the male high school students was higher than for the female high school students? *(Adapted from National Center for Educational Statistics)*

2. A science teacher suggests that the mean scores on a science assessment test for nine-year-old boys and girls are equal. If the mean score for 13 randomly selected boys is 230.9 with a standard deviation of 1.3 and the mean score for 15 randomly selected girls is 227.9 with a standard deviation of 1.1, can you reject the teacher's suggestion? Assume the populations are normally distributed and the variances are equal. Use $\alpha = 0.01$. *(Adapted from National Center for Educational Statistics)*

3. In a recent year, of 1,296,000 accidents involving drivers aged 21 to 24, 5% involved alcohol. That year, of 856,000 accidents involving drivers aged 65 years and older, 1% involved alcohol. At $\alpha = 0.10$, can you support the claim that the proportion of accidents involving alcohol is higher for drivers in the 21 to 24 age group than for drivers aged 65 and older? *(Source: U.S. National Highway Traffic Safety Administration)*

DATA

4. The table shows the scores for 12 randomly selected students the first and second times they took the mathematics SAT. At $\alpha = 0.05$, is there enough evidence to conclude that the students' SAT scores improved on the second test? Assume the populations are normally distributed.

Student	1	2	3	4	5	6
Score on first SAT	457	419	343	539	394	413
Score on second SAT	532	523	427	607	444	490

Student	7	8	9	10	11	12
Score on first SAT	392	421	439	340	493	339
Score on second SAT	428	524	532	397	550	357

1. (a) H_0: $\mu_1 \le \mu_2$
H_a: $\mu_1 > \mu_2$ (claim)

(b) Right tailed z-test because n_1 and n_2 are each greater than 30 and the samples are independent.

(c) $z_0 = 1.645$

Rejection region: $z > 1.645$

(d) 31.426 (e) Reject H_0.

(f) There is sufficient evidence at the 5% level to support the claim that the mean score on the science assessment for the male high school students was higher than for the female high school students.

2. (a) H_0: $\mu_1 = \mu_2$ (claim)
H_a: $\mu_1 \ne \mu_2$

(b) Two tailed t-test because n_1 and n_2 are less than 30, the samples are independent, and the populations are normally distributed.

(c) $-t_0 = -2.779$ and $t_0 = 2.779$

Rejection regions: $t < -2.779$, $t > 2.779$

(d) 6.617 (e) Reject H_0.

(f) There is sufficient evidence at the 1% level to reject the teacher's suggestion that the mean scores on the science assessment test are the same for nine-year-old boys and girls.

3. (a) H_0: $p_1 \le p_2$; H_a: $p_1 > p_2$ (claim)

(b) Right tailed z-test because you are testing proportions and $n_1\bar{p}$, $n_2\bar{p}$, $n_1\bar{q}$, and $n_2\bar{q} \ge 5$ and the samples are independent.

(c) $z_0 = 1.282$

Rejection region: $z > 1.282$

(d) 158.471 (e) Reject H_0.

(f) There is sufficient evidence at the 10% level to support the claim that the proportion of accidents involving alcohol is higher for drivers in the 21 to 24 age group than for drivers aged 65 and older.

4. See Odd Answers, page A76.

PUTTING IT ALL TOGETHER

Real Statistics ■ Real Decisions

You work for the statistics department of the National Center for Health Statistics. In an effort to determine if the mean length of hospital stays for patients diagnosed with pneumonia is different today from what it was 10 years ago, your department decides to analyze data from a random selection of hospital records. The results for several patients ages 17 and younger are shown in the histograms for 10 years ago and this year.

Your department believes that there is a difference in the mean length of hospital stays for patients diagnosed with pneumonia. As the department's research analyst, it is your job to work with the data and determine if this claim can be supported.

Exercises

1. *How Would You Do It?*

(a) What sampling technique would you use to select the sample for the study? Why? What sampling technique would you use if you divided hospital records according to four geographic regions (northeast, south, midwest, and west), randomly selected one region, and obtained records from each hospital in that region?

(b) Which technique in part (a) would be easiest to implement? Why?

(c) Identify possible flaws or biases in your study.

2. *Choosing a Test*

To test the claim that there is a difference in the mean length of hospital stays for patients (ages 17 and younger) diagnosed with pneumonia, should you use a z-test or a t-test? Are the samples independent or dependent? Do you need to know what each population's distribution is? Do you need to know anything about the population variances?

3. *Testing a Mean*

Test the claim that there is a difference in the mean length of hospital stays for patients (ages 17 and younger) diagnosed with pneumonia. Assume the populations are normal and the population variances are not equal. Use $\alpha = 0.05$. Write a paragraph that interprets the test's decision. Does the decision support your department's claim?

Patients 17 and Younger with Pneumonia (10 Years Ago)

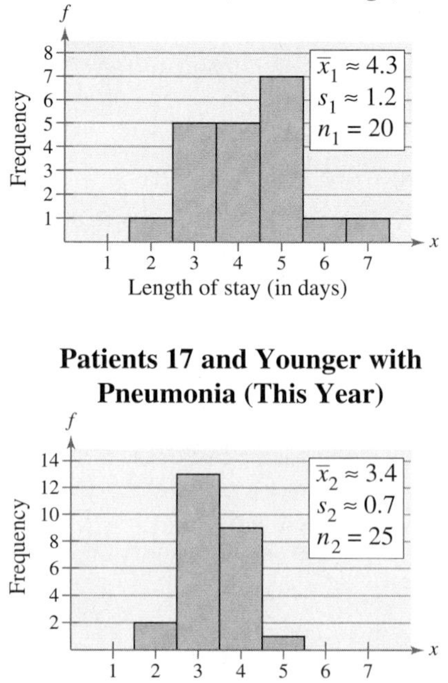

Patients 17 and Younger with Pneumonia (This Year)

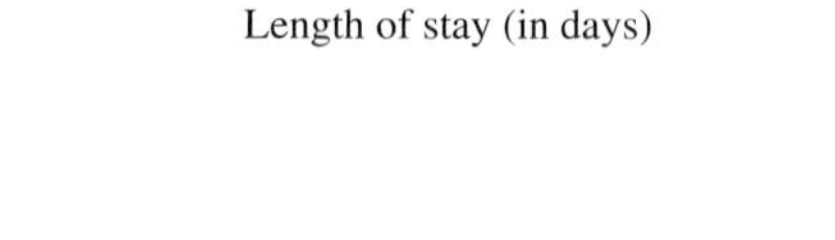

TECHNOLOGY | MINITAB | EXCEL | TI-83

Tails Over Heads

In the article "Tails over Heads" in the *Washington Post* (Oct. 13, 1996), journalist William Casey describes one of his hobbies—keeping track of every coin he finds on the street! From January 1, 1985 until the article was written, Casey found 11,902 coins.

As each coin is found, Casey records the time, date, location, value, mint location, and whether the coin is lying heads up or tails up. In the article, Casey notes that 6130 coins were found tails up and 5772 were found heads up. Of the 11,902 coins found, 43 were minted in San Francisco, 7133 were minted in Philadelphia, and 4726 were minted in Denver.

A simulation of Casey's experiment can be done in MINITAB as shown below. A frequency histogram of one simulation's results is shown at the right.

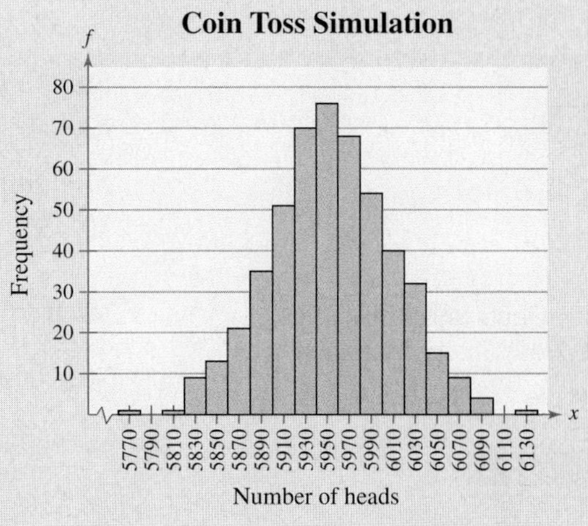

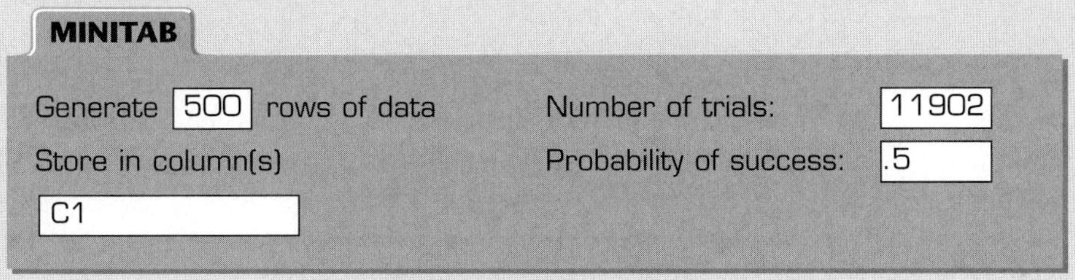

Calc, Random Data, Binomial Distribution

Exercises

1. Use a technology tool to perform a one-sample z-test to test the hypothesis that the probability that a "found coin" will be lying heads up is 0.5. Use $\alpha = 0.01$. Use Casey's data as your sample and write your conclusion as a sentence.

2. Do Casey's data differ significantly from chance? If so, what might be the reason?

3. In the simulation shown above, what percent of the trials had heads less than or equal to the number of heads in Casey's data? Use a technology tool to repeat the simulation. Are your results comparable?

In Exercises 4 and 5, use a technology tool to perform a two-sample z-test to decide whether there is a difference in the mint dates and in the values of coins found on a street from 1985 through 1996. Write your conclusion as a sentence. Use $\alpha = 0.05$.

4. Mint dates of coins (years)

Philadelphia:	Denver:
$\bar{x}_1 = 1984.8$	$\bar{x}_2 = 1983.4$
$s_1 = 8.6$	$s_2 = 8.4$

5. Value of coins (dollars)

Philadelphia:	Denver:
$\bar{x}_1 = \$0.034$	$\bar{x}_2 = \$0.033$
$s_1 = \$0.054$	$s_2 = \$0.052$

Extended solutions are given in the *Technology Supplement*.
Technical instruction is provided for MINITAB, Excel, and the TI-83.

8 Using Technology to Perform Two-Sample Hypothesis Tests

Here are some MINITAB and TI-83 printouts for several examples in this chapter. To duplicate the MINITAB results, you need the original data. For the TI-83, you can simply enter the descriptive statistics.

(See Example 1, page 418.)

Stopping Distances (in feet) for 10 Firestone Winterfire Tires

63 48 57 42 54 58 45 55 37 51

Stopping Distances (in feet) for 12 Michelin XM+S Alpin Tires

49 52 52.5 53 54.5 55 55.5 56 57 57.5 58.5 59.5

Display Descriptive Statistics...
Store Descriptive Statistics...
1-Sample Z...
1-Sample t...
2-Sample t...
Paired t...
1 Proportion...
2 Proportions...

MINITAB

Two-Sample T-Test and Confidence Interval

```
Two-sample T for C1 vs C2

      N     Mean    StDev    SE Mean
C1   10    51.00     8.00        2.5
C2   12    55.00     3.01       0.87

99% CI for difference: (-12.31, 4.31)
T-Test of difference = 0 (vs not = ): T-Value = -1.50
P-Value = 0.163  DF = 11
Difference = mu C1 - mu C2
Estimate for difference: -4.00
```

(See Example 2, page 428.)

Golf Scores, Old and New Club Design

Golfer	1	2	3	4	5	6	7	8
Score (old design)	89	84	96	82	74	92	85	91
Score (new design)	83	83	92	84	76	91	80	91

Display Descriptive Statistics...
Store Descriptive Statistics...
1-Sample Z...
1-Sample t...
2-Sample t...
Paired t...
1 Proportion...
2 Proportions...

MINITAB

Paired T-Test

```
Paired T for C1 - C2

             N     Mean    StDev    SE Mean
C1           8    86.63     6.89       2.43
C2           8    85.00     5.81       2.05
Difference   8     1.63     3.07       1.08

T-Test of mean difference = 0 (vs>0): T-Value = 1.50  P-Value = 0.089
```

(See Example 1, page 407.)

TI-83

```
EDIT CALC TESTS
1: Z-Test...
2: T-Test...
3: 2-SampZTest...
4: 2-SampTTest...
5: 1-PropZTest...
6: 2-PropZTest...
7↓ZInterval...
```

TI-83

```
2-SampZTest
 Inpt:Data Stats
 σ1: 12000
 σ2: 15000
 x̄1: 60900
 n1: 100
 x̄2: 64300
↓n2: 100
```

TI-83

```
2-SampZTest
↑σ2: 15000
 x̄1: 60900
 n1: 100
 x̄2: 64300
 n2: 100
 μ1: ≠μ2 <μ2 >μ2
 Calculate Draw
```

TI-83

```
2-SampZTest
 μ1 ≠ μ2
 z= −1.769969301
 p= .0767321594
 x̄1= 60900
 x̄2= 64300
↓n1= 100
```

(See Example 2, page 419.)

TI-83

```
EDIT CALC TESTS
1: Z-Test...
2: T-Test...
3: 2-SampZTest...
4: 2-SampTTest...
5: 1-PropZTest...
6: 2-PropZTest...
7↓ZInterval...
```

TI-83

```
2-SampTTest
 Inpt:Data Stats
 x̄1: 1275
 Sx1: 45
 n1: 14
 x̄2: 1250
 Sx2: 30
↓n2: 16
```

TI-83

```
2-SampTTest
↑n1: 14
 x̄2: 1250
 Sx2: 30
 n2: 16
 μ1: ≠μ2 <μ2 >μ2
 Pooled: No Yes
 Calculate Draw
```

TI-83

```
2-SampTTest
 μ1 > μ2
 t= 1.811358919
 p= .0404131295
 df= 28
 x̄1= 1275
↓x̄2= 1250
```

(See Example 1, page 439.)

TI-83

```
EDIT CALC TESTS
1: Z-Test...
2: T-Test...
3: 2-SampZTest...
4: 2-SampTTest...
5: 1-PropZTest...
6: 2-PropZTest...
7↓ZInterval...
```

TI-83

```
2-PropZTest
 x1: 60
 n1: 200
 x2: 95
 n2: 250
 p1: ≠p2 <p2 >p2
 Calculate Draw
```

TI-83

```
2-PropZTest
 p1 ≠ p2
 z= −1.774615984
 p= .0759612188
 p̂1= .3
 p̂2= .38
↓p̂= .3444444444
```

CHAPTER 9

Correlation and Regression

9.1 Correlation
9.2 Linear Regression
Case Study
9.3 Measures of Regression and Prediction Intervals
9.4 Multiple Regression
Uses and Abuses
Real Statistics–Real Decisions
Technology

The modern summer Olympic games began in Athens, Greece, in 1896. The winter Olympic games began in Chamonix, France, in 1924. The games have been held in the United States eight times: St. Louis (S 1904), Lake Placid (W 1932), Los Angeles (S 1932), Squaw Valley (W 1960), Lake Placid (W 1980), Los Angeles (S 1984), Atlanta (S 1996), and Salt Lake City (W 2002).

Where You've Been

In Chapters 1–8, you studied descriptive statistics, probability, and inferential statistics. One of the techniques you learned in descriptive statistics was graphing paired data with a scatter plot (Section 2.2). For instance, the winning times for the men's and women's 100-meter run in the summer Olympics from 1928 through 2004 are shown in tabular form at the right and in graphical form below.

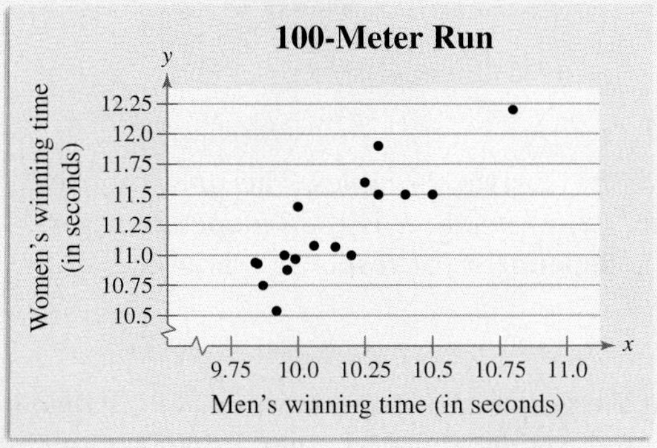

Winning Times for 100-Meter Run Summer Olympics

Year	Men (in seconds), x	Women (in seconds), y
1928	10.80	12.20
1932	10.30	11.90
1936	10.30	11.50
1948	10.30	11.90
1952	10.40	11.50
1956	10.50	11.50
1960	10.20	11.00
1964	10.00	11.40
1968	9.95	11.00
1972	10.14	11.07
1976	10.06	11.08
1980	10.25	11.60
1984	9.99	10.97
1988	9.92	10.54
1992	9.96	10.82
1996	9.84	10.94
2000	9.87	10.75
2004	9.85	10.93

Where You're Going

In this chapter, you will study how to describe and test the significance of relationships between two variables when data are presented as ordered pairs. For instance, in the scatter plot for the 100-meter run, it appears that women's fast times tend to correspond to men's fast times and women's slow times tend to correspond to men's slow times. This relationship is described by saying the women's times are positively correlated to the men's times. Graphically, the relationship can be described by drawing a line, called a regression line, that fits the points as closely as possible. The second scatter plot below shows a similar result—the women's running high jump heights are positively correlated to the men's running high jump heights.

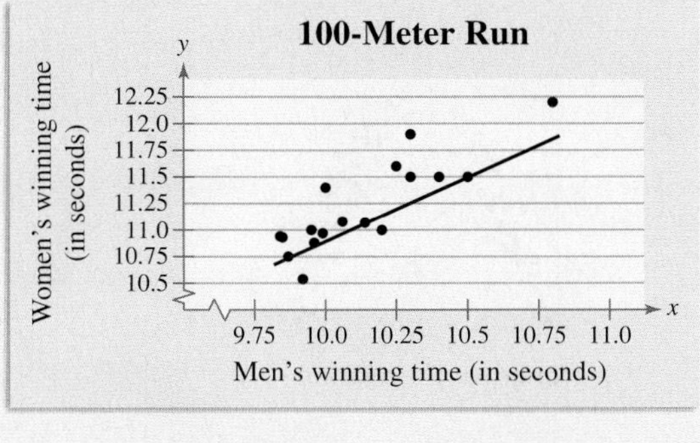

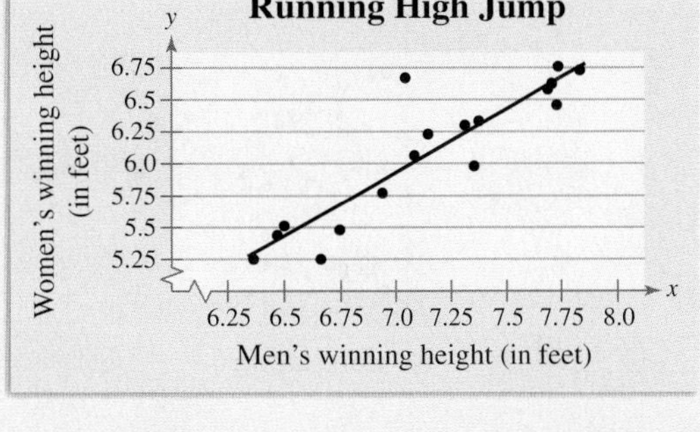

9.1 Correlation

What You Should Learn

- An introduction to linear correlation, independent and dependent variables, and the types of correlation
- How to find a correlation coefficient
- How to test a population correlation coefficient ρ using a table
- How to perform a hypothesis test for a population correlation coefficient ρ
- How to distinguish between correlation and causation

An Overview of Correlation • Correlation Coefficient • Using a Table to Test a Population Correlation Coefficient ρ • Hypothesis Testing for a Population Correlation Coefficient ρ • Correlation and Causation

An Overview of Correlation

Suppose a safety inspector wants to determine whether a relationship exists between the number of hours of training for an employee and the number of accidents involving that employee. Or suppose a psychologist wants to know whether a relationship exists between the number of hours a person sleeps each night and that person's reaction time. How would you help them determine if any relationship exists?

In this section, you will study how to describe what type of relationship, or correlation, exists between two variables and how to determine whether the correlation is significant.

DEFINITION

A **correlation** is a relationship between two variables. The data can be represented by the ordered pairs (x, y) where x is the **independent** (or **explanatory**) **variable** and y is the **dependent** (or **response**) **variable.**

Note to Instructor

You can begin Chapter 9 using an intuitive approach. Ask students what type of relationship exists between variables such as scores and college grades, height, and IQ. Also, throughout the chapter, emphasize interpretation rather than calculation. Many of the calculations can be performed using technology.

In Section 2.2, you learned that the graph of ordered pairs (x, y) is called a **scatter plot.** A scatter plot can be used to determine whether a linear (straight line) correlation exists between two variables. In a scatter plot, the ordered pairs (x, y) are graphed as points in a coordinate plane. The independent (explanatory) variable x is measured by the horizontal axis, and the dependent (response) variable y is measured by the vertical axis. The following scatter plots show several types of correlation.

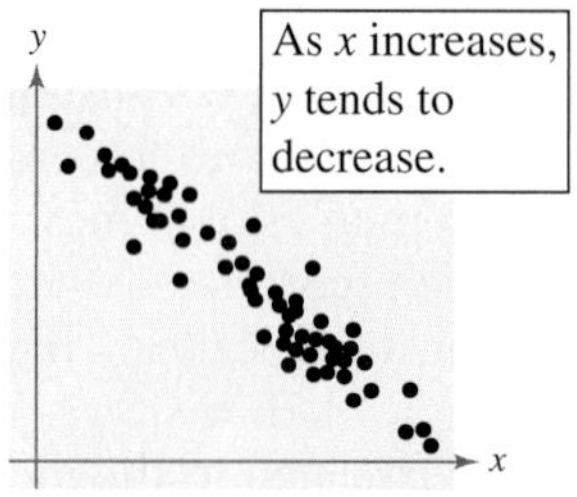

Negative Linear Correlation

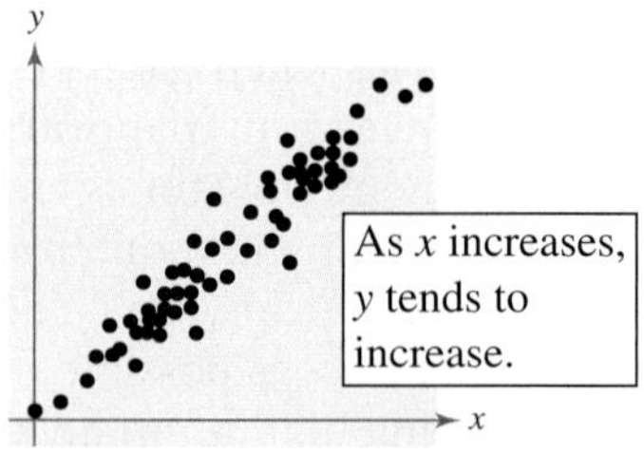

Positive Linear Correlation

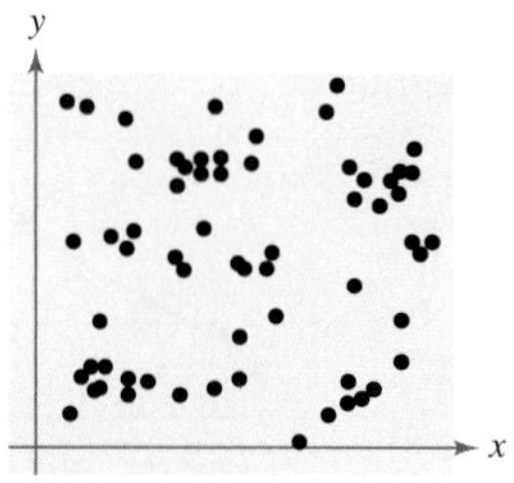

No Correlation

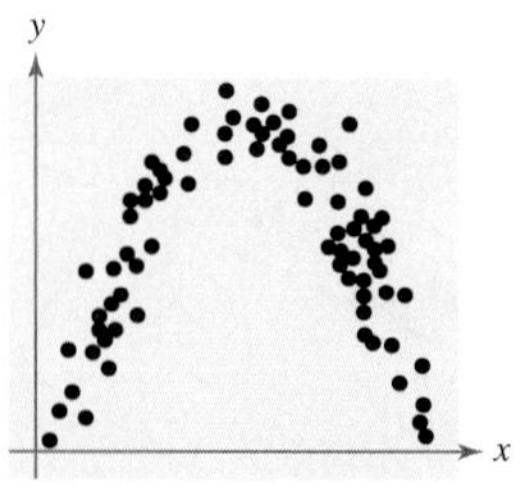

Nonlinear Correlation

EXAMPLE 1

Constructing a Scatter Plot

Advertising expenses (1000s of \$), *x*	Company sales (1000s of \$), *y*
2.4	225
1.6	184
2.0	220
2.6	240
1.4	180
1.6	184
2.0	186
2.2	215

A marketing manager conducted a study to determine whether there is a linear relationship between money spent on advertising and company sales. The data are shown in the table at the left. Display the data in a scatter plot and determine whether there appears to be a positive or negative linear correlation or no linear correlation.

SOLUTION The scatter plot is shown at the right. From the scatter plot, it appears that there is a positive linear correlation between the variables.

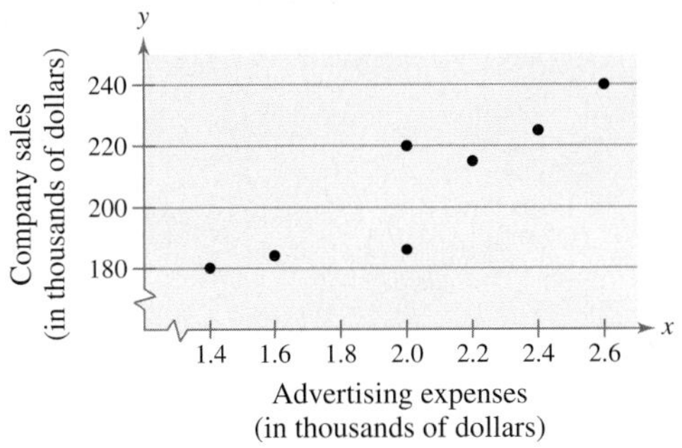

Interpretation Reading from left to right, as the advertising expenses increase, the sales tend to increase.

Try It Yourself 1

Income level (in 1000s of \$), *x*	Donating percent, *y*
42	9
48	10
50	8
59	5
65	6
72	3

A sociologist conducted a study to determine whether there is a linear relationship between family income level (in thousands of dollars) and percent of income donated to charities. The data are shown in the table at the left. Display the data in a scatter plot and determine the type of correlation.

a. *Draw* and *label* the x- and y-axes.
b. *Plot* each ordered pair on the graph.
c. Does there appear to be a linear correlation?

Answer: Page A42

Note to Instructor

Have students try to picture a line going through the points. If the line clearly has a positive slope, the correlation is positive. If the slope is clearly negative, the correlation is negative. If they have trouble picturing a line through the points, there is probably no linear correlation. Other types of correlation are introduced in the Extending Concepts of this section's exercise set.

EXAMPLE 2

Constructing a Scatter Plot

A student nurse conducts a study to determine whether there is a linear relationship between an individual's weight (in pounds) and daily water consumption (in ounces). The data are shown in the following table. Organize the data in a scatter plot and describe the type of correlation.

Weight, *x*	102	119	124	141	142	154	201	220
Water, *y*	50	32	82	64	54	21	86	39

SOLUTION The scatter plot is shown at the right. From the scatter plot, it appears that there is no linear correlation between the variables.

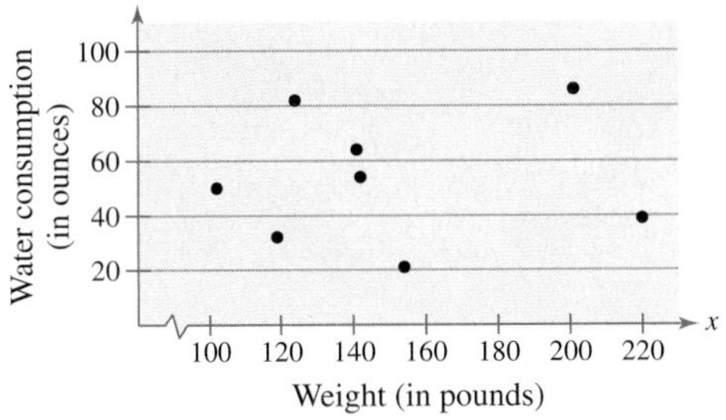

Interpretation A person's weight does not appear to be related to the amount of water that person consumes.

Try It Yourself 2

A marketing manager conducts a study to determine whether there is a linear relationship between a person's age and the number of magazines to which that person subscribes. The data are shown in the following table. Display the data in a scatter plot and determine the type of correlation.

Age, x	21	26	33	35	48	50	55	64
Subscriptions, y	4	0	3	1	3	0	2	6

a. *Draw* and *label* the x- and y-axes.
b. *Plot* each ordered pair on the graph.
c. Does there appear to be a linear correlation?

Answer: Page A42

Study Tip

Save any data put into a technology tool because these data will be used throughout the chapter.

EXAMPLE 3

Constructing a Scatter Plot Using Technology

Old Faithful, located in Yellowstone National Park, is the world's most famous geyser. The duration (in minutes) of several of Old Faithful's eruptions and the times (in minutes) until the next eruption are shown in the table at the left. Using a TI-83, display the data in a scatter plot. Determine the type of correlation.

Duration, x	Time, y	Duration, x	Time, y
1.80	56	3.78	79
1.82	58	3.83	85
1.88	60	3.87	81
1.90	62	3.88	80
1.92	60	4.10	89
1.93	56	4.27	90
1.98	57	4.30	84
2.03	60	4.30	89
2.05	57	4.43	84
2.13	60	4.43	89
2.30	57	4.47	86
2.35	57	4.47	80
2.37	61	4.53	89
2.82	73	4.55	86
3.13	76	4.60	88
3.27	77	4.60	92
3.65	77	4.63	91
3.70	82		

SOLUTION Begin by entering the x-values into List 1 and the y-values into List 2. Use *Stat Plot* to construct the scatter plot. The plot should look similar to the one shown below. From the scatter plot, it appears that the variables have a positive linear correlation.

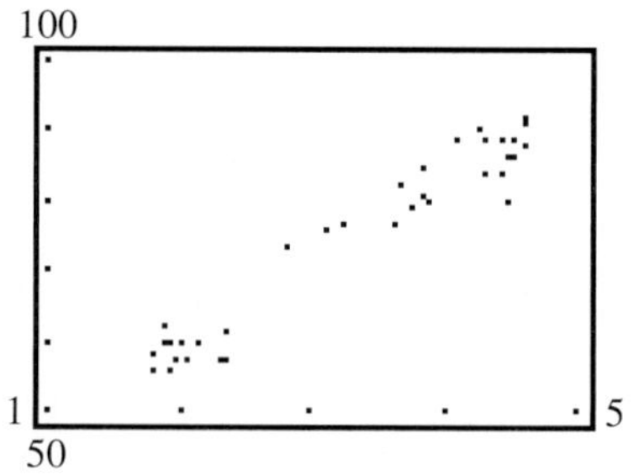

Interpretation You can conclude that the longer the duration of the eruption, the longer the time before the next eruption begins.

Try It Yourself 3

Consider the data from the Chapter Opener on men's and women's winning times in the 100-meter run on page 457. Use a technology tool to display the data in a scatter plot. Determine the type of correlation.

a. *Enter* the data into List 1 and List 2.
b. *Construct* the scatter plot.
c. Does there appear to be a linear correlation?

Answer: Page A42

Study Tip

You can also use MINITAB and Excel to construct scatter plots.

Correlation Coefficient

Interpreting correlation using a scatter plot can be subjective. A more precise way to measure the type and strength of a linear correlation between two variables is to calculate the correlation coefficient.

DEFINITION

The **correlation coefficient** is a measure of the strength and the direction of a linear relationship between two variables. The symbol r represents the sample correlation coefficient. The formula for r is

$$r = \frac{n\sum xy - (\sum x)(\sum y)}{\sqrt{n\sum x^2 - (\sum x)^2}\sqrt{n\sum y^2 - (\sum y)^2}}$$

where n is the number of pairs of data.

The population correlation coefficient is represented by ρ (the lowercase Greek letter rho, pronounced "row").

Insight

The formal name for r is the **Pearson product moment correlation coefficient.** It is named after the English statistician Karl Pearson (1857– 1936). (See page 29.)

The range of the correlation coefficient is -1 to 1. If x and y have a strong positive linear correlation, r is close to 1. If x and y have a strong negative linear correlation, r is close to -1. If there is no linear correlation or a weak linear correlation, r is close to 0. Several examples are shown below.

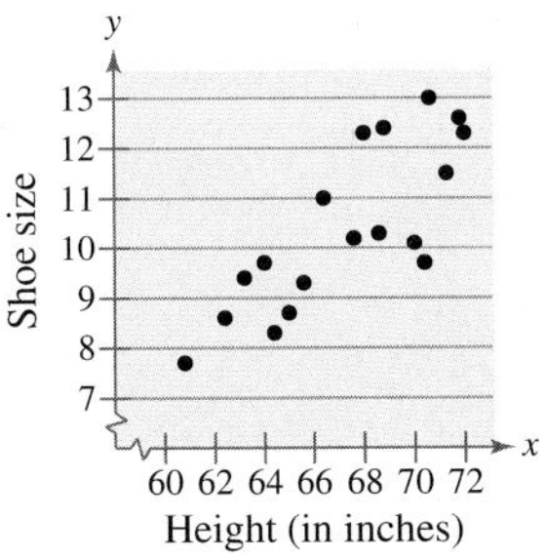

Strong positive correlation
$r = 0.81$

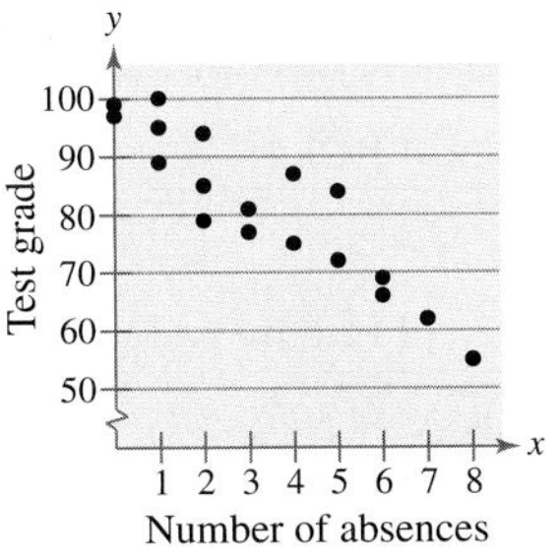

Strong negative correlation
$r = -0.92$

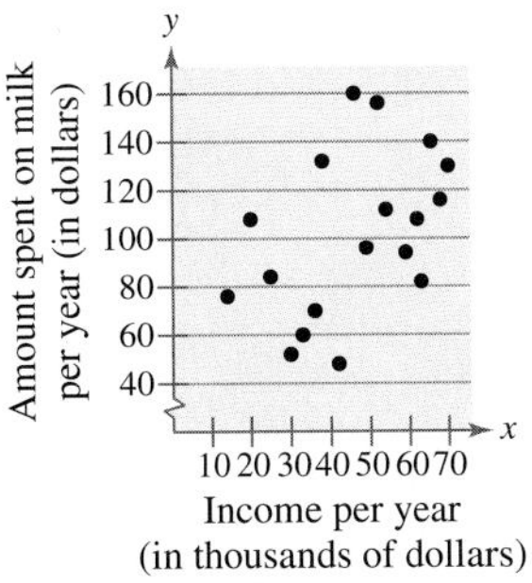

Weak positive correlation
$r = 0.45$

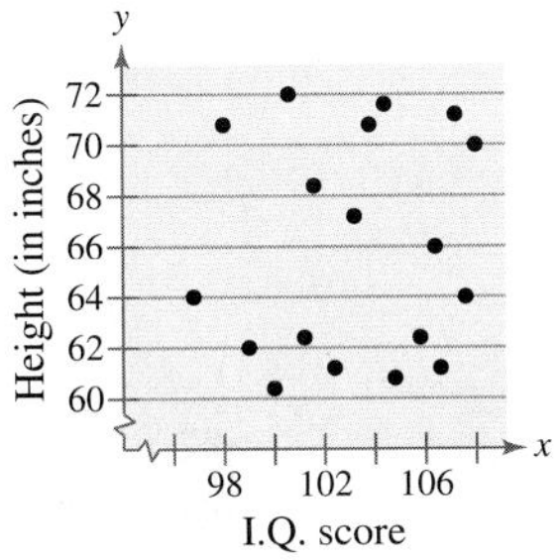

No correlation
$r = 0.04$

GUIDELINES

Calculating a Correlation Coefficient

In Words	*In Symbols*
1. Find the sum of the x-values.	$\sum x$
2. Find the sum of the y-values.	$\sum y$
3. Multiply each x-value by its corresponding y-value and find the sum.	$\sum xy$
4. Square each x-value and find the sum.	$\sum x^2$
5. Square each y-value and find the sum.	$\sum y^2$
6. Use these five sums to calculate the correlation coefficient.	$r = \frac{n\sum xy - (\sum x)(\sum y)}{\sqrt{n\sum x^2 - (\sum x)^2}\sqrt{n\sum y^2 - (\sum y)^2}}$

Study Tip

$\sum x^2$ means square each value and add the squares. $(\sum x)^2$ means add the values and square the sum.

Note to Instructor

It helps students if you emphasize the interpretation of the correlation coefficient *r* in each example.

EXAMPLE 4

Finding the Correlation Coefficient

Calculate the correlation coefficient for the advertising expenditures and company sales data given in Example 1. What can you conclude?

SOLUTION Use a table to help calculate the correlation coefficient.

Advertising expenses (1000s of \$), x	Company sales (1000s of \$), y	xy	x^2	y^2
2.4	225	540	5.76	50,625
1.6	184	294.4	2.56	33,856
2.0	220	440	4	48,400
2.6	240	624	6.76	57,600
1.4	180	252	1.96	32,400
1.6	184	294.4	2.56	33,856
2.0	186	372	4	34,596
2.2	215	473	4.84	46,225
$\Sigma x = 15.8$	$\Sigma y = 1634$	$\Sigma xy = 3289.8$	$\Sigma x^2 = 32.44$	$\Sigma y^2 = 337{,}558$

With these sums and $n = 8$, the correlation coefficient is

$$r = \frac{n\Sigma xy - (\Sigma x)(\Sigma y)}{\sqrt{n\Sigma x^2 - (\Sigma x)^2}\sqrt{n\Sigma y^2 - (\Sigma y)^2}}$$

$$= \frac{8(3289.8) - (15.8)(1634)}{\sqrt{8(32.44) - 15.8^2}\sqrt{8(337{,}558) - 1634^2}}$$

$$= \frac{501.2}{\sqrt{9.88}\sqrt{30{,}508}} \approx 0.913.$$

The result $r \approx 0.913$ suggests a strong positive linear correlation.

Interpretation As the amount spent on advertising increases, the company sales also increase.

Study Tip

Notice that the correlation coefficient *r* in Example 4 is rounded to three decimal places. This *round-off rule* will be used throughout the text.

Try It Yourself 4

Calculate the correlation coefficient for the income level and donating percent data given in Try It Yourself 1. What can you conclude?

Income level (in 1000s of \$), x	Donating percent, y
42	9
48	10
50	8
59	5
65	6
72	3

a. *Identify n.*
b. *Use a table* to calculate Σx, Σy, Σxy, Σx^2, and Σy^2.
c. *Use the resulting sums and n* to calculate r.
d. What can you conclude?

Answer: Page A42

EXAMPLE 5

Using Technology to Find a Correlation Coefficient

Use a technology tool to calculate the correlation coefficient for the Old Faithful data given in Example 3. What can you conclude?

SOLUTION MINITAB, Excel, and the TI-83 each have features that allow you to calculate a correlation coefficient for paired data sets. Try using this technology to find r. You should obtain results similar to the following.

MINITAB

Correlations: C1, C2

Pearson correlation of C1 and C2 = 0.970

EXCEL

	A	B	C
36	CORREL(A1:A35,B1:B35)		
37			0.969787

Before using the TI-83 to calculate r, you must enter the Diagnostic On command. To do so, enter the following keystrokes:

[2nd] [0]

cursor to *DiagnosticOn*

[ENTER] [ENTER].

The following screens describe how to find r using a TI-83 with the data stored in List 1 and List 2.

TI-83

EDIT CALC TESTS
1: 1–Var Stats
2: 2–Var Stats
3: Med–Med
4: LinReg(ax+b)
5: QuadReg
6: CubicReg
7↓Quart Reg

TI-83

LinReg(ax+b) L1,
L2

TI-83

LinReg
y=ax+b
a=11.8244078
b=35.30117105
r²=.9404868083
r=.9697869912

Correlation coefficient

The result, $r \approx 0.970$, suggests a strong positive linear correlation.

Try It Yourself 5

Calculate the correlation coefficient for the data from the Chapter Opener on men's and women's winning times in the 100-meter run on page 457. What can you conclude?

a. *Enter* the data.
b. Use the appropriate feature to *calculate r.*
c. What can you conclude?

Answer: Page A42

Using a Table to Test a Population Correlation Coefficient ρ

Once you have calculated r, the sample correlation coefficient, you will want to determine whether there is enough evidence to decide that the population correlation coefficient ρ is significant at a specified level of significance α. Remember that you are using sample data to make a decision about population data. It is always possible that your inference may be wrong. Using a level of significance of $\alpha = 0.05$ means that 5% of the time you will say the population correlation coefficient is significant when it is really not. For $\alpha = 0.01$, you will make this type of error 1% of the time.

One way to determine whether the population correlation coefficient ρ is significant is to use the critical values given in Table 11 in Appendix B. As you examine Table 11, notice that the first column represents the number n of pairs of data in the sample, and the second and third columns represent the critical values for a level of significance $\alpha = 0.05$ and $\alpha = 0.01$, respectively.

If $|r|$ is greater than the critical value, there is enough evidence to decide that the correlation is significant. Otherwise, there is *not* enough evidence to say that the correlation is significant. For instance, to determine whether ρ is significant for five pairs of data ($n = 5$) at a level of significance of $\alpha = 0.01$, you need to compare $|r|$ with a critical value of 0.959, as shown in the table.

Note to Instructor

Another way to determine whether the population correlation coefficient ρ is significant is to perform a hypothesis test (which is explained later in this section).

n	$\alpha = 0.05$	$\alpha = 0.01$
4	0.950	0.990
5	0.878	0.959
6	0.811	0.917

If $|r| > 0.959$, the correlation is significant. Otherwise, there is *not* enough evidence to support that the correlation is significant. The guidelines for this process are as follows.

Study Tip

If you determine that the linear correlation is significant, then you will be able to proceed to write the equation for the line that best describes the data. This line, called the regression line, can be used to predict the value of y when given a value of x. You will learn how to write this equation in the next section.

GUIDELINES

Using Table 11 for the Correlation Coefficient ρ

In Words	*In Symbols*
1. Determine the number of pairs of data in the sample.	Determine n.
2. Specify the level of significance.	Specify α.
3. Find the critical value.	Use Table 11 in Appendix B.
4. Decide if the correlation is significant.	If $\lvert r \rvert >$ critical value, the correlation is significant. Otherwise, there is *not* enough evidence to support that the correlation is significant.
5. Interpret the decision in the context of the original claim.	

EXAMPLE 6

Using Table 11 for a Correlation Coefficient

In Example 5, you used 35 pairs of data to find $r \approx 0.970$. Is the correlation coefficient significant? Use $\alpha = 0.05$.

SOLUTION

The number of pairs of data is 35, so $n = 35$. The level of significance is $\alpha = 0.05$. Using Table 11, find the critical value in the $\alpha = 0.05$ column that corresponds to the row with $n = 35$. The number in that column and row is 0.334.

n	$\alpha = 0.05$	$\alpha = 0.01$
4	0.950	0.990
5	0.878	0.959
6	0.811	0.917
7	0.754	0.875
8	0.707	0.834
9	0.666	0.798
10	0.632	0.765
11	0.602	0.735
12	0.576	0.708
13	0.553	0.684
14	0.532	0.661
23	0.413	0.526
24	0.404	0.515
25	0.396	0.505
26	0.388	0.496
27	0.381	0.487
28	0.374	0.479
29	0.367	0.471
30	0.361	0.463
35	0.334	0.430
40	0.312	0.403
45	0.294	0.380

Because $|r| \approx 0.970 > 0.334$, you can decide that the population correlation is significant.

Interpretation There is enough evidence at the 5% level of significance to conclude that there is a significant linear correlation between the duration of Old Faithful's eruptions and the time between eruptions.

Try It Yourself 6

In Try It Yourself 4, you calculated the correlation of the income level and donating percent data to be $r \approx -0.916$. Is the correlation coefficient significant? Use $\alpha = 0.01$.

a. *Determine* the number of pairs of data in the sample.
b. *Specify* the level of significance.
c. *Find* the critical value. Use Table 11 in Appendix B.
d. *Compare* $|r|$ with the critical value and *decide* if the correlation is significant.
e. *Interpret* the decision in the context of the original claim.

Answer: Page A42

Note to Instructor

The material on hypothesis testing requires previous coverage of Chapter 7. This material can be omitted if you cover correlation and regression before covering Chapters 7 and 8.

Hypothesis Testing for a Population Correlation Coefficient ρ

You can also use a hypothesis test to determine whether the sample correlation coefficient r provides enough evidence to conclude that the population correlation coefficient ρ is significant. A hypothesis test for ρ can be one tailed or two tailed. The null and alternative hypotheses for these tests are as follows.

$$\begin{cases} H_0\text{: } \rho \geq 0 \text{ (no significant negative correlation)} \\ H_a\text{: } \rho < 0 \text{ (significant negative correlation)} \end{cases}$$ **Left-tailed test**

$$\begin{cases} H_0\text{: } \rho \leq 0 \text{ (no significant positive correlation)} \\ H_a\text{: } \rho > 0 \text{ (significant positive correlation)} \end{cases}$$ **Right-tailed test**

$$\begin{cases} H_0\text{: } \rho = 0 \text{ (no significant correlation)} \\ H_a\text{: } \rho \neq 0 \text{ (significant correlation)} \end{cases}$$ **Two-tailed test**

In this text, you will consider hypothesis tests only for ρs that are two-tailed.

Note to Instructor

Remind students that this is the same family of distributions that was used for hypothesis testing of means with small and/or dependent samples. There are $n - 2$ degrees of freedom because one degree of freedom is lost for each variable.

The *t*-Test for the Correlation Coefficient

A ***t*-test** can be used to test whether the correlation between two variables is significant. The **test statistic** is r and the **standardized test statistic**

$$t = \frac{r}{\sigma_r} = \frac{r}{\sqrt{\dfrac{1 - r^2}{n - 2}}}$$

follows a t-distribution with $n - 2$ degrees of freedom.

GUIDELINES

Using the *t*-Test for the Correlation Coefficient ρ

In Words	*In Symbols*
1. State the null and alternative hypotheses.	State H_0 and H_a.
2. Specify the level of significance.	Identify α.
3. Identify the degrees of freedom.	d.f. $= n - 2$
4. Determine the critical value(s) and the rejection region(s).	Use Table 5 in Appendix B.
5. Find the standardized test statistic.	$t = \dfrac{r}{\sqrt{\dfrac{1 - r^2}{n - 2}}}$
6. Make a decision to reject or fail to reject the null hypothesis.	If t is in the rejection region, reject H_0. Otherwise, fail to reject H_0.
7. Interpret the decision in the context of the original claim.	

EXAMPLE 7

The t-Test for a Correlation Coefficient

In Example 4, you used eight pairs of data to find $r \approx 0.913$. Test the significance of this correlation coefficient. Use $\alpha = 0.05$.

SOLUTION The null and alternative hypotheses are

$$H_0: \rho = 0 \text{ (no correlation)} \quad \text{and} \quad H_a: \rho \neq 0 \text{ (significant correlation).}$$

Because there are eight pairs of data in the sample, there are $8 - 2 = 6$ degrees of freedom. Because the test is a two-tailed test, $\alpha = 0.05$, and d.f. $= 6$, the critical values are $-t_0 = -2.447$ and $t_0 = 2.447$. The rejection regions are $t < -2.447$ and $t > 2.447$. Using the t-test, the standardized test statistic is

$$t = \frac{r}{\sqrt{\frac{1-r^2}{n-2}}} = \frac{0.913}{\sqrt{\frac{1-(0.913)^2}{8-2}}} \approx 5.482.$$

The following graph shows the location of the rejection regions and the standardized test statistic.

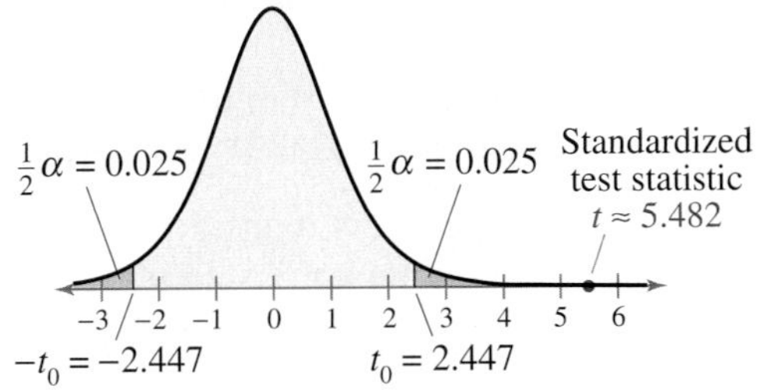

Because t is in the rejection region, you should decide to reject the null hypothesis.

Interpretation At the 5% level, there is enough evidence to conclude that there is a significant linear correlation between advertising expenses and company sales.

Try It Yourself 7

In Try It Yourself 5, you calculated the correlation coefficient of the Olympic 100-meter winning times data to be $r \approx 0.849$. Test the significance of this correlation coefficient. Use $\alpha = 0.01$.

a. *State* the null and alternative hypotheses.
b. *Specify* the level of significance.
c. *Identify* the degrees of freedom.
d. *Determine* the critical values and the rejection regions.
e. *Find* the standardized test statistic.
f. *Make a decision* to reject or fail to reject the null hypothesis.
g. *Interpret* the decision in the context of the original claim.

Answer: Page A42

Insight

In Example 7, you can use Table 11 in Appendix B to test the population correlation coefficient ρ. Given $n = 8$ and $\alpha = 0.05$, the critical value from Table 11 is 0.707. Because

$$|r| \approx 0.913 > 0.707,$$

the correlation is significant. Note that this is the same result you obtained using a t-test for the population correlation coefficient ρ.

Study Tip

Be sure you see in Example 7 that rejecting the null hypothesis means that there is sufficient evidence that the correlation is significant.

Correlation and Causation

The fact that two variables are strongly correlated does not in itself imply a cause-and-effect relationship between the variables. More in-depth study is usually needed to determine whether there is a causal relationship between the variables.

If there is a significant correlation between two variables, a researcher should consider the following possibilities.

1. Is there a direct cause-and-effect relationship between the variables?

That is, does x cause y? For instance, consider the relationship between advertising expenses and company sales that has been discussed throughout this section. It is reasonable to conclude that spending more money on advertising will result in more sales.

2. Is there a reverse cause-and-effect relationship between the variables?

That is, does y cause x? For instance, consider the Old Faithful data that have been discussed throughout this section. These variables have a positive linear correlation, and it is possible to conclude that the duration of an eruption affects the time before the next eruption. However, it is also possible that the time between eruptions affects the duration of the next eruption.

3. Is it possible that the relationship between the variables can be caused by a third variable or perhaps a combination of several other variables?

For instance, consider the men's and women's winning times in the 100-meter run listed in the Chapter Opener. Although these variables have a positive linear correlation, it is doubtful that just because the men's winning time decreases the women's winning time will also decrease. The relationship is probably due to several other variables, such as training techniques, improvements in athletic equipment, or climate conditions at the Olympic location.

4. Is it possible that the relationship between two variables may be a coincidence?

For instance, although it may be possible to find a significant correlation between the number of animal species living in certain regions and the number of people who own more than two cars in those regions, it is highly unlikely that the variables are directly related. The relationship is probably due to coincidence.

Determining which of the above cases is valid for a data set can be difficult. For instance, consider the following example. Suppose a person breaks out in a rash each time he eats shrimp at a certain restaurant. The natural conclusion is that the person is allergic to shrimp. However, upon further study by an allergist, it is found that the person is not allergic to shrimp, but to a type of seasoning the chef is putting into the shrimp.

Picturing the World

The following scatter plot shows the results of a survey conducted as a group project by students in a high school statistics class in the San Francisco area. In the survey, 125 high school students were asked their grade point average (GPA) and the number of caffeine drinks they consumed each day.

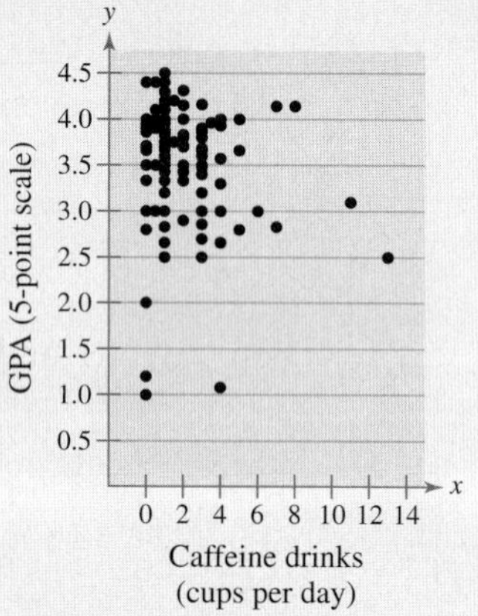

What type of correlation, if any, does the scatter plot show between caffeine consumption and GPA?

9.1 Exercises

Building Basic Skills and Vocabulary

1. Which value of r indicates a stronger correlation: $r = 0.731$ or $r = -0.845$? Explain your reasoning.

2. Which of the following values could not represent a correlation coefficient? Explain why.

   (a) $r = 0.926$ (b) $r = 1.054$ (c) $r = -0.730$ (d) $r = -0.053$

3. Explain how to perform a hypothesis test for a population correlation coefficient ρ.

4. Discuss the difference between r and ρ.

Graphical Analysis In Exercises 5–8, the scatter plots of paired data sets are shown. Determine whether there is a positive linear correlation, negative linear correlation, or no linear correlation between the variables.

5.

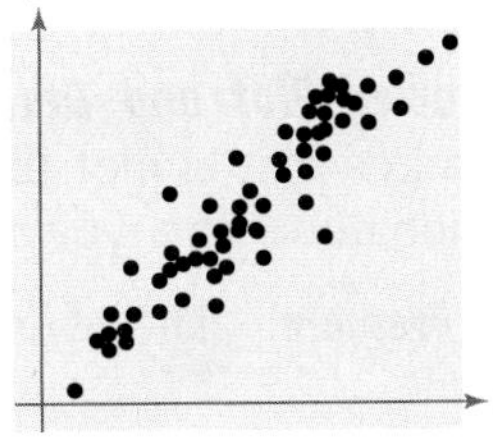

6.

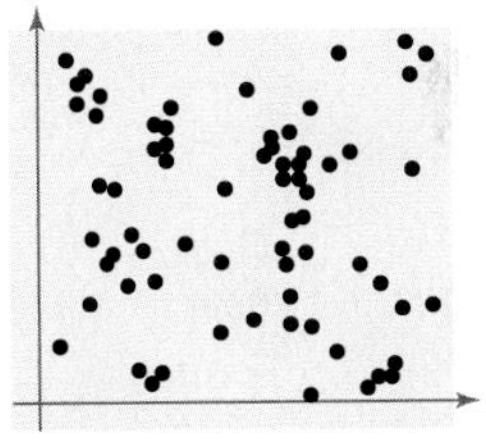

7.

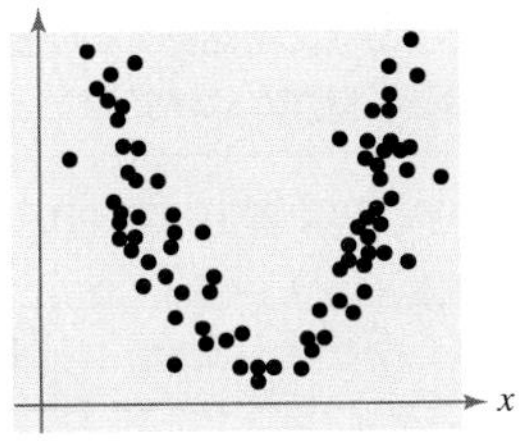

8.

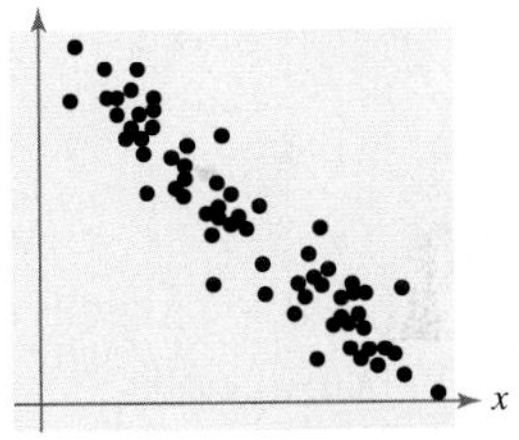

Graphical Analysis In Exercises 9–12, the scatter plots show the results of a survey of 20 randomly selected males ages 24–35. Using age as the explanatory variable, match each scatter plot to the appropriate description. Explain your reasoning.

(a) Age and body temperature

(b) Age and balance on student loans

(c) Age and income

(d) Age and height

9.

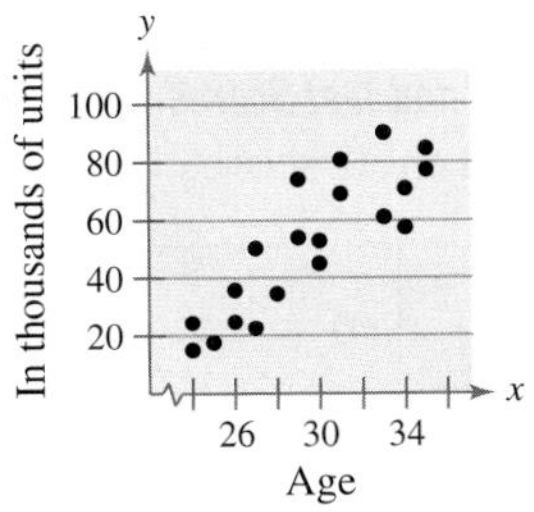

10.

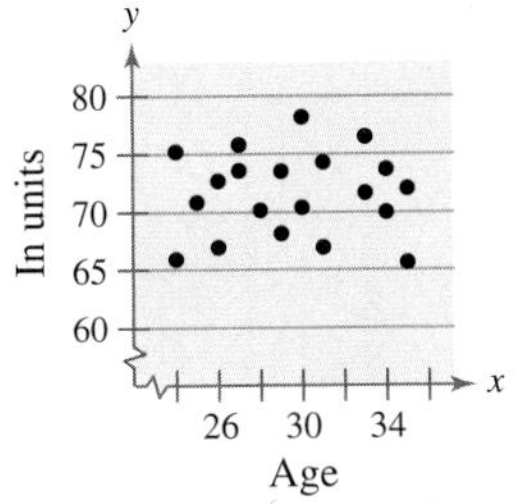

1. $r = -0.845$ represents a stronger correlation because it is closer to -1 than $r = 0.731$ is to $+1$.
2. b; correlation coefficients must be contained in the interval $[-1, 1]$.
3. State the null and alternative hypotheses. Specify the level of significance and determine the degrees of freedom. Identify the rejection regions and calculate the standardized test statistic. Make a decision and interpret in the context of the original claim.
4. r = sample correlation coefficient

   ρ = population correlation coefficient
5. Positive linear correlation
6. No linear correlation
7. No linear correlation (but there is a nonlinear correlation between the variables)
8. Negative linear correlation
9. c; You would expect a positive linear correlation between age and income.
10. d; You would not expect age and height to be correlated.

11. b; You would expect a negative linear correlation between age and balance on student loans.

12. a; You would expect the relationship between age and body temperature to be fairly constant.

13. Explanatory variable: Amount of water consumed

14. Explanatory variable: Hours of safety classes

Response variable: Number of accidents

Response variable: Weight loss

15. (a)

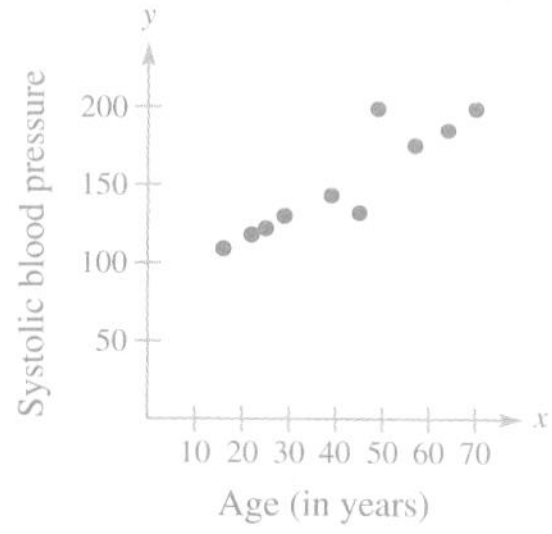

(b) 0.908

(c) Strong positive linear correlation

16. (a)

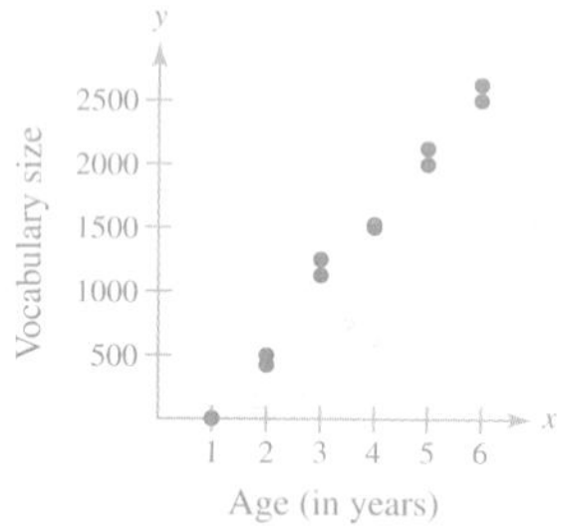

(b) 0.996

(c) Strong positive linear correlation

17. (a)

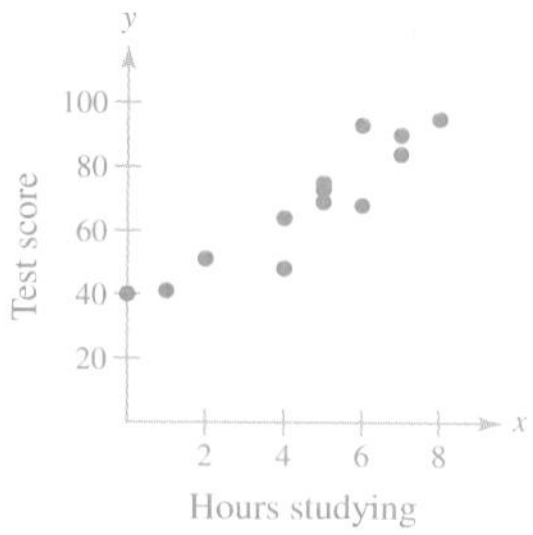

(b) 0.923

(c) Strong positive linear correlation

11.

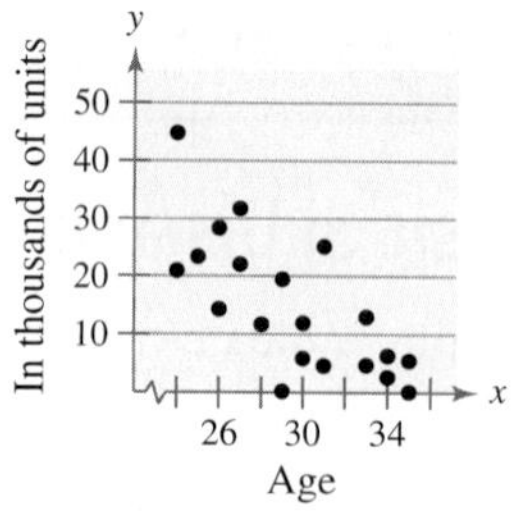

12.

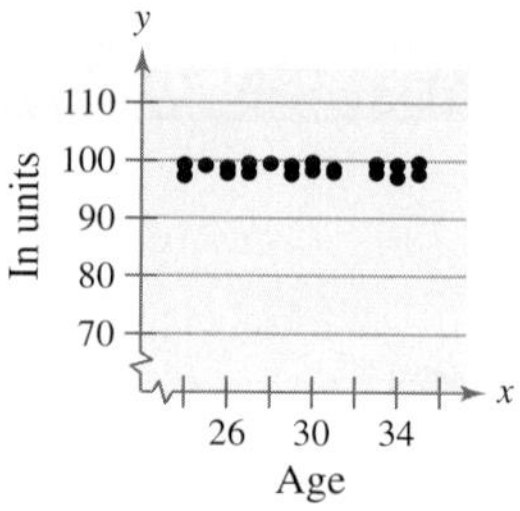

In Exercises 13 and 14, identify the explanatory variable and the response variable.

13. A nutritionist wants to determine if the amount of water consumed each day by persons of the same weight and on the same diet can be used to predict individual weight loss.

14. An insurance company hires an actuary to determine whether the number of hours of safety driving classes can be used to predict the number of driving accidents for each driver.

Using and Interpreting Concepts

Constructing a Scatter Plot and Determining Correlation In Exercises 15–22, (a) display the data in a scatter plot, (b) calculate the correlation coefficient r, and (c) make a conclusion about the type of correlation.

DATA **15. *Age and Blood Pressure*** The ages (in years) of 10 men and their systolic blood pressures

Age, x	16	25	39	45	49
Systolic blood pressure, y	109	122	143	132	199
Age, x	64	70	29	57	22
Systolic blood pressure, y	185	199	130	175	118

DATA **16. *Age and Vocabulary*** The ages (in years) of 11 children and the number of words in their vocabulary

Age, x	1	2	3	4	5	6
Vocabulary size, y	3	440	1200	1500	2100	2600

Age, x	3	5	2	4	6
Vocabulary size, y	1100	2000	500	1525	2500

DATA **17. *Hours Studying and Test Scores*** The number of hours 13 students spent studying for a test and their scores on that test

Hours spent studying, x	0	1	2	4	4	5	5	5	6	6	7	7	8
Test score, y	40	41	51	48	64	69	73	75	68	93	84	90	95

DATA **18.** ***TV and Test Scores*** The number of hours 12 students watched television during the weekend and the scores of each student who took a test the following Monday

Hours spent watching TV, x	0	1	2	3	3	5	5	5	6	7	7	10
Test score, y	96	85	82	74	95	68	76	84	58	65	75	50

19. ***Coffee Sales and Temperature*** The high outdoor temperature (in degrees Fahrenheit) and coffee sales (in hundreds of dollars) for a coffee shop for eight randomly selected days

Temperature, x	32	39	51	60	65	72	78	81
Coffee sales, y	26.2	24.8	19.7	20.0	13.3	13.9	11.4	11.2

20. ***Age and Hours Slept*** The age (in years) and the number of hours slept in a day by six infants

Age, x	0.1	0.2	0.5	0.7	0.8	0.9
Hours slept, y	14.9	14.5	13.4	14.1	13.4	13.7

DATA **21.** ***Earnings and Dividends*** The earnings per share and dividends per share for 10 electric utility companies in a recent year *(Source: The Value Line Investment Survey)*

Earnings per share, x	2.78	1.41	2.74	0.92	2.44	3.50	3.68	1.97	1.95	2.07
Dividends per share, y	2.16	0.88	1.04	1.10	0.96	2.18	1.54	1.39	1.41	2.88

DATA **22.** ***Crimes and Arrests*** The number of crimes reported (in millions) and the number of arrests reported (in millions) by the FBI for 11 years *(Source: Bureau of Justice Statistics)*

Crimes, x	1.66	1.65	1.61	1.55	1.44	1.41	1.32	1.23	1.22	1.23	1.22
Arrests, y	0.72	0.72	0.78	0.80	0.73	0.72	0.68	0.64	0.63	0.63	0.62

18. (a)

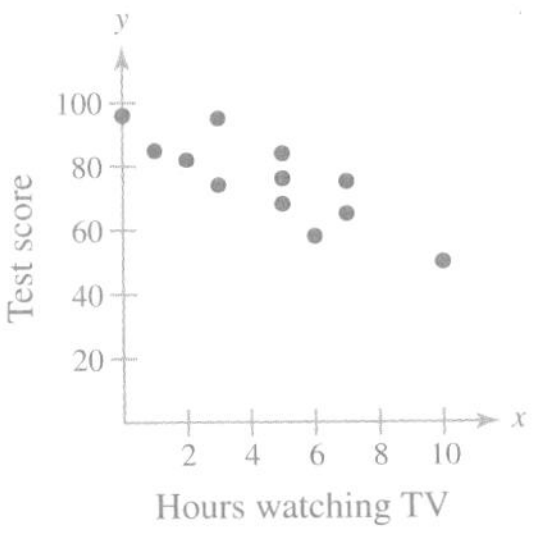

(b) −0.831

(c) Strong negative linear correlation

19. (a)

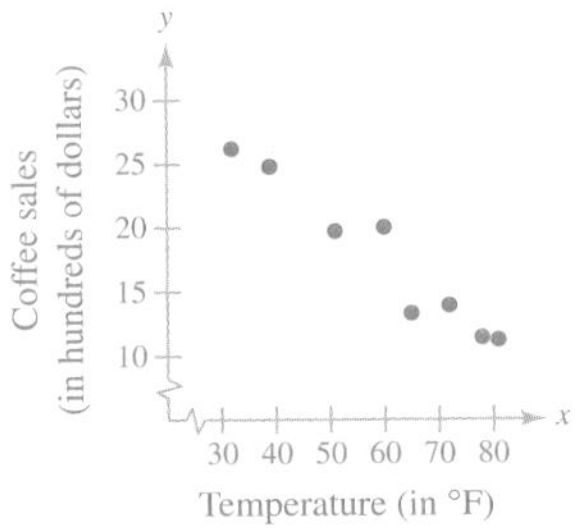

(b) −0.971

(c) Strong negative linear correlation

20. (a)

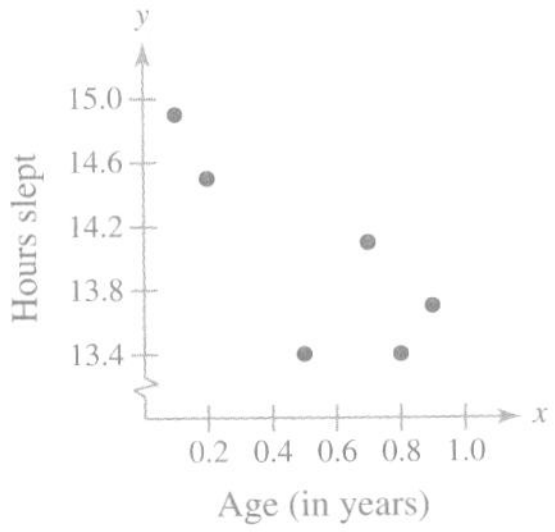

(b) −0.789

(c) Negative linear correlation

21. (a)

(b) 0.342

(c) No linear correlation

22. See Selected Answers, page A97.

23. There is enough evidence at the 1% level to conclude that there is no significant linear correlation between vehicle weight and the variability in braking distance.

24. There is enough evidence at the 5% level to conclude that there is a significant linear correlation between vehicle weight and variability in braking distance.

25. There is enough evidence at the 1% level to conclude that there is a significant linear correlation between the number of hours spent studying for a test and the score received on the test.

26. There is enough evidence at the 5% level to conclude that there is a significant linear correlation between the number of hours spent watching television before a test and the score received on the test.

Testing Claims In Exercises 23–28, use Table 11 in Appendix B as shown in Example 6 or perform a hypothesis test using Table 5 in Appendix B as shown in Example 7 to make a conclusion about the indicated correlation coefficient. Also, if convenient, use technology to solve the problem.

23. ***Braking Distances: Dry Surface*** The weights (in pounds) of eight vehicles and the variability of their braking distances (in feet) when stopping on a dry surface are shown in the table. Can you conclude that there is a significant linear correlation between vehicle weight and variability in braking distance on a dry surface? Use $\alpha = 0.01$. *(Adapted from National Highway Traffic Safety Administration)*

Weight, x	5940	5340	6500	5100	5850	4800	5600	5890
Variability in braking distance, y	1.78	1.93	1.91	1.59	1.66	1.50	1.61	1.70

24. ***Braking Distances: Wet Surface*** The weights (in pounds) of eight vehicles and the variability of their braking distances (in feet) when stopping on a wet surface are shown in the table. At $\alpha = 0.05$, can you conclude that there is a significant linear correlation between vehicle weight and variability in braking distance on a wet surface? *(Adapted from National Highway Traffic Safety Administration)*

Weight, x	5890	5340	6500	4800	5940	5600	5100	5850
Variability in braking distance, y	2.92	2.40	4.09	1.72	2.88	2.53	2.32	2.78

DATA **25.** ***Hours Studying and Test Scores*** The number of hours 13 students spent studying for a test and their scores on that test are shown in the table. Is there enough evidence to conclude that there is a significant linear correlation between the data? Use $\alpha = 0.01$. (Use the value of r found in Exercise 17.)

Hours spent studying, x	0	1	2	4	4	5	5	5	6	6	7	7	8
Test score, y	40	41	51	48	64	69	73	75	68	93	84	90	95

DATA **26.** ***TV and Test Scores*** An instructor wants to show students that there is a linear correlation between the number of hours they watch television during a certain weekend and their scores on a test taken the following Monday. The number of television viewing hours and the test scores for 12 randomly selected students are shown in the table. At $\alpha = 0.05$, is there enough evidence for the instructor to conclude that there is a significant linear correlation between the data? (Use the value of r found in Exercise 18.)

Hours spent watching TV, x	0	1	2	3	3	5	5	5	6	7	7	10
Test score, y	96	85	82	74	95	68	76	84	58	65	75	50

27. There is enough evidence at the 1% level to conclude that there is no significant linear correlation between earnings per share and dividends per share.

28. There is enough evidence at the 5% level to conclude that there is a significant linear correlation between crimes and arrests.

29. The correlation coefficient remains unchanged when the x-values and y-values are switched.

30. The correlation coefficient remains unchanged when the x-values and y-values are switched.

31. Answers will vary.

DATA **27. *Earnings and Dividends*** The following table shows the earnings per share and dividends per share for 10 electric utility companies in a recent year. At $\alpha = 0.01$, can you conclude that there is a significant linear correlation between earnings per share and dividends per share? (Use the value of r found in Exercise 21.) *(Source: The Value Line Investment Survey)*

Earnings per share, x	2.78	1.41	2.74	0.92	2.44
Dividends per share, y	2.16	0.88	1.04	1.10	0.96

Earnings per share, x	3.50	3.68	1.97	1.95	2.07
Dividends per share, y	2.18	1.54	1.39	1.41	2.88

DATA **28. *Crimes and Arrests*** The following table shows the number of crimes reported (in millions) and the number of arrests reported (in millions) by the FBI for 11 years. At $\alpha = 0.05$, can you conclude that there is a significant linear correlation between the number of crimes and the number of arrests? (Use the value of r found in Exercise 22.) *(Source: Bureau of Justice Statistics)*

Crimes, x	1.66	1.65	1.61	1.55	1.44	1.41
Arrests, y	0.72	0.72	0.78	0.80	0.73	0.72

Crimes, x	1.32	1.23	1.22	1.23	1.22
Arrests, y	0.68	0.64	0.63	0.63	0.62

Extending Concepts

Interchanging x and y In Exercises 29 and 30, calculate the correlation coefficient r, letting Row 1 represent the x-values and Row 2 the y-values. Then calculate the correlation coefficient r, letting Row 2 represent the x-values and Row 1 the y-values. What effect does switching the explanatory and response variables have on the correlation coefficient?

29.

Row 1	16	25	39	45	49	64	70
Row 2	109	122	143	132	199	185	199

30.

Row 1	0	1	2	3	3	5	5	5	6	7
Row 2	96	85	82	74	95	68	76	84	58	65

31. *Writing* Use your school's library, the Internet, or some other reference source to find a real-life data set with the indicated cause-and-effect relationship. Write a paragraph describing each variable and explain why you think the variables have the indicated cause-and-effect relationship.

(a) *Direct Cause-and-Effect:* Changes in one variable cause changes in the other variable.

(b) *Other Factors:* The relationship between the variables is caused by a third variable.

(c) *Coincidence:* The relationship between the variables is a coincidence.

9.2 Linear Regression

What You Should Learn

- How to find the equation of a regression line
- How to predict y-values using a regression equation

Regression Lines • Applications of Regression Lines

Regression Lines

After verifying that the linear correlation between two variables is significant, the next step is to determine the equation of the line that best models the data. This line is called a regression line, and its equation can be used to predict the value of y for a given value of x. Although many lines can be drawn through a set of points, a regression line is determined by specific criteria.

Consider the scatter plot and the line shown below. For each data point, d_i represents the difference between the observed y-value and the predicted y-value for a given x-value on the line. These differences are called **residuals** and can be positive, negative, or zero. When the point is above the line, d_i is positive. When the point is below the line, d_i is negative. If the observed y-value equals the predicted y-value, $d_i = 0$. Of all possible lines that can be drawn through a set of points, the regression line is the line for which the sum of the squares of all the residuals

$$\Sigma d_i^2$$

is a minimum.

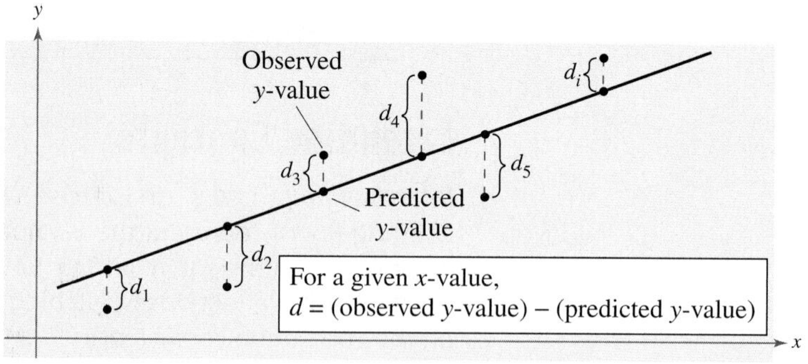

DEFINITION

A **regression line,** also called a **line of best fit,** is the line for which the sum of the squares of the residuals is a minimum.

In algebra, you learned that you can write an equation of a line by finding its slope m and y-intercept b. The equation has the form

$$y = mx + b.$$

Recall that the slope of a line is the ratio of its rise over its run and the y-intercept is the y-value of the point at which the line crosses the y-axis. It is the y-value when $x = 0$.

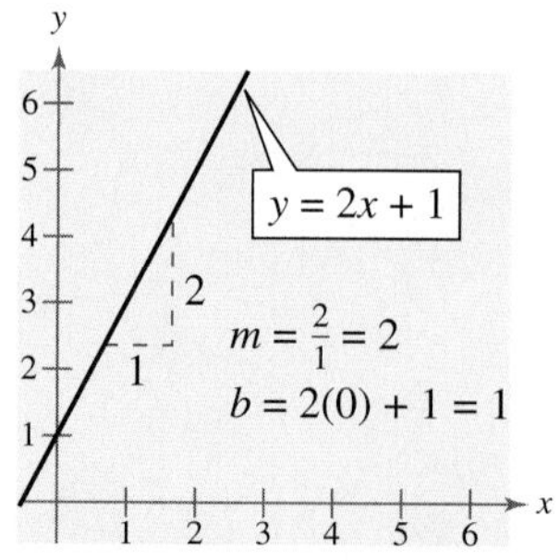

In algebra, you used two points to determine the equation of a line. In statistics, you will use every point in the data set to determine the equation of the regression line.

The equation of a regression line allows you to use the independent (explanatory) variable x to make predictions for the dependent (response) variable y.

Study Tip

Notice that both the slope m and the y-intercept b in Example 1 are rounded to three decimal places. This *round-off rule* will be used throughout the text.

The Equation of a Regression Line

The equation of a regression line for an independent variable x and a dependent variable y is

$$\hat{y} = mx + b$$

where $\hat{y}$ is the predicted y-value for a given x-value. The slope m and y-intercept b are given by

$$m = \frac{n\sum xy - (\sum x)(\sum y)}{n\sum x^2 - (\sum x)^2} \quad \text{and} \quad b = \bar{y} - m\bar{x} = \frac{\sum y}{n} - m\frac{\sum x}{n}$$

where $\bar{y}$ is the mean of the y-values in the data set and $\bar{x}$ is the mean of the x-values. The regression line always passes through the point $(\bar{x}, \bar{y})$.

EXAMPLE 1

Finding the Equation of a Regression Line

Find the equation of the regression line for the advertising expenditures and company sales data used in Section 9.1.

Advertising expenses (1000s of \$), x	Company sales (1000s of \$), y
2.4	225
1.6	184
2.0	220
2.6	240
1.4	180
1.6	184
2.0	186
2.2	215

SOLUTION In Example 4 of Section 9.1, you found that $n = 8$, $\sum x = 15.8$, $\sum y = 1634$, $\sum xy = 3289.8$, and $\sum x^2 = 32.44$. You can use these values to calculate the slope and y-intercept of the regression line as shown.

$$m = \frac{n\sum xy - (\sum x)(\sum y)}{n\sum x^2 - (\sum x)^2} = \frac{8(3289.8) - (15.8)(1634)}{8(32.44) - 15.8^2} = \frac{501.2}{9.88} \approx 50.7287$$

$$b = \bar{y} - m\bar{x} = \frac{1634}{8} - (50.7287)\frac{15.8}{8} = 204.25 - (50.7287)(1.975) \approx 104.0608$$

So, the equation of the regression line is

$$\hat{y} = 50.729x + 104.061.$$

The regression line and scatter plot of the data are shown at the right. If you plot the point $(\bar{x}, \bar{y}) = (1.975, 204.25)$, you will notice that the line passes through this point.

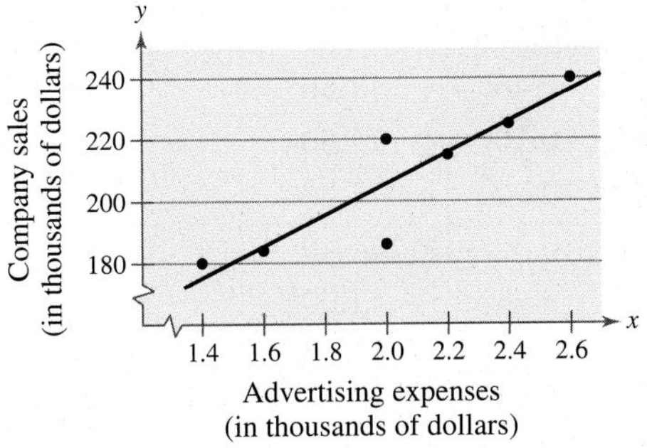

Try It Yourself 1

Find the equation of the regression line for the income level and donating percent data used in Section 9.1.

Income level (in 1000s of \$), x	Donating percent, y
42	9
48	10
50	8
59	5
65	6
72	3

a. *Identify* n, $\sum x$, $\sum y$, $\sum xy$, and $\sum x^2$ from Try It Yourself 4 of Section 9.1.
b. *Calculate* the slope m.
c. *Calculate* the y-intercept b.
d. *Write* the equation of the regression line.

Answer: Page A42

Duration, x	Time, y	Duration, x	Time, y
1.80	56	3.78	79
1.82	58	3.83	85
1.88	60	3.87	81
1.90	62	3.88	80
1.92	60	4.10	89
1.93	56	4.27	90
1.98	57	4.30	84
2.03	60	4.30	89
2.05	57	4.43	84
2.13	60	4.43	89
2.30	57	4.47	86
2.35	57	4.47	80
2.37	61	4.53	89
2.82	73	4.55	86
3.13	76	4.60	88
3.27	77	4.60	92
3.65	77	4.63	91
3.70	82		

Note to Instructor

Point out the different notation used by the TI-83, MINITAB, and Excel.

Men's times, x	Women's times, y
10.80	12.20
10.30	11.90
10.30	11.50
10.30	11.90
10.40	11.50
10.50	11.50
10.20	11.00
10.00	11.40
9.95	11.00
10.14	11.07
10.06	11.08
10.25	11.60
9.99	10.97
9.92	10.54
9.96	10.82
9.84	10.94
9.87	10.75
9.85	10.93

EXAMPLE 2

Using Technology to Find a Regression Equation

Use a technology tool to find the equation of the regression line for the Old Faithful data used in Section 9.1.

SOLUTION MINITAB, Excel, and the TI-83 each have features that automatically calculate a regression equation. Try using this technology to find the regression equation. You should obtain results similar to the following.

MINITAB

Regression Analysis: C2 versus C1

The regression equation is
C2 = 35.3 + 11.8 C1

Predictor	Coef	SE Coef	T	P
Constant	35.301	1.804	19.57	0.000
C1	11.8244	0.5178	22.84	0.000

S = 3.27704 R-Sq = 94.0% R-Sq(adj) = 93.9%

EXCEL

	A	B	C	D
1	Slope:			
2	INDEX(LINEST(known_y's,known_x's),1)			
3				11.82441
4				
5	Y-intercept:			
6	INDEX(LINEST(known_y's,known_x's),2)			
7				35.30117

TI-83

LinReg
y=ax+b
a=11.8244078
b=35.30117105
r^2=.9404868083
r=.9697869912

From the displays, you can see that the regression equation is

$$\hat{y} = 11.824x + 35.301.$$

The TI-83 display at the right shows the regression line and a scatter plot of the data.

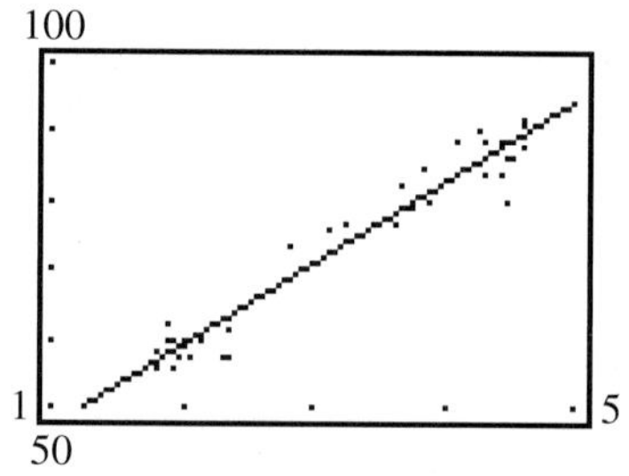

Try It Yourself 2

Use a technology tool to find the equation of the regression line for the men's and women's winning times in the 100-meter run given in the Chapter Opener on page 457.

a. *Enter* the data.
b. *Perform the necessary steps* to calculate the slope and y-intercept.
c. *Specify* the regression equation.

Answer: Page A43

Applications of Regression Lines

After finding the equation of a regression line, you can use the equation to predict y-values over the range of the data *if the correlation between x and y is significant*. For example, an advertising executive could forecast company sales on the basis of advertising expenditures. To predict y-values, substitute the given x-value into the regression equation, then calculate $\hat{y}$, the predicted y-value.

Note to Instructor

This Picturing the World is a good place to discuss outliers.

Picturing the World

The following scatter plot shows the relationship between the number of motor vehicles in a state and the number of motor vehicle-caused deaths. *(Source: U.S. National Highway Safety Administration)*

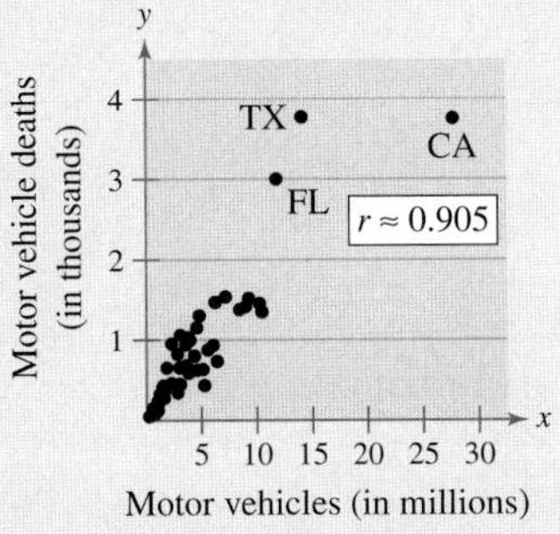

Describe the correlation between these two variables in words. Use the scatter plot to predict the number of motor vehicle-related deaths in a state that has 16 million motor vehicles. The regression line for this scatter plot is $\hat{y} = 0.159x + 0.130$. Use this equation to make your prediction. (Assume x and y have a significant linear correlation.) How does your algebraic prediction compare with your graphical one?

EXAMPLE 3

Predicting y-Values Using Regression Equations

The regression equation for the advertising expenses (in thousands of dollars) and company sales (in thousands of dollars) data is

$$\hat{y} = 50.729x + 104.061.$$

Use this equation to predict the *expected* company sales for the following advertising expenses. (Recall from Section 9.1, Example 7 that x and y have a significant linear correlation.)

1. 1.5 thousand dollars 2. 1.8 thousand dollars 3. 2.5 thousand dollars

SOLUTION To predict the expected company sales, substitute each advertising expenditure for x in the regression equation. Then calculate $\hat{y}$.

1. $\hat{y} = 50.729x + 104.061$
$= 50.729(1.5) + 104.061$
≈ 180.155
When the advertising expenses are \$1500, the company sales are about \$180,155.

2. $\hat{y} = 50.729x + 104.061$
$= 50.729(1.8) + 104.061$
≈ 195.373
When the advertising expenses are \$1800, the company sales are about \$195,373.

3. $\hat{y} = 50.729x + 104.061$
$= 50.729(2.5) + 104.061$
≈ 230.884
When the advertising expenses are \$2500, the company sales are about \$230,884.

Prediction values are meaningful only for x-values in (or close to) the range of the data. The x-values in the original data set range from 1.4 to 2.6. So, it would not be appropriate to use the regression line $\hat{y} = 50.729x + 104.061$ to predict company sales for advertising expenditures such as 0.5 (\$500) or 5.0 (\$5000).

Try It Yourself 3

The regression equation for the Old Faithful data is $\hat{y} = 11.824x + 35.301$. Use this to predict the time until the next eruption for each of the following eruption durations. (Recall from Section 9.1, Example 6 that x and y have a significant linear correlation.)

1. 2 minutes
2. 3.32 minutes

a. *Substitute* each value for x in the regression equation.
b. *Calculate* $\hat{y}$.
c. *Specify* the predicted time until the next eruption for each eruption duration.

Answer: Page A43

9.2 Exercises

Help

MyMathLab

Student Study Pack

Building Basic Skills and Vocabulary

In Exercises 1–8, match the description in the left column with a description in the right column.

1. Regression line

2. Residual

3. The y-value of a data point corresponding to x_i

4. The y-value for a point on the regression line corresponding to x_i

5. Slope

6. y-intercept

7. The mean of the y-values

8. The point a regression line always passes through

a. The difference between the observed y-value of a data point and the predicted y-value on the line for the same data point

b. $\hat{y}_i$

c. The line of best fit

d. y_i

e. b

f. $(\bar{x}, \bar{y})$

g. m

h. $\bar{y}$

Graphical Analysis In Exercises 9–12, match the regression equation with the appropriate graph. (Note that the x- and y-axes are broken.)

9. $\hat{y} = -1.04x + 50.3$

10. $\hat{y} = -0.902x + 90.5$

11. $\hat{y} = 0.00114x + 2.53$

12. $\hat{y} = -0.667x + 52.6$

a.

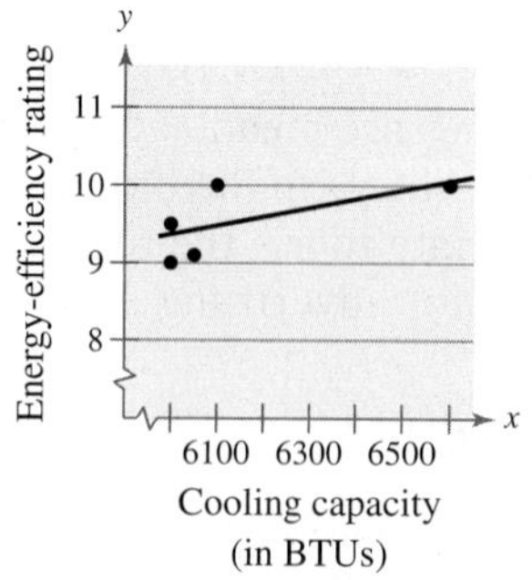

b.

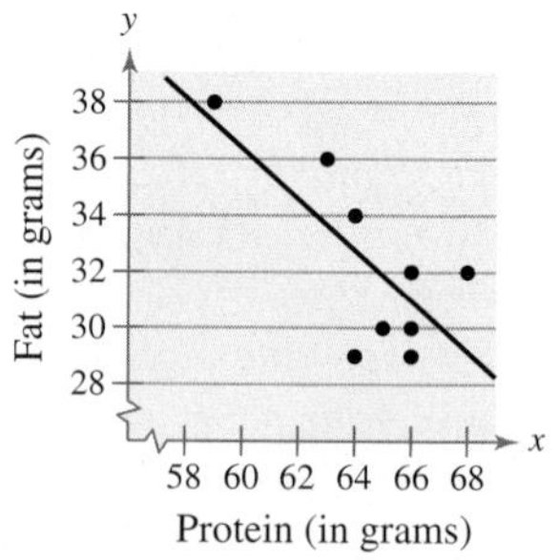

c.

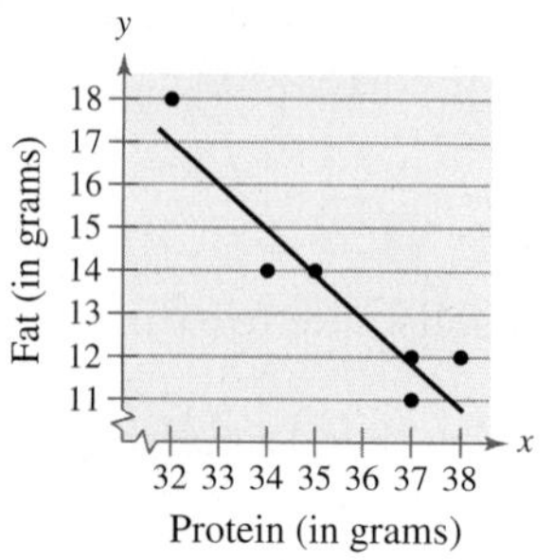

d.

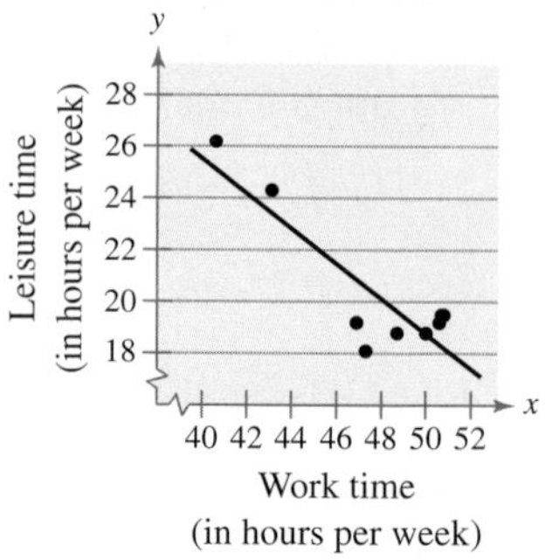

1. c
2. a
3. d
4. b
5. g
6. e
7. h
8. f
9. c
10. b
11. a
12. d
13. $\hat{y} = 1.662x + 83.338$

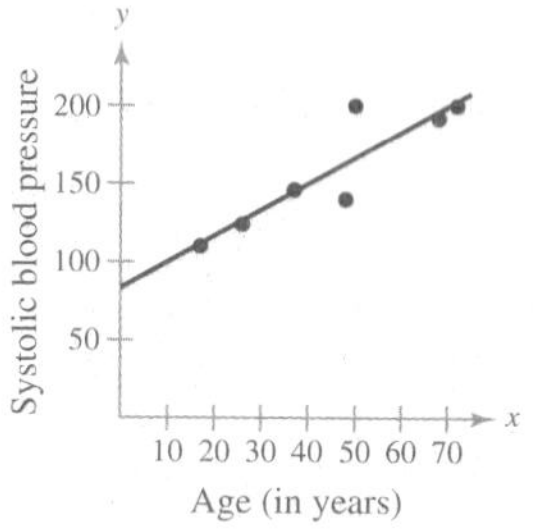

(a) 113.254
(b) 201.340
(c) 131.536
(d) 174.748

Using and Interpreting Concepts

Finding the Equation of a Regression Line In Exercises 13–22, find the equation of the regression line for the given data. Then construct a scatter plot of the data and draw the regression line. (Each pair of variables has a significant correlation.) Then use the regression equation to predict the value of y for each of the given x-values, if meaningful. If the x-value is not meaningful to predict the value of y, explain why not.

13. ***Age and Blood Pressure*** The ages (in years) of seven men and their systolic blood pressures

Age, x	17	26	37	48	50	68	72
Systolic blood pressure, y	110	124	146	140	200	192	200

(a) $x = 18$ years (b) $x = 71$ years
(c) $x = 29$ years (d) $x = 55$ years

14. ***Age and Vocabulary*** The ages (in years) of seven children and the number of words in their vocabulary

Age, x	3	4	4	5	6	2	3
Vocabulary size, y	1100	1300	1500	2100	2600	460	1200

(a) $x = 2$ years (b) $x = 3$ years
(c) $x = 6$ years (d) $x = 12$ years

DATA **15.** ***Hours Studying and Test Scores*** The number of hours 13 students spent studying for a test and their scores on that test

Hours spent studying, x	0	1	2	4	4	5	5
Test score, y	40	41	51	48	64	69	73

Hours spent studying, x	5	6	6	7	7	8
Test score, y	75	68	93	84	90	95

(a) $x = 3$ hours (b) $x = 6.5$ hours
(c) $x = 13$ hours (d) $x = 4.5$ hours

DATA **16.** ***TV and Test Scores*** The number of hours 12 students watched television during the weekend and the scores of each student who took a test the following Monday

Hours spent watching TV, x	0	1	2	3	3	5
Test score, y	96	85	82	74	95	68

Hours spent watching TV, x	5	5	6	7	7	10
Test score, y	76	84	58	65	75	50

(a) $x = 4$ hours (b) $x = 8$ hours
(c) $x = 9$ hours (d) $x = 15$ hours

14. $\hat{y} = 510.789x - 504.474$

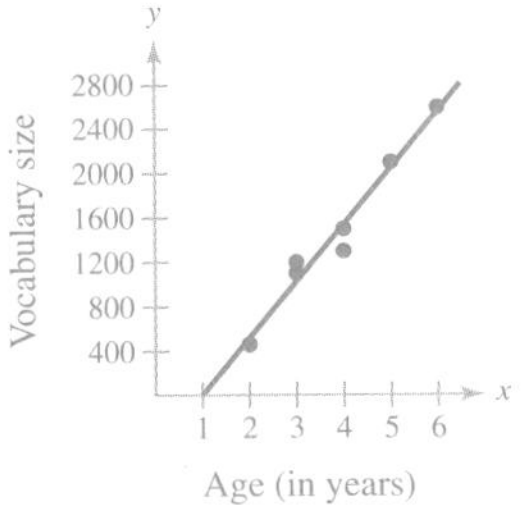

(a) 517
(b) 1028
(c) 2560
(d) It is not meaningful to predict the value of y for $x = 12$ because $x = 12$ is outside the range of the original data.

15. $\hat{y} = 7.350x + 34.617$

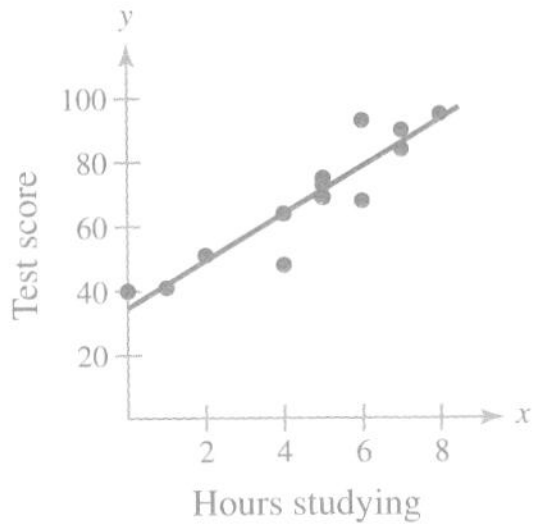

(a) 56.7
(b) 82.4
(c) It is not meaningful to predict the value of y for $x = 13$ because $x = 13$ is outside the range of the original data.
(d) 67.7

16. $\hat{y} = -4.067x + 93.970$

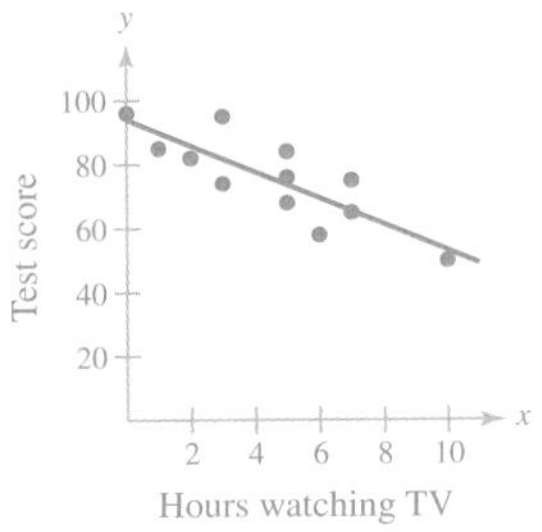

(a) 77.7
(b) 61.4
(c) 57.4
(d) It is not meaningful to predict the value of y for $x = 15$ because $x = 15$ is outside the range of the original data.

17. $\hat{y} = 4.347x - 299.482$

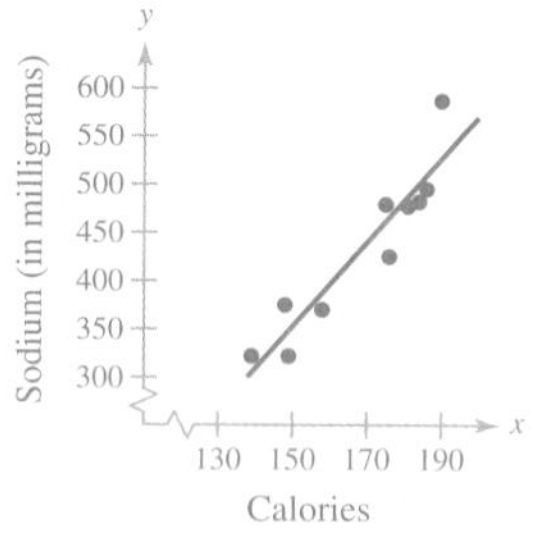

(a) 439.508 milligrams

(b) It is not meaningful to predict the value of y for $x = 136$ because $x = 136$ is outside the range of the original data.

(c) 330.833 milligrams

(d) 496.019 milligrams

18. $\hat{y} = 0.432x - 20.297$

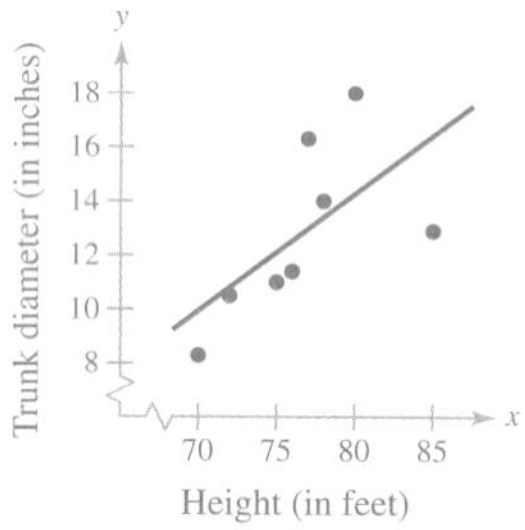

(a) 11.67 inches

(b) 14.70 inches

(c) It is not meaningful to predict the value of y for $x = 87$ because $x = 87$ is outside the range of the original data.

(d) 13.83 inches

19. $\hat{y} = 1.870x + 51.360$

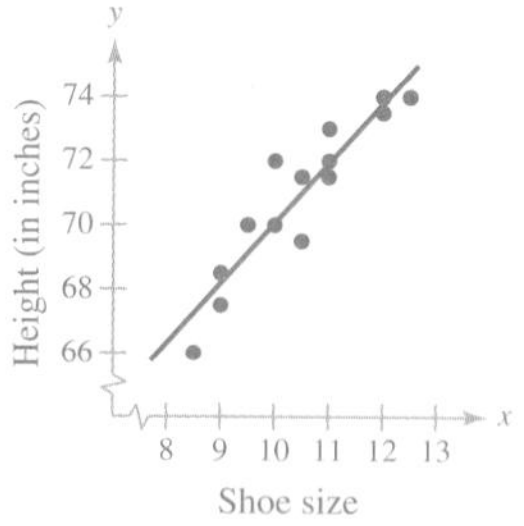

(a) 72.865 inches

(b) 66.320 inches

(c) It is not meaningful to predict the value of y for $x = 15.5$ because $x = 15.5$ is outside the range of the original data.

(d) 70.060 inches

20. See Selected Answers, page A97.

DATA **17. *Hot Dogs: Caloric and Sodium Content*** The caloric content and the sodium content (in milligrams) for 10 beef hot dogs *(Source: Consumer Reports)*

Calories, x	186	181	176	149	184
Sodium, y	495	477	425	322	482

Calories, x	190	158	139	175	148
Sodium, y	587	370	322	479	375

(a) $x = 170$ calories
(b) $x = 136$ calories
(c) $x = 145$ calories
(d) $x = 183$ calories

18. *Trees: Heights and Diameters* The height (in feet) and trunk diameters (in inches) of eight trees

Height, x	70	72	75	76
Trunk diameter, y	8.3	10.5	11.0	11.4

Height, x	85	78	77	80
Trunk diameter, y	12.9	14.0	16.3	18.0

(a) $x = 74$ feet
(b) $x = 81$ feet
(c) $x = 87$ feet
(d) $x = 79$ feet

DATA **19. *Shoe Size and Height*** The shoe sizes and heights (in inches) for 14 men

Shoe size, x	8.5	9.0	9.0	9.5	10.0	10.0	10.5
Height, y	66.0	68.5	67.5	70.0	70.0	72.0	71.5

Shoe size, x	10.5	11.0	11.0	11.0	12.0	12.0	12.5
Height, y	69.5	71.5	72.0	73.0	73.5	74.0	74.0

(a) $x =$ size 11.5
(b) $x =$ size 8.0
(c) $x =$ size 15.5
(d) $x =$ size 10.0

DATA **20. *Age and Hours Slept*** The age (in years) and the number of hours slept in a day by 10 infants

Age, x	0.1	0.2	0.4	0.7	0.6	0.9
Hours slept, y	14.9	14.5	13.9	14.1	13.9	13.7

Age, x	0.2	0.6	0.5	0.3
Hours slept, y	14.3	13.9	14.0	14.1

(a) $x = 0.3$ year
(b) $x = 3.9$ years
(c) $x = 0.6$ year
(d) $x = 0.8$ year

21. ***Braking Distances: Dry Surface*** The weights (in pounds) of eight vehicles and the variability of their braking distances (in feet) when stopping on a dry surface *(Adapted from National Highway Traffic Safety Administration)*

Weight, x	5720	4050	6130	5000	5010	4270	5500	5550
Variability in braking distance, y	2.19	1.36	2.58	1.74	1.78	1.69	1.80	1.87

(a) $x = 4500$ pounds (b) $x = 6000$ pounds
(c) $x = 7500$ pounds (d) $x = 5750$ pounds

22. ***Braking Distances: Wet Surface*** The weights (in pounds) of eight vehicles and the variability of their braking distances (in feet) when stopping on a wet surface *(Adapted from National Highway Traffic Safety Administration)*

Weight, x	5720	4050	6130	5000	5010	4270	5500	5550
Variability in braking distance, y	3.78	2.43	4.63	2.88	3.25	2.76	3.42	3.51

(a) $x = 4800$ pounds (b) $x = 5850$ pounds
(c) $x = 4075$ pounds (d) $x = 3000$ pounds

23. ***Writing*** Explain how to predict y-values using the equation of a regression line.

24. ***Writing*** Given a set of data and a corresponding regression line, describe all values of x that provide meaningful predictions for y.

Extending Concepts

Interchanging x and y In Exercise 25 and 26, do the following.

(a) Find the equation of the regression line for the given data, letting Row 1 represent the x-values and Row 2 the y-values. Graph a scatter plot of the data and draw the regression line with it.

(b) Find the equation of the regression line for the given data, letting Row 2 represent the x-values and Row 1 the y-values. Graph a scatter plot of the data and draw the regression line with it.

(c) What effect does switching the explanatory and response variables have on the regression line?

25.

Row 1	16	25	39	45	49	64	70
Row 2	109	122	143	132	199	185	199

26.

Row 1	0	1	2	3	3	5	5	5	6	7
Row 2	96	85	82	74	95	68	76	84	58	65

21. $\hat{y} = 0.000447x - 0.425$

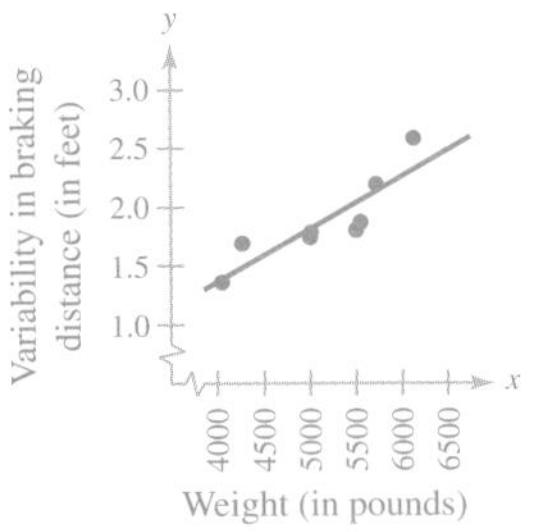

(a) 1.587 feet (b) 2.257 feet

(c) It is not meaningful to predict the value of y for $x = 7500$ because $x = 7500$ is outside the range of the original data.

(d) 2.145 feet

22. See Selected Answers, page A97.

23. Substitute a value x into the equation of a regression line and solve for y.

24. Prediction values are meaningful only for x-values in (or close to) the range of the original data.

25. (a) $\hat{y} = 1.724x + 79.733$

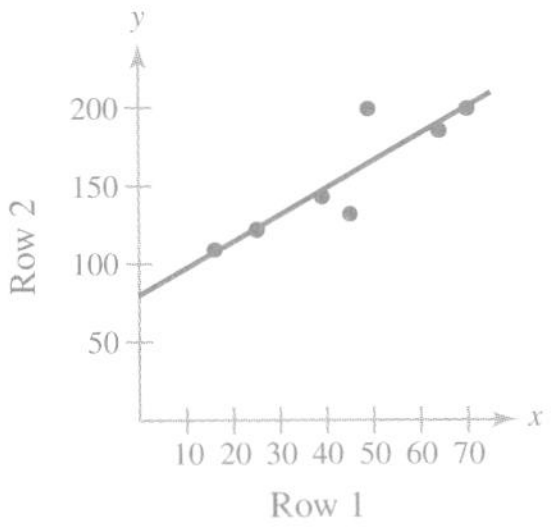

(b) $\hat{y} = 0.453x - 26.448$

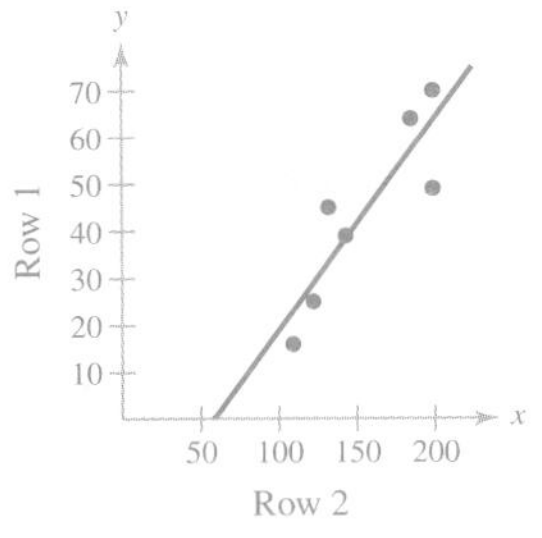

(c) The slope of the line keeps the same sign, but the values of m and b change.

26. See Selected Answers, page A97.

Number of subscribers (in millions), x	Average monthly bill (in dollars), y
3.5	89.30
5.3	80.90
7.6	72.74
11.0	68.68
16.0	61.48
24.1	56.21
33.8	51.00
44.0	47.70
55.3	42.78
69.2	39.43
86.0	41.24
109.5	45.27
128.4	47.37

27. Strong negative nonlinear correlation; as the number of subscribers increases, the average monthly bill decreases.

28. See Selected Answers, page A98.

29. \$17.73; This is not a valid prediction because $x = 180$ is outside the range of the original data.

30. $t \approx -3.831 < -3.106$ or $|r| \approx 0.756 > 0.684$

There is enough evidence at the 1% level to conclude that there is a significant correlation between the number of subscribers and the average monthly bill.

31. As the number of cellular phone subscribers increases, the average monthly bill decreases. Then at approximately 70 million subscribers, the average monthly bill starts to increase as subscribers increase. (Answers will vary.)

32. See Selected Answers, page A98.

33. The exponential equation seems to be the better fit because the line is curved.

34. See Selected Answers, page A98.

35. See Odd Answers, page A78.

36. The logarithmic equation is better because it curves with the data.

37. The logarithmic equation is a better model because the data do not follow a linear pattern.

Cellular Phones In Exercises 27–31, use the following information. You work for a cellular phone industry analyst and gather the data shown in the table. The table shows the number of cellular phone subscribers and the average monthly cellular phone bill in the United States for 13 years. *(Source: Cellular Telecommunications & Internet Association)*

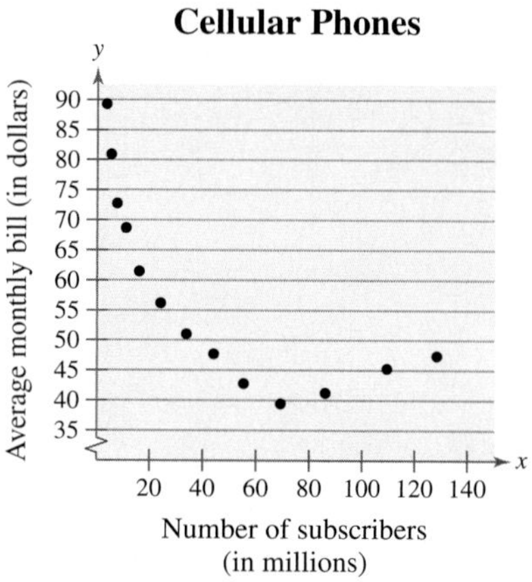

27. ***Correlation*** Using the scatter plot of the cellular phone data shown, what type of correlation, if any, do you think the data have? Explain.

DATA 28. ***Regression Line*** Find an equation of the regression line for the data. Graph a scatter plot of the data and the regression line.

29. ***Using the Regression Line*** The analyst uses the regression line you found in Exercise 28 to predict the average monthly bill for $x = 180$ million subscribers. Is this a valid prediction? Explain your reasoning.

30. ***Significant Correlation?*** The analyst claims that the data have a significant correlation for $\alpha = 0.01$. Verify this claim.

31. ***Cause and Effect*** Write a paragraph describing the cause-and-effect relationship between the number of cellular phone subscribers and the average monthly cellular phone bill.

Exponential Equation In Exercises 32 and 33, use the following information and a technology tool. The **exponential equation** is a nonlinear regression equation of the form

$$y = ab^x.$$

32. Find and graph the exponential equation for the data given in Exercises 27–31.

33. Compare your results in Exercise 32 with the equation of the regression line and its graph. Which equation is a better model for the data? Explain.

Logarithmic Equation In Exercises 34–37, use the following information and a technology tool. The **logarithmic equation** is a nonlinear regression equation of the form

$$y = a + b \ln x.$$

34. Find and graph the logarithmic equation for the data given in Exercise 19.

35. Find and graph the logarithmic equation for the data given in Exercise 20.

36. Compare your results in Exercise 34 with the equation of the regression line and its graph. Which equation is a better model for the data? Explain.

37. Compare your results in Exercise 35 with the equation of the regression line and its graph. Which equation is a better model for the data? Explain.

Case Study

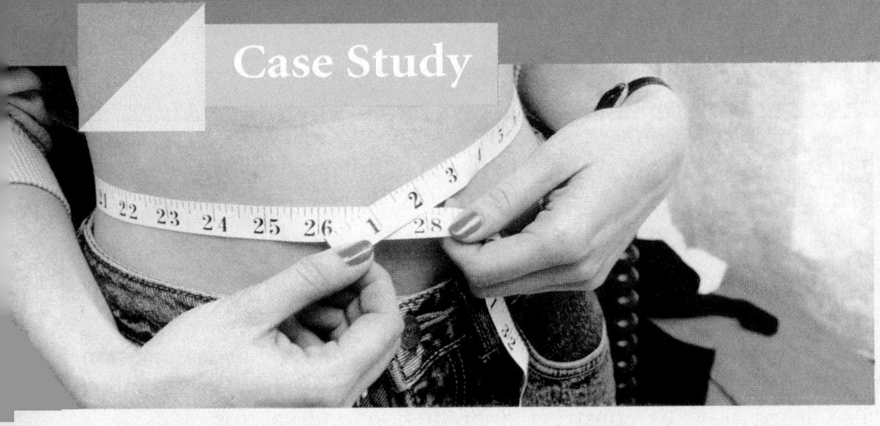

Correlation of Body Measurements

In a study published in *Medicine and Science in Sports and Exercise* (volume 17, no. 2, page 189) the measurements of 252 men (ages 22–81) are given. Of the 14 measurements taken of each man, some have significant correlations and others don't. For instance, the scatter plot at the right shows that the hip and abdomen circumferences of the men have a strong linear correlation ($r = 0.85$). The partial table shown here lists only the first nine rows of the data.

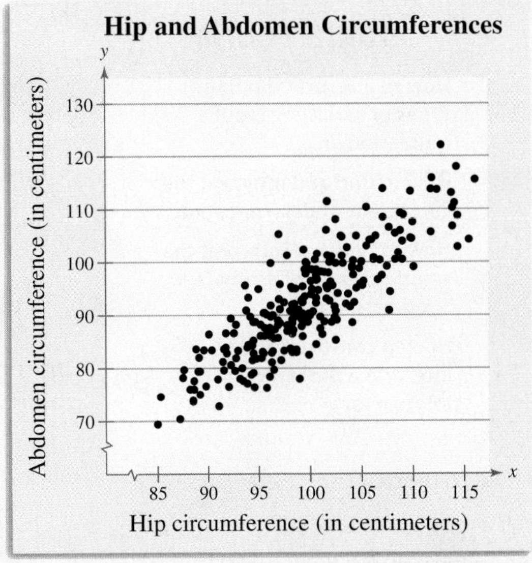

Age (yr)	Weight (lb)	Height (in.)	Neck (cm)	Chest (cm)	Abdom. (cm)	Hip (cm)	Thigh (cm)	Knee (cm)	Ankle (cm)	Bicep (cm)	Forearm (cm)	Wrist (cm)	Body fat %
22	173.25	72.25	38.5	93.6	83.0	98.7	58.7	37.3	23.4	30.5	28.9	18.2	6.1
22	154.00	66.25	34.0	95.8	87.9	99.2	59.6	38.9	24.0	28.8	25.2	16.6	25.3
23	154.25	67.75	36.2	93.1	85.2	94.5	59.0	37.3	21.9	32.0	27.4	17.1	12.3
23	198.25	73.50	42.1	99.6	88.6	104.1	63.1	41.7	25.0	35.6	30.0	19.2	11.7
23	159.75	72.25	35.5	92.1	77.1	93.9	56.1	36.1	22.7	30.5	27.2	18.2	9.4
23	188.15	77.50	38.0	96.6	85.3	102.5	59.1	37.6	23.2	31.8	29.7	18.3	10.3
24	184.25	71.25	34.4	97.3	100.0	101.9	63.2	42.2	24.0	32.2	27.7	17.7	28.7
24	210.25	74.75	39.0	104.5	94.4	107.8	66.0	42.0	25.6	35.7	30.6	18.8	20.9
24	156.00	70.75	35.7	92.7	81.9	95.3	56.4	36.5	22.0	33.5	28.3	17.3	14.2

Exercises

1. Using your intuition, classify the following (x, y) pairs as having a weak correlation ($0 < r < 0.5$), a moderate correlation ($0.5 < r < 0.8$), or a strong correlation ($0.8 < r < 1.0$).

(a) (weight, neck)
(b) (weight, height)
(c) (age, body fat)
(d) (chest, hip)
(e) (age, wrist)
(f) (ankle, wrist)
(g) (forearm, height)
(h) (bicep, forearm)
(i) (weight, body fat)
(j) (knee, thigh)
(k) (hip, abdomen)
(l) (abdomen, hip)

2. Now, use a technology tool to find the correlation coefficient for each pair in Exercise 1. Compare your results with those obtained by intuition.

3. Use a technology tool to find the regression line for each pair in Exercise 1 that has a strong correlation.

4. Use the results of Exercise 3 to predict the following.

(a) The neck circumference of a man whose weight is 180 pounds
(b) The abdomen circumference of a man whose hip circumference is 100 centimeters

5. Are there pairs of measurements that have stronger correlation coefficients than 0.85? Use a technology tool and intuition to reach a conclusion.

9.3 Measures of Regression and Prediction Intervals

What You Should Learn

- How to interpret the three types of variation about a regression line
- How to find and interpret the coefficient of determination
- How to find and interpret the standard error of estimate for a regression line
- How to construct and interpret a prediction interval for y

Note to Instructor

Tell students that without information about the regression line, the best predictor for any value of x is $\bar{y}$, the mean of the y-values.

Variation about a Regression Line • The Coefficient of Determination • The Standard Error of Estimate • Prediction Intervals

Variation about a Regression Line

In this section, you will study two measures used in correlation and regression studies—the coefficient of determination and the standard error of estimate. You will also learn how to construct a prediction interval for y using a regression line and a given value of x. Before studying these concepts, you need to understand the three types of variation about a regression line.

To find the total variation, the explained variation, and the unexplained variation about a regression line, you must first calculate the **total deviation,** the **explained deviation,** and the **unexplained deviation** for each ordered pair (x_i, y_i) in a data set. These deviations are shown in the graph.

Total deviation $= y_i - \bar{y}$

Explained deviation $= \hat{y}_i - \bar{y}$

Unexplained deviation $= y_i - \hat{y}_i$

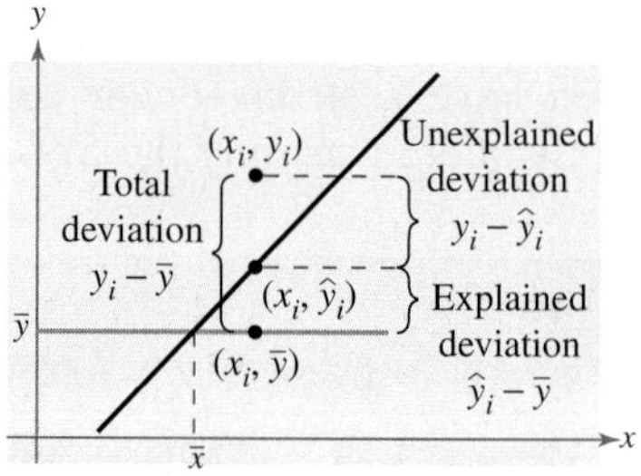

After calculating the deviations for each data point (x_i, y_i), you can find the total variation, the explained variation, and the unexplained variation.

Study Tip

Consider the advertising and sales data used throughout this chapter with a regression line of

$\hat{y} = 50.729x + 104.061.$

Using the data point (2.0, 220), you can find the total, explained, and unexplained deviations as follows.

Total deviation:

$y_i - \bar{y} = 220 - 204.25$
$= 15.75$

Explained deviation:

$\hat{y}_i - \bar{y} = 205.519 - 204.25$
$= 1.269$

Unexplained deviation:

$y_i - \hat{y}_i = 220 - 205.519$
$= 14.481$

DEFINITION

The **total variation** about a regression line is the sum of the squares of the differences between the y-value of each ordered pair and the mean of y.

$$\text{Total variation} = \Sigma(y_i - \bar{y})^2$$

The **explained variation** is the sum of the squares of the differences between each predicted y-value and the mean of y.

$$\text{Explained variation} = \Sigma(\hat{y}_i - \bar{y})^2$$

The **unexplained variation** is the sum of the squares of the differences between the y-value of each ordered pair and each corresponding predicted y-value.

$$\text{Unexplained variation} = \Sigma(y_i - \hat{y}_i)^2$$

The sum of the explained and unexplained variations is equal to the total variation.

$$\text{Total variation} = \text{Explained variation} + \text{Unexplained variation}$$

As its name implies, the *explained variation* can be explained by the relationship between x and y. The *unexplained variation* cannot be explained by the relationship between x and y and is due to chance or other variables.

Picturing the World

Janette Benson (Psychology Department, University of Denver) performed a study relating the age that infants crawl (in weeks after birth) with the average monthly temperature six months after birth. Her results are based on a sample of 414 infants. Janette Benson believes that the reason for the correlation in temperature to crawling age is that parents tend to bundle infants in more restrictive clothing and blankets during cold months. This bundling doesn't allow the infant as much opportunity to move and experiment with crawling.

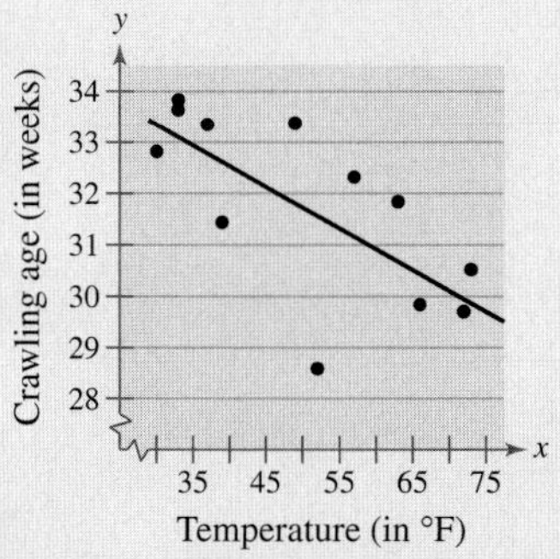

The correlation coefficient is $r = 0.70$. What percent of the variation in the data can be explained? What percent is due to chance, sampling error, or other factors?

The Coefficient of Determination

You already know how to calculate the correlation coefficient r. The square of this coefficient is called the coefficient of determination. It can be shown that the coefficient of determination is equal to the ratio of the explained variation to the total variation.

DEFINITION

The **coefficient of determination r^2** is the ratio of the explained variation to the total variation. That is,

$$r^2 = \frac{\text{Explained variation}}{\text{Total variation}}.$$

It is important that you interpret the coefficient of determination correctly. For instance, if the correlation coefficient is $r = 0.90$, then the coefficient of determination is

$$r^2 = 0.90^2$$
$$= 0.81.$$

This means that 81% of the variation of y can be explained by the relationship between x and y. The remaining 19% of the variation is unexplained and is due to other factors or to sampling error.

EXAMPLE 1

Finding the Coefficient of Determination

The correlation coefficient for the advertising expenses and company sales data given in Section 9.1 is $r \approx 0.913$, as shown in Example 4. Find the coefficient of determination. What does this tell you about the explained variation of the data about the regression line? About the unexplained variation?

SOLUTION The coefficient of determination is

$$r^2 = (0.913)^2$$
$$\approx 0.834.$$

Interpretation About 83.4% of the variation in the company sales can be explained by the variation in the advertising expenditures. About 16.6% of the variation is unexplained and is due to chance or other variables.

Try It Yourself 1

The correlation coefficient for the Old Faithful data given in Section 9.1 is $r \approx 0.970$, as shown in Example 5. Find the coefficient of determination. What does this tell you about the explained variation of the data about the regression line? About the unexplained variation?

a. *Identify* the correlation coefficient r.
b. *Square* r.
c. What percent of the variation in the times is explained? What percent is unexplained? *Answer: Page A43*

The Standard Error of Estimate

When a $\hat{y}$-value is predicted from an x-value, the prediction is a point estimate. You can construct an interval estimate for $\hat{y}$, but first you need to calculate the standard error of estimate.

DEFINITION

The **standard error of estimate s_e** is the standard deviation of the observed y_i-values about the predicted $\hat{y}$-value for a given x_i-value. It is given by

$$s_e = \sqrt{\frac{\Sigma(y_i - \hat{y}_i)^2}{n - 2}}$$

where n is the number of ordered pairs in the data set.

From this formula, you can see that the standard error of estimate is the square root of the unexplained variation divided by $n - 2$. So, the closer the observed y-values are to the predicted y-values, the smaller the standard error of estimate will be.

GUIDELINES

Finding the Standard Error of Estimate s_e

In Words	*In Symbols*
1. Make a table that includes the column headings shown at the right.	$x_i, y_i, \hat{y}_i, (y_i - \hat{y}_i), (y_i - \hat{y}_i)^2$
2. Use the regression equation to calculate the predicted y-values.	$\hat{y}_i = mx_i + b$
3. Calculate the sum of the squares of the differences between each observed y-value and the corresponding predicted y-value.	$\Sigma(y_i - \hat{y}_i)^2$
4. Find the standard error of estimate.	$s_e = \sqrt{\frac{\Sigma(y_i - \hat{y}_i)^2}{n - 2}}$

Shoe size, x	Men's heights (in inches), y
10.0	70.0
10.5	71.0
9.5	70.0
11.0	72.0
12.0	74.0
8.5	66.0
9.0	68.5
13.0	76.0
10.5	71.5
10.5	70.5
10.0	72.0
9.5	70.0
10.0	71.0
10.5	69.5
11.0	71.5
12.0	73.5

You can also find the standard error of estimate using the following formula.

$$s_e = \sqrt{\frac{\Sigma y^2 - b\Sigma y - m\Sigma xy}{n - 2}}$$

This formula is easy to use if you have already calculated the slope m, the y-intercept b, and several of the sums. For instance, the regression line for the data set given at the left is $\hat{y} = 1.84247x + 51.77413$, and the values of the sums are $\Sigma y^2 = 80{,}877.5$, $\Sigma y = 1137$, and $\Sigma xy = 11{,}940.25$. When the alternative formula is used, the standard error of estimate is

$$s_e = \sqrt{\frac{\Sigma y^2 - b\Sigma y - m\Sigma xy}{n - 2}}$$

$$= \sqrt{\frac{80{,}877.5 - 51.77413(1137) - 1.84247(11{,}940.25)}{16 - 2}} \approx 0.877.$$

EXAMPLE 2

Finding the Standard Error of Estimate

The regression equation for the advertising expenses and company sales data as calculated in Example 1 of Section 9.2 is

$$\hat{y} = 50.729x + 104.061.$$

Find the standard error of estimate.

SOLUTION Use a table to calculate the sum of the squared differences of each observed y-value and the corresponding predicted y-value.

x_i	y_i	$\hat{y}_i$	$(y_i - \hat{y}_i)^2$
2.4	225	225.81	0.6561
1.6	184	185.23	1.5129
2.0	220	205.52	209.6704
2.6	240	235.96	16.3216
1.4	180	175.08	24.2064
1.6	184	185.23	1.5129
2.0	186	205.52	381.0304
2.2	215	215.66	0.4356
			$\Sigma = 635.3463$

Unexplained variation

When $n = 8$ and $\Sigma(y_i - \hat{y}_i)^2 = 635.3463$ are used, the standard error of estimate is

$$s_e = \sqrt{\frac{\Sigma(y_i - \hat{y}_i)^2}{n - 2}}$$

$$= \sqrt{\frac{635.3463}{8 - 2}}$$

$$\approx 10.290.$$

Interpretation The standard deviation of the company sales for a specific advertising expense is about \$10,290.

Try It Yourself 2

Radio ad time	Weekly sales
15	26
20	32
20	38
30	56
40	54
45	78
50	80
60	88

A researcher collects the data shown at the left and concludes that there is a significant relationship between the amount of radio advertising time (in minutes per week) and the weekly sales of a product (in hundreds of dollars). Find the standard error of estimate. Use the regression equation

$$\hat{y} = 1.405x + 7.311.$$

a. *Use a table* to calculate the sum of the squared differences of each observed y-value and the corresponding predicted y-value.
b. *Identify* the number n of ordered pairs in the data set.
c. *Calculate* s_e.
d. *Interpret* the results.

Answer: Page A43

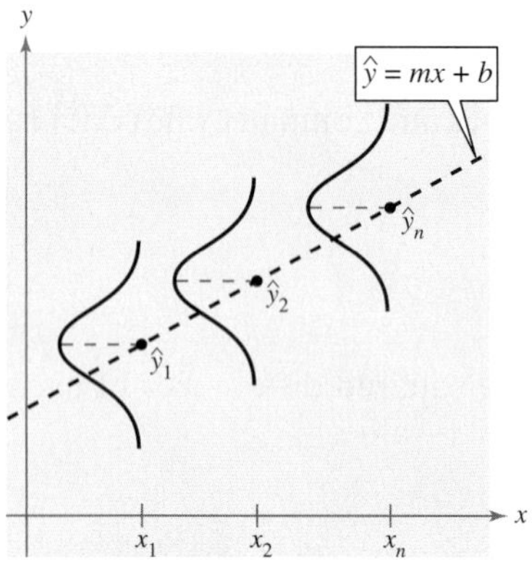

Bivariate Normal Distribution

Prediction Intervals

Two variables have a **bivariate normal distribution** if for any fixed values of x, the corresponding values of y are normally distributed and for any fixed values of y, the corresponding values of x are normally distributed. Because regression equations are determined using sample data and because x and y are assumed to have a bivariate normal distribution, you can construct a prediction interval for the true value of y. To construct the prediction interval, use a t-distribution with $n - 2$ degrees of freedom.

DEFINITION

Given a linear regression equation $\hat{y} = mx + b$ and x_0, a specific value of x, a ***c*-prediction interval** for y is

$$\hat{y} - E < y < \hat{y} + E$$

where

$$E = t_c s_e \sqrt{1 + \frac{1}{n} + \frac{n(x_0 - \bar{x})^2}{n\sum x^2 - (\sum x)^2}}.$$

The point estimate is $\hat{y}$ and the margin of error is E. The probability that the prediction interval contains y is c.

GUIDELINES

Construct a Prediction Interval for y for a Specific Value of x

In Words	*In Symbols*
1. Identify the number of ordered pairs in the data set n and the degrees of freedom.	d.f. $= n - 2$
2. Use the regression equation and the given x-value to find the point estimate $\hat{y}$.	$\hat{y}_i = mx_i + b$
3. Find the critical value t_c that corresponds to the given level of confidence c.	Use Table 5 in Appendix B.
4. Find the standard error of estimate s_e.	$s_e = \sqrt{\frac{\sum (y_i - \hat{y}_i)^2}{n - 2}}$
5. Find the margin of error E.	$E = t_c s_e \sqrt{1 + \frac{1}{n} + \frac{n(x_0 - \bar{x})^2}{n\sum x^2 - (\sum x)^2}}$
6. Find the left and right end-points and form the prediction interval.	Left endpoint: $\hat{y} - E$ Right endpoint: $\hat{y} + E$ Interval: $\hat{y} - E < y < \hat{y} + E$

Study Tip

The formulas for s_e and E use the quantities $\sum(y_i - \hat{y}_i)^2$, $(\sum x)^2$, and $\sum x^2$. Use a table to calculate these quantities.

EXAMPLE 3

Constructing a Prediction Interval

Construct a 95% prediction interval for the company sales when the advertising expenses are $2100 (see Example 2). What can you conclude?

SOLUTION Because $n = 8$, there are

$$8 - 2 = 6$$

degrees of freedom. Using the regression equation

$$\hat{y} = 50.729x + 104.061$$

and

$$x = 2.1$$

the point estimate is

$$\begin{aligned}\hat{y} &= 50.729x + 104.061\\ &= 50.729(2.1) + 104.061\\ &\approx 210.592.\end{aligned}$$

From Table 5, the critical value is $t_c = 2.447$ and from Example 2, $s_e = 10.290$. Using these values, the margin of error is

$$\begin{aligned}E &= t_c s_e \sqrt{1 + \frac{1}{n} + \frac{n(x_0 - \bar{x})^2}{n(\sum x^2) - (\sum x)^2}}\\ &= (2.447)(10.290)\sqrt{1 + \frac{1}{8} + \frac{8(2.1 - 1.975)^2}{8(32.44) - (15.8)^2}}\\ &\approx 26.857.\end{aligned}$$

Using $\hat{y} = 210.592$ and $E = 26.857$, the confidence interval is

Left Endpoint	Right Endpoint
$210.592 - 26.857 = 183.735$	$210.592 + 26.857 = 237.449$

$$183.735 < y < 237.449.$$

Interpretation You can be 95% confident that when advertising expenses are $2100, the company sales will be between $183,735 and $237,449.

Try It Yourself 3

Construct a 95% prediction interval for the company sales when the advertising expenses are $2500. What can you conclude?

a. *Specify* n, d.f., t_c, s_e.
b. *Calculate* $\hat{y}$ when $x = 2.5$.
c. *Calculate* the margin of error E.
d. *Construct* the prediction interval.
e. *Interpret* the results.

Answer: Page A43

Insight

The greater the difference between x and $\bar{x}$, the wider the prediction interval is. For instance, in Example 3 the 95% prediction intervals for $1.4 < x < 2.6$ are shown below.

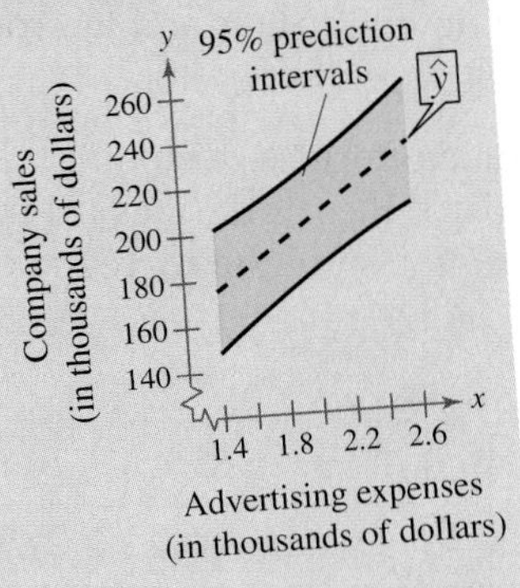

9.3 Exercises

Building Basic Skills and Vocabulary

Graphical Analysis In Exercises 1–3, use the graph to answer the question.

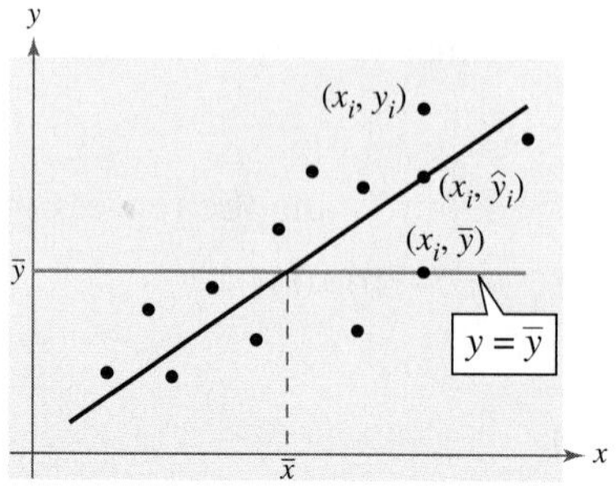

1. Use the graph to describe the total variation about a regression line in words and in symbols.

2. Use the graph to describe the explained variation about a regression line in words and in symbols.

3. Use the graph to describe the unexplained variation about a regression line in words and in symbols.

4. The coefficient of determination is the ratio of which two types of variations? What does the coefficient of determination measure?

In Exercises 5–8, use the value of the linear correlation coefficient to calculate the coefficient of determination. What does this tell you about the explained variation of the data about the regression line? About the unexplained variation?

5. $r = 0.250$

6. $r = -0.375$

7. $r = -0.891$

8. $r = 0.964$

Using and Interpreting Concepts

Finding Types of Variations and the Coefficient of Determination In Exercises 9–16, use the data to find the (a) coefficient of determination and interpret the result, and (b) standard error of estimate s_e and interpret the result.

DATA **9. *Stock Offerings*** The number of initial public offerings of stock issued in a recent 12-year period and the total proceeds of these offerings (in millions of U.S. dollars) are shown in the table. The equation of the regression line is $\hat{y} = 51.185x - 35.462$. *(Source: Securities Data Co. and Thomson Financial Securities Data)*

No. of issues, x	518	222	209	172	366	512
Proceeds, y	16,745.7	6111.7	6082.0	4519.0	16,283.2	23,379.8

No. of issues, x	667	571	575	865	432	267
Proceeds, y	34,461.1	22,771.9	29,270.8	48,789.8	31,701.0	34,628.0

Help

MyMathLab

Student Study Pack

1. Total variation $= \sum(y_i - \bar{y})^2$; the sum of the squares of the differences between the y-values of each ordered pair and the mean of the y-values of the ordered pairs.
2. Explained variation $= \sum(\hat{y}_i - \bar{y})^2$; the sum of the squares of the differences between the predicted y-values and the mean of the y-values of the ordered pairs.
3. Unexplained variation $= \sum(y_i - \hat{y}_i)^2$; the sum of the squares of the differences between the observed y-values and the predicted y-values.
4. The coefficient of determination $r^2 = \dfrac{\sum(\hat{y}_i - \bar{y})^2}{\sum(y_i - \bar{y})^2}$ is the ratio of the explained variation to the total variation and is the percent of variation of y that is explained by the relationship between x and y.
5. 0.063; 6.3% of the variation is explained. 93.7% of the variation is unexplained.
6. 0.141; 14.1% of the variation is explained, and 85.9% of the variation is unexplained.
7. 0.794; 79.4% of the variation is explained. 20.6% of the variation is unexplained.
8. 0.929; 92.9% of the variation is explained, and 7.1% of the variation is unexplained.

9. (a) 0.623; 62.3% of the variation in proceeds can be explained by the variation in the number of issues, and 37.7% of the variation is unexplained.

(b) 8775.838; the standard deviation of the proceeds for a specific number of issues is about \$8,775,838,000.

10. (a) 0.995; 99.5% of the variation in cigarettes exported from the United States can be explained by the variation in the number of cigarettes consumed in the United States, and 0.5% of the variation is unexplained.

(b) 3.066; the standard deviation of the number of cigarettes exported for a specific number of cigarettes consumed is about 3,066,000,000.

11. (a) 0.985; 98.5% of the variation in sales can be explained by the variation in the total square footage, and 1.5% of the variation is unexplained.

(b) 35.652; the standard deviation of the sales for a specific total square footage is about \$35,652,000,000.

12. (a) 0.578; 57.8% of the variation in the median number of leisure hours per week can be explained by the variation in the median number of work hours per week, and 42.2% of the variation is unexplained.

(b) 2.013; the standard deviation of the median number of leisure hours per week for a specific median number of work hours per week is about 2.013 hours.

13. (a) 0.982; 98.2% of the variation in the median weekly earnings of female workers can be explained by the variation in the median weekly earnings of male workers, and 1.8% of the variation is unexplained.

(b) 8.226; the standard deviation of the median weekly earnings of female workers for a specific median weekly earnings of male workers is about \$8.226.

10. ***Cigarettes*** The table shows the number of cigarettes consumed (in billions) in the United States and the number of cigarettes exported (in billions) from the United States for seven years. The equation of the regression line is $\hat{y} = 1.528x - 500.435$. *(Source: Alcohol and Tobacco Tax and Trade Bureau)*

Cigarettes consumed in U.S., *x*	480	465	435	430	425	415	410
Cigarettes exported by U.S., *y*	232	213	166	154	145	137	127

DATA **11.** ***Retail Space and Sales*** The table shows the total square footage (in billions) of retailing space at shopping centers and their sales (in billions of U.S. dollars) for 11 years. The equation of the regression line is $\hat{y} = 230.831x - 289.809$. *(Source: International Council of Shopping Centers)*

Total square footage, *x*	1.6	2.3	3.0	3.4	3.9	4.6
Sales, *y*	123.2	211.5	385.5	475.1	641.1	716.9

Total square footage, *x*	4.7	4.8	4.9	5.0	5.1
Sales, *y*	768.2	806.6	851.3	893.8	933.9

DATA **12.** ***Work and Leisure Time*** The median number of work hours per week and the median number of leisure hours per week for people in the United States for 10 recent years are shown in the table. The equation of the regression line is $\hat{y} = -0.646x + 50.734$. *(Source: Louis Harris & Associates)*

Median no. of work hrs. per week, *x*	40.6	43.1	46.9	47.3	46.8
Median no. of leisure hrs. per week, *y*	26.2	24.3	19.2	18.1	16.6

Median no. of work hrs. per week, *x*	48.7	50.0	50.7	50.6	50.8
Median no. of leisure hrs. per week, *y*	18.8	18.8	19.5	19.2	19.5

13. ***Earnings of Men and Women*** The table shows median weekly earnings (in U.S. dollars) of full-time male and female workers for five years. The equation of the regression line is $\hat{y} = 0.919x - 95.462$. *(Source: U.S. Bureau of Labor Statistics)*

Median weekly earnings of male workers, *x*	695	679	672	618	557
Median weekly earnings of female workers, *y*	552	529	511	473	418

14. (a) 0.888; 88.8% of the variation in the turnout for federal elections can be explained by the variation in the voting age population, and 11.2% of the variation is unexplained.
(b) 3.732; the standard deviation of the turnout in a federal election for a specified voting age population is about 3,732,000.

15. (a) 0.992; 99.2% of the variation in the money spent can be explained by the variation in the money raised, and 0.8% of the variation is unexplained.
(b) 16.079; the standard deviation of the money spent for a specified amount of money raised is about $16,079,000.

16. (a) 0.877; 87.7% of the variation in bond and income funds can be explained by the variation in equity funds, and 12.3% of the variation is unexplained.
(b) 114.237; the standard deviation of the bond and income funds for a specified equity fund is about $114,237,000,000.

17. $14,752,729,000 $< y <$ $58,063,787,000
You can be 95% confident that the proceeds will be between $14,752,729,000 and $58,063,787,000 when the number of initial offerings is 712.

18. 237,610,000,000 $< y <$ 258,960,000,000
You can be 95% confident that the number of cigarettes exported will be between 237,610,000,000 and 258,960,000,000 when the number of cigarettes consumed is 490 billion.

19. $679,992,000,000 $< y <$ $817,870,000,000
You can be 90% confident that the sales will be between $679,992,000,000 and $817,870,000,000 when the total square footage is 4.5 billion.

20. See Selected Answers, page A98.

14. *Voter Turnout* The U.S. voting age population (in millions) and the turnout of the voting age population (in millions) for federal elections for eight years are shown in the table. The data can be modeled by the regression equation $\hat{y} = 0.373x + 26.377$. *(Source: Federal Election Commission)*

Voting age population, x	120.3	140.8	152.3	164.6
Turnout in federal elections, y	73.2	77.7	81.6	86.5

Voting age population, x	174.5	182.8	189.5	196.5
Turnout in federal elections, y	92.7	91.6	104.4	96.5

15. *Campaign Money* The money raised and spent (both in millions of U.S. dollars) by all congressional campaigns for eight recent years are shown in the table. The data can be modeled by the regression equation $\hat{y} = 1.020x - 25.854$. *(Source: Federal Election Commission)*

Money raised, x	354.7	397.2	472.0	477.6	471.7	659.3	740.5	790.5
Money spent, y	342.4	374.1	450.9	459.0	446.3	680.2	725.2	765.3

16. *Fund Assets* The table shows the total assets (in billions of U.S. dollars) of equity funds and bond and income funds for nine years. The equation of the regression line is $\hat{y} = 0.689x + 68.861$. *(Source: Investment Company Institute)*

Equity funds, x	35.9	41.2	77.0	116.9	180.7
Bond and income funds, y	13.1	14.0	36.6	134.8	273.1

Equity funds, x	249.0	411.6	749.0	1269.0
Bond and income funds, y	304.8	441.4	761.1	798.3

Constructing and Interpreting Prediction Intervals In Exercises 17–24, construct the indicated prediction interval and interpret the results.

17. *Proceeds* Construct a 95% prediction interval for the proceeds from initial public offerings in Exercise 9 when the number of issues is 712.

18. *Cigarettes Consumed* Construct a 95% prediction interval for the cigarettes exported by the United States in Exercise 10 when the number of cigarettes consumed in the United States is 490 billion.

19. *Retail Sales* Using the results of Exercise 11, construct a 90% prediction interval for shopping center sales when the total square footage of shopping centers is 4.5 billion.

20. *Leisure Hours* Using the results of Exercise 12, construct a 90% prediction interval for the median number of leisure hours per week when the median number of work hours per week is 45.1.

21. *Earnings of Female Workers* When the median weekly earnings of male workers is \$500, find a 99% prediction interval for the median weekly earnings of female workers. Use the results of Exercise 13.

22. *Predicting Voter Turnout* When the voting age population is 190 million, construct a 99% prediction interval for the voter turnout in federal elections. Use the results of Exercise 14.

23. *Campaign Spending* A total of \$775.8 million is raised in one year for congressional campaigns. Construct a prediction interval for the money spent by the campaigns. Use the results of Exercise 15 and $c = 0.95$.

24. *Total Assets* The total assets in equity funds is \$900 billion. Construct a prediction interval for the total assets in bond and income funds. Use the results of Exercise 16 and $c = 0.90$.

Extending Concepts

Old Vehicles In Exercises 25–31, use the information shown at the right.

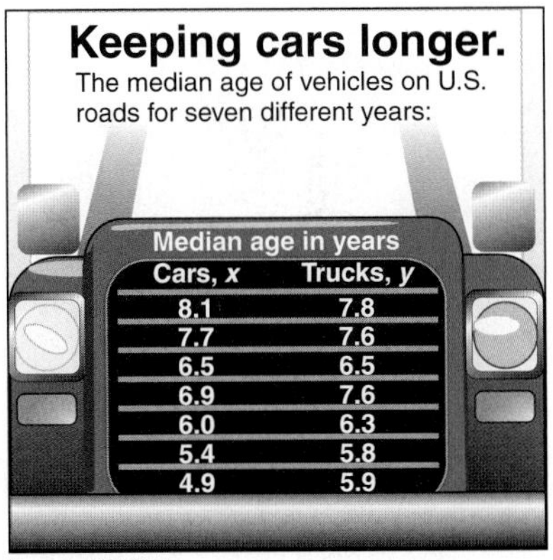

Cars, x	Trucks, y
8.1	7.8
7.7	7.6
6.5	6.5
6.9	7.6
6.0	6.3
5.4	5.8
4.9	5.9

(Source: Polk Co.)

25. *Scatter Plot* Construct a scatter plot of the data. Show $\bar{y}$ and $\bar{x}$ on the graph.

26. *Regression Line* Find and graph the regression line.

27. *Deviation* Calculate the explained deviation, the unexplained deviation, and the total deviation for each data point.

28. *Variation* Find the (a) explained variation, (b) unexplained variation, and (c) total variation.

29. *Coefficient of Determination* Find the coefficient of determination. What can you conclude?

30. *Error of Estimate* Find the standard error of estimate s_e and interpret the results.

31. *Prediction Interval* Construct a 95% prediction interval for the median age of trucks in use when the median age of cars in use is 7.3 years.

32. *Correlation Coefficient and Slope* Recall that the formula for the correlation coefficient r is

$$r = \frac{n\sum xy - (\sum x)(\sum y)}{\sqrt{n\sum x^2 - (\sum x)^2}\sqrt{n\sum y^2 - (\sum y)^2}}$$

and the formula for the slope m of a regression line is

$$m = \frac{n\sum xy - (\sum x)(\sum y)}{n\sum x^2 - (\sum x)^2}.$$

Given a set of data, why must the slope m of the data's regression line always have the same sign as the data's correlation coefficient r?

21. $\$283.386 < y < \$444.690$
You can be 99% confident that the median weekly earnings of female workers will be between \$283.386 and \$444.690 when the median weekly earnings of male workers is \$500.

22. 81.745 million $< y <$ 112.741 million
You can be 99% confident that the voter turnout in federal elections will be between 81.745 million and 112.741 million when the voting age population is 190 million.

23. \$718.708 million $< y <$ \$812.216 million
You can be 95% confident that the money spent for congressional campaigns will be between \$718.708 million and \$812.216 million when the money raised is \$775.8 million.

24. \$438.949 billion $< y <$ \$938.973 billion
You can be 90% confident that the total assets in bond and income funds will be between \$438.949 billion and \$938.973 billion when the total assets in equity funds is \$900 billion.

25.

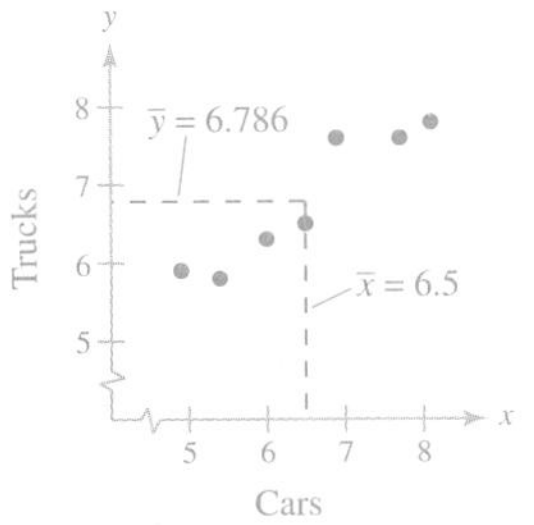

26. See Selected Answers, page A98.

27. See Odd Answers, page A79.

28. (a) 3.930 (b) 0.498 (c) 4.429

29. 0.887; About 88.7% of the variation in the median age of trucks can be explained by the variation in the median age of cars, and 11.3% of the variation is unexplained.

30. 0.316; 31.6% of the variation in the median age of trucks can be explained by the variation in the median age of cars, and 68.4% of the variation is unexplained.

31. (6.441, 8.237)

32. See Selected Answers, page A98.

9.4 Multiple Regression

What You Should Learn

- How to use technology to find a multiple regression equation, the standard error of estimate, and the coefficient of determination
- How to use a multiple regression equation to predict y-values

Finding a Multiple Regression Equation • Predicting y-Values

Finding a Multiple Regression Equation

In many instances, a better prediction model can be found for a dependent (response) variable by using more than one independent (explanatory) variable. For example, a more accurate prediction for the company sales discussed in previous sections might be made by considering the number of employees on the sales staff as well as the advertising expenses. Models that contain more than one independent variable are multiple regression models.

Insight

Because the mathematics associated with multiple regression is complicated, this section focuses on how to use technology to find a multiple regression equation and how to interpret the results.

DEFINITION

A **multiple regression equation** has the form

$$\hat{y} = b + m_1x_1 + m_2x_2 + m_3x_3 + \cdots + m_kx_k$$

where $x_1, x_2, x_3, \ldots, x_k$ are the independent variables, b is the y-intercept, and y is the dependent variable.

The y-intercept b is the value of y when all x_i are 0. Each coefficient m_i is the amount of change in y when the independent variable x_i is changed by one unit and all other independent variables are held constant.

EXAMPLE 1

Finding a Multiple Regression Equation

A researcher wants to determine how employee salaries at a certain company are related to the length of employment, previous experience, and education. The researcher selects eight employees from the company and obtains the following data.

Employee	Salary, y	Employment (in years), x_1	Experience (in years), x_2	Education (in years), x_3
A	37,310	10	2	16
B	37,380	5	6	16
C	34,135	3	1	12
D	36,985	6	5	14
E	38,715	8	8	16
F	40,620	20	0	12
G	39,200	8	4	18
H	40,320	14	6	17

Use MINITAB to find a multiple regression equation that models the data.

SOLUTION Enter the y-values in C1 and the x_1-, x_2-, and x_3-values in C2, C3, and C4, respectively. Select "Regression▶Regression . . . " from the *Stat* menu. Using the salaries as the response variable and the remaining data as the predictors, you should obtain results similar to the following.

MINITAB

Regression Analysis: Salary, y versus x1, x2, x3

```
The regression equation is
Salary, y = 29764 + 364 x1 + 228 x2 + 267 x3

Predictor      Coef          SE Coef       T        P
Constant      29764   (b)     1981      15.02    0.000
x1           364.41   (m1)    48.32      7.54    0.002
x2            227.6   (m2)    123.8      1.84    0.140
x3            266.9   (m3)    147.4      1.81    0.144

S = 659.490     R-Sq = 94.4%     R-Sq(adj) = 90.2%
```

The regression equation is $\hat{y} = 29{,}764 + 364x_1 + 228x_2 + 267x_3$.

Study Tip

In Example 1, it is important that you interpret the coefficients m_1, m_2, and m_3 correctly. For instance, if x_2 and x_3 are held constant and x_1 increases by 1, then y increases by \$364. Similarly, if x_1 and x_3 are held constant and x_2 increases by 1, then y increases by \$228. If x_1 and x_2 are held constant and x_3 increases by 1, then y increases by \$267.

Try It Yourself 1

A statistics professor wants to determine how students' final grades are related to the midterm exam grades and number of classes missed. The professor selects 10 students from her class and obtains the following data.

Student	Final grade, y	Midterm exam, x_1	Classes missed, x_2
1	81	75	1
2	90	80	0
3	86	91	2
4	76	80	3
5	51	62	6
6	75	90	4
7	44	60	7
8	81	82	2
9	94	88	0
10	93	96	1

Use technology to find a multiple regression equation that models the data.

a. *Enter* the data.
b. *Calculate* the regression line.

Answer: Page A43

MINITAB displays much more than the regression equation and the coefficients of the independent variables. For example, it also displays the standard error of estimate, denoted by S, and the coefficient of determination, denoted by $R\text{-}Sq$. In Example 1, $S = 659.490$ and $R\text{-}Sq = 94.4\%$. So, the standard error of estimate is \$659.49. The coefficient of determination tells you that 94.4% of the variation in y can be explained by the multiple regression model. The remaining 5.6% is unexplained and is due to other factors or chance.

Picturing the World

In a lake in Finland, 159 fish of seven species were caught and measured for weight G, length L, height H, and width W (H and W are percents of L). The regression equation for G and L is

$$G = -488 + 28.5L,$$
$$r = 0.925.$$

When all four variables are used, the regression equation is

$$G = -711 + 28.2L + 1.46H + 13.4W,$$
$$r = 0.931.$$

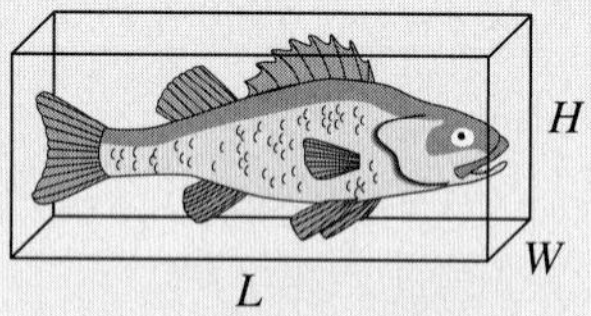

Predict the weight of a fish with the following measurements: $L = 40$, $H = 17$, and $W = 11$. How do your predictions vary when you use a single variable versus many variables? Which do you think is more accurate?

Predicting y-Values

After finding the equation of the multiple regression line, you can use the equation to predict y-values over the range of the data. To predict y-values, substitute the given value for each independent variable into the equation, then calculate $\hat{y}$.

EXAMPLE 2

Predicting y-Values Using Multiple Regression Equations

Use the regression equation found in Example 1 to predict an employee's salary given the following conditions.

1. 12 years of current employment, 5 years of experience, and 16 years of education
2. 4 years of current employment, 2 years of experience, and 12 years of education
3. 8 years of current employment, 7 years of experience, and 17 years of education

SOLUTION To predict each employee's salary, substitute the values for x_1, x_2, and x_3 into the regression equation. Then calculate $\hat{y}$.

1. $\hat{y} = 29{,}764 + 364x_1 + 228x_2 + 267x_3$
   $= 29{,}764 + 364(12) + 228(5) + 267(16)$
   $= 39{,}544$

   The employee's predicted salary is \$39,544.

2. $\hat{y} = 29{,}764 + 364x_1 + 228x_2 + 267x_3$
   $= 29{,}764 + 364(4) + 228(2) + 267(12)$
   $= 34{,}880$

   The employee's predicted salary is \$34,880.

3. $\hat{y} = 29{,}764 + 364x_1 + 228x_2 + 267x_3$
   $= 29{,}764 + 364(8) + 228(7) + 267(17)$
   $= 38{,}811$

   The employee's predicted salary is \$38,811.

Try It Yourself 2

Use the regression equation found in Try It Yourself 1 to predict a student's final grade given the following conditions.

1. A student has a midterm exam score of 89 and misses 1 class.
2. A student has a midterm exam score of 78 and misses 3 classes.
3. A student has a midterm exam score of 83 and misses 2 classes.

a. *Substitute* the midterm score for x_1 into the regression equation.
b. *Substitute* the corresponding number of missed classes for x_2 into the regression equation.
c. *Calculate* $\hat{y}$.
d. What is each student's final grade?

Answer: Page A43

9.4 Exercises

Help
MyMathLab
Student Study Pack

Building Basic Skills and Vocabulary

Predicting y-Values In Exercises 1–4, use the multiple regression equation to predict the y-values for the given values of the independent variables.

1. *Peanut Yield* The equation used to predict peanut yield (in pounds) is

$$\hat{y} = 6503 - 14.8x_1 + 12.2x_2$$

where x_1 is the number of acres planted (in thousands) and x_2 is the number of acres harvested (in thousands). *(Source: U.S. National Agricultural Statistics Service)*

(a) $x_1 = 1458,\ x_2 = 1450$ (b) $x_1 = 1500,\ x_2 = 1475$
(c) $x_1 = 1400,\ x_2 = 1385$ (d) $x_1 = 1525,\ x_2 = 1500$

2. *Rice Yield* To predict the annual rice yield (in pounds), use the equation

$$\hat{y} = 859 + 5.76x_1 + 3.82x_2$$

where x_1 is the number of acres planted (in thousands) and x_2 is the number of acres harvested (in thousands). *(Source: U.S. National Agricultural Statistics Service)*

(a) $x_1 = 2532,\ x_2 = 2255$ (b) $x_1 = 3581,\ x_2 = 3021$
(c) $x_1 = 3213,\ x_2 = 3065$ (d) $x_1 = 2758,\ x_2 = 2714$

3. *Black Cherry Tree Volume* The volume (in cubic feet) of black cherry trees can be modeled by the equation

$$\hat{y} = -52.2 + 0.3x_1 + 4.5x_2$$

where x_1 is the tree's height (in feet) and x_2 is the tree's diameter (in inches). *(Source: Journal of the Royal Statistical Society)*

(a) $x_1 = 70,\ x_2 = 8.6$ (b) $x_1 = 65,\ x_2 = 11.0$
(c) $x_1 = 83,\ x_2 = 17.6$ (d) $x_1 = 87,\ x_2 = 19.6$

4. *Earnings per Share* The earnings per share (in dollars) for McDonald's Corporation are given by the equation

$$\hat{y} = 0.668 + 0.046x_1 - 0.015x_2$$

where x_1 represents total revenue (in billions of dollars) and x_2 represents total net worth (in billions of dollars). *(Source: McDonald's Corporation)*

(a) $x_1 = 11.4,\ x_2 = 8.6$ (b) $x_1 = 8.1,\ x_2 = 6.2$
(c) $x_1 = 10.7,\ x_2 = 8.5$ (d) $x_1 = 7.3,\ x_2 = 5.1$

1. (a) 2614.6 pounds
(b) 2298 pounds
(c) 2680 pounds
(d) 2233 pounds
2. (a) 24,057.42 pounds
(b) 33,025.78 pounds
(c) 31,074.18 pounds
(d) 27,112.56 pounds
3. (a) 7.5 cubic feet
(b) 16.8 cubic feet
(c) 51.9 cubic feet
(d) 62.1 cubic feet
4. (a) \$1.063
(b) \$0.948
(c) \$1.033
(d) \$0.927

Using and Interpreting Concepts

Finding a Multiple Regression Equation In Exercises 5 and 6, use technology to find the multiple regression equation for the data shown in the table. Then answer the following.

(a) What is the standard error of estimate?
(b) What is the coefficient of determination?
(c) Interpret the results of (a) and (b).

5. $\hat{y} = -256.293 + 103.502x_1 + 14.649x_2$
   (a) 34.16
   (b) 0.988
   (c) The standard deviation of the predicted sales given a specific total square footage and number of shopping centers is \$34.16 billion. The multiple regression model explains 98.8% of the variation in y.
6. $\hat{y} = -1.661 + 0.156x_1 + 0.031x_2$
   (a) 0.549
   (b) 0.997
   (c) The standard deviation of the predicted equity given a specific net sales and total assets is \$0.549 billion. The multiple regression model explains 99.7% of the variation in y.
7. 0.985
8. 0.996

DATA **5. *Sales*** The total square footage (in billions) of retailing space at shopping centers, the number (in thousands) of shopping centers, and the sales (in billions of U.S. dollars) for shopping centers for a recent 11-year period are shown in the table. *(Source: International Council of Shopping Centers)*

Sales, y	Total square footage, x_1	Number of shopping centers, x_2
123.2	1.6	13.2
211.5	2.3	17.5
385.5	3.0	22.1
475.1	3.4	25.5
641.1	3.9	32.6
716.9	4.6	38.0
768.2	4.7	39.0
806.6	4.8	39.6
851.3	4.9	40.4
893.8	5.0	41.2
933.9	5.1	42.1

DATA **6. *Shareholder's Equity*** The following table shows the net sales (in billions of dollars), total assets (in billions of dollars), and shareholder's equity (in billions of dollars) for Wal-Mart for a recent six-year period. *(Source: Wal-Mart Annual Report)*

Shareholder's equity, y	Net sales, x_1	Total assets, x_2
18.5	118.0	45.4
21.1	137.6	50.0
25.8	165.0	70.3
31.3	191.3	78.1
35.1	217.8	83.5
39.3	244.5	94.7

Extending Concepts

Adjusted r^2 The calculation of r^2, the coefficient of determination, depends on the number of data pairs and the number of independent variables. An adjusted value of r^2 can be calculated, based on the number of degrees of freedom, as follows.

$$r^2_{adj} = 1 - \left[\frac{(1 - r^2)(n - 1)}{n - k - 1}\right]$$

where n is the number of data pairs and k is the number of independent variables.

In Exercises 7 and 8, after calculating r^2_{adj}, determine the percentage of the variation in y that can be explained by the relationships between variables according to r^2_{adj}. Compare this result with the one obtained with r^2.

7. Calculate r^2_{adj} for the data in Exercise 5.

8. Calculate r^2_{adj} for the data in Exercise 6.

Uses and Abuses
Statistics in the Real World

Uses

Correlation and Regression Correlation and regression analysis can be used to determine whether there is a significant relationship between two variables. If there is, you can use one of the variables to predict the value of the other variable. For example, educators have used correlation and regression analysis to determine that there is a significant correlation between a student's SAT score and the grade point average from a student's freshman year at college. Consequently, many colleges and universities use SAT scores of high school applicants as a predictor of the applicant's initial success at college.

Abuses

Confusing Correlation and Causation The most common abuse of correlation in studies is to confuse the concepts of correlation with those of causation (see page 468). Good SAT scores do not cause good college grades. Rather, there are other variables, such as good study habits and motivation, that contribute to both. When a strong correlation is found between two variables, look for other variables that are correlated with both.

Considering Only Linear Correlation The correlation studied in this chapter is linear correlation. When the correlation coefficient is close to 1 or close to −1, the data points can be modeled by a straight line. It is possible that a correlation coefficient is close to zero but there is still a strong correlation of a different type. Consider the data listed in the following table. The value of the correlation coefficient is 0; however, the data are perfectly correlated with the equation $x^2 + y^2 = 1$.

x	y
1	0
0	1
−1	0
0	−1

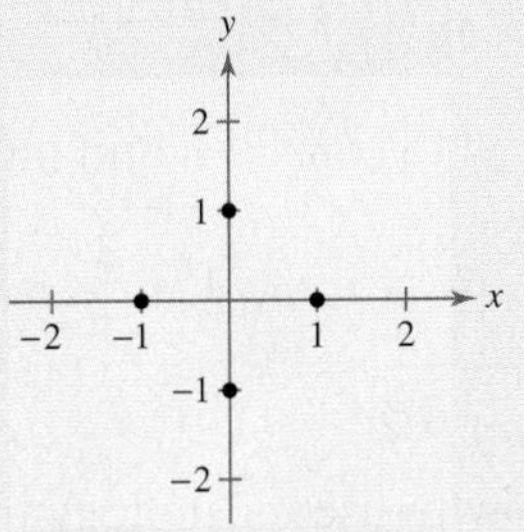

Exercises

1. ***Confusing Correlation and Causation*** Find an example of an article in a newspaper or magazine that confuses correlation and causation. Discuss other variables that could contribute to the relationship between the variables.

2. ***Considering Only Linear Correlation*** Find an example of two real-life variables that have a nonlinear correlation.

9 Chapter Summary

What did you learn?

Review Exercises

Section 9.1

- How to find a correlation coefficient — *1–4*

$$r = \frac{n\Sigma xy - (\Sigma x)(\Sigma y)}{\sqrt{n\Sigma x^2 - (\Sigma x)^2}\sqrt{n\Sigma y^2 - (\Sigma y)^2}}$$

- How to perform a hypothesis test for a population correlation coefficient ρ — *5–10*

$$t = \frac{r}{\sqrt{\dfrac{1 - r^2}{n - 2}}}$$

Section 9.2

- How to find the equation of a regression line, $\hat{y} = mx + b$ — *11–14*

$$m = \frac{n\Sigma xy - (\Sigma x)(\Sigma y)}{n\Sigma x^2 - (\Sigma x)^2}$$

$$b = \bar{y} - m\bar{x}$$

$$= \frac{\Sigma y}{n} - m\frac{\Sigma x}{n}$$

- How to predict y-values using a regression equation — *15–18*

Section 9.3

- How to find and interpret the coefficient of determination r^2 — *19–24*
- How to find and interpret the standard error of estimate for a regression line — *23, 24*

$$s_e = \sqrt{\frac{\Sigma(y_i - \hat{y}_i)^2}{n - 2}} = \sqrt{\frac{\Sigma y^2 - b\Sigma y - m\Sigma xy}{n - 2}}$$

- How to construct and interpret a prediction interval for y, $\hat{y} - E < y < \hat{y} + E$ — *25–30*

$$E = t_c s_e \sqrt{1 + \frac{1}{n} + \frac{n(x_0 - \bar{x})^2}{n\Sigma x^2 - (\Sigma x)^2}}$$

Section 9.4

- How to use technology to find a multiple regression equation, the standard error of estimate, and the coefficient of determination — *31, 32*

- How to use a multiple regression equation to predict y-values — *33, 34*

$$\hat{y} = b + m_1x_1 + m_2x_2 + m_3x_3 + \cdots + m_kx_k$$

9 Review Exercises

Section 9.1

In Exercises 1–4, display the data in a scatter plot. Then find the sample correlation coefficient r. Determine whether there is a positive linear correlation, a negative linear correlation, or no linear correlation between the variables. What can you conclude?

1. The ages of eight cows (in years) and their milk production (in gallons) per week

Age, x	4	4	6	7	7	8	10	11
Milk production, y	37.0	35.4	33.3	33.1	32.3	33.7	30.2	29.6

2. The number of convertibles sold at a car dealership in a month and the average monthly temperature (in degrees Fahrenheit)

Convertibles sold, x	0	3	6	5	7	8
Average monthly temperature, y	42	50	58	62	75	77

3. The weight (in pounds) and the daily caloric intake of nine adults

Weight, x	102	118	125	149	155	180	192	195	211
Calories, y	1500	1570	1800	1720	2000	1880	1900	2230	2520

4. The annual per capita sugar consumption (in kilograms) and the average number of cavities of 11- and 12-year-old children in seven countries

Sugar consumption, x	2.1	5.0	6.3	6.5	7.7	8.7	11.6
Cavities, y	0.59	1.51	1.55	1.70	2.18	2.10	2.43

In Exercises 5 and 6, use the given sample statistics to test the claim about the population correlation coefficient ρ at the indicated level of significance α.

5. Claim: $\rho = 0$, $\alpha = 0.01$. Sample statistics: $r = 0.24$, $n = 26$

6. Claim: $\rho \neq 0$, $\alpha = 0.05$. Sample statistics: $r = -0.55$, $n = 22$

In Exercises 7–10, test the claim about the population correlation coefficient ρ at the indicated level of significance α. Then interpret the decision in the context of the original claim.

7. Refer to the data in Exercise 1. At $\alpha = 0.05$, test the claim that there is a linear correlation between a cow's age and milk production.

8. Refer to the data in Exercise 2. Is there enough evidence to conclude that there is a linear correlation between the number of convertibles sold and the average monthly temperature? Use $\alpha = 0.05$.

1.
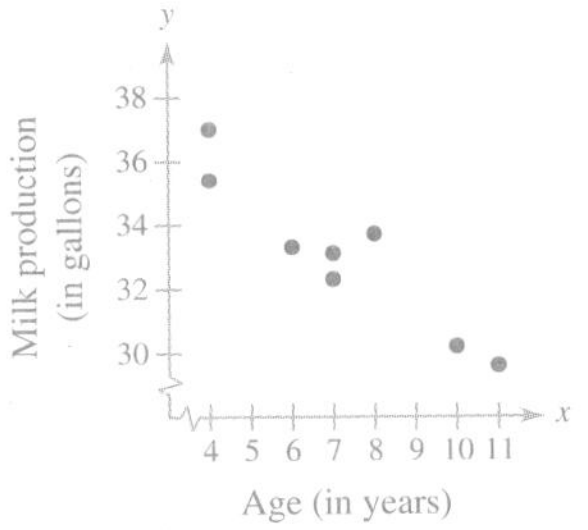

$r \approx -0.939$; negative linear correlation; milk production decreases with age.

2.
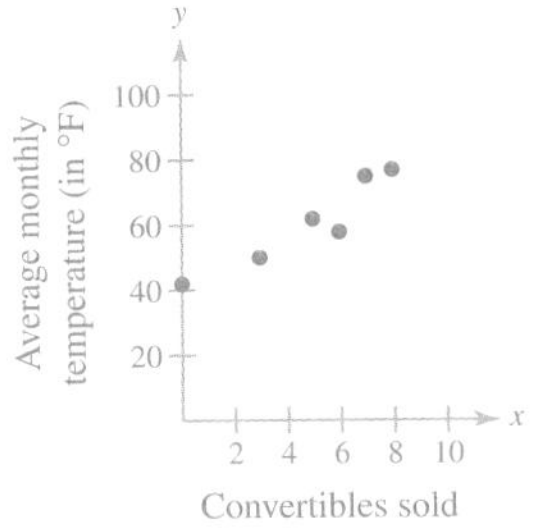

$r \approx 0.944$; positive linear correlation; the average monthly temperature increases as the number of convertibles sold increases.

3.
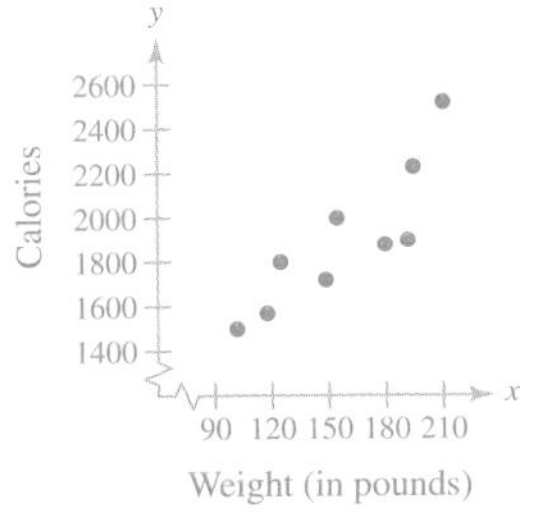

$r \approx 0.868$; positive linear correlation; caloric intake increases with weight.

4. See Selected Answers, page A98.
5. There is not enough evidence at the 1% level to conclude that there is a significant linear correlation.
6. There is enough evidence at the 5% level to conclude that there is a significant linear correlation.
7. There is enough evidence at the 5% level to conclude that there is a significant linear correlation between the age of a cow and its milk production.
8. See Selected Answers, page A98.

9. There is enough evidence at the 1% level to conclude that there is a significant linear correlation between weight and daily caloric intake.

10. There is enough evidence at the 1% level to conclude that there is a significant linear correlation between sugar consumption and tooth decay.

11. $\hat{y} = 0.757x + 21.5$

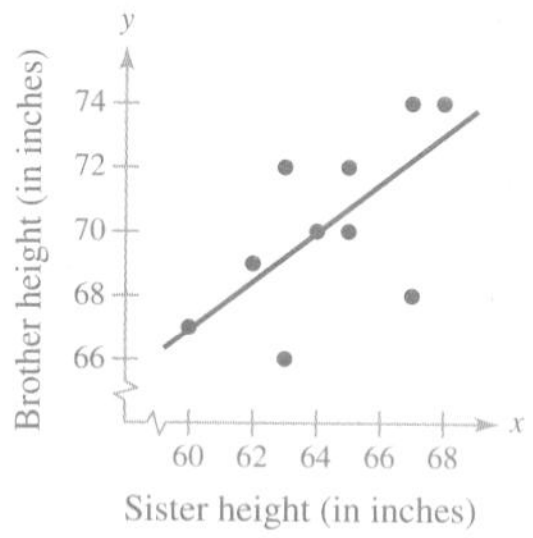

$r \approx 0.688$

12. See Selected Answers, page A98.

13. $\hat{y} = -0.086x + 10.450$

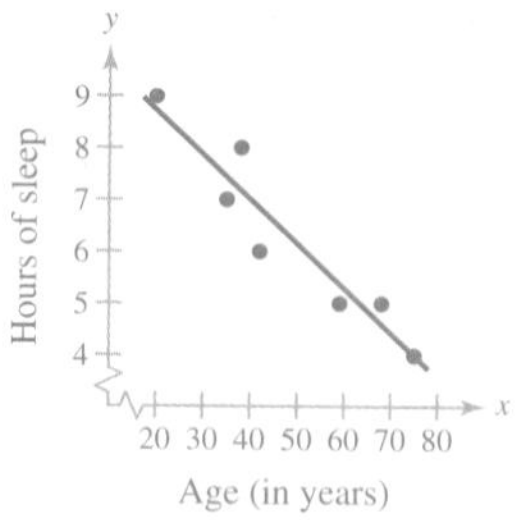

$r \approx -0.949$

14. See Selected Answers, page A98.

15. (a) 68 inches (b) 71 inches

(c) It is not meaningful to predict the value of y for $x = 71$ because $x = 71$ is outside the range of the original data.

(d) It is not meaningful to predict the value of y for $x = 50$ because $x = 50$ is outside the range of the original data.

16. See Selected Answers, page A98.

17. (a) It is not meaningful to predict the value of y for $x = 18$ because $x = 18$ is outside the range of the original data.

(b) 8.3 hours

9. Refer to the data in Exercise 3. At $\alpha = 0.01$, test the claim that there is a linear correlation between an adult's weight and daily caloric intake.

10. Refer to the data in Exercise 4. Is there enough evidence to conclude that there is a linear correlation between sugar consumption and tooth decay? Use $\alpha = 0.01$.

Section 9.2

In Exercises 11–14, use the data to find the equation of the regression line. Then construct a scatter plot of the data and draw the regression line. Can you make a guess about the sign and magnitude of r? Calculate r and check your guess.

DATA **11.** The heights (in inches) of adult brothers and sisters from 11 families

Sister, x	65	62	63	68	63	67	64	60	67	65	68
Brother, y	70	69	66	74	72	74	70	67	68	72	74

12. The weights (in pounds) of adult brothers and sisters from eight families

Sister, x	108	136	182	151	144	133	127	157
Brother, y	120	156	212	180	156	175	158	190

13. The ages (in years) and the number of hours of sleep in one night of seven adults

Age, x	35	20	59	42	68	38	75
Hours of sleep, y	7	9	5	6	5	8	4

14. The engine displacement (in cubic inches) and the fuel efficiency (in miles per gallon) of seven automobiles

Displacement, x	170	134	220	305	109	256	322
Fuel efficiency, y	29.5	34.5	23.0	17.0	33.5	23.0	15.5

In Exercises 15–18, use the regression equations found in Exercises 11–14 to predict the value of y for each value of x, if meaningful. If not, explain why not. (Correlation between x and y in Exercises 11–14 is significant for $\alpha = 0.05$.)

15. Refer to Exercise 11. What height would you predict for a male whose sister is (a) 61 inches? (b) 66 inches? (c) 71 inches? (d) 50 inches?

16. Refer to Exercise 12. What weight would you predict for a male whose sister weighs (a) 140 pounds? (b) 100 pounds? (c) 170 pounds? (d) 200 pounds?

17. Refer to Exercise 13. How many hours of sleep would you predict for an adult of age (a) 18 years? (b) 25 years? (c) 80 years? (d) 50 years?

18. Refer to Exercise 14. What fuel efficiency rating would you predict for a car with an engine displacement of (a) 86 cubic inches? (b) 198 cubic inches? (c) 289 cubic inches? (d) 407 cubic inches?

Section 9.3

In Exercises 19–22, use the value of the linear correlation coefficient r to find the coefficient of determination. Interpret the result.

19. $r = -0.553$

20. $r = -0.962$

21. $r = 0.181$

22. $r = 0.740$

In Exercises 23 and 24, use the data to find the (a) coefficient of determination r^2 and interpret the result with regard to the regression line, and (b) the standard error of estimate s_e and interpret the result.

23. The table shows the area of eight living spaces (in square feet) and the cooling capacity (in Btu per hour) of the air conditioners used in those spaces. The regression equation is $\hat{y} = 9.468x + 3002.991$. *(Adapted from Consumer Reports)*

Living area, x	730	485	205	420	550	590	385	630
Cooling capacity, y	10,200	7000	5300	6800	7250	9000	6900	9400

DATA **24.** The table shows the prices of 16 gas grills (in U.S. dollars) and their usable cooking area (in square inches). The regression equation is $\hat{y} = 1.305x - 209.506$. *(Source: Consumer Reports)*

Area, x	430	338	426	446	465	372	305	403
Price, y	480	360	570	450	350	250	175	200

Area, x	389	424	306	424	309	386	328	261
Price, y	270	270	200	200	180	200	190	150

In Exercises 25–30, construct the indicated prediction interval and interpret the results.

25. Construct a 90% prediction interval for the height of a male in Exercise 11 whose sister is 64 inches tall.

26. Construct a 90% prediction interval for the weight of a male in Exercise 12 whose sister weighs 125 pounds.

27. Construct a 95% prediction interval for the number of hours of sleep for an adult in Exercise 13 who is 45 years old.

28. Construct a 95% prediction interval for the fuel efficiency of an automobile in Exercise 14 that has an engine displacement of 265 cubic inches.

(c) It is not meaningful to predict the value of y for $x = 80$ because $x = 80$ is outside the range of the original data.
(d) 6.15 hours

18. (a) It is not meaningful to predict the value of y for $x = 86$ because $x = 86$ is outside the range of the original data.
(b) 26.855 miles per gallon
(c) 18.665 miles per gallon
(d) It is not meaningful to predict the value of y for $x = 407$ because $x = 407$ is outside the range of the original data.

19. 0.306; 30.6% of the variation in y is explained by the model, and 69.4% of the variation is unexplained.

20. 0.925; 92.5% of the variation in y is explained by the model, and 7.5% of the variation is unexplained.

21. 0.033; 3.3% of the variation in y is explained by the model, and 96.7% of the variation is unexplained.

22. 0.548; 54.8% of the variation in y is explained by the model, and 45.2% of the variation is unexplained.

23. (a) 0.897; 89.7% of the variation in the cooling capacity of an air conditioner can be explained by the variation in the living area, and 10.3% of the variation is unexplained.
(b) 568.011; the standard deviation of the cooling capacity for a specific living area is 568.0 Btu per hour.

24. (a) 0.397; 39.7% of the variation in the price of gas grills can be explained by the variation in the cooking area, and 60.3% of the variation is unexplained.
(b) 101.0; the standard deviation of the price of a gas grill for a specific cooking area is $101.

25. $65.71 < y < 74.19$

26. $125.284 < y < 171.232$

27. $4.865 < y < 8.295$

28. $16.668 < y < 24.982$

29. $7342.01 < y < 12{,}297.89$
30. $1.015 < y < 623.973$
31. $\hat{y} = 6.317 + 0.822x_1 + 0.031x_2 - 0.004x_3$
32. $s_e = 0.9139$; $r^2 = 0.942$
94.2% of the variation of y can be explained by the model.
33. (a) 21.705
(b) 25.210
(c) 30.100
(d) 25.860
34. (a) 16.116
(b) 8.450
(c) 15.734
(d) 9.847

29. Construct a 99% prediction interval for the cooling capacity of an air conditioner in Exercise 23 that is used in a living area of 720 square feet.

30. Construct a 99% prediction interval for the price of a gas grill in Exercise 24 with a usable cooking area of 400 square inches.

Section 9.4

In Exercises 31 and 32, use the data in the table, which shows the carbon monoxide, tar, nicotine content and weight, all in milligrams, of 13 brands of U.S. cigarettes. *(Source: Federal Trade Commission)*

Carbon monoxide, y	Tar, x_1	Nicotine, x_2	Weight, x_3
13.6	14.1	0.86	985.3
16.6	16.0	1.06	1093.8
10.2	8.0	0.67	928.0
5.4	4.1	0.40	946.2
15.0	15.0	1.04	888.5
9.0	8.8	0.76	1026.7
12.3	12.4	0.95	922.5
16.3	16.6	1.12	937.2
15.4	14.9	1.02	885.8
13.0	13.7	1.01	964.3
14.4	15.1	0.90	931.6
10.0	7.8	0.57	970.5
10.2	11.4	0.78	1124.0

DATA

31. Use technology to find the multiple regression equation from the data.

32. Find the standard error of estimate s_e and the coefficient of determination r^2. What percentage of the variation of y can be explained by the regression equation?

In Exercises 33 and 34, use the multiple regression equation to predict the y-values for the given values of the independent variables.

33. An equation that can be used to predict fuel economy (in miles per gallon) for automobiles is $\hat{y} = 41.3 - 0.004x_1 - 0.0049x_2$, where x_1 is the engine displacement (in cubic inches) and x_2 is the vehicle weight (in pounds).

(a) $x_1 = 305, x_2 = 3750$ (b) $x_1 = 225, x_2 = 3100$
(c) $x_1 = 105, x_2 = 2200$ (d) $x_1 = 185, x_2 = 3000$

34. Use the regression equation found in Exercise 31.

(a) $x_1 = 16.5, x_2 = 0.69, x_3 = 946.4$ (b) $x_1 = 7.2, x_2 = 0.55, x_3 = 950.5$
(c) $x_1 = 16.1, x_2 = 0.88, x_3 = 961.2$ (d) $x_1 = 9.7, x_2 = 0.99, x_3 = 1118.6$

9 Chapter Quiz

Take this quiz as you would take a quiz in class. After you are done, check your work against the answers given in the back of the book.

DATA For Exercises 1–8, use the data in the table, which shows the personal income and outlays (both in trillions of dollars) for Americans for 11 recent years. *(Source: U.S. Commerce Department, Bureau of Economic Analysis)*

Personal income, x	Personal outlays, y
5.6	4.6
5.8	4.9
6.2	5.2
6.5	5.5
6.9	5.8
7.4	6.1
7.8	6.5
8.4	7.0
8.7	7.3
8.9	7.7
9.2	8.0

1. Construct a scatter plot for the data. Do the data appear to have a positive linear correlation, a negative linear correlation, or no linear correlation? Explain.
2. Calculate the correlation coefficient r. What can you conclude?
3. Test the level of significance of the correlation coefficient r. Use $\alpha = 0.05$.
4. Find the equation of the regression line for the data. Include the regression line in the scatter plot.
5. Use the regression line to predict the personal outlays when the personal income is 5.3 trillion dollars.
6. Find the coefficient of determination and interpret the results.
7. Find the standard error of estimate s_e and interpret the results.
8. Construct a 95% prediction interval for personal outlays when personal income is 6.4 trillion dollars. Interpret the results.
9. The equation used to predict sunflower yield (in pounds) is

$$\hat{y} = 1257 - 1.34x_1 + 1.41x_2$$

where x_1 is the number of acres planted (in thousands) and x_2 is the number of acres harvested (in thousands). Use the regression equation to predict the y-values for the given values of the independent variables listed below. Then determine which variable has a greater influence on the value of y. *(Source: U.S. National Agricultural Statistics Service)*

(a) $x_1 = 2103, x_2 = 2037$ (b) $x_1 = 3387, x_2 = 3009$
(c) $x_1 = 2185, x_2 = 1980$ (d) $x_1 = 3485, x_2 = 3404$

---

1. 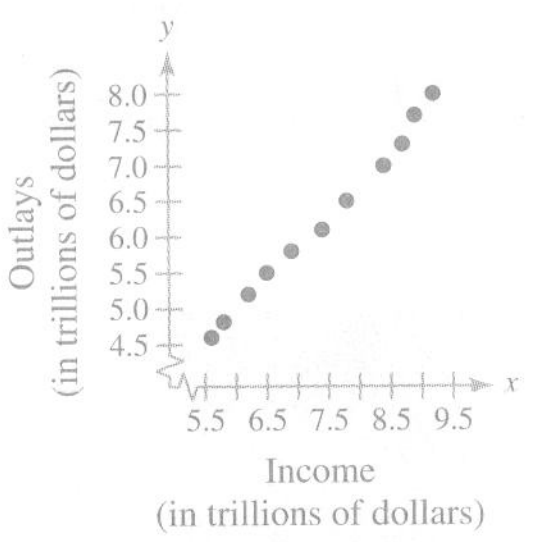

The data appear to have a positive correlation. The outlays increase as the incomes increases.

2. 0.996; strong positive linear correlation
3. There is enough evidence at the 5% level to conclude that a significant correlation exists between personal income and personal outlays for Americans.
4. $\hat{y} = 0.891x - 0.359$

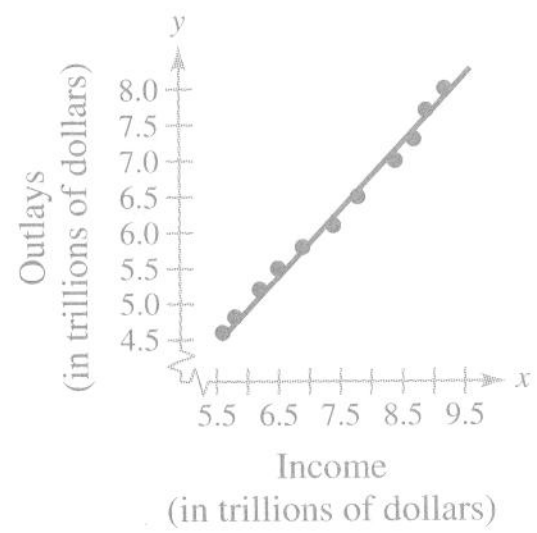

5. \$4.363 trillion
6. 0.992; 99.2% of the variation in personal outlays can be explained by the variation in personal income, and 0.8% of the variation is unexplained.
7. 0.110; the standard deviation of personal outlays for a specified personal income is \$0.110 trillion.
8. $\$5.076 \text{ trillion} < y < \$5.610 \text{ trillion}$

You can be 95% confident that the personal outlays will be between \$5.076 trillion and \$5.610 trillion when personal income is \$6.4 trillion.

9. (a) 1311.150 pounds
(b) 961.110 pounds
(c) 1120.900 pounds
(d) 1386.740 pounds
x_2 has the greater influence on y.

PUTTING IT ALL TOGETHER

Real Statistics ■ Real Decisions

The Center for Infant Development, part of the Department of Psychology at the University of Denver, is a research center that focuses on perceptual and cognitive development in infants from two months to three years of age. Graduate students, postdoctoral fellows, and members of the research staff can perform studies to see how infants think and behave under particular circumstances.

You are a member of the research staff at the Center for Infant Development at the University of Denver. You are interested in determining whether babies take longer to learn how to crawl during the cold winter months than in the warmer summer months. You decide to record the age (in weeks) of each child when the child is able to crawl and the temperature (in degrees Fahrenheit) when the child crawls. You want to determine if there is a significant correlation between the mean crawling age and the mean temperature.

UNIVERSITY OF DENVER 1864

www.du.edu

Mean monthly temperature, x	Mean crawling age, y
33	33.64
30	32.82
33	33.83
37	33.35
48	33.38
57	32.32
66	29.84
73	30.52
72	29.70
63	31.84
52	28.58
39	31.44

(Source: Janette Benson *Infant Behavior and Development*)

Exercises

1. *Analyzing the Data*

(a) The data in the table at the right show the average temperature (in degrees Fahrenheit) for each month and the average age (in weeks) of the infants that began to crawl during that month. Construct a scatter plot of the data. Use the graph to determine whether there is a positive linear correlation, negative linear correlation, or no linear correlation between the mean crawling age and the mean monthly temperature.

(b) Calculate the correlation coefficient r and verify your conclusion in part (a).

(c) Test the significance of the correlation coefficient found in part (b). Use $\alpha = 0.05$.

(d) Find the equation of the regression line for the temperature and the mean crawling age. Add the graph of the regression line to your scatter plot in part (a). Does the regression line appear to be a good fit?

(e) Can you use the equation of the regression line to predict the mean crawling age given the temperature? Why or why not?

(f) Find the coefficient of determination and the standard error of estimate s_e and interpret the results.

2. *What Do You Think?*

According to your analysis, would you agree or disagree with the following? Why or why not?

Because there is a significant linear correlation between temperature and the mean crawling age, parents can consider the temperature when determining when their baby will crawl.

TECHNOLOGY

MINITAB | EXCEL | TI-83

U.S. Food and Drug Administration

Nutrients in Breakfast Cereals

The U.S. Food and Drug Administration (FDA) requires nutrition labeling for most foods. Under FDA regulations, manufacturers are required to list the amounts of certain nutrients in their foods, such as calories, sugar, fat, and carbohydrates. This nutritional information is displayed in the "Nutrition Facts" panel on the food's package.

The table shows the following nutritional content for one cup of each of 21 different breakfast cereals.

C = calories

S = sugar in grams

F = fat in grams

R = carbohydrates in grams

Cereal name	C	S	F	R
Apple Jacks	130	16	0.5	30
Berry Burst Cheerios	110	11	1.5	24
Booberry	120	14	1	27
Cheerios	110	1	2	22
Cocoa Puffs	120	14	1	26
Cookie Crisp	120	13	1	26
Corn Chex	110	3	0	25
Corn Flakes	100	2	0	24
Corn Pops	120	14	0	28
Count Chocula	120	14	1	26
Crispix	110	3	0	25
Frankenberry	120	14	1	27
Froot Loops	120	15	1	28
Frosted Cheerios	120	13	1	25
Honey Nut Cheerios	120	11	1.5	24
Lucky Charms	120	13	1	25
Multi Grain Cheerios	110	6	1	24
Product 19	100	4	0	25
Special K	110	4	0	22
Trix	120	13	1	27
Wheaties	110	4	1	24

Exercises

1. Use a technology tool to draw a scatter plot of the following (x, y) pairs in the data set.

(a) (calories, sugar)

(b) (calories, fat)

(c) (calories, carbohydrates)

(d) (sugar, fat)

(e) (sugar, carbohydrates)

(f) (fat, carbohydrates)

2. From the scatter plots in Exercise 1, which pairs of variables appear to have a strong linear correlation?

3. Use a technology tool to find the correlation coefficient for each pair of variables in Exercise 1. Which has the strongest linear correlation?

4. Use a technology tool to find an equation of a regression line for the following variables.

(a) (calories, sugar)

(b) (calories, carbohydrates)

5. Use the results of Exercise 4 to predict the following.

(a) The sugar content of one cup of cereal that has a caloric content of 120 calories

(b) The carbohydrate content of one cup of cereal that has a caloric content of 120 calories

6. Use a technology tool to find the multiple regression equations of the following forms.

(a) $C = b + m_1S + m_2F + m_3R$

(b) $C = b + m_1S + m_2R$

7. Use the results of Exercise 6 to predict the caloric content of one cup of cereal that has 10 grams of sugar and 25 grams of carbohydrates.

Extended solutions are given in the *Technology Supplement.* Technical instruction is provided for MINITAB, Excel, and the TI-83.

CHAPTER

10

Chi-Square Tests and the *F*-Distribution

10.1 Goodness of Fit

10.2 Independence

Case Study

10.3 Comparing Two Variances

10.4 Analysis of Variance

Uses and Abuses

Real Statistics–Real Decisions

Technology

Federal law requires all passenger cars to pass a 30-mile per hour frontal crash test. These tests are conducted through the New Car Assessment Program, in which the federal government buys brand new vehicles directly off a car lot and crashes them. Results of these tests are classified using a one- to five-star rating, with one star indicating the least protection and five stars the most protection.

Where You've Been

In its New Car Assessment Program, the government buys new cars each year and crashes them into a wall at 35 miles per hour to compare how different vehicles protect front-seat passengers in a head-on collision. The forces and impacts that occur during a crash test are measured by equipping dummies with special instruments and placing them in the car. The crash test results include data on head, chest, and leg injuries. For a low crash test number, the injury potential in a 35-mile per hour frontal crash is low. If the crash test number is high, then the injury potential is high. Using the techniques of Chapter 8, you can determine if the mean chest injury potential is the same for pickups and vans. The sample statistics are as follows. *(Source: National Highway Traffic Safety Administration)*

Vehicle	Number	Mean chest injury	Standard deviation
Vans	$n_1 = 30$	$\bar{x}_1 = 54.2$	$s_1 = 12.1$
Pickups	$n_2 = 36$	$\bar{x}_2 = 50.6$	$s_2 = 8.94$

For the means of chest injury, the P-value for the hypothesis that $\mu_1 = \mu_2$ is about 0.177. At $\alpha = 0.05$, you fail to reject the null hypothesis. So, you do not have enough evidence to conclude that there is a significant difference in the means of the chest injury potential in a 35-mile per hour frontal crash for vans and pickups.

Where You're Going

In Chapter 8, you learned how to test a hypothesis that compares two populations by basing your decisions on sample statistics and their distributions. In this chapter, you will learn how to test a hypothesis that compares three or more populations.

For instance, in addition to the crash tests for vans and pickups, a third group of vehicles was also tested. The results for all three types of vehicles are as follows.

Vehicle	Number	Mean chest injury	Standard deviation
Vans	$n_1 = 30$	$\bar{x}_1 = 54.2$	$s_1 = 12.1$
Pickups	$n_2 = 36$	$\bar{x}_2 = 50.6$	$s_2 = 8.94$
Light	$n_3 = 35$	$\bar{x}_3 = 46.4$	$s_3 = 6.90$

From these three samples, is there evidence of a difference in chest injury potential among vans, pickups, and light vehicles in a 35-mile per hour frontal crash?

In this chapter, you will learn that you can answer this question by testing the hypothesis that the three means are equal. For the means of chest injury, the P-value for the hypothesis that $\mu_1 = \mu_2 = \mu_3$ is about 0.005. At $\alpha = 0.05$, you can reject the null hypothesis. So, you can conclude that for the three types of vehicles tested, at least one of the means of the chest injury potential in a 35-mile per hour frontal crash is different from the others.

10.1 Goodness of Fit

What You Should Learn

- How to use the chi-square distribution to test whether a frequency distribution fits a claimed distribution

The Chi-Square Goodness-of-Fit Test

The Chi-Square Goodness-of-Fit Test

Suppose a marketing executive is planning a new advertising campaign and wants to determine the proportions of radio music listeners in a specific broadcast region who prefer each of six types of music. To determine these proportions, the executive can perform a multinomial experiment. A **multinomial experiment** is a probability experiment consisting of a fixed number of trials in which there are more than two possible outcomes for each independent trial. The probability for each outcome is fixed, and each outcome is classified into **categories.** (Remember from Section 4.2 that a **binomial** experiment has only two possible outcomes.)

Now, suppose the marketing executive wants to test a radio station's claim concerning the distribution of proportions of music preferences. To do so, the executive could compare the distribution of proportions obtained in the multinomial experiment with the radio station's specified distribution. How can the executive compare the distributions? The answer is, perform a chi-square goodness-of-fit test.

DEFINITION

A **chi-square goodness-of-fit test** is used to test whether a frequency distribution fits an expected distribution.

Insight

The hypothesis tests described in Sections 10.1 and 10.2 can be used for qualitative data.

To begin a goodness-of-fit test, you must first state a null and an alternative hypothesis. Generally, the null hypothesis states that the frequency distribution fits the specified distribution and the alternative hypothesis states that the frequency distribution does not fit the specified distribution.

For example, suppose the radio station claims that the distribution of music preferences for listeners in the broadcast region is as shown below.

Distribution of music preferences			
Classical	4%	Oldies	2%
Country	36%	Pop	18%
Gospel	11%	Rock	29%

To test the radio station's claim, the executive can perform a chi-square goodness-of-fit test using the following null and alternative hypotheses.

H_0: The distribution of music preferences in the broadcast region is 4% classical, 36% country, 11% gospel, 2% oldies, 18% pop, and 29% rock. (Claim)

H_a: The distribution of music preferences differs from the claimed or expected distribution.

Picturing the World

How often do adults visit the doctor? According to the distribution in the following display, 45.4% of all U.S. adults visit the doctor one to three times in a year.

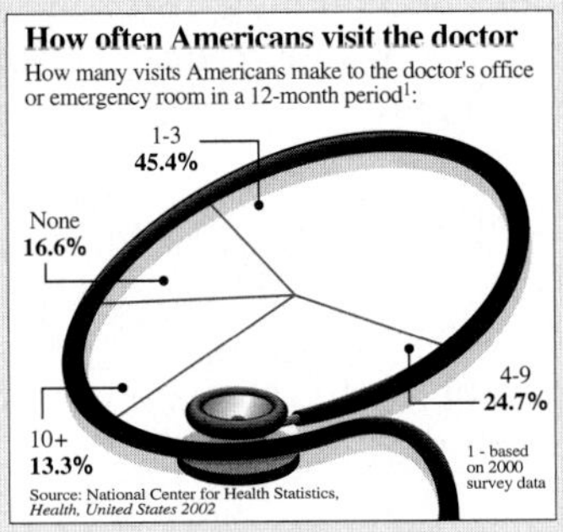

A sample of 274 adults is asked how many visits they make to the doctor in a year: 1–3, 4–9, 10 or more, or none. What is the expected frequency for each response?

Insight

The sum of the expected frequencies always equals the sum of the observed frequencies. For instance, in Example 1 the sum of the observed frequencies and the sum of the expected frequencies are both 500.

To calculate the test statistic for the chi-square goodness-of-fit test, you can use observed frequencies and expected frequencies. To calculate the expected frequencies, you must assume the null hypothesis is true.

DEFINITION

The **observed frequency** O of a category is the frequency for the category observed in the sample data.

The **expected frequency** E of a category is the *calculated* frequency for the category. Expected frequencies are obtained assuming the specified (or hypothesized) distribution. The expected frequency for the ith category is

$$E_i = np_i$$

where n is the number of trials (the sample size) and p_i is the assumed probability of the ith category.

EXAMPLE 1

Finding Observed Frequencies and Expected Frequencies

A marketing executive randomly selects 500 music listeners from the broadcast region and asks each whether he or she prefers classical, country, gospel, oldies, pop, or rock music. The results are shown at the right. Find the observed frequencies and the expected frequencies for each type of music.

Survey results ($n = 500$)	
Classical	8
Country	210
Gospel	72
Oldies	10
Pop	75
Rock	125

SOLUTION The observed frequency for each type of music is the number of music listeners naming a particular type of music. The expected frequency for each type of music is the product of the number of listeners in the survey and the probability that a listener will name a particular type of music. The observed frequencies and expected frequencies are shown in the following table.

Type of music	% of listeners	Observed frequency	Expected frequency
Classical	4%	8	500(0.04) = 20
Country	36%	210	500(0.36) = 180
Gospel	11%	72	500(0.11) = 55
Oldies	2%	10	500(0.02) = 10
Pop	18%	75	500(0.18) = 90
Rock	29%	125	500(0.29) = 145

Try It Yourself 1

Suppose the executive randomly selects 300 music listeners in the listening region. Find the expected frequencies for each type of music.

a. Multiply 300 by the probability that a listener will name each particular type of music.

Answer: Page A43

For the chi-square goodness-of-fit test to be used, the following must be true.

1. The observed frequencies must be obtained using a random sample.
2. Each expected frequency must be greater than or equal to 5.

If the expected frequency of a category is less than 5, it may be possible to combine it with another category to meet the requirements.

The Chi-Square Goodness-of-Fit Test

If the conditions listed above are satisfied, then the sampling distribution for the goodness-of-fit test is approximated by a chi-square distribution with $k - 1$ degrees of freedom, where k is the number of categories. The test statistic for the chi-square goodness-of-fit test is

$$\chi^2 = \Sigma \frac{(O - E)^2}{E}$$

where O represents the observed frequency of each category and E represents the expected frequency of each category.

When the observed frequencies closely match the expected frequencies, the differences between O and E will be small and the chi-square test statistic will be close to 0. As such, the null hypothesis is unlikely to be rejected. However, when there are large discrepancies between the observed frequencies and the expected frequencies, the differences between O and E will be large, resulting in a large chi-square test statistic. A large chi-square test statistic is evidence for rejecting the null hypothesis. So, the chi-square goodness-of-fit test is always a right-tailed test.

GUIDELINES

Performing a Chi-Square Goodness-of-Fit Test

In Words	*In Symbols*
1. Identify the claim. State the null and alternative hypotheses.	State H_0 and H_a.
2. Specify the level of significance.	Identify α.
3. Identify the degrees of freedom.	d.f. $= k - 1$
4. Determine the critical value.	Use Table 6 in Appendix B.
5. Determine the rejection region.	
6. Calculate the test statistic.	$\chi^2 = \Sigma \frac{(O - E)^2}{E}$
7. Make a decision to reject or fail to reject the null hypothesis.	If χ^2 is in the rejection region, reject H_0. Otherwise, fail to reject H_0.
8. Interpret the decision in the context of the original claim.	

EXAMPLE 2

Performing a Chi-Square Goodness-of-Fit Test

A radio station claims that the music preferences of the listeners in the station's broadcast region are distributed as shown in the table at the left below. You randomly select 500 radio music listeners from the broadcast region and ask each whether he or she prefers classical, country, gospel, oldies, pop, or rock music. The survey results are shown in the table at the right below. Using $\alpha = 0.01$, perform a chi-square goodness-of-fit test to test the claimed distribution. What can you conclude?

Distribution of music preferences			
Classical	4%	Oldies	2%
Country	36%	Pop	18%
Gospel	11%	Rock	29%

Survey results ($n = 500$)			
Classical	8	Oldies	10
Country	210	Pop	75
Gospel	72	Rock	125

Type of music	Observed frequency	Expected frequency
Classical	8	20
Country	210	180
Gospel	72	55
Oldies	10	10
Pop	75	90
Rock	125	145

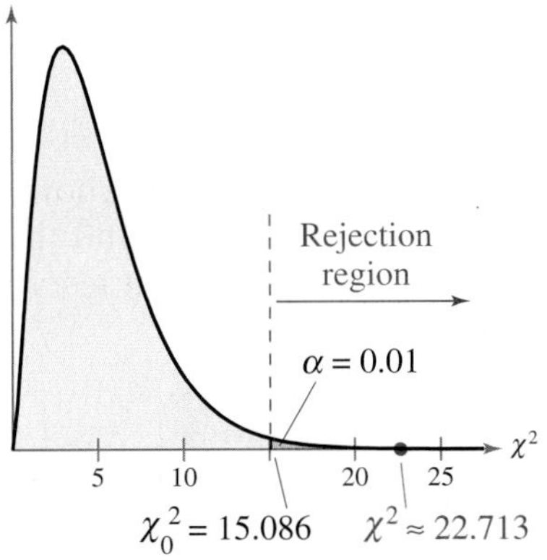

SOLUTION The observed and expected frequencies are shown in the table at the left. The expected frequencies were calculated in Example 1. Because the observed frequencies were obtained using a random sample and each expected frequency is at least 5, you can use the chi-square goodness-of-fit test to test the proposed distribution. The null and alternative hypotheses are as follows.

H_0: The distribution of music preferences in the broadcast region is 4% classical, 36% country, 11% gospel, 2% oldies, 18% pop, and 29% rock. (Claim)

H_a: The distribution of music preferences differs from the claimed or expected distribution.

Because there are six categories, the chi-square distribution has $k - 1 = 6 - 1 = 5$ degrees of freedom. With d.f. $= 5$ and $\alpha = 0.01$, the critical value is $\chi_0^2 = 15.086$. With the observed and expected frequencies, the chi-square test statistic is

$$\chi^2 = \Sigma\frac{(O - E)^2}{E}$$

$$= \frac{(8 - 20)^2}{20} + \frac{(210 - 180)^2}{180} + \frac{(72 - 55)^2}{55} + \frac{(10 - 10)^2}{10} + \frac{(75 - 90)^2}{90} + \frac{(125 - 145)^2}{145}$$

$$\approx 22.713.$$

The graph shows the location of the rejection region and the chi-square test statistic. Because χ^2 is in the rejection region, you should decide to reject the null hypothesis.

Interpretation At the 1% level, there is enough evidence to conclude that the distribution of music preferences differs from the radio station's claimed or expected distribution.

Ages	Claimed distribution	Survey results
0–9	16%	76
10–19	20%	84
20–29	8%	30
30–39	14%	60
40–49	15%	54
50–59	12%	40
60–69	10%	42
70+	5%	14

Try It Yourself 2

A sociologist claims that the age distribution for the residents of a certain city is different than it was 10 years ago. The distribution of ages 10 years ago is shown in the table at the left. You randomly select 400 residents and record the age of each. The survey results are shown in the table. At $\alpha = 0.05$, perform a chi-square goodness-of-fit test to determine whether the distribution has changed.

a. *Verify* that the expected frequency is at least 5 for each category.
b. *Identify* the claimed distribution and state H_0 and H_a.
c. *Specify* the level of significance α.
d. *Identify* the degrees of freedom.
e. *Determine* the critical value and the rejection region.
f. *Calculate* the chi-square test statistic.
g. *Decide* whether to reject the null hypothesis. Use a graph if necessary.
h. *Interpret* the decision in the context of the original claim. *Answer: Page A43*

EXAMPLE 3

Performing a Chi-Square Goodness-of-Fit Test

Men's survey results ($n = 300$)	
For irradiation	168
Against irradiation	71
No opinion	61

Irradiation is a controversial method for preserving meat. The display at the right shows two distributions describing attitudes toward the irradiation of red meat. You work for a meat-packing plant and want to test the distribution describing men's attitudes. To test the distribution, you randomly select 300 men and ask each whether he is in favor of irradiation, is against irradiation, or has no opinion. The results are shown in the table at the left. At $\alpha = 0.05$, test the claimed or expected distribution.

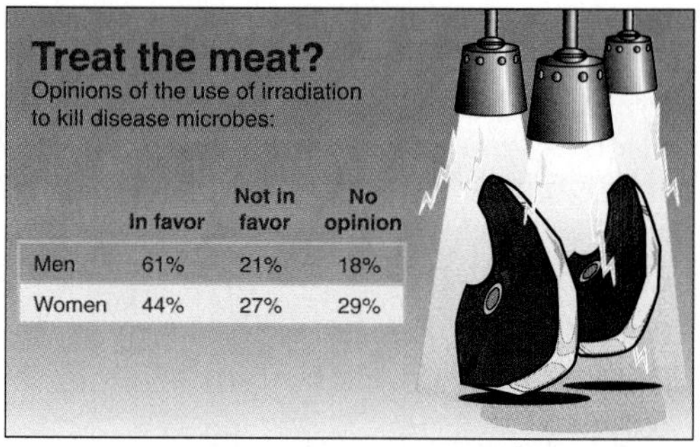

(Adapted from Peter D. Hart Research Associates for Grocery Manufacturers of America)

SOLUTION The observed frequencies and the expected frequencies are shown in the following table. Because each expected frequency is at least 5 and the men were randomly selected, you can use the chi-square goodness-of-fit test to test the claimed distribution.

Response	Observed frequency	Expected frequency
In favor of	168	$300(0.61) = 183$
Against	71	$300(0.21) = 63$
No opinion	61	$300(0.18) = 54$

The null and alternative hypotheses are as follows.

H_0: The distribution of men's attitudes toward the irradiation of red meat is 61% in favor, 21% against, and 18% no opinion. (Claim)

H_a: The distribution of men's attitudes toward the irradiation of red meat differs from the claimed or expected distribution.

Because there are three categories, the chi-square distribution has $k - 1 = 3 - 1 = 2$ degrees of freedom. Using d.f. $= 2$ and $\alpha = 0.05$, the critical value is $\chi_0^2 = 5.991$. With the observed and expected frequencies, the chi-square test statistic is as shown in the following table.

Study Tip

Another way to calculate the chi-square test statistic is to organize the calculations in a table as shown in Example 3.

O	E	$O - E$	$(O - E)^2$	$\frac{(O - E)^2}{E}$
168	183	−15	225	1.229508197
71	63	8	64	1.015873016
61	54	7	49	0.9074074074
				$\chi^2 = \Sigma \frac{(O - E)^2}{E} \approx 3.1528$

The graph shows the location of the rejection region and the chi-square test statistic. Because χ^2 is not in the rejection region, you should decide not to reject the null hypothesis.

Interpretation At the 5% level, there is not enough evidence to dispute the claimed or expected distribution of men's opinions.

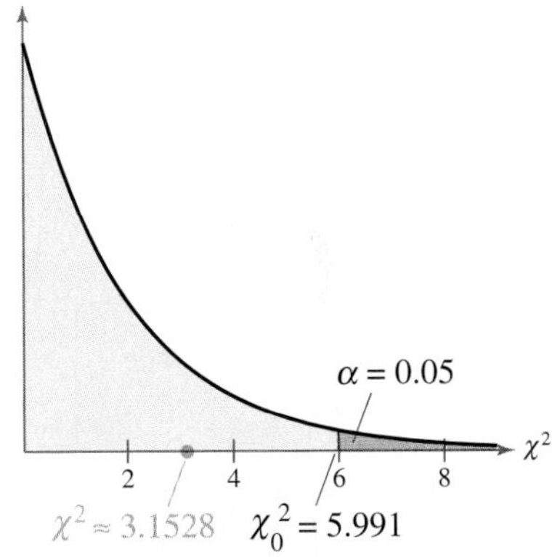

Try It Yourself 3

Women's survey results ($n = 200$)	
For irradiation	100
Against irradiation	48
No opinion	52

You also want to test the distribution describing women's attitudes. To test the distribution, you randomly select 200 women and ask each whether she is in favor of irradiation, is against irradiation, or has no opinion. The results are shown in the table at the left. At $\alpha = 0.01$, test the claimed or expected distribution. What can you conclude?

a. *Verify* that the expected frequency is at least 5 for each category.
b. *Identify* the claimed distribution and state H_0 and H_a.
c. *Specify* the level of significance α.
d. *Identify* the degrees of freedom.
e. *Determine* the critical value and the rejection region.
f. Use the observed and expected frequencies to *calculate the chi-square test statistic.*
g. *Decide* whether to reject the null hypothesis. Use a graph if necessary.
h. *Interpret* the decision in the context of the original claim.

Answer: Page A43

The chi-square goodness-of-fit test is often used to determine whether a distribution is uniform. For such tests, the expected frequencies of the categories are equal. When testing a uniform distribution, you can find the expected frequency of each category by dividing the sample size by the number of categories. For example, suppose a company believes that the number of sales made by its sales force is uniform throughout the five-day work week. If the sample consists of 1000 sales, then the expected value of the sales for each day will be $1000/5 = 200$.

10.1 Exercises

Building Basic Skills and Vocabulary

1. What is a multinomial experiment?

2. What conditions are necessary to use the chi-square goodness-of-fit test?

1. A multinomial experiment is a probability experiment consisting of a fixed number of trials in which there are more than two possible outcomes for each independent trial.

2. See Selected Answers, page A98.

Finding Expected Frequencies In Exercises 3–6, find the expected frequency for the given values of n and p_i.

3. $n = 150, p_i = 0.3$

4. $n = 500, p_i = 0.9$

5. $n = 230, p_i = 0.25$

6. $n = 415, p_i = 0.08$

3. 45 4. 450
5. 57.5 6. 33.2

Using and Interpreting Concepts

Performing a Chi-Square Goodness-of-Fit Test In Exercises 7–18, (a) identify the claim and state H_0 and H_a, (b) determine the critical value and the rejection region, (c) calculate the test statistic χ^2, and (d) decide whether to reject or fail to reject the null hypothesis. Then interpret the decision in the context of the original claim.

7. ***Employees' Well-Being*** Results from a survey three years ago asking employees if they agree or disagree that their boss is concerned about their personal well-being are shown in the graph. To determine whether this distribution has changed, you randomly select 425 employees and ask each if their boss is concerned about their personal well-being. The results are shown in the table. Can you conclude that there has been a change in the distribution? Use $\alpha = 0.05$.

Figure for Exercise 7

Survey results	
Response	**Frequency, f**
Agree	311
Disagree	84
Neither	30

8. ***Coffee*** Results from a survey five years ago asking where coffee drinkers typically drink their first cup of coffee are shown in the graph. To determine whether this distribution has changed, you randomly select 581 coffee drinkers and ask each where they typically drink their first cup of coffee. The results are shown in the table. Can you conclude that there has been a change in the claimed or expected distribution? Use $\alpha = 0.05$. *(Adapted from USA TODAY)*

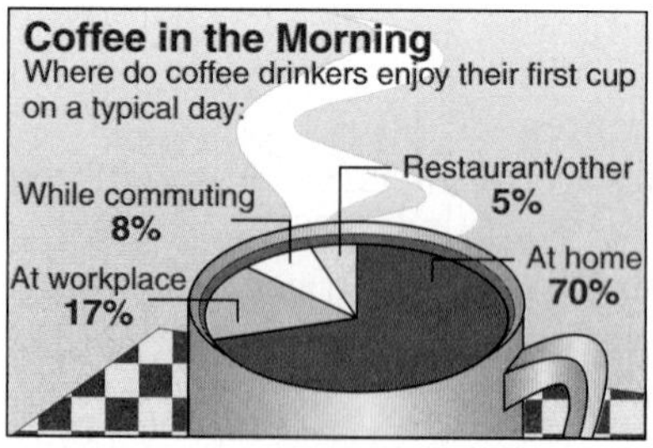

Figure for Exercise 8

Survey results	
Response	**Frequency, f**
At home	389
At workplace	110
While commuting	55
Restaurant/other	27

7. (a) Claimed distribution:

Response	Distribution
Agree	78%
Disagree	17%
Neither	5%

H_0: Distribution of responses is as shown in table above.

H_a: Distribution of responses differs from the claimed distribution.

(b) $\chi_0^2 = 5.991$

Rejection region: $\chi^2 > 5.991$

(c) 6.782

(d) Reject H_0. There is enough evidence at the 5% level to conclude that the distribution of responses differs from the claimed distribution.

8. See Selected Answers, page A98.

9. (a) Claimed distribution:

Response	Distribution
Keep	77%
Tax cuts	15%
Not sure	8%

H_0: Distribution of responses is as shown in table above.

H_a: Distribution of responses differs from the claimed distribution.

(b) $\chi_0^2 = 9.210$

Rejection region: $\chi^2 > 9.210$

(c) 9.375

(d) Reject H_0. There is enough evidence at the 1% level to conclude that the distribution of responses differs from the claimed distribution.

10. See Selected Answers, page A98.

11. See Odd Answers, page A81.

9. *Tax Cuts or Entitlements* An economist believes that the distribution of the opinions of U.S. adults on whether tax cuts or entitlements are more important is different from the one shown in the graph. The economist randomly selects 600 adults and asks each which is more important—keeping Medicare, Social Security, and Medicaid or cutting taxes. The results are shown in the table. At $\alpha = 0.01$, are the distributions different?

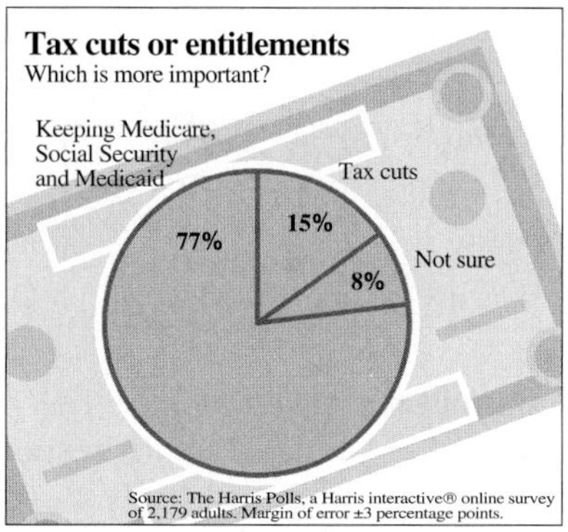

Survey results	
Response	**Frequency, *f***
Keeping Medicare, Social Security, and Medicaid	431
Tax cuts	107
Not sure	62

10. *Reasons Workers Leave* A personnel director believes that the distribution of the reasons workers leave their jobs is different from the one shown in the graph. The director randomly selects 200 workers who recently left their jobs and asks each his or her reason for doing so. The results are shown in the table. At $\alpha = 0.01$, are the distributions different? *(Adapted from USA TODAY)*

Survey results	
Response	**Frequency, *f***
Limited advancement potential	78
Lack of recognition	52
Low salary/benefits	30
Unhappy with mgmt.	25
Bored/don't know	15

11. *Bicycle Accidents: Day of Week* A bicycle safety organization claims that fatal bicycle accidents are uniformly distributed throughout the week. The table shows the day of the week for which 911 randomly selected fatal bicycle accidents occurred. At $\alpha = 0.10$, can you reject the claim that the distribution is uniform? *(Adapted from Insurance Institute for Highway Safety)*

Day	Frequency, *f*	Day	Frequency, *f*
Sunday	118	Thursday	129
Monday	119	Friday	146
Tuesday	127	Saturday	135
Wednesday	137		

12. See Selected Answers, page A99.

13. (a) Claimed distribution:

Object struck	Distribution
Tree	28%
Embankment	10%
Utility pole	10%
Guardrail	9%
Ditch	7%
Curb	6%
Culvert	5%
Sign/post/fence	10%
Other	15%

H_0: Distribution of objects struck is as shown in table above.

H_a: Distribution of objects struck differs from the claimed distribution.

(b) $\chi_0^2 = 20.090$

Rejection region: $\chi^2 > 20.090$

(c) 49.665

(d) Reject H_0. There is enough evidence at the 1% level to conclude that the distribution of objects struck changed after warning signs were erected.

14. (a) Claimed distribution:

Time of day	Distribution
Midnight–6 A.M.	34%
6 A.M.–Noon	15%
Noon–6 P.M.	22%
6 P.M.–Midnight	29%

H_0: Distribution of the time of day of roadside hazard crash deaths is as shown in the table above.

H_a: Distribution of the time of day of roadside hazard crash deaths differs from the claimed distribution.

(b) $\chi_0^2 = 11.345$

Rejection region: $\chi^2 > 11.345$

(c) 19.240

(d) Reject H_0. There is enough evidence at the 1% level to conclude that the distribution of the time of day of roadside hazard crash deaths has changed.

DATA

12. *Bicycle Accidents: Month of Year* A bicycle safety organization conducted a study of 996 randomly selected fatal bicycle accidents. The month each accident occurred is shown in the table. At $\alpha = 0.10$, can you conclude that fatal bicycle accidents are not uniformly distributed by month? *(Adapted from Insurance Institute for Highway Safety)*

Month	Frequency, f	Month	Frequency, f
January	50	July	129
February	48	August	122
March	81	September	89
April	72	October	87
May	90	November	73
June	101	December	54

13. *Crash Deaths: Object Struck* The pie chart shows the distribution of roadside hazard crash deaths with respect to the object hit. After highway warning signs were erected, a study was conducted to see if there was a change in the distribution. The results are shown in the table. Can you conclude that there is a change in the distribution? Use $\alpha = 0.01$. *(Adapted from Insurance Institute for Highway Safety)*

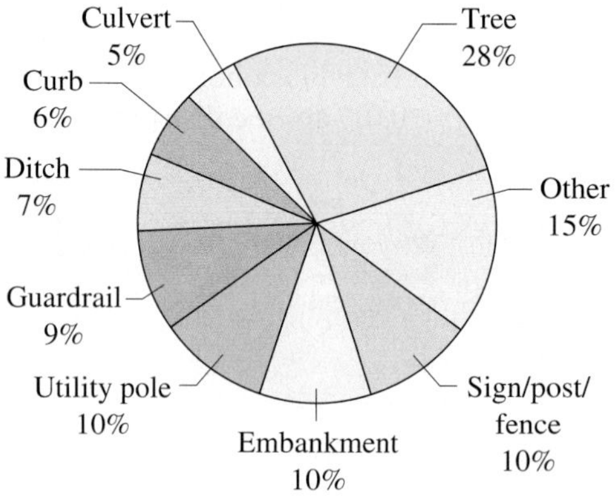

Survey results	
Object struck	**Frequency, f**
Tree	179
Embankment	100
Utility pole	107
Guardrail	57
Ditch	36
Curb	43
Culvert	28
Sign/post/fence	68
Other	73

14. *Crash Deaths: Time of Day* The pie chart shows the distribution of the time of day of roadside hazard crash deaths for a previous year. The results of a recent study of 627 randomly selected hazard crash deaths are shown in the table. At $\alpha = 0.01$, has the distribution changed? *(Adapted from Insurance Institute for Highway Safety)*

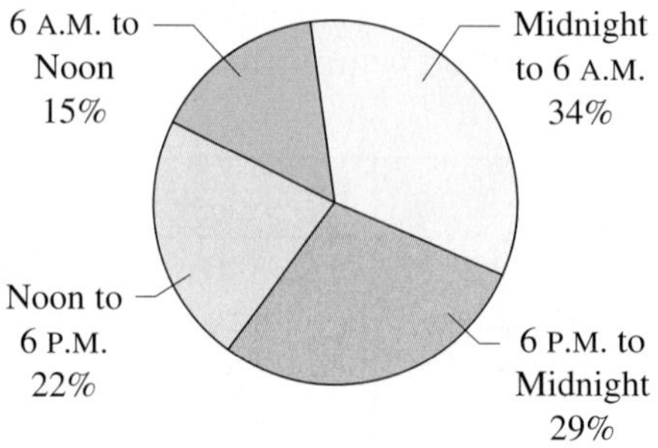

Survey results	
Time of day	**Frequency, f**
Midnight to 6 A.M.	224
6 A.M. to Noon	128
Noon to 6 P.M.	115
6 P.M. to Midnight	160

15. (a) Claimed distribution:

Response	Distribution
Not a HS grad	33.333%
HS graduate	33.333%
College (1yr+)	33.333%

H_0: Distribution of the responses is uniform as shown in table above.

H_a: Distribution of the responses is not uniform.

(b) $\chi_0^2 = 7.378$

Rejection region: $\chi^2 > 7.378$

(c) 5.637

(d) Fail to reject H_0. There is not enough evidence at the 2.5% level to conclude that the distribution of responses is not uniform.

16. See Selected Answers, page A99.

17. (a) Claimed distribution:

Cause	Distribution
Trans. accidents	41%
Assaults	20%
Objects/equipment	15%
Falls	10%
Exposure	10%
Other	4%

H_0: Distribution of the causes is as shown in table above.

H_a: Distribution of the causes differs from the claimed distribution.

(b) $\chi_0^2 = 11.071$

Rejection region: $\chi^2 > 11.071$

(c) 9.493

(d) Fail to reject H_0. There is not enough evidence at the 5% level to conclude that the distribution of causes in the western United States differs from the national distribution.

15. ***Educational Attainment*** A social service organization reports that the level of educational attainment of mothers receiving food stamps is uniformly distributed. To test this claim, you randomly select 99 mothers who currently receive food stamps and record the educational attainment of each. The results are shown in the following table. At $\alpha = 0.025$, can you reject the claim that the distribution is uniform? *(Adapted from U.S. Census Bureau)*

Response	Frequency, f
Not a high school graduate	37
High school graduate	40
College (1 year or more)	22

16. ***Marital Status*** A social service worker believes that the marital status of mothers receiving food stamps is uniformly distributed. To test this claim, you randomly select 101 mothers who currently receive food stamps and record the marital status of each. The results are shown below. At $\alpha = 0.025$, can you reject the claim that the distribution is uniform? *(Adapted from U.S. Census Bureau)*

Response	Frequency, f
Married, husband present	20
Married, husband absent	19
Widowed or divorced	23
Never married	39

17. ***Fatal Work Injuries*** The pie chart shows the national distribution of fatal work injuries in the United States. You believe that the distribution of fatal work injuries is different in the western United States and randomly select 6231 fatal work injuries occurring in that region and record how each occurred. The results are shown in the table. At $\alpha = 0.05$, can you conclude that the distribution of fatal work injuries in the western United States is different from the national distribution? *(Adapted from U.S. Bureau of Labor Statistics)*

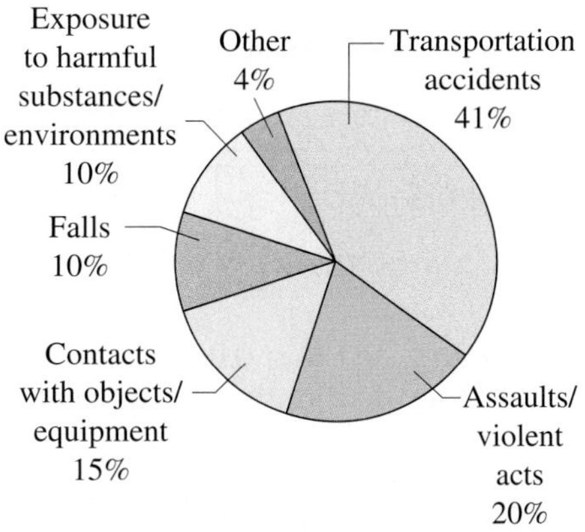

Study results: Western U.S.	
Cause	**Frequency, f**
Transportation accidents	2500
Assaults/violent acts	1300
Contacts with objects and equipment	985
Falls	620
Exposure to harmful substances or environments	602
Other	224

18. (a) Claimed distribution:

Response	Distribution
First marriage of bride & groom	50%
1st marriage bride remarriage groom	12%
1st marriage groom remarriage bride	14%
Remarriage of bride & groom	24%

H_0: Distribution of the responses is as shown in table above.

H_a: Distribution of the responses differs from the claimed distribution.

(b) $\chi_0^2 = 11.345$

Rejection region: $\chi^2 > 11.345$

(c) 0.675

(d) Fail to reject H_0. There is not enough evidence at the 1% level to reject the counselor's claim.

19. See Odd Answers, page A81.

20. See Selected Answers, page A99.

18. *Marriage* A marriage counselor says that 50% of all marriages are the first marriage for both the bride and the groom, 12% are the first for the bride and a remarriage for the groom, 14% are a remarriage for the bride and a first for the groom, and 24% are a remarriage for both. You randomly select 103 married couples and gather the results shown in the table. At $\alpha = 0.01$, can you reject the counselor's claim? *(Adapted from U.S. National Center for Health Statistics)*

Response	Frequency, *f*
First marriage of bride and groom	55
First marriage of bride, remarriage of groom	12
Remarriage of bride, first marriage of groom	12
Remarriage of bride and groom	24

Extending Concepts

Testing for Normality Using a chi-square goodness-of-fit test, you can decide, with some degree of certainty, whether a variable is normally distributed. In all chi-square tests for normality, the null and alternative hypotheses are as follows.

H_0: The variable has a normal distribution.

H_a: The variable does not have a normal distribution.

To determine the expected frequencies when performing a chi-square test for normality, first find the mean and standard deviation of the frequency distribution. Then use the mean and standard deviation to compute the z-score for each class boundary. Then use the z-scores to calculate the area under the standard normal curve for each class. Multiplying the resulting class areas by the sample size yields the expected frequency for each class.

In Exercises 19 and 20, (a) find the expected frequencies, (b) determine the critical value and the rejection region, (c) calculate the test statistic χ^2, and (d) decide whether to reject or fail to reject the null hypothesis. Then interpret the decision in the context of the original claim.

19. *Test Scores* The frequency distribution shows the results of 200 test scores. Are the test scores normally distributed? Use $\alpha = 0.01$.

Class boundaries	49.5–58.5	58.5–67.5	67.5–76.5	76.5–85.5	85.5–94.5
Frequency, *f*	19	61	82	34	4

20. *Test Scores* At $\alpha = 0.05$, test the claim that the 400 test scores shown in the frequency distribution are normally distributed.

Class boundaries	50.5–60.5	60.5–70.5	70.5–80.5	80.5–90.5	90.5–100.5
Frequency, *f*	28	106	151	97	18

10.2 Independence

What You Should Learn

- How to use a contingency table to find expected frequencies
- How to use a chi-square distribution to test whether two variables are independent

Contingency Tables • The Chi-Square Test for Independence

Contingency Tables

In Section 3.2, you learned that two events are **independent** if the occurrence of one event does not affect the probability of the occurrence of the other event. For instance, the outcomes of a roll of a die and a toss of a coin are independent. But, suppose a medical researcher wants to determine if there is a relationship between caffeine consumption and heart attack risk. Are these variables independent or are they dependent? In this section, you will learn how to use the chi-square test for independence to answer such a question. To perform a chi-square test for independence, you will use sample data that are organized in a contingency table.

DEFINITION

An $r \times c$ **contingency table** shows the observed frequencies for two variables. The observed frequencies are arranged in r rows and c columns. The intersection of a row and a column is called a **cell.**

For example, the following table is a 2×5 contingency table. It has two rows and five columns and shows the results of a random sample of 550 company CEOs classified by age and size of company. From the table, you can see that 108 of the CEOs between the ages of 50 and 59 direct small or midsize companies, and 85 of the CEOs in this age group direct large companies.

Study Tip

In a contingency table, the notation $E_{r,c}$ represents the expected frequency for the cell in row r, column c. For instance, in the table at the right, $E_{1,4}$ represents the expected frequency for the cell in row 1, column 4.

Company size	Age				
	39 and under	40–49	50–59	60–69	70 and over
Small/midsize	42	69	108	60	21
Large	5	18	85	120	22

(Adapted from Grant Thornton LLP, The Segal Company)

Assuming the two variables of study in a contingency table are independent, you can use the contingency table to find the expected frequency for each cell. The formula for calculating the expected frequency for each cell is given below.

Finding the Expected Frequency for Contingency Table Cells

The expected frequency for a cell $E_{r,c}$ in a contingency table is

$$\text{Expected frequency } E_{r,c} = \frac{(\text{Sum of row } r) \times (\text{Sum of column } c)}{\text{Sample size}}.$$

Insight

In Example 1, once the expected frequency for $E_{1,1}$ has been calculated to be 25.64, you can determine the expected frequency for $E_{2,1}$ to be $47 - 25.64 = 21.36$. That is, the expected frequency for the last cell in each row or column can be found by subtracting from the total.

EXAMPLE 1

Finding Expected Frequencies

Find the expected frequency for each cell in the contingency table. Assume that the variables, age and company size, are independent.

	Age					
Company size	**39 and under**	**40–49**	**50–59**	**60–69**	**70 and over**	**Total**
Small/midsize	42	69	108	60	21	300
Large	5	18	85	120	22	250
Total	47	87	193	180	43	550

SOLUTION Using the formula

$$\text{Expected frequency } E_{r,c} = \frac{(\text{Sum of row } r) \times (\text{Sum of column } c)}{\text{Sample size}}$$

you can find each expected frequency as shown.

$$E_{1,1} = \frac{300 \cdot 47}{550} \approx 25.64 \quad E_{1,2} = \frac{300 \cdot 87}{550} \approx 47.45 \quad E_{1,3} = \frac{300 \cdot 193}{550} \approx 105.27$$

$$E_{1,4} = \frac{300 \cdot 180}{550} \approx 98.18 \quad E_{1,5} = \frac{300 \cdot 43}{550} \approx 23.45 \quad E_{2,1} = \frac{250 \cdot 47}{550} \approx 21.36$$

$$E_{2,2} = \frac{250 \cdot 87}{550} \approx 39.55 \quad E_{2,3} = \frac{250 \cdot 193}{550} \approx 87.73 \quad E_{2,4} = \frac{250 \cdot 180}{550} \approx 81.82$$

$$E_{2,5} = \frac{250 \cdot 43}{550} \approx 19.55$$

Try It Yourself 1

The marketing consultant for a travel agency wants to determine whether certain travel concerns are related to travel purpose. A random sample of 300 travelers is selected and the results are classified as shown in the contingency table. Assuming that the variables travel concern and travel purpose are independent, find the expected frequency for each cell. *(Adapted from NPD Group for Embassy Suites)*

	Travel concern			
Travel purpose	**Hotel room**	**Leg room on plane**	**Rental car size**	**Other**
Business	36	108	14	22
Leisure	38	54	14	14

a. *Calculate* the sum of each row.
b. *Calculate* the sum of each column.
c. *Determine* the sample size.
d. *Use the formula* to find the expected frequency for each cell.

Answer: Page A44

The Chi-Square Test for Independence

After finding the expected frequencies, you can test whether the variables are independent using a chi-square independence test.

DEFINITION

A **chi-square independence test** is used to test the independence of two variables. Using a chi-square test, you can determine whether the occurrence of one variable affects the probability of the occurrence of the other variable.

For the chi-square independence test to be used, the following conditions must be true.

1. The observed frequencies must be obtained using a random sample.
2. Each expected frequency must be greater than or equal to 5.

The Chi-Square Independence Test

If the conditions listed above are satisfied, then the sampling distribution for the chi-square independence test is approximated by a chi-square distribution with

$$(r - 1)(c - 1)$$

degrees of freedom, where r and c are the number of rows and columns, respectively, of a contingency table. The test statistic for the chi-square independence test is

$$\chi^2 = \Sigma\frac{(O - E)^2}{E}$$

where O represents the observed frequencies and E represents the expected frequencies.

To begin the independence test, you must first state a null and an alternative hypothesis. For a chi-square independence test, the null and alternative hypotheses are always some variation of the following statements.

H_0: The variables are independent.

H_a: The variables are dependent.

The expected frequencies are calculated on the assumption that the two variables are independent. If the variables are independent, then you can expect little difference between the observed frequencies and the expected frequencies. When the observed frequencies closely match the expected frequencies, the differences between O and E will be small and the chi-square test statistic will be close to 0. As such, the null hypothesis is unlikely to be rejected.

However, if the variables are dependent, there will be large discrepancies between the observed frequencies and the expected frequencies. When the differences between O and E are large, the chi-square test statistic is also large. A large chi-square test statistic is evidence for rejecting the null hypothesis. So, the chi-square independence test is always a right-tailed test.

Picturing the World

A researcher wishes to determine if a relationship exists between job status (full- or part-time) and two different work arrangements. The results of a random sample of 772 workers are shown in the contingency table. *(Adapted from the Bureau of Labor Statistics)*

	Work arrangement	
Status	**Contract**	**Traditional**
Full-time	5	630
Part-time	1	136

Can the researcher use this sample to test for independence using a chi-square independence test? Why or why not?

Study Tip

A contingency table with three rows and four columns will have
$(3 - 1)(4 - 1) = (2)(3)$
$= 6$ d.f.

GUIDELINES

Performing a Chi-Square Test for Independence

In Words	*In Symbols*
1. Identify the claim. State the null and alternative hypotheses.	State H_0 and H_a.
2. Specify the level of significance.	Identify α.
3. Identify the degrees of freedom.	d.f. $= (r - 1)(c - 1)$
4. Determine the critical value.	Use Table 6 in Appendix B.
5. Determine the rejection region.	
6. Calculate the test statistic.	$\chi^2 = \Sigma \frac{(O - E)^2}{E}$
7. Make a decision to reject or fail to reject the null hypothesis.	If χ^2 is in the rejection region, reject H_0. Otherwise, fail to reject H_0.
8. Interpret the decision in the context of the original claim.	

EXAMPLE 2

Performing a Chi-Square Independence Test

The contingency table shows the results of a random sample of 550 company CEOs classified by age and size of company. The expected frequencies are displayed in parentheses. At $\alpha = 0.01$, can you conclude that the CEOs' ages are related to company size?

	Age of CEOs					
Company size	**39 and under**	**40–49**	**50–59**	**60–69**	**70 and over**	**Total**
Small/midsize	42 (25.64)	69 (47.45)	108 (105.27)	60 (98.18)	21 (23.45)	300
Large	5 (21.36)	18 (39.55)	85 (87.73)	120 (81.82)	22 (19.55)	250
Total	47	87	193	180	43	550

SOLUTION The expected frequencies were calculated in Example 1. Because each expected frequency is at least 5 and the CEOs were randomly selected, you can use the chi-square independence test to test whether the variables are independent. The null and alternative hypotheses are as follows.

H_0: The CEOs' ages are independent of the company size.

H_a: The CEOs' ages are dependent on the company size. (Claim)

Because the contingency table has two rows and five columns, the chi-square distribution has $(r - 1)(c - 1) = (2 - 1)(5 - 1) = 4$ degrees of freedom. Because d.f. = 4 and $\alpha = 0.01$, the critical value is $\chi_0^2 = 13.277$. With the observed and expected frequencies, the chi-square test statistic is as shown.

O	E	$O - E$	$(O - E)^2$	$\frac{(O - E)^2}{E}$
42	25.64	16.36	267.6496	10.4388
69	47.45	21.55	464.4025	9.7872
108	105.27	2.73	7.4529	0.0708
60	98.18	−38.18	1457.7124	14.8473
21	23.45	−2.45	6.0025	0.2560
5	21.36	−16.36	267.6496	12.5304
18	39.55	−21.55	464.4025	11.7422
85	87.73	−2.73	7.4529	0.0850
120	81.82	38.18	1457.7124	17.8161
22	19.55	2.45	6.0025	0.3070
			$\chi^2 = \sum\frac{(O - E)^2}{E}$	≈ 77.9

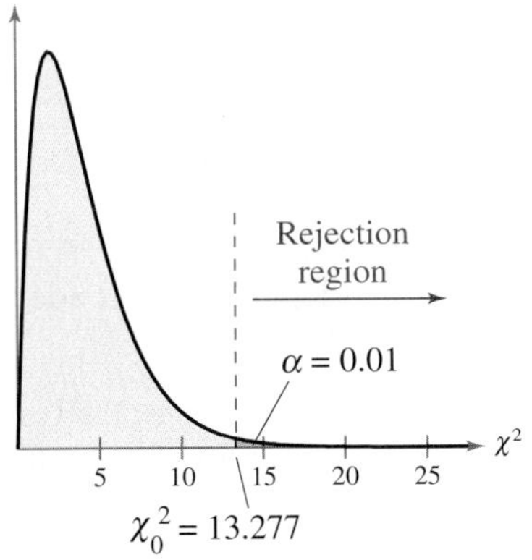

The graph shows the location of the rejection region. Because $\chi^2 \approx 77.9$ is in the rejection region, you should decide to reject the null hypothesis.

Interpretation There is enough evidence at the 1% level of significance to conclude that the CEOs' ages and the company size are dependent.

Try It Yourself 2

The marketing consultant for a travel agency wants to determine whether travel concerns are related to travel purpose. A random sample of 300 travelers is selected and each is asked his or her primary travel concern. The results are classified as shown in the contingency table. At $\alpha = 0.01$, can the consultant conclude that the travel concerns depend on the purpose of travel? (The expected frequencies are displayed in parentheses.) *(Adapted from NPD Group for Embassy Suites)*

	Travel concern				
Travel purpose	**Hotel room**	**Leg room on plane**	**Rental car size**	**Other**	**Total**
Business	36 (44.4)	108 (97.2)	14 (16.8)	22 (21.6)	180
Leisure	38 (29.6)	54 (64.8)	14 (11.2)	14 (14.4)	120
Total	74	162	28	36	300

a. *Identify* the claim and state H_0 and H_a.
b. *Specify* the level of significance α.
c. *Determine* the degrees of freedom.
d. *Determine* the critical value and the rejection region.
e. Use the observed and expected frequencies to *calculate the chi-square test statistic.*
f. *Decide* whether to reject the null hypothesis. Use a graph if necessary.
g. *Interpret* the decision in the context of the original claim. *Answer: Page A44*

EXAMPLE 3

Using Technology for a Chi-Square Independence Test

A health club manager wants to determine whether the number of days per week that college students spend exercising is related to gender. A random sample of 275 college students is selected and the results are classified as shown in the table. At $\alpha = 0.05$, is there enough evidence to conclude that the number of days spent exercising per week is related to gender?

	Days per week spent exercising				
Gender	**0–1**	**2–3**	**4–5**	**6–7**	**Total**
Male	40	53	26	6	125
Female	34	68	37	11	150
Total	74	121	63	17	275

Setup

X²–Test
Observed: [A]
Expected: [B]
Calculate Draw

Calculate

X²–Test
x²=3.493357223
p=.321624691
df=3

Draw

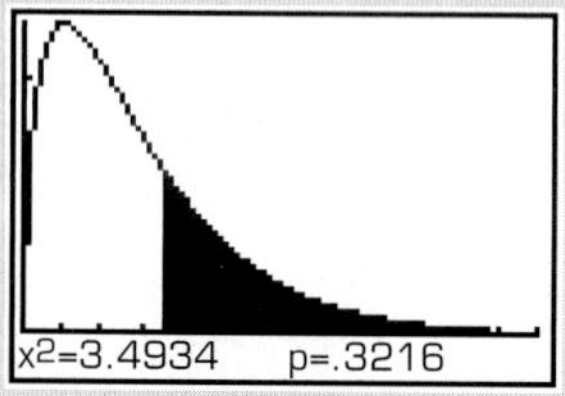

SOLUTION The null and alternative hypotheses can be stated as follows.

H_0: The number of days spent exercising per week is independent of gender.

H_a: The number of days spent exercising per week depends on gender. (Claim)

Enter the observed frequencies into matrix A. Then set up the chi-square test using a TI-83 as shown at the left. Display the expected value matrix to verify that the expected value for each cell is at least 5.

The displays at the left show the results of selecting *Calculate* or *Draw*. Because d.f. $= 3$ and $\alpha = 0.05$, the rejection region is $\chi^2 > 7.815$. The test statistic $\chi^2 \approx 3.49$ is not in the rejection region, so you should fail to reject the null hypothesis.

Interpretation There is not enough evidence to conclude that the number of days spent exercising per week is related to gender.

Study Tip

You can also use *P*-values to perform a chi-square test for independence. For instance, in Example 3, note that the TI-83 displays $P \approx 0.322$. Because $P > \alpha$, you should fail to reject the null hypothesis.

Try It Yourself 3

A researcher wants to determine whether the number of minutes adults spend online per day is related to gender. A random sample of 450 adults is selected and the results are classified as shown in the table. At $\alpha = 0.05$, is there enough evidence to conclude that the number of minutes spent online per day is related to gender?

	Minutes spent online per day					
Gender	**0–15**	**15–30**	**30–45**	**45–60**	**60 and over**	**Total**
Male	19	36	75	90	55	275
Female	21	72	45	19	18	175
Total	40	108	120	109	73	450

a. *Determine* the critical value and the rejection region.
b. *Enter* the observed frequencies.
c. *Use* a technology tool to find the χ^2 test statistic.
d. *Decide* whether to reject the null hypothesis. Use a graph if necessary.
e. *Interpret* the decision in the context of the original claim. *Answer: Page A44*

10.2 Exercises

Building Basic Skills and Vocabulary

1. Explain how to find the expected frequency for a cell in a contingency table.

2. Explain why the chi-square independence test is always a right-tailed test.

True or False? In Exercises 3 and 4, determine whether the statement is true or false. If it is false, rewrite it so that it is a true statement.

3. A contingency table with two rows and five columns will have $(2)(5) = 10$ degrees of freedom.

4. If the two variables of chi-square test for independence are dependent, then you can expect little difference between the observed frequencies and the expected frequencies.

Finding Expected Frequencies In Exercises 5–8, find the expected frequency for each cell in the contingency table. Assume that the variables are independent.

5.

	Athlete has		
Result	**Stretched**	**Not stretched**	**Total**
Injury	18	22	40
No injury	211	189	400
Total	229	211	440

6.

	Rating			
Size of restaurant	**Excellent**	**Fair**	**Poor**	**Total**
Seats 100 or fewer	182	203	165	550
Seats over 100	180	311	159	650
Total	362	514	324	1200

7.

	Treatment			
Result	**Brand-name**	**Generic**	**Placebo**	**Total**
Improvement	24	21	10	55
No change	12	13	45	70
Total	36	34	55	125

8.

	Type of car				
Gender	**Compact**	**Full size**	**SUV**	**Truck/van**	**Total**
Male	28	39	21	22	110
Female	24	32	20	14	90
Total	52	71	41	36	200

Help
MyMathLab
Student Study Pack

1. The expected frequency for a cell in a contingency table is the sum of the row the cell is in times the sum of the column the cell is in, all divided by the sample size.
2. A chi-square independence test is always a right-tailed test because if the variables are dependent, then the chi-square test statistic will be large and a large chi-square test statistic is evidence for rejecting the null hypothesis.
3. False. A contingency table with two rows and five columns will have $(2 - 1)(5 - 1) = 4$ degrees of freedom.
4. False. If the two variables of a chi-square test for independence are dependent, then you can expect a large difference between the observed frequencies and the expected frequencies.

5.

	Athlete has	
Result	Stretched	Not stretched
Injury	20.82	19.18
No injury	208.18	191.82

6.

	Rating		
Size of restaurant	Excellent	Fair	Poor
Seats 100 or less	165.92	235.58	148.5
Seats over 100	196.08	278.42	175.5

7. See Odd Answers, page A82.

8. See Selected Answers, page A99.

9. (a) H_0: Skill level in a subject is independent of location. (claim)

H_a: Skill level in a subject is dependent on location.

(b) d.f. = 2; $\chi_0^2 = 9.210$

Rejection region: $\chi^2 > 9.210$

(c) 0.297

(d) Fail to reject H_0. There is not enough evidence at the 1% level to conclude that skill level in a subject is dependent on location.

10. (a) H_0: Attitudes about safety are independent of the type of school.

H_a: Attitudes about safety are dependent on the type of school. (claim)

(b) d.f. = 1; $\chi_0^2 = 6.635$

Rejection region: $\chi^2 > 6.635$

(c) 8.691

(d) Reject H_0. There is enough evidence at the 1% level to conclude that attitudes about safety are dependent on the type of school.

11. (a) H_0: Grades are independent of the institution.

H_a: Grades are dependent on the institution. (claim)

(b) d.f. = 8; $\chi_0^2 = 15.507$

Rejection region: $\chi^2 > 15.507$

(c) 48.488

(d) Reject H_0. There is enough evidence at the 5% level to conclude that grades are dependent on the institution.

12. (a) H_0: Adults' ratings are independent of the type of school.

H_a: Adults' ratings are dependent on the type of school. (claim)

(b) d.f. = 3; $\chi_0^2 = 7.815$

Rejection region: $\chi^2 > 7.815$

(c) 148.389

(d) Reject H_0. There is enough evidence at the 5% level to conclude that adults' ratings are dependent on the type of school.

Using and Interpreting Concepts

Performing a Chi-Square Test for Independence In Exercises 9–19, perform the indicated chi-square test for independence by doing the following.

(a) Identify the claim and state the null and alternative hypotheses.

(b) Determine the degrees of freedom, find the critical value, and identify the rejection region.

(c) Calculate the test statistic. (If possible, use a technology tool.)

(d) Decide to reject or fail to reject the null hypothesis. Then interpret the decision in the context of the original claim.

9. ***Achievement and School Location*** Is achieving a basic skill level in a subject related to the location of the school? A random sample of students by the location of school and the number achieving basic skill levels in three subjects is shown in the table. At $\alpha = 0.01$, test the hypothesis that the variables are independent. *(Adapted from USA TODAY)*

	Subject		
Location of school	**Reading**	**Math**	**Science**
Urban	43	42	38
Suburban	63	66	65

10. ***Attitudes about Safety*** The results of a random sample of students by type of school and their attitudes on safety steps taken by the school staff are shown in the table. At $\alpha = 0.01$, can you conclude that attitudes about the safety steps taken by the school staff are related to the type of school? *(Adapted from USA TODAY)*

	School staff has	
Type of school	**Taken all steps necessary for student safety**	**Taken some steps toward student safety**
Public	40	51
Private	64	34

11. ***Grades for Our Leaders*** The contingency table shows how a random sample of college freshmen graded the leaders of three types of institutions. At $\alpha = 0.05$, can you conclude that the grades are related to the institution? *(Adapted from Louis Harris for Northwestern Mutual Life Insurance)*

	Grade				
Institution	**A**	**B**	**C**	**D**	**F**
Military	25	46	19	5	3
Religious	18	44	24	7	5
Media/press	5	23	37	21	12

12. ***Rating Public Schools*** The contingency table shows how a random sample of adults rated their local public schools and how they rated America's public schools. At $\alpha = 0.05$, can you conclude that the adults' ratings are related to the type of school? *(Adapted from USA TODAY)*

	Rating			
Type of school	**Excellent**	**Good**	**Fair**	**Poor**
Local	120	405	263	151
National	41	238	481	179

13. ***Obsessive-Compulsive Disorder*** The results of a random sample of patients with obsessive-compulsive disorder treated with a drug or with a placebo are shown in the contingency table. At $\alpha = 0.10$, can you conclude that the treatment is related to the result? On the basis of these results, would you recommend using the drug as part of a treatment for obsessive-compulsive disorder? *(Adapted from The Journal of the American Medical Association)*

	Treatment	
Result	**Drug**	**Placebo**
Improvement	39	25
No change	54	70

14. ***Chronic Fatigue Syndrome*** The contingency table shows a random sample of patients with chronic fatigue syndrome treated with a drug or with a placebo. At $\alpha = 0.10$, can you conclude that the variables treatment and result are dependent? On the basis of these results, would you recommend using the drug as part of a treatment for chronic fatigue syndrome? *(Adapted from The Journal of the American Medical Association)*

	Treatment	
Result	**Drug**	**Placebo**
Improvement	20	19
No change	10	16

15. ***Continuing Education*** You work for a college's continuing education department and want to determine whether the reasons given by workers for continuing their education is related to job type. In your study, you randomly collect the data shown in the contingency table. At $\alpha = 0.01$, can you conclude that the variables reason and type of worker are dependent? How could you use this information in your marketing efforts? *(Adapted from USA TODAY)*

	Reason		
Type of worker	**Professional**	**Personal**	**Professional and personal**
Technical	30	36	41
Other	47	25	30

13. (a) H_0: Results are independent of the type of treatment.
H_a: Results are dependent on the type of treatment. (claim)
(b) d.f. = 1; $\chi_0^2 = 2.706$
Rejection region: $\chi^2 > 2.706$
(c) 5.106
(d) Reject H_0. There is enough evidence at the 10% level to conclude that results are dependent on the type of treatment. I would recommend using the drug.

14. (a) H_0: Results are independent of the type of treatment.
H_a: Results are dependent on the type of treatment. (claim)
(b) d.f. = 1; $\chi_0^2 = 2.706$
Rejection region: $\chi^2 > 2.706$
(c) 1.032
(d) Fail to reject H_0. There is not enough evidence at the 10% level to conclude that results are dependent on the type of treatment. I would not recommend using the drug.

15. (a) H_0: Reasons are independent of the type of worker.
H_a: Reasons are dependent on the type of worker. (claim)
(b) d.f. = 2; $\chi_0^2 = 9.210$
Rejection region: $\chi^2 > 9.210$
(c) 7.326
(d) Fail to reject H_0. There is not enough evidence at the 1% level to conclude that reason(s) for continuing education are dependent on the type of worker. On the basis of these results, marketing strategies should not differ between technical and nontechnical audiences in regard to reason(s) for continuing education.

16. *Blood Alcohol Concentration* You are investigating the relationship between the ages and the blood alcohol concentration of pedestrians fatally injured at night. During your investigation, you randomly collect the data shown in the contingency table. At $\alpha = 0.01$, is there enough evidence to conclude that blood alcohol concentration is related to age in nighttime pedestrian deaths? *(Adapted from Insurance Institute for Highway Safety)*

	Blood Alcohol Concentration		
Age	**0.00**	**0.01–0.09**	**0.10 and greater**
16–34 years	439	85	696
35 years and older	513	98	622

17. *Vehicles and Crashes* You work for an insurance company and are studying the relationship between types of crashes and the vehicles involved. As part of your study, you randomly select 3207 vehicle crashes and organize the resulting data as shown in the contingency table. At $\alpha = 0.05$, can you conclude that the type of crash depends on the type of vehicle? *(Adapted from Insurance Institute for Highway Safety)*

	Vehicle		
Type of crash	**Car**	**Pickup truck/ utility vehicle**	**Cargo/large passenger van**
Single-vehicle	895	493	45
Multiple-vehicle	1400	336	38

18. *Alcohol-Related Accidents* The contingency table shows a random sample of fatally injured passenger vehicle drivers (with blood alcohol concentrations greater than or equal to 0.10) by age and gender. At $\alpha = 0.05$, can you conclude that age is related to gender in such alcohol-related accidents? *(Adapted from Insurance Institute for Highway Safety)*

	Age					
Gender	**16–20**	**21–30**	**31–40**	**41–50**	**51–60**	**61 and older**
Male	32	51	52	43	28	10
Female	13	22	33	21	10	6

19. *Coauthored Articles* The contingency table shows a random sample of engineering, psychology, and biology articles and if and how they were coauthored. At $\alpha = 0.10$, can you conclude that the subject matter and coauthorship are related? *(Adapted from CHI Research, Inc.)*

	Subject matter		
Coauthorship	**Engineering**	**Psychology**	**Biology**
Coauthored	47	44	50
Internationally coauthored	17	9	16
Not coauthored	36	47	34

16. (a) H_0: Blood alcohol concentration is independent of age.
H_a: Blood alcohol concentration is dependent on age. (claim)
(b) d.f. = 2; $\chi_0^2 = 9.210$
Rejection region: $\chi^2 > 9.210$
(c) 10.762
(d) Reject H_0. There is enough evidence at the 1% level to conclude that blood alcohol concentration is dependent on age.

17. (a) H_0: Type of crash is independent of the type of vehicle.
H_a: Type of crash is dependent on the type of vehicle. (claim)
(b) d.f. = 2; $\chi_0^2 = 5.991$
Rejection region: $\chi^2 > 5.991$
(c) 106.390
(d) Reject H_0. There is enough evidence at the 5% level to conclude that the type of crash is dependent on the type of vehicle.

18. See Selected Answers, page A99.

19. (a) H_0: Subject is independent of coauthorship.
H_a: Coauthorship and subject are dependent. (claim)
(b) d.f. = 4; $\chi_0^2 = 7.779$
Rejection region: $\chi^2 > 7.779$
(c) 5.610
(d) Fail to reject H_0. There is not enough evidence at the 10% level to conclude that subject matter and coauthorship are related.

20. Fail to reject H_0. There is not enough evidence at the 5% level to conclude that at least one of the proportions is different from the others.

21. Reject H_0. There is enough evidence at the 10% level to conclude that at least one of the proportions is different from the others.

Extending Concepts

Homogeneity of Proportions Test Another chi-square test that involves a contingency table is the **homogeneity of proportions test.** This test is used to determine if several proportions are equal when samples are taken from different populations. Before the populations are sampled and the contingency table is made, the sample sizes are determined. After randomly sampling different populations, you can test whether the proportion of elements in a category is the same for each population using the same guidelines in performing a chi-square independence test. The null and alternative hypotheses are always some variation of the following statements.

H_0: The proportions are equal.

H_a: At least one of the proportions is different from the others.

Performing a homogeneity of proportions test requires that the observed frequencies be obtained using a random sample, and each expected frequency must be greater than or equal to 5.

20. ***Motor Vehicle Deaths*** The table shows a random sample of motor vehicle deaths by age and gender. At $\alpha = 0.05$, perform a homogeneity of proportions test on the claim that the proportions of motor vehicle deaths involving males or females are the same for each age group. *(Adapted from Insurance Institute for Highway Safety)*

	Age			
Gender	**16–24**	**25–34**	**35–44**	**45–54**
Male	80	54	41	39
Female	35	21	18	17

	Age			
Gender	**55–64**	**65–74**	**75–84**	**85 and older**
Male	39	45	73	55
Female	20	27	40	19

21. ***Obsessive-Compulsive Disorder*** The contingency table shows the results of a random sample of patients with obsessive-compulsive disorder after being treated with a drug or with a placebo. At $\alpha = 0.10$, perform a homogeneity of proportions test on the claim that the proportions of the results for drug and placebo treatment are the same. *(Adapted from The Journal of the American Medical Association)*

	Treatment	
Result	**Drug**	**Placebo**
Improvement	39	25
No change	54	70

Case Study

Traffic Safety Facts

Each year, the National Highway Traffic Safety Administration (NHTSA) together with the National Center for Statistics and Analysis (NCSA) publishes *Traffic Safety Facts,* which summarizes the motor vehicle traffic crash experience for the United States. *Traffic Safety Facts 2001* includes trend data, crash data, vehicle data, and people data. Also, the NHTSA and NCSA publish a report summarizing the motor vehicle crash data of the 17 states in the NHTSA's State Data System.

In 2001, there were 42,116 fatalities in the United States as a result of motor vehicle crashes. The pie chart at the right shows the national distribution of traffic fatalities with respect to age group. For example, 24% of all motor vehicle fatalities were young adults aged 16–24. Using the data from the 17 states in the NHTSA's State Data System as a sample, the contingency table shows the number of motor vehicle fatalities according to age and geographic location within the United States.

U.S. Distribution of Traffic Fatalities

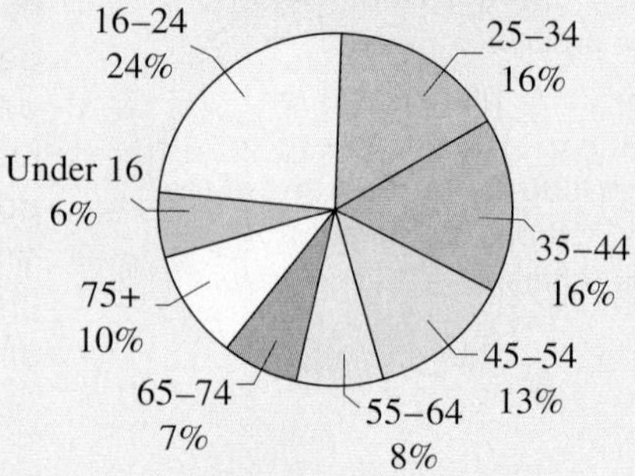

Motor Vehicle Fatalities

	Region		
Age	Eastern U.S.	Central U.S.	Western U.S.
Under 16	503	361	654
16–24	2175	1555	2373
25–34	1460	934	1613
35–44	1494	1008	1527
45–54	1206	776	1225
55–64	733	469	739
65–74	637	411	594
75+	973	603	808

Exercises

1. In 2001, how many people in the United States aged 16–24 died as a result of a motor vehicle crash?

2. Assuming the variables region and age are independent, in which region did the number of motor vehicle fatalities for the 16–24 age group exceed the expected number of fatalities?

3. Assuming the variables region and age are independent, in which region did the number of motor vehicle fatalities for the 25–34 age group exceed the expected number of fatalities?

4. At $\alpha = 0.05$, perform a chi-square test to determine whether the variables region and age are independent. What can you conclude?

In Exercises 5–7, perform a chi-square goodness-of-fit test to compare the national distribution of motor vehicle fatalities with the distribution of each region of the United States. Use the national distribution as the claimed distribution. Use $\alpha = 0.05$.

5. Compare the distribution of the sample of motor vehicle fatalities from the eastern United States with the national distribution. What can you conclude?

6. Compare the distribution of the sample of motor vehicle fatalities from the central United States with the national distribution. What can you conclude?

7. Compare the distribution of the sample of motor vehicle fatalities from the western United States with the national distribution. What can you conclude?

8. In addition to the variables used in this case study, what other variables do you think are important considerations when studying the distribution of motor vehicle fatalities?

10.3 Comparing Two Variances

What You Should Learn

- How to interpret the F-distribution and use an F-table to find critical values
- How to perform a two-sample F-test to compare two variances

The F-Distribution • The Two-Sample F-Test for Variances

The F-Distribution

In Chapter 8, you learned how to perform hypothesis tests to compare population means and population proportions. Recall from Section 8.2 that the t-test for the difference between two population means depends on whether the population variances are equal. To determine whether the population variances are equal, you can perform a two-sample F-test.

In this section, you will learn about the F-distribution and how to use the F-distribution to compare two variances.

Note to Instructor

If you prefer, you can cover this section before Section 8.2 so students can test whether variances of two populations are equal.

DEFINITION

Let s_1^2 and s_2^2 represent the sample variances of two different populations. If both populations are normal and the population variances σ_1^2 and σ_2^2 are equal, then the sampling distribution of

$$F = \frac{s_1^2}{s_2^2}$$

is called an ***F*-distribution.** Several properties of the F-distribution are as follows.

1. The F-distribution is a family of curves each of which is determined by two types of degrees of freedom: the degrees of freedom corresponding to the variance in the numerator, denoted by $\mathbf{d.f._N}$, and the degrees of freedom corresponding to the variance in the denominator, denoted by $\mathbf{d.f._D}$.
2. F-distributions are positively skewed.
3. The total area under each curve of an F-distribution is equal to 1.
4. F-values are always greater than or equal to zero.
5. For all F-distributions, the mean value of F is approximately equal to 1.

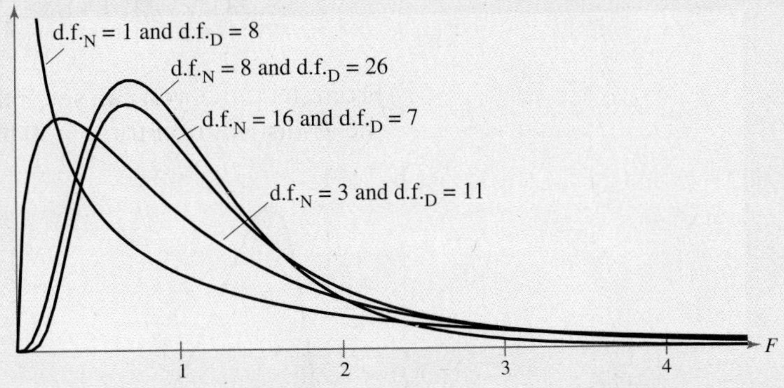

F-Distributions

Table 7 in Appendix B lists the critical values for the *F*-distribution for selected levels of significance α and degrees of freedom, $d.f._N$ and $d.f._D$.

GUIDELINES

Finding Critical Values for the *F*-Distribution

1. Specify the level of significance α.
2. Determine the degrees of freedom for the numerator, $d.f._N$.
3. Determine the degrees of freedom for the denominator, $d.f._D$.
4. Use Table 7 in Appendix B to find the critical value. If the hypothesis test is
   - **a.** one tailed, use the α *F*-table.
   - **b.** two tailed, use the $\frac{1}{2}\alpha$ *F*-table.

Study Tip

In the sampling distribution of $F = \frac{s_1^2}{s_2^2}$, the larger variance is always in the numerator. So, *F* is always greater than or equal to 1. As such, all one-tailed tests are right-tailed tests, and for all two-tailed tests, you need only to find the right-tailed critical value.

EXAMPLE 1

Finding Critical F-Values for a Right-Tailed Test

Find the critical *F*-value for a right-tailed test when $\alpha = 0.05$, $d.f._N = 6$, and $d.f._D = 29$.

SOLUTION A portion of Table 7 is shown below. Using the $\alpha = 0.05$ *F*-table with $d.f._N = 6$ and $d.f._D = 29$, you can find the critical value as shown by the highlighted areas in the table.

$d.f._D$: Degrees of freedom, denominator	$\alpha = 0.05$ $d.f._N$: Degrees of freedom, numerator							
	1	2	3	4	5	6	7	8
1	161.4	199.5	215.7	224.6	230.2	234.0	236.8	238.9
2	18.51	19.00	19.16	19.25	19.30	19.33	19.35	19.37
26	4.23	3.37	2.98	2.74	2.59	2.47	2.39	2.32
27	4.21	3.35	2.96	2.73	2.57	2.46	2.37	2.31
28	4.20	3.34	2.95	2.71	2.56	2.45	2.36	2.29
29	4.18	3.33	2.93	2.70	2.55	2.43	2.35	2.28
30	4.17	3.32	2.92	2.69	2.53	2.42	2.33	2.27

From the table, you can see that the critical value is $F_0 = 2.43$. The graph shows the *F*-distribution for $\alpha = 0.05$, $d.f._N = 6$, $d.f._D = 29$, and $F_0 = 2.43$.

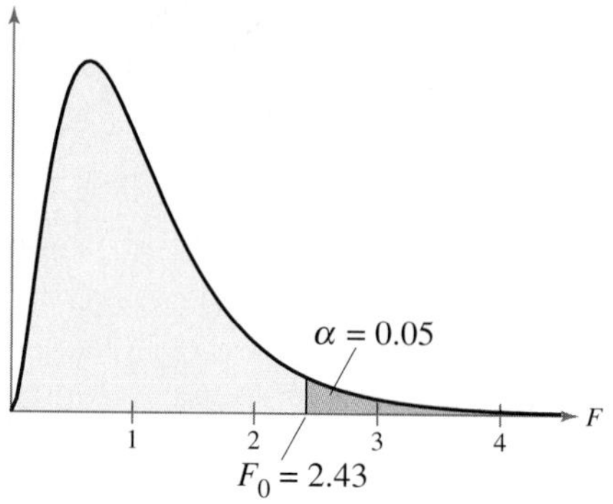

Try It Yourself 1

Find the critical F-value for a right-tailed test when $\alpha = 0.01$, $\text{d.f.}_N = 3$, and $\text{d.f.}_D = 15$.

a. *Specify* the level of significance α.
b. *Use* Table 7 in Appendix B to find the critical value.

Answer: Page A44

When performing a two-tailed hypothesis test using the F-distribution, you need only to find the right-tailed critical value. You must, however, remember to use the $\frac{1}{2}\alpha$ F-table.

EXAMPLE 2

Finding Critical F-Values for a Two-Tailed Test

Find the critical F-value for a two-tailed test when $\alpha = 0.05$, $\text{d.f.}_N = 4$, and $\text{d.f.}_D = 8$.

SOLUTION A portion of Table 7 is shown below. Using the

$$\frac{1}{2}\alpha = \frac{1}{2}(0.05) = 0.025$$

F-table with $\text{d.f.}_N = 4$, and $\text{d.f.}_D = 8$, you can find the critical value as shown by the highlighted areas in the table.

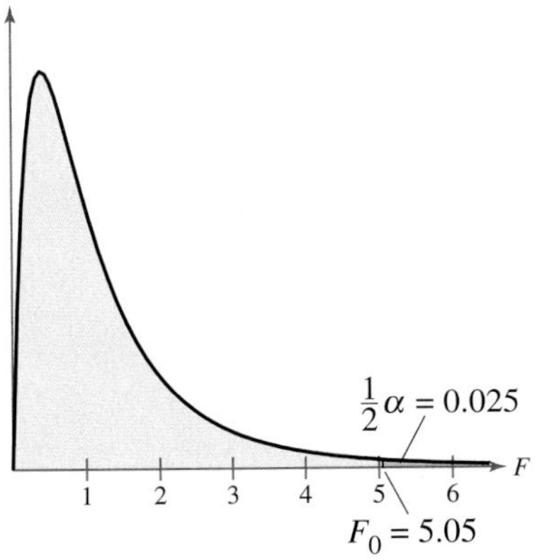

d.f.$_D$: Degrees of freedom, denominator	$\alpha = 0.025$							
	d.f.$_N$: Degrees of freedom, numerator							
	1	2	3	4	5	6	7	8
1	647.8	799.5	864.2	899.6	921.8	937.1	948.2	956.7
2	38.51	39.00	39.17	39.25	39.30	39.33	39.36	39.37
3	17.44	16.04	15.44	15.10	14.88	14.73	14.62	14.54
4	12.22	10.65	9.98	9.60	9.36	9.20	9.07	8.98
5	10.01	8.43	7.76	7.39	7.15	6.98	6.85	6.76
6	8.81	7.26	6.60	6.23	5.99	5.82	5.70	5.60
7	8.07	6.54	5.89	5.52	5.29	5.12	4.99	4.90
8	7.57	6.06	5.42	5.05	4.82	4.65	4.53	4.43
9	7.21	5.71	5.08	4.72	4.48	4.32	4.20	4.10

From the table, the critical value is $F_0 = 5.05$. The graph shows the F-distribution for $\frac{1}{2}\alpha = 0.025$, $\text{d.f.}_N = 4$, $\text{d.f.}_D = 8$, and $F_0 = 5.05$.

Try It Yourself 2

Find the critical F-value for a two-tailed test when $\alpha = 0.01$, $\text{d.f.}_N = 2$, and $\text{d.f.}_D = 5$.

a. *Specify* the level of significance α.
b. *Use* Table 7 in Appendix B with $\frac{1}{2}\alpha$ to find the critical value.

Answer: Page A44

The Two-Sample *F*-Test for Variances

In the remainder of this section, you will learn how to perform a two-sample F-test for comparing two population variances using a sample from each population. Such a test has three conditions that must be met.

1. The samples must be randomly selected.
2. The samples must be independent.
3. Each population must have a normal distribution.

If these requirements are met, you can use the F-test to compare the population variances σ_1^2 and σ_2^2.

Two-Sample *F*-Test for Variances

A two-sample F-test is used to compare two population variances σ_1^2 and σ_2^2 when a sample is randomly selected from each population. The populations must be independent and normally distributed. The test statistic is

$$F = \frac{s_1^2}{s_2^2}$$

where s_1^2 and s_2^2 represent the sample variances with $s_1^2 \geq s_2^2$. The degrees of freedom for the numerator is $\text{d.f.}_N = n_1 - 1$ and the degrees of freedom for the denominator is $\text{d.f.}_D = n_2 - 1$, where n_1 is the size of the sample having variance s_1^2 and n_2 is the size of the sample having variance s_2^2.

GUIDELINES

Using a Two-Sample *F*-Test to Compare σ_1^2 and σ_2^2

In Words	*In Symbols*
1. Identify the claim. State the null and alternative hypotheses.	State H_0 and H_a.
2. Specify the level of significance.	Identify α.
3. Identify the degrees of freedom.	$\text{d.f.}_N = n_1 - 1$ $\text{d.f.}_D = n_2 - 1$
4. Determine the critical value.	Use Table 7 in Appendix B.
5. Determine the rejection region.	
6. Calculate the test statistic.	$F = \frac{s_1^2}{s_2^2}$
7. Make a decision to reject or fail to reject the null hypothesis.	If F is in the rejection region, reject H_0. Otherwise, fail to reject H_0.
8. Interpret the decision in the context of the original claim.	

EXAMPLE 3

Performing a Two-Sample F-Test

A bank manager is designing a system that is intended to decrease the variance of the time customers wait in line for the ATM. Under the old system, a random sample of 10 customers had a variance of 144. Under the new system, a random sample of 21 customers had a variance of 100. At $\alpha = 0.10$, is there enough evidence to convince the manager to switch to the new system? Assume both populations are normally distributed.

SOLUTION Because $144 > 100$, $s_1^2 = 144$, and $s_2^2 = 100$. Therefore, s_1^2 and σ_1^2 represent the sample and population variances for the old system, respectively. With the claim "the variance of the waiting times under the new system is less than the variance of the waiting times under the old system," the null and alternative hypotheses are

H_0: $\sigma_1^2 \leq \sigma_2^2$ and H_a: $\sigma_1^2 > \sigma_2^2$. (Claim)

Because the test is a right-tailed test with $\alpha = 0.10$, d.f.$_N = n_1 - 1 = 10 - 1 = 9$, and d.f.$_D = n_2 - 1 = 21 - 1 = 20$, the critical value is $F_0 = 1.96$. With the F-test, the test statistic is

$$F = \frac{s_1^2}{s_2^2} = \frac{144}{100} = 1.44.$$

The graph shows the location of the rejection region and the test statistic. Because F is not in the rejection region, you should fail to reject the null hypothesis.

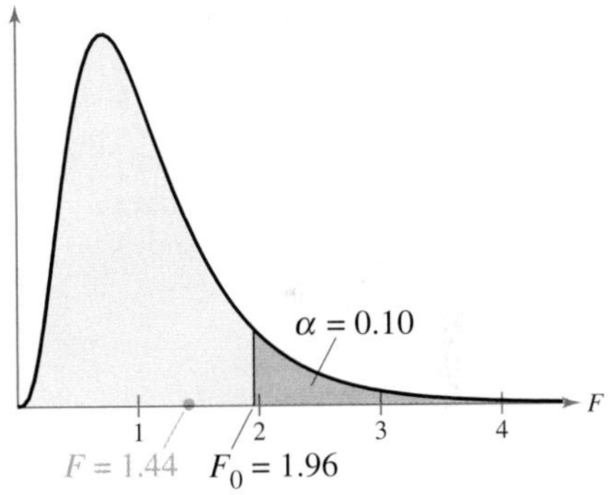

Interpretation There is not enough evidence to convince the manager to switch to the new system.

Try It Yourself 3

A medical researcher claims that a specially treated intravenous solution decreases the variance of the time required for nutrients to enter the bloodstream. Independent samples from each type of solution are randomly selected, and the results are shown in the table at the left. At $\alpha = 0.01$, is there enough evidence to support the researcher's claim? Assume the populations are normally distributed.

Normal solution	Treated solution
$n = 25$	$n = 20$
$s^2 = 180$	$s^2 = 56$

a. *Identify* the claim and state H_0 and H_a.
b. *Specify* the level of significance α.
c. *Identify* the degrees of freedom for the numerator and for the denominator.
d. *Determine* the critical value and the rejection region.
e. Use the F-test to *calculate the test statistic F.*
f. *Decide* whether to reject the null hypothesis. Use a graph if necessary.
g. *Interpret* the decision in the context of the original claim.

Answer: Page A44

Picturing the World

Does location have an effect on the variance of real estate selling prices? A random sample of selling prices (in thousands of dollars) of houses sold in Albuquerque, New Mexico, is shown in the table. The first column represents the selling prices of houses in northeastern Albuquerque, and the second column lists the selling prices of houses in the remaining Albuquerque regions. *(Source: Albuquerque Board of Realtors)*

Northeast section	Other sections
205.0	129.5
208.0	97.5
215.0	93.9
215.0	82.0
199.9	78.0
190.0	77.0
180.0	70.0
156.0	62.0
145.0	54.0
144.9	107.0

Assuming the population of selling prices is normally distributed, is it possible to use a two-sample F-test to compare the population variances?

EXAMPLE 4

Performing a Two-Sample F-Test

Stock A	Stock B
$n_2 = 30$	$n_1 = 31$
$s_2 = 3.5$	$s_1 = 5.7$

You want to purchase stock in a company and are deciding between two different stocks. Because a stock's risk can be associated with the standard deviation of its daily closing prices, you randomly select samples of the daily closing prices for each stock to obtain the results shown at the left. At $\alpha = 0.05$, can you conclude that one of the two stocks is a riskier investment? Assume the stock closing prices are normally distributed.

SOLUTION Because $5.7^2 > 3.5^2$, $s_1^2 = 5.7^2$, and $s_2^2 = 3.5^2$. Therefore, s_1^2 and σ_1^2 represent the sample and population variances for Stock B, respectively. With the claim "one of the stocks is a riskier investment," the null and alternative hypotheses are

$$H_0\colon \sigma_1^2 = \sigma_2^2 \quad \text{and} \quad H_a\colon \sigma_1^2 \neq \sigma_2^2. \text{ (Claim)}$$

Because the test is a two-tailed test with $\frac{1}{2}\alpha = \frac{1}{2}(0.05) = 0.025$, d.f.$_N = n_1 - 1 = 31 - 1 = 30$, and d.f.$_D = n_2 - 1 = 30 - 1 = 29$, the critical value is $F_0 = 2.09$. With the F-test, the test statistic is

$$F = \frac{s_1^2}{s_2^2}$$

$$= \frac{5.7^2}{3.5^2} \qquad \text{Square sample standard deviations to obtain sample variances.}$$

$$\approx 2.65.$$

Study Tip

You can also use technology and a *P*-value to perform a two-sample *F*-test. For instance, in Example 4, you can use the TI-83 to obtain a *P*-value of 0.0102. Because $P < \alpha$, you should decide to reject the null hypothesis.

The graph shows the location of the rejection region and the test statistic. Because F is in the rejection region, you should reject the null hypothesis.

Interpretation One of the two stocks is a riskier investment.

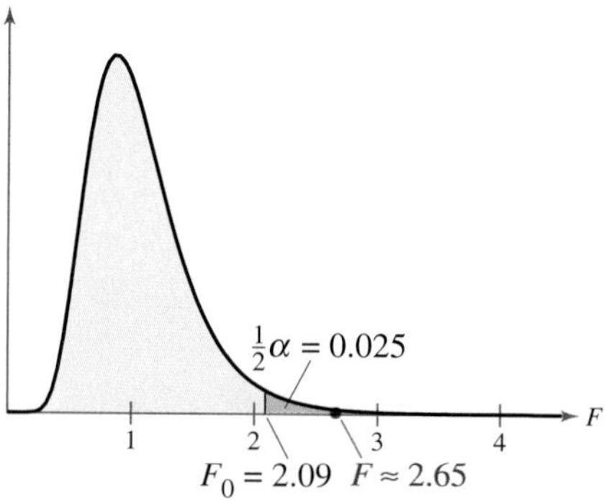

Try It Yourself 4

Location A	Location B
$n = 16$	$n = 22$
$s = 0.95$	$s = 0.78$

A biologist claims that the pH levels of the soil in two geographic locations have equal standard deviations. Independent samples from each location are randomly selected, and the results are shown at the left. At $\alpha = 0.01$, is there enough evidence to reject the biologist's claim? Assume the pH levels are normally distributed.

a. *Identify* the claim and state H_0 and H_a.
b. *Specify* the level of significance α.
c. *Identify* the degrees of freedom for the numerator and for the denominator.
d. *Determine* the critical value and the rejection region.
e. Use the F-test to *calculate the test statistic F.*
f. *Decide* whether to reject the null hypothesis. Use a graph if necessary.
g. *Interpret* the decision in the context of the original claim.

Answer: Page A44

10.3 Exercises

Building Basic Skills and Vocabulary

1. Explain how to find the critical value for an F-test.

2. List five properties of the F-distribution.

In Exercises 3–8, find the critical F-value for a right-tailed test using the indicated level of significance α and degrees of freedom d.f._{N} and d.f._{D}.

3. $\alpha = 0.05, \text{d.f.}_{\text{N}} = 4, \text{d.f.}_{\text{D}} = 18$

4. $\alpha = 0.01, \text{d.f.}_{\text{N}} = 3, \text{d.f.}_{\text{D}} = 24$

5. $\alpha = 0.01, \text{d.f.}_{\text{N}} = 5, \text{d.f.}_{\text{D}} = 11$

6. $\alpha = 0.05, \text{d.f.}_{\text{N}} = 6, \text{d.f.}_{\text{D}} = 19$

7. $\alpha = 0.10, \text{d.f.}_{\text{N}} = 10, \text{d.f.}_{\text{D}} = 15$

8. $\alpha = 0.025, \text{d.f.}_{\text{N}} = 7, \text{d.f.}_{\text{D}} = 3$

In Exercises 9–14, test the claim about the differences between two population variances σ_1^2 and σ_2^2 at the given level of significance α using the given sample statistics. Assume the sample statistics are from independent samples that are randomly selected and each population has a normal distribution.

9. Claim: $\sigma_1^2 > \sigma_2^2, \alpha = 0.10$.
Sample statistics: $s_1^2 = 773$,
$n_1 = 5; s_2^2 = 765, n_2 = 6$

10. Claim: $\sigma_1^2 = \sigma_2^2, \alpha = 0.05$.
Sample statistics: $s_1^2 = 310$,
$n_1 = 7; s_2^2 = 297, n_2 = 8$

11. Claim: $\sigma_1^2 \leq \sigma_2^2, \alpha = 0.01$.
Sample statistics: $s_1^2 = 842$,
$n_1 = 11; s_2^2 = 836, n_2 = 10$

12. Claim: $\sigma_1^2 \neq \sigma_2^2, \alpha = 0.05$.
Sample statistics: $s_1^2 = 141$,
$n_1 = 15; s_2^2 = 117, n_2 = 14$

13. Claim: $\sigma_1^2 = \sigma_2^2, \alpha = 0.05$.
Sample statistics: $s_1^2 = 5.2$,
$n_1 = 7; s_2^2 = 4.8, n_2 = 9$

14. Claim: $\sigma_1^2 > \sigma_2^2, \alpha = 0.05$.
Sample statistics: $s_1^2 = 44.6$,
$n_1 = 16; s_2^2 = 39.3, n_2 = 12$

Using and Interpreting Concepts

Comparing Two Variances In Exercises 15–22, (a) identify the claim and state H_0 and H_a, (b) determine the critical value and the rejection region, (c) calculate the test statistic, and (d) decide whether to reject or fail to reject the null hypothesis. Then interpret the decision in the context of the original claim. Assume the samples are independent and each population has a normal distribution.

15. ***Life of Appliances*** Company A claims that the variance of the life of its appliances is less than the variance of the life of Company B's appliances. A random sample of the lives of 20 of Company A's appliances has a variance of 2.6. A random sample of the lives of 23 of Company B's appliances has a variance of 2.8. At $\alpha = 0.05$, can you support Company A's claim?

16. ***Fuel Consumption*** An automobile manufacturer claims that the variance of the fuel consumption for its luxury sedans is less than the variance of the fuel consumption for the luxury sedans of a top competitor. A random sample of the fuel consumption of 19 of the manufacturer's sedans has a variance of 4.2. A random sample of the fuel consumption of 22 of its competitor's sedans has a variance of 4.5. At $\alpha = 0.05$, can you support the manufacturer's claim?

1. Specify the level of significance α. Determine the degrees of freedom for the numerator and denominator. Use Table 7 in Appendix B to find the critical value F.
2. See Selected Answers, page A99.
3. 2.93
4. 4.72
5. 5.32
6. 2.63
7. 2.06
8. 14.62
9. Fail to reject H_0. There is not enough evidence at the 10% level to support the claim.
10. Fail to reject H_0. There is not enough evidence at the 5% level to reject the claim.
11. Fail to reject H_0. There is not enough evidence at the 1% level to reject the claim.
12. Fail to reject H_0. There is not enough evidence at the 5% level to support the claim.
13. Fail to reject H_0. There is not enough evidence at the 5% level to reject the claim.
14. See Selected Answers, page A100.
15. (a) H_0: $\sigma_1^2 \leq \sigma_2^2$
H_a: $\sigma_1^2 > \sigma_2^2$ (claim)
(b) $F_0 = 2.13$
Rejection region: $F > 2.13$
(c) 1.077
(d) Fail to reject H_0. There is not enough evidence at the 5% level to support Company A's claim that the variance of the life of its appliances is less than the variance of the life of Company B's appliances.
16. See Selected Answers, page A100.

17. (a) H_0: $\sigma_1^2 = \sigma_2^2$ (claim)
H_a: $\sigma_1^2 \neq \sigma_2^2$
(b) $F_0 = 2.63$
Rejection region: $F > 2.63$
(c) 1.126
(d) Fail to reject H_0. There is not enough evidence at the 10% level to reject the claim that the standard deviation of physical science assessment test scores for eighth grade students is the same in Districts 1 and 2.

18. (a) H_0: $\sigma_1^2 = \sigma_2^2$ (claim)
H_a: $\sigma_1^2 \neq \sigma_2^2$
(b) $F_0 = 5.20$
Rejection region: $F > 5.20$
(c) 1.155
(d) Fail to reject H_0. There is not enough evidence at the 1% level to reject the claim that the standard deviation of eighth grade student test scores is the same for Districts 1 and 2.

19. (a) H_0: $\sigma_1^2 \leq \sigma_2^2$
H_a: $\sigma_1^2 > \sigma_2^2$ (claim)
(b) $F_0 = 1.77$
Rejection region: $F > 1.77$
(c) 1.96
(d) Reject H_0. There is enough evidence at the 10% level to support the claim that the standard deviation of waiting times has decreased.

20. (a) H_0: $\sigma_1^2 = \sigma_2^2$ (claim)
H_a: $\sigma_1^2 \neq \sigma_2^2$
(b) $F_0 = 4.77$
Rejection region: $F > 4.77$
(c) 1.170
(d) Fail to reject H_0. There is not enough evidence at the 1% level to reject the claim that the standard deviations of hotel room rates for the two cities are the same.

17. *Physical Science Assessment* In a recent interview, a state school administrator stated that the standard deviations of physical science assessment test scores for eighth grade students are the same in Districts 1 and 2. A random sample of 12 test scores from District 1 has a standard deviation of 27.7 points and a random sample of 14 test scores from District 2 has a standard deviation of 26.1 points. Can you reject the administrator's claim? Use $\alpha = 0.10$. *(Adapted from National Center for Educational Statistics)*

18. *Comparison of Test Scores* A school administrator reports that the standard deviation of the test scores for eighth grade students are the same in District 1 and 2. As proof, the administrator gives the results of a study of test scores in each district. The study shows that a random sample of 10 test scores from District 1 has a standard deviation of 28.8 points and a random sample of 13 test scores from District 2 has a standard deviation of 26.8 points. At $\alpha = 0.01$, can you reject the administrator's claim? *(Adapted from National Center for Educational Statistics)*

19. *Waiting Times* A random sample of 25 waiting times (in minutes) before patients saw a medical professional in a hospital's minor emergency department had a standard deviation of 0.7 minute. After a new admissions procedure was implemented, a random sample of 21 waiting times had a standard deviation of 0.5 minute. At $\alpha = 0.10$, can you support the hospital's claim that the standard deviation of the waiting times has decreased?

20. *Room Rates* A travel agency's marketing brochure indicates that the standard deviations of hotel room rates for two cities are the same. A random sample of 13 hotel room rates in one city has a standard deviation of \$27.50 and a random sample of 15 hotel room rates in the other city has a standard deviation of \$29.75. Can you reject the agency's claim? Use $\alpha = 0.01$. *(Adapted from Smith Travel Research)*

21. *Annual Salaries* The annual salaries for a random sample of 16 actuaries working in California have a standard deviation of \$13,900. The annual salaries for a random sample of 17 actuaries working in New York have a standard deviation of \$8800. Using this information, can you conclude that the standard deviation of the annual salaries for actuaries is greater in California than in New York? Use $\alpha = 0.05$. *(Adapted from America's Career InfoNet)*

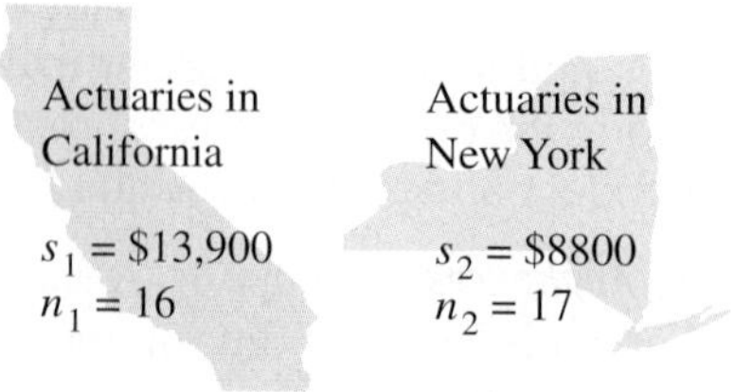

Figure for Exercise 21

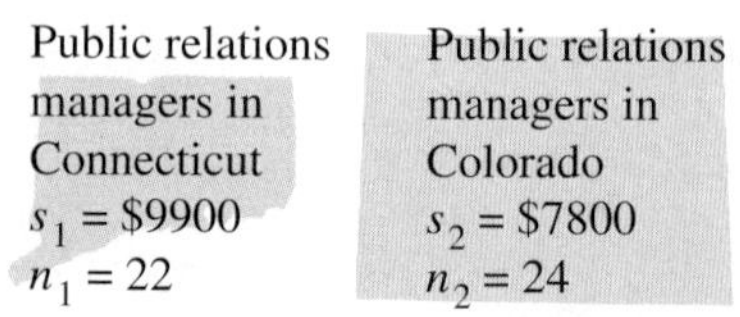

Figure for Exercise 22

22. *Annual Salaries* An employment information service claims the standard deviation of the annual salaries for public relations managers is greater in Connecticut than in Colorado. The annual salaries for a random sample of 22 public relations managers in Connecticut have a standard deviation of \$9900. The annual salaries for a random sample of 24 public relations managers in Colorado have a standard deviation of \$7800. At $\alpha = 0.05$, can you support the service's claim? *(Adapted from America's Career InfoNet)*

21. (a) H_0: $\sigma_1^2 \le \sigma_2^2$
H_a: $\sigma_1^2 > \sigma_2^2$ (claim)
(b) $F_0 = 2.35$
Rejection region: $F > 2.35$
(c) 2.495
(d) Reject H_0. There is enough evidence at the 5% level to conclude that the standard deviation of the annual salaries for actuaries is greater in California than in New York.

22. (a) H_0: $\sigma_1^2 \le \sigma_2^2$
H_a: $\sigma_1^2 > \sigma_2^2$ (claim)
(b) $F_0 = 2.04$
Rejection region: $F > 2.04$
(c) 1.611
(d) Fail to reject H_0. There is not enough evidence at the 5% level to support the claim that the standard deviation of the annual salaries for public relations managers is greater in Connecticut than in Colorado.

23. Right tailed: 14.73
Left tailed: 0.152

24. Right tailed: 2.33
Left tailed: 0.455

25. (0.366, 3.839)

26. (0.478, 5.019)

Extending Concepts

Finding Left-Tailed Critical F-values In this section you learned that if s_1^2 is larger than s_2^2, then you only need to calculate the right-tailed critical F-value for a two-tailed test. For other applications of the F-distribution, you will need to calculate the left-tailed critical F-value. To calculate the left-tailed critical F-value, do the following.

(1) Interchange the values for d.f._N and d.f._D.

(2) Find the corresponding F-value in Table 7.

(3) Calculate the reciprocal of the F-value to obtain the left-tailed critical F-value.

In Exercises 23 and 24, find the right- and left-tailed critical F-values for a two-tailed test with the given values of α, d.f._N, and d.f._D.

23. $\alpha = 0.05$, $\text{d.f.}_N = 6$, $\text{d.f.}_D = 3$

24. $\alpha = 0.10$, $\text{d.f.}_N = 20$, $\text{d.f.}_D = 15$

Confidence Interval for σ_1^2/σ_2^2 When s_1^2 and s_2^2 are the variances of randomly selected, independent samples from normally distributed populations, then a confidence interval for σ_1^2/σ_2^2 is

$$\frac{s_1^2}{s_2^2}F_L < \frac{\sigma_1^2}{\sigma_1^2} < \frac{s_1^2}{s_2^2}F_R$$

where F_L is the left-tailed critical F-value and F_R is the right-tailed critical F-value.

In Exercises 25 and 26, construct the indicated confidence interval for σ_1^2/σ_2^2. Assume the samples are independent and each population has a normal distribution.

25. ***Cholesterol Content*** In a recent study of the cholesterol content in grilled chicken sandwiches served at fast food restaurants, a nutritionist found that a random sample of sandwiches from Arby's and from McDonald's had the sample statistics shown in the table. Construct a 95% confidence interval for σ_1^2/σ_2^2, where σ_1^2 and σ_2^2 are the variances of the cholesterol content of grilled chicken sandwiches from Arby's and McDonald's, respectively. *(Adapted from Fast Food Facts, Minnesota Attorney General's Office)*

Cholesterol content for grilled chicken sandwiches		
Restaurant	Arby's	McDonald's
Sample variance	$s_1^2 = 9.61$	$s_2^2 = 8.41$
Sample size	$n_1 = 15$	$n_2 = 12$

Table for Exercise 25

Carbohydrate content for grilled chicken sandwiches		
Restaurant	Arby's	McDonald's
Sample variance	$s_1^2 = 4.84$	$s_2^2 = 3.24$
Sample size	$n_1 = 15$	$n_2 = 12$

Table for Exercise 26

26. ***Carbohydrate Content*** A fast food study found that the carbohydrate content of 15 randomly selected grilled chicken sandwiches from Arby's had a variance of 4.84. The study also found that the carbohydrate content of 12 randomly selected grilled chicken sandwiches from McDonald's had a variance of 3.24. Construct a 95% confidence interval for σ_1^2/σ_2^2, where σ_1^2 and σ_2^2 are the variances of the carbohydrate content of grilled chicken sandwiches from Arby's and McDonald's, respectively. *(Adapted from Fast Food Facts, Minnesota Attorney General's Office)*

10.4 Analysis of Variance

What You Should Learn

- How to use one-way analysis of variance to test claims involving three or more means
- An introduction to two-way analysis of variance

One-Way ANOVA • Two-Way ANOVA

One-Way ANOVA

Suppose a medical researcher is analyzing the effectiveness of three types of pain relievers and wants to determine whether there is a difference in the mean length of the time it takes each medication to provide relief. To determine whether such a difference exists, the researcher can use the F-distribution together with a technique called *analysis of variance.* Because one independent variable is being studied, the process is called *one-way analysis of variance.*

DEFINITION

One-way analysis of variance is a hypothesis-testing technique that is used to compare means from three or more populations. Analysis of variance is usually abbreviated as **ANOVA.**

To begin a one-way analysis of variance test, you should first state a null and an alternative hypothesis. For a one-way ANOVA test, the null and alternative hypotheses are always similar to the following statements.

H_0: $\mu_1 = \mu_2 = \mu_3 = \cdots = \mu_k$ (All population means are equal.)

H_a: At least one of the means is different from the others.

When you reject the null hypothesis in an ANOVA test, you can conclude that one of the means is different from the others. Without performing more statistical tests, however, you cannot determine which of the means is different.

In a one-way ANOVA test, the following conditions must be true.

1. Each sample must be randomly selected from a normal, or approximately normal, population.
2. The samples must be independent of each other.
3. Each population must have the same variance.

The test statistic for a one-way ANOVA test is the ratio of two variances: the variance between samples and the variance within samples.

$$\text{Test statistic} = \frac{\text{Variance between samples}}{\text{Variance within samples}}$$

1. The variance between samples MS_B measures the differences related to the treatment given to each sample and is sometimes called the **mean square between.**
2. The variance within samples MS_W measures the differences related to entries within the same sample. This variance, sometimes called the **mean square within,** is usually due to sampling error.

One-Way Analysis of Variance Test

If the conditions for a one-way analysis of variance test are satisfied, then the sampling distribution for the test is approximated by the F-distribution. The test statistic is

$$F = \frac{MS_B}{MS_W}.$$

The degrees of freedom for the F-test are

$$\text{d.f.}_N = k - 1$$

and

$$\text{d.f.}_D = N - k$$

where k is the number of samples and N is the sum of the sample sizes.

If there is little or no difference between the means, then MS_B will be approximately equal to MS_W and the test statistic will be approximately 1. Values of F close to 1 suggest that you should fail to reject the null hypothesis. However, if one of the means differs significantly from the others, MS_B will be greater than MS_W and the test statistic will be greater than 1. Values of F significantly greater than 1 suggest that you reject the null hypothesis. As such, all one-way ANOVA tests are right-tailed tests. That is, if the test statistic is greater than the critical value, H_0 will be rejected.

Study Tip

The notations n_i, $\bar{x}_i$, and s_i^2 represent the sample size, mean, and variance of the ith sample, respectively. $\bar{\bar{x}}$ is sometimes called the grand mean.

GUIDELINES

Finding the Test Statistic for a One-Way ANOVA Test

In Words	*In Symbols*
1. Find the mean and variance of each sample.	$\bar{x} = \frac{\Sigma x}{n}$ $\quad s^2 = \frac{\Sigma(x - \bar{x})^2}{n - 1}$
2. Find the mean of all entries in all samples (the grand mean).	$\bar{\bar{x}} = \frac{\Sigma x}{N}$
3. Find the sum of squares between the samples.	$SS_B = \Sigma n_i(\bar{x}_i - \bar{\bar{x}})^2$
4. Find the sum of squares within the samples.	$SS_W = \Sigma(n_i - 1)s_i^2$
5. Find the variance between the samples.	$MS_B = \frac{SS_B}{k - 1} = \frac{SS_B}{\text{d.f.}_N}$
6. Find the variance within the samples.	$MS_W = \frac{SS_W}{N - k} = \frac{SS_W}{\text{d.f.}_D}$
7. Find the test statistic.	$F = \frac{MS_B}{MS_W}$

The notation SS_B represents the sum of squares between groups.

$$SS_B = n_1(\bar{x}_1 - \bar{\bar{x}})^2 + n_2(\bar{x}_2 - \bar{\bar{x}})^2 + \cdots + n_k(\bar{x}_k - \bar{\bar{x}})^2$$
$$= \Sigma n_i(\bar{x}_i - \bar{\bar{x}})^2$$

The notation SS_W represents the sum of squares within groups.

$$SS_W = (n_1 - 1)s_1^2 + (n_2 - 1)s_2^2 + \cdots + (n_k - 1)s_k^2$$
$$= \Sigma(n_i - 1)s_i^2$$

GUIDELINES

Performing a One-Way Analysis of Variance Test

In Words	*In Symbols*
1. Identify the claim. State the null and alternative hypotheses.	State H_0 and H_a.
2. Specify the level of significance.	Identify α.
3. Identify the degrees of freedom.	$\text{d.f.}_N = k - 1$ $\text{d.f.}_D = N - k$
4. Determine the critical value.	Use Table 7 in Appendix B.
5. Determine the rejection region.	
6. Calculate the test statistic.	$F = \dfrac{MS_B}{MS_W}$
7. Make a decision to reject or fail to reject the null hypothesis.	If F is in the rejection region, reject H_0. Otherwise, fail to reject H_0.
8. Interpret the decision in the context of the original claim.	

Tables are a convenient way to summarize the results of a one-way analysis of variance test. ANOVA summary tables are set up as shown below.

ANOVA Summary Table

Variation	Sum of squares	Degrees of freedom	Mean squares	F
Between	SS_B	d.f._N	$MS_B = \dfrac{SS_B}{\text{d.f.}_N}$	$MS_B \div MS_W$
Within	SS_W	d.f._D	$MS_W = \dfrac{SS_W}{\text{d.f.}_D}$	

EXAMPLE 1

Performing a One-Way ANOVA Test

A medical researcher wants to determine whether there is a difference in the mean length of time it takes three types of pain relievers to provide relief from headache pain. Several headache sufferers are randomly selected and given one of the three medications. Each headache sufferer records the time (in minutes) it takes the medication to begin working. The results are shown in the table. At $\alpha = 0.01$, can you conclude that the mean times are different? Assume that each population of relief times is normally distributed and that the population variances are equal.

Medication 1	Medication 2	Medication 3
12	16	14
15	14	17
17	21	20
12	15	15
	19	
$\bar{x}_1 = \frac{56}{4} = 14$	$\bar{x}_2 = \frac{85}{5} = 17$	$\bar{x}_3 = \frac{66}{4} = 16.5$
$s_1^2 = 6$	$s_2^2 = 8.5$	$s_3^2 = 7$

SOLUTION The null and alternative hypotheses are as follows.

H_0: $\mu_1 = \mu_2 = \mu_3$

H_a: At least one mean is different from the others. (Claim)

Because there are $k = 3$ samples, $\text{d.f.}_N = k - 1 = 3 - 1 = 2$. The sum of the sample sizes is $N = n_1 + n_2 + n_3 = 4 + 5 + 4 = 13$. So, $\text{d.f.}_D = N - k = 13 - 3 = 10$. Using $\text{d.f.}_N = 2$, $\text{d.f.}_D = 10$, and $\alpha = 0.01$, the critical value is $F_0 = 7.56$. To find the test statistic, first calculate $\bar{\bar{x}}$, MS_B, and MS_W.

$$\bar{\bar{x}} = \frac{\Sigma x}{N} = \frac{56 + 85 + 66}{13} \approx 15.92$$

$$MS_B = \frac{SS_B}{\text{d.f.}_N} = \frac{\Sigma n_i(\bar{x}_i - \bar{\bar{x}})^2}{k - 1}$$

$$\approx \frac{4(14 - 15.92)^2 + 5(17 - 15.92)^2 + 4(16.5 - 15.92)^2}{3 - 1}$$

$$\approx \frac{21.92}{2} = 10.96$$

$$MS_W = \frac{SS_W}{\text{d.f.}_D} = \frac{\Sigma(n_i - 1)s_i^2}{N - k}$$

$$= \frac{(4 - 1)(6) + (5 - 1)(8.5) + (4 - 1)(7)}{13 - 3} = \frac{73}{10} = 7.3$$

Using $MS_B = 10.96$ and $MS_W = 7.3$, the test statistic is

$$F = \frac{MS_B}{MS_W} = \frac{10.96}{7.3} \approx 1.50.$$

The graph shows the location of the rejection region. Because F is not in the rejection region, you should decide not to reject the null hypothesis.

Interpretation There is not enough evidence at the 1% level of significance to conclude that there is a difference in the mean length of time it takes the three pain relievers to provide relief from headache pain.

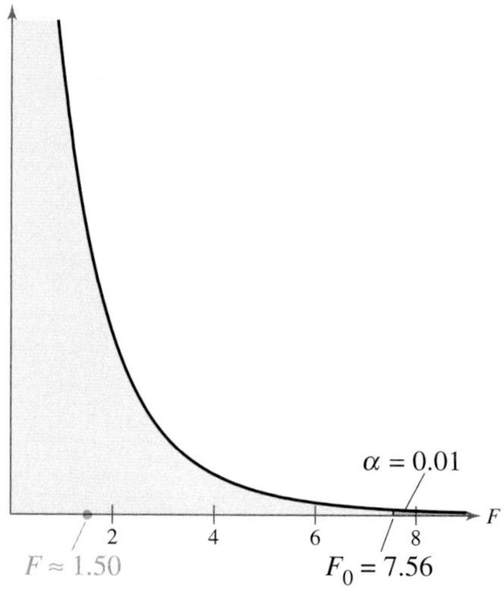

The ANOVA summary table for Example 1 is shown below.

Variation	Sum of squares	Degrees of freedom	Mean squares	F
Between	21.92	2	10.96	1.50
Within	73	10	7.3	

Try It Yourself 1

A sales analyst wants to determine whether there is a difference in the mean monthly sales of a company's four sales regions. Several salespersons from each region are randomly selected and each provides his or her sales amounts (in thousands of dollars) for the previous month. The results are shown in the table. At $\alpha = 0.05$, can the analyst conclude that there is a difference in the mean monthly sales among the sales regions? Assume that each population of sales is normally distributed and that the population variances are equal.

North	East	South	West
34	47	40	21
28	36	30	30
18	30	41	24
24	38	29	37
	44		23
$\bar{x}_1 = 26$	$\bar{x}_2 = 39$	$\bar{x}_3 = 35$	$\bar{x}_4 = 27$
$s_1^2 \approx 45.33$	$s_2^2 = 45$	$s_3^2 \approx 40.67$	$s_4^2 = 42.5$

a. *State* H_0 and H_a.
b. *Specify* the level of significance α.
c. *Identify* the degrees of freedom for the numerator and for the denominator.
d. *Determine* the critical value and the rejection region.
e. *Calculate* the test statistic F.
f. *Decide* whether to reject the null hypothesis. Use a graph if necessary.
g. *Interpret* the decision in the context of the original claim.

Answer: Page A44

Picturing the World

Adult common cuckoos do not care for their own eggs. Cuckoos lay their eggs in the nests of other birds, such as sparrows, robins, and wrens. These birds adopt the cuckoo eggs, hatching and caring for the eggs along with their own. A biologist is studying the lengths of cuckoo eggs found in sparrow, robin, and wren nests. The randomly selected egg lengths (in millimeters) are shown in the table.

Host nest		
Sparrow	**Robin**	**Wren**
24.08	22.66	20.89
23.40	22.51	20.97
22.95	21.44	22.31
22.82	22.70	21.54
23.98	22.15	21.19
24.59	22.75	20.19
22.95	23.02	21.38
25.16	21.72	20.50
23.39	21.49	20.99
23.74	22.15	21.16
21.15	22.15	20.83
23.81	22.28	20.74
22.40	22.32	21.40
24.17	22.98	21.90

At $\alpha = 0.05$, can you conclude that the mean length of the cuckoo eggs found in one type of nest is different from the others? Assume that each population of cuckoo egg lengths is normally distributed and that the population variances are equal.

Using technology greatly simplifies the one-way ANOVA process. When using a technology tool such as Excel, MINITAB, or the TI-83 to perform a one-way analysis of variance test, you can use P-values to decide whether to reject the null hypothesis. If the P-value is less than α, you should reject H_0.

EXAMPLE 2

Using Technology to Perform ANOVA Tests

Airline 1	Airline 2	Airline 3
122	119	120
135	133	158
126	143	155
131	149	126
125	114	147
116	124	164
120	126	134
108	131	151
142	140	131
113	136	141

Three airline companies offer flights between Corydon and Lincolnville. Several randomly selected flight times (in minutes) between the towns for each airline are shown at the left. Assume that the populations of flight times are normally distributed, the samples are independent, and the population variances are equal. At $\alpha = 0.01$, can you conclude that there is a difference in the means of the flight times? Use a technology tool.

SOLUTION The results obtained using the TI-83 are shown below. From the results, you can see that $P \approx 0.006$. Because $P < \alpha$, you should reject the null hypothesis.

TI-83

```
One-way ANOVA
 F=6.13023478
 p=.0063828069
 Factor
  df=2
  SS=1806.46667
↓ MS=903.233333
```

TI-83

```
One-way ANOVA
↑ MS=903.233333
 Error
  df=27
  SS=3978.2
  MS=147.340741
 Sxp=12.1383994
```

Interpretation You can conclude that there is a difference in the means of the flight times.

Try It Yourself 2

The data shown in the table represent the GPAs of randomly selected freshmen, sophomores, juniors, and seniors. At $\alpha = 0.05$, can you conclude that there is a difference in the means of the GPAs? Assume that the populations of GPAs are normally distributed and that the population variances are equal. Use a technology tool.

Freshmen	2.34	2.38	3.31	2.39	3.40	2.70	2.34			
Sophomores	3.26	2.22	3.26	3.29	2.95	3.01	3.13	3.59	2.84	3.00
Juniors	2.80	2.60	2.49	2.83	2.34	3.23	3.49	3.03	2.87	
Seniors	3.31	2.35	3.27	2.86	2.78	2.75	3.05	3.31		

a. *Enter* the data.
b. *Perform* the ANOVA test.
c. *Compare* the resulting P-value to the given level of significance α.
d. *Interpret* the results.

Answer: Page A44

Two-Way ANOVA

Insight

The conditions for a two-way ANOVA test are the same as those for a one-way ANOVA test with the additional condition that all samples must be of equal size.

When you want to test the effect of *two* independent variables, or factors, on one dependent variable, you can use a **two-way analysis of variance test.** For example, suppose a medical researcher wants to test the effect of gender *and* type of medication on the mean length of time it takes pain relievers to provide relief. To perform such an experiment, the researcher can use the following two-way ANOVA block design.

Type of medication	Gender: M	Gender: F
I	Males taking type I	Females taking type I
II	Males taking type II	Females taking type II
III	Males taking type III	Females taking type III

Insight

If gender and type of medication have no effect on the length of time it takes a pain reliever to provide relief, then there will be no significant difference in the means of the relief times.

A two-way ANOVA test has three null hypotheses—two for each main effect and one for the interaction effect. A **main effect** is the effect of one independent variable on the dependent variable, and the **interaction effect** is the effect of both independent variables on the dependent variable. For example, the hypotheses for the pain reliever experiment are as follows.

Hypotheses for main effects:

H_0: Gender has no effect on the mean length of time it takes a pain reliever to provide relief.

H_a: Gender has an effect on the mean length of time it takes a pain reliever to provide relief.

H_0: The type of medication has no effect on the mean length of time it takes a pain reliever to provide relief.

H_a: The type of medication has an effect on the mean length of time it takes a pain reliever to provide relief.

Hypotheses for interaction effect:

H_0: There is no interaction effect between gender and type of medication on the mean length of time it takes a pain reliever to provide relief.

H_a: There is an interaction effect between gender and type of medication on the mean length of time it takes a pain reliever to provide relief.

To test these hypotheses, you can perform a two-way ANOVA test. Using the *F*-distribution, a two-way ANOVA test calculates an *F*-test statistic for each hypothesis. As a result, it is possible to reject none, one, two, or all of the null hypotheses. The statistics involved with a two-way ANOVA test is beyond the scope of this course. You can, however, use a technology tool such as MINITAB to perform a two-way ANOVA test.

10.4 Exercises

Help

MyMathLab

Student Study Pack

Building Basic Skills and Vocabulary

1. State a null and alternative hypothesis for a one-way ANOVA test.

2. What conditions are necessary in order to use a one-way ANOVA test?

3. Describe the difference between the variance between samples MS_B and the variance within samples MS_W.

4. Describe the hypotheses for a two-way ANOVA test.

Using and Interpreting Concepts

Performing a One-Way ANOVA Test In Exercises 5–16, perform the indicated one-way ANOVA test by

(a) identifying the claim and stating H_0 and H_a.

(b) identifying the degrees of freedom for the numerator and for the denominator, determining the critical value, and determining the rejection region.

(c) calculating the test statistic.

(d) deciding to reject or fail to reject the null hypothesis and interpreting the decision in the context of the original claim.

Assume that each sample is drawn from a normal, or approximately normal, population, that the samples are independent of each other, and that the populations have the same variances. If convenient, use technology to solve the problem.

DATA

5. ***Toothpaste*** The table shows the cost per month (in dollars) for a random sample of toothpastes exhibiting moderate abrasiveness, low abrasiveness, or very low abrasiveness. At $\alpha = 0.05$, can you conclude that the mean costs per month are different? *(Source: Consumer Reports)*

Moderate abrasiveness	Low abrasiveness	Very low abrasiveness
0.75	0.74	0.44
0.46	0.96	1.40
3.16	0.65	4.00
0.97	0.70	0.91
0.72	0.72	
2.74	1.71	
0.62	1.46	
0.72	1.57	
1.00	0.77	
0.78	0.98	

6. ***Automobile Batteries*** The prices (in dollars) for 16 randomly selected automobile batteries are shown in the table. The prices are classified according to battery type. At $\alpha = 0.05$, is there enough evidence to conclude that at least one of the mean battery prices is different from the others? *(Source: Consumer Reports)*

Group size 24	50	65	75	64		
Group size 58	83	63	65	63	85	65
Group size 34/78	89	84	75	90	70	60

1. H_0: $\mu_1 = \mu_2 = \mu_3 = \ldots = \mu_k$
H_a: At least one of the means is different from the others.

2. Each sample must be randomly selected from a normal, or approximately normal, population. The samples must be independent of each other. Each population must have the same variance.

3. The MS_B measures the differences related to the treatment given to each sample. The MS_W measures the differences related to entries within the same sample.

4. See Selected Answers, page A100.

5. (a) H_0: $\mu_1 = \mu_2 = \mu_3$
H_a: At least one mean is different from the others. (claim)
(b) d.f.$_N = 2$; d.f.$_D = 21$; $F_0 = 3.47$
Rejection region: $F > 3.47$
(c) 0.77
(d) Fail to reject H_0. There is not enough evidence at the 5% level to conclude that the mean costs per month are different.

6. (a) H_0: $\mu_1 = \mu_2 = \mu_3$
H_a: At least one mean is different from the others. (claim)
(b) d.f.$_N = 2$; d.f.$_D = 13$; $F_0 = 3.81$
Rejection region: $F > 3.81$
(c) 2.15
(d) Fail to reject H_0. There is not enough evidence at the 5% level to conclude that at least one of the mean battery prices is different from the others.

7. (a) H_0: $\mu_1 = \mu_2 = \mu_3$ (claim)
   H_a: At least one mean is different from the others.
   (b) $d.f._N = 2$; $d.f._D = 12$; $F_0 = 2.81$
   Rejection region: $F > 2.81$
   (c) 1.77
   (d) Fail to reject H_0. There is not enough evidence at the 10% level to reject the claim that the mean prices are all the same for the three types of treatment.
8. (a) H_0: $\mu_1 = \mu_2 = \mu_3 = \mu_4$ (claim)
   H_a: At least one mean is different from the others.
   (b) $d.f._N = 3$; $d.f._D = 23$; $F_0 = 2.34$
   Rejection region: $F > 2.34$
   (c) 1.61
   (d) Fail to reject H_0. There is not enough evidence at the 10% level to reject the claim that the mean annual amounts are the same in all regions.
9. (a) H_0: $\mu_1 = \mu_2 = \mu_3 = \mu_4$ (claim)
   H_a: At least one mean is different from the others.
   (b) $d.f._N = 3$; $d.f._D = 29$; $F_0 = 4.54$
   Rejection region: $F > 4.54$
   (c) 4.98
   (d) Reject H_0. There is enough evidence at the 1% level to reject the claim that the mean number of days patients spend in the hospital is the same for all four regions.

7. *Exterior Deck Treatments* The table shows the price per gallon (in dollars) for a random sample of exterior deck treatments. At $\alpha = 0.10$, can you reject the claim that the mean prices are the same for the three types of treatments? *(Source: Consumer Reports)*

Semitransparent treatments	Lightly tinted treatments	Clear treatments
24	51	13
23	14	13
22	21	10
17	16	12
21		22
17		

DATA

8. *Reading Expenditures* The table shows the annual amount spent on reading (in dollars) for a random sample of consumers from four regions of the United States. At $\alpha = 0.10$, can you reject the claim that the mean annual amounts are the same in all regions? *(Adapted from U.S. Bureau of Labor Statistics)*

Northeast	Midwest	South	West
308	246	103	223
58	169	143	184
141	246	164	221
109	158	119	269
220	167	99	199
144	76	214	171
316		108	204

DATA

9. *Days Spent at a Hospital* In a recent study, a health insurance company investigated the number of days patients spend at a hospital. In part of the study, the company randomly selected patients from various parts of the United States and recorded the number of days each patient spent at a hospital. The results of the study are shown in the table. At $\alpha = 0.01$, can the company reject the claim that the mean number of days patients spend in the hospital is the same for all four regions? *(Adapted from U.S. National Center for Health Statistics)*

Northeast	Midwest	South	West
8	7	3	5
6	7	5	4
9	8	6	6
4	5	6	4
5	5	3	6
6	5	6	6
8	5	4	5
10	5		4
11	5		

10. (a) H_0: $\mu_1 = \mu_2 = \mu_3 = \mu_4$
H_a: At least one mean is different from the others. (claim)
(b) $d.f._N = 3$; $d.f._D = 48$; $F_0 = 2.20$
Rejection region: $F > 2.20$
(c) 2.29
(d) Reject H_0. There is enough evidence at the 10% level to conclude that the mean square footage for at least one of the four regions is different.

11. (a) H_0: $\mu_1 = \mu_2 = \mu_3 = \mu_4$ (claim)
H_a: At least one mean is different from the others.
(b) $d.f._N = 3$; $d.f._D = 43$; $F_0 = 2.22$
Rejection region: $F > 2.22$
(c) 3.04
(d) Reject H_0. There is enough evidence at the 10% level to reject the claim that the mean price is the same for all four cities.

12. (a) H_0: $\mu_1 = \mu_2 = \mu_3 = \mu_4$
H_a: At least one mean is different from the others. (claim)
(b) $d.f._N = 3$; $d.f._D = 16$; $F_0 = 3.24$
Rejection region: $F > 3.24$
(c) 1.870
(d) Fail to reject H_0. There is not enough evidence at the 5% level to conclude that the mean salary for an individual is different in at least one of the areas.

DATA **10. *Building Space*** The table shows the square footage (in thousands) for a random sample of buildings from four regions of the United States. At $\alpha = 0.10$, can you conclude that the mean square footage for one of the regions is different? *(Adapted from U.S. Energy Information Administration)*

Northeast	Midwest	South	West
13.9	10.4	14.1	13.0
18.0	11.6	16.4	15.0
15.7	5.3	15.6	11.4
11.7	10.8	3.1	10.4
24.6	12.3	16.6	11.5
12.2	14.5	17.0	16.0
18.1	3.8	13.2	5.5
10.0	15.2	8.6	20.0
8.7	18.8	25.5	7.0
17.1	13.6	9.6	2.8
17.4	11.4	12.8	15.3
16.3	7.9	15.4	14.6
19.4	15.2	16.2	11.0

DATA **11. *Housing Prices*** A realtor is comparing the prices of one-family houses in four cities. After randomly selecting one-family houses in the four cities and determining the price for each, the realtor organizes the prices (in thousands of dollars) in a table as shown below. At $\alpha = 0.10$, can the realtor reject the claim that the mean price is the same for all four cities? *(Adapted from National Association of Realtors)*

City A	City B	City C	City D
131.2	137.9	164.4	83.9
122.6	88.0	49.4	98.3
103.7	95.1	144.7	168.9
148.8	79.3	108.6	93.7
144.5	81.5	137.2	204.3
206.9	101.9	83.6	85.1

City A	City B	City C	City D
164.6	128.8	100.5	100.4
45.0	147.2	99.4	211.7
162.8	38.7	67.8	140.1
132.7	57.9	147.9	131.3
178.2	67.3	39.2	167.0
131.2		149.9	84.3

DATA **12. *Personal Income*** The table shows the salaries of randomly selected individuals from four large metropolitan areas. At $\alpha = 0.05$, can you conclude that the mean salary is different in at least one of the areas? *(Adapted from U.S. Bureau of Economic Analysis)*

Pittsburgh	Dallas	Chicago	Minneapolis
27,800	30,000	32,000	30,000
28,000	33,900	35,800	40,000
25,500	29,750	28,000	35,000
29,150	25,000	38,900	33,000
30,295	34,055	27,245	29,805

13. (a) H_0: $\mu_1 = \mu_2 = \mu_3 = \mu_4$
H_a: At least one mean is different from the others. (claim)
(b) d.f.$_N$ = 3; d.f.$_D$ = 26; $F_0 = 2.31$
Rejection region: $F > 2.31$
(c) 6.36
(d) Reject H_0. There is enough evidence at the 10% level to conclude that at least one of the mean prices is different.

14. (a) H_0: $\mu_1 = \mu_2 = \mu_3 = \mu_4$
H_a: At least one mean is different from the others. (claim)
(b) d.f.$_N$ = 3; d.f.$_D$ = 36; $F_0 = 4.38$
Rejection region: $F > 4.38$
(c) 8.46
(d) Reject H_0. There is enough evidence at the 1% level to conclude that the mean energy consumption for at least one region is different.

15. (a) H_0: $\mu_1 = \mu_2 = \mu_3 = \mu_4$(claim)
H_a: At least one mean is different from the others.
(b) d.f.$_N$ = 3; d.f.$_D$ = 19; $F_0 = 5.01$
Rejection region: $F > 5.01$
(c) 0.19
(d) Fail to reject H_0. There is not enough evidence at the 1% level to reject the claim that the mean numbers of female students are equal for all grades.

DATA **13. *Mobile Home Prices*** The table shows the prices (in dollars) for a random sample of new mobile homes from four manufacturers. At $\alpha = 0.10$, can you conclude that at least one of the mean prices is different? *(Adapted from U.S. Census Bureau)*

Manufacturer A	Manufacturer B	Manufacturer C	Manufacturer D
34,918	37,223	39,029	43,176
34,904	38,079	38,711	52,844
33,971	35,646	21,774	42,936
39,404	39,420	28,836	41,685
58,000	34,732	35,535	45,320
32,710	51,223	25,185	54,841
45,049	40,630	35,791	
48,373		32,607	
		27,728	

DATA **14. *Energy Consumption*** The table shows the energy consumed (in millions of Btu's) in one year for a random sample of households from four regions of the United States. At $\alpha = 0.01$, can you conclude that, for at least one region, the mean energy consumption is different? *(Adapted from U.S. Energy Information Administration)*

Northeast	Midwest	South	West
115.2	53.0	43.6	24.7
123.2	231.1	55.7	43.4
167.3	107.9	82.7	124.9
139.5	125.6	83.5	58.6
142.7	207.2	43.3	115.6
174.0	246.2	38.2	67.0
138.3	144.9	50.1	15.7
98.9	180.4	12.5	125.0
196.6	137.8	179.7	105.8
57.3	201.8		55.5
	84.2		

DATA **15. *Sports Team Involvement*** The table shows the number of students who played on a sports team for a random sample of female students in grades 9 through 12. At $\alpha = 0.01$, can you reject the claim that the mean numbers of female students who played on a sports team are equal for all grades? *(Adapted from U.S. Centers for Disease Control and Prevention)*

Grade 9	82	91	53	133	64	212
Grade 10	125	51	58	206	87	77
Grade 11	187	81	46	56	115	
Grade 12	42	77	165	65	49	102

16. (a) H_0: $\mu_1 = \mu_2 = \mu_3 = \mu_4$ (claim)

H_a: At least one mean is different from the others.

(b) d.f.$_N$ = 3; d.f.$_D$ = 33; $F_0 = 4.51$

Rejection region: $F > 4.51$

(c) 1.41

(d) Fail to reject H_0. There is not enough evidence at the 1% level to reject the claim that the mean amounts spent are equal for all regions.

17. Fail to reject all null hypotheses. The interaction between advertising medium and the length of the ad has no effect on the rating and therefore there is no significant difference in the means of the ratings.

18. Type of vehicle has an effect on the mean number of vehicles sold in a month. There is an interaction effect between type of vehicle and gender on the number of vehicles sold in a month.

DATA **16. *Amount Spent on Energy*** The table shows the amount spent (in dollars) on energy in one year for a random sample of households from four regions of the United States. At $\alpha = 0.01$, can you reject the claim that the mean amounts spent are equal for all regions? *(Adapted from U.S. Energy Information Administration)*

Northeast	Midwest	South	West
2623	1310	1252	947
573	940	1997	1801
2142	1342	1678	1706
1146	1294	1105	1244
853	1190	1593	1345
1073	1514	1667	768
2193	689	1367	1109
2140	1025	1777	963
826	1100		1598
2542			
801			

Extending Concepts

Using Technology to Perform a Two-Way ANOVA Test In Exercises 17–20, use a technology tool and the given block design to perform a two-way ANOVA test. Use $\alpha = 0.10$. Interpret the results.

DATA **17. *Advertising*** A study was conducted in which a random sample of 20 adults was asked to rate the effectiveness of advertisements. Each adult rated a radio or television advertisement that lasted 30 or 60 seconds. The block design shows these ratings (on a scale of 1 to 5, with 5 being extremely effective).

Length of ad	Advertising medium: Radio	Advertising medium: Television
30 sec	2, 3, 5, 1, 3	3, 5, 4, 1, 2
60 sec	1, 4, 2, 2, 5	2, 5, 3, 4, 4

DATA **18. *Vehicle Sales*** The owner of a car dealership wants to determine if the gender of a salesperson and the type of vehicle sold affect the number of vehicles sold in a month. The block design shows the number of cars, listed by type, sold in a month by a random sample of eight salespeople.

Gender	Type of vehicle: Car	Truck	Van/SUV
Male	6, 5, 4, 5	2, 2, 1, 3	4, 3, 4, 2
Female	5, 7, 8, 7	1, 0, 1, 2	4, 2, 0, 1

19. Fail to reject all null hypotheses. The interaction between age and gender has no effect on GPA and therefore there is no significant difference in the means of the GPAs.

20. There is an interaction effect between technician and brand on the mean time it takes to repair a disk drive.

21. $CV_{\text{Scheffé}} = 6.60$

$(1, 2) \rightarrow 4.927 \rightarrow$ No difference
$(1, 3) \rightarrow 0.650 \rightarrow$ No difference
$(1, 4) \rightarrow 4.426 \rightarrow$ No difference
$(2, 3) \rightarrow 1.998 \rightarrow$ No difference
$(2, 4) \rightarrow 0.013 \rightarrow$ No difference
$(3, 4) \rightarrow 1.684 \rightarrow$ No difference

22. $CV_{\text{Scheffé}} = 6.660$

$(1, 2) \rightarrow 7.212 \rightarrow$ Significant difference
$(1, 3) \rightarrow 3.519 \rightarrow$ No difference
$(1, 4) \rightarrow 0.260 \rightarrow$ No difference
$(2, 3) \rightarrow 0.724 \rightarrow$ No difference
$(2, 4) \rightarrow 4.782 \rightarrow$ No difference
$(3, 4) \rightarrow 1.866 \rightarrow$ No difference

23. $CV_{\text{Scheffé}} \approx 13.14$

$(1, 2) \rightarrow 0.966 \rightarrow$ No difference
$(1, 3) \rightarrow 9.587 \rightarrow$ No difference
$(1, 4) \rightarrow 7.897 \rightarrow$ No difference
$(2, 3) \rightarrow 16.979 \rightarrow$ Significant difference
$(2, 4) \rightarrow 14.892 \rightarrow$ Significant difference
$(3, 4) \rightarrow 0.130 \rightarrow$ No difference

24. $CV_{\text{Scheffé}} = 6.930$

$(1, 2) \rightarrow 0.148 \rightarrow$ No difference
$(1, 3) \rightarrow 7.842 \rightarrow$ Significant difference
$(1, 4) \rightarrow 2.581 \rightarrow$ No difference
$(2, 3) \rightarrow 5.311 \rightarrow$ No difference
$(2, 4) \rightarrow 3.678 \rightarrow$ No difference
$(3, 4) \rightarrow 17.877 \rightarrow$ Significant difference

DATA

19. ***Grade Point Average*** A study was conducted in which a random sample of 24 high school students was asked to give their grade point average (GPA). The block design shows the GPAs of male and female students from four different age groups.

	Age: Under 15	15–16	17–18	Over 18
Gender: Male	2.5, 2.1, 3.8	4.0, 1.4, 2.0	3.5, 2.2, 2.0	3.1, 0.7, 2.8
Gender: Female	4.0, 2.1, 1.9	3.5, 3.0, 2.1	4.0, 2.2, 1.7	1.6, 2.5, 3.6

DATA

20. ***Disk Drive Repairs*** The manager of a computer repair service wants to determine whether there is a difference in the time it takes four technicians to repair different brands of disk drives. The block design shows the times (in minutes) it took for each technician to repair three different brands of disk drives.

	Technician: Technician 1	Technician 2	Technician 3	Technician 4
Brand: Brand A	67, 82, 64	42, 56, 39	69, 47, 38	70, 44, 50
Brand: Brand B	44, 62, 55	47, 58, 62	55, 45, 66	47, 29, 40
Brand: Brand C	47, 36, 68	39, 74, 51	74, 80, 70	45, 62, 59

The Scheffé Test If the null hypothesis is rejected in a one-way ANOVA test of three or more means, a **Scheffé Test** can be performed to find which means have a significant difference. In a Scheffé Test, the means are compared two at a time. For example, with three means you would have the following comparisons: $\bar{x}_1$ versus $\bar{x}_2$, $\bar{x}_1$ versus $\bar{x}_3$, and $\bar{x}_2$ versus $\bar{x}_3$. For each comparison, calculate

$$\frac{(\bar{x}_a - \bar{x}_b)^2}{\dfrac{SS_w}{\Sigma(n_i - 1)}[(1/n_a) + (1/n_b)]}$$

where $\bar{x}_a$ and $\bar{x}_b$ are the means being compared and n_a and n_b are the corresponding sample sizes. Then compare the value with the critical value. Calculate the critical value using the same steps as in a one-way ANOVA test and multiply the result by $k - 1$. Use this information to solve Exercises 21–24.

21. Refer to the data in Exercise 10. At $\alpha = 0.10$, perform a Scheffé Test to determine which means have a significant difference.

22. Refer to the data in Exercise 11. At $\alpha = 0.10$, perform a Scheffé Test to determine which means have a significant difference.

23. Refer to the data in Exercise 14. At $\alpha = 0.01$, perform a Scheffé Test to determine which means have a significant difference.

24. Refer to the data in Exercise 13. At $\alpha = 0.01$, perform a Scheffé Test to determine which means have a significant difference.

Uses and Abuses
Statistics in the Real World

Uses

One-Way Analysis of Variance (ANOVA) ANOVA can help you make important decisions about the allocation of resources. For instance, suppose you work for a large manufacturing company and part of your responsibility is to determine the distribution of the company's sales throughout the world and decide where to focus the company's efforts. Because wrong decisions will cost your company money, you want to make sure that you make the right decisions.

Abuses

Preconceived Notions There are several ways that the tests presented in this chapter can be abused. For example, it is easy to allow preconceived notions to affect the results of a chi-square goodness-of-fit test and a test for independence. When testing to see whether a distribution has changed, do not let the existing distribution "cloud" the study results. Similarly, when determining whether two variables are independent, do not let your intuition "get in the way." As with any hypothesis test, you must properly gather appropriate data and perform the corresponding test before you can reach a logical conclusion

Incorrect Interpretation of Rejection of Null Hypothesis It is important to remember that when you reject the null hypothesis of an ANOVA test, you are simply stating that you have enough evidence to determine that at least one of the population means is different from the others. You are not finding them all to be different. One way to further test which of the population means differs from the others is explained in Extending Concepts of Section 10.4 Exercises.

Exercises

1. ***Preconceived Notions*** ANOVA depends on having independent variables. Describe an abuse that might occur by having dependent variables. Then describe how the abuse could be avoided.
2. ***Incorrect Interpretation of Rejection of Null Hypothesis*** Find an example of the use of ANOVA. In that use, describe what would be meant by "rejection of the null hypothesis." How should rejection of the null hypothesis be correctly interpreted?

10 Chapter Summary

What did you learn? | Review Exercises

Section 10.1

- How to use the chi-square distribution to test whether a frequency distribution fits a claimed distribution — *1–4*

$$\chi^2 = \Sigma \frac{(O - E)^2}{E}$$

Section 10.2

- How to use a contingency table to find expected frequencies and how to use a chi-square distribution to test whether two variables are independent — *5–8*

$$E_{r,c} = \frac{(\text{Sum of row } r) \times (\text{Sum of column } c)}{\text{Sample size}}$$

Section 10.3

- How to interpret the F-distribution and use an F-table to find critical values — *9–12*

$$F = \frac{s_1^2}{s_2^2}$$

- How to perform a two-sample F-test to compare two variances — *13–18*

Section 10.4

- How to use one-way analysis of variance to test claims involving three or more means — *19, 20*

$$F = \frac{MS_B}{MS_W}$$

10 Review Exercises

1. Reject H_0. There is enough evidence at the 5% level to conclude that the distribution of office visits differs from the claimed distribution.
2. Fail to reject H_0. There is not enough evidence at the 10% level to conclude that the distribution of Internet advertisers differs from the claimed distribution.
3. Fail to reject H_0. There is not enough evidence at the 10% level to conclude that the distribution of prices differs from the claimed distribution.
4. Reject H_0. There is enough evidence at the 1% level to support the claim that the distribution of ages of the jury differs from the age distribution of available jurors.

Section 10.1

In Exercises 1–4, use a χ^2 goodness-of-fit test to test the claim about the population distribution. Interpret the decision in the context of the original claim.

1. A health care investigator wishes to test the following claim: Of all doctor's office visits in the United States, 25% are from new patients, 25% are from old patients with a new problem, and the remainder are old patients with a recurring problem. A random sample of various doctors finds that 97 patients were new, 142 were old with a new problem, and 457 were old patients with a recurring problem. Test the claim at $\alpha = 0.05$. *(Adapted from U.S. National Center for Health Statistics)*

2. A website design company wants to test the following claim: Of all Internet advertisers in the United States, 26% are retail goods and services, 18% are financial services, 15% are web media, and the remainder are from other categories. A random sample of Internet advertisers finds that 88 are retail goods and services, 73 are financial services, 66 are web media, and 138 are from other categories. Test the claim at $\alpha = 0.10$. *(Adapted from Nielsen// Net Ratings)*

3. An economist believes that the distribution of the percent of adults who would opt for fuel-efficient cars when gas prices reach certain levels (in dollars per gallon) is different from the one shown at the left below. The economist randomly selects 800 adults and asks each at what gas price they would opt for a more fuel-efficient car. The results are shown at the right below. Test the economist's claim at $\alpha = 0.10$. *(Adapted from Harris Poll)*

Distribution	
Less than \$1.50	6%
\$1.50–\$1.99	22%
\$2.00	29%
\$2.01–\$2.49	5%
\$2.50 or more	38%

Survey Results	
Less than \$1.50	54
\$1.50–\$1.99	172
\$2.00	244
\$2.01–\$2.49	28
\$2.50 or more	302

4. A legal researcher is studying the age distribution of juries by comparing them with the overall age distribution of available jurors. The researcher claims that the jury distribution is different from the overall distribution; that is, there is a noticeable age bias in jury selection in this area. The table shows the number of jurors at a county court in one year and the percent of persons residing in that county, by age. Use the population distribution to find the expected juror frequencies. Test the researcher's claim at $\alpha = 0.01$.

	21–29	30–39	40–49	50–59	60 and above
Jury	45	128	244	224	359
Population	20.5%	21.7%	18.1%	17.3%	22.4%

5. See Odd Answers, page A84.
6. See Selected Answers, page A100.
7. See Odd Answers, page A84.
8. See Selected Answers, page A100.

Section 10.2

In Exercises 5–8, use the given contingency table to (a) find the expected frequencies of each cell in the table, (b) perform a chi-square test for independence, and (c) comment on the relationship between the two variables. Assume the variables are independent.

5. The table shows the highest level of education of people in the United States by age category in a recent year. The numbers listed are in thousands of persons. Use $\alpha = 0.10$. *(Source: U.S. Census Bureau)*

	H.S.—did not complete	H.S. completed	College 1–3 years	College 4 or more years
25–44	10,102	26,980	23,014	23,123
45 and older	18,782	30,914	20,169	20,636

6. The contingency table shows the results of a random sample of 480 individuals classified by gender and type of vehicle owned. Use $\alpha = 0.05$.

	Type of vehicle owned			
Gender	**Car**	**Truck**	**SUV**	**Van**
Male	85	96	45	6
Female	110	75	60	3

7. The table shows the age distribution of a random sample of fatally injured male and female passenger vehicle drivers whose blood alcohol content was at least 0.08 in a recent year. Use $\alpha = 0.05$. *(Adapted from Insurance Institute for Highway Safety)*

	Age Group					
Gender	**16–20**	**21–30**	**31–40**	**41–50**	**51–60**	**61+**
Male	155	280	265	225	155	50
Female	60	145	145	130	65	15

8. The table shows the distribution of a random sample of fatal pedestrian motor vehicle collisions by the time of day and gender in a recent year. Use $\alpha = 0.10$. *(Adapted from Insurance Institute for Highway Safety)*

	Time of day			
Gender	**12 A.M.–5:59 A.M.**	**6 A.M.–11:59 A.M.**	**12 P.M.–5:59 P.M.**	**6 P.M.–11:59 P.M.**
Male	654	490	654	1471
Female	308	231	308	692

9. 2.295

10. 4.71

11. 2.39

12. 2.01

13. Fail to reject H_0. There is not enough evidence at the 1% level to reject the claim.

14. Fail to reject H_0. There is not enough evidence at the 10% level to support the claim.

15. Fail to reject H_0. There is not enough evidence at the 10% level to support the claim that the variation in wheat production is greater in Garfield County than in Kay County.

16. Reject H_0. There is enough evidence at the 5% level to conclude that the yield strength of the nontempered couplings is more variable than that of the tempered couplings.

17. Fail to reject H_0. There is not enough evidence at the 1% level to support the claim that the test score variance for females is different from that for males.

Section 10.3

In Exercises 9–12, find the critical F-value for a right-tailed test using the indicated level of significance α and degrees of freedom d.f._N and d.f._D.

9. $\alpha = 0.05, \text{d.f.}_N = 6, \text{d.f.}_D = 50$

10. $\alpha = 0.01, \text{d.f.}_N = 12, \text{d.f.}_D = 10$

11. $\alpha = 0.10, \text{d.f.}_N = 5, \text{d.f.}_D = 12$

12. $\alpha = 0.05, \text{d.f.}_N = 20, \text{d.f.}_D = 25$

In Exercises 13 and 14, test the claim about the differences between two population variances σ_1^2 and σ_2^2 at the indicated level of significance α using the given sample statistics. Assume the sample statistics are from independent samples that are randomly selected and each population has a normal distribution.

13. Claim: $\sigma_1^2 \leq \sigma_2^2$, $\alpha = 0.01$. Sample statistics: $s_1^2 = 653$, $n_1 = 16$ and $s_2^2 = 270$, $n_2 = 21$

14. Claim: $\sigma_1^2 \neq \sigma_2^2$, $\alpha = 0.10$. Sample statistics: $s_1^2 = 112{,}676$, $n_1 = 6$ and $s_2^2 = 49{,}572$, $n_2 = 11$

In Exercises 15–18, test the claim about two population variances at the indicated level of significance α. Interpret the results in the context of the claim.

15. An agricultural analyst is comparing the wheat production in Oklahoma counties. The analyst claims that the variation in wheat production is greater in Garfield County than in Kay County. A random sample of 21 Garfield County farms yields a standard deviation of 0.76 bushel per acre; 16 Kay County farms are found to have a standard deviation of 0.58 bushel per acre. Diagnostics suggest that wheat production is normally distributed in both counties. Test the analyst's claim at $\alpha = 0.10$. *(Adapted from Environmental Verification and Analysis Center—University of Oklahoma)*

16. A steel pipe fittings company claims that the yield strength of its nontempered couplings is more variable than that of its tempered couplings. A random sample of nine tempered couplings has a standard deviation of 13.1 megapascals, and a similar sample of nine nontempered couplings has a standard deviation of 25.4 megapascals. From past data, it is known that the company's production process results in normally distributed yield strengths. Test the company's claim at $\alpha = 0.05$.

DATA

17. The table shows the SAT verbal test scores for nine randomly selected female students and 14 randomly selected male students. Assume that SAT verbal test scores are normally distributed. At $\alpha = 0.01$, test the claim that the test score variance for females is different from that for males.

Female	480	610	340	630	520	690	540
Male	560	680	360	530	380	460	630

Female	600	800					
Male	310	730	740	520	560	400	510

18. Reject H_0. There is enough evidence at the 5% level to support the claim that the new mold produces inserts that are less variable in diameter than the current mold.

19. Reject H_0. There is enough evidence at the 10% level to conclude that the mean residential energy expenditures are not the same for all four regions.

20. Fail to reject H_0. There is not enough evidence at the 5% level to conclude that the average annual income is different for the four regions.

DATA **18.** A plastics company that produces automobile dashboard inserts has just received a new injection mold that is supposedly more consistent than its current mold. A quality technician wishes to test whether this new mold will produce inserts that are less variable in diameter than those produced with the company's current mold. The table shows independent random samples (of size 12) of insert diameters (in centimeters) for both the current and new molds. Assume that the samples are from normally distributed populations. At $\alpha = 0.05$, test the claim that the new mold produces inserts that are less variable in diameter than the current mold produces.

New	9.611	9.618	9.594	9.580	9.611	9.597
Current	9.571	9.642	9.650	9.651	9.596	9.636

New	9.638	9.568	9.605	9.603	9.647	9.590
Current	9.570	9.537	9.641	9.625	9.626	9.579

Section 10.4

In Exercises 19 and 20, use the given sample data to perform a one-way ANOVA test using the indicated level of significance α. What can you conclude?

DATA **19.** The table at the right shows the residential energy expenditures (in dollars) in one year of a random sample of households in four regions of the United States Diagnostics suggest that the sample is drawn from a normally distributed population. Use $\alpha = 0.10$ to test for differences among the means for the four regions. *(Adapted from U.S. Energy Information Administration)*

Northeast	Midwest	South	West
2088	759	888	1115
1259	1346	773	898
1762	975	1141	1605
1783	1523	1232	857
1623	1233	1310	465
2175	1903	953	359
1265	1236	1005	909
1258	1176	1475	1193

DATA **20.** The table at the right shows the annual income (in dollars) of a random sample of households in four regions of the United States Diagnostics suggest that the sample is drawn from an approximately normally distributed population. Use $\alpha = 0.05$ to test for differ-ences among the means for the four regions. *(Adapted from U.S. Census Bureau)*

Northeast	Midwest	South	West
50,533	27,175	37,348	43,970
41,798	50,788	61,090	36,378
50,661	48,847	25,393	32,587
26,789	29,602	35,098	30,665
47,620	27,152	23,261	
	42,364	47,798	
	9,078		

10 Chapter Quiz

Take this quiz as you would take a quiz in class. After you are done, check your work against the answers given in the back of the book.

For each Exercise,

(a) state H_0 and H_a.
(b) specify the level of significance.
(c) determine the critical value.
(d) determine the rejection region.
(e) calculate the test statistic.
(f) make a decision.
(g) interpret the results in the context of the problem.

For Exercises 1 and 2, use the following data. The data list the annual wages (in thousands of dollars) for randomly selected individuals from three metropolitan areas. Assume the wages are normally distributed and that the samples are independent. *(Adapted from U.S. Bureau of Labor Statistics)*

San Jose, CA: 48.6, 61.9, 19.8, 35.2, 25.5, 58.4, 47.4, 48.6, 17.6, 57.6, 81.1, 26.1, 76.8

Dallas, TX: 32.9, 46.4, 19.7, 24.0, 19.7, 18.4, 25.2, 10.3, 26.9, 20.5, 52.6, 40.7, 24.4, 36.5, 33.9, 28.0, 35.3, 25.5

Ann Arbor, MI: 22.7, 36.0, 20.7, 26.8, 24.0, 49.2, 29.5, 28.8, 38.8, 35.8, 23.1, 27.6, 30.8, 33.5, 23.3, 38.7

DATA

1. At $\alpha = 0.01$, is there enough evidence to conclude that the variances in annual wages for San Jose, CA and Dallas, TX are different?

2. Are the mean annual wages equal for all three cities? Use $\alpha = 0.10$. Assume that the population variances are equal.

For Exercises 3 and 4, use the following table. The table lists the distribution of educational achievement for people in the United States aged 25 and older. It also lists the results of a random survey for two additional age categories. *(Adapted from U.S. Census Bureau)*

	25 and older	35–44	65–74
Not a H.S. graduate	16.6%	37	124
High school graduate	33.3%	102	148
Some college, no degree	17.3%	59	61
Associate degree	7.5%	27	15
Bachelor's degree	17.0%	51	37
Advanced degree	8.3%	25	22

3. Does the distribution for people in the United States aged 25 and older describe people in the United States aged 35–44? Use $\alpha = 0.01$.

4. Does the distribution for people in the United States aged 25 and older describe people in the United States aged 65–74? Use $\alpha = 0.05$.

1. (a) H_0: $\sigma_1^2 = \sigma_2^2$
H_a: $\sigma_1^2 \neq \sigma_2^2$ (claim)
(b) 0.01 (c) $F_0 = 3.97$
(d) Rejection region: $F > 3.97$
(e) 3.813 (f) Fail to reject H_0.
(g) There is not enough evidence at the 1% level to conclude that the variances in annual wages for San Jose, CA and Dallas, TX are different.

2. (a) H_0: $\mu_1 = \mu_2 = \mu_3$ (claim)
H_a: At least one mean is different from the others.
(b) 0.10 (c) $F_0 = 2.43$
(d) Rejection region: $F > 2.43$
(e) 7.39 (f) Reject H_0.
(g) There is enough evidence at the 10% level to conclude that the mean annual wages for the three cities are not all equal.

3. (a) Claimed distribution:

Education	25 & Over
Not a HS graduate	16.6%
HS graduate	33.3%
Some college, no degree	17.3%
Associate degree	7.5%
Bachelor's degree	17.0%
Advanced degree	8.2%

H_0: Distribution of educational achievement for people in the United States aged 35–44 is as shown in table above.

H_a: Distribution of educational achievement for people in the United States aged 35–44 differs from the claimed distribution.

(b) 0.01 (c) $\chi_0^2 = 15.086$
(d) Rejection region: $\chi^2 > 15.086$
(e) 5.189 (f) Fail to reject H_0.
(g) There is not enough evidence at the 1% level to conclude the distribution of educational achievement for people in the United States aged 35–44 differs from the claimed distribution.

4. See Odd Answers, page A84.

PUTTING IT ALL TOGETHER

Real Statistics ■ Real Decisions

The National Fraud Information Center (NFIC) was established in 1992 by the National Consumers League (NCL) to combat the growing problem of telemarketing and Internet fraud by improving prevention and enforcement. NCL works to protect and promote the economic and social interests of consumers in the United States, using education, research, science, and publications. The NCL was formed in 1899.

You work for the NFIC as a statistical analyst. You are studying data on telemarketing fraud. Part of your analysis involves testing the goodness of fit, testing for independence, comparing variances, and performing ANOVA.

NATIONAL FRAUD INFORMATION CENTER

www.fraud.org

Exercises

1. ***Goodness of Fit*** A claimed distribution for the ages of telemarketing fraud victims is shown in the table at the right. The results of a survey of 1000 randomly selected telemarketing fraud victims are also shown in the table. Using $\alpha = 0.01$, perform a chi-square goodness-of-fit test to test the claimed distribution. What can you conclude? Do you think the claimed distribution is valid? Why or why not?

Ages	Claimed distribution	Survey results
Under 20	1%	30
20–29	13%	200
30–39	16%	300
40–49	19%	270
50–59	16%	150
60–69	13%	40
70+	22%	10

Table for Exercise 1

2. ***Independence*** The following contingency table shows the results of a random sample of 2000 telemarketing fraud victims classified by age and type of fraud. The frauds were committed using bogus sweepstakes or credit card offers.

   (a) Calculate the expected frequency for each cell in the contingency table. Assume the variables age and type of fraud are independent.

   (b) Can you conclude that the age of the victims is related to the type of fraud? Use $\alpha = 0.01$.

Type of Fraud	Age								
	Under 20	20–29	30–39	40–49	50–59	60–69	70–79	80+	Total
Sweepstakes	10	60	70	130	90	160	280	200	1000
Credit cards	20	180	260	240	180	70	30	20	1000
Total	30	240	330	370	270	230	310	220	2000

TECHNOLOGY | MINITAB | EXCEL | TI-83

Teacher Salaries

In 1916, the American Federation of Teachers (AFT) was formed by teachers from three Chicago, Illinois unions and one Gary, Indiana union. Today, the AFT represents over 1.3 million teachers, higher education faculty and staff, school support staff, state and municipal employees, and health care professionals.

Each year, the AFT publishes the *Survey and Analysis of Teacher Salary Trends*. This report is divided into four sections—the first focuses on state salary comparisons, the second on trends in national salary averages over the past few decades, the third on beginning teachers' salaries, and the fourth on salary data for school districts in the nation's 100 largest cities.

The table at the right shows the salaries of a random sample of teachers from California, Ohio, and Wyoming.

Teacher Salaries		
California	Ohio	Wyoming
58,245	40,500	32,500
47,622	41,200	55,422
38,000	37,650	30,000
56,500	52,665	29,840
42,750	59,600	39,650
59,800	36,000	33,000
65,000	39,780	42,000
50,000	45,225	50,200
58,978	42,100	31,200
55,473	40,500	33,560
49,995	39,650	35,555
52,600	60,000	39,660
59,800	32,560	41,585
64,000	47,226	29,850
62,005	48,600	33,440
48,800	45,000	48,186

Exercises

In Exercises 1–3, refer to the following samples. Use $\alpha = 0.05$.

(a) California teachers

(b) Ohio teachers

(c) Wyoming teachers

1. Are the samples independent of each other? Explain.

2. Use a technology tool to determine whether each sample is from a normal, or approximately normal, population.

3. Use a technology tool to determine whether the samples were selected from populations having equal variances.

4. Using the results of Exercises 1–3, discuss whether the three conditions for a one-way ANOVA test are satisfied. If so, use a technology tool to test the claim that teachers from California, Ohio, and Wyoming have the same mean salary. Use $\alpha = 0.05$.

5. Repeat Exercises 1–4 using the data in the table below. The table displays the salaries of a random sample of teachers from Alaska, Nevada, and New York.

Teacher Salaries		
Alaska	Nevada	New York
58,750	67,620	70,006
49,875	50,950	79,000
66,500	32,000	81,000
48,900	40,756	28,000
31,560	62,000	39,650
43,268	40,074	31,681
56,985	40,005	28,560
50,285	36,652	46,000
48,632	44,878	85,000
44,652	49,659	30,050
43,669	58,284	82,000
34,560	40,665	32,008
36,985	30,899	37,614
47,895	42,705	75,750
44,662	40,202	29,999
77,270	36,587	40,002

Extended solutions are given in the *Technology Supplement*. Technical instruction is provided for MINITAB, Excel, and the TI-83.

CHAPTER 11

Nonparametric Tests

11.1 The Sign Test

11.2 The Wilcoxon Tests

Case Study

11.3 The Kruskal-Wallis Test

11.4 Rank Correlation

11.5 The Runs Test

Uses and Abuses

Real Statistics–Real Decisions

Technology

The space shuttle *Challenger* flew nine successful missions before it exploded on January 28, 1986. The explosion was caused by the failure of an O-ring seal.

Where You've Been

Up to this point in the text, you have studied dozens of different statistical formulas and tests that can help you in a decision-making process. Knowing how to apply various statistical tests is important. However, it is even more important that you always use critical thinking when using a statistical test. Using critical thinking when applying statistical techniques can help prevent costly mistakes and tragic errors.

The explosion of the space shuttle *Challenger* is a classic example of how tragic results might have been avoided by using correct statistical techniques. Seven astronauts died because two large rubber O-rings leaked during takeoff. The rings leaked because of the low temperature at the time of takeoff.

The correlation between O-ring failure and air temperature was known before the time of takeoff. In fact, an engineering group had recommended that the flight be delayed. In their argument, however, the engineers failed to present the data clearly. How would you present the following data?

Temperature (°C)	12	14	14	17	19	19	19	19	20	21	21	21
Damage Index	11	4	4	2	0	0	0	0	0	4	0	4

Temperature (°C)	21	21	22	23	24	24	24	24	26	26	27
Damage Index	0	0	0	0	4	0	0	0	0	0	0

Where You're Going

In this chapter, you will study additional statistical tests. Each of these has usefulness in real-life applications. Remember, however, that no statistical test, including those in this chapter, can be applied blindly. Moreover, the more important your decision, the more important it is that you look at the data from several perspectives.

With the O-ring data above, the air temperature T and the damage index D can be related by the regression line $D = 10.280 - 0.426T$. The correlation coefficient, however, is relatively weak ($r \approx -0.6$). The P-value is 0.001, which would indicate that the correlation is significant, but the D-values do not pass the normality requirement.

So, although a simple correlation test might indicate a relationship between air temperature and O-ring damage, one might question the results because the data do not fit the requirements for the test. Similar tests you will study in this chapter, such as Spearman's rank correlation test, will give you additional information. If you had seen the scatter plot at the right before the *Challenger's* takeoff, what decision would you have made? Is the relationship between air temperature and O-ring damage evident enough for you to have postponed the flight? (*Note:* The air temperature at takeoff on January 28, 1986 was 0° Celsius, and the O-ring temperature was about 6° colder.) *(Source: Tufte, E.R., Visual Explanations)*

Temperature and O-ring Damage for 23 Space Shuttle Flights

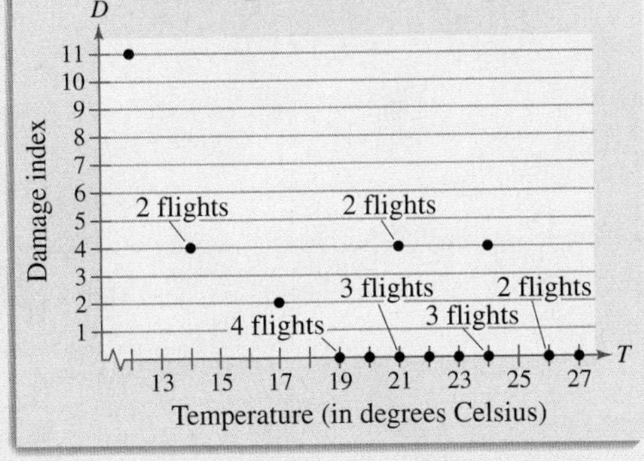

11.1 The Sign Test

What You Should Learn

- How to use the sign test to test a population median
- How to use the sign test to test the difference between two population medians (dependent samples)

The Sign Test for a Population Median • The Paired-Sample Sign Test

The Sign Test for a Population Median

Many of the hypothesis tests studied so far have imposed one or more requirements for a population distribution. For example, some tests require that a population must have a normal distribution, and other tests require that population variances be equal. What if, for a given test, such requirements cannot be met? For these cases, statisticians have developed hypothesis tests that are "distribution free." Such tests are called nonparametric tests.

DEFINITION

A **nonparametric test** is a hypothesis test that does not require any specific conditions concerning the shape of populations or the value of any population parameters.

Nonparametric tests are usually easier to perform than corresponding parametric tests. However, they are usually less efficient than parametric tests. Stronger evidence is required to reject a null hypothesis using the results of a nonparametric test. Consequently, whenever possible, you should use a parametric test. One of the easiest nonparametric tests to perform is the sign test.

Insight

For many nonparametric tests, statisticians test the median instead of the mean.

DEFINITION

The **sign test** is a nonparametric test that can be used to test a population median against a hypothesized value k.

The sign test for a population median can be left tailed, right tailed, or two tailed. The null and alternative hypotheses for each type of test are as follows.

Left-tailed test:

H_0: median $\geq k$ and H_a: median $< k$

Right-tailed test:

H_0: median $\leq k$ and H_a: median $> k$

Two-tailed test:

H_0: median $= k$ and H_a: median $\neq k$

To use the sign test, first compare each entry in the sample with the hypothesized median k. If the entry is below the median, assign it a $-$ sign; if the entry is above the median, assign it a $+$ sign; and if the entry is equal to the median, assign it a 0. Then compare the number of $+$ and $-$ signs. (The 0s are ignored.) If there is a large difference between the number of $+$ signs and the number of $-$ signs, it is likely that the median is different from the hypothesized value and the null hypothesis should be rejected.

Insight

Because the 0s are ignored, there are two possible outcomes when comparing a data entry with a hypothesized median: a + or a − sign. If the median is k, then about half of the values will be above k and half will be below. As such, the probability for each sign is 0.5. Table 8 in Appendix B is constructed using the binomial distribution where $p = 0.5$.

When $n > 25$, you can use the normal approximation (with a correction for continuity) for the binomial. In this case, use $\mu = np = 0.5n$ and $\sigma = \sqrt{npq} = \frac{\sqrt{n}}{2}$.

Table 8 in Appendix B lists the critical values for the sign test for selected levels of significance and sample sizes. When the sign test is used, the sample size n is the total number of + and − signs. If the sample size is greater than 25, you can use the standard normal distribution to find the critical values.

Test Statistic for the Sign Test

When $n \leq 25$, the test statistic x for the sign test is the smaller number of + or − signs.

When $n > 25$, the test statistic for the sign test is

$$z = \frac{(x + 0.5) - 0.5n}{\frac{\sqrt{n}}{2}}$$

where x is the smaller number of + or − signs and n is the sample size, i.e., the total number of + and − signs.

Because x is defined to be the smaller number of + or − signs, the rejection region is always in the left tail. Consequently, the sign test for a population median is always a left-tailed test or a two-tailed test. When the test is two tailed, use only the left-tailed critical value. (If x is defined to be the larger number of + or − signs, the rejection region is always in the right tail. Right-tailed sign tests are presented in the exercises.)

GUIDELINES

Performing a Sign Test for a Population Median

In Words	*In Symbols*
1. State the claim. Identify the null and alternative hypotheses.	State H_0 and H_a.
2. Specify the level of significance.	Identify α.
3. Determine the sample size n by assigning + signs and − signs to the sample data.	n = total number of + and − signs
4. Determine the critical value.	If $n \leq 25$, use Table 8 in App. B. If $n > 25$, use Table 4 in App. B.
5. Calculate the test statistic.	If $n \leq 25$, use x. If $n > 25$, use $z = \frac{(x + 0.5) - 0.5n}{\frac{\sqrt{n}}{2}}$.
6. Make a decision to reject or fail to reject the null hypothesis.	If x or z is less than or equal to the critical value, reject H_0. Otherwise, fail to reject H_0.
7. Interpret the decision in the context of the original claim.	

EXAMPLE 1

Using the Sign Test

A bank manager claims that the median number of customers per day is no more than 750. A teller doubts the accuracy of this claim. The number of bank customers per day for 16 randomly selected days are listed below. At $\alpha = 0.05$, can the teller reject the bank manager's claim?

775	765	801	742
754	753	739	751
745	750	777	769
756	760	782	789

SOLUTION The teller must disprove the bank manager's claim that "the median number of customers per day is no more than 750." So, the null and alternative hypotheses are

H_0: median ≤ 750 (Claim) and H_a: median > 750.

The table below shows the results of comparing each data entry with the hypothesized median 750.

+	+	+	−
+	+	−	+
−	0	+	+
+	+	+	+

From the table, you can see that there are 3 − signs and 12 + signs. So, $n = 12 + 3 = 15$. Using Table 8 with $\alpha = 0.05$ (one tailed) and $n = 15$, you find that the critical value is 3. Because $n \leq 25$, the test statistic x is the smaller number of + or − signs. So, $x = 3$. Because $x = 3$ is equal to the critical value, the teller should reject the null hypothesis.

Interpretation At the 5% level, the teller can reject the bank manager's claim and conclude that the median number of customers per day is more than 750.

Try It Yourself 1

A supermarket manager claims that the median number of customers per day is greater than 2500. A supplier wants to verify the accuracy of this claim. The number of customers per day for 24 randomly selected days are shown below. At $\alpha = 0.025$, can the supplier support the manager's claim?

Number of customers per day for 24 days							
2174	2491	2682	2510	2557	2418	2709	2562
2390	2467	2500	2205	2246	2054	2243	2627
1949	2500	2592	2567	2478	2348	2692	2580

a. *Identify* the claim and *state* H_0 and H_a.
b. *Specify* the level of significance α.
c. *Determine* the sample size n.
d. *Determine* the critical value.
e. *Calculate* the test statistic x.
f. *Decide* whether to reject the null hypothesis.
g. *Interpret* the decision in the context of the original claim.

Answer: Page A44

Picturing the World

The National Confectioners Association reports that people in the United States spend a lot of money on candy. In 2002, people in the United States spent a total of \$13.3 billion on candy. The U.S. Department of Commerce reported that in 2002, the average person in the United States ate about 22.2 pounds of candy.

Candy Consumption

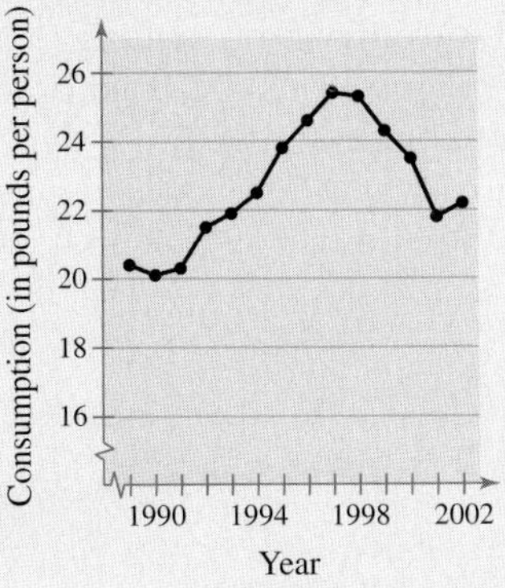

If you were to test the U.S. Department of Commerce's claim concerning per capita candy consumption, would you use a parametric test or a nonparametric test? What factors must you consider?

EXAMPLE 2

Using the Sign Test

A car dealership claims to give customers a median trade-in offer of at least \$6000. A random sample of 103 transactions revealed that the trade-in offer for 60 automobiles was less than \$6000 and the trade-in offer for 40 automobiles was more than \$6000. At $\alpha = 0.01$, can you reject the dealership's claim?

SOLUTION To reject the dealership's claim, you must disprove the claim that "the median trade-in offer is at least \$6000." The null and alternative hypotheses are

H_0: median ≥ 6000 (Claim) and H_a: median < 6000.

Because $n > 25$, you should use Table 4, the Standard Normal Table, to find the critical value. Because the test is a left-tailed test with $\alpha = 0.01$, the critical value is -2.33. Of the 103 transactions, there are 60 $-$ signs and 40 $+$ signs. When the zeros are ignored, the sample size is

$$n = 60 + 40 = 100$$

and

$$x = 40.$$

With these values, the test statistic is

$$z = \frac{(40 + 0.5) - 0.5(100)}{\sqrt{100}/2}$$

$$= \frac{-9.5}{5}$$

$$= -1.9.$$

The graph at the right shows the location of the rejection region and the test statistic z. Because z is greater than the critical value, it is not in the rejection region, and you should fail to reject the null hypothesis.

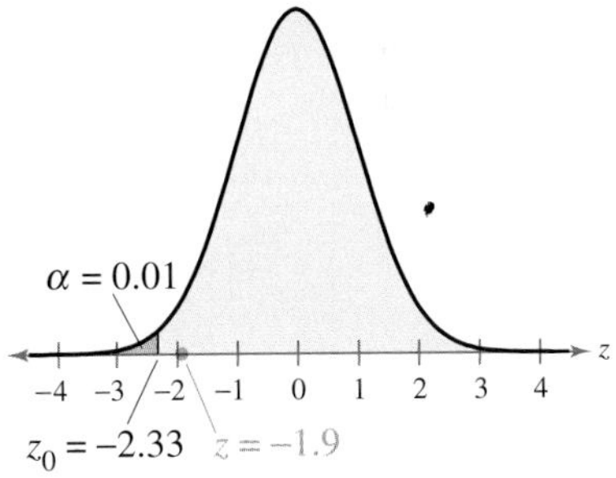

Interpretation At the 1% level of significance, you cannot reject the dealership's claim that the median trade-in offer is at least \$6000.

Try It Yourself 2

A realtor claims that the median sales price of houses sold in a certain region is \$134,500. A random sample of 85 house sales reveals that 30 houses were sold for less than \$134,500 and 51 houses were sold for more than \$134,500. At $\alpha = 0.10$, can you reject the realtor's claim?

a. *Identify* the claim and *state* H_0 and H_a.
b. *Specify* the level of significance α.
c. *Determine* the sample size n.
d. *Determine* the critical value.
e. *Calculate* the test statistic z.
f. *Decide* whether to reject the null hypothesis.
g. *Interpret* the decision in the context of the original claim. *Answer: Page A44*

Study Tip

When performing a two-tailed sign test, remember to use only the left-tailed critical value.

The Paired-Sample Sign Test

In Section 8.3, you learned how to use a t-test for the difference between means of dependent samples. That test required both populations to be normally distributed. If the parametric condition of normality cannot be satisfied, you can use the sign test to test the difference between two population medians. To perform the paired-sample sign test for the difference between two population medians, the following conditions must be met.

1. A sample must be randomly selected from each population.
2. The samples must be dependent (paired).

The paired-sample sign test can be left tailed, right tailed, or two tailed. This test is similar to the sign test for a single population median. However, instead of comparing each data entry with a hypothesized median and recording a $+$, $-$, or 0, you find the difference between corresponding data entries and record the sign of the difference. Generally, to find the difference, subtract the entry representing the second variable from the entry representing the first variable. Then compare the number of $+$ and $-$ signs. (The 0s are ignored.) If the number of $+$ signs is approximately equal to the number of $-$ signs, the null hypothesis should not be rejected. If, however, there is a significant difference between the number of $+$ signs and the number of $-$ signs, the null hypothesis should be rejected.

GUIDELINES

Performing a Paired-Sample Sign Test

In Words	*In Symbols*
1. Identify the claim. State the null and alternative hypotheses.	State H_0 and H_a.
2. Specify the level of significance.	Identify α.
3. Determine the sample size n by finding the difference for each data pair. Assign a $+$ sign for a positive difference, a $-$ sign for a negative difference, and a 0 for no difference.	$n =$ total number of $+$ and $-$ signs
4. Determine the critical value.	Use Table 8 in Appendix B.
5. Find the test statistic.	$x =$ lesser number of $+$ and $-$ signs
6. Make a decision to reject or fail to reject the null hypothesis.	If the test statistic is less than or equal to the critical value, reject H_0. Otherwise, fail to reject H_0.
7. Interpret the decision in the context of the original claim.	

EXAMPLE 3

Using the Paired-Sample Sign Test

A psychologist claims that the number of repeat offenders will decrease if first-time offenders complete a special rehabilitation course. You randomly select 10 prisons and record the number of repeat offenders during a two-year period. Then after first-time offenders complete the course, you record the number of repeat offenders at each prison for another two-year period. The results are shown in the following table. At $\alpha = 0.025$, can you support the psychologist's claim?

Prison	1	2	3	4	5	6	7	8	9	10
Before	21	34	9	45	30	54	37	36	33	40
After	19	22	16	31	21	30	22	18	17	21

SOLUTION To support the psychologist's claim, you could use the following null and alternative hypotheses.

H_0: The number of repeat offenders will not decrease.

H_a: The number of repeat offenders will decrease. (Claim)

The table below shows the sign of the differences between the "before" and "after" data.

Prison	1	2	3	4	5	6	7	8	9	10
Before	21	34	9	45	30	54	37	36	33	40
After	19	22	16	31	21	30	22	18	17	21
Sign	+	+	−	+	+	+	+	+	+	+

You can see that there is 1 − sign and there are 9 + signs. So, $n = 1 + 9 = 10$. In Table 8 with $\alpha = 0.025$ (one tailed) and $n = 10$, the critical value is 1. The test statistic x is the smaller number of + or − signs. So, $x = 1$. Because $x = 1$ is equal to the critical value, you should reject the null hypothesis.

Interpretation At the 2.5% level of significance, you can support the psychologist's claim that the number of repeat offenders will decrease.

Try It Yourself 3

A medical researcher claims that a new vaccine will decrease the number of colds in adults. You randomly select 14 adults and record the number of colds each has in a one-year period. After giving the vaccine to each adult, you again record the number of colds each has in a one-year period. The results are shown in the table at the left. At $\alpha = 0.05$, can you support the researcher's claim?

Adult	Before vaccine	After vaccine
1	3	2
2	4	1
3	2	0
4	1	1
5	3	1
6	6	3
7	4	3
8	5	2
9	2	2
10	0	2
11	2	3
12	5	4
13	3	3
14	3	2

a. *Identify* the claim and *state* H_0 and H_a.
b. *Specify* the level of significance α.
c. *Determine* the sample size n.
d. *Determine* the critical value.
e. *Find* the test statistic x.
f. *Decide* whether to reject the null hypothesis.
g. *Interpret* the decision in the context of the original claim. *Answer: Page A44*

11.1 Exercises

Help

MyMathLab

Student Study Pack

Building Basic Skills and Vocabulary

1. What is a nonparametric test? How does a nonparametric test differ from a parametric test? What are the advantages and disadvantages of using a nonparametric test?

2. Explain how to use the sign test to test a population median.

Using and Interpreting Concepts

Performing a Sign Test In Exercises 3–18, (a) write the claim mathematically and identify H_0 and H_a, (b) determine the critical value, (c) calculate the test statistic, (d) decide whether to reject or fail to reject the null hypothesis, and (e) interpret the decision in the context of the original claim.

3. ***Credit Card Charges*** In order to estimate the median amount of new credit card charges for the previous month, a financial service accountant randomly selects 12 credit card accounts and records the amount of new charges for each account for the previous month. The amounts are listed below. At $\alpha = 0.01$, can the accountant conclude that the median amount of new credit card charges for the previous month was more than \$200? *(Adapted from Board of Governors of the Federal Reserve System)*

\$246.71	\$282.59	\$155.03	\$102.17	\$209.80	\$165.88
\$199.41	\$170.83	\$196.54	\$216.46	\$145.92	\$209.47

4. ***Temperature*** A meteorologist estimates that the daily median temperature for the month of July in Pittsburgh is 73° Fahrenheit. The temperatures (in degrees Fahrenheit) for 15 randomly selected July days in Pittsburgh are listed below. At $\alpha = 0.01$, is there enough evidence to reject the meteorologist's hypothesis? *(Adapted from U.S. National Oceanic and Atmospheric Administration)*

64	72	72	75	76	70	66	73
67	65	74	74	67	77	66	

5. ***Sales Prices of Homes*** A real estate agent believes that the median sales price of new privately owned one-family homes sold in the past year is \$148,000 or less. The sales prices of eight randomly selected homes are listed below. At $\alpha = 0.05$, is there enough evidence to reject the agent's claim? *(Adapted from National Association of Realtors)*

\$178,000	\$141,450	\$97,500	\$152,600
\$83,800	\$169,000	\$157,000	\$168,000

6. ***Temperature*** During a weather report, a meteorologist states that the daily median temperature for the month of January in San Diego is 58° Fahrenheit. The temperatures (in degrees Fahrenheit) for 18 randomly selected January days in San Diego are listed below. At $\alpha = 0.01$, can you reject the meteorologist's claim? *(Adapted from U.S. National Oceanic and Atmospheric Administration)*

58	62	55	55	53	52	52	59	55
55	60	56	57	61	58	63	63	55

1. A nonparametric test is a hypothesis test that does not require any specific conditions concerning the shape of populations or the value of any population parameters.

   A nonparametric test is usually easier to perform than its corresponding parametric test, but the nonparametric test is usually less efficient.

2. Identify the claim and state H_0 and H_a. Identify the level of significance and sample size. Find the critical value using Table 8 (if $n \leq 25$) or Table 4 ($n > 25$). Calculate the test statistic. Make a decision and interpret in the context of the problem.

3. (a) H_0: median $\leq$ \$200

   H_a: median $>$ \$200 (claim)

   (b) 1 (c) 5 (d) Fail to reject H_0.

   (e) There is not enough evidence at the 1% level to support the claim that the median amount of new credit card charges for the previous month was more than \$200.

4. See Selected Answers, page A100.

5. (a) H_0: median $\leq$ \$148,000 (claim)

   H_a: median $>$ \$148,000

   (b) 1 (c) 3 (d) Fail to reject H_0.

   (e) There is not enough evidence at the 5% level to reject the claim that the median sales price of new privately owned one-family homes sold in the past year is \$148,000 or less.

6. See Selected Answers, page A100.

7. (a) H_0: median $\geq$ \$1900 (claim)
H_a: median $<$ \$1900
(b) -2.055 (c) -3.040
(d) Reject H_0.
(e) There is enough evidence at the 2% level to reject the claim that the median amount of credit card debt for families holding such debts is at least \$1900.

8. (a) H_0: median $\geq$ \$38,800
H_a: median $<$ \$38,800 (claim)
(b) -1.96 (c) -1.554
(d) Fail to reject H_0.
(e) There is not enough evidence at the 2.5% level to support the claim that the median amount of financial debt for families holding such debts is less than \$38,800.

9. (a) H_0: median $\leq$ 34
H_a: median $>$ 34 (claim)
(b) 3 (c) 8
(d) Fail to reject H_0.
(e) There is not enough evidence at the 1% level to support the claim that the median age of recipients of social science doctorates is greater than 34.

10. See Selected Answers, page A100.

11. (a) H_0: median $=$ 4 (claim)
H_a: median $\neq$ 4
(b) -1.96 (c) -2.334
(d) Reject H_0.
(e) There is enough evidence at the 5% level to reject the claim that the median number of rooms in renter-occupied units is 4.

12. See Selected Answers, page A100.

13. (a) H_0: median $=$ \$11.63 (claim)
H_a: median $\neq$ \$11.63
(b) -2.575 (c) -0.961
(d) Fail to reject H_0.
(e) There is not enough evidence at the 1% level to reject the claim that the median hourly earnings of male workers paid hourly rates is \$11.63.

7. *Credit Card Debt* A financial services institution reports that the median amount of credit card debt for families holding such debts is at least \$1900. In a random sample of 104 families holding debt, you see that the debt for 68 families is less than \$1900 and the debt for 36 families is greater than \$1900. At $\alpha = 0.02$, can you reject the institution's claim? *(Adapted from Board of Governors of the Federal Reserve System)*

8. *Financial Debt* A financial services accountant estimates that the median amount of financial debt for families holding such debts is less than \$38,800. In a random sample of 70 families holding debt, the debt for 28 families was less than \$38,800 and the debt for 42 families was greater than \$38,800. At $\alpha = 0.025$, can you support the accountant's estimate? *(Adapted from Board of Governors of the Federal Reserve System)*

9. *Social Science Doctorates* A social sciences association conducted a study to determine the median age of recipients of social science doctorates. In part of the study, the association randomly selected 20 social science doctorates and found that nine were conferred before age 34, eight were conferred after age 34, and three were conferred at age 34. Test the association's claim that the median age of recipients of social science doctorates is greater than 34 years. Use $\alpha = 0.01$. *(Adapted from U.S. National Science Foundation)*

10. *Physical Science Doctorates* A science association claims that the median age of recipients of physical science doctorates is less than 30 years. In a random sample of 24 physical science doctorates, 8 were conferred at less than 30 years, 12 were conferred at greater than 30 years, and 4 were conferred at 30 years. At $\alpha = 0.05$, can you support the association's claim? *(Adapted from U.S. National Science Foundation)*

11. *Unit Size* A renters' organization claims that the median number of rooms in renter-occupied units is four. You randomly select 50 renter-occupied units and obtain the results shown below. At $\alpha = 0.05$, can you reject the organization's claim? *(Adapted from U.S. Census Bureau)*

Unit size	Number of units
Fewer than 4 rooms	15
4 rooms	3
More than 4 rooms	32

Data for Exercise 11

Square footage	Number of units
Less than 1300	11
1300	2
More than 1300	9

Data for Exercise 12

12. *Square Footage* A renters' organization believes that the median square footage of renter-occupied units is 1300 square feet. To test this claim, you randomly select 22 renter-occupied units and obtain the results shown above. At $\alpha = 0.10$, can you reject the organization's claim? *(Adapted from U.S. Census Bureau)*

13. *Hourly Earnings* A labor organization estimates that the median hourly earnings of male workers paid hourly rates is \$11.63. In a random sample of 41 male workers paid hourly rates, 16 are paid less than \$11.63 per hour, 23 are paid more than \$11.63 per hour, and 2 are paid \$11.63 per hour. At $\alpha = 0.01$, can you reject the organization's claim? *(Adapted from U.S. Bureau of Labor Statistics)*

14. (a) H_0: median $\leq$ \$9.89 (claim)

H_a: median $>$ \$9.89

(b) 5

(c) 9

(d) Fail to reject H_0.

(e) There is not enough evidence at the 5% level to reject the claim that the median hourly earnings of female workers paid hourly rates is at most \$9.89.

15. (a) H_0: The headache hours have not decreased.

H_a: The headache hours have decreased. (claim)

(b) 1

(c) 3

(d) Fail to reject H_0.

(e) There is not enough evidence at the 5% level to support the claim that the daily headache hours were reduced after the soft-tissue therapy and spinal manipulation.

16. (a) H_0: The headache hours have not decreased.

H_a: The headache hours have decreased. (claim)

(b) 1

(c) 4

(d) Fail to reject H_0.

(e) There is not enough evidence at the 1% level to support the claim that daily headache hours have decreased.

17. (a) H_0: The SAT scores have not improved.

H_a: The SAT scores have improved. (claim)

(b) 2

(c) 4

(d) Fail to reject H_0.

(e) There is not enough evidence at the 5% level to support the claim that verbal SAT scores improved.

14. *Hourly Earnings* A labor organization claims that the median hourly earnings of female workers paid hourly rates is at most \$9.89. In a random sample of 23 female workers paid hourly rates, 9 are paid less than \$9.89 per hour, 11 are paid more than \$9.89 per hour, and 3 are paid \$9.89 per hour. At $\alpha = 0.05$, can you reject the organization's claim? *(Adapted from U.S. Bureau of Labor Statistics)*

15. *Headaches* The table shows the daily headache hours suffered by eight patients before and after receiving soft tissue therapy and spinal manipulation for seven weeks. At $\alpha = 0.05$, is there enough evidence to conclude that daily headache hours were reduced after the soft tissue therapy and spinal manipulation? *(Adapted from The Journal of the American Medical Association)*

Patient	1	2	3	4	5	6	7	8
Headache hours (before)	0.8	2.4	2.8	2.6	2.7	0.9	1.2	2.2
Headache hours (after)	1.6	1.3	1.6	1.4	1.5	1.6	1.7	1.8

DATA **16. *Headaches*** The table shows the daily headache hours suffered by 12 patients before and after receiving soft tissue therapy and laser treatment for seven weeks. At $\alpha = 0.01$, is there enough evidence to conclude that daily headache hours were reduced after receiving the soft tissue therapy and laser treatment? *(Adapted from The Journal of the American Medical Association)*

Patient	1	2	3	4	5	6
Headache hours (before)	2.1	3.9	3.8	2.5	2.4	3.6
Headache hours (after)	2.2	2.8	2.5	2.6	1.9	1.8

Patient	7	8	9	10	11	12
Headache hours (before)	3.4	2.4	3.5	2.0	2.7	2.4
Headache hours (after)	2.0	1.6	1.5	2.1	1.8	3.0

DATA **17. *Improving SAT Scores*** A tutoring agency believes that by completing a special course, students can improve their verbal SAT scores. In part of a study, 12 students take the verbal part of the SAT, complete the special course, then take the verbal part of the SAT again. The students' scores are shown below. At $\alpha = 0.05$, is there enough evidence to conclude that the students' verbal SAT scores improved?

Student	1	2	3	4	5	6
Score on first SAT	308	456	352	433	306	471
Score on second SAT	300	524	409	419	304	483

Student	7	8	9	10	11	12
Score on first SAT	538	207	205	351	360	251
Score on second SAT	708	253	399	350	480	303

18. (a) H_0: The SAT scores have not improved.

H_a: The SAT scores have improved. (claim)

(b) 1

(c) 3

(d) Fail to reject H_0.

(e) There is not enough evidence at the 1% level to support the claim that the verbal SAT scores have improved.

19. (a) Fail to reject H_0.

(b) There is not enough evidence at the 5% level to reject the claim that the proportion of companies that provide transportation home is equal to the proportion of companies that do not.

20. (a) Fail to reject H_0.

(b) There is not enough evidence at the 5% level to reject the claim that the proportion of credit card holders who almost always pay off their credit card balances is equal to the proportion of credit card holders who do not.

DATA

18. ***SAT Scores*** Students at a certain school are required to take the SAT twice. The table shows both verbal SAT scores for 12 students. At $\alpha = 0.01$, can you conclude that the students' scores improved the second time they took the SAT?

Student	1	2	3	4	5	6
Score on first SAT	445	510	429	452	629	453
Score on second SAT	446	571	517	478	610	453

Student	7	8	9	10	11	12
Score on first SAT	358	477	325	513	636	571
Score on second SAT	378	532	299	501	648	603

19. ***Office Parties*** An alcohol awareness organization conducted a survey by randomly selecting companies and asking them if they give their employees a holiday party at which alcohol is served. The organization asked companies that said yes if the companies provided transportation home or offered alternatives to driving home. The results are shown in the figure. *(Adapted from USA TODAY)*

(a) Use a sign test to test the null hypothesis that the proportion of companies that provide transportation home or offer alternatives to driving home is equal to the proportion of companies that do not. Assign a + sign to a company that does, assign a − sign to a company that does not, and assign a 0 to a company that does not know. Use $\alpha = 0.05$.

(b) What can you conclude?

20. ***Credit Cards*** A financial institution conducts a survey by randomly selecting credit cardholders and asking them if they almost always pay off their credit card balances. The results are shown in the figure. *(Adapted from Board of Governors of the Federal Reserve System)*

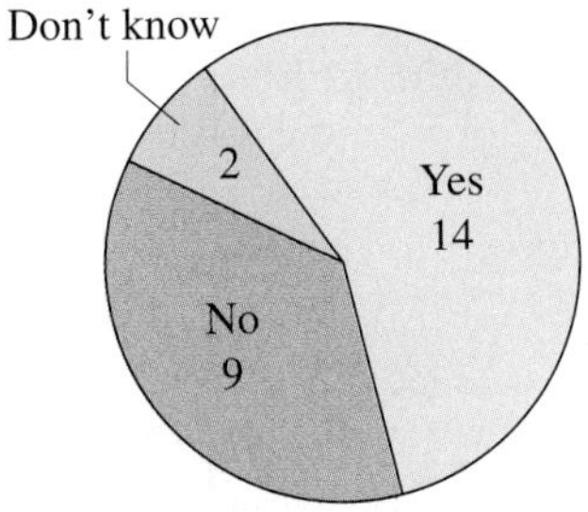

(a) Use a sign test to test the null hypothesis that the proportion of credit cardholders who almost always pay off their credit card balances is equal to the proportion of credit cardholders who do not. Assign a + sign to a credit cardholder who does, assign a − sign to a credit cardholder who does not, and assign a 0 to a credit cardholder who does not know. Use $\alpha = 0.05$.

(b) What can you conclude?

Extending Concepts

More on Sign Tests When you are using a sign test for $n > 25$ and the test is left tailed, you know you can reject the null hypothesis if the test statistic

$$z = \frac{(x + 0.5) - 0.5n}{\frac{\sqrt{n}}{2}}$$

is less than or equal to the *left-tailed* critical value, where x is the *smaller* number of + and − signs. For a right-tailed test, you can reject the null hypothesis if the test statistic

$$z = \frac{(x - 0.5) - 0.5n}{\frac{\sqrt{n}}{2}}$$

is greater than or equal to the *right-tailed* critical value, where x is the *larger* number of + and − signs.

In Exercises 21–24, (a) write the claim mathematically and identify H_0 and H_a, (b) determine the critical value, (c) calculate the test statistic, and (d) decide whether to reject or fail to reject the null hypothesis. Then interpret the decision in the context of the original claim.

21. ***Weekly Earnings*** A labor organization claims that the median weekly earnings of female workers is less than or equal to \$530. To test this claim, you randomly select 50 female workers and ask each to provide her weekly earnings. The results are shown in the table. At $\alpha = 0.01$, can you reject the organization's claim? *(Adapted from U.S. Bureau of Labor Statistics)*

Weekly earnings	Number of workers
Less than \$530	18
\$530	3
More than \$530	29

Data for Exercise 21

Weekly earnings	Number of workers
Less than \$680	21
\$680	1
More than \$680	48

Data for Exercise 22

22. ***Weekly Earnings*** A labor organization states that the median weekly earnings of male workers is greater than \$680. To test this claim, you randomly select 70 male workers and ask each to provide his weekly earnings. The results are shown in the table. At $\alpha = 0.01$, can you support the organization's claim? *(Adapted from U.S. Bureau of Labor Statistics)*

23. ***Ages of Brides*** A marriage counselor estimates that the median age of brides at the time of their first marriage is greater than 25.3 years. In a random sample of 65 brides, 22 are less than 25.3 years, 38 are more than 25.3 years, and 5 are 25.3 years. At $\alpha = 0.05$, can you support the counselor's claim? *(Adapted from U.S. Census Bureau)*

24. ***Ages of Grooms*** A marriage counselor estimates that the median age of grooms at the time of their first marriage is less than or equal to 26.9 years. In a random sample of 56 grooms, 23 are less than 26.9 years, 33 are more than 26.9 years, and none are 26.9 years. At $\alpha = 0.05$, is there enough evidence to reject the counselor's claim? *(Adapted from U.S. Census Bureau)*

21. (a) H_0: median $\leq$ \$530 (claim)
H_a: median $>$ \$530
(b) 2.33
(c) 1.459
(d) Fail to reject H_0. There is not enough evidence at the 1% level to reject the claim that the median weekly earnings of female workers is less than or equal to \$530.

22. (a) H_0: median $\leq$ \$680
H_a: median $>$ \$680 (claim)
(b) 2.33
(c) 3.130
(d) Reject H_0. There is enough evidence at the 1% level to support the claim that the median weekly earnings of male workers is greater than \$680.

23. (a) H_0: median $\leq$ 25.3
H_a: median $>$ 25.3 (claim)
(b) 1.645
(c) 1.936
(d) Reject H_0. There is enough evidence at the 5% level to support the claim that the median age of first-time brides is greater than 25.3 years.

24. (a) H_0: median $\leq$ 26.9 (claim)
H_a: median $>$ 26.9
(b) 1.645
(c) 1.203
(d) Fail to reject H_0. There is not enough evidence at the 5% level to reject the claim that the median age of first-time grooms is less than or equal to 26.9 years.

11.2 The Wilcoxon Tests

What You Should Learn

- How to use the Wilcoxon signed-rank test to determine if two dependent samples are selected from populations having the same distribution
- How to use the Wilcoxon rank sum test to determine if two independent samples are selected from populations having the same distribution

The Wilcoxon Signed-Rank Test • The Wilcoxon Rank Sum Test

The Wilcoxon Signed-Rank Test

In this section, you will study the Wilcoxon signed-rank test and the Wilcoxon rank sum test. Unlike the sign test, the strength of these two nonparametric tests is that each considers the magnitude, or size, of the data entries.

In Section 8.3, you used a t-test together with dependent samples to determine whether there was a difference between two populations. To use the t-test to test such a difference, you must assume (or know) that the dependent samples are randomly selected from populations having a normal distribution. But, what if this assumption cannot be made? Instead of using the two-sample t-test, you can use the Wilcoxon signed-rank test.

DEFINITION

The **Wilcoxon signed-rank test** is a nonparametric test that can be used to determine whether two *dependent* samples were selected from populations having the same distribution.

Study Tip

The absolute value of a number is its value, disregarding its sign. A pair of vertical bars, | |, is used to denote absolute value. For example, $|3| = 3$ and $|-7| = 7$.

GUIDELINES

Performing a Wilcoxon Signed-Rank Test

In Words	*In Symbols*
1. Identify the claim. State the null and alternative hypotheses.	State H_0 and H_a.
2. Specify the level of significance.	Identify α.
3. Determine the sample size n.	
4. Determine the critical value.	Use Table 9 in Appendix B.
5. Calculate the test statistic w_s. **a.** Complete a table using the headers listed at the right. **b.** Find the sum of the positive ranks and the sum of the negative ranks. **c.** Select the smaller of absolute values of the sums.	Headers: **Sample 1, Sample 2, Difference, Absolute value, Rank,** and **Signed rank.**
6. Make a decision to reject or fail to reject the null hypothesis.	If w_s is less than or equal to the critical value, reject H_0. Otherwise, fail to reject H_0.
7. Interpret the decision in the context of the original claim.	

EXAMPLE 1

Performing a Wilcoxon Signed-Rank Test

A sports psychologist believes that listening to music affects the length of athletes' workout sessions. The length of time (in minutes) of 10 athletes' workout sessions, while listening to music and while not listening to music, are shown in the table. At $\alpha = 0.05$, can you support the sports psychologist's claim?

Length of workout session, with music	45	38	28	39	41	47	62	54	33	44
Length of workout session, without music	38	40	33	36	42	41	54	47	28	35

SOLUTION The claim is "Music affects the length of athletes' workout sessions." To test this claim, use the following null and alternative hypotheses.

H_0: There is no difference in the length of the athletes' workout sessions.

H_a: There is a difference in the length of the athletes' workout sessions.

This Wilcoxon signed-rank test is a two-tailed test with $\alpha = 0.05$ and $n = 10$. From Table 9 in Appendix B, the critical value is 8. To find the test statistic w_s complete a table as shown below.

Length of session, with music	Length of session, without music	Difference	Absolute value	Rank	Signed rank
45	38	7	7	7.5	7.5
38	40	−2	2	2	−2
28	33	−5	5	4.5	−4.5
39	36	3	3	3	3
41	42	−1	1	1	−1
47	41	6	6	6	6
62	54	8	8	9	9
54	47	7	7	7.5	7.5
33	28	5	5	4.5	4.5
44	35	9	9	10	10

The sum of the negative ranks is

$$-1 + (-2) + (-4.5) = -7.5.$$

The sum of the positive ranks is

$$(+3) + (+4.5) + (+6) + (+7.5) + (+7.5) + (+9) + (+10) = 47.5.$$

The test statistic is the smaller of the absolute value of these two sums. Because $|-7.5| < |47.5|$, the test statistic is $w_s = 7.5$. Because the test statistic is less than the critical value, that is, $7.5 < 8$, you should decide to reject the null hypothesis.

Interpretation At the 5% level of significance, you have enough evidence to support the claim that music makes a difference in the length of athletes' workout sessions.

Study Tip

Do not assign a rank to any differences of zero. The sample size n should be the total number of nonzero differences. In the case of a tie between data entries, use the average of the corresponding ranks. For instance, if two data entries are tied for the fifth rank, use the average of 5 and 6, which is 5.5. If three entries are tied for the fifth rank, use the average of 5, 6, and 7, or 6.

Picturing the World

In an effort to determine whether smoking is a risk factor for pneumonia in patients with chicken pox, doctors recorded the carbon monoxide transfer factor levels in seven patients (all smokers) upon their admission to the hospital and one week later. *(Source: British Medical Journal)*

Patient	Level at admission	Level after one week
1	62	63
2	60	64
3	58	85
4	56	80
5	66	60
6	40	73
7	50	52

At $\alpha = 0.05$, can you conclude that the carbon monoxide transfer level changed significantly in one week?

Study Tip

Use the Wilcoxon signed-rank test for dependent samples and the Wilcoxon rank sum test for independent samples.

Try It Yourself 1

A quality control inspector wants to test the claim that a spray-on water repellent is effective. To test this claim, he selects 12 pieces of fabric, sprays water on each, and measures the amount of water repelled (in milliliters). He then applies the water repellent and repeats the experiment. The results are shown in the table. At $\alpha = 0.01$, can he conclude that the water repellent is effective?

No repellent	8	7	7	4	6	10
Repellent applied	15	12	11	6	6	8

No repellent	9	5	9	11	8	4
Repellent applied	8	6	12	8	14	8

a. *Identify* the claim and *state* H_0 and H_a.
b. *Specify* the level of significance α.
c. *Determine* the sample size n.
d. *Determine* the critical value.
e. *Calculate* the test statistic w_s by making a table, finding the sum of the positive ranks and the sum of the negative ranks, and finding the absolute value of each.
f. *Decide* whether to reject the null hypothesis. Use a graph if necessary.
g. *Interpret* the decision in the context of the original claim.

Answer: Page A44

The Wilcoxon Rank Sum Test

In Sections 8.1 and 8.2, you used a z-test or a t-test together with independent samples to determine whether there was a difference between two populations. To use these tests to test such a difference, you had to make several assumptions concerning the distribution of each population. But, what if these assumptions cannot be made? You can still compare the populations using the Wilcoxon rank sum test.

DEFINITION

The **Wilcoxon rank sum test** is a nonparametric test that can be used to determine whether two *independent* samples were selected from populations having the same distribution.

A requirement for the Wilcoxon rank sum test is that the sample size of both samples must be at least 10. When calculating the test statistic for the Wilcoxon rank sum test, let n_1 represent the sample size of the smaller sample and n_2 represent the sample size of the larger sample. If the two samples have the same size, it does not matter which one is n_1 or n_2.

When calculating the sum of the ranks R, use the ranks for the smaller of the two samples. If the two samples have the same size, you can use the ranks from either sample, but you must use the ranks from the sample you associate with n_1.

Test Statistic for the Wilcoxon Rank Sum Test

Given two independent samples, the test statistic z for the Wilcoxon rank sum test is

$$z = \frac{R - \mu_R}{\sigma_R}$$

where

$R =$ sum of the ranks for the smaller sample,

$$\mu_R = \frac{n_1(n_1 + n_2 + 1)}{2},$$

and

$$\sigma_R = \sqrt{\frac{n_1 n_2(n_1 + n_2 + 1)}{12}}.$$

GUIDELINES

Performing a Wilcoxon Rank Sum Test

In Words	*In Symbols*
1. Identify the claim. State the null and alternative hypotheses.	State H_0 and H_a.
2. Specify the level of significance.	Identify α.
3. Determine the critical value.	Use Table 4 in Appendix B.
4. Determine the sample sizes.	$n_1 \leq n_2$
5. Find the sum of the ranks for the smaller sample. **a.** List the combined data in ascending order. **b.** Rank the combined data. **c.** Add the sum of the ranks for the smaller sample.	R
6. Calculate the test statistic.	$z = \frac{R - \mu_R}{\sigma_R}$
7. Make a decision to reject or fail to reject the null hypothesis.	If z is in the rejection region, reject H_0. Otherwise, fail to reject H_0.
8. Interpret the decision in the context of the original claim.	

EXAMPLE 2

Performing a Wilcoxon Rank Sum Test

The table shows the earnings (in thousands of dollars) of a random sample of 10 male and 12 female salespersons. At $\alpha = 0.10$, can you conclude that there is a difference between the males' and females' earnings?

Male earnings	28	43	64	51	48	44	36	45	67	49		
Female earnings	36	27	51	43	35	48	41	37	34	47	50	40

SOLUTION The claim is "there is a difference between the males' and females' earnings." The null and alternative hypotheses for this test are as follows.

H_0: There is no difference between the males' and the females' earnings.

H_a: There is a difference between the males' and the females' earnings.

Because the test is a two-tailed test with $\alpha = 0.10$, the critical values are

$-z_0 = -1.645$ and $z_0 = 1.645.$

The rejection regions are

$z \leq -1.645$ and $z \geq 1.645.$

Before calculating the test statistic, you must find the values of R, μ_R, and σ_R. The table shows the combined data listed in ascending order and the corresponding ranks.

Ordered data	Sample	Rank
27	F	1
28	M	2
34	F	3
35	F	4
36	M	5.5
36	F	5.5
37	F	7
40	F	8
41	F	9
43	M	10.5
43	F	10.5

Ordered data	Sample	Rank
44	M	12
45	M	13
47	F	14
48	M	15.5
48	F	15.5
49	M	17
50	F	18
51	M	19.5
51	F	19.5
64	M	21
67	M	22

Because the smaller sample is the sample of males, R is the sum of the male rankings.

$$R = 2 + 5.5 + 10.5 + 12 + 13 + 15.5 + 17 + 19.5 + 21 + 22$$
$$= 138$$

Using $n_1 = 10$ and $n_2 = 12$, you can find μ_R and σ_R as follows.

$$\mu_R = \frac{n_1(n_1 + n_2 + 1)}{2} = \frac{10(10 + 12 + 1)}{2} = \frac{230}{2} = 115$$

$$\sigma_R = \sqrt{\frac{n_1 n_2(n_1 + n_2 + 1)}{12}}$$
$$= \sqrt{\frac{(10)(12)(10 + 12 + 1)}{12}}$$
$$= \sqrt{\frac{2760}{12}}$$
$$= \sqrt{230}$$
$$\approx 15.2$$

When $R = 138$, $\mu_R = 115$, and $\sigma_R = 15.2$, the test statistic is

$$z = \frac{R - \mu_R}{\sigma_R}$$
$$= \frac{138 - 115}{15.2}$$
$$\approx 1.51.$$

From the graph at the right, you can see that the test statistic z is not in the rejection region. So, you should decide to fail to reject the null hypothesis.

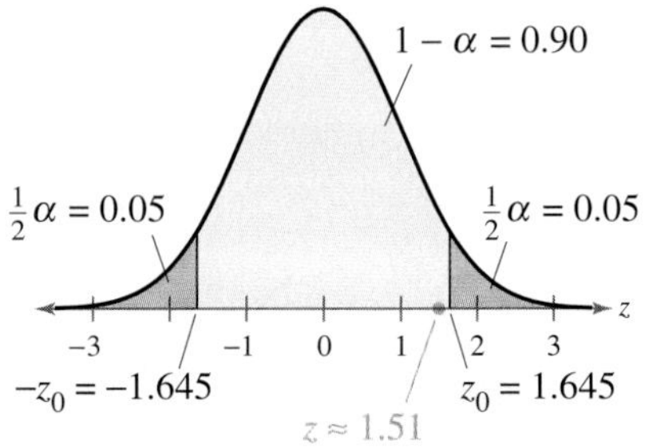

Interpretation At the 10% level of significance, you cannot conclude that there is a difference between the males' and females' earnings.

Try It Yourself 2

You are investigating the automobile insurance claims paid (in thousands of dollars) by two insurance companies. The table shows a random, independent sample of 12 claims paid by two insurance companies. At $\alpha = 0.05$, can you conclude that there is a difference in the claims paid by the companies?

Company A	6.2	10.6	2.5	4.5	6.5	7.4
Company B	7.3	5.6	3.4	1.8	2.2	4.7

Company A	9.9	3.0	5.8	3.9	6.0	6.3
Company B	10.8	4.1	1.7	3.0	4.4	5.3

a. *Identify* the claim and *state* H_0 and H_a.
b. *Specify* the level of significance α.
c. *Determine* the critical value.
d. *Determine* the sample sizes n_1 and n_2.
e. *List* the combined data in ascending order, *rank* the data, and *find* the sum of the ranks of the smaller sample.
f. *Calculate* the test statistic.
g. *Decide* whether to reject the null hypothesis. Use a graph if necessary.
h. *Interpret* the decision in the context of the original claim. *Answer: Page A45*

11.2 Exercises

Building Basic Skills and Vocabulary

1. How do you know whether to use a Wilcoxon signed-rank test or a Wilcoxon rank sum test?

2. What is the requirement for the sample size of both samples when using the Wilcoxon rank sum test?

Using and Interpreting Concepts

Performing a Wilcoxon Test In Exercises 3–8,

(a) write the claim mathematically and identify H_0 and H_a.

(b) decide whether to use a Wilcoxon signed-rank test or a Wilcoxon rank sum test.

(c) determine the critical value.

(d) calculate the test statistic.

(e) decide whether to reject or fail to reject the null hypothesis.

(f) interpret the decision in the context of the original claim.

DATA

3. ***Calcium Supplements and Blood Pressure*** In a study testing the effects of calcium supplements on blood pressure in men, 12 men were randomly chosen and given a calcium supplement for 12 weeks. The measurements shown in the table are for each subject's diastolic blood pressure taken before and after the 12-week treatment period. At $\alpha = 0.01$, can you reject the claim that there was no reduction in diastolic blood pressure? *(Adapted from The Journal of American Medicine)*

Patient	1	2	3	4	5	6
Before treatment	108	109	120	129	112	111
After treatment	99	115	105	116	115	117

Patient	7	8	9	10	11	12
Before treatment	117	135	124	118	130	115
After treatment	108	122	120	126	128	106

DATA

4. ***Comparing Salaries: Manufacturing and Construction*** A private industry analyst claims that there is no difference in the salaries earned by workers in the manufacturing and construction industries. A random sample of 10 manufacturing and 10 construction workers and their salaries is shown in the table. At $\alpha = 0.10$, can you reject the analyst's claim? *(Adapted from U.S. Bureau of Labor Statistics)*

Industry	Salary (in thousands of dollars)									
Manufacturing	31	38	33	33	35	47	33	29	38	45
Construction	31	30	27	32	28	34	30	33	26	35

1. If the samples are dependent, use a Wilcoxon signed-rank test. If the samples are independent, use a Wilcoxon rank sum test.
2. The sample size must be at least 10 for both samples to use the Wilcoxon rank sum test.
3. (a) H_0: There is no reduction in diastolic blood pressure. (claim)
   H_a: There is a reduction in diastolic blood pressure.
   (b) Wilcoxon signed-rank test
   (c) 10
   (d) 17
   (e) Fail to reject H_0.
   (f) There is not enough evidence at the 1% level to reject the claim that there was no reduction in diastolic blood pressure.
4. (a) H_0: There is no difference in salaries. (claim)
   H_a: There is a difference in salaries.
   (b) Wilcoxon rank sum test
   (c) ± 1.645
   (d) -2.306
   (e) Reject H_0.
   (f) There is enough evidence at the 10% level to reject the claim that there is no difference in the salaries.

5. (a) H_0: There is no difference in the earnings.

H_a: There is a difference in the earnings. (claim)

(b) Wilcoxon rank sum test

(c) ± 1.96 (d) -3.87

(e) Reject H_0.

(f) There is enough evidence at the 5% level to support the claim that there is a difference in the earnings.

6. (a) H_0: There is no difference in the number of months mothers breast-feed their babies. (claim)

H_a: There is a difference in the number of months mothers breast-feed their babies.

(b) Wilcoxon rank sum test

(c) ± 2.575 (d) -1.723

(e) Fail to reject H_0.

(f) There is not enough evidence at the 1% level to reject the claim that there is no difference in the number of months mothers breast-feed their babies.

7. (a) H_0: There is no difference in salaries.

H_a: There is a difference in salaries. (claim)

(b) Wilcoxon rank sum test

(c) ± 1.96 (d) -1.82

(e) Fail to reject H_0.

(f) There is not enough evidence at the 5% level to support the claim that there is a difference in salaries.

8. (a) H_0: The new drug does not affect the number of headache hours.

H_a: The new drug does affect the number of headache hours. (claim)

(b) Wilcoxon signed-rank test

(c) 2 (d) 6

(e) Fail to reject H_0.

(f) There is not enough evidence at the 5% level to conclude that the drug affects the number of headache hours.

5. *Earnings by Degree* A college administrator believes that there is a difference in the earnings of people with bachelor's degrees and those with associate degrees. The table shows the earnings (in thousands of dollars) of a random sample of 11 people with bachelor's degrees and 10 people with associate degrees. At $\alpha = 0.05$, is there enough evidence to support the administrator's belief? *(Adapted from U.S. Census Bureau)*

Level of highest degree	Salary (in thousands of dollars)										
Bachelor's	50	54	68	47	69	76	53	41	56	40	47
Associate	28	32	30	24	24	24	27	23	30	30	

6. *Mother's Age and Baby Care* A natal health care researcher conducts a study of a random selection of mothers under 20 years and 20–24 years. The number of weeks each mother breast-fed her baby is shown in the table. At $\alpha = 0.01$, can you reject the researcher's claim that there is no difference in the number of months mothers in these two age groups breast-feed their babies? *(Adapted from U.S. National Center for Health Statistics)*

Age of mother	Duration (in weeks)											
Under 20 years	21	24	24	30	11	17	10	28	21	12	15	
20–24 years	25	28	26	23	22	13	28	21	27	32	19	27

7. *Teacher Salaries* A teacher's union representative claims that there is a difference in the salaries earned by teachers in Ohio and Pennsylvania. A random sample of 12 Ohio and 12 Pennsylvania teachers and their salaries is shown in the table. At $\alpha = 0.05$, is there enough evidence to support the representative's claim? *(Adapted from National Education Association)*

State	Salary (in thousands of dollars)											
Ohio	42	45	50	47	44	51	42	51	45	37	59	36
Pennsylvania	47	66	44	52	49	54	69	65	56	51	44	34

8. *Headaches* A medical researcher wants to determine whether a new drug affects the number of headache hours experienced by headache sufferers. To do so, the researcher selects seven patients and asks each to give the number of headache hours (per day) each experiences before and after taking the drug. The results are shown in the table. At $\alpha = 0.05$, can the researcher conclude that the new drug affects the number of headache hours?

Patient	1	2	3	4	5	6	7
Headache hours (before)	0.8	2.4	2.8	2.6	2.7	0.9	1.2
Headache hours (after)	1.6	1.3	1.6	1.4	1.5	1.6	1.7

9. Reject H_0. There is enough evidence at the 10% level to conclude that the gas mileage is improved.

10. Reject H_0. There is enough evidence at the 5% level to conclude that the fuel additive improves gas mileage.

Extending Concepts

Wilcoxon Signed-Rank Test for n > 30 If you are performing a Wilcoxon signed-rank test and the sample size *n* is greater than 30, you can use the Standard Normal Table and the following formula to find the test statistic.

$$z = \frac{w_s - \dfrac{n(n+1)}{4}}{\sqrt{\dfrac{n(n+1)(2n+1)}{24}}}$$

In Exercises 9 and 10, perform the indicated Wilcoxon signed-rank test using the test statistic for $n > 30$.

DATA **9.** ***Fuel Additive*** A petroleum engineer wants to know whether a certain fuel additive improves a car's gas mileage. To decide, the engineer records the gas mileage (in miles per gallon) of 33 cars with and without the additive. The results are shown in the table. At $\alpha = 0.10$, can the engineer conclude that the gas mileage is improved?

Car	1	2	3	4	5	6	7	8	9	10	11
Without additive	36.4	36.4	36.6	36.6	36.8	36.9	37.0	37.1	37.2	37.2	36.7
With additive	36.7	36.9	37.0	37.5	38.0	38.1	38.4	38.7	38.8	38.9	36.3

Car	12	13	14	15	16	17	18	19	20	21	22
Without additive	37.5	37.6	37.8	37.9	37.9	38.1	38.4	40.2	40.5	40.9	35.0
With additive	38.9	39.0	39.1	39.4	39.4	39.5	39.8	40.0	40.0	40.1	36.3

Car	23	24	25	26	27	28	29	30	31	32	33
Without additive	32.7	33.6	34.2	35.1	35.2	35.3	35.5	35.9	36.0	36.1	37.2
With additive	32.8	34.2	34.7	34.9	34.9	35.3	35.9	36.4	36.6	36.6	38.3

DATA **10.** ***Fuel Additive*** A petroleum engineer claims that a fuel additive improves gas mileage. The table shows the gas mileage (in miles per gallon) of 32 cars measured with and without the fuel additive. Test the petroleum engineer's claim at $\alpha = 0.05$.

Car	1	2	3	4	5	6	7	8	9	10	11
Without additive	34.0	34.2	34.4	34.4	34.6	34.8	35.6	35.7	30.2	31.6	32.3
With additive	36.6	36.7	37.2	37.2	37.3	37.4	37.6	37.7	34.2	34.9	34.9

Car	12	13	14	15	16	17	18	19	20	21	22
Without additive	33.0	33.1	33.7	33.7	33.8	35.7	36.1	36.1	36.6	36.6	36.8
With additive	34.9	35.7	36.0	36.2	36.5	37.8	38.1	38.2	38.3	38.3	38.7

Car	23	24	25	26	27	28	29	30	31	32
Without additive	37.1	37.1	37.2	37.9	37.9	38.0	38.0	38.4	38.8	42.1
With additive	38.8	38.9	39.1	39.1	39.2	39.4	39.8	40.3	40.8	43.2

Earnings by College Degree

Each month, the U.S. Census Bureau conducts the *Current Population Survey* (*CPS*) for the U.S. Bureau of Labor Statistics. The survey provides information about the characteristics of the U.S. work force, including employment, unemployment, earnings, and hours of work. The survey also provides demographics on characteristics such as age, sex, race, marital status, and educational attainment. This information is essential for interpreting the economic condition of the entire country.

The data from the *CPS* are available in data files and reports issued by the U.S. Census Bureau. One such report discusses the earnings of the U.S. work force by the type of college degree earned. The table shows the annual earnings (in dollars) by the highest degree earned for randomly selected members of the U.S. work force in a recent year.

Earnings (in dollars) by Highest Degree			
All persons with a doctorate	Males with a doctorate	All persons with a bachelor's	Females with a bachelor's
96,890	111,363	55,589	41,162
69,554	86,022	63,943	47,165
105,718	67,763	59,523	37,423
68,622	108,783	50,626	19,161
81,569	81,354	53,633	37,196
77,117	93,291	41,097	28,373
94,662	84,342	67,245	49,228
75,322	87,295	31,128	34,078
100,724	68,840	51,349	28,297
86,570	170,225	32,100	47,050

Exercises

1. Construct a box-and-whisker plot for each group on the same graph. Do any of the median earnings appear to be the same? Different?
   (a) All persons with a doctorate degree
   (b) Males with a doctorate degree
   (c) All persons with a bachelor's degree
   (d) Females with a bachelor's degree

In Exercises 2–5, use the sign test to test the claim. What can you conclude? Use $\alpha = 0.05$.

2. The median earnings for all persons with a doctorate degree are less than or equal to \$85,000.
3. The median earnings for males with a doctorate degree are \$95,500.
4. For all persons with a bachelor's degree, the median earnings are greater than or equal to \$55,000.
5. The median earnings for females with a bachelor's degree are different from \$40,000.

In Exercises 6 and 7, use the Wilcoxon rank sum test to test the claim. Use $\alpha = 0.01$.

6. There is no difference in the median earnings for all persons with a doctorate degree and the median earnings for males with a doctorate degree.
7. There is a difference in the median earnings for all persons with a bachelor's degree and the median earnings for females with a bachelor's degree.

11.3 The Kruskal-Wallis Test

What You Should Learn

- How to use the Kruskal-Wallis test to determine whether three or more samples were selected from populations having the same distribution

The Kruskal-Wallis Test

The Kruskal-Wallis Test

In Section 10.4, you learned how to use one-way ANOVA techniques to compare the means of three or more populations. When using one-way ANOVA, you should verify that each independent sample is selected from a population that is normally, or approximately normally, distributed. If, however, you cannot verify that the populations are normal, you can still compare the distributions of three or more populations. To do so, you can use the Kruskal-Wallis test.

DEFINITION

The **Kruskal-Wallis test** is a nonparametric test that can be used to determine whether three or more independent samples were selected from populations having the same distribution.

The null and alternative hypotheses for the Kruskal-Wallis test are as follows.

H_0: There is no difference in the distribution of the populations.

H_a: There is a difference in the distribution of the populations.

Two conditions for using the Kruskal-Wallis test are that each sample must be randomly selected and the size of each sample must be at least 5. If these conditions are met, the sampling distribution for the Kruskal-Wallis test is approximated by a chi-square distribution with $k - 1$ degrees of freedom, where k is the number of samples. You can calculate the Kruskal-Wallis test statistic using the following formula.

Test Statistic for the Kruskal-Wallis Test

Given three or more independent samples, the test statistic H for the Kruskal-Wallis test is

$$H = \frac{12}{N(N+1)}\left(\frac{R_1^2}{n_1} + \frac{R_2^2}{n_2} + \cdots + \frac{R_k^2}{n_k}\right) - 3(N+1)$$

where

k represents the number of samples,

n_i is the size of the ith sample,

N is the sum of the sample sizes,

and

R_i is the sum of the ranks of the ith sample.

Performing a Kruskal-Wallis test consists of combining and ranking the sample data. The data are then separated according to sample and the sum of the ranks of each sample is calculated. These sums are then used to calculate the test statistic H, which is an approximation of the variance of the rank sums. If the samples are selected from populations having the same distribution, the sums of the ranks will be approximately equal, H will be small, and the null hypothesis should not be rejected. If, however, the sums of the ranks are quite different, H will be large, and the null hypothesis should be rejected.

Because the null hypothesis is rejected only when H is significantly large, the Kruskal-Wallis test is always a right-tailed test.

GUIDELINES

Performing a Kruskal-Wallis Test

In Words	*In Symbols*
1. Identify the claim. State the null and alternative hypotheses.	State H_0 and H_a.
2. Specify the level of significance.	Identify α.
3. Identify the degrees of freedom.	d.f. $= k - 1$
4. Determine the critical value and the rejection region.	Use Table 6 in Appendix B.
5. Find the sum of the ranks for each sample. **a.** List the combined data in ascending order. **b.** Rank the combined data.	
6. Calculate the test statistic.	$H = \frac{12}{N(N+1)} \cdot \left(\frac{R_1^2}{n_1} + \frac{R_2^2}{n_2} + \cdots + \frac{R_k^2}{n_k}\right) - 3(N+1)$
7. Make a decision to reject or fail to reject the null hypothesis.	If H is in the rejection region, reject H_0. Otherwise, fail to reject H_0.
8. Interpret the decision in the context of the original claim.	

EXAMPLE 1

Performing a Kruskal-Wallis Test

You want to compare the hourly pay rates of accountants who work in Michigan, New York, and Virginia. To do so, you randomly select several accountants in each state and record their hourly pay rate. The hourly pay rates are shown in the table. At $\alpha = 0.01$, can you conclude that the distributions of accountants' hourly pay rates in these three states are different?

Sample Hourly Pay Rates		
MI (Sample 1)	NY (Sample 2)	VA (Sample 3)
14.24	21.18	17.02
14.06	20.94	20.63
14.85	16.26	17.47
17.47	21.03	15.54
14.83	19.95	15.38
19.01	17.54	14.90
13.08	14.89	20.48
15.94	18.88	18.50
13.48	20.06	12.80
16.94		15.57

SOLUTION You want to test the claim that there is no difference in the hourly pay rates in Michigan, New York, and Virginia. The null and alternative hypotheses are as follows.

H_0: There is no difference in the hourly pay rates in the three states.

H_a: There is a difference in the hourly pay rates in the three states.

The test is a right-tailed test with $\alpha = 0.01$ and d.f. $= k - 1 = 3 - 1 = 2$. From Table 6, the critical value is $\chi_0^2 = 9.210$. Before calculating the test statistic, you must find the sum of the ranks for each sample. The table shows the combined data listed in ascending order and the corresponding ranks.

Ordered data	Sample	Rank
12.80	VA	1
13.08	MI	2
13.48	MI	3
14.06	MI	4
14.24	MI	5
14.83	MI	6
14.85	MI	7
14.89	NY	8
14.90	VA	9
15.38	VA	10

Ordered data	Sample	Rank
15.54	VA	11
15.57	VA	12
15.94	MI	13
16.26	NY	14
16.94	MI	15
17.02	VA	16
17.47	VA	17.5
17.47	MI	17.5
17.54	NY	19
18.50	VA	20

Ordered data	Sample	Rank
18.88	NY	21
19.01	MI	22
19.95	NY	23
20.06	NY	24
20.48	VA	25
20.63	VA	26
20.94	NY	27
21.03	NY	28
21.18	NY	29

Picturing the World

The following randomly collected data were used to compare the water temperature (in degrees Fahrenheit) of cities bordering the Gulf of Mexico. *(Adapted from The USA TODAY Weather Almanac)*

Port Mansfield, TX	Eugene Island, LA	Dauphin Island, AL
62	51	63
69	55	51
77	57	54
52	63	60
60	74	75
75	82	80
83	85	70
65	60	78
79	64	82
83	76	84
82	83	
	86	

At $\alpha = 0.05$, can you conclude that the temperature distributions of the three cities are different?

The sum of the ranks for each sample is as follows.

$$R_1 = 2 + 3 + 4 + 5 + 6 + 7 + 13 + 15 + 17.5 + 22 = 94.5$$

$$R_2 = 8 + 14 + 19 + 21 + 23 + 24 + 27 + 28 + 29 = 193$$

$$R_3 = 1 + 9 + 10 + 11 + 12 + 16 + 17.5 + 20 + 25 + 26 = 147.5$$

Using these sums and the values $n_1 = 10$, $n_2 = 9$, $n_3 = 10$, and $N = 29$, the test statistic is

$$H = \frac{12}{29(29 + 1)}\left(\frac{94.5^2}{10} + \frac{193^2}{9} + \frac{147.5^2}{10}\right) - 3(29 + 1) \approx 9.41.$$

From the graph at the right, you can see that the test statistic H is in the rejection region. So, you should decide to reject the null hypothesis.

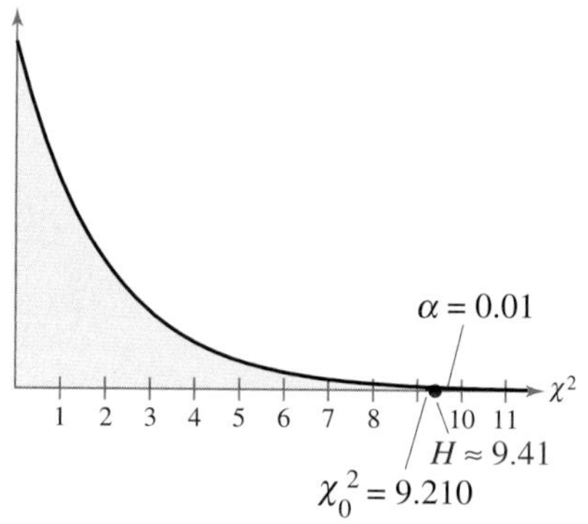

Interpretation At the 1% level of significance, you can conclude that there is a difference in accountants' hourly pay rates in Michigan, New York, and Virginia.

Try It Yourself 1

You want to compare the salaries of underwriters who work in California, Connecticut, and New Jersey. To compare the salaries, you randomly select 10 underwriters in each state and record their salary. The salaries (in thousands of dollars) are listed in the table. At $\alpha = 0.10$, can you conclude that the distributions of the underwriters' salaries in these three states are different?

Sample Salaries		
CA (Sample 1)	CT (Sample 2)	NJ (Sample 3)
26.42	25.57	50.16
48.46	29.86	46.72
33.68	51.03	29.73
37.18	57.07	44.57
36.55	45.04	36.35
41.33	37.39	37.24
38.36	30.00	43.26
40.31	33.68	34.29
46.55	45.29	33.91
47.17	61.46	43.89

a. *Identify* the claim and *state* H_0 and H_a.
b. *Specify* the level of significance α.
c. *Identify* the degrees of freedom.
d. *Determine* the critical value and the rejection region.
e. *List* the combined data in ascending order, *rank* the data, and *find* the sum of the ranks of each sample.
f. *Calculate* the test statistic.
g. *Decide* whether to reject the null hypothesis. Use a graph if necessary.
h. *Interpret* the decision in the context of the original claim.

Answer: Page A45

11.3 Exercises

Building Basic Skills and Vocabulary

1. What are the conditions for using a Kruskal-Wallis test?

2. Explain why the Kruskal-Wallis test is always a right-tailed test.

Using and Interpreting Concepts

Performing a Kruskal-Wallis Test In Exercises 3–6, (a) write the claim mathematically and identify H_0 and H_a, (b) determine the critical value, (c) find the sums of the ranks for each sample and calculate the test statistic, (d) decide whether to reject or fail to reject the null hypothesis, and (e) interpret the decision in the context of the original claim.

3. ***Home Insurance*** The table shows the annual premium for a random sample of high-risk home insurance policies in California, Florida, and Illinois. At $\alpha = 0.05$, can you conclude that the distributions of the annual premiums in these three states are different? *(Adapted from Consumer Reports)*

State	Annual Premium (in dollars)						
California	1654	553	839	777	884	753	643
Florida	1682	2067	1392	1237	1609	1550	1441
Illinois	756	496	381	437	480	566	647

4. ***Home Insurance*** An independent insurance representative wants to determine whether there is a difference in the annual premiums for high-risk home insurance in three states: New Jersey, New York, and Pennsylvania. To do so, the representative randomly selects several homes with high-risk insurance in each state and records the annual premium for each in the table shown. At $\alpha = 0.05$, can the representative conclude that the distributions of the annual premiums in these states are different? *(Adapted from Consumer Reports)*

State	Annual Premium (in dollars)						
New Jersey	441	420	474	411	371	470	
New York	753	684	869	719	1036	613	663
Pennsylvania	653	405	380	484	383	382	387

5. ***Annual Salaries*** The table shows the annual salaries for a random sample of workers in Maryland, Missouri, New Mexico, and Ohio. At $\alpha = 0.10$, can you conclude that the distributions of the annual salaries in these four states are different? *(Adapted from U.S. Bureau of Labor Statistics)*

State	Annual Salary (in thousands of dollars)						
Maryland	20.3	36.4	37.5	36.7	46.9	46.8	39.9
Missouri	31.3	51.5	34.2	31.3	29.6	49.0	30.7
New Mexico	30.3	33.9	31.3	40.1	40.2	28.2	22.8
Ohio	38.2	47.1	21.9	38.6	23.8	39.7	35.1

1. The conditions for using a Kruskal-Wallis test are that the sample must be randomly selected and the size of each sample must be at least 5.

2. See Selected Answers, page A100.

3. (a) H_0: There is no difference in the premiums.
H_a: There is a difference in the premiums. (claim)
(b) 5.991 (c) 14.05
(d) Reject H_0.
(e) There is enough evidence at the 5% level to support the claim that the distributions of the annual premiums in California, Florida, and Illinois are different.

4. (a) H_0: There is no difference in the premiums.
H_a: There is a difference in the premiums. (claim)
(b) 5.991 (c) 12.55
(d) Reject H_0.
(e) There is enough evidence at the 5% level to support the claim that the distributions of the annual premiums in the three states are different.

5. (a) H_0: There is no difference in the salaries.
H_a: There is a difference in the salaries. (claim)
(b) 6.251 (c) 1.45
(d) Fail to reject H_0.
(e) There is not enough evidence at the 10% level to support the claim that the distributions of the annual salaries in the four states are different.

6. (a) H_0: There is no difference in the annual salaries.
H_a: There is a difference in the annual salaries. (claim)
(b) 6.251
(c) 10.569
(d) Reject H_0.
(e) There is enough evidence at the 10% level to support the claim that there is a difference in the annual salaries.

7. (a) Fail to reject H_0.
(b) Fail to reject H_0.
(c) Both tests come to the same decision, which is, that there is not enough evidence to support the claim that there is a difference in the number of days spent in the hospital.

8. (a) Reject H_0.
(b) Reject H_0.
(c) Both tests come to the same decision, which is, that the mean energy consumptions are different.

DATA **6. *Annual Salaries*** You are writing an article concerning the annual salaries of workers in four U.S. states: Alabama, Colorado, Georgia, and Iowa. The annual salaries of randomly selected workers from each state are shown in the table. Can you write in your report that the distributions of the annual salaries in these four states are different? Use $\alpha = 0.10$. *(Adapted from U.S. Bureau of Labor Statistics)*

State	Annual Salary (in thousands of dollars)							
Alabama	30.8	33.8	26.1	33.9	28.4	19.4	32.3	31.7
Colorado	34.5	29.3	20.3	38.3	35.0	36.4	32.9	45.0
Georgia	42.0	32.7	47.8	47.4	30.4	39.8	44.1	40.8
Iowa	28.5	25.4	28.4	39.2	42.6	27.6	31.2	26.0

Extending Concepts

Comparing Two Tests In Exercises 7 and 8, perform the indicated test using

(a) a Kruskal-Wallis test.

(b) a one-way ANOVA test, assuming that each population is normally distributed and the population variances are equal.

(c) Compare the results.

If convenient, use technology to solve the problem.

DATA **7. *Hospital Patient Stays*** An insurance underwriter reports that the mean number of days patients spend in a hospital differs according to the region of the United States in which the patient lives. The table shows the number of days randomly selected patients spent in a hospital in four U.S. regions. At $\alpha = 0.01$, can you support the underwriter's claim? *(Adapted from U.S. National Center for Health Statistics)*

Region	Number of Days									
Northeast	8	6	6	3	5	11	3	8	1	6
Midwest	5	4	3	9	1	4	6	3	4	7
South	5	8	1	5	8	7	5	1		
West	2	3	6	6	5	4	3	6	5	

DATA **8. *Energy Consumption*** The table shows the energy consumed (in millions of Btus) in one year for a random sample of households from four U.S. regions. At $\alpha = 0.01$, can you conclude that the mean energy consumptions are different? *(Adapted from U.S. Energy Information Administration)*

Region	Energy Consumed (in millions of Btus)									
Northeast	72	106	151	138	104	108	95	134	100	174	
Midwest	84	183	194	165	120	212	148	129	113	62	97
South	91	40	72	91	147	74	70	67			
West	74	32	78	28	106	39	118	63	70	56	

11.4 Rank Correlation

What You Should Learn

- How to use the Spearman rank correlation coefficient to determine whether the correlation between two variables is significant

The Spearman Rank Correlation Coefficient

The Spearman Rank Correlation Coefficient

In Section 9.1, you learned how to measure the strength of the relationship between two variables using the Pearson correlation coefficient r. Two requirements for the Pearson correlation coefficient are that the variables are linearly related and that the population represented by each variable is normally distributed. If these requirements cannot be met, you can examine the relationship between two variables using the nonparametric equivalent to the Pearson correlation coefficient—the Spearman rank correlation coefficient.

The Spearman rank correlation coefficient has several advantages over the Pearson correlation coefficient. For instance, the Spearman rank correlation coefficient can be used to describe the relationship between linear or nonlinear data. The Spearman rank correlation coefficient can be used for data at the ordinal level. And, the Spearman rank correlation coefficient is easier to calculate by hand than the Pearson coefficient.

DEFINITION

The **Spearman rank correlation coefficient r_s** is a measure of the strength of the relationship between two variables. The Spearman rank correlation coefficient is calculated using the ranks of paired sample data entries. The formula for the Spearman rank correlation coefficient is

$$r_s = 1 - \frac{6\Sigma d^2}{n(n^2 - 1)}$$

where

n is the number of paired data entries

and

d is the difference between the ranks of a paired data entry.

The values of r_s range from -1 to $+1$, inclusive. If the ranks of corresponding data pairs are exactly identical, r_s is equal to $+1$. If the ranks are in "reverse" order, r_s is equal to -1. If the ranks of corresponding data pairs have no relationship, r_s is equal to 0.

After calculating the Spearman rank correlation coefficient, you can determine whether the correlation between the variables is significant. You can make this determination by performing a hypothesis test for the population correlation coefficient ρ_s. The null and alternative hypotheses for this test are as follows.

H_0: $\rho_s = 0$ (There is no correlation between the variables.)

H_a: $\rho_s \neq 0$ (There is a significant correlation between the variables.)

The critical values for the Spearman rank correlation coefficient are listed in Table 10 of Appendix B. Table 10 lists critical values for selected levels of significance and for sample sizes of 30 or less. The test statistic for the hypothesis test is the Spearman rank correlation coefficient r_s.

GUIDELINES

Testing the Significance of the Spearman Rank Correlation Coefficient

In Words	*In Symbols*
1. State the null and alternative hypotheses.	State H_0 and H_a.
2. Specify the level of significance.	Identify α.
3. Determine the critical value.	Use Table 10 in Appendix B.
4. Find the test statistic.	$r_s = 1 - \dfrac{6\Sigma d^2}{n(n^2 - 1)}$
5. Make a decision to reject or fail to reject the null hypothesis.	If $\lvert r_s \rvert$ is greater than the critical value, reject H_0. Otherwise, fail to reject H_0.
6. Interpret the decision in the context of the original claim.	

EXAMPLE 1

The Spearman Rank Correlation Coefficient

The table shows the prices (in dollars per 100 pounds) received by U.S. farmers for beef and lamb from 1995 to 2001. At $\alpha = 0.05$, can you conclude that there is a correlation between the beef and lamb prices? *(Source: U.S. Department of Agriculture)*

Year	Beef	Lamb
1995	61.8	78.2
1996	58.7	88.2
1997	63.1	90.3
1998	59.6	72.3
1999	63.4	74.5
2000	68.6	79.8
2001	71.3	66.9

SOLUTION The null and alternative hypotheses are as follows.

H_0: $\rho_s = 0$ (There is no correlation between the beef and lamb prices.)

H_a: $\rho_s \neq 0$ (There is a correlation between the beef and lamb prices.)

Picturing the World

The table lists the number of men and women (in thousands) who graduated from a U.S. college with a bachelor's degree from 1990 to 2003. *(Source: U.S. Department of Education)*

Year	Male	Female
1990	492	560
1991	504	590
1992	521	616
1993	533	632
1994	532	637
1995	526	634
1996	522	642
1997	521	652
1998	520	664
1999	519	682
2000	530	708

Does a correlation exist between the number of men and women who graduate with bachelor's degrees each year? Use $\alpha = 0.05$.

Each data set has seven entries. From Table 10 with $\alpha = 0.05$ and $n = 7$, the critical value is 0.786. Before calculating the test statistic, you must find d^2, the sum of the squares of the differences of the ranks of the data sets. You can use a table to calculate d^2, as shown below.

Beef prices	Rank	Lamb prices	Rank	d	d^2
61.8	3	78.2	4	−1	1
58.7	1	88.2	6	−5	25
63.1	4	90.3	7	−3	9
59.6	2	72.3	2	0	0
63.4	5	74.5	3	2	4
68.6	6	79.8	5	1	1
71.3	7	66.9	1	6	36
					$\Sigma = 76$

When $n = 7$ and $\Sigma d^2 = 76$, the test statistic is

$$\begin{aligned} r_s &= 1 - \frac{6\Sigma d^2}{n(n^2 - 1)} \\ &= 1 - \frac{6(76)}{7(7^2 - 1)} \\ &\approx -0.357. \end{aligned}$$

Because $|-0.357| < 0.786$, you should fail to reject the null hypothesis.

Interpretation At the 5% level, you cannot conclude that there is a significant correlation between beef and lamb prices between 1995 and 2001.

Try It Yourself 1

The table lists the prices (in dollars per bushel) received by U.S. farmers for oat and wheat from 1995 to 2003. At $\alpha = 0.05$. can you conclude that there is a correlation between the oat and wheat prices? *(Source: U.S. Department of Agriculture)*

Year	1995	1996	1997	1998	1999	2000	2001	2002	2003
Oat	1.67	1.96	1.60	1.10	1.12	1.10	1.59	1.81	1.45
Wheat	4.55	4.30	3.38	2.65	2.48	2.62	2.78	3.56	3.35

a. *State* the null and alternative hypotheses.
b. *Specify* the level of significance.
c. *Determine* the critical value.
d. *Use a table* to calculate d^2.
e. *Find* the standardized test statistic.
f. *Make a decision* to reject or fail to reject the null hypothesis.
g. *Interpret* the decision in the context of the original claim.

Answer: Page A45

11.4 Exercises

Help

MyMathLab

Student Study Pack

Building Basic Skills and Vocabulary

1. What are some advantages of the Spearman rank correlation coefficient over the Pearson correlation coefficient?

2. What does it mean when r_s is equal to 1? What does it mean when r_s is equal to -1? What does it mean when r_s is equal to 0?

Using and Interpreting Concepts

Testing a Claim In Exercises 3–6, (a) identify the claim and state H_0 and H_a, (b) determine the critical value using Table 10 in Appendix B, (c) find the standardized test statistic r_s, (d) decide whether to reject or fail to reject the null hypothesis, and (e) interpret the decision in the context of the original claim.

3. ***Farming: Debt and Income*** In an agricultural report, a commodities analyst suggests that there is a correlation between debt and income in the farming business. The table shows the total debts and total incomes for farms in seven states for a recent year. At $\alpha = 0.01$, is there enough evidence to support the analyst's claim? *(Source: U.S. Department of Agriculture)*

State	Debt (in millions of dollars)	Income (in millions of dollars)
California	18,578	25,510
Illinois	9,755	7,022
Iowa	13,613	10,774
Minnesota	9,271	7,522
Nebraska	9,595	8,952
North Carolina	3,805	7,410
Texas	12,341	13,344

DATA

4. ***Portable CD Players*** Is the price of a portable CD player related to its quality? To answer this question, you randomly select 11 portable CD players and determine the overall score and price of each. The overall score represents the error correction, locate speed, battery life, and headphone quality of a CD player. The results of the study are shown in the table. At $\alpha = 0.01$, can you conclude that there is a correlation between the overall score and the price? *(Adapted from Consumer Reports)*

Overall score	Price (in dollars)
82	150
78	100
68	120
67	140
61	145
60	100

Overall score	Price (in dollars)
60	150
58	80
57	200
55	80
49	75

1. The Spearman rank correlation coefficient can be used to describe the relationship between linear or nonlinear data. Also, it can be used for data at the ordinal level and it is easier to calculate by hand than the Pearson correlation coefficient.
2. The ranks of the corresponding data are exactly identical when r_s is equal to 1. The ranks are in "reverse" order when r_s is equal to -1. The ranks have no relationship when r_s is equal to 0.
3. (a) H_0: $\rho_s = 0$; H_a: $\rho_s \neq 0$ (claim)
   (b) 0.929 (c) 0.750
   (d) Fail to reject H_0.
   (e) There is not enough evidence at the 1% level to support the claim that there is a correlation between debt and income in the farming business.
4. (a) H_0: $\rho_s = 0$; H_a: $\rho_s \neq 0$ (claim)
   (b) 0.818 (c) 0.423
   (d) Fail to reject H_0.
   (e) There is not enough evidence at the 1% level to support the claim that there is a correlation between the overall score and price.

5. (a) H_0: $\rho_s = 0$; H_a: $\rho_s \neq 0$ (claim)
(b) 0.881
(c) 0.173
(d) Fail to reject H_0.
(e) There is not enough evidence at the 1% level to support the claim that there is a correlation between the overall score and price.

6. (a) H_0: $\rho_s = 0$; H_a: $\rho_s \neq 0$ (claim)
(b) 0.497
(c) 0.568
(d) Reject H_0.
(e) There is enough evidence at the 10% level to support the claim that there is a correlation between overall score and price.

7. Fail to reject H_0. There is not enough evidence at the 5% level to conclude that there is a correlation between science achievement scores and GNP.

8. Fail to reject H_0. There is not enough evidence at the 5% level to conclude that there is a correlation between mathematics achievement scores and GNP.

9. Fail to reject H_0. There is not enough evidence at the 5% level to conclude that there is a correlation between science and mathematics achievement scores.

10. At the 5% level, there is not enough evidence to conclude that there is a significant correlation between eighth grade science achievement scores and the GNP of a country, between eighth grade math achievement scores and the GNP of a country, or between eighth grade science and math achievement scores. But this also means that there is not enough evidence to conclude that there's not a significant correlation between the variables. More evidence is needed to state whether or not a significant correlation exists. (Answers will vary.)

5. ***Suitcases*** The table shows the overall scores and the prices for eight different suitcases. The overall score represents the ease of use, features, construction, and durability of a suitcase. At $\alpha = 0.01$, can you conclude that there is a correlation between the overall score and the price? *(Adapted from Consumer Reports)*

Overall score	90	85	81	78	72	68	64	61
Price (in dollars)	495	230	190	160	350	230	260	200

DATA

6. ***Air Conditioners*** The table shows the overall scores and the prices for 12 different models of air conditioners. The overall score represents the air conditioner's performance and quietness. At $\alpha = 0.10$, can you conclude that there is a correlation between the overall score and the price? *(Adapted from Consumer Reports)*

Overall score	83	82	79	78	78	74
Price (in dollars)	320	360	360	300	230	285

Overall score	71	67	64	63	61	33
Price (in dollars)	320	300	340	295	220	280

Test Scores and GNP In Exercises 7–10, use the following table. The table shows the average achievement score in eighth grade science and mathematics along with the gross national product (GNP) of nine countries for a recent year. (The GNP is a measure of a nation's total economic activity.) *(Source: IEA Third International Mathematics and Science Study, Boston College; U.S. Census Bureau)*

Country	Science average	Mathematics average	GNP (in billions of dollars)
Australia	545	530	342
Canada	531	527	542
Czech Republic	574	564	101
France	498	538	1521
Japan	571	605	5153
Portugal	480	454	103
Spain	517	487	554
Switzerland	522	545	316
USA	534	500	7247

DATA

7. ***Science and GNP*** At $\alpha = 0.05$, can you conclude that there is a correlation between science achievement scores and GNP?

8. ***Math and GNP*** At $\alpha = 0.05$, can you conclude that there is a correlation between mathematics achievement scores and GNP?

9. ***Science and Math*** At $\alpha = 0.05$, can you conclude that there is a correlation between science and mathematics achievement scores?

10. ***Writing a Summary*** Use the results from Exercises 7–9 to write a summary about the correlation (or lack of correlation) between test scores and GNP.

11. Fail to reject H_0. There is not enough evidence at the 5% level to support the claim that there is a correlation between average hours worked and the number of on-the-job injuries.

12. Fail to reject H_0. There is not enough evidence at the 5% level to support the claim that there is a correlation between average hours worked and the number of on-the-job injuries.

Extending Concepts

Testing the Rank Correlation Coefficient for n > 30 If you are testing the significance of the Spearman rank correlation coefficient and the sample size n is greater than 30, you can use the following formula to find the critical value.

$$\frac{\pm z}{\sqrt{n-1}}, \quad z \text{ corresponds to the level of significance}$$

In Exercises 11 and 12, perform the indicated test.

DATA **11.** ***Work Injuries*** The table shows the average hours worked per week and the number of on-the-job injuries for a random sample of U.S. industries in a recent year. At $\alpha = 0.05$, can you conclude that there is a correlation between average hours worked and the number of on-the-job injuries? *(Adapted from U.S. Bureau of Labor Statistics; National Safety Council)*

Hours worked	47.6	44.1	45.6	45.5	44.5	47.3	44.6	45.9	45.5	43.7	44.8	42.5
Injuries	16	33	25	33	18	20	21	18	21	28	15	26

Hours worked	46.5	42.3	45.5	41.8	43.1	44.4	44.5	43.7	44.9	47.8	46.6	45.5
Injuries	34	32	26	28	22	19	23	20	28	24	26	29

Hours worked	43.5	42.8	44.8	43.5	47.0	44.5	50.1	46.7	43.1
Injuries	21	28	23	26	24	20	28	26	25

DATA **12.** ***Work Injuries in Construction*** The table shows the average hours worked per week and the number of on-the-job injuries for a random sample of U.S. construction companies in a recent year. At $\alpha = 0.05$, can you conclude that there is a correlation between average hours worked and the number of on-the-job injuries? *(Adapted from U.S. Bureau of Labor Statistics; National Safety Council)*

Hours worked	40.5	38.3	37.8	38.2	38.6	41.2	39.0	41.0	40.6	44.1	39.7	41.2
Injuries	12	13	19	18	22	22	17	13	15	10	18	19

Hours worked	41.1	38.2	42.3	39.2	36.1	36.2	38.7	36.0	37.3	36.5	37.9	38.0
Injuries	13	24	12	12	13	15	18	11	24	16	13	23

Hours worked	36.7	40.1	35.5	38.2	42.3	39.0	39.6	39.1	39.6	39.1
Injuries	14	10	5	14	13	18	15	23	15	23

11.5 The Runs Test

What You Should Learn

- How to use the runs test to determine whether a data set is random

The Runs Test for Randomness

The Runs Test for Randomness

In obtaining a sample of data, it is important for the data to be selected randomly. But how do you know if the sample data are truly random? One way to test for randomness in a data set is to use a *runs test for randomness*.

Before using a runs test for randomness, you must first know how to determine the number of runs in a data set.

DEFINITION

A **run** is a sequence of data having the same characteristic. Each run is preceded by and followed by data with a different characteristic or by no data at all. The number of data in a run is called the **length** of the run.

EXAMPLE 1

Finding the Number of Runs

A liquid-dispensing machine has been designed to fill one-liter bottles. A quality control inspector decides whether each bottle is filled to an acceptable level and passes inspection (P) or fails inspection (F). Determine the number of runs for each sequence and find the length of each run.

1. $P\ P\ P\ P\ P\ P\ P\ P\ F\ F\ F\ F\ F\ F\ F\ F$
2. $P\ F\ P\ F\ P\ F\ P\ F\ P\ F\ P\ F\ P\ F\ P\ F$
3. $P\ P\ F\ F\ F\ F\ P\ F\ F\ F\ P\ P\ P\ P\ P\ P$

SOLUTION

1. There are two runs. The first eight Ps form a run of length 8 and the first eight Fs form another run of length 8, as shown below.

$$\underbrace{P\ P\ P\ P\ P\ P\ P\ P}_{\text{1st run}}\quad\underbrace{F\ F\ F\ F\ F\ F\ F\ F}_{\text{2nd run}}$$

2. There are 16 runs each of length 1, as shown below.

$$\underbrace{P}_{\text{1st run}}\ \underbrace{F}_{\text{2nd run...}}\ P\ F\ P\ F\ P\ F\ P\ F\ P\ F\ P\ F\ P\ \underbrace{F}_{\text{...16th run}}$$

3. There are five runs, the first of length 2, the second of length 4, the third of length 1, the fourth of length 3, and the fifth of length 6, as shown below.

$$\underbrace{P\ P}_{\text{1st run}}\quad\underbrace{F\ F\ F\ F}_{\text{2nd run}}\quad\underbrace{P}_{\text{3rd run}}\quad\underbrace{F\ F\ F}_{\text{4th run}}\quad\underbrace{P\ P\ P\ P\ P\ P}_{\text{5th run}}$$

Try It Yourself 1

A machine produces engine parts. An inspector measures the diameter of each engine part and determines if the part passes inspection (P) or fails inspection (F). The results are shown below. Determine the number of runs in the sequence and find the length of each run.

$P\ P\ P\ F\ P\ F\ P\ P\ P\ P\ F\ F\ P\ F\ P\ P\ F\ F\ F\ P\ P\ P\ F\ P\ P\ P$

a. *Separate the data* each time there is a change in the characteristic of the data.
b. *Count the number of groups* to determine the number of runs.
c. *Count the number of data* within each run to determine the length.

Answer: Page A46

When each value in a set of data can be categorized into one of two separate categories, you can use the runs test for randomness to determine whether the data are random.

DEFINITION

The **runs test for randomness** is a nonparametric test that can be used to determine whether a sequence of sample data is random.

The runs test for randomness considers the number of runs in a sequence of sample data in order to test whether a sequence is random. If a sequence has too few or too many runs, it is usually not random. For example, the sequence

$P\ P\ P\ P\ P\ P\ P\ P\ F\ F\ F\ F\ F\ F\ F\ F$

from Example 1, Part 1 has too few runs (only 2 runs). The sequence

$P\ F\ P\ F\ P\ F\ P\ F\ P\ F\ P\ F\ P\ F\ P\ F$

from Example 1, Part 2 has too many runs (16 runs). So, these sample data are probably not random.

You can use a hypothesis test to determine whether the number of runs in a sequence of sample data is too high or too low. The runs test is a two-tailed test, and the null and alternative hypotheses are as follows.

H_0: The sequence of data is random.

H_a: The sequence of data is not random.

When using the runs test, let n_1 represent the number of data that have one characteristic and let n_2 represent the number of data that have the second characteristic. It does not matter which characteristic you choose to be represented by n_1. Let G represent the number of runs.

n_1 = number of data with one characteristic

n_2 = number of data with other characteristic

G = number of runs

Table 12 in Appendix B lists the critical values for the runs test for selected values of n_1 and n_2 at the $\alpha = 0.05$ level of significance. (In this text, you will use only the $\alpha = 0.05$ level of significance when performing runs tests.) If n_1 or n_2 is greater than 20, you can use the standard normal distribution to find the critical values.

You can calculate the test statistic for the runs test as follows.

Test Statistic for the Runs Test

When $n_1 \leq 20$ and $n_2 \leq 20$, the test statistic for the runs test is G, the number of runs.

When $n_1 > 20$ or $n_2 > 20$, the test statistic for the runs test is

$$z = \frac{G - \mu_G}{\sigma_G}$$

where

$$\mu_G = \frac{2n_1n_2}{n_1 + n_2} + 1 \quad \text{and} \quad \sigma_G = \sqrt{\frac{2n_1n_2(2n_1n_2 - n_1 - n_2)}{(n_1 + n_2)^2(n_1 + n_2 - 1)}}.$$

GUIDELINES

Performing a Runs Test for Randomness

In Words	*In Symbols*
1. Identify the claim. State the null and alternative hypotheses.	State H_0 and H_a.
2. Specify the level of significance. (Use $\alpha = 0.05$ for the runs test.)	Identify α.
3. Determine the number of data that have each characteristic and the number of runs.	Determine n_1, n_2, and G.
4. Determine the critical values.	If $n_1 \leq 20$ and $n_2 \leq 20$, use Table 12 in Appendix B. If $n_1 > 20$ or $n_2 > 20$, use Table 4 in Appendix B.
5. Calculate the test statistic.	If $n_1 \leq 20$ and $n_2 \leq 20$, use G. If $n_1 > 20$ or $n_2 > 20$, use $z = \frac{G - \mu_G}{\sigma_G}$.
6. Make a decision to reject or fail to reject the null hypothesis.	If G is less than or equal to the lower critical value or greater than or equal to the upper critical value, reject H_0. Otherwise, fail to reject H_0. Or, if z is in the rejection region, reject H_0. Otherwise, fail to reject H_0.
7. Interpret the decision in the context of the original claim.	

EXAMPLE 2

Using the Runs Test

A foreman for a construction company records injuries reported by workers during his shift. The following sequence shows whether any injuries were reported during each month in a recent year. *I* represents a month in which at least one injury was reported and *N* represents a month in which no injuries were reported. At $\alpha = 0.05$, can you conclude that the occurrence of injuries each month is not random?

I I N N N I N I I N N N

SOLUTION The claim is "The occurrence of injuries is not random." To test this claim, use the following null and alternative hypotheses.

H_0: The occurrence of injuries is random.

H_a: The occurrence of injuries is not random. (Claim)

To find the critical values, first determine n_1, the number of *I*s, n_2, the number of *N*s, and G, the number of runs.

II	NNN	I	N	II	NNN
1st run	2nd run	3rd run	4th run	5th run	6th run

n_1 = number of *I*s = 5

n_2 = number of *N*s = 7

G = number of runs = 6

Because $n_1 \le 20$, $n_2 \le 20$, and $\alpha = 0.05$, use Table 12 to find the lower critical value 3 and the upper critical value 11. The test statistic is the number of runs $G = 6$. Because the test statistic G is between the critical values 3 and 11, you should fail to reject the null hypothesis.

Interpretation At the 5% level, you do not have enough evidence to support the claim that the occurrence of injuries is not random. So, it appears that the injuries reported by workers during the foreman's shift occur randomly.

Try It Yourself 2

The genders of 14 students as they enter a classroom are shown below, where *F* represents the females and *M* represents the males. At $\alpha = 0.05$, can you conclude that the sequence of genders is not random?

F F F M M F F M F M M F F F

a. *Identify* the claim and *state* H_0 and H_a.
b. *Specify* the level of significance α.
c. *Determine* n_1, n_2, and G.
d. *Determine* the critical values.
e. *Calculate* the test statistic G.
f. *Decide* whether to reject the null hypothesis.
g. *Interpret* the decision in the context of the original claim.

Answer: Page A46

EXAMPLE 3

Using the Runs Test

You want to determine whether the selection of recently hired employees in a large company is random with respect to gender. The genders of 36 recently hired employees are shown below. At $\alpha = 0.05$, can you conclude that the selection is not random?

$M\ M\ F\ F\ F\ F\ M\ M\ M\ M\ M\ M$
$F\ F\ F\ F\ F\ M\ M\ M\ M\ M\ M\ M$
$F\ F\ F\ M\ M\ M\ M\ F\ M\ M\ F\ M$

SOLUTION The claim is "The selection of employees is not random." To test this claim, use the following null and alternative hypotheses.

H_0: The selection of employees is random.

H_a: The selection of employees is not random. (Claim)

To find the critical values, first determine n_1, the number of Fs, n_2, the number of Ms, and G, the number of runs.

$$\underbrace{MM}_{\text{1st run}}\ \underbrace{FFFF}_{\text{2nd run}}\ \underbrace{MMMMMM}_{\text{3rd run}}$$

$$\underbrace{FFFFF}_{\text{4th run}}\ \underbrace{MMMMMMM}_{\text{5th run}}$$

$$\underbrace{FFF}_{\text{6th run}}\ \underbrace{MMMM}_{\text{7th run}}\ \underbrace{F}_{\text{8th run}}\ \underbrace{MM}_{\text{9th run}}\ \underbrace{F}_{\text{10th run}}\ \underbrace{M}_{\text{11th run}}$$

n_1 = number of Fs = 14

n_2 = number of Ms = 22

G = number of runs = 11

Because $n_2 > 20$, use Table 4 in Appendix B to find that the critical values at the $\alpha = 0.05$ level of significance are

$$-z_0 = -1.96$$

and

$$z_0 = 1.96.$$

Before calculating the test statistic, find the values of μ_G and σ_G, as follows.

$$\begin{aligned}\mu_G &= \frac{2n_1n_2}{n_1 + n_2} + 1 \\ &= \frac{2(14)(22)}{14 + 22} + 1 \\ &= \frac{616}{36} + 1 \\ &\approx 18.1\end{aligned}$$

Picturing the World

The table shows the NFL conference of each winning team from Super Bowl I to Super Bowl XXXVIII, where *A* represents the American Football Conference and *N* represents the National Football Conference. *(Source: NFL.com)*

Year	Winning Team	Conference
1967	Green Bay	*N*
1968	Green Bay	*N*
1969	N.Y. Jets	*A*
1970	Kansas City	*A*
1971	Baltimore	*A*
1972	Dallas	*N*
1973	Miami	*A*
1974	Miami	*A*
1975	Pittsburgh	*A*
1976	Pittsburgh	*A*
1977	Oakland	*A*
1978	Dallas	*N*
1979	Pittsburgh	*A*
1980	Pittsburgh	*A*
1981	Oakland	*A*
1982	San Francisco	*N*
1983	Washington	*N*
1984	L.A. Raiders	*A*
1985	San Francisco	*N*
1986	Chicago	*N*
1987	N.Y. Giants	*N*
1988	Washington	*N*
1989	San Francisco	*N*
1990	San Francisco	*N*
1991	N.Y. Giants	*N*
1992	Washington	*N*
1993	Dallas	*N*
1994	Dallas	*N*
1995	San Francisco	*N*
1996	Dallas	*N*
1997	Green Bay	*N*
1998	Denver	*A*
1999	Denver	*A*
2000	St. Louis	*N*
2001	Baltimore	*A*
2002	New England	*A*
2003	Tampa Bay	*N*
2004	New England	*A*

At $\alpha = 0.05$, can you conclude that the sequence is not random?

$$\sigma_G = \sqrt{\frac{2n_1n_2(2n_1n_2 - n_1 - n_2)}{(n_1 + n_2)^2(n_1 + n_2 - 1)}}$$

$$= \sqrt{\frac{2(14)(22)[2(14)(22) - 14 - 22]}{(14 + 22)^2(14 + 22 - 1)}}$$

$$\approx 2.8$$

You can find the test statistic as follows.

$$z = \frac{G - \mu_G}{\sigma_G}$$

$$= \frac{11 - 18.1}{2.8}$$

$$\approx -2.54$$

From the graph below, you can see that the test statistic z is in the rejection region. So, you should decide to reject the null hypothesis.

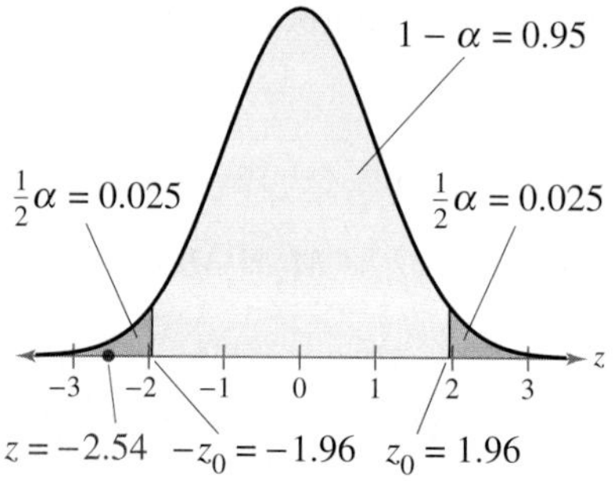

Interpretation You have enough evidence at the 5% level of significance to support the claim that the selection of employees with respect to gender is not random.

Try It Yourself 3

Let *S* represent a day in a small town in which it snowed and let *N* represent a day in the same town in which it did not snow. The following are the snowfall results for the entire month of January. At $\alpha = 0.05$, can you conclude that the sequence is not random?

N N N S S N N S N S N N N N N S
N S N S N N S N S S N N N N N

a. *Identify* the claim and *state* H_0 and H_a.
b. *Specify* the level of significance α.
c. *Determine* n_1, n_2, and G.
d. *Determine* the critical values.
e. *Calculate* the test statistic z.
f. *Decide* whether to reject the null hypothesis.
g. *Interpret* the decision in the context of the original claim.

Answer: Page A46

When n_1 or n_2 is greater than 20, you can also use a P-value to perform a hypothesis test for the randomness of the data. In Example 3, you can calculate the P-value to be 0.011. Because $P < \alpha$, you should reject the null hypothesis.

11.5 Exercises

Building Basic Skills and Vocabulary

Finding the Number of Runs In Exercises 1–4, determine the number of runs in the given sequence. Then find the length of each run.

1. $T\ F\ T\ F\ T\ T\ T\ F\ F\ F\ T\ F$

2. $U\ U\ D\ D\ U\ D\ U\ U\ D\ D\ U\ D\ U\ U$

3. $M\ F\ M\ F\ M\ F\ F\ F\ F\ F\ F\ M\ M\ M\ F\ F\ M\ M\ M\ M$

4. $A\ A\ A\ B\ B\ B\ A\ B\ B\ A\ A\ A\ A\ A\ A\ B\ A\ A\ B\ A\ B\ B$

5. Find the values for n_1 and n_2 in Exercise 1.

6. Find the values for n_1 and n_2 in Exercise 2.

7. Find the values for n_1 and n_2 in Exercise 3.

8. Find the values for n_1 and n_2 in Exercise 4.

Finding Critical Values In Exercises 9–12, use the given sequence and Table 12 in Appendix B to determine the number of runs that are considered too high and the number of runs that are considered too low for the data to be in random order.

9. $T\ F\ T\ F\ T\ F\ T\ F\ T\ F\ T\ F$

10. $M\ F\ M\ M\ M\ M\ M\ M\ F\ F\ M\ M$

11. $N\ S\ S\ S\ N\ N\ N\ N\ N\ S\ N\ S\ N\ S\ S\ N\ N\ N$

12. $X\ X\ X\ X\ X\ X\ X\ Y\ Y\ Y\ Y\ Y\ Y\ Y\ Y\ Y\ Y\ Y\ Y\ Y$

Using and Interpreting Concepts

Performing a Runs Test In Exercises 13–18, use the runs test to

(a) write the claim mathematically and identify H_0 and H_a.

(b) determine the critical values using Table 4 or Table 12 in Appendix B.

(c) calculate the test statistic.

(d) decide whether to reject or fail to reject the null hypothesis.

(e) interpret the decision in the context of the original claim. Use $\alpha = 0.05$.

13. ***Coin Toss*** A coach records the results of the coin toss at the beginning of each football game for a season. The results are shown, where H represents heads and T represents tails. The coach claimed the tosses were not random. Use the runs test to test the coach's claim.

$H\ T\ T\ T\ H\ T\ H\ H\ T\ T\ T\ T\ H\ T\ H\ H$

DATA **14.** ***Council Members*** The mayor called on council members at a town meeting in the sequence shown, where R represents a Republican council member and D represents a Democrat council member. Can you conclude that the selection of members was not random?

$R\ D\ D\ D\ R\ R\ D\ R\ D\ D\ R\ D\ D\ D\ R\ R\ D$
$R\ R\ R\ R\ D\ R\ R\ R\ D\ D\ D\ R\ D\ R\ D\ R\ R$

1. Number of runs: 8
Run lengths: 1, 1, 1, 1, 3, 3, 1, 1

2. Number of runs: 9
Run lengths: 2, 2, 1, 1, 2, 2, 1, 1, 2

3. Number of runs: 9
Run lengths: 1, 1, 1, 1, 1, 6, 3, 2, 4

4. Number of runs: 10
Run lengths: 3, 3, 1, 2, 6, 1, 2, 1, 1, 2

5. n_1 = number of Ts = 6
n_2 = number of Fs = 6

6. n_1 = number of Us = 8
n_2 = number of Ds = 6

7. n_1 = number of Ms = 10
n_2 = number of Fs = 10

8. n_1 = number of As = 13
n_2 = number of Bs = 9

9. too high: 11; too low: 3

10. too high: 8; too low: 2

11. too high: 14; too low: 5

12. too high: 15; too low: 5

13. (a) H_0: The coin tosses are random.
H_a: The coin tosses are not random. (claim)
(b) lower critical value = 4
upper critical value = 14
(c) 9
(d) Fail to reject H_0.
(e) There is not enough evidence at the 5% level to support the claim that the coin tosses are not random.

14. See Selected Answers, page A100.

15. (a) H_0: The sequence of digits are randomly generated.

H_a: The sequence of digits are not randomly generated. (claim)

(b) lower critical value = 11
upper critical value = 23

(c) 9 (d) Reject H_0.

(e) There is enough evidence at the 5% level to support the claim that the sequence of digits was not randomly generated.

16. (a) H_0: The births are random by gender. (claim)

H_a: The births are not random by gender.

(b) lower critical value = 8
upper critical value = 18

(c) 12 (d) Fail to reject H_0.

(e) There is not enough evidence at the 5% level to reject the claim that births are random by gender.

17. (a) H_0: The production of defective parts is random.

H_a: The production of defective parts is not random. (claim)

(b) ± 1.96 (c) -0.85

(d) Fail to reject H_0.

(e) There is not enough evidence at the 5% level to support the claim that the production of defective parts is not random.

18. (a) H_0: The sequence of arrivals is random.

H_a: The sequence of arrivals is not random. (claim)

(b) ± 1.96 (c) 2.6

(d) Reject H_0.

(e) There is enough evidence at the 5% level to support the claim that the sequence of arrivals is not random.

19. Fail to reject H_0. There is not enough evidence at the 5% level to support the claim that the daily high temperatures do not occur randomly.

20. Fail to reject H_0. There is not enough evidence at the 5% level to support the claim that the selection of students' GPAs is not random.

21. Answers will vary.

15. *Number Generator* DATA A number generator outputs the sequence of digits shown below, where O represents an odd digit and E represents an even digit. Test the claim that the digits were not randomly generated.

$O\ O\ O\ E\ E\ E\ E\ O\ O\ O\ O\ O\ E\ E\ E\ E$
$O\ O\ E\ E\ E\ E\ O\ O\ O\ O\ E\ E\ E\ E\ O\ O$

16. *Newborn Babies* DATA The gender of babies born in a hospital in one month was recorded in order of birth, where F represents a female and M represents a male. A doctor claims that the births are random by gender. Do you have enough evidence to reject the doctor's claim?

$M\ M\ F\ M\ F\ F\ F\ F\ F\ M\ M\ M\ F\ F\ F$
$F\ F\ F\ F\ M\ F\ M\ F\ F\ F\ M\ F\ F\ F$

17. *Defective Parts* DATA A machine produces parts that are either defective (D) or not defective (N). An inspector records whether each part produced in one day is defective or not defective and the results are shown. Can you conclude that the production of defective parts is not random?

$N\ N\ N\ N\ N\ D\ D\ N\ N\ N\ N\ N\ D\ N\ N\ N\ N\ N\ D$
$N\ N\ N\ N\ N\ N\ N\ N\ N\ D\ N\ N\ N\ N\ N\ N\ N\ N\ N$
$D\ N\ N\ N\ N\ N\ D\ N\ N\ D\ D$

18. *Tardiness* DATA Students in a class record whether their professor is late (L) or on time (T) for class each day. The results are shown. Can you conclude that the sequence is not random?

$T\ T\ L\ T\ T\ L\ T\ T\ L\ T\ T\ L\ T\ T\ L\ T\ T\ L\ T\ T\ T\ T\ T\ T$
$L\ T\ T\ L\ T\ T\ L\ T\ T\ L\ T\ T\ L\ T\ L\ T\ T\ L\ T\ T\ T\ T\ T\ L$
$T\ T\ T\ T\ T\ L\ T\ T\ T\ L$

Extending Concepts

Runs Test with Quantitative Data In Exercises 19 and 20, use the following information to perform a runs test. You can also use the runs test for randomness with quantitative data. First, calculate the median. Then assign + to those values above the median and − to those values below the median. Ignore any values that are equal to the median.

19. *Daily High Temperatures* DATA The sequence shows the daily high temperatures (in degrees Fahrenheit) for a city during the month of July. Test the claim that the daily high temperatures do not occur randomly. Use $\alpha = 0.05$.

84 87 92 93 95 84 82 83 81 87 92
98 99 93 84 85 86 92 91 95 84 92
83 81 87 92 98 89 93 84 85

20. *Grade Point Averages* The sequence shows the grade point averages (GPAs) of students selected from a class. Test the claim that the selection of students' GPAs was not random.

3.8 4.0 1.8 1.9 2.5 2.3 3.6 2.4 2.1
1.8 0.7 0.9 1.3 1.8 2.9 3.5 3.9 4.0

21. Use a technology tool to generate a sequence of 30 numbers between 1 and 99. Test the claim that the sequence of numbers is not random.

Uses and Abuses
Statistics in the Real World

Uses

Nonparametric Tests Before you could perform many of the hypothesis tests you learned about in previous chapters, you had to ensure that certain conditions about the population were satisfied. For example, before you could run a t-test, you had to verify that the population is normally distributed. One advantage of the nonparametric tests shown in this chapter is that they are distribution free. That is, they do not require any particular information about the population or populations being tested. Another advantage of nonparametric tests is that they are easier to perform than their parametric counterparts. This means that they are easier to understand and quicker to use. Nonparametric tests can often be used when data are at the nominal or ordinal level.

Abuses

Insufficient Evidence Stronger evidence is needed to reject a null hypothesis in a nonparametric test than it is to reject a null hypothesis in a corresponding parametric test. That is, when you are trying to support a claim represented by the alternative hypothesis, you might need a larger sample when performing a nonparametric test. If the outcome of a nonparametric test results in failing to reject the null hypothesis, you should investigate the sample size used. It may be that a larger sample will produce different results.

Using an Inappropriate Test In general, when information about the populations (such as the condition of normality) is known, it is more efficient to use a parametric test. However, if information about the population is not known, nonparametric tests can be helpful.

Exercises

1. ***Insufficient Evidence*** Give an example of a nonparametric test in which there is not enough evidence to reject the null hypothesis.
2. ***Using an Inappropriate Test*** Discuss the nonparametric tests described in this chapter and match each test with its parametric counterpart, which you studied in earlier chapters.

11 Chapter Summary

What did you learn?

Review Exercises

Section 11.1

- How to use the sign test to test a population median and to test the difference between two population medians (dependent samples) — *1–6*

$$z = \frac{(x + 0.5) - 0.5n}{\frac{\sqrt{n}}{2}}$$

Section 11.2

- How to use the Wilcoxon signed-rank test and Wilcoxon rank sum test to test the difference between two population distributions — *7, 8*

$$z = \frac{R - \mu_R}{\sigma_R}, \quad \mu_R = \frac{n_1(n_1 + n_2 + 1)}{2}, \quad \sigma_R = \sqrt{\frac{n_1 n_2(n_1 + n_2 + 1)}{12}}$$

Section 11.3

- How to use the Kruskal-Wallis test to test for differences among three or more population distributions — *9, 10*

$$H = \frac{12}{N(N + 1)}\left(\frac{R_1^2}{n_1} + \frac{R_2^2}{n_2} + \cdots + \frac{R_k^2}{n_k}\right) - 3(N + 1)$$

Section 11.4

- How to use the Spearman rank correlation coefficient to determine whether the correlation between two variables is significant — *11, 12*

$$r_s = 1 - \frac{6\Sigma d^2}{n(n^2 - 1)}$$

Section 11.5

- How to use the runs test to determine whether a data set is random — *13, 14*

$$G = \text{number of runs}, \quad z = \frac{G - \mu_G}{\sigma_G}, \quad \mu_G = \frac{2n_1 n_2}{n_1 + n_2} + 1, \quad \sigma_G = \sqrt{\frac{2n_1 n_2(2n_1 n_2 - n_1 - n_2)}{(n_1 + n_2)^2(n_1 + n_2 - 1)}}$$

The table summarizes parametric and nonparametric tests. Always use the parametric test if the conditions for that test are satisfied.

Test application	Parametric test	Nonparametric test
One-sample tests	*z*-test for a population mean *t*-test for a population mean	Sign test for a population median
Two-sample tests		
Dependent samples	*t*-test for the difference between means	Paired-sample sign test Wilcoxon signed-rank test
Independent samples	*z*-test for the difference between means *t*-test for the difference between means	Wilcoxon rank sum test
Tests involving three or more samples	One-way ANOVA	Kruskal-Wallis test
Correlation	Pearson correlation coefficient	Spearman rank correlation coefficient
Randomness	(No parametric test)	Runs test

11 Review Exercises

1. (a) H_0: median = \$34,300 (claim)
   H_a: median ≠ \$34,300
   (b) 2
   (c) 7
   (d) Fail to reject H_0.
   (e) There is not enough evidence at the 1% level to reject the claim that the median value of stock among families that own stock is \$34,300.
2. (a) H_0: median ≤ \$1200
   H_a: median > \$1200 (claim)
   (b) 1
   (c) 6
   (d) Fail to reject H_0.
   (e) There is not enough evidence at the 1% level to support the claim that the median credit card debt is more than \$1200.
3. (a) H_0: median ≤ 6 (claim)
   H_a: median > 6
   (b) −1.28
   (c) −2.032
   (d) Reject H_0.
   (e) There is enough evidence at the 10% level to reject the claim that the median turnover time is no more than 6 hours.
4. (a) H_0: There is no reduction in diastolic blood pressure. (claim)
   H_a: There is a reduction in diastolic blood pressure.
   (b) 1
   (c) 4
   (d) Fail to reject H_0.
   (e) There is not enough evidence at the 5% level to reject the claim that there is no reduction in diastolic blood pressure.

Section 11.1

In Exercises 1–6, use a sign test to test the claim by doing the following.

(a) Write the claim mathematically and identify H_0 and H_a.

(b) Determine the critical value.

(c) Calculate the test statistic.

(d) Decide whether to reject or fail to reject the null hypothesis.

(e) Interpret the decision in the context of the original claim.

1. A financial services institution estimates that the median value of stock among families that own stock is \$34,300. The stock values (in thousands of dollars) among 17 randomly selected families that own stock are listed below. At $\alpha = 0.01$, can you reject the institution's claim? *(Adapted from Board of Governors of the Federal Reserve System)*

24.89	38.89	29.32	40.10	35.27	30.62	33.26	21.75	41.01
30.01	30.09	30.74	31.64	29.94	34.87	38.13	34.54	

2. A financial services institution claims that the median credit card debt among families with annual earnings of \$10,000 to \$24,999 is more than \$1200. The credit card debts (in thousands of dollars) among 13 randomly selected families are listed below. At $\alpha = 0.01$, can you support the institution's claim? *(Adapted from Board of Governors of the Federal Reserve System)*

1.97	1.10	1.02	1.05	0.98	1.36	1.74
1.69	1.48	0.99	1.29	0.92	1.45	

3. A mail-order company believes that the median turnover time between receipt of a telephone order and packing of that order is six hours or less. Over a five-day period, 78 orders are randomly selected and their turnover time is recorded in half-hour increments. Eight orders took six hours, 26 orders took under six hours, and 44 orders took over six hours. At $\alpha = 0.10$, can you reject the company's claim?

DATA **4.** In a study testing the effects of calcium supplements on blood pressure in men, 10 randomly selected men were given a calcium supplement for 12 weeks. The following measurements are for each subject's diastolic blood pressure taken before and after the 12-week treatment period. At $\alpha = 0.05$, can you reject the claim that there was no reduction in diastolic blood pressure? *(Adapted from The Journal of American Medicine)*

Patient	1	2	3	4	5	6	7
Before treatment	107	110	123	129	112	111	107
After treatment	100	114	105	112	115	116	106

Patient	8	9	10
Before treatment	112	136	102
After treatment	102	125	104

5. (a) H_0: There is no reduction in diastolic blood pressure. (claim)

   H_a: There is a reduction in diastolic blood pressure.

   (b) 2

   (c) 3

   (d) Fail to reject H_0.

   (e) There is not enough evidence at the 5% level to reject the claim that there was no reduction in diastolic blood pressure.

6. (a) H_0: median = \$27,900 (claim)

   H_a: median $\neq$ \$27,900

   (b) ± 1.96

   (c) -1.497

   (d) Fail to reject H_0.

   (e) There is not enough evidence at the 5% level to reject the claim that the median starting salary is \$27,900.

7. (a) Independent; Wilcoxon rank sum test

   (b) H_0: There is no difference in the amount of time that it takes to earn a doctorate.

   H_a: There is a difference in the amount of time that it takes to earn a doctorate. (claim)

   (c) ± 2.575

   (d) -3.175

   (e) Reject H_0.

   (f) There is enough evidence at the 1% level to support the claim that there is a difference in the amount of time that it takes to earn a doctorate.

DATA

5. In a study testing the effects of an herbal supplement on blood pressure in men, 11 randomly selected men were given an herbal supplement for 15 weeks. The following measurements are for each subject's diastolic blood pressure taken before and after the 15-week treatment period. At $\alpha = 0.05$, can you reject the claim that there was no reduction in diastolic blood pressure? *(Adapted from The Journal of American Medicine)*

Patient	1	2	3	4	5	6	7
Before treatment	123	109	112	102	98	114	119
After treatment	124	97	113	105	95	119	114

Patient	8	9	10	11
Before treatment	112	110	117	130
After treatment	114	121	118	133

6. The career placement office at a large university claims that the median starting salary of graduates with a bachelor's degree in marketing is \$27,900. Of last year's graduates in marketing, 57 were randomly surveyed about their salaries. Of the 54 graduates who were currently employed, 21 were paid less than \$27,900 annually, and 33 were paid more than \$27,900. At $\alpha = 0.05$, can you reject the office's claim? *(Adapted from National Association of Colleges and Employers)*

Section 11.2

In Exercises 7 and 8, use a Wilcoxon test to test the claim by doing the following.

(a) Decide whether the samples are dependent or independent; then choose the appropriate Wilcoxon test.

(b) Write the claim mathematically and identify H_0 and H_a.

(c) Determine the critical value.

(d) Calculate the test statistic.

(e) Decide whether to reject or fail to reject the null hypothesis.

(f) Interpret the decision in the context of the original claim.

DATA

7. A career placement advisor estimates that there is a difference in the total time to earn a doctorate degree by female and male graduate students. A random sample of 12 female and 12 male graduate students and their total time to earn a doctorate degree is shown in the table. At $\alpha = 0.01$, can you support the advisor's claim? *(Adapted from National Research Council)*

Gender	**Total Time (in years)**											
Female	15	14	15	15	15	11	13	16	9	9	11	13
Male	11	8	9	11	10	8	8	10	11	9	10	8

8. A medical researcher claims that a new drug affects the number of headache hours experienced by headache sufferers. The number of headache hours (per day) experienced by eight randomly selected patients before and after taking the drug are shown in the table. At $\alpha = 0.05$, can you support the researcher's claim?

Patient	1	2	3	4	5	6	7	8
Headache hours (before)	0.9	2.3	2.7	2.4	2.9	1.9	1.3	3.1
Headache hours (after)	1.4	1.5	1.4	1.8	1.3	0.6	0.7	1.9

Section 11.3

In Exercises 9 and 10, use the Kruskal-Wallis test to test the claim by doing the following.

(a) Write the claim mathematically and identify H_0 and H_a.
(b) Determine the critical value.
(c) Find the sums of the ranks for each sample and calculate the test statistic.
(d) Decide whether to reject or fail to reject the null hypothesis.
(e) Interpret the decision in the context of the original claim.

DATA **9.** The table shows the starting salaries for a random sample of college graduates in three fields of study. At $\alpha = 0.05$, can you conclude that the distributions of the starting salaries in these three fields of study are different? *(Adapted from U.S. Department of Education)*

Field of Study	Starting Salary (in thousands of dollars)									
Education	20.4	20.5	20.4	20.2	20.4	20.2	20.1	20.1	19.4	19.9
Humanities	21.1	21.3	21.1	20.8	20.9	21.3	20.7	21.5	20.6	21.0
Natural science	22.2	22.1	22.2	21.0	21.1	21.8	22.6	22.0	22.5	21.2

DATA **10.** The table shows the total time to earn a doctorate degree for a random sample of college graduates in three fields of study. At $\alpha = 0.05$, can you conclude that the distributions of the total times in these three fields of study are different? *(Adapted from National Research Council)*

Field of Study	Total Time (in years)										
Computer science	9	10	9	7	11	9	8	10	9	9	10
Natural science	8	8	7	8	7	7	8	8	8	8	7
Social science	12	13	11	11	10	9	14	10	10	11	12

8. (a) Dependent; Wilcoxon signed-rank test
(b) H_0: The new drug does not affect the number of headache hours experienced.
H_a: The new drug does affect the number of headache hours experienced. (claim)
(c) 4
(d) 1
(e) Reject H_0.
(f) There is enough evidence at the 5% level to support the claim that the new drug does affect the number of headache hours experienced.

9. (a) H_0: There is no difference in salaries between the fields of study.
H_a: There is a difference in salaries between the fields of study. (claim)
(b) 5.991
(c) 22.98
(d) Reject H_0.
(e) There is enough evidence at the 5% level to conclude that there is a difference in salaries between the fields of study.

10. (a) H_0: There is no difference in the amount of time to earn a doctorate between the fields of study.
H_a: There is a difference in the amount of time to earn a doctorate between the fields of study. (claim)
(b) 5.991
(c) 22.28
(d) Reject H_0.
(e) There is enough evidence at the 5% level to conclude that there is a difference in the amount of time to earn a doctorate between the fields of study.

11. (a) H_0: $\rho_s = 0$; H_a: $\rho_s \neq 0$ (claim)
(b) 0.881
(c) −0.429
(d) Fail to reject H_0.
(e) There is not enough evidence at the 1% level to conclude that there is a correlation between overall score and price.

12. (a) H_0: $\rho_s = 0$; H_a: $\rho_s \neq 0$ (claim)
(b) 0.700
(c) 0.238
(d) Fail to reject H_0.
(e) There is not enough evidence at the 5% level to support the claim that there is a correlation between overall score and price.

13. (a) H_0: The traffic stops are random by gender.
H_a: The traffic stops are not random by gender. (claim)
(b) lower critical value = 8
upper critical value = 19
(c) 14
(d) Fail to reject H_0.
(e) There is not enough evidence at the 5% level to support the claim that the traffic stops are not random by gender.

14. (a) H_0: The departure status of buses is random.
H_a: The departure status of buses is not random. (claim)
(b) lower critical value = 5
upper critical value = 14
(c) 5
(d) Reject H_0.
(e) There is enough evidence at the 5% level to support the claim that the departure status of buses is not random.

Section 11.4

In Exercises 11 and 12, use the Spearman rank correlation coefficient to test the claim by doing the following.

(a) Write the claim mathematically and identify H_0 and H_a.
(b) Determine the critical value using Table 10 in Appendix B.
(c) Find the standardized test statistic r_s.
(d) Decide whether to reject the null hypothesis.
(e) Interpret the decision in the context of the original claim.

11. The table shows the overall scores and the prices for eight randomly selected televisions. The overall score represents the television's picture quality, sound quality, ease of use, and cable performance. At $\alpha = 0.01$, can you conclude that there is a correlation between overall score and price? *(Adapted from Consumer Reports)*

Overall score	80	79	70	65	63	62	61	58
Price (in dollars)	580	740	630	660	600	590	700	750

12. The table shows the overall scores and the prices for nine randomly selected cordless phones. The overall score represents the phone's speech clarity and handset convenience. At $\alpha = 0.05$, can you conclude that there is a correlation between overall score and price? *(Adapted from Consumer Reports)*

Overall score	81	78	72	68	66	65	67	63	57
Price (in dollars)	130	200	70	80	105	130	100	150	60

Section 11.5

In Exercises 13 and 14, use the runs test to (a) write the claim mathematically and identify H_0 and H_a, (b) determine the critical values using Table 12 in Appendix B, (c) calculate the test statistic, (d) decide whether to reject or fail to reject the null hypothesis, and (e) interpret the decision in the context of the original claim.

DATA **13.** A highway patrol officer stops speeding vehicles on an interstate highway. The following shows the gender of the last 25 drivers who were stopped, where F represents a female driver and M represents a male driver. At $\alpha = 0.05$, can you conclude that the stops are not random by gender?

F M M M F M F M F F F M M
F F F M M M F M M F F M

14. The following data represent the departure status of the last 18 buses to leave a bus station, where T represents a bus departed on time and L represents a bus that was late. At $\alpha = 0.05$, can you conclude that the departure status of the buses is not random?

T T T T L L L L T L L L T T T T T T

11 Chapter Quiz

1. (a) H_0: There is no difference in the salaries between genders.
   H_a: There is a difference in the salaries between genders. (claim)
   (b) Wilcoxon rank sum test
   (c) ± 1.645 (d) -1.722
   (e) Reject H_0.
   (f) There is enough evidence at the 10% level to support the claim that there is a difference in the salaries between genders.
2. (a) H_0: median $= 28$ (claim)
   H_a: median $\neq 28$
   (b) Sign test (c) 6 (d) 10
   (e) Fail to reject H_0.
   (f) There is not enough evidence at the 5% level to reject the claim that the median age in Puerto Rico is 28 years.
3. (a) H_0: There is no difference in the annual premiums between the states.
   H_a: There is a difference in the annual premiums between the states. (claim)
   (b) Kruskal-Wallis test
   (c) 5.991 (d) 1.43
   (e) Fail to reject H_0.
   (f) There is not enough evidence at the 5% level to conclude that there is a difference in the annual premiums between the states.
4. (a) H_0: The days with rain are random.
   H_a: The days with rain are not random. (claim)
   (b) Runs test
   (c) lower critical value = 10
   upper critical value = 22
   (d) 16 (e) Fail to reject H_0.
   (f) There is not enough evidence at the 5% level to support the claim that days with rain are not random.

Take this quiz as you would take a quiz in class. After you are done, check your work against the answers given in the back of the book.

For this quiz, do the following.

(a) Write the claim mathematically and identify H_0 and H_a.
(b) Decide which test to use.
(c) Determine the critical value(s).
(d) Calculate the test statistic.
(e) Decide whether to reject or fail to reject the null hypothesis.
(f) Interpret the decision in the context of the original claim.

1. A women's organization claims that there is a difference in the salaries earned by female and male employees of state and local governments. A random sample of nine female and nine male state and local government employees and their salaries is listed in the table. At $\alpha = 0.10$, can you support the organization's claim? *(Adapted from U.S. Equal Employment Opportunity Commission)*

Gender	Salary (in thousands of dollars)								
Female	29.2	28.8	28.4	32.0	23.8	34.3	24.8	25.6	22.8
Male	43.9	35.3	20.4	31.3	38.8	30.7	31.8	27.3	32.4

2. A government official believes that the median age in Puerto Rico is 28 years. In a random sample of 25 Puerto Ricans, 10 were younger than 28 years, 13 were older than 28 years, and 2 were 28 years. At $\alpha = 0.05$, can you reject the official's claim? *(Adapted from U.S. Census Bureau)*

DATA **3.** An independent insurance representative wants to determine whether there is a difference in the annual premiums for low-risk homeowner's insurance in three states: California, Florida, and Texas. To do so, the representative randomly selects several homes with low-risk insurance in each state and determines the annual homeowner's insurance premium for each. At $\alpha = 0.05$, can the representative conclude that the distributions of the annual premiums in these states are different? *(Adapted from Consumer Reports)*

State	Annual Premium (in dollars)						
California	756	429	476	427	501	512	492
Florida	677	701	573	405	604	506	458
Texas	591	490	735	546	633	502	447

DATA **4.** A meteorologist wants to determine whether days with rain occur randomly in April in his home town. To do so, the meteorologist records whether it rains for each day in April. The results are shown, where R represents a day with rain and N represents a day with no rain. At $\alpha = 0.05$, can the meteorologist conclude that days with rain are not random?

N R R N N N N R N R R N R R R
R N N N N R N R R R N R N N R

PUTTING IT ALL TOGETHER

Real Statistics ■ Real Decisions

In a recent year, according to the Bureau of Labor Statistics, the median number of years that wage and salary workers had been with their current employer (called employee tenure) was 3.5 years. Information on employee tenure has been gathered since the early 1950s using the *Current Population Survey* (*CPS*), a monthly survey of 50,000 households that provides information on employment, unemployment, earnings, demographics, and other characteristics of the U.S. population. With respect to employee tenure, the questions measure how long workers have been with their current employer, not how long they plan to stay with their employer.

stats.bls.gov

Employee Tenure of 20 Workers		
4.1	2.1	2.8
2.3	1.0	1.4
3.5	4.5	3.4
4.6	3.2	4.9
3.1	3.4	5.7
1.2	4.1	2.6
3.9	3.1	

Table for Exercise 2

Employee tenure for a sample of male workers	Employee tenure for a sample of female workers
3.3	3.7
3.9	4.2
4.1	2.7
3.3	3.6
4.4	3.3
3.3	1.1
3.1	4.4
4.1	4.4
2.7	2.6
4.9	1.5
0.9	4.5
4.6	2.0
	4.0
	3.0

Table for Exercise 3

Exercises

1. *How Would You Do It?*

(a) What sampling technique would you use to select the sample for the *CPS*?

(b) Do you think the technique in part (a) will give you a sample that is representative of the U.S. population? Why or why not?

(c) Identify possible flaws or biases in the survey on the basis of the technique you chose in part (a).

2. *Is There a Difference?*

A congressional representative claims that the median tenure for workers from the representative's district is less than the national median tenure of 3.5 years. The claim is based on the representative's data and is shown in the table at the above right. (Assume that the employees were randomly selected.)

(a) Is it possible that the claim is true? What questions should you ask about how the data were collected?

(b) How would you test the representative's claim? Can you use a parametric test, or do you need to use a nonparametric test?

(c) State the null hypothesis and the alternative hypothesis.

(d) Test the claim using $\alpha = 0.05$. What can you conclude?

3. *Comparing Male and Female Employee Tenures*

A congressional representative claims that the median tenure for male workers is greater than the median tenure for female workers. The claim is based on the representative's data and is shown in the table at the right. (Assume that the employees were randomly selected from the representative's district.)

(a) How would you test the representative's claim? Can you use a parametric test, or do you need to use a nonparametric test?

(b) State the null hypothesis and the alternative hypothesis.

(c) Test the claim using $\alpha = 0.05$. What can you conclude?

TECHNOLOGY

MINITAB | EXCEL | TI-83

U.S. Income and Economic Research

The National Bureau of Economic Research (NBER) is a private, nonprofit economic research organization. The NBER provides information to better understand how the U.S. economy works. Researchers at the NBER concentrate on four types of empirical research: developing new statistical measurements, estimating quantitative models of economic behavior, assessing the effects of public policies on the U.S. economy, and projecting the effects of alternative policy proposals.

One of the NBER's interests is the median income of people in different regions of the United States. The table at the right shows the annual incomes (in dollars) of a random sample of people (15 years and over) in a recent year in four U.S. regions: Northeast, Midwest, South, and West.

Annual Income of People (in dollars)			
Northeast	**Midwest**	**South**	**West**
44,501	26,000	18,755	34,000
32,645	28,500	32,668	29,118
28,997	26,775	31,005	22,500
25,000	34,625	32,000	30,000
26,000	32,669	17,000	31,444
32,900	31,475	22,700	20,500
30,928	30,500	23,000	26,900
30,000	42,000	29,798	38,415
27,450	19,665	34,445	22,228
22,600	40,178	31,500	20,000
39,600	29,370	20,400	26,776
47,450	27,660	23,600	33,564

Exercises

In Exercises 1–5, refer to the annual income of people in the table. Use $\alpha = 0.05$ for all tests.

1. Construct a box-and-whisker plot for each region. Do the median annual incomes appear to differ between regions?

2. Use a technology tool to perform a sign test to test the claim that the median annual income in the Midwest is at least \$25,000.

3. Use a technology tool to perform a Wilcoxon rank sum test to test the claim that the median annual incomes in the Northeast and South are the same.

4. Use a technology tool to perform a Kruskal-Wallis test to test the claim that the median annual incomes for all four regions are the same.

5. Use a technology tool to perform a one-way ANOVA to test the claim that the average annual incomes for all four regions are the same. How do your results compare with those in Exercise 4?

6. Repeat Exercises 1, 3, 4, and 5 using the data in the following table. The table shows the annual incomes (in dollars) of a random sample of families in a recent year in four U.S. regions: Northeast, Midwest, South, and West.

Annual Income of Families (in dollars)			
Northeast	**Midwest**	**South**	**West**
50,410	50,420	46,000	50,478
99,665	76,000	48,787	58,795
68,000	29,800	40,200	39,652
102,800	45,000	85,000	56,000
56,128	64,558	46,009	42,500
42,100	42,788	26,000	67,890
38,566	60,500	52,600	49,778
58,950	54,576	29,556	74,800
41,000	55,660	55,421	75,625
65,720	50,000	43,600	30,400
36,995	55,000	70,222	54,720
57,850	52,336	41,885	36,580
50,500	63,250	50,000	51,034
65,000	54,200	35,900	55,550
49,999	70,100	47,862	35,692

Extended solutions are given in the *Technology Supplement.* Technical instruction is provided for MINITAB, Excel, and the TI-83.

Appendix A

In this appendix, we use a 0-to-*z* table as an alternative development of the standard normal distribution. It is intended that this appendix be used after completing the "Properties of a Normal Distribution" subsection of Section 5.1 in the text. If used, this appendix should replace the material in the "Standard Normal Distribution" subsection of Section 5.1 except for the exercises.

Standard Normal Distribution (0-to-*z*)

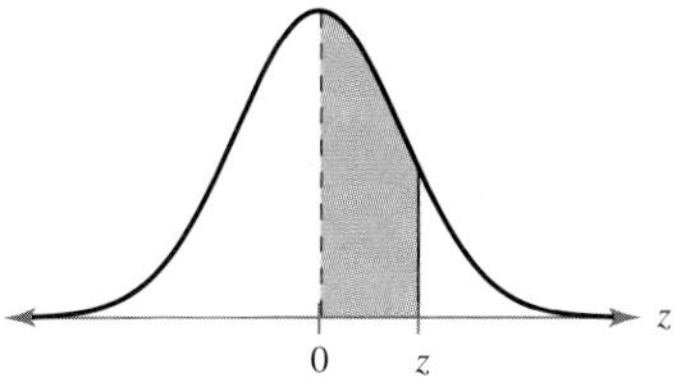

z	.00	.01	.02	.03	.04	.05	.06	.07	.08	.09
0.0	.0000	.0040	.0080	.0120	.0160	.0199	.0239	.0279	.0319	.0359
0.1	.0398	.0438	.0478	.0517	.0557	.0596	.0636	.0675	.0714	.0753
0.2	.0793	.0832	.0871	.0910	.0948	.0987	.1026	.1064	.1103	.1141
0.3	.1179	.1217	.1255	.1293	.1331	.1368	.1406	.1443	.1480	.1517
0.4	.1554	.1591	.1628	.1664	.1700	.1736	.1772	.1808	.1844	.1879
0.5	.1915	.1950	.1985	.2019	.2054	.2088	.2123	.2157	.2190	.2224
0.6	.2257	.2291	.2324	.2357	.2389	.2422	.2454	.2486	.2517	.2549
0.7	.2580	.2611	.2642	.2673	.2704	.2734	.2764	.2794	.2823	.2852
0.8	.2881	.2910	.2939	.2967	.2995	.3023	.3051	.3078	.3106	.3133
0.9	.3159	.3186	.3212	.3238	.3264	.3289	.3315	.3340	.3365	.3389
1.0	.3413	.3438	.3461	.3485	.3508	.3531	.3554	.3577	.3599	.3621
1.1	.3643	.3665	.3686	.3708	.3729	.3749	.3770	.3790	.3810	.3830
1.2	.3849	.3869	.3888	.3907	.3925	.3944	.3962	.3980	.3997	.4015
1.3	.4032	.4049	.4066	.4082	.4099	.4115	.4131	.4147	.4162	.4177
1.4	.4192	.4207	.4222	.4236	.4251	.4265	.4279	.4292	.4306	.4319
1.5	.4332	.4345	.4357	.4370	.4382	.4394	.4406	.4418	.4429	.4441
1.6	.4452	.4463	.4474	.4484	.4495	.4505	.4515	.4525	.4535	.4545
1.7	.4554	.4564	.4573	.4582	.4591	.4599	.4608	.4616	.4625	.4633
1.8	.4641	.4649	.4656	.4664	.4671	.4678	.4686	.4693	.4699	.4706
1.9	.4713	.4719	.4726	.4732	.4738	.4744	.4750	.4756	.4761	.4767
2.0	.4772	.4778	.4783	.4788	.4793	.4798	.4803	.4808	.4812	.4817
2.1	.4821	.4826	.4830	.4834	.4838	.4842	.4846	.4850	.4854	.4857
2.2	.4861	.4864	.4868	.4871	.4875	.4878	.4881	.4884	.4887	.4890
2.3	.4893	.4896	.4898	.4901	.4904	.4906	.4909	.4911	.4913	.4916
2.4	.4918	.4920	.4922	.4925	.4927	.4929	.4931	.4932	.4934	.4936
2.5	.4938	.4940	.4941	.4943	.4945	.4946	.4948	.4949	.4951	.4952
2.6	.4953	.4955	.4956	.4957	.4959	.4960	.4961	.4962	.4963	.4964
2.7	.4965	.4966	.4967	.4968	.4969	.4970	.4971	.4972	.4973	.4974
2.8	.4974	.4975	.4976	.4977	.4977	.4978	.4979	.4979	.4980	.4981
2.9	.4981	.4982	.4982	.4983	.4984	.4984	.4985	.4985	.4986	.4986
3.0	.4987	.4987	.4987	.4988	.4988	.4989	.4989	.4989	.4990	.4990
3.1	.4990	.4991	.4991	.4991	.4992	.4992	.4992	.4992	.4993	.4993
3.2	.4993	.4993	.4994	.4994	.4994	.4994	.4994	.4995	.4995	.4995
3.3	.4995	.4995	.4995	.4996	.4996	.4996	.4996	.4996	.4996	.4997
3.4	.4997	.4997	.4997	.4997	.4997	.4997	.4997	.4997	.4997	.4998

Alternative Presentation of the Standard Normal Distribution

What You Should Learn

- How to find areas under the standard normal curve

The Standard Normal Distribution

The Standard Normal Distribution

There are infinitely many normal distributions, each with its own mean and standard deviation. The normal distribution with a mean of 0 and a standard deviation of 1 is called the **standard normal distribution.** The horizontal scale of the graph of the standard normal distribution corresponds to z-scores. In Section 2.5, you learned that a z-score is a measure of position that indicates the number of standard deviations a value lies from the mean. Recall that you can transform an x-value to a z-score using the formula

$$z = \frac{\text{value} - \text{mean}}{\text{standard deviation}} = \frac{x - \mu}{\sigma}.$$

DEFINITION

The **standard normal distribution** is a normal distribution with a mean of 0 and a standard deviation of 1.

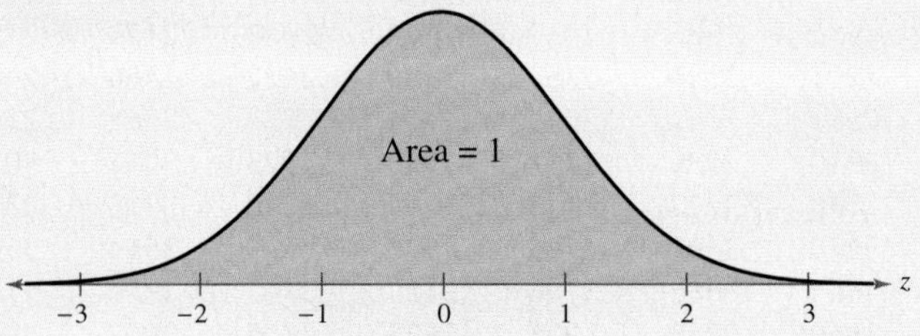

Standard Normal Distribution

If each data value of a normally distributed random variable x is transformed into a z-score, the result will be the standard normal distribution. When this transformation takes place, the area that falls in the interval under the nonstandard normal curve is the *same* as that under the standard normal curve within the corresponding z-boundaries.

In Section 5.1, you learned to approximate areas under a normal curve when values of the random variable x corresponded to −3, −2, −1, 0, 1, 2, or 3 standard deviations from the mean. In this section, you will learn to calculate areas corresponding to other x-values. After you transform an x-value to a z-score, you can use the Standard Normal Table (0-to-z) on page A1. The table lists the area under the standard normal curve between 0 and the given z-score. As you examine the table, notice the following properties.

Properties of the Standard Normal Distribution

1. The distribution is symmetric about the mean ($z = 0$).
2. The area under the standard normal curve to the left of $z = 0$ is 0.5 and the area to the right of $z = 0$ is 0.5.
3. The area under the standard normal curve increases as the distance between 0 and z increases.

Insight

Because every normal distribution can be transformed to the standard normal distribution, you can use z-scores and the standard normal curve to find areas (and therefore probability) under any normal curve.

Note to Instructor

Mention that the formula for a normal probability density function on page 206 is greatly simplified when $\mu = 0$ and $\sigma = 1$.

$$y = \frac{e^{-x^2/2}}{\sqrt{2\pi}}$$

Study Tip

It is important that you know the difference between x and z. The random variable x is sometimes called a raw score and represents values in a *nonstandard* normal distribution, whereas z represents values in the *standard* normal distribution.

At first glance, the table on page A1 appears to give areas for positive z-scores only. However, because of the symmetry of the standard normal curve, the table also gives areas for negative z-scores (see Example 1).

Study Tip

When the z-score is not in the table, use the entry closest to it. If the given z-score is exactly midway between two z-scores, then use the area midway between the corresponding areas.

EXAMPLE 1

Using the Standard Normal Table (0-to-z)

1. Find the area under the standard normal curve between $z = 0$ and $z = 1.15$.
2. Find the z-scores that correspond to an area of 0.0948.

SOLUTION

1. Find the area that corresponds to $z = 1.15$ by finding 1.1 in the left column and then moving across the row to the column under 0.05. The number in that row and column is 0.3749. So, the area between $z = 0$ and $z = 1.15$ is 0.3749.

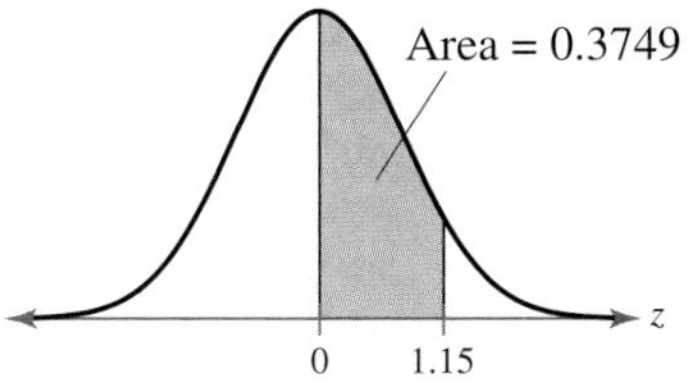

z	.00	.01	.02	.03	.04	.05	.06
0.0	.0000	.0040	.0080	.0120	.0160	.0199	.0239
0.1	.0398	.0438	.0478	.0517	.0557	.0596	.0636
0.2	.0793	.0832	.0871	.0910	.0948	.0987	.1026
0.3	.1179	.1217	.1255	.1293	.1331	.1368	.1406
0.9	.3159	.3186	.3212	.3238	.3264	.3289	.3315
1.0	.3413	.3438	.3461	.3485	.3508	.3531	.3554
1.1	.3643	.3665	.3686	.3708	.3729	.3749	.3770
1.2	.3849	.3869	.3888	.3907	.3925	.3944	.3962
1.3	.4032	.4049	.4066	.4082	.4099	.4115	.4131
1.4	.4192	.4207	.4222	.4236	.4251	.4265	.4279

2. Find the z-scores that correspond to an area of 0.0948 by locating 0.0948 in the table. The values at the beginning of the corresponding row and at the top of the corresponding column give the z-score. For an area of 0.0948, the row value is 0.2 and the column value is 0.04. So, the z-scores are $z = -0.24$ and $z = 0.24$.

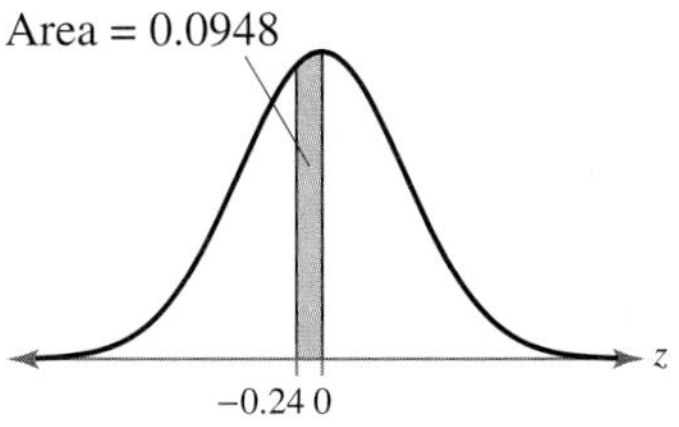

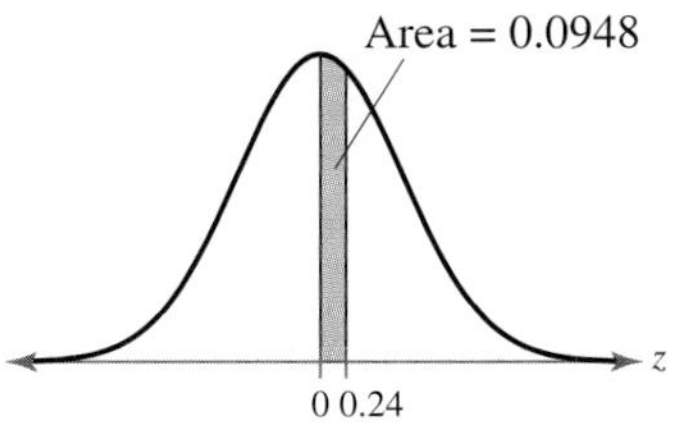

z	.00	.01	.02	.03	.04	.05	.06
0.0	.0000	.0040	.0080	.0120	.0160	.0199	.0239
0.1	.0398	.0438	.0478	.0517	.0557	.0596	.0636
0.2	.0793	.0832	.0871	.0910	.0948	.0987	.1026
0.3	.1179	.1217	.1255	.1293	.1331	.1368	.1406
0.4	.1554	.1591	.1628	.1664	.1700	.1736	.1772
0.5	.1915	.1950	.1985	.2019	.2054	.2088	.2123

Try It Yourself 1

1. Find the area under the standard normal curve between $z = 0$ and $z = 2.19$.
   a. Locate the given z-score and *find the corresponding area* in the Standard Normal Table (0-to-z) on page A1.
2. Find the z-scores that correspond to an area of 0.4850.
   a. Locate the given area in the Standard Normal Table (0-to-z) on page A1 and *find the corresponding z-score.*

Answer: Page A46

Use the following guidelines to find various types of areas under the standard normal curve.

GUIDELINES

Finding Areas under the Standard Normal Curve

1. Sketch the standard normal curve and shade the appropriate area under the curve.

2. Use the Standard Normal Table (0-to-z) on page A1 to find the area that corresponds to the given z-score(s).

3. Find the desired area by following the directions for each case shown.

a. Area to the left of z

i. When $z < 0$, *subtract* the area from 0.5.

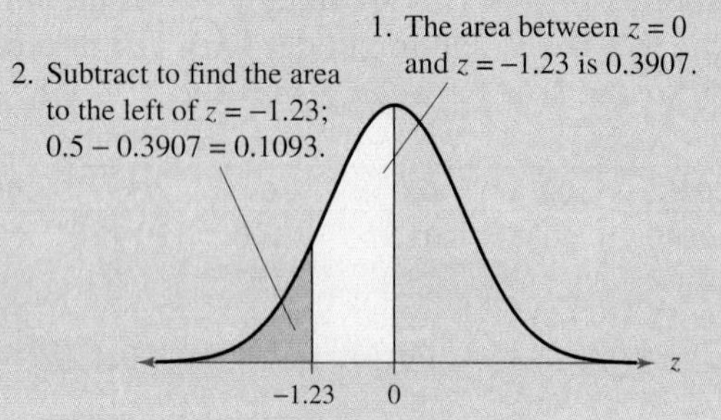

ii. When $z > 0$, *add* 0.5 to the area.

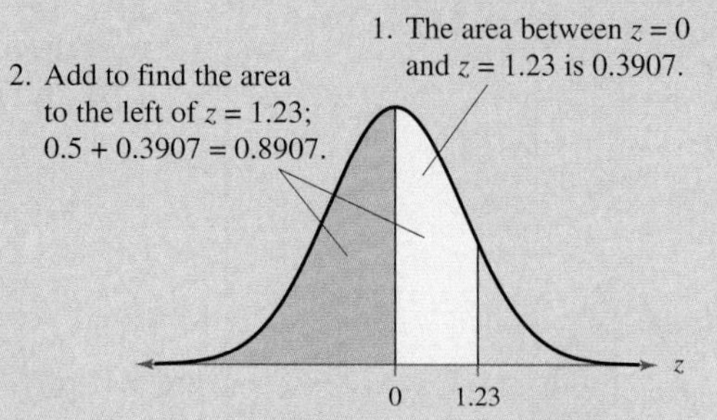

b. Area to the right of z

i. When $z < 0$, *add* 0.5 to the area.

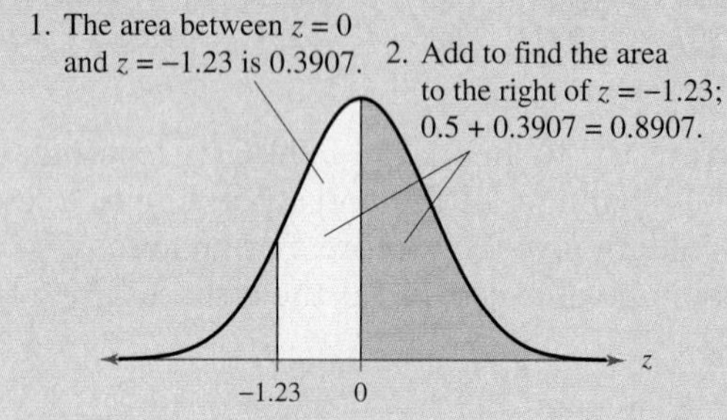

ii. When $z > 0$, *subtract* the area from 0.5.

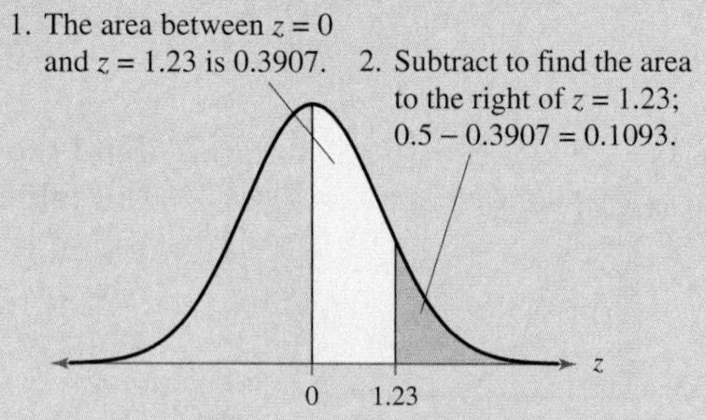

c. Area between two z-scores

i. When the two z-scores have the same sign (both positive or both negative), *subtract* the smaller area from the larger area.

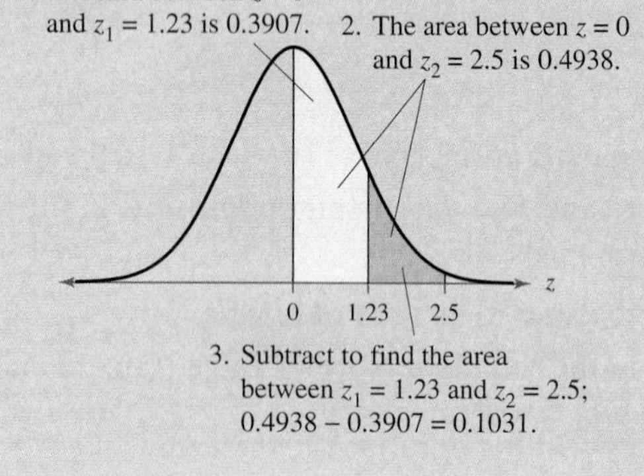

ii. When the two z-scores have opposite signs (one negative and one positive), *add* the areas.

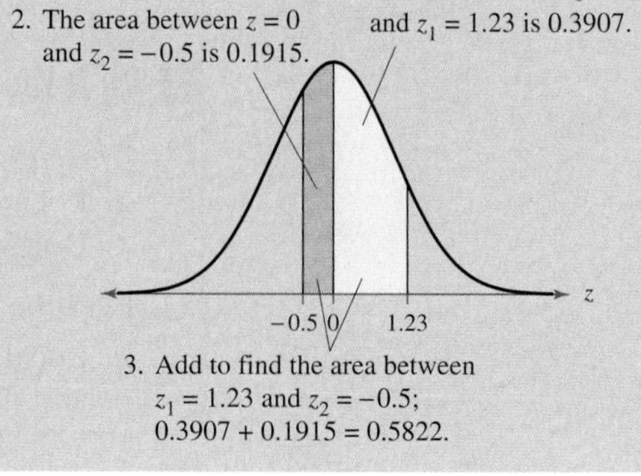

EXAMPLE 2

Finding Area under the Standard Normal Curve

Find the area under the standard normal curve to the left of $z = -0.99$.

SOLUTION The area under the standard normal curve to the left of $z = -0.99$ is shown.

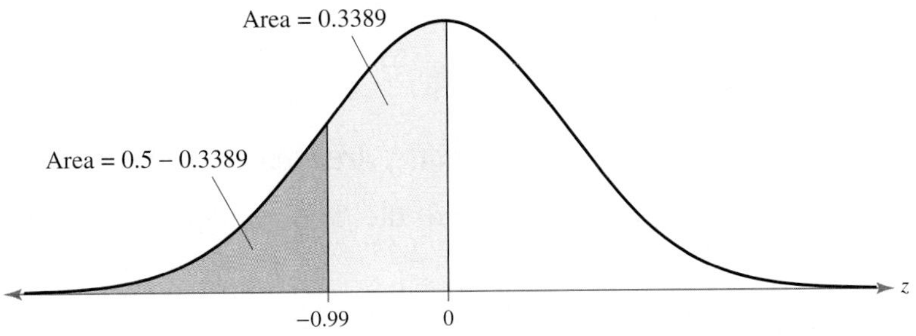

From the Standard Normal Table (0-to-z), the area corresponding to $z = -0.99$ is 0.3389. Because the area to the left of $z = 0$ is 0.5, the area to the left of $z = -0.99$ is $0.5 - 0.3389 = 0.1611$.

> **Insight**
>
> Because the normal distribution is a continuous probability distribution, the area under the standard normal curve to the left of a z-score gives the probability that z is less than that z-score. For instance, in Example 2, the area to the left of $z = -0.99$ is 0.1611. So, $P(z < -0.99) = 0.1611$, which is read as "the probability that z is less than -0.99 is 0.1611."

Try It Yourself 2

Find the area under the standard normal curve to the left of $z = 2.13$.

a. *Draw* the standard normal curve and shade the area under the curve and to the left of $z = 2.13$.
b. Use the Standard Normal Table (0-to-z) on page A1 to *find the area* that corresponds to $z = 2.13$.
c. *Add* 0.5 to the resulting area.

Answer: Page A46

EXAMPLE 3

Finding Area under the Standard Normal Curve

Find the area under the standard normal curve to the right of $z = 1.06$.

SOLUTION The area under the standard normal curve to the right of $z = 1.06$ is shown.

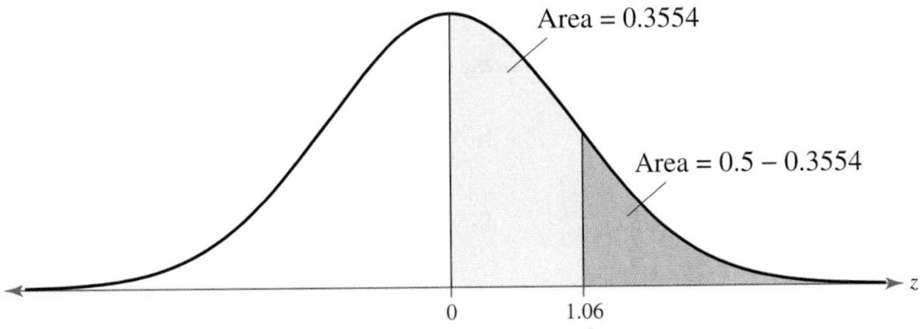

From the Standard Normal Table (0-to-z), the area corresponding to $z = 1.06$ is 0.3554. Because the area to the right of $z = 0$ is 0.5, the area to the right of $z = 1.06$ is $0.5 - 0.3554 = 0.1446$.

Picturing the World

Each year the Centers for Disease Control and Prevention and the National Center for Health Statistics jointly publish a report summarizing the vital statistics from the previous year. According to one publication, the number of births in a recent year was 4,021,726. The weights of the newborns can be approximated by a normal distribution, as shown by the following graph.

Weights of Newborns

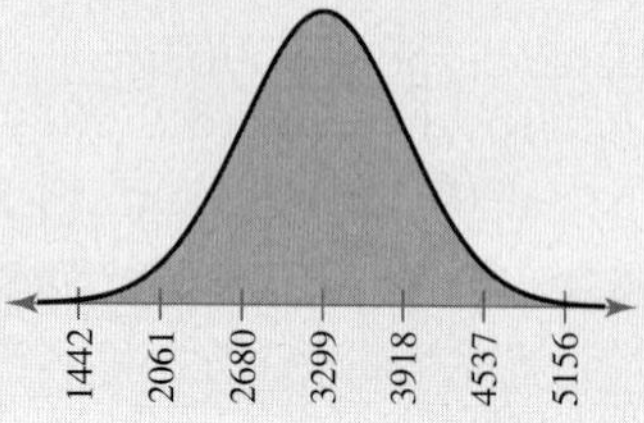

The weights of three newborns are 2000 grams, 3000 grams, and 4000 grams. Find the z-score that corresponds to each weight. Are any of these unusually heavy or light?

Try It Yourself 3

Find the area under the standard normal curve to the right of $z = -2.16$.

a. *Draw* the standard normal curve and shade the area below the curve and to the right of $z = -2.16$.
b. Use the Standard Normal Table (0-to-z) on page A1 to *find the area* that corresponds to $z = -2.16$.
c. *Add* 0.5 to the resulting area.

Answer: Page A46

EXAMPLE 4

Finding Area under the Standard Normal Curve

Find the area under the standard normal curve between $z = -1.5$ and $z = 1.25$.

SOLUTION The area under the standard normal curve between $z = -1.5$ and $z = 1.25$ is shown.

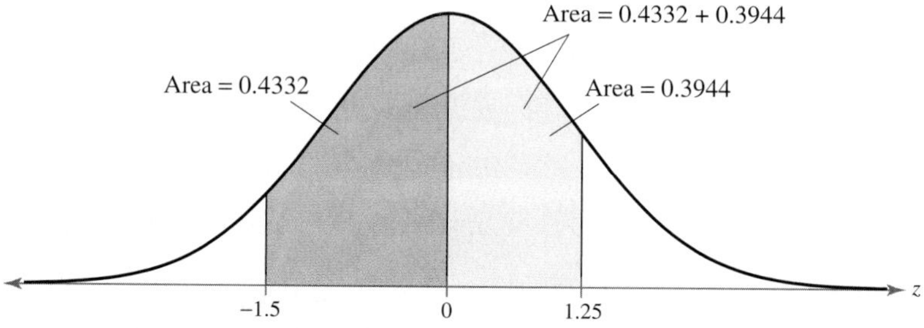

From the Standard Normal Table, the area corresponding to $z = -1.5$ is 0.4332 and the area corresponding to $z = 1.25$ is 0.3944. To find the area between these two z-scores, add the resulting areas.

$$\text{Area} = 0.4332 + 0.3944 = 0.8276$$

In other words, 82.76% of the area under the curve falls between $z = -1.5$ and $z = 1.25$.

Try It Yourself 4

Find the area under the standard normal curve between $z = -2.16$ and $z = -1.35$.

a. *Draw* the standard normal curve and shade the area below the curve that is between $z = -2.16$ and $z = -1.35$.
b. Use the Standard Normal Table (0-to-z) on page A1 to *find the areas* that correspond to $z = -2.16$ and to $z = -1.35$.
c. *Subtract* the smaller area from the larger area.

Answer: Page A46

Recall in Section 2.5 you learned, using the Empirical Rule, that values lying more than two standard deviations from the mean are considered unusual. Values lying more than three standard deviations from the mean are considered *very* unusual. So if a z-score is greater than 2 or less than -2, it is unusual. If a z-score is greater than 3 or less than -3, it is *very* unusual.

Appendix B

Table 1—Random Numbers

92630	78240	19267	95457	53497	23894	37708	79862	76471	66418
79445	78735	71549	44843	26104	67318	00701	34986	66751	99723
59654	71966	27386	50004	05358	94031	29281	18544	52429	06080
31524	49587	76612	39789	13537	48086	59483	60680	84675	53014
06348	76938	90379	51392	55887	71015	09209	79157	24440	30244
28703	51709	94456	48396	73780	06436	86641	69239	57662	80181
68108	89266	94730	95761	75023	48464	65544	96583	18911	16391
99938	90704	93621	66330	33393	95261	95349	51769	91616	33238
91543	73196	34449	63513	83834	99411	58826	40456	69268	48562
42103	02781	73920	56297	72678	12249	25270	36678	21313	75767
17138	27584	25296	28387	51350	61664	37893	05363	44143	42677
28297	14280	54524	21618	95320	38174	60579	08089	94999	78460
09331	56712	51333	06289	75345	08811	82711	57392	25252	30333
31295	04204	93712	51287	05754	79396	87399	51773	33075	97061
36146	15560	27592	42089	99281	59640	15221	96079	09961	05371
29553	18432	13630	05529	02791	81017	49027	79031	50912	09399
23501	22642	63081	08191	89420	67800	55137	54707	32945	64522
57888	85846	67967	07835	11314	01545	48535	17142	08552	67457
55336	71264	88472	04334	63919	36394	11196	92470	70543	29776
10087	10072	55980	64688	68239	20461	89381	93809	00796	95945
34101	81277	66090	88872	37818	72142	67140	50785	21380	16703
53362	44940	60430	22834	14130	96593	23298	56203	92671	15925
82975	66158	84731	19436	55790	69229	28661	13675	99318	76873
54827	84673	22898	08094	14326	87038	42892	21127	30712	48489
25464	59098	27436	89421	80754	89924	19097	67737	80368	08795
67609	60214	41475	84950	40133	02546	09570	45682	50165	15609
44921	70924	61295	51137	47596	86735	35561	76649	18217	63446
33170	30972	98130	95828	49786	13301	36081	80761	33985	68621
84687	85445	06208	17654	51333	02878	35010	67578	61574	20749
71886	56450	36567	09395	96951	35507	17555	35212	69106	01679
00475	02224	74722	14721	40215	21351	08596	45625	83981	63748
25993	38881	68361	59560	41274	69742	40703	37993	03435	18873
92882	53178	99195	93803	56985	53089	15305	50522	55900	43026
25138	26810	07093	15677	60688	04410	24505	37890	67186	62829
84631	71882	12991	83028	82484	90339	91950	74579	03539	90122
34003	92326	12793	61453	48121	74271	28363	66561	75220	35908
53775	45749	05734	86169	42762	70175	97310	73894	88606	19994
59316	97885	72807	54966	60859	11932	35265	71601	55577	67715
20479	66557	50705	26999	09854	52591	14063	30214	19890	19292
86180	84931	25455	26044	02227	52015	21820	50599	51671	65411
21451	68001	72710	40261	61281	13172	63819	48970	51732	54113
98062	68375	80089	24135	72355	95428	11808	29740	81644	86610
01788	64429	14430	94575	75153	94576	61393	96192	03227	32258
62465	04841	43272	68702	01274	05437	22953	18946	99053	41690
94324	31089	84159	92933	99989	89500	91586	02802	69471	68274
05797	43984	21575	09908	70221	19791	51578	36432	33494	79888
10395	14289	52185	09721	25789	38562	54794	04897	59012	89251
35177	56986	25549	59730	64718	52630	31100	62384	49483	11409
25633	89619	75882	98256	02126	72099	57183	55887	09320	73463
16464	48280	94254	45777	45150	68865	11382	11782	22695	41988

Table 2—Binomial Distribution

This table shows the probability of *x* successes in *n* independent trials, each with probability of success *p*.

		p																			
n	*x*	.01	.05	.10	.15	.20	.25	.30	.35	.40	.45	.50	.55	.60	.65	.70	.75	.80	.85	.90	.95
2	0	.980	.902	.810	.723	.640	.563	.490	.423	.360	.303	.250	.203	.160	.123	.090	.063	.040	.023	.010	.002
	1	.020	.095	.180	.255	.320	.375	.420	.455	.480	.495	.500	.495	.480	.455	.420	.375	.320	.255	.180	.095
	2	.000	.002	.010	.023	.040	.063	.090	.123	.160	.203	.250	.303	.360	.423	.490	.563	.640	.723	.810	.902
3	0	.970	.857	.729	.614	.512	.422	.343	.275	.216	.166	.125	.091	.064	.043	.027	.016	.008	.003	.001	.000
	1	.029	.135	.243	.325	.384	.422	.441	.444	.432	.408	.375	.334	.288	.239	.189	.141	.096	.057	.027	.007
	2	.000	.007	.027	.057	.096	.141	.189	.239	.288	.334	.375	.408	.432	.444	.441	.422	.384	.325	.243	.135
	3	.000	.000	.001	.003	.008	.016	.027	.043	.064	.091	.125	.166	.216	.275	.343	.422	.512	.614	.729	.857
4	0	.961	.815	.656	.522	.410	.316	.240	.179	.130	.092	.062	.041	.026	.015	.008	.004	.002	.001	.000	.000
	1	.039	.171	.292	.368	.410	.422	.412	.384	.346	.300	.250	.200	.154	.112	.076	.047	.026	.011	.004	.000
	2	.001	.014	.049	.098	.154	.211	.265	.311	.346	.368	.375	.368	.346	.311	.265	.211	.154	.098	.049	.014
	3	.000	.000	.004	.011	.026	.047	.076	.112	.154	.200	.250	.300	.346	.384	.412	.422	.410	.368	.292	.171
	4	.000	.000	.000	.001	.002	.004	.008	.015	.026	.041	.062	.092	.130	.179	.240	.316	.410	.522	.656	.815
5	0	.951	.774	.590	.444	.328	.237	.168	.116	.078	.050	.031	.019	.010	.005	.002	.001	.000	.000	.000	.000
	1	.048	.204	.328	.392	.410	.396	.360	.312	.259	.206	.156	.113	.077	.049	.028	.015	.006	.002	.000	.000
	2	.001	.021	.073	.138	.205	.264	.309	.336	.346	.337	.312	.276	.230	.181	.132	.088	.051	.024	.008	.001
	3	.000	.001	.008	.024	.051	.088	.132	.181	.230	.276	.312	.337	.346	.336	.309	.264	.205	.138	.073	.021
	4	.000	.000	.000	.002	.006	.015	.028	.049	.077	.113	.156	.206	.259	.312	.360	.396	.410	.392	.328	.204
	5	.000	.000	.000	.000	.000	.001	.002	.005	.010	.019	.031	.050	.078	.116	.168	.237	.328	.444	.590	.774
6	0	.941	.735	.531	.377	.262	.178	.118	.075	.047	.028	.016	.008	.004	.002	.001	.000	.000	.000	.000	.000
	1	.057	.232	.354	.399	.393	.356	.303	.244	.187	.136	.094	.061	.037	.020	.010	.004	.002	.000	.000	.000
	2	.001	.031	.098	.176	.246	.297	.324	.328	.311	.278	.234	.186	.138	.095	.060	.033	.015	.006	.001	.000
	3	.000	.002	.015	.042	.082	.132	.185	.236	.276	.303	.312	.303	.276	.236	.185	.132	.082	.042	.015	.002
	4	.000	.000	.001	.006	.015	.033	.060	.095	.138	.186	.234	.278	.311	.328	.324	.297	.246	.176	.098	.031
	5	.000	.000	.000	.000	.002	.004	.010	.020	.037	.061	.094	.136	.187	.244	.303	.356	.393	.399	.354	.232
	6	.000	.000	.000	.000	.000	.000	.001	.002	.004	.008	.016	.028	.047	.075	.118	.178	.262	.377	.531	.735
7	0	.932	.698	.478	.321	.210	.133	.082	.049	.028	.015	.008	.004	.002	.001	.000	.000	.000	.000	.000	.000
	1	.066	.257	.372	.396	.367	.311	.247	.185	.131	.087	.055	.032	.017	.008	.004	.001	.000	.000	.000	.000
	2	.002	.041	.124	.210	.275	.311	.318	.299	.261	.214	.164	.117	.077	.047	.025	.012	.004	.001	.000	.000
	3	.000	.004	.023	.062	.115	.173	.227	.268	.290	.292	.273	.239	.194	.144	.097	.058	.029	.011	.003	.000
	4	.000	.000	.003	.011	.029	.058	.097	.144	.194	.239	.273	.292	.290	.268	.227	.173	.115	.062	.023	.004
	5	.000	.000	.000	.001	.004	.012	.025	.047	.077	.117	.164	.214	.261	.299	.318	.311	.275	.210	.124	.041
	6	.000	.000	.000	.000	.000	.001	.004	.008	.017	.032	.055	.087	.131	.185	.247	.311	.367	.396	.372	.257
	7	.000	.000	.000	.000	.000	.000	.000	.001	.002	.004	.008	.015	.028	.049	.082	.133	.210	.321	.478	.698
8	0	.923	.663	.430	.272	.168	.100	.058	.032	.017	.008	.004	.002	.001	.000	.000	.000	.000	.000	.000	.000
	1	.075	.279	.383	.385	.336	.267	.198	.137	.090	.055	.031	.016	.008	.003	.001	.000	.000	.000	.000	.000
	2	.003	.051	.149	.238	.294	.311	.296	.259	.209	.157	.109	.070	.041	.022	.010	.004	.001	.000	.000	.000
	3	.000	.005	.033	.084	.147	.208	.254	.279	.279	.257	.219	.172	.124	.081	.047	.023	.009	.003	.000	.000
	4	.000	.000	.005	.018	.046	.087	.136	.188	.232	.263	.273	.263	.232	.188	.136	.087	.046	.018	.005	.000
	5	.000	.000	.000	.003	.009	.023	.047	.081	.124	.172	.219	.257	.279	.279	.254	.208	.147	.084	.033	.005
	6	.000	.000	.000	.000	.001	.004	.010	.022	.041	.070	.109	.157	.209	.259	.296	.311	.294	.238	.149	.051
	7	.000	.000	.000	.000	.000	.000	.001	.003	.008	.016	.031	.055	.090	.137	.198	.267	.336	.385	.383	.279
	8	.000	.000	.000	.000	.000	.000	.000	.000	.001	.002	.004	.008	.017	.032	.058	.100	.168	.272	.430	.663
9	0	.914	.630	.387	.232	.134	.075	.040	.021	.010	.005	.002	.001	.000	.000	.000	.000	.000	.000	.000	.000
	1	.083	.299	.387	.368	.302	.225	.156	.100	.060	.034	.018	.008	.004	.001	.000	.000	.000	.000	.000	.000
	2	.003	.063	.172	.260	.302	.300	.267	.216	.161	.111	.070	.041	.021	.010	.004	.001	.000	.000	.000	.000
	3	.000	.008	.045	.107	.176	.234	.267	.272	.251	.212	.164	.116	.074	.042	.021	.009	.003	.001	.000	.000
	4	.000	.001	.007	.028	.066	.117	.172	.219	.251	.260	.246	.213	.167	.118	.074	.039	.017	.005	.001	.000
	5	.000	.000	.001	.005	.017	.039	.074	.118	.167	.213	.246	.260	.251	.219	.172	.117	.066	.028	.007	.001
	6	.000	.000	.000	.001	.003	.009	.021	.042	.074	.116	.164	.212	.251	.272	.267	.234	.176	.107	.045	.008
	7	.000	.000	.000	.000	.000	.001	.004	.010	.021	.041	.070	.111	.161	.216	.267	.300	.302	.260	.172	.063
	8	.000	.000	.000	.000	.000	.000	.000	.001	.004	.008	.018	.034	.060	.100	.156	.225	.302	.368	.387	.299
	9	.000	.000	.000	.000	.000	.000	.000	.000	.000	.001	.002	.005	.010	.021	.040	.075	.134	.232	.387	.630

Table 2—Binomial Distribution *(continued)*

		p																			
n	*x*	.01	.05	.10	.15	.20	.25	.30	.35	.40	.45	.50	.55	.60	.65	.70	.75	.80	.85	.90	.95
10	0	.904	.599	.349	.197	.107	.056	.028	.014	.006	.003	.001	.000	.000	.000	.000	.000	.000	.000	.000	.000
	1	.091	.315	.387	.347	.268	.188	.121	.072	.040	.021	.010	.004	.002	.000	.000	.000	.000	.000	.000	.000
	2	.004	.075	.194	.276	.302	.282	.233	.176	.121	.076	.044	.023	.011	.004	.001	.000	.000	.000	.000	.000
	3	.000	.010	.057	.130	.201	.250	.267	.252	.215	.166	.117	.075	.042	.021	.009	.003	.001	.000	.000	.000
	4	.000	.001	.011	.040	.088	.146	.200	.238	.251	.238	.205	.160	.111	.069	.037	.016	.006	.001	.000	.000
	5	.000	.000	.001	.008	.026	.058	.103	.154	.201	.234	.246	.234	.201	.154	.103	.058	.026	.008	.001	.000
	6	.000	.000	.000	.001	.006	.016	.037	.069	.111	.160	.205	.238	.251	.238	.200	.146	.088	.040	.011	.001
	7	.000	.000	.000	.000	.001	.003	.009	.021	.042	.075	.117	.166	.215	.252	.267	.250	.201	.130	.057	.010
	8	.000	.000	.000	.000	.000	.000	.001	.004	.011	.023	.044	.076	.121	.176	.233	.282	.302	.276	.194	.075
	9	.000	.000	.000	.000	.000	.000	.000	.000	.002	.004	.010	.021	.040	.072	.121	.188	.268	.347	.387	.315
	10	.000	.000	.000	.000	.000	.000	.000	.000	.000	.000	.001	.003	.006	.014	.028	.056	.107	.197	.349	.599
11	0	.895	.569	.314	.167	.086	.042	.020	.009	.004	.001	.000	.000	.000	.000	.000	.000	.000	.000	.000	.000
	1	.099	.329	.384	.325	.236	.155	.093	.052	.027	.013	.005	.002	.001	.000	.000	.000	.000	.000	.000	.000
	2	.005	.087	.213	.287	.295	.258	.200	.140	.089	.051	.027	.013	.005	.002	.001	.000	.000	.000	.000	.000
	3	.000	.014	.071	.152	.221	.258	.257	.225	.177	.126	.081	.046	.023	.010	.004	.001	.000	.000	.000	.000
	4	.000	.001	.016	.054	.111	.172	.220	.243	.236	.206	.161	.113	.070	.038	.017	.006	.002	.000	.000	.000
	5	.000	.000	.002	.013	.039	.080	.132	.183	.221	.236	.226	.193	.147	.099	.057	.027	.010	.002	.000	.000
	6	.000	.000	.000	.002	.010	.027	.057	.099	.147	.193	.226	.236	.221	.183	.132	.080	.039	.013	.002	.000
	7	.000	.000	.000	.000	.002	.006	.017	.038	.070	.113	.161	.206	.236	.243	.220	.172	.111	.054	.016	.001
	8	.000	.000	.000	.000	.000	.001	.004	.010	.023	.046	.081	.126	.177	.225	.257	.258	.221	.152	.071	.014
	9	.000	.000	.000	.000	.000	.000	.001	.002	.005	.013	.027	.051	.089	.140	.200	.258	.295	.287	.213	.087
	10	.000	.000	.000	.000	.000	.000	.000	.000	.001	.002	.005	.013	.027	.052	.093	.155	.236	.325	.384	.329
	11	.000	.000	.000	.000	.000	.000	.000	.000	.000	.000	.000	.001	.004	.009	.020	.042	.086	.167	.314	.569
12	0	.886	.540	.282	.142	.069	.032	.014	.006	.002	.001	.000	.000	.000	.000	.000	.000	.000	.000	.000	.000
	1	.107	.341	.377	.301	.206	.127	.071	.037	.017	.008	.003	.001	.000	.000	.000	.000	.000	.000	.000	.000
	2	.006	.099	.230	.292	.283	.232	.168	.109	.064	.034	.016	.007	.002	.001	.000	.000	.000	.000	.000	.000
	3	.000	.017	.085	.172	.236	.258	.240	.195	.142	.092	.054	.028	.012	.005	.001	.000	.000	.000	.000	.000
	4	.000	.002	.021	.068	.133	.194	.231	.237	.213	.170	.121	.076	.042	.020	.008	.002	.001	.000	.000	.000
	5	.000	.000	.004	.019	.053	.103	.158	.204	.227	.223	.193	.149	.101	.059	.029	.011	.003	.001	.000	.000
	6	.000	.000	.000	.004	.016	.040	.079	.128	.177	.212	.226	.212	.177	.128	.079	.040	.016	.004	.000	.000
	7	.000	.000	.000	.001	.003	.011	.029	.059	.101	.149	.193	.223	.227	.204	.158	.103	.053	.019	.004	.000
	8	.000	.000	.000	.000	.001	.002	.008	.020	.042	.076	.121	.170	.213	.237	.231	.194	.133	.068	.021	.002
	9	.000	.000	.000	.000	.000	.000	.001	.005	.012	.028	.054	.092	.142	.195	.240	.258	.236	.172	.085	.017
	10	.000	.000	.000	.000	.000	.000	.000	.001	.002	.007	.016	.034	.064	.109	.168	.232	.283	.292	.230	.099
	11	.000	.000	.000	.000	.000	.000	.000	.000	.000	.001	.003	.008	.017	.037	.071	.127	.206	.301	.377	.341
	12	.000	.000	.000	.000	.000	.000	.000	.000	.000	.000	.000	.001	.002	.006	.014	.032	.069	.142	.282	.540
15	0	.860	.463	.206	.087	.035	.013	.005	.002	.000	.000	.000	.000	.000	.000	.000	.000	.000	.000	.000	.000
	1	.130	.366	.343	.231	.132	.067	.031	.013	.005	.002	.000	.000	.000	.000	.000	.000	.000	.000	.000	.000
	2	.009	.135	.267	.286	.231	.156	.092	.048	.022	.009	.003	.001	.000	.000	.000	.000	.000	.000	.000	.000
	3	.000	.031	.129	.218	.250	.225	.170	.111	.063	.032	.014	.005	.002	.000	.000	.000	.000	.000	.000	.000
	4	.000	.005	.043	.116	.188	.225	.219	.179	.127	.078	.042	.019	.007	.002	.001	.000	.000	.000	.000	.000
	5	.000	.001	.010	.045	.103	.165	.206	.212	.186	.140	.092	.051	.024	.010	.003	.001	.000	.000	.000	.000
	6	.000	.000	.002	.013	.043	.092	.147	.191	.207	.191	.153	.105	.061	.030	.012	.003	.001	.000	.000	.000
	7	.000	.000	.000	.003	.014	.039	.081	.132	.177	.201	.196	.165	.118	.071	.035	.013	.003	.001	.000	.000
	8	.000	.000	.000	.001	.003	.013	.035	.071	.118	.165	.196	.201	.177	.132	.081	.039	.014	.003	.000	.000
	9	.000	.000	.000	.000	.001	.003	.012	.030	.061	.105	.153	.191	.207	.191	.147	.092	.043	.013	.002	.000
	10	.000	.000	.000	.000	.000	.001	.003	.010	.024	.051	.092	.140	.186	.212	.206	.165	.103	.045	.010	.001
	11	.000	.000	.000	.000	.000	.000	.001	.002	.007	.019	.042	.078	.127	.179	.219	.225	.188	.116	.043	.005
	12	.000	.000	.000	.000	.000	.000	.000	.000	.002	.005	.014	.032	.063	.111	.170	.225	.250	.218	.129	.031
	13	.000	.000	.000	.000	.000	.000	.000	.000	.000	.001	.003	.009	.022	.048	.092	.156	.231	.286	.267	.135
	14	.000	.000	.000	.000	.000	.000	.000	.000	.000	.000	.000	.002	.005	.013	.031	.067	.132	.231	.343	.366
	15	.000	.000	.000	.000	.000	.000	.000	.000	.000	.000	.000	.000	.000	.002	.005	.013	.035	.087	.206	.463

Table 2—Binomial Distribution *(continued)*

		p																			
n	*x*	.01	.05	.10	.15	.20	.25	.30	.35	.40	.45	.50	.55	.60	.65	.70	.75	.80	.85	.90	.95
16	0	.851	.440	.185	.074	.028	.010	.003	.001	.000	.000	.000	.000	.000	.000	.000	.000	.000	.000	.000	.000
	1	.138	.371	.329	.210	.113	.053	.023	.009	.003	.001	.000	.000	.000	.000	.000	.000	.000	.000	.000	.000
	2	.010	.146	.275	.277	.211	.134	.073	.035	.015	.006	.002	.001	.000	.000	.000	.000	.000	.000	.000	.000
	3	.000	.036	.142	.229	.246	.208	.146	.089	.047	.022	.009	.003	.001	.000	.000	.000	.000	.000	.000	.000
	4	.000	.006	.051	.131	.200	.225	.204	.155	.101	.057	.028	.011	.004	.001	.000	.000	.000	.000	.000	.000
	5	.000	.001	.014	.056	.120	.180	.210	.201	.162	.112	.067	.034	.014	.005	.001	.000	.000	.000	.000	.000
	6	.000	.000	.003	.018	.055	.110	.165	.198	.198	.168	.122	.075	.039	.017	.006	.001	.000	.000	.000	.000
	7	.000	.000	.000	.005	.020	.052	.101	.152	.189	.197	.175	.132	.084	.044	.019	.006	.001	.000	.000	.000
	8	.000	.000	.000	.001	.006	.020	.049	.092	.142	.181	.196	.181	.142	.092	.049	.020	.006	.001	.000	.000
	9	.000	.000	.000	.000	.001	.006	.019	.044	.084	.132	.175	.197	.189	.152	.101	.052	.020	.005	.000	.000
	10	.000	.000	.000	.000	.000	.001	.006	.017	.039	.075	.122	.168	.198	.198	.165	.110	.055	.018	.003	.000
	11	.000	.000	.000	.000	.000	.000	.001	.005	.014	.034	.067	.112	.162	.201	.210	.180	.120	.056	.014	.001
	12	.000	.000	.000	.000	.000	.000	.000	.001	.004	.011	.028	.057	.101	.155	.204	.225	.200	.131	.051	.006
	13	.000	.000	.000	.000	.000	.000	.000	.000	.001	.003	.009	.022	.047	.089	.146	.208	.246	.229	.142	.036
	14	.000	.000	.000	.000	.000	.000	.000	.000	.000	.001	.002	.006	.015	.035	.073	.134	.211	.277	.275	.146
	15	.000	.000	.000	.000	.000	.000	.000	.000	.000	.000	.000	.001	.003	.009	.023	.053	.113	.210	.329	.371
	16	.000	.000	.000	.000	.000	.000	.000	.000	.000	.000	.000	.000	.000	.001	.003	.010	.028	.074	.185	.440
20	0	.818	.358	.122	.039	.012	.003	.001	.000	.000	.000	.000	.000	.000	.000	.000	.000	.000	.000	.000	.000
	1	.165	.377	.270	.137	.058	.021	.007	.002	.000	.000	.000	.000	.000	.000	.000	.000	.000	.000	.000	.000
	2	.016	.189	.285	.229	.137	.067	.028	.010	.003	.001	.000	.000	.000	.000	.000	.000	.000	.000	.000	.000
	3	.001	.060	.190	.243	.205	.134	.072	.032	.012	.004	.001	.000	.000	.000	.000	.000	.000	.000	.000	.000
	4	.000	.013	.090	.182	.218	.190	.130	.074	.035	.014	.005	.001	.000	.000	.000	.000	.000	.000	.000	.000
	5	.000	.002	.032	.103	.175	.202	.179	.127	.075	.036	.015	.005	.001	.000	.000	.000	.000	.000	.000	.000
	6	.000	.000	.009	.045	.109	.169	.192	.171	.124	.075	.036	.015	.005	.001	.000	.000	.000	.000	.000	.000
	7	.000	.000	.002	.016	.055	.112	.164	.184	.166	.122	.074	.037	.015	.005	.001	.000	.000	.000	.000	.000
	8	.000	.000	.000	.005	.022	.061	.114	.161	.180	.162	.120	.073	.035	.014	.004	.001	.000	.000	.000	.000
	9	.000	.000	.000	.001	.007	.027	.065	.116	.160	.177	.160	.119	.071	.034	.012	.003	.000	.000	.000	.000
	10	.000	.000	.000	.000	.002	.010	.031	.069	.117	.159	.176	.159	.117	.069	.031	.010	.002	.000	.000	.000
	11	.000	.000	.000	.000	.000	.003	.012	.034	.071	.119	.160	.177	.160	.116	.065	.027	.007	.001	.000	.000
	12	.000	.000	.000	.000	.000	.001	.004	.014	.035	.073	.120	.162	.180	.161	.114	.061	.022	.005	.000	.000
	13	.000	.000	.000	.000	.000	.000	.001	.005	.015	.037	.074	.122	.166	.184	.164	.112	.055	.016	.002	.000
	14	.000	.000	.000	.000	.000	.000	.000	.001	.005	.015	.037	.075	.124	.171	.192	.169	.109	.045	.009	.000
	15	.000	.000	.000	.000	.000	.000	.000	.000	.001	.005	.015	.036	.075	.127	.179	.202	.175	.103	.032	.002
	16	.000	.000	.000	.000	.000	.000	.000	.000	.000	.001	.005	.014	.035	.074	.130	.190	.218	.182	.090	.013
	17	.000	.000	.000	.000	.000	.000	.000	.000	.000	.000	.001	.004	.012	.032	.072	.134	.205	.243	.190	.060
	18	.000	.000	.000	.000	.000	.000	.000	.000	.000	.000	.000	.001	.003	.010	.028	.067	.137	.229	.285	.189
	19	.000	.000	.000	.000	.000	.000	.000	.000	.000	.000	.000	.000	.000	.002	.007	.021	.058	.137	.270	.377
	20	.000	.000	.000	.000	.000	.000	.000	.000	.000	.000	.000	.000	.000	.000	.001	.003	.012	.039	.122	.358

Table 3—Poisson Distribution

	μ									
x	**0.1**	**0.2**	**0.3**	**0.4**	**0.5**	**0.6**	**0.7**	**0.8**	**0.9**	**1.0**
0	.9048	.8187	.7408	.6703	.6065	.5488	.4966	.4493	.4066	.3679
1	.0905	.1637	.2222	.2681	.3033	.3293	.3476	.3595	.3659	.3679
2	.0045	.0164	.0333	.0536	.0758	.0988	.1217	.1438	.1647	.1839
3	.0002	.0011	.0033	.0072	.0126	.0198	.0284	.0383	.0494	.0613
4	.0000	.0001	.0003	.0007	.0016	.0030	.0050	.0077	.0111	.0153
5	.0000	.0000	.0000	.0001	.0002	.0004	.0007	.0012	.0020	.0031
6	.0000	.0000	.0000	.0000	.0000	.0000	.0001	.0002	.0003	.0005
7	.0000	.0000	.0000	.0000	.0000	.0000	.0000	.0000	.0000	.0001

	μ									
x	**1.1**	**1.2**	**1.3**	**1.4**	**1.5**	**1.6**	**1.7**	**1.8**	**1.9**	**2.0**
0	.3329	.3012	.2725	.2466	.2231	.2019	.1827	.1653	.1496	.1353
1	.3662	.3614	.3543	.3452	.3347	.3230	.3106	.2975	.2842	.2707
2	.2014	.2169	.2303	.2417	.2510	.2584	.2640	.2678	.2700	.2707
3	.0738	.0867	.0998	.1128	.1255	.1378	.1496	.1607	.1710	.1804
4	.0203	.0260	.0324	.0395	.0471	.0551	.0636	.0723	.0812	.0902
5	.0045	.0062	.0084	.0111	.0141	.0176	.0216	.0260	.0309	.0361
6	.0008	.0012	.0018	.0026	.0035	.0047	.0061	.0078	.0098	.0120
7	.0001	.0002	.0003	.0005	.0008	.0011	.0015	.0020	.0027	.0034
8	.0000	.0000	.0001	.0001	.0001	.0002	.0003	.0005	.0006	.0009
9	.0000	.0000	.0000	.0000	.0000	.0000	.0001	.0001	.0001	.0002

	μ									
x	**2.1**	**2.2**	**2.3**	**2.4**	**2.5**	**2.6**	**2.7**	**2.8**	**2.9**	**3.0**
0	.1225	.1108	.1003	.0907	.0821	.0743	.0672	.0608	.0550	.0498
1	.2572	.2438	.2306	.2177	.2052	.1931	.1815	.1703	.1596	.1494
2	.2700	.2681	.2652	.2613	.2565	.2510	.2450	.2384	.2314	.2240
3	.1890	.1966	.2033	.2090	.2138	.2176	.2205	.2225	.2237	.2240
4	.0992	.1082	.1169	.1254	.1336	.1414	.1488	.1557	.1622	.1680
5	.0417	.0476	.0538	.0602	.0668	.0735	.0804	.0872	.0940	.1008
6	.0146	.0174	.0206	.0241	.0278	.0319	.0362	.0407	.0455	.0504
7	.0044	.0055	.0068	.0083	.0099	.0118	.0139	.0163	.0188	.0216
8	.0011	.0015	.0019	.0025	.0031	.0038	.0047	.0057	.0068	.0081
9	.0003	.0004	.0005	.0007	.0009	.0011	.0014	.0018	.0022	.0027
10	.0001	.0001	.0001	.0002	.0002	.0003	.0004	.0005	.0006	.0008
11	.0000	.0000	.0000	.0000	.0000	.0001	.0001	.0001	.0002	.0002
12	.0000	.0000	.0000	.0000	.0000	.0000	.0000	.0000	.0000	.0001

	μ									
x	**3.1**	**3.2**	**3.3**	**3.4**	**3.5**	**3.6**	**3.7**	**3.8**	**3.9**	**4.0**
0	.0450	.0408	.0369	.0334	.0302	.0273	.0247	.0224	.0202	.0183
1	.1397	.1304	.1217	.1135	.1057	.0984	.0915	.0850	.0789	.0733
2	.2165	.2087	.2008	.1929	.1850	.1771	.1692	.1615	.1539	.1465
3	.2237	.2226	.2209	.2186	.2158	.2125	.2087	.2046	.2001	.1954
4	.1734	.1781	.1823	.1858	.1888	.1912	.1931	.1944	.1951	.1954
5	.1075	.1140	.1203	.1264	.1322	.1377	.1429	.1477	.1522	.1563
6	.0555	.0608	.0662	.0716	.0771	.0826	.0881	.0936	.0989	.1042
7	.0246	.0278	.0312	.0348	.0385	.0425	.0466	.0508	.0551	.0595
8	.0095	.0111	.0129	.0148	.0169	.0191	.0215	.0241	.0269	.0298
9	.0033	.0040	.0047	.0056	.0066	.0076	.0089	.0102	.0116	.0132
10	.0010	.0013	.0016	.0019	.0023	.0028	.0033	.0039	.0045	.0053
11	.0003	.0004	.0005	.0006	.0007	.0009	.0011	.0013	.0016	.0019
12	.0001	.0001	.0001	.0002	.0002	.0003	.0003	.0004	.0005	.0006
13	.0000	.0000	.0000	.0000	.0001	.0001	.0001	.0001	.0002	.0002
14	.0000	.0000	.0000	.0000	.0000	.0000	.0000	.0000	.0000	.0001

Reprinted with permission from W. H. Beyer, *Handbook of Tables for Probability and Statistics*, 2e, CRC Press, Boca Raton, Florida, 1986.

Table 3—Poisson Distribution *(continued)*

	μ									
x	**4.1**	**4.2**	**4.3**	**4.4**	**4.5**	**4.6**	**4.7**	**4.8**	**4.9**	**5.0**
0	.0166	.0150	.0136	.0123	.0111	.0101	.0091	.0082	.0074	.0067
1	.0679	.0630	.0583	.0540	.0500	.0462	.0427	.0395	.0365	.0337
2	.1393	.1323	.1254	.1188	.1125	.1063	.1005	.0948	.0894	.0842
3	.1904	.1852	.1798	.1743	.1687	.1631	.1574	.1517	.1460	.1404
4	.1951	.1944	.1933	.1917	.1898	.1875	.1849	.1820	.1789	.1755
5	.1600	.1633	.1662	.1687	.1708	.1725	.1738	.1747	.1753	.1755
6	.1093	.1143	.1191	.1237	.1281	.1323	.1362	.1398	.1432	.1462
7	.0640	.0686	.0732	.0778	.0824	.0869	.0914	.0959	.1002	.1044
8	.0328	.0360	.0393	.0428	.0463	.0500	.0537	.0575	.0614	.0653
9	.0150	.0168	.0188	.0209	.0232	.0255	.0280	.0307	.0334	.0363
10	.0061	.0071	.0081	.0092	.0104	.0118	.0132	.0147	.0164	.0181
11	.0023	.0027	.0032	.0037	.0043	.0049	.0056	.0064	.0073	.0082
12	.0008	.0009	.0011	.0014	.0016	.0019	.0022	.0026	.0030	.0034
13	.0002	.0003	.0004	.0005	.0006	.0007	.0008	.0009	.0011	.0013
14	.0001	.0001	.0001	.0001	.0002	.0002	.0003	.0003	.0004	.0005
15	.0000	.0000	.0000	.0000	.0001	.0001	.0001	.0001	.0001	.0002

	μ									
x	**5.1**	**5.2**	**5.3**	**5.4**	**5.5**	**5.6**	**5.7**	**5.8**	**5.9**	**6.0**
0	.0061	.0055	.0050	.0045	.0041	.0037	.0033	.0030	.0027	.0025
1	.0311	.0287	.0265	.0244	.0225	.0207	.0191	.0176	.0162	.0149
2	.0793	.0746	.0701	.0659	.0618	.0580	.0544	.0509	.0477	.0446
3	.1348	.1293	.1239	.1185	.1133	.1082	.1033	.0985	.0938	.0892
4	.1719	.1681	.1641	.1600	.1558	.1515	.1472	.1428	.1383	.1339
5	.1753	.1748	.1740	.1728	.1714	.1697	.1678	.1656	.1632	.1606
6	.1490	.1515	.1537	.1555	.1571	.1584	.1594	.1601	.1605	.1606
7	.1086	.1125	.1163	.1200	.1234	.1267	.1298	.1326	.1353	.1377
8	.0692	.0731	.0771	.0810	.0849	.0887	.0925	.0962	.0998	.1033
9	.0392	.0423	.0454	.0486	.0519	.0552	.0586	.0620	.0654	.0688
10	.0200	.0220	.0241	.0262	.0285	.0309	.0334	.0359	.0386	.0413
11	.0093	.0104	.0116	.0129	.0143	.0157	.0173	.0190	.0207	.0225
12	.0039	.0045	.0051	.0058	.0065	.0073	.0082	.0092	.0102	.0113
13	.0015	.0018	.0021	.0024	.0028	.0032	.0036	.0041	.0046	.0052
14	.0006	.0007	.0008	.0009	.0011	.0013	.0015	.0017	.0019	.0022
15	.0002	.0002	.0003	.0003	.0004	.0005	.0006	.0007	.0008	.0009
16	.0001	.0001	.0001	.0001	.0001	.0002	.0002	.0002	.0003	.0003
17	.0000	.0000	.0000	.0000	.0000	.0000	.0001	.0001	.0001	.0001

Table 3—Poisson Distribution *(continued)*

	μ									
x	**6.1**	**6.2**	**6.3**	**6.4**	**6.5**	**6.6**	**6.7**	**6.8**	**6.9**	**7.0**
0	.0022	.0020	.0018	.0017	.0015	.0014	.0012	.0011	.0010	.0009
1	.0137	.0126	.0116	.0106	.0098	.0090	.0082	.0076	.0070	.0064
2	.0417	.0390	.0364	.0340	.0318	.0296	.0276	.0258	.0240	.0223
3	.0848	.0806	.0765	.0726	.0688	.0652	.0617	.0584	.0552	.0521
4	.1294	.1249	.1205	.1162	.1118	.1076	.1034	.0992	.0952	.0912
5	.1579	.1549	.1519	.1487	.1454	.1420	.1385	.1349	.1314	.1277
6	.1605	.1601	.1595	.1586	.1575	.1562	.1546	.1529	.1511	.1490
7	.1399	.1418	.1435	.1450	.1462	.1472	.1480	.1486	.1489	.1490
8	.1066	.1099	.1130	.1160	.1188	.1215	.1240	.1263	.1284	.1304
9	.0723	.0757	.0791	.0825	.0858	.0891	.0923	.0954	.0985	.1014
10	.0441	.0469	.0498	.0528	.0558	.0588	.0618	.0649	.0679	.0710
11	.0245	.0265	.0285	.0307	.0330	.0353	.0377	.0401	.0426	.0452
12	.0124	.0137	.0150	.0164	.0179	.0194	.0210	.0227	.0245	.0264
13	.0058	.0065	.0073	.0081	.0089	.0098	.0108	.0119	.0130	.0142
14	.0025	.0029	.0033	.0037	.0041	.0046	.0052	.0058	.0064	.0071
15	.0010	.0012	.0014	.0016	.0018	.0020	.0023	.0026	.0029	.0033
16	.0004	.0005	.0005	.0006	.0007	.0008	.0010	.0011	.0013	.0014
17	.0001	.0002	.0002	.0002	.0003	.0003	.0004	.0004	.0005	.0006
18	.0000	.0001	.0001	.0001	.0001	.0001	.0001	.0002	.0002	.0002
19	.0000	.0000	.0000	.0000	.0000	.0000	.0000	.0001	.0001	.0001

	μ									
x	**7.1**	**7.2**	**7.3**	**7.4**	**7.5**	**7.6**	**7.7**	**7.8**	**7.9**	**8.0**
0	.0008	.0007	.0007	.0006	.0006	.0005	.0005	.0004	.0004	.0003
1	.0059	.0054	.0049	.0045	.0041	.0038	.0035	.0032	.0029	.0027
2	.0208	.0194	.0180	.0167	.0156	.0145	.0134	.0125	.0116	.0107
3	.0492	.0464	.0438	.0413	.0389	.0366	.0345	.0324	.0305	.0286
4	.0874	.0836	.0799	.0764	.0729	.0696	.0663	.0632	.0602	.0573
5	.1241	.1204	.1167	.1130	.1094	.1057	.1021	.0986	.0951	.0916
6	.1468	.1445	.1420	.1394	.1367	.1339	.1311	.1282	.1252	.1221
7	.1489	.1486	.1481	.1474	.1465	.1454	.1442	.1428	.1413	.1396
8	.1321	.1337	.1351	.1363	.1373	.1382	.1388	.1392	.1395	.1396
9	.1042	.1070	.1096	.1121	.1144	.1167	.1187	.1207	.1224	.1241
10	.0740	.0770	.0800	.0829	.0858	.0887	.0914	.0941	.0967	.0993
11	.0478	.0504	.0531	.0558	.0585	.0613	.0640	.0667	.0695	.0722
12	.0283	.0303	.0323	.0344	.0366	.0388	.0411	.0434	.0457	.0481
13	.0154	.0168	.0181	.0196	.0211	.0227	.0243	.0260	.0278	.0296
14	.0078	.0086	.0095	.0104	.0113	.0123	.0134	.0145	.0157	.0169
15	.0037	.0041	.0046	.0051	.0057	.0062	.0069	.0075	.0083	.0090
16	.0016	.0019	.0021	.0024	.0026	.0030	.0033	.0037	.0041	.0045
17	.0007	.0008	.0009	.0010	.0012	.0013	.0015	.0017	.0019	.0021
18	.0003	.0003	.0004	.0004	.0005	.0006	.0006	.0007	.0008	.0009
19	.0001	.0001	.0001	.0002	.0002	.0002	.0003	.0003	.0003	.0004
20	.0000	.0000	.0001	.0001	.0001	.0001	.0001	.0001	.0001	.0002
21	.0000	.0000	.0000	.0000	.0000	.0000	.0000	.0000	.0001	.0001

Table 3—Poisson Distribution *(continued)*

	μ									
x	**8.1**	**8.2**	**8.3**	**8.4**	**8.5**	**8.6**	**8.7**	**8.8**	**8.9**	**9.0**
0	.0003	.0003	.0002	.0002	.0002	.0002	.0002	.0002	.0001	.0001
1	.0025	.0023	.0021	.0019	.0017	.0016	.0014	.0013	.0012	.0011
2	.0100	.0092	.0086	.0079	.0074	.0068	.0063	.0058	.0054	.0050
3	.0269	.0252	.0237	.0222	.0208	.0195	.0183	.0171	.0160	.0150
4	.0544	.0517	.0491	.0466	.0443	.0420	.0398	.0377	.0357	.0337
5	.0882	.0849	.0816	.0784	.0752	.0722	.0692	.0663	.0635	.0607
6	.1191	.1160	.1128	.1097	.1066	.1034	.1003	.0972	.0941	.0911
7	.1378	.1358	.1338	.1317	.1294	.1271	.1247	.1222	.1197	.1171
8	.1395	.1392	.1388	.1382	.1375	.1366	.1356	.1344	.1332	.1318
9	.1256	.1269	.1280	.1290	.1299	.1306	.1311	.1315	.1317	.1318
10	.1017	.1040	.1063	.1084	.1104	.1123	.1140	.1157	.1172	.1186
11	.0749	.0776	.0802	.0828	.0853	.0878	.0902	.0925	.0948	.0970
12	.0505	.0530	.0555	.0579	.0604	.0629	.0654	.0679	.0703	.0728
13	.0315	.0334	.0354	.0374	.0395	.0416	.0438	.0459	.0481	.0504
14	.0182	.0196	.0210	.0225	.0240	.0256	.0272	.0289	.0306	.0324
15	.0098	.0107	.0116	.0126	.0136	.0147	.0158	.0169	.0182	.0194
16	.0050	.0055	.0060	.0066	.0072	.0079	.0086	.0093	.0101	.0109
17	.0024	.0026	.0029	.0033	.0036	.0040	.0044	.0048	.0053	.0058
18	.0011	.0012	.0014	.0015	.0017	.0019	.0021	.0024	.0026	.0029
19	.0005	.0005	.0006	.0007	.0008	.0009	.0010	.0011	.0012	.0014
20	.0002	.0002	.0002	.0003	.0003	.0004	.0004	.0005	.0005	.0006
21	.0001	.0001	.0001	.0001	.0001	.0002	.0002	.0002	.0002	.0003
22	.0000	.0000	.0000	.0000	.0001	.0001	.0001	.0001	.0001	.0001

	μ									
x	**9.1**	**9.2**	**9.3**	**9.4**	**9.5**	**9.6**	**9.7**	**9.8**	**9.9**	**10.0**
0	.0001	.0001	.0001	.0001	.0001	.0001	.0001	.0001	.0001	.0000
1	.0010	.0009	.0009	.0008	.0007	.0007	.0006	.0005	.0005	.0005
2	.0046	.0043	.0040	.0037	.0034	.0031	.0029	.0027	.0025	.0023
3	.0140	.0131	.0123	.0115	.0107	.0100	.0093	.0087	.0081	.0076
4	.0319	.0302	.0285	.0269	.0254	.0240	.0226	.0213	.0201	.0189
5	.0581	.0555	.0530	.0506	.0483	.0460	.0439	.0418	.0398	.0378
6	.0881	.0851	.0822	.0793	.0764	.0736	.0709	.0682	.0656	.0631
7	.1145	.1118	.1091	.1064	.1037	.1010	.0982	.0955	.0928	.0901
8	.1302	.1286	.1269	.1251	.1232	.1212	.1191	.1170	.1148	.1126
9	.1317	.1315	.1311	.1306	.1300	.1293	.1284	.1274	.1263	.1251
10	.1198	.1210	.1219	.1228	.1235	.1241	.1245	.1249	.1250	.1251
11	.0991	.1012	.1031	.1049	.1067	.1083	.1098	.1112	.1125	.1137
12	.0752	.0776	.0799	.0822	.0844	.0866	.0888	.0908	.0928	.0948
13	.0526	.0549	.0572	.0594	.0617	.0640	.0662	.0685	.0707	.0729
14	.0342	.0361	.0380	.0399	.0419	.0439	.0459	.0479	.0500	.0521
15	.0208	.0221	.0235	.0250	.0265	.0281	.0297	.0313	.0330	.0347
16	.0118	.0127	.0137	.0147	.0157	.0168	.0180	.0192	.0204	.0217
17	.0063	.0069	.0075	.0081	.0088	.0095	.0103	.0111	.0119	.0128
18	.0032	.0035	.0039	.0042	.0046	.0051	.0055	.0060	.0065	.0071
19	.0015	.0017	.0019	.0021	.0023	.0026	.0028	.0031	.0034	.0037
20	.0007	.0008	.0009	.0010	.0011	.0012	.0014	.0015	.0017	.0019
21	.0003	.0003	.0004	.0004	.0005	.0006	.0006	.0007	.0008	.0009
22	.0001	.0001	.0002	.0002	.0002	.0002	.0003	.0003	.0004	.0004
23	.0000	.0001	.0001	.0001	.0001	.0001	.0001	.0001	.0002	.0002
24	.0000	.0000	.0000	.0000	.0000	.0000	.0000	.0001	.0001	.0001

Table 3—Poisson Distribution *(continued)*

	μ									
x	**11**	**12**	**13**	**14**	**15**	**16**	**17**	**18**	**19**	**20**
0	.0000	.0000	.0000	.0000	.0000	.0000	.0000	.0000	.0000	.0000
1	.0002	.0001	.0000	.0000	.0000	.0000	.0000	.0000	.0000	.0000
2	.0010	.0004	.0002	.0001	.0000	.0000	.0000	.0000	.0000	.0000
3	.0037	.0018	.0008	.0004	.0002	.0001	.0000	.0000	.0000	.0000
4	.0102	.0053	.0027	.0013	.0006	.0003	.0001	.0001	.0000	.0000
5	.0224	.0127	.0070	.0037	.0019	.0010	.0005	.0002	.0001	.0001
6	.0411	.0255	.0152	.0087	.0048	.0026	.0014	.0007	.0004	.0002
7	.0646	.0437	.0281	.0174	.0104	.0060	.0034	.0018	.0010	.0005
8	.0888	.0655	.0457	.0304	.0194	.0120	.0072	.0042	.0024	.0013
9	.1085	.0874	.0661	.0473	.0324	.0213	.0135	.0083	.0050	.0029
10	.1194	.1048	.0859	.0663	.0486	.0341	.0230	.0150	.0095	.0058
11	.1194	.1144	.1015	.0844	.0663	.0496	.0355	.0245	.0164	.0106
12	.1094	.1144	.1099	.0984	.0829	.0661	.0504	.0368	.0259	.0176
13	.0926	.1056	.1099	.1060	.0956	.0814	.0658	.0509	.0378	.0271
14	.0728	.0905	.1021	.1060	.1024	.0930	.0800	.0655	.0514	.0387
15	.0534	.0724	.0885	.0989	.1024	.0992	.0906	.0786	.0650	.0516
16	.0367	.0543	.0719	.0866	.0960	.0992	.0963	.0884	.0772	.0646
17	.0237	.0383	.0550	.0713	.0847	.0934	.0963	.0936	.0863	.0760
18	.0145	.0256	.0397	.0554	.0706	.0830	.0909	.0936	.0911	.0844
19	.0084	.0161	.0272	.0409	.0557	.0699	.0814	.0887	.0911	.0888
20	.0046	.0097	.0177	.0286	.0418	.0559	.0692	.0798	.0866	.0888
	μ									
x	**11**	**12**	**13**	**14**	**15**	**16**	**17**	**18**	**19**	**20**
21	.0024	.0055	.0109	.0191	.0299	.0426	.0560	.0684	.0783	.0846
22	.0012	.0030	.0065	.0121	.0204	.0310	.0433	.0560	.0676	.0769
23	.0006	.0016	.0037	.0074	.0133	.0216	.0320	.0438	.0559	.0669
24	.0003	.0008	.0020	.0043	.0083	.0144	.0226	.0328	.0442	.0557
25	.0001	.0004	.0010	.0024	.0050	.0092	.0154	.0237	.0336	.0446
26	.0000	.0002	.0005	.0013	.0029	.0057	.0101	.0164	.0246	.0343
27	.0000	.0001	.0002	.0007	.0016	.0034	.0063	.0109	.0173	.0254
28	.0000	.0000	.0001	.0003	.0009	.0019	.0038	.0070	.0117	.0181
29	.0000	.0000	.0001	.0002	.0004	.0011	.0023	.0044	.0077	.0125
30	.0000	.0000	.0000	.0001	.0002	.0006	.0013	.0026	.0049	.0083
31	.0000	.0000	.0000	.0000	.0001	.0003	.0007	.0015	.0030	.0054
32	.0000	.0000	.0000	.0000	.0001	.0001	.0004	.0009	.0018	.0034
33	.0000	.0000	.0000	.0000	.0000	.0001	.0002	.0005	.0010	.0020
34	.0000	.0000	.0000	.0000	.0000	.0000	.0001	.0002	.0006	.0012
35	.0000	.0000	.0000	.0000	.0000	.0000	.0000	.0001	.0003	.0007
36	.0000	.0000	.0000	.0000	.0000	.0000	.0000	.0001	.0002	.0004
37	.0000	.0000	.0000	.0000	.0000	.0000	.0000	.0000	.0001	.0002
38	.0000	.0000	.0000	.0000	.0000	.0000	.0000	.0000	.0000	.0001
39	.0000	.0000	.0000	.0000	.0000	.0000	.0000	.0000	.0000	.0001

Table 4—Standard Normal Distribution

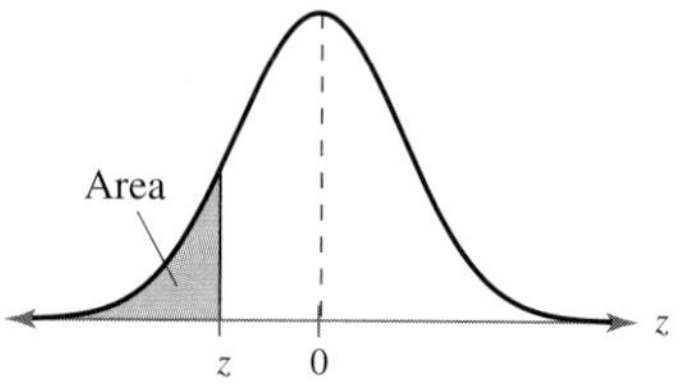

z	.09	.08	.07	.06	.05	.04	.03	.02	.01	.00
−3.4	.0002	.0003	.0003	.0003	.0003	.0003	.0003	.0003	.0003	.0003
−3.3	.0003	.0004	.0004	.0004	.0004	.0004	.0004	.0005	.0005	.0005
−3.2	.0005	.0005	.0005	.0006	.0006	.0006	.0006	.0006	.0007	.0007
−3.1	.0007	.0007	.0008	.0008	.0008	.0008	.0009	.0009	.0009	.0010
−3.0	.0010	.0010	.0011	.0011	.0011	.0012	.0012	.0013	.0013	.0013
−2.9	.0014	.0014	.0015	.0015	.0016	.0016	.0017	.0018	.0018	.0019
−2.8	.0019	.0020	.0021	.0021	.0022	.0023	.0023	.0024	.0025	.0026
−2.7	.0026	.0027	.0028	.0029	.0030	.0031	.0032	.0033	.0034	.0035
−2.6	.0036	.0037	.0038	.0039	.0040	.0041	.0043	.0044	.0045	.0047
−2.5	.0048	.0049	.0051	.0052	.0054	.0055	.0057	.0059	.0060	.0062
−2.4	.0064	.0066	.0068	.0069	.0071	.0073	.0075	.0078	.0080	.0082
−2.3	.0084	.0087	.0089	.0091	.0094	.0096	.0099	.0102	.0104	.0107
−2.2	.0110	.0113	.0116	.0119	.0122	.0125	.0129	.0132	.0136	.0139
−2.1	.0143	.0146	.0150	.0154	.0158	.0162	.0166	.0170	.0174	.0179
−2.0	.0183	.0188	.0192	.0197	.0202	.0207	.0212	.0217	.0222	.0228
−1.9	.0233	.0239	.0244	.0250	.0256	.0262	.0268	.0274	.0281	.0287
−1.8	.0294	.0301	.0307	.0314	.0322	.0329	.0336	.0344	.0351	.0359
−1.7	.0367	.0375	.0384	.0392	.0401	.0409	.0418	.0427	.0436	.0446
−1.6	.0455	.0465	.0475	.0485	.0495	.0505	.0516	.0526	.0537	.0548
−1.5	.0559	.0571	.0582	.0594	.0606	.0618	.0630	.0643	.0655	.0668
−1.4	.0681	.0694	.0708	.0721	.0735	.0749	.0764	.0778	.0793	.0808
−1.3	.0823	.0838	.0853	.0869	.0885	.0901	.0918	.0934	.0951	.0968
−1.2	.0985	.1003	.1020	.1038	.1056	.1075	.1093	.1112	.1131	.1151
−1.1	.1170	.1190	.1210	.1230	.1251	.1271	.1292	.1314	.1335	.1357
−1.0	.1379	.1401	.1423	.1446	.1469	.1492	.1515	.1539	.1562	.1587
−0.9	.1611	.1635	.1660	.1685	.1711	.1736	.1762	.1788	.1814	.1841
−0.8	.1867	.1894	.1922	.1949	.1977	.2005	.2033	.2061	.2090	.2119
−0.7	.2148	.2177	.2206	.2236	.2266	.2296	.2327	.2358	.2389	.2420
−0.6	.2451	.2483	.2514	.2546	.2578	.2611	.2643	.2676	.2709	.2743
−0.5	.2776	.2810	.2843	.2877	.2912	.2946	.2981	.3015	.3050	.3085
−0.4	.3121	.3156	.3192	.3228	.3264	.3300	.3336	.3372	.3409	.3446
−0.3	.3483	.3520	.3557	.3594	.3632	.3669	.3707	.3745	.3783	.3821
−0.2	.3859	.3897	.3936	.3974	.4013	.4052	.4090	.4129	.4168	.4207
−0.1	.4247	.4286	.4325	.4364	.4404	.4443	.4483	.4522	.4562	.4602
−0.0	.4641	.4681	.4721	.4761	.4801	.4840	.4880	.4920	.4960	.5000

Critical Values

Level of Confidence c	z_c
0.80	1.28
0.90	1.645
0.95	1.96
0.99	2.575

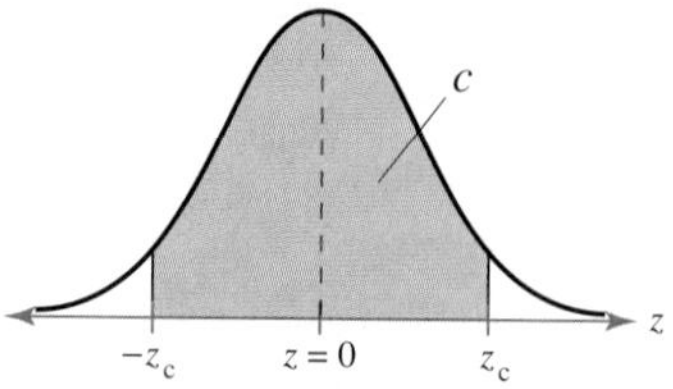

Table 4—Standard Normal Distribution *(continued)*

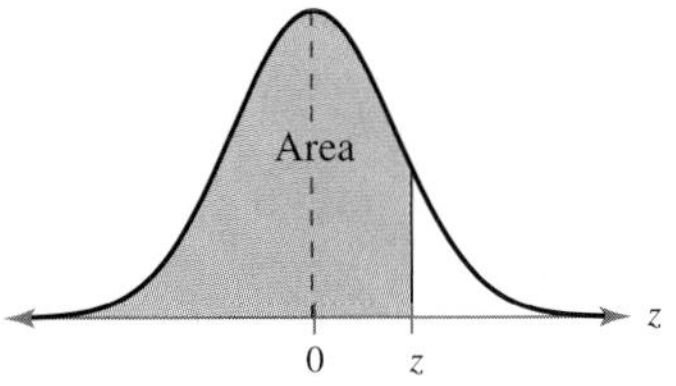

z	.00	.01	.02	.03	.04	.05	.06	.07	.08	.09
0.0	.5000	.5040	.5080	.5120	.5160	.5199	.5239	.5279	.5319	.5359
0.1	.5398	.5438	.5478	.5517	.5557	.5596	.5636	.5675	.5714	.5753
0.2	.5793	.5832	.5871	.5910	.5948	.5987	.6026	.6064	.6103	.6141
0.3	.6179	.6217	.6255	.6293	.6331	.6368	.6406	.6443	.6480	.6517
0.4	.6554	.6591	.6628	.6664	.6700	.6736	.6772	.6808	.6844	.6879
0.5	.6915	.6950	.6985	.7019	.7054	.7088	.7123	.7157	.7190	.7224
0.6	.7257	.7291	.7324	.7357	.7389	.7422	.7454	.7486	.7517	.7549
0.7	.7580	.7611	.7642	.7673	.7704	.7734	.7764	.7794	.7823	.7852
0.8	.7881	.7910	.7939	.7967	.7995	.8023	.8051	.8078	.8106	.8133
0.9	.8159	.8186	.8212	.8238	.8264	.8289	.8315	.8340	.8365	.8389
1.0	.8413	.8438	.8461	.8485	.8508	.8531	.8554	.8577	.8599	.8621
1.1	.8643	.8665	.8686	.8708	.8729	.8749	.8770	.8790	.8810	.8830
1.2	.8849	.8869	.8888	.8907	.8925	.8944	.8962	.8980	.8997	.9015
1.3	.9032	.9049	.9066	.9082	.9099	.9115	.9131	.9147	.9162	.9177
1.4	.9192	.9207	.9222	.9236	.9251	.9265	.9279	.9292	.9306	.9319
1.5	.9332	.9345	.9357	.9370	.9382	.9394	.9406	.9418	.9429	.9441
1.6	.9452	.9463	.9474	.9484	.9495	.9505	.9515	.9525	.9535	.9545
1.7	.9554	.9564	.9573	.9582	.9591	.9599	.9608	.9616	.9625	.9633
1.8	.9641	.9649	.9656	.9664	.9671	.9678	.9686	.9693	.9699	.9706
1.9	.9713	.9719	.9726	.9732	.9738	.9744	.9750	.9756	.9761	.9767
2.0	.9772	.9778	.9783	.9788	.9793	.9798	.9803	.9808	.9812	.9817
2.1	.9821	.9826	.9830	.9834	.9838	.9842	.9846	.9850	.9854	.9857
2.2	.9861	.9864	.9868	.9871	.9875	.9878	.9881	.9884	.9887	.9890
2.3	.9893	.9896	.9898	.9901	.9904	.9906	.9909	.9911	.9913	.9916
2.4	.9918	.9920	.9922	.9925	.9927	.9929	.9931	.9932	.9934	.9936
2.5	.9938	.9940	.9941	.9943	.9945	.9946	.9948	.9949	.9951	.9952
2.6	.9953	.9955	.9956	.9957	.9959	.9960	.9961	.9962	.9963	.9964
2.7	.9965	.9966	.9967	.9968	.9969	.9970	.9971	.9972	.9973	.9974
2.8	.9974	.9975	.9976	.9977	.9977	.9978	.9979	.9979	.9980	.9981
2.9	.9981	.9982	.9982	.9983	.9984	.9984	.9985	.9985	.9986	.9986
3.0	.9987	.9987	.9987	.9988	.9988	.9989	.9989	.9989	.9990	.9990
3.1	.9990	.9991	.9991	.9991	.9992	.9992	.9992	.9992	.9993	.9993
3.2	.9993	.9993	.9994	.9994	.9994	.9994	.9994	.9995	.9995	.9995
3.3	.9995	.9995	.9995	.9996	.9996	.9996	.9996	.9996	.9996	.9997
3.4	.9997	.9997	.9997	.9997	.9997	.9997	.9997	.9997	.9997	.9998

Table 5—*t*-Distribution

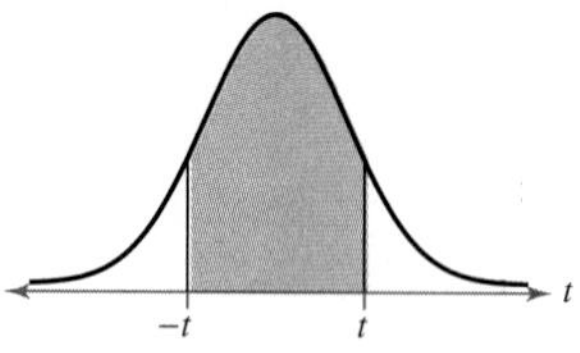

c-confidence interval

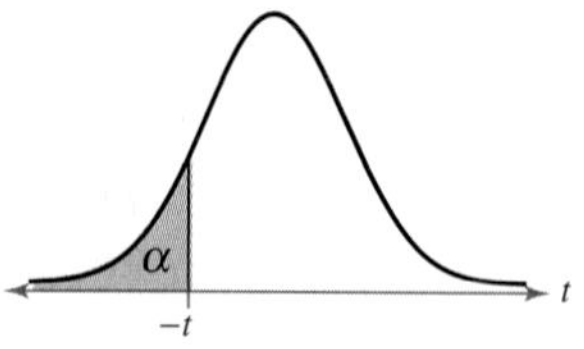

Left-tailed test

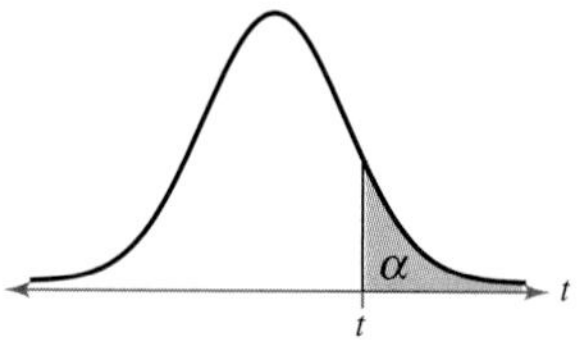

Right-tailed test

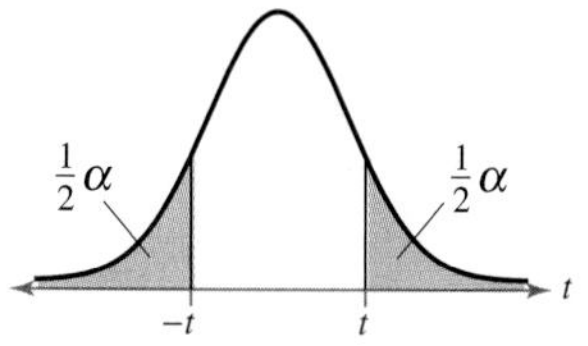

Two-tailed test

	Level of confidence, *c*	0.50	0.80	0.90	0.95	0.98	0.99
	One tail, α	**0.25**	**0.10**	**0.05**	**0.025**	**0.01**	**0.005**
d.f.	**Two tails, α**	**0.50**	**0.20**	**0.10**	**0.05**	**0.02**	**0.01**
1		1.000	3.078	6.314	12.706	31.821	63.657
2		.816	1.886	2.920	4.303	6.965	9.925
3		.765	1.638	2.353	3.182	4.541	5.841
4		.741	1.533	2.132	2.776	3.747	4.604
5		.727	1.476	2.015	2.571	3.365	4.032
6		.718	1.440	1.943	2.447	3.143	3.707
7		.711	1.415	1.895	2.365	2.998	3.499
8		.706	1.397	1.860	2.306	2.896	3.355
9		.703	1.383	1.833	2.262	2.821	3.250
10		.700	1.372	1.812	2.228	2.764	3.169
11		.697	1.363	1.796	2.201	2.718	3.106
12		.695	1.356	1.782	2.179	2.681	3.055
13		.694	1.350	1.771	2.160	2.650	3.012
14		.692	1.345	1.761	2.145	2.624	2.977
15		.691	1.341	1.753	2.131	2.602	2.947
16		.690	1.337	1.746	2.120	2.583	2.921
17		.689	1.333	1.740	2.110	2.567	2.898
18		.688	1.330	1.734	2.101	2.552	2.878
19		.688	1.328	1.729	2.093	2.539	2.861
20		.687	1.325	1.725	2.086	2.528	2.845
21		.686	1.323	1.721	2.080	2.518	2.831
22		.686	1.321	1.717	2.074	2.508	2.819
23		.685	1.319	1.714	2.069	2.500	2.807
24		.685	1.318	1.711	2.064	2.492	2.797
25		.684	1.316	1.708	2.060	2.485	2.787
26		.684	1.315	1.706	2.056	2.479	2.779
27		.684	1.314	1.703	2.052	2.473	2.771
28		.683	1.313	1.701	2.048	2.467	2.763
29		.683	1.311	1.699	2.045	2.462	2.756
∞		.674	1.282	1.645	1.960	2.326	2.576

Table 6—Chi-Square Distribution

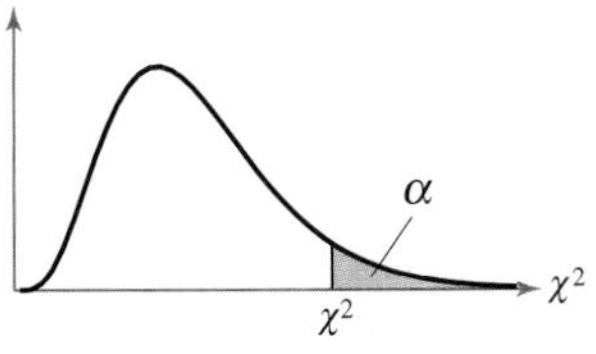

Right tail

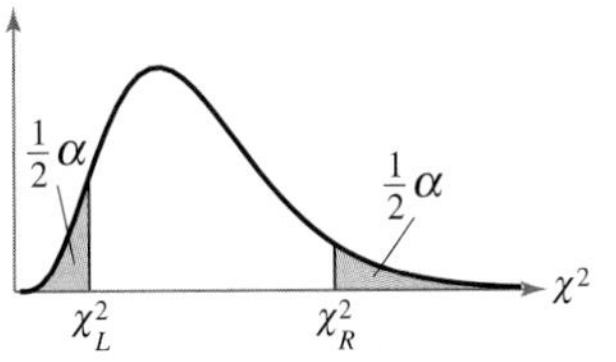

Two tails

Degrees of freedom	α									
	0.995	**0.99**	**0.975**	**0.95**	**0.90**	**0.10**	**0.05**	**0.025**	**0.01**	**0.005**
1	—	—	0.001	0.004	0.016	2.706	3.841	5.024	6.635	7.879
2	0.010	0.020	0.051	0.103	0.211	4.605	5.991	7.378	9.210	10.597
3	0.072	0.115	0.216	0.352	0.584	6.251	7.815	9.348	11.345	12.838
4	0.207	0.297	0.484	0.711	1.064	7.779	9.488	11.143	13.277	14.860
5	0.412	0.554	0.831	1.145	1.610	9.236	11.071	12.833	15.086	16.750
6	0.676	0.872	1.237	1.635	2.204	10.645	12.592	14.449	16.812	18.548
7	0.989	1.239	1.690	2.167	2.833	12.017	14.067	16.013	18.475	20.278
8	1.344	1.646	2.180	2.733	3.490	13.362	15.507	17.535	20.090	21.955
9	1.735	2.088	2.700	3.325	4.168	14.684	16.919	19.023	21.666	23.589
10	2.156	2.558	3.247	3.940	4.865	15.987	18.307	20.483	23.209	25.188
11	2.603	3.053	3.816	4.575	5.578	17.275	19.675	21.920	24.725	26.757
12	3.074	3.571	4.404	5.226	6.304	18.549	21.026	23.337	26.217	28.299
13	3.565	4.107	5.009	5.892	7.042	19.812	22.362	24.736	27.688	29.819
14	4.075	4.660	5.629	6.571	7.790	21.064	23.685	26.119	29.141	31.319
15	4.601	5.229	6.262	7.261	8.547	22.307	24.996	27.488	30.578	32.801
16	5.142	5.812	6.908	7.962	9.312	23.542	26.296	28.845	32.000	34.267
17	5.697	6.408	7.564	8.672	10.085	24.769	27.587	30.191	33.409	35.718
18	6.265	7.015	8.231	9.390	10.865	25.989	28.869	31.526	34.805	37.156
19	6.844	7.633	8.907	10.117	11.651	27.204	30.144	32.852	36.191	38.582
20	7.434	8.260	9.591	10.851	12.443	28.412	31.410	34.170	37.566	39.997
21	8.034	8.897	10.283	11.591	13.240	29.615	32.671	35.479	38.932	41.401
22	8.643	9.542	10.982	12.338	14.042	30.813	33.924	36.781	40.289	42.796
23	9.262	10.196	11.689	13.091	14.848	32.007	35.172	38.076	41.638	44.181
24	9.886	10.856	12.401	13.848	15.659	33.196	36.415	39.364	42.980	45.559
25	10.520	11.524	13.120	14.611	16.473	34.382	37.652	40.646	44.314	46.928
26	11.160	12.198	13.844	15.379	17.292	35.563	38.885	41.923	45.642	48.290
27	11.808	12.879	14.573	16.151	18.114	36.741	40.113	43.194	46.963	49.645
28	12.461	13.565	15.308	16.928	18.939	37.916	41.337	44.461	48.278	50.993
29	13.121	14.257	16.047	17.708	19.768	39.087	42.557	45.722	49.588	52.336
30	13.787	14.954	16.791	18.493	20.599	40.256	43.773	46.979	50.892	53.672
40	20.707	22.164	24.433	26.509	29.051	51.805	55.758	59.342	63.691	66.766
50	27.991	29.707	32.357	34.764	37.689	63.167	67.505	71.420	76.154	79.490
60	35.534	37.485	40.482	43.188	46.459	74.397	79.082	83.298	88.379	91.952
70	43.275	45.442	48.758	51.739	55.329	85.527	90.531	95.023	100.425	104.215
80	51.172	53.540	57.153	60.391	64.278	96.578	101.879	106.629	112.329	116.321
90	59.196	61.754	65.647	69.126	73.291	107.565	113.145	118.136	124.116	128.299
100	67.328	70.065	74.222	77.929	82.358	118.498	124.342	129.561	135.807	140.169

Table 7—*F*-Distribution

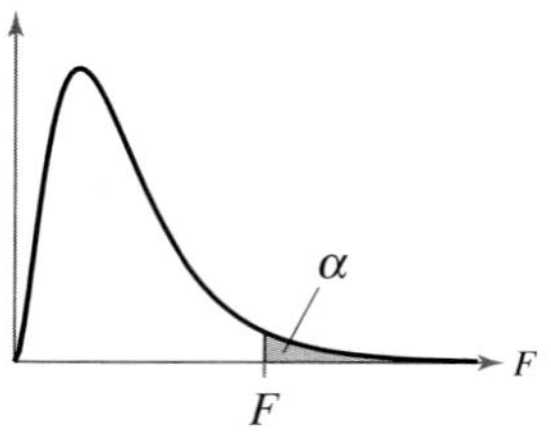

d.f.$_D$: Degrees of freedom, denominator	$\alpha = 0.005$																		
	d.f.$_N$: Degrees of freedom, numerator																		
	1	2	3	4	5	6	7	8	9	10	12	15	20	24	30	40	60	120	∞
1	16211	20000	21615	22500	23056	23437	23715	23925	24091	24224	24426	24630	24836	24940	25044	25148	25253	25359	25465
2	198.5	199.0	199.2	199.2	199.3	199.3	199.4	199.4	199.4	199.4	199.4	199.4	199.4	199.5	199.5	199.5	199.5	199.5	199.5
3	55.55	49.80	47.47	46.19	45.39	44.84	44.43	44.13	43.88	43.69	43.39	43.08	42.78	42.62	42.47	42.31	42.15	41.99	41.83
4	31.33	26.28	24.26	23.15	22.46	21.97	21.62	21.35	21.14	20.97	20.70	20.44	20.17	20.03	19.89	19.75	19.61	19.47	19.32
5	22.78	18.31	16.53	15.56	14.94	14.51	14.20	13.96	13.77	13.62	13.38	13.15	12.90	12.78	12.66	12.53	12.40	12.27	12.14
6	18.63	14.54	12.92	12.03	11.46	11.07	10.79	10.57	10.39	10.25	10.03	9.81	9.59	9.47	9.36	9.24	9.12	9.00	8.88
7	16.24	12.40	10.88	10.05	9.52	9.16	8.89	8.68	8.51	8.38	8.18	7.97	7.75	7.65	7.53	7.42	7.31	7.19	7.08
8	14.69	11.04	9.60	8.81	8.30	7.95	7.69	7.50	7.34	7.21	7.01	6.81	6.61	6.50	6.40	6.29	6.18	6.06	5.95
9	13.61	10.11	8.72	7.96	7.47	7.13	6.88	6.69	6.54	6.42	6.23	6.03	5.83	5.73	5.62	5.52	5.41	5.30	5.19
10	12.83	9.43	8.08	7.34	6.87	6.54	6.30	6.12	5.97	5.85	5.66	5.47	5.27	5.17	5.07	4.97	4.86	4.75	4.64
11	12.73	8.91	7.60	6.88	6.42	6.10	5.86	5.68	5.54	5.42	5.24	5.05	4.86	4.76	4.65	4.55	4.44	4.34	4.23
12	11.75	8.51	7.23	6.52	6.07	5.76	5.52	5.35	5.20	5.09	4.91	4.72	4.53	4.43	4.33	4.23	4.12	4.01	3.90
13	11.37	8.19	6.93	6.23	5.79	5.48	5.25	5.08	4.94	4.82	4.64	4.46	4.27	4.17	4.07	3.97	3.87	3.76	3.65
14	11.06	7.92	6.68	6.00	5.56	5.26	5.03	4.86	4.72	4.60	4.43	4.25	4.06	3.96	3.86	3.76	3.66	3.55	3.44
15	10.80	7.70	6.48	5.80	5.37	5.07	4.85	4.67	4.54	4.42	4.25	4.07	3.88	3.79	3.69	3.58	3.48	3.37	3.26
16	10.58	7.51	6.30	5.64	5.21	4.91	4.69	4.52	4.38	4.27	4.10	3.92	3.73	3.64	3.54	3.44	3.33	3.22	3.11
17	10.38	7.35	6.16	5.50	5.07	4.78	4.56	4.39	4.25	4.14	3.97	3.79	3.61	3.51	3.41	3.31	3.21	3.10	2.98
18	10.22	7.21	6.03	5.37	4.96	4.66	4.44	4.28	4.14	4.03	3.86	3.68	3.50	3.40	3.30	3.20	3.10	2.99	2.87
19	10.07	7.09	5.92	5.27	4.85	4.56	4.34	4.18	4.04	3.93	3.76	3.59	3.40	3.31	3.21	3.11	3.00	2.89	2.78
20	9.94	6.99	5.82	5.17	4.76	4.47	4.26	4.09	3.96	3.85	3.68	3.50	3.32	3.22	3.12	3.02	2.92	2.81	2.69
21	9.83	6.89	5.73	5.09	4.68	4.39	4.18	4.01	3.88	3.77	3.60	3.43	3.24	3.15	3.05	2.95	2.84	2.73	2.61
22	9.73	6.81	5.65	5.02	4.61	4.32	4.11	3.94	3.81	3.70	3.54	3.36	3.18	3.08	2.98	2.88	2.77	2.66	2.55
23	9.63	6.73	5.58	4.95	4.54	4.26	4.05	3.88	3.75	3.64	3.47	3.30	3.12	3.02	2.92	2.82	2.71	2.60	2.48
24	9.55	6.66	5.52	4.89	4.49	4.20	3.99	3.83	3.69	3.59	3.42	3.25	3.06	2.97	2.87	2.77	2.66	2.55	2.43
25	9.48	6.60	5.46	4.84	4.43	4.15	3.94	3.78	3.64	3.54	3.37	3.20	3.01	2.92	2.82	2.72	2.61	2.50	2.38
26	9.41	6.54	5.41	4.79	4.38	4.10	3.89	3.73	3.60	3.49	3.33	3.15	2.97	2.87	2.77	2.67	2.56	2.45	2.33
27	9.34	6.49	5.36	4.74	4.34	4.06	3.85	3.69	3.56	3.45	3.28	3.11	2.93	2.83	2.73	2.63	2.52	2.41	2.25
28	9.28	6.44	5.32	4.70	4.30	4.02	3.81	3.65	3.52	3.41	3.25	3.07	2.89	2.79	2.69	2.59	2.48	2.37	2.29
29	9.23	6.40	5.28	4.66	4.26	3.98	3.77	3.61	3.48	3.38	3.21	3.04	2.86	2.76	2.66	2.56	2.45	2.33	2.24
30	9.18	6.35	5.24	4.62	4.23	3.95	3.74	3.58	3.45	3.34	3.18	3.01	2.82	2.73	2.63	2.52	2.42	2.30	2.18
40	8.83	6.07	4.98	4.37	3.99	3.71	3.51	3.35	3.22	3.12	2.95	2.78	2.60	2.50	2.40	2.30	2.18	2.06	1.93
60	8.49	5.79	4.73	4.14	3.76	3.49	3.29	3.13	3.01	2.90	2.74	2.57	2.39	2.29	2.19	2.08	1.96	1.83	1.69
120	8.18	5.54	4.50	3.92	3.55	3.28	3.09	2.93	2.81	2.71	2.54	2.37	2.19	2.09	1.98	1.87	1.75	1.61	1.43
∞	7.88	5.30	4.28	3.72	3.35	3.09	2.90	2.74	2.62	2.52	2.36	2.19	2.00	1.90	1.79	1.67	1.53	1.36	1.00

Table 7—*F*-Distribution *(continued)*

d.f.$_D$: Degrees of freedom, denominator	$\alpha = 0.01$																		
	d.f.$_N$: Degrees of freedom, numerator																		
	1	**2**	**3**	**4**	**5**	**6**	**7**	**8**	**9**	**10**	**12**	**15**	**20**	**24**	**30**	**40**	**60**	**120**	**∞**
1	4052	4999.5	5403	5625	5764	5859	5928	5982	6022	6056	6106	6157	6209	6235	6261	6287	6313	6339	6366
2	98.50	99.00	99.17	99.25	99.30	99.33	99.36	99.37	99.39	99.40	99.42	99.43	99.45	99.46	99.47	99.47	99.48	99.49	99.50
3	34.12	30.82	29.46	28.71	28.24	27.91	27.67	27.49	27.35	27.23	27.05	26.87	26.69	26.60	26.50	26.41	26.32	26.22	26.13
4	21.20	18.00	16.69	15.98	15.52	15.21	14.98	14.80	14.66	14.55	4.37	14.20	14.02	13.93	13.84	13.75	13.65	13.56	13.46
5	16.26	13.27	12.06	11.39	10.97	10.67	10.46	10.29	10.16	10.05	9.89	9.72	9.55	9.47	9.38	9.29	9.20	9.11	9.02
6	13.75	10.92	9.78	9.15	8.75	8.47	8.26	8.10	7.98	7.87	7.72	7.56	7.40	7.31	7.23	7.14	7.06	6.97	6.88
7	12.25	9.55	8.45	7.85	7.46	7.19	6.99	6.84	6.72	6.62	6.47	6.31	6.16	6.07	5.99	5.91	5.82	5.74	5.65
8	11.26	8.65	7.59	7.01	6.63	6.37	6.18	6.03	5.91	5.81	5.67	5.52	5.36	5.28	5.20	5.12	5.03	4.95	4.86
9	10.56	8.02	6.99	6.42	6.06	5.80	5.61	5.47	5.35	5.26	5.11	4.96	4.81	4.73	4.65	4.57	4.48	4.40	4.31
10	10.04	7.56	6.55	5.99	5.64	5.39	5.20	5.06	4.94	4.85	4.71	4.56	4.41	4.33	4.25	4.17	4.08	4.00	3.91
11	9.65	7.21	6.22	5.67	5.32	5.07	4.89	4.74	4.63	4.54	4.40	4.25	4.10	4.02	3.94	3.86	3.78	3.69	3.60
12	9.33	6.93	5.95	5.41	5.06	4.82	4.64	4.50	4.39	4.30	4.16	4.01	3.86	3.78	3.70	3.62	3.54	3.45	3.36
13	9.07	6.70	5.74	5.21	4.86	4.62	4.44	4.30	4.19	4.10	3.96	3.82	3.66	3.59	3.51	3.43	3.34	3.25	3.17
14	8.86	6.51	5.56	5.04	4.69	4.46	4.28	4.14	4.03	3.94	3.80	3.66	3.51	3.43	3.35	3.27	3.18	3.09	3.00
15	8.68	6.36	5.42	4.89	4.56	4.32	4.14	4.00	3.89	3.80	3.67	3.52	3.37	3.29	3.21	3.13	3.05	2.96	2.87
16	8.53	6.23	5.29	4.77	4.44	4.20	4.03	3.89	3.78	3.69	3.55	3.41	3.26	3.18	3.10	3.02	2.93	2.84	2.75
17	8.40	6.11	5.18	4.67	4.34	4.10	3.93	3.79	3.68	3.59	3.46	3.31	3.16	3.08	3.00	2.92	2.83	2.75	2.65
18	8.29	6.01	5.09	4.58	4.25	4.01	3.84	3.71	3.60	3.51	3.37	3.23	3.08	3.00	2.92	2.84	2.75	2.66	2.57
19	8.18	5.93	5.01	4.50	4.17	3.94	3.77	3.63	3.52	3.43	3.30	3.15	3.00	2.92	2.84	2.76	2.67	2.58	2.49
20	8.10	5.85	4.94	4.43	4.10	3.87	3.70	3.56	3.46	3.37	3.23	3.09	2.94	2.86	2.78	2.69	2.61	2.52	2.42
21	8.02	5.78	4.87	4.37	4.04	3.81	3.64	3.51	3.40	3.31	3.17	3.03	2.88	2.80	2.72	2.64	2.55	2.46	2.36
22	7.95	5.72	4.82	4.31	3.99	3.76	3.59	3.45	3.35	3.26	3.12	2.98	2.83	2.75	2.67	2.58	2.50	2.40	2.31
23	7.88	5.66	4.76	4.26	3.94	3.71	3.54	3.41	3.30	3.21	3.07	2.93	2.78	2.70	2.62	2.54	2.45	2.35	2.26
24	7.82	5.61	4.72	4.22	3.90	3.67	3.50	3.36	3.26	3.17	3.03	2.89	2.74	2.66	2.58	2.49	2.40	2.31	2.21
25	7.77	5.57	4.68	4.18	3.85	3.63	3.46	3.32	3.22	3.13	2.99	2.85	2.70	2.62	2.54	2.45	2.36	2.27	2.17
26	7.72	5.53	4.64	4.14	3.82	3.59	3.42	3.29	3.18	3.09	2.96	2.81	2.66	2.58	2.50	2.42	2.33	2.23	2.13
27	7.68	5.49	4.60	4.11	3.78	3.56	3.39	3.26	3.15	3.06	2.93	2.78	2.63	2.55	2.47	2.38	2.29	2.20	2.10
28	7.64	5.45	4.57	4.07	3.75	3.53	3.36	3.23	3.12	3.03	2.90	2.75	2.60	2.52	2.44	2.35	2.26	2.17	2.06
29	7.60	5.42	4.54	4.04	3.73	3.50	3.33	3.20	3.09	3.00	2.87	2.73	2.57	2.49	2.41	2.33	2.23	2.14	2.03
30	7.56	5.39	4.51	4.02	3.70	3.47	3.30	3.17	3.07	2.98	2.84	2.70	2.55	2.47	2.39	2.30	2.21	2.11	2.01
40	7.31	5.18	4.31	3.83	3.51	3.29	3.12	2.99	2.89	2.80	2.66	2.52	2.37	2.29	2.20	2.11	2.02	1.92	1.80
60	7.08	4.98	4.13	3.65	3.34	3.12	2.95	2.82	2.72	2.63	2.50	2.35	2.20	2.12	2.03	1.94	1.84	1.73	1.60
120	6.85	4.79	3.95	3.48	3.17	2.96	2.79	2.66	2.56	2.47	2.34	2.19	2.03	1.95	1.86	1.76	1.66	1.53	1.38
∞	6.63	4.61	3.78	3.32	3.02	2.80	2.64	2.51	2.41	2.32	2.18	2.04	1.88	1.79	1.70	1.59	1.47	1.32	1.00

Table 7—*F*-Distribution *(continued)*

d.f.$_D$: Degrees of freedom, denominator	$\alpha = 0.025$ d.f.$_N$: Degrees of freedom, numerator																		
	1	2	3	4	5	6	7	8	9	10	12	15	20	24	30	40	60	120	∞
1	647.8	799.5	864.2	899.6	921.8	937.1	948.2	956.7	963.3	968.6	976.7	984.9	993.1	997.2	1001	1006	1010	1014	1018
2	38.51	39.00	39.17	39.25	39.30	39.33	39.36	39.37	39.39	39.40	39.41	39.43	39.45	39.46	39.46	39.47	39.48	39.49	39.50
3	17.44	16.04	15.44	15.10	14.88	14.73	14.62	14.54	14.47	14.42	14.34	14.25	14.17	14.12	14.08	14.04	13.99	13.95	13.90
4	12.22	10.65	9.98	9.60	9.36	9.20	9.07	8.98	8.90	8.84	8.75	8.66	8.56	8.51	8.46	8.41	8.36	8.31	8.26
5	10.01	8.43	7.76	7.39	7.15	6.98	6.85	6.76	6.68	6.62	6.52	6.43	6.33	6.28	6.23	6.18	6.12	6.07	6.02
6	8.81	7.26	6.60	6.23	5.99	5.82	5.70	5.60	5.52	5.46	5.37	5.27	5.17	5.12	5.07	5.01	4.96	4.90	4.85
7	8.07	6.54	5.89	5.52	5.29	5.12	4.99	4.90	4.82	4.76	4.67	4.57	4.47	4.42	4.36	4.31	4.25	4.20	4.14
8	7.57	6.06	5.42	5.05	4.82	4.65	4.53	4.43	4.36	4.30	4.20	4.10	4.00	3.95	3.89	3.84	3.78	3.73	3.67
9	7.21	5.71	5.08	4.72	4.48	4.32	4.20	4.10	4.03	3.96	3.87	3.77	3.67	3.61	3.56	3.51	3.45	3.39	3.33
10	6.94	5.46	4.83	4.47	4.24	4.07	3.95	3.85	3.78	3.72	3.62	3.52	3.42	3.37	3.31	3.26	3.20	3.14	3.08
11	6.72	5.26	4.63	4.28	4.04	3.88	3.76	3.66	3.59	3.53	3.43	3.33	3.23	3.17	3.12	3.06	3.00	2.94	2.88
12	6.55	5.10	4.47	4.12	3.89	3.73	3.61	3.51	3.44	3.37	3.28	3.18	3.07	3.02	2.96	2.91	2.85	2.79	2.72
13	6.41	4.97	4.35	4.00	3.77	3.60	3.48	3.39	3.31	3.25	3.15	3.05	2.95	2.89	2.84	2.78	2.72	2.66	2.60
14	6.30	4.86	4.24	3.89	3.66	3.50	3.38	3.29	3.21	3.15	3.05	2.95	2.84	2.79	2.73	2.67	2.61	2.55	2.49
15	6.20	4.77	4.15	3.80	3.58	3.41	3.29	3.20	3.12	3.06	2.98	2.86	2.76	2.70	2.64	2.59	2.52	2.46	2.40
16	6.12	4.69	4.08	3.73	3.50	3.34	3.22	3.12	3.05	2.99	2.89	2.79	2.68	2.63	2.57	2.51	2.45	2.38	2.32
17	6.04	4.62	4.01	3.66	3.44	3.28	3.16	3.06	2.98	2.92	2.82	2.72	2.62	2.56	2.50	2.44	2.38	2.32	2.25
18	5.98	4.56	3.95	3.61	3.38	3.22	3.10	3.01	2.93	2.87	2.77	2.67	2.56	2.50	2.44	2.38	2.32	2.26	2.19
19	5.92	4.51	3.90	3.56	3.33	3.17	3.05	2.96	2.88	2.82	2.72	2.62	2.51	2.45	2.39	2.33	2.27	2.20	2.13
20	5.87	4.46	3.86	3.51	3.29	3.13	3.01	2.91	2.84	2.77	2.68	2.57	2.46	2.41	2.35	2.29	2.22	2.16	2.09
21	5.83	4.42	3.82	3.48	3.25	3.09	2.97	2.87	2.80	2.73	2.64	2.53	2.42	2.37	2.31	2.25	2.18	2.11	2.04
22	5.79	4.38	3.78	3.44	3.22	3.05	2.93	2.84	2.76	2.70	2.60	2.50	2.39	2.33	2.27	2.21	2.14	2.08	2.00
23	5.75	4.35	3.75	3.41	3.18	3.02	2.90	2.81	2.73	2.67	2.57	2.47	2.36	2.30	2.24	2.18	2.11	2.04	1.97
24	5.72	4.32	3.72	3.38	3.15	2.99	2.87	2.78	2.70	2.64	2.54	2.44	2.33	2.27	2.21	2.15	2.08	2.01	1.94
25	5.69	4.29	3.69	3.35	3.13	2.97	2.85	2.75	2.68	2.61	2.51	2.41	2.30	2.24	2.18	2.12	2.05	1.98	1.91
26	5.66	4.27	3.67	3.33	3.10	2.94	2.82	2.73	2.65	2.59	2.49	2.39	2.25	2.22	2.16	2.09	2.03	1.95	1.88
27	5.63	4.24	3.65	3.31	3.08	2.92	2.80	2.71	2.63	2.57	2.47	2.36	2.25	2.19	2.13	2.07	2.00	1.93	1.85
28	5.61	4.22	3.63	3.29	3.06	2.90	2.78	2.69	2.61	2.55	2.45	2.34	2.23	2.17	2.11	2.05	1.98	1.91	1.83
29	5.59	4.20	3.61	3.27	3.04	2.88	2.76	2.67	2.59	2.53	2.43	2.32	2.21	2.15	2.09	2.03	1.96	1.89	1.81
30	5.57	4.18	3.59	3.25	3.03	2.87	2.75	2.65	2.57	2.51	2.41	2.31	2.20	2.14	2.07	2.01	1.94	1.87	1.79
40	5.42	4.05	3.46	3.13	2.90	2.74	2.62	2.53	2.45	2.39	2.29	2.18	2.07	2.01	1.94	1.88	1.80	1.72	1.64
60	5.29	3.93	3.34	3.01	2.79	2.63	2.51	2.41	2.33	2.27	2.17	2.06	1.94	1.88	1.82	1.74	1.67	1.58	1.48
120	5.15	3.80	3.23	2.89	2.67	2.52	2.39	2.30	2.22	2.16	2.05	1.94	1.82	1.76	1.69	1.61	1.53	1.43	1.31
∞	5.02	3.69	3.12	2.79	2.57	2.41	2.29	2.19	2.11	2.05	1.94	1.83	1.71	1.64	1.57	1.48	1.39	1.27	1.00

Table 7—*F*-Distribution *(continued)*

d.f.$_D$: Degrees of freedom, denominator	$\alpha = 0.05$																		
	d.f.$_N$: Degrees of freedom, numerator																		
	1	2	3	4	5	6	7	8	9	10	12	15	20	24	30	40	60	120	∞
1	161.4	199.5	215.7	224.6	230.2	234.0	236.8	238.9	240.5	241.9	243.9	245.9	248.0	249.1	250.1	251.1	252.2	253.3	254.3
2	18.51	19.00	19.16	19.25	19.30	19.33	19.35	19.37	19.38	19.40	19.41	19.43	19.45	19.45	19.46	19.47	19.48	19.49	19.50
3	10.13	9.55	9.28	9.12	9.01	8.94	8.89	8.85	8.81	8.79	8.74	8.70	8.66	8.64	8.62	8.59	8.57	8.55	8.53
4	7.71	6.94	6.59	6.39	6.26	6.16	6.09	6.04	6.00	5.96	5.91	5.86	5.80	5.77	5.75	5.72	5.69	5.66	5.63
5	6.61	5.79	5.41	5.19	5.05	4.95	4.88	4.82	4.77	4.74	4.68	4.62	4.56	4.53	4.50	4.46	4.43	4.40	4.36
6	5.99	5.14	4.76	4.53	4.39	4.28	4.21	4.15	4.10	4.06	4.00	3.94	3.87	3.84	3.81	3.77	3.74	3.70	3.67
7	5.59	4.74	4.35	4.12	3.97	3.87	3.79	3.73	3.68	3.64	3.57	3.51	3.44	3.41	3.38	3.34	3.30	3.27	3.23
8	5.32	4.46	4.07	3.84	3.69	3.58	3.50	3.44	3.39	3.35	3.28	3.22	3.15	3.12	3.08	3.04	3.01	2.97	2.93
9	5.12	4.26	3.86	3.63	3.48	3.37	3.29	3.23	3.18	3.14	3.07	3.01	2.94	2.90	2.86	2.83	2.79	2.75	2.71
10	4.96	4.10	3.71	3.48	3.33	3.22	3.14	3.07	3.02	2.98	2.91	2.85	2.77	2.74	2.70	2.66	2.62	2.58	2.54
11	4.84	3.98	3.59	3.36	3.20	3.09	3.01	2.95	2.90	2.85	2.79	2.72	2.65	2.61	2.57	2.53	2.49	2.45	2.40
12	4.75	3.89	3.49	3.26	3.11	3.00	2.91	2.85	2.80	2.75	2.69	2.62	2.54	2.51	2.47	2.43	2.38	2.34	2.30
13	4.67	3.81	3.41	3.18	3.03	2.92	2.83	2.77	2.71	2.67	2.60	2.53	2.46	2.42	2.38	2.34	2.30	2.25	2.21
14	4.60	3.74	3.34	3.11	2.96	2.85	2.76	2.70	2.65	2.60	2.53	2.46	2.39	2.35	2.31	2.27	2.22	2.18	2.13
15	4.54	3.68	3.29	3.06	2.90	2.79	2.71	2.64	2.59	2.54	2.48	2.40	2.33	2.29	2.25	2.20	2.16	2.11	2.07
16	4.49	3.63	3.24	3.01	2.85	2.74	2.66	2.59	2.54	2.49	2.42	2.35	2.28	2.24	2.19	2.15	2.11	2.06	2.01
17	4.45	3.59	3.20	2.96	2.81	2.70	2.61	2.55	2.49	2.45	2.38	2.31	2.23	2.19	2.15	2.10	2.06	2.01	1.96
18	4.41	3.55	3.16	2.93	2.77	2.66	2.58	2.51	2.46	2.41	2.34	2.27	2.19	2.15	2.11	2.06	2.02	1.97	1.92
19	4.38	3.52	3.13	2.90	2.74	2.63	2.54	2.48	2.42	2.38	2.31	2.23	2.16	2.11	2.07	2.03	1.98	1.93	1.88
20	4.35	3.49	3.10	2.87	2.71	2.60	2.51	2.45	2.39	2.35	2.28	2.20	2.12	2.08	2.04	1.99	1.95	1.90	1.84
21	4.32	3.47	3.07	2.84	2.68	2.57	2.49	2.42	2.37	2.32	2.25	2.18	2.10	2.05	2.01	1.96	1.92	1.87	1.81
22	4.30	3.44	3.05	2.82	2.66	2.55	2.46	2.40	2.34	2.30	2.23	2.15	2.07	2.03	1.98	1.94	1.89	1.84	1.78
23	4.28	3.42	3.03	2.80	2.64	2.53	2.44	2.37	2.32	2.27	2.20	2.13	2.05	2.01	1.96	1.91	1.86	1.81	1.76
24	4.26	3.40	3.01	2.78	2.62	2.51	2.42	2.36	2.30	2.25	2.18	2.11	2.03	1.98	1.94	1.89	1.84	1.79	1.73
25	4.24	3.39	2.99	2.76	2.60	2.49	2.40	2.34	2.28	2.24	2.16	2.09	2.01	1.96	1.92	1.87	1.82	1.77	1.71
26	4.23	3.37	2.98	2.74	2.59	2.47	2.39	2.32	2.27	2.22	2.15	2.07	1.99	1.95	1.90	1.85	1.80	1.75	1.69
27	4.21	3.35	2.96	2.73	2.57	2.46	2.37	2.31	2.25	2.20	2.13	2.06	1.97	1.93	1.88	1.84	1.79	1.73	1.67
28	4.20	3.34	2.95	2.71	2.56	2.45	2.36	2.29	2.24	2.19	2.12	2.04	1.96	1.91	1.87	1.82	1.77	1.71	1.65
29	4.18	3.33	2.93	2.70	2.55	2.43	2.35	2.28	2.22	2.18	2.10	2.03	1.94	1.90	1.85	1.81	1.75	1.70	1.64
30	4.17	3.32	2.92	2.69	2.53	2.42	2.33	2.27	2.21	2.16	2.09	2.01	1.93	1.89	1.84	1.79	1.74	1.68	1.62
40	4.08	3.23	2.84	2.61	2.45	2.34	2.25	2.18	2.12	2.08	2.00	1.92	1.84	1.79	1.74	1.69	1.64	1.58	1.51
60	4.00	3.15	2.76	2.53	2.37	2.25	2.17	2.10	2.04	1.99	1.92	1.84	1.75	1.70	1.65	1.59	1.53	1.47	1.39
120	3.92	3.07	2.68	2.45	2.29	2.17	2.09	2.02	1.96	1.91	1.83	1.75	1.66	1.61	1.55	1.50	1.43	1.35	1.25
∞	3.84	3.00	2.60	2.37	2.21	2.10	2.01	1.94	1.88	1.83	1.75	1.67	1.57	1.52	1.46	1.39	1.32	1.22	1.00

Table 7—*F*-Distribution *(continued)*

$d.f._D$: Degrees of freedom, denominator	$\alpha = 0.10$																		
	$d.f._N$: Degrees of freedom, numerator																		
	1	2	3	4	5	6	7	8	9	10	12	15	20	24	30	40	60	120	∞
1	39.86	49.50	53.59	55.83	57.24	58.20	58 91	59.44	59.86	60.19	60.71	61.22	61.74	62.00	62.26	62.53	62.79	63.06	63.33
2	8.53	9.00	9.16	9.24	9.29	9.33	9.35	9.37	9.38	9.39	9.41	9.42	9.44	9.45	9.46	9.47	9.47	9.48	9.49
3	5.54	5.46	5.39	5.34	5.31	5.28	5.27	5.25	5.24	5.23	5.22	5.20	5.18	5.18	5.17	5.16	5.15	5.14	5.13
4	4.54	4.32	4.19	4.11	4.05	4.01	3.98	3.95	3.94	3.92	3.90	3.87	3.84	3.83	3.82	3.80	3.79	3.78	3.76
5	4.06	3.78	3.62	3.52	3.45	3.40	3.37	3.34	3.32	3.30	3.27	3.24	3.21	3.19	3.17	3.16	3.14	3.12	3.10
6	3.78	3.46	3.29	3.18	3.11	3.05	3.01	2.98	2.96	2.94	2.90	2.87	2.84	2.82	2.80	2.78	2.76	2.74	2.72
7	3.59	3.26	3.07	2.96	2.88	2.83	2.78	2.75	2.72	2.70	2.67	2.63	2.59	2.58	2.56	2.54	2.51	2.49	2.47
8	3.46	3.11	2.92	2.81	2.73	2.67	2.62	2.59	2.56	2.54	2.50	2.46	2.42	2.40	2.38	2.36	2.34	2.32	2.29
9	3.36	3.01	2.81	2.69	2.61	2.55	2.51	2.47	2.44	2.42	2.38	2.34	2.30	2.28	2.25	2.23	2.21	2.18	2.16
10	3.29	2.92	2.73	2.61	2.52	2.46	2.41	2.38	2.35	2.32	2.28	2.24	2.20	2.18	2.16	2.13	2.11	2.08	2.06
11	3.23	2.86	2.66	2.54	2.45	2.39	2.34	2.30	2.27	2.25	2.21	2.17	2.12	2.10	2.08	2.05	2.03	2.00	1.97
12	3.18	2.81	2.61	2.48	2.39	2.33	2.28	2.24	2.21	2.19	2.15	2.10	2.06	2.04	2.01	1.99	1.96	1.93	1.90
13	3.14	2.76	2.56	2.43	2.35	2.28	2.23	2.20	2.16	2.14	2.10	2.05	2.01	1.98	1.96	1.93	1.90	1.88	1.85
14	3.10	2.73	2.52	2.39	2.31	2.24	2.19	2.15	2.12	2.10	2.05	2.01	1.96	1.94	1.91	1.89	1.86	1.83	1.80
15	3.07	2.70	2.49	2.36	2.27	2.21	2.16	2.12	2.09	2.06	2.02	1.97	1.92	1.90	1.87	1.85	1.82	1.79	1.76
16	3.05	2.67	2.46	2.33	2.24	2.18	2.13	2.09	2.06	2.03	1.99	1.94	1.89	1.87	1.84	1.81	1.78	1.75	1.72
17	3.03	2.64	2.44	2.31	2.22	2.15	2.10	2.06	2.03	2.00	1.96	1.91	1.86	1.84	1.81	1.78	1.75	1.72	1.69
18	3.01	2.62	2.42	2.29	2.20	2.13	2.08	2.04	2.00	1.98	1.93	1.89	1.84	1.81	1.78	1.75	1.72	1.69	1.66
19	2.99	2.61	2.40	2.27	2.18	2.11	2.06	2.02	1.98	1.96	1.91	1.86	1.81	1.79	1.76	1.73	1.70	1.67	1.63
20	2.97	2.59	2.38	2.25	2.16	2.09	2.04	2.00	1.96	1.94	1.89	1.84	1.79	1.77	1.74	1.71	1.68	1.64	1.61
21	2.96	2.57	2.36	2.23	2.14	2.08	2.02	1.98	1.95	1.92	1.87	1.83	1.78	1.75	1.72	1.69	1.66	1.62	1.59
22	2.95	2.56	2.35	2.22	2.13	2.06	2.01	1.97	1.93	1.90	1.86	1.81	1.76	1.73	1.70	1.67	1.64	1.60	1.57
23	2.94	2.55	2.34	2.21	2.11	2.05	1.99	1.95	1.92	1.89	1.84	1.80	1.74	1.72	1.69	1.66	1.62	1.59	1.55
24	2.93	2.54	2.33	2.19	2.10	2.04	1.98	1.94	1.91	1.88	1.83	1.78	1.73	1.70	1.67	1.64	1.61	1.57	1.53
25	2.92	2.53	2.32	2.18	2.09	2.02	1.97	1.93	1.89	1.87	1.82	1.77	1.72	1.69	1.66	1.63	1.59	1.56	1.52
26	2.91	2.52	2.31	2.17	2.08	2.01	1.96	1.92	1.88	1.86	1.81	1.76	1.71	1.68	1.65	1.61	1.58	1.54	1.50
27	2.90	2.51	2.30	2.17	2.07	2.00	1.95	1.91	1.87	1.85	1.80	1.75	1.70	1.67	1.64	1.60	1.57	1.53	1.49
28	2.89	2.50	2.29	2.16	2.06	2.00	1.94	1.90	1.87	1.84	1.79	1.74	1.69	1.66	1.63	1.59	1.56	1.52	1.48
29	2.89	2.50	2.28	2.15	2.06	1.99	1.93	1.89	1.86	1.83	1.78	1.73	1.68	1.65	1.62	1.58	1.55	1.51	1.47
30	2.88	2.49	2.28	2.14	2.05	1.98	1.93	1.88	1.85	1.82	1.77	1.72	1.67	1.64	1.61	1.57	1.54	1.50	1.46
40	2.84	2.44	2.23	2.09	2.00	1.93	1.87	1.83	1.79	1.76	1.71	1.66	1.61	1.57	1.54	1.51	1.47	1.42	1.38
60	2.79	2.39	2.18	2.04	1.95	1.87	1.82	1.77	1.74	1.71	1.66	1.60	1.54	1.51	1.48	1.44	1.40	1.35	1.29
120	2.75	2.35	2.13	1.99	1.90	1.82	1.77	1.72	1.68	1.65	1.60	1.55	1.48	1.45	1.41	1.37	1.32	1.26	1.19
∞	2.71	2.30	2.08	1.94	1.85	1.77	1.72	1.67	1.63	1.60	1.55	1.49	1.42	1.38	1.34	1.30	1.24	1.17	1.00

From M. Merrington and C. M. Thompson (1943). Table of Percentage Points of the Inverted Beta (F) Distribution. *Biometrika* 33, pp. 74–87. Reprinted with permission from *Biometrika*.

Table 8—Critical Values for the Sign Test

Reject the null hypothesis if the test statistic x is less than or equal to the value in the table.

	One-tailed, $\alpha = 0.005$	$\alpha = 0.01$	$\alpha = 0.025$	$\alpha = 0.05$
n	**Two-tailed, $\alpha = 0.01$**	**$\alpha = 0.02$**	**$\alpha = 0.05$**	**$\alpha = 0.10$**
8	0	0	0	1
9	0	0	1	1
10	0	0	1	1
11	0	1	1	2
12	1	1	2	2
13	1	1	2	3
14	1	2	3	3
15	2	2	3	3
16	2	2	3	4
17	2	3	4	4
18	3	3	4	5
19	3	4	4	5
20	3	4	5	5
21	4	4	5	6
22	4	5	5	6
23	4	5	6	7
24	5	5	6	7
25	5	6	6	7

Note: Table 8 is for one-tailed or two-tailed tests. The sample size n represents the total number of + and − signs. The test value x is the smaller number of + or − signs..

From *Journal of American Statistical Association* Vol. 41 (1946), pp. 557–66. W. J. Dixon and A. M. Mood. Reprinted with permission.

Table 9—Critical Values for the Wilcoxon Signed-Rank Test

Reject the null hypothesis if the value of the test statistic w_s is less than or equal to the value given in the table.

	One-tailed, $\alpha = 0.05$	$\alpha = 0.025$	$\alpha = 0.01$	$\alpha = 0.005$
n	**Two-tailed, $\alpha = 0.10$**	$\alpha = 0.05$	$\alpha = 0.02$	$\alpha = 0.01$
5	1	—	—	—
6	2	1	—	—
7	4	2	0	—
8	6	4	2	0
9	8	6	3	2
10	11	8	5	3
11	14	11	7	5
12	17	14	10	7
13	21	17	13	10
14	26	21	16	13
15	30	25	20	16
16	36	30	24	19
17	41	35	28	23
18	47	40	33	28
19	54	46	38	32
20	60	52	43	37
21	68	59	49	43
22	75	66	56	49
23	83	73	62	55
24	92	81	69	61
25	101	90	77	68
26	110	98	85	76
27	120	107	93	84
28	130	117	102	92
29	141	127	111	100
30	152	137	120	109

Table 10—Critical Values for the Spearman Rank Correlation

Reject H_0: $\rho_s = 0$ if the absolute value of r_s is greater than the value given in the table.

n	$\alpha = 0.10$	$\alpha = 0.05$	$\alpha = 0.01$
5	0.900	—	—
6	0.829	0.886	—
7	0.714	0.786	0.929
8	0.643	0.738	0.881
9	0.600	0.700	0.833
10	0.564	0.648	0.794
11	0.536	0.618	0.818
12	0.497	0.591	0.780
13	0.475	0.566	0.745
14	0.457	0.545	0.716
15	0.441	0.525	0.689
16	0.425	0.507	0.666
17	0.412	0.490	0.645
18	0.399	0.476	0.625
19	0.388	0.462	0.608
20	0.377	0.450	0.591
21	0.368	0.438	0.576
22	0.359	0.428	0.562
23	0.351	0.418	0.549
24	0.343	0.409	0.537
25	0.336	0.400	0.526
26	0.329	0.392	0.515
27	0.323	0.385	0.505
28	0.317	0.377	0.496
29	0.311	0.370	0.487
30	0.305	0.364	0.478

From N. L. Johnson and F. C. Leone, *Statistical and Experimental Design,* Vol. I (1964), p. 412. Reprinted with permission from the Institute of Mathematical Statistics.

Table 11—Critical Values for the Pearson Correlation Coefficient

Reject H_0: $\rho = 0$ if the absolute value of r is greater than the value given in the table.

n	$\alpha = 0.05$	$\alpha = 0.01$
4	0.950	0.990
5	0.878	0.959
6	0.811	0.917
7	0.754	0.875
8	0.707	0.834
9	0.666	0.798
10	0.632	0.765
11	0.602	0.735
12	0.576	0.708
13	0.553	0.684
14	0.532	0.661
15	0.514	0.641
16	0.497	0.623
17	0.482	0.606
18	0.468	0.590
19	0.456	0.575
20	0.444	0.561
21	0.433	0.549
22	0.423	0.537
23	0.413	0.526
24	0.404	0.515
25	0.396	0.505
26	0.388	0.496
27	0.381	0.487
28	0.374	0.479
29	0.367	0.471
30	0.361	0.463
35	0.334	0.430
40	0.312	0.403
45	0.294	0.380
50	0.279	0.361
55	0.266	0.345
60	0.254	0.330
65	0.244	0.317
70	0.235	0.306
75	0.227	0.296
80	0.220	0.286
85	0.213	0.278
90	0.207	0.270
95	0.202	0.263
100	0.197	0.256

The critical values in Table 11 were generated using Excel.

Table 12—Critical Values for the Number of Runs

Reject the null hypothesis if the test statistic G is less than or equal to the smaller entry or greater than or equal to the larger entry.

Value of n_1	Value of n_2																		
	2	**3**	**4**	**5**	**6**	**7**	**8**	**9**	**10**	**11**	**12**	**13**	**14**	**15**	**16**	**17**	**18**	**19**	**20**
2	1	1	1	1	1	1	1	1	1	1	2	2	2	2	2	2	2	2	2
	6	6	6	6	6	6	6	6	6	6	6	6	6	6	6	6	6	6	6
3	1	1	1	1	2	2	2	2	2	2	2	2	2	3	3	3	3	3	3
	6	8	8	8	8	8	8	8	8	8	8	8	8	8	8	8	8	8	8
4	1	1	1	2	2	2	3	3	3	3	3	3	3	3	4	4	4	4	4
	6	8	9	9	9	10	10	10	10	10	10	10	10	10	10	10	10	10	10
5	1	1	2	2	3	3	3	3	3	4	4	4	4	4	4	4	5	5	5
	6	8	9	10	10	11	11	12	12	12	12	12	12	12	12	12	12	12	12
6	1	2	2	3	3	3	3	4	4	4	4	5	5	5	5	5	5	6	6
	6	8	9	10	11	12	12	13	13	13	13	14	14	14	14	14	14	14	14
7	1	2	2	3	3	3	4	4	5	5	5	5	5	6	6	6	6	6	6
	6	8	10	11	12	13	13	14	14	14	14	15	15	15	16	16	16	16	16
8	1	2	3	3	3	4	4	5	5	5	6	6	6	6	6	7	7	7	7
	6	8	10	11	12	13	14	14	15	15	15	16	16	16	16	17	17	17	17
9	1	2	3	3	4	4	5	5	5	6	6	6	7	7	7	7	8	8	8
	6	8	10	12	13	14	14	15	16	16	16	17	17	18	18	18	18	18	18
10	1	2	3	3	4	5	5	5	6	6	7	7	7	7	8	8	8	8	9
	6	8	10	12	13	14	15	16	16	17	17	18	18	18	18	18	18	20	20
11	1	2	3	4	4	5	5	6	6	7	7	7	8	8	8	9	9	9	9
	6	8	10	12	13	14	15	16	17	17	18	19	19	19	20	20	20	21	21
12	2	2	3	4	4	5	6	6	7	7	7	8	8	8	9	9	9	10	10
	6	8	10	12	13	14	16	16	17	18	19	19	20	20	21	21	21	22	22
13	2	2	3	4	5	5	6	6	7	7	8	8	9	9	9	10	10	10	10
	6	8	10	12	14	15	16	17	18	19	19	20	20	21	21	22	22	23	23
14	2	2	3	4	5	5	6	7	7	8	8	9	9	9	10	10	10	11	11
	6	8	10	12	14	15	16	17	18	19	20	20	21	22	22	23	23	23	24
15	2	3	3	4	5	6	6	7	7	8	8	9	9	10	10	11	11	11	12
	6	8	10	12	14	15	16	17	17	18	29	21	22	22	23	23	24	24	25
16	2	3	4	4	5	6	6	7	8	8	9	9	10	10	11	11	11	12	12
	6	8	10	12	14	16	17	18	19	20	21	21	22	23	23	24	25	25	25
17	2	3	4	4	5	6	7	7	8	9	9	10	10	11	11	11	12	12	13
	6	8	10	12	14	16	17	18	18	20	21	22	23	23	24	25	25	26	26
18	2	3	4	5	5	6	7	8	8	9	9	10	10	11	11	12	12	13	13
	6	8	10	12	14	16	17	18	19	20	21	22	23	24	25	25	26	26	27
19	2	3	4	5	6	6	7	8	8	9	10	10	11	11	12	12	13	13	13
	6	8	10	12	14	16	17	18	20	21	22	23	23	24	25	26	26	27	27
20	2	3	4	5	6	6	7	8	9	9	10	10	11	12	12	13	13	13	14
	6	8	10	12	14	16	17	18	20	21	22	23	24	25	25	26	27	27	28

Note: Table 12 is for a two-tailed test with $\alpha = 0.05$.

From "Tables for testing randomness of groupings in a sequence of alternatives," *The Annals of Mathematical Statistics*, Vol. 14, No. 1. Reprinted with permission from the Institute of Mathematical Statistics.

Try It Yourself Answers

CHAPTER 1

Section 1.1

1a. The population consists of the prices per gallon of regular gasoline at all gasoline stations in the United States.

b. The sample consists of the prices per gallon of regular gasoline at the 900 surveyed stations.

c. The data set consists of the 900 prices.

2a. Population **b.** Parameter

3a. Descriptive statistics involve the statement "76% of women and 60% of men had a physical examination within the previous year."

b. An inference drawn from the study is that a higher percentage of women had a physical examination within the previous year.

Section 1.2

1a. City names and city population

b. City name: Nonnumerical
City population: Numerical

c. City name: Qualitative
City population: Quantitative

2. (1a) The final standings represent a ranking of hockey teams.

(1b) Ordinal, because the data can be put in order.

(2a) The collection of phone numbers represents labels. No mathematical computations can be made.

(2b) Nominal, because you cannot make calculations on the data.

3. (1a) The data set is the collection of body temperatures.

(1b) Interval, because the data can be ordered and meaningful differences can be calculated, but it does not make sense to write a ratio using the temperatures.

(2a) The data set is the collection of heart rates.

(2b) Ratio, because the data can be ordered, you can calculate meaningful differences, can be written as a ratio, and the data set contains an inherent zero.

Section 1.3

1. (1a) Focus: Effect of exercise on senior citizens.

(1b) Population: Collection of all senior citizens.

(1c) Experiment

(2a) Focus: Effect of radiation fallout on senior citizens.

(2b) Population: Collection of all senior citizens.

(2c) Sampling

2a. Answers will vary. Example: start with the first digits 92630782 . . .

b. 92|63|07|82|40|19|26 **c.** 63, 7, 40, 19, 26

3. (1a) The sample was selected by using only available students.

(1b) Convenience sampling

(2a) The sample was selected by numbering each student in the school, randomly choosing a starting number, and selecting students at regular intervals from the starting number.

(2b) Systematic sampling

CHAPTER 2

Section 2.1

1a. 8 classes

b. Min = 0; Max = 72; Class width = 10

c.

Lower limit	Upper limit
0	9
10	19
20	29
30	39
40	49
50	59
60	69
70	79

d. See part (e).

e.

Class	Frequency, *f*
0–9	15
10–19	19
20–29	14
30–39	7
40–49	14
50–59	6
60–69	4
70–79	1

2a. See part (b).

b.

Class	Frequency, *f*	Midpoint	Relative frequency	Cumulative frequency
0–9	15	4.5	0.1875	15
10–19	19	14.5	0.2375	34
20–29	14	24.5	0.1750	48
30–39	7	34.5	0.0875	55
40–49	14	44.5	0.1750	69
50–59	6	54.5	0.0750	75
60–69	4	64.5	0.0500	79
70–79	1	74.5	0.0125	80
	$\sum f = 80$		$\sum \frac{f}{n} = 1$	

c. 42.5% of the population is under 20 years old. 6.25% of the population is over 59 years old.

3a.

Class boundaries
−0.5–9.5
9.5–19.5
19.5–29.5
29.5–39.5
39.5–49.5
49.5–59.5
59.5–69.5
69.5–79.5

b. Use class midpoints for the horizontal scale and frequency for the vertical scale.

c. Ages of Akhiok Residents

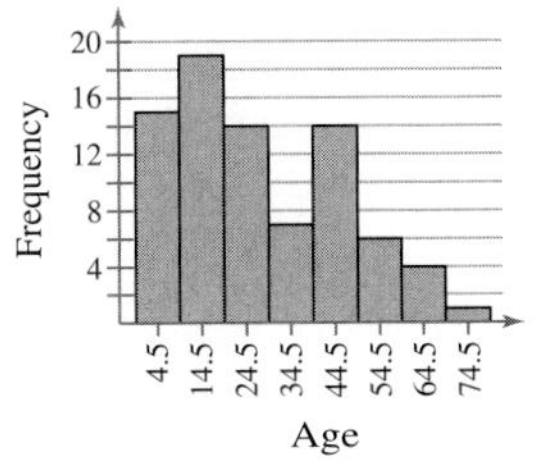

d. Same as 2c.

4a. Same as 3b. **b.** See part (c).

c. Ages of Ahkiok Residents

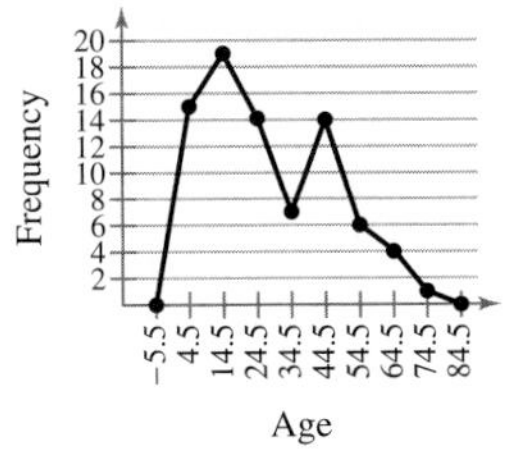

d. The population increases up to the age of 14.5 and then decreases. Population increases again between the ages of 34.5 and 44.5, but then after 44.5, the population decreases.

5abc. Ages of Akhiok Residents

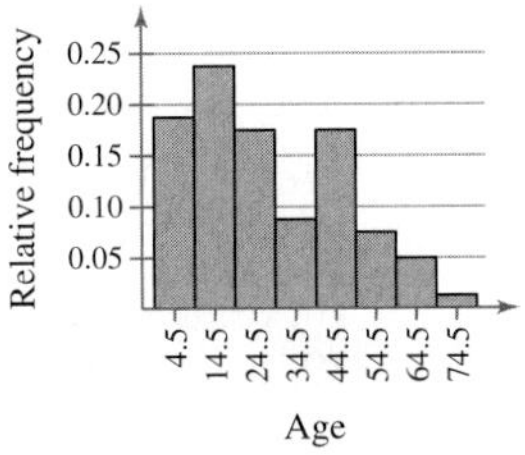

6a. Use upper class boundaries for the horizontal scale and cumulative frequency for the vertical scale.

b. See part (c).

c. Ages of Akhiok Residents

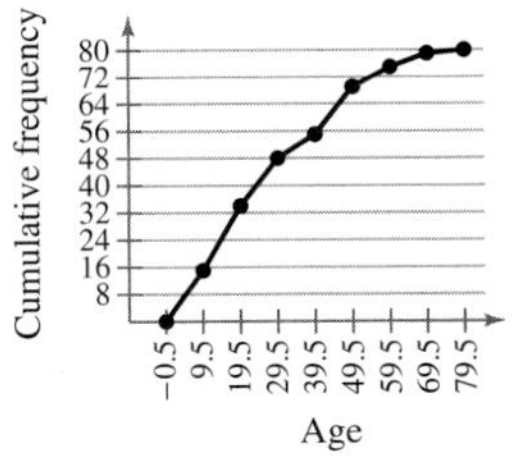

d. Approximately 69 residents are 49 years old or younger.

7a. Enter data.

b.

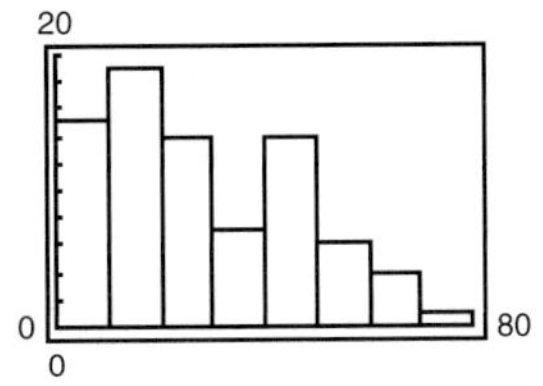

Section 2.2

1a.

0	
1	
2	
3	
4	
5	
6	
7	

b. Key: 3|3 = 33

0	5 2 7 1 5 3 1 0 1 3 3 9 0 4 5
1	8 2 5 6 3 3 7 3 0 7 8 2 3 8 9 3 6 9 9
2	5 4 2 0 3 3 4 0 1 5 9 6 6 6
3	9 6 9 7 9 9 3
4	4 2 4 7 1 8 0 0 1 9 9 5 1 9
5	8 3 1 6 8 9
6	0 8 7 8
7	2

c. Key: 3|3 = 33

0	0 0 1 1 1 2 3 3 3 4 5 5 5 7 9
1	0 2 2 3 3 3 3 3 5 6 6 7 7 8 8 8 9 9 9
2	0 0 1 2 3 3 4 4 5 5 6 6 6 9
3	3 6 7 9 9 9 9
4	0 0 1 1 1 2 4 4 5 7 8 9 9 9
5	1 3 6 8 8 9
6	0 7 8 8
7	2

d. It seems that most of the residents are under 40.

2ab. Key: 3|3 = 33

0	0 0 1 1 1 2 3 3 3 4
0	5 5 5 7 9
1	0 2 2 3 3 3 3 3
1	5 6 6 7 7 8 8 8 9 9 9
2	0 0 1 2 3 3 4 4
2	5 5 6 6 6 9
3	3
3	6 7 9 9 9 9
4	0 0 1 1 1 2 4 4
4	5 7 8 9 9 9
5	1 3
5	6 8 8 9
6	0
6	7 8 8
7	2
7	

3a. Use ages for the horizontal axis.

b. **Ages of Akhiok Residents**

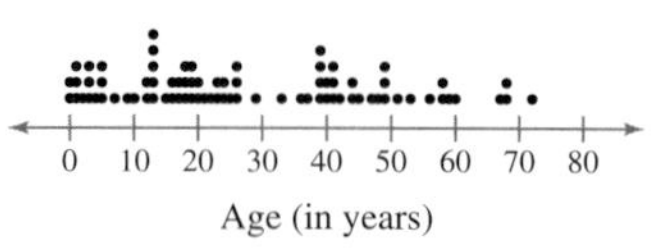

c. A large percentage of the residents are under 40 years old.

4a.

Vehicle type	Killed (frequency)	Relative frequency	Central angle
Cars	22,385	0.6556	236°
Trucks	8,457	0.2477	89°
Motorcycles	2,806	0.0822	30°
Other	497	0.0146	5°
	$\Sigma f = 34{,}145$	$\Sigma \frac{f}{n} \approx 1$	$\sum = 360°$

b. **Motor Vehicle Occupants Killed in 1991**

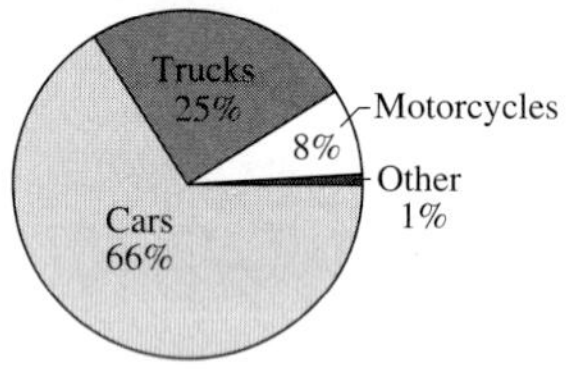

c. As a percentage of total motor vehicle deaths, car deaths decreased by 10%, truck deaths increased by 9%, and motorcycle deaths stayed about the same.

5a.

Cause	Frequency, *f*
Auto Dealers	14,668
Auto Repair	9,728
Home Furnishing	7,792
Computer Sales	5,733
Dry Cleaning	4,649

b. **Causes of BBB Complaints**

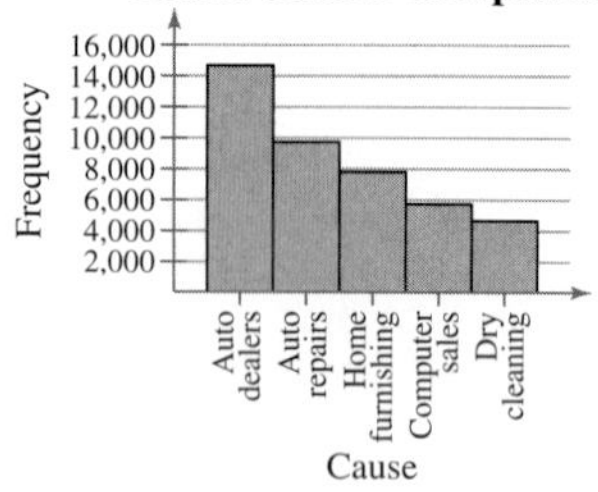

c. It appears that the auto industry (dealers and repair shops) account for the largest portion of complaints filed at the BBB.

6ab. **Salaries**

c. It appears that the longer an employee is with the company, the larger his or her salary will be.

7ab. **Cellular Phone Bills**

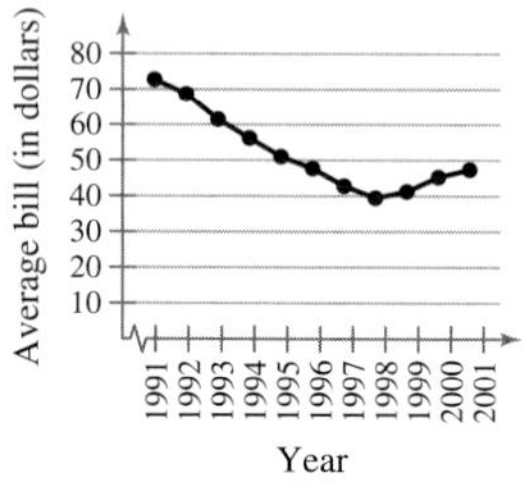

c. From 1991 to 1998, the average bill decreased significantly. From 1998 until 2001, the average bill increased slightly.

Section 2.3

1a. 578 **b.** 41.3

c. The typical age of an employee in a department store is 41.3 years old.

2a. 0, 0, 1, 1, 1, 2, 3, 3, 4, 5, 5, 5, 9, 10, 12, 12, 13, 13, 13, 13, 13, 15, 16, 16, 17, 17, 18, 18, 18, 19, 19, 19, 20, 20, 21, 22, 23, 23, 24, 24, 25, 25, 26, 26, 26, 29, 36, 39, 39, 39, 39, 40, 40, 41, 41, 41, 42, 44, 44, 45, 47, 48, 49, 49, 49, 51, 53, 56, 58, 58, 60, 67, 68, 68, 72

b. 23

3a. 0, 0, 1, 1, 1, 2, 3, 3, 3, 4, 5, 5, 5, 7, 9, 10, 12, 12, 13, 13, 13, 13, 13, 15, 16, 16, 17, 17, 18, 18, 18, 19, 19, 19, 20, 20, 21, 22, 23, 23, 24, 24, 25, 25, 26, 26, 26, 29, 33, 36, 37, 39, 39, 39, 39, 40, 40, 41, 41, 41, 42, 44, 44, 45, 47, 48, 49, 49, 49, 51, 53, 56, 58, 58, 59, 60, 67, 68, 68, 72

b. 23.5

c. Half of the residents of Akhiok are younger than 23.5 years old and half are older than 23.5 years old.

4a. 0, 0, 1, 1, 1, 2, 3, 3, 3, 4, 5, 5, 5, 7, 9, 10, 12, 12, 13, 13, 13, 13, 13, 15, 16, 16, 17, 17, 18, 18, 18, 19, 19, 19, 20, 20, 21, 22, 23, 23, 24, 24, 25, 25, 26, 26, 26, 29, 33, 36, 37, 39, 39, 39, 39, 40, 40, 41, 41, 41, 42, 44, 44, 45, 47, 48, 49, 49, 49, 51, 53, 56, 58, 58, 59, 60, 67, 68, 68, 72

b. 13 **c.** The mode of the ages is 13 years old.

5a. Yes **b.** The mode of the responses to the survey is "Yes."

6a. 21.6; 21; 20

b. The mean in Example 6 ($\bar{x} \approx 23.8$) was heavily influenced by the age 65. Neither the median nor the mode was affected as much by the age 65.

7ab.

Source	Score, x	Weight, w	$x \cdot w$
Test Mean	86	0.50	43.0
Midterm	96	0.15	14.4
Final	98	0.20	19.6
Computer Lab	98	0.10	9.8
Homework	100	0.05	5.0
		$\Sigma w = 1.00$	$\Sigma(x \cdot w) = 91.8$

c. 91.8 **d.** The weighted mean for the course is 91.8.

8abc.

Class	Midpoint, x	Frequency, f	$x \cdot f$
0–9	4.5	15	67.50
10–19	14.5	19	275.50
20–29	24.5	14	343.00
30–39	34.5	7	241.50
40–49	44.5	14	623.00
50–59	54.5	6	327.00
60–69	64.5	4	258.00
70–79	74.5	1	74.50
		$N = 80$	$\Sigma(x \cdot f) = 2210$

d. 27.6

Section 2.4

1a. Min = 23, or $23,000; Max = 58, or $58,000

b. 35, or $35,000

c. The range of the starting salaries for Corporation B is 35, or $35,000 (much larger than the range of Corporation A).

2a. 41.5, or $41,500

b.

Salary, x (1000s of dollars)	Deviation, $x - \mu$ (1000s of dollars)
23	−18.5
29	−12.5
32	−9.5
40	−1.5
41	−0.5
41	−0.5
49	7.5
50	8.5
52	10.5
58	16.5
$\Sigma x = 415$	$\Sigma(x - \mu) = 0$

3ab. $\mu = 41.5$, or $41,500

Salary, x	$x - \mu$	$(x - \mu)^2$
23	−18.5	342.25
29	−12.5	156.25
32	−9.5	90.25
40	−1.5	2.25
41	−0.5	0.25
41	−0.5	0.25
49	7.5	56.25
50	8.5	72.25
52	10.5	110.25
58	16.5	272.25
$\Sigma x = 415$	$\Sigma(x - \mu) = 0$	$\Sigma(x - \mu)^2 = 1102.5$

c. 110.3 **d.** 10.5, or $10,500

e. The population variance is 110.3 and the population standard deviation is 10.5, or $10,500.

4a. See 3ab. **b.** 122.5 **c.** 11.1, or $11,100

5a. Enter data. **b.** 37.89; 3.98

6a. 7, 7, 7, 7, 7, 13, 13, 13, 13, 13 **b.** 3

7a. 1 standard deviation **b.** 34%

c. The estimated percent of the heights that are between 61.25 and 64 inches is 34%.

8a. 0 **b.** 70.6

c. At least 75% of the data lie within 2 standard deviations of the mean. At least 75% of the population of Alaska is between 0 and 70.6 years old.

9a.

x	f	xf
0	10	0
1	19	19
2	7	14
3	7	21
4	5	20
5	1	5
6	1	6
	$n = 50$	$\Sigma xf = 85$

b. 1.7

c.

$x - \bar{x}$	$(x - \bar{x})^2$	$(x - \bar{x})^2 \cdot f$
−1.70	2.8900	28.90
−0.70	0.4900	9.31
0.30	0.0900	0.63
1.30	1.6900	11.83
2.30	5.2900	26.45
3.30	10.8900	10.89
4.30	18.4900	18.49
		$\Sigma(x - \bar{x})^2 f = 106.5$

d. 1.5

10a.

Class	x	f	xf
0–99	49.5	380	18,810
100–199	149.5	230	34,385
200–299	249.5	210	52,395
300–399	349.5	50	17,475
400–499	449.5	60	26,970
500+	650.0	70	45,500
		$n = 1000$	$\Sigma xf = 195{,}535$

b. 195.5

c.

$x - \bar{x}$	$(x - \bar{x})^2$	$(x - \bar{x})^2 f$
−146.04	21,327.68	8,104,518.4
−46.04	2,119.68	487,526.4
53.96	2,911.68	611,452.8
153.96	23,703.68	1,185,184.0
253.96	64,495.68	3,869,740.8
454.46	206,533.89	14,457,372.3
		$\Sigma(x - \bar{x})^2 f = 28{,}715{,}794.7$

d. 169.5

Section 2.5

1a. 0, 0, 1, 1, 1, 2, 3, 3, 3, 4, 5, 5, 5, 7, 9, 10, 12, 12, 13, 13, 13, 13, 13, 15, 16, 16, 17, 17, 18, 18, 18, 19, 19, 19, 20, 20, 21, 22, 23, 23, 24, 24, 25, 25, 26, 26, 26, 29, 33, 36, 37, 39, 39, 39, 39, 40, 40, 41, 41, 41, 42, 44, 44, 45, 47, 48, 49, 49, 49, 51, 53, 56, 58, 58, 59, 60, 67, 68, 68, 72

b. 23.5 **c.** 13, 41.5

2a. Enter data. **b.** 17, 23, 28.5

c. One quarter of the tuition costs is $17,000 or less, one half is $23,000 or less, and three quarters is $28,500 or less.

3a. 13, 41.5 **b.** 28.5

c. The ages in the middle half of the data set vary by 28.5 years.

4a. 0, 13, 23.5, 41.5, 72

bc. Ages of Akhiok Residents

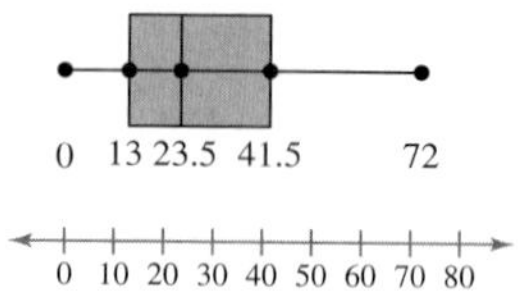

d. It appears that half of the ages are between 13 and 41.5 years.

5a. 80th percentile

b. 80% of the ages are 45 years or younger.

6a. $\mu = 70,\ \sigma = 8$

b. $z_1 = \dfrac{60 - 70}{8} = -1.25$

$z_2 = \dfrac{71 - 70}{8} = 0.125$

$z_3 = \dfrac{92 - 70}{8} = 2.75$

c. From the z-score, $60 is 1.25 standard deviations below the mean, $71 is 0.125 standard deviation above the mean, and $92 is 2.75 standard deviations above the mean.

7a. NFL: $\mu = 23.6,\ \sigma = 6.0$

AFL: $\mu = 11.7,\ \sigma = 4.6$

b. Kansas City: $z = -1.27$

Tampa Bay: $z = 0.07$

c. The number of field goals scored by Kansas City is 1.27 standard deviations below the mean and the number of field goals scored by Tampa Bay is 0.07 standard deviations above the mean. Comparing the two measures of position indicates that Tampa Bay has a higher position within the AFL than Kansas City has in the NFL.

CHAPTER 3

Section 3.1

1ab. (1)

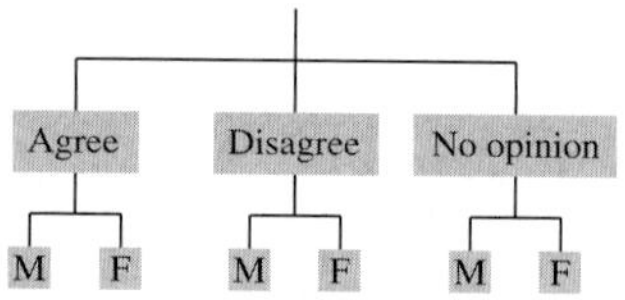

(2)

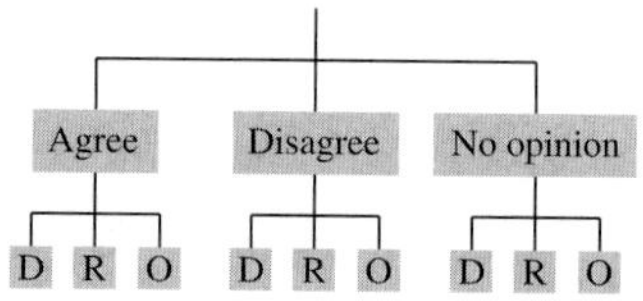

c. (1) 6 (2) 9

d. (1) Let A = Agree, D = Disagree, N = No Opinion, M = Male, F = Female

Sample Space = $\{AM, AF, DM, DF, NM, NF\}$

(2) Let A = Agree, D = Disagree, N = No Opinion, De = Democrat, R = Republican, O = Other

Sample Space = $\{ADe, AR, AO, DDe, DR, DO, NDe, NR, NO\}$

2a. (1) 6 (2) 1

b. (1) Not a simple event (2) Simple event

3a. (1) 52 (2) 1 (3) 0.019

b. (1) 52 (2) 13 (3) 0.25

c. (1) 52 (2) 52 (3) 1

4a. The event is "the next claim processed is fraudulent." The frequency is 4.

b. 100 **c.** 0.04

5a. 54 **b.** 1000 **c.** 0.054

6a. The event is "salmon successfully passing through a dam on the Columbia River."

b. Estimated **c.** Empirical probability

7a. 0.425 **b.** 0.575 **c.** $\frac{23}{40}$ or 0.575

Section 3.2

1a. (1) 30 and 102 (2) 11 and 50

b. (1) 0.294 (2) 0.22

2a. (1) No (2) Yes

b. (1) Independent (2) Dependent

3a. (1) Independent (2) Dependent

b. (1) 0.723 (2) 0.059

4a. (1) Event (2) Complement

b. (1) 0.729 (2) 0.999

Section 3.3

1a. (1) None are true. (2) None are true.

(3) Events A and B cannot occur at the same time.

b. (1) Not mutually exclusive (2) Not mutually exclusive

(3) Mutually exclusive

2a. (1) Mutually exclusive (2) Not mutually exclusive

b. (1) $\frac{1}{6}, \frac{1}{2}$ (2) $\frac{12}{52}, \frac{13}{52}, \frac{3}{52}$

c. (1) 0.667 (2) 0.423

3a. $A = \{\text{sales between \$0 and \$24,999}\}$

$B = \{\text{sales between \$25,000 and \$49,000}\}$

b. A and B cannot occur at the same time.
A and B are mutually exclusive.

c. $\frac{3}{36}, \frac{5}{36}$ **d.** 0.222

4a. (1) Mutually exclusive (2) Not mutually exclusive

b. (1) 0.149 (2) 0.910

5a. 0.188 **b.** 0.812

Section 3.4

1a. Manufacturer: 4 **b.** 72
Size: 3
Color: 6

2a. (1) Each letter is an event (26 choices).

(2) Each letter is an event (26, 25, 24, 23, 22, and 21 choices).

b. (1) 308,915,776 (2) 165,765,600

3a. 6 **b.** 720

4a. 336

b. There are 336 possible ways that three horses can finish in first, second, and third place.

5a. $n = 12, r = 4$ **b.** 11,880

6a. $n = 20, n_1 = 6, n_2 = 9, n_3 = 5$ **b.** 77,597,520

7a. $n = 16, r = 3$ **b.** 560

c. There are 560 different possible three-person committees that can be selected from 16 employees.

8a. 1 outcome and 180 distinguishable permutations

b. 0.006

9a. 10 **b.** 220 **c.** 0.045

CHAPTER 4

Section 4.1

1a. (1) Measured (2) Counted

b. (1) The random variable is continuous because x can be any amount of time needed to complete a test.

(2) The random variable is discrete because x can be counted.

2ab.

x	f	$P(x)$
0	16	0.16
1	19	0.19
2	15	0.15
3	21	0.21
4	9	0.09
5	10	0.10
6	8	0.08
7	2	0.02
	$n = 100$	$\Sigma P(x) = 1$

c.

3a. Each $P(x)$ is between 0 and 1. **b.** $\Sigma P(x) = 1$

c. Because both conditions are met, the distribution is a probability distribution.

4a. (1) Yes, each outcome is between 0 and 1.

(2) Yes, each outcome is between 0 and 1.

b. (1) No (2) Yes

c. (1) Not a probability distribution

(2) A probability distribution

5ab.

x	$P(x)$	$xP(x)$
0	0.16	0.00
1	0.19	0.19
2	0.15	0.30
3	0.21	0.63
4	0.09	0.36
5	0.10	0.50
6	0.08	0.48
7	0.02	0.14
	$\Sigma P(x) = 1$	$\Sigma xP(x) = 2.60$

c. $\mu = 2.6$
On average, 2.6 sales are made per day.

6ab.

x	$P(x)$	$x - \mu$	$(x - \mu)^2$	$P(x)(x - \mu)^2$
0	0.16	−2.6	6.76	1.0816
1	0.19	−1.6	2.56	0.4864
2	0.15	−0.6	0.36	0.0540
3	0.21	0.4	0.16	0.0336
4	0.09	1.4	1.96	0.1764
5	0.10	2.4	5.76	0.5760
6	0.08	3.4	11.56	0.9248
7	0.02	4.4	19.36	0.3872
	$\Sigma P(x) = 1$			$\Sigma P(x)(x - \mu)^2 = 3.72$

c. 1.9

d. Most of the data values differ from the mean by no more than 1.9 sales per day.

7ab.

x	f	$P(x)$	$xP(x)$
0	25	0.111	0.000
1	48	0.213	0.213
2	60	0.267	0.533
3	45	0.200	0.600
4	20	0.089	0.356
5	10	0.044	0.222
6	8	0.036	0.213
7	5	0.022	0.156
8	3	0.013	0.107
9	1	0.004	0.040
	$n = 225$	$\Sigma P(x) \approx 1$	$\Sigma xP(x) = 2.440$

c. 2.4

d. You can expect an average of 2.4 sales per day.

Section 4.2

1a. Trial: answering a question

Success: question answered correctly

b. Yes

c. $n = 10, p = 0.25, q = 0.75, x = 0, 1, 2, \ldots, 9, 10$

2a. Trial: drawing a card with replacement
Success: card drawn is a club
Failure: card drawn is not a club

b. $n = 5, p = 0.25, q = 0.75, x = 3$

c. $P(3) = \frac{5!}{2!\,3!}(0.25)^3(0.75)^2 \approx 0.088$

3a. Trial: selecting a worker and asking a question
Success: selecting a worker who will rely on pension
Failure: selecting a worker who will not rely on pension

b. $n = 7, p = 0.34, q = 0.66, x = 0, 1, 2, \ldots, 6, 7$

c. $P(0) = {}_7C_0\,(0.34)^0\,(0.66)^7 = 0.0546$
$P(1) = {}_7C_1\,(0.34)^1\,(0.66)^6 = 0.197$
$P(2) = {}_7C_2\,(0.34)^2\,(0.66)^5 = 0.304$
$P(3) = {}_7C_3\,(0.34)^3\,(0.66)^4 = 0.261$
$P(4) = {}_7C_4\,(0.34)^4\,(0.66)^3 = 0.134$
$P(5) = {}_7C_5\,(0.34)^5\,(0.66)^2 = 0.0416$
$P(6) = {}_7C_6\,(0.34)^6\,(0.66)^1 = 0.00714$
$P(7) = {}_7C_7\,(0.34)^7\,(0.66)^0 = 0.000525$

d.

x	$P(x)$
0	0.0546
1	0.197
2	0.304
3	0.261
4	0.134
5	0.0416
6	0.00714
7	0.000525
	$\Sigma P(x) \approx 1$

4a. $n = 250, p = 0.71, x = 178$ **b.** 0.056

c. The probability that exactly 178 people in the United States will use more than one topping on their hotdog is about 0.056.

5a. (1) $x = 2$ (2) $x = 2, 3, 4,$ or 5 (3) $x = 0$ or 1

b. (1) 0.217 (2) 0.217, 0.058, 0.008, 0.0004; 0.283
(3) 0.308, 0.409; 0.717

c. (1) The probability that exactly two men consider fishing their favorite leisure-time activity is about 0.217.

(2) The probability that at least two men consider fishing their favorite leisure-time activity is about 0.283.

(3) The probability that fewer than two men consider fishing their favorite leisure-time activity is about 0.717.

6a. Trial: selecting a business and asking if it has a website

Success: selecting a business with a website

Failure: selecting a business without a website

b. $n = 10, p = 0.30, x = 4$ **c.** 0.200

d. The probability that exactly four of the 10 small business have a website is 0.200.

7a. 0.006, 0.050, 0.167, 0.294, 0.293, 0.155, 0.034

b.

x	$P(x)$
0	0.006
1	0.050
2	0.167
3	0.294
4	0.293
5	0.155
6	0.034

c. Owning a Computer

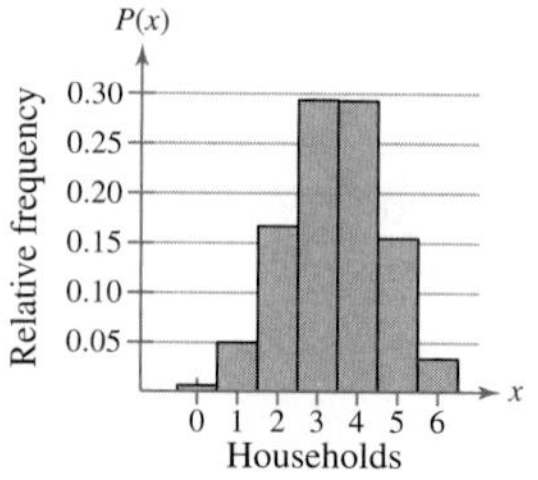

8a. Success: selecting a clear day

$n = 31,\ p = 0.44,\ q = 0.56$

b. 13.6 **c.** 7.6 **d.** 2.8

e. On average, there are about 14 clear days during the month of May. The standard deviation is about 3 days.

Section 4.3

1a. 0.23, 0.177, 0.136 **b.** 0.543

c. The probability that your first sale will occur before your fourth sales call is 0.543.

2a. $P(0) \approx 0.050$

$P(1) \approx 0.149$

$P(2) \approx 0.224$

$P(3) \approx 0.224$

$P(4) \approx 0.168$

b. 0.815 **c.** 0.185

d. The probability that more than four accidents will occur in any given month at the intersection is 0.185.

3a. 0.10 **b.** 0.10, 3 **c.** 0.0002

d. The probability of finding three brown trout in any given cubic meter of the lake is 0.0002.

CHAPTER 5

Section 5.1

1a. A: 45, B: 60, C: 45; B has the greatest mean.

b. Curve C is more spread out, so curve C has the greatest standard deviation.

2a. 3.5 feet **b.** 3.3, 3.7; 0.2 foot

3a. (1) 0.0143 (2) 0.9850

4a.

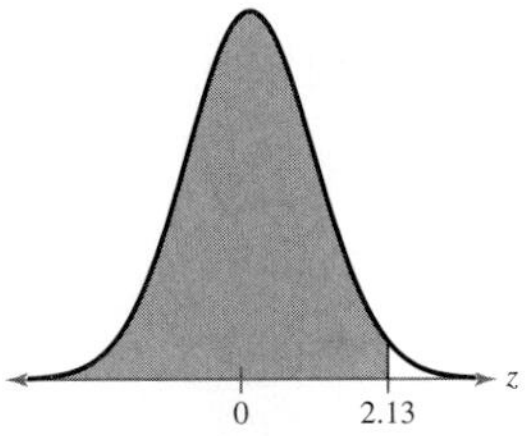

b. 0.9834

5a.

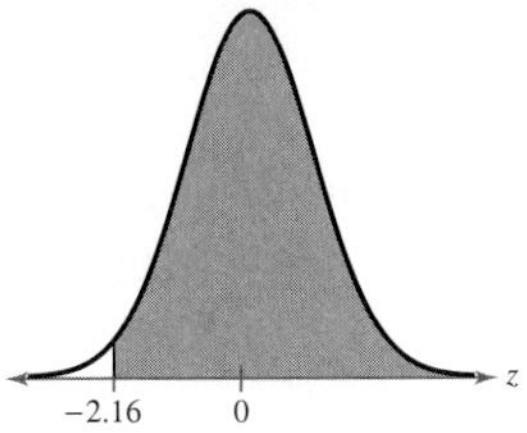

b. 0.0154

c. 0.9846

6a. 0.0885 **b.** 0.0154 **c.** 0.0731

Section 5.2

1a.

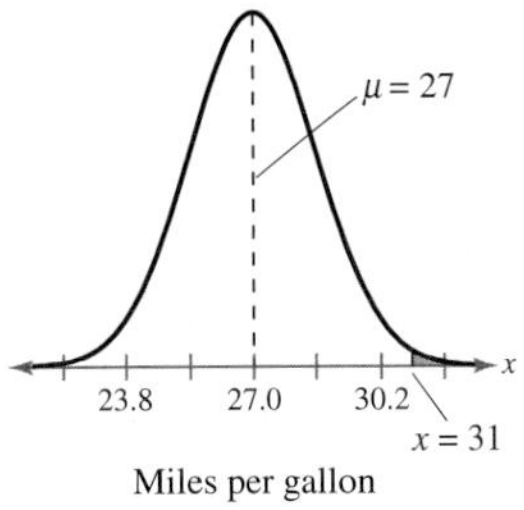

b. 2.50

c. 0.0062

d. The probability that a randomly selected manual transmission Focus will get more than 31 miles per gallon in city driving is 0.0062.

2a.

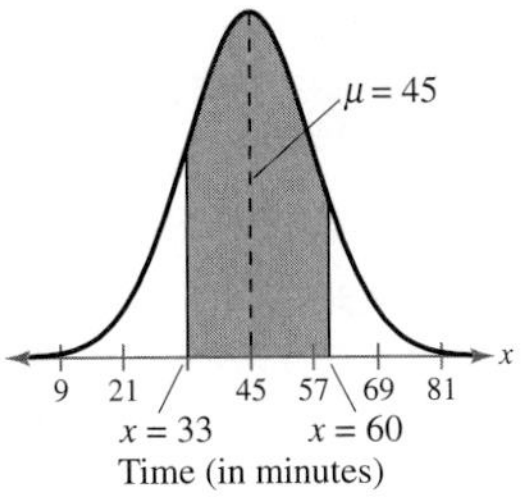

b. −1, 1.25

c. 0.1587; 0.8944

d. 0.7357

3a. Read user's guide for the technology tool.

b. Enter the data.

c. The probability that a randomly selected U.S. man's cholesterol is between 190 and 225 is about 0.4968.

Section 5.3

1a. (1) 0.0384 (2) 0.0250 and 0.9750

bc. (1) −1.77 (2) ±1.96

2a. (1) Area = 0.10 (2) Area = 0.20

(3) Area = 0.99

bc. (1) −1.28 (2) −0.84 (3) 2.33

3a. $\mu = 70,\ \sigma = 8$ **b.** 64; 104.32; 55.44

c. 64 is below the mean, 104.32 is above the mean, and 55.44 is below the mean.

4ab.

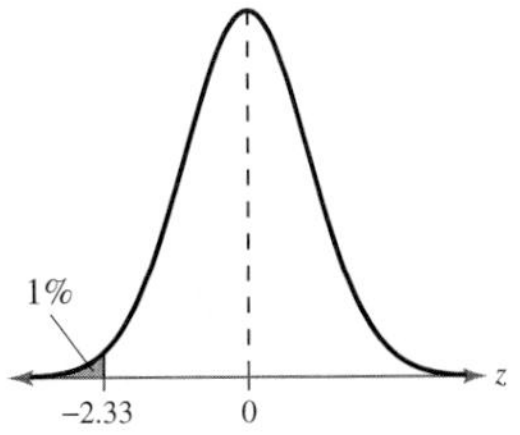

c. 142.83

d. So, the longest braking distance a Ford F-150 could have and still be in the top 1% is 143 feet.

5ab.

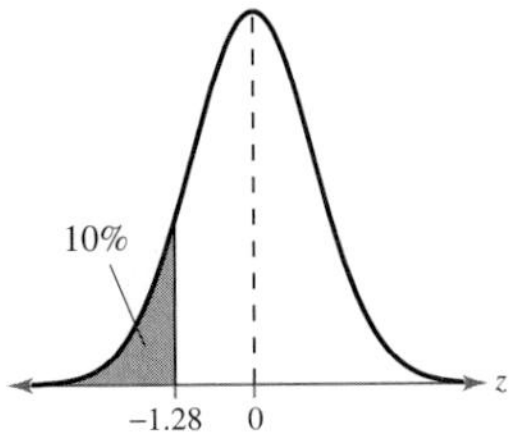

c. 8.512

d. So, the maximum length of time an employee could have worked and still be laid off is 8 years.

Section 5.4

1a.

Sample	Mean	Sample	Mean	Sample	Mean
1, 1, 1	1	3, 3, 5	3.67	5, 7, 1	4.33
1, 1, 3	1.67	3, 3, 7	4.33	5, 7, 3	5
1, 1, 5	2.33	3, 5, 1	3	5, 7, 5	5.67
1, 1, 7	3	3, 5, 3	3.67	5, 7, 7	6.33
1, 3, 1	1.67	3, 5, 5	4.33	7, 1, 1	3
1, 3, 3	2.33	3, 5, 7	5	7, 1, 3	3.67
1, 3, 5	3	3, 7, 1	3.67	7, 1, 5	4.33
1, 3, 7	3.67	3, 7, 3	4.33	7, 1, 7	5
1, 5, 1	2.33	3, 7, 5	5	7, 3, 1	3.67
1, 5, 3	3	3, 7, 7	5.67	7, 3, 3	4.33
1, 5, 5	3.67	5, 1, 1	2.33	7, 3, 5	5
1, 5, 7	4.33	5, 1, 3	3	7, 3, 7	5.67
1, 7, 1	3	5, 1, 5	3.67	7, 5, 1	4.33
1, 7, 3	3.67	5, 1, 7	4.33	7, 5, 3	5
1, 7, 5	4.33	5, 3, 1	3	7, 5, 5	5.67
1, 7, 7	5	5, 3, 3	3.67	7, 5, 7	6.33
3, 1, 1	1.67	5, 3, 5	4.33	7, 7, 1	5
3, 1, 3	2.33	5, 3, 7	5	7, 7, 3	5.67
3, 1, 5	3	5, 5, 1	3.67	7, 7, 5	6.33
3, 1, 7	3.67	5, 5, 3	4.33	7, 7, 7	7
3, 3, 1	2.33	5, 5, 5	5		
3, 3, 3	3	5, 5, 7	5.67		

b.

$\bar{x}$	f	Probability
1	1	0.0156
1.67	3	0.0469
2.33	6	0.0938
3	10	0.1563
3.67	12	0.1875
4.33	12	0.1875
5	10	0.1563
5.67	6	0.0938
6.33	3	0.0469
7	1	0.0156

$\mu_{\bar{x}} = 4$
$(\sigma_{\bar{x}})^2 = 1.667$
$\sigma_{\bar{x}} = 1.291$

c. $\mu_{\bar{x}} = \mu = 4$

$$(\sigma_{\bar{x}})^2 = \frac{\sigma^2}{n} = \frac{5}{3} = 1.667; \sigma_{\bar{x}} = \frac{\sigma}{\sqrt{n}} = \frac{\sqrt{5}}{\sqrt{3}} = 1.291$$

2a. 64, 0.9

b. $n = 100$

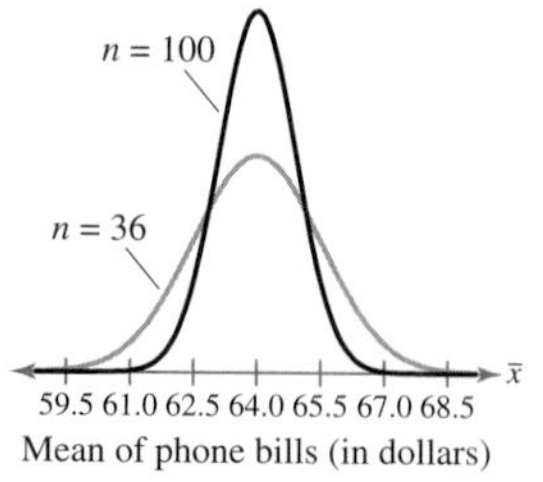

c. With a larger sample size, the mean stays the same but the standard deviation decreases.

3a. 3.5, 0.05

b.

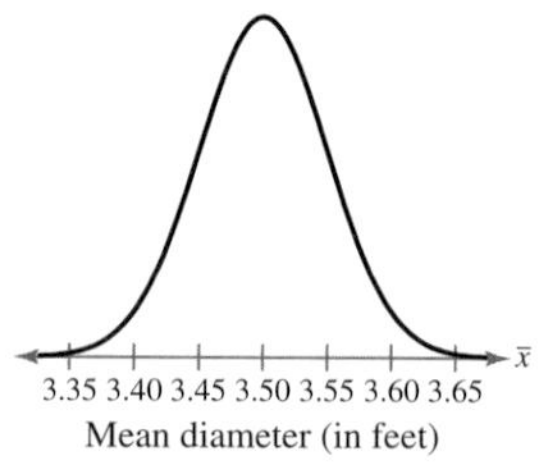

4a. $\mu_{\bar{x}} = 25$ **b.** −2, 3.33 **c.** 0.9768

$\sigma_{\bar{x}} = 0.15$

5a. 243,756; 12,701.71 **b.** −3.44 **c.** 0.9997

6a. 0.5, 1.58 **b.** 0.6915, 0.9429

c. There is a 69% chance an *individual receiver* will cost less than $700. There is a 94% chance that the *mean of a sample of 10 receivers* is less than $700.

Section 5.5

1a. 70, 0.61, 0.39 **b.** 42.7, 27.3

c. Normal distribution can be used.

d. 42.7, 4.08

2a. (1) 57, 58, . . . , 83 (2) . . . , 52, 53, 54

b. (1) $56.5 < x < 83.5$ (2) $x < 54.5$

3a. Normal distribution can be used. **b.** 42.7, 4.08

c.

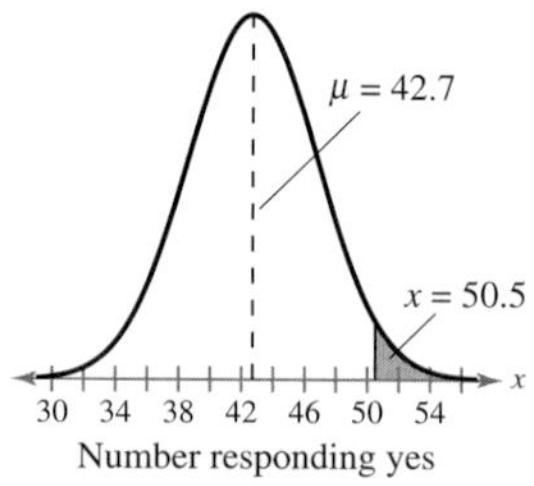

d. 1.91 **e.** 0.0281

4a. Normal distribution can be used. **b.** 76, 6.86

c. $P(x \leq 85.5)$ **d.** 1.38 **e.** 0.9162

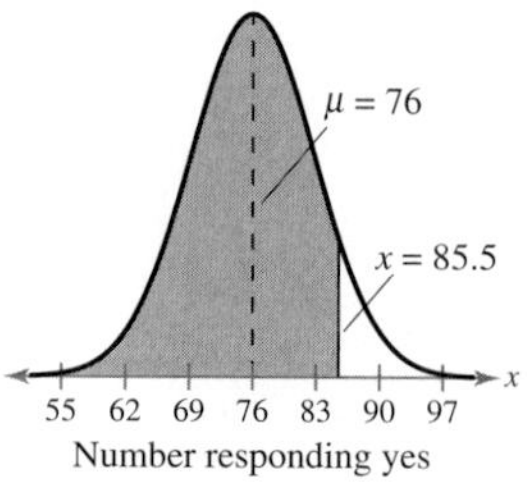

5a. Normal distribution can be used. **b.** 190, 3.08

c. $P(190.5 < x < 191.5)$ **d.** 0.16, 0.49 **e.** 0.1243

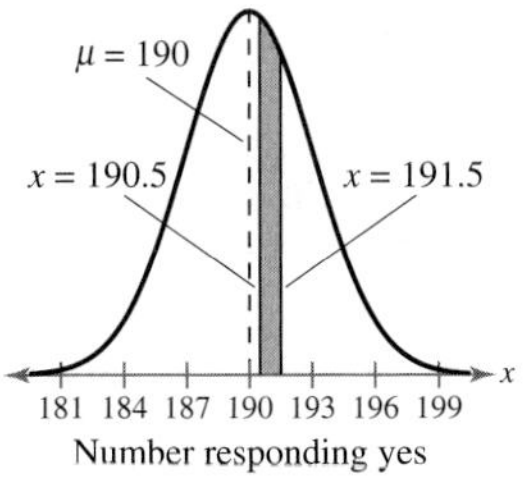

CHAPTER 6

Section 6.1

1a. 14.767

b. The mean number of sentences per magazine advertisement is 14.767.

2a. $z_c = 1.96, n = 30, s \approx 16.536$ **b.** 5.917

c. You are 95% confident that the maximum error of the estimate is about 5.917 sentences per magazine advertisement.

3a. 14.767, 5.917 **b.** 8.850, 20.684

c. You are 95% confident that the mean number of sentences per magazine advertisement is between 8.850 and 20.684.

4a. Enter the data. **b.** (11.6, 13.2); (11.4, 13.4)

c. As the confidence level increases, so does the width of the interval.

5a. $n = 30, \bar{x} = 22.9, \sigma = 1.5, z_c = 1.645$ **b.** 22.4; 23.4

c. You are 90% confident that the mean age of the students is between 22.4 and 23.4 years.

6a. $z_c = 1.96, E = 2, s \approx 5.0$ **b.** 25

c. You should have at least 25 magazine advertisements in your sample.

Section 6.2

1a. d.f. = 21 **b.** $c = 0.90$ **c.** 1.721

2a. 1.753; 4.383; 2.947; 7.368

b. (157.6, 166.4); (154.6, 169.4)

c. You are 90% confident that the mean temperature of coffee sold is between 157.6°F and 166.4°F.

You are 99% confident that the mean temperature of coffee sold is between 154.6°F and 169.4°F.

3a. 1.729; 0.162; 2.093; 0.197

b. (6.77, 7.09); (6.73, 7.13)

c. You are 90% confident that the mean mortgage interest rate is contained between 6.77% and 7.09%.

You are 95% confident that the mean mortgage interest rate is contained between 6.73% and 7.13%.

4a. No; Yes; No; Use t-distribution.

Section 6.3

1a. $x = 1177, n = 3859$ **b.** 0.305

2a. 0.305, 0.695

b. $n\hat{p} = 1176.995 > 5$ and $n\hat{q} = 2682.005 > 5$

c. 1.645; 0.012 **d.** 0.293, 0.317

e. You are 90% confident that the proportion of adults who thought Franklin D. Roosevelt was the greatest president of the 20th century is contained between 29.3% and 31.7%.

3a. $n = 900, \hat{p} = 0.33$ **b.** 0.67

c. $n\hat{p} = 297 > 5$ and $n\hat{q} = 603 > 5$

d. $z_c = 2.575$ **e.** 0.290, 0.370

f. You are 99% confident that the proportion of adults who think that people over 75 are the more dangerous drivers is contained between 29.0% and 37.0%.

4a. (1) $\hat{p} = 0.5, \hat{q} = 0.5, z_c = 1.645, E = 0.02$

(2) $\hat{p} = 0.04, \hat{q} = 0.96, z_c = 1.645, E = 0.02$

b. (1) 1691.27 (2) 259.78 **c.** (1) 1692 (2) 260

Section 6.4

1a. d.f. = 24, $c = 0.95$ **b.** 0.025, 0.975 **c.** 39.364, 12.401

2a. 42.557, 17.708; 45.722, 16.047

b. (0.98, 2.36); (0.91, 2.60) **c.** (0.99, 1.54); (0.96, 1.61)

d. You are 90% confident that the population variance is between 0.98 and 2.36 and that the population standard deviation is between 0.99 and 1.54. You are 95% confident that the population variance is between 0.91 and 2.60 and that the population standard deviation is between 0.96 and 1.61.

CHAPTER 7

Section 7.1

1a. (1) The mean ... is not 74 months.
$\mu \neq 74$

(2) The variance ... is less than or equal to 3.5.
$\sigma^2 \leq 3.5$

(3) The proportion ... is greater than 39%.
$p > 0.39$

b. (1) $\mu = 74$ (2) $\sigma^2 > 3.5$ (3) $p \leq 0.39$

c. (1) H_0: $\mu = 74$; H_a: $\mu \neq 74$ (claim)

(2) H_0: $\sigma^2 \leq 3.5$ (claim); H_a: $\sigma^2 > 3.5$

(3) H_0: $p \leq 0.39$; H_a: $p > 0.39$ (claim)

2a. H_0: $p \leq 0.01$; H_a: $p > 0.01$

b. Type I error will occur if the actual proportion is less than or equal to 0.01, but you decided to reject H_0.

Type II error will occur if the actual proportion is greater than 0.01, but you do not reject H_0.

c. Type II error is more serious because you would be misleading the consumer, possibly causing serious injury or death.

3a. (1) H_0: $\mu = 74$; H_a: $\mu \neq 74$

(2) H_0: $p \leq 0.39$; H_a: $p > 0.39$

b. (1) Two-tailed (2) Right-tailed

c. (1) 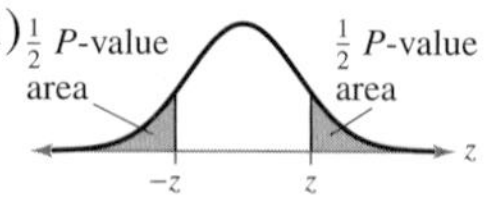

(2) 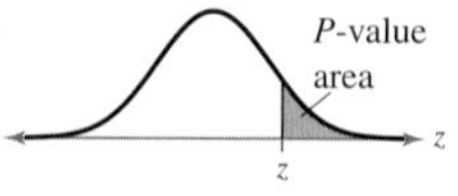

4a. There is enough evidence to support the radio station's claim.

b. There is not enough evidence to support the radio station's claim.

5a. No answer required.

b. (1) Support claim. (2) Reject claim.

c. (1) H_0: $\mu \geq 650$; H_a: $\mu < 650$ (claim)

(2) H_0: $\mu = 98.6$ (claim); H_a: $\mu \neq 98.6$

Section 7.2

1a. (1) $0.01 < 0.0347$ (2) $0.0347 < 0.05$

b. (1) Fail to reject H_0 because $0.0347 > 0.01$.

(2) Reject H_0 because $0.0347 < 0.05$.

2ab. 0.0526

c. Fail to reject H_0 because $0.0526 > 0.05$.

3a. 0.9896 **b.** 0.0208

c. Fail to reject H_0 because $0.0208 > 0.01$.

4a. The claim is "the mean speed is greater than 35 miles per hour."

H_0: $\mu \leq 35$; H_a: $\mu > 35$ (claim)

b. 0.05 **c.** 2.5 **d.** 0.0062

e. Reject H_0 because $0.0062 < 0.05$.

f. There is enough evidence to conclude the average speed limit is greater than 35 miles per hour.

5a. H_0: $\mu = 150$ (claim); H_a: $\mu \neq 150$

b. 0.01 **c.** -2.76 **d.** 0.0058

e. Reject H_0 because $0.0058 < 0.01$.

f. There is enough evidence at the 1% level to reject the claim that the distributorship averages 150 sales per day.

6a. $0.039 > 0.01$ **b.** Fail to reject H_0.

7a. 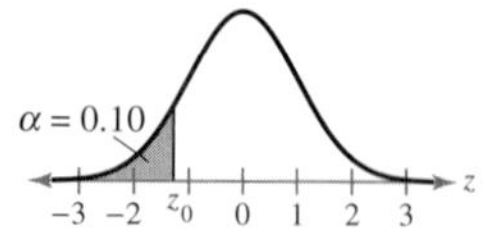

b. 0.1003

c. -1.28

d. $z < -1.28$

8a. 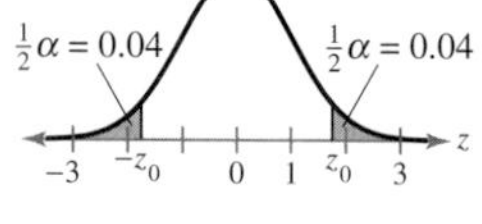

b. 0.0401 and 0.9599

c. $-z_0 = -1.75$ and $z_0 = 1.75$

d. $z < -1.75$, $z > 1.75$

9a. H_0: $\mu \geq 8.5$; H_a: $\mu < 8.5$ (claim) **b.** 0.01

c. $z_0 = -2.33$; Rejection region: $z < -2.33$

d. -3.550

e. 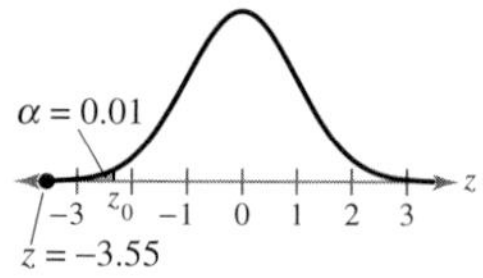

Because $-3.55 < -2.33$, reject H_0.

f. There is enough evidence at the 1% level to support the claim that the mean work day is less than 8.5 hours.

10a. 0.01

b. $-z_0 = -2.575$ and $z_0 = 2.575$

Rejection regions: $z < -2.575$, $z > 2.575$

c. Fail to reject H_0.

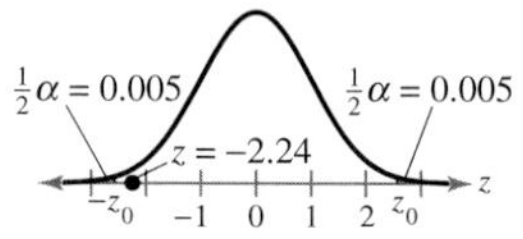

d. There is not enough evidence to support the claim that the mean cost is significantly different from \$9405 at the 1% level of significance.

Section 7.3

1a. 2.650 **b.** -2.650

2a. 1.860 **b.** 1.860

3a. 2.947 **b.** ± 2.947

4a. H_0: $\mu \geq \$1240$ (claim); H_a: $\mu < \$1240$

b. $\alpha = 0.01$, d.f. $= 8$

c. $t_0 = -2.896$; Rejection region: $t < -2.896$ **d.** -2.419

e. Fail to reject H_0.

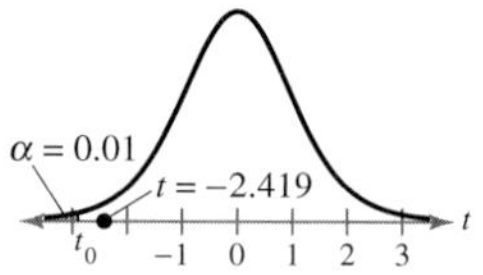

f. There is not enough evidence at the 1% level to reject the claim that the mean cost of insuring a 2002 Ford F-150 Suber Cab is at least \$1240.

5a. H_0: $\mu = 1890$ (claim); H_a: $\mu \neq 1890$

b. $\alpha = 0.01$, d.f. $= 18$

c. $-t_0 = -2.878$ and $t_0 = 2.878$

Rejection regions: $t < -2.878$, $t > 2.878$

d. 3.798

e. Reject H_0.

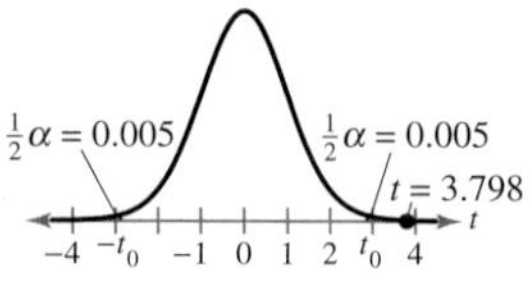

f. There is enough evidence at the 1% level to reject the company's claim that the mean conductivity of the river is 1890 milligrams per liter.

6a. 0.0484 **b.** $0.0484 < 0.05$ **c.** Reject H_0.

d. There is enough evidence at the 5% level to reject the claim that the mean nightly lodging rate is at least \$136.

Section 7.4

1a. $np = 25.8 > 5$, $nq = 60.2 > 5$

b. H_0: $p \geq 0.30$; H_a: $p < 0.30$ (claim) **c.** 0.05

d. $z_0 = -1.645$; Rejection region: $z < -1.645$

e. -2.024 **f.** Reject H_0.

g. There is enough evidence at the 5% level to support the claim that less than 30% of adults in the United States are allergic to trees, weeds, flowers, and grasses.

2a. $np = 12.5 > 5$, $nq = 237.5 > 5$

b. H_0: $p = 0.05$ (claim); H_a: $p \neq 0.05$ **c.** 0.01

d. $-z_0 = -2.575$ and $z_0 = 2.575$
Rejection regions: $z < -2.575$, $z > 2.575$

e. 2.176 **f.** Fail to reject H_0.

g. There is not enough evidence at the 1% level to reject the claim that 5% of U.S. adults have seen an extraterrestrial being.

3a. $np = 28.5 > 5$, $nq = 46.5 > 5$

b. H_0: $p \leq 0.38$; H_a: $p > 0.38$ (claim) **c.** 0.01

d. $z_0 = 2.33$; Rejection region: $z > 2.33$ **e.** 1.071

f. Fail to reject H_0.

g. There is not enough evidence at the 1% level to support the claim that more than 38% of U.S. adults regularly watch a cable news broadcast.

Section 7.5

1a. 33.409 **2a.** 17.708 **3a.** 31.526 **b.** 8.231

4a. H_0: $\sigma^2 \leq 0.40$ (claim); H_a: $\sigma^2 > 0.40$

b. $\alpha = 0.01$, d.f. $= 30$

c. $\chi_0^2 = 50.892$; Rejection region: $\chi^2 > 50.892$

d. 56.250 **e.** Reject H_0.

f. There is enough evidence at the 1% level to reject the claim that the variance of the amount of sports drink in a 12-ounce bottle is no more than 0.40.

5a. H_0: $\sigma \geq 3.7$; H_a: $\sigma < 3.7$ (claim) **b.** $\alpha = 0.05$, d.f. $= 8$

c. $\chi_0^2 = 2.733$; Rejection region: $\chi^2 < 2.733$

d. 5.259 **e.** Fail to reject H_0.

f. There is not enough evidence at the 5% level to support the claim that the standard deviation in the length of response times is less than 3.7 minutes.

6a. H_0: $\sigma^2 = 8.6$ (claim); H_a: $\sigma^2 \neq 8.6$

b. $\alpha = 0.01$, d.f. $= 9$

c. $\chi_L^2 = 1.735$, $\chi_R^2 = 23.589$
Rejection regions: $\chi^2 > 23.589$, $\chi^2 < 1.735$

d. 4.50 **e.** Fail to reject H_0.

f. There is not enough evidence at the 1% level to reject the claim that the variance of the diameters in a certain tire model is 8.6.

CHAPTER 8

Section 8.1

1a. H_0: $\mu_1 = \mu_2$; H_a: $\mu_1 \neq \mu_2$ (claim) **b.** 0.01

c. $-z_0 = -2.575$ and $z_0 = 2.575$
Rejection regions: $z < -2.575$, $z > 2.575$

d. 2.747 **e.** Reject H_0.

f. There is enough evidence at the 1% level to support the claim that there is a difference in the mean credit card charges.

2a. $z \approx 4.542$; P-value ≈ 0.00000278 **b.** Reject H_0.

Section 8.2

1a. H_0: $\mu_1 = \mu_2$; H_a: $\mu_1 \neq \mu_2$ (claim) **b.** 0.05 **c.** 9

d. $-t_0 = -2.262$ and $t_0 = 2.262$
Rejection regions: $t < -2.262$, $t > 2.262$

e. 2.376 **f.** Reject H_0.

g. There is enough evidence at the 5% level to support the claim that the mean stopping distances are different.

2a. H_0: $\mu_1 \geq \mu_2$; H_a: $\mu_1 < \mu_2$ (claim) **b.** 0.10 **c.** 25

d. $t_0 = -1.316$; Rejection region: $t < -1.316$

e. -0.898 **f.** Fail to reject H_0.

g. There is not enough evidence at the 10% level to support the claim that the watt usage is less than that of its leading competitor.

Section 8.3

1a. (1) Independent (2) Dependent

2a. H_0: $\mu_d \geq 0$; H_a: $\mu_d < 0$ (claim)

b. $\alpha = 0.05$, d.f. $= 11$

c. $t_0 \approx -1.796$; Rejection region: $t < -1.796$

d. $\bar{d} = -1$; $s_d \approx 2.374$ **e.** -1.459

f. Fail to reject H_0.

g. There is not enough evidence at the 5% level to support the claim that experimental medication increases an individual's heart rate.

3a. H_0: $\mu_d = 0$; H_a: $\mu_d \neq 0$ (claim)

b. $\alpha = 0.05$, d.f. $= 6$

c. $-t_0 = -2.447$ and $t_0 = 2.447$
Rejection regions: $t < -2.447$, $t > 2.447$

d. $\bar{d} \approx 0.557$; $s_d \approx 0.924$

e. 1.595 **f.** Fail to reject H_0.

g. There is not enough evidence at the 5% level to conclude that the drug changes the body's temperature at the specified level of significance.

Section 8.4

1a. H_0: $p_1 = p_2$; H_a: $p_1 \neq p_2$ (claim) **b.** 0.05

c. $-z_0 = -1.96$ and $z_0 = 1.96$

Rejection regions: $z < -1.96$, $z > 1.96$

d. $\overline{p} \approx 0.229$; $\overline{q} \approx 0.771$

e. $n_1\overline{p} \approx 1551 > 5$, $n_1\overline{q} \approx 5220 > 5$, $n_2\overline{p} \approx 1550 > 5$, and $n_2\overline{q} \approx 5217 > 5$.

f. 4.71 **g.** Reject H_0.

h. There is enough evidence at the 5% level to support the claim that there is a difference between the proportion of male high school students who smoke cigarettes and the proportion of female high school students who smoke cigarettes.

2a. H_0: $p_1 \leq p_2$; H_a: $p_1 > p_2$ (claim)

b. 0.05 **c.** $z_0 = 1.645$; Rejection region: $z > 1.645$

d. $\overline{p} \approx 0.116$; $\overline{q} \approx 0.884$

e. $n_1\overline{p} \approx 785 > 5$, $n_1\overline{q} \approx 5986 > 5$, $n_2\overline{p} \approx 785 > 5$, and $n_2\overline{q} \approx 5982 > 5$

f. 19.45 **g.** Reject H_0.

h. There is enough evidence at the 5% level to support the claim that the proportion of male high school students who smoke cigars is greater than the proportion of female high school students who smoke cigars.

CHAPTER 9

Section 9.1

1ab.

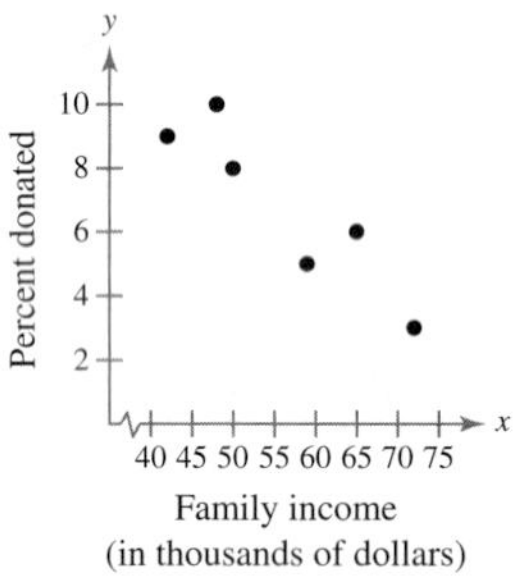

c. Yes, it appears that there is a negative linear correlation. As family income increases, the percent of income donated to charity decreases.

2ab.

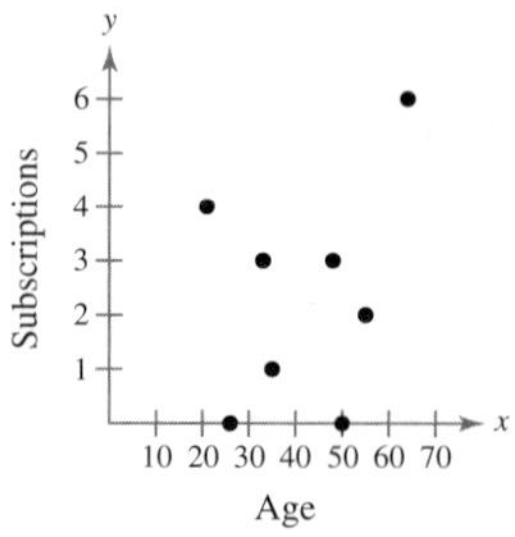

c. No, it appears that there is no correlation between age and subscriptions.

3ab.

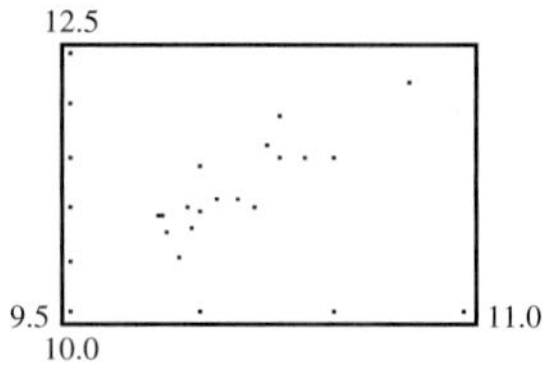

c. Yes, there appears to be a positive linear relationship between men's winning time and women's winning time.

4a. $n = 6$

b.

x	y	xy	x^2	y^2
42	9	378	1764	81
48	10	480	2304	100
50	8	400	2500	64
59	5	295	3481	25
65	6	390	4225	36
72	3	216	5184	9
$\Sigma x = 336$	$\Sigma y = 41$	$\Sigma xy = 2159$	$\Sigma x^2 = 19{,}458$	$\Sigma y^2 = 315$

c. -0.916

d. Because r is close to -1, there appears to be a strong negative linear correlation between income level and donating percent.

5a. Enter data. **b.** 0.849

c. Because r is close to 1, there appears to be a strong positive linear correlation between men's winning time and women's winning time.

6a. 6 **b.** 0.01 **c.** 0.917

d. $|r| \approx 0.916 < 0.917$; the correlation is not significant.

e. There is not enough evidence at the 1% level to conclude that there is a significant correlation between the income level and the donating percent.

7a. H_0: $\rho = 0$ and H_a: $\rho \neq 0$ **b.** 0.01 **c.** 16

d. $-t_0 = -2.921$ and $t_0 = 2.921$

Rejection regions: $t < -2.921$, $t > 2.921$

e. 6.427 **f.** Reject H_0.

g. There is enough evidence at the 1% level to conclude that there is a significant linear correlation between men's winning time and women's winning time.

Section 9.2

1a. $n = 6$

x	y	xy	x^2
42	9	378	1764
48	10	480	2304
50	8	400	2500
59	5	295	3481
65	6	390	4225
72	3	216	5184
$\Sigma x = 336$	$\Sigma y = 41$	$\Sigma xy = 2159$	$\Sigma x^2 = 19{,}458$

b. $m \approx -0.2134$ **c.** $b \approx 18.7837$

d. $\hat{y} = -0.213x + 18.784$

2a. Enter data. **b.** $m \approx 1.486$; $b \approx -3.819$

c. $\hat{y} = 1.486x - 3.819$

3a. (1) $\hat{y} = 11.824(2) + 35.301$

(2) $\hat{y} = 11.824(3.32) + 35.301$

b. (1) 58.949 (2) 74.557

c. (1) 58.949 minutes (2) 74.557 minutes

Section 9.3

1a. 0.970 **b.** 0.941

c. 94.1% of the variation in the times is explained.

5.9% of the variation is unexplained.

2a.

x_i	y_i	$\hat{y}_i$	$(y_i - \hat{y}_i)^2$
15	26	28.386	5.693
20	32	35.411	11.635
20	38	35.411	6.703
30	56	49.461	42.759
40	54	63.511	90.459
45	78	70.536	55.711
50	80	77.561	5.949
60	88	91.611	13.039
			$\Sigma = 231.948$

b. 8 **c.** 6.218

d. The standard deviation of the weekly sales for a specific radio ad time is about $621.80.

3a. $n = 8$, d.f. $= 6$, $t_c = 2.447$, $s_e \approx 10.290$

b. 230.884 **c.** 29.236 **d.** $201.648 < y < 260.120$

e. You can be 95% confident that the company sales will be between $201,648 and $260,120 when advertising expenses are $2500.

Section 9.4

1a. Enter data.

b. $\hat{y} = 46.385 + 0.540x_1 - 4.897x_2$

2ab. (1) $\hat{y} = 46.385 + 0.540(89) - 4.897(1)$

(2) $\hat{y} = 46.385 + 0.540(78) - 4.897(3)$

(3) $\hat{y} = 46.385 + 0.540(83) - 4.897(2)$

c. (1) $\hat{y} = 89.548$ (2) $\hat{y} = 73.814$ (3) $\hat{y} = 81.411$

d. (1) 90 (2) 74 (3) 81

CHAPTER 10

Section 10.1

1a.

Music	% of listeners	Expected frequency
Classical	4%	12
Country	36%	108
Gospel	11%	33
Oldies	2%	6
Pop	18%	54
Rock	29%	87

2a. The expected frequencies are 64, 80, 32, 56, 60, 48, 40, and 20, all of which are at least 5.

b. Claimed distribution:

Ages	Distribution
0–9	16%
10–19	20%
20–29	8%
30–39	14%
40–49	15%
50–59	12%
60–69	10%
70+	5%

H_0: Distribution of ages is as shown in table above.

H_a: Distribution of ages differs from the claimed distribution.

c. 0.05 **d.** 7

e. $\chi_0^2 = 14.067$; Rejection region: $\chi^2 > 14.067$

f. 6.694 **g.** Fail to reject H_0.

h. There is not enough evidence at the 5% level to conclude that the distribution of ages differs from the claimed distribution.

3a. The expected frequencies are 88, 54, and 58, all of which are at least 5.

b. Claimed distribution:

Response	Distribution
In favor of	44%
Against	27%
No opinion	29%

H_0: Distribution of responses is as shown in table above.

H_a: Distribution of responses differs from the claimed distribution.

c. 0.01 **d.** 2

e. $\chi_0^2 = 9.210$; Rejection region: $\chi^2 > 9.210$

f. 2.924 **g.** Fail to reject H_0.

h. There is not enough evidence at the 1% level to conclude that the distribution of responses differs from the claimed distribution.

Section 10.2

1ab.

	Travel concern				
Travel purpose	Hotel	Leg room	Rental size	Other	Total
Business	36	108	14	22	180
Leisure	38	54	14	14	120
Total	74	162	28	36	300

c. 300

d.

	Travel concern			
Travel purpose	Hotel	Leg room	Rental size	Other
Business	44.4	97.2	16.8	21.6
Leisure	29.6	64.8	11.2	14.4

2a. H_0: Travel concern is independent of travel purpose.
H_a: Travel concern is dependent on travel purpose. (claim)
b. 0.01 **c.** 3
d. $\chi_0^2 = 11.345$; Rejection region: $\chi^2 > 11.345$
e. 8.158 **f.** Fail to reject H_0.
g. There is not enough evidence to conclude that travel concern is dependent on travel purpose.

3a. $\chi_0^2 = 9.488$; Rejection region: $\chi^2 > 9.488$
b. Enter the data. **c.** 65.619 **d.** Reject H_0.
e. There is enough evidence at the 1% level to conclude that the number of minutes spent online per day is dependent on gender.

Section 10.3

1a. 0.01 **b.** 5.42

2a. 0.01 **b.** 18.31

3a. H_0: $\sigma_1^2 \leq \sigma_2^2$; H_a: $\sigma_1^2 > \sigma_2^2$ (claim)
b. 0.01 **c.** $\text{d.f.}_N = 24$, $\text{d.f.}_D = 19$
d. $F_0 = 2.92$; Rejection region: $F > 2.92$
e. 3.214 **f.** Reject H_0.
g. There is enough evidence at the 1% level to support the claim that a specially treated intravenous solution decreases the variance of the time required for nutrients to enter the bloodstream.

4a. H_0: $\sigma_1 = \sigma_2$ (claim); H_a: $\sigma_1 \neq \sigma_2$
b. 0.01 **c.** $\text{d.f.}_N = 15$, $\text{d.f.}_D = 21$
d. $F_0 = 3.43$; Rejection region: $F > 3.43$
e. 1.483 **f.** Fail to reject H_0.
g. There is not enough evidence at the 1% level to reject the claim that the pH levels of the soil in two geographic locations have equal standard deviations.

Section 10.4

1a. H_0: $\mu_1 = \mu_2 = \mu_3 = \mu_4$
H_a: At least one mean is different from the others.
b. 0.05 **c.** $\text{d.f.}_N = 3$, $\text{d.f.}_D = 14$
d. $F_0 = 3.34$; Rejection region: $F > 3.34$
e. 4.22 **f.** Reject H_0.
g. There is enough evidence at the 5% level to conclude that at least one mean is different from the others.

2a. Enter the data.
b. $F = 1.34$; P-value $= 0.280$
c. Fail to reject H_0.
d. There is not enough evidence at the 5% level to conclude that at least one mean is different from the others.

CHAPTER 11

Section 11.1

1a. H_0: median ≤ 2500; H_a: median > 2500 (claim)
b. 0.025 **c.** 22 **d.** 5 **e.** 10 **f.** Fail to reject H_0.
g. There is not enough evidence at the 2.5% level to support the claim that the median number of customers per day is greater than 2500.

2a. H_0: median $= \$134{,}500$ (claim); H_a: median $\neq \$134{,}500$
b. 0.10 **c.** 81 **d.** -1.645 **e.** -2.22 **f.** Reject H_0.
g. There is enough evidence at the 10% level to reject the claim that the median sales price of houses sold in a certain region is $134,500.

3a. H_0: The number of colds will not decrease.
H_a: The number of colds will decrease. (claim)
b. 0.05 **c.** 11 **d.** 2 **e.** 2 **f.** Reject H_0.
g. There is enough evidence at the 5% level to support the claim that a new vaccine will decrease the number of colds in adults.

Section 11.2

1a. H_0: The water repellent is not effective.
H_a: The water repellent is effective. (claim)
b. 0.01 **c.** 11 **d.** 7

e.

No repellent	Repellent applied	Difference	Absolute value	Rank	Signed rank
8	15	−7	7	11	−11
7	12	−5	5	9	−9
7	11	−4	4	7.5	−7.5
4	6	−2	2	3.5	−3.5
6	6	0	0		
10	8	2	2	3.5	3.5
9	8	1	1	1.5	1.5
5	6	−1	1	1.5	−1.5
9	12	−3	3	5.5	−5.5
11	8	3	3	5.5	5.5
8	14	−6	6	10	−10
4	8	−4	4	7.5	−7.5

$w_s = 10.5$

f. Reject H_0.

g. There is enough evidence at the 1% level to support the claim that the water repellent is effective.

2a. H_0: There is no difference in the claims paid by the companies.

H_a: There is a difference in the claims paid by the companies. (claim)

b. 0.05 **c.** ± 1.96 **d.** $n_1 = 12$ and $n_2 = 12$

e.

Ordered data	Sample	Rank	Ordered data	Sample	Rank
1.7	B	1	5.3	B	13
1.8	B	2	5.6	B	14
2.2	B	3	5.8	A	15
2.5	A	4	6.0	A	16
3.0	A	5.5	6.2	A	17
3.0	B	5.5	6.3	A	18
3.4	B	7	6.5	A	19
3.9	A	8	7.3	B	20
4.1	B	9	7.4	A	21
4.4	B	10	9.9	A	22
4.5	A	11	10.6	A	23
4.7	B	12	10.8	B	24

$R = 120.5$

f. -1.703 **g.** Fail to reject H_0.

h. There is not enough evidence at the 5% level to conclude that there is a difference in the claims paid by the companies.

Section 11.3

1a. H_0: There is no difference in the salaries in the three states.

H_a: There is a difference in the salaries in the three states. (claim)

b. 0.10 **c.** 2

d. $\chi_0^2 = 4.605$; Rejection region: $\chi^2 > 4.605$

e.

Ordered data	State	Rank	Ordered data	State	Rank
25.57	CT	1	40.31	CA	16
26.42	CA	2	41.33	CA	17
29.73	NJ	3	43.26	NJ	18
29.86	CT	4	43.89	NJ	19
30.00	CT	5	44.57	NJ	20
33.68	CA	6.5	45.04	CT	21
33.68	CT	6.5	45.29	CT	22
33.91	NJ	8	46.55	CA	23
34.29	NJ	9	46.72	NJ	24
36.35	NJ	10	47.17	CA	25
36.55	CA	11	48.46	CA	26
37.18	CA	12	50.16	NJ	27
37.24	NJ	13	51.03	CT	28
37.39	CT	14	57.07	CT	29
38.36	CA	15	61.46	CT	30

$R_1 = 153.5$

$R_2 = 160.5$

$R_3 = 151$

f. 0.063 **g.** Fail to reject H_0.

h. There is not enough evidence at the 10% level to support the claim that the distributions of the underwriters' salaries in the three states are different.

Section 11.4

1a. H_0: $\rho_s = 0$; H_a: $\rho_s \neq 0$

b. 0.05 **c.** 0.700

d.

Oat	Rank	Wheat	Rank	d	d^2
1.67	7	4.55	9	−2	4
1.90	9	4.30	8	1	1
1.60	6	3.38	6	0	0
1.10	1.5	2.65	3	−1.5	2.25
1.12	3	2.48	1	2	4
1.10	1.5	2.62	2	−0.5	0.25
1.59	5	2.78	4	1	1
1.81	8	3.56	7	1	1
1.45	4	3.35	5	−1	1
					$\Sigma = 14.5$

$\Sigma d^2 = 14.5$

e. 0.879 **f.** Reject H_0.

g. There is enough evidence at the 5% level to conclude that a significant correlation exists between oat and wheat prices.

Section 11.5

1a. *P P P F P F P P P P F F P F P P F F F P P P F P P P*

b. 13

c. 3, 1, 1, 1, 4, 2, 1, 1, 2, 3, 3, 1, 3

2a. H_0: The sequence of genders is random.

H_a: The sequence of genders is not random. (claim)

b. 0.05

c. n_1 = number of *F*s = 9

n_2 = number of *M*s = 5

G = number of runs = 7

d. lower critical value = 3

upper critical value = 12

e. 7 **f.** Fail to reject H_0.

g. There is not enough evidence at the 5% level to support the claim that the sequence of genders is not random.

3a. H_0: The sequence of weather conditions is random.

H_a: The sequence of weather conditions is not random. (claim)

b. 0.05

c. n_1 = number of *N*s = 21

n_2 = number of *S*s = 10

G = number of runs = 17

d. ±1.96 **e.** 1.04 **f.** Fail to reject H_0.

g. There is not enough evidence at the 5% level to support the claim that the sequence of weather conditions is not random.

APPENDIX A

1a. (1) 0.4857

(2) $z = \pm 2.17$

2a.

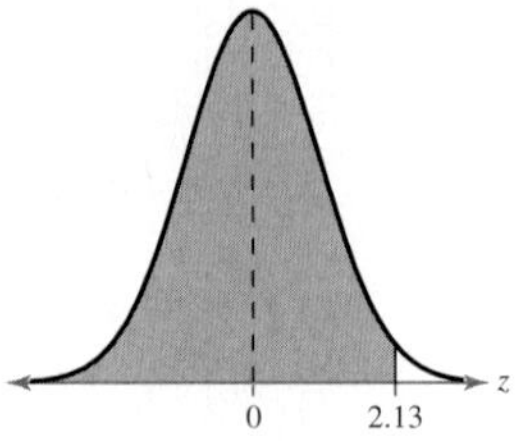

b. 0.4834 **c.** 0.9834

3a.

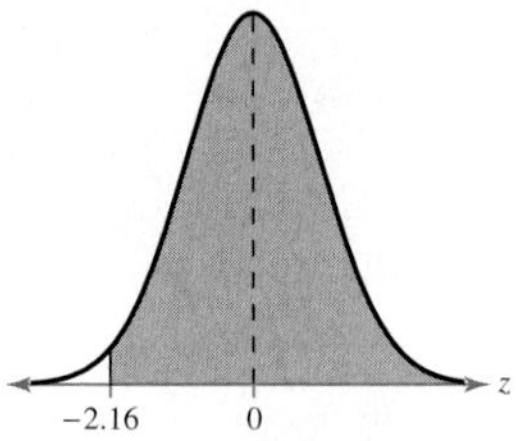

b. 0.4846 **c.** 0.9846

4a.

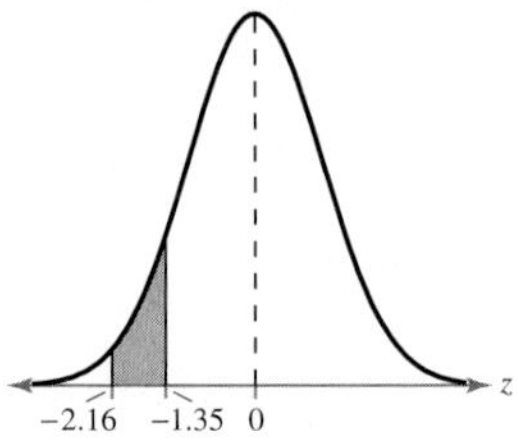

b. 0.4846; 0.4115 **c.** 0.0731

Odd Answers

CHAPTER 1

Section 1.1 *(page 6)*

1. A sample is a subset of a population.
3. A parameter is a numerical description of a population characteristic. A statistic is a numerical description of a sample characteristic.
5. False. A statistic is a measure that describes a sample characteristic.
7. True
9. False. A population is the collection of *all* outcomes, responses, measurements, or counts that are of interest.
11. Population, because it is a collection of the ages of all the state governors.
13. Sample, because the collection of the 500 students is a subset within the population.
15. Sample, because the collection of the 20 patients is a subset within the population.
17. Population: Party of registered voters in Bucks County

    Sample: Party of Bucks County voters responding to phone survey
19. Population: Ages of adults in the U.S. who own computers

    Sample: Ages of adults in the U.S. who own Dell computers
21. Population: Collection of all infants in Italy

    Sample: Collection of the 33,043 infants in the study
23. Population: Collection of all women in the U.S.

    Sample: Collection of the 546 U.S. women surveyed
25. Statistic. The value $68,000 is a numerical description of a sample of annual salaries.
27. Statistic. 8% is a numerical description of a sample of computer users.
29. Statistic. 47% is a numerical description of a sample of people in the United States.
31. The statement "56% are the primary investor in their household" is an application of descriptive statistics.

    An inference drawn from the sample is that an association exists between U.S. women and being the primary investor in their household.
33. Answers will vary.
35. (a) An inference drawn from the sample is that senior citizens who own luxury cars have more of their natural teeth than senior citizens who do not own luxury cars.

    (b) This inference may incorrectly imply that if you purchase a luxury car, you will keep more of your natural teeth.
37. Answers will vary.

Section 1.2 *(page 13)*

1. Nominal and ordinal
3. False. Data at the ordinal level can be qualitative or quantitative.
5. False. More types of calculations can be performed with data at the interval level than with data at the nominal level.
7. Qualitative **9.** Quantitative
11. Ordinal. Data can be arranged in order, but the differences between data entries make no sense.
13. Nominal. No mathematical computations can be made and data are categorized by region.
15. Ordinal. Data can be arranged in order, but the differences between data entries are not meaningful.
17. Ordinal **19.** Nominal
21. An inherent zero is a zero that implies "none." Answers will vary.

Section 1.3 *(page 21)*

1. False. Using stratified sampling guarantees that members of each group within a population will be sampled.
3. False. A systematic sample is selected by ordering a population in some way and then selecting members of the population at regular intervals.
5. Perform an experiment because you want to measure the effect of a treatment on the human digestive system.
7. Use a simulation because the situation is impractical.
9. Simple random sampling is used because each telephone number has an equal chance of being dialed, and all samples of 1599 phone numbers have an equal chance of being selected. The sample may be biased because only homes with telephones will be sampled.
11. Convenience sampling is used because the students were chosen owing to their convenience of location. Bias may enter into the sample because the students sampled may not be representative of the population of students.
13. Simple random sampling is used because each outpatient had an equal chance of being contacted, and all samples of 1210 outpatients had an equal chance of being selected.
15. Stratified sampling is used because a sample is taken from each one-acre subplot.
17. Systematic sampling is used because every ninth name on a list is being selected.
19. Census, because it is relatively easy to obtain the salaries of the 50 employees.

21. The question is biased because it already suggests that drinking fruit juice is good for you. The question might be rewritten as "How does drinking fruit juice affect your health?"

23. The survey question is unbiased.

25. The households sampled represent various locations, ethnic groups, and income brackets. Each of these variables is considered a stratum.

27. Open Question

Advantage: Allows respondent to express some depth and shades of meaning in the answer.

Disadvantage: Not easily quantified and difficult to compare surveys.

Closed Question

Advantage: Easy to analyze results.

Disadvantage: May not provide appropriate alternatives and may influence the opinion of the respondent.

Uses and Abuses for Chapter 1 *(page 23)*

1. Answers will vary. **2.** Answers will vary.

Review Answers for Chapter 1 *(page 25)*

1. Population: Collection of all U.S. adult VCR owners

Sample: Collection of the 898 U.S. adult VCR owners that were sampled

3. Population: Collection of all U.S. ATMs

Sample: Collection of 82,188 ATMs that were sampled

5. Parameter **7.** Parameter

9. The statement "the average surcharge for withdrawals from a competing bank was $1.48" is an application of descriptive statistics.

An inference drawn from the sample is that all ATMs charge an average of $1.48 for withdrawals from a competing bank.

11. Quantitative, because monthly salaries are numerical measurements.

13. Quantitative, because ages are numerical measurements.

15. Interval. It makes no sense saying that 100 degrees is twice as hot as 50 degrees.

17. Nominal. The data are categorical and cannot be arranged in a meaningful order.

19. Take a census because judges keep accurate records of charitable donations.

21. Perform an experiment because you want to measure the effect of a treatment on chrysanthemums.

23. Simple random sampling is used because random telephone numbers were generated and called.

25. Cluster sampling is used because each community is considered a cluster and every pregnant woman in a selected community is surveyed.

27. Stratified sampling is used because 25 students are randomly selected from each grade level.

29. Telephone sampling samples only individuals who have telephones, are available, and are willing to respond.

31. The selected communities may not be representative of the entire area.

Chapter Quiz for Chapter 1 *(page 27)*

1. Population: Collection of all individuals with sleep disorders

Sample: Collection of 254 patients in study

2. (a) Statistic (b) Parameter

3. (a) Qualitative (b) Quantitative

4. (a) Nominal, because no mathematical computations can be made.

(b) Ratio, because one data value can be expressed as a multiple of another.

(c) Interval, because meaningful differences between entries can be calculated but a zero entry is not an inherent zero.

5. (a) Perform an experiment because you want to measure the effect of a treatment on lead levels in adults.

(b) Use sampling because it would be impossible to question everyone in the population.

6. (a) Convenience sampling, because all of the people sampled are in one convenient location.

(b) Systematic sampling, because every fifth part is sampled.

(c) Stratified sampling, because the population is first stratified and then a sample is collected from each stratum.

7. Convenience

8. (a) False. A statistic is a numerical measure that describes a sample characteristic.

(b) False. Ratio data represent the highest level of measurement.

Real Statistics–Real Decisions for Chapter 1
(page 28)

1. (a) Answers will vary.

(b) Yes (c) Use sampling and surveys.

(d) You may take too large of a percentage of your sample from a subgroup of the population that is relatively small.

2. (a) Response: qualitative; Percent responding: quantitative

(b) Response: nominal; Percent responding: ratio
Method: nominal; Percent responding: ratio

(c) Sample (d) Statistics

3. (a) Answers will vary.

(b) Answers will vary

CHAPTER 2

Section 2.1 *(page 43)*

1. Organizing the data into a frequency distribution may make patterns within the data more evident.

3. Class limits determine which numbers can belong to that class.

Class boundaries are the numbers that separate classes without forming gaps between them.

5. False. The midpoint of a class is the sum of the lower and upper limits of the class divided by two.

7. True

9. (a) 10

(b) and (c)

Class	Midpoint	Class boundaries
20–29	24.5	19.5–29.5
30–39	34.5	29.5–39.5
40–49	44.5	39.5–49.5
50–59	54.5	49.5–59.5
60–69	64.5	59.5–69.5
70–79	74.5	69.5–79.5
80–89	84.5	79.5–89.5

11.

Class	Frequency, f	Midpoint	Relative frequency	Cumulative frequency
20–29	10	24.5	0.01	10
30–39	132	34.5	0.13	142
40–49	284	44.5	0.29	426
50–59	300	54.5	0.30	726
60–69	175	64.5	0.18	901
70–79	65	74.5	0.07	966
80–89	25	84.5	0.03	991
	$\Sigma f = 991$		$\Sigma \frac{f}{n} = 1$	

13. (a) Number of classes = 7 (b) Least frequency ≈ 10
(c) Greatest frequency ≈ 300 (d) Class width = 10

15. (a) 50 (b) 12.5–13.5 pounds

17. (a) 24 (b) 19.5 pounds

19. (a) Class with greatest relative frequency: 8–9 inches
Class with least relative frequency: 17–18 inches
(b) Greatest relative frequency ≈ 0.195
Least relative frequency ≈ 0.005
(c) Approximately 0.015

21. Class with greatest frequency: 500–550
Classes with least frequency: 250–300 and 700–750

23.

Class	Frequency, f	Midpoint	Relative frequency	Cumulative frequency
0–7	8	3.5	0.32	8
8–15	8	11.5	0.32	16
16–23	3	19.5	0.12	19
24–31	3	27.5	0.12	22
32–39	3	35.5	0.12	25
	$\Sigma f = 25$		$\Sigma \frac{f}{n} = 1$	

25.

Class	Frequency, f	Midpoint	Relative frequency	Cumulative frequency
1000–2019	12	1509.5	0.5455	12
2020–3039	3	2529.5	0.1364	15
3040–4059	2	3549.5	0.0909	17
4060–5079	3	4569.5	0.1364	20
5080–6099	1	5589.5	0.0455	21
6100–7119	1	6609.5	0.0455	22
	$\Sigma f = 22$		$\Sigma \frac{f}{N} \approx 1$	

July Sales for Representatives

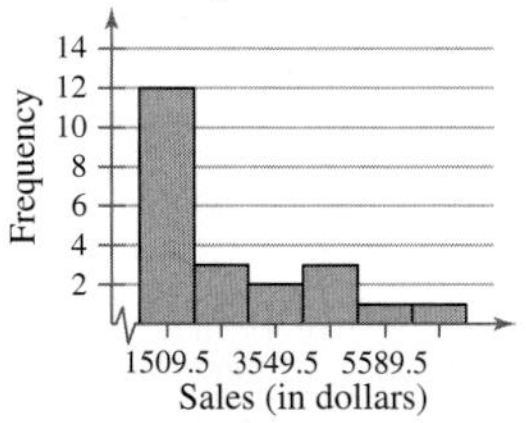

Class with greatest frequency: 1000–2019
Classes with least frequency: 5080–6099 and 6100–7119

27.

Class	Frequency, f	Midpoint	Relative frequency	Cumulative frequency
291–318	5	304.5	0.1667	5
319–346	4	332.5	0.1333	9
347–374	3	360.5	0.1000	12
375–402	5	388.5	0.1667	17
403–430	6	416.5	0.2000	23
431–458	4	444.5	0.1333	27
459–486	1	472.5	0.0333	28
487–514	2	500.5	0.0667	30
	$\Sigma f = 30$		$\Sigma \frac{f}{n} = 1$	

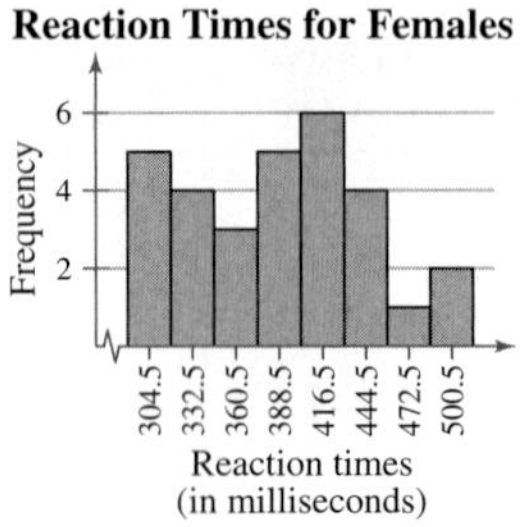

Class with greatest frequency: 403–430

Class with least frequency: 459–486

29.

Class	Frequency, f	Mid-point	Relative frequency	Cumulative frequency
146–169	6	157.5	0.2308	6
170–193	9	181.5	0.3462	15
194–217	3	205.5	0.1154	18
218–241	6	229.5	0.2308	24
242–265	2	253.5	0.0769	26
	$\sum f = 26$		$\sum \frac{f}{n} \approx 1$	

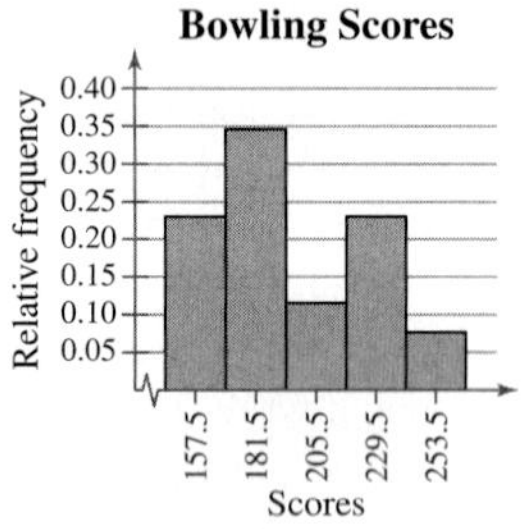

Class with greatest relative frequency: 170–193

Class with least relative frequency: 242–265

31.

Class	Frequency, f	Mid-point	Relative frequency	Cumulative frequency
33–36	8	34.5	0.3077	8
37–40	6	38.5	0.2308	14
41–44	5	42.5	0.1923	19
45–48	2	46.5	0.0769	21
49–52	5	50.5	0.1923	26
	$\sum f = 26$		$\sum \frac{f}{n} \approx 1$	

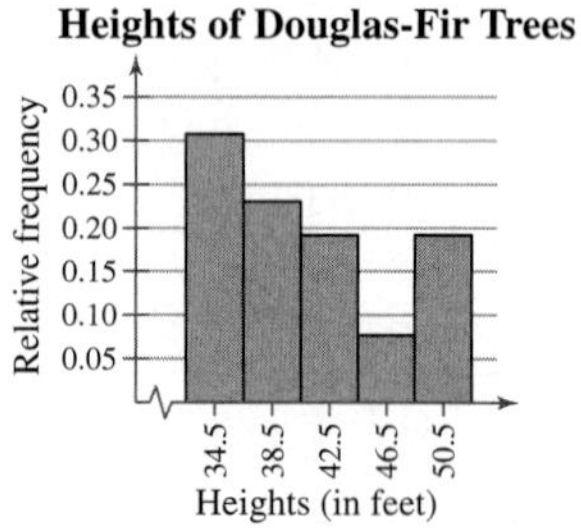

Class with greatest relative frequency: 33–36

Class with least relative frequency: 45– 48

33.

Class	Frequency, f	Relative frequency	Cumulative frequency
50–53	1	0.0417	1
54–57	0	0.0000	1
58–61	4	0.1667	5
62–65	9	0.3750	14
66–69	7	0.2917	21
70–73	3	0.1250	24
	$\sum f = 24$	$\sum \frac{f}{n} \approx 1$	

Location of the greatest increase in frequency: 62–65

35.

Class	Frequency, f	Relative frequency	Cumulative frequency
2–4	9	0.3214	9
5–7	6	0.2143	15
8–10	7	0.2500	22
11–13	3	0.1071	25
14–16	2	0.0714	27
17–19	1	0.0357	28
	$\sum f = 28$	$\sum \frac{f}{n} \approx 1$	

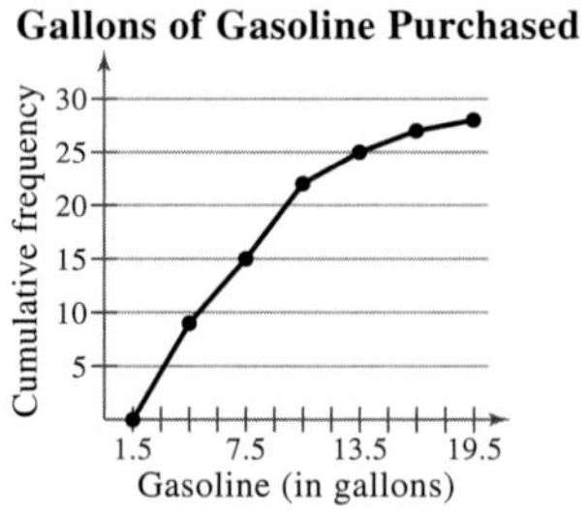

Location of the greatest increase in frequency: 2–4

37.

Class	Frequency, f	Midpoint	Relative frequency	Cumulative frequency
47–57	1	52	0.05	1
58–68	1	63	0.05	2
69–79	5	74	0.25	7
80–90	8	85	0.40	15
91–101	5	96	0.25	20
	$\Sigma f = 20$		$\Sigma \frac{f}{N} = 1$	

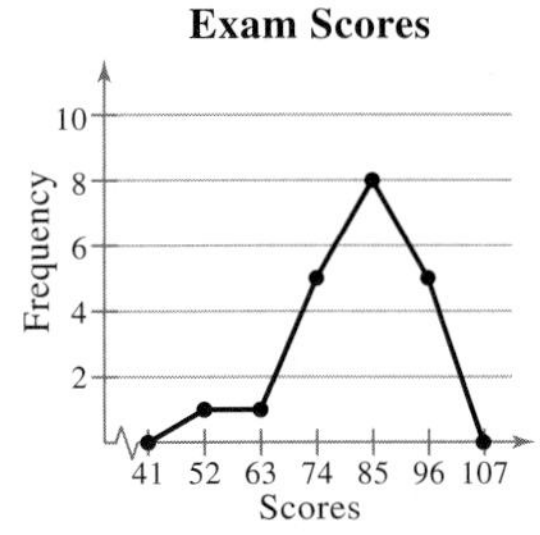

Class with greatest frequency: 80–90

Classes with least frequency: 47–57 and 58–68

39. (a)

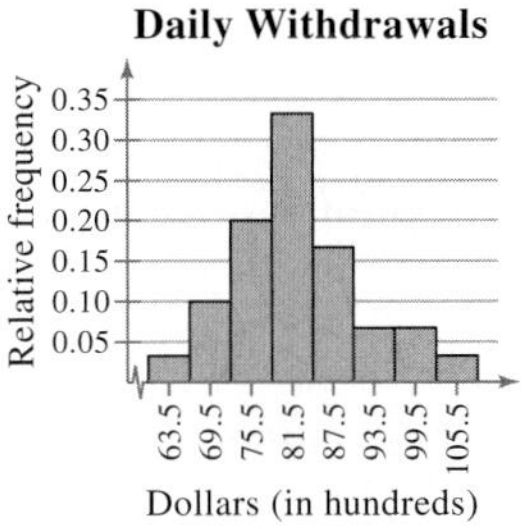

(b) 16.7%, because the sum of the relative frequencies for the last three classes is 0.167.

(c) $9600, because the sum of the relative frequencies for the last two classes is 0.10.

41.

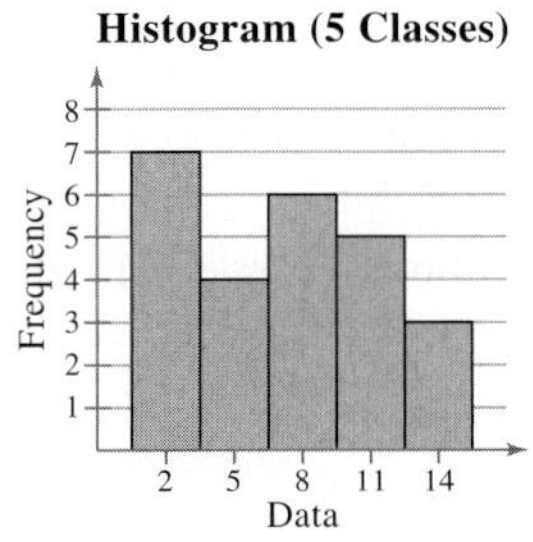

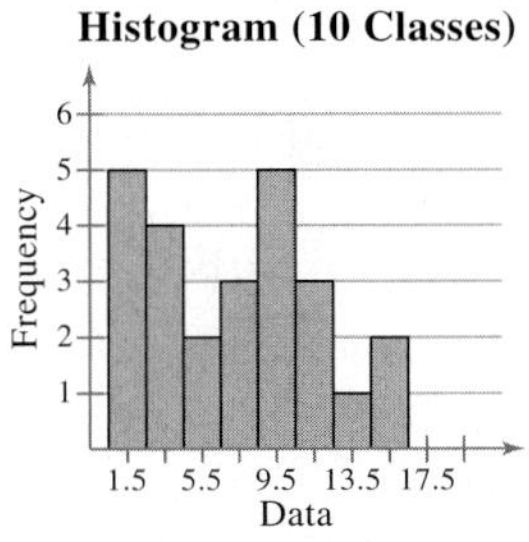

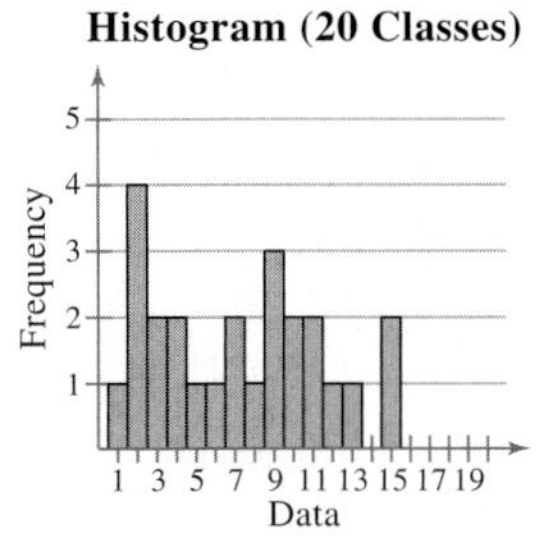

In general, a greater number of classes better preserves the actual values of the data set but is not as helpful for observing general trends and making conclusions. In choosing the number of classes, an important consideration is the size of the data set. For instance, you would not want to use 20 classes if your data set contained 20 entries. In this particular example, as the number of classes increases, the histogram shows more fluctuation. The histograms with 10 and 20 classes have classes with zero frequencies. Not much is gained by using more than five classes. Therefore, it appears that five classes would be best.

Section 2.2 *(page 56)*

1. Quantitative: stem-and-leaf plot, dot plot, histogram, scatter plot, time series chart

Qualitative: pie chart, Pareto chart

3. a **4.** d **5.** b **6.** c

7. 27, 32, 41, 43, 43, 44, 47, 47, 48, 50, 51, 51, 52, 53, 53, 53, 54, 54, 54, 54, 55, 56, 56, 58, 59, 68, 68, 68, 73, 78, 78, 85

Max: 85; Min: 27

9. 13, 13, 14, 14, 14, 15, 15, 15, 15, 15, 16, 17, 17, 18, 19

Max: 19; Min: 13

11. Anheuser-Busch spends the most on advertising and Honda spends the least. (Answers will vary.)

13. Tailgaters irk drivers the most, and too-cautious drivers irk drivers the least. (Answers will vary.)

15. Key: 3|3 = 33

3	2 3 3 4 5 9
4	0 1 1 3 4 5 5 6 6 7 8
5	1 3 3
6	0 0 6 9

It appears that most elephants tend to drink less than 55 gallons of water per day. (Answers will vary.)

17. Key: 17|5 = 17.5

16	4 8
17	1 1 3 4 5 5 6 7 9
18	1 3 4 4 6 6 6 9
19	0 0 2 3 3 5 6
20	1 8

It appears that most farmers charge 17 to 19 cents per pound of apples. (Answers will vary.)

19.

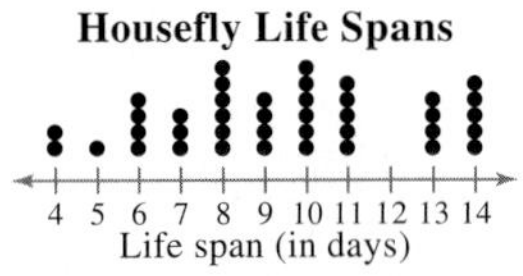

It appears that the life span of a housefly tends to be between 4 and 14 days. (Answers will vary.)

21.

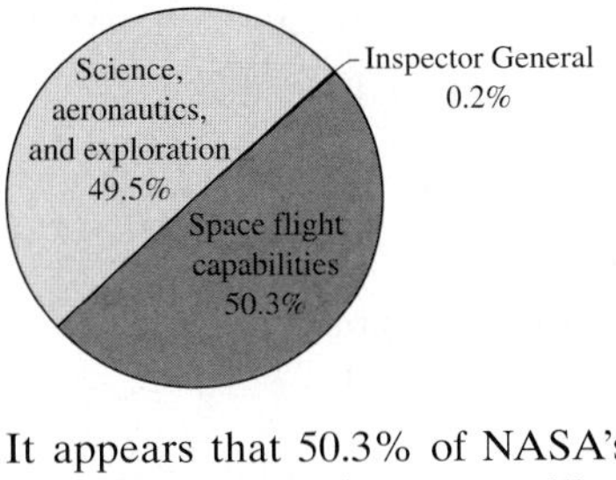

It appears that 50.3% of NASA's budget went to space flight capabilities. (Answers will vary).

23. 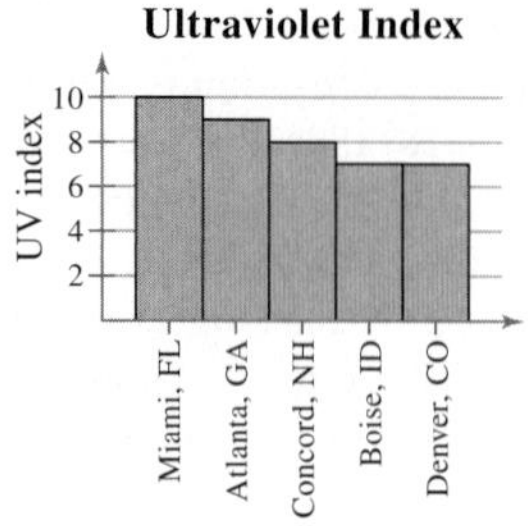

It appears that Boise, ID, and Denver, CO, have the same UV index. (Answers will vary.)

25. 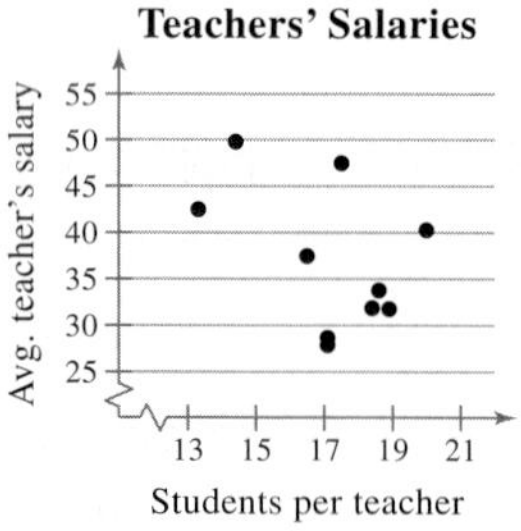

It appears that a teacher's average salary decreases as the number of students per teacher increases. (Answers will vary.)

27. 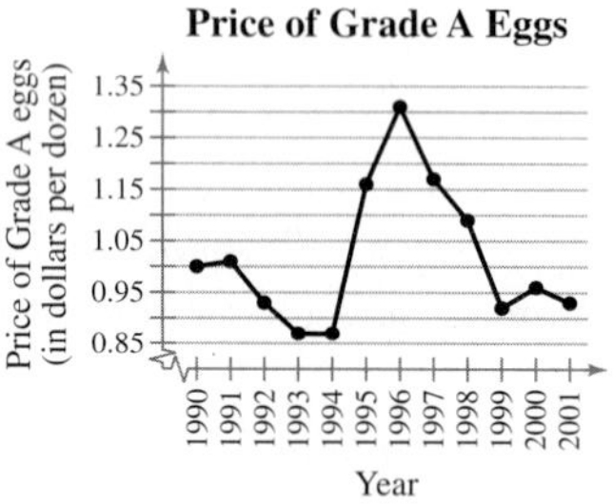

It appears the price of eggs peaked in 1996. (Answers will vary.)

29. (a) When data are taken at regular intervals over a period of time, a time series chart should be used. (Answers will vary.)

(b) 

Section 2.3 *(page 67)*

1. False. The mean is the measure of central tendency most likely to be affected by an extreme value (or outlier).

3. False. All quantitative data sets have a median.

5. A data set with an outlier within it would be an example. (Answers will vary.)

7. The shape of the distribution is skewed right because the bars have a "tail" to the right.

9. The shape of the distribution is uniform because the bars are approximately the same height.

11. (9), because the distribution of values ranges from 1 to 12 and has (approximately) equal frequencies.

13. (10), because the distribution has a maximum value of 90 and is skewed left owing to a few students' scoring much lower than the majority of the students.

15. (a) $\bar{x} \approx 6.2$

median $= 6$; mode $= 5$

(b) Median, because the distribution is skewed.

17. (a) $\bar{x} \approx 4.57$

median $= 4.8$; mode $= 4.8$

(b) Median, because there are no outliers.

19. (a) $\bar{x} \approx 93.81$

median $= 92.9$; mode $= 90.3, 91.8$

(b) Median, because the distribution is skewed.

21. (a) $\bar{x} =$ not possible

median $=$ not possible; mode $=$ "Worse"

(b) Mode, because the data are at the nominal level of measurement.

23. (a) $\bar{x} \approx 170.63$

median $= 169.3$; mode $=$ none

(b) Mean, because there are no outliers.

25. (a) $\bar{x} = 22.6$

median $= 19$; mode $= 14$

(b) Median, because the distribution is skewed.

27. (a) $\bar{x} \approx 14.11$

median $= 14.25$; mode $= 2.5$

(b) Mean, because there are no outliers.

29. (a) $\bar{x} = 41.3$

median $= 39.5$; mode $= 45$

(b) Median, because the distribution is skewed.

31. (a) $\bar{x} \approx 19.5$

median $= 20$; mode $= 15$

(b) Median, because the distribution is skewed.

33. A $=$ mode, because it's the data entry that occurred most often.

B $=$ median, because the distribution is skewed right.

C $=$ mean, because the distribution is skewed right.

35. Mode, because the data are at the nominal level of measurement.

37. Mean, because there are no outliers.

39. 89.3 **41.** 2.8 **43.** 65.5 **45.** 35.0

47.

Class	Frequency, f	Midpoint
3–4	3	3.5
5–6	8	5.5
7–8	4	7.5
9–10	2	9.5
11–12	2	11.5
13–14	1	13.5
	$\Sigma f = 20$	

Hospitalization Positively skewed

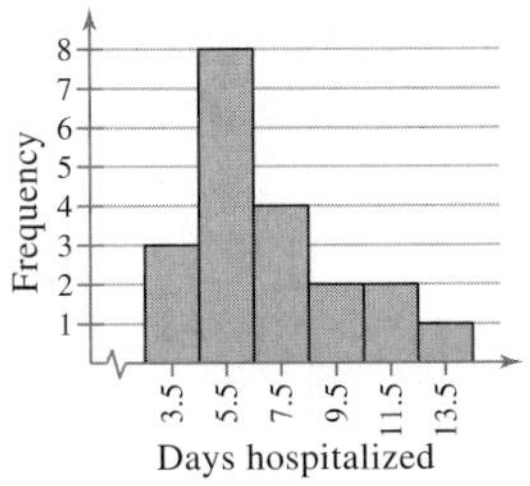

49.

Class	Frequency, f	Midpoint
62–64	3	63
65–67	7	66
68–70	9	69
71–73	8	72
74–76	3	75
	$\Sigma f = 30$	

Heights of Males Symmetric

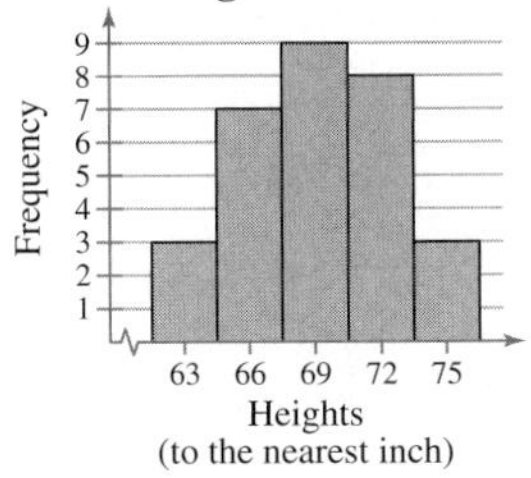

51. (a) $\bar{x} = 6.005$, median $= 6.01$ (b) $\bar{x} = 5.945$, median $= 6.01$

(c) Mean

53. (a) Mean, because Car A has the highest mean of the three.

(b) Median, because Car B has the highest median of the three.

(c) Mode, because Car C has the highest mode of the three.

55. (a) $\bar{x} \approx 49.2$ (b) median $= 46.5$

(c) Key: $3|6 = 36$ (d) Positively skewed

1	1 3	
2	2 8	
3	6 6 6 7 7 7 8	
4	1 3 4 6 7	← mean
5	1 1 1 3	
6	1 2 3 4	median (7 in row 4)
7	2 2 4 6	
8	5	
9	0	

57. Two different symbols are needed because they describe a measure of central tendency for two different sets of data (sample is a subset of the population).

Section 2.4 *(page 84)*

1. Range $= 7$, mean $= 8.1$, variance ≈ 5.7, standard deviation ≈ 2.4

3. Range $= 14$, mean ≈ 11.1, variance ≈ 21.6, standard deviation ≈ 4.6

5. 73

7. The range is the difference between the maximum and minimum values of a data set. The advantage of the range is that it is easy to calculate. The disadvantage is that it uses only two entries from the data set.

9. The units of variance are squared. Its units are meaningless. (Example: dollars^2)

11. (a) Range $= 25.1$ (b) Range $= 45.1$

(c) Changing the maximum value of the data set greatly affects the range.

13. (a) has a standard deviation of 24 and (b) has a standard deviation of 16, because the data in (a) have more variability.

15. When calculating the population standard deviation, you divide the sum of the squared deviations by n, then take the square root of that value. When calculating the sample standard deviation, you divide the sum of the squared deviations by $n - 1$, then take the square root of that value.

17. Company B

19. (a) Los Angeles: 17.6, 37.35, 6.11

Long Beach: 8.7, 8.71, 2.95

(b) It appears from the data that the annual salaries in Los Angeles are more variable than the salaries in Long Beach.

21. (a) Males: 405; 16,225.3; 127.4

Females: 552; 34,575.1; 185.9

(b) It appears from the data that the SAT scores for females are more variable than the SAT scores for males.

23. (a) Greatest sample standard deviation: (ii)

Data set (ii) has more entries that are farther away from the mean.

Least sample standard deviation: (iii)

Data set (iii) has more entries that are close to the mean.

(b) The three data sets have the same mean but have different standard deviations.

25. (a) Greatest sample standard deviation: (ii)

Data set (ii) has more entries that are farther away from the mean.

Least sample standard deviation: (iii)

Data set (iii) has more entries that are close to the mean.

(b) The three data sets have the same mean, median, and mode but have different standard deviations.

27. Similarity: Both estimate proportions of the data contained within k standard deviations of the mean.

Difference: The Empirical Rule assumes the distribution is bell shaped; Chebychev's Theorem makes no such assumption.

29. 68% **31.** (a) 51 (b) 17

33. \$1450, \$550 **35.** 24

37. Sample mean ≈ 2.1

Sample standard deviation ≈ 1.3

39. Class width $= \dfrac{\text{Max} - \text{Min}}{5} = \dfrac{14-4}{5} = 2 \Rightarrow 3$

Class	Midpoint, x	f	xf
4–6	5	13	65
7–9	8	6	48
10–12	11	11	121
13–15	14	2	28
16–18	17	0	0
		$N = 32$	$\Sigma xf = 262$

$x - \mu$	$(x-\mu)^2$	$(x-\mu)^2 f$
−3.2	10.24	133.12
−0.2	0.04	0.24
2.8	7.84	86.24
5.8	33.64	67.28
8.8	77.44	0
		$\Sigma(x-\mu)^2 f = 286.88$

$$\mu = \frac{\Sigma xf}{N} = \frac{262}{32} \approx 8.2$$

$$\sigma = \sqrt{\frac{\Sigma(x-\mu)^2 f}{N}} = \sqrt{\frac{286.88}{32}} \approx 3.0$$

41.

f	Midpoint, x	xf	$x - \bar{x}$	$(x-\bar{x})^2$	$(x-\bar{x})^2 f$
1	70.5	70.5	−44	1936	1936
12	92.5	1110.0	−22	484	5808
25	114.5	2862.5	0	0	0
10	136.5	1365.0	22	484	4840
2	158.5	317.0	44	1936	3872
$n = 50$		$\Sigma xf = 5725$		$\Sigma(x-\bar{x})^2 f = 16{,}456$	

$$\bar{x} = \frac{\Sigma xf}{n} = \frac{5725}{50} = 114.5$$

$$s = \sqrt{\frac{\Sigma(x-\bar{x})^2 f}{n-1}} = \sqrt{\frac{16{,}456}{49}} \approx 18.33$$

43.

Class	f	Midpoint, x	xf
0–4	19.9	2.0	39.80
5–13	35.2	9.0	316.80
14–17	16.9	15.5	261.95
18–24	29.8	21.0	625.80
25–34	38.3	29.5	1129.85
35–44	40.0	39.5	1580.00
45–64	78.3	54.5	4267.35
65+	39.0	70.0	2730.00
	$n = 297.4$		$\Sigma xf = 10{,}951.55$

$x - \bar{x}$	$(x-\bar{x})^2$	$(x-\bar{x})^2 f$
−34.82	1212.43	24,127.36
−27.82	773.95	27,243.04
−21.32	454.54	7,681.73
−15.82	250.27	7,458.05
−7.32	53.58	2,052.11
2.68	7.18	287.20
17.68	312.58	24,475.01
33.18	1100.91	42,935.49
	$\Sigma(x-\bar{x})^2 f = 136{,}259.99$	

$$\bar{x} = \frac{\Sigma xf}{n} = \frac{10{,}951.55}{297.4} \approx 36.82$$

$$s = \sqrt{\frac{\Sigma(x-\bar{x})^2 f}{n-1}} = \sqrt{\frac{136{,}259.99}{296.4}} \approx 21.44$$

45. $CV_{\text{heights}} = \dfrac{3.44}{72.75} \cdot 100 \approx 4.73$

$CV_{\text{weights}} = \dfrac{18.47}{187.83} \cdot 100 \approx 9.83$

It appears that weight is more variable than height.

47. (a) $\bar{x} = 550,\ s \approx 302.8$

(b) $\bar{x} = 5500,\ s \approx 3028$

(c) $\bar{x} = 55,\ s \approx 30.28$

(d) When each entry is multiplied by a constant k, the new sample mean is $k \cdot \bar{x}$, and the new sample standard deviation is $k \cdot s$.

49. 10

Set $1 - \dfrac{1}{k^2} = 0.99$ and solve for k.

Section 2.5 *(page 100)*

1. (a) $Q_1 = 4.5,\ Q_2 = 6,\ Q_3 = 7.5$

(b)

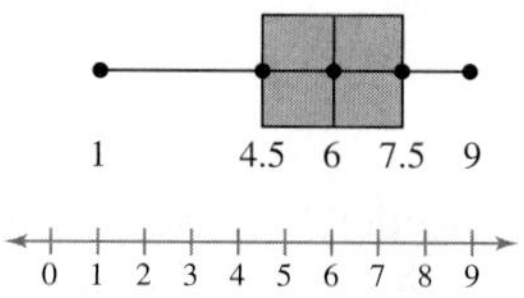

3. The basketball team scored more points per game than 75% of the teams in the league.

5. The student scored above 63% of the students who took the ACT placement test.

7. True

9. False. The 50th percentile is equivalent to Q_2.

11. (a) Min = 10 (b) Max = 20
(c) $Q_1 = 13$ (d) $Q_2 = 15$
(e) $Q_3 = 17$ (f) IQR = 4

13. (a) Min = 900 (b) Max = 2100
(c) $Q_1 = 1250$ (d) $Q_2 = 1500$
(e) $Q_3 = 1950$ (f) IQR = 700

15. (a) Min = −1.9 (b) Max = 2.1
(c) $Q_1 = -0.5$ (d) $Q_2 = 0.1$
(e) $Q_3 = 0.7$ (f) IQR = 1.2

17. $Q_1 = B$, $Q_2 = A$, $Q_3 = C$, because about one quarter of the data fall on or below 17, 18.5 is the median of the entire data set, and about three quarters of the data fall on or below 20.

19. (a) $Q_1 = 2, Q_2 = 4, Q_3 = 5$
(b)

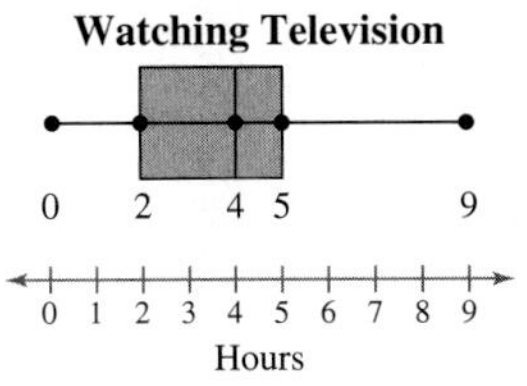

21. (a) $Q_1 = 3.2, Q_2 = 3.65, Q_3 = 3.9$
(b)

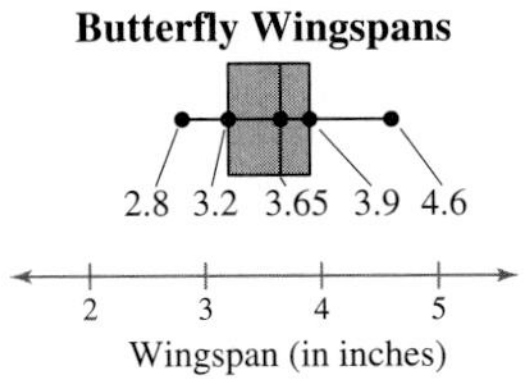

23. (a) 5 (b) 50% (c) 25%

25. A → $z = -1.43$
B → $z = 0$
C → $z = 2.14$
A z-score of 2.14 would be unusual.

27. (a) Statistics: $z = \dfrac{73 - 63}{7} \approx 1.43$

Biology: $z = \dfrac{26 - 23}{3.9} \approx 0.77$

(b) The student did better on the statistics test.

29. (a) Statistics: $z = \dfrac{78 - 63}{7} \approx 2.14$

Biology: $z = \dfrac{29 - 23}{3.9} \approx 1.54$

(b) The student did better on the statistics test.

31. (a) $z_1 = \dfrac{34{,}000 - 35{,}000}{2250} \approx -0.44$

$z_2 = \dfrac{37{,}000 - 35{,}000}{2250} \approx 0.89$

$z_3 = \dfrac{31{,}000 - 35{,}000}{2250} \approx -1.78$

None of the selected tires have unusual life spans.

(b) For 30,500, 2.5th percentile
For 37,250, 84th percentile
For 35,000, 50th percentile

33. About 67 inches; 20% of the heights are below 67 inches.

35. $z_1 = \dfrac{74 - 69.2}{2.9} \approx 1.66$

$z_2 = \dfrac{62 - 69.2}{2.9} \approx -2.48$

$z_3 = \dfrac{80 - 69.2}{2.9} \approx 3.72$

The heights that are 62 and 80 inches are unusual.

37. $z = \dfrac{71.1 - 69.2}{2.9} \approx 0.66$

About the 70th percentile

39. (a) $Q_1 = 42, Q_2 = 49, Q_3 = 56$
(b)

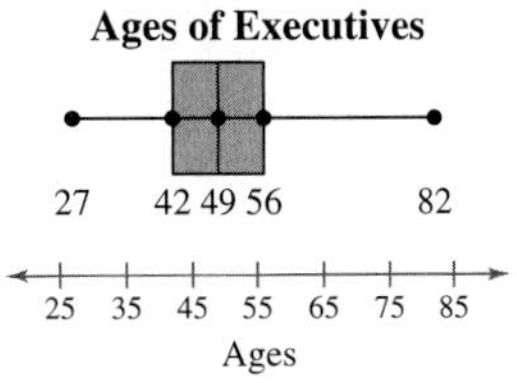

(c) Half of the ages are between 42 and 56 years.
(d) 49, because half of the executives are older and half are younger.

41. 33.75

43. 19.8

Uses and Abuses for Chapter 2 *(page 105)*

1. Answers will vary.

2. The salaries of employees at a business could contain an outlier.

The median is not affected by an outlier because the median does not take into account the outlier's numerical value.

Review Answers for Chapter 2 *(page 107)*

1.

Class	Mid-point	Boundaries	Frequency, f	Rel freq	Cum freq
20–23	21.5	19.5–23.5	1	0.05	1
24–27	25.5	23.5–27.5	2	0.10	3
28–31	29.5	27.5–31.5	6	0.30	9
32–35	33.5	31.5–35.5	7	0.35	16
36–39	37.5	35.5–39.5	4	0.20	20
			$\Sigma f = 20$	$\Sigma \frac{f}{n} = 1$	

3.

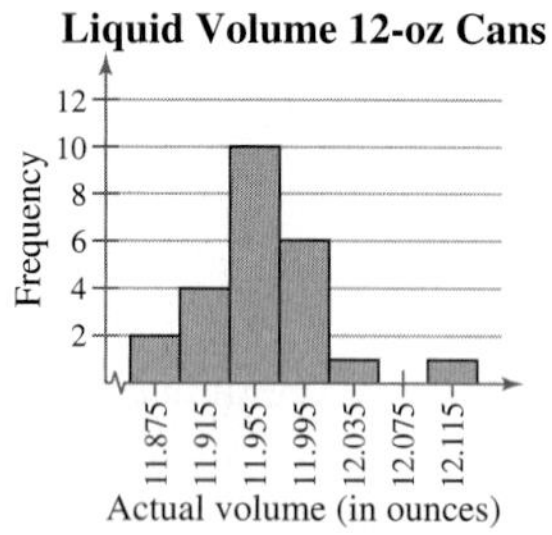

5.

Class	Midpoint	Frequency, f
79–93	86	9
94–108	101	12
109–123	116	5
124–138	131	3
139–153	146	2
154–168	161	1
		$\Sigma f = 32$

7.

1	3 7 8 9
2	0 1 2 3 3 3 4 4 5 5 5 7 8 8 9
3	1 1 2 3 4 5 7 8
4	3 4 7
5	1

9.

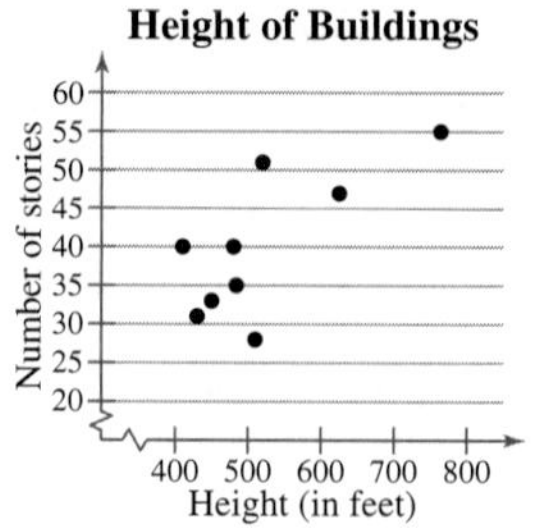

The number of stories appears to increase with height.

11.

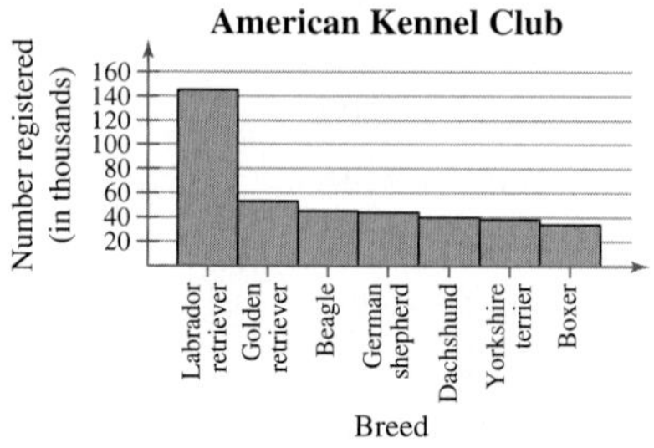

13. Mean = 8.6; Median = 9; Mode = 9

15. 31.7 **17.** 79.5 **19.** Skewed **21.** Skewed left

23. Median **25.** 2.8

27. Population mean = 9; Standard deviation ≈ 3.2

29. Sample mean = 2453.4; Standard deviation ≈ 306.1

31. Between \$21.50 and \$36.50 **33.** 30

35. Sample mean ≈ 2.5; Standard deviation ≈ 1.2

37. 56 **39.** 14 **41.** 4

43. 23% scored higher than 68.

45. $z = 2.33$, unusual **47.** $z = 1.25$, not unusual

Chapter Quiz for Chapter 2 *(page 111)*

1. (a)

Class	Mid-point	Class boundaries	Frequency, f	Rel freq	Cum freq
101–112	106.5	100.5–112.5	3	0.12	3
113–124	118.5	112.5–124.5	11	0.44	14
125–136	130.5	124.5–136.5	7	0.28	21
137–148	142.5	136.5–148.5	2	0.08	23
149–160	154.5	148.5–160.5	2	0.08	25

(b) Frequency histogram and polygon

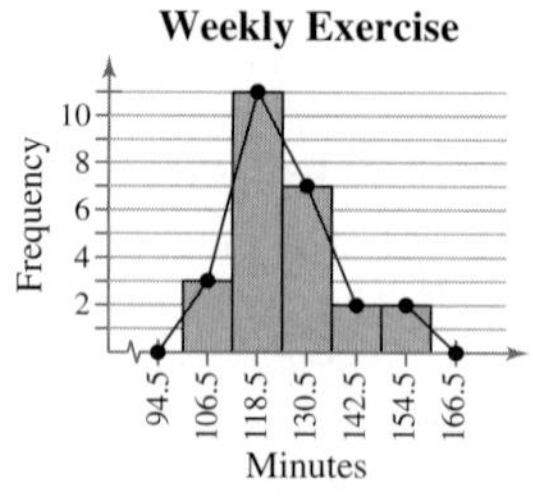

(c) Relative frequency histogram

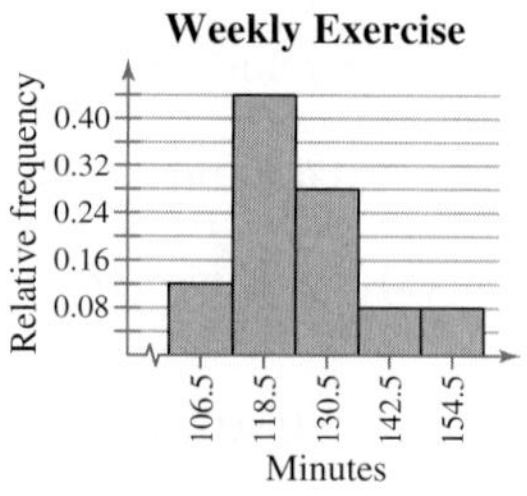

(d) Skewed

(e)

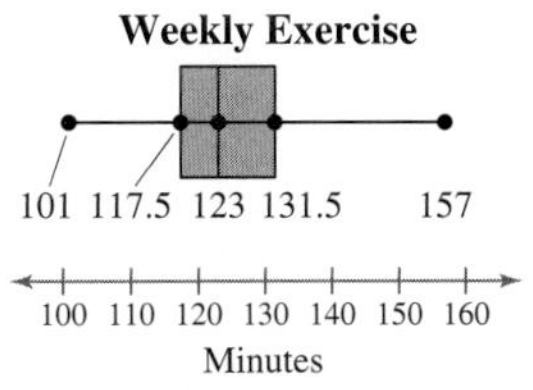

(f)

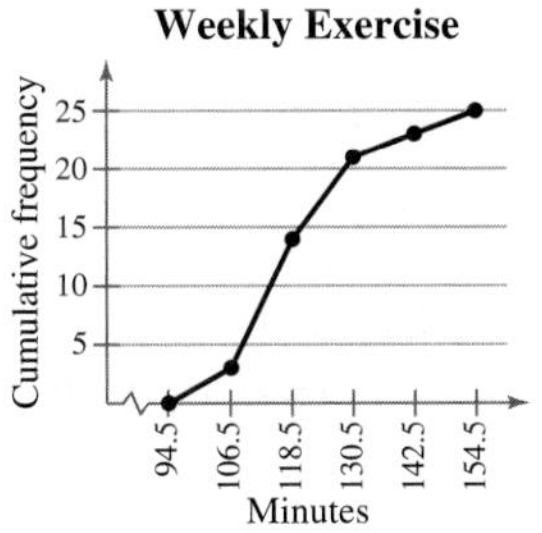

2. 125.2, 13.0

3. (a)

(b)

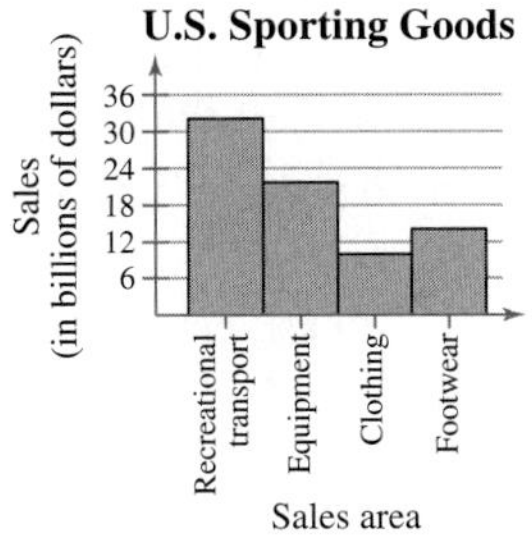

4. (a) 751.6, 784.5, none

The mean best describes a typical salary because there are no outliers.

(b) 575; 48,135.1; 219.4

5. Between \$125,000 and \$185,000

6. (a) $z = 3.0$, unusual

(b) $z \approx -6.67$, very unusual

(c) $z \approx 1.33$

(d) $z = -2.2$, unusual

7. (a) 71, 84.5, 90 (b) 19

(c)

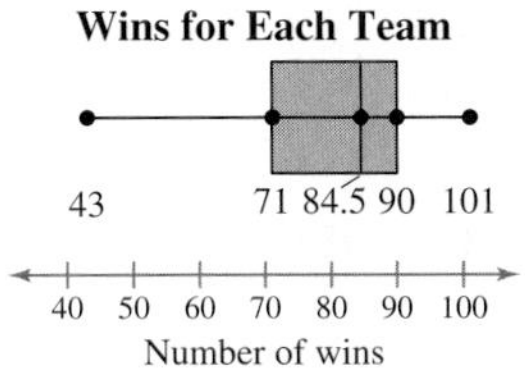

Real Statistics–Real Decisions for Chapter 2

(page 112)

1. (a) Find the average price of automobile insurance for each city and do a comparison.

(b) Find the mean, range, and population standard deviation for each city.

2. (a) Construct a Pareto chart because the data in use are quantitative and a Pareto chart positions data in order of decreasing height, with the tallest bar positioned at the left.

(b)

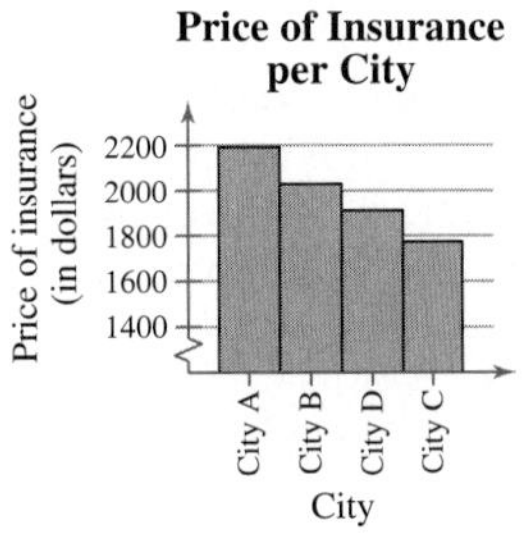

(c) Yes. From the Pareto chart you can see that City A has the highest average automobile insurance premium followed by City B, City D, and City C.

3. (a) Find the mean, range, and population standard deviation for each city.

(b)

City A	*City B*
$\bar{x} = \$2191.00$	$\bar{x} = \$2029.20$
$\sigma \approx \$351.86$	$\sigma \approx \$437.54$
range $= \$1015.00$	range $= \$1336.00$
City C	*City D*
$\bar{x} = \$1772.00$	$\bar{x} = \$1909.30$
$\sigma \approx \$418.52$	$\sigma \approx \$361.14$
range $= \$1347.00$	range $= \$1125.00$

(c) Yes. City A has the highest mean and lowest range and standard deviation.

4. (a) Tell your readers that on average, the price of automobile insurance premiums is higher in this city than in other cities.

(b) Location, weather, population

CHAPTER 3

Section 3.1 *(page 125)*

1. (a) Yes. The probability of an event occurring must be contained in the interval [0, 1] or [0%, 100%].

(b) No. The probability of an event occurring cannot be greater than 1.

(c) No. The probability of an event occurring cannot be less than 0.

(d) Yes. The probability of an event occurring must be contained in the interval [0, 1] or [0%, 100%].

(e) Yes. The probability of an event occurring must be contained in the interval [0, 1] or [0%, 100%].

3. {0, 1, 2, 3, 4, 5, 6, 7, 8, 9}

5. 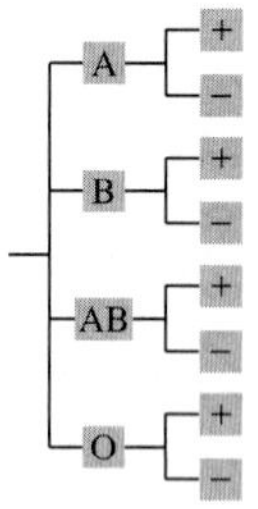

$\{(P, A), (P, B), (P, AB), (P, O), (N, A), (N, B), (N, AB), (N, O)\}$, where (P, A) represents positive Rh-factor with A-blood type and (N, A) represents negative Rh-factor with A-blood type.

7. Simple event because it is an event that consists of a single outcome.

9. Not a simple event because it is an event that consists of more than a single outcome.

11. Empirical probability because company records were used to calculate the frequency of a washing machine breaking down.

13. 0.159 **15.** 0.000953 **17.** 0.474 **19.** 0.056

21. 0.963 **23.** 0.090 **25.** 0.236

27. (a) 0.5 (b) 0.25 (c) 0.25

29. 0.761 **31.** 0.239

33. The probability of choosing a tea drinker who does not have a college degree.

35. (a) 0.55 (b) 0.175 (c) 0.013

37. (a)

Sum	Probability
2	0.028
3	0.056
4	0.083
5	0.111
6	0.139
7	0.167
8	0.139
9	0.111
10	0.083
11	0.056
12	0.028

(b) Answers will vary.
(c) Answers will vary.

39. 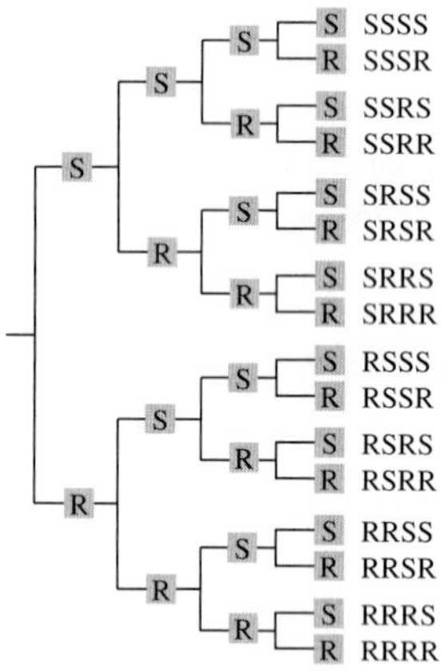

41. No. The odds of winning a prize are 1 : 6 (one winning cap and six losing caps). So, the statement should read, "one in seven game pieces wins a prize."

43. 13 : 39 = 1 : 3

Section 3.2 *(page 134)*

1. Two events are independent if the occurrence of one of the events does not affect the probability of the occurrence of the other event.
If $P(B|A) = P(B)$ or $P(A|B) = P(A)$, then events A and B are independent.

3. False. If two events are independent, then $P(A|B) = P(A)$.

5. Independent. The outcome of the 1st card drawn does not affect the outcome of the 2nd card drawn.

7. Dependent. The outcome of the 1st ball drawn affects the outcome of the 2nd ball drawn.

9. (a) 0.8 (b) 0.003
(c) Dependent. The probability that a woman develops breast cancer depends on whether she carries the gene mutation.

11. (a) 0.807 (b) 0.315
(c) Dependent. The probability that the family takes a summer vacation depends on whether they own a computer.

13. (a) 0.017 (b) 0.93
(c) Yes, this would be unusual because the probability of a pregnancy and a multiple birth is only 1.7%.

15. 0.291

17. (a) 0.109 (b) 0.382 (c) 0.618

19. (a) 0.481 (b) 0.465 (c) 0.449
(d) Dependent. Whether a person has at least one month's income saved depends on whether the person is male.

21. (a) 0.0000000243 (b) 0.859 (c) 0.141

23. (a) 0.2 (b) 0.04 (c) 0.008 (d) 0.512 (e) 0.488

25. 0.011 **27.** 0.444 **29.** 0.167

31. (a) 0.074 (b) 0.999 **33.** 0.954

Section 3.3 *(page 145)*

1. $P(A \text{ and } B) = 0$ because A and B cannot occur at the same time.

3. True

5. False. The probability that event A or event B will occur is $P(A \text{ or } B) = P(A) + P(B) - P(A \text{ and } B)$.

7. Not mutually exclusive. The student can be taking both economics and statistics.

9. Not mutually exclusive. The worker can be female and have a college degree.

11. Mutually exclusive. The person cannot be in both age classes.

13. (a) Not mutually exclusive. For five weeks the events overlap.
(b) 0.423

15. (a) Not mutually exclusive. A carton can have a puncture and a smashed corner.
(b) 0.126

17. (a) 0.308 (b) 0.538 (c) 0.308

19. (a) 0.067 (b) 0.868 (c) 0.231

21. (a) 0.110 (b) 0.720

23. (a) 0.014 (b) 0.774 (c) 0.226

25. Answers will vary. 27. 0.54

Section 3.4 *(page 157)*

1. You are counting the number of ways two or more events can occur in sequence.

3. False. A permutation is an ordered arrangement of objects.

5. 5040 7. 35

9. Permutation. The order of the 15 people in line matters.

11. 6240 13. 4500 15. 40,320 17. 3,628,800

19. 1320 21. 24,024 23. 9,189,180

25. 5,586,853,480 27. 84

29. (a) 70 (b) 16 (c) 0.086

31. (a) 67,600,000 (b) 19,656,000 (c) 0.000000015

33. (a) 120 (b) 12 (c) 12 (d) 0.4

35. 0.000012 37. 6.00×10^{-20}

39. (a) 658,008 (b) 0.00000152 41. 1001; 1000

43.

Team (worst team first)	1	2	3	4	5	6	7
Probability	0.25	0.20	0.157	0.12	0.089	0.064	0.044

Team (worst team first)	8	9	10	11	12	13
Probability	0.029	0.018	0.011	0.007	0.006	0.005

45. 0.566

Uses and Abuses for Chapter 3 *(page 161)*

1. (a) 0.000001 (b) 0.001 (c) 0.001

2. Answers will vary.

Review Answers for Chapter 3 *(page 163)*

1. Sample space:
{HHHH, HHHT, HHTH, HHTT, HTHH, HTHT, HTTH, HTTT, THHH, THHT, THTH, THTT, TTHH, TTHT, TTTH, TTTT}
Event: Getting three heads
{HHHT, HHTH, HTHH, THHH}

3. Sample space:
$\{0, 1, 2, 3, 4, 5, 6, 7, 8, 9\}$
Event: Choosing a number less than 2
$\{0, 1\}$

5. Empirical probability 7. Subjective probability

9. Classical probability

11. 0.72 13. 0.92 15. Independent 17. 0.025

19. Mutually exclusive 21. 0.60 23. 0.538

25. 0.583 27. 0.463 29. 0.23 31. 144 33. 2730

35. 2380 37. 0.00000923

39. (a) 0.955 (b) 0.000000761 (c) 0.045
(d) 0.999999239

Chapter Quiz for Chapter 3 *(page 167)*

1. (a) 0.536 (b) 0.527 (c) 0.550 (d) 0.792
(e) 0.025 (f) 0.672 (g) 0.094 (h) 0.582

2. Not mutually exclusive. A golfer can score the best round in a four-round tournament and still lose the tournament.
Dependent. One event can affect the occurrence of the second event.

3. (a) 518,665 (b) 1 (c) 551,299

4. (a) 0.941 (b) 0.00000181 (c) 0.999998

5. 4500 6. 303,600

Real Statistics–Real Decisions for Chapter 3
(page 168)

1. (a) Answers will vary.
(b) Use the Multiplication Rule, Fundamental Counting Principle, and Combinations.

2. If you played only the red ball, the probability of matching it is $\frac{1}{42}$. However, because you must pick 5 white balls, you must get the white balls wrong. So, using the Multiplication Rule, you get
P(matching only the red ball and not matching any of the 5 white balls)
$$= \frac{1}{42} \cdot \frac{48}{53} \cdot \frac{47}{52} \cdot \frac{46}{51} \cdot \frac{45}{50} \cdot \frac{44}{49}$$
$$\approx 0.014$$
$$\approx \frac{1}{70}.$$

3. The overall probability of winning a prize is determined by calculating the number of ways to win and dividing by the total number of outcomes.
To calculate the number of ways to win something, you must use combinations.

CHAPTER 4

Section 4.1 *(page 179)*

1. A random variable represents a numerical value assigned to an outcome of a probability experiment.
Examples: Answers will vary.

3. False. In most applications, discrete random variables represent counted data, whereas continuous random variables represent measured data.

5. True

7. Discrete because home attendance is a random variable that is countable.

9. Continuous because annual vehicle-miles driven is a random variable that cannot be counted.

11. Discrete because number of highway fatalities in one year in Texas is a random variable that is countable.

13. Continuous because volume of blood drawn for a blood test is a random variable that cannot be counted.

15. Discrete because number of books sold per quarter at a book store is a random variable that is countable.

17. Continuous because amount of snow that fell in Alaska last winter is a random variable that cannot be counted.

19. (a) 0.35 (b) 0.90 **21.** 0.22 **23.** Yes

25. No, $\Sigma P(x) = 0.95$ and $P(5) < 0$.

27. (a)

x	f	$P(x)$	$xP(x)$
0	1491	0.686	0
1	425	0.195	0.195
2	168	0.077	0.154
3	48	0.022	0.066
4	29	0.013	0.052
5	14	0.006	0.030
	$n = 2175$	$\Sigma P(x) \approx 1$	$\Sigma xP(x) = 0.497$

x	$(x - \mu)^2 P(x)$
0	0.169
1	0.049
2	0.173
3	0.138
4	0.160
5	0.122
	$\Sigma(x - \mu)^2 P(x) = 0.811$

(b) 0.5 (c) 0.8 (d) 0.9

(e) A household on average has 0.5 dogs with a standard deviation of 0.9 dogs.

29. (a)

x	f	$P(x)$	$xP(x)$
0	300	0.432	0
1	280	0.403	0.403
2	95	0.137	0.274
3	20	0.029	0.087
	$n = 695$	$\Sigma P(x) \approx 1$	$\Sigma xP(x) = 0.764$

x	$(x - \mu)^2 P(x)$
0	0.252
1	0.022
2	0.209
3	0.145
	$\Sigma(x - \mu)^2 P(x) = 0.629$

(b) 0.8 (c) 0.6 (d) 0.8

(e) A household on average has 0.8 computers with a standard deviation of 0.8 computers.

31. (a)

x	f	$P(x)$	$xP(x)$
0	6	0.031	0
1	12	0.063	0.063
2	29	0.151	0.302
3	57	0.297	0.891
4	42	0.219	0.876
5	30	0.156	0.780
6	16	0.083	0.498
	$n = 192$	$\Sigma P(x) = 1$	$\Sigma xP(x) = 3.410$

x	$(x - \mu)^2 P(x)$
0	0.360
1	0.366
2	0.300
3	0.050
4	0.076
5	0.394
6	0.557
	$\Sigma(x - \mu)^2 P(x) = 2.103$

(b) 3.4 (c) 2.1 (d) 1.4

(e) An employee works an average of 3.4 overtime hours per week with a standard deviation of 1.4 hours.

33. (a) 5.3 (b) 3.3 (c) 1.8 (d) 5.3

(e) The expected number of questions answered correctly is 5.3 with a standard deviation of 1.8.

35. (a) 2.1 (b) 1.1 (c) 1.1 (d) 2.1

(e) The average number of hurricanes that hit the U.S. is 2.1 with a standard deviation of 1.1.

37. (a) 2.6 (b) 1.9 (c) 1.4 (d) 2.6

(e) The average household size is 2.6 with a standard deviation of 1.4.

39. (a) 0.881 (b) 0.119 (c) 0.112

41. A household with three dogs is unusual because the probability of this event is 0.022.

43. −\$0.05

Section 4.2 *(page 193)*

1. (a) $p = 0.50$ (b) $p = 0.20$ (c) $p = 0.80$

3. (a) $n = 12$ (b) $n = 4$ (c) $n = 8$

As n increases, the distribution becomes more symmetric.

5. Binomial experiment

Success: baby recovers

$n = 5, p = 0.80, q = 0.20, x = 0, 1, 2, \ldots, 5$

7. Not a binomial experiment because there are more than two possible outcomes for each trial.

9. 40; 24; 4.9 **11.** 22.08; 18.547; 4.307

13. (a) 0.088 (b) 0.104 (c) 0.896

15. (a) 0.069 (b) 0.089 (c) 0.911

17. (a) 0.213 (b) 0.139 (c) 0.861

19. (a) $n = 6, p = 0.36$ (b)

x	$P(x)$
0	0.069
1	0.232
2	0.326
3	0.245
4	0.103
5	0.023
6	0.002

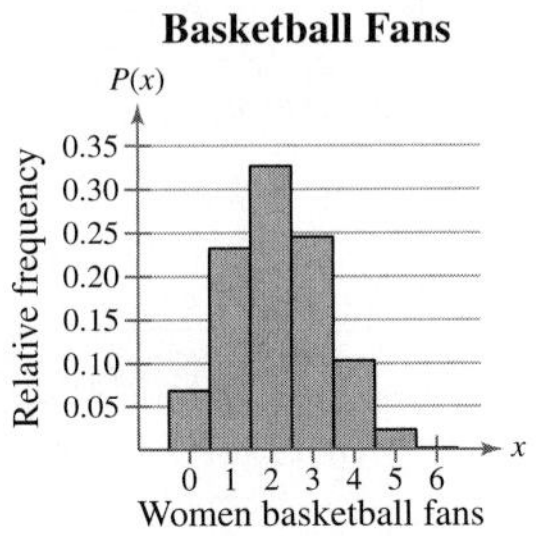

(c) Symmetric (d) 2.2 (e) 1.4 (f) 1.2

(g) On average 2.2, out of 6, women would consider themselves basketball fans. The standard deviation is 1.2 women.

$x = 0$, 5, or 6 would be uncommon because of their low probabilities.

21. (a) $n = 4, p = 0.05$ (b)

x	$P(x)$
0	0.814506
1	0.171475
2	0.013537
3	0.000475
4	0.000006

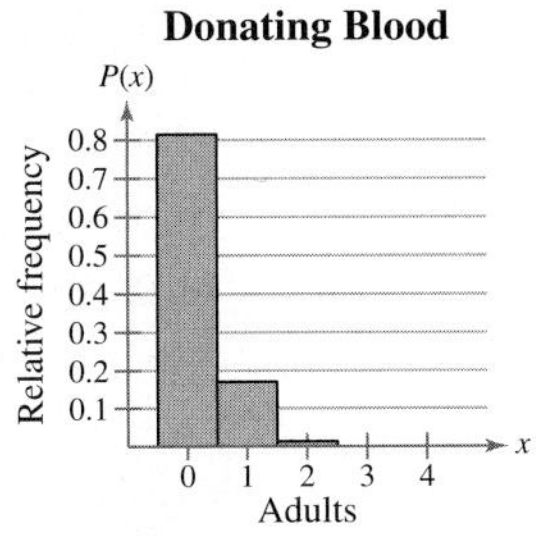

(c) Skewed right (d) 0.2 (e) 0.2 (f) 0.4

(g) On average 0.2 eligible adults, out of every 4, give blood. The standard deviation is 0.4 adults.

$x = 2$, 3, or 4 would be uncommon because of their low probabilities.

23. (a) $n = 6, p = 0.61$ (b) 0.129 (c) 0.249

x	$P(x)$
0	0.004
1	0.033
2	0.129
3	0.269
4	0.316
5	0.198
6	0.052

25. 3.7; 1.2

If six drivers are randomly selected, on average 3.7 drivers will name talking on cell phones as the most annoying habit of other drivers.

None of the six randomly selected drivers naming talking on cell phones as the most annoying habit of other drivers is unusual because $P(0) = 0.004$.

27. 0.033

Section 4.3 *(page 202)*

1. 0.032 **3.** 0.056 **5.** 0.09 **7.** 0.251

9. Geometric. You are interested in counting the number of trials until the first success.

11. Poisson. You are interested in counting the number of occurrences that take place within a given unit of space.

13. Binomial. You are interested in counting the number of successes out of n trials.

15. (a) 0.082 (b) 0.469 (c) 0.531

17. (a) 0.009 (b) 0.030 (c) 0.904

19. (a) 0.057 (b) 0.958 (c) 0.900

21. (a) 0.348 (b) 0.844 (c) 0.156

23. (a) 0.1453996179

(b) 0.1453688667; The results are approximately the same.

25. (a) 1000; 999,000; 999.5

On average you would have to play 1000 times until you won the lottery. The standard deviation is 999.5.

(b) 1000 times

Lose money. On average you would win \$500 every 1000 times you play the lottery. So, the net gain would be −\$500.

27. (a) $\sigma^2 = 3.9$, $\sigma \approx 2.0$

The standard deviation is 2.0 strokes.

(b) 0.352

Uses and Abuses for Chapter 4 *(page 205)*

1. 40, 0.081

2. 0.739; Answers will vary.

3. The probability of finding 36 adults out of 100 who prefer Brand A is 0.059. So, the manufacturer's claim is hard to believe.

4. The probability of finding 25 adults out of 100 who prefer Brand A is 0.000627. So, the manufacturer's claim is not believable.

Review Answers for Chapter 4 *(page 207)*

1. Discrete **3.** Continuous

5. No, $\Sigma P(x) \neq 1$. **7.** Yes

9. Yes

11. (a)

x	f	$P(x)$
2	3	0.005
3	12	0.018
4	72	0.111
5	115	0.177
6	169	0.260
7	120	0.185
8	83	0.128
9	48	0.074
10	22	0.034
11	6	0.009
	$n = 650$	$\Sigma P(x) \approx 1$

(b)

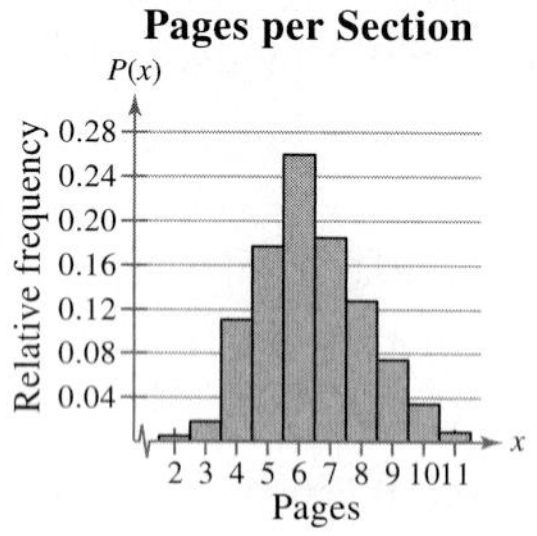

(c) 6.4, 2.9, 1.7

13. (a)

x	f	$P(x)$
0	3	0.015
1	38	0.190
2	83	0.415
3	52	0.260
4	18	0.090
5	5	0.025
6	1	0.005
	$n = 200$	$\Sigma P(x) = 1$

(b)

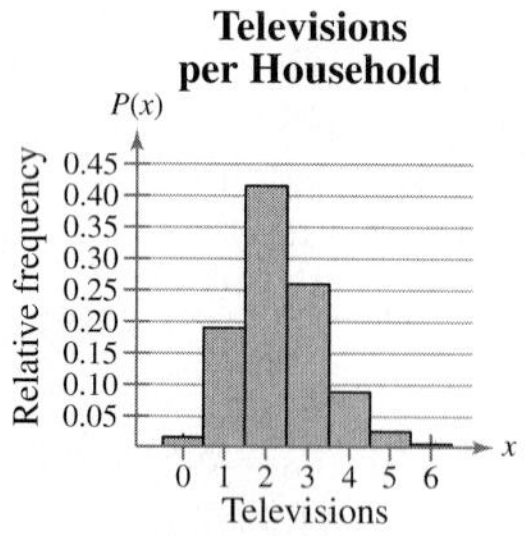

(c) 2.3, 1.1, 1.0

15. 3.4

17. Yes, $n = 12$, $p = 0.24$, $q = 0.76$, $x = 0, 1, \ldots, 12$

19. (a) 0.208 (b) 0.322 (c) 0.114

21. (a) 0.294 (b) 0.644 (c) 0.350

23. (a)

x	$P(x)$
0	0.007
1	0.059
2	0.201
3	0.342
4	0.291
5	0.099

(b)

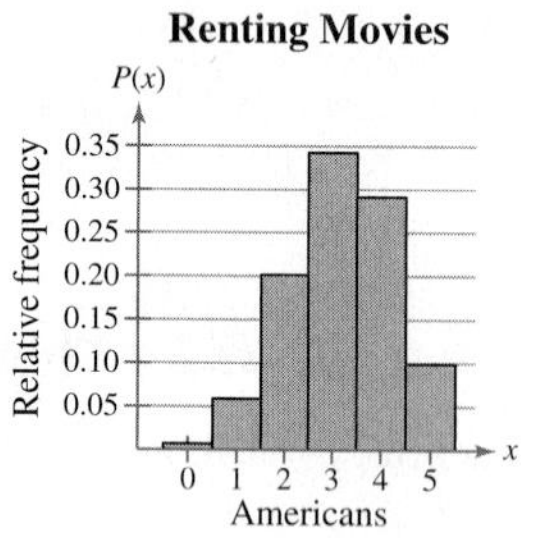

(c) 3.2, 1.2, 1.1

25. (a)

x	$P(x)$
0	0.012
1	0.096
2	0.293
3	0.397
4	0.202

(b)

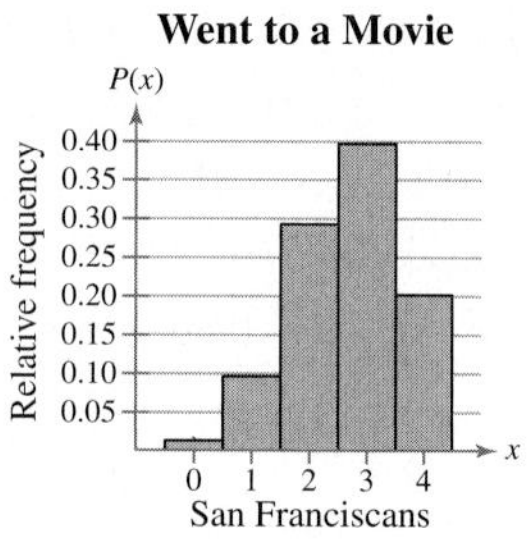

(c) 2.7, 0.9, 0.9

27. (a) 0.096 (b) 0.518 (c) 0.579

29. (a) 0.810 (b) 0.171 (c) 0.019

Chapter Quiz for Chapter 4 *(page 211)*

1. (a) Discrete because the random variable is countable.

(b) Continuous because the random variable has an infinite number of possible outcomes and cannot be counted.

2. (a)

x	f	$P(x)$
1	61	0.370
2	39	0.236
3	48	0.291
4	14	0.085
5	3	0.018
	$n = 165$	$\Sigma P(x) = 1$

(b)

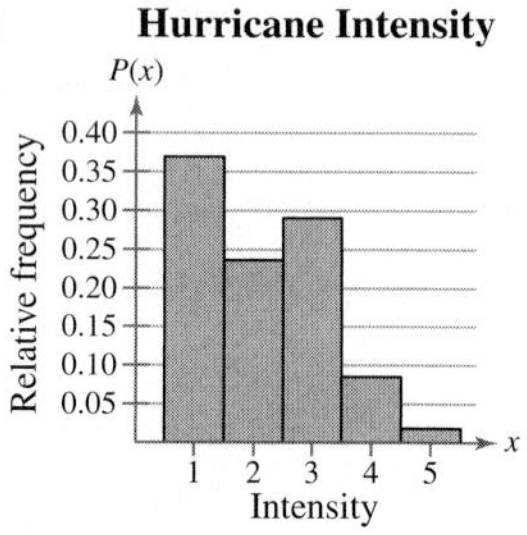

(c) 2.1, 1.1, 1.1

On average, the intensity of a hurricane will be 2.1 with a standard deviation of 1.1.

(d) 0.103

3. (a) $n = 8$, $p = 0.80$

x	$P(x)$
0	0.000003
1	0.000082
2	0.001147
3	0.009175
4	0.045875
5	0.146801
6	0.293601
7	0.335544
8	0.167772

(b)

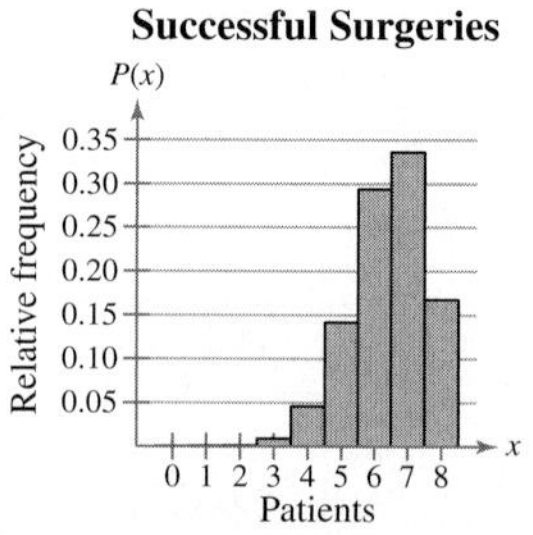

(c) 6.4, 1.3, 1.1 (d) 0.001 (e) 0.000085

4. (a) 0.176 (b) 0.440 (c) 0.007

Real Statistics–Real Decisions for Chapter 4

(page 212)

1. (a) Answers will vary. For example, calculate the probability of obtaining zero clinical pregnancies out of 10 randomly selected ART cycles.

(b) Binomial. The distribution is discrete because the number of clinical pregnancies is countable.

2. $n = 10$, $p = 0.328$, $P(0) = 0.0188$

x	P(x)
0	0.0188
1	0.0917
2	0.2013
3	0.2621
4	0.2238
5	0.1311
6	0.0533
7	0.0149
8	0.0027
9	0.0003
10	0.00001

3. (a) Suspicious, because the probability is very small.

(b) Not suspicious, because the probability is not that small.

CHAPTER 5

Section 5.1 *(page 224)*

1. Answers will vary.

3. Answers will vary.

Similarities: The two curves will have the same line of symmetry.

Differences: One curve will be more spread out than the other.

5. $\mu = 0, \sigma = 1$

7. "The" standard normal distribution is used to describe one specific normal distribution ($\mu = 0, \sigma = 1$). "A" normal distribution is used to describe a normal distribution with any mean and standard deviation.

9. No, the graph crosses the x-axis.

11. Yes, the graph fulfills the properties of the normal distribution.

13. No, the graph is skewed right.

15. It is normal because it is bell shaped and symmetric.

17. 0.3849 **19.** 0.6247 **21.** 0.9382 **23.** 0.975

25. 0.8289 **27.** 0.1003 **29.** 0.005 **31.** 0.05

33. 0.475 **35.** 0.437 **37.** 0.95 **39.** 0.2006

41. (a)

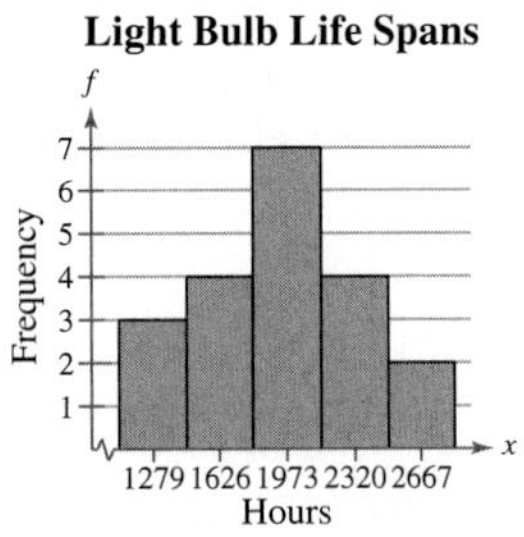

It is reasonable to assume that the life span is normally distributed because the histogram is nearly symmetric and bell shaped.

(b) 1941.35, 432.385

(c) The sample mean of 1941.35 hours is less than the claimed mean, so, on average, the bulbs in the sample lasted for a shorter time. The sample standard deviation of 432 hours is greater than the claimed standard deviation, so the bulbs in the sample had a greater variation in life span than the manufacturer's claim.

43. (a) A = 2.97; B = 2.98; C = 3.01; D = 3.05

(b) 0.5; −1.5; −1; 2.5

(c) $x = 3.05$ is unusual owing to a relatively large z-score (2.5).

45. (a) A = 801; B = 950; C = 1250; D = 1467

(b) −0.36; 1.07; 2.11; −1.08

(c) $x = 1467$ is unusual owing to a relatively large z-score (2.11).

47. 0.6915 **49.** 0.05 **51.** 0.5328 **53.** 0.9265

55. 0.9744 **57.** 0.3133 **59.** 0.901 **61.** 0.0098

63.

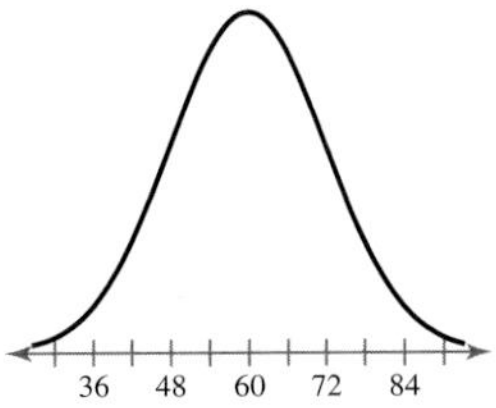

The normal distribution curve is centered at its mean (60) and has 2 points of inflection (48 and 72) representing $\mu \pm \sigma$.

65. (a) Area under curve = area of rectangle = (1)(1) = 1

(b) 0.25 (c) 0.4

Section 5.2 *(page 232)*

1. 0.1151 **3.** 0.1151 **5.** 0.1144

7. 0.3022 **9.** 0.2742 **11.** 0.0566

13. (a) 0.1357 (b) 0.6983 (c) 0.1660

15. (a) 0.1711 (b) 0.7018 (c) 0.1271

17. (a) 0.0062 (b) 0.9876 (c) 0.0062

19. (a) 0.0073 (b) 0.806 (c) 0.1867

21. (a) 79.95% (b) 348 **23.** (a) 64.8% (b) 18

25. (a) 30.85% (b) 31 **27.** (a) 99.87% (b) 0.798

29. 1.5%; It is unusual for a battery to have a life span that is more than 2065 hours because of the relatively large z-score (2.17).

31. Out of control, because there is a point more than 3 standard deviations beyond the mean.

33. Out of control, because there are nine consecutive points below the mean, and two out of three consecutive points lie more than 2 standard deviations from the mean.

Section 5.3 *(page 242)*

1. −2.05 **3.** 0.85 **5.** −0.16 **7.** 2.39

9. −1.645 **11.** 0.995 **13.** −2.33 **15.** −0.25

17. 1.175 **19.** −0.67 **21.** 0.67 **23.** −0.39

25. −0.38 **27.** −0.58 **29.** −1.645, 1.645

31. 0.325 **33.** −0.33 **35.** 1.28 **37.** ±0.06

39. (a) 68.52 (b) 62.14 **41.** (a) 12.28 (b) 20.08

43. (a) 139.22 (b) 96.92 **45.** 19.89

47. Tires that wear out by 26,800 miles will be replaced free of charge.

49. 7.93

Section 5.4 *(page 254)*

1. 100, 2.12 **3.** 100, 0.949

5. False. As the size of a sample increases, the mean of the distribution of sample means does not change.

7. False. The shape of a sampling distribution is normal if either $n \geq 30$ or the shape of the population is normal.

9.

Sample	Mean	Sample	Mean	Sample	Mean
0, 0, 0	0	2, 2, 4	2.67	4, 8, 0	4
0, 0, 2	0.67	2, 2, 8	4	4, 8, 2	4.67
0, 0, 4	1.33	2, 4, 0	2	4, 8, 4	5.33
0, 0, 8	2.67	2, 4, 2	2.67	4, 8, 8	6.67
0, 2, 0	0.67	2, 4, 4	3.33	8, 0, 0	2.67
0, 2, 2	1.33	2, 4, 8	4.67	8, 0, 2	3.33
0, 2, 4	2	2, 8, 0	3.33	8, 0, 4	4
0, 2, 8	3.33	2, 8, 2	4	8, 0, 8	5.33
0, 4, 0	1.33	2, 8, 4	4.67	8, 2, 0	3.33
0, 4, 2	2	2, 8, 8	6	8, 2, 2	4
0, 4, 4	2.67	4, 0, 0	1.33	8, 2, 4	4.67
0, 4, 8	4	4, 0, 2	2	8, 2, 8	6
0, 8, 0	2.67	4, 0, 4	2.67	8, 4, 0	4
0, 8, 2	3.33	4, 0, 8	4	8, 4, 2	4.67
0, 8, 4	4	4, 2, 0	2	8, 4, 4	5.33
0, 8, 8	5.33	4, 2, 2	2.67	8, 4, 8	6.67
2, 0, 0	0.67	4, 2, 4	3.33	8, 8, 0	5.33
2, 0, 2	1.33	4, 2, 8	4.67	8, 8, 2	6
2, 0, 4	2	4, 4, 0	2.67	8, 8, 4	6.67
2, 0, 8	3.33	4, 4, 2	3.33	8, 8, 8	8
2, 2, 0	1.33	4, 4, 4	4		
2, 2, 2	2	4, 4, 8	5.33		

$\mu_{\bar{x}} = 3.5, \sigma_{\bar{x}} = 1.708$

$\mu = 3.5, \sigma = 2.958$

11. (c), because $\mu = 16.5$, $\sigma = 1.19$, and the graph approximates a normal curve.

13. 87.5, 1.804

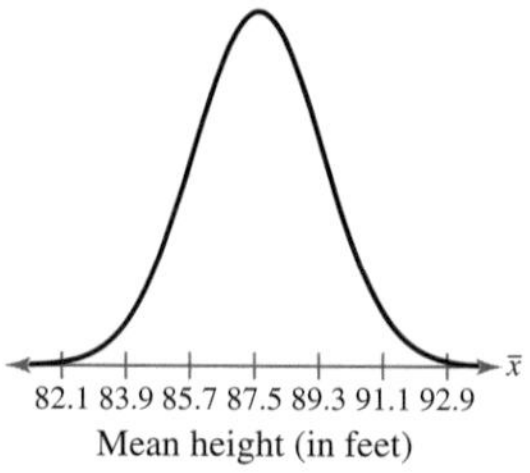

15. 349, 1.26

17. 113.5, 8.61

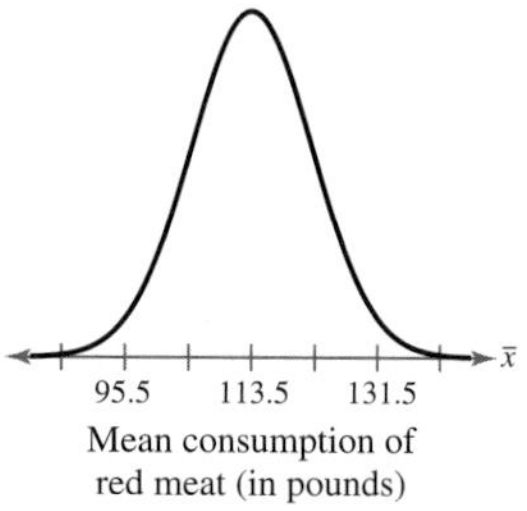

19. 87.5, 1.276; 87.5, 1.042

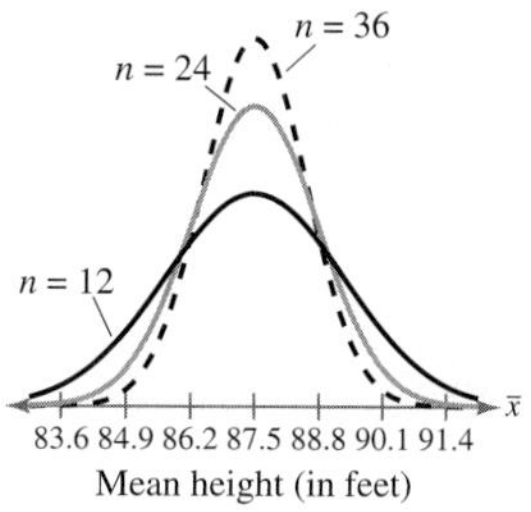

As the sample size increases, the standard error decreases, while the mean of the sample means remains constant.

21. 0.0019 **23.** 0.6319 **25.** ≈ 0

27. It is more likely to select a sample of 20 women with a mean height less than 70 inches because the sample of 20 has a higher probability.

29. Yes, it is very unlikely that you would have randomly sampled 40 cans with a mean equal to 127.9 ounces.

31. (a) 0.0008 (b) Claim is inaccurate.

(c) No, assuming the manufacturer's claim is true, because 96.25 is within 1 standard deviation of the mean for an individual board.

33. (a) 0.0002 (b) Claim is inaccurate.

(c) No, assuming the manufacturer's claim is true, because 49,721 is within 1 standard deviation of the mean for an individual tire.

35. Yes, because of the relatively large z-score (2.12).

37. ≈ 0

Section 5.5 *(page 265)*

1. Cannot use normal distribution.

3. Can use normal distribution.

5. Cannot use normal distribution because $np < 5$.

7. Cannot use normal distribution because $nq < 5$.

9. d **10.** b **11.** a **12.** c **13.** a **14.** d

15. c **16.** b **17.** Binomial: 0.549; Normal: 0.5463

19. Cannot use normal distribution because $np < 5$.

(a) 0.0000199 (b) 0.000023

(c) 0.999977 (d) 0.1635

21. Can use normal distribution.

(a) 0.1174

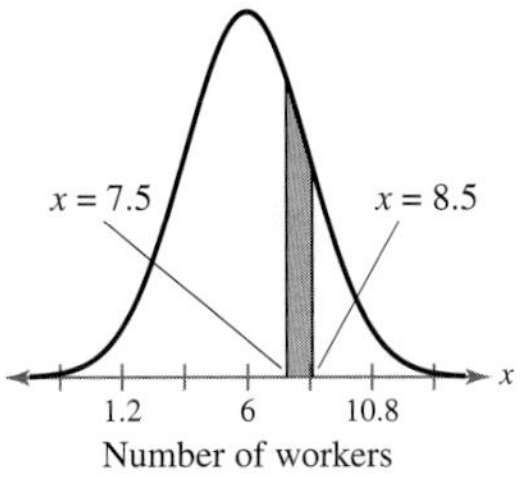

(b) 0.2643

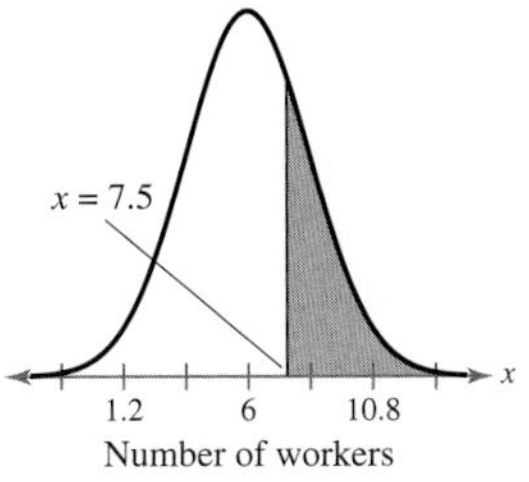

(c) 0.7357

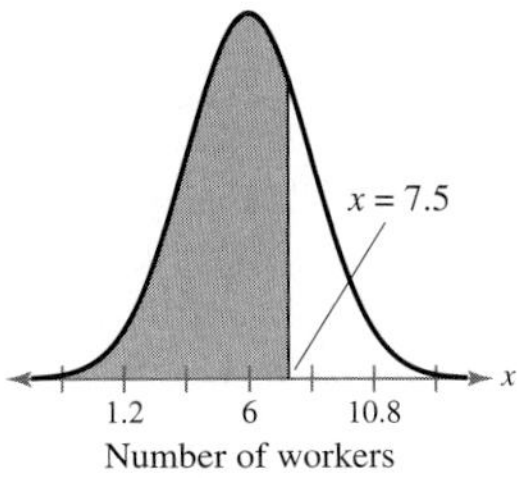

(d) 0.7190

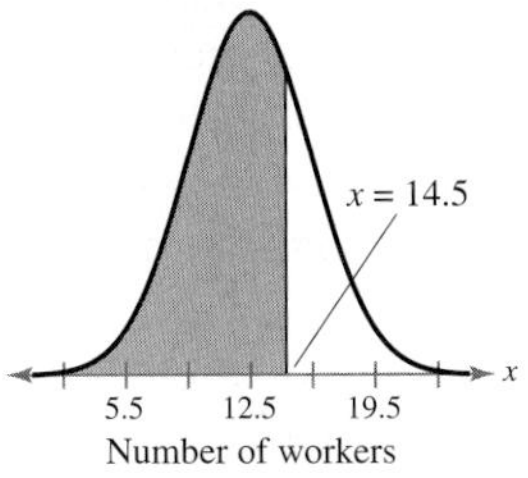

23. Can use normal distribution.

(a) 0.0465

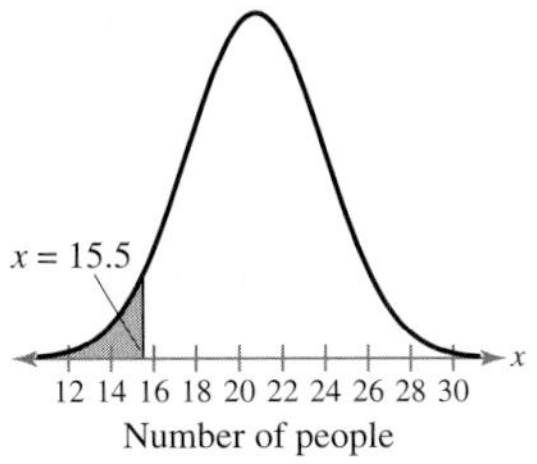

(b) 0.9767

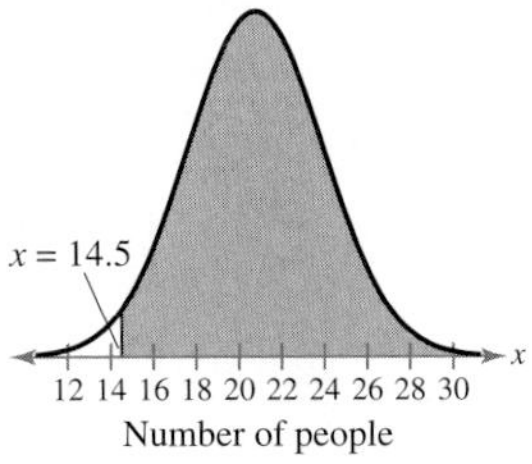

(c) 0.9535

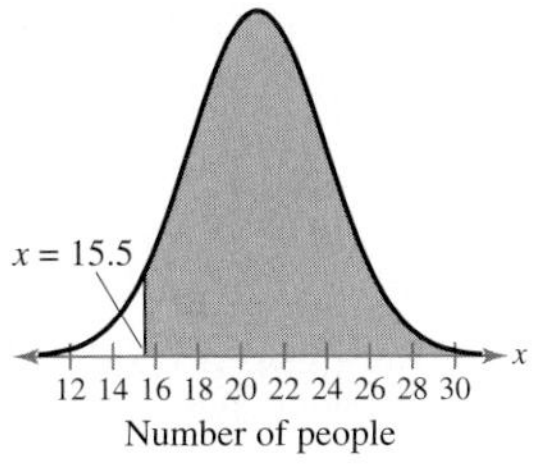

(d) 0.1635

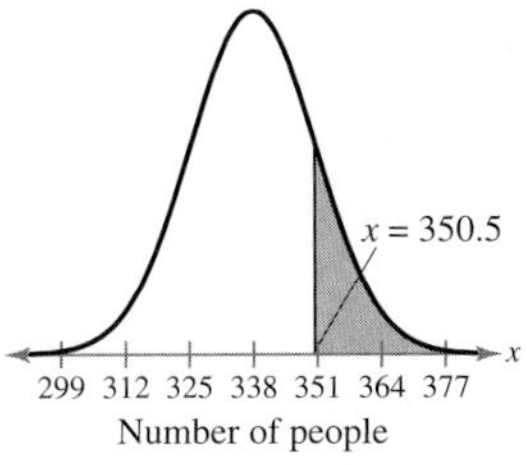

25. (a) $np = 6 \geq 5$

$nq = 19 \geq 5$

(b) 0.121

(c) No, because the z-score is within one standard deviation of the mean.

27. Highly unlikely. Answers will vary.

29. 0.1020

Uses and Abuses for Chapter 5 *(page 269)*

1. No. Answers will vary.

2. It is more likely that all 10 people lie within 2 standard deviations of the mean. This can be shown by using the Empirical Rule and the Multiplication Rule.

(a) By the Empirical Rule, the probability of lying within 2 standard deviations of the mean is 0.95. Let x = number of people selected who lie within 2 standard deviations of the mean.

$P(x = 10) = (0.95)^{10} \approx 0.599$

(b) P (at least one person does not lie within 2 standard deviations of the mean) $= 1 - P(x = 10) \approx 1 - 0.599 = 0.401$

Review Answers for Chapter 5 *(page 271)*

1. $\mu = 15, \sigma = 3$ **3.** $-2.25; 0.5; 2; 3.5$ **5.** 0.2005

7. 0.3936 **9.** 0.0465 **11.** 0.4495 **13.** 0.3519

15. 0.1336 **17.** 0.8997 **19.** 0.9236 **21.** 0.0124

23. (a) 0.3156 (b) 0.3099 (c) 0.3446

25. -0.07 **27.** 1.13 **29.** 1.04 **31.** 43.9 meters

33. 45.9 meters **35.** 45.74 meters

37. {0 0 0, 0 0 200, 0 0 40, 0 0 600, 0 0 80, 0 200 0, 0 200 200, 0 200 40, 0 200 600, 0 200 80, 0 40 0, 0 40 200, 0 40 40, 0 40 600, 0 40 80, 0 600 0, 0 600 200, 0 600 40, 0 600 600, 0 600 80, 0 80 0, 0 80 200, 0 80 40, 0 80 600, 0 80 80, 200 0 0, 200 0 200, 200 0 40, 200 0 600, 200 0 80, 200 200 0, 200 200 200, 200 200 40, 200 200 600, 200 200 80, 200 40 0, 200 40 200, 200 40 40, 200 40 600, 200 40 80, 200 600 0, 200 600 200, 200 600 40, 200 600 600, 200 600 80, 200 80 0, 200 80 200, 200 80 40, 200 80 600, 200 80 80, 40 0 0, 40 0 200, 40 0 40, 40 0 600, 40 0 80, 40 200 0, 40 200 200, 40 200 40, 40 200 600, 40 200 80, 40 40 0, 40 40 200, 40 40 40, 40 40 600, 40 40 80, 40 600 0, 40 600 200, 40 600 40, 40 600 600, 40 600 80, 40 80 0, 40 80 200, 40 80 40, 40 80 600, 40 80 80, 600 0 0, 600 0 200, 600 0 40, 600 0 600, 600 0 80, 600 200 0, 600 200 200, 600 200 40, 600 200 600, 600 200 80, 600 40 0, 600 40 200, 600 40 40, 600 40 600, 600 40 80, 600 600 0, 600 600 200, 600 600 40, 600 600 600, 600 600 80, 600 80 0, 600 80 200, 600 80 40, 600 80 600, 600 80 80, 80 0 0, 80 0 200, 80 0 40, 80 0 600, 80 0 80, 80 200 0, 80 200 200, 80 200 40, 80 200 600, 80 200 80, 80 40 0, 80 40 200, 80 40 40, 80 40 600, 80 40 80, 80 600 0, 80 600 200, 80 600 40, 80 600 600, 80 600 80, 80 80 0, 80 80 200, 80 80 40, 80 80 600, 80 80 80}

$\mu = 184, \sigma \approx 218.504$

$\mu_{\bar{x}} = 184, \sigma_{\bar{x}} \approx 126.153$

39. 152.7, 8.7

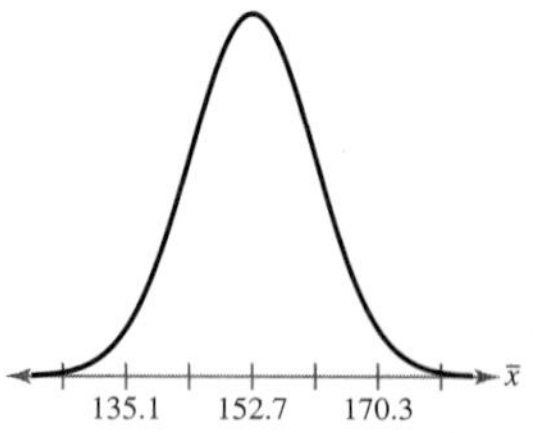

41. (a) 0.0485 (b) 0.8180 (c) 0.0823

(a) and (c) are smaller, (b) is larger. This is to be expected because the standard error of the sample means is smaller.

43. (a) ≈ 0 (b) ≈ 0 **45.** 0.0019

47. Cannot use normal distribution because $nq < 5$.

49. $P(x > 24.5)$ **51.** $P(44.5 < x < 45.5)$

53. Can use normal distribution.
0.0032

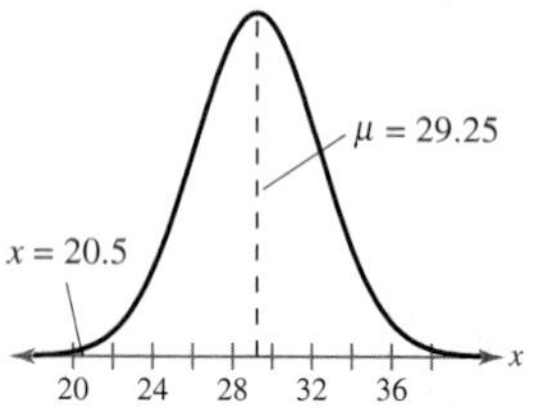

Chapter Quiz for Chapter 5 *(page 275)*

1. (a) 0.9821 (b) 0.9994 (c) 0.9802 (d) 0.8135

2. (a) 0.9198 (b) 0.1940 (c) 0.0456

3. 0.1611 **4.** 0.5739 **5.** 81.59% **6.** 1417.6

7. 337.588 **8.** 257.952 **9.** ≈ 0

10. More likely to select one student with a test score greater than 300 because the standard error of the mean is less than the standard deviation.

11. Can use normal distribution.
$\mu = 16.32, \sigma \approx 2.285$

12. 0.3594

Real Statistics–Real Decisions for Chapter 5 *(page 276)*

1. (a) 0.8413 (b) 0.9999999997

2. (a) 0.9772 (b) 0.9999881476 (c) mean

3. Answers will vary.

CHAPTER 6

Section 6.1 *(page 287)*

1. You are more likely to be correct using an interval estimate because it is unlikely that a point estimate will equal the population mean exactly.

3. d; As the level of confidence increases, z_c increases therefore causing wider intervals.

5. 1.28 **7.** 1.15 **9.** 0.47 **11.** 1.76 **13.** 0.60

15. 0.685 **17.** 0.197 **19.** c **20.** d **21.** b **22.** a

23. (14.8, 15.6) **25.** (4.18, 4.36)

27. 126 **29.** 7 **31.** 0.7, 2.8

33. (614.41, 647.39); (611.25, 650.55)

The 95% CI is wider.

35. (25.7, 27.9); (25.5, 28.1) **37.** (94.58, 105.42)

The 95% CI is wider.

39. (96.17, 103.84)

$n = 40$ CI is wider because a smaller sample is taken, giving less information about the population.

41. (9.719, 11.185)

43. (8.687, 12.217)

$s = 5.130$ CI is wider because of the increased variability within the population.

45. (a) An increase in the level of confidence will widen the confidence interval.

(b) An increase in the sample size will narrow the confidence interval.

(c) An increase in the standard deviation will widen the confidence interval.

47. (8.4, 9.7); (8.1, 10.1) **49.** 89

The 99% CI is wider.

51. (a) 121 (b) 208

99% CI requires larger sample because more information is needed from the population to be 99% confident.

53. (a) 32 (b) 87

$E = 0.15$ requires a larger sample size. As the error size decreases, a larger sample must be taken to obtain enough information from the population to ensure desired accuracy.

55. (a) 16 (b) 62

$E = 0.0625$ requires a larger sample size. As the error size decreases, a larger sample must be taken to obtain enough information from the population to ensure desired accuracy.

57. (a) 42 (b) 60

$\sigma = 0.30$ requires a larger sample size. Owing to the increased variability in the population, a larger sample size is needed to ensure the desired accuracy.

59. (a) An increase in the level of confidence will increase the minimum sample size required.

(b) An increase (larger E) in the error tolerance will decrease the minimum sample size required.

(c) An increase in the population standard deviation will increase the minimum sample size required.

61. (303.5, 311.3) **63.** (13.680, 15.200)

65. (a) 0.707 (b) 0.949 (c) 0.962 (d) 0.975

(e) The finite population correction factor approaches 1 as the sample size decreases while the population size remains the same.

67. $E = \dfrac{z_c \sigma}{\sqrt{n}} \Rightarrow E\sqrt{n} = z_c \sigma \Rightarrow \sqrt{n} = \dfrac{z_c \sigma}{E} \Rightarrow n = \left(\dfrac{z_c \sigma}{E}\right)^2$

Section 6.2 *(page 300)*

1. 1.833 **3.** 2.947 **5.** (a) 2.450 (b) 2.664

7. (a) (10.9, 14.1) (b) t-CI is wider.

9. (a) (4.1, 4.5) (b) t-CI is wider.

11. (59.48, 90.52); 15.518

13. (61.85, 88.15); 13.148

The t-CI is wider.

15. (a) (3.8, 5.2) (b) (4.4, 4.6); t-CI is wider.

17. (a) 4174.75 (b) 100.341 (c) (4071.63, 4277.87)

19. (a) 909.1 (b) 305.3 (c) (635.4, 1182.8)

21. Use normal distribution because $n \geq 30$.

(1.248, 1.252)

23. Use t-distribution because $n < 30$, the miles per gallon are normally distributed, and σ is unknown.

(23, 25)

25. Cannot use normal or t-distribution because $n < 30$ and the times are not normally distributed.

27. $n = 25, \bar{x} = 56.0, s = 0.25$

$\pm t_{0.99} \Rightarrow 99\%$ t-CI

$$\bar{x} \pm t_c \frac{s}{\sqrt{n}} = 56.0 \pm 2.797 \frac{0.25}{\sqrt{25}} \approx (55.9, 56.1)$$

They are not making good tennis balls because desired bounce height of 55.5 inches is not contained between 55.9 and 56.1 inches.

Section 6.3 *(page 308)*

1. False. To estimate the value of p, the population proportion of successes, use the point estimate $\hat{p} = x/n$.

3. 0.119, 0.881 **5.** 0.64, 0.36 **7.** 0.12, 0.88

9. 0.694, 0.306 **11.** 0.451, 0.549 **13.** (0.45, 0.51)

15. (0.099, 0.139); (0.093, 0.145)

The 99% CI is wider.

17. (0.625, 0.655); (0.621, 0.659)

The 99% CI is wider.

19. (0.098, 0.142); (0.091, 0.149)

The 99% CI is wider.

21. (a) 1068 (b) 822

(c) Having an estimate of the proportion reduces the minimum sample size needed.

23. (a) 1690 (b) 1267

(c) Having an estimate of the proportion reduces the minimum sample size needed.

25. (a) (0.554, 0.666)

(b) (0.383, 0.497)

It is unlikely that the two proportions are equal because the confidence intervals estimating the proportions do not overlap.

27. (a) (0.195, 0.285) (b) (0.252, 0.348) (c) (0.31, 0.43)

29. (30.4%, 32.4%) is approximately a 95.2% CI.

31. If $n\hat{p} < 5$ or $n\hat{q} < 5$, the sampling distribution of $\hat{p}$ may not be normally distributed, therefore preventing the use of z_c when calculating the confidence interval.

33.

p	$q = 1 - p$	pq
0.0	1.0	0.00
0.1	0.9	0.09
0.2	0.8	0.16
0.3	0.7	0.21
0.4	0.6	0.24
0.5	0.5	0.25
0.6	0.4	0.24
0.7	0.3	0.21
0.8	0.2	0.16
0.9	0.1	0.09
1.0	0.0	0.00

p	$q = 1 - p$	pq
0.45	0.55	0.2475
0.46	0.54	0.2484
0.47	0.53	0.2491
0.48	0.52	0.2496
0.49	0.51	0.2499
0.50	0.50	0.2500
0.51	0.49	0.2499
0.52	0.48	0.2496
0.53	0.47	0.2491
0.54	0.46	0.2484
0.55	0.45	0.2475

$\hat{p} = 0.5$ gives the maximum value of $\hat{p}\hat{q}$.

Section 6.4 *(page 316)*

1. 16.919, 3.325 **3.** 35.479, 10.283

5. 52.336, 13.121

7. (a) (0.0000413, 0.000157) (b) (0.00643, 0.0125)

9. (a) (0.031, 0.191) (b) (0.175, 0.437)

11. (a) (4.34, 44.64) (b) (2.08, 6.68)

13. (a) (360, 1830) (b) (19, 43)

15. (a) (128, 492) (b) (11, 22)

17. (a) (6622, 24,422) (b) (81, 156)

19. (a) (0.08, 0.61) (b) (0.28, 0.78)

21. Yes, because the confidence interval is below 0.015.

Uses and Abuses for Chapter 6 *(page 319)*

1. Answers will vary.

2. Answers will vary.

Review Answers for Chapter 6 *(page 321)*

1. (a) 103.5 (b) 9.016 **3.** (10.2, 10.4)

5. 47 **7.** 49 **9.** 1.415 **11.** 2.624 **13.** 10.255

15. 0.706 **17.** (42.5, 63.1) **19.** (6.1, 7.5)

21. (74, 86) **23.** 0.65, 0.35 **25.** 0.591, 0.409

27. 0.18, 0.82 **29.** 0.4, 0.6 **31.** (0.630, 0.670)

33. (0.566, 0.616) **35.** (0.141, 0.219)
37. (0.374, 0.426) **39.** 273 **41.** 23.337, 4.404
43. 14.067, 2.167 **45.** (0.0029, 0.0127); (0.0539, 0.1127)
47. (0.82, 2.20); (0.91, 1.48)

Chapter Quiz for Chapter 6 *(page 325)*

1. (a) 98.110 (b) 8.847
(c) (89.263, 106.957)
You are 95% confident that the mean repair cost is between \$89.26 and \$106.96.

2. 34

3. (a) 6.61 (b) 3.38 (c) (4.65, 8.57)
(d) (4.79, 8.43)
The t-CI is wider because less information is available.

4. (4259, 5229)

5. (a) 0.620 (b) (0.595, 0.645) (c) 977

6. (a) (387.650, 1104.514) (b) (19.689, 33.234)

Real Statistics – Real Decisions for Chapter 6
(page 326)

1. (a) No, there has not been a change in the mean concentration levels because the confidence interval for year 1 overlaps with the confidence interval for year 2.
(b) No, there has not been a change in the mean concentration levels because the confidence interval for year 2 overlaps with the confidence interval for year 3.
(c) Yes, there has been a change in the mean concentration levels because the confidence interval for year 1 does not overlap with the confidence interval for year 3.

2. The efforts to reduce the acetylene concentration are significant over the three-year period.

3. (a) A point estimate was used because it is the most unbiased estimate of the population mean.
(b) No, because typically σ is unknown. They could have used the sample standard deviation.

CHAPTER 7

Section 7.1 *(page 343)*

1. The two types of hypotheses used in a hypothesis test are the null hypothesis and the alternative hypothesis.
The alternative hypothesis is the complement of the null hypothesis.

3. False. In a hypothesis test, you assume the null hypothesis is true.

5. True

7. False. A small P-value in a test will favor a rejection of the null hypothesis.

9. H_0: $\mu \leq 645$; H_a: $\mu > 645$

11. H_0: $\sigma = 5$; H_a: $\sigma \neq 5$

13. H_0: $p \geq 0.45$: H_a: $p < 0.45$

15. c; H_0: $\mu \leq 3$

16. d; H_0: $\mu \geq 3$

17. b; H_0: $\mu = 3$

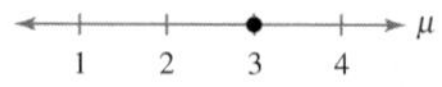

18. a; H_0: $\mu \leq 2$

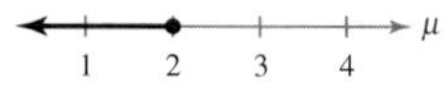

19. Right-tailed **21.** Two-tailed

23. $\mu > 750$
H_0: $\mu \leq 750$; H_a: $\mu > 750$ (claim)

25. $\sigma \leq 1220$
H_0: $\sigma \leq 1220$ (claim); H_a: $\sigma > 1220$

27. $p = 0.84$
H_0: $p = 0.84$ (claim); H_a: $p \neq 0.84$

29. Type I: Rejecting H_0: $p \geq 0.24$ when actually $p \geq 0.24$
Type II: Not rejecting H_0: $p \geq 0.24$ when actually $p < 0.24$

31. Type I: Rejecting H_0: $\sigma \leq 23$ when actually $\sigma \leq 23$
Type II: Not rejecting H_0: $\sigma \leq 23$ when actually $\sigma > 23$

33. Type I: Rejecting H_0: $p = 0.88$ when actually $p = 0.88$
Type II: Not rejecting H_0: $p = 0.88$ when actually $p \neq 0.88$

35. Left-tailed because the alternative hypothesis contains $<$.

37. Two-tailed because the alternative hypothesis contains $\neq$.

39. Two-tailed because the alternative hypothesis contains $\neq$.

41. (a) There is enough evidence to support the company's claim.
(b) There is not enough evidence to support the company's claim.

43. (a) There is enough evidence to support the Department of Labor's claim.
(b) There is not enough evidence to decide that the Department of Labor's claim is true.

45. (a) There is enough evidence to reject the manufacturer's claim.
(b) There is not enough evidence to reject the manufacturer's claim.

47. H_0: $\mu = 10$; H_a: $\mu \neq 10$

49. (a) H_0: $\mu \leq 15$; H_a: $\mu > 15$
(b) H_0: $\mu \geq 15$; H_a: $\mu < 15$

51. If you decrease α, you are decreasing the probability that you will reject H_0. Therefore, you are increasing the probability of failing to reject H_0. This could increase β, the probability of failing to reject H_0 when H_0 is false.

53. (a) Reject H_0 because the confidence interval is located below 70.
(b) Do not reject H_0 because the confidence interval includes values larger than 70.
(c) Do not reject H_0 because the confidence interval includes values larger than 70.

55. (a) Reject H_0 because the confidence interval is located above 0.20.

(b) Do not reject H_0 because the confidence interval includes values less than 0.20.

(c) Do not reject H_0 because the confidence interval includes values less than 0.20.

Section 7.2 *(page 357)*

1. $P = 0.1151$; fail to reject H_0.

3. $P = 0.0096$; reject H_0.

5. $P = 0.1188$; fail to reject H_0.

7. c 8. d 9. e 10. f 11. b 12. a

13. (a) Fail to reject H_0. (b) Reject H_0.

15. 1.645 17. −1.88 19. ±2.33

21. Right-tailed ($\alpha = 0.01$) 23. Two-tailed ($\alpha = 0.10$)

25. (a) Fail to reject H_0 because $-1.645 < z < 1.645$.

(b) Reject H_0 because $z > 1.645$.

(c) Fail to reject H_0 because $-1.645 < z < 1.645$.

(d) Reject H_0 because $z < -1.645$.

27. (a) Fail to reject H_0 because $z < 1.285$.

(b) Fail to reject H_0 because $z < 1.285$.

(c) Fail to reject H_0 because $z < 1.285$.

(d) Reject H_0 because $z > 1.285$.

29. Reject H_0. There is not enough evidence at the 5% level to support the claim.

31. Reject H_0. There is enough evidence at the 1% level to support the claim.

33. (a) H_0: $\mu \leq 260$; H_a: $\mu > 260$ (claim)

(b) 0.838; 0.7995 (c) 0.2005 (d) Fail to reject H_0.

(e) There is insufficient evidence at the 4% level of significance to support the claim that the mean score for Illinois' eighth graders is more than 260.

35. (a) H_0: $\mu \leq 7$; H_a: $\mu > 7$ (claim)

(b) 2.996; 0.9987 (c) 0.0013 (d) Reject H_0.

(e) There is sufficient evidence at the 7% level to support the claim that the mean consumption of tea by a person in the United States is more than 7 gallons per year.

37. (a) H_0: $\mu = 15$ (claim); H_a: $\mu \neq 15$

(b) −0.219; 0.4129 (c) 0.8258 (d) Fail to reject H_0.

(e) There is insufficient evidence at the 5% level to reject the claim that the mean time it takes smokers to quit smoking permanently is 15 years.

39. (a) H_0: $\mu = 40$ (claim); H_a: $\mu \neq 40$

(b) $-z_0 = -2.575$ and $z_0 = 2.575$

Rejection regions: $z < -2.575$, $z > 2.575$

(c) −0.584 (d) Fail to reject H_0.

(e) There is insufficient evidence at the 1% level to reject the claim that the mean caffeine content per one 12-ounce bottle of cola is 40 milligrams.

41. (a) H_0: $\mu \geq 750$ (claim); H_a: $\mu < 750$

(b) $z_0 = -2.05$; Rejection region: $z < -2.05$

(c) −0.500 (d) Fail to reject H_0.

(e) There is insufficient evidence at the 2% level to reject the claim that the mean life of the bulb is at least 750 hours.

43. (a) H_0: $\mu \leq 28$; H_a: $\mu > 28$ (claim)

(b) $z_0 = 1.55$; Rejection region: $z > 1.55$

(c) 1.318 (d) Fail to reject H_0.

(e) There is insufficient evidence at the 6% level to support the claim that the mean nitrogen dioxide level in West London is greater than 28 parts per billion.

45. (a) H_0: $\mu \geq 10$ (claim); H_a: $\mu < 10$

(b) $z_0 = -1.88$; Rejection region: $z < -1.88$

(c) −0.510 (d) Fail to reject H_0.

(e) There is insufficient evidence at the 3% level to reject the claim that the mean weight loss after one month is at least 10 pounds.

47. Fail to reject H_0 because the standardized test statistic $z = -1.96$ is greater than the critical value $z_0 = -2.33$.

49. b, c, d

51. In the classical z-test, the test statistic is compared with critical values. The z-test using a P-value compares the P-value with the level of significance α.

Section 7.3 *(page 370)*

1. Identify the level of significance α and the degrees of freedom, d.f. $= n - 1$. Find the critical value(s) using the t-distribution table in the row with $n - 1$ d.f. If the hypothesis test is

(1) left-tailed, use "One Tail, α" column with a negative sign.

(2) right-tailed, use "One Tail, α" column with a positive sign.

(3) two-tailed, use "Two Tail, α" column with a negative and a positive sign.

3. 1.717 5. −2.101 7. ±2.779

9. 1.328 11. −2.473 13. ±3.747

15. (a) Fail to reject H_0 because $t > -2.086$.

(b) Fail to reject H_0 because $t > -2.086$.

(c) Fail to reject H_0 because $t > -2.086$.

(d) Reject H_0 because $t < -2.086$.

17. (a) Fail to reject H_0 because $-2.602 < t < 2.602$.

(b) Fail to reject H_0 because $-2.602 < t < 2.602$.

(c) Reject H_0 because $t > 2.602$.

(d) Reject H_0 because $t < -2.602$.

19. Fail to reject H_0. There is not enough evidence at the 1% level to reject the claim.

21. Reject H_0. There is enough evidence at the 1% level to reject the claim.

23. (a) H_0: $\mu \geq 100$; H_a: $\mu < 100$ (claim)
(b) $t_0 = -3.747$; Rejection region: $t < -3.747$
(c) -4.472 (d) Reject H_0.
(e) There is sufficient evidence at the 1% level to support the claim that the mean repair cost for damaged microwave ovens is less than \$100.

25. (a) H_0: $\mu \leq 1$; H_a: $\mu > 1$ (claim)
(b) $t_0 = 1.796$; Rejection region: $t > 1.796$
(c) 2.309 (d) Reject H_0.
(e) There is sufficient evidence at the 5% level to support the claim that the mean waste recycled by adults in the United States is more than 1 pound per person per day.

27. (a) H_0: $\mu = \$22{,}300$ (claim); H_a: $\mu \neq \$22{,}300$
(b) $-t_0 = -2.262$ and $t_0 = 2.262$
Rejection regions: $t < -2.262$, $t > 2.262$
(c) -0.572 (d) Fail to reject H_0.
(e) There is insufficient evidence at the 5% level to reject the claim that the mean salary for full-time male workers over age 25 without a high school diploma is \$22,300.

29. (a) H_0: $\mu \geq 3.0$; H_a: $\mu < 3.0$ (claim)
(b) 0.130 (c) Fail to reject H_0.
(d) There is insufficient evidence at the 5% level to support the claim that teenage males drink fewer than three 12-ounce servings of soda per day.

31. (a) H_0: $\mu \geq 32$; H_a: $\mu < 32$ (claim)
(b) 0.034 (c) Fail to reject H_0.
(d) There is insufficient evidence at the 1% level to support the claim that the mean class size for full-time faculty is fewer than 32 students.

33. (a) H_0: $\mu = \$2116$ (claim); H_a: $\mu \neq \$2116$
(b) 0.210 (c) Fail to reject H_0.
(d) There is sufficient evidence at the 2% level to reject the claim that the typical household in the United States spends a mean amount of \$2116 per year on food away from home.

35. Because the P-value $= 0.096 > 0.01$, fail to reject H_0.

37. Use the t-distribution because the population is normal, $n < 30$, and σ is unknown.
Fail to reject H_0. There is insufficient evidence at the 5% level to reject the claim that the mean gas mileage for the luxury sedan is at least 21 miles per gallon.

Section 7.4 *(page 378)*

1. Verify that $np \geq 5$ and $nq \geq 5$. State H_0 and H_a. Specify the level of significance α. Determine the critical value(s) and rejection region(s). Find the standardized test statistic. Make a decision and interpret in the context of the original claim.

3. Use normal distribution.
Fail to reject H_0. There is not enough evidence at the 5% level to support the claim.

5. Cannot use normal distribution.

7. Use normal distribution.
Reject H_0. There is enough evidence at the 10% level to reject the claim.

9. (a) H_0: $p \geq 0.23$ (claim); H_a: $p < 0.23$
(b) $z_0 = -2.33$; Rejection region: $z < -2.33$
(c) -0.168 (d) Fail to reject H_0.
(e) There is insufficient evidence at the 1% level to reject the claim that at least 23% of U.S. adults are smokers.

11. (a) H_0: $p \leq 0.30$; H_a: $p > 0.30$ (claim)
(b) $z_0 = 1.88$; Rejection region: $z > 1.88$
(c) 1.414 (d) Fail to reject H_0.
(e) There is insufficient evidence at the 3% level to support the claim that more than 30% of U.S. consumers have stopped buying the product because the manufacturing of the product pollutes the environment.

13. (a) H_0: $p = 0.44$ (claim); H_a: $p \neq 0.44$
(b) $-z_0 = -2.33$ and $z_0 = 2.33$
Rejection regions: $z < -2.33$, $z > 2.33$
(c) -2.537 (d) Reject H_0.
(e) There is sufficient evidence at the 2% level to reject the claim that 44% of home buyers found their real estate agent through a friend.

15. Fail to reject H_0. There is not enough evidence at the 5% level to reject the claim that at least 52% of the adults are more likely to buy a product when there are free samples.

17. Reject H_0.
The results are the same.

Section 7.5 *(page 386)*

1. Specify the level of significance α. Determine the degrees of freedom. Determine the critical values using the χ^2 distribution. If (a) right-tailed test, use the value that corresponds to d.f. and α; (b) left-tailed test, use the value that corresponds to d.f. and $1 - \alpha$; (c) two-tailed test, use the value that corresponds to d.f. and $\frac{1}{2}\alpha$ and $1 - \frac{1}{2}\alpha$.

3. 38.885 **5.** 0.872 **7.** $\chi^2_L = 7.261$, $\chi^2_R = 24.996$

9. (a) Fail to reject H_0. (b) Fail to reject H_0.
(c) Fail to reject H_0. (d) Reject H_0.

11. (a) Fail to reject H_0. (b) Reject H_0.
(c) Reject H_0. (d) Fail to reject H_0.

13. Fail to reject H_0. There is not enough evidence at the 5% level to reject the claim.

15. (a) H_0: $\sigma^2 = 3$ (claim); H_a: $\sigma^2 \neq 3$
(b) $\chi^2_L = 13.844$, $\chi^2_R = 41.923$
Rejection regions: $\chi^2 > 41.923$, $\chi^2 < 13.844$

(c) 24.267 (d) Fail to reject H_0.

(e) There is insufficient evidence at the 5% level of significance to reject the claim that the variance of the life of the appliances is 3.

17. (a) H_0: $\sigma \geq 29$; H_a: $\sigma < 29$ (claim)

(b) $\chi_0^2 = 13.240$; Rejection region: $\chi^2 < 13.240$

(c) 19.159 (d) Fail to reject H_0.

(e) There is insufficient evidence at the 10% level of significance to support the claim that the standard deviation for eighth graders on the examination is less than 29.

19. (a) H_0: $\sigma \leq 0.5$ (claim); H_a: $\sigma > 0.5$

(b) $\chi_0^2 = 33.196$; Rejection region: $\chi^2 > 33.196$

(c) 47.040 (d) Reject H_0.

(e) There is sufficient evidence at the 10% level to reject the claim that the standard deviation of the waiting times is no more than 0.5 minute.

21. (a) H_0: $\sigma \geq \$18{,}000$; H_a: $\sigma < \$18{,}000$ (claim)

(b) $\chi_0^2 = 18.114$; Rejection region: $\chi^2 < 18.114$

(c) 28.521 (d) Fail to reject H_0.

(e) There is insufficient evidence at the 10% level to support the claim that the standard deviation of the total charge for patients involved in a crash in which the vehicle struck a wall is less than $18,000.

23. (a) H_0: $\sigma \leq 20{,}000$; H_a: $\sigma > 20{,}000$ (claim)

(b) $\chi_0^2 = 24.996$; Rejection region: $\chi^2 > 24.996$

(c) 16.011 (d) Fail to reject H_0.

(e) There is insufficient evidence at the 5% level to support the claim that the standard deviation of the annual salaries for actuaries is more than $20,000.

25. P-value = 0.615
Fail to reject H_0.

27. P-value = 0.381
Fail to reject H_0.

Uses and Abuses for Chapter 7 *(page 389)*

1. Answers will vary.
2. H_0: $p = 0.47$; Answers will vary.
3. Answers will vary.
4. Answers will vary.

Review Answers for Chapter 7 *(page 393)*

1. H_0: $\mu \leq 1593$ (claim); H_a: $\mu > 1593$

3. H_0: $p \geq 0.205$; H_a: $p < 0.205$ (claim)

5. H_0: $\sigma \leq 4.5$; H_a: $\sigma > 4.5$ (claim)

7. (a) H_0: $p = 0.73$ (claim); H_a: $p \neq 0.73$

(b) Type I error will occur if H_0 is rejected when the actual proportion of American adults who use nonprescription pain relievers is 0.73.

Type II error will occur if H_0 is not rejected when the actual proportion of American adults who use nonprescription pain relievers is not 0.73.

(c) Two-tailed because the alternative hypothesis contains $\neq$.

(d) There is enough evidence to reject the claim.

(e) There is not enough evidence to reject the claim.

9. (a) H_0: $\mu \leq 50$ (claim); H_a: $\mu > 50$

(b) Type I error will occur if H_0 is rejected when the actual standard deviation of sodium content is no more than 50 milligrams.

Type II error will occur if H_0 is not rejected when the actual standard deviation of sodium content is more than 50 milligrams.

(c) Right-tailed because the alternative hypothesis contains $>$.

(d) There is enough evidence to reject the claim.

(e) There is not enough evidence to reject the claim.

11. -2.05 **13.** 1.96

15. Reject H_0. There is enough evidence at the 5% level to reject the claim.

17. Fail to reject H_0. There is not enough evidence at the 1% level to support the claim.

19. H_0: $\mu \leq 0.05$ (claim); H_a: $\mu > 0.05$

$z \approx 2.200$; P-value = 0.0139

$\alpha = 0.10 \Rightarrow$ Reject H_0.

$\alpha = 0.05 \Rightarrow$ Reject H_0.

$\alpha = 0.01 \Rightarrow$ Fail to reject H_0.

21. Reject H_0. There is sufficient evidence to reject the claim that mean daily cost of meals and lodging is $284.

23. ± 2.093 **25.** -1.345

27. Fail to reject H_0. There is not enough evidence at the 5% level to support the claim.

29. Fail to reject H_0. There is not enough evidence at the 10% level to reject the claim.

31. Reject H_0. There is enough evidence at the 1% level to reject the claim.

33. Fail to reject H_0. There is not enough evidence at the 10% level to reject the claim that the average monthly cost of joining a health club is $25.

35. Fail to reject H_0. There is not enough evidence at the 1% level to reject the claim that the mean number of hours of sleep high school students get each night is at least 6 hours.

37. Fail to reject H_0. There is not enough evidence at the 5% level to reject the claim.

39. Because $np = 3.6$ is less than 5, the normal distribution cannot be used to approximate the binomial distribution.

41. Because $np = 1.2$ is less than 5, the normal distribution cannot be used to approximate the binomial distribution.

43. Fail to reject H_0. There is not enough evidence at the 1% level to support the claim.

45. Reject H_0. There is not enough evidence at the 10% level to support the claim that over 75% of people in the United States prepare and file their taxes before April 15th.

47. 30.144 **49.** 33.196

51. Fail to reject H_0. There is not enough evidence at the 10% level to support the claim.

53. Fail to reject H_0. There is not enough evidence at the 5% level to reject the claim.

55. Reject H_0. There is enough evidence at the 0.5% level to reject the claim that the variance is at most 0.01.

Chapter Quiz for Chapter 7 *(page 397)*

1. (a) H_0: $\mu \geq 94$ (claim); H_a: $\mu < 94$

(b) Left-tailed because the alternative hypothesis contains $<$; z-test because $n \geq 30$.

(c) $z_0 = -2.05$; Rejection region: $z < -2.05$

(d) -0.169

(e) Fail to reject H_0. There is insufficient evidence at the 2% level to reject the claim that the mean consumption of fresh citrus fruits by people in the United States is at least 94 pounds per year.

2. (a) H_0: $\mu \geq 25$ (claim); H_a: $\mu < 25$

(b) Left-tailed because the alternative hypothesis contains $<$; t-test because the population is normal, $n < 30$, and σ is unknown.

(c) $t_0 = -1.895$; Rejection region: $t < -1.895$

(d) -1.131

(e) Fail to reject H_0. There is insufficient evidence at the 5% level to reject the claim that the mean gas mileage is at least 25 miles per gallon.

3. (a) H_0: $p \leq 0.10$ (claim); H_a: $p > 0.10$

(b) Right-tailed because the alternative hypothesis contains $>$; z-test because np, $nq \geq 5$.

(c) $z_0 = 1.75$; Rejection region: $z > 1.75$ (d) 0.755

(e) Fail to reject H_0. There is insufficient evidence at the 4% level to reject the claim that no more than 10% of the microwaves need repair during the first five years of use.

4. (a) H_0: $\sigma = 105$ (claim); H_a: $\sigma \neq 105$

(b) Two-tailed because the alternative hypothesis contains $\neq$; χ^2 test because the population is normal.

(c) $\chi^2_L = 3.565$, $\chi^2_R = 29.819$

Rejection regions: $\chi^2 < 3.565$, $\chi^2 > 29.819$

(d) 15.056

(e) Fail to reject H_0. There is insufficient evidence at the 1% level to reject the claim that the standard deviation of the SAT verbal scores for the state is 105.

5. (a) H_0: $\mu = \$59{,}482$ (claim); H_a: $\mu \neq \$59{,}482$

(b) Two-tailed because the alternative hypothesis contains $\neq$; t-test because the population is normal, $n < 30$, and σ is unknown.

(c) Not applicable. (d) $t \approx -0.480$; P-value $= 0.640$

(e) Fail to reject H_0. There is insufficient evidence at the 5% level to reject the claim that the mean annual salary for full-time male workers ages 25 to 34 with a bachelor's degree is \$59,482.

6. (a) H_0: $\mu = \$276$ (claim); H_a: $\mu \neq \$276$

(b) Two-tailed because the alternative hypothesis contains $\neq$; z-test because $n \geq 30$.

(c) Not applicable. (d) $z \approx 1.775$; P-value $= 0.0759$

(e) Fail to reject H_0. There is insufficient evidence at the 5% level to reject the claim that the mean daily cost of meals and lodging for a family of four traveling in Massachusetts is \$276.

Real Statistics–Real Decisions for Chapter 7
(page 398)

1. (a) Take a random sample to get a diverse group of people. Stratified random sampling.

(b) Random sample (c) Answers will vary.

2. Fail to reject H_0. There is not enough evidence to support the claim that less than 40% of people in the United States think the Social Security system will have money available to provide the benefits they expect for their retirements.

3. Reject H_0. There is enough evidence to reject the claim that the mean age of people in the United States who would say yes to the survey question is 60 years or older.

4. Answers will vary.

CHAPTER 8

Section 8.1 *(page 409)*

1. State the hypotheses and identify the claim. Specify the level of significance. Find the critical value(s) and identify the rejection region(s). Find the standardized test statistic. Make a decision and interpret in the context of the claim.

3. (a) 2 (b) 7.603 (c) z is in the rejection region.

(d) Reject H_0. There is enough evidence at the 5% level to reject the claim.

5. (a) 30 (b) 1.84 (c) z is not in the rejection region.

(d) Fail to reject H_0. There is not enough evidence at the 1% level to support the claim.

7. Fail to reject H_0. There is enough evidence at the 1% level to support the claim.

9. (a) H_0: $\mu_1 = \mu_2$; H_a: $\mu_1 \neq \mu_2$ (claim)

(b) $-z_0 = -1.645$ and $z_0 = 1.645$

Rejection regions: $z < -1.645$, $z > 1.645$

(c) -2.786 (d) Reject H_0.

(e) There is sufficient evidence at the 10% level to support the claim that the mean braking distance is different for the two types of tires.

11. (a) H_0: $\mu_1 \geq \mu_2$; H_a: $\mu_1 < \mu_2$ (claim)
(b) $z_0 = -2.33$; Rejection region: $z < -2.33$
(c) -1.536 (d) Fail to reject H_0.
(e) There is insufficient evidence at the 1% level to conclude that the repair costs for Model A are lower than for Model B.

13. (a) H_0: $\mu_1 = \mu_2$ (claim); H_a: $\mu_1 \neq \mu_2$
(b) $-z_0 = -2.575$ and $z_0 = 2.575$
Rejection regions: $z < -2.575$, $z > 2.575$
(c) 0.202 (d) Fail to reject H_0.
(e) There is insufficient evidence at the 1% level to reject the claim that male and female high school students have equal ACT scores.

15. (a) H_0: $\mu_1 = \mu_2$ (claim); H_a: $\mu_1 \neq \mu_2$
(b) $-z_0 = -1.645$ and $z_0 = 1.645$
Rejection regions: $z < -1.645$, $z > 1.645$
(c) -0.606 (d) Fail to reject H_0.
(e) There is insufficient evidence at the 10% level to reject the claim that the lodging cost for a family traveling in California is the same as in Florida.

17. (a) H_0: $\mu_1 = \mu_2$ (claim); H_a: $\mu_1 \neq \mu_2$
(b) $-z_0 = -1.645$ and $z_0 = 1.645$
Rejection regions: $z < -1.645$, $z > 1.645$
(c) 0.645 (d) Fail to reject H_0.
(e) There is insufficient evidence at the 10% level to reject the claim that the lodging cost for a family traveling in California is the same as in Florida.
The new samples do not lead to a different conclusion.

19. (a) H_0: $\mu_1 \leq \mu_2$; H_a: $\mu_1 > \mu_2$ (claim)
(b) $z_0 = 1.96$; Rejection region: $z > 1.96$
(c) 4.988 (d) Reject H_0.
(e) At the 2.5% level of significance, there is sufficient evidence to support the claim that children spent more time watching television in 1981 than children do today.

21. (a) H_0: $\mu_1 = \mu_2$ (claim); H_a: $\mu_1 \neq \mu_2$
(b) $-z_0 = -2.575$ and $z_0 = 2.575$
Rejection regions: $z < -2.575$, $z > 2.575$
(c) 66.172 (d) Reject H_0.
(e) At the 1% level of significance, there is sufficient evidence to reject the claim that there is no difference in mean washer diameters.

23. They are equivalent through algebraic manipulation of the equation.
$\mu_1 = \mu_2 \Rightarrow \mu_1 - \mu_2 = 0$

25. H_0: $\mu_1 - \mu_2 = -9$ (claim); H_a: $\mu_1 - \mu_2 \neq -9$
Fail to reject H_0. There is not enough evidence at the 1% level to reject the claim that children spend 9 hours a week more in day care or preschool.

27. H_0: $\mu_1 - \mu_2 \leq 6000$; H_a: $\mu_1 - \mu_2 > 6000$ (claim)
Fail to reject H_0. There is not enough evidence at the 10% level to support the claim that the difference in mean annual salaries of statisticians is more than \$6000.

29. $-2.45 < \mu_1 - \mu_2 < 0.65$

31. There is not enough evidence at the 5% level to support the claim. You should not recommend using the herbal supplement with a high-fiber, low-calorie diet to lose weight because there was not a significant difference between the weight loss it produced and that of the placebo.

33. The 95% CI for $\mu_1 - \mu_2$ in Exercise 29 contained values less than or equal to zero and, as found in Exercise 31, there was not enough evidence at the 5% level of significance to support the claim.
If zero is contained in the CI for $\mu_1 - \mu_2$, you fail to reject H_0 because the null hypothesis states that $\mu_1 - \mu_2$ is less than or equal to zero.

Section 8.2 *(page 420)*

1. State hypotheses and identify the claim. Specify the level of significance. Determine the degrees of freedom. Find the critical value(s) and identify the rejection region(s). Find the standardized test statistic. Make a decision and interpret in the context of the original claim.

3. (a) $t_0 = \pm 1.725$ (b) $t_0 = \pm 1.833$

5. (a) $t_0 = -2.074$ (b) $t_0 = -2.306$

7. (a) $t_0 = 1.729$ (b) $t_0 = 1.895$

9. (a) $t_0 = \pm 2.771$ (b) $t_0 = \pm 3.106$

11. (a) -1.8 (b) -1.199
(c) t is not in the rejection region.
(d) Fail to reject H_0. There is not enough evidence at the 1% level to reject the claim.

13. There is no need to run the test because this is a right-tailed test and the test statistic is negative. It is obvious that the standardized test statistic will also be negative and fall outside of the rejection region. So, the decision is to fail to reject H_0.

15. (a) H_0: $\mu_1 = \mu_2$ (claim); H_a: $\mu_1 \neq \mu_2$
(b) $-t_0 = -1.706$ and $t_0 = 1.706$
Rejection regions: $t < -1.706$, $t > 1.706$
(c) 0.876 (d) Fail to reject H_0.
(e) There is not enough evidence at the 10% level to reject the claim that the mean footwell intrusions for small and midsize cars are equal.

17. (a) H_0: $\mu_1 \geq \mu_2$; H_a: $\mu_1 < \mu_2$ (claim)
(b) $t_0 = -1.350$; Rejection region: $t < -1.350$
(c) -4.038 (d) Reject H_0.
(e) There is enough evidence at the 10% level to support the claim that mean bumper repair cost is less for small cars than it is for midsize cars.

19. (a) H_0: $\mu_1 \le \mu_2$; H_a: $\mu_1 > \mu_2$ (claim)
(b) $t_0 = 1.345$; Rejection region: $t > 1.345$
(c) 1.440 (d) Reject H_0.
(e) There is enough evidence at the 10% level to support the claim that the mean annual wage is greater in Allegheny County than it is in Erie County.

21. (a) H_0: $\mu_1 = \mu_2$; H_a: $\mu_1 \ne \mu_2$ (claim)
(b) $-t_0 = -2.831$ and $t_0 = 2.831$
Rejection regions: $t < -2.831$, $t > 2.831$
(c) -6.410 (d) Reject H_0.
(e) There is enough evidence at the 1% level to support the claim that the new treatment makes a difference in the tensile strength of steel bars.

23. (a) H_0: $\mu_1 \ge \mu_2$; H_a: $\mu_1 < \mu_2$ (claim)
(b) $t_0 = -1.282$; Rejection region: $t < -1.282$
(c) -4.295 (d) Reject H_0.
(e) There is enough evidence at the 10% level to support the claim and to recommend changing to the new method.

25. $4.336 < \mu_1 - \mu_2 < 15.664$

27. $-11.580 < \mu_1 - \mu_2 < -8.420$

Section 8.3 *(page 431)*

1. Two samples are dependent if each member of one sample corresponds to a member of the other sample. Example: The weights of 22 people before starting an exercise program and the weights of the same 22 people 6 weeks after starting the exercise program.

Two samples are independent if the sample selected from one population is not related to the sample selected from the second population. Example: The weights of 25 cats and the weights of 20 dogs.

3. Independent because the scores are from different students.

5. Dependent because the same adults were sampled.

7. Independent because different boats were sampled.

9. Dependent because the same tire sets were sampled.

11. Left-tailed test; Fail to reject H_0.

13. Right-tailed test; Reject H_0.

15. Left-tailed test; Reject H_0.

17. (a) H_0: $\mu_d \ge 0$; H_a: $\mu_d < 0$ (claim)
(b) $t_0 = -2.650$; Rejection region: $t < -2.650$
(c) $\bar{d} \approx -33.714$; $s_d \approx 42.034$
(d) -3.001 (e) Reject H_0.
(f) There is enough evidence at the 1% level to support the claim that the second SAT scores are improved.

19. (a) H_0: $\mu_d \ge 0$; H_a: $\mu_d < 0$ (claim)
(b) $t_0 = -1.415$; Rejection region: $t < -1.415$
(c) $\bar{d} \approx -1.125$; $s_d \approx 0.871$
(d) -3.653 (e) Reject H_0.
(f) There is enough evidence at the 10% level to support the claim that the fuel additive improved gas mileage.

21. (a) H_0: $\mu_d \le 0$; H_a: $\mu_d > 0$ (claim)
(b) $t_0 = 1.363$; Rejection region: $t > 1.363$
(c) $\bar{d} = 3.75$; $s_d \approx 7.841$
(d) 1.657 (e) Reject H_0.
(f) There is enough evidence at the 10% level to support the claim that the exercise program helps participants lose weight.

23. (a) H_0: $\mu_d \le 0$; H_a: $\mu_d > 0$ (claim)
(b) $t_0 = 2.764$; Rejection region: $t > 2.764$
(c) $\bar{d} \approx 1.255$; $s_d \approx 0.441$
(d) 9.438 (e) Reject H_0.
(f) There is enough evidence at the 1% level to support the claim that soft tissue therapy and spinal manipulation help reduce the length of time patients suffer from headaches.

25. (a) H_0: $\mu_d \le 0$; H_a: $\mu_d > 0$ (claim)
(b) $t_0 = 1.895$; Rejection region: $t > 1.895$
(c) $\bar{d} = 14.75$; $s_d \approx 6.861$
(d) 6.082 (e) Reject H_0.
(f) There is enough evidence at the 5% level to support the claim that the new drug reduces systolic blood pressure.

27. (a) H_0: $\mu_d = 0$; H_a: $\mu_d \ne 0$ (claim)
(b) $-t_0 = -2.365$ and $t_0 = 2.365$
Rejection regions: $t < -2.365$, $t > 2.365$
(c) $\bar{d} = -1$; $s_d \approx 1.309$
(d) -2.159 (e) Fail to reject H_0.
(f) There is not enough evidence at the 5% level to support the claim that the product ratings have changed from last year to this year.

29. $-1.763 < \mu_d < -1.287$

Section 8.4 *(page 441)*

1. State the hypotheses and identify the claim. Specify the level of significance. Find the critical value(s) and rejection region(s). Find $\bar{p}$ and $\bar{q}$. Find the standardized test statistic. Make a decision and interpret in the context of the claim.

3. Two-tailed test; Fail to reject H_0.

5. Right-tailed test; Fail to reject H_0.

7. (a) H_0: $p_1 = p_2$ (claim); H_a: $p_1 \ne p_2$
(b) $-z_0 = -1.96$ and $z_0 = 1.96$
Rejection regions: $z < -1.96$, $z > 1.96$
(c) -5.060 (d) Reject H_0.
(e) There is sufficient evidence at the 5% level to reject the claim that the proportion of adults using alternative medicines has not changed since 1991.

9. (a) H_0: $p_1 = p_2$ (claim); H_a: $p_1 \neq p_2$

(b) $-z_0 = -1.645$ and $z_0 = 1.645$

Rejection regions: $z < -1.645$, $z > 1.645$

(c) 4.351 (d) Reject H_0.

(e) There is sufficient evidence at the 10% level to reject the claim that the proportions of male and female senior citizens who eat the daily recommended number of servings of vegetables are the same.

11. (a) H_0: $p_1 \geq p_2$; H_a: $p_1 < p_2$ (claim)

(b) $z_0 = -2.33$; Rejection region: $z < -2.33$

(c) -1.596 (d) Fail to reject H_0.

(e) There is insufficient evidence at the 1% level to support the claim that the proportion of adults who are smokers is lower in Alabama than in Missouri.

13. (a) H_0: $p_1 \geq p_2$; H_a: $p_1 < p_2$ (claim)

(b) $z_0 = -2.33$; Rejection region: $z < -2.33$

(c) 2.395 (d) Fail to reject H_0.

(e) There is insufficient evidence at the 1% level to support the claim that the proportion of twelfth grade males who said they had smoked in the last 30 days is less than the proportion of twelfth grade females.

15. (a) H_0: $p_1 = p_2$ (claim); H_a: $p_1 \neq p_2$

(b) $-z_0 = -1.96$ and $z_0 = 1.96$

Rejection regions: $z < -1.96$, $z > 1.96$

(c) 26.135 (d) Reject H_0.

(e) There is sufficient evidence at the 5% level to reject the claim that the proportions of adults who have chronic backaches are the same for the two groups.

17. There is insufficient evidence at the 1% level to support the claim that the proportions of teens who watch less than 1 hour of TV per week is greater for girls than for boys.

19. There is sufficient evidence at the 5% level to reject the organization's claim that the proportions of teens who watch 1 to 5 hours of TV per week is the same for boys and for girls.

21. There is sufficient evidence at the 5% level to support the claim that the proportions of men age 18 to 24 living in parents' homes was more in 1992 than in 2002.

23. There is sufficient evidence at the 1% level to reject the claim that the proportions of 18- to 24-year-olds living in parents' homes in 1992 was the same for men and women.

25. $0.0312 < p_1 - p_2 < 0.0328$

Uses and Abuses for Chapter 8 *(page 445)*

1. Answers will vary.

2. Blind: The patients do not know which group (medicine or placebo) they belong to.

Double Blind: Both the researcher and patient do not know which group (medicine or placebo) that the patient belongs to.

Review Answers for Chapter 8 *(page 447)*

1. Reject H_0. There is enough evidence at the 5% level to reject the claim.

3. Reject H_0. There is enough evidence at the 10% level to support the claim.

5. (a) H_0: $\mu_1 \leq \mu_2$; H_a: $\mu_1 > \mu_2$ (claim)

(b) $z_0 = 1.645$; Rejection region: $z > 1.645$

(c) 1.750 (d) Reject H_0.

(e) There is sufficient evidence at the 5% level to conclude that the McDonald's sandwich has fewer calories than the Long John Silver's sandwich.

7. Fail to reject H_0. There is not enough evidence at the 5% level to reject the claim.

9. Fail to reject H_0. There is not enough evidence at the 10% level to reject the claim.

11. Reject H_0. There is enough evidence at the 1% level to support the claim.

13. (a) H_0: $\mu_1 \leq \mu_2$; H_a: $\mu_1 > \mu_2$ (claim)

(b) $t_0 = 1.645$; Rejection region: $t > 1.645$

(c) 2.266 (d) Reject H_0.

(e) There is sufficient evidence at the 5% level to support the claim that third graders taught with the directed reading activities scored higher than those taught without the activities.

15. Independent because the two samples of laboratory mice are different groups.

17. Two-tailed test; Reject H_0.

19. Right-tailed test; Reject H_0.

21. (a) H_0: $\mu_d \leq 0$; H_a: $\mu_d > 0$ (claim)

(b) $t_0 = 1.383$; Rejection region: $t > 1.383$

(c) $\bar{d} = 5$; $s_d \approx 8.743$ (d) 1.808 (e) Reject H_0.

(f) There is enough evidence at the 10% level to support the claim that the men's diastolic blood pressure decreased.

23. Two-tailed test; Fail to reject H_0.

25. Right-tailed test; Fail to reject H_0.

27. (a) H_0: $p_1 = p_2$ (claim); H_a: $p_1 \neq p_2$

(b) $-z_0 = -1.645$ and $z_0 = 1.645$

Rejection regions: $z < -1.645$, $z > 1.645$

(c) -0.718 (d) Fail to reject H_0.

(e) There is not enough evidence at the 10% level to reject the claim that the proportions of Canadians with a college degree were the same for the two years.

Chapter Quiz for Chapter 8 *(page 451)*

1. (a) H_0: $\mu_1 \leq \mu_2$; H_a: $\mu_1 > \mu_2$ (claim)

(b) Right tailed z-test because n_1 and n_2 are each greater than 30 and the samples are independent.

(c) $z_0 = 1.645$; Rejection region: $z > 1.645$

(d) 31.426 (e) Reject H_0.

(f) There is sufficient evidence at the 5% level to support the claim that the mean score on the science assessment for the male high school students was higher than for the female high school students.

2. (a) H_0: $\mu_1 = \mu_2$ (claim); H_a: $\mu_1 \neq \mu_2$

(b) Two tailed t-test because n_1 and n_2 are less than 30, the samples are independent, and the populations are normally distributed.

(c) $-t_0 = -2.779$ and $t_0 = 2.779$

Rejection regions: $t < -2.779$, $t > 2.779$

(d) 6.617 (e) Reject H_0.

(f) There is sufficient evidence at the 1% level to reject the teacher's suggestion that the mean scores on the science assessment test are the same for nine-year-old boys and girls.

3. (a) H_0: $p_1 \leq p_2$; H_a: $p_1 > p_2$ (claim)

(b) Right tailed z-test because you are testing proportions and $n_1\bar{p}$, $n_2\bar{p}$, $n_1\bar{q}$, and $n_2\bar{q} \geq 5$ and the samples are independent.

(c) $z_0 = 1.282$; Rejection region: $z > 1.282$

(d) 158.471 (e) Reject H_0.

(f) There is sufficient evidence at the 10% level to support the claim that the proportion of accidents involving alcohol is higher for drivers in the 21 to 24 age group than for drivers aged 65 and older.

4. (a) H_0: $\mu_d \geq 0$; H_a: $\mu_d < 0$ (claim)

(b) Dependent t-test because both populations are normally distributed and the samples are dependent.

(c) $t_0 = -1.796$; Rejection region: $t < -1.796$

(d) -9.016 (e) Reject H_0.

(f) There is sufficient evidence at the 5% level to conclude that the students' SAT scores improved on the second test.

Real Statistics–Real Decisions for Chapter 8

(page 452)

1. (a) Take a simple random sample of records today and 10 years ago from a random sample of hospitals.

Use a cluster sample.

(b) Answers will vary.

(c) Answers will vary.

2. Use a t-test; independent; yes, you need to know if the population distributions are normal or not; yes, you need to know if the population variances are equal or not.

3. There is enough evidence at the 5% level to support the claim that there is a difference in the mean length of hospital stays for patients (ages 17 and younger) diagnosed with pneumonia.

CHAPTER 9

Section 9.1 *(page 469)*

1. $r = -0.845$ represents a stronger correlation because it is closer to -1 than $r = 0.731$ is to $+1$.

3. State the null and alternative hypotheses. Specify the level of significance and determine the degrees of freedom. Identify the rejection regions and calculate the standardized test statistic. Make a decision and interpret in the context of the original claim.

5. Positive linear correlation

7. No linear correlation (but there is a nonlinear correlation between the variables)

9. c; You would expect a positive linear correlation between age and income.

10. d; You would not expect age and height to be correlated.

11. b; You would expect a negative linear correlation between age and balance on student loans.

12. a; You would expect the relationship between age and body temperature to be fairly constant.

13. Explanatory variable: Amount of water consumed

Response variable: Weight loss

15. (a)

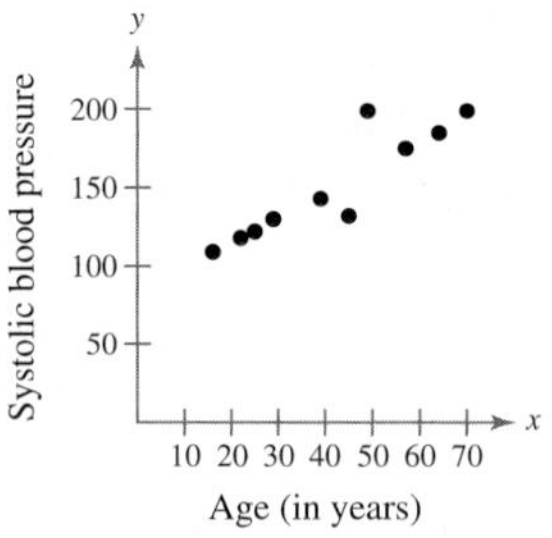

(b) 0.908 (c) Strong positive linear correlation

17. (a)

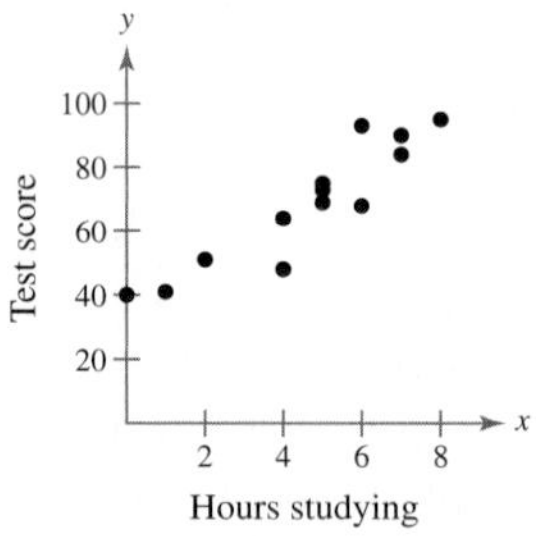

(b) 0.923 (c) Strong positive linear correlation

19. (a)

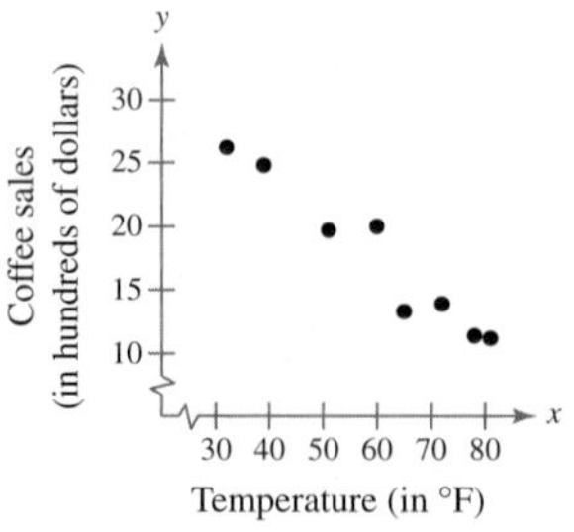

(b) -0.971 (c) Strong negative linear correlation

21. (a)

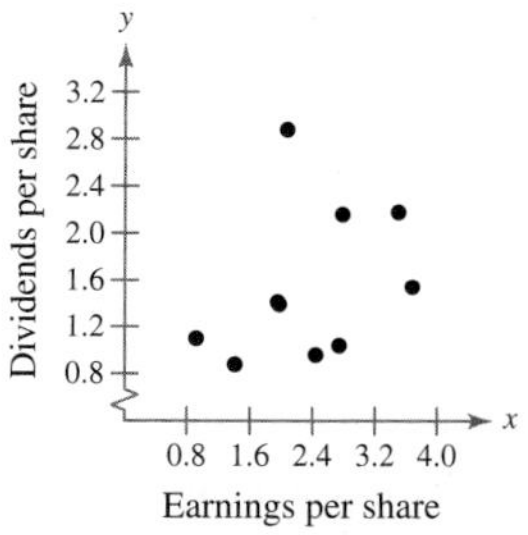

(b) 0.342 (c) No linear correlation

23. There is enough evidence at the 1% level to conclude that there is no significant linear correlation between vehicle weight and the variability in braking distance.

25. There is enough evidence at the 1% level to conclude that there is a significant linear correlation between the number of hours spent studying for a test and the score received on the test.

27. There is enough evidence at the 1% level to conclude that there is no significant linear correlation between earnings per share and dividends per share.

29. The correlation coefficient remains unchanged when the x-values and y-values are switched.

31. Answers will vary.

Section 9.2 *(page 478)*

1. c **2.** a **3.** d **4.** b **5.** g **6.** e **7.** h
8. f **9.** c **10.** b **11.** a **12.** d

13. $\hat{y} = 1.662x + 83.338$

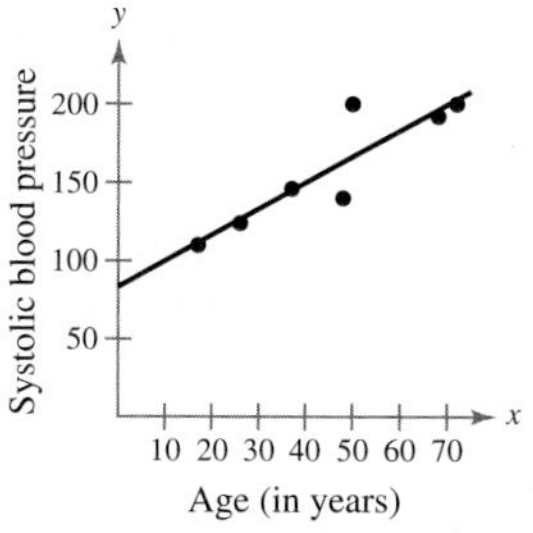

(a) 113.254 (b) 201.340 (c) 131.536 (d) 174.748

15. $\hat{y} = 7.350x + 34.617$

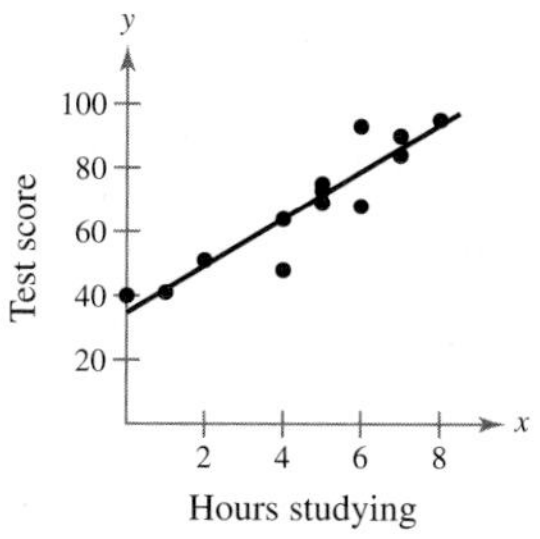

(a) 56.7 (b) 82.4

(c) It is not meaningful to predict the value of y for $x = 13$ because $x = 13$ is outside the range of the original data.

(d) 67.7

17. $\hat{y} = 4.347x - 299.482$

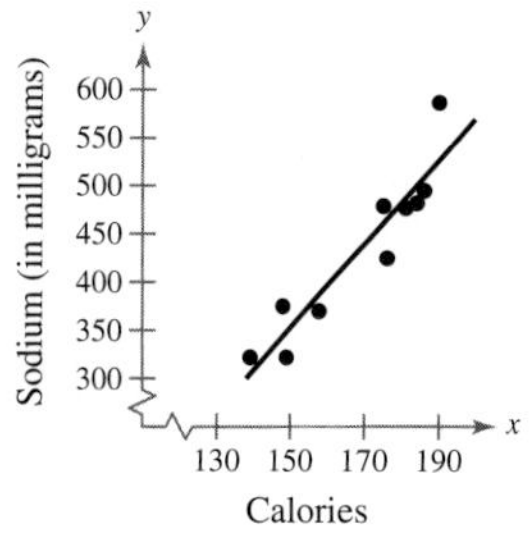

(a) 439.508 milligrams

(b) It is not meaningful to predict the value of y for $x = 136$ because $x = 136$ is outside the range of the original data.

(c) 330.833 milligrams (d) 496.019 milligrams

19. $\hat{y} = 1.870x + 51.360$

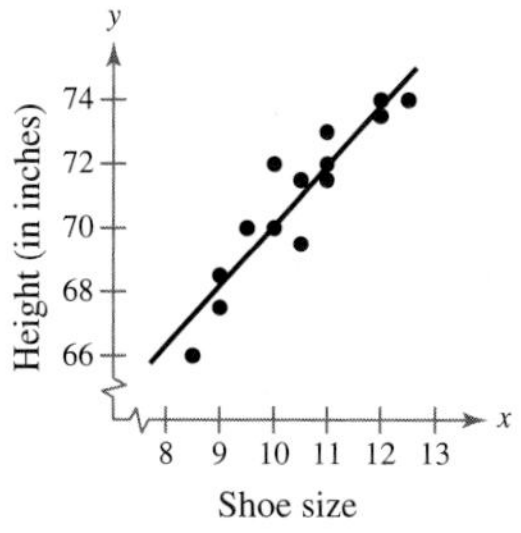

(a) 72.865 inches (b) 66.320 inches

(c) It is not meaningful to predict the value of y for $x = 15.5$ because $x = 15.5$ is outside the range of the original data.

(d) 70.060 inches

21. $\hat{y} = 0.000447x - 0.425$

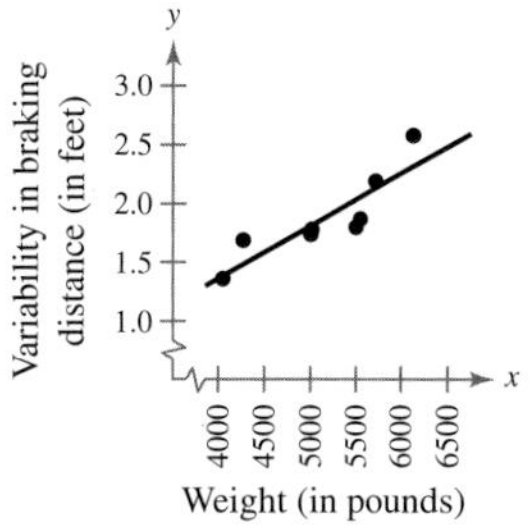

(a) 1.587 feet (b) 2.257 feet

(c) It is not meaningful to predict the value of y for $x = 7500$ because $x = 7500$ is outside the range of the original data.

(d) 2.145 feet

23. Substitute a value x into the equation of a regression line and solve for y.

25. (a) $\hat{y} = 1.724x + 79.733$

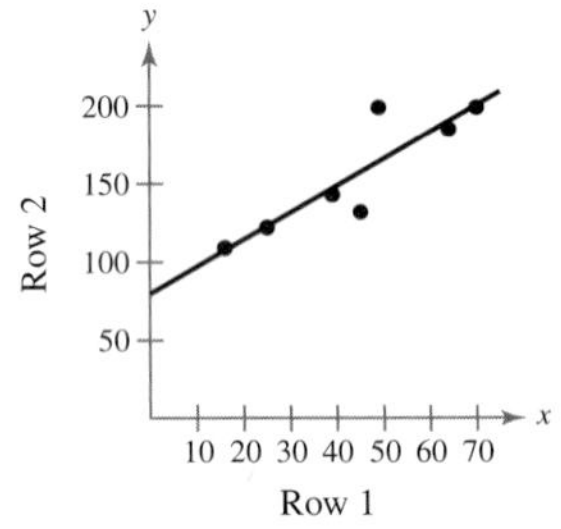

(b) $\hat{y} = 0.453x - 26.448$

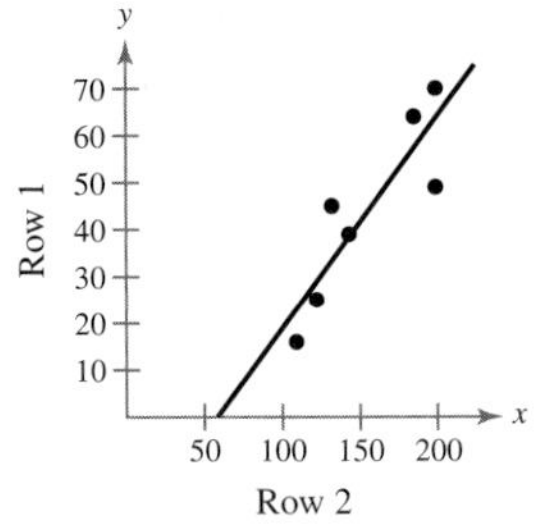

(c) The slope of the line keeps the same sign, but the values of m and b change.

27. Strong negative nonlinear correlation; as the number of subscribers increases, the average monthly bill decreases.

29. $17.73; This is not a valid prediction because $x = 180$ is outside the range of the original data.

31. As the number of cellular phone subscribers increases, the average monthly bill decreases. Then at approximately 70 million subscribers, the average monthly bill starts to increase as subscribers increase. (Answers will vary.)

33. The exponential equation seems to be the better fit because the line is curved.

35. $\hat{y} = 13.667 - 0.471 \ln x$

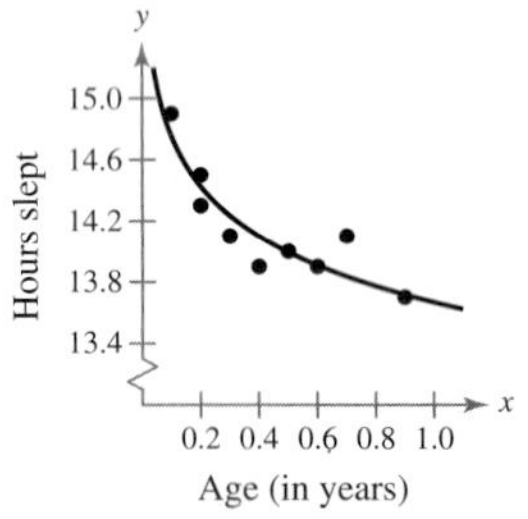

37. The logarithmic equation is a better model because the data do not follow a linear pattern.

Section 9.3 *(page 490)*

1. Total variation $= \Sigma(y_i - \bar{y})^2$; the sum of the squares of the differences between the y-values of each ordered pair and the mean of the y-values of the ordered pairs.

3. Unexplained variation $= \Sigma(y_i - \hat{y}_i)^2$; the sum of the squares of the differences between the observed y-values and the predicted y-values.

5. 0.063; 6.3% of the variation is explained. 93.7% of the variation is unexplained.

7. 0.794; 79.4% of the variation is explained. 20.6% of the variation is unexplained.

9. (a) 0.623; 62.3% of the variation in proceeds can be explained by the variation in the number of issues, and 37.7% of the variation is unexplained.

(b) 8775.838; the standard deviation of the proceeds for a specific number of issues is about $8,775, 838,000.

11. (a) 0.985; 98.5% of the variation in sales can be explained by the variation in the total square footage, and 1.5% of the variation is unexplained.

(b) 35.652; the standard deviation of the sales for a specific total square footage is about $35,652,000,000.

13. (a) 0.982; 98.2% of the variation in the median weekly earnings of female workers can be explained by the variation in the median weekly earnings of male workers, and 1.8% of the variation is unexplained.

(b) 8.226; the standard deviation of the median weekly earnings of female workers for a specific median weekly earnings of male workers is about $8.226.

15. (a) 0.992; 99.2% of the variation in the money spent can be explained by the variation in the money raised, and 0.8% of the variation is unexplained.

(b) 16.079; the standard deviation of the money spent for a specified amount of money raised is about $16,079,000.

17. $14,752,729,000 $< y <$ $58,063,787,000

You can be 95% confident that the proceeds will be between $14,752,729,000 and $58,063,787,000 when the number of initial offerings is 712.

19. $679,992,000,000 $< y <$ $817,870,000,000

You can be 90% confident that the sales will be between $679,992,000,000 and $817,870,000,000 when the total square footage is 4.5 billion.

21. $283.386 $< y <$ $444.690

You can be 99% confident that the median weekly earnings of female workers will be between $283.386 and $444.690 when the median weekly earnings of male workers is $500.

23. $718.708 million $< y <$ $812.216 million

You can be 95% confident that the money spent for congressional campaigns will be between $718.708 million and $812.216 million when the money raised is $775.8 million.

25.

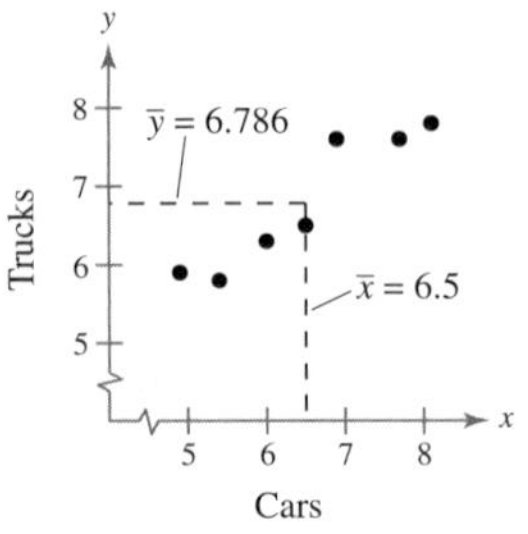

27.

x_i	y_i	$\hat{y}_i$	$\hat{y}_i - \bar{y}$	$y_i - \hat{y}_i$	$y_i - \bar{y}$
8.1	7.8	7.893	1.107	−0.093	1.014
7.7	7.6	7.616	0.830	−0.016	0.814
6.5	6.5	6.785	−0.001	−0.285	−0.286
6.9	7.6	7.062	0.276	0.538	0.814
6.0	6.3	6.438	−0.348	−0.138	−0.486
5.4	5.8	6.022	−0.764	−0.222	−0.986
4.9	5.9	5.676	−1.110	0.224	−0.886

29. 0.887; About 88.7% of the variation in the median age of trucks can be explained by the variation in the median age of cars, and 11.3% of the variation is unexplained.

31. (6.441, 8.237)

Section 9.4 *(page 497)*

1. (a) 2614.6 pounds (b) 2298 pounds
(c) 2680 pounds (d) 2233 pounds

3. (a) 7.5 cubic feet (b) 16.8 cubic feet
(c) 51.9 cubic feet (d) 62.1 cubic feet

5. $\hat{y} = -256.293 + 103.502x_1 + 14.649x_2$
(a) 34.16 (b) 0.988
(c) The standard deviation of the predicted sales given a specific total square footage and number of shopping centers is \$34.16 billion. The multiple regression model explains 98.8% of the variation in y.

7. 0.985

Uses and Abuses for Chapter 9 *(page 499)*

1. Answers will vary. **2.** Answers will vary.

Review Answers for Chapter 9 *(page 501)*

1.

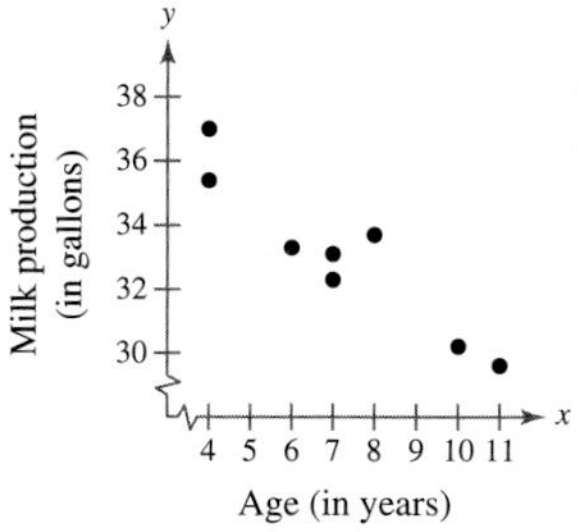

$r \approx -0.939$; negative linear correlation; milk production decreases with age.

3.

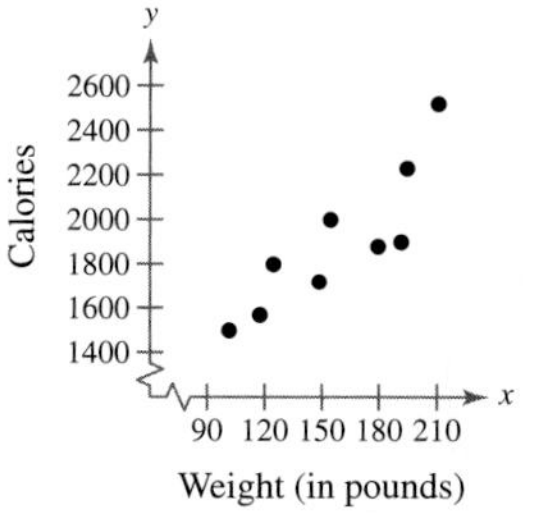

$r \approx 0.868$; positive linear correlation; caloric intake increases with weight.

5. There is not enough evidence at the 1% level to conclude that there is a significant linear correlation.

7. There is enough evidence at the 5% level to conclude that there is a significant linear correlation between the age of a cow and its milk production.

9. There is enough evidence at the 1% level to conclude that there is a significant linear correlation between weight and daily caloric intake.

11. $\hat{y} = 0.757x + 21.5$

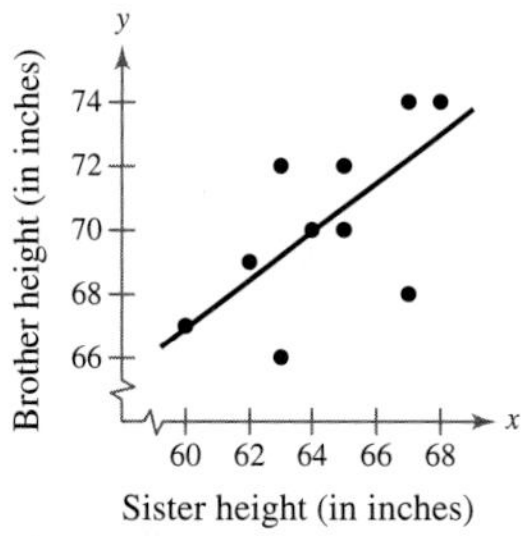

$r \approx 0.688$

13. $\hat{y} = -0.086x + 10.450$

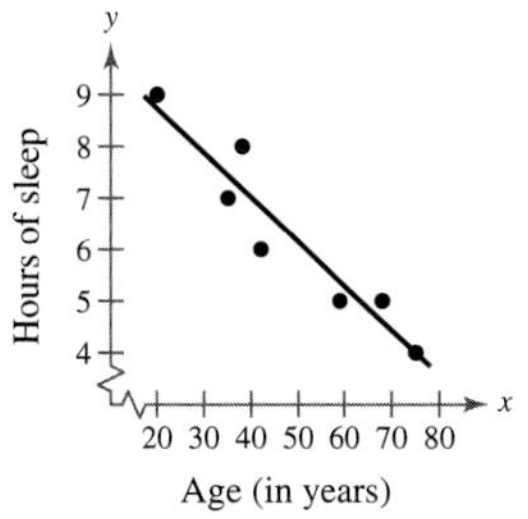

$r \approx -0.949$

15. (a) 68 inches (b) 71 inches
(c) It is not meaningful to predict the value of y for $x = 71$ because $x = 71$ is outside the range of the original data.
(d) It is not meaningful to predict the value of y for $x = 50$ because $x = 50$ is outside the range of the original data.

17. (a) It is not meaningful to predict the value of y for $x = 18$ because $x = 18$ is outside the range of the original data.
(b) 8.3 hours
(c) It is not meaningful to predict the value of y for $x = 80$ because $x = 80$ is outside the range of the original data.
(d) 6.15 hours

19. 0.306; 30.6% of the variation in y is explained by the model, and 69.4% of the variation is unexplained.

21. 0.033; 3.3% of the variation in y is explained by the model, and 96.7% of the variation is unexplained.

23. (a) 0.897; 89.7% of the variation in the cooling capacity of an air conditioner can be explained by the variation in the living area, and 10.3% of the variation is unexplained.
(b) 568.011; the standard deviation of the cooling capacity for a specific living area is 568.0 Btu per hour.

25. $65.71 < y < 74.19$ **27.** $4.865 < y < 8.295$

29. $7342.01 < y < 12{,}297.89$

31. $\hat{y} = 6.317 + 0.822x_1 + 0.031x_2 - 0.004x_3$

33. (a) 21.705 (b) 25.210 (c) 30.100 (d) 25.860

Chapter Quiz for Chapter 9 *(page 505)*

1.

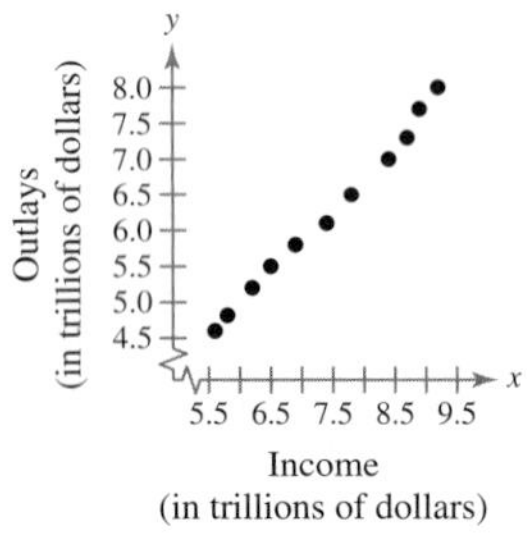

The data appear to have a positive correlation. The outlays increase as the incomes increases.

2. 0.996; strong positive linear correlation

3. There is enough evidence at the 5% level to conclude that a significant correlation exists between personal income and personal outlays for Americans.

4. $\hat{y} = 0.891x - 0.359$

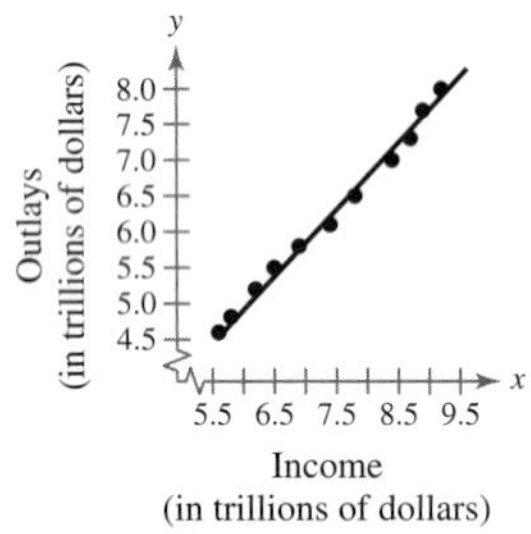

5. \$4.363 trillion

6. 0.992; 99.2% of the variation in personal outlays can be explained by the variation in personal income, and 0.8% of the variation is unexplained.

7. 0.110; the standard deviation of personal outlays for a specified personal income is \$0.110 trillion.

8. \$5.076 trillion $< y <$ \$5.610 trillion

You can be 95% confident that the personal outlays will be between \$5.076 trillion and \$5.610 trillion when personal income is \$6.4 trillion.

9. (a) 1311.150 pounds (b) 961.110 pounds
(c) 1120.900 pounds (d) 1386.740 pounds

x_2 has the greater influence on y.

Real Statistics–Real Decisions for Chapter 9
(page 506)

1. (a)

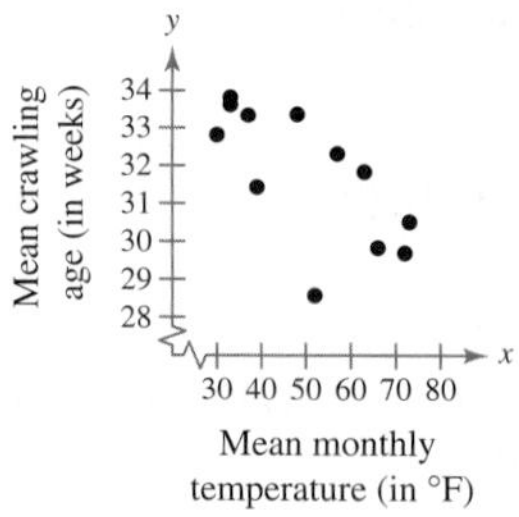

There appears to be a negative linear correlation between mean crawling age and the mean monthly temperature.

(b) −0.700

(c) There is enough evidence at the 5% level to conclude that there is a significant linear correlation between mean crawling age and the mean monthly temperature.

(d) $\hat{y} = -0.078x + 35.678$

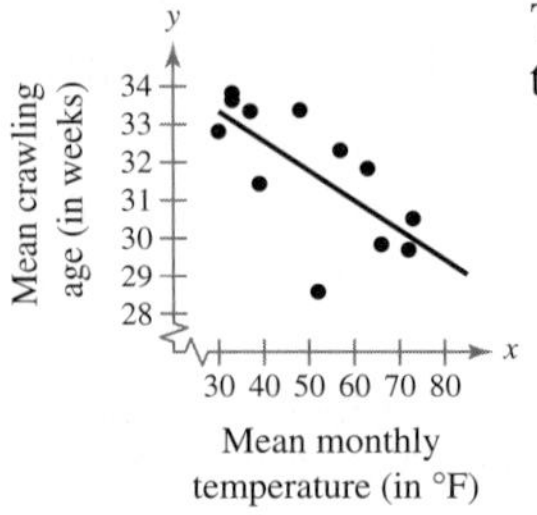

The regression line appears to be a good fit.

(e) Answers will vary.

(f) 0.490; 49.0% of the variation in mean crawling age can be explained by the variation in the mean monthly temperature, and 51.0% of the variation is unexplained; the standard deviation of mean crawling age for a specific mean monthly temperature is 1.319 weeks.

2. Answers will vary.

CHAPTER 10

Section 10.1 *(page 516)*

1. A multinomial experiment is a probability experiment consisting of a fixed number of trials in which there are more than two possible outcomes for each independent trial.

3. 45 **5.** 57.5

7. (a) Claimed distribution:

Response	Distribution
Agree	78%
Disagree	17%
Neither	5%

H_0: Distribution of responses is as shown in table above.

H_a: Distribution of responses differs from the claimed distribution.

(b) $\chi_0^2 = 5.991$; Rejection region: $\chi^2 > 5.991$ (c) 6.782

(d) Reject H_0. There is enough evidence at the 5% level to conclude that the distribution of responses differs from the claimed distribution.

9. (a) Claimed distribution:

Response	Distribution
Keep	77%
Tax cuts	15%
Not sure	8%

H_0: Distribution of responses is as shown in table above.

H_a: Distribution of responses differs from the claimed distribution.

(b) $\chi_0^2 = 9.210$; Rejection region: $\chi^2 > 9.210$ (c) 9.375

(d) Reject H_0. There is enough evidence at the 1% level to conclude that the distribution of responses differs from the claimed distribution.

11. (a) Claimed distribution:

Day	Distribution
Sunday	14.286%
Monday	14.286%
Tuesday	14.286%
Wednesday	14.286%
Thursday	14.286%
Friday	14.286%
Saturday	14.286%

H_0: The distribution of fatal bicycle accidents throughout the week is uniform as shown in the table above.

H_a: The distribution of fatal bicycle accidents throughout the week is not uniform.

(b) $\chi_0^2 = 10.645$; Rejection region: $\chi^2 > 10.645$

(c) 4.648

(d) Fail to reject H_0. There is not enough evidence at the 10% level to conclude that the distribution of fatal bicycle accidents throughout the week is not uniform.

13. (a) Claimed distribution:

Object struck	Distribution
Tree	28%
Embankment	10%
Utility pole	10%
Guardrail	9%
Ditch	7%
Curb	6%
Culvert	5%
Sign/post/fence	10%
Other	15%

H_0: Distribution of objects struck is as shown in table above.

H_a: Distribution of objects struck differs from the claimed distribution.

(b) $\chi_0^2 = 20.090$; Rejection region: $\chi^2 > 20.090$

(c) 49.665

(d) Reject H_0. There is enough evidence at the 1% level to conclude that the distribution of objects struck changed after warning signs were erected.

15. (a) Claimed distribution:

Response	Distribution
Not a HS grad	33.333%
HS graduate	33.333%
College (1yr+)	33.333%

H_0: Distribution of the responses is uniform as shown in table above.

H_a: Distribution of the responses is not uniform.

(b) $\chi_0^2 = 7.378$; Rejection region: $\chi^2 > 7.378$

(c) 5.637

(d) Fail to reject H_0. There is not enough evidence at the 2.5% level to conclude that the distribution of responses is not uniform.

17. (a) Claimed distribution:

Cause	Distribution
Trans. accidents	41%
Assaults	20%
Objects/equipment	15%
Falls	10%
Exposure	10%
Other	4%

H_0: Distribution of the causes is as shown in table above.

H_a: Distribution of the causes differs from the claimed distribution.

(b) $\chi_0^2 = 11.071$; Rejection region: $\chi^2 > 11.071$

(c) 9.493

(d) Fail to reject H_0. There is not enough evidence at the 5% level to conclude that the distribution of causes in the western United States differs from the national distribution.

19. (a) Frequency distribution: $\mu = 69.435$; $\sigma \approx 8.337$

Lower boundary	Upper boundary	Lower z-score	Upper z-score	Area
49.5	58.5	−2.39	−1.31	0.0867
58.5	67.5	−1.31	−0.23	0.3139
67.5	76.5	−0.23	0.85	0.3933
76.5	85.5	0.85	1.93	0.1709
85.5	94.5	1.93	3.01	0.0255

Class boundaries	Distribution	Freq. f	Expected	$\frac{(O-E)^2}{E}$
49.5–58.5	8.67%	19	17	0.235
58.5–67.5	31.39%	61	63	0.063
67.5–76.5	39.33%	82	79	0.114
76.5–85.5	17.09%	34	34	0
85.5–94.5	2.55%	4	5	0.2
		200		0.612

H_0: Variable has a normal distribution.

H_a: Variable does not have a normal distribution.

(b) $\chi_0^2 = 13.277$; Rejection region: $\chi^2 > 13.277$

(c) 0.612

(d) Fail to reject H_0. There is not enough evidence at the 1% level to conclude that the distribution of test scores is not normal.

Section 10.2 *(page 527)*

1. The expected frequency for a cell in a contingency table is the sum of the row the cell is in times the sum of the column the cell is in, all divided by the sample size.

3. False. A contingency table with two rows and five columns will have $(2 - 1)(5 - 1) = 4$ degrees of freedom.

5.

	Athlete has	
Result	**Stretched**	**Not stretched**
Injury	20.82	19.18
No injury	208.18	191.82

7.

	Treatment		
Result	**Brand-name**	**Generic**	**Placebo**
Improvement	15.84	14.96	24.20
No change	20.16	19.04	30.80

9. (a) H_0: Skill level in a subject is independent of location. (claim)

H_a: Skill level in a subject is dependent on location.

(b) d.f. = 2; $\chi_0^2 = 9.210$; Rejection region: $\chi^2 > 9.210$

(c) 0.297

(d) Fail to reject H_0. There is not enough evidence at the 1% level to conclude that skill level in a subject is dependent on location.

11. (a) H_0: Grades are independent of the institution.

H_a: Grades are dependent on the institution. (claim)

(b) d.f. = 8; $\chi_0^2 = 15.507$; Rejection region: $\chi^2 > 15.507$

(c) 48.488

(d) Reject H_0. There is enough evidence at the 5% level to conclude that grades are dependent on the institution.

13. (a) H_0: Results are independent of the type of treatment.

H_a: Results are dependent on the type of treatment. (claim)

(b) d.f. = 1; $\chi_0^2 = 2.706$; Rejection region: $\chi^2 > 2.706$

(c) 5.106

(d) Reject H_0. There is enough evidence at the 10% level to conclude that results are dependent on the type of treatment. I would recommend using the drug.

15. (a) H_0: Reasons are independent of the type of worker.

H_a: Reasons are dependent on the type of worker. (claim)

(b) d.f. = 2; $\chi_0^2 = 9.210$; Rejection region: $\chi^2 > 9.210$

(c) 7.326

(d) Fail to reject H_0. There is not enough evidence at the 1% level to conclude that reason(s) for continuing education are dependent on the type of worker. On the basis of these results, marketing strategies should not differ between technical and nontechnical audiences in regard to reason(s) for continuing education.

17. (a) H_0: Type of crash is independent of the type of vehicle.

H_a: Type of crash is dependent on the type of vehicle. (claim)

(b) d.f. = 2; $\chi_0^2 = 5.991$; Rejection region: $\chi^2 > 5.991$

(c) 106.390

(d) Reject H_0. There is enough evidence at the 5% level to conclude that the type of crash is dependent on the type of vehicle.

19. (a) H_0: Subject is independent of coauthorship.

H_a: Coauthorship and subject are dependent. (claim)

(b) d.f. = 4; $\chi_0^2 = 7.779$; Rejection region: $\chi^2 > 7.779$

(c) 5.610

(d) Fail to reject H_0. There is not enough evidence at the 10% level to conclude that subject matter and coauthorship are related.

21. Reject H_0. There is enough evidence at the 10% level to conclude that at least one of the proportions is different from the others.

Section 10.3 *(page 539)*

1. Specify the level of significance α. Determine the degrees of freedom for the numerator and denominator. Use Table 7 in Appendix B to find the critical value F.

3. 2.93 **5.** 5.32 **7.** 2.06

9. Fail to reject H_0. There is not enough evidence at the 10% level to support the claim.

11. Fail to reject H_0. There is not enough evidence at the 1% level to reject the claim.

13. Fail to reject H_0. There is not enough evidence at the 5% level to reject the claim.

15. (a) H_0: $\sigma_1^2 \leq \sigma_2^2$; H_a: $\sigma_1^2 > \sigma_2^2$ (claim)

(b) $F_0 = 2.13$; Rejection region: $F > 2.13$

(c) 1.077

(d) Fail to reject H_0. There is not enough evidence at the 5% level to support Company A's claim that the variance of the life of its appliances is less than the variance of the life of Company B's appliances.

17. (a) H_0: $\sigma_1^2 = \sigma_2^2$ (claim); H_a: $\sigma_1^2 \neq \sigma_2^2$

(b) $F_0 = 2.63$; Rejection region: $F > 2.63$

(c) 1.126

(d) Fail to reject H_0. There is not enough evidence at the 10% level to reject the claim that the standard deviation of physical science assessment test scores for eighth grade students is the same in Districts 1 and 2.

19. (a) H_0: $\sigma_1^2 \leq \sigma_2^2$; H_a: $\sigma_1^2 > \sigma_2^2$ (claim)

(b) $F_0 = 1.77$; Rejection region: $F > 1.77$

(c) 1.96

(d) Reject H_0. There is enough evidence at the 10% level to support the claim that the standard deviation of waiting times has decreased.

21. (a) H_0: $\sigma_1^2 \leq \sigma_2^2$; H_a: $\sigma_1^2 > \sigma_2^2$ (claim)

(b) $F_0 = 2.35$; Rejection region: $F > 2.35$ (c) 2.495

(d) Reject H_0. There is enough evidence at the 5% level to conclude that the standard deviation of the annual salaries for actuaries is greater in California than in New York.

23. Right tailed: 14.73 **25.** (0.366, 3.839)

Left tailed: 0.152

Section 10.4 *(page 549)*

1. H_0: $\mu_1 = \mu_2 = \mu_3 = \ldots = \mu_k$

H_a: At least one of the means is different from the others.

3. The MS_B measures the differences related to the treatment given to each sample. The MS_W measures the differences related to entries within the same sample.

5. (a) H_0: $\mu_1 = \mu_2 = \mu_3$

H_a: At least one mean is different from the others. (claim)

(b) $\text{d.f.}_N = 2$; $\text{d.f.}_D = 21$; $F_0 = 3.47$

Rejection region: $F > 3.47$

(c) 0.77

(d) Fail to reject H_0. There is not enough evidence at the 5% level to conclude that the mean costs per month are different.

7. (a) H_0: $\mu_1 = \mu_2 = \mu_3$ (claim)

H_a: At least one mean is different from the others.

(b) $\text{d.f.}_N = 2$; $\text{d.f.}_D = 12$; $F_0 = 2.81$

Rejection region: $F > 2.81$

(c) 1.77

(d) Fail to reject H_0. There is not enough evidence at the 10% level to reject the claim that the mean prices are all the same for the three types of treatment.

9. (a) H_0: $\mu_1 = \mu_2 = \mu_3 = \mu_4$ (claim)

H_a: At least one mean is different from the others.

(b) $\text{d.f.}_N = 3$; $\text{d.f.}_D = 29$; $F_0 = 4.54$

Rejection region: $F > 4.54$

(c) 4.98

(d) Reject H_0. There is enough evidence at the 1% level to reject the claim that the mean number of days patients spend in the hospital is the same for all four regions.

11. (a) H_0: $\mu_1 = \mu_2 = \mu_3 = \mu_4$ (claim)

H_a: At least one mean is different from the others.

(b) $\text{d.f.}_N = 3$; $\text{d.f.}_D = 43$; $F_0 = 2.22$

Rejection region: $F > 2.22$

(c) 3.04

(d) Reject H_0. There is enough evidence at the 10% level to reject the claim that the mean price is the same for all four cities.

13. (a) H_0: $\mu_1 = \mu_2 = \mu_3 = \mu_4$

H_a: At least one mean is different from the others. (claim)

(b) $\text{d.f.}_N = 3$; $\text{d.f.}_D = 26$; $F_0 = 2.31$

Rejection region: $F > 2.31$

(c) 6.36

(d) Reject H_0. There is enough evidence at the 10% level to conclude that at least one of the mean prices is different.

15. (a) H_0: $\mu_1 = \mu_2 = \mu_3 = \mu_4$ (claim)

H_a: At least one mean is different from the others.

(b) $\text{d.f.}_N = 3$; $\text{d.f.}_D = 19$; $F_0 = 5.01$

Rejection region: $F > 5.01$

(c) 0.19

(d) Fail to reject H_0. There is not enough evidence at the 1% level to reject the claim that the mean numbers of female students are equal for all grades.

17. Fail to reject all null hypotheses. The interaction between advertising medium and the length of the ad has no effect on the rating and therefore there is no significant difference in the means of the ratings.

19. Fail to reject all null hypotheses. The interaction between age and gender has no effect on GPA and therefore there is no significant difference in the means of the GPAs.

21. $CV_{\text{Scheffé}} = 6.60$

$(1, 2) \rightarrow 4.927 \rightarrow$ No difference

$(1, 3) \rightarrow 0.650 \rightarrow$ No difference

$(1, 4) \rightarrow 4.426 \rightarrow$ No difference

$(2, 3) \rightarrow 1.998 \rightarrow$ No difference

$(2, 4) \rightarrow 0.013 \rightarrow$ No difference

$(3, 4) \rightarrow 1.684 \rightarrow$ No difference

23. $CV_{\text{Scheffé}} \approx 13.14$

$(1, 2) \rightarrow 0.966 \rightarrow$ No difference

$(1, 3) \rightarrow 9.587 \rightarrow$ No difference

$(1, 4) \rightarrow 7.897 \rightarrow$ No difference

$(2, 3) \rightarrow 16.979 \rightarrow$ Significant difference

$(2, 4) \rightarrow 14.892 \rightarrow$ Significant difference

$(3, 4) \rightarrow 0.130 \rightarrow$ No difference

Uses and Abuses for Chapter 10 *(page 555)*

1. Answers will vary.

2. Answers will vary.

Review Answers for Chapter 10 *(page 557)*

1. Reject H_0. There is enough evidence at the 5% level to conclude that the distribution of office visits differs from the claimed distribution.

3. Fail to reject H_0. There is not enough evidence at the 10% level to conclude that the distribution of prices differs from the claimed distribution.

5. (a) Expected frequencies:

	HS—did not complete	HS complete	College 1–3 years
25–44	13,836.62	27,733.60	20,686.43
45+	15,047.38	30,160.40	22,496.57
Total	28,884	57,894	43,183

	College 4+ years	Total
25–44	20,962.35	83,219
45+	22,796.65	90,501
Total	43,759	173,720

(b) Reject H_0.

(c) There is enough evidence at the 10% level to conclude that the education level of people in the United States and their age are dependent.

7. (a) Expected frequencies:

	Age Group			
Gender	16–20	21–30	31–40	41–50
Male	143.76	284.17	274.14	237.37
Female	71.24	140.83	135.86	117.63
Total	215	425	410	355

	Age Group		
Gender	51–60	61+	Total
Male	147.10	43.46	1130
Female	72.90	21.54	560
Total	220	65	1690

(b) Fail to reject H_0.

(c) There is not enough evidence at the 5% level to conclude that age group is dependent on gender.

9. 2.295 **11.** 2.39

13. Fail to reject H_0. There is not enough evidence at the 1% level to reject the claim.

15. Fail to reject H_0. There is not enough evidence at the 10% level to support the claim that the variation in wheat production is greater in Garfield County than in Kay County.

17. Fail to reject H_0. There is not enough evidence at the 1% level to support the claim that the test score variance for females is different from that for males.

19. Reject H_0. There is enough evidence at the 10% level to conclude that the mean residential energy expenditures are not the same for all four regions.

Chapter Quiz for Chapter 10 *(page 561)*

1. (a) H_0: $\sigma_1^2 = \sigma_2^2$; H_a: $\sigma_1^2 \neq \sigma_2^2$ (claim)

(b) 0.01 (c) $F_0 = 3.97$

(d) Rejection region: $F > 3.97$

(e) 3.813 (f) Fail to reject H_0.

(g) There is not enough evidence at the 1% level to conclude that the variances in annual wages for San Jose, CA and Dallas, TX are different.

2. (a) H_0: $\mu_1 = \mu_2 = \mu_3$ (claim)

H_a: At least one mean is different from the others.

(b) 0.10 (c) $F_0 = 2.43$

(d) Rejection region: $F > 2.43$

(e) 7.39 (f) Reject H_0.

(g) There is enough evidence at the 10% level to conclude that the mean annual wages for the three cities are not all equal.

3. (a) Claimed distribution:

Education	25 & Over
Not a HS graduate	16.6%
HS graduate	33.3%
Some college, no degree	17.3%
Associate degree	7.5%
Bachelor's degree	17.0%
Advanced degree	8.2%

H_0: Distribution of educational achievement for people in the United States aged 35–44 is as shown in table above.

H_a: Distribution of educational achievement for people in the United States aged 35–44 differs from the claimed distribution.

(b) 0.01 (c) $\chi_0^2 = 15.086$

(d) Rejection region: $\chi^2 > 15.086$

(e) 5.189 (f) Fail to reject H_0.

(g) There is not enough evidence at the 1% level to conclude the distribution of educational achievement for people in the United States aged 35–44 differs from the claimed distribution.

4. (a) Claimed distribution:

Education	25 & Over
Not a HS graduate	16.6%
HS graduate	33.3%
Some college, no degree	17.3%
Associate degree	7.5%
Bachelor's degree	17.0%
Advanced degree	8.2%

H_0: Distribution of educational achievement for people in the United States aged 65–74 is as shown in table above.

H_a: Distribution of educational achievement for people in the United States aged 65–74 differs from the claimed distribution.

(b) 0.05 (c) $\chi_0^2 = 11.071$

(d) Rejection region: $\chi^2 > 11.071$

(e) 76.532 (f) Reject H_0.

(g) There is enough evidence at the 5% level to conclude that the distribution of educational achievement for people in the United States aged 65–74 differs from the claimed distribution.

Real Statistics–Real Decisions for Chapter 10

(page 562)

1. Claimed distribution:

Response	Distribution
Under 20	1%
20–29	13%
30–39	16%
40–49	19%
50–59	16%
60–69	13%
70+	22%

H_0: Distribution of responses is as shown in table above. (claim)

H_a: Distribution of responses differs from the claimed distribution.

$\chi_0^2 = 16.812$

$\chi^2 = 497.26$

Reject H_0. There is enough evidence at the 1% level to conclude that the distribution of the ages of telemarketing fraud victims differs from the survey.

2. (a)

Type of Fraud	Age				
	Under 20	20–29	30–39	40–49	50–59
Sweepstakes	10 (15)	60 (120)	70 (165)	130 (185)	90 (135)
Credit Card	20 (15)	180 (120)	260 (165)	240 (185)	180 (135)
Total	30	240	330	370	270

Type of Fraud	Age			
	60–69	70–79	80+	Total
Sweepstakes	160 (115)	280 (155)	200 (110)	1000
Credit Card	70 (115)	30 (155)	20 (110)	1000
Total	230	310	220	2000

(b) H_0: The type of fraud is independent of the victim's age.

H_a: The type of fraud is dependent on the victim's age.

$\chi_0^2 = 18.475$; $\chi^2 \approx 619.533$

Reject H_0. There is enough evidence at the 1% level to conclude that the victims' ages and the type of fraud are dependent.

CHAPTER 11

Section 11.1 *(page 572)*

1. A nonparametric test is a hypothesis test that does not require any specific conditions concerning the shape of populations or the value of any population parameters.

A nonparametric test is usually easier to perform than its corresponding parametric test, but the nonparametric test is usually less efficient.

3. (a) H_0: median $\leq$ \$200; H_a: median $>$ \$200 (claim)

(b) 1 (c) 5 (d) Fail to reject H_0.

(e) There is not enough evidence at the 1% level to support the claim that the median amount of new credit card charges for the previous month was more than \$200.

5. (a) H_0: median $\leq$ \$148,000 (claim)

H_a: median $>$ \$148,000

(b) 1 (c) 3 (d) Fail to reject H_0.

(e) There is not enough evidence at the 5% level to reject the claim that the median sales price of new privately owned one-family homes sold in the past year is \$148,000 or less.

7. (a) H_0: median $\geq$ \$1900 (claim); H_a: median $<$ \$1900
(b) -2.055 (c) -3.040 (d) Reject H_0.
(e) There is enough evidence at the 2% level to reject the claim that the median amount of credit card debt for families holding such debts is at least \$1900.

9. (a) H_0: median $\leq$ 34; H_a: median $>$ 34 (claim)
(b) 3 (c) 8 (d) Fail to reject H_0.
(e) There is not enough evidence at the 1% level to support the claim that the median age of recipients of social science doctorates is greater than 34.

11. (a) H_0: median $=$ 4 (claim); H_a: median $\neq$ 4
(b) -1.96 (c) -2.334 (d) Reject H_0.
(e) There is enough evidence at the 5% level to reject the claim that the median number of rooms in renter-occupied units is 4.

13. (a) H_0: median $=$ \$11.63 (claim); H_a: median $\neq$ \$11.63
(b) -2.575 (c) -0.961 (d) Fail to reject H_0.
(e) There is not enough evidence at the 1% level to reject the claim that the median hourly earnings of male workers paid hourly rates is \$11.63.

15. (a) H_0: The headache hours have not decreased.
H_a: The headache hours have decreased. (claim)
(b) 1 (c) 3 (d) Fail to reject H_0.
(e) There is not enough evidence at the 5% level to support the claim that the daily headache hours were reduced after the soft-tissue therapy and spinal manipulation.

17. (a) H_0: The SAT scores have not improved.
H_a: The SAT scores have improved. (claim)
(b) 2 (c) 4 (d) Fail to reject H_0.
(e) There is not enough evidence at the 5% level to support the claim that verbal SAT scores improved.

19. (a) Fail to reject H_0.
(b) There is not enough evidence at the 5% level to reject the claim that the proportion of companies that provide transportation home is equal to the proportion of companies that do not.

21. (a) H_0: median $\leq$ \$530 (claim); H_a: median $>$ \$530
(b) 2.33 (c) 1.459
(d) Fail to reject H_0. There is not enough evidence at the 1% level to reject the claim that the median weekly earnings of female workers is less than or equal to \$530.

23. (a) H_0: median $\leq$ 25.3; H_a: median $>$ 25.3 (claim)
(b) 1.645 (c) 1.936
(d) Reject H_0. There is enough evidence at the 5% level to support the claim that the median age of first-time brides is greater than 25.3 years.

Section 11.2 *(page 583)*

1. If the samples are dependent, use a Wilcoxon signed-rank test. If the samples are independent, use a Wilcoxon rank sum test.

3. (a) H_0: There is no reduction in diastolic blood pressure. (claim)
H_a: There is a reduction in diastolic blood pressure.
(b) Wilcoxon signed-rank test
(c) 10 (d) 17 (e) Fail to reject H_0.
(f) There is not enough evidence at the 1% level to reject the claim that there was no reduction in diastolic blood pressure.

5. (a) H_0: There is no difference in the earnings.
H_a: There is a difference in the earnings. (claim)
(b) Wilcoxon rank sum test
(c) ± 1.96 (d) -3.87 (e) Reject H_0.
(f) There is enough evidence at the 5% level to support the claim that there is a difference in the earnings.

7. (a) H_0: There is no difference in salaries.
H_a: There is a difference in salaries. (claim)
(b) Wilcoxon rank sum test
(c) ± 1.96 (d) -1.82 (e) Fail to reject H_0.
(f) There is not enough evidence at the 5% level to support the claim that there is a difference in salaries.

9. Reject H_0. There is enough evidence at the 10% level to conclude that the gas mileage is improved.

Section 11.3 *(page 591)*

1. The conditions for using a Kruskal-Wallis test are that the sample must be randomly selected and the size of each sample must be at least 5.

3. (a) H_0: There is no difference in the premiums.
H_a: There is a difference in the premiums. (claim)
(b) 5.991 (c) 14.05 (d) Reject H_0.
(e) There is enough evidence at the 5% level to support the claim that the distributions of the annual premiums in California, Florida, and Illinois are different.

5. (a) H_0: There is no difference in the salaries.
H_a: There is a difference in the salaries. (claim)
(b) 6.251 (c) 1.45 (d) Fail to reject H_0.
(e) There is not enough evidence at the 10% level to support the claim that the distributions of the annual salaries in the four states are different.

7. (a) Fail to reject H_0.
(b) Fail to reject H_0.
(c) Both tests come to the same decision, which is, that there is not enough evidence to support the claim that there is a difference in the number of days spent in the hospital.

Section 11.4 *(page 596)*

1. The Spearman rank correlation coefficient can be used to describe the relationship between linear or nonlinear data. Also, it can be used for data at the ordinal level and it is easier to calculate by hand than the Pearson correlation coefficient.

3. (a) H_0: $\rho_s = 0$; H_a: $\rho_s \neq 0$ (claim)

(b) 0.929 (c) 0.750 (d) Fail to reject H_0.

(e) There is not enough evidence at the 1% level to support the claim that there is a correlation between debt and income in the farming business.

5. (a) H_0: $\rho_s = 0$; H_a: $\rho_s \neq 0$ (claim)

(b) 0.881 (c) 0.173 (d) Fail to reject H_0.

(e) There is not enough evidence at the 1% level to support the claim that there is a correlation between the overall score and price.

7. Fail to reject H_0. There is not enough evidence at the 5% level to conclude that there is a correlation between science achievement scores and GNP.

9. Fail to reject H_0. There is not enough evidence at the 5% level to conclude that there is a correlation between science and mathematics achievement scores.

11. Fail to reject H_0. There is not enough evidence at the 5% level to support the claim that there is a correlation between average hours worked and the number of on-the-job injuries.

Section 11.5 *(page 605)*

1. Number of runs: 8

Run lengths: 1, 1, 1, 1, 3, 3, 1, 1

3. Number of runs: 9

Run lengths: 1, 1, 1, 1, 1, 6, 3, 2, 4

5. n_1 = number of Ts = 6

n_2 = number of Fs = 6

7. n_1 = number of Ms = 10

n_2 = number of Fs = 10

9. too high: 11
too low: 3

11. too high: 14
too low: 5

13. (a) H_0: The coin tosses are random.

H_a: The coin tosses are not random. (claim)

(b) lower critical value = 4

upper critical value = 14

(c) 9 (d) Fail to reject H_0.

(e) There is not enough evidence at the 5% level to support the claim that the coin tosses are not random.

15. (a) H_0: The sequence of digits are randomly generated.

H_a: The sequence of digits are not randomly generated. (claim)

(b) lower critical value = 11

upper critical value = 23

(c) 9 (d) Reject H_0.

(e) There is enough evidence at the 5% level to support the claim that the sequence of digits was not randomly generated.

17. (a) H_0: The production of defective parts is random.

H_a: The production of defective parts is not random. (claim)

(b) ± 1.96 (c) -0.85 (d) Fail to reject H_0.

(e) There is not enough evidence at the 5% level to support the claim that the production of defective parts is not random.

19. Fail to reject H_0. There is not enough evidence at the 5% level to support the claim that the daily high temperatures do not occur randomly.

21. Answers will vary.

Uses and Abuses for Chapter 11 *(page 607)*

1. Answers will vary.

2. Sign test → z-test

Paired-sample sign test → t-test

Wilcoxon signed-rank test → t-test

Wilcoxon rank sum test → z or t-test

Kruskal-Wallis test → one-way ANOVA

Spearman rank correlation coefficient → Pearson correlation coefficient

Review Answers for Chapter 11 *(page 609)*

1. (a) H_0: median = \$34,300 (claim)

H_a: median ≠ \$34,300

(b) 2 (c) 7 (d) Fail to reject H_0.

(e) There is not enough evidence at the 1% level to reject the claim that the median value of stock among families that own stock is \$34,300.

3. (a) H_0: median ≤ 6 (claim); H_a: median > 6

(b) -1.28 (c) -2.032 (d) Reject H_0.

(e) There is enough evidence at the 10% level to reject the claim that the median turnover time is no more than 6 hours.

5. (a) H_0: There is no reduction in diastolic blood pressure. (claim)

H_a: There is a reduction in diastolic blood pressure.

(b) 2 (c) 3 (d) Fail to reject H_0.

(e) There is not enough evidence at the 5% level to reject the claim that there was no reduction in diastolic blood pressure.

7. (a) Independent; Wilcoxon rank sum test

(b) H_0: There is no difference in the amount of time that it takes to earn a doctorate.

H_a: There is a difference in the amount of time that it takes to earn a doctorate. (claim)

(c) ± 2.575 (d) -3.175 (e) Reject H_0.

(f) There is enough evidence at the 1% level to support the claim that there is a difference in the amount of time that it takes to earn a doctorate.

9. (a) H_0: There is no difference in salaries between the fields of study.

H_a: There is a difference in salaries between the fields of study. (claim)

(b) 5.991 (c) 22.98 (d) Reject H_0.

(e) There is enough evidence at the 5% level to conclude that there is a difference in salaries between the fields of study.

11. (a) H_0: $\rho_s = 0$; H_a: $\rho_s \neq 0$ (claim)

(b) 0.881 (c) -0.429 (d) Fail to reject H_0.

(e) There is not enough evidence at the 1% level to conclude that there is a correlation between overall score and price.

13. (a) H_0: The traffic stops are random by gender.

H_a: The traffic stops are not random by gender. (claim)

(b) lower critical value = 8

upper critical value = 19

(c) 14 (d) Fail to reject H_0.

(e) There is not enough evidence at the 5% level to support the claim that the traffic stops are not random by gender.

Chapter Quiz for Chapter 11 *(page 613)*

1. (a) H_0: There is no difference in the salaries between genders.

H_a: There is a difference in the salaries between genders. (claim)

(b) Wilcoxon rank sum test

(c) ± 1.645 (d) -1.722 (e) Reject H_0.

(f) There is enough evidence at the 10% level to support the claim that there is a difference in the salaries between genders.

2. (a) H_0: median = 28 (claim); H_a: median $\neq$ 28

(b) Sign test (c) 6 (d) 10 (e) Fail to reject H_0.

(f) There is not enough evidence at the 5% level to reject the claim that the median age in Puerto Rico is 28 years.

3. (a) H_0: There is no difference in the annual premiums between the states.

H_a: There is a difference in the annual premiums between the states. (claim)

(b) Kruskal-Wallis test

(c) 5.991 (d) 1.43 (e) Fail to reject H_0.

(f) There is not enough evidence at the 5% level to conclude that there is a difference in the annual premiums between the states.

4. (a) H_0: The days with rain are random.

H_a: The days with rain are not random. (claim)

(b) Runs test

(c) lower critical value = 10

upper critical value = 22

(d) 16 (e) Fail to reject H_0.

(f) There is not enough evidence at the 5% level to support the claim that days with rain are not random.

Real Statistics–Real Decisions for Chapter 11
(page 614)

1. (a) Random sample

(b) Answers will vary.

(c) Answers will vary.

2. (a) Answers will vary.

(b) Sign test; You need to use the nonparametric test because nothing is known about the shape of the population.

(c) H_0: median $\geq$ 3.5; H_a: median $<$ 3.5 (claim)

(d) Fail to reject H_0. There is not enough evidence at the 5% level to support the claim that the median tenure is less than 3.5 years.

3. (a) Wilcoxon rank sum test

(b) H_0: median male tenure $\leq$ median female tenure

H_a: median male tenure $>$ median female tenure (claim)

(c) Fail to reject H_0. There is not enough evidence at the 5% level to conclude that the median tenure for male workers is greater than the median tenure for female workers.

Selected Answers

CHAPTER 1

Section 1.1

28. Parameter. 12% is a numerical description of all new magazines.

30. Parameter. 20.7 is a numerical description of ACT scores for all graduates.

36. (a) An inference drawn from the sample is that the number of people who have strokes has increased every year for the past 15 years.

(b) This inference may incorrectly imply that the same trend will continue for the next 15 years.

Section 1.3

2. False. A census is a count of an entire population.

6. Use sampling because it would be impossible to ask every consumer whether he or she would still buy a product with a warning label.

8. Take a census because the U.S. Congress keeps records on the ages of its members.

10. Stratified sampling is used because the persons are divided into strata and a sample is selected from each stratum.

12. Cluster sampling is used because the disaster area was divided into grids and 30 grids were then entirely selected. Certain grids may have been much more severely damaged than others, so this is a possible source of bias.

14. Systematic sampling is used because every twentieth engine part is sampled. It is possible for bias to enter into the sample if, for some reason, the assembly line performs differently on a consistent basis.

18. Simple random sampling is used because each telephone has an equal chance of being dialed and all samples of 1012 phone numbers have an equal chance of being selected. The sample may be biased because only homes with telephones have a chance of being sampled.

20. Sampling. The population of cars is too large to easily record their color. Cluster sampling is advised because it would be easy to randomly select car dealerships then record the color for every car sold at the selected dealerships.

26. Stratified sampling ensures that each segment of the population is represented.

28. (a) Advantage: Usually results in a savings in the survey cost.

(b) Disadvantage: There tends to be a lower response rate and this can introduce a bias into the sample.

Sampling technique: Convenience sampling

Review Answers for Chapter 1

28. Convenience sampling is used because of the convenience of surveying people leaving one restaurant.

30. Because of the convenience sample taken, the study may be biased toward the opinions of the student's friends.

32. In heavy interstate traffic, it may be difficult to identify every tenth car that passed the law enforcement official.

CHAPTER 2

Section 2.1

10. (a) 5

(b) and (c)

Class	Midpoint	Class boundaries
16–20	18	15.5–20.5
21–25	23	20.5–25.5
26–30	28	25.5–30.5
31–35	33	30.5–35.5
36–40	38	35.5–40.5
41–45	43	40.5–45.5
46–50	48	45.5–50.5

12.

Class	Frequency, f	Midpoint	Relative frequency	Cumulative frequency
16–20	100	18	0.03	100
21–25	122	23	0.04	222
26–30	900	28	0.30	1122
31–35	207	33	0.07	1329
36–40	795	38	0.26	2124
41–45	568	43	0.19	2692
46–50	322	48	0.11	3014
	$\Sigma f = 3014$		$\Sigma \frac{f}{n} = 1$	

24.

Class	Frequency, f	Midpoint	Relative frequency	Cumulative frequency
30–113	5	71.5	0.1724	5
114–197	7	155.5	0.2414	12
198–281	8	239.5	0.2759	20
282–365	2	323.5	0.0690	22
366–449	3	407.5	0.1034	25
450–533	4	491.5	0.1379	29
	$\Sigma f = 29$		$\Sigma \frac{f}{n} = 1$	

26.

Class	Frequency, f	Midpoint	Relative frequency	Cumulative frequency
32–35	3	33.5	0.1250	3
36–39	9	37.5	0.3750	12
40–43	8	41.5	0.3333	20
44–47	3	45.5	0.1250	23
48–51	1	49.5	0.0417	24
	$\Sigma f = 24$		$\Sigma \frac{f}{n} = 1$	

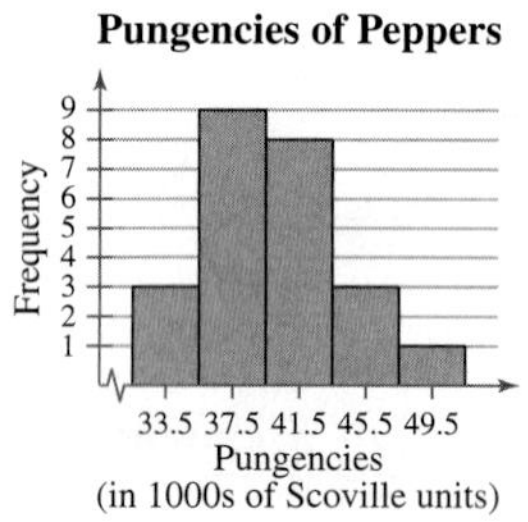

Class with greatest frequency: 36–39

Class with least frequency: 48–51

28.

Class	Frequency, f	Midpoint	Relative frequency	Cumulative frequency
2456–2542	7	2499	0.28	7
2543–2629	3	2586	0.12	10
2630–2716	2	2673	0.08	12
2717–2803	4	2760	0.16	16
2804–2890	9	2847	0.36	25
	$\Sigma f = 25$		$\Sigma \frac{f}{n} = 1$	

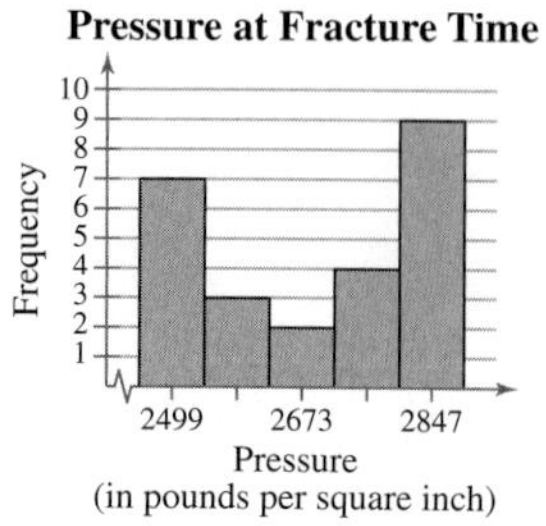

Class with greatest frequency: 2804–2890

Class with least frequency: 2630–2716

30.

Class	Frequency, f	Midpoint	Relative frequency	Cumulative frequency
10–23	11	16.5	0.3438	11
24–37	9	30.5	0.2813	20
38–51	6	44.5	0.1875	26
52–65	2	58.5	0.0625	28
66–80	4	72.5	0.1250	32
	$\Sigma f = 32$		$\Sigma \frac{f}{n} \approx 1$	

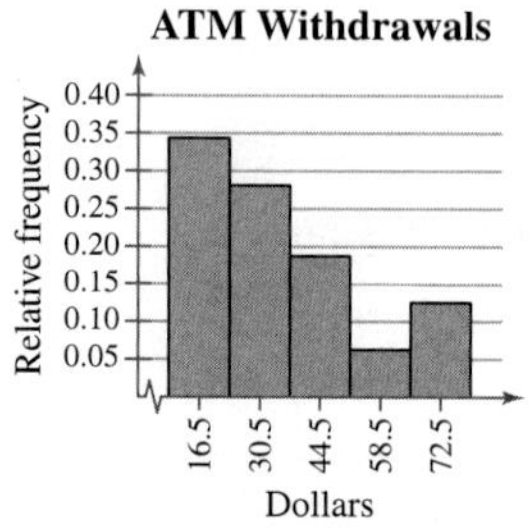

Class with greatest relative frequency: 10–23

Class with least relative frequency: 52–65

32.

Class	Frequency, f	Midpoint	Relative frequency	Cumulative frequency
6–7	3	6.5	0.12	3
8–9	10	8.5	0.38	13
10–11	6	10.5	0.23	19
12–13	6	12.5	0.23	25
14–15	1	14.5	0.04	26
	$\Sigma f = 26$		$\Sigma \frac{f}{n} = 1$	

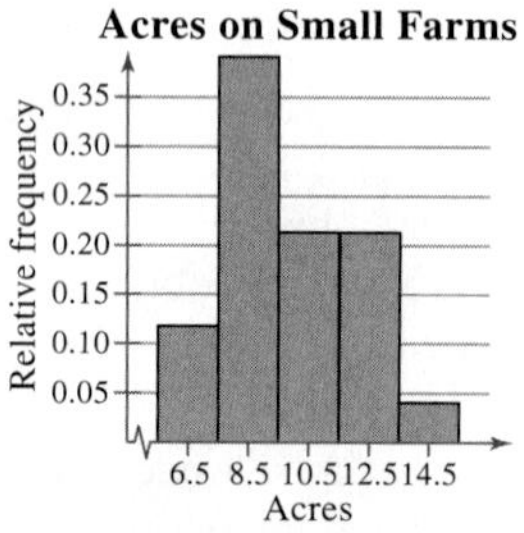

Class with greatest relative frequency: 8–9

Class with least relative frequency: 14–15

34.

Class	Frequency, f	Relative frequency	Cumulative frequency
16–22	2	0.10	2
23–29	3	0.15	5
30–36	8	0.40	13
37–43	5	0.25	18
44–50	0	0.00	18
51–57	2	0.10	20
	$\Sigma f = 20$	$\Sigma \frac{f}{n} = 1$	

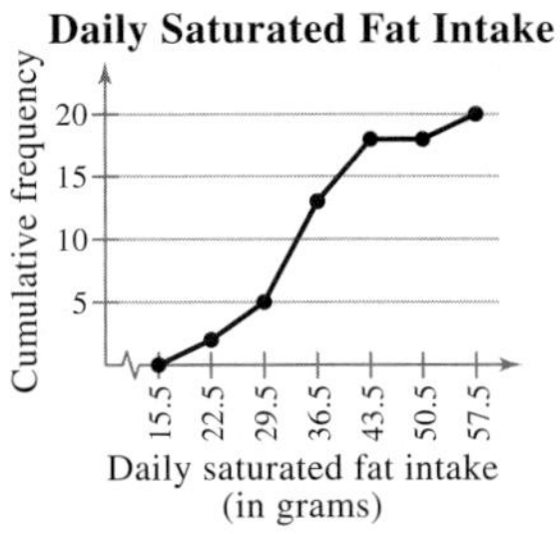

Location of the greatest increase in frequency: 30–36

36.

Class	Frequency, f	Relative frequency	Cumulative frequency
1–5	5	0.2083	5
6–10	9	0.3750	14
11–15	3	0.1250	17
16–20	4	0.1667	21
21–25	2	0.0833	23
26–30	1	0.0417	24
	$\Sigma f = 24$	$\Sigma \frac{f}{n} = 1$	

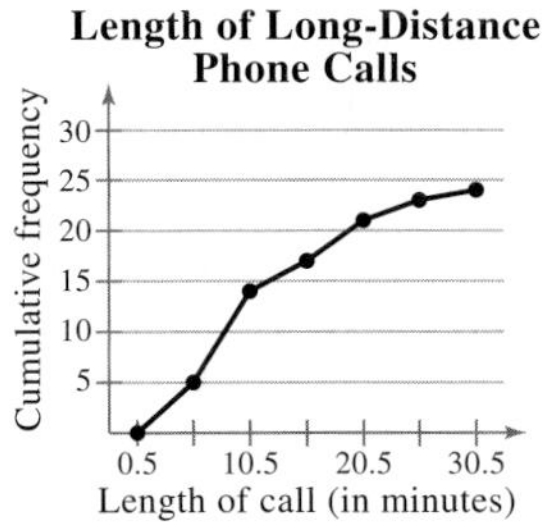

Location of the greatest increase in frequency: 6–10

38.

Class	Frequency, f	Mid-point	Relative frequency	Cumulative frequency
0–2	17	1	0.4048	17
3–5	16	4	0.3810	33
6–8	7	7	0.1667	40
9–11	1	10	0.0238	41
12–14	0	13	0.0000	41
15–17	1	16	0.0238	42
	$\Sigma f = 42$		$\Sigma \frac{f}{n} \approx 1$	

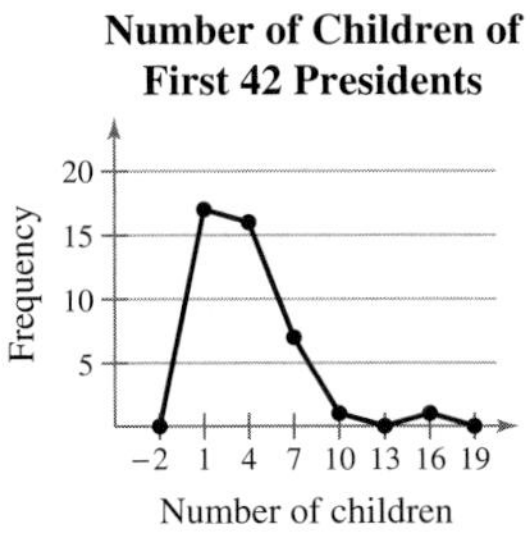

Class with greatest frequency: 0–2

Class with least frequency: 12–14

40. (a)

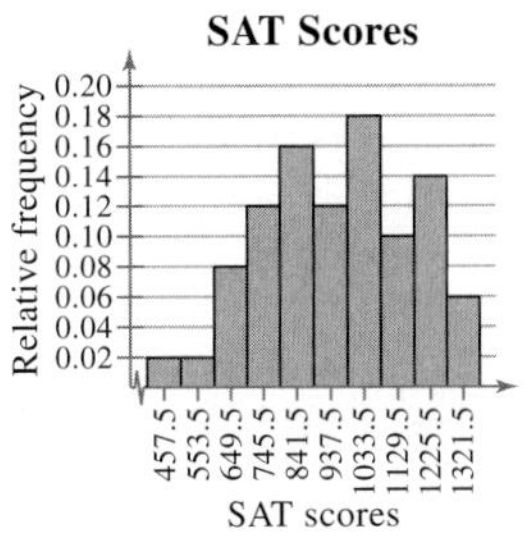

(b) 48%, because the sum of the relative frequencies for the last four classes is 0.48.

(c) 698, because the sum of the relative frequencies for the last seven classes is 0.88.

Section 2.2

18.

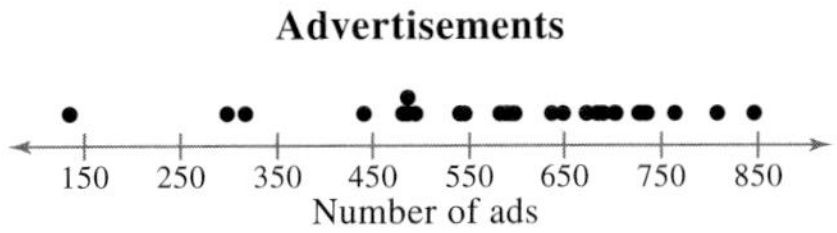

It appears that most of the 30 people from the United States see or hear between 450 and 750 advertisements per week. (Answers will vary.)

22.

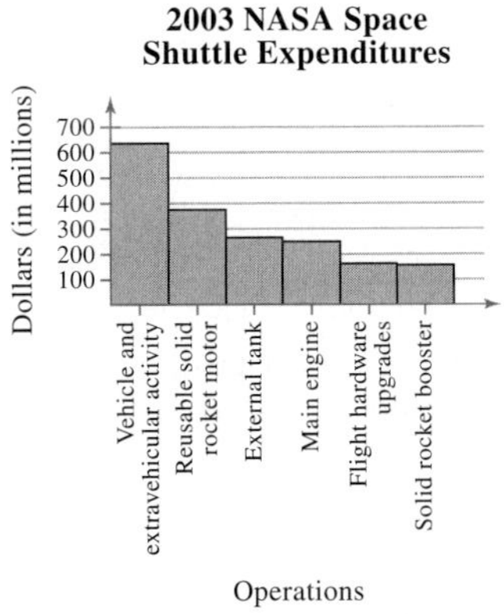

The greatest NASA space shuttle operations expenditures in 2003 were for vehicle and extravehicular activity; the least were for solid rocket booster. (Answers will vary.)

26.

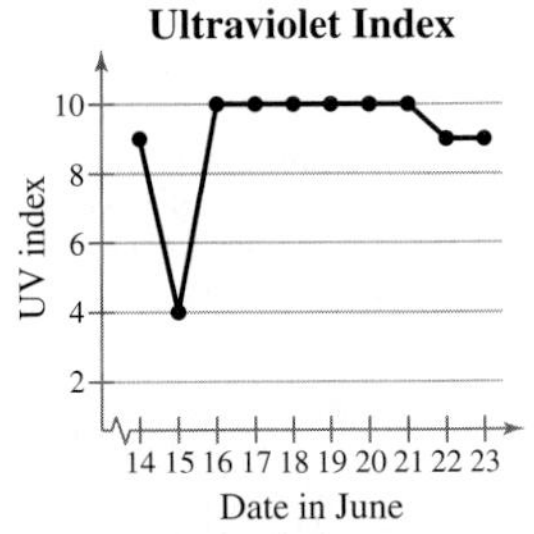

Of the period from June 14 to 23, the ultraviolet index was highest from June 16 to 21 in Memphis, TN. (Answers will vary.)

28.

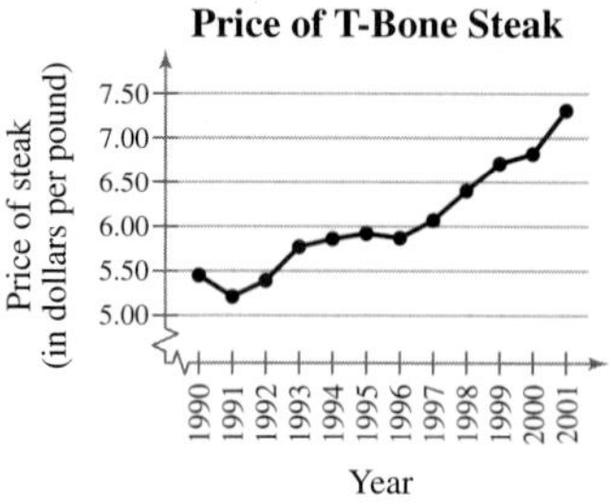

It appears that the price of a T-bone steak steadily increased from 1991 to 2001.

30. (a) The pie chart should be displaying all four quarters, not just the first three.

(b)

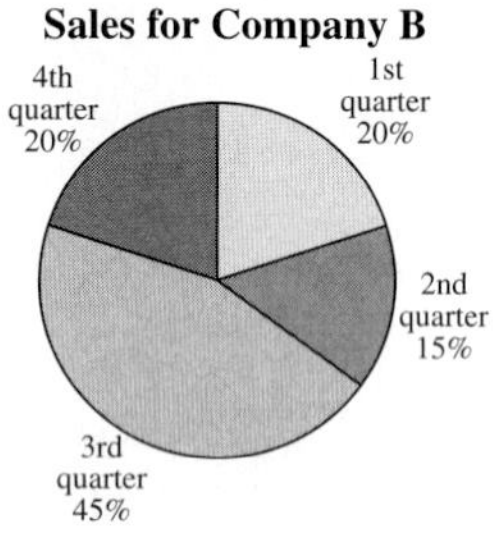

Section 2.3

10. The shape of the distribution is skewed left because the bars have a "tail" to the left.

12. (7), the distribution of values ranges from 20,000 to 100,000 and the distribution is skewed right owing to a few executives' having much higher salaries.

14. (8), the distribution of values ranges from 80 to 160 and the distribution is basically symmetric.

32. (a) $\bar{x} \approx 213.4$

median = 214

mode = 217

(b) Median, because the distribution is skewed.

34. A = mean, because the distribution is skewed left.

B = median, because the distribution is skewed left.

C = mode, because it's the data entry that occurred most often.

50.

Class	Frequency, f
1	6
2	5
3	4
4	6
5	4
6	5
	$\Sigma f = 30$

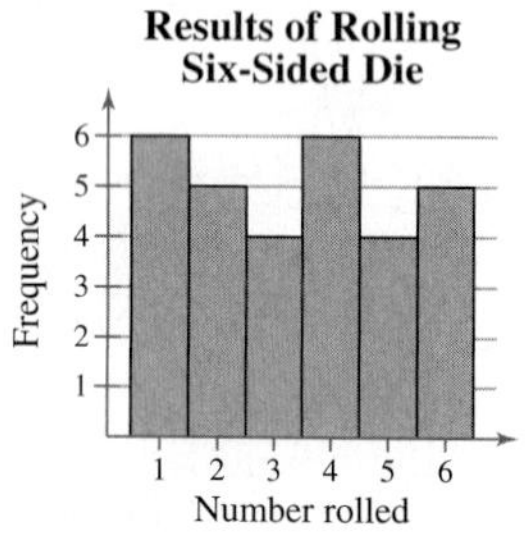

Uniform

Section 2.4

40.

Class	f	Midpoint, x	xf
145–164	8	154.5	1236.0
165–184	7	174.5	1221.5
185–204	3	194.5	583.5
205–224	1	214.5	214.5
225–244	1	234.5	234.5
	$N = 20$		$\Sigma xf = 3490.0$

$x - \mu$	$(x - \mu)^2$	$(x - \mu)^2 f$
−20	400	3200
0	0	0
20	400	1200
40	1600	1600
60	3600	3600
		$\Sigma(x - \mu)^2 f = 9600$

$$\mu = \frac{\Sigma xf}{N} = \frac{3490}{20} = 174.5$$

$$\sigma = \sqrt{\frac{\Sigma(x - \mu)^2 f}{N}} = \sqrt{\frac{9600}{20}} \approx 21.9$$

42.

Class	f	xf	$x - \bar{x}$	$(x - \bar{x})^2$	$(x - \bar{x})^2 f$
0	1	0	−1.93	3.72	3.72
1	9	9	−0.93	0.86	7.74
2	13	26	0.07	0.00	0.00
3	5	15	1.07	1.14	5.70
4	2	8	2.07	4.28	8.56
	$n = 30$	$\Sigma xf = 58$			$\Sigma(x - \bar{x})^2 f = 25.27$

$$\bar{x} = \frac{\Sigma xf}{n} = \frac{58}{30} \approx 1.9$$

$$s = \sqrt{\frac{\Sigma(x - \bar{x})^2 f}{n - 1}} = \sqrt{\frac{25.72}{29}} \approx 0.9$$

44.

Class	f	Midpoint, x	xf
0.5–9.5	11.9	5	59.5
10.5–19.5	12.1	15	181.5
20.5–29.5	14.0	25	350.0
30.5–39.5	18.5	35	647.5
40.5–49.5	16.6	45	747.0
50.5–59.5	16.3	55	896.5
60.5–69.5	17.8	65	1157.0
70.5–79.5	12.4	75	930.0
80.5–89.5	6.3	85	535.5
90.5–99.5	1.3	95	123.5
	$n = 127.2$		$\Sigma xf = 5628$

$x - \bar{x}$	$(x - \bar{x})^2$	$(x - \bar{x})^2 f$
−39.25	1540.5625	18,332.69
−29.25	855.5625	10,352.31
−19.25	370.5625	5,187.88
−9.25	85.5625	1,582.91
0.75	0.5625	9.34
10.75	115.5625	1,883.67
20.75	430.5625	7,664.01
30.75	945.5625	11,724.98
40.75	1660.5625	10,461.54
50.75	2575.5625	3,348.23
		$\Sigma(x - \bar{x})^2 f = 70{,}547.56$

$$\bar{x} = \frac{\Sigma xf}{n} = \frac{5628}{127.2} \approx 44.25$$

$$s = \sqrt{\frac{\Sigma(x - \bar{x})^2 f}{n - 1}} = \sqrt{\frac{70{,}547.56}{126.2}} \approx 23.64$$

Section 2.5

22. (a) $Q_1 = 15.125$, $Q_2 = 15.8$, $Q_3 = 17.65$

(b) **Railroad Equipment Manufacturers**

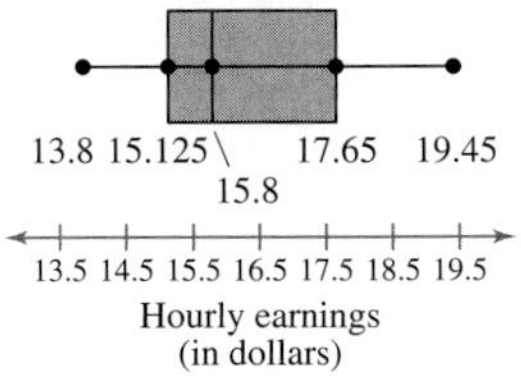

Review Answers for Chapter 2

2. **Income of Employees**

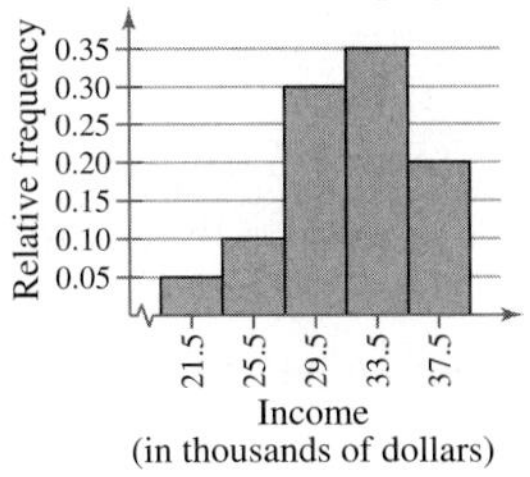

The class with the greatest relative frequency is 32–35 and that with the least is 20–23.

4. **Liquid Volume 12-oz Cans**

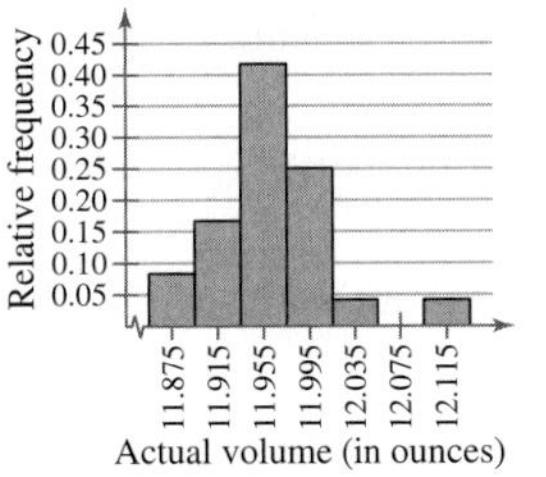

6. **Meals Purchased**

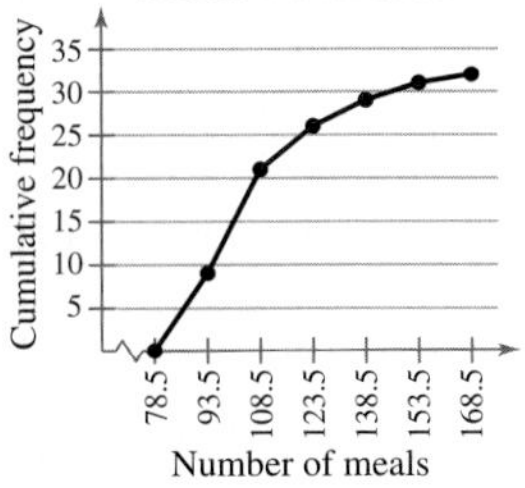

8. **Average Daily Highs**

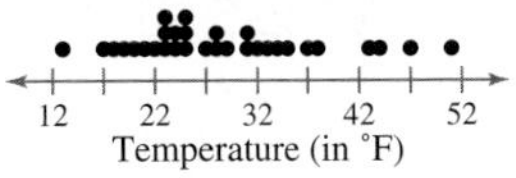

CHAPTER 3

4.

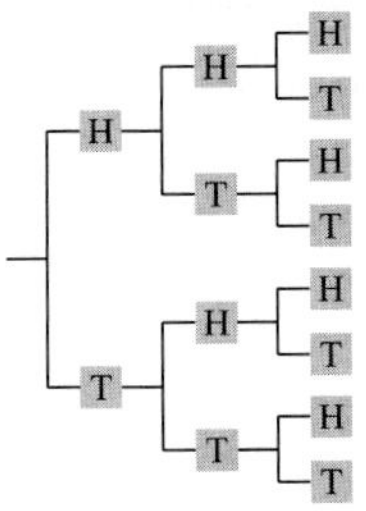

{HHH HHT, HTH, HTT, THH, THT, TTH, TTT}

6.

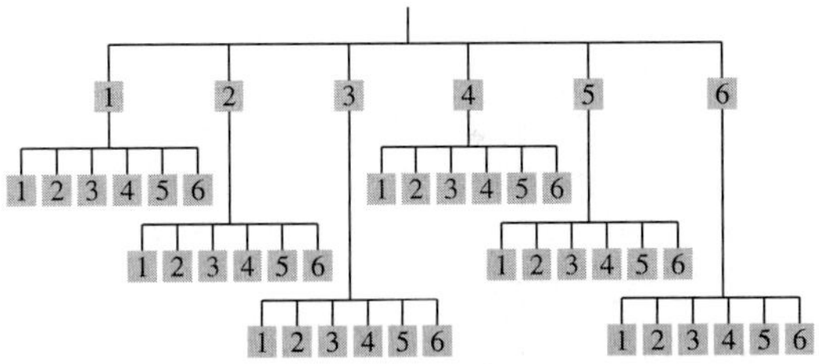

{(1, 1), (1, 2), (1, 3), (1, 4), (1, 5), (1, 6), (2, 1), (2, 2), (2, 3), (2, 4), (2, 5), (2, 6), (3, 1), (3, 2), (3, 3), (3, 4), (3, 5), (3, 6), (4, 1), (4, 2), (4, 3), (4, 4), (4, 5), (4, 6), (5, 1), (5, 2), (5, 3), (5, 4), (5, 5), (5, 6), (6, 1), (6, 2), (6, 3), (6, 4), (6, 5), (6, 6)}

12. Classical probability because it is possible to generate the sample space and count the total number of outcomes.

CHAPTER 4

Section 4.1

2. A random variable is discrete if it has a finite or countable number of possible outcomes that can be listed.

Condition 1: $0 \le P(x) \le 1$ Condition 2: $\Sigma P(x) = 1$

6. False. The expected value of a discrete random variable is equal to the mean of the random variable.

8. Continuous because length of time is a random variable that cannot be counted.

10. Discrete because the number of fatalities is a random variable that is countable.

12. Continuous because length of time is a random variable that cannot be counted.

14. Discrete because the number of rainy days in a month is a random variable that is countable.

16. Continuous because weight is a random variable that cannot be counted.

18. Discrete because number of blackouts is a random variable that is countable.

28. (a)

x	f	P(x)	xP(x)
0	1941	0.728	0
1	349	0.131	0.131
2	203	0.076	0.152
3	78	0.029	0.087
4	57	0.021	0.084
5	40	0.015	0.075
	$n = 2668$	$\Sigma P(x) = 1$	$\Sigma xP(x) = 0.529$

x	$(x - \mu)^2 P(x)$
0	0.204
1	0.029
2	0.164
3	0.177
4	0.253
5	0.300
	$\Sigma(x - \mu)^2 P(x) = 1.127$

(b) 0.5 (c) 1.1 (d) 1.0

(e) A household on average has 0.5 cats with a standard deviation of 1.0 cat.

30. (a)

x	f	P(x)	xP(x)
0	260	0.152	0.000
1	500	0.292	0.292
2	425	0.249	0.498
3	305	0.178	0.534
4	175	0.102	0.408
5	45	0.026	0.130
	$n = 1710$	$\Sigma P(x) \approx 1$	$\Sigma xP(x) = 1.862$

x	$(x - \mu)^2 P(x)$
0	0.527
1	0.217
2	0.005
3	0.231
4	0.466
5	0.256
	$\Sigma(x - \mu)^2 P(x) = 1.701$

(b) 1.9 (c) 1.7 (d) 1.3

(e) The average number of accidents per student is 1.9 accidents with a standard deviation of 1.3 accidents.

32. (a)

x	f	P(x)	xP(x)
0	19	0.059	0
1	39	0.122	0.122
2	52	0.163	0.325
3	57	0.178	0.534
4	68	0.213	0.850
5	41	0.128	0.641
6	27	0.084	0.506
7	17	0.053	0.372
	$n = 320$	$\Sigma P(x) = 1$	$\Sigma xP(x) = 3.350$

x	$(x - \mu)^2 P(x)$
0	0.666
1	0.673
2	0.296
3	0.022
4	0.090
5	0.349
6	0.590
7	0.706
	$\Sigma(x - \mu)^2 P(x) = 3.392$

(b) 3.4

(c) 3.4

(d) 1.8

(e) A student has an average of 3.4 school-related extracurricular activities with a standard deviation of 1.8 activities.

Section 4.2

18. (a) 1.627×10^{-9} (b) ≈ 1 (c) 1.662×10^{-9}

20. (a) $n = 5, p = 0.25$ (b)

x	P(x)
0	0.237
1	0.396
2	0.264
3	0.088
4	0.015
5	0.001

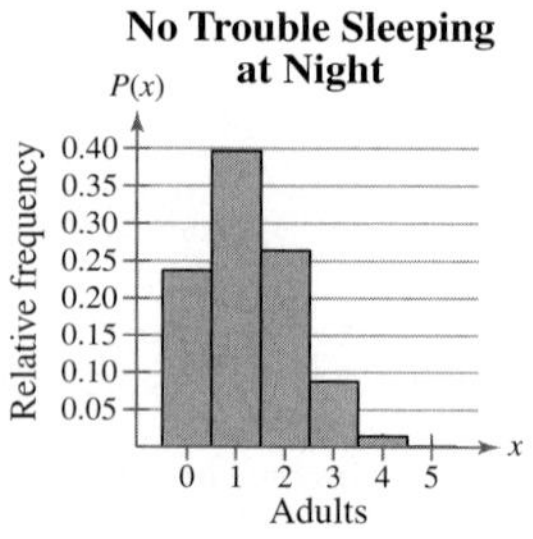

(c) Skewed right (d) 1.3 (e) 0.9 (f) 0.9

(g) On average 1.3 adults, out of every 5, have difficulty sleeping at night. The standard deviation is 0.9 adults.

$x = 4$ or 5 would be uncommon because of their low probabilities.

22. (a) $n = 5, p = 0.38$ (b)

x	$P(x)$
0	0.092
1	0.281
2	0.344
3	0.211
4	0.065
5	0.008

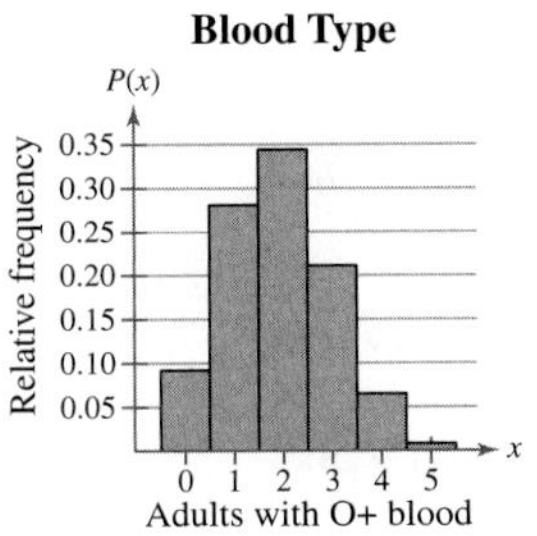

(c) Skewed right (d) 1.9 (e) 1.2 (f) 1.1

(g) On average 1.9 adults, out of every 5, has O^+ blood. The standard deviation is 1.1 adults.

$x = 4$ or 5 would be uncommon because of their low probabilities.

Review Answers for Chapter 4

12. (a)

x	f	$P(x)$
0	7	0.219
1	8	0.250
2	10	0.313
3	3	0.094
4	3	0.094
5	1	0.031
	$n = 32$	$\Sigma P(x) \approx 1$

(b)

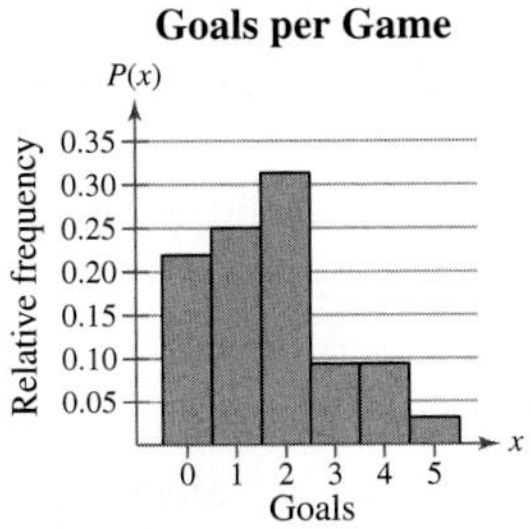

(c) 1.7, 1.8, 1.3

26. (a)

x	$P(x)$
0	0.394
1	0.403
2	0.165
3	0.0338
4	0.00347
5	0.000142

(b)

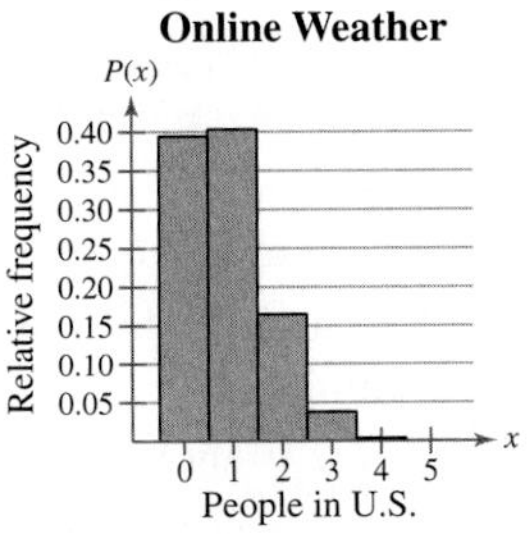

(c) 0.9, 0.7, 0.8

CHAPTER 5

Section 5.4

12. (b), because $\mu = 5.8$, $\sigma = 0.23$, and the graph approximates a normal curve.

14. 800, 25.820

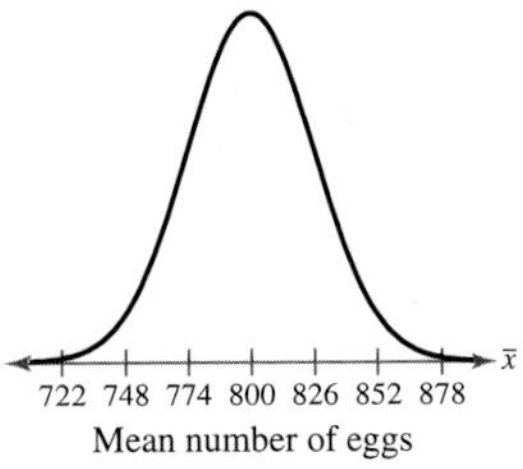

16. 47.2, 0.6

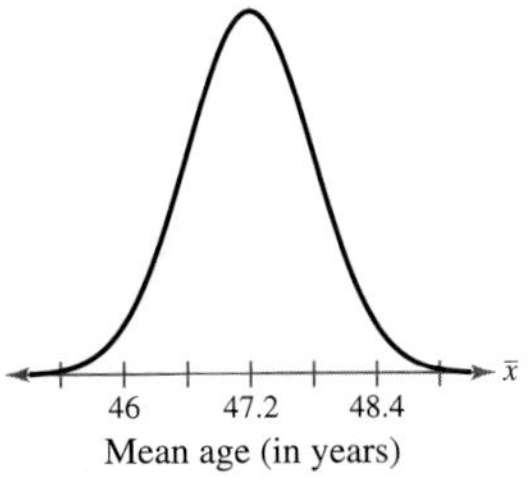

18. 49.3, 3.42

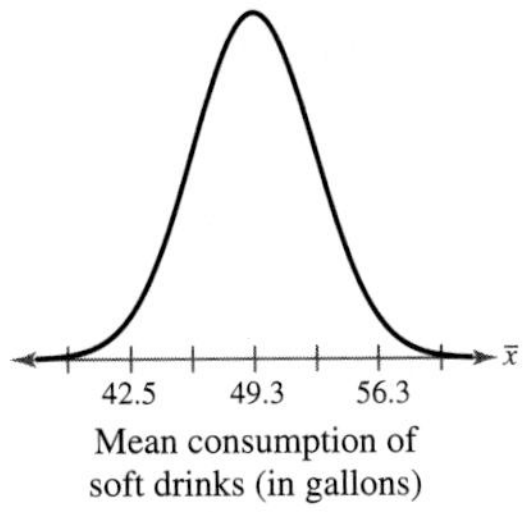

20. 800, 18.257; 800, 14.907

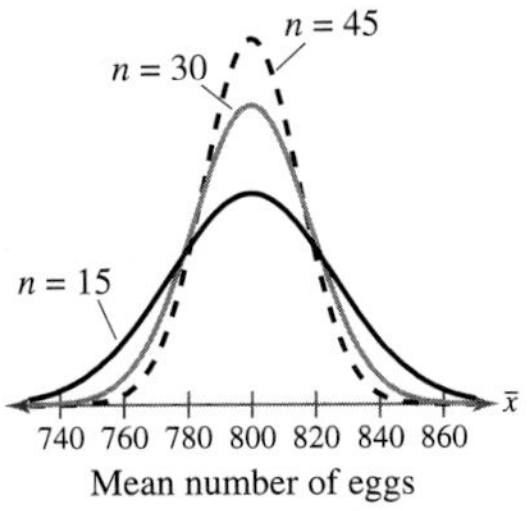

As the sample size increases, the standard error decreases.

Section 5.5

20. Can use normal distribution.

(a) 0.1347

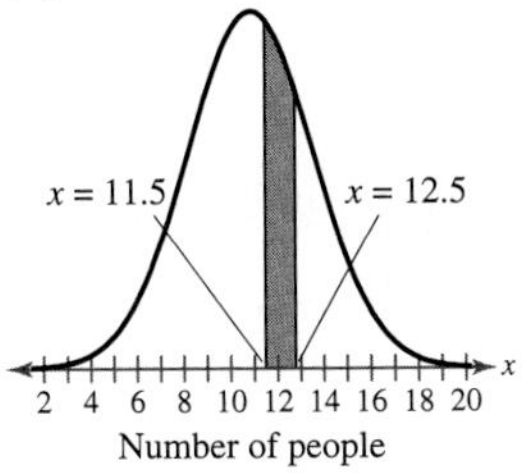

(b) 0.4090

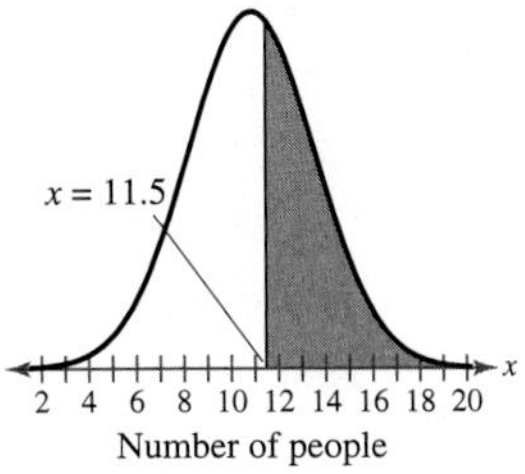

(c) 0.5910

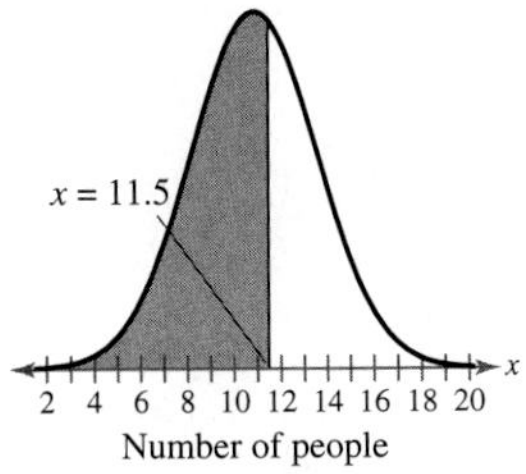

(d) 0.9292

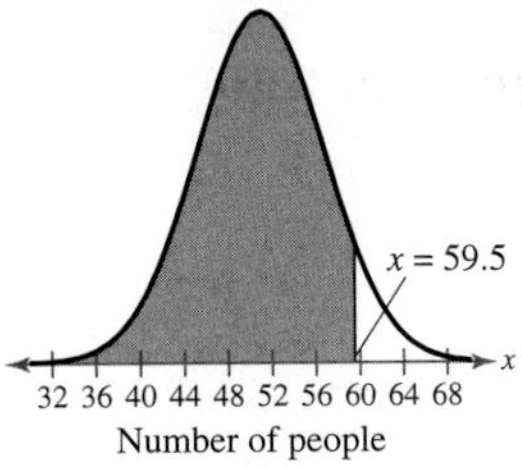

22. Can use normal distribution.

(a) 0.1074

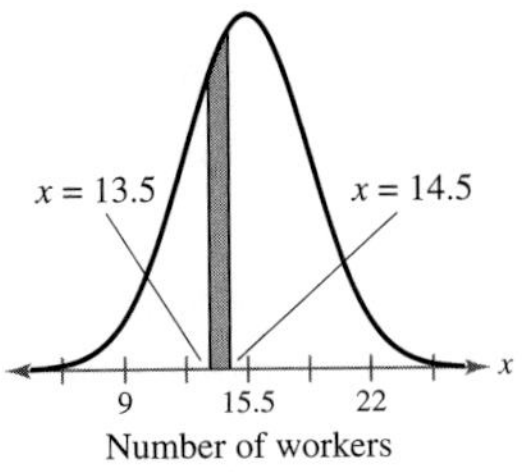

(b) 0.7291

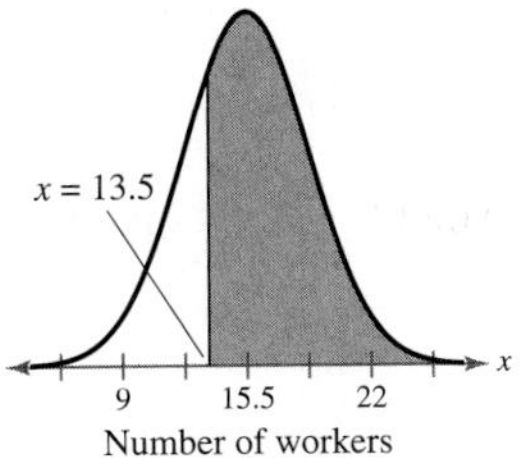

(c) 0.2709

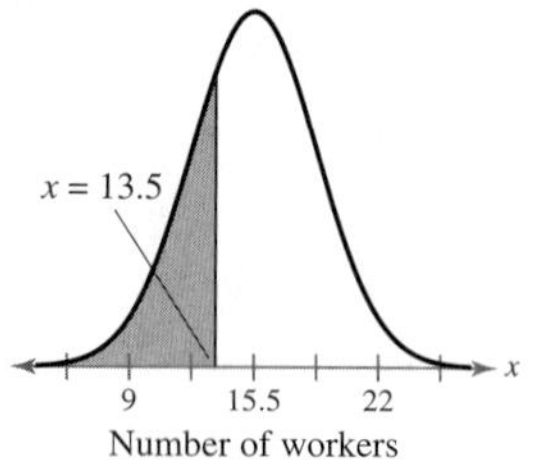

(d) 0.0013

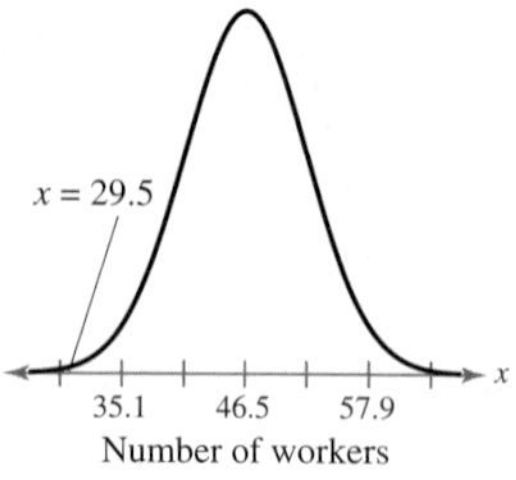

CHAPTER 7

Section 7.1

8. False. If you want to support a claim, write it as your alternative hypothesis.

Section 7.3

2. Identify the claim. State H_0 and H_a. Specify the level of significance. Identify the degrees of freedom and sketch the sampling distribution. Determine the critical value(s) and rejection region(s). Find standardized test statistic. Make a decision and interpret it in the context of the original claim. The population distribution must be normal or nearly normal.

16. (a) Fail to reject H_0 because $-1.372 < t < 1.372$.
(b) Reject H_0 because $t < -1.372$.
(c) Reject H_0 because $t > 1.372$.
(d) Fail to reject H_0 because $-1.372 < t < 1.372$.

18. (a) Fail to reject H_0 because $-1.725 < t < 1.725$.
(b) Reject H_0 because $t < -1.725$.
(c) Fail to reject H_0 because $-1.725 < t < 1.725$.
(d) Reject H_0 because $t > 1.725$.

Section 7.4

12. (a) H_0: $p \le 0.58$; H_a: $p > 0.58$ (claim)
(b) $z_0 = 1.28$; Rejection region: $z > 1.28$
(c) 0.608 (d) Fail to reject H_0.
(e) There is insufficient evidence at the 10% level to support the claim that more than 58% of British consumers want supermarkets to stop selling genetically modified foods.

14. (a) H_0: $p = 0.44$ (claim); H_a: $p \ne 0.44$
(b) $-z_0 = -1.96$ and $z_0 = 1.96$
Rejection regions: $z < -1.96$, $z > 1.96$
(c) 2.544 (d) Reject H_0.
(e) There is sufficient evidence at the 5% level to reject the claim that 44% of adults in the United States do volunteer work.

16. $$z = \frac{\hat{p} - p}{\sqrt{\frac{pq}{n}}} = \frac{(x/n - p)}{\sqrt{\frac{pq}{n}}} \Rightarrow \frac{(x/n) - p}{\frac{\sqrt{pq}}{\sqrt{n}}} \Rightarrow$$
$$\frac{[(x/n) - p]\sqrt{n}}{\sqrt{pq}} \cdot \frac{\sqrt{n}}{\sqrt{n}} \Rightarrow \frac{[(x/n) - p]n}{\sqrt{pqn}} = \frac{x - np}{\sqrt{pqn}}$$

Section 7.5

18. (a) H_0: $\sigma \ge 30$; H_a: $\sigma < 30$ (claim)
(b) $\chi_0^2 = 2.088$; Rejection region: $\chi^2 < 2.088$
(c) 8.294 (d) Fail to reject H_0.
(e) There is insufficient evidence at the 1% level to support the claim that the standard deviation of test scores for eighth grade students who took a life science assessment test is less than 30.

20. (a) H_0: $\sigma = 6.14$ (claim); H_a: $\sigma \ne 6.14$
(b) $\chi_L^2 = 8.907$, $\chi_R^2 = 32.852$
Rejection regions: $\chi^2 > 32.852$, $\chi^2 < 8.907$
(c) 21.293 (d) Fail to reject H_0.
(e) There is insufficient evidence at the 5% level to reject the claim that the standard deviation of the lengths of stay for patients involved in a crash in which the vehicle struck a tree is 6.14 days.

Review Answers for Chapter 7

8. (a) H_0: $\mu \ge 30{,}000$ (claim); H_a: $\mu < 30{,}000$
(b) Type I error will occur if H_0 is rejected when the actual mean tire life is at least 30,000 miles.
Type II error will occur if H_0 is not rejected when the actual mean tire life is less than 30,000 miles.
(c) Left-tailed because the alternative hypothesis contains $<$.
(d) There is enough evidence to reject the claim.
(e) There is not enough evidence to reject the claim.

10. (a) H_0: $\mu \ge 20$; H_a: $\mu < 20$ (claim)
(b) Type I error will occur if H_0 is rejected when the actual mean number of fat calories is at least 20.
Type II error will occur if H_0 is not rejected when the actual mean number of fat calories is less than 20.
(c) Left-tailed because the alternative hypothesis contains $<$.
(d) There is enough evidence to support the claim.
(e) There is not enough evidence to support the claim.

CHAPTER 8

Section 8.1

16. (a) H_0: $\mu_1 \ge \mu_2$; H_a: $\mu_1 < \mu_2$ (claim)
(b) $z_0 = -1.645$; Rejection region: $z < -1.645$
(c) -8.03 (d) Reject H_0.

(e) There is sufficient evidence at the 5% level to support the claim that households in the United States headed by people under the age of 25 spend less on food away from home than households headed by people ages 55–64.

18. (a) H_0: $\mu_1 \geq \mu_2$; H_a: $\mu_1 < \mu_2$ (claim)

(b) $z_0 = -1.645$; Rejection region: $z < -1.645$

(c) -6.26 (d) Reject H_0.

(e) There is sufficient evidence at the 5% level to support the claim that households in the United States headed by people under the age of 25 spend less on food away from home than households headed by people ages 55–64.

The new samples do not lead to a different conclusion.

Section 8.2

18. (a) H_0: $\mu_1 \leq \mu_2$; H_a: $\mu_1 > \mu_2$ (claim)

(b) $t_0 = 1.363$; Rejection region: $t > 1.363$

(c) 2.723 (d) Reject H_0.

(e) There is enough evidence at the 10% level to support the claim that the mean bumper repair cost is greater for small pickups than for small utility vehicles.

Section 8.4

12. (a) H_0: $p_1 \geq p_2$; H_a: $p_1 < p_2$ (claim)

(b) $z_0 = -1.645$; Rejection region: $z < -1.645$

(c) -3.762 (d) Reject H_0.

(e) There is sufficient evidence at the 5% level to support the claim that the proportion of adults who are smokers is lower in California than in Oregon.

16. (a) H_0: $p_1 = p_2$ (claim); H_a: $p_1 \neq p_2$

(b) $-z_0 = -1.96$ and $z_0 = 1.96$

Rejection regions: $z < -1.96$, $z > 1.96$

(c) 22.862 (d) Reject H_0.

(e) There is sufficient evidence at the 5% level to reject the claim that the proportions of adults who have high cholesterol are the same for the two groups.

20. There is insufficient evidence at the 5% level to reject the claim that the proportion of teens who watch 5 to 10 hours of TV per week is the same for boys and for girls.

CHAPTER 9

Section 9.1

22. (a)

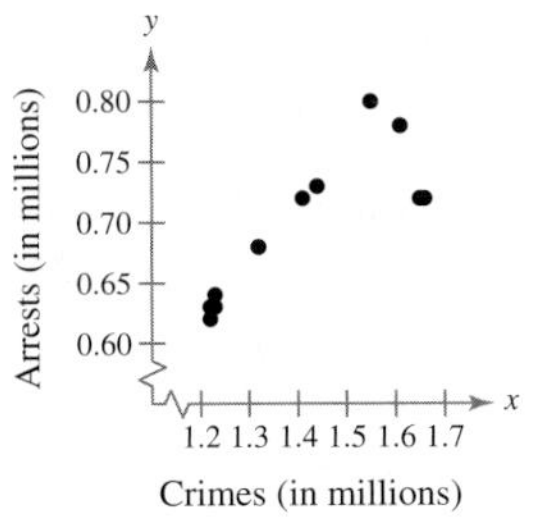

(b) 0.843

(c) Strong positive linear correlation

Section 9.2

20. $\hat{y} = -1.137x + 14.642$

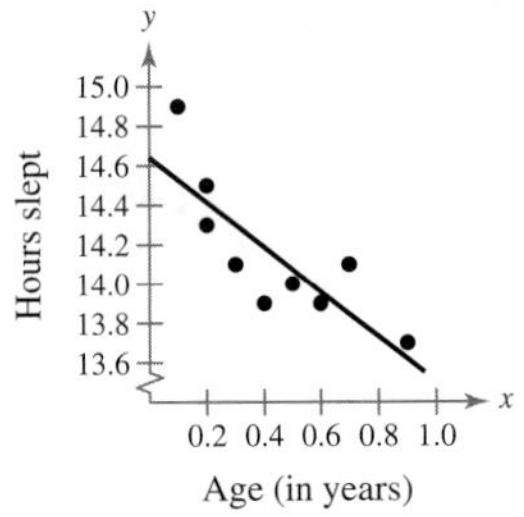

(a) 14.30 hours

(b) It is not meaningful to predict the value of y for $x = 3.9$ because $x = 3.9$ is outside the range of the original data.

(c) 13.96 hours (d) 13.73 hours

22. $\hat{y} = 0.000892x - 1.262$

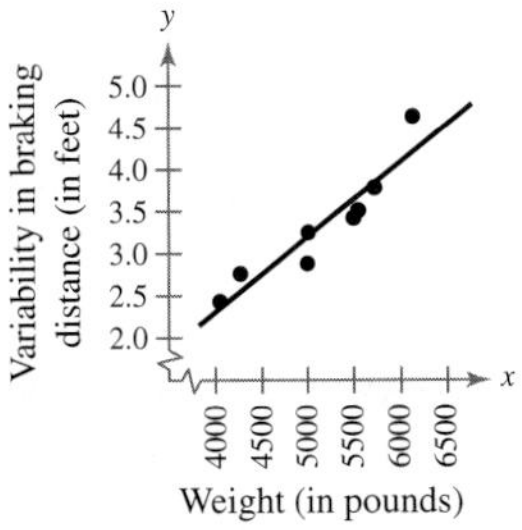

(a) 3.020 feet (b) 3.956 feet (c) 2.373 feet

(d) It is not meaningful to predict the value of y for $x = 3000$ because $x = 3000$ is outside the range of the original data.

26. (a) $\hat{y} = -4.297x + 94.200$

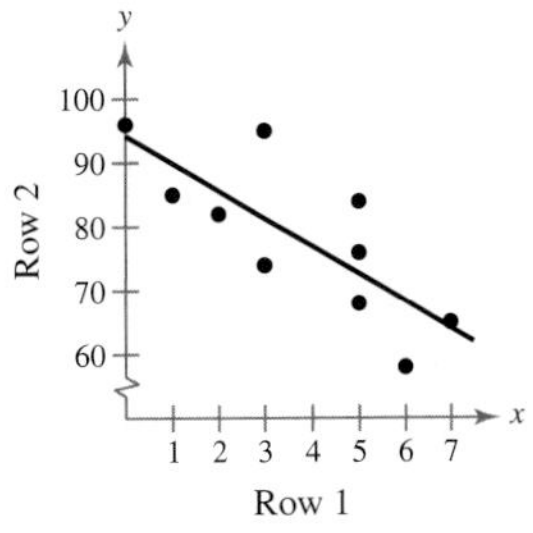

(b) $\hat{y} = -0.141x + 14.763$

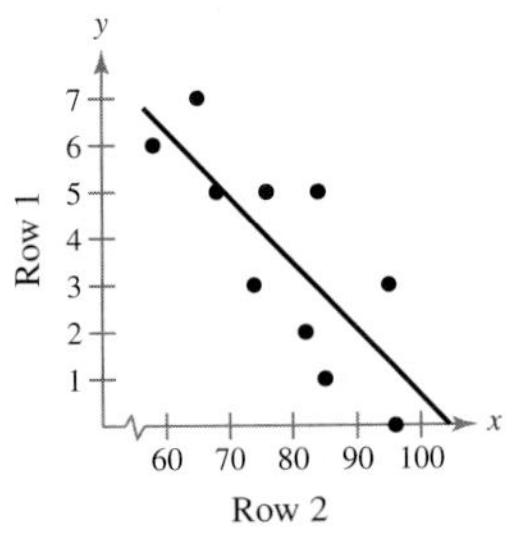

(c) The slope of the line keeps the same sign, but the values of m and b change.

28. $\hat{y} = -0.294x + 70.652$

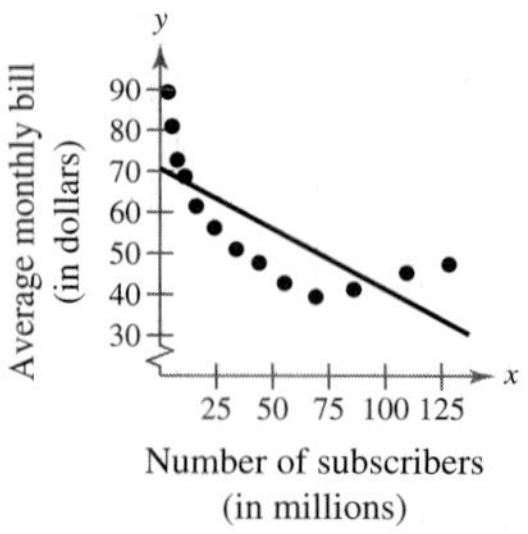

32. $\hat{y} = (69.481)0.995^x$

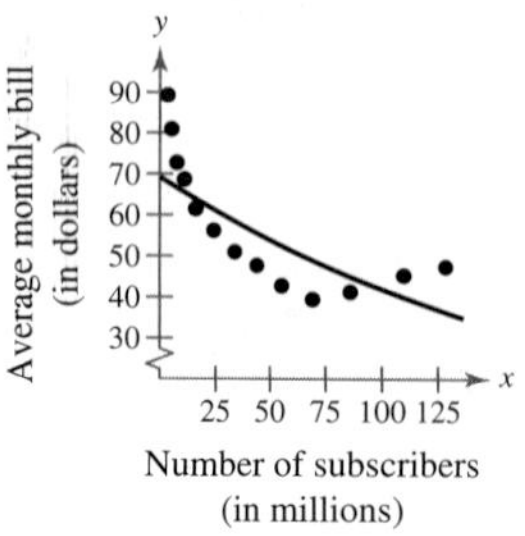

34. $\hat{y} = 25.035 + 19.599 \ln x$

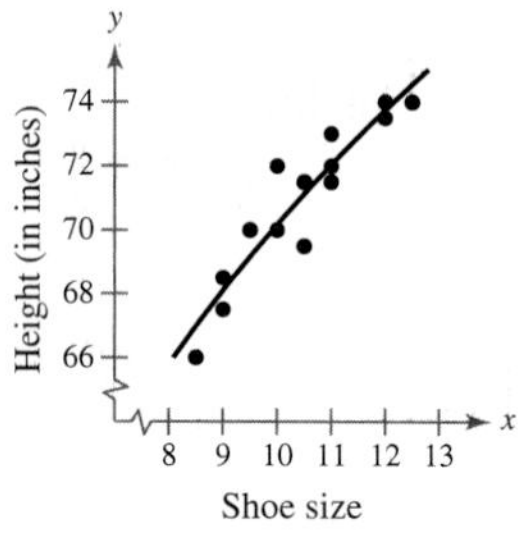

Section 9.3

20. $17.573 < y < 25.625$

You can be 90% confident that the median number of leisure hours per week will be between 17.573 and 25.625 when the median number of work hours per week is 45.1.

26. $\hat{y} = 0.693x + 2.280$

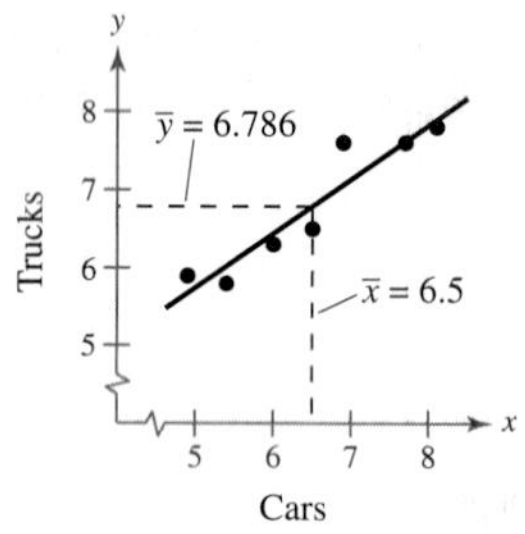

32. The slope and correlation coefficient will have the same sign because the denominators of both formulas are always positive while the numerators are always equal.

Review Answers for Chapter 9

4.

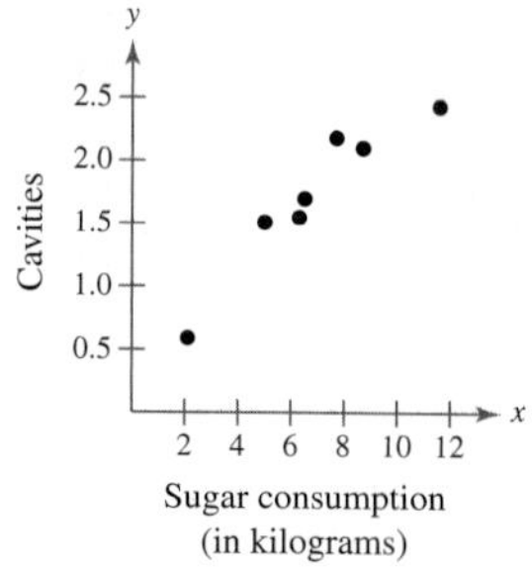

$r \approx 0.953$; positive linear correlation; the average number of cavities increases as the annual per capita sugar consumption increases.

8. There is enough evidence at the 5% level to conclude that there is a significant linear correlation between convertibles sold and average monthly temperature.

12. $\hat{y} = 1.163x + 2.883$

$r \approx 0.932$

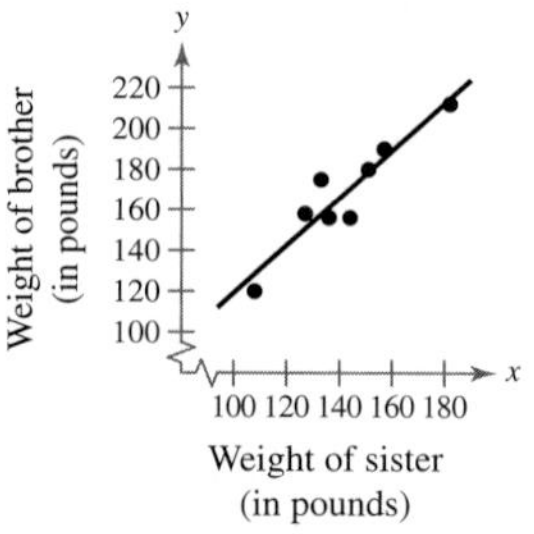

14. $\hat{y} = -0.090x + 44.675$

$r \approx -0.984$

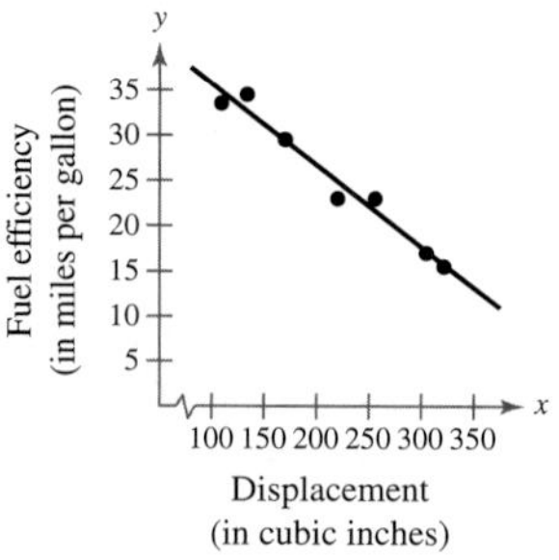

16. (a) 165.7 pounds

(b) It is not meaningful to predict the value of y for $x = 100$ because $x = 100$ is outside the range of the original data.

(c) 200.6 pounds

(d) It is not meaningful to predict the value of y for $x = 200$ because $x = 200$ is outside the range of the original data.

CHAPTER 10

Section 10.1

2. The observed frequencies must be obtained using a random sample, and each expected frequency must be greater than or equal to 5.

8. (a) Claimed distribution:

Response	Distribution
Home	70%
Work	17%
Commuting	8%
Other	5%

H_0: Distribution of responses is as shown in table above.

H_a: Distribution of responses differs from the claimed distribution.

(b) $\chi_0^2 = 7.815$; Rejection region: $\chi^2 > 7.815$

(c) 3.754

(d) Fail to reject H_0. There is not enough evidence at the 5% level to conclude that the distribution of responses differs from the claimed distribution.

10. (a) Claimed distribution:

Response	Distribution
Limited advancement	41%
Lack of recognition	25%
Low salary	15%
Unhappy with mgmt.	10%
Bored/don't know	9%

H_0: Distribution of responses is as shown in table above.

H_a: Distribution of responses differs from the claimed distribution.

(b) $\chi_0^2 = 13.277$; Rejection region: $\chi^2 > 13.277$

(c) 2.025

(d) Fail to reject H_0. There is not enough evidence at the 5% level to conclude that the distribution of responses differs from the claimed distribution.

12. (a) Claimed distribution:

Month	Distribution	Month	Distribution
January	8.333%	July	8.333%
February	8.333%	August	8.333%
March	8.333%	September	8.333%
April	8.333%	October	8.333%
May	8.333%	November	8.333%
June	8.333%	December	8.333%

H_0: The distribution of fatal bicycle accidents throughout the year is uniform as shown in table above.

H_a: The distribution of fatal bicycle accidents throughout the year is not uniform.

(b) $\chi_0^2 = 17.275$; Rejection region: $\chi^2 > 17.275$

(c) 89.663

(d) Reject H_0. There is enough evidence at the 10% level to conclude that the distribution of fatal bicycle accidents throughout the year is not uniform.

16. (a) Claimed distribution:

Response	Distribution
Married, husband present	25%
Married, husband absent	25%
Widowed/divorced	25%
Never married	25%

H_0: Distribution of the responses is uniform as shown in table above.

H_a: Distribution of the responses is not uniform.

(b) $\chi_0^2 = 9.348$; Rejection region: $\chi^2 > 9.348$

(c) 10.327

(d) Reject H_0. There is enough evidence at the 2.5% level to reject the social service worker's claim that the distribution is uniform.

20. (a) Frequency distribution: $\mu = 74.775$; $\sigma = 9.822$

Lower boundary	Upper boundary	Lower *z*-score	Upper *z*-score	Area
50.5	60.5	−2.47	−1.45	0.0668
60.5	70.5	−1.45	−0.44	0.2564
70.5	80.5	−0.44	0.58	0.3891
80.5	90.5	0.58	1.60	0.2262
90.5	100.5	1.60	2.62	0.0504

Class boundaries	Distribution	Freq. *f*	Expected	$\frac{(O-E)^2}{E}$
50.5–60.5	6.68%	28	27	0.037
60.5–70.5	25.64%	106	103	0.087
70.5–80.5	38.91%	151	156	0.160
80.5–90.5	22.62%	97	90	0.544
90.5–100.5	5.04%	18	20	0.2
		400		1.028

H_0: Variable has a normal distribution.

H_a: Variable does not have a normal distribution.

(b) $\chi_0^2 = 9.488$; Rejection region: $\chi^2 > 9.488$

(c) 1.028

(d) Fail to reject H_0. There is not enough evidence at the 5% level to conclude that the distribution of test scores is not normal.

Section 10.2

8.

	Type of car			
Gender	**Compact**	**Full size**	**SUV**	**Truck/van**
Male	28.6	39.05	22.55	19.8
Female	23.4	31.95	18.45	16.2

18. (a) H_0: Age is independent of gender.

H_a: Age is dependent on gender. (claim)

(b) d.f. $= 5$; $\chi_0^2 = 11.071$; Rejection region: $\chi^2 > 11.071$

(c) 2.834

(d) Fail to reject H_0. There is not enough evidence at the 5% level to conclude that age is dependent on gender in such alcohol-related accidents.

Section 10.3

2. (1) The *F*-distribution is a family of curves determined by two types of degrees of freedom, d.f._N and d.f._D.

(2) *F*-distributions are positively skewed.

(3) The area under the F-distribution curve is equal to 1.

(4) F-values are always greater than or equal to 0.

(5) For all F-distributions, the mean value of F is approximately equal to 1.

14. Fail to reject H_0. There is not enough evidence at the 5% level to support the claim.

16. (a) H_0: $\sigma_1^2 \leq \sigma_2^2$; H_a: $\sigma_1^2 > \sigma_2^2$ (claim)

(b) $F_0 = 2.18$; Rejection region: $F > 2.18$

(c) 1.071

(d) Fail to reject H_0. There is not enough evidence at the 5% level to conclude that the variance of the fuel consumption for the company's luxury sedan is less than that of the competitor's luxury sedan.

Section 10.4

4. There are three null hypothesis for a two-way ANOVA test. There is one for each main effect and one for the interaction effect.

Review Answers for Chapter 10

6. (a) Expected frequencies:

	Type of vehicle owned				
Gender	**Car**	**Truck**	**SUV**	**Van**	**Total**
Male	94.25	82.65	50.75	4.35	232
Female	100.75	88.35	54.25	4.65	248
Total	195	171	105	9	480

(b) Reject H_0.

(c) There is enough evidence at the 5% level to conclude that type of vehicle owned is dependent on gender.

8. (a) Expected frequencies:

	Time of day				
Gender	**12 A.M.– 5:59 A.M.**	**6 A.M.– 11:59 A.M.**	**12 P.M.– 5:59 P.M.**	**6 P.M.– 11:59 P.M.**	**Total**
Male	654.07	490.21	654.07	1470.64	3269
Female	307.93	230.79	307.93	692.36	1539
Total	962	721	962	2163	4808

(b) Fail to reject H_0.

(c) There is not enough evidence at the 10% level to conclude that time of day is dependent on gender.

CHAPTER 11

Section 11.1

4. (a) H_0: median = 73 (claim); H_a: median ≠ 73

(b) 2 (c) 5 (d) Fail to reject H_0.

(e) There is not enough evidence at the 1% level to reject the claim that the daily median temperature for the month of July in Pittsburgh is 73°F.

6. (a) H_0: median = 58 (claim); H_a: median ≠ 58

(b) 2 (c) 6 (d) Fail to reject H_0.

(e) There is not enough evidence at the 1% level to reject the claim that the daily median temperature for the month of January in San Diego is 58°F.

10. (a) H_0: median ≥ 30; H_a: median < 30 (claim)

(b) 5 (c) 8 (d) Fail to reject H_0.

(e) There is not enough evidence at the 5% level to support the claim that the median age of recipients of physical science doctorates is less than 30 years.

12. (a) H_0: median = 1300 (claim); H_a: median ≠ 1300

(b) 5 (c) 9 (d) Fail to reject H_0.

(e) There is not enough evidence at the 10% level to reject the claim that the median square footage of renter-occupied units is 1300 square feet.

Section 11.3

2. The Kruskal-Wallis test is always a right-tailed test because the null hypothesis is rejected only when H is significantly large.

Section 11.5

14. (a) H_0: The selection of members was random.

H_a: The selection of members was not random. (claim)

(b) lower critical value = 11

upper critical value = 25

(c) 19 (d) Fail to reject H_0.

(e) There is not enough evidence at the 5% level to support the claim that the selection of members was not random.

Index

A

Addition Rule, 171
  for the probability of *A* or *B*, 141, 144
alternative formula
  for the standardized test statistic for a proportion, 379
  for variance and standard deviation, 91
alternative hypothesis
  one-sample, 333
  two-sample, 405
analysis of variance (ANOVA) test
  one-way, 542
  two-way, 548
approximating binomial probabilities, 262
area of a region
  under a probability curve, 219
  under a Standard Normal Curve, 219

B

Bayes's Theorem, 138
biased sample, 18
bimodal, 62
binomial distribution
  normal approximation to, 259
  population parameters of, 188
binomial experiment, 184, 510
  notation for, 184
binomial probabilities, using the normal distribution to approximate, 259
binomial probability distribution, 187, 201
binomial probability formula, 186
bivariate normal distribution, 488
block, 16
box-and-whisker plot, 95

C

calculating a correlation coefficient, 461
categories, 510
c-confidence interval
  for the population mean, 283
  for the population proportion, 304
cell, 521
census, 17
central angle, 52
Central Limit Theorem, 248
central tendency, measure of, 60
chart
  control, 236
  Pareto, 53
  pie, 52
  time series, 55
Chebychev's Theorem, 81
chi-square distribution, 312
chi-square goodness-of-fit test, 510, 512
chi-square independence test, 523
chi-square test,
  for independence, 523
  for standard deviation, 382, 391
  for variance, 382, 391
class, 34
  boundaries, 38
  mark, 36
  width, 34
class limit
  lower, 34
  upper, 34
classical probability, 120, 144
cluster sample, 19
clusters, 19
coefficient
  of correlation, 461
    t-test for, 464
  of determination, 485
  Spearman rank correlation, 593
  of variation, 90
combination, 154, 155
  of *n* objects taken *r* at a time, 154, 155
complement of event *E*, 124
complementary events, 144
conditional probability, 130
confidence interval(s), 283
  for a population mean, finding a, 283
  for a population proportion, constructing a, 304
  for a population standard deviation, 314
  for a population variance, 314
confidence, level of, 281
constructing
  a confidence interval for the mean: *t*-distribution, 297
  a discrete probability distribution, 174
  a frequency distribution from a data set, 34
  an ogive, 41
  a prediction interval for *y* for a specific value of *x*, 488
contingency table, 521
contingency table cells, finding the expected frequency for, 521
continuity, correction for, 261
continuous probability distribution, 216
continuous random variable, 172, 216
control chart, 236
convenience sample, 20
correction factor
  finite, 258
  finite population, 293
correction for continuity, 261
correlation, 458
correlation coefficient, 461
  Pearson product moment, 461
  Spearman rank, 593
  t-test for, 466
  using a table for, 464
counting principle
  fundamental, 150, 155
c-prediction interval, 488
critical region, 352
critical value, 281, 352
  in a normal distribution, finding, 352
  in a *t*-distribution, finding, 364
cumulative frequency, 36
  graph, 40
curve, normal, 216

D

data (datum), 2
  qualitative, 9
  quantitative, 9
data sets
  paired, 54
decision rule
  based on *P*-value, 340, 347
  based on rejection region, 354
degrees of freedom, 295
dependent
  event, 131
  sample, 425
  variable, 458
descriptive statistics, 5
designing a statistical study, 16
determination, coefficient of, 485
deviation, 75
  explained, 484
  total, 484
  unexplained, 484
diagram, tree, 118
discrete probability distribution, 173
discrete random variable, 172
  expected value of, 178
  mean of a, 176
  standard deviation of, 177
  variance of, 177
distinguishable permutations, 153
distribution
  binomial probability, 187, 201
  bivariate normal, 488
  chi-square, 312
  continuous probability, 216
  discrete probability, 173
  frequency, 34
  F-, 533
  geometric, 198, 201
  sampling of sample means, 246
  normal, 216
    finding critical values in, 352
  Poisson, 199, 201
  sampling, 246
  standard normal, 219, A1, A2
    properties of, A2
  uniform, 228
dot plot, 51
drawing a box-and-whisker plot, 96

E

e, 199
effect
  interaction, 548
  main, 548
empirical probability, 121, 144
Empirical Rule (or 68-95-99.7 Rule), 80
equation
  multiple regression, 494
  of a regression line, 475
error
  of estimate
    standard, 486
  margin of, 282
  of the mean
    standard, 246
  sampling, 282

type I, 335
type II, 335
estimate
interval, 281
point, 280
pooled, of the standard deviation, 416
standard error of, 486
estimating p by minimum sample size, 307
event, 118
complement of an, 124
dependent, 131
independent, 131, 144
mutually exclusive, 140, 144
simple, 119
expected frequency, 511
finding for contingency table cells, 521
expected value, 178
of a discrete random variable, 178
experiment, 16
binomial, 184, 510
multinomial, 196, 510
probability, 118
explained deviation, 484
explained variation, 484
explanatory variable, 458
exploratory data analysis (EDA), 49
exponential equation, 482

F

factorial, 151
false positive, 139
F-distribution, 533
finding a confidence interval for a population mean, 283
finding a minimum sample size
to estimate μ, 286
to estimate p, 307
finding areas under the standard normal curve, 221, A4
finding critical values
for the chi-square test, 380
for the F-distribution, 534
in a normal distribution, 352
in a t-distribution, 364
finding the expected frequency for contingency table cells, 521
finding the P-value for hypothesis test, 347
finding the standard error of estimate, 486
finding the test statistic for the one-way ANOVA test, 543
finite correction factor, 258
finite population correction factor, 293
first quartile, 93
five-number summary, 96
formula, binomial probability, 186
fractiles, 93
frequency, 34
cumulative, 36
expected, 511
observed, 511
relative, 36
frequency distribution, 34
mean of, 65
rectangular, 66
skewed left (negatively skewed), 66
skewed right (positively skewed), 66
symmetric, 66
uniform, 66
frequency histogram, 38
relative, 40
frequency polygon, 39
F-test for variances, two-sample, 536
fundamental counting principle, 150, 155

G

geometric distribution, 198, 201
goodness-of-fit test, chi-square, 510, 512
graph, cumulative frequency, 40

H

histogram
frequency, 38
relative frequency, 40
history of statistics timeline, 29
homogeneity of proportions test, 531
hypergeometric distribution, 204
hypothesis
alternative, 333, 405
null, 333, 405
statistical, 333
hypothesis test, 332
finding the P-value for, 347
hypothesis testing
steps for, 341
summary of, 389, 390

I

independence test, chi-square, 523
independent, 521
event, 131, 144
sample, 405, 425
variable, 458
inferential statistics, 5
inflection points, 216
inherent zero, 11
interaction effect, 548
interquartile range (IQR), 95
interval estimate, 281
interval level of measurement, 11, 12
interval, c-prediction, 488
intervals, 34

K

Kruskal-Wallis test, 587
test statistic for, 587

L

law of large numbers, 122
leaf, 49
left, skewed, 66
left-tailed test, 338
for a population correlation coefficient, 466
length of a run, 599
level of confidence, 281
level of significance, 337
levels of measurement
interval, 11, 12
nominal, 10, 12
ordinal, 10, 12
ratio, 11, 12
limit
lower class, 34
upper class 34
line
of best fit, 474
regression, 474
logarithmic equation, 482
lower class limit, 34

M

main effect, 548
making an interval estimate, 281
margin of error, 282
matched samples, 425
maximum error of estimate, 282
mean, 60
of a binomial distribution, 192
difference between
t-test for, 366
two-sample t-test for, 416
two-sample z-test for, 405
of a discrete random variable, 176
of a frequency distribution, 65
standard error, 246, 248
weighted, 64
mean square
between, 542
within, 542
means, sampling distribution of sample, 246
measure of central tendency, 60
measurement
interval level of, 11, 12
nominal level of, 10, 12
ordinal level of, 10, 12
ratio level of, 11, 12
median, 61
midpoint, 36
midquartile, 104
midrange, 73
minimum sample size
to estimate μ, 286
to estimate p, 307
mode, 62
multinomial experiment, 196, 510
multiple regression equation, 494
Multiplication Rule for the probability of A and B, 132, 144
mutually exclusive, 140, 144

N

n factorial, 151
negative linear correlation, 458
negatively skewed, 66
no correlation, 458
nominal level of measurement, 10, 12
nonlinear correlation, 458
nonparametric test, 566
normal approximation to a binomial distribution, 259
normal curve, 216
normal distribution, 216
bivariate, 488
finding critical values in, 352
properties of, 216
standard, 219,
finding areas under, 221, A4
properties of, 219, A2
notation for binomial experiments, 184
null hypothesis,
one-sample, 333
two-sample, 405

O

observational study, 16
observed frequency, 511
odds, 129
ogive, 40
one-way analysis of variance, 542
test, 543, 544
finding the test statistic for, 543
ordered stem-and-leaf plot, 49
ordinal level of measurement, 10, 12
outcome, 118
outlier, 49, 63

P

paired data sets, 54
paired samples, 425
  sign test, performing a, 570
parameter, 4
  population
    binomial distribution, 192
Pareto chart, 53
Pearson product moment correlation coefficient, 461
Pearson's index of skewness, 91
Performing a chi-square goodness-of-fit test, 512
Performing a chi-square test for independence, 524
Performing a Kruskal-Wallis test, 588
Performing a one-way analysis of variance test, 544
Performing a paired-sample sign test, 570
Performing a runs test for randomness, 601
Performing a sign test for a population median, 567
Performing a Wilcoxon rank sum test, 580
Performing a Wilcoxon signed-rank test, 577
permutation, 151, 155
  distinguishable, 153
  of n objects taken r at a time, 152
pie chart, 52
plot
  dot, 51
  scatter, 54, 458
  stem-and-leaf, 49
point estimate, 280
  for σ, 312
  for σ^2, 312
  for ρ, 303
Poisson distribution, 199, 210
polygon, frequency, 39
pooled estimate of the standard deviation, 416
population, 3
  correlation coefficient
    using a table for the, 464
    using a t-test for the, 466
  mean, finding a confidence interval for, 283
  parameters
    of a binomial distribution, 192
  proportion
    constructing confidence intervals for, 304
  standard deviation, 76
  variance, 75, 76
positive linear correlation, 458
positively skewed, 66
principle, fundamental counting, 150, 155
probability
  Addition Rule for, 141, 144
  classical, 120, 144
  conditional, 130
  curve, area of region under, 219
  empirical, 121, 144
  experiment, 118
  formula, binomial, 186
  Multiplication Rule for, 132, 144
  rule, range of, 123
  statistical, 121
  subjective, 122
  that the first success will occur on trial number x, 198
  theoretical, 120
  value, 337
probability distribution
  binomial, 187, 201
  chi-square, 312
  continuous, 216
  discrete, 173
  geometric, 198, 201
  normal
    properties of a, 216
  Poisson, 199, 201
  sampling, 246
  standard normal, 219
properties
  of a normal distribution, 216
  of sampling distributions of sample means, 246
  of the standard normal distribution, 219, A2
proportion
  population, 303
    confidence interval for, 304
    z-test for, 374
proportions test
  homogeneity of, 531
P-value, 337
  decision rule based on, 340, 347
  for hypothesis test, finding the, 347

Q

qualitative data, 9
quantitative data, 9
quartile, 93
  first, 93
  second, 93
  third, 93

R

random sample, simple, 18
random variable, 172
  continuous, 216
  discrete, 172
    expected value of a, 178
    mean of a, 176
    standard deviation of a, 177
    variance of a, 177
randomness, runs test for, 600
range, 34, 74
  interquartile, 95
  of probability rule, 123, 144
rank correlation coefficient, Spearman, 593
rank sum test, Wilcoxon, 579
ratio level of measurement, 11, 12
rectangular, frequency distribution, 66
region
  critical, 352
  rejection, 352
regression line, 474
  deviation about, 484
  equation of, 475
  variation about, 484
regression, multiple, 494
rejection region, 352
  decision rule based on, 354
relative frequency, 36
  histogram, 40
replacement
  with, 19
  without, 19
residuals, 474
response variable, 458
right, skewed, 66
right-tailed test, 338
  for a population correlation coefficient, 466
rule
  addition, 141, 144
  decision
    based on P-value, 340, 347
    based on rejection region, 354
  empirical, 80
  multiplication, 132, 144
  range of probability, 123, 144
run, 599
runs test for randomness, 600

S

sample, 3
  biased, 18
  cluster, 19
  convenience, 20
  dependent, 425
  independent, 425
  matched, 425
  paired, 425
  random, 18
    simple, 18
  stratified, 19
  systematic, 19
sample mean, sampling distribution of, 246
sample size
  minimum to estimate ρ, 309
  minimum to estimate μ, 286
sample space, 118
sample standard deviation, 77
  for grouped data, 82
sample variance, 77
sampling, 17
sampling distribution, 246
  of sample means, 246
  properties of, 246
sampling error, 282
sampling process
  with replacement, 19
  without replacement, 19
scatter plot, 54, 458
Scheffé Test, 554
scores, standard, 98
second quartile, 93
sigma, 35
sign test, 566
  performing a paired sample, 570
  test statistic for, 567
signed-rank test, Wilcoxon, 577
significance, level of, 337
simple event, 119
simple random sample, 18
simulation, 17
skewed
  left, 66
  negatively, 66
  positively, 66
  right, 66
Spearman rank correlation coefficient, 593
standard deviation
  of a binomial distribution, 192
  chi-square test for, 382
  confidence interval for, 314
  of a discrete random variable, 177
  point estimate for, 312
  pooled estimate of, 416
  population, 76
  sample, 77
standard error
  of estimate, 486
  of the mean, 246, 248
standard normal curve, finding areas under, 221, A4
standard normal distribution, 219, A1, A2
  properties of, 219, A2
standard score, 98
standardized test statistic,
  for a chi-square test
    for standard deviation, 382, 391
    for variance, 382, 391

for the correlation coefficient
t-test, 466
for the difference between means
t-test, 427
z-test, 406
for the difference between proportions
z-test, 438
for a *t*-test
for a mean, 366, 391
for a *z*-test
for a mean, 349, 391
for a proportion, 374, 391
two-sample, 406
statistic, 4
statistical hypothesis, 333
statistical probability, 121
statistical process control (SPC), 236
statistical study, designing a, 16
statistics, 2, 3
descriptive, 5
history of, timeline, 29
inferential, 5
status, 2
stem, 49
stem-and-leaf plot, 49
steps for hypothesis testing, 341
strata, 19
stratified sample, 19
study
observational, 16
statistical, designing a, 16
subjective probability, 122
success, population proportion of, 303
sum of squares, 76, 77
sum test, Wilcoxon rank, 579
summary
of discrete probability distributions, 201
five-number, 96
of four levels of measurement, 12
of hypothesis testing, 389, 390
of probability, 144
survey, 17
symmetric, frequency distribution, 66
systematic sample, 19

T

table, contingency, 521
t-distribution, 295
constructing a confidence interval for the mean, 297
finding critical values in, 364
test
chi-square
goodness-of-fit, 510
chi-square independence, 523
for randomness, runs, 600
hypothesis, 332
Kruskal-Wallis, 587
left-tailed, 338
nonparametric, 566
one-way analysis of variance, 543
paired-sample sign, 570
right-tailed, 338
sign, 566
two-tailed, 338
two-way analysis of variance, 548
Wilcoxon rank sum, 579
Wilcoxon signed-rank, 577
test statistic, 337
for a chi-square test, 382, 391
for the correlation coefficient, 466
for the difference between means, 427
for the difference between proportions, 438
for the Kruskal-Wallis test, 587
for a mean
large sample, 349, 391
small sample, 366, 391
for a proportion, 374, 391
for the runs test, 601
for the sign test, 567
for a two-sample *z*-test, 406
for the Wilcoxon rank sum test, 580
testing the significance of the Spearman rank correlation coefficient, 594
theorem
Central Limit, 248
Chebychev's, 81
theoretical probability, 120
third quartile, 93
time series, 55
chart, 55
timeline, history of statistics, 29
total deviation, 484
total variation, 484
tree diagram, 118
trimmed mean, 73
t-test
for the correlation coefficient, 466
for the difference between means, 427
for a mean, 366, 391
two-sample
for the difference between means, 416
two-sample
F-test for variances, 536
t-test, 416
z-test
for the difference between
proportions, 438
means, 406
two-tailed test, 338
for a population correlation coefficient, 466
two-way analysis of variance test, 548
type I error, 335
type II error, 335

U

unexplained deviation, 484
unexplained variation, 484
uniform distribution, 228
uniform frequency distribution, 66
unordered stem-and-leaf plot, 49
upper class limit, 34
using the chi-square test for a variance or standard deviation, 382
using the normal distribution to approximate binomial probabilities, 262
using *P*-values for a *z*-test for a mean, 349
using rejection regions for a *z*-test for a mean, 354
using Table 11 for the correlation coefficient, 461
using the *t*-test
for the correlation coefficient, 466
for the difference between means, 427
for a mean, 366
using a two-sample *F*-test to compare σ_1^2 and σ_2^2, 536
using a two-sample *t*-test for the difference between means, 417
using a two-sample *z*-test
for the difference between means, 406
for the difference between proportions, 438
using a *z*-test for a proportion, 374

V

value
critical, 281, 352
expected, 178
probability, 337
variable
dependent, 458
explanatory, 458
independent, 458
random, 172
continuous, 216
response, 458
variance
of a binomial distribution, 192
chi-square test for, 382, 391
confidence intervals for, 314
of a discrete random variable, 177
mean square
between, 542
within, 542
one-way analysis of, 542
point estimate for, 312
population, 75, 76
sample, 77
two-sample *F*-test for, 536
two-way analysis of, 548
variation
explained, 484
total, 484
unexplained, 484

W

weighted mean, 64
Wilcoxon rank sum test, 579
test statistic for, 580
Wilcoxon signed-rank test, 577
with replacement, 19
without replacement, 19

X

x, random variable, 172

Z

z-score, 98
z-test
for a mean, 349, 391
test statistic for, 349, 391
using *P*-values for, 349
using rejection regions for, 352
for a proportion, 374, 391
test statistic for, 374, 391
two-sample
difference between means, 406
difference between proportions, 438
zero, inherent, 11

Credits

TOC

Ch. 1 Aerial view of Philadelphia, Pennsylvania/ Lien/Nibauer/Getty Images Inc.-Liaison; **Ch. 2** Akhiok child/Ray Corral Photography; **Ch. 3** Bob Barker on the set of "The Price is Right"/Kevin Fleming/Corbis /Bettman; **Ch. 4** Hay (Reeds)/Superstock, Inc.; **Ch. 5** Woman jogging/Sean Thompson/Getty Images, Inc./ PhotoDisc; **Ch. 6** Auto mechanic at auto dialysis computer/Corbis/Bettmann; **Ch. 7** Students using computers at a library/Dennis Degnan/Corbis/ Bettmann; **Ch. 8** Teens smoking cigarettes at pool next to "No smoking" sign/Michael Newman/PhotoEdit; **Ch. 9** Athens 2004, Athletics—Men's 100 mm Finals. Justin Gatlin of the U.S. celebrates his gold medal/Mulvey/Dallas Morning News/Corbis/Bettmann; **Ch. 10** Crash-test dummies inside car/Wayne Eastep/ Getty Images Inc.–Stone Allstock; **Ch. 11** Kennedy Space Center, Florida/NASA Headquarters

CHAPTER 1 p. 0 Aerial view of Philadelphia, Pennsylvania /Lien/Nibauer. Getty Images Inc.-Liaison; **p. 15** Family watching television/Lisette LeBon/ SuperStock, Inc.; **p. 29** Karl Pearson/University College London; Sir Ronald Aylmer Fisher/Library of Congress; John W. Tukey/Princeton University-Stanhope Hall

CHAPTER 2 p. 32 Akhiok child/Ray Corral Photography; **p. 61** Fishing village, Akhiok, Alaska/Roy Corral Photography; **p. 92** Two African-American female twins wearing sunglasses/Digital Vision/Robert Harding World Imagery; **p. 112** Runzheimer International Logo/Runzheimer International; **p. 113** Dairy Farmers of America Logo/DFA/Dairy Farmers of America

CHAPTER 3 p. 116 Bob Barker on the set of "The Price is Right"/Kevin Fleming/Corbis/Bettman; **p. 117** The Price is Right Million Dollar Game with Bob Barker/Getty Images, Inc.-Liaison; **p. 149** Aerial view of parking lot, Wareham, MA/Michael J. Doolittle/The Image Works; **p. 161** Kristen Helm rehearsing for the Tennessee Lottery Cash 3 drawing/AP Wide World Photos; **p. 168** Powerball Logo run by MUSL/Multi-State Lottery Association

CHAPTER 4 p. 170 Hay (Reeds)/Superstock Inc.; **p. 197** Rescue truck and plane crash in Grapevine, Texas/AP Wide World Photos

CHAPTER 5 p. 214 Woman jogging/Sean Thompson/ Getty Images, Inc./Photodisc; **p. 245** Baby being weighed by a doctor/Photolibrary.Com; **p. 277** Census Bureau Logo/U.S. Department of Commerce.

CHAPTER 6 p. 268 Auto mechanic at auto dialysis computer/Corbis/Bettmann; **p. 294** Appalachian black bear/Lynn Rogers. Northart/Lynn Rogers; **p. 295** William Sealy Gosset/The Granger Collection; **p. 326** EPA Logo/U.S. Environmental Protection Agency Headquarters; **p. 326** Female lab technician in laboratory/Alamy/Getty Images, Inc.-Comstock Images

CHAPTER 7 p. 330 Students using computers at a library/Dennis Degnan/Corbis/Bettmann; **p. 333** A man in bed taking medicine/R.B. Studio/Corbis/Stock Market; **p. 355** New Honda Civic Hybrid/AP Wide World Photos; **p. 363** Mother taking son's temperature/ Ariel Skelley/Corbis/Bettmann; **p. 398** Social Security Administration Logo/National Archives and Records Administration

CHAPTER 8 p. 402 Teens smoking cigarettes at pool next to "No smoking" sign/Michael Newman/PhotoEdit; **p. 415** Home Health Care date photographed: 1996/Don Mason/Corbis/Bettmann; **p. 417** Music sheets and notes/Masahiro Sano/Corbis/Stock Market; **p. 438** Reflexology/Sauze Francoise/Photo Researchers, Inc.; **p. 445** Female with arthritis in her hand/ZEFA/ Masterfile Corporation

CHAPTER 9 p. 456 Athens 2004, Athletics—Men's 100 mm Finals. Justin Gatlin of the U.S. celebrates his gold medal/Mulvey/Dallas Morning News/Corbis/ Bettmann; **p. 483** Woman measuring her waist/ Dion Ogust/The Image Works; **p. 506** University of Denver Logo/University of Denver; **p. 507** FDA Food and Drug Administration Logo/U.S. Food and Drug Administration

CHAPTER 10 p. 508 Crash-test dummies inside car/Wayne Eastep/Getty Images Inc.–Stone Allstock; **p. 532** Firefighters at motor vehicle accident/Corbis RF; **p. 562** National Fraud Information Center Logo/ National Archives and Records Administration

CHAPTER 11 p. 564 Kennedy Space Center, Florida/ NASA Headquarters; **p. 586** Portrait of a group of graduates/Bill Losh/Getty Images, Inc.-Taxi; **p. 614** Department of Labor USA Logo/U.S. Department of Labor

COVER

Commuter reading newspaper—©Taxi/Getty Images, Inc.; crowd applauding—©Lisette Le Bon/Superstock; research chemist—©Kim Steele/Photodisc/Getty Images, Inc.; soccer players—©Royalty-Free/CORBIS

Ron Larson/Elizabeth Farber
Elementary Statistics: Picturing the World, Instructor-to-Instructor CD
0-13-186162-X

Pearson Prentice Hall
Pearson Education, Inc.
Upper Saddle River, NJ 07458

YOU SHOULD CAREFULLY READ THE TERMS AND CONDITIONS BEFORE USING THE DISKETTE PACKAGE. USING THIS CD-ROM PACKAGE INDICATES YOUR ACCEPTANCE OF THESE TERMS AND CONDITIONS.

Prentice-Hall, Inc. provides this program and licenses its use. You assume responsibility for the selection of the program to achieve your intended results, and for the installation, use, and results obtained from the program. This license extends only to use of the program in the United States or countries in which the program is marketed by authorized distributors.

LICENSE GRANT

You hereby accept a nonexclusive, nontransferable, permanent license to install and use the program on an unlimited number of computers at one site. A site is a single campus or branch of an educational institution at one geographic location. You may copy the program solely for backup or archival purposes in support of your use of the program. You may not modify, translate, disassemble, decompile, or reverse engineer the program, in whole or in part.

TERM

The License is effective until terminated. Prentice-Hall, Inc. reserves the right to terminate this License automatically if any provision of the License is violated. You may terminate the License at any time. To terminate this License, you must return the program, including documentation, along with a written warranty stating that all copies in your possession have been returned or destroyed.

LIMITED WARRANTY

THE PROGRAM IS PROVIDED "AS IS" WITHOUT WARRANTY OF ANY KIND, EITHER EXPRESSED OR IMPLIED, INCLUDING, BUT NOT LIMITED TO, THE IMPLIED WARRANTIES OR MERCHANTABILITY AND FITNESS FOR A PARTICULAR PURPOSE. THE ENTIRE RISK AS TO THE QUALITY AND PERFORMANCE OF THE PROGRAM IS WITH YOU. SHOULD THE PROGRAM PROVE DEFECTIVE, YOU (AND NOT PRENTICE-HALL, INC. OR ANY AUTHORIZED DEALER) ASSUME THE ENTIRE COST OF ALL NECESSARY SERVICING, REPAIR, OR CORRECTION. NO ORAL OR WRITTEN INFORMATION OR ADVICE GIVEN BY PRENTICE-HALL, INC., ITS DEALERS, DISTRIBUTORS, OR AGENTS SHALL CREATE A WARRANTY OR INCREASE THE SCOPE OF THIS WARRANTY.

SOME STATES DO NOT ALLOW THE EXCLUSION OF IMPLIED WARRANTIES, SO THE ABOVE EXCLUSION MAY NOT APPLY TO YOU. THIS WARRANTY GIVES YOU SPECIFIC LEGAL RIGHTS AND YOU MAY ALSO HAVE OTHER LEGAL RIGHTS THAT VARY FROM STATE TO STATE.

Prentice-Hall, Inc. does not warrant that the functions contained in the program will meet your requirements or that the operation of the program will be uninterrupted or error-free.

However, Prentice-Hall, Inc. warrants the CD-ROM(s) on which the program is furnished to be free from defects in material and workmanship under normal use for a period of ninety (90) days from the date of delivery to you as evidenced by a copy of your receipt.

The program should not be relied on as the sole basis to solve a problem whose incorrect solution could result in injury to person or property. If the program is employed in such a manner, it is at the user's own risk and Prentice-Hall, Inc. explicitly disclaims all liability for such misuse.

LIMITATION OF REMEDIES

Prentice-Hall, Inc.'s entire liability and your exclusive remedy shall be:

1. the replacement of any CD-ROM not meeting Prentice-Hall, Inc.'s "LIMITED WARRANTY" and that is returned to Prentice-Hall, or
2. if Prentice-Hall is unable to deliver a replacement CD-ROM that is free of defects in materials or workmanship, you may terminate this agreement by returning the program.

IN NO EVENT WILL PRENTICE-HALL, INC. BE LIABLE TO YOU FOR ANY DAMAGES, INCLUDING ANY LOST PROFITS, LOST SAVINGS, OR OTHER INCIDENTAL OR CONSEQUENTIAL DAMAGES ARISING OUT OF THE USE OR INABILITY TO USE SUCH PROGRAM EVEN IF PRENTICE-HALL, INC. OR AN AUTHORIZED DISTRIBUTOR HAS BEEN ADVISED OF THE POSSIBILITY OF SUCH DAMAGES, OR FOR ANY CLAIM BY ANY OTHER PARTY.

SOME STATES DO NOT ALLOW FOR THE LIMITATION OR EXCLUSION OF LIABILITY FOR INCIDENTAL OR CONSEQUENTIAL DAMAGES, SO THE ABOVE LIMITATION OR EXCLUSION MAY NOT APPLY TO YOU.

GENERAL

You may not sublicense, assign, or transfer the license of the program. Any attempt to sublicense, assign or transfer any of the rights, duties, or obligations hereunder is void.

This Agreement will be governed by the laws of the State of New York.

Should you have any questions concerning this Agreement, you may contact Prentice-Hall, Inc. by writing to:

Director of Multimedia Development
Higher Education Division
Prentice-Hall, Inc.
1 Lake Street
Upper Saddle River, NJ 07458

YOU ACKNOWLEDGE THAT YOU HAVE READ THIS AGREEMENT, UNDERSTAND IT, AND AGREE TO BE BOUND BY ITS TERMS AND CONDITIONS. YOU FURTHER AGREE THAT IT IS THE COMPLETE AND EXCLUSIVE STATEMENT OF THE AGREEMENT BETWEEN US THAT SUPERSEDES ANY PROPOSAL OR PRIOR AGREEMENT, ORAL OR WRITTEN, AND ANY OTHER COMMUNICATIONS BETWEEN US RELATING TO THE SUBJECT MATTER OF THIS AGREEMENT.

**Mac users, please note: MINITAB files are viewable only on PC platforms.